Progress in Scientific Computing
Vol. 4

Edited by
S. Abarbanel
R. Glowinski
G. Golub
P. Henrici
H.-O. Kreiss

Birkhäuser
Boston · Basel · Stuttgart

Jane K. Cullum
Ralph A. Willoughby

Lanczos Algorithms for Large Symmetric Eigenvalue Computations Vol. II Programs

1985

Birkhäuser
Boston · Basel · Stuttgart

Authors:
Jane K. Cullum
Ralph A. Willoughby
IBM T. J. Watson Research Center
Yorktown Heights, NY 10598 (USA)

Library of Congress Cataloging in Publication Data

(Revised for volume 2)
Cullum, Jan K., 1938–
  Lanczos algorithms for large symmetric eigenvalue
computations.
  (Progress in Scientific computing ; v. 3–4)
  Includes bibliographies and indexes.
  Contents: v. 1. Theory – – v. 2. Programs.
  1. Symmetric matrices – – Data processing.
2. Eigenvalues – – Data processing.   I. Willoughby,
Ralph A.   II. Title.   III. Series.
QA193.C84   512.9'434   81-38450

CIP-Kurztitelaufnahme der Deutschen Bibliothek

Cullum, Jane K.:
Lanczos algorithms for large symmetric eigenvalue
computations / Jane K. Cullum ; Ralph A. Willoughby.
– Boston ; Basel ; Stuttgart : Birkhäuser

NE: Willoughby, Ralph A.:
Vol. 2. Programs. – 1985.
  (Progress in Scientific computing ; Vol. 4)

NE: GT

ISBN 978-1-4684-9180-7   ISBN 978-1-4684-9178-4 (eBook)
DOI 10.1007/978-1-4684-9178-4

Table of Contents

CHAPTER 5

REAL, SYMMETRIC GENERALIZED PROBLEMS

CHAPTER 6

REAL RECTANGULAR MATRICES

CHAPTER 7

NONDEFECTIVE, COMPLEX SYMMETRIC MATRICES

CHAPTER 8

REAL SYMMETRIC MATRICES, BLOCK LANCZOS CODE

CHAPTER 9

FACTORED INVERSES, REAL SYMMETRIC MATRICES, BLOCK LANCZOS CODE

REFERENCES

INDEX TO FORTRAN PROGRAMS

CHAPTER 1

LANCZOS PROCEDURES

SECTION 1.1   INTRODUCTION

The FORTRAN codes contained in this volume are designed for computing eigenvalues and eigenvectors or singular values and singular vectors of large, sparse matrices. Large means of order several hundred to perhaps 10,000. The largest matrix which we tested was real symmetric and had order 4900. This book is divided into 9 chapters. In this first chapter we give a brief description of Lanczos eigenelement procedures and then make some comments about what the Lanczos codes in this book can and cannot be expected to compute. Detailed analyses of the ideas used in these procedures are contained in Volume 1 of this book.

Chapters 2 through 7 contain procedures which are based upon the single-vector Lanczos recursion with no reorthogonalization of any kind. Six different classes of problems are addressed in these 6 chapters: Eigenelement computations for real symmetric matrices (Chapter 2); for Hermitian matrices (Chapter 3); for factored inverses of real symmetric matrices (Chapter 4); for real symmetric, generalized problems (Chapter 5) and for nondefective, complex symmetric matrices (Chapter 7); and Singular value and vector computations for real, rectangular matrices (Chapter 6). Chapters 8 and 9 contain Lanczos procedures which are based upon a 'block' version of the Lanczos recursions. These iterative block procedures include some reorthogonalization within each iteration, but this reorthogonalization is limited to reorthogonalizations w.r.t. certain vectors in each first Lanczos block.

The single-vector procedures can be used to compute anywhere from a very few to very many eigenvalues (singular values). These eigenvalues (singular values) need not be at the extremes of the spectrum. For some matrices it is even possible to compute all of the eigenvalues. The iterative block procedures can only be used to compute a few extreme eigenvalues of the specified matrix. The single vector codes consist of two phases. First eigenvalues or singular values are computed and then corresponding eigenvectors or singular vectors are computed. The iterative 'block' codes compute eigenvalues and corresponding eigenvector approximations simultaneously. Block codes for computing singular values are not included in this book. See for example, Golub, Luk, and Overton [1981] for an example of such a block algorithm.

With three exceptions which are given below, each Chapter 2 - 9 contains the following types of information for the particular class of problems considered in that chapter: documentation; main program(s); LANCZS subroutine for computing Lanczos matrices; sample matrix-vector multiply and/or solve subroutines; other subroutines needed by the codes in that chapter; and definitions of the files used by the programs together with sample input files. Because of

1

the similarities between the variables, flags, etc., the documentation for the codes contained in Chapters 2,3,4, and 5 was combined and is contained in Section 2.2 of Chapter 2. The codes in Chapters 2, 3, 4 and 5 use essentially (with 2 exceptions) the same set of 'other or additional subroutines' so these subroutines were combined and are given only in Chapter 2, Section 2.6. Similarly, the block codes in Chapters 8 and 9 use the same set of additional subroutines and these are given only in Section 8.5. Some additional optional, preprocessing codes are also provided, and again each of these is included in only one of the chapters and not in each of the ones where it might be useful.

Each set of codes contains many write statements. These write statements serve two major functions: to provide consistency checks on the information supplied by the user, and to provide running commentary on the progress of the computations. Much of the code has been modularized to help make the program logic more transparent to the user. These codes are not designed as efficiently as they could be. Many internal comments have been included. Numerous consistency checks have been used to verify that the user has set up the procedure properly. Basically, we have compromised some efficiency for safety and robustness.

Each LANCZS subroutine together with the corresponding sample matrix-vector multiply and solve subroutines are in files labelled as *MULT. For example in Chapter 2 where real symmetric matrices are discussed this file is labelled LEMULT. The user should note that within a given *MULT file, each sample USPEC* and *MATV subroutine has been given two names so that these subroutines can co-exist with similar subroutines for other test matrices. However, two different *MULT files cannot co-exist because subroutine names are reused in going from one category of matrices to another category. In particular for the codes in Chapters 2,3, and 7, the matrix-vector multiply subroutine is called CMATV. Moreover, in all of the chapters, the matrix specification subroutines are called USPEC*. This reuse of names makes it easier for the user to pass from one set of codes to another. Furthermore, from category to category, subroutines with similar function were typically given the same name. For example, all of the subroutines which generate families of Lanczos matrices are named LANCZS. There are two BISEC bisection subroutines for computing eigenvalues of real symmetric tridiagonal matrices, one for Chapters 2,3,4, and 5 and the other one is for Chapter 6. If these sets of codes had to co-exist in one computer file, then it would be necessary for the user to devise a scheme for renaming those subroutines which have the same names.

With respect to portability, each of these programs and subroutines has been individually checked for portability by the PFORT Verifier [1981], but the communications between these subroutines have not been checked. Obvious problems with portability like non-Fortran items in the format statements have all been removed. However, certain nonportable constructions have been retained because they make the programs somewhat easier to use. The header of each of the programs contains a list of those constructions in that program which were identified by the PFORT verifier as being nonportable. These headers can be used to locate the nonportable items so that if necessary they can be modified. A list of most of the nonportable items and the reasons for retaining them are given in Table 1.1.1.

TABLE 1.1.1   Nonportable Constructions Used in the Codes

Nonportable Construction	Where Used	Why Used
Entry	Passes storage locations of arrays and parameters needed to define user-specified matrix from subroutine USPEC where arrays are dimensioned and initialized to the corresponding matrix-vector multiply or solve subroutine.	Codes do not need to 'see' the user-specified matrix. Codes need only output from matrix-vector multiply or solve subroutines for the matrix being used. User does not have to alter the calling sequences to these subroutines every time the number or kind of arrays needed to define the given matrix is changed.
Formats (20A4) and (4Z20)	(20A4) is used to read and write explanatory comments within the main programs and in sample USPEC subroutines. Machine format (4Z20) is used to read in and write out the Lanczos tridiagonal matrices generated and other quantities for which conversion errors could cause numerical problems.	Allows the user to easily modify headers describing the matrix and code being used. Prevents format conversion errors incurred in input/output conversions.
Free Format Read (5, *)	Used in main program and in sample USPEC subroutines on read-ins of user-specified parameters from input file 5.	Ease of input. User does not have to have the input values properly aligned in the input file.
Complex*16 Variables	Used only in the Hermitian and in the complex symmetric Lanczos codes.	Computations require double precision complex arithmetic.
Specification of Machine Epsilon	Used in main programs	Required to define tolerances used at various points in the computations.

The single vector Lanczos codes in Chapters 2 - 7 are essentially self-contained. The user must provide the matrix-vector multiply and/or solve subroutines which are required by these codes, together with a matrix specification subroutine which defines, dimensions and initializes the matrix which will be used by the Lanczos procedure. The sample matrix-specification subroutines and sample matrix-vector multiply and solve subroutines contained with these codes can be modified and used if appropriate or they can be replaced completely. All of these procedures require a random number generator subroutine, inner product subroutines, and a subroutine to mask underflow. These procedures assume that each time the random number generator is called that the seed for this generator is automatically reset to a different value.

The iterative 'block' Lanczos codes in Chapters 8 and 9 require matrix specification and matrix-vector multiply and solve subroutines very similar to those used in the single vector codes, plus the same type of random number generating subroutine, inner product subroutine, and mask subroutine. However, as implemented here the block codes are not self-contained. These codes call two subroutines from the EISPACK Library [1976,1977], TRED2 and IMTQL2, which are used repeatedly to compute the eigenvalues and eigenvectors of the small Lanczos matrices generated on each iteration of the block procedures. The user can of course replace these calls by calls to subroutines which perform similar functions, if the EISPACK Library is not available.

The optional preprocessing programs in Sections 2.7, 4.5, 6.7, and 7.7 are stand-alone (if one includes the programs which must be supplied by the user), except for the subroutine PERMUT given in Section 4.5. PERMUT can be used in conjunction with the procedures in Chapters 4, 5 and 9. It calls the SPARSPAK Library [1979] (A. George, J. Liu, E. Ng, U. Waterloo) to try to determine a reordering of the given sparse matrix for which the sparsity of the given matrix translates into a sparse factorization of the reordered matrix.

SECTION 1.2   WHAT ARE LANCZOS PROCEDURES?

Lanczos procedures for computing eigenvalues and eigenvectors of real symmetric matrices are based upon one or more variants of the basic single-vector Lanczos recursion for tridiagonalizing a real symmetric matrix A. Given a starting vector v_1 which is typically-generated randomly, the Lanczos recursion implements a Gram-Schmidt orthogonalization of the matrix-vector products Av_i corresponding to the Lanczos vectors v_i generated by the recursion. See for example Bjorck [1967]. Specifically, we have that for $i = 2,...,m$,

$$\beta_{i+1}v_{i+1} = Av_i - \alpha_i v_i - \beta_i v_{i-1} \qquad (1.2.1)$$

where $\alpha_i \equiv v_i^T Av_i$ and $\beta_{i+1} \equiv v_{i+1}^T Av_i$. By definition $\alpha_i v_i$ and $\beta_i v_{i-1}$ are the projections of Av_i onto the two most recently-generated Lanczos vectors v_i and v_{i-1}. In practice to improve the numerical stability of this recursion, the above formulas are replaced by the following ones.

$$\alpha_i \equiv v_i^T(Av_i - \beta_i v_{i-1}) \quad \text{and} \quad \beta_{i+1} \equiv \| Av_i - \alpha_i v_i - \beta_i v_{i-1} \| . \qquad (1.2.2)$$

The α_i as defined in Eqn(1.2.2) correspond to a modified Gram-Schmidt orthogonalization procedure. The formula for β_{i+1} given in Eqn(1.2.2) is theoretically equivalent to the one given in Eqn(1.2.1). However, it is superior numerically because this choice directly controls the sizes of the Lanczos vectors. See Paige [1976].

Rewriting Eqn(1.2.1) in matrix form, we obtain

$$AV_j = V_jT_j + \beta_{j+1}v_{j+1}e_j^T \qquad (1.2.3)$$

where T_j denotes the real symmetric tridiagonal Lanczos matrix of order j whose diagonal entries are the scalars α_i, $1 \le i \le j$, and whose subdiagonal (superdiagonal) entries are the scalars β_{i+1}, $1 \le i \le j - 1$, generated by the Lanczos recursion. In Eqn(1.2.3), $V_j = \{v_1, v_2, ..., v_j\}$, the matrix whose columns are the Lanczos vectors generated by the recursion, and e_j is the coordinate vector whose jth component is 1 and whose other components are 0.

It is easy to demonstrate by induction that in exact arithmetic each set of vectors V_j generated by the recursion in Eqns(1.2.1) - (1.2.2) is an orthonormal set. Therefore for any A-matrix with n distinct eigenvalues and any starting vector v_1 which has a projection on every eigenspace of A, we have that for each $j \le n$,

$$T_j = V_j^TAV_j. \qquad (1.2.4)$$

Thus the symmetric tridiagonal matrices T_j are representations of the projections of the given matrix A onto the subspaces spanned by the corresponding sets of Lanczos vectors V_j. The eigenvalues of these matrices are the eigenvalues of the A-matrix restricted to these subspaces. Since the Lanczos vectors are obtained by orthogonalizing vectors of the form $\{v_1, Av_1, A^2v_1, ...,\}$, we expect the eigenvalues of the T_j to provide good approximations to some of the eigenvalues of A, if j is sufficiently large. Clearly, at least theoretically, if we extend the recursion to $j = n$, then the eigenvalues of T_n will be the eigenvalues of A. T_n is simply an orthogonal transformation of A and must therefore have the same eigenvalues as A. Moreover, any Ritz vector V_ju obtained from an eigenvector u of some T_j is an approximation to a corresponding eigenvector of A.

Basic steps in any Lanczos procedure for computing eigenvalues and eigenvectors of 'symmetric' matrices are the following. (1) Use a variant of the Lanczos recursion to transform the given 'symmetric' matrix A into a family of 'symmetric' tridiagonal matrices of varying sizes. (2) Compute eigenvalues and eigenvectors of certain members of this family. Because of the real symmetric tridiagonal structure this is a much simpler problem than computing the eigenvalues and eigenvectors of A directly. (3) Take some or all of these eigenvalues as approximations to eigenvalues of A and map the corresponding eigenvectors of the tridiagonal matrix into Ritz vectors for the matrix A. (4) Use these Ritz vectors as approximations to the eigenvectors of A.

The Lanczos recursion in Eqn(1.2.1) has several properties which make it particularly attractive for dealing with large but sparse matrices. First the given matrix enters the recursion only through the matrix-vector multiply terms Av_i. Thus contrary to what is done in the standard methods for solving small or medium size eigenvalue problems, see for example EISPACK [1976,1977], the given matrix is not explicitly modified. The user must provide only a subroutine which computes Ax for any given vector x. If the matrix A is sparse, this computation can be done using an amount of storage that is only linear in the size of the matrix instead of quadratic. Second, the recursion uses only the two most recently-generated Lanczos vectors. The Gram-Schmidt orthogonalization of an arbitrary set of vectors would require that at any given stage in the process that all of the vectors which have already been orthogonalized be available for orthogonalizing each additional vector as it is considered. Thus, the storage requirements for implementing the basic Lanczos recursion are minimal. If we use Eqns(1.2.1) - (1.2.2) then only 2 n-vectors are needed for the two most recently-generated Lanczos vectors plus storage for the α and β arrays.

There are however numerical problems if only a simple direct implementation of this recursion is programmed. In general such an implementation yields Lanczos matrices which have extra eigenvalues in addition to the 'good' eigenvalues which are approximations to eigenvalues of A. These extraneous or 'spurious' eigenvalues are caused by the losses in the orthogonality of the Lanczos vectors which in turn are caused by the combination of the roundoff errors resulting from the finite computer arithmetic and the convergence (as j is increased) of eigenvalues of the Lanczos matrices to eigenvalues of the original matrix A. This interaction between the computer arithmetic and the convergence of eigenvalues is discussed in Paige [1971,1980].

During the past 5-10 years many different types of Lanczos eigenelement algorithms have been proposed. See Volume 1, Chapter 2 of this book for a brief survey of the literature. Most of these procedures incorporate modifications to the basic Lanczos recursion in Eqns(1.2.1) - (1.2.2) which force the Lanczos vectors to stay nearly orthonormal. These approaches require either the repeated computation of Ritz vectors or the repeated reorthogonalization of the Lanczos vectors as they are generated or some combination of these two computations. In either case as the size of the Lanczos matrix generated is increased to be able to compute more eigenvalues, the associated Ritz vectors or the Lanczos vectors needed for the reorthogonalizations require more and more storage. These modifications often work well but destroy much of the simplicity of the basic procedure, and because of the added storage requirements resulting from the reorthogonalizations they limit the number of eigenelements which can be computed.

The approach which we have chosen and which is implemented in the enclosed FORTRAN programs in Chapters 2 - 7 is not to force the orthogonality of the Lanczos vectors by reorthogonalizing, but to work directly with the basic Lanczos recursion, accepting the losses in orthogonality, and then unraveling the effects of these losses. This approach allows us to retain the basic simplicity of the Lanczos recursion, to minimize the storage requirements, and to therefore maximize the number of eigenvalues of A which can be computed. In our approach in the single-vector algorithms in Chapters 2 - 7, Ritz vectors are not computed until after the eigenvalues have been computed accurately. Consequently, the basic storage requirements for our eigenvalue (singular value) algorithms are only a small multiple of the size of the largest Lanczos matrix used in the computations. Thus, we can compute many eigenvalues of very large but sparse matrices. Depending upon what is to be computed and upon the eigenvalue distribution in the given matrix A, the sizes of the Lanczos matrices used in these computations may be much smaller or considerably larger than the original A-matrix. However the Lanczos matrices generated by the procedures in Chapters 2 - 6 are real symmetric and tridiagonal so that these matrices can be very large and still not present insurmountable computational problems. Eigenvalue and eigenvector computations for such matrices require minimal amounts of storage and fairly reasonable numbers of arithmetic operations.

The computational problems which arise from not maintaining near orthogonality of the Lanczos vectors and which we must address in our single-vector codes are of two types. First and most importantly, we must deal with the question of sorting the eigenvalues of the Lanczos matrices into 2 classes, one corresponding to the 'good' eigenvalues which are approximations to the eigenvalues of A and the other corresponding to the extra or 'spurious' eigenvalues caused by the losses in orthogonality. The identification test used for doing this is discussed in Volume 1, Chapter 4, Section 4.5 of this book. For the procedures discussed in Chapters 2 - 6, this identification test is an integral and inexpensive part of the eigenvalue (singular value) computations. For the complex symmetric procedure discussed in Chapter 7 this test is handled in a considerably less eloquent manner and is expensive.

The second but much less serious difficulty we must address is the question of false multiplicities. The multiplicity of a particular 'good' eigenvalue as an eigenvalue of the Lanczos matrices is not related to the multiplicity of that eigenvalue as an eigenvalue of the A-matrix. 'Good' eigenvalues may replicate many times as eigenvalues of a Lanczos matrix, but be only simple eigenvalues of the original A-matrix. Thus, these single-vector procedures cannot directly determine the true multiplicities of the computed 'good' eigenvalues. Of course, this latter comment is also applicable to any single-vector Lanczos procedure not just to our procedures. Theoretically, at most one eigenvector for each distinct eigenvalue of the A-matrix can be obtained using the single-vector Lanczos recursion given in Eqns(1.2.1) and (1.2.2). (This of course is not true for iterative block Lanczos procedures.) It is interesting to note however that if the Lanczos recursion is used without any reorthogonalization, then it can yield sets of linearly independent eigenvectors for eigenvalues which are multiple in the A-matrix. The amount of work required to compute these additional eigenvectors depends upon the particular matrix in question and upon the particular eigenvalue. The codes provided in Chapters 2 - 7 of this book do not however incorporate this capability.

The iterative 'block' Lanczos procedures for real symmetric matrices given in Chapters 8 -

9 are based upon a block version of the Lanczos recursion

$$Q_{j+1}B_{j+1} = AQ_j - Q_jA_j - Q_{j-1}B_j^T \tag{1.2.5}$$

for $j=1,2,...,s$ where Q_1 is nxq and the coefficient matrices A_j and B_{j+1} are block analogs of the scalar coefficients in the single-vector Lanczos recursion in Eqns(1.2.1) and (1.2.2). The number of blocks s used on each iteration is chosen such that $qs<<n$, where n is the order of the given A-matrix and q is chosen such that $q \geq q'$, the number of eigenvalues and eigenvectors desired. The Lanczos matrices are real symmetric, block tridiagonal matrices. In Eqn(1.2.5) we used Q_j instead of V_j because in our block Lanczos procedures we maintain near-orthogonality of the blocks generated within each iteration by incorporating reorthogonalization of the blocks of Lanczos vectors with respect to certain vectors in the first Lanczos block.

The 'block' procedures provided in Chapters 8 and 9 are really hybrid algorithms, something between a true block Lanczos procedure, see for example Cullum and Donath [1974], and the single-vector Lanczos procedures given in Chapters 2 - 7. The sequence of 'blocks' generated on each iteration of this hybrid method has the property that the first Q-block contains at least as many vectors as the user is trying to compute, but the second and succeeding blocks each contain only one vector. The corresponding resulting Lanczos matrices are not block tridiagonal. Each Lanczos matrix has a border of blocks in the first q rows and columns and is tridiagonal below this border.

At the beginning of each chapter, a brief description is given of the particular variant of the Lanczos recursion used in the Lanczos codes included in that chapter, along with some additional comments relevant to the particular types of problems being considered in that chapter.

SECTION 1.3   COMMENTS AND DISCLAIMERS

The single-vector Lanczos procedures contained in Chapters 2 - 7 do not behave like standard eigenelement procedures. Their behavior is both non-classical and somewhat unorthodox. If one of these codes were run on two different kinds of computers but with the same original matrix and the same initial specifications, the computed results could be quite different. A primary cause for such differences can of course be a difference in the starting vector caused by a difference in the random number generators. However even if the same starting vector were read in, the results would almost surely differ due to the differences in the computer arithmetic. In practice, the Lanczos matrices generated on two different kinds of computers may agree for a certain number of Lanczos steps but will begin to diverge upon the convergence of one or more of the eigenvalues of these Lanczos matrices to eigenvalues of the A-matrix. If after a reasonable number of steps in the Lanczos recursion we were to compare the entries in the Lanczos matrices generated by the two different computers, the values would probably be very different.

Furthermore, if we were to compute the eigenvalues of the two sets of Lanczos matrices for various sizes and 'spurious' eigenvalues were present, then these spurious eigenvalues would be different and even appear in different portions of the spectrum. In fact, prior to the convergence of a particular 'good' eigenvalue, the values of that good eigenvalue, in terms of how accurate it is at any given stage in the computations, may differ. However once a 'good' eigenvalue in either set has converged, that 'good' eigenvalue will agree with a true eigenvalue of the original user-specified matrix to as many digits as can be expected.

Therefore, if the user carries out the sample eigenvalue computation provided in Chapter 2, he/she should not be alarmed or surprised if the output from the computer being used does not agree with what is shown in the sample, as long as the converged 'good' eigenvalues agree. Actually one may observe different rates of convergence on different kinds of computers, depending upon the computer arithmetic. With increased arithmetic precision in all of the computations, these procedures may converge more rapidly. With decreased precision, they will converge less rapidly. All of our codes use double precision arithmetic (for an IBM 3083) and any precision less than that is not recommended.

Each of these procedures requires the user to supply either a matrix-vector multiply subroutine or a matrix-vector solve subroutine. (Both types of subroutines are required for the codes in Chapter 5.) Such subroutines should perform the required computations rapidly and accurately, taking advantage of any special properties or structure in the given matrix. Our Lanczos programs see the original matrix as the outputs of these subroutines. The codes provided include sample matrix-vector multiply subroutines for a general sparse 'symmetric' matrix given in a particular sparse format. These are available for the user to use or modify as desired. Note that similar programs are also provided for the singular value/vector computations. Accuracy is important in these subroutines because consistency must be maintained in the information being provided to the LANCZS subroutine which is generating the Lanczos matrices. There is no built-in mechanism for preserving symmetry. Therefore, the matrix-vector multiply and solve subroutines must be coded with care. Without such consistency the Lanczos codes will not function properly.

The convergence characteristics of the two types of Lanczos procedures considered are quite different. These differences are discussed in Chapters 4 and 7 of Volume 1 of this book. However, in both cases, the degree of difficulty in computing the desired eigenvalues depends upon the eigenvalue gaps. For the single-vector procedures the primary factor in determining whether or not it is feasible to compute either large numbers of eigenvalues or the eigenvalues with the smallest gaps, is the gap ratio, the ratio of the largest gap between two neighboring eigenvalues to the smallest such gap. The smaller this ratio, the easier it is to compute all of the eigenvalues of the given matrix. The larger this ratio, the harder it is to compute those eigenvalues with the smallest gaps. The locations of the desired eigenvalues in the spectrum of the given matrix also play a significant role in the rate of convergence of individual eigenvalues.

Both types of Lanczos procedures favor extreme eigenvalues. The iterative block codes, in fact, can only compute a few extreme eigenvalues. However for the single-vector codes, it is possible for interior eigenvalues which have gaps which are significantly larger than the gaps for some of the extreme eigenvalues to converge prior to the convergence of those extreme eigenvalues. Examples of the convergence achievable are given in Volume 1, Chapter 4 of this book.

The convergence of the iterative block procedures depends primarily upon the gaps between the eigenvalues being computed and the closest eigenvalue not being approximated, the spread of the matrix, and the overall eigenvalue distribution. The block procedures discussed in Chapters 8 and 9 are iterative and the codes track the rate of convergence. If the observed rate is too slow (as specified by the user), these block procedures will terminate without achieving convergence. The user then has the option of restarting the block procedure with a different choice of parameters and using the current approximation to the basis for the desired eigenspace as the starting vectors.

Thus, the amount of work required for a particular eigenelement computation for a given matrix using a particular method depends directly upon the eigenvalue distribution in that matrix and upon which portion of the spectrum is being computed. Some problems are 'easy', others are hard. Therefore failure can occur, in the sense that these procedures may not be able to compute the information desired by the user within the computational bounds specified by the user. However the single-vector Lanczos procedures, even in 'failure', provide a great deal of information about the eigenvalue spectrum of the given matrix.

In deciding which procedure to use on a given problem, our preference is a single-vector procedure, although the iterative block procedures can often quickly provide simultaneously the desired eigenvalues and eigenvectors. If the user wants extreme eigenvalues and the user knows or suspects that one or more of these is multiple, then the block procedure is probably preferable. More details about the Lanczos procedures contained in this book can be found in Volume 1. Any questions about these programs including the question of obtaining copies of these codes or of problems with these codes, should be addressed directly to the authors. We hope that these codes will prove useful in many different applications in the engineering and scientific community.

CHAPTER 2

REAL SYMMETRIC MATRICES

SECTION 2.1   INTRODUCTION

The FORTRAN codes in this chapter address the question of computing distinct eigenvalues and corresponding eigenvectors of real symmetric matrices, using a single-vector Lanczos procedure.   For a given real symmetric matrix A, these codes compute real scalars λ and corresponding real vectors $x \neq 0$, such that

$$Ax = \lambda x. \tag{2.1.1}$$

DEFINITION 2.1.1 A real nxn matrix $A \equiv (a_{ij})$, $1 \leq i,j \leq n$, is a real symmetric matrix if and only if for every i and j, $a_{ij} = a_{ji}$.

Real symmetric matrices are discussed in detail in Stewart [1973].   Properties which we use are: (1) Real symmetric matrices have complete eigensystems.   That is, the dimension of the eigenspace corresponding to any given eigenvalue of the given matrix A is the same as the multiplicity of that eigenvalue as a root of the characteristic polynomial of A.   (2) For any two distinct eigenvalues of A, λ and μ, and corresponding eigenvectors x and y, $x^T y = 0$. Thus, eigenvectors corresponding to different eigenvalues are orthogonal, and we can construct an eigenvector basis which is orthonormal.   Vectors are orthonormal if they are orthogonal and each has a Euclidean norm of 1.   (The Euclidean norm of a vector is just the square root of the sum of the squares of its components.)   (3) Small perturbations in the matrix cause only small perturbations in the eigenvalues.   Of the classes of matrices which we consider, the class of real symmetric matrices is the most well-behaved and thus the 'easiest'.

The Lanczos codes contained in this chapter correspond to the most straight-forward implementation of the Lanczos recursion included in this book.   These codes can be used to compute either a very few or very many of the distinct eigenvalues of the given real symmetric matrix.   As the documentation in the next section indicates, the A-multiplicity of a given computed 'good' Lanczos eigenvalue can be obtained only with additional computation, and the modifications required to do this additional computation are not included in these versions of the codes.   This implementation uses the basic Lanczos recursion given in Eqns(1.1.2) - (1.2.1) in Section 1.2 of Chapter 1 to generate a family of real symmetric, tridiagonal matrices (T-matrices) whose sizes are specified by the user.   There is no reorthogonalization of the Lanczos vectors at any stage in any of the computations.

LEVAL, the main program for the real symmetric eigenvalue computations, calls the subroutine BISEC to compute eigenvalues of the user-specified Lanczos tridiagonal matrices on the user-specified intervals.   BISEC simultaneously computes these T-eigenvalues with their

T-multiplicities and sorts the computed T-cigenvalues into two classes, the 'good' T-eigenvalues and the 'spurious' T-eigenvalues. The 'good' T-eigenvalues are accepted as approximations to eigenvalues of the user-specified matrix A. The accuracy of these 'good' T-eigenvalues as eigenvalues of A is then estimated using error estimates computed by subroutine INVERR. Error estimates are computed only for isolated 'good' T-eigenvalues. All other 'good' T-eigenvalues are assumed to have converged. If convergence has not yet occurred and a larger Lanczos matrix has been specified by the user, the program will continue on to a larger Lanczos matrix, repeating the above procedure on this larger matrix.

Once the eigenvalues have been computed accurately enough, the user can select a subset of the 'converged' eigenvalues for which eigenvectors are to be computed. The main program LEVEC, for computing eigenvectors of real symmetric matrices, is then used to compute these desired eigenvectors.

All computations are in double precision real arithmetic. The user must supply a subroutine USPEC which defines and initializes the user-specified matrix A and a subroutine CMATV which computes matrix-vector multiplies Ax for any given vector x. These subroutines must be constructed in such a way as to take advantage of the sparsity (and/or structure) of the user-supplied A-matrix and such that these computations are done accurately. More details about these real symmetric single-vector Lanczos procedures are given in Chapter 4 of Volume 1 of this book.

SECTION 2.2   DOCUMENTATION FOR PROGRAMS IN CHAPTERS 2 - 5

```
C-----DOCUMENTATION FOR SINGLE-VECTOR--------------------------------------LEV00010
C     LANCZOS EIGENVALUE/EIGENVECTOR PROGRAMS FOR                          LEV00020
C     (1) REAL SYMMETRIC MATRICES                                          LEV00030
C     (2) HERMITIAN MATRICES                                               LEV00040
C     (3) FACTORED INVERSES OF REAL SYMMETRIC MATRICES                     LEV00050
C     (4) REAL SYMMETRIC, GENERALIZED PROBLEMS WHERE ONE OF THE            LEV00060
C         MATRICES IS POSITIVE DEFINITE AND ITS CHOLESKY FACTORS ARE       LEV00070
C         AVAILABLE                                                        LEV00080
C                                                                          LEV00090
C                                                                          LEV00100
C     AUTHORS:   JANE CULLUM AND RALPH A. WILLOUGHBY, IBM RESEARCH,        LEV00110
C                YORKTOWN HEIGHTS, NY 10598.  PHONE: 914-945-2227          LEV00120
C                                                                          LEV00130
C                                                                          LEV00140
C     REAL SYMMETRIC MATRICES:                                             LEV00150
C                                                                          LEV00160
C     GIVEN A REAL SYMMETRIC MATRIX A OF ORDER N THE THREE SETS OF         LEV00170
C     FORTRAN FILES LABELLED LEVAL, LESUB, AND LEMULT CAN BE USED TO       LEV00180
C     COMPUTE DISTINCT EIGENVALUES OF THE USER-SPECIFIED MATRIX            LEV00190
C     IN USER-SPECIFIED INTERVALS.                                         LEV00200
C                                                                          LEV00210
C     CORRESPONDING EIGENVECTORS FOR SELECTED, COMPUTED EIGENVALUES CAN    LEV00220
C     BE COMPUTED USING THE SETS OF FILES LABELLED LEVEC, LESUB, AND       LEV00230
C     LEMULT.                                                              LEV00240
C                                                                          LEV00250
C                                                                          LEV00260
C     HERMITIAN MATRICES:                                                  LEV00270
C                                                                          LEV00280
C     GIVEN A HERMITIAN MATRIX A OF ORDER N THE THREE SETS OF              LEV00290
C     FORTRAN FILES LABELLED HLEVAL, LESUB, AND HLEMULT CAN BE USED        LEV00300
C     TO COMPUTE DISTINCT EIGENVALUES IN USER-SPECIFIED INTERVALS.         LEV00310
C                                                                          LEV00320
C     CORRESPONDING EIGENVECTORS FOR SELECTED, COMPUTED EIGENVALUES        LEV00330
C     CAN BE COMPUTED USING THE SETS OF PROGRAMS LABELLED HLEVEC,          LEV00340
C     LESUB, AND HLEMULT.                                                  LEV00350
C                                                                          LEV00360
C                                                                          LEV00370
C     FACTORED INVERSES OF REAL SYMMETRIC MATRICES:                        LEV00380
C                                                                          LEV00390
C     GIVEN A REAL SYMMETRIC MATRIX A, THE LANCZOS RECURSION IS            LEV00400
C     APPLIED TO THE INVERSE OF A, USING A FACTORIZATION                   LEV00410
C     OF A.  THE SETS OF FILES LIVAL, LESUB, AND LIMULT                    LEV00420
C     CAN BE USED TO COMPUTE THE DISTINCT EIGENVALUES OF THE               LEV00430
C     INVERSE OF THE A-MATRIX AND OF A IN  USER-SPECIFIED                  LEV00440
C     INTERVALS.  THE PROGRAMS ACTUALLY ALLOW ONE TO WORK WITH            LEV00450
C     ANY MATRIX B = PCP' WHERE  C = SO*A + SHIFT*I, WHERE                 LEV00460
C     SO AND SHIFT ARE SCALARS CHOSEN BY THE USER AND P IS A               LEV00470
C     PERMUTATION MATRIX CHOSEN SUCH THAT THE FACTORIZATION                LEV00480
C     OF THE B-MATRIX RETAINS SPARSITY.  IN THE                            LEV00490
C     SAMPLE LIMULT SUBROUTINES PROVIDED, SO AND SHIFT MUST BE             LEV00500
C     CHOSEN SO THAT THE RESULTING B-MATRIX IS POSITIVE DEFINITE,          LEV00510
C     AND THE CHOLESKY FACTORS ARE USED TO SOLVE B*U = V.                  LEV00520
C     HOWEVER, THE USER CAN EASILY REPLACE THE SAMPLE USPEC AND            LEV00530
C     BSOLV SUBROUTINES PROVIDED BY SUBROUTINES THAT ALLOW THE             LEV00540
C     GENERAL FACTORIZATION L*D*(L-TRANSPOSE).  THESE LANCZOS              LEV00550
C     PROGRAMS APPLY THE LANCZOS RECURSION TO B-INVERSE, USING             LEV00560
C     THE FACTORIZATION PROVIDED.  OPTIONAL PREPROCESSING PROGRAMS         LEV00570
C     PERMUT, LORDER, LFACT, AND LTEST ARE PROVIDED FOR SET-UP PURPOSES.LEV00580
C     PERMUT USES THE SPARSPAK PACKAGE OF A. GEORGE, J. LIU AND            LEV00590
C     E. NG TO OBTAIN A REORDERING OF THE GIVEN MATRIX THAT                LEV00600
C     PRESERVES SPARSENESS ON SUBEQUENT FACTORIZATION.  LORDER             LEV00610
C     CAN BE USED TO REORDER A GIVEN MATRIX, USING A GIVEN                 LEV00620
C     PERMUTATION. LFACT CAN BE USED TO COMPUTE THE CHOLESKY               LEV00630
C     FACTORS OF A GIVEN POSITIVE DEFINITE B-MATRIX.  LTEST CAN            LEV00640
```

```
C     BE USED TO ESTIMATE THE NUMERICAL CONDITION OF THE              LEV00650
C     B-MATRIX.                                                       LEV00660
C                                                                     LEV00670
C     CORRESPONDING EIGENVECTORS FOR SELECTED, COMPUTED               LEV00680
C     EIGENVALUES CAN BE COMPUTED USING THE SETS OF FILES             LEV00690
C     LABELLED LIVEC, LESUB, AND LIMULT.                              LEV00700
C                                                                     LEV00710
C     GENERALIZED REAL SYMMETRIC PROBLEMS:                            LEV00720
C                                                                     LEV00730
C     GIVEN 2 REAL SYMMETRIC MATRICES A AND B WHERE IN ADDITION B IS  LEV00740
C     POSITIVE DEFINITE AND ITS CHOLESKY FACTORS ARE AVAILABLE,       LEV00750
C     THE SETS OF FILES LGVAL, LGMULT, AND LESUB CAN BE USED          LEV00760
C     TO COMPUTE THE DISTINCT EIGENVALUES OF THE GENERALIZED          LEV00770
C     PROBLEM A*X = EVAL*B*X.                                         LEV00780
C                                                                     LEV00790
C     CORRESPONDING EIGENVECTORS CAN BE COMPUTED USING THE PROGRAMS   LEV00800
C     LGVEC, LGMULT, AND LESUB.  NOTE THAT THE PREPROCESSING PROGRAMS LEV00810
C     AVAILABLE FOR USE IN CASE (3) (PERMUT, LORDER, LFACT, AND LTEST)LEV00820
C     CAN ALSO BE USED IN THIS CASE TO OBTAIN A SUITABLE PERMUTATION, LEV00830
C     AND A FACTORIZATION OF THE RESULTING B-MATRIX.  THE A-MATRIX    LEV00840
C     CAN THEN BE PERMUTED USING LORDER.                             LEV00850
C                                                                     LEV00860
C                                                                     LEV00870
C     THESE PROGRAMS ALL USE LANCZOS TRIDIAGONALIZATION WITHOUT       LEV00880
C     REORTHOGONALIZATION TO GENERATE REAL SYMMETRIC TRIDIAGONAL      LEV00890
C     MATRICES, T(1,MEV), OF ORDER MEV.  SUBSETS OF THE EIGENVALUES OF LEV00900
C     THESE T-MATRICES, LABELLED AS THE 'GOOD EIGENVALUES', YIELD     LEV00910
C     APPROXIMATIONS TO THE DESIRED EIGENVALUES.  CORRESPONDING       LEV00920
C     RITZ VECTORS ARE APPROXIMATIONS TO THE DESIRED EIGENVECTORS.    LEV00930
C     NOTE THAT FOR CASE (4) THE GENERALIZED LANCZOS RECURSION        LEV00940
C     B*V(I+1)*BETA(I+1) = A*V(I) - B*V(I)*ALPHA(I) - B*V(I-1)*BETA(I) LEV00950
C     IS USED, ALONG WITH THE B-NORM.                                 LEV00960
C                                                                     LEV00970
C     THE IDEAS USED IN THESE PROGRAMS ARE DISCUSSED IN THE FOLLOWING LEV00980
C     REFERENCES.                                                     LEV00990
C                                                                     LEV01000
C     1.   JANE CULLUM AND RALPH A. WILLOUGHBY, LANCZOS ALGORITHMS    LEV01010
C          FOR LARGE SYMMETRIC MATRICES, VOLUME ?, PROGRESS IN        LEV01020
C          SCIENTIFIC COMPUTING, EDITORS, G. GOLUB, H.O. KREISS,      LEV01030
C          S. ARBARBANEL, AND R. GLOWINSKI,  BIRKHAUSER BOSTON INC.,  LEV01040
C          CAMBRIDGE, MASSACHUSETTS, 1983.                            LEV01050
C                                                                     LEV01060
C     2.   JANE CULLUM AND RALPH A. WILLOUGHBY, COMPUTING EIGENVECTORS LEV01070
C          (AND EIGENVALUES) OF LARGE, SYMMETRIC MATRICES USING       LEV01080
C          LANCZOS TRIDIAGONALIZATION, LECTURE NOTES IN MATHEMATICS,  LEV01090
C          773, NUMERICAL ANALYSIS PROCEEDINGS, DUNDEE 1979, EDITED BY LEV01100
C          G. A. WATSON, SPRINGER-VERLAG, (1980), BERLIN, PP.46-63.   LEV01110
C                                                                     LEV01120
C     3. IBID, LANCZOS AND THE COMPUTATION IN SPECIFIED INTERVALS OF  LEV01130
C          THE SPECTRUM OF LARGE SPARSE, REAL SYMMETRIC MATRICES, SPARSE LEV01140
C          MATRIX PROCEEDINGS 1978, ED. I.S. DUFF AND G. W. STEWART,  LEV01150
C          SIAM, PHILADELPHIA, PP.220-255, 1979.                      LEV01160
C                                                                     LEV01170
C     4. IBID, COMPUTING EIGENVALUES OF VERY LARGE SYMMETRIC MATRICES- LEV01180
C          AN IMPLEMENTATION OF A LANCZOS ALGORITHM WITHOUT           LEV01190
C          REORTHOGONALIZATION, J. COMPUT. PHYS. 44(1981), 329-358.   LEV01200
C                                                                     LEV01210
C                                                                     LEV01220
C-----PORTABILITY--------------------------------------------------------LEV01230
C                                                                     LEV01240
C                                                                     LEV01250
C     PROGRAMS WERE TESTED FOR PORTABILITY USING THE PFORT VERIFIER.  LEV01260
C     FOR DETAILS OF THE VERIFIER SEE FOR EXAMPLE, B. G. RYDER AND    LEV01270
C     A. D. HALL, "THE PFORT VERIFIER", COMPUTING SCIENCE TECHNICAL   LEV01280
C     REPORT 12, BELL LABORATORIES, MURRAY HILL, NEW JERSEY 07974,    LEV01290
C     (REVISED), JANUARY 1981.                                        LEV01300
C                                                                     LEV01310
```

```
C     WITH THE EXCEPTION OF THE PROGRAMS FOR HERMITIAN MATRICES WHICH    LEVO1320
C     ARE NOT PORTABLE BECAUSE OF THEIR USE OF COMPLEX*16 VARIABLES,     LEVO1330
C     THE OTHER PROGRAMS INCLUDED ARE PORTABLE EXCEPT FOR A FEW          LEVO1340
C     CONSTRUCTIONS WHICH, IF NECESSARY, WILL HAVE TO BE MODIFIED        LEVO1350
C     BY THE USER FOR THE PARTICULAR COMPUTER BEING USED.               LEVO1360
C                                                                        LEVO1370
C     NONPORTABLE CONSTRUCTIONS:                                         LEVO1380
C                                                                        LEVO1390
C     REAL SYMMETRIC MATRICES:                                           LEVO1400
C     IN LEVAL AND IN LEVEC                                              LEVO1410
C        1.  DATA/MACHEP STATEMENT                                       LEVO1420
C        2.  ALL READ(5,*) STATEMENTS (FREE FORMAT)                      LEVO1430
C        3.  FORMAT(20A4) USED FOR THE EXPLANATORY HEADER ARRAY, EXPLANLEVO1440
C        4.  FORMAT(4Z20) USED TO READ AND WRITE ALPHA/BETA FILES.       LEVO1450
C     IN LEMULT                                                          LEVO1460
C        1.  IN CMATV AND  USPEC THE ENTRY THAT PASSES THE STORAGE       LEVO1470
C            LOCATIONS OF THE ARRAYS DEFINING THE USER-SPECIFIED         LEVO1480
C            MATRIX.                                                     LEVO1490
C        2.  IN THE SAMPLE USPEC PROVIDED:  FREE FORMAT (8,*),           LEVO1500
C            THE FORMAT (20A4), AND DATA/MACHEP STATEMENT.               LEVO1510
C                                                                        LEVO1520
C     HERMITIAN MATRICES:                                                LEVO1530
C     IN HLEVAL AND IN HLEVEC                                            LEVO1540
C        1.  DATA/MACHEP STATEMENT                                       LEVO1550
C        2.  ALL READ(5,*) STATEMENTS (FREE FORMAT)                      LEVO1560
C        3.  FORMAT(20A4) USED FOR THE EXPLANATORY HEADER ARRAY, EXPLANLEVO1570
C        4.  COMPLEX*16 VARIABLES AND FUNCTIONS SUCH AS DCMPLX.          LEVO1580
C        5.  FORMAT (4Z20) USED TO READ AND WRITE ALPHA/BETA FILES.      LEVO1590
C     IN HLEMULT                                                         LEVO1600
C        1.  IN CMATV AND  USPEC THE ENTRY THAT PASSES THE STORAGE       LEVO1610
C            LOCATIONS OF THE ARRAYS DEFINING THE USER-SPECIFIED         LEVO1620
C            MATRIX.                                                     LEVO1630
C        2.  COMPLEX*16 VARIABLES AND FUNCTIONS SUCH AS DCMPLX.          LEVO1640
C        3.  IN THE SAMPLE USPEC PROVIDED:  FREE FORMAT (8,*),           LEVO1650
C            THE FORMAT (20A4), AND DATA/MACHEP STATEMENT.               LEVO1660
C                                                                        LEVO1670
C     FACTORED INVERSES OF REAL SYMMETRIC MATRICES:                      LEVO1680
C     IN LIVAL AND IN LIVEC                                              LEVO1690
C        1.  DATA/MACHEP STATEMENT                                       LEVO1700
C        2.  ALL READ(5,*) STATEMENTS (FREE FORMAT)                      LEVO1710
C        3.  FORMAT(20A4) USED FOR THE EXPLANATORY HEADER ARRAY, EXPLANLEVO1720
C        4.  FORMAT(4Z20) USED TO READ AND WRITE ALPHA/BETA FILES.       LEVO1730
C     IN LIMULT                                                          LEVO1740
C        1.  IN USPEC AND BSOLV, THE ENTRIES THAT PASS                   LEVO1750
C            THE STORAGE LOCATIONS OF THE ARRAYS DEFINING THE            LEVO1760
C            USER-SPECIFIED MATRIX.                                      LEVO1770
C        2.  IN THE SAMPLE USPEC SUBROUTINES PROVIDED:                   LEVO1780
C            FORMATS (20A4) AND (4Z20), FREE FORMAT (8,*), AND           LEVO1790
C            DATA/MACHEP STATEMENTS.                                     LEVO1800
C                                                                        LEVO1810
C                                                                        LEVO1820
C     GENERALIZED SYMMETRIC PROBLEM, B-MATRIX POSITIVE                   LEVO1830
C     DEFINITE AND CHOLESKY FACTORS AVAILABLE:                           LEVO1840
C     IN LGVAL AND IN LGVEC                                              LEVO1850
C        1.  DATA/MACHEP STATEMENT                                       LEVO1860
C        2.  ALL READ(5,*) STATEMENTS (FREE FORMAT)                      LEVO1870
C        3.  FORMAT(20A4) USED FOR THE EXPLANATORY HEADER ARRAY, EXPLANLEVO1880
C        4.  FORMAT(4Z20) USED TO READ AND WRITE ALPHA/BETA FILES.       LEVO1890
C     IN LGMULT                                                          LEVO1900
C        1.  IN USPECA, USPECB, AMATV AND LSOLV THE ENTRIES             LEVO1910
C            THAT PASS THE STORAGE LOCATIONS OF THE ARRAYS DEFINING      LEVO1920
C            THE USER-SPECIFIED MATRICES.                                LEVO1930
C        2.  IN THE SAMPLE USPECA AND USPECB SUBROUTINES PROVIDED:       LEVO1940
C            FORMATS (20A4) AND (4Z20), FREE FORMAT (8,*), AND           LEVO1950
C            DATA/MACHEP STATEMENTS.                                     LEVO1960
C                                                                        LEVO1970
C     ALL 4 CASES USE THE FORTRAN FILE LESUB:                           LEVO1980
```

```
C     IN LESUB ALL STATEMENTS ARE PORTABLE EXCEPT FOR:              LEVO1990
C        (1)  THE ENTRY IN SUBROUTINE LPERM THAT PASSES THE         LEVO2000
C             PERMUTATION FROM THE USPEC SUBROUTINE TO LPERM.       LEVO2010
C             (THIS IS USED ONLY IN CASES (3) AND (4)).             LEVO2020
C        (2)  THE COMPLEX*16 VARIABLES AND FUNCTIONS USED IN        LEVO2030
C             SUBROUTINE CINPRD. (THIS IS USED ONLY IN CASE (2)).   LEVO2040
C                                                                   LEVO2050
C     IN THE COMMENTS BELOW:                                       LEVO2060
C                                                                   LEVO2070
C     COMPLEX*16 = COMPLEX VARIABLE, 16 BYTES OF STORAGE            LEVO2080
C     REAL*8 = REAL VARIABLE, 8 BYTES OF STORAGE                   LEVO2090
C     REAL*4 = REAL VARIABLE, 4 BYTES OF STORAGE                   LEVO2100
C     INTEGER*4 = INTEGER VARIABLE, 4 BYTES                        LEVO2110
C                                                                   LEVO2120
C                                                                   LEVO2130
C-----MATRIX SPECIFICATION---------------------------------------------LEVO2140
C                                                                   LEVO2150
C                                                                   LEVO2160
C     IN CASES (1) AND (2), SUBROUTINE USPEC IS USED TO SPECIFY THE LEVO2170
C     USER-SUPPLIED A-MATRIX.  SIMILARLY, IN CASE (4) SUBROUTINES   LEVO2180
C     USPECA AND USPECB DEFINE THE USER-SUPPLIED A-MATRIX AND B-MATRIX. LEVO2190
C     IN CASE (3) ((4)), SUBROUTINE USPECB DEFINES THE FACTORIZATION LEVO2200
C     OF THE MATRIX (B-MATRIX) USED BY THE LANCZOS PROCEDURE.       LEVO2210
C     (IN CASE (3) THE A-MATRIX IS NOT USED DIRECTLY.)             LEVO2220
C                                                                   LEVO2230
C     IN CASES (1) AND (2), SUBROUTINE CMATV IS A CORRESPONDING     LEVO2240
C     MATRIX-VECTOR MULTIPLY SUBROUTINE WHICH SHOULD BE DESIGNED    LEVO2250
C     TO TAKE ADVANTAGE OF ANY SPECIAL PROPERTIES OF THE GIVEN      LEVO2260
C     MATRIX.  IN CASE (4) THIS SUBROUTINE IS NEEDED FOR THE        LEVO2270
C     A-MATRIX AND THUS IS CALLED AMATV.  IN CASES (3) AND (4)      LEVO2280
C     SUBROUTINES THAT CAN SOLVE B*U = V, USING A SPARSE           LEVO2290
C     FACTORIZATION OF B ARE NEEDED.  THESE SUBROUTINES ARE         LEVO2300
C     CALLED RESPECTIVELY, BSOLV AND LSOLV.  IN ALL CASES,         LEVO2310
C     ANY MATRIX-VECTOR MULTIPLY AND SOLVE SUBROUTINES USED         LEVO2320
C     MUST BE DESIGNED TO COMPUTE RAPIDLY AND ACCURATELY.          LEVO2330
C                                                                   LEVO2340
C     IN ALL CASES:                                                LEVO2350
C     SUBROUTINE USPEC(A OR B) HAS THE CALLING SEQUENCE            LEVO2360
C                                                                   LEVO2370
C          CALL USPEC(N,MATNO)                                     LEVO2380
C                                                                   LEVO2390
C     WHERE N IS THE ORDER OF THE USER-SUPPLIED MATRIX A, AND       LEVO2400
C     MATNO IS A <= 8 DIGIT INTEGER USED AS A MATRIX AND           LEVO2410
C     TEST IDENTIFICATION NUMBER.  IN ALL CASES THIS (THESE)        LEVO2420
C     SUBROUTINE(S) DEFINES (DIMENSIONS) THE ARRAYS REQUIRED        LEVO2430
C     TO SPECIFY THE MATRIX (MATRICES IN CASE (4)) THAT WILL BE     LEVO2440
C     USED BY THE LANCZS SUBROUTINE.  IN CASES (1) AND (2)          LEVO2450
C     THIS IS THE A-MATRIX; IN CASE (3) THIS IS THE FACTORIZATION   LEVO2460
C     OF A SCALED, SHIFTED AND PERMUTED VERSION OF THE              LEVO2470
C     USER-SPECIFIED A-MATRIX.  IN CASE (4) THE A-MATRIX            LEVO2480
C     IS SPECIFIED AS WELL AS THE FACTORIZATION OF THE              LEVO2490
C     B-MATRIX.  THIS SUBROUTINE ALSO INITIALIZES THE ARRAYS        LEVO2500
C     AND ANY OTHER PARAMETERS NEEDED TO DEFINE THE MATRIX          LEVO2510
C     (MATRICES).  THE STORAGE LOCATIONS OF THESE PARAMETERS        LEVO2520
C     AND ARRAYS ARE THEN PASSED TO THE MATRIX-VECTOR MULTIPLY      LEVO2530
C     SUBROUTINE CMATV IN CASES (1) AND (2), TO THE SUBROUTINE      LEVO2540
C     BSOLV IN CASE (3), AND TO THE SUBROUTINES AMATV              LEVO2550
C     AND LSOLV IN CASE (4) VIA ENTRY CALLS.  IN CASES (3) AND (4)  LEVO2560
C     WHENEVER A MATRIX HAS BEEN PERMUTED, THERE IS ALSO AN         LEVO2570
C     ENTRY INTO THE SUBROUTINE LPERM TO PASS THE LOCATIONS OF      LEVO2580
C     THE PERMUTATIONS IPR AND IPRT USED.  SAMPLE USPECS, CMATV,    LEVO2590
C     AMATV, BSOLV AND LSOLV SUBROUTINES ARE INCLUDED              LEVO2600
C     IN THE RELEVANT FILES.  THESE SAMPLE PROGRAMS ASSUME THAT     LEVO2610
C     THE USER-SUPPLIED A-MATRIX IS STORED ON FILE 8 IN CASES (1),  LEVO2620
C     (2), AND (4), AND THAT THE FACTORIZATION OF THE B-MATRIX      LEVO2630
C     IS ON FILE 7 IN CASES (3) AND (4).  THE USER SHOULD SEE       LEVO2640
C     THE INDIVIDUAL SAMPLE SUBROUTINES FOR MORE DETAILS.          LEVO2650
```

```
C                                                                      LEVO2660
C       IN CASES (1) AND (2):                                          LEVO2670
C       SUBROUTINE CMATV HAS THE CALLING SEQUENCE                      LEVO2680
C                                                                      LEVO2690
C            CALL CMATV(W,U,SUM)                                       LEVO2700
C                                                                      LEVO2710
C       IN THE REAL SYMMETRIC CASE, U AND W ARE REAL*8 VECTORS         LEVO2720
C       AND SUM IS A REAL*8 SCALAR.  IN THE HERMITIAN CASE, U          LEVO2730
C       AND W ARE COMPLEX*16 VECTORS AND SUM IS A REAL*8 SCALAR.       LEVO2740
C       CMATV CALCULATES U = A*W - SUM*U FOR THE USER-SPECIFIED        LEVO2750
C       MATRIX A. ONE OF THE SAMPLE CMATV SUBROUTINES INCLUDED         LEVO2760
C       COMPUTES MATRIX-VECTOR MULTIPLIES FOR AN ARBITRARY SPARSE,     LEVO2770
C       SYMMETRIC MATRIX STORED IN THE SPARSE FORMAT SPECIFIED IN THE  LEVO2780
C       CORRESPONDING SAMPLE USPEC SUBROUTINE.  FOR CASES (1) AND      LEVO2790
C       (2) CMATV IS THE SUBROUTINE USED BY THE LANCZS SUBROUTINE      LEVO2800
C       THAT GENERATES THE T-MATRICES.  IN CASE (4) SUBROUTINE         LEVO2810
C       AMATV HAS THE SAME CALLING SEQUENCE AS CMATV IN CASE (1).      LEVO2820
C                                                                      LEVO2830
C       IN CASES (3) AND (4):                                          LEVO2840
C       ALPHA/BETA HISTORY IS GENERATED USING SPARSE MATRIX INVERSION. LEVO2850
C       IN CASE (3), AT EACH ITERATION OF THE LANCZOS RECURSION        LEVO2860
C       GIVEN A FACTORIZATION OF THE MATRIX BEING USED, THE            LEVO2870
C       SUBROUTINE BSOLV FOR A GIVEN V, COMPUTES U SUCH THAT B*U = V.  LEVO2880
C       THE CALLING SEQUENCE OF BSOLV IS                              LEVO2890
C                                                                      LEVO2900
C            CALL BSOLV(V,U,IBSOLV)                                    LEVO2910
C                                                                      LEVO2920
C       WHEN IBSOLV = 2,  U = (B-INVERSE)*V IS RETURNED.  IN CASE (4), LEVO2930
C       AT EACH ITERATION OF THE GENERALIZED LANCZOS RECURSION BOTH THE LEVO2940
C       SUBROUTINE AMATV AND THE SUBROUTINE LSOLV ARE USED.  THE       LEVO2950
C       CALLING SEQUENCE OF LSOLV IS                                  LEVO2960
C                                                                      LEVO2970
C            CALL LSOLV(V,U,ISOLV)                                    LEVO2980
C                                                                      LEVO2990
C       WHERE U AND V ARE REAL*8 VECTORS.  LSOLV PERFORMS 4 FUNCTIONS. LEVO3000
C       LET L DENOTE THE CHOLESKY FACTOR OF THE B-MATRIX USED IN LANCZS. LEVO3010
C       WHEN ISOLV = 1, LSOLV COMPUTES U = L*V.  WHEN ISOLV = 2,       LEVO3020
C       LSOLV COMPUTES U = (L-TRANSPOSE)*V.  WHEN ISOLV = 3, LSOLV     LEVO3030
C       COMPUTES U = (L-INVERSE)*V.  WHEN ISOLV = 4, LSOLV            LEVO3040
C       COMPUTES   U = ((L-TRANSPOSE)-INVERSE)*V.                      LEVO3050
C                                                                      LEVO3060
C       SAMPLE PROGRAMS ASSUME THAT THE A-MATRIX (CASES (1),(2),(4))   LEVO3070
C       IS ON FILE 8 AND STORED IN THE FOLLOWING SPARSE FORMAT:        LEVO3080
C       ICOL(K), K = 1,NZL, NUMBER OF SUBDIAGONAL NONZEROS IN COLUMN K. LEVO3090
C       IROW(K), K = 1,NZS, ROW INDEX OF ASD(K).                      LEVO3100
C       AD(K), K=1,N, CONTAINS THE DIAGONAL ELEMENTS OF THE A-MATRIX.  LEVO3110
C       ASD(K), K=1,NZS  CONTAINS THE SUBDIAGONAL ELEMENTS OF A BY COLUMN. LEVO3120
C       NZS = NUMBER OF NONZERO ELEMENTS BELOW THE DIAGONAL OF A       LEVO3130
C       NZL = INDEX OF LAST COLUMN WITH NONZERO SUBDIAGONAL ENTRIES    LEVO3140
C       N = ORDER OF THE A-MATRIX.                                    LEVO3150
C                                                                      LEVO3160
C       NOTE THAT THE OPTIONAL PREPROCESSING PROGRAMS PERMUT AND       LEVO3170
C       LORDER ASSUME THAT THE GIVEN MATRIX IS ON FILE 8.  CASES (3)   LEVO3180
C       AND (4) ASSUME THAT THE SPARSE FACTORIZATION OF B IS STORED ON LEVO3190
C       FILE 7.  THE SAMPLE BSOLV SUBROUTINE SUPPLIED ASSUMES          LEVO3200
C       THAT THE B-MATRIX IS POSITIVE DEFINITE AND THAT ITS CHOLESKY   LEVO3210
C       FACTOR IS PROVIDED ON FILE 7, STORED IN SPARSE FORMAT IN       LEVO3220
C       ARRAYS BD AND BSD.  THE USER CAN EASILY REPLACE THIS SAMPLE    LEVO3230
C       BSOLV SUBROUTINE AND THE CORRESPONDING SAMPLE USPEC            LEVO3240
C       SUBROUTINE BY SUBROUTINES THAT DEFINE AND USE A GENERAL        LEVO3250
C       FACTORIZATION L*D*(L-TRANSPOSE).                              LEVO3260
C                                                                      LEVO3270
C       THE SAMPLE USPEC, CMATV (CASES (1) AND (2)), AMATV (CASE (4)), LEVO3280
C       BSOLV (CASE (3)), AND LSOLV (CASE(4)) MUST BE MODIFIED BY      LEVO3290
C       THE USER TO ACCOMODATE THE USER-SPECIFIED MATRIX OR MATRICES.  LEVO3300
C                                                                      LEVO3310
C                                                                      LEVO3320
```

```
C-----MACHEP-----------------------------------------------------------LEV03330
C                                                                       LEV03340
C                                                                       LEV03350
C       MACHEP IS A MACHINE DEPENDENT PARAMETER SPECIFYING THE RELATIVE  LEV03360
C       PRECISION OF THE FLOATING POINT ARITHMETIC USED.                LEV03370
C       MACHEP = 2.2 ± 10**-16 FOR DOUBLE PRECISION ARITHMETIC ON       LEV03380
C       IBM 370-3081.                                                   LEV03390
C                                                                       LEV03400
C       THE USER WILL HAVE TO RESET THIS PARAMETER TO                   LEV03410
C       THE CORRESPONDING VALUE FOR THE MACHINE BEING USED.  NOTE THAT  LEV03420
C       IF A MACHINE WITH A MACHINE EPSILON THAT IS MUCH LARGER THAN THE LEV03430
C       VALUE GIVEN HERE IS BEING USED, THEN THERE COULD BE             LEV03440
C       PROBLEMS WITH THE TOLERANCES.                                   LEV03450
C                                                                       LEV03460
C                                                                       LEV03470
C-----SUBROUTINES AND FUNCTIONS USER MUST SUPPLY------------------------LEV03480
C                                                                       LEV03490
C                                                                       LEV03500
C       GENRAN, FINPRO, MASK, USPEC, AND                               LEV03510
C       CASES (1) AND (2), CMATV:  CASE (3), BSOLV:                     LEV03520
C       CASE (4),  AMATV AND LSOLV.                                     LEV03530
C                                                                       LEV03540
C       GENRAN = COMPUTES K PSEUDO-RANDOM NUMBERS AND STORES THEM IN    LEV03550
C                THE REAL*4 ARRAY, G.  THIS SUBROUTINE IS USED TO       LEV03560
C                GENERATE A STARTING VECTOR FOR THE LANCZOS PROCEDURE   LEV03570
C                IN THE SUBROUTINE LANCZS AND A STARTING RIGHT-HAND SIDE LEV03580
C                FOR INVERSE ITERATION IN THE SUBROUTINE INVERR.        LEV03590
C                                                                       LEV03600
C                TESTS REPORTED IN THE REFERENCES USED EITHER GGL1 OR   LEV03610
C                GGL2 FROM THE IBM LIBRARY SLMATH.                      LEV03620
C                THE EXISTING CALLING SEQUENCE IS:                      LEV03630
C                                                                       LEV03640
C                       CALL GENRAN(IIX,G,K).                           LEV03650
C                                                                       LEV03660
C                WHERE IIX =INTEGER SEED, G = REAL*4 ARRAY WHOSE        LEV03670
C                DIMENSION MUST BE >= K.  K RANDOM NUMBERS ARE GENERATED LEV03680
C                AND PLACED IN G.                                       LEV03690
C                                                                       LEV03700
C       FINPRO = DOUBLE PRECISION FUNCTION WHICH COMPUTES THE INNER     LEV03710
C                PRODUCT OF 2 DOUBLE PRECISION VECTORS OF DIMENSION  N. LEV03720
C                TESTS REPORTED IN THE REFERENCES USED THE HARWELL      LEV03730
C                LIBRARY SUBROUTINE FM02AD.                             LEV03740
C                EXISTING CALLING SEQUENCE IS                           LEV03750
C                                                                       LEV03760
C                       CALL FINPRO(N,V,J,W,K).                         LEV03770
C                                                                       LEV03780
C                COMPUTES THE INNER PRODUCT OF DIMENSION N OF THE VECTORS LEV03790
C                V AND W.  SUCCESSIVE COMPONENTS OF V AND OF W ARE STORED LEV03800
C                AT LOCATIONS THAT ARE ,RESPECTIVELY, J AND K UNITS APART.LEV03810
C                                                                       LEV03820
C       MASK = MASKS OVERFLOW AND UNDERFLOW.                            LEV03830
C              USER MUST SUPPLY OR COMMENT OUT CALL.                    LEV03840
C                                                                       LEV03850
C       USPEC = DIMENSIONS AND INITIALIZES ARRAYS NEEDED TO SPECIFY     LEV03860
C               MATRIX THAT WILL BE USED BY LANCZS SUBROUTINE.          LEV03870
C               IN CASE (4) A-MATRIX AND B-MATRIX MUST BOTH BE SPECIFIED.LEV03880
C                                                                       LEV03890
C       CMATV = MATRIX-VECTOR MULTIPLY FOR USER-SUPPLIED MATRIX.        LEV03900
C               CASES (1) AND (2).  SEE MATRIX SPECIFICATION SECTION.   LEV03910
C                                                                       LEV03920
C       AMATV = MATRIX-VECTOR MULTIPLY FOR USER-SUPPLIED A-MATRIX.      LEV03930
C               CASES (4) ONLY.  SEE MATRIX SPECIFICATION SECTION.      LEV03940
C                                                                       LEV03950
C       BSOLV = GIVEN A VECTOR V COMPUTES U SUCH THAT B*U = V,          LEV03960
C               USING THE FACTORIZATION OF B.  USED IN CASE (3) ONLY.   LEV03970
C               SEE MATRIX SPECIFICATION SECTION.                       LEV03980
C                                                                       LEV03990
```

```
C    LSOLV = PERFORMS 4 FUNCTIONS.  GIVEN A VECTOR V COMPUTES        LEV04000
C            U = L*V, U = (L-TRANSPOSE)*V, U = (L-INVERSE)*V OR       LEV04010
C            U = (L-TRANSPOSE)-INVERSE*V, USING THE CHOLESKY          LEV04020
C            FACTORS OF B.  USED ONLY IN CASE (4).  SEE MATRIX        LEV04030
C            SPECIFICATION SECTION.                                   LEV04040
C                                                                     LEV04050
C                                                                     LEV04060
C---------------------------------------------------------------------LEV04070
C                                                                     LEV04080
C         COMMENTS FOR EIGENVALUE COMPUTATIONS                        LEV04090
C                                                                     LEV04100
C---------------------------------------------------------------------LEV04110
C                                                                     LEV04120
C                                                                     LEV04130
C-----PARAMETER CONTROLS FOR EIGENVALUE PROGRAMS----------------------LEV04140
C                                                                     LEV04150
C                                                                     LEV04160
C    PARAMETER CONTROLS ARE INTRODUCED TO ALLOW SEGMENTATION OF THE   LEV04170
C    EIGENVALUE COMPUTATIONS AND TO ALLOW VARIOUS COMBINATIONS OF     LEV04180
C    READ/WRITES.                                                     LEV04190
C                                                                     LEV04200
C    THE FLAG ISTART CONTROLS THE T-MATRIX (ALPHA/BETA HISTORY)       LEV04210
C    GENERATION.                                                      LEV04220
C                                                                     LEV04230
C    ISTART = (0,1)  MEANS                                            LEV04240
C                                                                     LEV04250
C           (0) THERE IS NO EXISTING ALPHA/BETA HISTORY AND ONE       LEV04260
C               MUST BE GENERATED.                                    LEV04270
C                                                                     LEV04280
C           (1) THERE IS AN EXISTING ALPHA/BETA HISTORY AND IT IS     LEV04290
C               TO BE READ IN FROM FILE 2 AND EXTENDED IF NECESSARY.  LEV04300
C                                                                     LEV04310
C    THE FLAG ISTOP CAN BE USED IN CONJUNCTION WITH THE FLAG ISTART TO LEV04320
C    ALLOW SEGMENTATION OF THE EIGENVALUE COMPUTATIONS.              LEV04330
C                                                                     LEV04340
C    ISTOP  = (0,1)  MEANS                                            LEV04350
C                                                                     LEV04360
C           (0) PROGRAM COMPUTES ONLY THE REQUESTED ALPHAS/BETAS,     LEV04370
C               STORES THEM AND THE LAST 2 LANCZOS VECTORS GENERATED  LEV04380
C               IN FILE 1 AND THEN TERMINATES.  IN CASE (4) THERE     LEV04390
C               ARE ACTUALLY 3 VECTORS TO BE SAVED.                   LEV04400
C                                                                     LEV04410
C           (1) PROGRAM COMPUTES REQUESTED ALPHAS/BETAS AND THEN      LEV04420
C               USES THE BISEC SUBROUTINE TO CALCULATE EIGENVALUES    LEV04430
C               OF THE TRIDIAGONAL MATRICES GENERATED FOR THE ORDERS  LEV04440
C               SPECIFIED BY THE USER AND ON THE USER-SPECIFIED       LEV04450
C               INTERVALS.  PROGRAM THEN USES THE SUBROUTINE INVERR   LEV04460
C               TO COMPUTE ERROR ESTIMATES FOR THE ISOLATED GOOD      LEV04470
C               T-EIGENVALUES WHICH ARE USED TO CHECK THE            LEV04480
C               CONVERGENCE OF THESE T-EIGENVALUES.                   LEV04490
C                                                                     LEV04500
C    CONTROL PARAMETERS FOR WRITES                                    LEV04510
C                                                                     LEV04520
C    IHIS  = (0,1)  MEANS                                             LEV04530
C                                                                     LEV04540
C           (0) IF ISTOP .GT. 0 THEN ALPHA/BETAS ARE NOT SAVED ON     LEV04550
C               FILE 1.                                               LEV04560
C                                                                     LEV04570
C           (1) PROGRAM WRITES ALPHAS/BETAS AND LAST 2 LANCZOS        LEV04580
C               VECTORS TO FILE 1 SO THAT THE T-MATRIX GENERATION     LEV04590
C               MAY BE REUSED OR CONTINUED LATER IF NECESSARY.        LEV04600
C               TYPICALLY ONE WOULD ALWAYS DO THIS ON ANY RUN WHERE   LEV04610
C               A HISTORY FILE IS BEING GENERATED.  HISTORY MUST BE   LEV04620
C               SAVED IN MACHINE FORMAT ((4Z20) FOR IBM 3081) SO      LEV04630
C               THAT NO ERRORS ARE INTRODUCED BY FORMAT CONVERSIONS.  LEV04640
C                                                                     LEV04650
C    IDIST = (0,1)  MEANS                                             LEV04660
```

```
C                                                                    LEV04670
C                 (0) DISTINCT EIGENVALUES OF T-MATRICES ARE NOT SAVED.  LEV04680
C                                                                    LEV04690
C                 (1) PROGRAM WRITES COMPUTED DISTINCT EIGENVALUES OF  LEV04700
C                     T-MATRICES ALONG WITH THEIR T-MULTIPLICITIES   LEV04710
C                     TO FILE 11.                                    LEV04720
C                                                                    LEV04730
C    IWRITE = (0,1)  MEANS                                           LEV04740
C                                                                    LEV04750
C                 (0) NO EXTENDED OUTPUT FROM SUBROUTINES BISEC AND INVERR LEV04760
C                     IS SENT TO FILE 6.                             LEV04770
C                                                                    LEV04780
C                 (1) INDIVIDUAL COMPUTED T-EIGENVALUES AND CORRESPONDING  LEV04790
C                     ERROR ESTIMATES FROM THE SUBROUTINES BISEC AND INVERRLEV04800
C                     ARE PRINTED OUT TO FILE 6 AS THEY ARE COMPUTED.  LEV04810
C                                                                    LEV04820
C    THE PROGRAM ALWAYS MAKES A SEPARATE LIST OF THE COMPUTED GOOD   LEV04830
C    T-EIGENVALUES ALONG WITH THEIR MINIMAL GAPS AND WRITES THEM OUT LEV04840
C    TO FILE 3.  CORRESPONDING ERROR ESTIMATES FOR ANY ISOLATED      LEV04850
C    GOOD T-EIGENVALUES ARE ALWAYS WRITTEN TO FILE 4.                LEV04860
C                                                                    LEV04870
C                                                                    LEV04880
C-----INPUT/OUTPUT FILES FOR EIGENVALUE PROGRAMS----------------------LEV04890
C                                                                    LEV04900
C    ANY INPUT DATA OTHER THAN THE ALPHA/BETA HISTORY SHOULD BE STORED LEV04910
C    ON FILE 5. SEE SAMPLE INPUT/OUTPUT FROM TYPICAL RUN.            LEV04920
C    THE READ STATEMENTS IN THE GIVEN FORTRAN PROGRAM ASSUME THAT    LEV04930
C    THE DATA STORED ON FILE 5 IS IN FREE FORMAT.  USER SHOULD NOTE  LEV04940
C    THAT 'FREE FORMAT' IS NOT CLASSIFIED AS PORTABLE BY PFORT SO THAT LEV04950
C    THE USER MAY HAVE TO MODIFY THE READ STATEMENTS FROM FILE 5 TO  LEV04960
C    CONFORM TO WHAT IS PERMISSIBLE ON THE MACHINE BEING USED.       LEV04970
C                                                                    LEV04980
C    FILE 6 WAS USED AS THE INTERACTIVE TERMINAL OUTPUT FILE.        LEV04990
C    THIS FILE PROVIDES A RUNNING ACCOUNT OF THE PROGRESS OF THE     LEV05000
C    COMPUTATIONS.  THE AMOUNT OF INFORMATION PRINTED OUT IS         LEV05010
C    CONTROLLED BY THE PARAMETER IWRITE.                             LEV05020
C                                                                    LEV05030
C DESCRIPTION OF OTHER I/O FILES                                     LEV05040
C                                                                    LEV05050
C FILE (K)    CONTAINS:                                              LEV05060
C                                                                    LEV05070
C    (1)      OUTPUT FILE:                                           LEV05080
C             HISTORY FILE OF NEWLY-GENERATED T-MATRIX (ALPHA AND    LEV05090
C             BETA VECTORS) AND LAST 2 LANCZOS VECTORS USED          LEV05100
C             IN THE T-MATRIX GENERATION.  NOTE THAT IN CASE (4)     LEV05110
C             THREE 'LANCZOS' VECTORS ARE WRITTEN TO FILE 1.         LEV05120
C             IF IHIS = 0 AND ISTOP = 1, FILE 1 IS NOT WRITTEN.      LEV05130
C                                                                    LEV05140
C    (2)      INPUT FILE:                                            LEV05150
C             SAME AS FILE 1 EXCEPT THAT IT CONTAINS A               LEV05160
C             PREVIOUSLY-GENERATED T-MATRIX (IF ANY).  IF ISTART = 1, LEV05170
C             PROGRAM ASSUMES THAT THERE IS A HISTORY FILE OF ALPHAS LEV05180
C             AND BETAS ON FILE 2.  THESE ALPHAS AND BETAS ARE       LEV05190
C             READ IN ALONG WITH THE LAST 2 LANCZOS VECTORS THAT     LEV05200
C             WERE GENERATED.  IN CASE (4) THREE 'LANCZOS' VECTORS   LEV05210
C             ARE READ IN FROM FILE 2.                               LEV05220
C                                                                    LEV05230
C    (3)      OUTPUT FILE:                                           LEV05240
C             COMPUTED GOOD EIGENVALUES OF THE T-MATRICES USED. ALSO LEV05250
C             CONTAINS T-MULTIPLICITIES OF THESE EIGENVALUES AS      LEV05260
C             EIGENVALUES OF THE T-MATRIX, AND THEIR GAPS AS         LEV05270
C             EIGENVALUES IN THE A-MATRIX AND IN THE T-MATRIX.       LEV05280
C             FILE 3 IS ALWAYS WRITTEN.  IN CASE (3) THIS OUTPUT     LEV05290
C             CONTAINS THE EIGENVALUES OF THE B-INVERSE MATRIX       LEV05300
C             SINCE IN THIS CASE THE T-MATRICES CORRESPOND TO        LEV05310
C             THE B-INVERSE MATRIX AND NOT TO THE A-MATRIX.  IN      LEV05320
C             THIS CASE, 3 SETS OF GAPS ARE GIVEN, THOSE IN          LEV05330
```

```
C                    THE T-MATRIX, IN THE B-INVERSE MATRIX AND THOSE          LEV05340
C                    FOR THE CORRESPONDING EIGENVALUES IN THE A-MATRIX.        LEV05350
C                                                                             LEV05360
C      (4)    OUTPUT FILE:                                                    LEV05370
C                    ERROR ESTIMATES FOR THE ISOLATED GOOD T-EIGENVALUES      LEV05380
C                    WHICH ARE OBTAINED USING THE SUBROUTINE INVERR. THESE    LEV05390
C                    ESITMATES USE THE LAST COMPONENTS OF THE ASSOCIATED      LEV05400
C                    T-EIGENVECTORS WHICH ARE COMPUTED USING INVERSE          LEV05410
C                    ITERATION.  FILE 4 IS ALWAYS WRITTEN.                    LEV05420
C                                                                             LEV05430
C                                                                             LEV05440
C      (7)    INPUT FILE:                                                     LEV05450
C                    USED ONLY IN CASES (3) AND (4), FACTORED INVERSES        LEV05460
C                    OF REAL SYMMETRIC MATRICES AND GENERALIZED EIGENVALUE    LEV05470
C                    PROBLEM.  CONTAINS THE REQUIRED FACTORIZATION OF THE     LEV05480
C                    B-MATRIX.                                                LEV05490
C                                                                             LEV05500
C      (8)    INPUT FILE:                                                     LEV05510
C                    SAMPLE USPEC SUBROUTINE ASSUMES THAT THE ARRAYS          LEV05520
C                    REQUIRED TO SPECIFY THE USER'S-MATRIX ARE STORED ON      LEV05530
C                    FILE 8.  USERS MUST MAKE WHATEVER DEFINITIONS ARE        LEV05540
C                    APPROPRIATE FOR THEIR MATRICES.  NOTE THAT IN CASE       LEV05550
C                    (3) THE LANCZS SUBROUTINE DOES NOT USE THE MATRIX        LEV05560
C                    ON FILE 8 IN THE T-MATRIX GENERATION, RATHER IT          LEV05570
C                    USES THE FACTORIZATION OF AN ASSOC.ATED                  LEV05580
C                    B-MATRIX WHICH IS STORED ON FILE 7.  IN CASE (4),        LEV05590
C                    THE INFORMATION STORED ON BOTH FILES 7 AND 8 IS USED.    LEV05600
C                                                                             LEV05610
C      (9)    INPUT AND OUTPUT FILE:                                          LEV05620
C                    CAN BE USED TO STORE THE TRUE EIGENVALUES OF THE         LEV05630
C                    GIVEN PROBLEM,  WHEN THE PROCEDURE                       LEV05640
C                    IS BEING EXERCISED ON A TEST MATRIX.                     LEV05650
C                                                                             LEV05660
C      (11)   OUTPUT FILE:                                                    LEV05670
C                    COMPUTED DISTINCT EIGENVALUES OF T-MATRICES USED.        LEV05680
C                    ALSO CONTAINS THEIR T-MULTIPLICITIES AND T-GAPS TO       LEV05690
C                    NEAREST DISTINCT EIGENVALUES, AND THE T-MULTIPLICITY     LEV05700
C                    PATTERN OF THE GOOD AND THE SPURIOUS T-EIGENVALUES.      LEV05710
C                    FILE 11 IS WRITTEN ONLY IF IDIST = 1.                    LEV05720
C                                                                             LEV05730
C                                                                             LEV05740
C-----PARAMETERS SET BY THE EIGENVALUE PROGRAMS----------------------- LEV05750
C                                                                             LEV05760
C                                                                             LEV05770
C      THESE PARAMETERS ARE SET INTERNALLY IN THE PROGRAM                     LEV05780
C                                                                             LEV05790
C      SCALEK     K = 1,2,3,4                                                  LEV05800
C                                                                             LEV05810
C                    THE SCALING FACTORS SCALEK HAVE BEEN INTRODUCED IN AN    LEV05820
C                    ATTEMPT TO MAKE THE TOLERANCES USED IN THE               LEV05830
C                    T-MULTIPLICITY, SPURIOUS, ISOLATION AND PRTESTS ADJUST   LEV05840
C                    TO THE SCALE OF THE GIVEN MATRIX.  THESE FACTORS MUST    LEV05850
C                    NOT BE MODIFIED.  THESE TOLERANCES OCCUR IN LUMP,        LEV05860
C                    ISOEV, AND PRTEST.                                       LEV05870
C                                                                             LEV05880
C      NOTE:    THE USER SHOULD NOTE THAT IF THE MATRIX BEING                 LEV05890
C      PROCESSED IS VERY STIFF, THAT IS THE RATIO OF THE LARGEST              LEV05900
C      EIGENVALUE IN MAGNITUDE TO THE SMALLEST IN MAGNITUDE IS VERY           LEV05910
C      LARGE, THEN THE TOLERANCES BEING USED IN BISEC, LUMP, ISOEV            LEV05920
C      AND PRTEST MAY NOT TREAT THE SMALL END (SMALL IN MAGNITUDE)            LEV05930
C      VERY WELL.  IN SOME SUCH CASES A USER-INTRODUCED REDUCTION             LEV05940
C      IN THE SIZE OF TKMAX AND THE SUBSEQUENT RECOMPUTATION OF               LEV05950
C      THE T-MATRIX EIGENVALUES IN (ONLY) THE LOWER END OF THE                LEV05960
C      SPECTRUM WITH THIS TKMAX MAY RESULT IN IMPROVED COMPUTATIONS           LEV05970
C      AT THE LOW END.                                                        LEV05980
C                                                                             LEV05990
C      THE LUMP, ISOEV, AND PRTEST TOLERANCES THAT WERE USED                  LEV06000
```

22

```
C      MOST IN  THE TESTING OF THESE ALGORITHMS WERE NOT            LEV06010
C      SCALE INVARIANT BUT SEEMED TO WORK WELL ON MATRICES THAT     LEV06020
C      HAD EIGENVALUES WITH MAGNITUDES BOTH GREATER THAN AND LESS   LEV06030
C      THAN 1.  THESE TOLERANCES ARE ALSO INCLUDED IN THESE THREE   LEV06040
C      SUBROUTINES BUT AS COMMENTED OUT STATEMENTS.  THEY CAN BE    LEV06050
C      REVIVED BY COMMENTING OUT THE CORRESPONDING TOLERANCES       LEV06060
C      SPECIFIED IN THE STATEMENT ABOVE EACH OF THESE.              LEV06070
C                                                                   LEV06080
C      IMPORTANT TOLERANCES OR SCALES THAT ARE USED REPEATEDLY      LEV06090
C      THROUGHOUT THE LANCZOS EIGENVALUE PROGRAMS ARE THE FOLLOWING: LEV06100
C      SCALED MACHINE EPSILON:  TTOL = TKMAX*EPSM WHERE             LEV06110
C      EPSM = 2*MACHINE EPSILON AND                                 LEV06120
C      TKMAX = MAX(|ALPHA(J)|,BETA(J), J = 1,MEV)                   LEV06130
C      BISEC CONVERGENCE TOLERANCE:  BISTOL = DSQRT(1000+MEV)*TTOL  LEV06140
C      BISEC T-MULTIPLICITY TOLERANCE:  MULTOL = (1000+MEV)*TTOL    LEV06150
C      LANCZOS CONVERGENCE TOLERANCE:   CONTOL = BETA(MEV+1)*1.D-10 LEV06160
C                                                                   LEV06170
C                                                                   LEV06180
C      BTOL = RELATIVE TOLERANCE USED TO ESTIMATE ANY LOSS OF LOCAL LEV06190
C             ORTHOGONALITY OF THE LANCZOS VECTORS AFTER THE T-MATRIX LEV06200
C             HAS BEEN GENERATED.  THE LANCZOS PROCEDURE WORKS WELL LEV06210
C             ONLY IF LOCAL ORTHOGONALITY BETWEEN SUCCESSIVE LANCZOS LEV06220
C             VECTORS IS MAINTAINED.  THE TNORM SUBROUTINE TESTS    LEV06230
C             WHETHER OR NOT                                        LEV06240
C                                                                   LEV06250
C                  MINIMUM  |BETA(I)|/||A|| > BTOL.                 LEV06260
C                  I=2,KMAX                                         LEV06270
C                                                                   LEV06280
C             IF THIS TEST IS VIOLATED BY SOME BETA AND A T-MATRIX THAT LEV06290
C             WOULD INCLUDE SUCH A BETA IS REQUESTED, THEN THE LANCZOS LEV06300
C             PROCEDURE WILL TERMINATE FOR THE USER TO DECIDE WHAT TO LEV06310
C             DO.  THE USER CAN OVER-RIDE THIS TEST BY SIMPLY DECREASING LEV06320
C             THE SIZE OF BTOL, BUT THEN CONVERGENCE IS NOT AS CERTAIN. LEV06330
C             THE PROGRAM SETS BTOL = 1.D-8 WHICH IS A VERY CONSERVATIVE LEV06340
C             CHOICE. THE || A || IS ESTIMATED BY USING AN ESTIMATE LEV06350
C             OF THE NORM OF THE T-MATRIX, T(1,KMAX).               LEV06360
C                                                                   LEV06370
C      GAPTOL = RELATIVE TOLERANCE USED IN THE SUBROUTINE ISOEV     LEV06380
C               TO DETERMINE WHICH OF THE GOOD T-EIGENVALUES NEED   LEV06390
C               ERROR ESTIMATES.  THE PROGRAM SETS GAPTOL = 1.D-8.  LEV06400
C               IF FOR A GIVEN 'GOOD' T-EIGENVALUE THE COMPUTED GAP LEV06410
C               IS TOO SMALL AND IS DUE TO A 'SPURIOUS' T-EIGENVALUE LEV06420
C               THEN THE 'GOOD' T-EIGENVALUE IS ASSUMED TO HAVE CONVERGEDLEV06430
C               AND NO ERROR ESTIMATES ARE COMPUTED.               LEV06440
C                                                                   LEV06450
C                                                                   LEV06460
C-----USER-SPECIFIED PARAMETERS FOR EIGENVALUE PROGRAMS------------------LEV06470
C                                                                   LEV06480
C                                                                   LEV06490
C      RELTOL = RELATIVE TOLERANCE USED IN 'COMBINING' COMPUTED     LEV06500
C               EIGENVALUES OF T(1,MEV) PRIOR TO COMPUTING ERROR    LEV06510
C               ESTIMATES.                                          LEV06520
C                                                                   LEV06530
C      THE LUMPING OF T-EIGENVALUES OCCURS IN SUBROUTINE LUMP.      LEV06540
C      LUMPING IS NECESSARY BECAUSE IT IS IMPOSSIBLE TO ACCURATELY  LEV06550
C      PREDICT THE ACCURACY OF THE BISEC SUBROUTINE.  LUMP 'COMBINES' LEV06560
C      T-EIGENVALUES THAT HAVE SLIPPED BY THE TOLERANCE THAT WAS USED LEV06570
C      IN THE T-MULTIPLICITY TESTS.  IN PARTICULAR IF FOR SOME J,   LEV06580
C                                                                   LEV06590
C      |EVALUE(J)-EVALUE(J-1)| < DMAX1(RELTOL*|EVALUE(J)|,SCALE2*MULTOL) LEV06600
C                                                                   LEV06610
C      THEN THESE T-EIGENVALUES ARE 'COMBINED'.  MULTOL IS THE TOLERANCE LEV06620
C      THAT WAS USED IN THE T-MULTIPLICITY TEST IN BISEC.  SEE THE HEADERLEV06630
C      ON THE LUMP SUBROUTINE FOR MORE DETAILS.                     LEV06640
C                                                                   LEV06650
C      RELTOL IS SET TO 1.D-10.                                     LEV06660
C                                                                   LEV06670
```

```
C       MXINIT = MAXIMUM NUMBER OF INVERSE ITERATIONS ALLOWED IN        LEVO6680
C                SUBROUTINE INVERR FOR EACH ISOLATED GOOD T-EIGENVALUE.  LEVO6690
C                TYPICALLY ONLY ONE ITERATION IS REQUIRED.              LEVO6700
C                                                                       LEVO6710
C       SEEDS FOR RANDOM NUMBER GENERATORS = INTEGER*4 SCALARS.         LEVO6720
C                                                                       LEVO6730
C                 (1) SVSEED = SEED FOR STARTING VECTOR USED IN         LEVO6740
C                     T-MATRIX GENERATION IN LANCZS SUBROUTINE          LEVO6750
C                                                                       LEVO6760
C                 (2) RHSEED = SEED FOR RIGHT-HAND SIDE USED IN         LEVO6770
C                     INVERSE ITERATION COMPUTATIONS IN INVERR.         LEVO6780
C                                                                       LEVO6790
C       BISEC DATA                                                      LEVO6800
C                                                                       LEVO6810
C       (1) NINT  =  NUMBER OF SUBINTERVALS ON WHICH EIGENVALUES ARE    LEVO6820
C                    TO BE COMPUTED.                                    LEVO6830
C                                                                       LEVO6840
C       (2) LB(J) = (J = 1,NINT)  =  LEFT END POINTS OF THESE INTERVALS. LEVO6850
C                   MUST BE PROVIDED IN INCREASING ORDER.  THAT IS,     LEVO6860
C                   LB(J) < LB(J+1) FOR J = 1,NINT.                     LEVO6870
C                                                                       LEVO6880
C       (3) UB(J) = (J = 1,NINT)  =  RIGHT END POINTS OF THESE INTERVALS. LEVO6890
C                   MUST BE PROVIDED IN INCREASING ORDER.  THAT IS,     LEVO6900
C                   UB(J) < UB(J+1) FOR J = 1,NINT.                     LEVO6910
C                                                                       LEVO6920
C       (4) MXSTUR  =  MAXIMUM NUMBER OF STURM ITERATIONS ALLOWED FOR   LEVO6930
C                   ENTIRE SET OF EIGENVALUE CALCULATIONS OVER ALL      LEVO6940
C                   SPECIFIED SIZE T-MATRICES.  PROGRAM WILL            LEVO6950
C                   TERMINATE IF THIS LIMIT IS EXCEEDED.                LEVO6960
C                                                                       LEVO6970
C       T-MATRICES                                                      LEVO6980
C                                                                       LEVO6990
C       SIZES OF T-MATRICES                                             LEVO7000
C                                                                       LEVO7010
C            (1) KMAX= MAXIMUM ORDER FOR T-MATRIX THAT USER IS WILLING  LEVO7020
C                   TO CONSIDER.                                        LEVO7030
C                                                                       LEVO7040
C            (2) NMEVS = MAXIMUM NUMBER OF T-MATRICES THAT WILL BE      LEVO7050
C                   CONSIDERED.                                         LEVO7060
C                                                                       LEVO7070
C            (3) NMEV(J)  (J=1,NMEVS)  = SIZES OF T-MATRIX TO BE        LEVO7080
C                                   CONSIDERED SEQUENTIALLY.            LEVO7090
C                                                                       LEVO7100
C       T-MATRIX-GENERATION                                             LEVO7110
C                                                                       LEVO7120
C       USER SHOULD NOTE THAT THIS PROGRAM FIRST COMPUTES A T-MATRIX    LEVO7130
C       OF ORDER KMAX AND THEN CYCLES THROUGH THE T-MATRICES SPECIFIED  LEVO7140
C       A PRIORI BY THE USER, USING THE SUBROUTINE BISEC TO COMPUTE THE LEVO7150
C       EIGENVALUES OF THE T-MATRICES ON THE INTERVALS SPECIFIED BY     LEVO7160
C       THE USER.                                                       LEVO7170
C                                                                       LEVO7180
C       IDEALLY, ONE WOULD COMPUTE THE EIGENVALUE APPROXIMATIONS AT A   LEVO7190
C       REASONABLE SIZE T-MATRIX, LOOK AT THE ACCURACY OF THE COMPUTED  LEVO7200
C       RESULTS AND USE THAT TO DETERMINE AN APPROPRIATE                LEVO7210
C       INCREMENT FOR THE SIZE OF THE T-MATRIX BASED UPON WHAT          LEVO7220
C       HAS ALREADY CONVERGED AND UPON THE SIZES OF THE ERROR ESTIMATES LEVO7230
C       ON THOSE EIGENVALUES THAT ARE DESIRED BUT THAT HAVE NOT YET     LEVO7240
C       CONVERGED. HOWEVER, IN THE INTERESTS OF GENERALITY AND          LEVO7250
C       SIMPLICITY WE DID NOT DO THAT HERE.                            LEVO7260
C                                                                       LEVO7270
C                                                                       LEVO7280
C-----CONVERGENCE TESTS FOR THE EIGENVALUE PROGRAMS-------------------- LEVO7290
C                                                                       LEVO7300
C                                                                       LEVO7310
C       THE CONVERGENCE TEST INCORPORATED IN THIS PROGRAM IS           LEVO7320
C       BASED UPON THE ASSUMPTION THAT THOSE T-EIGENVALUES AND THEIR    LEVO7330
C       ASSOCIATED T-EIGENVECTORS WHICH CORRESPOND TO THE              LEVO7340
```

```
C      EIGENVALUES AND RITZVECTORS WHICH WE WISH TO COMPUTE            LEVO7350
C      CONVERGE AS THE T-SIZE IS INCREASED.                            LEVO7360
C                                                                      LEVO7370
C      AS CURRENTLY PROGRAMMED, CONVERGENCE IS CHECKED BY EXAMINING    LEVO7380
C      THE SIZES OF ALL OF THE COMPUTED ERROR ESTIMATES ON ALL OF THE  LEVO7390
C      INTERVALS SPECIFIED BY THE USER.  IDEALLY CONVERGENCE SHOULD     LEVO7400
C      BE CHECKED ONLY ON THOSE EIGENVALUES OF INTEREST AND            LEVO7410
C      ONCE THE EIGENVALUES ON SUB-INTERVALS OF THESE INTERVALS HAVE   LEVO7420
C      CONVERGED, ANY SUBSEQUENT EIGENVALUE COMPUTATIONS SHOULD BE     LEVO7430
C      MADE ONLY ON THE UNCONVERGED PORTIONS.  OBVIOUSLY, IT WOULD BE   LEVO7440
C      DIFFICULT TO INCORPORATE CODE TO DO THE ABOVE WITHOUT KNOWING   LEVO7450
C      A PRIORI PRECISELY WHAT THE USER IS TRYING TO COMPUTE.          LEVO7460
C      THEREFORE, WE DID NOT ATTEMPT TO DO THIS.  IF ONE WISHES TO     LEVO7470
C      MAKE SUCH A MODIFICATION THEN ONE MUST ALSO MODIFY THE PROGRAM  LEVO7480
C      SO THAT IT CREATES AN OVERALL LIST OF THE CONVERGED 'GOOD'      LEVO7490
C      T-EIGENVALUES AS THEY ARE COMPUTED, SINCE CONVERGED 'GOOD'      LEVO7500
C      T-EIGENVALUES WHICH WERE COMPUTED AT A PARTICULAR VALUE OF MEV  LEVO7510
C      WOULD NO LONGER BE RECOMPUTED AT LARGER VALUES OF MEV.          LEVO7520
C                                                                      LEVO7530
C      IF ONLY A FEW EIGENVALUES ARE TO BE COMPUTED THEN SUCH CHANGES  LEVO7540
C      WOULD NOT MAKE MUCH DIFFERENCE IN THE RUNNING TIME.             LEVO7550
C                                                                      LEVO7560
C                                                                      LEVO7570
C-----ARRAYS REQUIRED BY THE EIGENVALUE PROGRAMS------------------------LEVO7580
C                                                                      LEVO7590
C                                                                      LEVO7600
C      ALL 4 CASES                                                     LEVO7610
C                                                                      LEVO7620
C      ALPHA(J) = REAL*8 ARRAY. ITS DIMENSION MUST BE AT LEAST KMAX,   LEVO7630
C                 THE LENGTH OF THE LARGEST T-MATRIX ALLOWED.  THIS    LEVO7640
C                 ARRAY CONTAINS THE DIAGONAL ENTRIES OF THE T-MATRICES. LEVO7650
C                                                                      LEVO7660
C      BETA(J) = REAL*8 ARRAY.  ITS DIMENSION MUST BE AT LEAST KMAX+1. LEVO7670
C                THIS ARRAY CONTAINS THE SUBDIAGONAL ENTRIES OF THE    LEVO7680
C                T-MATRICES.                                           LEVO7690
C                                                                      LEVO7700
C                THE ALPHA AND BETA VECTORS ARE NOT ALTERED           LEVO7710
C                DURING THE CALCULATIONS.                              LEVO7720
C                                                                      LEVO7730
C      V1(J),V2(J),VS(J) = REAL*8 ARRAYS IN REAL SYMMETRIC CASES.      LEVO7740
C                          V1 AND V2 ARE COMPLEX*16 IN HERMITIAN CASE. LEVO7750
C                          IN CASES (1) AND (2) VS MUST BE OF          LEVO7760
C                          DIMENSION AT LEAST KMAX.  IN CASES (3) AND  LEVO7770
C                          (4) VS MUST BE OF DIMENSION AT LEAST        LEVO7780
C                          MAX(N,KMAX).  IN REAL SYMMETRIC CASES       LEVO7790
C                          V1 MUST BE OF DIMENSION AT LEAST            LEVO7800
C                          MAX(KMAX+1,N) AND V2 MUST BE OF DIMENSION   LEVO7810
C                          MAX(KMAX,N).  IN HERMITIAN CASES V1         LEVO7820
C                          MUST BE OF DIMENSION MAX(N,(KMAX+1)/2)      LEVO7830
C                          AND V2 OF DIMENSION AT LEAST MAX(N,KMAX/2). LEVO7840
C                          IN ALL CASES HOWEVER, THE ABOVE DIMENSIONS  LEVO7850
C                          FOR V2 ARE VALID ONLY IF NO MORE            LEVO7860
C                          THAN KMAX/2 EIGENVALUES OF THE GIVEN        LEVO7870
C                          T-MATRICES ARE TO BE COMPUTED IN ANY GIVEN  LEVO7880
C                          SUBINTERVAL. V2 IS USED IN THE SUBROUTINE   LEVO7890
C                          BISEC TO HOLD THE UPPER AND LOWER           LEVO7900
C                          ENDPOINTS OF THE SUBINTERVALS GENERATED     LEVO7910
C                          DURING THE BISECTIONS.  THEREFORE, ITS      LEVO7920
C                          REAL*8 DIMENSION MUST ALWAYS BE AT LEAST    LEVO7930
C                          2*Q WHERE Q IS THE MAXIMUM NUMBER OF        LEVO7940
C                          EIGENVALUES OF THE SPECIFIED T-MATRIX IN ANY LEVO7950
C                          ONE OF THE SPECIFIED INTERVALS.             LEVO7960
C                          NOTE THAT IN THE HERMITIAN CASE, V1 AND V2  LEVO7970
C                          ARE DEFINED AS COMPLEX*16 IN THE MAIN PROGRAM LEVO7980
C                          AND IN THE LANCZS SUBROUTINE BUT ARE        LEVO7990
C                          REDEFINED AS REAL*8 IN OTHER SUBROUTINES.   LEVO8000
C                                                                      LEVO8010
```

```
C     LB(J),UB(J) = REAL*8 ARRAYS. EACH MUST BE OF DIMENSION AT LEAST   LEV08020
C                   NINT, THE NUMBER OF SUBINTERVALS TO BE CONSIDERED.   LEV08030
C                   LB CONTAINS THE LEFT-END POINTS OF THE INTERVALS     LEV08040
C                   ON WHICH EIGENVALUES ARE TO BE COMPUTED.  UB         LEV08050
C                   CONTAINS THE RIGHT-END POINTS.                       LEV08060
C                                                                        LEV08070
C     EXPLAN(J) = REAL*4 ARRAY.  ITS DIMENSION IS 20.  THIS ARRAY IS     LEV08080
C                 USED TO ALLOW EXPLANATORY COMMENTS IN THE INPUT FILES. LEV08090
C                                                                        LEV08100
C     G(J) = REAL*4 ARRAY.  ITS DIMENSION MUST BE >= MAX(2*KMAX,N)       LEV08110
C            IT IS USED FOR HOLDING THE RANDOM VECTORS GENERATED,        LEV08120
C            HOLDING THE COMPUTED ERROR ESTIMATES AND THE COMPUTED       LEV08130
C            MINIMAL GAPS FOR THE GOOD T-EIGENVALUES.                    LEV08140
C                                                                        LEV08150
C     MP(J) = INTEGER*4 ARRAY.  ITS DIMENSION MUST BE AT LEAST KMAX,     LEV08160
C             THE MAXIMUM SIZE OF THE T-MATRICES ALLOWED.  IT CONTAINS   LEV08170
C             THE T-MULTIPLICITIES OF THE COMPUTED EIGENVALUES. NOTE     LEV08180
C             THAT 'SPURIOUS' T-EIGENVALUES ARE DENOTED BY A            LEV08190
C             T-MULTIPLICITY OF 0.  T-EIGENVALUES THAT THE SUBROUTINE    LEV08200
C             PRTEST HAS IDENTIFIED AS 'GOOD' BUT HIDDEN ARE IDENTIFIED  LEV08210
C             BY A T-MULTIPLICITY OF -10.                                LEV08220
C                                                                        LEV08230
C     NMEV(J) = INTEGER*4 ARRAY.  ITS DIMENSION MUST BE AT LEAST THE     LEV08240
C               NUMBER OF T-MATRICES ALLOWED.  IT CONTAINS THE ORDERS    LEV08250
C               OF THE T-MATRICES TO BE CONSIDERED.                      LEV08260
C                                                                        LEV08270
C                                                                        LEV08280
C     FOR CASE (3) ONLY:                                                 LEV08290
C     GR(J),GC(J) = REAL*8 ARRAYS. USED ONLY IN THE HERMITIAN CASE.      LEV08300
C                   GR AND GC MUST EACH BE OF DIMENSION AT LEAST N.      LEV08310
C                   BOTH ARE USED IN THE RANDOM VECTOR GENERATION.       LEV08320
C                   GR IS ALSO USED TO STORE MINIMAL GAPS BETWEEN        LEV08330
C                   'GOOD' T-EIGENVALUES.                                LEV08340
C                                                                        LEV08350
C     FOR CASES (3) AND (4) FOR THE PERMUTATION:                         LEV08360
C                                                                        LEV08370
C     IPR(J), IPT(J) = INTEGER*4 ARRAYS.  EACH OF DIMENSION AT LEAST N.  LEV08380
C                      USED TO STORE THE REORDERING OF THE GIVEN MATRIX  LEV08390
C                      OR MATRICES.                                      LEV08400
C                                                                        LEV08410
C                                                                        LEV08420
C     OTHER ARRAYS                                                       LEV08430
C                                                                        LEV08440
C     THE USER MUST SPECIFY IN THE SUBROUTINE USPEC (A OR B) WHATEVER    LEV08450
C     ARRAYS ARE REQUIRED TO DEFINE THE MATRIX OR MATRICES BEING USED.   LEV08460
C     ALSO IN CASES (3) AND (4) ONLY, WHEN WORKING WITH INVERSES         LEV08470
C     OF SPARSE SYMMETRIC MATRICES, IN THE OPTIONAL, PREPROCESSING       LEV08480
C     PROGRAMS PERMUT, LFACT, LORDER, AND LTEST IT IS NECESSARY TO       LEV08490
C     SPECIFY ADDITIONAL ARRAYS JUST FOR THESE COMPUTATIONS.  THE USER   LEV08500
C     IS REFERRED TO THOSE PROGRAMS FOR DETAILS.                         LEV08510
C                                                                        LEV08520
C                                                                        LEV08530
C-----SUBROUTINES INCLUDED-----------------------------------------------LEV08540
C                                                                        LEV08550
C                                                                        LEV08560
C     LANCZS = COMPUTES THE ALPHA/BETA HISTORY. IN CASES (1) AND (2)     LEV08570
C              REAL SYMMETRIC AND HERMITIAN MATRICES, USES SUBROUTINES   LEV08580
C              FINPRO, GENRAN AND CMATV.  IN CASE (3), INVERSES OF       LEV08590
C              REAL SYMMETRIC MATRICES, USES SUBROUTINES FINPRO,         LEV08600
C              GENRAN AND BSOLV.  IN CASE (4), GENERALIZED EIGENVALUE    LEV08610
C              PROBLEM, USES SUBROUTINES FINPRO, GENRAN, AMATV AND       LEV08620
C              LSOLV.                                                    LEV08630
C                                                                        LEV08640
C     BISEC = COMPUTES EIGENVALUES OF THE SPECIFIED T-MATRIX             LEV08650
C             USING STURM SEQUENCING, ON SEQUENCE OF INTERVALS           LEV08660
C             SPECIFIED BY THE USER.  EACH SUBINTERVAL IS TREATED        LEV08670
C             AS OPEN ON THE LEFT AND CLOSED ON THE RIGHT.               LEV08680
```

```
C          EIGENVALUES ARE COMPUTED WITH SIMULTANEOUS DETERMINATION   LEV08690
C          OF THE T-MULTIPLICITIES AND OF SPURIOUS T-EIGENVALUES.      LEV08700
C                                                                      LEV08710
C  INVERR = USES INVERSE ITERATION ON T-MATRICES TO COMPUTE ERROR      LEV08720
C          ESTIMATES ON COMPUTED GOOD T-EIGENVALUES. (USES GENRAN)     LEV08730
C                                                                      LEV08740
C  LUMP = 'COMBINES' EIGENVALUES OF T-MATRIX USING THE RELATIVE        LEV08750
C          TOLERANCE RELTOL.                                           LEV08760
C                                                                      LEV08770
C  ISOEV = CALCULATES GAPS BETWEEN DISTINCT EIGENVALUES OF T-MATRIX    LEV08780
C          AND THEN USES THESE GAPS TO LABEL THOSE 'GOOD'              LEV08790
C          T-EIGENVALUES FOR WHICH ERROR ESTIMATES ARE NOT COMPUTED.   LEV08800
C                                                                      LEV08810
C  TNORM = COMPUTES THE SCALE TKMAX USED IN DETERMINING THE            LEV08820
C          TOLERANCES FOR THE SPURIOUS, T-MULTIPLICITY AND PRTESTS.    LEV08830
C          IT ALSO CHECKS FOR LOCAL ORTHOGONALITY OF THE LANCZOS       LEV08840
C          VECTORS BY TESTING THE RELATIVE SIZE OF THE BETAS USING     LEV08850
C          THE RELATIVE TOLERANCE BTOL.                                LEV08860
C                                                                      LEV08870
C  PRTEST = LOOKS FOR GOOD T-EIGENVALUES THAT HAVE BEEN MISLABELLED    LEV08880
C          BY THE SPURIOUS TEST BECAUSE THEY HAD 'TOO SMALL' A         LEV08890
C          PROJECTION ON THE STARTING LANCZOS VECTOR.                  LEV08900
C          (LESS THAN SINGLE PRECISION)                                LEV08910
C          TESTS INDICATE THAT SUCH EIGENVALUES ARE RARE.              LEV08920
C          PRTEST SHOULD BE CALLED ONLY AFTER CONVERGENCE              LEV08930
C          HAS BEEN ESTABLISHED.                                       LEV08940
C                                                                      LEV08950
C  INVERM = USED TO COMPUTE ERROR ESTIMATES FOR ANY T-EIGENVALUES      LEV08960
C          WHICH PRTEST INDICATES MAY HAVE BEEN MISLABELLED.           LEV08970
C          SUCH T-EIGENVALUES ARE RELABELLED ONLY IF THEIR ERROR       LEV08980
C          ESTIMATES ARE SUFFICIENTLY SMALL.  PRIMARY USE OF           LEV08990
C          INVERM IS IN THE CORRESPONDING EIGENVECTOR COMPUTATIONS.    LEV09000
C                                                                      LEV09010
C  CASES (3) AND (4) ONLY, FACTORED INVERSES:                          LEV09020
C                                                                      LEV09030
C  FOR OPTIONAL, PRELIMINARY PROCESSING:                               LEV09040
C  PERMUT  (PROGRAM CALLS SPARSPAK PACKAGE) :                          LEV09050
C  USES THE NONZERO STRUCTURE OF A GIVEN MATRIX A.                     LEV09060
C  CAN BE USED TO OBTAIN A REORDERING OF A THAT PRESERVES              LEV09070
C  SPARSITY UNDER FACTORIZATION.  PERMUT CALLS                         LEV09080
C  THE SPARSPAK PROGRAMS, (A. GEORGE, J. LIU, E.NG,                    LEV09090
C  U. WATERLOO).  PERMUT ALSO TAKES THE USER-SPECIFIED MATRIX,         LE/09100
C  APPLIES THE SCALE SO AND THE SHIFT TO IT, AND THEN WRITES           LEV09110
C  OUT THE CORRESPONDING SPARSE MATRIX DATA FILE FOR THE               LEV09120
C  RESULTING MATRIX C = SO*A + SHIFT*I.  SEE THE PERMUT FORTRAN        LEV09130
C  CODE FOR DETAILS.                                                   LEV09140
C                                                                      LEV09150
C  LORDER (STAND-ALONE PROGRAM):                                       LEV09160
C  GIVEN A MATRIX C IN SPARSE FORMAT AND A PERMUTATION P,              LEV09170
C  COMPUTES THE REORDERED MATRIX B = P*C*P' AND WRITES IT              LEV09180
C  TO FILE 9 IN SPARSE FORMAT.  SEE THE LORDER FORTRAN CODE            LEV09190
C  FOR DETAILS.                                                        LEV09200
C                                                                      LEV09210
C  LFACT (STAND-ALONE PROGRAM) :                                       LEV09220
C  GIVEN A POSITIVE DEFINITE MATRIX B IN SPARSE FORMAT                 LEV09230
C  COMPUTES THE SPARSE CHOLESKY FACTOR L OF B AND WRITES IT            LEV09240
C  TO FILE 7 IN SPARSE FORMAT.  THUS, B = L*L'.                        LEV09250
C  SEE THE LFACT FORTRAN CODE FOR DETAILS.                             LEV09260
C                                                                      LEV09270
C  LTEST (CALLS 3 USER-SUPPLIED PROGRAMS CMATV, CMATS, AND BSOLV):     LEV09280
C  GIVEN THE FACTORIZATION OF A MATRIX B, LTEST COMPUTES               LEV09290
C  THE SOLUTION OF THE EQUATION B*U = B*V1 FOR A SPECIFIC RANDOMLY-    LEV09300
C  GENERATED V1, WITH AND WITHOUT ITERATIVE REFINEMENT, TO            LEV09310
C  OBTAIN A ROUGH CHECK ON THE NUMERICAL CONDITION OF THE MATRIX B.    LEV09320
C  THIS PROGRAM USES 3 SUBROUTINES CMATV, CMATS, AND BLSOLV.          LEV09330
C  SEE THE LTEST FORTRAN PROGRAM FOR DETAILS.                          LEV09340
C                                                                      LEV09350
```

```
C                                                                    LEV09360
C-----OTHER PROGRAMS PROVIDED-----------------------------------------LEV09370
C                                                                    LEV09380
C     LECOMPAC (STAND ALONE PROGRAM):                               LEV09390
C                 TRANSLATES A REAL SYMMETRIC MATRIX PROVIDED IN THE LEV09400
C                 FORMAT I, J, A(I,J) INTO THE SPARSE MATRIX         LEV09410
C                 FORMAT USED IN THE SAMPLE USPEC, CMATV, BSOLV AND  LEV09420
C                 LSOLV SUBROUTINES PROVIDED. IT ASSUMES THAT THE    LEV09430
C                 MATRIX ENTRIES ARE GIVEN EITHER COLUMN BY COLUMN OR LEV09440
C                 ROW BY ROW.  THE DATA SET CREATED IS WRITTEN TO    LEV09450
C                 FILE 8.                                            LEV09460
C                                                                    LEV09470
C                                                                    LEV09480
C-----COMMENTS ON THE STORAGE REQUIRED FOR EIGENVALUE PROGRAMS-------LEV09490
C                                                                    LEV09500
C                                                                    LEV09510
C     CASE (1), REAL SYMMETRIC MATRICES:                            LEV09520
C                                                                    LEV09530
C     THE ARRAYS IN THE REAL SYMMETRIC EIGENVALUE PROGRAM REQUIRE    LEV09540
C     APPROXIMATELY THE EQUIVALENT OF ONE REAL*8 ARRAY OF DIMENSION  LEV09550
C                                                                    LEV09560
C         3.5*KMAX + 2*MAX(KMAX,N) + .5* MAX(2*KMAX,N)               LEV09570
C                                                                    LEV09580
C     PLUS WHATEVER IS NEEDED TO GENERATE A*X FOR THE GIVEN MATRIX A. LEV09590
C     THE ARRAYS ALPHA, BETA, VS AND MP CONSUME 3.5*KMAX*8 BYTES.    LEV09600
C     THE ARRAYS V1 AND V2 CONSUME 2*MAXIMUM(KMAX,N)*8 BYTES, WITH THE LEV09610
C     QUALIFICATION STATED ABOVE WHERE V2 IS DEFINED.  THE G-ARRAY   LEV09620
C     CONSUMES .5*MAX(2*KMAX,N)*8 BYTES.                            LEV09630
C                                                                    LEV09640
C     CASE (2), HERMITIAN MATRICES:                                 LEV09650
C                                                                    LEV09660
C     THE ARRAYS IN THE HERMITIAN EIGENVALUE PROGRAMS REQUIRE        LEV09670
C     THE EQUIVALENT OF ONE REAL*8 ARRAY OF DIMENSION               LEV09680
C                                                                    LEV09690
C         3.5*KMAX + 4*MAX(KMAX/2,N) + .5*MAX(2*KMAX,N) + 2*N        LEV09700
C                                                                    LEV09710
C     PLUS WHATEVER IS NEEDED TO GENERATE A*X FOR THE GIVEN MATRIX A. LEV09720
C     THE ARRAYS ALPHA, BETA, VS, AND MP CONSUME 3.5*KMAX*8 BYTES.   LEV09730
C     THE ARRAYS V1 AND V2 CONSUME 4*MAXIMUM(KMAX/2,N)*8 BYTES, WITH THELEV09740
C     QUALIFICATION STATED ABOVE WHERE V2 IS DEFINED.  THE G-ARRAY   LEV09750
C     CONSUMES .5*MAX(2*KMAX,N)*8 BYTES.  GR REQUIRES               LEV09760
C     AND GC REQUIRE 2*N*8BYTES.                                    LEV09770
C                                                                    LEV09780
C                                                                    LEV09790
C     CASE (3), INVERSES OF REAL SYMMETRIC MATRICES:                LEV09800
C                                                                    LEV09810
C     THE ARRAYS IN THE EIGENVALUE PROGRAMS DESIGNED FOR            LEV09820
C     CASE (3), INVERSES OF REAL SYMMETRIC MATRICES USING           LEV09830
C     REORDERING AND FACTORIZATION, REQUIRE                         LEV09840
C     THE EQUIVALENT OF ONE REAL*8 ARRAY OF DIMENSION               LEV09850
C                                                                    LEV09860
C         3*KMAX + 3*MAX(KMAX,N) + .5*MAX(2*KMAX,N)                 LEV09870
C                                                                    LEV09880
C     PLUS WHATEVER IS NEEDED TO GENERATE B(INVERSE)*X  FOR THE      LEV09890
C     SCALED, SHIFTED AND PERMUTED VERSION OF A WHICH WE DENOTE      LEV09900
C     BY B.  THE ARRAYS ALPHA, BETA, MP, AND MP2 CONSUME 3*KMAX*8   LEV09910
C     BYTES.  THE ARRAYS V1, V2, AND VS CONSUME 3*MAX(KMAX,N)*8 BYTES, LEV09920
C     WITH THE QUALIFICATION STATED ABOVE WHERE V2 IS DEFINED.      LEV09930
C     THE G ARRAY CONSUMES .5*MAX(2*KMAX,N)*8 BYTES. THESE NUMBERS  LEV09940
C     DO NOT INCLUDE THE STORAGE REQUIRED BY THE PREPROCESSING PROGRAMS LEV09950
C     PERMUT, LORDER, LFACT, AND LTEST.                            LEV09960
C                                                                    LEV09970
C                                                                    LEV09980
C     A SYMMETRIC, SPARSE MATRIX OF ORDER N WITH NZS NONZERO ELEMENTS LEV09990
C     BELOW THE MAIN DIAGONAL WOULD REQUIRE THE EQUIVALENT OF ONE    LEV10000
C     REAL*8 ARRAY OF DIMENSION 1.5*(NZS + N) IF THE POINTERS USED   LEV10010
C     ARE INTEGER*4.                                               LEV10020
```

```
C                                                                      LEV10030
C     SOME OF THE ARRAY STORAGE IS NOT ESSENTIAL AND COULD BE          LEV10040
C     ELIMINATED IF STORAGE IS A PROBLEM.                              LEV10050
C     THE FOLLOWING COMMENTS APPLY DIRECTLY ONLY TO CASE (1),          LEV10060
C     THE PROGRAMS FOR REAL SYMMETRIC MATRICES, HOWEVER, SIMILAR       LEV10070
C     STATEMENTS COULD BE MADE ABOUT THE OTHER CASES.                  LEV10080
C                                                                      LEV10090
C     CASE (1), REAL SYMMETRIC PROGRAMS:                               LEV10100
C     THE G ARRAY COULD BE REMOVED IF THE USER IS WILLING TO           LEV10110
C                                                                      LEV10120
C             (1) REGENERATE THE RANDOM STARTING VECTOR IN INVERR      LEV10130
C                 FOR EACH ERROR ESTIMATE                              LEV10140
C             (2) WRITE OUT THE ERROR ESTIMATES AND VARIOUS GAPS AS    LEV10150
C                 THEY ARE GENERATED RATHER THAN STORING THEM IN G FOR LEV10160
C                 LATER PRINTOUT                                       LEV10170
C             (3) CHECK CONVERGENCE WITHIN INVERR                      LEV10180
C                                                                      LEV10190
C     CLEARLY THE INDEX VECTOR MP COULD BE AN INTEGER*2 ARRAY AS COULD LEV10200
C     THE POINTERS USED TO DEFINE THE USER'S MATRIX.                   LEV10210
C                                                                      LEV10220
C     THE USER SHOULD NOTE THAT WITH AN EIGENVALUE SUBROUTINE THAT     LEV10230
C     USES BISECTION (LIKE BISEC) IF MORE THAN 25% OF THE              LEV10240
C     EIGENVALUES ARE TO BE COMPUTED, THEN IT MAY BE MORE             LEV10250
C     ECONOMICAL TO USE THE EISPACK SUBROUTINE IMTQL1.                 LEV10260
C     (SEE MATRIX EIGENSYTEM ROUTINES-EISPACK GUIDE (2ND EDITION)      LEV10270
C     B.T. SMITH ET AL, SPRINGER-VERLAG, NEW YORK, 1976, P213.).       LEV10280
C     HOWEVER, IF THE SUBROUTINE IMTQL1 IS TO BE USED IN PLACE         LEV10290
C     OF BISEC, THEN NONTRIVIAL CHANGES IN THE LANCZOS CODE MUST BE    LEV10300
C     MADE. FOR DETAILS OF ONE SUCH IMPLEMENTATION SEE                 LEV10310
C     IBM RESEARCH REPORT 8298, COMPUTING                             LEV10320
C     EIGENVALUES OF LARGE SYMMETRIC MATRICES - AN IMPLEMENTATION OF A LEV10330
C     LANCZOS ALGORITHM WITH NO REORTHOGONALIZATION.  PART II. COMPUTER LEV10340
C     PROGRAMS., DECEMBER 1980, WHICH CONTAINS A GENERAL              LEV10350
C     LANCZOS CODE WHICH INCLUDES AN IMTQL1 OPTION OR                  LEV10360
C     PREFERABLY CONTACT THE AUTHORS.                                  LEV10370
C                                                                      LEV10380
C     THE BISEC SUBROUTINE WHICH IS INCLUDED IS A MODIFIED FORM OF     LEV10390
C     THE BISECT SUBROUTINE IN EISPACK. BISEC ASSUMES THAT THE         LEV10400
C     VECTOR V2 IS LONG ENOUGH TO HOLD BOTH THE UPPER AND THE          LEV10410
C     LOWER BOUNDS ON THE BISECTION INTERVALS USED TO COMPUTE          LEV10420
C     THE EIGENVALUES OF THE T-MATRICES. THEREFORE, IF THE             LEV10430
C     LENGTH OF V2 IS ONLY KMAX, BISEC CAN COMPUTE ONLY AT MOST        LEV10440
C     KMAX/2 EIGENVALUES OF THE GIVEN T-MATRIX IN ANY GIVEN            LEV10450
C     SUBINTERVAL.                                                     LEV10460
C                                                                      LEV10470
C     AS PROGRAMMED BISEC USES THE ARRAYS ALPHA,BETA,V1,V2,VS AND MP.  LEV10480
C     HOWEVER, V1 IS USED ONLY TO STORE BETA(J)**2 SO THAT THEY DO NOT LEV10490
C     HAVE TO BE REGENERATED ON EACH STURM. IF THE USER IS WILLING TO  LEV10500
C     COMPUTE THE BETA(J)**2 AS NEEDED, THEN V1 COULD BE ELIMINATED    LEV10510
C     FROM BISEC. BISEC STORAGE THEN BECOMES A REAL*8 ARRAY OF DIMENSION LEV10520
C     4.25*KMAX IF WE ALSO REDUCE MP TO INTEGER*2.                     LEV10530
C     IF ONE KNEW THAT ONLY Q*MEV EIGENVALUES OF T(1,MEV) WERE TO BE   LEV10540
C     COMPUTED AT EACH STAGE FOR SOME Q<.5 THEN FURTHER REDUCTIONS IN  LEV10550
C     STORAGE COULD BE MADE IN BISEC.                                  LEV10560
C                                                                      LEV10570
C     AS PROGRAMMED INVERR USES ALPHA, BETA,V1,V2,VS,G AND MP.         LEV10580
C     VS CONTAINS THE COMPUTED EIGENVALUES OF T(1,MEV). MP GIVES       LEV10590
C     THEIR T-MULTIPLICITIES AND FLAGS WHICH EIGENVALUES ARE TO HAVE   LEV10600
C     ERROR ESTIMATES COMPUTED. V2 IS USED FOR THE SOLUTION            LEV10610
C     VECTOR IN THE INVERSE ITERATION AND V1 FOR THE FACTORIZATION.    LEV10620
C     G CONTAINS THE RANDOMLY-GENERATED STARTING VECTOR FOR THE        LEV10630
C     INVERSE ITERATION. THE BASIC STORAGE FOR INVERR IS THEREFORE     LEV10640
C     A REAL*8 ARRAY OF DIMENSION 4*KMAX PLUS THE STORAGE NEEDED FOR   LEV10650
C     THE COMPUTED T-EIGENVALUES AND THEIR T-MULTIPLICITIES.           LEV10660
C                                                                      LEV10670
C     VS COULD BE USED TO STORE ONLY THOSE COMPUTED EIGENVALUES OF     LEV10680
C     T(1,MEV) THAT ARE OF INTEREST. IN THAT CASE THE DIMENSIONS OF VS LEV10690
```

```
C       AND OF MP NEED NOT BE ANY LONGER THAN THE NUMBER OF SUCH          LEV10700
C       EIGENVALUES. AS PROGRAMMED, ALL THE COMPUTED DISTINCT EIGENVALUES LEV10710
C       OF T(1,MEV) ARE STORED IN VS.  THEREFORE TO TAKE ADVANTAGE OF     LEV10720
C       SUCH A REDUCTION IN STORAGE THE USER WOULD HAVE TO MODIFY THAT    LEV10730
C       PART OF THE PROGRAM AND ALSO COMMENT OUT THE CALL TO THE          LEV10740
C       SUBROUTINE PRTEST.                                               LEV10750
C                                                                        LEV10760
C                                                                        LEV10770
C-----------------------------------------------------------------------LEV10780
C                                                                        LEV10790
C       COMMENTS FOR EIGENVECTOR COMPUTATIONS                            LEV10800
C                                                                        LEV10810
C-----------------------------------------------------------------------LEV10820
C                                                                        LEV10830
C                                                                        LEV10840
C       THE EIGENVALUES WHOSE EIGENVECTORS ARE TO BE COMPUTED MUST       LEV10850
C       HAVE BEEN COMPUTED USING THE CORRESPONDING LANCZOS EIGENVALUE    LEV10860
C       PROGRAMS BECAUSE THE EIGENVECTOR PROGRAMS WILL USE THE SAME      LEV10870
C       FAMILY OF LANCZOS TRIDIAGONAL MATRICES THAT WAS USED IN THE      LEV10880
C       CORRESPONDING EIGENVALUE COMPUTATIONS.                          LEV10890
C                                                                        LEV10900
C       THESE PROGRAMS ASSUME THAT THE EIGENVALUES SUPPLIED TO IT        LEV10910
C       HAVE BEEN COMPUTED ACCURATELY, AS MEASURED BY THE               LEV10920
C       ERROR ESTIMATES COMPUTED IN THE CORRESPONDING LANCZOS           LEV10930
C       EIGENVALUE COMPUTATIONS, ALTHOUGH THESE ESTIMATES ARE           LEV10940
C       TYPICALLY CONSERVATIVE.  IN CASES (1), (2) AND (4), THE          LEV10950
C       EIGENVALUES OF INTEREST ARE STORED IN THE ARRAY GOODEV(J),       LEV10960
C       J=1,NGOOD.  IN CASE (3) THE PROGRAM WORKS WITH THE              LEV10970
C       EIGENVALUES OF B(INVERSE) WHICH ARE STORED IN THE ARRAY          LEV10980
C       GOODBI(J), J=1,NGOOD.  THE CORRESPONDING EIGENVALUES            LEV10990
C       OF A ARE STORED IN GOODA(J), J=1,NGOOD.                         LEV11000
C                                                                        LEV11010
C       FOR EACH GOODEV(J), THE SUBROUTINE STURMI COMPUTES THE           LEV11020
C       SMALLEST SIZE LANCZOS TRIDIAGONAL MATRIX, T(1,M1(J)), FOR        LEV11030
C       WHICH GOODEV(J) IS AN EIGENVALUE TO WITHIN A SPECIFIED           LEV11040
C       TOLERANCE.  IT ALSO ATTEMPTS TO COMPUTE THE SIZE, M2(J),         LEV11050
C       BY WHICH THE GIVEN EIGENVALUE BECOMES A DOUBLE EIGENVALUE        LEV11060
C       TO WITHIN THE GIVEN TOLERANCE.  THESE VALUES ARE USED            LEV11070
C       TO DETERMINE 1ST GUESSES AT SIZES FOR THE T-EIGENVECTORS         LEV11080
C       THAT WILL BE USED IN THE RITZ VECTOR COMPUTATIONS.              LEV11090
C       SUBROUTINE INVERM SUCCESSIVELY COMPUTES CORRESPONDING            LEV11100
C       EIGENVECTORS OF ENLARGED T-MATRICES UNTIL A SUITABLE            LEV11110
C       SIZE T-MATRIX IS DETERMINED FOR EACH J.  UP TO 10 SUCH          LEV11120
C       EIGENVECTOR COMPUTATIONS ARE ALLOWED FOR EACH EIGENVALUE.        LEV11130
C                                                                        LEV11140
C       AFTER APPROPRIATE T-EIGENVECTORS HAVE BEEN COMPUTED,             LEV11150
C       RITZ VECTOR CORRESPONDING TO THESE T-EIGENVECTORS ARE THEN       LEV11160
C       COMPUTED AND TAKEN AS APPROXIMATE EIGENVECTORS OF A FOR THE      LEV11170
C       GIVEN EIGENVALUES, GOODEV(J), J = 1, ..., NGOOD.               LEV11180
C                                                                        LEV11190
C       THIS IMPLEMENTATION FIRST COMPUTES ALL OF THE RELEVANT          LEV11200
C       EIGENVECTORS OF THE SYMMETRIC TRIDIAGONAL MATRICES              LEV11210
C       IN THE VECTOR, TVEC.                                            LEV11220
C                                                                        LEV11230
C       THEN, AS EACH OF THE LANCZOS VECTORS IS REGENERATED, ALL        LEV11240
C       OF THE RITZ VECTORS CORRESPONDING TO THESE                      LEV11250
C       T-EIGENVECTORS ARE UPDATED USING THE CURRENTLY-GENERATED        LEV11260
C       LANCZOS VECTOR.  LANCZOS VECTORS ARE GENERATED (NOTE            LEV11270
C       THAT THEY ARE NOT BEING KEPT), UNTIL ENOUGH HAVE                LEV11280
C       BEEN GENERATED TO MAP THE LONGEST T-EIGENVECTOR INTO ITS        LEV11290
C       CORRESPONDING RITZ VECTOR.  THE ARRAY RITVEC CONTAINS THE       LEV11300
C       SUCCESSIVE RITZ VECTORS WHICH ARE THE APPROXIMATE              LEV11310
C       EIGENVECTORS OF A.                                              LEV11320
C                                                                        LEV11330
C                                                                        LEV11340
C-----PARAMETER CONTROLS FOR EIGENVECTOR PROGRAMS----------------------LEV11350
C                                                                        LEV11360
```

```
C                                                                    LEV11370
C     IN CASES (3) AND (4) WHERE A SPARSE FACTORIZATION OF A         LEV11380
C     SPECIFIED MATRIX IS USED, THE USER SPECIFIES USING THE FLAG    LEV11390
C     JPERM WHETHER OR NOT THE FACTORIZATION SUPPLIED CORRESPONDS    LEV11400
C     TO THE ORIGINAL MATRIX OR TO A PERMUTATION OF THE ORIGINAL     LEV11410
C     MATRIX.                                                        LEV11420
C                                                                    LEV11430
C     JPERM = (0,1) MEANS                                            LEV11440
C          0    NO PERMUTATION                                       LEV11450
C          1    MATRIX HAS BEEN PERMUTED.  NOTE THAT IN              LEV11460
C               CASE (4) THE PROGRAM WILL ASSUME THAT THE            LEV11470
C               DATA SUPPLIED FOR THE A-MATRIX CORRESPONDS TO THE    LEV11480
C               CORRESPONDING PERMUTATION OF THE ORIGINAL A-MATRIX.  LEV11490
C               IN BOTH CASES THE LANCZS CODES WILL WORK WITH THE    LEV11500
C               PERMUTED MATRICES AND THE PERMUTATION WILL BE        LEV11510
C               UNDONE ONLY IN THE EIGENVECTOR PROGRAM AFTER         LEV11520
C               THE RITZ VECTORS FOR THE PERMUTED PROBLEM HAVE       LEV11530
C               BEEN COMPUTED.                                       LEV11540
C                                                                    LEV11550
C     OTHER PARAMETER CONTROLS ARE INTRODUCED TO ALLOW SEGMENTATION  LEV11560
C     OF THE EIGENVECTOR COMPUTATIONS AND TO ALLOW VARIOUS COMBINATIONS LEV11570
C     OF READ/WRITES.                                                LEV11580
C                                                                    LEV11590
C     THE FLAG MBOUND ALLOWS THE USER TO DETERMINE A FIRST GUESS ON THE LEV11600
C     STORAGE THAT WILL BE REQUIRED BY THE T-EIGENVECTORS FOR THE    LEV11610
C     EIGENVALUES WHOSE EIGENVECTORS ARE TO BE COMPUTED.             LEV11620
C     THIS CAN BE USED TO ESTIMATE THE REQUIRED SIZE OF THE TVEC ARRAY. LEV11630
C                                                                    LEV11640
C     MBOUND = (0,1) MEANS                                           LEV11650
C                                                                    LEV11660
C          (0)  PROGRAM COMPUTES FIRST GUESSES AT THE SIZES          LEV11670
C               OF THE T-MATRICES REQUIRED BY EACH OF THE            LEV11680
C               EIGENVALUES SUPPLIED AND THEN CONTINUES WITH         LEV11690
C               THE CORRESPONDING T-EIGENVECTOR COMPUTATIONS.        LEV11700
C                                                                    LEV11710
C          (1)  PROGRAM COMPUTES FIRST GUESSES AT THE SIZES          LEV11720
C               OF THE T-MATRICES REQUIRED BY EACH OF THE            LEV11730
C               EIGENVALUES SUPPLIED, STORES THESE IN FILE 10        LEV11740
C               AND THEN TERMINATES.  THE USER CAN USE THESE         LEV11750
C               SIZES TO ESTIMATE THE SIZE TVEC ARRAY NEEDED         LEV11760
C               FOR THE DESIRED T-EIGENVECTOR COMPUTATIONS.          LEV11770
C                                                                    LEV11780
C     THE FLAGS NTVCON, TVSTOP, LVCONT, AND ERCONT CONTROL THE STOPPING LEV11790
C     CRITERIA FOR INTERMEDIATE POINTS IN THE LANCZOS PROCEDURE.  THEY LEV11800
C     TERMINATE THE PROCEDURE IF VARIOUS QUANTITIES COULD NOT BE     LEV11810
C     COMPUTED AS DESIRED.                                           LEV11820
C                                                                    LEV11830
C     NTVCON = (0,1) MEANS                                           LEV11840
C                                                                    LEV11850
C          (0)  IF THE ESTIMATED STORAGE FOR THE T-EIGENVECTORS      LEV11860
C               EXCEEDS THE USER-SPECIFIED DIMENSION OF THE          LEV11870
C               TVEC ARRAY PROGRAM DOES NOT CONTINUE WITH THE        LEV11880
C               T-EIGENVECTOR COMPUTATIONS.  TERMINATION OCCURS.     LEV11890
C                                                                    LEV11900
C          (1)  CONTINUE WITH THE T-EIGENVECTOR COMPUTATIONS         LEV11910
C               EVEN IF THE ESTIMATED STORAGE FOR TVEC EXCEEDS       LEV11920
C               THE USER-SPECIFIED DIMENSION OF THE TVEC ARRAY.      LEV11930
C               IN THIS SITUATION THE PROGRAM COMPUTES AS MANY       LEV11940
C               T-EIGENVECTORS AS IT HAS ROOM FOR, IN THE SAME       LEV11950
C               ORDER IN WHICH THE EIGENVALUES ARE PROVIDED.         LEV11960
C                                                                    LEV11970
C     SVTVEC = (0,1) MEANS                                           LEV11980
C                                                                    LEV11990
C          (0)  DO NOT STORE THE COMPUTED T-EIGENVECTORS ON          LEV12000
C               FILE 11 UNLESS ALSO HAVE THE FLAG TVSTOP = 1,        LEV12010
C               IN WHICH CASE THE T-EIGENVECTORS ARE ALWAYS          LEV12020
C               WRITTEN TO FILE 11.                                  LEV12030
```

```
C                                                                       LEV12040
C                  (1)  STORE THE COMPUTED T-EIGENVECTORS ON FILE 11.   LEV12050
C                                                                       LEV12060
C          TVSTOP = (0,1) MEANS                                         LEV12070
C                                                                       LEV12080
C                  (0)  ATTEMPT TO CONTINUE ON TO THE COMPUTATION       LEV12090
C                       OF THE RITZVECTORS AFTER COMPLETING THE         LEV12100
C                       COMPUTATION OF THE T-EIGENVECTORS.              LEV12110
C                                                                       LEV12120
C                  (1)  TERMINATE AFTER COMPUTING THE                   LEV12130
C                       T-EIGENVECTORS AND STORING THEM ON FILE 11.     LEV12140
C                                                                       LEV12150
C          LVCONT = (0,1) MEANS                                         LEV12160
C                                                                       LEV12170
C                  (0)  IF SOME OF THE T-EIGENVECTORS THAT WERE         LEV12180
C                       REQUESTED WERE NOT COMPUTED, EXIT               LEV12190
C                       FROM THE PROGRAM WITHOUT COMPUTING THE          LEV12200
C                       CORRESPONDING RITZ VECTORS.                     LEV12210
C                                                                       LEV12220
C                  (1)  CONTINUE ON TO THE RITZ VECTOR COMPUTATIONS     LEV12230
C                       EVEN IF NOT ALL OF THE T-EIGENVECTORS           LEV12240
C                       REQUESTED WERE COMPUTED.                        LEV12250
C                                                                       LEV12260
C          ERCONT = (0,1) MEANS                                         LEV12270
C                                                                       LEV12280
C                  (0)  PROCEDURE WILL NOT COMPUTE A RITZ VECTOR FOR    LEV12290
C                       ANY EIGENVALUE FOR WHICH NO T-EIGENVECTOR WHICH LEV12300
C                       SATISFIES THE ERROR ESTIMATE TEST (ERTOL) HAS   LEV12310
C                       BEEN IDENTIFIED.                                LEV12320
C                                                                       LEV12330
C                  (1)  A RITZ VECTOR WILL BE COMPUTED FOR EVERY        LEV12340
C                       EIGENVALUE FOR WHICH A T-EIGENVECTOR HAS BEEN   LEV12350
C                       COMPUTED REGARDLESS OF WHETHER OR NOT THAT      LEV12360
C                       T-EIGENVECTOR SATISFIED THE ERROR ESTIMATE TEST.LEV12370
C                                                                       LEV12380
C                                                                       LEV12390
C-----INPUT/OUTPUT FILES FOR THE EIGENVECTOR COMPUTATIONS---------------LEV12400
C                                                                       LEV12410
C                                                                       LEV12420
C      ANY INPUT DATA OTHER THAN THE T-MATRIX HISTORY FILE AND THE      LEV12430
C      PREVIOUSLY COMPUTED EIGENVALUES AND CORRESPONDING ERROR          LEV12440
C      ESTIMATES SHOULD BE STORED ON FILE 5 IN FREE FORMAT.             LEV12450
C      SEE SAMPLE INPUT/OUTPUT FOR TYPICAL INPUT FILE.                  LEV12460
C                                                                       LEV12470
C      FILE 6 WAS USED AS THE INTERACTIVE TERMINAL OUTPUT FILE.         LEV12480
C      THIS FILE PROVIDES A RUNNING ACCOUNT OF THE PROGRESS OF THE      LEV12490
C      COMPUTATIONS.  ADDITIONAL PRINTOUT IS GENERATED WHEN             LEV12500
C      THE FLAG IWRITE = 1.                                             LEV12510
C                                                                       LEV12520
C                                                                       LEV12530
C DESCRIPTION OF OTHER I/O FILES                                        LEV12540
C                                                                       LEV12550
C FILE (K)     CONTAINS:                                                LEV12560
C                                                                       LEV12570
C      (2)     INPUT FILE:                                              LEV12580
C              PREVIOUSLY-GENERATED T-MATRICES (ALPHA/BETA ARRAYS)      LEV12590
C              AND THE FINAL TWO LANCZOS VECTORS USED ON THAT           LEV12600
C              COMPUTATION.  THIS PROGRAM ALLOWS ENLARGEMENT            LEV12610
C              OF ANY T-MATRICES PROVIDED ON FILE 2.  NOTE THAT         LEV12620
C              IN CASE (4), THREE 'LANCZOS' VECTORS ARE ON FILE 2.      LEV12630
C                                                                       LEV12640
C      (3)     INPUT FILE:                                              LEV12650
C              THE GOOD EIGENVALUES OF THE T-MATRIX  T(1,MEV)           LEV12660
C              FOR WHICH RITZ VECTORS ARE REQUESTED.                    LEV12670
C              FILE 3 ALSO CONTAINS THE T-MULTIPLICITIES OF THESE       LEV12680
C              EIGENVALUES AND THEIR COMPUTED GAPS IN THE               LEV12690
C              T-MATRICES AND IN THE USER-SUPPLIED MATRIX.  IN          LEV12700
```

32

```
C                  CASE (3) FILE 3 CONTAINS THE EIGENVALUES OF THE           LEV12710
C                  B-INVERSE MATRIX AND 3 SETS OF CORRESPONDING GAPS.         LEV12720
C                  THIS FILE IS CREATED IN THE LANCZOS EIGENVALUE            LEV12730
C                  COMPUTATIONS.                                             LEV12740
C                                                                           LEV12750
C        (4)       INPUT FILE:                                              LEV12760
C                  ERROR ESTIMATES FOR THE ISOLATED GOOD T-EIGENVALUES      LEV12770
C                  ON FILE 3.  THIS FILE IS CREATED DURING THE LANCZOS      LEV12780
C                  EIGENVALUE COMPUTATIONS.                                 LEV12790
C                                                                           LEV12800
C        (7)       INPUT FILE:                                              LEV12810
C                  IN CASE (3) ((4)),                                       LEV12820
C                  CONTAINS SPARSE MATRIX REPRESENTATION OF FACTORIZATION   LEV12830
C                  OF MATRIX (B-MATRIX) USED BY LANCZS SUBROUTINE.          LEV12840
C                                                                           LEV12850
C        (8)       INPUT FILE:                                              LEV12860
C                  IN CASES (1),(2) AND (4), USPEC SUBROUTINE ASSUMES       LEV12870
C                  THAT USER-SUPPLIED A-MATRIX IS ON FILE 8.  IN CASE (3)   LEV12880
C                  A-MATRIX CAN BE STORED ON FILE 8, BUT IT IS NOT          LEV12890
C                  USED BY THE LANCZOS PROGRAMS.                            LEV12900
C                                                                           LEV12910
C        (9)       OUTPUT FILE:                                             LEV12920
C                  ERROR ESTIMATES FOR THE COMPUTED RITZ VECTORS CONSIDERED LEV12930
C                  AS EIGENVECTORS OF THE MATRIX USED BY THE LANCZS         LEV12940
C                  SUBROUTINE.   THESE ESTIMATES ARE OF THE FORM            LEV12950
C                       AERROR = || A*RITVEC - EVAL*RITVEC ||               LEV12960
C                  WHERE A DENOTES THE MATRIX USED BY LANCZS, EVAL DENOTES  LEV12970
C                  THE EIGENVALUE BEING CONSIDERED AND RITVEC DENOTES       LEV12980
C                  THE COMPUTED RITZ VECTOR.                                LEV1299C
C                                                                           LEV13000
C        (10)      OUTPUT FILE:                                             LEV13010
C                  GUESSES AT APPROPRIATE SIZE T-MATRICES FOR THE           LEV13020
C                  T-EIGENVECTORS FOR EACH SUPPLIED EIGENVALUE, GOODEV(J).  LEV13030
C                                                                           LEV13040
C        (11)      OUTPUT FILE:                                             LEV13050
C                  COMPUTED T-EIGENVECTORS CORRESPONDING TO EIGENVALUES     LEV13060
C                  IN THE GOODEV ARRAY.  NOTE THAT IT IS POSSIBLE IN        LEV13070
C                  CERTAIN SITUATIONS THAT FOR SOME EIGENVALUES IN THE      LEV13080
C                  GOODEV ARRAY A T-EIGENVECTOR WILL NOT BE COMPUTED.       LEV13090
C                                                                           LEV13100
C        (12)      OUTPUT FILE:                                             LEV13110
C                  CONTAINS COMPUTED RITZ VECTORS CORRESPONDING TO          LEV13120
C                  THE T-EIGENVECTORS ON FILE 11. NOTE THAT IN             LEV13130
C                  SOME SITUATIONS THAT FOR SOME EIGENVALUES IN            LEV13140
C                  THE GOODEV ARRAY FOR WHICH T-EIGENVECTORS HAVE          LEV13150
C                  BEEN COMPUTED NO RITZ VECTOR WILL HAVE BEEN             LEV13160
C                  COMPUTED.                                               LEV13170
C                                                                           LEV13180
C        (13)      OUTPUT FILE:                                             LEV13190
C                  ADDITIONAL INFORMATION ABOUT THE BOUNDS AND ERROR        LEV13200
C                  ESTIMATES OBTAINED.                                      LEV13210
C                                                                           LEV13220
C                                                                           LEV13230
C-----SEEDS FOR EIGENVECTOR PROGRAMS-------------------------------------LEV13240
C                                                                           LEV13250
C     SEEDS FOR RANDOM NUMBER GENERATOR GENRAN                              LEV13260
C                  (1) SVSEED = INTEGER*4 SCALAR USED IN THE SUBROUTINE     LEV13270
C                              GENRAN TO GENERATE THE STARTING VECTOR FORLEV13280
C                              THE REGENERATION OF THE LANCZOS VECTORS.     LEV13290
C                                                                           LEV13300
C                  (2) RHSEED = INTEGER*4 SCALAR USED IN THE SUBROUTINE     LEV13310
C                              GENRAN TO GENERATE A RANDOM VECTOR FOR       LEV13320
C                              USE IN SUBROUTINE INVERM.                    LEV13330
C                                                                           LEV13340
C     USER SHOULD NOTE THAT SVSEED MUST BE THE SAME SEED THAT               LEV13350
C     WAS USED TO GENERATE THE T-MATRICES THAT WERE USED TO                 LEV13360
C     COMPUTE THE EIGENVALUES WHOSE EIGENVECTORS ARE TO BE COMPUTED.        LEV13370
```

```
C      SVSEED IS READ IN FROM FILE 3.                              LEV13380
C                                                                  LEV13390
C                                                                  LEV13400
C-----USER-SPECIFIED PARAMETERS FOR THE EIGENVECTOR PROGRAMS-----------LEV13410
C                                                                  LEV13420
C                                                                  LEV13430
C      NGOOD    = NUMBER OF EIGENVALUES READ INTO THE GOODEV ARRAY LEV13440
C                 READ FROM FILE 3.                                LEV13450
C                                                                  LEV13460
C      N        = SIZE OF THE USER-SUPPLIED MATRIX.                LEV13470
C                                                                  LEV13480
C      MEV      = SIZE OF THE T-MATRIX THAT WAS USED TO COMPUTE    LEV13490
C                 THE EIGENVALUES WHOSE EIGENVECTORS ARE REQUESTED. LEV13500
C                 MEV IS READ IN FROM FILE 3.                      LEV13510
C                                                                  LEV13520
C      KMAX     =  SIZE OF THE T-MATRIX PROVIDED ON FILE 2.        LEV13530
C                                                                  LEV13540
C      MDIMTV  = MAXIMUM CUMULATIVE SIZE OF THE TVEC ARRAY ALLOWED LEV13550
C                 FOR ALL OF THE T-EIGENVECTORS REQUIRED.  MDIMTV  LEV13560
C                 MUST NOT EXCEED THE USER-SPECIFIED DIMENSION OF  LEV13570
C                 THE TVEC ARRAY.  PROGRAM CAN BE RUN WITH THE FLAG LEV13580
C                 MBOUND = 1 TO DETERMINE AN EDUCATED GUESS ON AN  LEV13590
C                 APPROPRIATE DIMENSION FOR THE TVEC ARRAY.        LEV13600
C                                                                  LEV13610
C      MDIMRV = MAXIMUM CUMULATIVE SIZE OF THE RITVEC ARRAY ALLOWED LEV13620
C                 FOR ALL OF THE RITZ VECTORS TO BE COMPUTED. MDIMRV LEV13630
C                 MUST NOT EXCEED THE USER-SPECIFIED DIMENSION OF  LEV13640
C                 THE RITVEC ARRAY.  MUST BE SELECTED SO THAT      LEV13650
C                 THERE IS ENOUGH ROOM FOR A RITZ VECTOR FOR EVERY LEV13660
C                 GOODEV(J) READ INTO PROGRAM.  (>= NGOOD*N)       LEV13670
C                                                                  LEV13680
C                                                                  LEV13690
C-----ARRAYS REQUIRED BY THE EIGENVECTOR PROGRAMS--------------------- LEV13700
C                                                                  LEV13710
C                                                                  LEV13720
C      ALL 4 CASES                                                 LEV13730
C                                                                  LEV13740
C      ALPHA(J) = REAL*8 ARRAY.  ITS DIMENSION MUST BE AT LEAST    LEV13750
C                 KMAXN, THE LARGEST SIZE T-MATRIX CONSIDERED BY   LEV13760
C                 THE PROGRAM.  NOTE THAT KMAXN IS THE LARGER OF   LEV13770
C                 THE SIZE OF THE ALPHA, BETA HISTORY PROVIDED     LEV13780
C                 ON FILE 2 (IF ANY ) AND THE SIZE WHICH THE PROGRAM LEV13790
C                 SPECIFIES INTERNALLY, THIS LATTER IS ALWAYS      LEV13800
C                 < = 11*MEV / 8  +  12, WHERE MEV IS THE SIZE     LEV13810
C                 T-MATRIX THAT WAS USED IN THE CORRESPONDING EIGENVALUE LEV13820
C                 COMPUTATIONS.  ALPHA CONTAINS THE DIAGONAL ENTRIES LEV13830
C                 OF THE LANCZOS T-MATRICES.  ALPHA IS NOT DESTROYED LEV13840
C                 IN THE COMPUTATIONS.                             LEV13850
C                                                                  LEV13860
C      BETA(J) = REAL*8 ARRAY.  ITS DIMENSION MUST BE AT LEAST 1   LEV13870
C                 MORE THAN THAT OF ALPHA.  DIMENSION COMMENTS ABOVE LEV13880
C                 ABOUT ALPHA APPLY ALSO TO THE BETA ARRAY.  BETA  LEV13890
C                 CONTAINS THE SUBDIAGONAL ENTRIES OF THE T-MATRICES. LEV13900
C                 BETA IS NOT DESTROYED IN THE COMPUTATIONS.       LEV13910
C                                                                  LEV13920
C      RITVEC(J) = REAL*8 ARRAY IN REAL SYMMETRIC AND INVERSE OF   LEV13930
C                 REAL SYMMETRIC CASES.  COMPLEX*16 IN CASE (2),   LEV13940
C                 HERMITIAN MATRICES.  IN EACH CASE ITS DIMENSION >= LEV13950
C                 NGOOD*N WHERE N IS THE ORDER OF THE USER-SUPPLIED LEV13960
C                 MATRIX AND NGOOD IS THE NUMBER OF EIGENVALUES WHOSE LEV13970
C                 EIGENVECTORS ARE TO BE COMPUTED.  IT CONTAINS THE LEV13980
C                 COMPUTED APPROXIMATE EIGENVECTORS OF A.  THESE   LEV13990
C                 COMPUTED RITZ VECTORS ARE STORED ON FILE 12.     LEV14000
C                                                                  LEV14010
C      TVEC(J)  = REAL*8 ARRAY.  ITS DIMENSION MUST BE AT LEAST    LEV14020
C                 MTOL = |MA(1)| + |MA(2)| + ... + |MA(NGOOD)|     LEV14030
C                 WHERE NGOOD IS THE NUMBER OF EIGENVALUES BEING   LEV14040
```

```
C                CONSIDERED AND |MA(J)| IS THE SIZE OF THE        LEV14050
C                T-MATRIX BEING USED FOR THE RITZ VECTOR          LEV14060
C                COMPUTATION FOR GOODEV(J). THESE SIZES           LEV14070
C                ARE COMPUTED BY THE PROGRAM.  AN ESTIMATE OF     LEV14080
C                MTOL CAN BE OBTAINED BY SETTING MBOUND = 1,      LEV14090
C                RUNNING THE PROGRAM, AND MULTIPLYING THE RESULTING LEV14100
C                TOTAL OF THE T-SIZES SPECIFIED BY 5/4.  THE ARRAY LEV14110
C                TVEC CONTAINS THE COMPUTED T-EIGENVECTORS.  IF THE LEV14120
C                FLAG SVTVEC = 1 OR THE FLAG TVSTOP = 1, THEN     LEV14130
C                THESE VECTORS ARE SAVED ON FILE 11.              LEV14140
C                                                                 LEV14150
C     V1(J)      = REAL*8 ARRAY IN REAL SYMMETRIC AND INVERSE OF  LEV14160
C                REAL SYMMETRIC CASES.  COMPLEX*16 IN CASE (2),   LEV14170
C                HERMITIAN MATRICES.  IN THE REAL CASES ITS       LEV14180
C                DIMENSION MUST BE THE MAXIMUM OF KMAX AND N.     LEV14190
C                IN THE HERMITIAN CASE ITS DIMENSION MUST BE      LEV14200
C                THE MAXIMUM OF KMAX/2 AND N WHERE KMAX IS THE    LEV14210
C                LARGEST SIZE T-MATRIX THAT IS TO BE CONSIDERED   LEV14220
C                IN THE T-EIGENVECTOR COMPUTATIONS.  V1 IS USED   LEV14230
C                IN THE SUBROUTINE INVERM AND IN THE REGENERATION LEV14240
C                OF THE LANCZOS VECTORS.                          LEV14250
C                                                                 LEV14260
C     V2(J)      = REAL*8 ARRAY IN THE REAL SYMMETRIC AND INVERSE LEV14270
C                OF REAL SYMMETRIC CASES.  COMPLEX*16 IN CASE (2),LEV14280
C                HERMITIAN MATRICES.  IN CASES (1),(3) AND (4), ITS LEV14290
C                DIMENSION MUST BE > = MAX(KMAX,N); IN CASE (2)   LEV14300
C                > = MAX(KMAX/2,N).  IT IS USED IN THE REGENERATION LEV14310
C                OF THE LANCZOS VECTORS AND IN SUBROUTINE INVERM. LEV14320
C                                                                 LEV14330
C   GOODEV(J),   = REAL*8 ARRAYS EACH OF DIMENSION AT LEAST NGOOD.LEV14340
C   EVNEW(J)       CONTAIN THE EIGENVALUES FOR WHICH EIGENVECTORS LEV14350
C                  ARE REQUESTED.  EIGENVALUES IN GOODEV ARE READ LEV14360
C                  IN FROM FILE 3.  IN CASE (3) GOODEV IS REPLACED LEV14370
C                  BY GOODA AND GOODBI ARRAYS, SEE BELOW.         LEV14380
C                                                                 LEV14390
C   AMINGP(J),   = REAL*4 ARRAYS OF DIMENSION AT LEAST NGOOD.     LEV14400
C   TMINGP(J)      CONTAIN, RESPECTIVELY, THE MINIMAL GAPS FOR    LEV14410
C                  CORRESPONDING EIGENVALUES IN GOODEV ARRAY IN   LEV14420
C                  A-MATRIX AND IN T-MATRIX.                      LEV14430
C                                                                 LEV14440
C   TERR(J), ERR(J),   = REAL*4 ARRAYS (EXCEPT TLAST WHICH IS     LEV14450
C   ERRDGP(J), TLAST(J)    REAL*8).  EACH MUST BE OF DIMENSION    LEV14460
C   RNORM(J), TBETA(J)     AT LEAST NGOOD.  USED TO STORE QUANTITIES LEV14470
C                          GENERATED DURING THE COMPUTATIONS FOR  LEV14480
C                          LATER PRINTOUT.                        LEV14490
C                                                                 LEV14500
C     G(J)      = REAL*4 ARRAY WHOSE DIMENSION MUST BE AT LEAST   LEV14510
C                MAX(KMAX,N).  USED IN SUBROUTINE GENRAN TO HOLD  LEV14520
C                RANDOM NUMBERS NEEDED FOR THE LANCZOS VECTOR     LEV14530
C                REGENERATION AND FOR THE INVERSE ITERATION       LEV14540
C                COMPUTATIONS IN THE SUBROUTINE INVERM.           LEV14550
C                                                                 LEV14560
C   MP(J) = INTEGER*4 ARRAY WHOSE DIMENSION IS AT LEAST NGOOD.    LEV14570
C           INITIALLY CONTAINS THE MULTIPLICITY OF THE EIGENVALUE LEV14580
C           GOODEV(J) AS AN EIGENVALUE OF THE T-MATRIX T(1,MEV).  LEV14590
C           USED TO FLAG EIGENVALUES FOR WHICH NO T-EIGENVECTOR   LEV14600
C           OR NO RITZ VECTOR IS TO BE COMPUTED.                  LEV14610
C                                                                 LEV14620
C   MA(J)     = INTEGER*4 ARRAYS EACH OF WHOSE DIMENSIONS         LEV14630
C               IS AT LEAST NGOOD.  USED IN DETERMINING           LEV14640
C               AN APPROPRIATE T-MATRIX FOR EACH EIGENVALUE       LEV14650
C               IN GOODEV ARRAY.                                  LEV14660
C                                                                 LEV14670
C   MINT(J),MFIN(J) = INTEGER*4 ARRAYS WHOSE DIMENSIONS MUST BE AT LEV14680
C                     LEAST NGOOD.  USED TO POINT TO THE BEGINNINGS LEV14690
C                     AND THE ENDS OF THE COMPUTED EIGENVECTOR    LEV14700
C                     OF THE T-MATRIX, T(1,|MA(J)|).              LEV14710
```

```
C                                                                    LEV14720
C     IDELTA(J)  = INTEGER*4 ARRAY WHOSE DIMENSION MUST BE AT         LEV14730
C                  LEAST NGOOD.  CONTAINS INCREMENTS USED IN LOOPS    LEV14740
C                  ON APPROPRIATE SIZE T-MATRIX FOR THE T-EIGENVECTOR LEV14750
C                  COMPUTATIONS.                                      LEV14760
C                                                                     LEV14770
C                                                                     LEV14780
C     CASE (2) ONLY, HERMITIAN MATRICES:                             LEV14790
C                                                                     LEV14800
C     GR(J),GC(J)    = REAL*8 ARRAYS USED ONLY IN CASE (2),          LEV14810
C                      HERMITIAN MATRICES.  EACH MUST BE AT          LEV14820
C                      LEAST MAX(N,KMAX).  USED TO HOLD              LEV14830
C                      STARTING VECTORS FOR LANCZS                   LEV14840
C                      COMPUTATIONS AND FOR INVERM SUBROUTINES.      LEV14850
C                                                                     LEV14860
C     CASES (3) AND (4) ONLY, FACTORED INVERSES OF REAL SYMMETRIC    LEV14870
C     MATRICES AND GENERALIZED EIGENVALUE PROBLEMS:                  LEV14880
C                                                                     LEV14890
C     VS(J) =  REAL*8 ARRAY WHOSE DIMENSION MUST BE AT LEAST N.      LEV14900
C              USED IN REGENERATION OF THE LANCZOS VECTORS.          LEV14910
C                                                                     LEV14920
C     IPR(J), IPT(J) = INTEGER*4 ARRAYS.  EACH MUST BE OF DIMENSION  LEV14930
C                      AT LEAST N, THE ORDER OF A. USED TO STORE     LEV14940
C                      THE REORDERING OF THE GIVEN MATRIX.           LEV14950
C                                                                     LEV14960
C     CASE (3) ONLY, INVERSES OF REAL SYMMETRIC MATRICES:           LEV14970
C                                                                     LEV14980
C     GCODA(J), GOODBI(J) = REAL*8 ARRAYS.  EACH MUST BE OF DIMENSION LEV14990
C                      AT LEAST NGOOD, THE NUMBER OF EIGENVALUES     LEV15000
C                      BEING CONSIDERED.  GOODA CONTAINS THE         LEV15010
C                      EIGENVALUES OF A AND GOODBI CONTAINS THE      LEV15020
C                      EIGENVALUES OF B(INVERSE).  THE PROGRAM       LEV15030
C                      WORKS DIRECTLY WITH THE GOODBI ARRAY.         LEV15040
C                                                                     LEV15050
C                                                                     LEV15060
C-----SUBROUTINES INCLUDED FOR THE EIGENVECTOR COMPUTATIONS------------LEV15070
C                                                                     LEV15080
C                                                                     LEV15090
C     STURMI = FOR EACH GIVEN EIGENVALUE GOODEV(J) DETERMINES        LEV15100
C              THE SMALLEST SIZE T-MATRIX FOR WHICH GOODEV(J) IS     LEV15110
C              AN EIGENVALUE (TO WITHIN A GIVEN TOLERANCE) AND IF    LEV15120
C              POSSIBLE THE SMALLEST SIZE T-MATRIX FOR WHICH         LEV15130
C              IT IS A DOUBLE EIGENVALUE (TO WITHIN THE SAME         LEV15140
C              TOLERANCE).  THE SIZE T-MATRIX USED IN THE            LEV15150
C              EIGENVECTOR COMPUTATIONS IS THEN DETERMINED BY        LEV15160
C              STARTING WITH AN INITIAL GUESS BASED UPON THE         LEV15170
C              INFORMATION FROM STURMI, AND LOOPING ON THE SIZE      LEV15180
C              OF THE T-EIGENVECTOR COMPUTATIONS.                    LEV15190
C                                                                     LEV15200
C     LBISEC = RECOMPUTES THE VALUE OF THE GIVEN EIGENVALUE AT THE   LEV15210
C              T-SIZE SPECIFIED FOR THE T-EIGENVECTOR COMPUTATION.   LEV15220
C              LBISEC IS A SIMPLIFICATION OF THE BISEC SUBROUTINE    LEV15230
C              USED IN THE LANCZOS EIGENVALUE COMPUTATIONS.          LEV15240
C                                                                     LEV15250
C     INVERM = FOR THE T-SIZES CONSIDERED BY THE PROGRAM COMPUTES    LEV15260
C              THE CORRESPONDING EIGENVECTORS OF THESE T-MATRICES    LEV15270
C              CORRESPONDING TO THE USER-SUPPLIED EIGENVALUES IN     LEV15280
C              THE GOODEV ARRAY.                                     LEV15290
C                                                                     LEV15300
C     THE LANCZS, TNORM , AND CINPRD (CASE (2) ONLY) SUBROUTINES     LEV15310
C     ARE USED HERE AS WELL AS IN THE EIGENVALUE COMPUTATIONS.       LEV15320
C                                                                     LEV15330
C     IN CASES (3) AND (4) ONLY AND THEN ONLY IF THE ORIGINAL MATRIX LEV15340
C     (MATRICES) HAS (HAVE) BEEN PERMUTED:                          LEV15350
C                                                                     LEV15360
C     LPERM  = (USED IN CASE (3) AND (4) ONLY).  GIVEN A B-MATRIX AND LEV15370
C              A PERMUTATION P DEFINED IN THE VECTORS IPR AND IPT,   LEV15380
```

```
C              AND A VECTOR X COMPUTE EITHER (P-TRANSPOSE)*X OR PX.     LEV15390
C                                                                      LEV15400
C----------------------------------------------------------------------LEV15410
```

SECTION 2.3   MAIN PROGRAM, EIGENVALUE CALCULATIONS

```
C-----LEVAL  (EIGENVALUES OF REAL SYMMETRIC MATRICES)------------------LEV00010
C                                                                       LEV00020
C       CONTAINS MAIN PROGRAM FOR COMPUTING DISTINCT EIGENVALUES OF      LEV00030
C       A REAL SYMMETRIC MATRIX USING LANCZOS TRIDIAGONALIZATION         LEV00040
C       WITHOUT REORTHOGONALIZATION.                                     LEV00050
C                                                                       LEV00060
C       PFORT VERIFIER IDENTIFIED THE FOLLOWING NONPORTABLE             LEV00070
C       CONSTRUCTIONS                                                    LEV00080
C                                                                       LEV00090
C       1.  DATA/MACHEP/ STATEMENT                                       LEV00100
C       2.  ALL READ(5,*) STATEMENTS (FREE FORMAT)                       LEV00110
C       3.  FORMAT(20A4) USED WITH EXPLANATORY HEADER EXPLAN.            LEV00120
C       4.  HEXADECIMAL FORMAT (4Z20) USED IN ALPHA/BETA FILES 1 AND 2.  LEV00130
C                                                                       LEV00140
C----------------------------------------------------------------------LEV00150
C                                                                       LEV00160
        DOUBLE PRECISION   ALPHA(5000),BETA(5001)                        LEV00170
        DOUBLE PRECISION   V1(5000),V2(5000),VS(5000)                    LEV00180
        DOUBLE PRECISION   LB(20),UB(20)                                 LEV00190
        DOUBLE PRECISION   BTOL,GAPTOL,TTOL,MACHEP,EPSM,RELTOL           LEV00200
        DOUBLE PRECISION   SCALE1,SCALE2,SCALE3,SCALE4,BISTOL,CONTOL,MULTOLLEV00210
        DOUBLE PRECISION   ONE,ZERO,TEMP,TKMAX,BETAM,BKMIN,TO,T1         LEV00220
        REAL  G(5000),EXPLAN(20)                                         LEV00230
        INTEGER  MP(5000),NMEV(20)                                       LEV00240
        INTEGER  SVSEED,RHSEED,SVSOLD                                    LEV00250
        INTEGER  IABS                                                    LEV00260
        REAL  ABS                                                        LEV00270
        DOUBLE PRECISION   DABS, DSQRT, DFLOAT                           LEV00280
        EXTERNAL CMATV                                                   LEV00290
C                                                                       LEV00300
C----------------------------------------------------------------------LEV00310
        DATA MACHEP/Z3410000000000000/                                   LEV00320
        EPSM = 2.0D0*MACHEP                                              LEV00330
C----------------------------------------------------------------------LEV00340
C                                                                       LEV00350
C       ARRAYS MUST BE DIMENSIONED AS FOLLOWS:                           LEV00360
C       DIMENSION OF V2 ASSUMES THAT NO MORE THAN KMAX/2 EIGENVALUES     LEV00370
C       OF THE T-MATRICES ARE BEING COMPUTED IN ANY ONE OF THE          LEV00380
C       SUB-INTERVALS BEING CONSIDERED.  V2 CONTAINS THE UPPER AND LOWER LEV00390
C       BOUNDS FOR EACH T-EIGENVALUE BEING COMPUTED BY BISEC IN ANY ONE  LEV00400
C       GIVEN INTERVAL.                                                  LEV00410
C                                                                       LEV00420
C       1.  ALPHA: >= KMAX,   BETA: >= (KMAX+1)                          LEV00430
C       2.  V1:  >= MAX(N,KMAX+1)                                        LEV00440
C       3.  V2:  >= MAX(N,KMAX)                                          LEV00450
C       4.  VS:  >= KMAX                                                 LEV00460
C       5.  G:  >= MAX(N,2*KMAX)                                         LEV00470
C       6.  MP:  >= KMAX                                                 LEV00480
C       7.  LB,UB: >= NUMBER OF SUBINTERVALS SUPPLIED TO BISEC.          LEV00490
C       8.  NMEV: >= NUMBER OF T-MATRICES ALLOWED.                       LEV00500
C       9.  EXPLAN: DIMENSION IS 20.                                     LEV00510
C                                                                       LEV00520
C                                                                       LEV00530
C       IMPORTANT TOLERANCES OR SCALES THAT ARE USED REPEATEDLY          LEV00540
C       THROUGHOUT THIS PROGRAM ARE THE FOLLOWING:                       LEV00550
C       SCALED MACHINE EPSILON:  TTOL = TKMAX*EPSM WHERE                 LEV00560
C       EPSM = 2*MACHINE EPSILON AND                                     LEV00570
C       TKMAX = MAX(|ALPHA(J)|,BETA(J), J = 1,MEV)                       LEV00580
C       BISEC CONVERGENCE TOLERANCE:  BISTOL = DSQRT(1000+MEV)*TTOL      LEV00590
C       BISEC T-MULTIPLICITY TOLERANCE:   MULTOL = (1000+MEV)*TTOL       LEV00600
C       LANCZOS CONVERGENCE TOLERANCE:    CONTOL = BETA(MEV+1)*1.D-10    LEV00610
C----------------------------------------------------------------------LEV00620
C       OUTPUT HEADER                                                    LEV00630
        WRITE(6,10)                                                      LEV00640
```

```
   10 FORMAT(/' LANCZOS PROCEDURE FOR REAL SYMMETRIC MATRICES'/)      LEV00650
C                                                                      LEV00660
C     SET PROGRAM PARAMETERS                                           LEV00670
C     SCALEK ARE USED IN TOLERANCES NEEDED IN SUBROUTINES LUMP,        LEV00680
C     ISOEV AND PRTEST.  USER MUST NOT MODIFY THESE SCALES.            LEV00690
      SCALE1 = 5.0D2                                                   LEV00700
      SCALE2 = 5.0D0                                                   LEV00710
      SCALE3 = 5.0D0                                                   LEV00720
      SCALE4 = 1.0D4                                                   LEV00730
      ONE  = 1.0D0                                                     LEV00740
      ZERO = 0.0D0                                                     LEV00750
      BTOL = 1.0D-8                                                    LEV00760
C     BTOL = EPSM                                                      LEV00770
      GAPTOL = 1.0D-8                                                  LEV00780
      ICONV = 0                                                        LEV00790
      MOLD = 0                                                         LEV00800
      MOLD1 = 1                                                        LEV00810
      ICT = 0                                                          LEV00820
      MMB = 0                                                          LEV00830
      IPROJ = 0                                                        LEV00840
C----------------------------------------------------------------------LEV00850
C     READ USER-SPECIFIED PARAMETERS FROM INPUT FILE 5 (FREE FORMAT)   LEV00860
C                                                                      LEV00870
C     READ USER-PROVIDED HEADER FOR RUN                               LEV00880
      READ(5,20) EXPLAN                                                LEV00890
      WRITE(6,20) EXPLAN                                               LEV00900
      READ(5,20) EXPLAN                                                LEV00910
      WRITE(6,20) EXPLAN                                               LEV00920
   20 FORMAT(20A4)                                                     LEV00930
C                                                                      LEV00940
C     READ ORDER OF MATRICES (N) , MAXIMUM ORDER OF T-MATRIX (KMAX),   LEV00950
C     NUMBER OF T-MATRICES ALLOWED (NMEVS), AND MATRIX IDENTIFICATION  LEV00960
C     NUMBERS (MATNO)                                                  LEV00970
      READ(5,20) EXPLAN                                                LEV00980
      READ(5,*) N,KMAX,NMEVS,MATNO                                     LEV00990
C                                                                      LEV01000
C     READ SEEDS FOR LANCZS AND INVERR SUBROUTINES (SVSEED AND RHSEED) LEV01010
C     READ MAXIMUM NUMBER OF ITERATIONS ALLOWED FOR EACH INVERSE       LEV01020
C     ITERATION (MXINIT) AND MAXIMUM NUMBER OF STURM SEQUENCES         LEV01030
C     ALLOWED (MXSTUR)                                                 LEV01040
      READ(5,20) EXPLAN                                                LEV01050
      READ(5,*) SVSEED,RHSEED,MXINIT,MXSTUR                            LEV01060
C                                                                      LEV01070
C     ISTART = (0,1):  ISTART = 0 MEANS ALPHA/BETA FILE IS NOT         LEV01080
C     AVAILABLE.  ISTART = 1 MEANS ALPHA/BETA FILE IS AVAILABLE ON     LEV01090
C     FILE 2.                                                          LEV01100
C     ISTOP = (0,1):  ISTOP = 0 MEANS PROCEDURE GENERATES ALPHA/BETA   LEV01110
C     FILE AND THEN TERMINATES.  ISTOP = 1 MEANS PROCEDURE GENERATES   LEV01120
C     ALPHAS/BETAS IF NEEDED AND THEN COMPUTES EIGENVALUES AND ERROR   LEV01130
C     ESTIMATES AND THEN TERMINATES.                                  LEV01140
      READ(5,20) EXPLAN                                                LEV01150
      READ(5,*) ISTART,ISTOP                                           LEV01160
C                                                                      LEV01170
C     IHIS = (0,1):  IHIS = 0 MEANS ALPHA/BETA FILE IS NOT WRITTEN     LEV01180
C     TO FILE 1.  IHIS = 1 MEANS ALPHA/BETA FILE IS WRITTEN TO FILE 1. LEV01190
C     IDIST = (0,1):  IDIST = 0 MEANS DISTINCT T-EIGENVALUES           LEV01200
C     ARE NOT WRITTEN TO FILE 11.  IDIST = 1 MEANS DISTINCT            LEV01210
C     T-EIGENVALUES ARE WRITTEN TO FILE 11.                           LEV01220
C     IWRITE = (0,1):  IWRITE = 0 MEANS NO INTERMEDIATE OUTPUT         LEV01230
C     FROM THE COMPUTATIONS IS WRITTEN TO FILE 6.  IWRITE = 1 MEANS    LEV01240
C     T-EIGENVALUES AND ERROR ESTIMATES ARE WRITTEN TO FILE 6          LEV01250
C     AS THEY ARE COMPUTED.                                           LEV01260
      READ(5,20) EXPLAN                                                LEV01270
      READ(5,*) IHIS,IDIST,IWRITE                                      LEV01280
C                                                                      LEV01290
C     READ IN THE RELATIVE TOLERANCE (RELTOL) FOR USE IN THE           LEV01300
C     SPURIOUS, T-MULTIPLICITY, AND PRTESTS.                           LEV01310
```

```
      READ(5,20) EXPLAN                                            LEVO1320
      READ(5,*) RELTOL                                             LEVO1330
C                                                                  LEVO1340
C     READ IN THE SIZES OF THE T-MATRICES TO BE CONSIDERED.        LEVO1350
      READ(5,20) EXPLAN                                            LEVO1360
      READ(5,*) (NMEV(J), J=1,NMEVS)                               LEVO1370
C                                                                  LEVO1380
C     READ IN THE NUMBER OF SUBINTERVALS TO BE CONSIDERED.         LEVO1390
      READ(5,20) EXPLAN                                            LEVO1400
      READ(5,*) NINT                                               LEVO1410
C                                                                  LEVO1420
C     READ IN THE LEFT-END POINTS OF THE SUBINTERVALS TO BE CONSIDERED. LEVO1430
C     THESE MUST BE IN ALGEBRAICALLY-INCREASING ORDER             LEVO1440
      READ(5,20) EXPLAN                                            LEVO1450
      READ(5,*) (LB(J), J=1,NINT)                                  LEVO1460
C                                                                  LEVO1470
C     READ IN THE RIGHT-END POINTS OF THE SUBINTERVALS TO BE CONSIDERED.LEVO1480
C     THESE MUST BE IN ALGEBRAICALLY-INCREASING ORDER             LEVO1490
      READ(5,20) EXPLAN                                            LEVO1500
      READ(5,*) (UB(J), J=1,NINT)                                  LEVO1510
C                                                                  LEVO1520
C------------------------------------------------------------------LEVO1530
C     INITIALIZE THE ARRAYS FOR THE USER-SPECIFIED MATRIX          LEVO1540
C     AND PASS THE STORAGE LOCATIONS OF THESE ARRAYS TO THE        LEVO1550
C     MATRIX-VECTOR MULTIPLY SUBROUTINE CMATV.                     LEVO1560
C                                                                  LEVO1570
      CALL USPEC(N,MATNO)                                          LEVO1580
C                                                                  LEVO1590
C------------------------------------------------------------------LEVO1600
C     MASK UNDERFLOW AND OVERFLOW                                  LEVO1610
C                                                                  LEVO1620
      CALL MASK                                                    LEVO1630
C                                                                  LEVO1640
C------------------------------------------------------------------LEVO1650
C                                                                  LEVO1660
C     WRITE TO FILE 6, A SUMMARY OF THE PARAMETERS FOR THIS RUN    LEVO1670
C                                                                  LEVO1680
      WRITE(6,30) MATNO,N,KMAX                                     LEVO1690
   30 FORMAT(/3X,'MATRIX ID',4X,'ORDER OF A',4X,'MAX ORDER OF T'/  LEVO1700
     1 I12,I14,I18/)                                               LEVO1710
C                                                                  LEVO1720
      WRITE(6,40) ISTART,ISTOP                                     LEVO1730
   40 FORMAT(/2X,'ISTART',3X,'ISTOP'/2I8/)                         LEVO1740
C                                                                  LEVO1750
      WRITE(6,50) IHIS,IDIST,IWRITE                                LEVO1760
   50 FORMAT(/4X,'IHIS',3X,'IDIST',2X,'IWRITE'/3I8/)               LEVO1770
C                                                                  LEVO1780
      WRITE(6,60) SVSEED,RHSEED                                    LEVO1790
   60 FORMAT(/' SEEDS FOR RANDOM NUMBER GENERATOR'//               LEVO1800
     1 4X,'LANCZS SEED',4X,'INVERR SEED'/2I15/)                    LEVO1810
C                                                                  LEVO1820
      WRITE(6,70) (NMEV(J), J=1,NMEVS)                             LEVO1830
   70 FORMAT(/' SIZES OF T-MATRICES TO BE CONSIDERED'/(6I12))      LEVO1840
C                                                                  LEVO1850
      WRITE(6,80) RELTOL,GAPTOL,BTOL                               LEVO1860
   80 FORMAT(/' RELATIVE TOLERANCE USED TO COMBINE COMPUTED T-EIGENVALUELEVO1870
     1S'/E15.3/' RELATIVE GAP TOLERANCES USED IN INVERSE ITERATION'/ LEVO1880
     1E15.3/' RELATIVE TOLERANCE FOR CHECK ON SIZE OF BETAS'/E15.3/) LEVO1890
C                                                                  LEVO1900
      WRITE(6,90) (J,LB(J),UB(J), J=1,NINT)                        LEVO1910
   90 FORMAT(/' BISEC WILL BE USED ON THE FOLLOWING INTERVALS'/    LEVO1920
     1 (I6,2E20.6))                                                LEVO1930
C                                                                  LEVO1940
      IF (ISTART.EQ.0) GO TO 140                                   LEVO1950
C                                                                  LEVO1960
C     READ IN ALPHA BETA HISTORY                                  LEVO1970
C                                                                  LEVO1980
```

```
      READ(2,100)MOLD,NOLD,SVSOLD,MATOLD                              LEV01990
  100 FORMAT(2I6,I12,I8)                                              LEV02000
C                                                                     LEV02010
      IF (KMAX.LT.MOLD) KMAX = MOLD                                   LEV02020
      KMAX1 = KMAX + 1                                                LEV02030
C                                                                     LEV02040
C     CHECK THAT ORDER N, MATRIX ID MATNO, AND RANDOM SEED SVSEED     LEV02050
C     AGREE WITH THOSE IN THE HISTORY FILE.  IF NOT PROCEDURE STOPS.  LEV02060
C                                                                     LEV02070
      ITEMP = (NOLD-N)**2+(MATNO-MATOLD)**2+(SVSEED-SVSOLD)**2         LEV02080
C                                                                     LEV02090
      IF (ITEMP.EQ.0) GO TO 120                                       LEV02100
C                                                                     LEV02110
      WRITE(6,110)                                                    LEV02120
  110 FORMAT(' PROGRAM TERMINATES'/    ' READ FROM FILE 2 CORRESPONDS TOLEV02130
     1 DIFFERENT MATRIX THAN MATRIX SPECIFIED'/)                      LEV02140
      GO TO 640                                                       LEV02150
C                                                                     LEV02160
  120 CONTINUE                                                        LEV02170
      MOLD1 = MOLD+1                                                  LEV02180
C                                                                     LEV02190
      READ(2,130)(ALPHA(J), J=1,MOLD)                                 LEV02200
      READ(2,130)(BETA(J), J=1,MOLD1)                                 LEV02210
  130 FORMAT(4Z20)                                                    LEV02220
C                                                                     LEV02230
      IF (KMAX.EQ.MOLD) GO TO 160                                     LEV02240
C                                                                     LEV02250
      READ(2,130)(V1(J), J=1,N)                                       LEV02260
      READ(2,130)(V2(J), J=1,N)                                       LEV02270
C                                                                     LEV02280
  140 CONTINUE                                                        LEV02290
      IIX = SVSEED                                                    LEV02300
C                                                                     LEV02310
C---------------------------------------------------------------------LEV02320
C                                                                     LEV02330
      CALL LANCZS(CMATV,ALPHA,BETA,V1,V2,G,KMAX,MOLD1,N,IIX)          LEV02340
C                                                                     LEV02350
C---------------------------------------------------------------------LEV02360
C                                                                     LEV02370
      KMAX1 = KMAX + 1                                                LEV02380
C                                                                     LEV02390
      IF (IHIS.EQ.0.AND.ISTOP.GT.0) GO TO 160                         LEV02400
C                                                                     LEV02410
      WRITE(1,150) KMAX,N,SVSEED,MATNO                                LEV02420
  150 FORMAT(2I6,I12,I8,' = KMAX,N,SVSEED,MATNO')                     LEV02430
C                                                                     LEV02440
      WRITE(1,130)(ALPHA(I), I=1,KMAX)                                LEV02450
      WRITE(1,130)(BETA(I), I=1,KMAX1)                                LEV02460
C                                                                     LEV02470
      WRITE(1,130)(V1(I), I=1,N)                                      LEV02480
      WRITE(1,130)(V2(I), I=1,N)                                      LEV02490
C                                                                     LEV02500
      IF (ISTOP.EQ.0) GO TO 540                                       LEV02510
C                                                                     LEV02520
  160 CONTINUE                                                        LEV02530
      BKMIN = BTOL                                                    LEV02540
      WRITE(6,170)                                                    LEV02550
  170 FORMAT(/' T-MATRICES (ALPHA AND BETA) ARE NOW AVAILABLE'/)      LEV02560
C                                                                     LEV02570
C---------------------------------------------------------------------LEV02580
C     SUBROUTINE TNORM CHECKS MIN(BETA)/(ESTIMATED NORM(A)) > BTOL .  LEV02590
C     IF THIS IS VIOLATED IB IS SET EQUAL TO THE NEGATIVE OF THE INDEX LEV02600
C     OF THE MINIMAL BETA.  IF(IB < 0) THEN SUBROUTINE TNORM IS       LEV02610
C     CALLED FOR EACH VALUE OF MEV TO DETERMINE WHETHER OR NOT THERE  LEV02620
C     IS A BETA IN THE T-MATRIX SPECIFIED THAT VIOLATES THIS TEST.    LEV02630
C     IF THERE IS SUCH A BETA THE PROGRAM TERMINATES FOR THE USER     LEV02640
C     TO DECIDE WHAT TO DO.  THIS TEST CAN BE OVER-RIDDEN BY          LEV02650
```

```
C      SIMPLY MAKING BTOL SMALLER, BUT THEN THERE IS THE POSSIBILITY       LEVO2660
C      THAT LOSSES IN THE LOCAL ORTHOGONALITY MAY HURT THE COMPUTATIONS. LEVO2670
C      BTOL = 1.D-8 IS HOWEVER A CONSERVATIVE CHOICE FOR BTOL.             LEVO2680
C                                                                          LEVO2690
C      TNORM ALSO COMPUTES TKMAX = MAX(|ALPHA(K)|,BETA(K), K=1,KMAX).      LEVO2700
C      TKMAX IS USED TO SCALE THE TOLERANCES USED IN THE                   LEVO2710
C      T-MULTIPLICITY AND SPURIOUS TESTS IN BISEC. TKMAX IS ALSO USED IN LEVO2720
C      THE PROJECTION TEST FOR HIDDEN EIGENVALUES THAT HAD 'TOO SMALL'     LEVO2730
C      A PROJECTION ON THE STARTING VECTOR.                               LEVO2740
C                                                                          LEVO2750
       CALL TNORM(ALPHA,BETA,BKMIN,TKMAX,KMAX,IB)                         LEVO2760
C                                                                          LEVO2770
C---------------------------------------------------------------------LEVO2780
C                                                                          LEVO2790
       TTOL = EPSM*TKMAX                                                   LEVO2800
C                                                                          LEVO2810
C      LOOP ON THE SIZE OF THE T-MATRIX                                    LEVO2820
C                                                                          LEVO2830
   180 CONTINUE                                                            LEVO2840
       MMB = MMB + 1                                                       LEVO2850
       MEV = NMEV(MMB)                                                     LEVO2860
C      IS MEV TOO LARGE ?                                                  LEVO2870
       IF(MEV.LE.KMAX) GO TO 200                                          LEVO2880
       WRITE(6,190) MMB, MEV, KMAX                                        LEVO2890
   190 FORMAT(/' TERMINATE PRIOR TO CONSIDERING THE',I6,'TH T-MATRIX'/    LEVO2900
      1' BECAUSE THE SIZE REQUESTED',I6,' IS GREATER THAN THE MAXIMUM SIZLEVO2910
      1E ALLOWED',I6/)                                                    LEVO2920
       GO TO 540                                                          LEVO2930
C                                                                          LEVO2940
   200 MP1 = MEV + 1                                                      LEVO2950
       BETAM = BETA(MP1)                                                  LEVO2960
C                                                                          LEVO2970
       IF (IB.GE.0) GO TO 210                                            LEVO2980
C                                                                          LEVO2990
       TO = BTOL                                                          LEVO3000
C                                                                          LEVO3010
C---------------------------------------------------------------------LEVO3020
C                                                                          LEVO3030
       CALL TNORM(ALPHA,BETA,TO,T1,MEV,IBMEV)                            LEVO3040
C                                                                          LEVO3050
C---------------------------------------------------------------------LEVO3060
C                                                                          LEVO3070
       TEMP = TO/TKMAX                                                    LEVO3080
       IBMEV = IABS(IBMEV)                                               LEVO3090
       IF (TEMP.GE.BTOL) GO TO 210                                       LEVO3100
       IBMEV = -IBMEV                                                    LEVO3110
       GO TO 600                                                         LEVO3120
C                                                                          LEVO3130
   210 CONTINUE                                                           LEVO3140
       IC = MXSTUR-ICT                                                   LEVO3150
C                                                                          LEVO3160
C---------------------------------------------------------------------LEVO3170
C      BISEC LOOP. THE SUBROUTINE BISEC INCORPORATES DIRECTLY THE          LEVO3180
C      T-MULTIPLICITY AND SPURIOUS TESTS. T-EIGENVALUES WILL BE            LEVO3190
C      CALCULATED BY BISEC SEQUENTIALLY ON INTERVALS                       LEVO3200
C      (LB(J),UB(J)), J = 1,NINT).                                        LEVO3210
C                                                                          LEVO3220
C      ON RETURN FROM BISEC                                                LEVO3230
C      NDIS = NUMBER OF DISTINCT EIGENVALUES OF T(1,MEV) ON UNION          LEVO3240
C             OF THE (LB,UB) INTERVALS                                    LEVO3250
C      VS = DISTINCT T-EIGENVALUES IN ALGEBRAICALLY INCREASING ORDER       LEVO3260
C      MP = MULTIPLICITIES OF THE T-EIGENVALUES IN VS                      LEVO3270
C      MP(I) = (0,1,MI), MI>1, I=1,NDIS  MEANS:                           LEVO3280
C        (0)  VS(I) IS SPURIOUS                                           LEVO3290
C        (1)  VS(I) IS T-SIMPLE AND GOOD                                  LEVO3300
C        (MI) VS(I) IS MULTIPLE AND IS THEREFORE NOT ONLY GOOD BUT         LEVO3310
C             ALSO A CONVERGED GOOD T-EIGENVALUE.                         LEVO3320
```

```
C                                                                  LEV03330
C                                                                  LEV03340
      CALL BISEC(ALPHA,BETA,V1,V2,VS,LB,UB,EPSM,TTOL,MP,NINT,      LEV03350
     1 MEV,NDIS,IC,IWRITE)                                         LEV03360
C                                                                  LEV03370
C------------------------------------------------------------------LEV03380
C                                                                  LEV03390
      IF (NDIS.EQ.0) GO TO 620                                     LEV03400
C                                                                  LEV03410
C     COMPUTE THE TOTAL NUMBER OF STURM SEQUENCES USED TO DATE     LEV03420
C     COMPUTE THE BISEC CONVERGENCE AND T-MULTIPLICITY TOLERANCES USED. LEV03430
C     COMPUTE THE CONVERGENCE TOLERANCE FOR EIGENVALUES OF A.      LEV03440
      ICT = ICT + IC                                               LEV03450
      TEMP = DFLOAT(MEV+1000)                                      LEV03460
      MULTOL = TEMP*TTOL                                           LEV03470
      TEMP = DSQRT(TEMP)                                           LEV03480
      BISTOL = TTOL*TEMP                                           LEV03490
      CONTOL = BETAM*1.D-10                                        LEV03500
C                                                                  LEV03510
C------------------------------------------------------------------LEV03520
C     SUBROUTINE LUMP 'COMBINES' T-EIGENVALUES THAT ARE 'TOO CLOSE'. LEV03530
C     NOTE HOWEVER THAT CLOSE SPURIOUS T-EIGENVALUES ARE NOT AVERAGED LEV03540
C     WITH GOOD ONES. HOWEVER, THEY MAY BE USED TO INCREASE THE    LEV03550
C     MULTIPLICITY OF A GOOD T-EIGENVALUE.                         LEV03560
C                                                                  LEV03570
      LOOP = NDIS                                                  LEV03580
      CALL LUMP(VS,RELTOL,MULTOL,SCALE2,MP,LOOP)                   LEV03590
C                                                                  LEV03600
C------------------------------------------------------------------LEV03610
C                                                                  LEV03620
      IF(NDIS.EQ.LOOP) GO TO 230                                   LEV03630
C                                                                  LEV03640
      WRITE(6,220) NDIS, MEV, LOOP                                 LEV03650
  220 FORMAT(/I6,' DISTINCT T-EIGENVALUES WERE COMPUTED IN BISEC AT MEV LEV03660
     1',I6/ 2X,' LUMP SUBROUTINE REDUCES NUMBER OF DISTINCT EIGENVALUES LEV03670
     10',I6)                                                       LEV03680
C                                                                  LEV03690
  230 CONTINUE                                                     LEV03700
      NDIS = LOOP                                                  LEV03710
      BETA(MP1) = BETAM                                            LEV03720
C                                                                  LEV03730
C------------------------------------------------------------------LEV03740
C     THE SUBROUTINE ISOEV LABELS THOSE SIMPLE EIGENVALUES OF T(1,MEV) LEV03750
C     WITH VERY SMALL GAPS BETWEEN NEIGHBORING EIGENVALUES OF T(1,MEV) LEV03760
C     TO AVOID COMPUTING ERROR ESTIMATES FOR ANY SIMPLE GOOD       LEV03770
C     T-EIGENVALUE THAT IS TOO CLOSE TO A SPURIOUS EIGENVALUE.     LEV03780
C     ON RETURN FROM ISOEV, G CONTAINS CODED MINIMAL GAPS          LEV03790
C     BETWEEN THE DISTINCT EIGENVALUES OF T(1,MEV). (G IS REAL).   LEV03800
C     G(I) < 0 MEANS MINGAP IS DUE TO LEFT GAP G(I) > 0 MEANS DUE TO LEV03810
C     RIGHT GAP. MP(I) = -1 MEANS THAT THE GOOD T-EIGENVALUE IS SIMPLE LEV03820
C     AND HAS A VERY SMALL MINGAP IN T(1,MEV) DUE TO A SPURIOUS    LEV03830
C     EIGENVALUE.  NG = NUMBER OF GOOD T-EIGENVALUES.              LEV03840
C     NISO = NUMBER OF ISOLATED GOOD T-EIGENVALUES.                LEV03850
C                                                                  LEV03860
      CALL ISOEV(VS,GAPTOL,MULTOL,SCALE1,G,MP,NDIS,NG,NISO)        LEV03870
C                                                                  LEV03880
C------------------------------------------------------------------LEV03890
C                                                                  LEV03900
      WRITE(6,240)NG,NISO,NDIS                                     LEV03910
  240 FORMAT(/I6,' GOOD T-EIGENVALUES HAVE BEEN COMPUTED'/         LEV03920
     1 I6,' OF THESE ARE T-ISOLATED'/                              LEV03930
     2 I6,' = NUMBER OF DISTINCT T-EIGENVALUES COMPUTED'/)         LEV03940
C                                                                  LEV03950
C     DO WE WRITE DISTINCT EIGENVALUES OF T-MATRIX TO FILE 11?     LEV03960
      IF (IDIST.EQ.0) GO TO 280                                    LEV03970
C                                                                  LEV03980
      WRITE(11,250) NDIS,NISO,MEV,N,SVSEED,MATNO                   LEV03990
```

```
  250 FORMAT(/4I6,I12,I8,' = NDIS,NISO,MEV,N,SVSEED,MATNO'/)              LEVO4000
C                                                                        LEVO4010
      WRITE(11,260) (MP(I),VS(I),G(I), I=1,NDIS)                         LEVO4020
  260 FORMAT(2(I3,E25.16,E12.3))                                         LEVO4030
C                                                                        LEVO4040
      WRITE(11,270) NDIS, (MP(I), I=1,NDIS)                              LEVO4050
  270 FORMAT(/I6,' = NDIS, T-MULTIPLICITIES (O MEANS  SPURIOUS)'/(20I4)) LEVO4060
C                                                                        LEVO4070
  280 CONTINUE                                                           LEVO4080
C                                                                        LEVO4090
      IF (NISO.NE.O) GO TO 310                                           LEVO4100
C                                                                        LEVO4110
      WRITE(4,290) MEV                                                   LEVO4120
  290 FORMAT(/' AT MEV = ',I6,' THERE ARE NO ISOLATED T-EIGENVALUES'/    LEVO4130
     1' SO NO ERROR ESTIMATES WERE COMPUTED'/')                         LEVO4140
C                                                                        LEVO4150
      WRITE(6,300)                                                       LEVO4160
  300 FORMAT(/' ALL COMPUTED GOOD T-EIGENVALUES ARE MULTIPLE'/           LEVO4170
     1 ' THEREFORE ALL SUCH EIGENVALUES ARE ASSUMED TO HAVE CONVERGED') LEVO4180
C                                                                        LEVO4190
      ICONV = 1                                                          LEVO4200
      GO TO 350                                                          LEVO4210
C                                                                        LEVO4220
  310 CONTINUE                                                           LEVO4230
C                                                                        LEVO4240
C-----------------------------------------------------------------------LEVO4250
C     SUBROUTINE INVERR COMPUTES ERROR ESTIMATES FOR ISOLATED GOOD       LEVO4260
C     T-EIGENVALUES USING INVERSE ITERATION ON T(1,MEV). ON RETURN       LEVO4270
C     G(J) = MINIMUM GAP IN T(1,MEV) FOR EACH VS(J), J=1,NDIS            LEVO4280
C     G(MEV+I) = BETAM*|U(MEV)| = ERROR ESTIMATE FOR ISOLATED GOOD       LEVO4290
C              T-EIGENVALUES, WHERE I = 1, NISO AND  BETAM = BETA(MEV+1) LEVO4300
C              U(MEV) IS MEVTH COMPONENT OF THE UNIT EIGENVECTOR OF T    LEVO4310
C              CORRESPONDING TO THE ITH ISOLATED GOOD T-EIGENVALUE.      LEVO4320
C     A NEGATIVE ERROR ESTIMATE MEANS THAT FOR THAT PARTICULAR           LEVO4330
C     EIGENVALUE THE INVERSE ITERATION DID NOT CONVERGE IN <= MXINIT     LEVO4340
C     STEPS AND THAT THE CORRESPONDING ERROR ESTIMATE IS QUESTIONABLE.   LEVO4350
C                                                                        LEVO4360
C     V2 CONTAINS THE ISOLATED GOOD T-EIGENVALUES                        LEVO4370
C     V1 CONTAINS THE MINGAPS TO THE NEAREST DISTINCT  EIGENVALUE        LEVO4380
C        OF T(1,MEV) FOR EACH ISOLATED GOOD T-EIGENVALUE IN V2.          LEVO4390
C     VS CONTAINS THE NDIS DISTINCT EIGENVALUES OF T(1,MEV)              LEVO4400
C     MP CONTAINS THE CORRESPONDING CODED T-MULTIPLICITIES               LEVO4410
C                                                                        LEVO4420
      IT = MXINIT                                                        LEVO4430
      CALL INVERR(ALPHA,BETA,V1,V2,VS,EPSM,G,MP,MEV,MMB,NDIS,NISO,N,     LEVO4440
     1   RHSEED,IT,IWRITE)                                              LEVO4450
C                                                                        LEVO4460
C-----------------------------------------------------------------------LEVO4470
C                                                                        LEVO4480
C     SIMPLE CHECK FOR CONVERGENCE. CHECKS TO SEE IF ALL OF THE ERROR    LEVO4490
C     ESTIMATES ARE SMALLER THAN CONTOL = BETAM*1.D-10.                  LEVO4500
C     IF THIS TEST IS SATISFIED, THEN CONVERGENCE FLAG, ICONV IS SET     LEVO4510
C     TO 1.  TYPICALLY ERROR ESTIMATES ARE VERY CONSERVATIVE.           LEVO4520
C                                                                        LEVO4530
      WRITE(6,320) CONTOL                                                LEVO4540
  320 FORMAT(/' CONVERGENCE IS TESTED USING THE CONVERGENCE TOLERANCE',  LEVO4550
     1E13.4/)                                                           LEVO4560
C                                                                        LEVO4570
      II = MEV +1                                                        LEVO4580
      IF = MEV+NISO                                                      LEVO4590
      DO 330 I = II,IF                                                   LEVO4600
      IF (ABS(G(I)).GT.CONTOL) GO TO 350                                 LEVO4610
  330 CONTINUE                                                           LEVO4620
      ICONV = 1                                                          LEVO4630
      MMB = NMEVS                                                        LEVO4640
C                                                                        LEVO4650
      WRITE(6,340) CONTOL                                                LEVO4660
```

```
  340 FORMAT(' ALL COMPUTED ERROR ESTIMATES WERE LESS THAN',E15.4/      LEV04670
     1 ' THEREFORE PROCEDURE TERMINATES'/)                             LEV04680
C                                                                       LEV04690
  350 CONTINUE                                                          LEV04700
C                                                                       LEV04710
C     IF CONVERGENCE IS INDICATED, THAT IS ICONV = 1 ,THEN             LEV04720
C     THE SUBROUTINE PRTEST IS CALLED TO CHECK FOR ANY CONVERGED       LEV04730
C     T-EIGENVALUES THAT HAVE BEEN MISLABELLED AS SPURIOUS BECAUSE     LEV04740
C     THE PROJECTION OF THEIR EIGENVECTOR(S) ON THE STARTING          LEV04750
C     VECTOR WERE TOO SMALL.                                          LEV04760
C     NUMERICAL TESTS INDICATE THAT SUCH EIGENVALUES ARE RARE.        LEV04770
C     IF FOR SOME REASON MANY OF THESE HIDDEN EIGENVALUES APPEAR      LEV04780
C     ON SOME RUN, YOU CAN BE CERTAIN THAT SOMETHING IS FOULED UP.    LEV04790
C                                                                       LEV04800
      IF (ICONV.EQ.0) GO TO 480                                        LEV04810
C                                                                       LEV04820
C----------------------------------------------------------------------LEV04830
C                                                                       LEV04840
      CALL PRTEST (ALPHA,BETA,VS,TKMAX,EPSM,RELTOL,SCALE3,SCALE4,      LEV04850
     1 MP,NDIS,MEV,IPROJ)                                             LEV04860
C                                                                       LEV04870
C----------------------------------------------------------------------LEV04880
C                                                                       LEV04890
      IF(IPROJ.EQ.0) GO TO 470                                         LEV04900
C                                                                       LEV04910
      IF(IDIST.EQ.1) WRITE(11,360) IPROJ                              LEV04920
  360 FORMAT(' SUBROUTINE PRTEST WANTS TO RELABEL',I6,' SPURIOUS T-EIGENLEV04930
     1VALUES'/' WE ACCEPT RELABELLING ONLY IF LAST COMPONENT OF T-EIGENVLEV04940
     1ECTOR IS L.T. 1.D-10'/)                                         LEV04950
C                                                                       LEV04960
      IIX = RHSEED                                                     LEV04970
C                                                                       LEV04980
C----------------------------------------------------------------------LEV04990
C                                                                       LEV05000
      CALL GENRAN(IIX,G,MEV)                                           LEV05010
C                                                                       LEV05020
C----------------------------------------------------------------------LEV05030
C                                                                       LEV05040
      ITEN = -10                                                      LEV05050
      NISOM = NISO + MEV                                              LEV05060
      IWRITO = IWRITE                                                 LEV05070
      IWRITE = 0                                                      LEV05080
C                                                                       LEV05090
      DO 390 J = 1,NDIS                                               LEV05100
      IF(MP(J).NE.ITEN) GO TO 390                                     LEV05110
      TO = VS(J)                                                      LEV05120
C                                                                       LEV05130
C----------------------------------------------------------------------LEV05140
C                                                                       LEV05150
      IT = MXINIT                                                     LEV05160
      CALL INVERM(ALPHA,BETA,V1,V2,TO,TEMP,T1,EPSM,G,MEV,IT,IWRITE)   LEV05170
C                                                                       LEV05180
C----------------------------------------------------------------------LEV05190
C                                                                       LEV05200
      IF(TEMP.LE.1.D-10) GO TO 380                                    LEV05210
C     ERROR ESTIMATE WAS NOT SMALL REJECT RELABELLING OF THIS EIGENVALUELEV05220
      IF(IDIST.EQ.1) WRITE(11,370) J,TO,TEMP                         LEV05230
  370 FORMAT(/' LAST COMPONENT FOR',I6,'TH EIGENVALUE',E20.12/' IS TOO LLEV05240
     1ARGE = ',E15.6,' SO DO NOT ACCEPT PRTEST RELABELLING'/)        LEV05250
      MP(J) = 0                                                       LEV05260
      IPROJ = IPROJ - 1                                               LEV05270
      GO TO 390                                                       LEV05280
C     RELABELLING ACCEPTED                                            LEV05290
  380 NISOM = NISOM + 1                                               LEV05300
      G(NISOM) = BETAM*TEMP                                           LEV05310
  390 CONTINUE                                                        LEV05320
```

```
      IWRITE = IWRIT0                                                    LEV05330
C                                                                       LEV05340
      IF(IPROJ.EQ.0) GO TO 430                                          LEV05350
      WRITE(6,400) IPROJ                                                LEV05360
  400 FORMAT(/I6,' T-EIGENVALUES WERE RECLASSIFIED AS GOOD.'/           LEV05370
     1' THESE ARE IDENTIFIED IN FILE 3 BY A T-MULTIPLICITY OF -10'/' USELEV05380
     2R SHOULD INSPECT EACH TO MAKE SURE NEIGHBORS HAVE CONVERGED'/)    LEV05390
C                                                                       LEV05400
      IF(IDIST.EQ.1) WRITE(11,410) IPROJ                                LEV05410
  410 FORMAT(/I6,' T-EIGENVALUES WERE RELABELLED AS GOOD'/              LEV05420
     1' BELOW IS CORRECTED T-MULTIPLICITY PATTERN'/)                    LEV05430
C                                                                       LEV05440
      WRITE(6,420) NDIS, (MP(I), I=1,NDIS)                              LEV05450
      IF(IDIST.EQ.1) WRITE(11,420) NDIS, (MP(I), I=1,NDIS)             LEV05460
  420 FORMAT(/I6,' = NDIS, T-MULTIPLICITIES (0 MEANS   SPURIOUS)'/      LEV05470
     1 6X, ' (-10) MEANS SPURIOUS T-EIGENVALUE RELABELLED AS GOOD'/(20I4LEV05480
     1))                                                                LEV05490
C                                                                       LEV05500
C     RECALCULATE MINGAPS FOR DISTINCT T(1,MEV) EIGENVALUES.            LEV05510
  430 NM1 = NDIS - 1                                                    LEV05520
      G(NDIS) = VS(NM1)-VS(NDIS)                                        LEV05530
      G(1) = VS(2)-VS(1)                                                LEV05540
C                                                                       LEV05550
      DO 440 J = 2,NM1                                                  LEV05560
      T0 = VS(J)-VS(J-1)                                                LEV05570
      T1 = VS(J+1)-VS(J)                                                LEV05580
      G(J) = T1                                                         LEV05590
      IF (T0.LT.T1) G(J) = -T0                                          LEV05600
  440 CONTINUE                                                          LEV05610
      IF(IPROJ.EQ.0) GO TO 470                                          LEV05620
C     WRITE TO FILE 4 ERROR ESTIMATES FOR THOSE T-EIGENVALUES RELABELLEDLEV05630
      NGOOD = 0                                                         LEV05640
      DO 450 J = 1,NDIS                                                 LEV05650
      IF(MP(J).EQ.0) GO TO 450                                          LEV05660
      NGOOD = NGOOD + 1                                                 LEV05670
      IF(MP(J).NE.ITEN) GO TO 450                                       LEV05680
      T0 = VS(J)                                                        LEV05690
      NISO = NISO + 1                                                   LEV05700
      NISOM = MEV + NISO                                                LEV05710
      WRITE(4,460) NGOOD,T0,G(NISOM),G(J)                               LEV05720
  450 CONTINUE                                                          LEV05730
  460 FORMAT(I10,E25.16,2E14.3)                                         LEV05740
C                                                                       LEV05750
  470 CONTINUE                                                          LEV05760
C                                                                       LEV05770
C     WRITE THE GOOD T-EIGENVALUES TO FILE 3.  FIRST TRANSFER THEM      LEV05780
C     TO V2 AND THEIR T-MULTIPLICITIES TO THE CORRESPONDING POSITIONS   LEV05790
C     IN MP AND COMPUTE THE A-MINGAPS, THE MINIMAL GAPS BETWEEN THE     LEV05800
C     GOOD T-EIGENVALUES.  THESE GAPS WILL BE PUT IN THE ARRAY G.       LEV05810
C     SINCE G CURRENTLY CONTAINS THE MINIMAL GAPS BETWEEN THE DISTINCT  LEV05820
C     EIGENVALUES OF THE T-MATRIX, THESE GAPS WILL FIRST BE             LEV05830
C     TRANSFERRED TO V1.  NOTE THAT V1<0 MEANS THAT THAT MINIMAL GAP    LEV05840
C     IN THE T-MATRIX IS DUE TO A SPURIOUS T-EIGENVALUE.                LEV05850
C     ALL THIS INFORMATION IS PRINTED TO FILE 3                         LEV05860
C                                                                       LEV05870
  480 CONTINUE                                                          LEV05880
C                                                                       LEV05890
      NG = 0                                                            LEV05900
      DO 490 I = 1,NDIS                                                 LEV05910
      IF (MP(I).EQ.0) GO TO 490                                         LEV05920
      NG = NG+1                                                         LEV05930
      MP(NG) = MP(I)                                                    LEV05940
      V2(NG) = VS(I)                                                    LEV05950
      TEMP = G(I)                                                       LEV05960
      TEMP = DABS(TEMP)                                                 LEV05970
      J = I+1                                                           LEV05980
      IF (G(I).LT.ZERO) J = I-1                                         LEV05990
```

```
      IF (MP(J).EQ.0) TEMP = -TEMP                               LEV06000
      V1(NG) = TEMP                                              LEV06010
  490 CONTINUE                                                   LEV06020
C                                                                LEV06030
      WRITE(6,500)MEV                                            LEV06040
  500 FORMAT(//' T-EIGENVALUE CALCULATION AT MEV = ',I6,'    IS COMPLETELEV06050
     1')                                                         LEV06060
C                                                                LEV06070
C     NG = NUMBER OF COMPUTED DISTINCT GOOD T-EIGENVALUES.  NEXT LEV06080
C     GENERATE GAPS BETWEEN GOOD T-EIGENVALUES (AMINGAPS) AND PUT THEM LEV06090
C     IN G.  G(J) < 0 MEANS THE AMINGAP IS DUE TO THE LEFT-HAND GAP. LEV06100
C                                                                LEV06110
      NGM1 = NG - 1                                              LEV06120
      G(NG) = V2(NGM1)-V2(NG)                                    LEV06130
      G(1) = V2(2)-V2(1)                                         LEV06140
C                                                                LEV06150
      DO 510 J = 2,NGM1                                          LEV06160
      TO = V2(J)-V2(J-1)                                         LEV06170
      T1 = V2(J+1)-V2(J)                                         LEV06180
      G(J) = T1                                                  LEV06190
      IF (TO.LT.T1) G(J) = -TO                                   LEV06200
  510 CONTINUE                                                   LEV06210
C                                                                LEV06220
C     WRITE GOOD T-EIGENVALUES OUT TO FILE 3.                    LEV06230
C                                                                LEV06240
      WRITE(3,520)NG,NDIS,MEV,N,SVSEED,MATNO,MULTOL,IB,BTOL      LEV06250
  520 FORMAT(4I6,I12,I8,' = NG,NDIS,MEV,N,SVEED,MATNO'/          LEV06260
     1 E20.12,I6,E13.4,' = MUTOL,INDEX MINIMAL BETA,BTOL'/       LEV06270
     1' EV NO',1X,'TMULT',10X,'GOOD EIGENVALUE',7X,'TMINGAP',7X,'AMINGAPLEV06280
     1')                                                         LEV06290
C                                                                LEV06300
      WRITE(3,530)(I,MP(I),V2(I),V1(I),G(I), I=1,NG)             LEV06310
  530 FORMAT(2I6,E25.16,2E14.3)                                  LEV06320
C                                                                LEV06330
C     IF CONVERGENCE FLAG ICONV.NE.1 AND NUMBER OF T-MATRICES    LEV06340
C     CONSIDERED TO DATE IS LESS THAN NUMBER ALLOWED, INCREMENT MEV. LEV06350
C     AND LOOP BACK TO 210 TO REPEAT COMPUTATIONS.  RESTORE BETA(MEV+1).LEV06360
C                                                                LEV06370
      BETA(MP1) = BETAM                                          LEV06380
C                                                                LEV06390
      IF (MMB.LT.NMEVS.AND.ICONV.NE.1) GO TO 180                 LEV06400
C                                                                LEV06410
C     END OF LOOP ON DIFFERENT SIZE T-MATRICES ALLOWED.          LEV06420
C                                                                LEV06430
  540 CONTINUE                                                   LEV06440
C                                                                LEV06450
      IF(ISTOP.EQ.0) WRITE(6,550)                                LEV06460
  550 FORMAT(/' T-MATRICES (ALPHA AND BETA) ARE NOW AVAILABLE, TERMINATELEV06470
     1')                                                         LEV06480
      IF (IHIS.EQ.1.AND.KMAX.NE.MOLD) WRITE(1,560)               LEV06490
  560 FORMAT(/' ABOVE ARE THE FOLLOWING VECTORS '/               LEV06500
     1 '   ALPHA(I), I = 1,KMAX'/                                LEV06510
     2 '   BETA(I), I = 1,KMAX+1'/                               LEV06520
     3 ' FINAL TWO LANCZOS VECTORS OF ORDER N FOR I = KMAX,KMAX+1'/ LEV06530
     4 ' ALL VECTORS IN THIS FILE HAVE HEX FORMAT 4Z20 '/        LEV06540
     5 ' ----- END OF FILE 1 NEW ALPHA, BETA HISTORY---------------'////)LEV06550
C                                                                LEV06560
      IF (ISTOP.EQ.0) GO TO 640                                  LEV06570
C                                                                LEV06580
      WRITE(3,570)                                               LEV06590
  570 FORMAT(/' ABOVE ARE COMPUTED GOOD T-EIGENVALUES'/          LEV06600
     1 ' NG = NUMBER OF GOOD T-EIGENVALUES COMPUTED'/            LEV06610
     2 ' NDIS = NUMBER OF COMPUTED DISTINCT EIGENVALUES OF T(1,MEV)'/ LEV06620
     3 ' N = ORDER OF A,  MATNO = MATRIX IDENT'/                 LEV06630
     4 ' MULTOL = T-MULTIPLICITY TOLERANCE FOR T-EIGENVALUES IN BISEC'/ LEV06640
     4 ' TMULT IS THE T-MULTIPLICITY OF GOOD T-EIGENVALUE'/      LEV06650
     5 ' TMULT = -1 MEANS SPURIOUS T-EIGENVALUE TOO CLOSE'/      LEV06660
```

```
      6 ' DO NOT COMPUTE ERROR ESTIMATES FOR SUCH EIGENVALUES'/        LEV06670
      7 ' AMINGAP = MINIMAL GAP BETWEEN THE COMPUTED A-EIGENVALUES'/    LEV06680
      8 ' AMINGAP .LT. 0. MEANS MINIMAL GAP IS DUE TO LEFT-HAND GAP'/   LEV06690
      9 ' TMINGAP= MINIMAL GAP W.R.T.  DISTINCT EIGENVALUES IN T(1,MEV)'/LEV06700
      1 ' TMINGAP .LT. 0. MEANS MINGAP IS DUE TO SPURIOUS T-EIGENVALUE'/ LEV06710
      2 ' ----- END OF FILE 3 GOODEIGENVALUES----------------------'///)LEV06720
C                                                                       LEV06730
      IF (IDIST.EQ.1) WRITE(11,580)                                     LEV06740
  580 FORMAT(/' ABOVE ARE THE DISTINCT EIGENVALUES OF T(1,MEV).'/       LEV06750
      2 ' THE FORMAT IS       T-MULTIPLICITY    T-EIGENVALUE   TMINGAP'/ LEV06760
      3 '         THIS FORMAT IS REPEATED TWICE ON EACH LINE.'/         LEV06770
      4 ' T-MULTIPLICITY = -1 MEANS THAT THE SUBROUTINE ISOEV HAS TAGGED'LEV06780
      5/'   THIS SIMPLE T-EIGENVALUE AS HAVING A VERY CLOSE SPURIOUS'/  LEV06790
      6 '   T-EIGENVALUE SO THAT NO ERROR ESTIMATE WILL BE COMPUTED'/   LEV06800
      7 '   FOR THAT EIGENVALUE IN SUBROUTINE INVERR.'/                 LEV06810
      8 ' TMINGAP .LT. 0, TMINGAP IS DUE TO LEFT GAP .GT. 0, RIGHT GAP.'/LEV06820
      9 ' EACH OF THE DISTINCT T-EIGENVALUE TABLES IS FOLLOWED'/        LEV06830
      9 ' BY THE T-MULTIPLICITY PATTERN.'/                              LEV06840
      1 ' NDIS = NUMBER OF COMPUTED DISTINCT EIGENVALUES OF T(1,MEV).'/ LEV06850
      2 ' NG = NUMBER OF GOOD T-EIGENVALUES. '/                         LEV06860
      3 ' NISO = NUMBER OF ISOLATED GOOD T-EIGENVALUES. '/             LEV06870
      4 ' NISO ALSO IS THE COUNT OF +1 ENTRIES IN T-MULTIPLICITY PATTERN.LEV06880
      5 '/' ----- END OF FILE 11 DISTINCT T-EIGENVALUES--------------'///LEV06890
      6 )                                                               LEV06900
C                                                                       LEV06910
      IF(NISO.NE.0)  WRITE(4,590)                                       LEV06920
  590 FORMAT(/' ABOVE ARE THE ERROR ESTIMATES OBTAINED FOR THE ISOLATED LEV06930
      1GOOD T-EIGENVALUES'/                                             LEV06940
      1' OBTAINED VIA INVERSE ITERATION IN THE SUBROUTINE INVERR.'/     LEV06950
      1' ALL OTHER GOOD T-EIGENVALUES HAVE CONVERGED.'/                 LEV06960
      2' ERROR ESTIMATE = BETAM*ABS(UM)'/                               LEV06970
      2' WHERE BETAM = BETA(MEV+1) AND UM = U(MEV).'/                   LEV06980
      3' U = UNIT EIGENVECTOR OF T WHERE T*U = EV*U AND EV = ISOLATED GOOLEV06990
      3D T-EIGENVALUE.'/                                                LEV07000
      4' TMINGAP = GAP TO NEAREST DISTINCT EIGENVALUE OF T(1,MEV).'/    LEV07010
      5' TMINGAP .LT. 0. MEANS MINGAP IS DUE TO A LEFT NEIGHBOR.'/      LEV07020
      6' ERROR ESTIMATE L.T. 0 MEANS INVERSE ITERATION DID NOT CONVERGE'/LEV07030
      7' ------ END OF FILE 4 ERRINV -----------------------------'//)  LEV07040
      GO TO 640                                                         LEV07050
C                                                                       LEV07060
  600 CONTINUE                                                          LEV07070
C                                                                       LEV07080
      IBB = IABS(IBMEV)                                                 LEV07090
      IF (IBMEV.LT.0) WRITE(6,610) MEV,IBB,BETA(IBB)                    LEV07100
  610 FORMAT(/' PROGRAM TERMINATES BECAUSE MEV REQUESTED = ',I6,' IS .GTLEV07110
      1',I6/' AT WHICH AN ABNORMALLY SMALL BETA = ' , E13.4,' OCCURRED'/)LEV07120
      GO TO 640                                                         LEV07130
C                                                                       LEV07140
  620 IF (NDIS.EQ.0.AND.ISTOP.GT.0) WRITE(6,630)                        LEV07150
  630 FORMAT(/' INTERVALS SPECIFIED FOR BISECT DID NOT CONTAIN ANY T-EIGLEV07160
      1ENVALUES'/' PROGRAM TERMINATES')                                 LEV07170
C                                                                       LEV07180
  640 CONTINUE                                                          LEV07190
C                                                                       LEV07200
      STOP                                                              LEV07210
C-----END OF MAIN PROGRAM FOR LANCZOS EIGENVALUE COMPUTATIONS----------LEV07220
      END                                                               LEV07230
```

SECTION 2.4   MAIN PROGRAM, EIGENVECTOR COMPUTATIONS

```
C-----LEVEC (EIGENVECTORS OF REAL SYMMETRIC MATRICES)-------------------LEV00010
C                                                                       LEV00020
C      CONTAINS MAIN PROGRAM FOR COMPUTING AN EIGENVECTOR CORRESPONDING  LEV00030
C      TO EACH OF A SET OF EIGENVALUES WHICH HAVE BEEN COMPUTED          LEV00040
C      ACCURATELY BY THE CORRESPONDING LANCZOS EIGENVALUE PROGRAM        LEV00050
C      (LEVAL) FOR REAL SYMMETRIC MATRICES.  THIS PROGRAM COULD BE       LEV00060
C      MODIFIED TO COMPUTE ADDITIONAL EIGENVECTORS FOR ANY EIGENVALUE    LEV00070
C      WHICH IS A MULTIPLE EIGENVALUE OF THE GIVEN A-MATRIX.  THE        LEV00080
C      AMOUNT OF ADDITIONAL COMPUTATION REQUIRED WOULD DEPEND UPON       LEV00090
C      THE GIVEN A-MATRIX AND UPON WHAT PART OF THE SPECTRUM OF          LEV00100
C      A IS INVOLVED.                                                    LEV00110
C                                                                       LEV00120
C      THE LANCZOS EIGENVECTOR COMPUTATIONS ASSUME THAT EACH            LEV00130
C      EIGENVALUE THAT IS BEING CONSIDERED HAS CONVERGED AS AN          LEV00140
C      EIGENVALUE OF THE LANCZOS TRIDIAGONAL MATRICES.                  LEV00150
C                                                                       LEV00160
C      PFORT VERIFIER IDENTIFIED THE FOLLOWING NONPORTABLE             LEV00170
C      CONSTRUCTIONS                                                    LEV00180
C                                                                       LEV00190
C      1.  DATA/MACHEP/ STATEMENT                                      LEV00200
C      2.  ALL READ(5,*) STATEMENTS (FREE FORMAT)                      LEV00210
C      3.  FORMAT(20A4) USED WITH THE EXPLANATORY HEADER, EXPLAN       LEV00220
C      4.  HEXADECIMAL FORMAT (4Z20) USED IN ALPHA/BETA FILES 1 AND 2. LEV00230
C.                                                                      LEV00240
C      IMPORTANT NOTE:  THIS PROGRAM ALLOWS ENLARGEMENT OF THE ALPHA,  LEV00250
C      BETA ARRAYS.  IN PARTICULAR, IF ANY ONE OF THE EIGENVALUES      LEV00260
C      SUPPLIED IS T-SIMPLE AND NOT CLOSE TO A SPURIOUS EIGENVALUE,    LEV00270
C      THE PROGRAM REQUIRES THAT KMAX BE AT LEAST 11*MEV/8 + 12.  IF   LEV00280
C      KMAX IS NOT THIS LARGE, THEN THE PROGRAM RESETS KMAX TO THIS    LEV00290
C      SIZE AND EXTENDS THE ALPHA, BETA HISTORY IF REQUIRED.           LEV00300
C      THUS, THE DIMENSIONS OF THE ALPHA AND BETA ARRAYS MUST BE       LEV00310
C      LARGE ENOUGH TO ALLOW FOR THIS POSSIBILITY.                     LEV00320
C      REMEMBER THAT THE BETA ARRAY, BETA(J), IS SUCH THAT            LEV00330
C      J = 1,..., KMAX+1.  SO IF THE KMAX USED BY THE PROGRAM         LEV00340
C      IS TO BE 3000, THEN BETA MUST BE OF LENGTH AT LEAST 3001.       LEV00350
C                                                                       LEV00360
C----------------------------------------------------------------------LEV00370
       DOUBLE PRECISION   ALPHA(5000),BETA(5001)                        LEV00380
       DOUBLE PRECISION   V1(5000),V2(5000)                             LEV00390
       DOUBLE PRECISION   RITVEC(30000),TVEC(30000),GOODEV(50),EVNEW(50) LEV00400
       DOUBLE PRECISION   EVAL,EVALN,TOLN,TTOL,ERTOL,ALFA,BATA          LEV00410
       DOUBLE PRECISION   MULTOL,SCALEO,STUTOL,BTOL,LB,UB               LEV00420
       DOUBLE PRECISION   ONE,ZERO,MACHEP,EPSM,TEMP,SUM,ERRMIN,BKMIN    LEV00430
       DOUBLE PRECISION   RELTOL,ERROR,TERROR,TLAST(50)                 LEV00440
       REAL   G(10000),AMINGP(50),TMINGP(50),EXPLAN(20)                 LEV00450
       REAL   TERR(50),ERR(50),ERRDGP(50),RNORM(50),TBETA(50)           LEV00460
       INTEGER   MP(50),M1(50),M2(50),MA(50),ML(50),MINT(50),MFIN(50)   LEV00470
       INTEGER   SVSEED,SVSOLD,RHSEED,IDELTA(50)                        LEV00480
       INTEGER   MBOUND,NTVCON,SVTVEC,TVSTOP,LVCONT,ERCONT,TFLAG        LEV00490
       DOUBLE PRECISION   FINPRO                                        LEV00500
       DOUBLE PRECISION   DABS, DMAX1, DSQRT, DFLOAT                    LEV00510
       REAL   ABS                                                       LEV00520
       INTEGER   IABS                                                   LEV00530
C----------------------------------------------------------------------LEV00540
       EXTERNAL CMATV                                                   LEV00550
       DATA MACHEP/Z3410000000000000/                                   LEV00560
       EPSM = 2.D0*MACHEP                                               LEV00570
C----------------------------------------------------------------------LEV00580
C                                                                       LEV00590
C      ARRAYS MUST BE DIMENSIONED AS FOLLOWS:                          LEV00600
C      1.  ALPHA:  >= KMAXN,  BETA: >= (KMAXN+1) WHERE KMAXN, THE      LEV00610
C                 LARGEST SIZE T-MATRIX CONSIDERED BY THE PROGRAM,     LEV00620
C                 IS THE LARGER OF THE SIZE OF THE ALPHA, BETA HISTORY LEV00630
C                 PROVIDED ON FILE 2 (IF ANY ) AND THE SIZE WHICH THE  LEV00640
```

```
C                    PROGRAM SPECIFIES INTERNALLY, THIS LATTER IS ALWAYS    LEV00650
C                    < = 11*MEV / 8 +  12, WHERE MEV IS THE SIZE            LEV00660
C                    T-MATRIX THAT WAS USED IN THE CORRESPONDING EIGENVALUE LEV00670
C                    COMPUTATIONS.                                          LEV00680
C        2.  V1:  >= MAX(N,KMAX)                                            LEV00690
C        3.  V2:  >= N                                                      LEV00700
C        4.  G:  >= MAX(N,KMAX)                                             LEV00710
C        5.  RITVEC: >= N*NGOOD, WHERE NGOOD IS NUMBER OF EIGENVALUES       LEV00720
C                    SUPPLIED TO THIS PROGRAM.                              LEV00730
C        6.  TVEC:  >= CUMULATIVE LENGTH OF ALL THE T-EIGENVECTORS          LEV00740
C                    NEEDED TO GENERATE THE DESIRED RITZ VECTORS.  AN EDUCATED LEV00750
C                    GUESS AT AN APPROPRIATE LENGTH CAN BE OBTAINED BY RUNNING THE LEV00760
C                    PROGRAM WITH THE FLAG MBOUND = 1 AND MULTIPLYING THE   LEV00770
C                    RESULTING SIZE BY 5/4.                                 LEV00780
C        7.  GCODEV, AMINGP, TMINGP, TERR, ERR, ERRGDP, RNORM, TBETA,       LEV00790
C            TLAST, EVNEW, MP, MA, M1, M2, MINT, MFIN AND IDELTA ALL MUST   LEV00800
C            BE >= NGOOD.                                                   LEV00810
C                                                                           LEV00820
C--------------------------------------------------------------------LEV00830
C        OUTPUT HEADER                                                      LEV00840
         WRITE(6,10)                                                        LEV00850
      10 FORMAT(/' LANCZOS EIGENVECTOR PROCEDURE FOR REAL SYMMETRIC MATRICELEV00860
     1S'/)                                                                  LEV00870
C                                                                           LEV00880
C        SET PROGRAM PARAMETERS                                             LEV00890
C        USER MUST NOT MODIFY SCALE0                                        LEV00900
         SCALE0 = 5.0D0                                                     LEV00910
         ZERO = 0.0D0                                                       LEV00920
         ONE = 1.0D0                                                        LEV00930
         MPMIN = -1000                                                      LEV00940
C        SET CONVERGENCE CRITERION FOR T-EIGENVECTORS.                      LEV00950
         ERTOL = 1.D-10                                                     LEV00960
C                                                                           LEV00970
C        READ USER-SPECIFIED PARAMETER FROM INPUT FILE 5 (FREE FORMAT)      LEV00980
C                                                                           LEV00990
C        READ USER-PROVIDED HEADER FOR RUN                                  LEV01000
         READ(5,20) EXPLAN                                                  LEV01010
         WRITE(6,20) EXPLAN                                                 LEV01020
      20 FORMAT(20A4)                                                       LEV01030
C                                                                           LEV01040
C        READ IN THE MAXIMUM PERMISSIBLE DIMENSIONS FOR THE TVEC ARRAY      LEV01050
C        (MDIMTV), FOR THE RITVEC ARRAY (MDIMRV), AND FOR THE BETA          LEV01060
C        ARRAY (MBETA).                                                     LEV01070
         READ(5,20) EXPLAN                                                  LEV01080
         READ(5,*) MDIMTV, MDIMRV, MBETA                                    LEV01090
C                                                                           LEV01100
C        READ IN RELATIVE TOLERANCE (RELTOL) USED IN DETERMINING            LEV01110
C        APPROPRIATE SIZES FOR THE T-MATRICES TO BE USED IN THE RITZ        LEV01120
C        VECTOR COMPUTATIONS.                                               LEV01130
         READ(5,20) EXPLAN                                                  LEV01140
         READ(5,*) RELTOL                                                   LEV01150
C                                                                           LEV01160
C        SET FLAGS TO 0 OR 1:                                               LEV01170
C        MBOUND = 1:  PROGRAM TERMINATES AFTER COMPUTING 1ST GUESSES        LEV01180
C                     ON APPROPRIATE T-SIZES FOR USE IN THE RITZ VECTOR     LEV01190
C                     COMPUTATIONS                                          LEV01200
C        NTVCON = 0:  PROGRAM TERMINATES IF THE TVEC ARRAY IS NOT           LEV01210
C                     LARGE ENOUGH TO HOLD ALL THE T-EIGENVECTORS REQUIRED.LEV01220
C        SVTVEC = 0:  THE T-EIGENVECTORS ARE NOT WRITTEN TO FILE 11         LEV01230
C                     UNLESS TVSTOP = 1                                     LEV01240
C        SVTVEC = 1:  WRITE THE T-EIGENVECTORS TO FILE 11.                  LEV01250
C        TVSTOP = 1:  PROGRAM TERMINATES AFTER COMPUTING THE                LEV01260
C                     T-EIGENVECTORS                                        LEV01270
C        LVCONT = 0:  PROGRAM TERMINATES IF THE NUMBER OF T-EIGENVECTORS    LEV01280
C                     COMPUTED IS NOT EQUAL TO THE NUMBER OF RITZ           LEV01290
C                     VECTORS REQUESTED.                                    LEV01300
C        ERCONT = 0:  MEANS FOR ANY GIVEN EIGENVALUE, A RITZ VECTOR         LEV01310
```

```
C                      WILL NOT BE COMPUTED FOR THAT EIGENVALUE UNLESS      LEV01320
C                      A T-EIGENVECTOR HAS BEEN IDENTIFIED WITH A LAST      LEV01330
C                      COMPONENT WHICH SATISFIES THE SPECIFIED              LEV01340
C                      CONVERGENCE CRITERION.                               LEV01350
C      ERCONT = 1:     MEANS FOR ANY GIVEN EIGENVALUE, A RITZ VECTOR        LEV01360
C                      WILL BE COMPUTED.  IF A T-EIGENVECTOR CANNOT         LEV01370
C                      BE IDENTIFIED WHICH SATISFIES THE LAST              LEV01380
C                      COMPONENT CRITERION, THEN THE PROGRAM WILL          LEV01390
C                      USE THE T-VECTOR THAT CAME CLOSEST TO               LEV01400
C                      SATISFYING THE CRITERION.                          LEV01410
C      IWRITE = 1:     EXTENDED OUTPUT OF INTERMEDIATE COMPUTATIONS        LEV01420
C                      IS WRITTEN TO FILE 6                                LEV01430
C      IREAD = 0:      ALPHA/BETA FILE IS REGENERATED.                     LEV01440
C      IREAD = 1:      ALPHA/BETA FILE USED IN EIGENVALUE COMPUTATIONS     LEV01450
C                      IS READ IN AND EXTENDED IF NECESSARY.  IN BOTH      LEV01460
C                      CASES IREAD = 0 OR 1, THE LANCZOS VECTORS ARE       LEV01470
C                      ALWAYS REGENERATED FOR THE RITZ VECTOR             LEV01480
C                      COMPUTATIONS                                        LEV01490
C                                                                          LEV01500
       READ(5,20) EXPLAN                                                   LEV01510
       READ(5,*) MBOUND,NTVCON,SVTVEC,IREAD                               LEV01520
C                                                                          LEV01530
       READ(5,20) EXPLAN                                                   LEV01540
       READ(5,*) TVSTOP,LVCONT,ERCONT,IWRITE                             LEV01550
       IF (TVSTOP.EQ.1) SVTVEC = 1                                        LEV01560
C                                                                          LEV01570
C      READ IN SEED (RHSEED) FOR GENERATING RANDOM STARTING VECTOR         LEV01580
C      FOR INVERSE ITERATION ON THE T-MATRICES.                           LEV01590
       READ(5,20) EXPLAN                                                   LEV01600
       READ(5,*) RHSEED                                                    LEV01610
C                                                                          LEV01620
C      READ IN MATNO = MATRIX/RUN IDENTIFICATION NUMBER AND               LEV01630
C      N = ORDER OF A-MATRIX                                               LEV01640
       READ(5,20) EXPLAN                                                   LEV01650
       READ(5,*) MATNO,N                                                   LEV01660
C                                                                          LEV01670
C------------------------------------------------------------------LEV01680
C      INITIALIZE THE ARRAYS FOR THE USER-SPECIFIED MATRIX                LEV01690
C      AND PASS THE STORAGE LOCATIONS OF THESE ARRAYS TO THE              LEV01700
C      MATRIX-VECTOR MULTIPLY SUBROUTINE CMATV.                           LEV01710
C                                                                          LEV01720
       CALL USPEC(N,MATNO)                                                 LEV01730
C                                                                          LEV01740
C------------------------------------------------------------------LEV01750
C      MASK UNDERFLOW AND OVERFLOW                                         LEV01760
       CALL MASK                                                           LEV01770
C                                                                          LEV01780
C------------------------------------------------------------------LEV01790
C      WRITE RUN PARAMETERS OUT TO FILE 6                                 LEV01800
C                                                                          LEV01810
       WRITE(6,30) MATNO,N                                                 LEV01820
   30 FORMAT(/' MATRIX IDENTIFICATION NO. = ',I10,' ORDER OF A = ',I5)     LEV01830
C                                                                          LEV01840
       WRITE(6,40) MBOUND,NTVCON,SVTVEC,IREAD                             LEV01850
   40 FORMAT(/3X,'MBOUND',3X,'NTVCON',3X,'SVTVEC',3X,'IREAD'/319,I8)      LEV01860
C                                                                          LEV01870
       WRITE(6,50) TVSTOP,LVCONT,ERCONT,IWRITE                           LEV01880
   50 FORMAT(/3X,'TVSTOP',3X,'LVCONT',3X,'ERCONT',3X,'IWRITE'/419)        LEV01890
C                                                                          LEV01900
       WRITE(6,60) MDIMTV,MDIMRV,MBETA                                     LEV01910
   60 FORMAT(/3X,'MDIMTV',3X,'MDIMRV',3X,'MBETA'/219,I8)                   LEV01920
C                                                                          LEV01930
       WRITE(6,70) RELTOL,RHSEED                                           LEV01940
   70 FORMAT(/7X,'RELTOL',3X,'RHSEED'/E13.4,I9)                           LEV01950
C                                                                          LEV01960
C                                                                          LEV01970
C      FROM FILE 3 READ IN THE NUMBER OF EIGENVALUES (NGOOD) FOR WHICH     LEV01980
```

```
C       EIGENVECTORS ARE REQUESTED, THE ORDER (MEV) OF THE LANCZOS          LEV01990
C       TRIDIAGONAL MATRIX USED IN COMPUTING THESE EIGENVALUES, THE         LEV02000
C       ORDER (NOLD) OF THE USER-SPECIFIED MATRIX USED IN THE EIGENVALUE    LEV02010
C       COMPUTATIONS, THE SEED (SVSEED) USED FOR GENERATING THE STARTING    LEV02020
C       VECTOR THAT WAS USED IN THOSE LANCZOS EIGENVALUE COMPUTATIONS,      LEV02030
C       AND THE MATRIX/RUN IDENTIFICATION NUMBER (MATOLD) USED IN THOSE     LEV02040
C       COMPUTATIONS.  ALSO READ IN THE NUMBER (NDIS) OF DISTINCT           LEV02050
C       EIGENVALUES OF T(1,MEV) THAT WERE COMPUTED BUT THIS VALUE IS        LEV02060
C       NOT USED IN THE EIGENVECTOR COMPUTATIONS.                          LEV02070
C                                                                          LEV02080
        READ(3,80) NGOOD,NDIS,MEV,NOLD,SVSEED,MATOLD                       LEV02090
     80 FORMAT(4I6,I12,I8)                                                 LEV02100
C                                                                          LEV02110
C       READ IN THE T-MULTIPLICITY TOLERANCE USED IN THE BISEC SUBROUTINE  LEV02120
C       DURING THE COMPUTATION OF THE GIVEN EIGENVALUES.                   LEV02130
C       ALSO READ IN THE FLAG IB.  IF IB < 0, THEN SOME BETA(I) IN THE     LEV02140
C       T-MATRIX FILE PROVIDED ON FILE 2 FAILED THE ORTHOGONALITY          LEV02150
C       TEST IN THE TNORM SUBROUTINE.  USER SHOULD NOTE THAT THIS VECTOR   LEV02160
C       PROGRAM PROCEEDS INDEPENDENTLY OF THE SIZE OF THE BETA USED.       LEV02170
C                                                                          LEV02180
        READ(3,90) MULTOL,IB,BTOL                                         LEV02190
     90 FORMAT(E20.12,I6,E13.4)                                           LEV02200
C                                                                          LEV02210
        TEMP = DFLOAT(MEV+1000)                                           LEV02220
        TTOL = MULTOL/TEMP                                                LEV02230
        WRITE(6,100) MULTOL,TTOL                                          LEV02240
    100 FORMAT(/' T-MULTIPLICITY TOLERANCE USED IN THE EIGENVALUE COMPUTATLEV02250
       1IONS WAS',E13.4/' SCALED MACHINE EPSILON IS',E13.4)              LEV02260
C                                                                          LEV02270
C       CONTINUE WRITE TO FILE 6 OF THE PARAMETERS FOR THIS RUN           LEV02280
C                                                                          LEV02290
        WRITE(6,110)NGOOD,NDIS,MEV,NOLD,MATOLD,SVSEED,MULTOL,IB,BTOL      LEV02300
    110 FORMAT(/' EIGENVALUES SUPPLIED ARE READ IN FROM FILE 3'/' FILE 3  LEV02310
       1HEADER IS'/4X,'NG',2X,'NDIS',3X,'MEV',2X,'NOLD',2X,'MATOLD',4X,   LEV02320
       1'SVSEED',6X,'MULTOL',6X,'IB',9X,'BTOL'/4I6,I8,I10,E12.3,I8,E13.4/)LEV02330
C                                                                          LEV02340
C       IS THE ARRAY RITVEC LONG ENOUGH TO HOLD ALL OF THE DESIRED        LEV02350
C       RITZ VECTORS (APPROXIMATE EIGENVECTORS)?                          LEV02360
        NMAX = NGOOD*N                                                    LEV02370
        IF(MBOUND.NE.0) GO TO 120                                         LEV02380
        IF(TVSTOP.NE.1.AND.NMAX.GT.MDIMRV) GO TO 1310                     LEV02390
C                                                                          LEV02400
C       CHECK THAT THE ORDER N AND THE MATRIX IDENTIFICATION NUMBER       LEV02410
C       MATNO SPECIFIED BY THE USER AGREE WITH THOSE READ IN FROM         LEV02420
C       FILE 3.                                                          LEV02430
    120 ITEMP = (NOLD-N)**2+(MATOLD-MATNO)**2                            LEV02440
        IF (ITEMP.NE.0) GO TO 1330                                        LEV02450
C                                                                          LEV02460
C       READ IN FROM FILE 3, THE T-MULTIPLICITIES OF THE EIGENVALUES      LEV02470
C       WHOSE EIGENVECTORS ARE TO BE COMPUTED, THE VALUES OF THESE        LEV02480
C       EIGENVALUES AND THEIR MINIMAL GAPS AS EIGENVALUES OF THE          LEV02490
C       USER-SPECIFIED MATRIX AND AS EIGENVALUES OF THE T-MATRIX.         LEV02500
C                                                                          LEV02510
        READ(3,20) EXPLAN                                                 LEV02520
        READ(3,130) (MP(J),GOODEV(J),TMINGP(J),AMINGP(J), J=1,NGOOD)      LEV02530
    130 FORMAT(6X,I6,E25.16,2E14.3)                                       LEV02540
C                                                                          LEV02550
        WRITE(6,140) (J,GOODEV(J),MP(J),TMINGP(J),AMINGP(J), J=1,NGOOD)   LEV02560
    140 FORMAT(/' EIGENVALUES READ IN, T-MULTIPLICITIES, T-GAPS AND A-GAPSLEV02570
       1 '/4X,' J ',5X,'GOOD EIGENVALUE',5X,'MULT',4X,' TMINGAP ',4X,     LEV02580
       1' AMINGAP '/(I6,E25.16,I4,2E15.4))                               LEV02590
C                                                                          LEV02600
C       READ IN ERROR ESTIMATES                                          LEV02610
        WRITE(6,150) MEV,SVSEED                                           LEV02620
    150 FORMAT(/' THESE EIGENVALUES WERE COMPUTED USING A T-MATRIX OF     LEV02630
```

```
      1ORDER ',I5/' AND SEED FOR RANDOM NUMBER GENERATOR =',I12)     LEVO2640
C     CHECK WHETHER OR NOT THERE ARE ANY T-ISOLATED EIGENVALUES IN   LEVO2650
C     THE EIGENVALUES PROVIDED                                       LEVO2660
      DO 160 J=1,NGOOD                                               LEVO2670
      IF(MP(J).EQ.1) GO TO 170                                       LEVO2680
  160 CONTINUE                                                       LEVO2690
      GO TO 190                                                      LEVO2700
  170 READ(4,20) EXPLAN                                              LEVO2710
      READ(4,20) EXPLAN                                              LEVO2720
      READ(4,20) EXPLAN                                              LEVO2730
      READ(4,180) NISO                                               LEVO2740
  180 FORMAT(18X,I6)                                                 LEVO2750
      READ(4,20) EXPLAN                                              LEVO2760
      READ(4,20) EXPLAN                                              LEVO2770
      READ(4,20) EXPLAN                                              LEVO2780
  190 DO 220 J=1,NGOOD                                               LEVO2790
      ERR(J) = 0.D0                                                  LEVO2800
      IF(MP(J).NE.1) GO TO 220                                       LEVO2810
      READ(4,200) EVAL, ERR(J)                                       LEVO2820
  200 FORMAT(10X,E25.16,E14.3)                                       LEVO2830
      IF(DABS(EVAL - GOODEV(J)).LT.1.D-10) GO TO 220                 LEVO2840
      WRITE(6,210) EVAL,GOODEV(J)                                    LEVO2850
  210 FORMAT(' PROBLEM WITH READ IN OF ERROR ESTIMATES'/' EIGENVALUE REALEVO2860
      1D IN',E20.12,' DOES NOT MATCH GOODEV(J) ='/E20.12)            LEVO2870
      GO TO 1550                                                     LEVO2880
C                                                                    LEVO2890
  220 CONTINUE                                                       LEVO2900
C                                                                    LEVO2910
      WRITE(6,230) (J,GOODEV(J),ERR(J), J=1,NGOOD)                   LEVO2920
  230 FORMAT(' ERROR ESTMATES ='/4X,' J',5X,'EIGENVALUE',10X,' ESTIMATE LEVO2930
      1'/(I6,E20.12,E14.3))                                          LEVO2940
C                                                                    LEVO2950
      IF(IREAD.EQ.0)  GO TO 330                                      LEVO2960
C                                                                    LEVO2970
C     READ IN THE SIZE OF THE T-MATRIX PROVIDED ON FILE 2.  READ IN  LEVO2980
C     THE ORDER OF THE USER-SPECIFIED MATRIX , THE SEED FOR THE      LEVO2990
C     RANDOM NUMBER GENERATOR, AND THE MATRIX/TEST IDENTIFICATION    LEVO3000
C     NUMBER THAT WERE USED IN THE LANCZOS EIGENVALUE COMPUTATIONS.  LEVO3010
C     THESE ARE USED IN A CONSISTENCY CHECK                          LEVO3020
C     IF FLAG IREAD = 0 REGENERATE ALPHA, BETA                       LEVO3030
C                                                                    LEVO3040
      READ(2,240) KMAX,NOLD,SVSOLD,MATOLD                            LEVO3050
  240 FORMAT(2I6,I12,I8)                                             LEVO3060
C                                                                    LEVO3070
      WRITE(6,250) KMAX,NOLD,SVSOLD,MATOLD                           LEVO3080
  250 FORMAT(/' READ IN THE T-MATRICES STORED ON FILE 2'/' FILE 2 HEADERLEVO3090
      1 IS'/2X,'KMAX',2X,'NOLD',6X,'SVSOLD',2X,'MATOLD'/2I6,I12,I8/) LEVO3100
C                                                                    LEVO3110
C     CHECK THAT THE ORDER, THE MATRIX/TEST IDENTIFICATION NUMBER    LEVO3120
C     AND THE SEED FOR THE RANDOM NUMBER GENERATOR USED IN THE       LEVO3130
C     LANCZOS COMPUTATIONS THAT GENERATED THE ALPHA,BETA FILE        LEVO3140
C     BEING USED AGREE WITH WHAT THE USER HAS SPECIFIED.             LEVO3150
      IF (NOLD.NE.N.OR.MATOLD.NE.MATNO.OR.SVSOLD.NE.SVSEED) GO TO 1350 LEVO3160
C                                                                    LEVO3170
      KMAX1 = KMAX + 1                                               LEVO3180
C                                                                    LEVO3190
C     READ IN THE T-MATRICES FROM FILE 2.  THESE ARE USED TO GENERATE LEVO3200
C     THE T-EIGENVECTORS THAT WILL BE USED IN THE RITZ VECTOR        LEVO3210
C     COMPUTATIONS.  HISTORY MUST BE STORED IN MACHINE FORMAT        LEVO3220
C     ((4Z20) FOR IBM/3081)                                          LEVO3230
C                                                                    LEVO3240
      READ(2,260) (ALPHA(J), J=1,KMAX)                               LEVO3250
      READ(2,260) (BETA(J), J=1,KMAX1)                               LEVO3260
  260 FORMAT(4Z20)                                                   LEVO3270
C                                                                    LEVO3280
      READ(2,260) (V1(J), J=1,N)                                     LEVO3290
```

```
       READ(2,260) (V2(J), J=1,N)                                    LEVO3300
C                                                                    LEVO3310
C    KMAX MAY BE ENLARGED IF THE SIZE AT WHICH THE EIGENVALUE        LEVO3320
C    COMPUTATIONS WERE PERFORMED IS ESSENTIALLY KMAX AND             LEVO3330
C    THERE IS AT LEAST ONE EIGENVALUE THAT IS T-SIMPLE AND           LEVO3340
C    T-ISOLATED, IN THE SENSE THAT IF ITS NEAREST NEIGHBOR IS TOO    LEVO3350
C    CLOSE THAT NEIGHBOR IS A 'GOOD' T-EIGENVALUE.                   LEVO3360
       DO 270 J = 1,NGOOD                                            LEVO3370
       IF(MP(J).EQ.1) GO TO 290                                      LEVO3380
   270 CONTINUE                                                      LEVO3390
       WRITE(6,280)                                                  LEVO3400
   280 FORMAT(/' ALL EIGENVALUES USED ARE T-MULTIPLE OR CLOSE TO SPURIOUSLEVO3410
      1 T-EIGENVALUES'/' SO KMAX IS NOT INCREASED')                  LEVO3420
       IF(KMAX.LT.MEV) GO TO 1370                                    LEVO3430
       GO TO 310                                                     LEVO3440
C                                                                    LEVO3450
   290 KMAXN= i1*MEV/8 + 12                                          LEVO3460
       IF(MBETA.LE.KMAXN) GO TO 1530                                 LEVO3470
       IF(KMAX.GE.KMAXN ) GO TO 310                                  LEVO3480
       WRITE(6,300) KMAX, KMAXN                                      LEVO3490
   300 FORMAT(' ENLARGE KMAX FROM ',I6,' TO ',I6)                    LEVO3500
       MOLD1 = KMAX + 1                                              LEVO3510
       KMAX = KMAXN                                                  LEVO3520
       GO TO 380                                                     LEVO3530
C                                                                    LEVO3540
   310 WRITE(6,320) KMAX                                             LEVO3550
   320 FORMAT(/' T-MATRICES HAVE BEEN READ IN FROM FILE 2'/' THE LARGEST LEVO3560
      1SIZE T-MATRIX ALLOWED IS',I6/)                                LEVO3570
C                                                                    LEVO3580
       IF(IREAD.EQ.1) GO TO 400                                      LEVO3590
C                                                                    LEVO3600
C    REGENERATE THE ALPHA AND BETA                                   LEVO3610
C                                                                    LEVO3620
   330 MOLD1 = 1                                                     LEVO3630
C                                                                    LEVO3640
       DO 340 J = 1,NGOOD                                            LEVO3650
       IF(MP(J).EQ.1) GO TO 360                                      LEVO3660
   340 CONTINUE                                                      LEVO3670
       KMAX = MEV + 12                                               LEVO3680
       WRITE(6,350) KMAX                                             LEVO3690
   350 FORMAT(/' ALL EIGENVALUES FOR WHICH EIGENVECTORS ARE TO BE COMPUTELEVO3700
      1D ARE EITHER T-MULTIPLE OR CLOSE TO'/' A SPURIOUS T-EIGENVALUE. THLEVO3710
      1EREFORE SET KMAX = MEV + 12 = ',I7)                           LEVO3720
       GO TO 380                                                     LEVO3730
C                                                                    LEVO3740
   360 KMAXN = 11*MEV/8 + 12                                         LEVO3750
       IF(MBETA.LE.KMAXN) GO TO 1530                                 LEVO3760
       WRITE(6,370) KMAXN                                            LEVO3770
   370 FORMAT(' SET KMAX EQUAL TO ',I6)                              LEVO3780
       KMAX = KMAXN                                                  LEVO3790
C                                                                    LEVO3800
   380 WRITE(6,390) MOLD1,KMAX                                       LEVO3810
   390 FORMAT(/' LANCZS SUBROUTINE GENERATES ALPHA(J), BETA(J+1), J =', LEVO3820
      1 I6,' TO ', I6/)                                              LEVO3830
C                                                                    LEVO3840
C-------------------------------------------------------------------LEVO3850
C                                                                    LEVO3860
       IIX = SVSEED                                                  LEVO3870
       CALL LANCZS(CMATV,ALPHA,BETA,V1,V2,G,KMAX,MOLD1,N,IIX)        LEVO3880
C                                                                    LEVO3890
C-------------------------------------------------------------------LEVO3900
C                                                                    LEVO3910
   400 CONTINUE                                                      LEVO3920
C                                                                    LEVO3930
C    THE SUBROUTINE STURMI DETERMINES THE SMALLEST SIZE T-MATRIX FOR LEVO3940
C    WHICH THE EIGENVALUE IN QUESTION IS A T-EIGENVALUE (TO WITHIN A  LEVO3950
C    GIVEN TOLERANCE) AND IF POSSIBLE THE SMALLEST SIZE T-MATRIX     LEVO3960
```

```
C      FOR WHICH IT IS A DOUBLE T-EIGENVALUE (TO WITHIN THE SAME        LEV03970
C      TOLERANCE).  THE SIZE T-MATRIX USED IN THE RITZ VECTOR          LEV03980
C      COMPUTATIONS IS THEN DETERMINED BY LOOPING ON SIZE OF THE       LEV03990
C      T-EIGENVECTORS STARTING WITH A T-SIZE DETERMINED FROM THE       LEV04000
C      OUTPUT FROM STURMI.                                             LEV04010
C                                                                      LEV04020
C                                                                      LEV04030
       STUTOL = SCALEO*MULTOL                                          LEV04040
       IF(IWRITE.EQ.1) WRITE(6,410)                                    LEV04050
   410 FORMAT(' FROM STURMI')                                          LEV04060
       DO 450 J = 1,NGOOD                                              LEV04070
       EVAL = GOODEV(J)                                                LEV04080
C      COMPUTE THE TOLERANCES USED BY STURMI TO DETERMINE AN INTERVAL  LEV04090
C      CONTAINING THE EIGENVALUE EVAL.                                 LEV04100
       TEMP = DABS(EVAL)*RELTOL                                        LEV04110
       TOLN = DMAX1(TEMP,STUTOL)                                       LEV04120
C                                                                      LEV04130
C-----------------------------------------------------------------LEV04140
C                                                                      LEV04150
       CALL STURMI(ALPHA,BETA,EVAL,TOLN,EPSM,KMAX,MK1,MK2,IC,IWRITE)   LEV04160
C                                                                      LEV04170
C-----------------------------------------------------------------LEV04180
C                                                                      LEV04190
C      STORE THE COMPUTED ORDERS OF T-MATRICES FOR LATER PRINTOUT      LEV04200
       M1(J) = MK1                                                     LEV04210
       M2(J) = MK2                                                     LEV04220
       ML(J) = (MK1 + 3*MK2)/4                                         LEV04230
       IF(MK2.EQ.KMAX)  ML(J) = KMAX                                   LEV04240
C                                                                      LEV04250
       IF(IC.GT.0) GO TO 430                                           LEV04260
C      IC = 0 MEANS THERE WAS NO EIGENVALUE IN THE DESIGNATED INTERVAL LEV04270
C      BY T-SIZE KMAX.  THIS MEANS THAT THE EIGENVALUE PROVIDED HAS    LEV04280
C      NOT YET CONVERGED SO ITS EIGENVECTOR WILL NOT BE COMPUTED.      LEV04290
       WRITE(6,420) J,GOODEV(J),MK1,MK2                                LEV04300
   420 FORMAT(I6,'TH EIGENVALUE',E20.12,' HAS NOT CONVERGED '/         LEV04310
      1' SO DO NOT COMPUTE ANY T-EIGENVECTOR OR RITZ VECTOR FOR IT'    LEV04320
      1/' MK1 AND MK2 FOR THIS EIGENVALUE WERE',2I6)                   LEV04330
       MP(J) = MPMIN                                                   LEV04340
       MA(J) = -2*KMAX                                                 LEV04350
       GO TO 450                                                       LEV04360
C      COMPUTE AN APPROPRIATE SIZE T-MATRIX FOR THE GIVEN EIGENVALUE.  LEV04370
   430 IF(M2(J).EQ.KMAX) GO TO 440                                     LEV04380
C      M1 AND M2 WERE BOTH DETERMINED                                  LEV04390
       MA(J) = (3*M1(J) + M2(J))/4  + 1                                LEV04400
       GO TO 450                                                       LEV04410
C      M2 NOT DETERMINED                                               LEV04420
   440 MA(J) = (5*M1(J))/4  + 1                                        LEV04430
C                                                                      LEV04440
   450 CONTINUE                                                        LEV04450
C                                                                      LEV04460
       IF (IWRITE.EQ.1) WRITE(6,460) (MA(JJ), JJ=1,NGOOD)              LEV04470
   460 FORMAT(/' 1ST GUESS AT APPROPRIATE SIZE T-MATRICES'/            LEV04480
      1 ' ACTUAL VALUES WILL PROBABLY BE 1/4 AGAIN AS MUCH'/(13I6))    LEV04490
C                                                                      LEV04500
C      PRINT OUT TO FILE 10 1ST GUESSES AT SIZES OF THE T-MATRICES TO  LEV04510
C      BE USED IN THE EIGENVECTOR COMPUTATIONS.                        LEV04520
C      PROGRAM LOOPS ON T-SIZE TO DETERMINE APPROPRIATE SIZE T-MATRIX. LEV04530
       WRITE(10,470) N,KMAX                                            LEV04540
   470 FORMAT(2I8,' = ORDER OF USER MATRIX AND MAX ORDER OF T(1,MEV)') LEV04550
C                                                                      LEV04560
       WRITE(10,480)                                                   LEV04570
   480 FORMAT(/' 1ST GUESS AT APPROPRIATE SIZE T-MATRICES'/            LEV04580
      1 ' ACTUAL VALUES WILL PROBABLY BE 1/4 AGAIN AS MUCH'/)          LEV04590
C                                                                      LEV04600
       WRITE(10,490)                                                   LEV04610
   490 FORMAT(4X,'J',4X,'A-EIGENVALUE',4X,'M1(J)',1X,'M2(J)',1X,'MA(J)') LEV04620
C                                                                      LEV04630
```

```
          WRITE(10,500) (J,GOODEV(J),M1(J),M2(J), MA(J), J=1,NGOOD)        LEV04640
  500 FORMAT(I5,E19.12,3I6)                                               LEV04650
C                                                                         LEV04660
          IF(MBOUND.EQ.1) WRITE(10,510)                                   LEV04670
  510 FORMAT(/'  EV = GOODEV(J) IS A GOOD EIGENVALUE OF T(1,MEV)'/        LEV04680
     1'  M1 = SMALLEST VALUE OF M SUCH THAT T(1,M) HAS AT LEAST'/         LEV04690
     1 '     ONE EIGENVALUE IN THE INTERVAL (EV-TOLN,EV+TOLN)'/           LEV04700
     1 '       TOLN(J) = DMAX1(GOODEV(J)*RELTOL, SCALE0*MULTOL)'/         LEV04710
     1 '  M2 = SMALLEST M (IF ANY) SUCH THAT IN THE ABOVE INTERVAL'/      LEV04720
     1 '       T(1,M) HAS AT LEAST TWO EIGENVALUES '/                     LEV04730
     1 '  IABS(MA(J)) = APPROPRIATE SIZE T-MATRIX FOR GOODEV(J)'/         LEV04740
     1 '  INITIAL VALUE OF MA(J) IS CHOSEN HEURISTICALLY'/                LEV04750
     1 '  PROGRAM LOOPS ON SIZE OF T-MATRIX TO GET BETTER SIZE'/          LEV04760
     1 '  END OF SIZES OF T-MATRICES FILE 10'///)                         LEV04770
C                                                                         LEV04780
C     TERMINATE AFTER COMPUTING 1ST GUESSES AT SIZES OF THE              LEV04790
C     T-MATRICES REQUIRED FOR THE GIVEN EIGENVALUES?                     LEV04800
          IF(MBOUND.EQ.1) GO TO 1390                                      LEV04810
C                                                                         LEV04820
C     IS THERE ROOM FOR ALL OF THE REQUESTED T-EIGENVECTORS?             LEV04830
          MTOL = 0                                                        LEV04840
          DO 520 J = 1,NGOOD                                             LEV04850
          IF(MP(J).EQ.MPMIN) GO TO 520                                    LEV04860
          MTOL = MTOL + IABS(MA(J))                                       LEV04870
  520 CONTINUE                                                            LEV04880
          MTOL = (5*MTOL)/4                                               LEV04890
          IF(MTOL.GT.MDIMTV.AND.NTVCON.EQ.0) GO TO 1410                   LEV04900
C                                                                         LEV04910
C----------------------------------------------------------------------LEV04920
C     GENERATE A RANDOM VECTOR TO BE USED REPEATEDLY BY                  LEV04930
C     SUBROUTINE INVERM                                                  LEV04940
C                                                                         LEV04950
          CALL GENRAN(RHSEED,G,KMAX)                                      LEV04960
C                                                                         LEV04970
C---------------------------------------------------------------------  LEV04980
C                                                                         LEV04990
C     LOOP ON GIVEN EIGENVALUES TO COMPUTE THE CORRESPONDING             LEV05000
C     T-EIGENVECTOR.                                                     LEV05010
C                                                                         LEV05020
          MTOL = 0                                                        LEV05030
          NTVEC = 0                                                       LEV05040
          ILBIS = 0                                                       LEV05050
          DO 710 J = 1,NGOOD                                             LEV05060
          ICOUNT = 0                                                      LEV05070
          ERRMIN = 10.D0                                                  LEV05080
          MABEST = MPMIN                                                  LEV05090
          IF(MP(J).EQ.MPMIN) GO TO 710                                    LEV05100
          TFLAG = 0                                                       LEVC5110
          EVAL = GOODEV(J)                                                LEV05120
          TEMP = DABS(EVAL)*RELTOL                                        LEV05130
          UB = EVAL + DMAX1(STUTOL,TEMP)                                  LEV05140
          LB = EVAL - DMAX1(STUTOL,TEMP)                                  LEV05150
  530 KMAXU = IABS(MA(J))                                                 LEV05160
C                                                                         LEV05170
C     SELECT A SUITABLE INCREMENT FOR THE ORDERS OF THE T-MATRICES       LEV05180
C     TO BE CONSIDERED IN DETERMINING APPROPRIATE SIZES FOR THE RITZ     LEV05190
C     VECTOR COMPUTATIONS.                                               LEV05200
          IF(ICOUNT.GT.0) GO TO 550                                       LEV05210
C     SELECT IDELTA(J) BASED UPON THE T-MULTIPLICITY OBTAINED            LEV05220
          IF(M2(J).EQ.KMAX) GO TO 540                                     LEV05230
C     M2 DETERMINED                                                      LEV05240
          IDELTA(J) = ((3*M1(J) + 5*M2(J))/8  + 1 - IABS(MA(J)))/10  + 1  LEV05250
          GO TO 550                                                       LEV05260
C     M2 NOT DETERMINED                                                  LEV05270
  540 MAMAX = MINO((11*MEV)/8 + 12, (13*M1(J))/8 + 1)                     LEV05280
          IDELTA(J) = (MAMAX - IABS(MA(J)))/10  + 1                       LEV05290
```

```
      550 ICOUNT = ICOUNT + 1                                           LEV05300
C                                                                       LEV05310
C-----------------------------------------------------------------------LEV05320
C     TO MIMIMIZE THE EFFECT OF THE ONE-SIDED ACCEPTANCE TEST FOR       LEV05330
C     EIGENVALUES IN THE BISEC SUBROUTINE, RECOMPUTE THE GIVEN          LEV05340
C     EIGENVALUE AT THE SPECIFIED KMAXU                                 LEV05350
C                                                                       LEV05360
      CALL LBISEC(ALPHA,BETA,EPSM,EVAL,EVALN,LB,UB,TTOL,KMAXU,NEVT)     LEV05370
C                                                                       LEV05380
C-----------------------------------------------------------------------LEV05390
C                                                                       LEV05400
C     CHECK WHETHER OR NOT GIVEN T-MATRIX HAS AN EIGENVALUE IN THE      LEV05410
C     SPECIFIED INTERVAL AND IF SO WHAT ITS T-MULTIPLICITY IS.          LEV05420
C                                                                       LEV05430
      IF(NEVT.EQ.1) GO TO 590                                           LEV05440
      IF(NEVT.NE.0) GO TO 570                                           LEV05450
      ILBIS = 1                                                         LEV05460
      WRITE(6,560) EVAL,KMAXU                                           LEV05470
  560 FORMAT(/' PROBLEM ENCOUNTERED IN RECOMPUTATION OF USER-SUPPLIED EILEV05480
     1GENVALUE',E20.12/' THE SIZE T-MATRIX SPECIFIED',I6,' DOES NOT     LEV05490
     1HAVE AN EIGENVALUE IN THE INTERVAL SPECIFIED'/' THEREFORE NO EIGENLEV05500
     1VECTOR WILL BE COMPUTED FOR THIS PARTICULAR EIGENVALUE'/)         LEV05510
      GO TO 610                                                         LEV05520
C                                                                       LEV05530
  570 IF(NEVT.GT.1)  WRITE(6,580) EVAL,KMAXU                            LEV05540
  580 FORMAT(/' PROBLEM ENCOUNTERED IN RECOMPUTATION OF USER-SUPPLIED   LEV05550
     1EIGENVALUE',E20.12/' FOR THE SIZE T-MATRIX SPECIFIED =',I6,' THE  LEV05560
     1GIVEN EIGENVALUE IS T-MULTIPLE IN THE INTERVAL SPECIFIED'/' SOMETHLEV05570
     1ING IS WRONG, THEREFORE NO EIGENVECTOR WILL BE COMPUTED FOR THIS ELEV05580
     1IGENVALUE'/)                                                      LEV05590
C                                                                       LEV05600
      MP(J) = MPMIN                                                     LEV05610
      MA(J) = -2*KMAX                                                   LEV05620
      GO TO 710                                                         LEV05630
C                                                                       LEV05640
  590 CONTINUE                                                          LEV05650
      ILBIS = 0                                                         LEV05660
C                                                                       LEV05670
      EVNEW(J) = EVALN                                                  LEV05680
      EVAL = EVALN                                                      LEV05690
      MTOL = MTOL+KMAXU                                                 LEV05700
C                                                                       LEV05710
C     IS THERE ROOM IN TVEC ARRAY FOR THE NEXT T-EIGENVECTOR?          LEV05720
C     IF NOT, SKIP TO RITZ VECTOR COMPUTATIONS.                        LEV05730
      IF (MTOL.GT.MDIMTV) GO TO 720                                     LEV05740
C                                                                       LEV05750
      IT = 3                                                            LEV05760
      KINT = MTOL - KMAXU +1                                            LEV05770
C                                                                       LEV05780
C     RECORD THE BEGINNING AND END OF THE T-EIGENVECTOR BEING COMPUTED  LEV05790
      MINT(J) = KINT                                                    LEV05800
      MFIN(J) = MTOL                                                    LEV05810
C                                                                       LEV05820
C-----------------------------------------------------------------------LEV05830
C     SUBROUTINE INVERM DOES INVERSE ITERATION, I.E. SOLVES            LEV05840
C     (T(1,KMAXU) - EVAL)*U = RHS  FOR EACH EIGENVALUE TO OBTAIN THE    LEV05850
C     DESIRED T-EIGENVECTOR.                                           LEV05860
C                                                                       LEV05870
      IF(IWRITE.EQ.1)  WRITE(6,600) J                                   LEV05880
  600 FORMAT(/I6,'TH EIGENVALUE')                                       LEV05890
C                                                                       LEV05900
      CALL INVERM(ALPHA,BETA,V1,TVEC(KINT),EVAL,ERROR,TERROR,EPSM,      LEV05910
     1 G,KMAXU,IT,IWRITE)                                               LEV05920
C                                                                       LEV05930
C-----------------------------------------------------------------------LEV05940
C                                                                       LEV05950
      TERR(J) = TERROR                                                  LEV05960
```

```
      TLAST(J) = ERROR                                             LEV05970
      KMAXU1 = KMAXU + 1                                           LEV05980
      TBETA(J) = BETA(KMAXU1)*ERROR                               LEV05990
C                                                                  LEV06000
C     AFTER EACH OF THE T-EIGENVECTORS IS COMPUTED, THE            LEV06010
C     SIZE OF THE ERROR ESTIMATE, ERROR IS CHECKED.               LEV06020
C     IF THIS ESTIMATE IS NOT AS SMALL AS DESIRED AND             LEV06030
C     |MA(J)| < ML(J), PROGRAM ATTEMPTS TO INCREASE THE SIZE OF |MA(J)| LEV06040
C     AND REPEAT THE T-EIGENVECTOR COMPUTATIONS.                  LEV06050
C                                                                  LEV06060
      IF(ERROR.LT.ERTOL.OR.TFLAG.EQ.1)  GO TO 700                 LEV06070
C                                                                  LEV06080
      IF(ERROR.GE.ERRMIN) GO TO 610                               LEV06090
C     LAST COMPONENT IS LESS THAN MINIMAL TO DATE                 LEV06100
      ERRMIN = ERROR                                               LEV06110
      MABEST = MA(J)                                               LEV06120
  610 CONTINUE                                                     LEV06130
C                                                                  LEV06140
      IF(MA(J).GT.0)  ITEST = MA(J) + IDELTA(J)                   LEV06150
      IF(MA(J).LT.0)  ITEST = -(IABS(MA(J)) + IDELTA(J))          LEV06160
      IF(IABS(ITEST).LE.ML(J).AND.ICOUNT.LE.10) GO TO 630         LEV06170
C     NEW MA(J) IS GREATER THAN MAXIMUM ALLOWED.                  LEV06180
      IF(ERCONT.EQ.0.OR.MABEST.EQ.MPMIN) GO TO 650                LEV06190
      TFLAG = 1                                                    LEV06200
      MA(J) = MABEST                                               LEV06210
      IF(ILBIS.EQ.0)  MTOL = MTOL - KMAXU                         LEV06220
      WRITE(6,620) MA(J)                                          LEV06230
  620 FORMAT(' 10 ORDERS WERE CONSIDERED.  NONE SATISFIED THE ERROR TESTLEV06240
     1'/' THEREFORE USE THE BEST ORDER OBTAINED FOR THE EIGENVECTORS'  LEV06250
     1,I6)                                                        LEV06260
      GO TO 530                                                    LEV06270
C                                                                  LEV06280
  630 MA(J) = ITEST                                                LEV06290
C                                                                  LEV06300
      MT = IABS(MA(J))                                             LEV06310
      IF(IWRITE.EQ.1) WRITE(6,640)  MT                            LEV06320
  640 FORMAT(/' CHANGE SIZE OF T-MATRIX TO ',I6,' RECOMPUTE T-EIGENVECTOLEV06330
     1R')                                                         LEV06340
C                                                                  LEV06350
      IF(ILBIS.EQ.0)  MTOL = MTOL - KMAXU                         LEV06360
C                                                                  LEV06370
      GO TO 530                                                    LEV06380
C                                                                  LEV06390
C     APPROPRIATE SIZE T-MATRIX WAS NOT OBTAINED                  LEV06400
  650 CONTINUE                                                     LEV06410
      WRITE(10,660) J,EVAL,MP(J)                                  LEV06420
  660 FORMAT(/' ON 10 INCREMENTS NOT ABLE TO IDENTIFY APPROPRIATE SIZE  LEV06430
     1T-MATRIX FOR'/                                              LEVC6440
     1' EIGENVALUE(',I4,') = ',E20.12,' T-MULTIPLICITY =',I4/)    LEV06450
      IF(M2(J).EQ.KMAX) WRITE(10,670)                             LEV06460
      IF(M2(J).LT.KMAX) WRITE(10,680)                             LEV06470
  670 FORMAT(/' ORDERS TESTED RANGED FROM 5*M1(J)/4 TO APPROXIMATELY    LEV06480
     1 '/'  MIN(11*MEV/8,13*M1(J)/8)'/)                           LEV06490
  680 FORMAT(/' ORDERS TESTED RANGED FROM (3*M1(J)+M2(J))/4 TO APPROXIMALEV06500
     1TELY'/'  (3*M1(J) + 5*M2(J))/8.'/)                          LEV06510
      WRITE(10,690)                                               LEV06520
  690 FORMAT(' ALLOWING LARGER ORDERS FOR THE T-MATRICES MAY RESULT IN  LEV06530
     1 SUCCESS'/' BUT PROBABLY WILL NOT.  PROBLEM IS PROBABLY DUE TO'   LEV06540
     1 /' LACK OF CONVERGENCE OF GIVEN EIGENVALUE, CHECK THE ERROR ESTIMLEV06550
     1ATE'/)                                                      LEV06560
      MP(J) = MPMIN                                                LEV06570
      IF(ILBIS.EQ.0)  MTOL = MTOL - KMAXU                         LEV06580
      GO TO 710                                                    LEV06590
  700 NTVEC = NTVEC + 1                                            LEV06600
C                                                                  LEV06610
  710 CONTINUE                                                     LEV06620
      NGOODC = NGOOD                                               LEV06630
```

```
      GO TO 740                                                    LEV06640
C                                                                  LEV06650
C     COME HERE IF THERE IS NOT ENOUGH ROOM FOR ALL OF T-EIGENVECTORS  LEV06660
  720 NGOODC = J-1                                                 LEV06670
      WRITE(6,730)  J, MTOL, MDIMTV                                LEV06680
  730 FORMAT(/' NOT ENOUGH ROOM IN TVEC FOR ',I4,'TH T-VECTOR'/' T-DIMLEV06690
     1ENSION REQUESTED = ',I6,' BUT TVEC HAS DIMENSION = ',I6/)    LEV06700
      IF(NGOODC.EQ.0)  GO TO 1430                                  LEV06710
      MTOL = MTOL-KMAXU                                            LEV06720
C                                                                  LEV06730
  740 CONTINUE                                                     LEV06740
C                                                                  LEV06750
C     THE LOOP ON T-EIGENVECTOR COMPUTATIONS IS COMPLETE.          LEV06760
C     WRITE OUT THE SIZE T-MATRICES THAT WILL BE USED FOR          LEV06770
C     THE RITZ VECTOR COMPUTATIONS.                                LEV06780
C                                                                  LEV06790
      WRITE(10,750)                                                LEV06800
  750 FORMAT(/' SIZES OF T-MATRICES THAT WILL BE USED IN THE RITZ COMPUTLEV06810
     1ATIONS'/5X,'J',16X,'GOODEV(J)',1X,'MA(J)')                   LEV06820
C                                                                  LEV06830
      WRITE(10,760)   (J,GOODEV(J),MA(J), J=1,NGOOD)               LEV06840
  760 FORMAT(I6,E25.14,I6)                                         LEV06850
      WRITE(10,510)                                                LEV06860
C                                                                  LEV06870
      WRITE(6,770) MTOL                                            LEV06880
  770 FORMAT(/' THE CUMULATIVE LENGTH OF THE T-EIGENVECTORS IS',I18)  LEV06890
C                                                                  LEV06900
      WRITE(6,780) NTVEC,NGOOD                                     LEV06910
  780 FORMAT(/I6,' T-EIGENVECTORS OUT OF',I6,' REQUESTED WERE COMPUTED')LEV06920
C                                                                  LEV06930
C     SAVE THE T-EIGENVECTORS ON FILE 11?                          LEV06940
      IF(TVSTOP.NE.1.AND.SVTVEC.EQ.0) GO TO 840                    LEV06950
C                                                                  LEV06960
      WRITE(11,790) NTVEC,MTOL,MATNO,SVSEED                        LEV06970
  790 FORMAT(I6,3I12,' = NTVEC,MTOL,MATNO,SVSEED')                 LEV06980
C                                                                  LEV06990
      DO 820 J=1,NGOODC                                            LEV07000
C     IF MP(J) = MPMIN THEN NO SUITABLE T-EIGENVECTOR IS AVAILABLE LEV07010
C     FOR THAT EIGENVALUE.                                         LEV07020
      IF(MP(J).EQ.MPMIN) WRITE(11,800) J,MA(J),GOODEV(J),MP(J)     LEV07030
  800 FORMAT(2I6,E20.12,I6/' TH EIGVAL,T-SIZE,EVALUE,FLAG,NO EIGVEC')  LEV07040
      IF(MP(J).NE.MPMIN) WRITE(11,810) J,MA(J),GOODEV(J),MP(J)     LEV07050
  810 FORMAT(I6,I6,E20.12,I6/' T-EIGVEC,SIZE T,EVALUE OF A,MP(J)') LEV07060
      IF(MP(J).EQ.MPMIN) GO TO 820                                 LEV07070
      KI = MINT(J)                                                 LEV07080
      KF = MFIN(J)                                                 LEV07090
C                                                                  LEV07100
      WRITE(11,260) (TVEC(K), K=KI,KF)                             LEV07110
C                                                                  LEV07120
  820 CONTINUE                                                     LEV07130
C                                                                  LEV07140
      IF(TVSTOP.NE.1) GO TO 840                                    LEV07150
C                                                                  LEV07160
      WRITE(6,830) TVSTOP, NTVEC,NGOOD                             LEV07170
  830 FORMAT(/' USER SET TVSTOP = ',I1/                            LEV07180
     1' THEREFORE PROGRAM TERMINATES AFTER T-EIGENVECTOR COMPUTATIONS'/ LEV07190
     1' T-EIGENVECTORS THAT WERE COMPUTED ARE SAVED ON FILE 11'/   LEV07200
     1I18,' T-EIGENVECTORS WERE COMPUTED OUT OF',I7,' REQUESTED'/) LEV07210
C                                                                  LEV07220
      GO TO 1550                                                   LEV07230
C                                                                  LEV07240
  840 CONTINUE                                                     LEV07250
C     IF NOT ABLE TO COMPUTE ALL THE REQUESTED T-EIGENVECTORS,     LEV07260
C     CONTINUE WITH THE LANCZOS VECTOR COMPUTATIONS ANYWAY?        LEV07270
      IF(NTVEC.NE.NGOOD.AND.LVCONT.EQ.0) GO TO 1450                LEV07280
C                                                                  LEV07290
C     COMPUTE THE MAXIMUM SIZE OF THE T-MATRIX USED FOR THOSE      LEV07300
```

```
C     EIGENVALUES WITH GOOD ERROR ESTIMATES.                        LEV07310
C                                                                   LEV07320
      KMAXU = 0                                                     LEV07330
      DO 850 J = 1,NGOODC                                           LEV07340
      MT = IABS(MA(J))                                              LEV07350
      IF(MT.LT.KMAXU.OR.MP(J).EQ.MPMIN) GO TO 850                   LEV07360
      KMAXU = MT                                                    LEV07370
  850 CONTINUE                                                      LEV07380
C                                                                   LEV07390
      IF(KMAXU.EQ.0) GO TO 1490                                     LEV07400
C                                                                   LEV07410
      WRITE(6,860) KMAXU                                            LEV07420
  860 FORMAT(/I6,' = LARGEST SIZE T-MATRIX TO BE USED IN THE RITZ VECTORLEV07430
     1 COMPUTATIONS')                                               LEV07440
C                                                                   LEV07450
C     COUNT THE NUMBER OF RITZ VECTORS NOT BEING COMPUTED           LEV07460
      MREJEC = 0                                                    LEV07470
      DO 870 J=1,NGOODC                                             LEV07480
  870 IF(MP(J).EQ.MPMIN)  MREJEC = MREJEC + 1                       LEV07490
      MREJET = MREJEC + (NGOOD-NGOODC)                              LEV07500
      IF(MREJET.NE.0) WRITE(6,880) MREJET                           LEV07510
  880 FORMAT(/' RITZ VECTORS ARE NOT COMPUTED FOR',I6,' OF THE EIGENVALUELEV07520
     1ES'/)                                                         LEV07530
      NACT = NGOODC - MREJEC                                        LEV07540
      WRITE(6,890) NGOOD,NTVEC,NACT                                 LEV07550
  890 FORMAT(/I6,' RITZ VECTORS WERE REQUESTED'/I6,' T-EIGENVECTORS WERELEV07560
     1 COMPUTED'/I6,' RITZ VECTORS WILL BE COMPUTED'/)              LEV07570
C     CHECK IF THERE ARE ANY RITZ VECTORS TO COMPUTE               LEV07580
      IF(MREJEC.EQ.NGOODC) GO TO 1470                               LEV07590
C                                                                   LEV07600
C     CONTINUE WITH THE LANCZOS VECTOR COMPUTATIONS?                LEV07610
      IF(LVCONT.EQ.0.AND.MREJEC.NE.0) GO TO 1450                    LEV07620
C                                                                   LEV07630
C     NOW COMPUTE THE RITZ VECTORS.  REGENERATE THE                 LEV07640
C     LANCZOS VECTORS.                                              LEV07650
C                                                                   LEV07660
      DO 900 I = 1,NMAX                                             LEV07670
  900 RITVEC(I) = ZERO                                              LEV07680
C                                                                   LEV07690
C-------------------------------------------------------------------LEV07700
C     REGENERATE THE STARTING VECTOR. THIS MUST BE GENERATED AND    LEV07710
C     NORMALIZED PRECISELY THE WAY IT WAS DONE IN THE EIGENVALUE    LEV07720
C     COMPUTATIONS, OTHERWISE THERE WILL BE A MISMATCH BETWEEN      LEV07730
C     THE T-EIGENVECTORS THAT HAVE BEEN COMPUTED FROM THE T-MATRICES LEV07740
C     READ IN FROM FILE 2 AND THE LANCZOS VECTORS THAT ARE          LEV07750
C     BEING REGENERATED.                                           LEV07760
C                                                                   LEV07770
      IIL = SVSEED                                                  LEV07780
      CALL GENRAN(IIL,G,N)                                          LEV07790
C                                                                   LEV07800
C-------------------------------------------------------------------LEV07810
C                                                                   LEV07820
      DO 910 J = 1,N                                                LEV07830
  910 V2(J) = G(J)                                                  LEV07840
C                                                                   LEV07850
      SUM = FINPRO(N,V2(1),1,V2(1),1)                               LEV07860
      SUM = ONE/DSQRT(SUM)                                          LEV07870
C                                                                   LEV07880
      DO 920 J = 1,N                                                LEV07890
      V1(J) = ZERO                                                  LEV07900
  920 V2(J) = V2(J)*SUM                                             LEV07910
C                                                                   LEV07920
C     LOOP FOR GENERATING RITZ VECTORS  (IVEC = 1,KMAXU)            LEV07930
      IVEC = 1                                                      LEV07940
      BATA = ZERO                                                   LEV07950
C                                                                   LEV07960
```

```
      GO TO 980                                                  LEV07970
C                                                                LEV07980
  930 CONTINUE                                                   LEV07990
C                                                                LEV08000
C     COMPUTE V1 = A*V2 - BATA*V1                                LEV08010
C                                                                LEV08020
C------------------------------------------------------------------LEV08030
C                                                                LEV08040
      CALL CMATV(V2,V1,BATA)                                     LEV08050
C                                                                LEV08060
C------------------------------------------------------------------LEV08070
C                                                                LEV08080
      ALFA = FINPRO(N,V1(1),1,V2(1),1)                           LEV08090
C                                                                LEV08100
      DO 940 J = 1,N                                             LEV08110
  940 V1(J) = V1(J)-ALFA*V2(J)                                   LEV08120
C                                                                LEV08130
      BATA = FINPRO(N,V1(1),1,V1(1),1)                           LEV08140
      BATA = DSQRT(BATA)                                         LEV08150
      SUM = ONE/BATA                                             LEV08160
C                                                                LEV08170
      TEMP = BETA(IVEC)                                          LEV08180
      TEMP = DABS(BATA - TEMP)/TEMP                              LEV08190
      IF (TEMP.LT.1.0D-10)GO TO 960                              LEV08200
C                                                                LEV08210
C     THE BETA BEING REGENERATED DO NOT MATCH THE BETA IN FILE 2. LEV08220
C     SOMETHING IS WRONG IN THE LANCZOS VECTOR GENERATION.       LEV08230
C     PROGRAM TERMINATES FOR USER TO CORRECT THE PROBLEM         LEV08240
C     WHICH MUST BE IN THE STARTING VECTOR GENERATION OR IN      LEV08250
C     THE MATRIX-VECTOR MULTIPLY SUBROUTINE CMATV SUPPLIED.      LEV08260
C     THIS SUBROUTINE MUST BE THE SAME ONE USED IN THE           LEV08270
C     EIGENVALUE COMPUTATIONS OR A MISMATCH WILL ENSUE.          LEV08280
C                                                                LEV08290
      WRITE(6,950) IVEC,BATA,BETA(IVEC),TEMP                     LEV08300
  950 FORMAT(/2X,'IVEC',16X,'BATA',10X,'BETA(IVEC)',14X,'RELDIF'/16, LEV08310
     13E20.12/' IN LANCZOS VECTOR REGENERATION THE ENTRIES OF THE TRIDIALEV08320
     1GONAL MATRICES BEING'/' GENERATED ARE NOT THE SAME AS THOSE IN THELEV08330
     1 MATRIX SUPPLIED ON FILE 2.'/' THEREFORE SOMETHING IS BEING INITIALEV08340
     1LIZED OR COMPUTED DIFFERENTLY FROM THE WAY'/' IT WAS COMPUTED IN TLEV08350
     1HE EIGENVALUE COMPUTATIONS'/' THE PROGRAM TERMINATES FOR THE USER LEV08360
     1TO DETERMINE WHAT THE PROBLEM IS'/)                        LEV08370
      GO TO 1550                                                 LEV08380
C                                                                LEV08390
C                                                                LEV08400
  960 CONTINUE                                                   LEV08410
      DO 970 J = 1,N                                             LEV08420
      TEMP = SUM*V1(J)                                           LEV08430
      V1(J) = V2(J)                                              LEV08440
  970 V2(J) = TEMP                                               LEV08450
C                                                                LEV08460
  980 CONTINUE                                                   LEV08470
C                                                                LEV08480
      LFIN = 0                                                   LEV08490
      DO 1000 J = 1,NGOODC                                       LEV08500
      LL = LFIN                                                  LEV08510
      LFIN = LFIN + N                                            LEV08520
C                                                                LEV08530
      IF(IABS(MA(J)).LT.IVEC.OR.MP(J).EQ.MPMIN) GO TO 1000       LEV08540
      II = IVEC + MINT(J) - 1                                    LEV08550
      TEMP = TVEC(II)                                            LEV08560
C     II IS THE (IVEC)TH COMPONENT OF THE T-EIGENVECTOR CONTAINED LEV08570
C     IN TVEC(MINT(J)).                                          LEV08580
C                                                                LEV08590
      DO 990 K = 1,N                                             LEV08600
      LL = LL + 1                                                LEV08610
  990 RITVEC(LL) = TEMP*V2(K) + RITVEC(LL)                       LEV08620
C                                                                LEV08630
```

```
 1000 CONTINUE                                                       LEV08640
C                                                                    LEV08650
      IVEC = IVEC + 1                                                LEV08660
      IF (IVEC.LE.KMAXU) GO TO 930                                   LEV08670
C                                                                    LEV08680
C                                                                    LEV08690
C     RITZVECTOR GENERATION IS COMPLETE. NORMALIZE EACH RITZVECTOR.  LEV08700
C     NOTE THAT IF CERTAIN RITZ VECTORS WERE NOT COMPUTED THEN THAT  LEV08710
C     PORTION OF THE RITVEC ARRAY WAS NOT UTILIZED.                  LEV08720
C                                                                    LEV08730
      LFIN = 0                                                       LEV08740
      DO 1050 J = 1,NGOODC                                           LEV08750
C                                                                    LEV08760
      KK = LFIN                                                      LEV08770
      LFIN = LFIN + N                                                LEV08780
      IF(MP(J).EQ.MPMIN) GO TO 1050                                  LEV08790
C                                                                    LEV08800
      DO 1010 K = 1,N                                                LEV08810
      KK = KK + 1                                                    LEV08820
 1010 V2(K) = RITVEC(KK)                                             LEV08830
C                                                                    LEV08840
      SUM = FINPRO(N,V2(1),1,V2(1),1)                                LEV08850
      SUM = DSQRT(SUM)                                               LEV08860
      RNORM(J) = SUM                                                 LEV08870
      TEMP = DABS(ONE-SUM)                                           LEV08880
      SUM = ONE/SUM                                                  LEV08890
C                                                                    LEV08900
      KK = LFIN - N                                                  LEV08910
      DO 1020 K = 1,N                                                LEV08920
      KK = KK + 1                                                    LEV08930
      V2(K) = SUM*V2(K)                                              LEV08940
 1020 RITVEC(KK) = V2(K)                                             LEV08950
C                                                                    LEV08960
      IF (IWRITE.NE.0) WRITE(6,1030) J,GOODEV(J)                     LEV08970
 1030 FORMAT(/I5,' TH EIGENVALUE CONSIDERED = ',E20.12/)             LEV08980
C                                                                    LEV08990
      IF (IWRITE.NE.0) WRITE(6,1040) TERR(J),TBETA(J),TEMP           LEV09000
 1040 FORMAT(' NORM OF ERROR IN T-EIGENVECTOR = ',E14.3/             LEV09010
     1 ' BETA(MA(J)+1)*U(MA(J)) = ',E14.3/                           LEV09020
     1 ' ABS(NORM(RITVEC) - 1.0)  = ',E14.3/)                        LEV09030
C                                                                    LEV09040
      LINT = LFIN - N + 1                                            LEV09050
      EVAL = EVNEW(J)                                                LEV09060
C                                                                    LEV09070
C-------------------------------------------------------------------LEV09080
C                                                                    LEV09090
      CALL CMATV(RITVEC(LINT),V2,EVAL)                               LEV09100
C                                                                    LEV09110
C-------------------------------------------------------------------LEV09120
C                                                                    LEV09130
C     COMPUTE ERROR IN RITZ VECTOR CONSIDERED AS A EIGENVECTOR OF A. LEV09140
C     V2 = A*RITVEC - EVAL*RITVEC                                    LEV09150
C                                                                    LEV09160
      SUM = FINPRO(N,V2(1),1,V2(1),1)                                LEV09170
      SUM = DSQRT(SUM)                                               LEV09180
      ERR(J) = SUM                                                   LEV09190
      GAP = ABS(AMINGP(J))                                           LEV09200
      ERRDGP(J) = SUM/GAP                                            LEV09210
C                                                                    LEV09220
 1050 CONTINUE                                                       LEV09230
C                                                                    LEV09240
C                                                                    LEV09250
C     RITZVECTORS ARE NORMALIZED AND ERROR ESTIMATES ARE IN ERR ARRAY LEV09260
C     AND IN ERRDGP ARRAY. STORE EVERYTHING                         LEV09270
C                                                                    LEV09280
C                                                                    LEV09290
      WRITE(9,1060)                                                  LEV09300
```

```
 1060 FORMAT(3X,'A-EIGENVALUE',2X,'MA(J)',3X,'A-MINGAP',6X,'AERROR',2X,  LEVO9310
     1 'AERROR/GAP',6X,'TERROR')                                         LEVO9320
C                                                                        LEVO9330
      WRITE(13,1070)                                                     LEVO9340
 1070 FORMAT(16X,'GOODEV(J)',5X,'RITZNORM',6X,'AMINGAP',5X,              LEVO9350
     1 'TBETA(J)',5X,'TLAST(J)')                                         LEVO9360
C                                                                        LEVO9370
      DO 1100 J=1,NGOODC                                                 LEVO9380
C                                                                        LEVO9390
      IF(MP(J).EQ.MPMIN) GO TO 1100                                      LEVO9400
C                                                                        LEVO9410
      WRITE(9,1080)EVNEW(J),MA(J),AMINGP(J),ERR(J),ERRDGP(J),TERR(J)     LEVO9420
 1080 FORMAT(E15.8,I6,4E12.4)                                            LEVO9430
C                                                                        LEVO9440
      WRITE(13,1090) EVNEW(J),RNORM(J),AMINGP(J),TBETA(J),TLAST(J)       LEVO9450
 1090 FORMAT(E25.14,4E13.5)                                              LEVO9460
C                                                                        LEVO9470
 1100 CONTINUE                                                           LEVO9480
C                                                                        LEVO9490
      IF(MREJEC.EQ.0) GO TO 1180                                         LEVO9500
      WRITE(9,1110)                                                      LEVO9510
 1110 FORMAT(/' RITZ VECTORS WERE NOT COMPUTED FOR THE FOLLOWING EIGENVALEVO9520
     1LUES'/' EITHER BECAUSE THEY HAD NOT CONVERGED OR BECAUSE THE ERRORLEVO9530
     1 ESTIMATE'/'  WAS NOT AS SMALL AS DESIRED'/)                       LEVO9540
C                                                                        LEVO9550
      DO 1170 J = 1,NGOODC                                               LEVO9560
      IF(MP(J).NE.MPMIN) GO TO 1170                                      LEVO9570
C     WRITE OUT MESSAGE FOR EACH EIGENVALUE FOR WHICH NO EIGENVECTOR     LEVO9580
C     WAS COMPUTED.                                                      LEVO9590
C                                                                        LEVO9600
      WRITE(9,1120)                                                      LEVO9610
 1120 FORMAT(6X,'GOODEV(J)',3X,'MA(J)',5X,'AMINGP(J)',6X,'TLAST(J)',3X,  LEVO9620
     1'MP(J)')                                                           LEVO9630
      WRITE(9,1130) GOODEV(J),MA(J),AMINGP(J),TBETA(J),MP(J)             LEVO9640
 1130 FORMAT(E15.8,I8,2E14.4,I8)                                         LEVO9650
C                                                                        LEVO9660
      WRITE(13,1140)                                                     LEVO9670
 1140 FORMAT(/' RITZ VECTORS WERE NOT COMPUTED FOR THE FOLLOWING EIGENVALEVO9680
     1LUES'/' BECAUSE THEY HAD NOT CONVERGED'/)                          LEVO9690
C                                                                        LEVO9700
      WRITE(13,1150)                                                     LEVO9710
 1150 FORMAT(6X,'GOODEV(J)',3X,'MA(J)',3X,'M1(J)',3X,'M2(J)',3X,'MP(J)'  LEVO9720
     1/)                                                                 LEVO9730
      WRITE(13,1160) GOODEV(J),MA(J),M1(J),M2(J),MP(J)                   LEVO9740
 1160 FORMAT(E15.8,4I8)                                                  LEVO9750
C                                                                        LEVO9760
 1170 CONTINUE                                                           LEVO9770
 1180 CONTINUE                                                           LEVO9780
C                                                                        LEVO9790
      WRITE(9,1190)                                                      LEVO9800
 1190 FORMAT(/' ABOVE ARE ERROR ESTIMATES FOR THE A AND T EIGENVECTORS'/LEVO9810
     1 ' ASSOCIATED WITH THE GOODEV LISTED IN COLUMN 1'/                 LEVO9820
     1 ' AERROR = NORM(A*X - EV*X)   TERROR = NORM(T*Y - EV*Y) '/        LEVO9830
     1 ' WHERE T = T(1,MA(J))   X = RITZ VECTOR = V*Y  V = SUCCESSIVE'/LEVO9840
     1 ' LANCZOS VECTORS. AMINGAP = GAP TO NEAREST A-EIGENVALUE'//)      LEVO9850
C                                                                        LEVO9860
      WRITE(13,1200)                                                     LEVO9870
 1200 FORMAT(/' ABOVE ARE ERROR ESTIMATES ASSOCIATED WITH THE GOODEV'/   LEVO9880
     1 ' RITZNORM = NORM(COMPUTED RITZ VECTOR)'/                         LEVO9890
     1 ' TBETA(J) = BETA(MA(J)+1)*Y(MA(J)),   T*Y = EVAL*Y'/             LEVO9900
     1 ' TLAST(J) = Y(MA(J))'/                                           LEVO9910
     1 ' AMINGAP = GAP TO NEAREST A-EIGENVALUE'/)                        LEVO9920
C                                                                        LEVO9930
C     NUMBER OF RITZ VECTORS COMPUTED                                    LEVO9940
      NCOMPU = NGOODC - MREJEC                                           LEVO9950
      WRITE(12,1210) N,NCOMPU,NGOODC,MATNO                               LEVO9960
```

```
 1210 FORMAT(316,112,' SIZE A, NO.RITZVECS, NO.EVALUES,MATNO')       LEV09970
C                                                                    LEV09980
      LFIN = 0                                                       LEV09990
      DO 1270 J = 1,NGOODC                                           LEV10000
      LINT = LFIN + 1                                                LEV10010
      LFIN = LFIN + N                                                LEV10020
C                                                                    LEV10030
      IF(MP(J).EQ.MPMIN) GO TO 1250                                  LEV10040
C     RITZ VECTOR WAS COMPUTED                                       LEV10050
      WRITE(12,1220) J, GOODEV(J), MP(J)                             LEV10060
 1220 FORMAT(16,4X,E20.12,16,' J, EIGENVAL, MP(J)')                  LEV10070
C                                                                    LEV10080
      WRITE(12,1230) ERR(J),ERRDGP(J)                                LEV10090
 1230 FORMAT(2E15.5,' = NORM(A*Z-EVAL*Z) AND  NORM(A*Z-EVAL*Z)/MINGAP') LEV10100
C                                                                    LEV10110
      WRITE(12,1240) (RITVEC(LL), LL=LINT,LFIN)                      LEV10120
 1240 FORMAT(4E20.12)                                                LEV10130
      GO TO 1270                                                     LEV10140
C     NO RITZ VECTOR WAS COMPUTED FOR THIS EIGENVALUE                LEV10150
 1250 WRITE(12,1260) J,GOODEV(J),MP(J)                               LEV10160
 1260 FORMAT(16,4X,E20.12,16,' J,EIGVALUE,NO RITZ VECTOR COMPUTED')  LEV10170
C                                                                    LEV10180
 1270 CONTINUE                                                       LEV10190
C                                                                    LEV10200
C     DID ANY T-MATRICES INCLUDE OFF-DIAGONAL ENTRIES SMALLER THAN   LEV10210
C     DESIRED, AS SPECIFIED BY BTOL?                                 LEV10220
C                                                                    LEV10230
      IF('B.GT.0) GO TO 1300                                         LEV10240
C                                                                    LEV10250
      WRITE(6,1280) KMAXU                                            LEV10260
 1280 FORMAT(/' FOR LARGEST T-MATRIX CONSIDERED',17,' CHECK THE SIZE OF LEV10270
     1BETAS')                                                        LEV10280
C                                                                    LEV10290
C--------------------------------------------------------------------LEV10300
C                                                                    LEV10310
      CALL TNORM(ALPHA,BETA,BKMIN,TEMP,KMAXU,IBMT)                   LEV10320
C                                                                    LEV10330
C--------------------------------------------------------------------LEV10340
C                                                                    LEV10350
      IF(IBMT.LT.0) WRITE (6,1290)                                   LEV10360
 1290 FORMAT(/' WARNING THE T-MATRICES FOR ONE OR MORE OF THE EIGENVALUELEV10370
     1S CONSIDERED'/' HAD AN OFF-DIAGONAL ENTRY THAT WAS SMALLER THAN THLEV10380
     1E BETA TOLERANCE THAT WAS SPECIFIED'/)                         LEV10390
 1300 CONTINUE                                                       LEV10400
C                                                                    LEV10410
      GO TO 1550                                                     LEV10420
C                                                                    LEV10430
 1310 WRITE(6,1320) NGOOD,NMAX,MDIMRV                                LEV10440
 1320 FORMAT(/14,' RITZ VECTORS WERE REQUESTED BUT THE REQUIRED DIMENSIOLEV10450
     1N',16/' IS LARGER THAN THE USER-SPECIFIED DIMENSION OF RITVEC',16 LEV10460
     1/' THEREFORE, THE EIGENVECTOR PROCEDURE TERMINATES FOR THE USER TOLEV10470
     1 INTERVENE')                                                   LEV10480
C                                                                    LEV10490
      GO TO 1550                                                     LEV10500
C                                                                    LEV10510
 1330 WRITE(6,1340) NOLD,N,MATOLD,MATNO                              LEV10520
 1340 FORMAT(//' PARAMETERS READ FROM FILE 3 DO NOT AGREE WITH THOSE SPELEV10530
     1CIFIED BY THE USER'/' N,NOLD,MATOLD,MATNO = ',216,2112/' PROGRAM TLEV10540
     1ERMINATES FOR USER TO RESOLVE PROBLEM'/)                       LEV10550
C                                                                    LEV10560
      GO TO 1550                                                     LEV10570
C                                                                    LEV10580
 1350 WRITE(6,1360)                                                  LEV10590
 1360 FORMAT(//' PARAMETERS IN THE ALPHA,BETA FILE HEADER DO NOT AGREE WLEV10600
     11TH PARAMTERS'/' SPECIFIED BY THE USER.  THEREFORE THE PROGRAM TERLEV10610
     1MINATES FOR THE USER'/' TO RESOLVE THE PROBLEM'/)              LEV10620
C                                                                    LEV10630
```

```
      GO TO 1550                                                   LEV10640
C                                                                  LEV10650
 1370 WRITE(6,1380) KMAX,MEV                                       LEV10660
 1380 FORMAT(/' ALPHA,BETA FILE HEADER GIVES KMAX =',I6/           LEV10670
     1' BUT EIGENVALUES WERE COMPUTED AT MEV = ',I6,' PROGRAM STOPS'/) LEV10680
C                                                                  LEV10690
      GO TO 1550                                                   LEV10700
C                                                                  LEV10710
 1390 WRITE(6,1400)                                                LEV10720
 1400 FORMAT(//' PROGRAM COMPUTED 1ST GUESSES AT T-MATRIX SIZES AND READLEV10730
     1 THEM TO'/' FILE 10, THEN TERMINATED AS REQUESTED.')         LEV10740
      GO TO 1550                                                   LEV10750
C                                                                  LEV10760
 1410 WRITE(6,'420) MTOL, MDIMTV                                   LEV10770
 1420 FORMAT(/ PROGRAM TERMINATES BECAUSE THE TVEC DIMENSION ANTICIPATELEV10780
     1D',I7/' IS LARGER THAN THE TVEC DIMENSION',I7,' SPECIFIED BY THE LEV10790
     1USER.'/' USER MAY RESET THE TVEC DIMENSION AND RESTART THE PROGRALEV10800
     1M')                                                          LEV10810
      GO TO 1550                                                   LEV10820
C                                                                  LEV10830
 1430 WRITE(6,1440)                                                LEV10840
 1440 FORMAT(/' PROGRAM TERMINATES BECAUSE NO SUITABLE T-EIGENVECTORS WELEV10850
     1RE IDENTIFIED'/' FOR ANY OF THE EIGENVALUES SUPPLIED.  PROBLEM COLEV10860
     1ULD BE CAUSED'/' BY TOO SMALL A TVEC DIMENSION OR SIMPLY THAT SUILEV10870
     1TABLE T-VECTORS COULD'/' NOT BE IDENTIFIED.  USER SHOULD CHECK OULEV10880
     1TPUT'/)                                                      LEV10890
      GO TO 1550                                                   LEV10900
C                                                                  LEV10910
 1450 WRITE(6,1460) LVCONT,NTVEC,NGOOD                             LEV10920
 1460 FORMAT(/' LVCONT FLAG =',I2,' AND NUMBER ',I5,' OF T-EIGENVECTORS LEV10930
     1 COMPUTED N.E.'/' NUMBER',I5,' REQUESTED SO PROGRAM TERMINATES'/) LEV10940
      GO TO 1550                                                   LEV10950
C                                                                  LEV10960
 1470 WRITE(6,1480)                                                LEV10970
 1480 FORMAT(//' PROGRAM TERMINATES WITHOUT COMPUTING RITZ VECTORS'/ LEV10980
     1' BECAUSE ALL T-EIGENVECTORS WERE REJECTED AS NOT SUITABLE FOR THELEV10990
     1 RITZ VECTOR'/' COMPUTATIONS.  PROBABLE CAUSE IS LACK OF CONVERGENLEV11000
     1CE OF THE EIGENVALUES SUPPLIED'/)                            LEV11010
      GO TO 1550                                                   LEV11020
C                                                                  LEV11030
 1490 WRITE(6,1500)                                                LEV11040
 1500 FORMAT(/' PROGRAM INDICATES THAT IT IS NOT POSSIBLE TO COMPUTE ANYLEV11050
     1 OF THE'/' REQUESTED EIGENVECTORS. THEREFORE PROGRAM TERMINATES') LEV11060
      DO 1510 J=1,NGOODC                                           LEV11070
 1510 WRITE(6,1520)  J,GOODEV(J),MP(J)                             LEV11080
 1520 FORMAT(/4X,' J',11X,'GOODEV(J)',4X,'MP(J)'/I6,E20.12,I9)     LEV11090
      GO TO 1550                                                   LEV11100
C                                                                  LEV11110
 1530 WRITE(6,1540) MBETA,KMAXN                                    LEV11120
 1540 FORMAT(/' PROGRAM TERMINATES BECAUSE THE STORAGE ALLOTTED FOR THE LEV11130
     1BETA ARRAY',I8/' IS NOT SUFFICIENT FOR THE ENLARGED KMAX =',I8,' TLEV11140
     1HAT THE PROGRAM WANTS'/' USER CAN ENLARGE THE DIMENSIONS OF THE ALLEV11150
     1PHA AND BETA ARRAYS'/' AND RERUN THE PROGRAM'/)              LEV11160
C                                                                  LEV11170
 1550 CONTINUE                                                     LEV11180
C                                                                  LEV11190
      STOP                                                         LEV11200
CC-----END OF MAIN PROGRAM FOR LANCZOS EIGENVECTORS--------------------LEV11210
      END                                                          LEV11220
```

SECTION 2.5   LANCZS AND SAMPLE MATRIX-VECTOR MULTIPLY SUBROUTINES

```
C-----LEMULT--------------------------------------------------------LEM00010
C                                                                    LEM00020
C     CONTAINS SUBROUTINES  LANCZS, USPEC, AND CMATV                 LEM00030
C     TO BE USED WITH THE REAL SYMMETRIC VERSION OF THE LANCZOS      LEM00040
C     EIGENVALUE/EIGENVECTOR PROCEDURES.                             LEM00050
C     ALSO CONTAINS SUBROUTINES FOR POISSON TEST MATRICES THAT ALLOW LEM00060
C     COMPUTATION OF TRUE ERRORS IN COMPUTED EIGENVALUES AND         LEM00070
C     IN CORRESPONDING EIGENVECTORS.                                 LEM00080
C                                                                    LEM00090
C     NONPORTABLE CONSTRUCTIONS:                                     LEM00100
C     1.  THE ENTRY MECHANISM USED TO PASS THE STORAGE              LEM00110
C         LOCATIONS OF THE USER-SPECIFIED MATRIX FROM THE           LEM00120
C         SUBROUTINE USPEC TO THE MATRIX-VECTOR SUBROUTINE          LEM00130
C         CMATV.                                                    LEM00140
C     2.  IN THE SAMPLE USPEC AND CMATV FOR DIAGONAL TEST MATRICES:  LEM00150
C         FREE FORMAT (8,*) AND THE FORMAT (20A4).                  LEM00160
C     3.  IN THE POISSON SUBROUTINES PROVIDED, THE DATA MACHEP      LEM00170
C         DEFINITION AND MANY OF THE INDICES FOR ARRAYS ARE NOT     LEM00180
C         IN A PORTABLE CONSTRUCTION.  THESE PROGRAMS SHOULD BE     LEM00190
C         REMOVED FROM THE LEMULT FILE IF THE USER IS NOT USING THEM. LEM00200
C                                                                    LEM00210
C-----LANCZS-COMPUTE THE LANCZOS TRIDIAGONAL MATRICES----------------LEM00220
C                                                                    LEM00230
      SUBROUTINE LANCZS(MATVEC,ALPHA,BETA,V1,V2,G,KMAX,MOLD1,N,IIX)    LEM00240
C                                                                    LEM00250
C------------------------------------------------------------------LEM00260
      DOUBLE PRECISION  ALPHA(1),BETA(1),V1(1),V2(1),SUM,TEMP,ONE,ZERO LEM00270
      REAL  G(1)                                                     LEM00280
      DOUBLE PRECISION FINPRO, DSQRT                                 LEM00290
      EXTERNAL MATVEC                                                LEM00300
C------------------------------------------------------------------LEM00310
C                                                                    LEM00320
      ZERO = 0.D0                                                    LEM00330
      ONE = 1.D0                                                     LEM00340
C                                                                    LEM00350
      IF(MOLD1.GT.1)GO TO 30                                         LEM00360
C                                                                    LEM00370
C     ALPHA/BETA GENERATION STARTS AT I = 1                         LEM00380
C     MOLD1 = 1 SET V1 = 0. AND V2 = RANDOM UNIT VECTOR             LEM00390
      BETA(1) = ZERO                                                LEM00400
      IIL=IIX                                                       LEM00410
C                                                                    LEM00420
C------------------------------------------------------------------LEM00430
      CALL GENRAN(IIL,G,N)                                           LEM00440
C------------------------------------------------------------------LEM00450
C                                                                    LEM00460
      DO 10 I = 1,N                                                 LEM00470
   10 V2(I) = G(I)                                                  LEM00480
C                                                                    LEM00490
C------------------------------------------------------------------LEM00500
      SUM = FINPRO(N,V2(1),1,V2(1),1)                               LEM00510
C------------------------------------------------------------------LEM00520
C                                                                    LEM00530
      SUM = ONE/DSQRT(SUM)                                          LEM00540
      DO 20 I = 1,N                                                 LEM00550
      V1(I) = ZERO                                                  LEM00560
   20 V2(I) = V2(I)*SUM                                             LEM00570
C                                                                    LEM00580
C     ALPHA BETA GENERATION LOOP                                    LEM00590
   30 CONTINUE                                                      LEM00600
C                                                                    LEM00610
      DO 60 I=MOLD1,KMAX                                            LEM00620
      SUM = BETA(I)                                                 LEM00630
C     MATVEC(V2,V1,SUM) CALCULATES   V1 = A*V2 - SUM*V1             LEM00640
```

```
C                                                                         LEM00650
C------------------------------------------------------------------------LEM00660
      CALL MATVEC(V2,V1,SUM)                                             LEM00670
C------------------------------------------------------------------------LEM00680
C                                                                         LEM00690
C------------------------------------------------------------------------LEM00700
      SUM = FINPRO(N,V1(1),1,V2(1),1)                                    LEM00710
C------------------------------------------------------------------------LEM00720
C                                                                         LEM00730
      ALPHA(I) = SUM                                                     LEM00740
      DO 40 J=1,N                                                        LEM00750
   40 V1(J) = V1(J)-SUM*V2(J)                                            LEM00760
C                                                                         LEM00770
C------------------------------------------------------------------------LEM00780
      SUM = FINPRO(N,V1(1),1,V1(1),1)                                    LEM00790
C------------------------------------------------------------------------LEM00800
C                                                                         LEM00810
      IN = I+1                                                           LEM00820
      BETA(IN) = DSQRT(SUM)                                              LEM00830
      SUM = ONE/BETA(IN)                                                 LEM00840
      DO 50 J=1,N                                                        LEM00850
      TEMP = SUM*V1(J)                                                   LEM00860
      V1(J) = V2(J)                                                      LEM00870
   50 V2(J) = TEMP                                                       LEM00880
   60 CONTINUE                                                           LEM00890
C                                                                         LEM00900
C     END ALPHA, BETA GENERATION LOOP                                    LEM00910
C                                                                         LEM00920
      RETURN                                                             LEM00930
C-----END OF LANCZS------------------------------------------------------LEM00940
      END                                                               LEM00950
C                                                                         LEM00960
C-----USPEC (GENERAL SYMMETRIC SPARSE MATRICES)-------------------------LEM00970
C                                                                         LEM00980
C     SUBROUTINE USPEC(N,MATNO)                                          LEM00990
      SUBROUTINE GUSPEC(N,MATNO)                                         LEM01000
C                                                                         LEM01010
C------------------------------------------------------------------------LEM01020
      DOUBLE PRECISION  A(10000),AD(5010)                               LEM01030
      INTEGER  IROW(10000),ICOL(5010)                                   LEM01040
C------------------------------------------------------------------------LEM01050
C     USPEC DIMENSIONS AND INITIALIZES THE ARRAYS NEEDED TO DEFINE       LEM01060
C     THE USER-SPECIFIED MATRIX AND THEN PASSES THE STORAGE LOCATIONS    LEM01070
C     OF THESE ARRAYS TO THE MULTIPLY SUBROUTINE CMATV.                  LEM01080
C                                                                         LEM01090
C     MATRIX IS STORED IN FOLLOWING SPARSE MATRIX FORMAT:                LEM01100
C     N = ORDER OF A-MATRIX,                                             LEM01110
C     NZS = NUMBER OF NONZERO SUBDIAGONAL ENTRIES,                       LEM01120
C     NZL = INDEX OF LAST COLUMN CONTAINING NONZERO SUBDIAGONAL ENTRIES, LEM01130
C     ICOL(J), J=1,NZL IS THE NUMBER OF NONZERO SUBDIAGONAL ELEMENTS     LEM01140
C             IN COLUMN J.                                               LEM01150
C     IROW(K), K = 1,NZS IS THE CORRESPONDING ROW INDEX FOR A(K).        LEM01160
C     AD(I), I=1,N CONTAINS DIAGONAL ENTRIES (INCLUDING ANY 0            LEM01170
C            DIAGONAL ENTRIES).                                          LEM01180
C     A(K), K=1,NZS CONTAINS NONZERO SUBDIAGONAL ENTRIES, BY COLUMN      LEM01190
C     FOR J > NZL THERE ARE NO NONZERO SUBDIAGONAL ELEMENTS IN COLUMN J. LEM01200
C     ICOL(J) = 0 IS ALLOWED                                             LEM01210
C                                                                         LEM01220
C------------------------------------------------------------------------LEM01230
C     ARRAYS THAT DEFINE THE MATRIX ARE READ IN FROM FILE 8              LEM01240
C                                                                         LEM01250
      READ(8,10) NZS,NOLD,NZL,MATOLD                                     LEM01260
   10 FORMAT(I10,2I6,I8)                                                 LEM01270
C                                                                         LEM01280
      WRITE(6,20) NZS,NOLD,NZL,MATOLD                                    LEM01290
```

```
      20 FORMAT(I10,2I6,I8,' = NZS,NOLD,NZL,MATOLD'/)                    LEM01300
C                                                                       LEM01310
C        TEST OF PARAMETER CORRECTNESS                                  LEM01320
         ITEMP = (NOLD-N)**2 + (MATNO-MATOLD)**2                        LEM01330
C                                                                       LEM01340
         IF(ITEMP.EQ.0) GO TO 40                                        LEM01350
C                                                                       LEM01360
         WRITE(6,30)                                                    LEM01370
      30 FORMAT(' PROGRAM TERMINATES BECAUSE EITHER ORDERS OF OR LABELS FORLEM01380
        1 MATRIX DISAGREE')                                             LEM01390
         GO TO 70                                                       LEM01400
C                                                                       LEM01410
      40 CONTINUE                                                       LEM01420
C                                                                       LEM01430
C        NUMBER OF NONZERO SUBDIAGONAL ENTRIES IN EACH COLUMN IS READ   LEM01440
C        THEN THE CORRESPONDING ROW INDEX FOR EACH SUCH ENTRY IS READ   LEM01450
         READ(8,50) (ICOL(K), K=1,NZL)                                  LEM01460
         READ(8,50) (IROW(K), K=1,NZS)                                  LEM01470
      50 FORMAT(13I6)                                                   LEM01480
C                                                                       LEM01490
C        DIAGONAL IS READ FIRST, THEN NONZERO BELOW DIAGONAL ENTRIES    LEM01500
         READ(8,60) (AD(K), K=1,N)                                      LEM01510
         READ(8,60) (A(K), K=1,NZS)                                     LEM01520
      60 FORMAT(4E19.10)                                                LEM01530
C                                                                       LEM01540
C-----------------------------------------------------------------------LEM01550
C        PASS STORAGE LOCATIONS OF ARRAYS THAT DEFINE THE MATRIX TO      LEM01560
C        THE MATRIX-VECTOR MULTIPLY SUBROUTINE CMATV                     LEM01570
C                                                                       LEM01580
         CALL CMATVE(A,AD,ICOL,IROW,N,NZL)                              LEM01590
C-----------------------------------------------------------------------LEM01600
C                                                                       LEM01610
         RETURN                                                         LEM01620
      70 STOP                                                           LEM01630
C-----END OF USPEC------------------------------------------------------LEM01640
         END                                                           LEM01650
C                                                                       LEM01660
C-----MATRIX-VECTOR MULTIPLY FOR REAL SPARSE SYMMETRIC MATRICES---------LEM01670
C                                                                       LEM01680
C        SUBROUTINE CMATV(W,U,SUM)                                      LEM01690
         SUBROUTINE GCMATV(W,U,SUM)                                     LEM01700
C                                                                       LEM01710
C-----------------------------------------------------------------------LEM01720
         DOUBLE PRECISION   U(1),W(1),A(1),AD(1),SUM                    LEM01730
         INTEGER   IROW(1),ICOL(1)                                      LEM01740
C-----------------------------------------------------------------------LEM01750
C        SPARSE MATRIX-VECTOR MULTIPLY FOR LANCZS   U = A*W - SUM*U      LEM01760
C        SEE USPEC SUBROUTINE FOR DESCRIPTION OF THE ARRAYS THAT DEFINE  LEM01770
C        THE MATRIX                                                     LEM01780
C-----------------------------------------------------------------------LEM01790
C                                                                       LEM01800
C        COMPUTE THE DIAGONAL TERMS                                     LEM01810
         DO 10 I = 1,N                                                  LEM01820
      10 U(I) = AD(I)*W(I)-SUM*U(I)                                     LEM01830
C                                                                       LEM01840
C        COMPUTE BY COLUMN                                              LEM01850
         LLAST = 0                                                      LEM01860
         DO 30 J = 1,NZL                                                LEM01870
C                                                                       LEM01880
         IF (ICOL(J).EQ.0) GO TO 30                                     LEM01890
         LFIRST = LLAST + 1                                             LEM01900
         LLAST = LLAST + ICOL(J)                                        LEM01910
C                                                                       LEM01920
         DO 20 L = LFIRST,LLAST                                         LEM01930
         I = IROW(L)                                                    LEM01940
C                                                                       LEM01950
         U(I) = U(I) + A(L)*W(J)                                        LEM01960
```

```
      U(J) = U(J) + A(L)*W(I)                                         LEM01970
C                                                                     LEM01980
   20 CONTINUE                                                        LEM01990
C                                                                     LEM02000
   30 CONTINUE                                                        LEM02010
C                                                                     LEM02020
      RETURN                                                          LEM02030
C                                                                     LEM02040
C--------------------------------------------------------------------LEM02050
C     STORAGE LOCATIONS OF ARRAYS ARE PASSED TO CMATV FROM USPEC      LEM02060
C                                                                     LEM02070
      ENTRY CMATVE(A,AD,ICOL,IROW,N,NZL)                              LEM02080
C--------------------------------------------------------------------LEM02090
C                                                                     LEM02100
      RETURN                                                          LEM02110
C-----END OF CMATV---------------------------------------------------LEM02120
      END                                                             LEM02130
C                                                                     LEM02140
C-----MATRIX-VECTOR MULTIPLY FOR DIAGONAL TEST MATRICES--------------LEM02150
C                                                                     LEM02160
      SUBROUTINE CMATV(W,U,SUM)                                       LEM02170
C     SUBROUTINE DCMATV(W,U,SUM)                                      LEM02180
C                                                                     LEM02190
C     CMATV COMPUTES  U = (DIAGONAL MATRIX) * W - SUM * U             LEM02200
C--------------------------------------------------------------------LEM02210
      DOUBLE PRECISION  W(1),U(1),SUM                                 LEM02220
      DOUBLE PRECISION  D(1)                                          LEM02230
C--------------------------------------------------------------------LEM02240
C                                                                     LEM02250
      DO 10 I=1,N                                                     LEM02260
   10 U(I)= D(I)*W(I) - SUM*U(I)                                      LEM02270
      RETURN                                                          LEM02280
C                                                                     LEM02290
C--------------------------------------------------------------------LEM02300
      ENTRY MVDIAE(D,N)                                               LEM02310
C--------------------------------------------------------------------LEM02320
C                                                                     LEM02330
      RETURN                                                          LEM02340
C-----END OF DIAGONAL TEST MATRIX  MULTIPLY-------------------------LEM02350
      END                                                             LEM02360
C                                                                     LEM02370
C                                                                     LEM02380
C-----START OF USPEC FOR DIAGONAL TEST MATRIX-----------------------LEM02390
C                                                                     LEM02400
      SUBROUTINE USPEC(N,MATNO)                                       LEM02410
C     SUBROUTINE DUSPEC(N,MATNO)                                      LEM02420
C                                                                     LEM02430
C--------------------------------------------------------------------LEM02440
      DOUBLE PRECISION  D(1000), SHIFT, SPACE                         LEM02450
      DOUBLE PRECISION  DABS, DFLOAT                                  LEM02460
      REAL  EXPLAN(20)                                                LEM02470
C--------------------------------------------------------------------LEM02480
C                                                                     LEM02490
      READ(8,10) EXPLAN                                               LEM02500
   10 FORMAT(20A4)                                                    LEM02510
      READ(8,*) NOLD,NUNIF,SPACE,D(1),SHIFT                           LEM02520
      NNUNIF = NOLD - NUNIF                                           LEM02530
      WRITE(6,20) NOLD,SPACE,NNUNIF,D(1),SHIFT                        LEM02540
   20 FORMAT(/' DIAGONAL TEST MATRIX, SIZE = ',I4/' MOST ENTRIES ARE ',LEM02550
     1E10.3,' UNITS APART.',I3,' ENTRIES'/' ARE IRREGULARLY SPACED. FIRSLEM02560
     1T ENTRY IS ',E10.3,' SHIFT = ',E10.3/)                         LEM02570
C                                                                     LEM02580
      IF(N.NE.NOLD) GO TO 90                                          LEM02590
C     COMPUTE THE UNIFORM PORTION OF THE SPECTRUM                     LEM02600
      DO 30 J=2,NUNIF                                                 LEM02610
   30 D(J) = D(1) - DFLOAT(J-1)*SPACE                                 LEM02620
      NUNIF1=NUNIF + 1                                                LEM02630
```

```
      READ(8,10) EXPLAN                                               LEM02640
      DO 40 J=NUNIF1,N                                                LEM02650
   40 READ(8,*) D(J)                                                  LEM02660
      NB = NUNIF - 2                                                  LEM02670
C                                                                     LEM02680
      IF(SHIFT.EQ.0.) GO TO 60                                        LEM02690
      DO 50 J=1,N                                                     LEM02700
   50 D(J) = D(J) + SHIFT                                             LEM02710
C                                                                     LEM02720
C     PRINT OUT THE EIGENVALUES OF INTEREST                           LEM02730
   60 WRITE(6,70) (D(I), I=1,10 )                                     LEM02740
      WRITE(6,80) (D(I), I = NB,N)                                    LEM02750
   70 FORMAT(/' REAL SYMMETRIC LANCZOS TEST, 1ST 10 ENTRIES OF DIAGONAL LEM02760
     1TEST MATRIX = '/(3E22.14))                                      LEM02770
   80 FORMAT(/' MIDDLE UNIFORM PORTION OF MATRIX IS NOT PRINTED OUT'/  LEM02780
     1' END OF UNIFORM PLUS NONUNIFORM SECTION = '/(3E25.16))         LEM02790
C                                                                     LEM02800
C     DIAGONAL GENERATION COMPLETE                                    LEM02810
C                                                                     LEM02820
C-------------------------------------------------------------------LEM02830
C     CALL ENTRY TO MATRIX-VECTOR MULTIPLY SUBROUTINE  TO PASS        LEM02840
C     STORAGE LOCATION OF D-ARRAY AND ORDER OF A-MATRIX.              LEM02850
      CALL MVDIAE(D,N)                                                LEM02860
C-------------------------------------------------------------------LEM02870
C                                                                     LEM02880
      RETURN                                                          LEM02890
   90 WRITE(6,100) NOLD,N                                             LEM02900
  100 FORMAT(' PROGRAM TERMINATES BECAUSE NOLD = ',I5,'DOES NOT EQUAL N LEM02910
     1 =',I5)                                                         LEM02920
C-----END OF USPEC SUBROUTINE FOR DIAGONAL TEST MATRICES-------------LEM02930
      STOP                                                            LEM02940
      END                                                             LEM02950
C                                                                     LEM02960
C-----POISSON TEST MATRICES-----------------------------------------LEM02970
C                                                                     LEM02980
C     CONTAINS SUBROUTINES USPEC, CMATV, EXEVG, EXERR AND  EXVEC      LEM02990
C                                                                     LEM03000
C-----START OF USPEC------------------------------------------------LEM03010
C                                                                     LEM03020
C     SUBROUTINE USPEC(N,MATNO)                                       LEM03030
      SUBROUTINE PUSPEC(N,MATNO)                                      LEM03040
C                                                                     LEM03050
C-------------------------------------------------------------------LEM03060
      DOUBLE PRECISION  C0,C1,C2,HALF,ONE                             LEM03070
      REAL  EXPLAN(20)                                                LEM03080
C------------------------------------------------------------------- LEM03090
      HALF = 0.5D0                                                    LEM03100
      ONE  = 1.0D0                                                    LEM03110
C                                                                     LEM03120
C     READ USER-SPECIFIED PARAMETERS FROM INPUT FILE 8 (FREE FORMAT)  LEM03130
C                                                                     LEM03140
      READ(8,10) EXPLAN                                               LEM03150
      WRITE(6,10) EXPLAN                                              LEM03160
   10 FORMAT(20A4)                                                    LEM03170
C                                                                     LEM03180
      READ(8,10) EXPLAN                                               LEM03190
      READ(8,*) KX,KY,C0                                              LEM03200
      N = KX*KY                                                       LEM03210
      C1 = HALF-C0                                                    LEM03220
      C2 = ONE                                                        LEM03230
C                                                                     LEM03240
      WRITE(6,20) N,KX,KY,C2,C0,C1                                    LEM03250
   20 FORMAT(/5X,'N',4X,'KX',4X,'KY',7X,'DIAGONAL',3X,'X-CODIAGONAL',  LEM03260
     1 3X,'Y-CODIAGONAL'/3I6,3E15.8/)                                 LEM03270
C                                                                     LEM03280
C-------------------------------------------------------------------LEM03290
      CALL PMATVE(C0,C1,C2,KX,KY)                                     LEM03300
```

```
      CALL EXEVE(CO,C1,C2,KX,KY)                                    LEM03310
      CALL EXERRP(CO,C1,C2,KX,KY)                                   LEM03320
      CALL EXVECP(CO,C1,C2,KX,KY)                                   LEM03330
C------------------------------------------------------------------LEM03340
C                                                                   LEM03350
      RETURN                                                        LEM03360
C-----END OF USPEC-------------------------------------------------LEM03370
      END                                                           LEM03380
C                                                                   LEM03390
C-----START OF CMATV----------------------------------------------LEM03400
C     CALCULATE U = A*W - SUM*U FOR REAL POISSON MATRICES           LEM03410
C                                                                   LEM03420
C     SUBROUTINE CMATV(W,U,SUM)                                     LEM03430
      SUBROUTINE PMATV(W,U,SUM)                                     LEM03440
C                                                                   LEM03450
C------------------------------------------------------------------LEM03460
      DOUBLE PRECISION  U(1),W(1)                                   LEM03470
      DOUBLE PRECISION  CO,C1,C2,CCO,CC1,SUM                        LEM03480
C------------------------------------------------------------------LEM03490
C                                                                   LEM03500
      N = KX*KY                                                     LEM03510
      KX1 = KX-1                                                    LEM03520
      KY1 = KY-1                                                    LEM03530
C                                                                   LEM03540
      KK = 1                                                        LEM03550
      U(KK) = (C2*W(KK)+CO*W(KK+1)+C1*W(KK+KX)) - SUM*U(KK)         LEM03560
      KK = KX                                                       LEM03570
      U(KK) = (C2*W(KK)+CO*W(KK-1)+C1*W(KK+KX)) - SUM*U(KK)         LEM03580
      KK = N - KX + 1                                               LEM03590
      U(KK) = (C2*W(KK)+CO*W(KK+1)+C1*W(KK-KX)) - SUM*U(KK)         LEM03600
      KK = N                                                        LEM03610
      U(KK) = (C2*W(KK)+CO*W(KK-1)+C1*W(KK-KX)) - SUM*U(KK)         LEM03620
C                                                                   LEM03630
      DO 10 J = 2,KX1                                               LEM03640
      KK = J                                                        LEM03650
      U(KK) = (C2*W(KK)+CO*W(KK-1)+CO*W(KK+1)+C1*W(KK+KX)) - SUM*U(KK) LEM03660
      KK = J+N-KX                                                   LEM03670
      U(KK) = (C2*W(KK)+CO*W(KK-1)+CO*W(KK+1)+C1*W(KK-KX))-SUM*U(KK) LEM03680
   10 CONTINUE                                                      LEM03690
C                                                                   LEM03700
      DO 30 J = 2,KY1                                               LEM03710
      KK = (J-1)*KX + 1                                             LEM03720
      U(KK) = (C2*W(KK)+CO*W(KK+1)+C1*W(KK-KX)+C1*W(KK+KX)) - SUM*U(KK) LEM03730
      DO 20 I = 2,KX1                                               LEM03740
      KK = KK + 1                                                   LEM03750
      U(KK) = (C2*W(KK)+CO*W(KK-1)+CO*W(KK+1)+C1*W(KK-KX)           LEM03760
     1 +C1*W(KK+KX)) - SUM*U(KK)                                    LEM03770
   20 CONTINUE                                                      LEM03780
      KK = KK + 1                                                   LEM03790
      U(KK) = (C2*W(KK)+CO*W(KK-1)+C1*W(KK-KX)+C1*W(KK+KX)) - SUM*U(KK) LEM03800
   30 CONTINUE                                                      LEM03810
C                                                                   LEM03820
      RETURN                                                        LEM03830
C                                                                   LEM03840
C------------------------------------------------------------------LEM03850
      ENTRY  PMATVE(CCO,CC1,C2,KX,KY)                               LEM03860
C------------------------------------------------------------------LEM03870
C                                                                   LEM03880
      CO = -CCO                                                     LEM03890
      C1 = -CC1                                                     LEM03900
C                                                                   LEM03910
      RETURN                                                        LEM03920
C-----END OF CMATV------------------------------ ------------------LEM03930
      END                                                           LEM03940
C                                                                   LEM03950
C-----START OF EXEVG----------------------------------------------LEM03960
C                                                                   LEM03970
```

```
C     COMPUTES TRUE EIGENVALUES OF POISSON MATRIX, GAPS BETWEEN        LEM03980
C     TRUE EIGENVALUES, AND MULTIPLICITIES OF TRUE EIGENVALUES         LEM03990
C     AND STORE THESE VALUES, RESPECTIVELY, IN U, G, AND MP.          LEM04000
C     THESE QUANTITIES ARE WRITTEN OUT TO FILE 9                       LEM04010
C                                                                      LEM04020
      SUBROUTINE EXEVG(U,G,MP)                                         LEM04030
C                                                                      LEM04040
C------------------------------------------------------------------LEM04050
      DOUBLE PRECISION  U(1)                                           LEM04060
      DOUBLE PRECISION MACHEP,EPSM,CO,C1,C2,TO,T1,PIK,PIL,ONE,TWO      LEM04070
      DOUBLE PRECISION ATOLN,EE                                        LEM04080
      REAL G(1)                                                        LEM04090
      INTEGER MP(1)                                                    LEM04100
C------------------------------------------------------------------LEM04110
      DATA MACHEP/Z3410000000000000/                                  LEM04120
      EPSM = 2.0D0*MACHEP                                              LEM04130
C------------------------------------------------------------------LEM04140
      N = KX*KY                                                        LEM04150
      ONE  = 1.0D0                                                     LEM04160
      TWO  = 2.0D0                                                     LEM04170
      TO = DARCOS(-ONE)                                                LEM04180
      T1 = DFLOAT(KX+1)                                                LEM04190
      PIK = TO/T1                                                      LEM04200
      T1 = DFLOAT(KY+1)                                                LEM04210
      PIL = TO/T1                                                      LEM04220
C     GENERATE TRUE EIGENVALUES                                       LEM04230
      KP = 0                                                           LEM04240
      DO 20 J = 1,KY                                                   LEM04250
      T1 = PIL*DFLOAT(J)                                               LEM04260
      TO = C2 - TWO*C1*DCOS(T1)                                        LEM04270
      DO 10 I = 1,KX                                                   LEM04280
      KP = KP+1                                                        LEM04290
      T1 = PIK*DFLOAT(I)                                               LEM04300
   10 U(KP) = TO - TWO*CO*DCOS(T1)                                     LEM04310
   20 CONTINUE                                                         LEM04320
C                                                                      LEM04330
C     ORDER U VECTOR BY INCREASING ALGEBRAIC SIZE                     LEM04340
      DO 40 K = 2,N                                                    LEM04350
      KM1 = K-1                                                        LEM04360
      DO 30 L = 1,KM1                                                  LEM04370
      JJ = K-L                                                         LEM04380
      IF (U(JJ+1).GE.U(JJ)) GO TO 40                                   LEM04390
      TO = U(JJ)                                                       LEM04400
      U(JJ) = U(JJ+1)                                                  LEM04410
   30 U(JJ+1) = TO                                                     LEM04420
   40 CONTINUE                                                         LEM04430
      ATOLN = DMAX1(DABS(U(1)),DABS(U(N)))*EPSM                        LEM04440
C                                                                      LEM04450
      WRITE(9,50)                                                      LEM04460
   50 FORMAT(' TRUE EIGENVALUES FOR POISSON'/)                        LEM04470
C                                                                      LEM04480
      WRITE(9,60)N,KX,KY,C2,CO,C1,ATOLN                               LEM04490
      WRITE(6,60) N,KX,KY,C2,CO,C1,ATOLN                              LEM04500
   60 FORMAT(1X,'A-SIZE',2X,'X-DIM',2X,'Y-DIM'/3I7/                   LEM04510
     1 5X,'A-DIAGONAL',3X,'X-CODIAGONAL',3X,'Y-CODIAGONAL',10X,'ATOLN'/ LEM04520
     2 4E15.8)                                                         LEM04530
C                                                                      LEM04540
C     DETERMINE MULTIPLICITIES FOR TRUE EIGENVALUES                   LEM04550
      I = 1                                                            LEM04560
      IDEX = 1                                                         LEM04570
      J = 1                                                            LEM04580
      NEXACT = 0                                                       LEM04590
   70 J = J+1                                                          LEM04600
      IF (J.GT.N) GO TO 80                                             LEM04610
      EE = DABS(U(J)-U(I))                                             LEM04620
      IF (EE.GT.ATOLN) GO TO 80                                        LEM04630
      IDEX = IDEX+1                                                    LEM04640
```

```
           GO TO 70                                                   LEM04650
     80 NEXACT = NEXACT+1                                             LEM04660
           U(NEXACT) = U(I)                                           LEM04670
           MP(NEXACT) = IDEX                                          LEM04680
  C        MP(K) = MULTIPLICITY OF KTH EIGENVALUE CLUSTER FOR A       LEM04690
           IDEX = 1                                                   LEM04700
           I = J                                                      LEM04710
           IF (I.GT.N) GO TO 90                                       LEM04720
           GO TO 70                                                   LEM04730
     90 CONTINUE                                                      LEM04740
  C                                                                   LEM04750
  C        MULTIPLICITIES HAVE BEEN DETERMINED                        LEM04760
  C        NEXACT = NUMBER OF DISTINCT A-EIGENVALUES                  LEM04770
  C                                                                   LEM04780
           WRITE(9,100)NEXACT                                         LEM04790
           WRITE(6,100)NEXACT                                         LEM04800
    100 FORMAT(I6,' = NUMBER OF TRUE A-EIGENVALUES WHICH ARE DISTINCT'/)  LEM04810
  C                                                                   LEM04820
  C        MINGAP CALCULATION FOR DISTINCT A-EIGENVALUES              LEM04830
           NM1 = NEXACT - 1                                           LEM04840
           G(NEXACT) = U(NM1)-U(NEXACT)                               LEM04850
           G(1) = U(2)-U(1)                                           LEM04860
  C                                                                   LEM04870
           DO 110 J = 2,NM1                                           LEM04880
           TO = U(J)-U(J-1)                                           LEM04890
           T1 = U(J+1)-U(J)                                           LEM04900
           G(J) = T1                                                  LEM04910
           IF (TO.LT.T1) G(J) = -TO                                   LEM04920
    110 CONTINUE                                                      LEM04930
  C                                                                   LEM04940
  C        NEXACT DISTINCT A-EIGENVALUES ARE IN U IN ASCENDING ORDER  LEM04950
  C        MP = MULTIPLICITIES OF THE DISTINCT EIGENVALUES OF A       LEM04960
  C        G = TRUE MINIMUM GAP IN A FOR EACH OF THESE EIGENVALUES    LEM04970
  C        G < 0 INDICATES THE LEFT-HAND GAP WAS MINIMAL.             LEM04980
  C        OUTPUT MULTIPLICITIES, DISTINCT EVS, AND MINGAPS TO FILE 9 LEM04990
  C                                                                   LEM05000
           WRITE(9,120)                                               LEM05010
    120 FORMAT(5X,'I',1X,'AMULT',5X,'TRUE A-EIGENVALUE(I)',           LEM05020
       1 3X,'A-MINGAP(I)')                                            LEM05030
  C                                                                   LEM05040
           WRITE(9,130)(J,MP(J),U(J),G(J), J=1,NEXACT)                LEM05050
    130 FORMAT(2I6,E25.16,E14.3)                                      LEM05060
  C                                                                   LEM05070
           WRITE(9,140)                                               LEM05080
    140 FORMAT(' NEXACT DISTINCT A-EIGENVALUES ARE IN ASCENDING ORDER'/  LEM05090
       1 ' AMULT = MULTIPLICITIES OF THE DISTINCT EIGENVALUES OF A.'/ LEM05100
       2 ' A-MINGAP(I) = TRUE MINIMUM GAP IN A FOR EACH EIGENVALUE.'/ LEM05110
       3 ' A-MINGAP(I) LT 0 INDICATES THE LEFT-HAND GAP WAS MINIMAL.'//) LEM05120
  C                                                                   LEM05130
  C        WE ORDER U VECTOR BY INCREASING SIZE OF THE GAPS           LEM05140
  C                                                                   LEM05150
           DO 150 K = 1,NEXACT                                        LEM05160
    150 MP(K) = K                                                     LEM05170
  C                                                                   LEM05180
           DO 170 K = 2,NEXACT                                        LEM05190
           KM1 = K-1                                                  LEM05200
  C                                                                   LEM05210
           DO 160 L = 1,KM1                                           LEM05220
           JJ = K - L                                                 LEM05230
           IF (ABS(G(JJ+1)).GE.ABS(G(JJ))) GO TO 170                  LEM05240
           EE = U(JJ)                                                 LEM05250
           U(JJ) = U(JJ+1)                                            LEM05260
           U(JJ+1) = EE                                               LEM05270
           GG = G(JJ)                                                 LEM05280
           G(JJ) = G(JJ+1)                                            LEM05290
           G(JJ+1) = GG                                               LEM05300
           IEE = MP(JJ)                                               LEM05310
```

```
      MP(JJ) = MP(JJ+1)                                          LEM05320
  160 MP(JJ+1) = IEE                                             LEM05330
C                                                                LEM05340
  170 CONTINUE                                                   LEM05350
C                                                                LEM05360
      WRITE(9,180)                                               LEM05370
  180 FORMAT(5X,'K',6X,'A-MINGAP',5X,'TRUE A-EIGENVALUE(I)',2X,'A-EVNO')LEM05380
C                                                                LEM05390
      WRITE(9,190)(J,G(J),U(J),MP(J), J=1,NEXACT)                LEM05400
  190 FORMAT(I6,E14.3,E25.16,I8)                                 LEM05410
C                                                                LEM05420
      WRITE(9,200)                                               LEM05430
  200 FORMAT(' NEXACT DISTINCT A-EIGENVALUES. GAPS IN ASCENDING ORDER'/ LEM05440
     2 ' A-MINGAP(I) = TRUE MINIMUM GAP IN A FOR EACH EIGENVALUE.'/    LEM05450
     3 ' A-MINGAP(I) LT 0 INDICATES THE LEFT-HAND GAP WAS MINIMAL.'/   LEM05460
     3 ' A-MATRIX IS BLOCK TRIDIAGONAL AND EACH DIAGONAL BLOCK IS OF ORDLEM05470
     3ER NX.'/                                                   LEM05480
     4 ' NX = NUMBER OF POINTS ON EACH X-LINE. THERE ARE NY DIAGONAL BLOLEM05490
     4CKS.'/                                                     LEM05500
     5 ' NY = NUMBER OF POINTS ON EACH Y-LINE.'/                 LEM05510
     5 ' A-DIAGONAL   = A(K,K)'/                                 LEM05520
     6 ' X-CODIAGONAL = A(I,I+1)'/                               LEM05530
     7 ' Y-CODIAGONAL = A(I,I+NX)'/                              LEM05540
     8 ' ----- END OF FILE 9 TRUEEV----------------------------'//)    LEM05550
C                                                                LEM05560
      RETURN                                                     LEM05570
C                                                                LEM05580
C---------------------------------------------------------------LEM05590
      ENTRY EXEVE(CO,C1,C2,KX,KY)                                LEM05600
C---------------------------------------------------------------LEM05610
C                                                                LEM05620
      RETURN                                                     LEM05630
C-----END OF EXEVG----------------------------------------------LEM05640
      END                                                        LEM05650
C                                                                LEM05660
C-----START OF EXERR--------------------------------------------LEM05670
C                                                                LEM05680
C     FOR GIVEN COMPUTED EIGENVALUES, V(I), I=1,2,...,NG         LEM05690
C     COMPUTES THE CLOSEST TRUE EIGENVALUES AND THE ERROR IN THE LEM05700
C     COMPUTED EIGENVALUES, AND STORES THESE RESPECTIVELY        LEM05710
C     IN U(I) AND IN G(MEV+I). THESE QUANTITIES ARE WRITTEN      LEM05720
C     TO FILE 10.                                                LEM05730
C                                                                LEM05740
      SUBROUTINE EXERR(V,U,G,MP,MEV,NG,NEXACT,IWRITE)            LEM05750
C                                                                LEM05760
C---------------------------------------------------------------LEM05770
      DOUBLE PRECISION U(1),V(1)                                 LEM05780
      DOUBLE PRECISION EV,EE,TO,T1,CO,C1,C2,PIK,PIL              LEM05790
      DOUBLE PRECISION ATOLN,EPSM,MACHEP,ZERO,ONE,TWO            LEM05800
      REAL G(1)                                                  LEM05810
      INTEGER MP(1)                                              LEM05820
C---------------------------------------------------------------LEM05830
      DATA MACHEP/Z3410000000000000/                            LEM05840
      EPSM = 2.0D0*MACHEP                                        LEM05850
C---------------------------------------------------------------LEM05860
C     ON ENTRY V CONTAINS NG GOOD EIGENVALUES OF T(1,MEV)        LEM05870
C     MP CONTAINS THE MULTIPLICITIES OF THESE EIGENVALUES.       LEM05880
C     U(I) = GAP TO NEAREST DISTINCT TMEV I=1,NG                 LEM05890
C     U < 0. MEANS GAP IS DUE TO SPURIOUS EV                     LEM05900
C                                                                LEM05910
C     ON EXIT G(MEV+I) = ERROR FOR V(I) < 0. IF MULT EV > 1      LEM05920
C     K = MP(I) MEANS |V(I) - U(K)| = MIN                        LEM05930
C     MP < 0 MEANS MORE THAN ONE I USES SAME K                   LEM05940
C                                                                LEM05950
C     TO = C2 - 2*C1*COS(PIL*J)                                  LEM05960
C     U(KP) = TO - 2*CO*COS(PIK*I)                               LEM05970
C     KP = (J-1)*KX + I                                          LEM05980
```

```
C     C2 = ONE                                                          LEM05990
C     CO = X-CODIAGONAL = INPUT                                         LEM06000
C     C1 = Y-CODIAGONAL = HALF - CO                                     LEM06010
C-----------------------------------------------------------------------LEM06020
      N = KX*KY                                                         LEM06030
      ZERO = 0.0D0                                                      LEM06040
      ONE  = 1.0D0                                                      LEM06050
      TWO  = 2.0D0                                                      LEM06060
C                                                                       LEM06070
C     SET G(I) = GAP FROM GOOD T(MEV) TO NEAREST DISTINCT TMEV I=1,NG   LEM06080
      DO 10 I = 1,NG                                                    LEM06090
      G(I) = U(I)                                                       LEM06100
   10 CONTINUE                                                          LEM06110
C                                                                       LEM06120
C     REGENERATE A-EIGENVALUES                                          LEM06130
      TO = DARCOS(-ONE)                                                 LEM06140
      T1 = DFLOAT(KX+1)                                                 LEM06150
      PIK = TO/T1                                                       LEM06160
      T1 = DFLOAT(KY+1)                                                 LEM06170
      PIL = TO/T1                                                       LEM06180
      KP = 0                                                            LEM06190
C                                                                       LEM06200
      DO 30 J = 1,KY                                                    LEM06210
      T1 = PIL*DFLOAT(J)                                                LEM06220
      TO = C2 - TWO*C1*DCOS(T1)                                         LEM06230
      DO 20 I = 1,KX                                                    LEM06240
      KP = KP+1                                                         LEM06250
      T1 = PIK*DFLOAT(I)                                                LEM06260
   20 U(KP) = TO - TWO*CO*DCOS(T1)                                      LEM06270
   30 CONTINUE                                                          LEM06280
C                                                                       LEM06290
C     ORDER U VECTOR BY INCREASING ALGEBRAIC SIZE                       LEM06300
      DO 50 K = 2,N                                                     LEM06310
      KM1 = K-1                                                         LEM06320
      DO 40 L = 1,KM1                                                   LEM06330
      JJ = K-L                                                          LEM06340
      IF (U(JJ+1).GE.U(JJ)) GO TO 50                                    LEM06350
      TO = U(JJ)                                                        LEM06360
      U(JJ) = U(JJ+1)                                                   LEM06370
   40 U(JJ+1) = TO                                                      LEM06380
   50 CONTINUE                                                          LEM06390
C                                                                       LEM06400
      ATOLN = DMAX1(DABS(U(1)),DABS(U(N)))*EPSM                         LEM06410
C                                                                       LEM06420
C     DETERMINE MULTIPLICITIES FOR TRUE EIGENVALUES                     LEM06430
      I = 1                                                             LEM06440
      J = 1                                                             LEM06450
      NEXACT = 0                                                        LEM06460
   60 J = J+1                                                           LEM06470
      IF (J.GT.N) GO TO 70                                              LEM06480
      EE = DABS(U(J)-U(I))                                              LEM06490
      IF (EE.GT.ATOLN) GO TO 70                                         LEM06500
      IDEX = IDEX+1                                                     LEM06510
      GO TO 60                                                          LEM06520
   70 NEXACT = NEXACT+1                                                 LEM06530
      U(NEXACT) = U(I)                                                  LEM06540
      I = J                                                             LEM06550
      IF (I.GT.N) GO TO 80                                              LEM06560
      GO TO 60                                                          LEM06570
   80 CONTINUE                                                          LEM06580
C                                                                       LEM06590
C     NEXACT = NUMBER OF DISTINCT A-EIGENVALUES                         LEM06600
C     U CONTAINS TRUE DISTINCT A-EV ORDERED BY INCREASING SIZE          LEM06610
C                                                                       LEM06620
      IF ( IWRITE.EQ.1) WRITE(6,90)MEV,NG,NEXACT                        LEM06630
   90 FORMAT(/3I6,' = MEV, NG, NEXACT, POISZ CASE'/                     LEM06640
```

```
      1 ' TRUE ERRORS FOR GOOD EIGENVALUES'/)                    LEM06650
C                                                                LEM06660
C     WRITE(6,61) (K,U(K), K=1,NEXACT)                           LEM06670
C  61 FORMAT(4(I5,E15.8))                                        LEM06680
C                                                                LEM06690
C     CALCULATION OF THE TRUE ERRORS.                            LEM06700
      KL = 1                                                     LEM06710
      DO 110 ITEV = 1,NG                                         LEM06720
      EV = V(ITEV)                                               LEM06730
      K = KL                                                     LEM06740
      T1 = DABS(EV - U(KL))                                      LEM06750
C                                                                LEM06760
      DO 100 KP = KL,NEXACT                                      LEM06770
      TO = DABS(EV - U(KP))                                      LEM06780
      IF (TO.GE.T1) GO TO 100                                    LEM06790
      K = KP                                                     LEM06800
      T1 = TO                                                    LEM06810
  100 CONTINUE                                                   LEM06820
C                                                                LEM06830
      IF (K.EQ.KL.AND.ITEV.GT.1) T1 = -T1                        LEM06840
      KL = K                                                     LEM06850
      MP(ITEV) = K                                               LEM06860
      G(MEV+ITEV) = T1                                           LEM06870
  110 CONTINUE                                                   LEM06880
C                                                                LEM06890
C     TRUE ERRORS HAVE BEEN COMPUTED OUTPUT THEM TO FILE 10      LEM06900
C     FORM HEADER FOR ERREXACT FILE 10                           LEM06910
      WRITE(10,120)N,KX,KY,C2,CO,C1                              LEM06920
  120 FORMAT(' POISSONZ TRUE ERROR FOR GOOD EIGENVALUES'/        LEM06930
     1 5X,'N',4X,'NX',4X,'NY'/3I6//                              LEM06940
     2 5X,'A-DIAGONAL',3X,'X-CODIAGONAL',3X,'Y-CODIAGONAL'/3E15.8//) LEM06950
C                                                                LEM06960
      WRITE(10,130)MEV,NG,NEXACT                                 LEM06970
  130 FORMAT(/3I6,' = MEV,NG,NEXACT'/1X,'T-EV NO',1X,'A-EV NO',  LEM06980
     1 10X,'GOOD EIGENVALUE',5X,'TRUEERROR',7X,'TMINGAP')        LEM06990
C                                                                LEM07000
      WRITE(10,140)(I,MP(I),V(I),G(MEV+I),G(I), I=1,NG)          LEM07010
  140 FORMAT(2I8,E25.16,2E14.3)                                  LEM07020
C                                                                LEM07030
      WRITE(10,150)                                              LEM07040
  150 FORMAT(' ABOVE ARE THE TRUE ERRORS FOR POISSON GOODEV'/    LEM07050
     1 ' IF A-EV NO LT 0 THEN GOODEV HAS MULTIPLICITY GT 1'/     LEM07060
     1 ' IF TRUE ERROR LT 0 THEN MORE THAN ONE GOODEV APPROXIMATES'/ LEM07070
     1 ' THE SAME TRUE POISSON EIGENVALUE'/                      LEM07080
     1 ' IF TMINGAP LT 0 THE MINGAP IS DUE TO SPURIOUS EIGENVALUE'//) LEM07090
C                                                                LEM07100
      RETURN                                                     LEM07110
C                                                                LEM07120
C----------------------------------------------------------------LEM07130
      ENTRY EXERRP(CO,C1,C2,KX,KY)                               LEM07140
C----------------------------------------------------------------LEM07150
C                                                                LEM07160
C-----END OF EXERR-----------------------------------------------LEM07170
      RETURN                                                     LEM07180
      END                                                        LEM07190
C                                                                LEM07200
C-----START OF EXVEC---------------------------------------------LEM07210
C                                                                LEM07220
C     (JVEC = 1): FOR A GIVEN RITZ VECTOR V AND EIGENVALUE X1, COMPUTES LEM07230
C     THE CLOSEST EIGENVALUE Y1 AND CORRESPONDING TRUE EIGENVECTOR U, LEM07240
C     AND THEN CALCULATES THE NORM OF THE DIFFERENCE BETWEEN     LEM07250
C     V AND U AND THE MAXIMAL DIFFERENCE BETWEEN THE COMPONENTS. LEM07260
C     THESE QUANTITIES ARE WRITTEN TO FILE 6.                    LEM07270
C                                                                LEM07280
C     (JVEC = 2):  COMPUTES THE PROJECTION OF EACH               LEM07290
C     OF THE TRUE EIGENVECTORS ON THE LANCZOS STARTING VECTOR    LEM07300
```

```
C     USED BY THE LANCZS SUBROUTINE AND WRITES THEM TO FILE 12.          LEM07310
C                                                                        LEM07320
      SUBROUTINE EXVEC(U,V,X1,Y1,G,MP,IIX,JVEC,ICOUNT)                   LEM07330
C                                                                        LEM07340
C-----------------------------------------------------------------------LEM07350
      DOUBLE PRECISION U(1),V(1)                                         LEM07360
      DOUBLE PRECISION WI(110),WJ(110),WII(110)                         LEM07370
      DOUBLE PRECISION X1,Y1,EV,EE,WS,PIK,PIL,SUM,PROJ,TEMP,S            LEM07380
      DOUBLE PRECISION ATOLN,EPSM,MACHEP,ZERO,HALF,ONE,TWO               LEM07390
      DOUBLE PRECISION CO,C1,C2,TO,T1,T2                                 LEM07400
      REAL G(1),GG                                                       LEM07410
      INTEGER MP(1)                                                      LEM07420
      DOUBLE PRECISION FINPRO                                            LEM07430
C-----------------------------------------------------------------------LEM07440
C     THIS PROGRAM CALCULATES THE TRUE EIGENVALUES AND EIGENVECTORS      LEM07450
C     OF THE POISSON MATRIX A OF ORDER  N = KX*KY                        LEM07460
C     A CONSISTS OF KY TRIDIAGONAL BLOCKS OF ORDER KX                    LEM07470
C     KX = X-DIMENSION    KY = Y-DIMENSION.                              LEM07480
C                                                                        LEM07490
C     IIX = SEED FOR RANDOM NUMBER GENERATOR USED TO CALCULATE           LEM07500
C           STARTING LANCZOS VECTOR IN LANCZS                            LEM07510
C     V = RANDOM UNIT STARTING VECTOR FOR LANCZS                         LEM07520
C     A*U = EV*U   ||U|| = ONE                                           LEM07530
C                                                                        LEM07540
C     C2 = DIAGONAL OF KX BY KX MATRIX                                   LEM07550
C    -CO = CO-DIAGONAL OF THE KX BY KX MATRIX.                           LEM07560
C    -C1 = Y-CODIAGONAL.                                                 LEM07570
C                                                                        LEM07580
C     NOTE THAT THE VECTORS WI,WJ,WII ARE DIMENSIONED INTERNALLY         LEM07590
C     THEY ARE USED JUST TO KEEP FROM REGENERATING INFORMATION.          LEM07600
C     WI,WII = REAL*8 ARRAYS OF DIMENSION AT LEAST KX                    LEM07610
C     WJ     = REAL*8 ARRAY  OF DIMENSION AT LEAST KY.                   LEM07620
C                                                                        LEM07630
C     NOTATION USED IN PROGRAM                                           LEM07640
C                                                                        LEM07650
C     PIK = ARCOS(-1)/(KX+1)    PIL = ARCOS(-1)/(KY+1)                   LEM07660
C     WI(I) = PIK*I        WJ(J) = PIL*J                                 LEM07670
C                                                                        LEM07680
C     U(K) IS A-EV ORDERED BY INCREASING SIZE, K = 1,N                   LEM07690
C     LATER U IS USED TO STORE THE TRUE EIGENVECTOR                      LEM07700
C     TO = C2 - 2*C1*COS(PIL*J)    EV(I,J) = TO - 2*CO*COS(PIK*I)        LEM07710
C     I = 1,KX    J = 1,KY    KP = (J-1)*KX + I                          LEM07720
C                                                                        LEM07730
C     U(KV) = SIN(PIK*I*IK)*SIN(PIL*J*JK)                                LEM07740
C     IK = 1,KX    JK = 1,KY    KV = (JK-1)*KX + IK                      LEM07750
C     U IS UNSCALED EIGENVECTOR FOR EV(I,J) = Y1                         LEM07760
C     WS = 1/||U|| : ||U|| = .5*DSQRT(T2*T3)  T2 = KX+1   T3 = KY+1      LEM07770
C                                                                        LEM07780
C     JVEC  = (1,2) FLAGS COMPUTATIONS TO BE PERFORMED.                  LEM07790
C                                                                        LEM07800
C           = (1) MEANS GIVEN X1 FIND Y1 AND KVEC SUCH THAT             LEM07810
C                 Y1 = EV(KVEC) AND |X1-Y1| = MIN                        LEM07820
C              ALSO GIVEN UNIT RITZ VECTOR ASSOCIATED WITH X1            LEM07830
C                 CALCULATE UNIT EIGENVECTOR U, A*U = Y1*U               LEM07840
C                 T2 = ||V-U||   T1 = MAX(|V(K)-U(K)|, K= 1,N)           LEM07850
C                 MAX OCCURS FIRST AT K = KK                             LEM07860
C                                                                        LEM07870
C           = (2) MEANS CALCULATION OF THE PROJECTION OF THE STARTING    LEM07880
C                 LANCZOS VECTOR ON EACH EIGENVECTOR OF A.               LEM07890
C                                                                        LEM07900
C-----------------------------------------------------------------------LEM07910
      DATA MACHEP/Z3410000000000000/                                     LEM07920
      EPSM = 2.0D0*MACHEP                                                 LEM07930
C----------------------------------------------------------------------- LEM07940
C     SPECIFY PARAMETERS                                                 LEM07950
      N = KX*KY                                                          LEM07960
      ZERO = 0.0D0                                                       LEM07970
```

```
      HALF = 0.5DO                                                LEM07980
      ONE  = 1.0DO                                                LEM07990
      TWO  = 2.0DO                                                LEM08000
      TO = DARCOS(-ONE)                                           LEM08010
      T1 = DFLOAT(KX+1)                                           LEM08020
      PIK = TO/T1                                                 LEM08030
      T2 = DFLOAT(KY+1)                                           LEM08040
      PIL = TO/T2                                                 LEM08050
      WS = TWO/DSQRT(T1*T2)                                       LEM08060
C                                                                 LEM08070
C     GENERATE WI WJ VECTORS                                      LEM08080
      KP = 0                                                      LEM08090
      DO 20 J = 1,KY                                              LEM08100
      T1 = PIL*DFLOAT(J)                                          LEM08110
      WJ(J) = T1                                                  LEM08120
      TO = C2 - TWO*C1*DCOS(T1)                                   LEM08130
      DO 10 I = 1,KX                                              LEM08140
      KP = KP+1                                                   LEM08150
      T1 = PIK*DFLOAT(I)                                          LEM08160
      WI(I) = T1                                                  LEM08170
   10 U(KP) = TO - TWO*CO*DCOS(T1)                                LEM08180
   20 CONTINUE                                                    LEM08190
C     U(KP) = EV(I,J) = C2 - 2*C1*COS(PIL*J) - 2*CO*COS(PIK*I)    LEM08200
C                                                                 LEM08210
C     INITIALIZE MP VECTOR                                        LEM08220
      DO 30 K = 1,N                                               LEM08230
   30 MP(K) = K                                                   LEM08240
C                                                                 LEM08250
C     WE ORDER U VECTOR BY INCREASING SIZE OF THE EVS             LEM08260
      DO 50 K = 2,N                                               LEM08270
      KM1 = K-1                                                   LEM08280
C                                                                 LEM08290
      DO 40 L = 1,KM1                                             LEM08300
      JJ = K - L                                                  LEM08310
      IF (U(JJ+1).GE.U(JJ)) GO TO 50                              LEM08320
      EE = U(JJ)                                                  LEM08330
      U(JJ) = U(JJ+1)                                             LEM08340
      U(JJ+1) = EE                                                LEM08350
      IEE = MP(JJ)                                                LEM08360
      MP(JJ) = MP(JJ+1)                                           LEM08370
   40 MP(JJ+1) = IEE                                              LEM08380
C                                                                 LEM08390
   50 CONTINUE                                                    LEM08400
C                                                                 LEM08410
      ATOLN = DMAX1(DABS(U(1)),DABS(U(N)))*EPSM                   LEM08420
C                                                                 LEM08430
      IF (ICOUNT.EQ.1) WRITE(6,60) N,KX,KY,JVEC,C2,CO,C1,ATOLN    LEM08440
   60 FORMAT(/' TRUE ERRORS FOR CONVERGED GOODEV'/                LEM08450
     1 4I6,' = N KX KY JVEC'//                                    LEM08460
     1 4E12.5,' = C2 CO C1 ATOLN'//)                              LEM08470
C                                                                 LEM08480
C     KP = MP(K) MEANS EIGENVALUE U(K) CORRESPONDS TO EIGENVECTOR W(KP) LEM08490
C     COMPUTE TOLERANCE USED IN COMPUTING TRUE MULTIPLICITIES     LEM08500
C                                                                 LEM08510
      IF (JVEC.EQ.1) GO TO 180                                    LEM08520
C                                                                 LEM08530
C     JVEC = 2 SO CALCULATE PROJECTIONS AND WRITE IN FILE 12      LEM08540
C                                                                 LEM08550
      WRITE(12,70)                                                LEM08560
   70 FORMAT(' PROJECTIONS OF LANCZOS STARTING VECTOR ON A-EIGENVECS') LEM08570
C                                                                 LEM08580
      WRITE(12,80)N,KX,KY,IIX,C2,CO,C1,ATOLN                      LEM08590
   80 FORMAT(1X,'A-SIZE',2X,'X-DIM',2X,'Y-DIM',6X,'SVSEED'/3I7,I12/ LEM08600
     1 5X,'A-DIAGONAL',3X,'X-CODIAGONAL',3X,'Y-CODIAGONAL',5X,'ATOLN'/ LEM08610
     2 3E15.8,E10.3)                                              LEM08620
C                                                                 LEM08630
      WRITE(12,90)                                                LEM08640
```

```
   90 FORMAT(5X,'PROJECTION',8X,'TRUE A-EIGENVALUE',1X,'EV NO'         LEM08650
     1,2X,'VEC NO')                                                    LEM08660
C                                                                      LEM08670
C     GENERATE SAME RANDOM UNIT VECTOR USED IN THE LANCZS RECURSIONS.  LEM08680
      IIL=IIX                                                          LEM08690
C                                                                      LEM08700
C----------------------------------------------------------------------LEM08710
      CALL GENRAN(IIL,G,N)                                             LEM08720
C----------------------------------------------------------------------LEM08730
C                                                                      LEM08740
      DO 100 I = 1,N                                                   LEM08750
  100 V(I) = G(I)                                                      LEM08760
C                                                                      LEM08770
C----------------------------------------------------------------------LEM08780
      SUM = FINPRO(N,V(1),1,V(1),1)                                    LEM08790
C----------------------------------------------------------------------LEM08800
C                                                                      LEM08810
      SUM = 1.D0/DSQRT(SUM)                                            LEM08820
C                                                                      LEM08830
      DO 110 I = 1,N                                                   LEM08840
  110 V(I) = V(I)*SUM                                                  LEM08850
C                                                                      LEM08860
C     DETERMINE UNIT EIGENVECTOR W ASSOCIATED WITH EACH EV(I,J) = Y1   LEM08870
C     AND CALCULATE THE PROJECTION G(K) OF U ON THE STARTING VECTOR V  LEM08880
C     A*U = EV*U       WS = 1/||WU||: WU = UNSCALED EIGENVECTOR        LEM08890
C                                                                      LEM08900
      DO 160 K =1,N                                                    LEM08910
C     DETERMINE I J FROM K: MP(K) = KP = (J-1)*KX+I                    LEM08920
      KP = MP(K)                                                       LEM08930
      I = MOD(KP,KX)                                                   LEM08940
      IF (I.EQ.0) I = KX                                               LEM08950
      T1 = WI(I)                                                       LEM08960
      J = 1 + (KP-1)/KX                                                LEM08970
      TO = WJ(J)                                                       LEM08980
      TO = WJ(J)                                                       LEM08990
C                                                                      LEM09000
      Y1 = C2 - TWO*C1*DCOS(WJ(J)) - TWO*CO*DCOS(WI(I))                LEM09010
C     Y1 = EV(I,J)                                                     LEM09020
C                                                                      LEM09030
      DO 120 II = 1,KX                                                 LEM09040
      T2 = T1*DFLOAT(II)                                               LEM09050
  120 WII(II) = WS*DSIN(T2)                                            LEM09060
C                                                                      LEM09070
      KV = 0                                                           LEM09080
      DO 140 JJ = 1,KY                                                 LEM09090
      T2 = TO*DFLOAT(JJ)                                               LEM09100
      T2 = DSIN(T2)                                                    LEM09110
C                                                                      LEM09120
      DO 130 II = 1,KX                                                 LEM09130
      KV = KV + 1                                                      LEM09140
  130 U(KV) = T2*WII(II)                                               LEM09150
C                                                                      LEM09160
  140 CONTINUE                                                         LEM09170
C                                                                      LEM09180
C     U IS UNIT EIGENVECTOR OF A ASSOCIATED WITH EV(I,J) = Y1          LEM09190
C     G(K) IS THE PROJECTION OF U ON V  FOR Y1                         LEM09200
C                                                                      LEM09210
C----------------------------------------------------------------------LEM09220
      PROJ = FINPRO(N,U(1),1,V(1),1)                                   LEM09230
C----------------------------------------------------------------------LEM09240
C                                                                      LEM09250
      TEMP = DABS(PROJ)                                                LEM09260
      G(K) = TEMP                                                      LEM09270
C                                                                      LEM09280
C     DESIRED PROJECTION HAS BEEN COMPUTED OUTPUT IT TO FILE 12.       LEM09290
      WRITE(12,150) G(K),Y1,K,MP(K)                                    LEM09300
```

```
     150 FORMAT(E15.8,E25.16,I6,I8)                                       LEM09310
C                                                                         LEM09320
     160 CONTINUE                                                         LEM09330
C                                                                         LEM09340
         WRITE(12,170)                                                    LEM09350
     170 FORMAT(' ----- END OF FILE 12 PROJECT ----------------------'//) LEM09360
C                                                                         LEM09370
         GO TO 310                                                        LEM09380
C                                                                         LEM09390
C        JVEC = 1                                                         LEM09400
C                                                                         LEM09410
C        X1 IS AN INPUT PARAMETER. WE CALCULATE TRUE                      LEM09420
C        A-EIGENVALUE WHICH IS CLOSEST TO X1, LABEL IT Y1 AND CALCULATE   LEM09430
C        UNIT EIGENVECTOR OF A ASSOCIATED WITH Y1. A*U = Y1*U, ||U|| = 1. LEM09440
C        Y1 = EJ(I,J). EIGENVALUES OF A ARE ORDERED BY INCREASING SIZE.   LEM09450
C        V = RITZ VECTOR ASSOCIATED WITH GOODEV X1                        LEM09460
C                                                                         LEM09470
     180 CONTINUE                                                         LEM09480
         KX1 = 0                                                          LEM09490
         IF (X1.LE.U(1)) KX1 = 1                                          LEM09500
         IF (X1.GE.U(N)) KX1 = N                                          LEM09510
         NM1 = N-1                                                        LEM09520
         IF (KX1.NE.0) GO TO 200                                          LEM09530
C                                                                         LEM09540
         DO 190 KVEC = 2,N                                                LEM09550
         IF (X1.GE.U(KVEC)) GO TO 190                                     LEM09560
C        U(KVEC-1).LE.X1.LT.U(KVEC)                                       LEM09570
         T1 = X1 - U(KVEC-1)                                              LEM09580
         T2 = U(KVEC) - X1                                                LEM09590
         KX1 = KVEC - 1                                                   LEM09600
         IF (T1.GT.T2) KX1 = KVEC                                         LEM09610
         GO TO 200                                                        LEM09620
     190 CONTINUE                                                         LEM09630
C                                                                         LEM09640
     200 Y1 = U(KX1)                                                      LEM09650
C                                                                         LEM09660
         IF (KX1.EQ.1) EE = U(2) - U(1)                                   LEM09670
         IF (KX1.EQ.N) EE = U(N) - U(NM1)                                 LEM09680
         IF (KX1.EQ.1.OR.KX1.EQ.N) GO TO 210                              LEM09690
         EE = DMIN1(U(KX1+1)-U(KX1),U(KX1)-U(KX1-1))                      LEM09700
     210 CONTINUE                                                         LEM09710
C                                                                         LEM09720
         TO = DABS(ONE - X1/Y1)                                           LEM09730
C                                                                         LEM09740
         WRITE(6,220) N,KX1,ICOUNT,Y1,X1,TO,EE                            LEM09750
     220 FORMAT(3I8,' = N, A-EV NUMBER,RITZ NUMBER'//                     LEM09760
        1 18X,' TRUEEV',19X,'GOODEV',4X,'RELERROR',4X,'A-MINGAP'/         LEM09770
        1 2E25.16,2E12.3/)                                               LEM09780
C                                                                         LEM09790
         IF (EE.GT.ATOLN) GO TO 240                                       LEM09800
C                                                                         LEM09810
         WRITE(6,230)                                                     LEM09820
     230 FORMAT(' Y1 IS A MULTIPLE EIGENVALUE OF A SO WE EXIT'/)          LEM09830
C                                                                         LEM09840
         GO TO 310                                                        LEM09850
C                                                                         LEM09860
C        Y1 IS TOEPLITZ EIGENVALUE CLOSEST TO X1.                         LEM09870
C        CALCULATION OF EIGENVECTOR ASSOCIATED WITH EIGENVALUE Y1         LEM09880
C        A*U = Y1*U                                                       LEM09890
C        DETERMINE I J FROM K: MP(K) = KP = (J-1)*KX+I                    LEM09900
     240 CONTINUE                                                         LEM09910
         K = KX1                                                          LEM09920
         KP = MP(K)                                                       LEM09930
         I = MOD(KP,KX)                                                   LEM09940
         IF (I.EQ.0) I = KX                                               LEM09950
         T1 = WI(I)                                                       LEM09960
         J = 1 + (KP-1)/KX                                                LEM09970
```

```
      T2 = WJ(J)                                              LEM09980
C                                                             LEM09990
      DO 250 II = 1,KX                                        LEM10000
      TO = T1*DFLOAT(II)                                      LEM10010
  250 WII(II) = WS*DSIN(TO)                                   LEM10020
C                                                             LEM10030
      KV = 0                                                  LEM10040
      DO 270 JJ = 1,KY                                        LEM10050
      TO = T2*DFLOAT(JJ)                                      LEM10060
      TO = DSIN(TO)                                           LEM10070
C                                                             LEM10080
      DO 260 II = 1,KX                                        LEM10090
      KV = KV + 1                                             LEM10100
  260 U(KV) = TO*WII(II)                                      LEM10110
C                                                             LEM10120
  270 CONTINUE                                                LEM10130
C                                                             LEM10140
C     U IS UNIT TRUE EIGENVECTOR OF A ASSOCIATED WITH Y1      LEM10150
C     V IS UNIT RITZVECTOR OF A ASSOCIATED WITH X1            LEM10160
C                                                             LEM10170
      KK = 0                                                  LEM10180
      S = ONE                                                 LEM10190
      T1 = ZERO                                               LEM10200
C                                                             LEM10210
      DO 280 K = 1,N                                          LEM10220
      IF (DABS(U(K)).LE.T1) GO TO 280                         LEM10230
      T1 = DABS(U(K))                                         LEM10240
      KK = K                                                  LEM10250
  280 CONTINUE                                                LEM10260
      IF (U(KK)*V(KK).LT.ZERO) S = - ONE                      LEM10270
C                                                             LEM10280
      KK = 0                                                  LEM10290
      T1 = ZERO                                               LEM10300
      T2 = ZERO                                               LEM10310
      DO 290 K = 1,N                                          LEM10320
      TEMP = DABS(S*U(K) - V(K))                              LEM10330
      T2 = T2 + TEMP**2                                       LEM10340
      IF (TEMP.LE.T1) GO TO 290                               LEM10350
      KK = K                                                  LEM10360
      T1 = TEMP                                               LEM10370
  290 CONTINUE                                                LEM10380
C                                                             LEM10390
      T2 = DSQRT(T2)                                          LEM10400
      WRITE(6,300) KK,T1,T2                                   LEM10410
  300 FORMAT(' EIGENVECTOR ERROR. MAX ERROR AT COMPONENT = ',I6/ LEM10420
     1 ' MAX DABS(TRUEVEC(K)-RITZVEC(K)) = ',E12.5/           LEM10430
     1 ' NORM(TRUEVEC-RITZVEC)  = ',E12.5/)                   LEM10440
C                                                             LEM10450
  310 CONTINUE                                                LEM10460
C                                                             LEM10470
      RETURN                                                  LEM10480
C                                                             LEM10490
C-------------------------------------------------------------LEM10500
      ENTRY EXVECP(CO,C1,C2,KX,KY)                            LEM10510
C-------------------------------------------------------------LEM10520
C                                                             LEM10530
      RETURN                                                  LEM10540
C-----END OF EXVEC--------------------------------------------LEM10550
      END                                                     LEM10560
```

SECTION 2.6   OTHER SUBROUTINES USED BY THE PROGRAMS IN CHAPTERS 2,3,4,5

```
C-----LESUB  (1) REAL SYMMETRIC----------------------------------------LES00010
C            (2) HERMITIAN MATRICES                                     LES00020
C            (3) FACTORED INVERSES OF REAL SYMMETRIC MATRICES AND       LES00030
C            (4) REAL SYMMETRIC GENERALIZED, A*X = EVAL*B*X WHERE       LES00040
C                B IS POSITIVE DEFINITE, CHOLESKY FACTOR AVAILABLE      LES00050
C                                                                       LES00060
C       ACCORDING TO PFORT THESE SUBROUTINES ARE PORTABLE EXCEPT FOR:   LES00070
C       (1) THE COMPLEX*16 VARIABLES AND THE CORRESPONDING FUNCTIONS    LES00080
C           FOR COMPLEX VARIABLES, DCMPLX, DREAL AND DCONJG USED IN     LES00090
C           THE SUBROUTINE CINPRD (USED ONLY IN CASE (2), HERMITIAN)    LES00100
C       (2) THE ENTRY IN THE SUBROUTINE LPERM USED TO PASS THE         LES00110
C           PERMUTATION FROM THE UPSEC SUBROUTINE TO LPERM. (USED       LES00120
C           ONLY IN CASES (3) AND (4), INVERSE AND GENERALIZED).        LES00130
C                                                                       LES00140
C       SUBROUTINES   BISEC, INVERR, TNORM, LUMP, ISOEV, PRTEST, AND    LES00150
C                     INVERM ARE USED WITH THE LANCZOS EIGENVALUE       LES00160
C                     PROGRAMS LEVAL, HLEVAL, LIVAL AND LGVAL. STURMI,  LES00170
C                     INVERM, LBISEC, AND TNORM ARE USED WITH THE       LES00180
C                     EIGENVECTOR PROGRAMS LEVEC, HLEVEC, LIVEC AND     LES00190
C                     LGVEC.  LPERM IS USED WITH LIVEC AND LGVEC.       LES00200
C                     IN THE HERMITIAN CASE, THE SUBROUTINE CINPRD      LES00210
C                     IS ALSO USED.                                     LES00220
C                                                                       LES00230
C-----COMPUTE T-EIGENVALUES BY BISECTION-------------------------------LES00240
C                                                                       LES00250
C       SUBROUTINE BISEC(ALPHA,BETA,BETA2,VB,VS,LBD,UBD,EPS,TTOL,MP,    LES00260
C      1 NINT,MEV,NDIS,IC,IWRITE)                                       LES00270
C                                                                       LES00280
C---------------------------------------------------------------------LES00290
C       DOUBLE PRECISION   ALPHA(1),BETA(1),BETA2(1),VB(1),VS(1)        LES00300
C       DOUBLE PRECISION   LBD(1),UBD(1),EPS,EPT,EPO,EP1,TEMP,TTOL      LES00310
C       DOUBLE PRECISION   ZERO,ONE,HALF,YU,YV,LB,UB,XL,XU,X1,X0,XS,BETAM LES00320
C       INTEGER   MP(1),IDEF(10)                                        LES00330
C       DOUBLE PRECISION   DABS, DSQRT, DMAX1, DMIN1, DFLOAT            LES00340
C---------------------------------------------------------------------LES00350
C       COMPUTES EIGENVALUES OF T(1,MEV) BY LOOPING INTERNALLY ON THE   LES00360
C       USER-SPECIFIED INTERVALS, (LB(J),UB(J)), J = 1,NINT. INTERVALS  LES00370
C       ARE TREATED AS OPEN ON THE LEFT AND CLOSED ON THE RIGHT.        LES00380
C       THE BISEC SUBROUTINE SIMULTANEOUSLY LABELS SPURIOUS T-EIGENVALUES LES00390
C       AND DETERMINES THE T-MULTIPLICITIES OF EACH GOOD T-EIGENVALUE.  LES00400
C       SPURIOUS T-EIGENVALUES ARE LABELLED BY A T-MULTIPLICITY = 0.    LES00410
C       ANY T-EIGENVALUE WITH A T-MULTIPLICITY >= 1 IS 'GOOD'.          LES00420
C                                                                       LES00430
C       IF IWRITE = 0 THEN MOST OF THE WRITES TO FILE 6 ARE NOT         LES00440
C       ACTIVATED.                                                      LES00450
C                                                                       LES00460
C       NOTE THAT PROGRAM ASSUMES THAT NO MORE THAN MMAX/2 EIGENVALUES  LES00470
C       OF T(1,MEV) ARE TO BE COMPUTED IN ANY ONE OF THE SUBINTERVALS   LES00480
C       CONSIDERED, WHERE MMAX = DIMENSION OF VB SPECIFIED BY THE USER  LES00490
C       IN THE MAIN PROGRAM LEVAL.                                      LES00500
C                                                                       LES00510
C       ON ENTRY                                                        LES00520
C       BETA2(J) IS SET = BETA(J)*BETA(J).  THE STORAGE FOR BETA2 COULD LES00530
C       BE ELIMINATED BY RECOMPUTING THE BETA(J)**2 FOR EACH STURM      LES00540
C       SEQUENCE.                                                       LES00550
C                                                                       LES00560
C       EPS = 2*MACHEP =  4.4 * 10**-16 ON IBM 3081.                    LES00570
C       TTOL = EPS*TKMAX WHERE                                          LES00580
C       TKMAX = MAX(|ALPHA(K)|,BETA(K), K=1,KMAX)                       LES00590
C                                                                       LES00600
C       ON EXIT                                                         LES00610
C       NDIS = TOTAL NUMBER OF COMPUTED DISTINCT EIGENVALUES OF         LES00620
C              T(1,MEV) ON THE UNION OF THE (LB,UB) INTERVALS.          LES00630
C       VS = COMPUTED DISTINCT EIGENVALUES OF T(1,MEV) IN ALGEBRAICALLY- LES00640
```

```
C          INCREASING ORDER                                            LES00650
C      MP = CORRESPONDING T-MULTIPLICITIES OF THESE EIGENVALUES         LES00660
C      MP(I) = (0,1,MI), MI>1, I=1,NDIS  MEANS:                         LES00670
C         (0)  V(I) IS SPURIOUS                                         LES00680
C         (1)  V(I) IS ISOLATED AND GOOD                                LES00690
C         (MI) V(I) IS MULTIPLE AND HENCE A CONVERGED GOOD T-EIGENVALUE. LES00700
C      IC = TOTAL NUMBER OF STURMS USED                                 LES00710
C                                                                       LES00720
C      DEFAULTS                                                         LES00730
C      ISKIP = 0 INITIALLY. IF DEFAULT OCCURS ON J-TH SUB-INTERVAL, SET LES00740
C             ISKIP=ISKIP+1 AND IDEF(ISKIP) = J                         LES00750
C             DEFAULTS OCCUR IF THERE ARE NO T-EIGENVALUES  IN THE      LES00760
C             SUBINTERVAL SPECIFIED OR IF THE NUMBER                    LES00770
C             OF STURMS SEQUENCES REQUIRED EXCEEDS MXSTUR.              LES00780
C             WHEN A DEFAULT OCCURS THE PROGRAM                         LES00790
C             SKIPS THE INTERVAL INVOLVED AND GOES ON TO THE NEXT       LES00800
C             INTERVAL.                                                 LES00810
C                                                                       LES00820
C-----------------------------------------------------------------------LES00830
C      SPECIFY PARAMETERS                                               LES00840
       ZERO = 0.0D0                                                     LES00850
       ONE  = 1.0D0                                                     LES00860
       HALF = 0.5D0                                                     LES00870
       MXSTUR = IC                                                      LES00880
       NDIS = 0                                                         LES00890
       IC = 0                                                           LES00900
       ISKIP = 0                                                        LES00910
       MP1 = MEV+1                                                      LES00920
C      SAVE THEN SET BETA(MEV+1) = 0. GENERATE BETA**2                  LES00930
       BETAM = BETA(MP1)                                                LES00940
       BETA(MP1) = ZERO                                                 LES00950
C                                                                       LES00960
       DO 10 I = 1,MP1                                                  LES00970
    10 BETA2(I) = BETA(I)*BETA(I)                                       LES00980
C                                                                       LES00990
C      EPO IS USED IN T-MULTIPLICITY AND SPURIOUS TESTS                 LES01000
C      EP1 AND EPS ARE USED IN THE BISEC CONVERGENCE TEST               LES01010
C                                                                       LES01020
       TEMP = DFLOAT(MEV+1000)                                          LES01030
       EPO  = TEMP*TTOL                                                 LES01040
       EP1  = DSQRT(TEMP)*TTOL                                          LES01050
C                                                                       LES01060
       WRITE(6,20)MEV,NINT                                              LES01070
    20 FORMAT(/' BISEC CALCULATION'/' ORDER OF T IS',I6/                LES01080
      1' NUMBER OF INTERVALS IS',I6/)                                   LES01090
C                                                                       LES01100
       WRITE(6,30) EPO,EP1                                              LES01110
    30 FORMAT(/' MULTOL, TOLERANCE USED IN T-MULTIPLICITY AND SPURIOUS TELES01120
      1STS = ',E10.3/' BISTOL, TOLERANCE USED IN BISEC CONVERGENCE TEST =LES01130
      1 ',E10.3/)                                                       LES01140
C                                                                       LES01150
C      LOOP ON THE NINT INTERVALS  (LB(J),UB(J)), J=1,NINT              LES01160
       DO 430 JIND = 1,NINT                                             LES01170
       LB = LBD(JIND)                                                   LES01180
       UB = UBD(JIND)                                                   LES01190
C                                                                       LES01200
       WRITE(6,40)JIND,LB,UB                                            LES01210
    40 FORMAT(//1X,'BISEC INTERVAL NO',2X,'LOWER BOUND',2X,'UPPER BOUND'/LES01220
      1I18,2E13.5/)                                                     LES01230
C                                                                       LES01240
C      INITIALIZATION AND PARAMETER SPECIFICATION                       LES01250
C      ICT IS TOTAL STURM COUNT ON (LB,UB)                              LES01260
C                                                                       LES01270
       NA = 0                                                           LES01280
       MD = 0                                                           LES01290
       NG = 0                                                           LES01300
```

```
      ICT = 0                                                          LES01310
C                                                                      LES01320
C     START OF T-EIGENVALUE CALCULATIONS                               LES01330
      X1 = UB                                                          LES01340
      ISTURM = 1                                                       LES01350
      GO TO 330                                                        LES01360
C     FORWARD STURM CALCULATION TO DETERMINE NA = NO. T-EIGENVALUES > UBLES01370
   50 NA = NEV                                                         LES01380
C                                                                      LES01390
      X1 = LB                                                          LES01400
      ISTURM = 2                                                       LES01410
      GO TO 330                                                        LES01420
C     FORWARD STURM CALC TO DETERMINE MT = NO. T-EIGENVALUES ON (LB,UB) LES01430
   60 CONTINUE                                                         LES01440
      MT=NEV                                                           LES01450
      ICT = !CT +2                                                     LES01460
C                                                                      LES01470
      WRITE(6,70)MT,NA                                                 LES01480
   70 FORMAT(/2I6,' = NO. TMEV ON (LB,UB) AND NO. .GT. UB'/)           LES01490
C                                                                      LES01500
C     DEFAULT TEST: IS ESTIMATED NUMBER OF STURMS > MXSTUR?            LES01510
      IEST = 30*MT                                                     LES01520
      IF (IEST.LT.MXSTUR) GO TO 90                                     LES01530
C                                                                      LES01540
      WRITE(6,80)                                                      LES01550
   80 FORMAT(//' ESTIMATED NUMBER OF STURMS REQUIRED EXCEEDS USER LIMIT'LES01560
     1/' SKIP THIS SUBINTERVAL')                                       LES01570
      GO TO 110                                                        LES01580
C                                                                      LES01590
   90 CONTINUE                                                         LES01600
C                                                                      LES01610
      IF (MT.GE.1) GO TO 120                                           LES01620
C                                                                      LES01630
      WRITE(6,100)                                                     LES01640
  100 FORMAT(//' THERE ARE NO T-EIGENVALUES ON THIS INTERVAL)'/)       LES01650
C                                                                      LES01660
  110 ISKIP = ISKIP+1                                                  LES01670
      IDEF(ISKIP) = JIND                                               LES01680
      GO TO 430                                                        LES01690
C                                                                      LES01700
C     REGULAR CASE.                                                    LES01710
  120 CONTINUE                                                         LES01720
C                                                                      LES01730
      IF (IWRITE.NE.0) WRITE(6,130)                                    LES01740
  130 FORMAT(/' DISTINCT T-EIGENVALUES COMPUTED USING BISEC'/          LES01750
     1 13X,'T-EIGENVALUE',2X,'TMULT',3X,'MD',4X,'NG')                  LES01760
C                                                                      LES01770
C     SET UP INITIAL UPPER AND LOWER BOUNDS FOR T-EIGENVALUES          LES01780
      DO 140 I=1,MT                                                    LES01790
      VB(I) = LB                                                       LES01800
      MTI = MT + I                                                     LES01810
  140 VB(MTI) = UB                                                     LES01820
C                                                                      LES01830
C     CALCULATE T-EIGENVALUES FROM LB UP TO UB  K = MT,...,1           LES01840
C     MAIN LOOP FOR FINDING KTH T-EIGENVALUE                           LES01850
C                                                                      LES01860
      K = MT                                                           LES01870
  150 CONTINUE                                                         LES01880
      ICO = 0                                                          LES01890
      XL = VB(K)                                                       LES01900
      MTK = MT+K                                                       LES01910
      XU = VB(MTK)                                                     LES01920
C                                                                      LES01930
      ISTURM = 3                                                       LES01940
      X1 = XU                                                          LES01950
      ICO = ICO + 1                                                    LES01960
```

```
        GO TO 330                                                   LES01970
C       FORWARD STURM CALCULATION AT XU                             LES01980
  160 NU=NEV                                                        LES01990
C                                                                   LES02000
C       BISECTION LOOP FOR KTH T-EIGENVALUE. TEST  X1=MIDPOINT OF (XL,XU) LES02010
        ISTURM = 4                                                  LES02020
  170 CONTINUE                                                      LES02030
        X1 = (XL+XU)*HALF                                           LES02040
        XS = DABS(XL)+DABS(XU)                                      LES02050
        XO = XU-XL                                                  LES02060
        EPT = EPS*XS+EP1                                            LES02070
C                                                                   LES02080
C       EPT IS CONVERGENCE TOLERANCE FOR KTH T-EIGENVALUE           LES02090
C                                                                   LES02100
        IF (XO.LE.EPT) GO TO 230                                    LES02110
C                                                                   LES02120
C       T-EIGENVALUE HAS NOT YET CONVERGED                          LES02130
C                                                                   LES02140
        ICO = ICO + 1                                               LES02150
        GO TO 330                                                   LES02160
C       FORWARD STURM CALCULATION AT CURRENT T-EIGENVALUE APPROXIMATION. LES02170
  180 CONTINUE                                                      LES02180
C                                                                   LES02190
C       UPDATE T-EIGENVALUE INTERVAL (XL,XU)                        LES02200
C                                                                   LES02210
        IF (NEV.LT.K) GO TO 190                                     LES02220
C                                                                   LES02230
C       NUMBER OF T-EIGENVALUES NEV = K                             LES02240
        XL = X1                                                     LES02250
        GO TO 170                                                   LES02260
  190 CONTINUE                                                      LES02270
C       NUMBER OF T-EIGENVALUES NEV<K                               LES02280
        XU = X1                                                     LES02290
        NU = NEV                                                    LES02300
C                                                                   LES02310
C       UPDATE OF T-EIGENVALUE BOUNDS                               LES02320
C                                                                   LES02330
        IF (NEV.EQ.0) GO TO 210                                     LES02340
C                                                                   LES02350
        DO 200 I = 1,NEV                                            LES02360
  200 VB(I) = DMAX1(X1,VB(I))                                       LES02370
C                                                                   LES02380
  210 NEV1 = NEV+1                                                  LES02390
C                                                                   LES02400
        DO 220 II = NEV1,K                                          LES02410
        I = MT+II                                                   LES02420
  220 VB(I) = DMIN1(X1,VB(I))                                       LES02430
C                                                                   LES02440
        GO TO 170                                                   LES02450
C                                                                   LES02460
C       END (XL,XU) BISECTION LOOP FOR KTH T-EIGENVALUE ON (LB,UB)  LES02470
C       TEST FOR T-MULTIPLICITY AND IF SIMPLE THEN TEST FOR SPURIOUSNESS LES02480
C                                                                   LES02490
  230 CONTINUE                                                      LES02500
        NDIS = NDIS+1                                               LES02510
        MD = MD+1                                                   LES02520
        VS(NDIS) = X1                                               LES02530
C                                                                   LES02540
        JSTURM = 1                                                  LES02550
        X1 = XL-EPO                                                 LES02560
        GO TO 370                                                   LES02570
C       BACKWARD STURM CALCULATION                                  LES02580
  240 KL = KEV                                                      LES02590
        JL = JEV                                                    LES02600
C                                                                   LES02610
        JSTURM = 2                                                  LES02620
        ICO = ICO + 2                                               LES02630
```

```
      X1 = XU+EPO                                                LES02640
      GO TO 370                                                  LES02650
C     BACKWARD STURM CALCULATION                                 LES02660
  250 JU = JEV                                                   LES02670
      KU = KEV                                                   LES02680
C                                                                LES02690
C     FOR T(1,MEV)                                               LES02700
C     NU - KU = NO. T-EIGENVALUES ON (XU, XU + EPO)              LES02710
C     KL - KU = NO. T-EIGENVALUES ON (XL - EPO, XU + EPO)        LES02720
C                                                                LES02730
C     FOR T(2,MEV)                                               LES02740
C     JL -JU = NO. T-EIGENVALUES ON (XL - EPO, XU + EPO)         LES02750
C                                                                LES02760
C     IS THIS A SIMPLE T-EIGENVALUE?                             LES02770
C                                                                LES02780
      IF (KL-KU-1.EQ.0) GO TO 290                                LES02790
C                                                                LES02800
C     VS(NDIS) = KTH-T-EIGENVALUE OF (LB,UB) IS MULTIPLE AND HENCE GOOD LES02810
      IF (KU.EQ.NU) GO TO 280                                    LES02820
C     CONTINUE TO CHECK FOR T-MULTIPLICITY                       LES02830
  260 CONTINUE                                                   LES02840
      ISTURM = 5                                                 LES02850
      X1 = X1+EPO                                                LES02860
      ICO = ICO + 1                                              LES02870
      GO TO 330                                                  LES02880
C     FORWARD STURM CALCULATION                                  LES02890
  270 KNE = KU-NEV                                               LES02900
      KU = NEV                                                   LES02910
      IF (KNE.NE.0) GO TO 260                                    LES02920
C     SPECIFY T-MULTIPLICITY = MP(NDIS)                          LES02930
  280 MPEV = KL-KU                                               LES02940
      KNEW = KU                                                  LES02950
      GO TO 300                                                  LES02960
C     END MULTIPLE CASE                                          LES02970
C                                                                LES02980
C     T-EIGENVALUE IS SIMPLE   CHECK IF IT IS SPURIOUS           LES02990
  290 CONTINUE                                                   LES03000
      MPEV = 1                                                   LES03010
      IF (JU.LT.JL) MPEV=0                                       LES03020
      KNEW = K-1                                                 LES03030
C                                                                LES03040
C     X1 >= XU+EPO                                               LES03050
C     SPURIOUS TEST AND T-SIMPLE CASE COMPLETED                  LES03060
C     START OF NEXT T-EIGENVALUE COMPUTATION                     LES03070
C                                                                LES03080
  300 K = KNEW                                                   LES03090
      MP(NDIS) = MPEV                                            LES03100
      IF (MPEV.GE.1) NG = NG + 1                                 LES03110
C                                                                LES03120
      IF (IWRITE.NE.0) WRITE(6,310) VS(NDIS),MPEV,MD,NG          LES03130
  310 FORMAT(E25.16,3I6)                                         LES03140
C                                                                LES03150
C     UPDATE STURM COUNT. ICO = STURM COUNT FOR KTH T-EIGENVALUE LES03160
      ICT = ICT + ICO                                            LES03170
C                                                                LES03180
C     EXIT TEST FOR K DO LOOP                                    LES03190
C                                                                LES03200
      IF (K.LE.0) GO TO 410                                      LES03210
C                                                                LES03220
C     UPDATE LOWER BOUNDS                                        LES03230
      DO 320 I=1,KNEW                                            LES03240
  320 VB(I) = DMAX1(X1,VB(I))                                    LES03250
C                                                                LES03260
      GO TO 150                                                  LES03270
C     END OF BISECTION LOOP FOR KTH T-EIGENVALUE                 LES03280
C                                                                LES03290
C     FORWARD STURM CALCULATION                                  LES03300
```

```
  330 NEV = -NA                                                      LES03310
      YU  = ONE                                                      LES03320
C                                                                    LES03330
      DO 360 I = 1,MEV                                               LES03340
      IF (YU.NE.ZERO) GO TO 340                                      LES03350
      YV = BETA(I)/EPS                                               LES03360
      GO TO 350                                                      LES03370
  340 YV = BETA2(I)/YU                                               LES03380
  350 YU = X1 - ALPHA(I) - YV                                        LES03390
      IF (YU.GE.ZERO) GO TO 360                                      LES03400
      NEV = NEV + 1                                                  LES03410
  360 CONTINUE                                                       LES03420
C     NEV = NUMBER OF T-EIGENVALUES ON (X1,UB)                       LES03430
C                                                                    LES03440
      GO TO (50,60,160,180,270), ISTURM                             LES03450
C                                                                    LES03460
C     BACKWARD STURM CALCULATION FOR T(1,MEV) AND T(2,MEV)          LES03470
  370 KEV = -NA                                                      LES03480
      YU  = ONE                                                      LES03490
C                                                                    LES03500
      DO 400 II = 1,MEV                                              LES03510
      I = MP1-II                                                     LES03520
      IF (YU.NE.ZERO) GO TO 380                                      LES03530
      YV = BETA(I+1)/EPS                                             LES03540
      GO TO 390                                                      LES03550
  380 YV = BETA2(I+1)/YU                                             LES03560
  390 YU = X1-ALPHA(I)-YV                                            LES03570
      JEV = C                                                        LES03580
      IF (YU.GE.ZERO) GO TO 400                                      LES03590
      KEV = KEV+1                                                    LES03600
      JEV = 1                                                        LES03610
  400 CONTINUE                                                       LES03620
      JEV = KEV-JEV                                                  LES03630
C                                                                    LES03640
      GO TO (240,250), JSTURM                                       LES03650
C                                                                    LES03660
C     KEV = -NA + (NUMBER OF T(1,MEV) EIGENVALUES) > X1             LES03670
C     JEV = -NA + (NUMBER OF T(2,MEV) EIGENVALUES) > X1             LES03680
C     SET PARAMETERS FOR NEXT INTERVAL                              LES03690
  410 CONTINUE                                                       LES03700
      IC = ICT+IC                                                    LES03710
      MXSTUR = MXSTUR-ICT                                            LES03720
C                                                                    LES03730
      WRITE(6,420) JIND,NG,MD                                        LES03740
  420 FORMAT(/' T-EIGENVALUE CALCULATION ON INTERVAL',I6,'  IS COMPLETE'LES03750
     1 /3X,'NO. GOOD',3X,'NO. DISTINCT'/I10,I13)                     LES03760
C                                                                    LES03770
  430 CONTINUE                                                       LES03780
C                                                                    LES03790
C     END LOOP ON THE SUBINTERVALS (LB(J),UB(J)), J=1,NINT          LES03800
C     ISKIP OUTPUT                                                   LES03810
C                                                                    LES03820
      IF (ISKIP.GT.0) WRITE(6,440)ISKIP                             LES03830
  440 FORMAT(' BISEC DEFAULTED ON',I3,3X,'INTERVALS'/                LES03840
     1 ' DEFAULTS OCCUR IF AN INTERVAL HAS NO T-EIGENVALUES'/        LES03850
     2 ' OR THE STURM ESTIMATE EXCEEDS THE USER-SPECIFIED LIMIT'/)   LES03860
C                                                                    LES03870
      IF (ISKIP.GT.0) WRITE(6,450)(IDEF(I), I=1,ISKIP)              LES03880
  450 FORMAT(' BISEC DEFAULTED ON INTERVALS'/(10I8))                LES03890
C                                                                    LES03900
C     RESET BETA AT I = MP1                                          LES03910
      BETA(MP1) = BETAM                                              LES03920
C-----END OF BISEC----------------------------------------------------LES03930
      RETURN                                                         LES03940
      END                                                            LES03950
C                                                                    LES03960
```

```
C-----INVERSE ITERATION ON T(1,MEV)-------------------------------------LES03970
C                                                                       LES03980
      SUBROUTINE INVERR(ALPHA,BETA,V1,V2,VS,EPS,G,MP,MEV,MMB,NDIS,NISO,  LES03990
     1 N,IKL,IT,IWRITE)                                                  LES04000
C                                                                       LES04010
C----------------------------------------------------------------------LES04020
      DOUBLE PRECISION ALPHA(1),BETA(1),V1(1),V2(1),VS(1)               LES04030
      DOUBLE PRECISION X1,U,Z,EST,TEMP,TO,T1,RATIO,SUM,XU,NORM,TSUM     LES04040
      DOUBLE PRECISION  BETAM,EPS,EPS3,EPS4,ZERO,ONE                    LES04050
      REAL G(1)                                                        LES04060
      INTEGER  MP(1)                                                   LES04070
      DOUBLE PRECISION FINPRO                                          LES04080
      REAL  ABS                                                        LES04090
      DOUBLE PRECISION  DABS, DMIN1, DSQRT, DFLOAT                     LES04100
C----------------------------------------------------------------------LES04110
C     COMPUTES ERROR ESTIMATES FOR COMPUTED ISOLATED GOOD T-EIGENVALUES LES04120
C     IN VS AND WRITES THESE T-EIGENVALUES AND ESTIMATES TO FILE 4.     LES04130
C     BY DEFINITION A GOOD T-EIGENVALUE IS ISOLATED IF ITS             LES04140
C     CLOSEST T-NEIGHBOR IS ALSO GOOD, OR ITS CLOSEST NEIGHBOR IS       LES04150
C     SPURIOUS, BUT THAT NEIGHBOR IS FAR ENOUGH AWAY.  SO              LES04160
C     IN PARTICULAR, WE COMPUTE ESTIMATES FOR GOOD T-EIGENVALUES        LES04170
C     THAT ARE IN CLUSTERS OF GOOD T-EIGENVALUES.                      LES04180
C                                                                       LES04190
C     USES INVERSE ITERATION ON T(1,MEV) SOLVING THE EQUATION           LES04200
C     (T - X1*J)V2 = RIGHT-HAND SIDE (RANDOMLY-GENERATED)              LES04210
C     FOR EACH SUCH GOOD T-EIGENVALUE X1.                             LES04220
C                                                                       LES04230
C     PROGRAM REFACTORS T-X1*I ON EACH ITERATION OF INVERSE ITERATION. LES04240
C     TYPICALLY ONLY ONE ITERATION IS NEEDED PER EIGENVALUE X1.        LES04250
C                                                                       LES04260
C     POSSIBLE STORAGE COMPRESSION                                     LES04270
C     G STORAGE COULD BE ELIMINATED BY REGENERATING THE RANDOM         LES04280
C     RIGHT-HAND SIDE ON EACH ITERATION AND PRINTING OUT THE           LES04290
C     ERROR ESTIMATES AS THEY ARE GENERATED.                          LES04300
C                                                                       LES04310
C     ON ENTRY AND EXIT                                               LES04320
C     MEV = ORDER OF T                                                LES04330
C     ALPHA, BETA CONTAIN THE NONZERO ENTRIES OF THE T-MATRIX          LES04340
C     VS = COMPUTED DISTINCT EIGENVALUES OF T(1,MEV)                   LES04350
C     MP = T-MULTIPLICITY OF EACH T-EIGENVALUE IN VS. MP(I) = -1 MEANS LES04360
C          VS(I) IS A GOOD T-EIGENVALUE BUT THAT IT IS SITTING CLOSE TO LES04370
C          A SPURIOUS T-EIGENVALUE.  MP(I) = 0 MEANS VS(I) IS SPURIOUS. LES04380
C          ESTIMATES ARE COMPUTED ONLY FOR THOSE T-EIGENVALUES         LES04390
C          WITH MP(I) = 1. FLAGGING WAS DONE IN SUBROUTINE ISOEV       LES04400
C          PRIOR TO ENTERING INVERR.                                  LES04410
C     NISO = NUMBER OF ISOLATED GOOD T-EIGENVALUES CONTAINED IN VS     LES04420
C     NDIS =  NUMBER OF DISTINCT T-EIGENVALUES IN VS                   LES04430
C     IKL = SEED FOR RANDOM NUMBER GENERATOR                          LES04440
C     EPS = 2. * MACHINE EPSILON                                      LES04450
C                                                                       LES04460
C     IN PROGRAM:                                                     LES04470
C     ITER = MAXIMUM NUMBER OF INVERSE ITERATION STEPS ALLOWED FOR EACH LES04480
C          X1.  ITER = IT ON ENTRY.                                   LES04490
C     G = ARRAY OF DIMENSION AT LEAST MEV + NISO.  USED TO STORE       LES04500
C          RANDOMLY-GENERATED RIGHT-HAND SIDE.  THIS IS NOT            LES04510
C          REGENERATED FOR EACH X1. G IS ALSO USED TO STORE ERROR      LES04520
C          ESTIMATES AS THEY ARE COMPUTED FOR LATER PRINTOUT.          LES04530
C     V1,V2 = WORK SPACES USED IN THE FACTORIZATION OF T(1,MEV).       LES04540
C     AT THE END OF THE INVERSE ITERATION COMPUTATION FOR X1, V2       LES04550
C     CONTAINS THE UNIT EIGENVECTOR OF T(1,MEV) CORRESPONDING TO X1.   LES04560
C     V1 AND V2 MUST BE OF DIMENSION AT LEAST MEV.                     LES04570
C                                                                       LES04580
C     ON EXIT                                                         LES04590
C     G(J) = MINIMUM GAP IN T(1,MEV) FOR EACH VS(J), J=1,NDIS         LES04600
C     G(MEV+I) = BETAM*|V2(MEV)| = ERROR ESTIMATE FOR ISOLATED GOOD    LES04610
C          T-EIGENVALUES, WHERE I = 1,NISO  AND  BETAM = BETA(MEV+1)LES04620
C          V2(MEV) IS LAST COMPONENT OF THE UNIT EIGENVECTOR OF        LES04630
```

```
C              T(1,MEV) CORRESPONDING TO ITH ISOLATED GOOD T-EIGENVALUE.LES04640
C                                                                        LES04650
C     IF FOR SOME X1 IT.GT.ITER THEN THE ERROR ESTIMATE IN G IS MARKED   LES04660
C     WITH A - SIGN.                                                     LES04670
C                                                                        LES04680
C     V2 = ISOLATED GOOD T-EIGENVALUES                                   LES04690
C     V1 = MINIMAL T-GAPS FOR THE T-EIGENVALUES IN V2.                   LES04700
C     THESE ARE CONSTRUCTED FOR WRITE-OUT PURPOSES ONLY AND NOT          LES04710
C     NEEDED ELSEWHERE IN THE PROGRAM.                                   LES04720
C-----------------------------------------------------------------------LES04730
C                                                                        LES04740
C     LABEL OUTPUT FILE  4                                               LES04750
      IF (MMB.EQ.1) WRITE(4,10)                                          LES04760
   10 FORMAT(' INVERSE ITERATION ERROR ESTIMATES'/)                      LES04770
C                                                                        LES04780
C     FILE 6 (TERMINAL) OUTPUT OF ERROR ESTIMATES                        LES04790
      IF (IWRITE.NE.0.AND.NISO.NE.0) WRITE(6,20)                         LES04800
   20 FORMAT(/' INVERSE ITERATION ERROR ESTIMATES'/'  JISO',' JDIST',8X  LES04810
     1,'GOOD T-EIGENVALUE',4X,'BETAM*UM',5X,'TMINGAP')                   LES04820
C                                                                        LES04830
C     INITIALIZATION AND PARAMETER SPECIFICATION                         LES04840
      ZERO = 0.0D0                                                       LES04850
      ONE = 1.0D0                                                        LES04860
      NG = 0                                                             LES04870
      NISO = 0                                                           LES04880
      ITER = IT                                                          LES04890
      MP1 = MEV+1                                                        LES04900
      MM1 = MEV-1                                                        LES04910
      BETAM = BETA(MP1)                                                  LES04920
      BETA(MP1) = ZERO                                                   LES04930
C                                                                        LES04940
C     CALCULATE SCALE AND TOLERANCES                                     LES04950
      TSUM = DABS(ALPHA(1))                                              LES04960
      DO 30 I = 2,MEV                                                    LES04970
   30 TSUM = TSUM + DABS(ALPHA(I)) + BETA(I)                             LES04980
C                                                                        LES04990
      EPS3 = EPS*TSUM                                                    LES05000
      EPS4 = DFLOAT(MEV)*EPS3                                            LES05010
C                                                                        LES05020
C     GENERATE SCALED RANDOM RIGHT-HAND SIDE                             LES05030
      ILL = IKL                                                          LES05040
C                                                                        LES05050
C-----------------------------------------------------------------------LES05060
      CALL GENRAN(ILL,G,MEV)                                             LES05070
C-----------------------------------------------------------------------LES05080
C                                                                        LES05090
      GSUM = ZERO                                                        LES05100
      DO 40 I = 1,MEV                                                    LES05110
   40 GSUM = GSUM+ABS(G(I))                                              LES05120
      GSUM = EPS4/GSUM                                                   LES05130
C                                                                        LES05140
      DO 50 I = 1,MEV                                                    LES05150
   50 G(I) = GSUM*G(I)                                                   LES05160
C                                                                        LES05170
C     LOOP ON ISOLATED GOOD T-EIGENVALUES IN VS (MP(I) = 1) TO           LES05180
C     CALCULATE CORRESPONDING UNIT EIGENVECTOR OF T(1,MEV)               LES05190
C                                                                        LES05200
      DO 180 JEV = 1,NDIS                                                LES05210
C                                                                        LES05220
      IF (MP(JEV).EQ.0) GO TO 180                                        LES05230
      NG = NG + 1                                                        LES05240
      IF (MP(JEV).NE.1) GO TO 180                                        LES05250
C                                                                        LES05260
      IT = 1                                                             LES05270
      NISO = NISO + 1                                                    LES05280
```

```
      X1 = VS(JEV)                                              LES05290
C                                                               LES05300
C     INITIALIZE RIGHT HAND SIDE FOR INVERSE ITERATION         LES05310
      DO 60 I = 1,MEV                                           LES05320
   60 V2(I) = G(I)                                              LES05330
C                                                               LES05340
C     TRIANGULAR FACTORIZATION WITH NEAREST NEIGHBOR PIVOT      LES05350
C     STRATEGY. INTERCHANGES ARE LABELLED BY SETTING BETA < 0.  LES05360
C                                                               LES05370
   70 CONTINUE                                                  LES05380
      U = ALPHA(1)-X1                                           LES05390
      Z = BETA(2)                                               LES05400
C                                                               LES05410
      DO 90 I = 2,MEV                                           LES05420
      IF (BETA(I).GT.DABS(U)) GO TO 80                          LES05430
C     NO INTERCHANGE                                            LES05440
      V1(I-1) = Z/U                                             LES05450
      V2(I-1) = V2(I-1)/U                                       LES05460
      V2(I) = V2(I)-BETA(I)*V2(I-1)                             LES05470
      RATIO = BETA(I)/U                                         LES05480
      U = ALPHA(I)-X1-Z*RATIO                                   LES05490
      Z = BETA(I+1)                                             LES05500
      GO TO 90                                                  LES05510
   80 CONTINUE                                                  LES05520
C     INTERCHANGE CASE                                          LES05530
      RATIO = U/BETA(I)                                         LES05540
      BETA(I) = -BETA(I)                                        LES05550
      V1(I-1) = ALPHA(I)-X1                                     LES05560
      U = Z-RATIO*V1(I-1)                                       LES05570
      Z = -RATIO*BETA(I+1)                                      LES05580
      TEMP = V2(I-1)                                            LES05590
      V2(I-1) = V2(I)                                           LES05600
      V2(I) = TEMP-RATIO*V2(I)                                  LES05610
   90 CONTINUE                                                  LES05620
      IF (U.EQ.ZERO) U = EPS3                                   LES05630
C                                                               LES05640
C     SMALLNESS TEST AND DEFAULT VALUE FOR LAST COMPONENT       LES05650
C     PIVOT(I-1) = |BETA(I)| FOR INTERCHANGE CASE               LES05660
C     (I-1,I+1) ELEMENT IN RIGHT FACTOR = BETA(I+1)             LES05670
C     END OF FACTORIZATION AND FORWARD SUBSTITUTION             LES05680
C                                                               LES05690
C     BACK SUBSTITUTION                                         LES05700
      V2(MEV) = V2(MEV)/U                                       LES05710
      DO 110 II = 1,MM1                                         LES05720
      I = MEV-II                                                LES05730
      IF (BETA(I+1).LT.ZERO) GO TO 100                          LES05740
C     NO INTERCHANGE                                            LES05750
      V2(I) = V2(I)-V1(I)*V2(I+1)                               LES05760
      GO TO 110                                                 LES05770
C     INTERCHANGE CASE                                          LES05780
  100 BETA(I+1) = -BETA(I+1)                                    LES05790
      V2(I) = (V2(I)-V1(I)*V2(I+1)-BETA(I+2)*V2(I+2))/BETA(I+1) LES05800
  110 CONTINUE                                                  LES05810
C                                                               LES05820
C     TESTS FOR CONVERGENCE OF INVERSE ITERATION                LES05830
C     IF SUM |V2| COMPS. LE. 1 AND IT. LE. ITER DO ANOTHER INVIT STEP  LES05840
C                                                               LES05850
      NORM = DABS(V2(MEV))                                      LES05860
      DO 120 II = 1,MM1                                         LES05870
      I = MEV-II                                                LES05880
  120 NORM = NORM+DABS(V2(I))                                   LES05890
C                                                               LES05900
      IF (NORM.GE.ONE) GO TO 140                                LES05910
      IT = IT+1                                                 LES05920
      IF (IT.GT.ITER) GO TO 140                                LES05930
      XU = EPS4/NORM                                            LES05940
C                                                               LES05950
```

```
      DO 130 I = 1,MEV                                          LES05960
  130 V2(I) = V2(I)*XU                                          LES05970
C                                                               LES05980
      GO TO 70                                                  LES05990
C     ANOTHER INVERSE ITERATION STEP                           LES06000
C                                                               LES06010
C     INVERSE ITERATION FINISHED                               LES06020
C     NORMALIZE COMPUTED T-EIGENVECTOR : V2 = V2/||V2||         LES06030
  140 CONTINUE                                                  LES06040
      SUM = FINPRO(MEV,V2(1),1,V2(1),1)                         LES06050
      SUM = ONE/DSQRT(SUM)                                      LES06060
C                                                               LES06070
      DO 150 II = 1,MEV                                         LES06080
  150 V2(II) = SUM*V2(II)                                       LES06090
C                                                               LES06100
C     SAVE ERROR ESTIMATE FOR LATER OUTPUT                     LES06110
      EST = BETAM*DABS(V2(MEV))                                 LES06120
      IF (IT.GT.ITER) EST = -EST                                LES06130
      MEVPNI = MEV + NISO                                       LES06140
      G(MEVPNI) = EST                                           LES06150
      IF (IWRITE.EQ.0) GO TO 180                                LES06160
C                                                               LES06170
C     FILE 6 (TERMINAL) OUTPUT OF ERROR ESTIMATES.             LES06180
      IF (JEV.EQ.1) GAP = VS(2) - VS(1)                         LES06190
      IF (JEV.EQ.MEV) GAP = VS(MEV) - VS(MEV-1)                 LES06200
      IF (JEV.EQ.MEV.OR.JEV.EQ.1) GO TO 160                     LES06210
      TEMP = DMIN1(VS(JEV+1)-VS(JEV),VS(JEV)-VS(JEV-1))         LES06220
      GAP = TEMP                                                LES06230
  160 CONTINUE                                                  LES06240
C                                                               LES06250
      WRITE(6,170) NISO,JEV,X1,EST,GAP                          LES06260
  170 FORMAT(2I6,E25.16,2E12.3)                                 LES06270
C                                                               LES06280
  180 CONTINUE                                                  LES06290
C                                                               LES06300
C     END ERROR ESTIMATE LOOP ON ISOLATED GOOD T-EIGENVALUES.  LES06310
C     GENERATE DISTINCT MINGAPS FOR T(1,MEV).  THIS IS USEFUL AS AN   LES06320
C     INDICATOR OF THE GOODNESS OF THE INVERSE ITERATION ESTIMATES.   LES06330
C     TRANSFER ISOLATED GOOD T-EIGENVALUES AND CORRESPONDING TMINGAPS LES06340
C     TO V2 AND V1 FOR OUTPUT PURPOSES ONLY.                   LES06350
C                                                               LES06360
      NM1 = NDIS - 1                                            LES06370
      G(NDIS) = VS(NM1)-VS(NDIS)                                LES06380
      G(1) = VS(2)-VS(1)                                        LES06390
C                                                               LES06400
      DO 190 J = 2,NM1                                          LES06410
      T0 = VS(J)-VS(J-1)                                        LES06420
      T1 = VS(J+1)-VS(J)                                        LES06430
      G(J) = T1                                                 LES06440
      IF (T0.LT.T1) G(J)=-T0                                    LES06450
  190 CONTINUE                                                  LES06460
      ISO = 0                                                   LES06470
      DO 200 J = 1,NDIS                                         LES06480
      IF (MP(J).NE.1) GO TO 200                                 LES06490
      ISO = ISO+1                                               LES06500
      V1(ISO) = G(J)                                            LES06510
      V2(ISO) = VS(J)                                           LES06520
  200 CONTINUE                                                  LES06530
C                                                               LES06540
      IF(NISO.EQ.0) GO TO 250                                   LES06550
C                                                               LES06560
C     ERROR ESTIMATES ARE WRITTEN TO FILE 4                    LES06570
      WRITE(4,210)MEV,NDIS,NG,NISO,N,IKL,ITER,BETAM             LES06580
  210 FORMAT(1X,'TSIZE',2X,'NDIS',1X,'NGOOD',2X,'NISO',1X,'ASIZE'/5I6/   LES06590
     1 4X,'RHSEED',2X,'MXINIT',5X,'BETAM'/I10,I8,E10.3/         LES06600
     2 2X,'GOODEVNO',8X,'GOOD T-EIGENVALUE',6X,'BETAM*UM',7X,'TMINGAP')  LES06610
C                                                               LES06620
```

```
      ISPUR = 0                                              LES06630
      I = 0                                                  LES06640
      DO 240 J = 1,NDIS                                      LES06650
      IF(MP(J).NE.0) GO TO 220                              LES06660
      ISPUR = ISPUR + 1                                     LES06670
      GO TO 240                                             LES06680
  220 IF(MP(J).NE.1) GO TO 240                              LES06690
      I = I + 1                                             LES06700
      MEVI = MEV + I                                         LES06710
      IGOOD = J - ISPUR                                      LES06720
      WRITE(4,230) IGOOD,V2(I),G(MEVI),V1(I)                LES06730
  230 FORMAT(I10,E25.16,2E14.3)                             LES06740
  240 CONTINUE                                              LES06750
      GO TO 270                                             LES06760
C                                                            LES06770
  250 WRITE(4,260)                                          LES06780
  260 FORMAT(/' THERE ARE NO ISOLATED T-EIGENVALUES SO NO ERROR ESTIMATELES06790
     1S WERE COMPUTED')                                     LES06800
C                                                            LES06810
C     RESTORE BETA(MEV+1) = BETAM                           LES06820
  270 BETA(MP1) = BETAM                                     LES06830
C-----END OF INVERR-------------------------------------------------LES06840
      RETURN                                                LES06850
      END                                                   LES06860
C                                                            LES06870
C-----START OF TNORM-----------------------------------------------LES06880
C                                                            LES06890
      SUBROUTINE TNORM(ALPHA,BETA,BMIN,TMAX,MEV,IB)          LES06900
C                                                            LES06910
C-----------------------------------------------------------------LES06920
      DOUBLE PRECISION  ALPHA(1),BETA(1)                    LES06930
      DOUBLE PRECISION  TMAX,BMIN,BMAX,BSIZE,BTOL           LES06940
      DOUBLE PRECISION  DABS, DMAX1                         LES06950
C-----------------------------------------------------------------LES06960
C     COMPUTE SCALING FACTOR USED IN THE T-MULTIPLICITY, SPURIOUS AND  LES06970
C     PRTESTS.  CHECK RELATIVE SIZE OF THE BETA(K), K=1,MEV LES06980
C     AS A TEST ON THE LOCAL ORTHOGONALITY OF THE LANCZOS VECTORS.   LES06990
C                                                            LES07000
C          TMAX = MAX (|ALPHA(I)|, BETA(I),  I=1,MEV)       LES07010
C          BMIN = MIN (BETA(I) I=2,MEV)                     LES07020
C          BSIZE = BMIN/TMAX                                LES07030
C          |IB| = INDEX OF MINIMAL(BETA)                    LES07040
C          IB < 0 IF BMIN/TMAX < BTOL                       LES07050
C-----------------------------------------------------------------LES07060
C     SPECIFY PARAMETERS                                    LES07070
      IB = 2                                                LES07080
      BTOL = BMIN                                           LES07090
      BMIN = BETA(2)                                        LES07100
      BMAX = BETA(2)                                        LES07110
      TMAX = DABS(ALPHA(1))                                 LES07120
C                                                            LES07130
      DO 20 I = 2,MEV                                        LES07140
      IF (BETA(I).GE.BMIN) GO TO 10                         LES07150
      IB = I                                                LES07160
      BMIN = BETA(I)                                        LES07170
   10 TMAX = DMAX1(TMAX,DABS(ALPHA(I)))                     LES07180
      BMAX = DMAX1(BETA(I),BMAX)                            LES07190
   20 CONTINUE                                              LES07200
      TMAX = DMAX1(BMAX,TMAX)                               LES07210
C                                                            LES07220
C     TEST OF LOCAL ORTHOGONALITY USING SCALED BETAS        LES07230
      BSIZE = BMIN/TMAX                                     LES07240
      IF (BSIZE.GE.BTOL) GO TO 40                           LES07250
C                                                            LES07260
C     DEFAULT.  BSIZE IS SMALLER THAN TOLERANCE BTOL SPECIFIED IN MAIN  LES07270
C     PROGRAM.  PROGRAM TERMINATES FOR USER TO DECIDE WHAT TO DO    LES07280
C     BECAUSE LOCAL ORTHOGONALITY OF THE LANCZOS VECTORS COULD BE   LES07290
```

```
C      LOST.                                                            LES07300
C                                                                       LES07310
       IB = -IB                                                         LES07320
       WRITE(6,30) MEV                                                  LES07330
    30 FORMAT(/' BETA TEST INDICATES POSSIBLE LOSS OF LOCAL ORTHOGONALITYLES07340
      1OVER 1ST',I6,' LANCZOS VECTORS'/)                                LES07350
C                                                                       LES07360
    40 CONTINUE                                                         LES07370
C                                                                       LES07380
       WRITE(6,50) IB                                                   LES07390
    50 FORMAT(/' MINIMUM BETA RATIO OCCURS AT',I6,' TH BETA'/)          LES07400
C                                                                       LES07410
       WRITE(6,60) MEV,BMIN,TMAX,BSIZE                                  LES07420
    60 FORMAT(/1X,'TSIZE',6X,'MIN BETA',5X,'TKMAX',6X,'MIN RATIO'/      LES07430
      1 I6,E14.3,E10.3,E15.3/)                                          LES07440
C                                                                       LES07450
C-----END OF TNORM----------------------------------------------------LES07460
       RETURN                                                           LES07470
       END                                                             LES07480
C                                                                       LES07490
C                                                                       LES07500
C-----START OF LUMP--------------------------------------------------LES07510
C                                                                       LES07520
       SUBROUTINE LUMP(V1,RELTOL,MULTOL,SCALE2,LINDEX,LOOP)             LES07530
C                                                                       LES07540
C--------------------------------------------------------------------LES07550
       DOUBLE PRECISION  V1(1),SUM,RELTOL,MULTOL,THOLD,ZERO,SCALE2      LES07560
       INTEGER  LINDEX(1)                                               LES07570
       DOUBLE PRECISION  DABS, DFLOAT, DMAX1                            LES07580
C--------------------------------------------------------------------LES07590
C      LINDEX(J) = T-MULTIPLICITY OF JTH DISTINCT T-EIGENVALUE          LES07600
C      LOOP = NUMBER OF DISTINCT T-EIGENVALUES                          LES07610
C      LUMP 'COMBINES' COMPUTED 'GOOD' T-EIGENVALUES THAT ARE           LES07620
C      'TOO CLOSE'.                                                     LES07630
C      VALUE OF RELTOL IS 1.D-10.                                       LES07640
C                                                                       LES07650
C      IF IN A SET OF T-EIGENVALUES TO BE COMBINED THERE IS AN EIGENVALUELES07660
C      WITH LINDEX=1, THEN THE VALUE OF THE COMBINED EIGENVALUES IS SET  LES07670
C      EQUAL TO THE VALUE OF THAT EIGENVALUE.  NOTE THAT IF A SPURIOUS   LES07680
C      T-EIGENVALUE IS TO BE 'COMBINED' WITH A GOOD T-EIGENVALUE, THEN   LES07690
C      THIS IS DONE ONLY BY INCREASING THE INDEX, LINDEX, FOR THAT       LES07700
C      T-EIGENVALUE.  NUMERICAL VALUES OF SPURIOUS EIGENVALUES ARE NEVER LES07710
C      COMBINE WITH THOSE OF GOOD T-EIGENVALUES.                         LES07720
C--------------------------------------------------------------------LES07730
       ZERO = 0.0D0                                                     LES07740
       NLOOP = 0                                                        LES07750
       J = 0                                                            LES07760
       ICOUNT = 1                                                       LES07770
       JI = 1                                                           LES07780
       THOLD = DMAX1(RELTOL*DABS(V1(1)),SCALE2*MULTOL)                  LES07790
C      THOLD = DMAX1(RELTOL*DABS(V1(1)),RELTOL)                         LES07800
C                                                                       LES07810
    10 J = J+1                                                          LES07820
       IF (J.EQ.LOOP) GO TO 20                                          LES07830
       SUM = DABS(V1(J)-V1(J+1))                                        LES07840
       IF (SUM.LT.THOLD) GO TO 60                                       LES07850
    20 JF = JI + ICOUNT - 1                                             LES07860
       INDSUM = 0                                                       LES07870
       ISPUR = 0                                                        LES07880
C                                                                       LES07890
       DO 30 KK = JI,JF                                                 LES07900
       IF (LINDEX(KK).NE.0) GO TO 30                                    LES07910
       ISPUR = ISPUR + 1                                                LES07920
       INDSUM = INDSUM + 1                                              LES07930
    30 INDSUM = INDSUM + LINDEX(KK)                                     LES07940
C                                                                       LES07950
C      IF (JF-JI.GE.1) WRITE(6,40) (V1(KKK), KKK=JI,JF)                 LES07960
```

```
   40 FORMAT(/' LUMP LUMPS THE T-EIGENVALUES'/(4E20.13))              LES07970
C                                                                     LES07980
C     COMPUTE THE 'COMBINED' T-EIGENVALUE AND THE RESULTING           LES07990
C     T-MULTIPLICITY                                                  LES08000
      K = JI - 1                                                      LES08010
   50 K = K+1                                                         LES08020
      IF (K.GT.JF) GO TO 70                                           LES08030
      IF (LINDEX(K) .NE.1) GO TO 50                                   LES08040
      NLOOP = NLOOP + 1                                               LES08050
      V1(NLOOP) = V1(K)                                               LES08060
      GO TO 100                                                       LES08070
   60 ICOUNT = ICOUNT + 1                                             LES08080
      GO TO 10                                                        LES08090
C                                                                     LES08100
C     ALL INDICES WERE 0 OR >1                                        LES08110
   70 NLOOP = NLOOP + 1                                               LES08120
      IDIF = INDSUM - ISPUR                                           LES08130
      IF (IDIF.EQ.0) GO TO 90                                         LES08140
C                                                                     LES08150
      SUM = ZERO                                                      LES08160
      DO 80 KK = JI,JF                                                LES08170
   80 SUM = SUM + V1(KK) * DFLOAT(LINDEX(KK))                         LES08180
C                                                                     LES08190
      V1(NLOOP) = SUM/DFLOAT(IDIF)                                    LES08200
      GO TO 100                                                       LES08210
   90 V1(NLOOP) = V1(JI)                                              LES08220
  100 LINDEX(NLOOP) = INDSUM                                          LES08230
      IDIF = INDSUM - ISPUR                                           LES08240
      IF (IDIF.EQ.0.AND.ISPUR.EQ.1) LINDEX(NLOOP) = 0                 LES08250
      IF (J.EQ.LOOP) GO TO 110                                        LES08260
      ICOUNT = 1                                                      LES08270
      JI= J+1                                                         LES08280
      THOLD = DMAX1(RELTOL*DABS(V1(JI)),SCALE2*MULTOL)                LES08290
C     THOLD = DMAX1(RELTOL*DABS(V1(JI)),RELTOL)                       LES08300
      IF (JI.LT.LOOP) GO TO 10                                        LES08310
      NLOOP = NLOOP + 1                                               LES08320
      V1(NLOOP)= V1(JI)                                               LES08330
      LINDEX(NLOOP) = LINDEX(JI)                                      LES08340
  110 CONTINUE                                                        LES08350
C                                                                     LES08360
C     ON RETURN V1 CONTAINS THE DISTINCT T-EIGENVALUES                LES08370
C     LINDEX CONTAINS THE CORRESPONDING T-MULTIPLICITIES              LES08380
C                                                                     LES08390
      LOOP = NLOOP                                                    LES08400
      RETURN                                                          LES08410
C-----END OF LUMP---------------------------------------------------- LES08420
      END                                                             LES08430
C                                                                     LES08440
C                                                                     LES08450
C-----START OF ISOEV------------------------------------------------- LES08460
C                                                                     LES08470
      SUBROUTINE ISOEV(VS,GAPTOL,MULTOL,SCALE1,G,MP,NDIS,NG,NISO)      LES08480
C                                                                     LES08490
C------------------------------------------------------------------- LES08500
      DOUBLE PRECISION  VS(1),T0,T1,MULTOL,GAPTOL,SCALE1,TEMP         LES08510
      REAL   G(1),GAP                                                 LES08520
      INTEGER  MP(1)                                                  LES08530
      REAL  ABS                                                       LES08540
      DOUBLE PRECISION  DABS, DMAX1                                   LES08550
C------------------------------------------------------------------- LES08560
C     GENERATE DISTINCT TMINGAPS AND USE THEM TO LABEL THE ISOLATED   LES08570
C     GOOD T-EIGENVALUES THAT ARE VERY CLOSE TO SPURIOUS ONES.        LES08580
C     ERROR ESTIMATES WILL NOT BE COMPUTED FOR THESE T-EIGENVALUES.   LES08590
C                                                                     LES08600
C     ON ENTRY AND EXIT                                               LES08610
C     VS CONTAINS THE COMPUTED DISTINCT EIGENVALUES OF T(1,MEV)       LES08620
C     MP CONTAINS THE CORRESPONDING T-MULTIPLICITIES                  LES08630
```

```
C       NDIS = NUMBER OF DISTINCT EIGENVALUES                             LES08640
C       GAPTOL = RELATIVE GAP TOLERANCE SET IN MAIN                       LES08650
C                                                                         LES08660
C       ON EXIT                                                           LES08670
C       G CONTAINS THE TMINGAPS.                                          LES08680
C       G(I) < O MEANS MINGAP IS DUE TO LEFT GAP                          LES08690
C       MP(I) IS NOT CHANGED EXCEPT THAT  MP(I)=-1, IF MP(I)=1,           LES08700
C       TMINGAP WAS TOO SMALL AND DUE TO A SPURIOUS T-EIGENVALUE.         LES08710
C                                                                         LES08720
C       IF MP(I)=-1 THAT SIMPLE GOOD T-EIGENVALUE WILL BE SKIPPED         LES08730
C       IN THE SUBSEQUENT ERROR ESTIMATE COMPUTATIONS IN INVERR           LES08740
C       THAT IS, WE COMPUTE ERROR ESTIMATES ONLY FOR THOSE GOOD           LES08750
C       T-EIGENVALUES WITH MP(I)=1.                                       LES08760
C------------------------------------------------------------------------LES08770
C       CALCULATE MINGAPS FOR DISTINCT T(1,MEV) EIGENVALUES.             LES08780
        NM1 = NDIS - 1                                                    LES08790
        G(NDIS) = VS(NM1)-VS(NDIS)                                        LES08800
        G(1) = VS(2)-VS(1)                                                LES08810
C                                                                         LES08820
        DO 10 J = 2,NM1                                                   LES08830
        TO = VS(J)-VS(J-1)                                                LES08840
        T1 = VS(J+1)-VS(J)                                                LES08850
        G(J) = T1                                                         LES08860
        IF (TO.LT.T1) G(J) = -TO                                          LES08870
     10 CONTINUE                                                          LES08880
C                                                                         LES08890
C       SET MP(I)=-1 FOR SIMPLE GOOD T-EIGENVALUES WHOSE MINGAPS  ARE     LES08900
C       'TOO SMALL' AND DUE TO SPURIOUS T-EIGENVALUES.                    LES08910
C                                                                         LES08920
        NISO = 0                                                          LES08930
        NG = 0                                                            LES08940
        DO 20 J = 1,NDIS                                                  LES08950
        IF (MP(J).EQ.0) GO TO 20                                          LES08960
        NG = NG+1                                                         LES08970
        IF (MP(J).NE.1) GO TO 20                                          LES08980
C       VS(J) IS NEXT SIMPLE GOOD T-EIGENVALUE                           LES08990
        NISO = NISO + 1                                                   LES09000
        I = J+1                                                           LES09010
        IF (G(J).LT.0.0) I = J-1                                          LES09020
        IF (MP(I).NE.0) GO TO 20                                          LES09030
        GAP = ABS(G(J))                                                   LES09040
        TO = DMAX1(SCALE1*MULTOL,GAPTOL*DABS(VS(J)))                      LES09050
C       TO = DMAX1(GAPTOL,GAPTOL*DABS(VS(J)))                            LES09060
        TEMP = TO                                                         LES09070
        IF (GAP.GT.TEMP) GO TO 20                                         LES09080
        MP(J) = -MP(J)                                                    LES09090
        NISO = NISO-1                                                     LES09100
     20 CONTINUE                                                          LES09110
C                                                                         LES09120
C-----END OF ISOEV------------------------------------------------------LES09130
        RETURN                                                            LES09140
        END                                                              LES09150
C                                                                         LES09160
C-----START OF PRTEST---------------------------------------------------LES09170
C                                                                         LES09180
        SUBROUTINE PRTEST(ALPHA,BETA,TEIG,TKMAX,EPSM,RELTOL,SCALE3,SCALE4,LES09190
       1 TMULT,NDIST,MEV,IPROJ)                                           LES09200
C                                                                         LES09210
C------------------------------------------------------------------------LES09220
        DOUBLE PRECISION  ALPHA(1), BETA(1),TEIG(1),SIGMA(10)             LES09230
        DOUBLE PRECISION  EPSM,RELTOL,PRTOL,TKMAX,LRATIO,URATIO           LES09240
        DOUBLE PRECISION  EPS,EPS1,BETAM,LBD,UBD,SIG,YU,YV,LRATS,URATS    LES09250
        DOUBLE PRECISION ZERO,ONE,TEN,BISTOL,SCALE3,SCALE4,AEV,TEMP       LES09260
        INTEGER  TMULT(1),ISIGMA(10)                                      LES09270
        DOUBLE PRECISION  DABS, DMAX1, DSQRT, DFLOAT                      LES09280
C------------------------------------------------------------------------LES09290
C       AFTER CONVERGENCE HAS BEEN ESTABLISHED, SUBROUTINE PRTEST         LES09300
```

```
C         TESTS COMPUTED EIGENVALUES OF T(1,MEV) THAT HAVE BEEN LABELLED      LES09310
C         SPURIOUS TO DETERMINE IF ANY EIGENVALUES OF A HAVE BEEN             LES09320
C         MISSED BY LANCZOS PROCEDURE.  AN EIGENVALUE WITH A VERY SMALL       LES09330
C         PROJECTION ON THE STARTING VECTOR (< SINGLE PRECISION)             LES09340
C         CAN BE MISSED BECAUSE IT IS ALSO AN EIGENVALUE OF T(2,MEV) TO       LES09350
C         WITHIN THE SQUARE OF THIS ORIGINAL PROJECTION.                      LES09360
C         OUR EXPERIENCE IS THAT SUCH SMALL PROJECTIONS OCCUR ONLY            LES09370
C         VERY INFREQUENTLY.                                                  LES09380
C                                                                             LES09390
C         THIS SUBROUTINE IS CALLED ONLY AFTER CONVERGENCE HAS BEEN           LES09400
C         ESTABLISHED. ONCE CONVERGENCE HAS BEEN OBSERVED ON THE              LES09410
C         OTHER EIGENVALUES THEN ONE CAN EXPECT TO ALSO HAVE CONVERGENCE      LES09420
C         ON ANY SUCH HIDDEN EIGENVALUES.(IF THERE ARE ANY).  THIS            LES09430
C         PROCEDURE CONSIDERS ONLY SPURIOUS T-EIGENVALUES AND ONLY THOSE      LES09440
C         SPURIOUS T-EIGENVALUES THAT ARE ISOLATED FROM GOOD T-EIGENVALUES.   LES09450
C         FOR EACH SUCH T-EIGENVALUE IT DOES 2 STURM SEQUENCES                LES09460
C         AND A FEW SCALAR MULTIPLICATIONS.  UPON RETURN TO MAIN              LES09470
C         PROGRAM ERROR ESTIMATES WILL BE COMPUTED FOR ANY EIGENVALUES        LES09480
C         THAT HAVE BEEN LABELLED AS 'HIDDEN'.  SUCH T-EIGENVALUES            LES09490
C         WILL BE RELABELLED AS 'GOOD' ONLY IF THESE ERROR ESTIMATES          LES09500
C         ARE SUFFICIENTLY SMALL.                                             LES09510
C---------------------------------------------------------------------------LES09520
          ZERO = 0.0D0                                                        LES09530
          ONE  = 1.0D0                                                        LES09540
          TEN  = 10.0D0                                                       LES09550
          PRTOL = 1.D-6                                                       LES09560
          TEMP = DFLOAT(MEV+1000)                                             LES09570
          TEMP = DSQRT(TEMP)                                                  LES09580
          BISTOL = TKMAX*EPSM*TEMP                                            LES09590
          NSIGMA = 4                                                          LES09600
          SIGMA(1) = TEN*TKMAX                                                LES09610
C                                                                             LES09620
          DO 10 J = 2,NSIGMA                                                  LES09630
       10 SIGMA(J) = TEN*SIGMA(J-1)                                           LES09640
C                                                                             LES09650
          IFIN = 0                                                            LES09660
          MF = 1                                                              LES09670
          ML = MEV                                                            LES09680
          BETAM = BETA(MF)                                                    LES09690
          BETA(MF) = ZERO                                                     LES09700
          IPROJ = 0                                                           LES09710
          J = 1                                                               LES09720
C                                                                             LES09730
          IF (TMULT(1).NE.0) GO TO 110                                        LES09740
C                                                                             LES09750
          AEV = DABS(TEIG(1))                                                 LES09760
          TEMP = PRTOL*AEV                                                    LES09770
          EPS1 = DMAX1(TEMP,SCALE4*BISTOL)                                    LES09780
C         EPS1 = DMAX1(TEMP,PRTOL)                                            LES09790
          TEMP = RELTOL*AEV                                                   LES09800
          EPS  = DMAX1(TEMP,SCALE3*BISTOL)                                    LES09810
C         EPS  = DMAX1(TEMP,RELTOL)                                           LES09820
C                                                                             LES09830
          IF (TEIG(2)-TEIG(1).LT.EPS1.AND.TMULT(2).NE.0) GO TO 110            LES09840
C                                                                             LES09850
       20 LBD = TEIG(J) - EPS                                                 LES09860
          UBD = TEIG(J) + EPS                                                 LES09870
          MEVL = 0                                                            LES09880
          IL = 0                                                              LES09890
          YU = ONE                                                            LES09900
C                                                                             LES09910
          DO 50 I=MF,ML                                                       LES09920
          IF (YU.NE.ZERO) GO TO 30                                            LES09930
          YV = BETA(I)/EPSM                                                   LES09940
          GO TO 40                                                            LES09950
       30 YV = BETA(I)*BETA(I)/YU                                             LES09960
       40 YU = ALPHA(I)-LBD-YV                                                LES09970
```

```
      IF (YU.GE.ZERO) GO TO 50                              LES09980
C     MEVL INCREMENTED                                      LES09990
      MEVL = MEVL + 1                                       LES10000
      IL = 1                                                LES10010
   50 CONTINUE                                              LES10020
C                                                           LES10030
      LRATIO = YU                                           LES10040
      MEV1L = MEVL                                          LES10050
      IF (IL.EQ.ML) MEV1L=MEVL-1                            LES10060
C                                                           LES10070
C     MEVL = NUMBER OF EVS OF T(1,MEV) WHICH ARE < LBD      LES10080
C     MEV1L = NUMBER OF EVS OF T(1,MEV-1) WHICH ARE < LBD   LES10090
C     LRATIO = DET(T(1,MEV)-LBD)/DET(T(1,MEV-1)-LBD):       LES10100
C                                                           LES10110
      MEVU = 0                                              LES10120
      IL = 0                                                LES10130
      YU = ONE                                              LES10140
C                                                           LES10150
      DO 80 I=MF,ML                                         LES10160
      IF (YU.NE.ZERO) GO TO 60                              LES10170
      YV = BETA(I)/EPSM                                     LES10180
      GO TO 70                                              LES10190
   60 YV = BETA(I)*BETA(I)/YU                               LES10200
   70 YU = ALPHA(I)-UBD-YV                                  LES10210
      IF (YU.GE.ZERO) GO TO 80                              LES10220
C     MEVU INCREMENTED                                      LES10230
      MEVU = MEVU + 1                                       LES10240
      IL = 1                                                LES10250
   80 CONTINUE                                              LES10260
C                                                           LES10270
      URATIO = YU                                           LES10280
      MEV1U = MEVU                                          LES10290
      IF (IL.EQ.ML) MEV1U=MEVU-1                            LES10300
C                                                           LES10310
C     MEVU = NUMBER OF EVS OF T(MEV) WHICH ARE < UBD        LES10320
C     MEV1U = NUMBER OF EVS OF T(MEV-1) WHICH ARE < UBD     LES10330
C     URATIO = DET(TM-UBD)/DET(T(M-1)-UBD): TM=T(MF,ML)     LES10340
C                                                           LES10350
      NEV1 = MEV1U-MEV1L                                    LES10360
C                                                           LES10370
      DO 90 K=1,NSIGMA                                      LES10380
      SIG = SIGMA(K)                                        LES10390
      LRATS = LRATIO-SIG                                    LES10400
      URATS = URATIO-SIG                                    LES10410
C     NOTE THE INCREMENT IS ON NUMBER OF EVALUES OF T(M-1)  LES10420
      MEVLS = MEV1L                                         LES10430
      IF (LRATS.LT.0.) MEVLS=MEV1L+1                        LES10440
      MEVUS = MEV1U                                         LES10450
      IF (URATS.LT.0.) MEVUS=MEV1U+1                        LES10460
      ISIGMA(K) = MEVUS - MEVLS                             LES10470
   90 CONTINUE                                              LES10480
C                                                           LES10490
      ICOUNT = 0                                            LES10500
      DO 100 K=1,NSIGMA                                     LES10510
  100 IF (ISIGMA(K).EQ.1) ICOUNT=ICOUNT + 1                 LES10520
C                                                           LES10530
      IF (ICOUNT.LT.2.OR.NEV1.EQ.0) GO TO 110               LES10540
      TMULT(J) = -10                                        LES10550
      IPROJ=IPROJ+1                                         LES10560
C                                                           LES10570
  110 J=J+1                                                 LES10580
C                                                           LES10590
      IF (J.GE.NDIST) GO TO 120                             LES10600
      IF (TMULT(J).NE.0) GO TO 110                          LES10610
C                                                           LES10620
      AEV = DABS(TEIG(J))                                   LES10630
      TEMP = PRTOL*AEV                                      LES10640
```

```
      EPS1 = DMAX1(TEMP,SCALE4*BISTOL)                              LES10650
C     EPS1 = DMAX1(TEMP,PRTOL)                                      LES10660
      TEMP = RELTOL*AEV                                             LES10670
      EPS  = DMAX1(TEMP,SCALE3*BISTOL)                              LES10680
C     EPS  = DMAX1(TEMP,RELTOL)                                     LES10690
C                                                                   LES10700
      IF (TEIG(J)-TEIG(J-1).LT.EPS1.AND.TMULT(J-1).NE.0) GO TO 110  LES10710
      IF (TEIG(J+1)-TEIG(J).LT.EPS1.AND.TMULT(J+1).NE.0) GO TO 110  LES10720
C                                                                   LES10730
      GO TO 20                                                      LES10740
C                                                                   LES10750
  120 IF (IFIN.EQ.1) GO TO 130                                      LES10760
      IF (TMULT(NDIST).NE.0) GO TO 130                              LES10770
C                                                                   LES10780
      AEV = DABS(TEIG(NDIST))                                       LES10790
      TEMP = PRTOL*AEV                                              LES10800
      EPS1 = DMAX1(TEMP,SCALE4*BISTOL)                              LES10810
C     EPS1 = DMAX1(TEMP,PRTOL)                                      LES10820
      TEMP = RELTOL*AEV                                             LES10830
      EPS  = DMAX1(TEMP,SCALE3*BISTOL)                              LES10840
C     EPS  = DMAX1(TEMP,RELTOL)                                     LES10850
C                                                                   LES10860
      NDIST1=NDIST -1                                               LES10870
      TEMP = TEIG(NDIST)-TEIG(NDIST1)                               LES10880
      IF (TEMP.LT.EPS1.AND.TMULT(NDIST1).NE.0) GO TO 130            LES10890
      IFIN = 1                                                      LES10900
C                                                                   LES10910
      GO TO 20                                                      LES10920
C                                                                   LES10930
  130 BETA(MF) = BETAM                                              LES10940
C                                                                   LES10950
C-----END OF PRTEST--------------------------------------------------LES10960
      RETURN                                                        LES10970
      END                                                           LES10980
C                                                                   LES10990
C------START OF STURMI----------------------------------------------LES11000
C                                                                   LES11010
      SUBROUTINE STURMI(ALPHA,BETA,X1,TOLN,EPSM,MMAX,MK1,MK2,IC,IWRITE) LES11020
C                                                                   LES11030
C------------------------------------------------------------------LES11040
      DOUBLE PRECISION  ALPHA(1),BETA(1)                            LES11050
      DOUBLE PRECISION  EPSM,X1,TOLN,EVL,EVU,BETA2                  LES11060
      DOUBLE PRECISION  U1,U2,V1,V2,ZERO,ONE                        LES11070
      INTEGER I,IC,ICD,ICO,IC1,IC2,MK1,MK2,MMAX                     LES11080
C------------------------------------------------------------------LES11090
C                                                                   LES11100
C     FOR ANY EIGENVALUE OF A THAT HAS CONVERGED AS AN EIGENVALUE   LES11110
C     OF THE T-MATRICES THIS SUBROUTINE CALCULATES                  LES11120
C     THE SMALLEST SIZE OF THE T-MATRIX, T(1,MK1) DEFINED           LES11130
C     BY THE ALPHA AND BETA ARRAYS SUCH THAT MK1.LE.MMAX            LES11140
C     AND THE INTERVAL (X1-TOLN,X1+TOLN) CONTAINS AT LEAST ONE      LES11150
C     EIGENVALUE OF T(1,MK1). IT ALSO CALCULATES MK2 <= MMAX        LES11160
C     AS THE SMALLEST SIZE T-MATRIX (IF ANY) SUCH THAT THIS INTERVAL LES11170
C     CONTAINS AT LEAST TWO EIGENVALUES OF T(1,MK2).                LES11180
C     IF NO T-MATRIX OF ORDER < MMAX SATISFIES THIS REQUIREMENT     LES11190
C     THEN MK2 IS SET EQUAL TO MMAX.  THE EIGENVECTOR PROGRAM       LES11200
C     USES THESE VALUES TO DETERMINE AN APPROPRIATE 1ST GUESS AT    LES11210
C     AN APPROPRIATE SIZE T-MATRIX FOR THE EIGENVALUE X1.           LES11220
C                                                                   LES11230
C     ON EXIT IC = NUMBER OF EIGENVALUES OF T(1,MK2) IN THIS INTERVAL LES11240
C                                                                   LES11250
C     STURMI REGENERATES THE QUANTITIES BETA(I)**2 EACH TIME IT IS  LES11260
C     CALLED, OBVIOUSLY FOR THE PRICE OF ANOTHER VECTOR OF LENGTH   LES11270
C     MMAX THIS GENERATION COULD BE DONE ONCE IN THE MAIN           LES11280
C     PROGRAM BEFORE THE LOOP ON THE CALLS TO SUBROUTINE STURMI.    LES11290
C                                                                   LES11300
C     IF ANY OF THE EIGENVALUES BEING CONSIDERED WERE MULTIPLE      LES11310
```

```
C       AS EIGENVALUES OF THE USER-SPECIFIED MATRIX, THEN            LES11320
C       THIS SUBROUTINE COULD BE MODIFIED TO COMPUTE ADDITIONAL      LES11330
C       SIZES MKJ, J = 3, ...  WHICH COULD THEN BE USED IN THE       LES11340
C       MAIN LANCZOS EIGENVECTOR PROGRAM TO COMPUTE ADDITIONAL       LES11350
C       EIGENVECTORS CORRESPONDING TO THESE MULTIPLE EIGENVALUES.    LES11360
C       THE MAIN PROGRAM PROVIDED DOES NOT INCLUDE THIS OPTION.      LES11370
C                                                                    LES11380
C--------------------------------------------------------------------LES11390
C       INITIALIZATION OF PARAMETERS                                 LES11400
        MK1 = 0                                                      LES11410
        MK2 = 0                                                      LES11420
        ZERO = 0.0D0                                                 LES11430
        ONE  = 1.0D0                                                 LES11440
        BETA(1) = ZERO                                               LES11450
        EVL = X1-TOLN                                                LES11460
        EVU = X1+TOLN                                                LES11470
        U1 = ONE                                                     LES11480
        U2 = ONE                                                     LES11490
        ICO = 0                                                      LES11500
        IC1 = 0                                                      LES11510
        IC2 = 0                                                      LES11520
C                                                                    LES11530
C       MAIN LOOP FOR CALCULATING THE SIZES MK1,MK2                  LES11540
        DO 60 I = 1,MMAX                                             LES11550
        BETA2 = BETA(I)*BETA(I)                                      LES11560
        IF (U1.NE.ZERO) GO TO 10                                     LES11570
        V1 = BETA(I)/EPSM                                            LES11580
        GO TO 20                                                     LES11590
     10 V1 = BETA2/U1                                                LES11600
     20 U1 = EVL - ALPHA(I) - V1                                     LES11610
        IF (U1.LT.ZERO) IC1 = IC1+1                                  LES11620
        IF (U2.NE.ZERO) GO TO 30                                     LES11630
        V2 = BETA(I)/EPSM                                            LES11640
        GO TO 40                                                     LES11650
     30 V2 = BETA2/U2                                                LES11660
     40 U2 = EVU - ALPHA(I) - V2                                     LES11670
        IF (U2.LT.ZERO) IC2 = IC2+1                                  LES11680
C       TEST FOR CHANGE IN NUMBER OF T-EIGENVALUES ON (EVL,EVU)      LES11690
        ICD = IC1-IC2                                                LES11700
        IC = ICD-ICO                                                 LES11710
        IF (IC.GE.1) GO TO 50                                        LES11720
        GO TO 60                                                     LES11730
     50 CONTINUE                                                     LES11740
        IF (ICO.EQ.0) MK1 = I                                        LES11750
        ICO = ICO+1                                                  LES11760
        IF (ICO.GT.1) GO TO 70                                       LES11770
     60 CONTINUE                                                     LES11780
C                                                                    LES11790
        I = I-1                                                      LES11800
        IF (ICO.EQ.0) MK1 = MMAX                                     LES11810
     70 MK2 = I                                                      LES11820
        IC = ICD                                                     LES11830
C                                                                    LES11840
        IF (IWRITE.EQ.1) WRITE(6,80) X1,MK1,MK2,IC                   LES11850
     80 FORMAT(' EVAL =',E20.12,' MK1 =',I6,' MK2 =',I6,' IC =',I3/) LES11860
C                                                                    LES11870
        RETURN                                                       LES11880
C-----END OF STURMI--------------------------------------------------LES11890
        END                                                          LES11900
C                                                                    LES11910
C                                                                    LES11920
C-----START OF INVERM------------------------------------------------LES11930
C                                                                    LES11940
        SUBROUTINE INVERM(ALPHA,BETA,V1,V2,X1,ERROR,ERRORV,EPS,G,MEV,IT, LES11950
       1 IWRITE)                                                     LES11960
C                                                                    LES11970
C--------------------------------------------------------------------LES11980
```

```
      DOUBLE PRECISION  ALPHA(1),BETA(1),V1(1),V2(1)                    LES11990
      DOUBLE PRECISION  X1,U,Z,TEMP,RATIO,SUM,XU,NORM,TSUM,BETAM        LES12000
      DOUBLE PRECISION  EPS,EPS3,EPS4,ERROR,ERRORV,ZERO,ONE            LES12010
      REAL  G(1)                                                       LES12020
      DOUBLE PRECISION  DABS, DSQRT, DFLOAT                            LES12030
      DOUBLE PRECISION FINPRO                                          LES12040
      REAL  ABS                                                        LES12050
C------------------------------------------------------------------LES12060
C                                                                       LES12070
C     COMPUTES T-EIGENVECTORS FOR ISOLATED GOOD T-EIGENVALUES X1       LES12080
C     USING INVERSE ITERATION ON T(1,MEV(X1)) SOLVING EQUATION         LES12090
C     (T - X1*I)V2 = RIGHT-HAND SIDE (RANDOMLY-GENERATED) .            LES12100
C     PROGRAM REFACTORS T- X1*I ON EACH ITERATION OF INVERSE ITERATION. LES12110
C     TYPICALLY ONLY ONE ITERATION IS NEEDED PER T-EIGENVALUE X1.      LES12120
C                                                                       LES12130
C     IF IWRITE = 1 THEN THERE ARE EXTENDED WRITES TO FILE 6 (TERMINAL) LES12140
C                                                                       LES12150
C     ON ENTRY G CONTAINS A REAL*4 RANDOM VECTOR WHICH WAS GENERATED   LES12160
C     IN MAIN PROGRAM.                                                 LES12170
C                                                                       LES12180
C     ON ENTRY AND EXIT                                               LES12190
C     MEV = ORDER OF T                                                 LES12200
C     ALPHA, BETA CONTAIN THE DIAGONAL AND OFFDIAGONAL ENTRIES OF T.   LES12210
C     EPS = 2. * MACHINE EPSILON                                       LES12220
C                                                                       LES12230
C     IN PROGRAM:                                                     LES12240
C     ITER = MAXIMUM NUMBER STEPS ALLOWED FOR INVERSE ITERATION        LES12250
C     ITER = IT ON ENTRY.                                             LES12260
C     V1,V2 = WORK SPACES USED IN THE FACTORIZATION OF T(1,MEV).       LES12270
C     V1 AND V2 MUST BE OF DIMENSION AT LEAST MEV.                    LES12280
C                                                                       LES12290
C     ON EXIT                                                         LES12300
C     V2 = THE UNIT EIGENVECTOR OF T(1,MEV) CORRESPONDING TO X1.       LES12310
C     ERROR =  |V2(MEV)| = ERROR ESTIMATE FOR CORRESPONDING           LES12320
C             RITZ VECTOR FOR X1.                                      LES12330
C                                                                       LES12340
C     ERRORV = || T*V2 - X1*V2 || = ERROR ESTIMATE ON T-EIGENVECTOR.  LES12350
C     IF IT.GT.ITER THEN ERRORV = -ERRORV                             LES12360
C     IT = NUMBER OF ITERATIONS ACTUALLY REQUIRED                     LES12370
C------------------------------------------------------------------LES12380
C     INITIALIZATION AND PARAMETER SPECIFICATION                      LES12390
      ONE  = 1.0D0                                                     LES12400
      ZERO = 0.0D0                                                     LES12410
      ITER = IT                                                        LES12420
      MP1 = MEV+1                                                      LES12430
      MM1 = MEV-1                                                      LES12440
      BETAM = BETA(MP1)                                                LES12450
      BETA(MP1) = ZERO                                                 LES12460
C                                                                       LES12470
C     CALCULATE SCALE AND TOLERANCES                                  LES12480
      TSUM = DABS(ALPHA(1))                                            LES12490
      DO 10 I = 2,MEV                                                  LES12500
   10 TSUM = TSUM + DABS(ALPHA(I)) + BETA(I)                          LES12510
C                                                                       LES12520
      EPS3 = EPS*TSUM                                                  LES12530
      EPS4 = DFLOAT(MEV)*EPS3                                          LES12540
C                                                                       LES12550
C     GENERATE SCALED RANDOM RIGHT-HAND SIDE                          LES12560
      GSUM = ZERO                                                      LES12570
      DO 20 I = 1,MEV                                                  LES12580
   20 GSUM = GSUM+ABS(G(I))                                           LES12590
      GSUM = EPS4/GSUM                                                 LES12600
C                                                                       LES12610
C     INITIALIZE RIGHT HAND SIDE FOR INVERSE ITERATION                LES12620
      DO 30 I = 1,MEV                                                  LES12630
   30 V2(I) = GSUM*G(I)                                               LES12640
```

```
         IT = 1                                                        LES12650
C                                                                      LES12660
C        CALCULATE UNIT EIGENVECTOR OF T(1,MEV) FOR ISOLATED GOOD      LES12670
C        T-EIGENVALUE X1.                                              LES12680
C                                                                      LES12690
C        TRIANGULAR FACTORIZATION WITH NEAREST NEIGHBOR PIVOT          LES12700
C        STRATEGY. INTERCHANGES ARE LABELLED BY SETTING BETA < 0.      LES12710
C                                                                      LES12720
      40 CONTINUE                                                      LES12730
         U = ALPHA(1)-X1                                               LES12740
         Z = BETA(2)                                                   LES12750
C                                                                      LES12760
         DO 60 I=2,MEV                                                 LES12770
         IF (BETA(I).GT.DABS(U)) GO TO 50                              LES12780
C        NO PIVOT INTERCHANGE                                          LES12790
         V1(I-1) = Z/U                                                 LES12800
         V2(I-1) = V2(I-1)/U                                           LES12810
         V2(I) = V2(I)-BETA(I)*V2(I-1)                                 LES12820
         RATIO = BETA(I)/U                                             LES12830
         U = ALPHA(I)-X1-Z*RATIO                                       LES12840
         Z = BETA(I+1)                                                 LES12850
         GO TO 60                                                      LES12860
C        PIVOT INTERCHANGE                                             LES12870
      50 CONTINUE                                                      LES12880
         RATIO = U/BETA(I)                                             LES12890
         BETA(I) = -BETA(I)                                            LES12900
         V1(I-1) = ALPHA(I)-X1                                         LES12910
         U = Z-RATIO*V1(I-1)                                           LES12920
         Z = -RATIO*BETA(I+1)                                          LES12930
         TEMP = V2(I-1)                                                LES12940
         V2(I-1) = V2(I)                                               LES12950
         V2(I) = TEMP-RATIO*V2(I)                                      LES12960
      60 CONTINUE                                                      LES12970
C                                                                      LES12980
         IF (U.EQ.ZERO) U=EPS3                                         LES12990
C                                                                      LES13000
C        SMALLNESS TEST AND DEFAULT VALUE FOR LAST COMPONENT           LES13010
C        PIVOT(I-1) = |BETA(I)| FOR INTERCHANGE CASE                   LES13020
C        (I-1,I+1) ELEMENT IN RIGHT FACTOR = BETA(I+1)                 LES13030
C        END OF FACTORIZATION AND FORWARD SUBSTITUTION                 LES13040
C                                                                      LES13050
C        BACK SUBSTITUTION                                             LES13060
         V2(MEV) = V2(MEV)/U                                           LES13070
         DO 80 II = 1,MM1                                              LES13080
         I = MEV-II                                                    LES13090
         IF (BETA(I+1).LT.ZERO) GO TO 70                               LES13100
C        NO PIVOT INTERCHANGE                                          LES13110
         V2(I) = V2(I)-V1(I)*V2(I+1)                                   LES13120
         GO TO 80                                                      LES13130
C        PIVOT INTERCHANGE                                             LES13140
      70 BETA(I+1) = -BETA(I+1)                                        LES13150
         V2(I) = (V2(I)-V1(I)*V2(I+1)-BETA(I+2)*V2(I+2))/BETA(I+1)     LES13160
      80 CONTINUE                                                      LES13170
C                                                                      LES13180
C                                                                      LES13190
C        TESTS FOR CONVERGENCE OF INVERSE ITERATION                    LES13200
C        IF SUM |V2| COMPS. LE. 1 AND IT. LE. ITER DO ANOTHER INVIT STEP  LES13210
C                                                                      LES13220
         NORM = DABS(V2(MEV))                                          LES13230
         DO 90 II = 1,MM1                                              LES13240
         I = MEV-II                                                    LES13250
      90 NORM = NORM+DABS(V2(I))                                       LES13260
C                                                                      LES13270
C        IS DESIRED GROWTH IN VECTOR ACHIEVED ?                        LES13280
C        IF NOT, DO ANOTHER INVERSE ITERATION STEP UNLESS NUMBER ALLOWED ISLES13290
C        EXCEEDED.                                                     LES13300
```

```
      IF (NORM.GE.ONE) GO TO 110                                   LES13310
C                                                                  LES13320
      IT=IT+1                                                      LES13330
      IF (IT.GT.ITER) GO TO 110                                   LES13340
C                                                                  LES13350
      XU = EPS4/NORM                                               LES13360
      DO 100 I=1,MEV                                               LES13370
  100 V2(I) = V2(I)*XU                                             LES13380
C                                                                  LES13390
      GO TO 40                                                     LES13400
C                                                                  LES13410
C     NORMALIZE COMPUTED T-EIGENVECTOR : V2 = V2/||V2||            LES13420
C                                                                  LES13430
  110 CONTINUE                                                     LES13440
C                                                                  LES13450
      SUM = FINPRO(MEV,V2(1),1,V2(1),1)                           LES13460
      SUM = ONE/DSQRT(SUM)                                         LES13470
      DO 120 II = 1,MEV                                            LES13480
  120 V2(II) = SUM*V2(II)                                          LES13490
C                                                                  LES13500
C     SAVE ERROR ESTIMATE FOR LATER OUTPUT                         LES13510
      ERROR = DABS(V2(MEV))                                        LES13520
C                                                                  LES13530
C     GENERATE ERRORV = ||T*V2 - X1*V2||.                          LES13540
      V1(MEV) = ALPHA(MEV)*V2(MEV)+BETA(MEV)*V2(MEV-1)-X1*V2(MEV)  LES13550
      DO 130 J = 2,MM1                                             LES13560
      JM = MP1 - J                                                 LES13570
      V1(JM) = ALPHA(JM)*V2(JM) + BETA(JM)*V2(JM-1) + BETA(JM+1)*V2(JM+1LES13580
     1) - X1*V2(JM)                                                LES13590
  130 CONTINUE                                                     LES13600
C                                                                  LES13610
      V1(1) = ALPHA(1)*V2(1) + BETA(2)*V2(2) - X1*V2(1)           LES13620
      ERRORV = FINPRO(MEV,V1(1),1,V1(1),1)                         LES13630
      ERRORV = DSQRT(ERRORV)                                       LES13640
      IF (IT.GT.ITER) ERRORV = -ERRORV                             LES13650
      IF (IWRITE.EQ.0) GO TO 150                                   LES13660
C                                                                  LES13670
C     FILE 6 (TERMINAL) OUTPUT OF ERROR ESTIMATES.                 LES13680
      WRITE(6,140) MEV,X1,ERROR,ERRORV                             LES13690
  140 FORMAT(2X,'TSIZE',15X,'EIGENVALUE',11X,'U(M)',9X,'ERRORV'/   LES13700
     1 I6,E25.16,2E15.5)                                           LES13710
C                                                                  LES13720
C     RESTORE BETA(MEV+1) = BETAM                                  LES13730
  150 CONTINUE                                                     LES13740
      BETA(MP1) = BETAM                                            LES13750
C-----END OF INVERM---------------------------------------------------LES13760
      RETURN                                                       LES13770
      END                                                          LES13780
C                                                                  LES13790
C-----START OF LBISEC-------------------------------------------------LES13800
C                                                                  LES13810
      SUBROUTINE LBISEC(ALPHA,BETA,EPSM,EVAL,EVALN,LB,UB,TTOL,M,NEVT) LES13820
C                                                                  LES13830
C-------------------------------------------------------------------LES13840
      DOUBLE PRECISION   ALPHA(1),BETA(1),X0,X1,XL,XU,YU,YV,LB,UB  LES13850
      DOUBLE PRECISION   EPSM,EP1,EVAL,EVALN,EVD,EPT               LES13860
      DOUBLE PRECISION   ZERO,ONE,HALF,TTOL,TEMP                   LES13870
      DOUBLE PRECISION   DABS,DSQRT,DFLOAT                         LES13880
C-------------------------------------------------------------------LES13890
C     SPECIFY PARAMETERS                                           LES13900
      ZERO = 0.0D0                                                 LES13910
      HALF = 0.5D0                                                 LES13920
      ONE  = 1.0D0                                                 LES13930
      XL = LB                                                      LES13940
      XU = UB                                                      LES13950
C                                                                  LES13960
C     EP1 = DSQRT(1000+M)*TTOL     TTOL = EPSM*TKMAX               LES13970
```

```
C       TKMAX = MAX(|ALPHA(K)|,BETA(K), K= 1,KMAX)                      LES13980
C                                                                       LES13990
        TEMP = DFLOAT(1000+M)                                          LES14000
        EP1 = DSQRT(TEMP)*TTOL                                         LES14010
C                                                                       LES14020
        NA = 0                                                         LES14030
        X1 = XU                                                        LES14040
        JSTURM = 1                                                     LES14050
        GO TO 60                                                       LES14060
C       FORWARD STURM CALCULATION                                      LES14070
   10 NA = NEV                                                         LES14080
        X1 = XL                                                        LES14090
        JSTURM = 2                                                     LES14100
        GO TO 60                                                       LES14110
C       FORWARD STURM CALCULATION                                      LES14120
   20 NEVT = NEV                                                       LES14130
C                                                                       LES14140
C       WRITE(6,30) M,EVAL,NEVT,EP1                                    LES14150
   30 FORMAT(/3X,'TSIZE',23X,'EV',9X/I8,E25.16/                        LES14160
      1 I6,' = NUMBER OF T(1,M) EIGENVALUES ON TEST INTERVAL'/         LES14170
      1 E12.3,' =  CONVERGENCE TOLERANCE'/)                            LES14180
C                                                                       LES14190
        IF (NEVT.NE.1) GO TO 120                                       LES14200
C                                                                       LES14210
C       BISECTION LOOP                                                 LES14220
        JSTURM = 3                                                     LES14230
   40 X1 = HALF*(XL+XU)                                                LES14240
        XO = XU-XL                                                     LES14250
        EPT = EPSM*(DABS(XL) + DABS(XU)) + EP1                         LES14260
C       CONVERGENCE TEST                                               LES14270
        IF (XO.LE.EP1) GO TO 100                                       LES14280
        GO TO 60                                                       LES14290
C       FORWARD STURM CALCULATION                                      LES14300
   50 CONTINUE                                                         LES14310
        IF(NEV.EQ.0) XU = X1                                           LES14320
        IF(NEV.EQ.1) XL = X1                                           LES14330
        GO TO 40                                                       LES14340
C       NEV = NUMBER OF T-EIGENVALUES OF T(1,M) ON (X1,XU)             LES14350
C       THERE IS EXACTLY ONE T-EIGENVALUE OF T(1,M) ON (XL,XU)         LES14360
C                                                                       LES14370
C       FORWARD STURM CALCULATION                                      LES14380
   60 NEV = -NA                                                        LES14390
        YU = ONE                                                       LES14400
        DO 90 I = 1,M                                                  LES14410
        IF (YU.NE.ZERO) GO TO 70                                       LES14420
        YV = BETA(I)/EPSM                                              LES14430
        GO TO 80                                                       LES14440
   70 YV = BETA(I)*BETA(I)/YU                                          LES14450
   80 YU = X1 - ALPHA(I) - YV                                          LES14460
        IF (YU.GE.ZERO) GO TO 90                                       LES14470
        NEV = NEV+1                                                    LES14480
   90 CONTINUE                                                         LES14490
        GO TO (10,20,50), JSTURM                                       LES14500
C                                                                       LES14510
  100 CONTINUE                                                         LES14520
C                                                                       LES14530
        EVALN = X1                                                     LES14540
        EVD = DABS(EVALN-EVAL)                                         LES14550
C       WRITE(6,110) EVALN,EVAL,EVD                                    LES14560
  110 FORMAT(/20X,'EVALN',21X,'EVAL',6X,'CHANGE'/2E25.16,E12.3/)       LES14570
C                                                                       LES14580
  120 CONTINUE                                                         LES14590
        RETURN                                                         LES14600
C-----END OF LBISEC----------------------------------------------------LES14610
        END                                                           LES14620
C-----START OF COMPLEX INNER PRODUCT-----------------------------------LES14630
C                                                                       LES14640
```

```
C      COMPLEX INNER PRODUCT                                          LES14650
C                                                                     LES14660
       SUBROUTINE CINPRD(V2,V1,SUM,N)                                 LES14670
C----------------------------------------------------------------LES14680
       DOUBLE PRECISION  ZERO,SUM                                     LES14690
       COMPLEX*16 V2(1),V1(1),SUMC                                    LES14700
C----------------------------------------------------------------LES14710
C                                                                     LES14720
C      NOTE THAT THE ORDER MATTERS HERE                               LES14730
C      COMPUTES THE INNER PRODUCT OF THE CONJUGATE OF V2 WITH V1.     LES14740
       ZERO = 0.D0                                                    LES14750
       SUMC = DCMPLX(ZERO,ZERO)                                       LES14760
       DO 10 J=1,N                                                    LES14770
   10  SUMC = SUMC + DCONJG(V2(J))*V1(J)                              LES14780
       SUM = DREAL(SUMC)                                              LES14790
C                                                                     LES14800
       RETURN                                                         LES14810
C-----END OF COMPLEX INNER PRODUCT SUBROUTINE--------------------LES14820
       END                                                            LES14830
C                                                                     LES14340
C-----LPERM-PERMUTES VECTORS-------------------------------------LES14850
C                                                                     LES14860
       SUBROUTINE LPERM(W,U,IPERM)                                    LES14870
C                                                                     LES14880
C----------------------------------------------------------------LES14890
       DOUBLE PRECISION  U(1),W(1)                                    LES14900
       INTEGER  IPR(1),IPT(1)                                         LES14910
C----------------------------------------------------------------LES14920
C      SUBROUTINE HAS 2 BRANCHES:  IPERM = 1,  CALCULATES            LES14930
C      U = P*W   WHERE  P IS THE PERMUTATION REPRESENTED BY IPR      LES14940
C      LET J = IPR(K) THEN U(K) = W(J), K = 1,N. WE SET W(K)=U(K), K=1,N LES14950
C      IPERM = 2, USING THE PERMUTATION IPT (P-TRANSPOSE) U = P'*W, W=U LES14960
C      LET J = IPT(K) THEN U(K) = W(J), K=1,N. WE SET W(K) = U(K), K=1,N LES14970
C----------------------------------------------------------------LES14980
C                                                                     LES14990
       IF(IPERM.EQ.2) GO TO 30                                        LES15000
C      IPERM = 1                                                      LES15010
       DO 10 K = 1,N                                                  LES15020
       J = IPR(K)                                                     LES15030
   10  U(K) = W(J)                                                    LES15040
       DO 20 K = 1,N                                                  LES15050
   20  W(K) = U(K)                                                    LES15060
       GO TO 60                                                       LES15070
C      IPERM = 2                                                      LES15080
   30  DO 40 K = 1,N                                                  LES15090
       J = IPT(K)                                                     LES15100
   40  U(K) = W(J)                                                    LES15110
       DO 50 K = 1,N                                                  LES15120
   50  W(K) = U(K)                                                    LES15130
   60  CONTINUE                                                       LES15140
C                                                                     LES15150
       RETURN                                                         LES15160
C                                                                     LES15170
C----------------------------------------------------------------LES15180
       ENTRY LPERME(IPR,IPT,N)                                        LES15190
C----------------------------------------------------------------LES15200
C                                                                     LES15210
C-----END OF LPERM-----------------------------------------       LES15220
       RETURN                                                         LES15230
       END                                                            LES15240
```

SECTION 2.7 OPTIONAL PREPROCESSING PROGRAM, FILE DEFINITIONS, SAMPLE INPUT/OUTPUT FILES

```
C-----LECOMPAC-(STAND-ALONE PROGRAM)-------------------------------------LEC00010
C                                                                         LEC00020
C     THIS PROGRAM TRANSLATES A SPARSE SYMMETRIC N X N MATRIX A,          LEC00030
C     GIVEN AS I, J, A(I,J), INTO THE SPARSE MATRIX FORMAT                LEC00040
C     REQUIRED BY THE SAMPLE USPEC AND CMATV PROGRAMS PROVIDED            LEC00050
C     FOR USE WITH THE LANCZOS EIGENVALUE/EIGENVECTOR PROCEDURES.         LEC00060
C     THIS PROGRAM ASSUMES THAT THE MATRIX ENTRIES ARE PROVIDED           LEC00070
C     EITHER COLUMN BY COLUMN OR ROW BY ROW.                             LEC00080
C     NOTE THAT THIS PROGRAM DOES NOT DIRECTLY APPLY TO THE               LEC00090
C     HERMITIAN CASE BECAUSE FOR HERMITIAN MATRICES THE DIAGONALS         LEC00100
C     ARE REAL AND THE OFF-DIAGONAL ENTRIES ARE COMPLEX VARIABLES.        LEC00110
C                                                                         LEC00120
C     NONPORTABLE STATEMENTS:  PFORT VERIFIER INDICATES THAT THIS         LEC00130
C                          ' IS PORTABLE.                                 LEC00140
C                                                                         LEC00150
C----------------------------------------------------------------------LEC00160
      DOUBLE PRECISION A(15000), AD(2000)                                LEC00170
      DOUBLE PRECISION ZERO                                              LEC00180
      INTEGER IROW(15000),ICOL(15000)                                    LEC00190
C----------------------------------------------------------------------LEC00200
C   INPUT FILE 7 CONTAINS THE SPARSE SYMMETRIC NXN MATRIX STORED AS:      LEC00210
C                                                                         LEC00220
C                NZ,M,N,MATNO                                             LEC00230
C                I(K)  J(K)  A(K)  K = 1,NZ                               LEC00240
C                                                                         LEC00250
C     WHERE NZ IS THE TOTAL NUMBER OF NONZEROS IN THE MATRIX A,           LEC00260
C     N IS THE ROW AND COLUMN DIMENSION OF A,                            LEC00270
C     AND A(K) ARE THE NONZERO ENTRIES STORED ROW BY ROW OR              LEC00280
C     COLUMN BY COLUMN.  PROGRAM READS THIS IN AS IROW(K) = I(K),        LEC00290
C     ICOL(K) = J(K), AND A(K) = A(K).                                   LEC00300
C                                                                         LEC00310
C     OUTPUT FILE = 8 CONTAINS THE A-MATRIX IN SPARSE FORMAT              LEC00320
C                                                                         LEC00330
C                NZS,N,NZL,MATNO                                          LEC00340
C              ' ICOL(K)    K = 1,NZL                                     LEC00350
C                IROW(K)    K = 1,NZS                                     LEC00360
C                AD(K)      K = 1,N                                       LEC00370
C                A(K)       K = 1,NZS                                     LEC00380
C                                                                         LEC00390
C     WHERE N IS THE ORDER OF THE INPUT MATRIX A,                         LEC00400
C     NZ EQUALS THE NUMBER OF NONZERO ELEMENTS IN A WHICH ARE ON          LEC00410
C     OR BELOW THE MAIN DIAGONAL.  NZL EQUALS THE NUMBER OF THE           LEC00420
C     LAST COLUMN HAVING NONZEROES BELOW THE DIAGONAL IN A.               LEC00430
C     NZS EQUALS THE NUMBER OF NONZERO ELEMENTS BELOW THE MAIN            LEC00440
C     DIAGONAL.   AD(K), K=1,N, CONTAINS THE DIAGONAL ELEMENTS OF A.      LEC00450
C     A(K), K=1,NZS, CONTAINS THE KTH NONZERO SUB-DIAGONAL ELEMENT        LEC00460
C     OF THE INPUT MATRIX.  A IS STORED COLUMN BY COLUMN.                 LEC00470
C     IROW(K), K=1,NZS, CONTAINS THE ROW INDEX OF THE  NONZERO            LEC00480
C     STRICTLY LOWER TRIANGULAR ELEMENT A(K).                            LEC00490
C     ICOL(K), K=1,NZL, EQUALS THE NUMBER OF STRICTLY LOWER               LEC00500
C     TRIANGULAR NONZEROES IN COLUMN K OF THE INPUT MATRIX.               LEC00510
C                                                                         LEC00520
C----------------------------------------------------------------------LEC00530
      ZERO = 0.0D0                                                       LEC00540
C                                                                         LEC00550
      READ(7,10) NZ,N,MATNO,IIROW                                        LEC00560
   10 FORMAT(2I6,I8,I4)                                                  LEC00570
C                                                                         LEC00580
      WRITE(6,20) NZ,N,MATNO,IIROW                                       LEC00590
   20 FORMAT(I10,I6,I10,' = NO. NONZERO AIJ J.GE.I, ORDER OF A, MATNO'/  LEC00600
     1 I6,' = IIROW IF IIROW=0 ORDERING IS BY COLS IIROW=1 BY ROWS'/)    LEC00610
C                                                                         LEC00620
      DO 30 K = 1,N                                                      LEC00630
```

```
   30 AD(K) = ZERO                                             LEC00640
C                                                              LEC00650
      IF (IIROW.EQ.0) READ(7,40) (IROW(K),ICOL(K),A(K), K=1,NZ) LEC00660
C                                                              LEC00670
      IF (IIROW.EQ.1) READ(7,40) (ICOL(K),IROW(K),A(K), K=1,NZ) LEC00680
   40 FORMAT(2I5,E14.7)                                        LEC00690
C                                                              LEC00700
      LCOUNT = 0                                               LEC00710
      K = 1                                                    LEC00720
C                                                              LEC00730
C     START OF A NEW COLUMN                                    LEC00740
   50 CONTINUE                                                 LEC00750
      J = ICOL(K)                                              LEC00760
      ICOL(J) = 0                                              LEC00770
   60 CONTINUE                                                 LEC00780
C                                                              LEC00790
      IF (J.NE.IROW(K)) GO TO 70                               LEC00800
C                                                              LEC00810
C     DIAGONAL CASE                                            LEC00820
      AD(J) = A(K)                                             LEC00830
      GO TO 80                                                 LEC00840
C                                                              LEC00850
C     SUB-DIAGONAL NONZERO                                     LEC00860
   70 CONTINUE                                                 LEC00870
      NZL = J                                                  LEC00880
      LCOUNT = LCOUNT + 1                                      LEC00890
      A(LCOUNT) = A(K)                                         LEC00900
      IROW(LCOUNT) = IROW(K)                                   LEC00910
      ICOL(J) = ICOL(J) + 1                                    LEC00920
C                                                              LEC00930
   80 CONTINUE                                                 LEC00940
      K = K+1                                                  LEC00950
C                                                              LEC00960
      IF(K.GT.NZ) GO TO 90                                     LEC00970
C                                                              LEC00980
      IF(ICOL(K).GT.J) GO TO 50                                LEC00990
C                                                              LEC01000
      GO TO 60                                                 LEC01010
C                                                              LEC01020
   90 CONTINUE                                                 LEC01030
      NZS = LCOUNT                                             LEC01040
C                                                              LEC01050
      WRITE(8,100) NZS,N,NZL,MATNO                             LEC01060
      WRITE(6,100) NZS,N,NZL,MATNO                             LEC01070
  100 FORMAT(I10,2I6,I8,' = NZS N NZL MATNO')                  LEC01080
C                                                              LEC01090
      WRITE(8,110) (ICOL(I), I=1,NZL)                          LEC01100
      WRITE(8,110) (IROW(K), K=1,NZS)                          LEC01110
  110 FORMAT(13I6)                                             LEC01120
C                                                              LEC01130
      WRITE(8,120) (AD(K), K=1,N)                              LEC01140
      WRITE(8,120) (A(K), K=1,NZS)                             LEC01150
  120 FORMAT(4E19.10)                                          LEC01160
C                                                              LEC01170
C-----END LECOMPAC----------------------------------------------LEC01180
      STOP                                                     LEC01190
      END                                                      LEC01200
```

-----INPUT/OUTPUT FILE DEFINITIONS FOR LEVAL AND LEVEC-----

Below is a listing of the input/output files which are accessed by the
real symmetric Lanczos eigenvalue program, LEVAL. Included also is a
sample of the input file which LEVAL requires on file 5. The parameters
supplied in this file are in free format. LEVAL computes
eigenvalues of real symmetric matrices on user-specified intervals.

SAMPLE DEFINITIONS OF THE INPUT/OUTPUT FILES FOR LEVAL
----
*LEVAL EXEC LANCZOS EIGENVALUE CALCULATION REAL SYMMETRIC MATRICES
FI 06 TERM
FILEDEF   1 DISK &1   NHISTORY   A (RECFM F LRECL 80 BLOCK 80
FILEDEF   2 DISK &1   HISTORY   A (RECFM F LRECL 80 BLOCK 80
FILEDEF   3 DISK &1   GOODEV   A (RECFM F LRECL 80 BLOCK 80
FILEDEF   4 DISK &1   ERRINV   A (RECFM F LRECL 80 BLOCK 80
FILEDEF   5 DISK LEVAL   INPUT   A (RECFM F LRECL 80 BLOCK 80
FILEDEF   8 DISK &1   INPUT   A (RECFM F LRECL 80 BLOCK 80
FILEDEF 11 DISK &1   DISTINCT   A (RECFM F LRECL 80 BLOCK 80
LOAD   LEVAL   LESUB   LEMULT

SAMPLE INPUT FILE FOR LEVAL

  LANCZOS EIGENVALUE COMPUTATIONS, NO REORTHOGONALIZATION
  TEST MATRIX
LINE 1   N   KMAX   NMEVS   MATNO
          143   429   1   706830
LINE 2   SVSEED   RHSEED   MXINIT   MXSTUR
         7892713   147935   5   100000
LINE 3   ISTART   ISTOP
             0   1
LINE 4   IHIS   IDIST   IWRITE
              1   0   1
LINE 5   RELTOL (RELATIVE TOLERANCE IN 'COMBINING' GOODEV)
        .0000000001
LINE 6   MB(1)   MB(2)   MB(3)   MB(4) (ORDERS OF $T(1,MEV)$)
            190
LINE 7   NINT   (NUMBER OF SUB-INTERVALS FOR BISEC)
              1
LINE 8   LB(1)   LB(2)   LB(3)   (INTERVAL LOWER BOUNDS)
             0.0
LINE 9   UB(1)   UB(2)   UB(3)   (INTERVAL UPPER BOUNDS)
           1.001

Below is a listing of the input/output files which are accessed by the
real symmetric Lanczos eigenvector program, LEVEC. Included also is a
sample of the input file which LEVEC requires on file 5. The parameters
are supplied in free format. LEVEC computes eigenvectors for each of
a user-specified subset of the eigenvalues computed by the companion
program LEVAL.

SAMPLE DEFINITIONS OF THE INPUT/OUTPUT FILES FOR LEVEC
----
*LEVEC EXEC TO RUN LANCZOS EIGENVECTOR PROGRAM, REAL SYMMETRIC MATRICES
FI 06 TERM
FILEDEF 2 DISK &1   HISTORY   A (RECFM F LRECL 80 BLOCK 80
FILEDEF 3 DISK &1   GOODEV   A (RECFM F LRECL 80 BLOCK 80
FILEDEF 4 DISK &1   ERRINV   A (RECFM F LRECL 80 BLOCK 80
FILEDEF 5 DISK LEVEC   INPUT   A (RECFM F LRECL 80 BLOCK 80
FILEDEF 8 DISK &1   INPUT   A (RECFM F LRECL 80 BLOCK 80
FILEDEF 9 DISK &1   ERREST   A (RECFM F LRECL 80 BLOCK 80
FILEDEF 10 DISK &1   BOUNDS   A (RECFM F LRECL 80 BLOCK 80
FILEDEF 11 DISK &1   TEIGVECS A (RECFM F LRECL 80 BLOCK 80
FILEDEF 12 DISK &1   RITZVECS A (RECFM F LRECL 80 BLOCK 80
FILEDEF 13 DISK &1   PAIGE   A (RECFM F LRECL 80 BLOCK 80
LOAD LEVEC LESUB LEMULT

SAMPLE INPUT FILE FOR LEVEC

 LEVEC REAL SYMMETRIC EIGENVECTOR COMPUTATIONS, NO REORTHOGONALIZATION
LINE 1 MDIMTV   MDIMRV MBETA(MAX.DIMENSIONS, TVEC, RITVEC AND BETA
        10000   10000   2000
LINE 2   RELTOL
        .0000000001
LINE 3 MBOUND   NTVCON SVTVEC IREAD (FLAGS
        0   1   0   1
LINE 4 TVSTOP   LVCONT ERCONT IWRITE (FLAGS
        0   1   1   1
LINE 5   RHSEED (RANDOM GENERATOR SEED FOR STARTING VECTOR IN INVERM)
        45329517
LINE 6 MATNO   N
        100   100

-----SAMPLE OUTPUT FOR REAL SYMMETRIC EIGENVALUE COMPUTATIONS-----

Below is a sample of the output from the LEVAL eigenvalue program.
The output which is printed to file 6 is given first, followed by the
outputs which are generated and written to other files. For this
particular test matrix the true eigenvalues are known. Therefore,
the true errors in the computed eigenvalue approximations are also
listed below.

LANCZOS PROCEDURE FOR REAL SYMMETRIC MATRICES

LEVAL INPUT LANCZOS EIGENVALUE COMPUTATION, NO REORTHOGONALIZATION
POISSON MATRIX N = 528 KX = 33   KY = 16
      POISSON INPUT

     N   KX   KY   DIAGONAL   X-CODIAGONAL   Y-CODIAGONAL
    528   33   16 0.10000000D+01 0.25000000D+00 0.25000000D+00

A-SIZE X-DIM Y-DIM
   528   33   16
    A-DIAGONAL   X-CODIAGONAL   Y-CODIAGONAL   ATOLN
 0.10000000D+01 0.25000000D+00 0.25000000D+00 0.88345049D-15
    401 = NUMBER OF TRUE A-EIGENVALUES WHICH ARE DISTINCT

    MATRIX ID   ORDER OF A   MAX ORDER OF T
     607840   528   1320

   ISTART   ISTOP
       0   1

    IHIS   IDIST   IWRITE
      1   0   0

SEEDS FOR RANDOM NUMBER GENERATOR

    LANCZS SEED   INVERR SEED
      7892713   147935

SIZES OF T-MATRICES TO BE CONSIDERED
       792   1056

RELATIVE TOLERANCE USED TO COMBINE COMPUTED T-EIGENVALUES
      0.100D-09
RELATIVE GAP TOLERANCES USED IN INVERSE ITERATION
      0.100D-07
RELATIVE TOLERANCE FOR CHECK ON SIZE OF BETAS
      0.100D-07

BISEC WILL BE USED ON THE FOLLOWING INTERVALS
    1   0.900000D+00   0.100001D+01

T-MATRICES (ALPHA AND BETA) ARE NOW AVAILABLE

MINIMUM BETA RATIO OCCURS AT   616 TH BETA

TSIZE   MIN BETA   TKMAX   MIN RATIO

1320   0.211D+00 0.153D+01   0.138D+00

BISEC CALCULATION
ORDER OF T IS   792
NUMBER OF INTERVALS IS   1

MULTOL, TOLERANCE USED IN T-MULTIPLICITY AND SPURIOUS TESTS = 0.122D-11
BISTOL, TOLERANCE USED IN BISEC CONVERGENCE TEST = 0.288D-13

BISEC INTERVAL NO LOWER BOUND UPPER BOUND
              1 0.90000D+00 0.10000D+01

  33   396 = NO. TMEV ON (LB,UB) AND NO. .GT. UB

T-EIGENVALUE CALCULATION ON INTERVAL   1 IS COMPLETE
  NO. GOOD   NO. DISTINCT
      33   33

  33 GOOD T-EIGENVALUES HAVE BEEN COMPUTED
  33 OF THESE ARE T-ISOLATED
  33 = NUMBER OF DISTINCT T-EIGENVALUES COMPUTED

CONVERGENCE IS TESTED USING THE CONVERGENCE TOLERANCE   0.4172D-10

T-EIGENVALUE CALCULATION AT MEV =   792   IS COMPLETE

BISEC CALCULATION
ORDER OF T IS 1056
NUMBER OF INTERVALS IS   1

MULTOL, TOLERANCE USED IN T-MULTIPLICITY AND SPURIOUS TESTS = 0.140D-11
BISTOL, TOLERANCE USED IN BISEC CONVERGENCE TEST = 0.308D-13

BISEC INTERVAL NO LOWER BOUND UPPER BOUND
              1 0.90000D+00 0.10000D+01

  40   528 = NO. TMEV ON (LB,UB) AND NO. .GT. UB

T-EIGENVALUE CALCULATION ON INTERVAL   1 IS COMPLETE
  NO. GOOD   NO. DISTINCT
      36   40

  36 GOOD T-EIGENVALUES HAVE BEEN COMPUTED
  36 OF THESE ARE T-ISOLATED
  40 = NUMBER OF DISTINCT T-EIGENVALUES COMPUTED

CONVERGENCE IS TESTED USING THE CONVERGENCE TOLERANCE   0.4755D-10

```
T-EIGENVALUE CALCULATION AT MEV =    1056    IS COMPLETE
RUNTIME = 1298 HNDRETHS OF A SECOND


    33    33    792    528      7892713  607840 = NG,NDIS,MEV,N,SVEED,MATNO
  0.121889836108D-11    616    0.1000D-07 = MUTOL,INDEX MINIMAL BETA,BTOL

EV NO TMULT          GOOD EIGENVALUE       TMINGAP        AMINGAP
    1     1    0.9023087045498001D+00     0.960D-03     0.960E-03
    2     1    0.9032683439082962D+00     0.960D-03    -0.960E-03
    3     1    0.9075220662489024D+00     0.210D-03     0.210E-03
    4     1    0.9077316472575748D+00     0.210D-03    -0.210E-03
    5     1    0.9093026847308840D+00     0.157D-02    -0.157E-02
    6     1    0.9117392539653555D+00     0.222D-02     0.222E-02
    7     1    0.9139623171524736D+00     0.222D-02    -0.222E-02
    8     1    0.9215518597033493D+00     0.371D-03     0.371E-03
    9     1    0.9219228129503947D+00     0.371D-03    -0.371E-03
   10     1    0.9272414797174763D+00     0.457D-02     0.457E-02
   11     1    0.9318128595969843D+00     0.960D-03     0.960E-03
   12     1    0.9327725002595535D+00     0.850D-03     0.850E-03
   13     1    0.9336220180364069D+00     0.850D-03    -0.850E-03
   14     1    0.9441957465130982D+00     0.200D-03     0.200E-03
   15     1    0.9443958908053349D+00     0.200D-03    -0.200E-03
   16     1    0.9538733953927414D+00     0.108D-02     0.108E-02
   17     1    0.9549583603516555D+00     0.108D-02    -0.108E-02
   18     1    0.9561549058563865D+00     0.120D-02    -0.120E-02
   19     1    0.9582577195407838D+00     0.750D-03     0.750E-03
   20     1    0.9590079809845515D+00     0.750D-03    -0.750E-03
   21     1    0.9618498555646015D+00     0.259D-02     0.259E-02
   22     1    0.9644445008375758D+00     0.259D-02    -0.259E-02
   23     1    0.9672303320363761D+00     0.113D-02     0.113E-02
   24     1    0.9683565650437763D+00     0.113D-02    -0.113E-02
   25     1    0.9704879822455046D+00     0.213D-02    -0.213E-02
   26     1    0.9738976059208907D+00     0.852D-03     0.852E-03
   27     1    0.9747491588235540D+00     0.852D-03    -0.852E-03
   28     1    0.9775269056442103D+00     0.278D-02    -0.278E-02
   29     1    0.9813455306889490D+00     0.382D-02    -0.382E-02
   30     1    0.9853232930778423D+00     0.398D-02    -0.398E-02
   31     1    0.9894262717440089D+00     0.410D-02    -0.410E-02
   32     1    0.9936194616944066D+00     0.419D-02    -0.419E-02
   33     1    0.9999999999999889D+00     0.638D-02    -0.638E-02
    36    40  1056    528      7892713  607840 = NG,NDIS,MEV,N,SVEED,MATNO
  0.139846820891D-11    616    0.1000D-07 = MUTOL,INDEX MINIMAL BETA,BTOL
EV NO TMULT          GOOD EIGENVALUE       TMINGAP        AMINGAP
    1     1    0.9023087045495045D+00     0.960D-03     0.960E-03
    2     1    0.9032683439081598D+00     0.960D-03    -0.960E-03
    3     1    0.9075220637981604D+00     0.210D-03     0.210E-03
    4     1    0.9077316405366920D+00     0.210D-03    -0.210E-03
    5     1    0.9093026846956148D+00    -0.332D-03    -0.157E-02
    6     1    0.9117392539584653D+00     0.210D-02     0.222E-02
    7     1    0.9139623171477664D+00     0.222D-02    -0.222E-02
    8     1    0.9215518596986422D+00     0.371D-03     0.371E-03
    9     1    0.9219228129327942D+00     0.371D-03    -0.371E-03
   10     1    0.9272414797172943D+00    -0.301D-04     0.457E-02
   11     1    0.9318128595792929D+00     0.960D-03     0.960E-03
   12     1    0.9327724989379482D+00     0.850D-03     0.850E-03
   13     1    0.9336220180228539D+00     0.850D-03    -0.850E-03
   14     1    0.9441957462784023D+00     0.200D-03     0.200E-03
   15     1    0.9443958907455297D+00     0.200D-03    -0.200E-03
   16     1    0.9538658202683405D+00     0.394D-03     0.394E-03
   17     1    0.9542594208233769D+00     0.394D-03    -0.394E-03
   18     1    0.9550432638722495D+00     0.784D-03    -0.784E-03
   19     1    0.9560950958355893D+00     0.116D-03     0.116E-03
   20     1    0.9562106619424611D+00     0.116D-03    -0.116E-03
   21     1    0.9577516552053169D+00     0.112D-02     0.112E-02
   22     1    0.9588724531626193D+00     0.781D-03     0.781E-03
```

```
23    1    0.9596530964495994D+00    0.781D-03    -0.781E-03
24    1    0.9618987632490549D+00    0.225D-02    -0.225E-02
25    1    0.9644694963663571D+00    0.257D-02    -0.257E-02
26    1    0.9673433632129476D+00    0.103D-02     0.103E-02
27    1    0.9683690265546639D+00    0.103D-02    -0.103E-02
28    1    0.9704958449702006D+00    0.213D-02    -0.213E-02
29    1    0.9739000457753179D+00    0.850D-03     0.850E-03
30    1    0.9747495648602239D+00    0.850D-03    -0.850E-03
31    1    0.9775269221872765D+00   -0.163D-02    -0.278E-02
32    1    0.9813455309753548D+00   -0.219D-02    -0.382E-02
33    1    0.9853232931157719D+00    0.398D-02    -0.398E-02
34    1    0.9894262717444636D+00    0.410D-02    -0.410E-02
35    1    0.9936194616944293D+00   -0.311D-04    -0.419E-02
36    1    0.9999999999999889D+00    0.638D-02    -0.638E-02
```

ABOVE ARE COMPUTED GOOD T-EIGENVALUES
NG = NUMBER OF GOOD T-EIGENVALUES COMPUTED
NDIS = NUMBER OF COMPUTED DISTINCT EIGENVALUES OF T(1,MEV)
N = ORDER OF A, MATNO = MATRIX IDENT
MULTOL = T-MULTIPLICITY TOLERANCE FOR T-EIGENVALUES IN BISEC
TMULT IS THE T-MULTIPLICITY OF GOOD T-EIGENVALUE
TMULT = -1 MEANS SPURIOUS T-EIGENVALUE TOO CLOSE
DO NOT COMPUTE ERROR ESTIMATES FOR SUCH EIGENVALUES
AMINGAP = MINIMAL GAP BETWEEN THE COMPUTED A-EIGENVALUES
AMINGAP .LT. O. MEANS MINIMAL GAP IS DUE TO LEFT-HAND GAP
TMINGAP= MINIMAL GAP W.R.T. DISTINCT EIGENVALUES IN T(1,MEV)
TMINGAP .LT. O. MEANS MINGAP IS DUE TO SPURIOUS T-EIGENVALUE
----- END OF FILE 3 GOODEIGENVALUES----------------------

INVERSE ITERATION ERROR ESTIMATES

TSIZE NDIS NGOOD NISO ASIZE
 792   33   33   33   528
  RHSEED   MXINIT   BETAM
  147935   5 0.417D+00
GOODEVNO   GOOD T-EIGENVALUE   BETAM*UM   TMINGAP
        1   0.9023087045498001D+00   0.797E-06   0.960D-03
        2   0.9032683439082962D+00   0.515E-06   -0.960D-03
        3   0.9075220662489024D+00   0.590E-04   0.210D-03
        4   0.9077316472575748D+00   0.970E-04   -0.210D-03
        5   0.9093026847308840D+00   0.667E-05   -0.157D-03
        6   0.9117392539653555D+00   0.273E-05   0.222D-02
        7   0.9139623171524736D+00   0.212E-05   -0.222D-02
        8   0.9215518597033493D+00   0.173E-05   0.371D-03
        9   0.9219228129503947D+00   0.331E-05   -0.371D-03
       10   0.9272414797174763D+00   0.297E-06   0.457D-02
       11   0.9318128595969843D+00   0.255E-05   0.960D-03
       12   0.9327725002595535D+00   0.215E-04   0.850D-03
       13   0.9336220180364069D+00   0.212E-05   -0.850D-03
       14   0.9441957465130982D+00   0.606E-05   0.200D-03
       15   0.9443958908053349D+00   0.303E-05   -0.200D-03
       16   0.9538733953927414D+00   0.282E-03   0.108D-02
       17   0.9549583603516555D+00   0.149E-02   -0.108D-02
       18   0.9561549058563865D+00   0.158E-02   -0.120D-02
       19   0.9582577195407838D+00   0.360E-02   0.750D-03
       20   0.9590079809845515D+00   0.169E-02   -0.750D-03
       21   0.9618498555646015D+00   0.140E-02   0.259D-02
       22   0.9644445008375758D+00   0.135E-02   -0.259D-02
       23   0.9672303320363761D+00   0.341E-02   0.113D-02
       24   0.9683565650437763D+00   0.113E-02   -0.113D-02
       25   0.9704879822455046D+00   0.102E-02   -0.213D-02
       26   0.9738976059208907D+00   0.647E-03   0.852D-03
       27   0.9747491588235540D+00   0.271E-03   -0.852D-03
       28   0.9775269056442103D+00   0.593E-04   -0.278D-02
```
```

```
        29   0.9813455306889490D+00      0.858E-05    -0.382D-02
        30   0.9853232930778423D+00      0.341E-05    -0.398D-02
        31   0.9894262717440089D+00      0.403E-06    -0.410D-02
        32   0.9936194616944066D+00      0.112E-06    -0.419D-02
        33   0.9999999999998889D+00      0.363E-07    -0.638D-02
TSIZE  NDIS NGOOD   NISO ASIZE
 1056    40    36     36   528
   RHSEED  MXINIT    BETAM
   147935       5 0.476D+00
  GOODEVNO        GOOD T-EIGENVALUE        BETAM*UM       TMINGAP
         1   0.9023087045495045D+00      0.391E-13     0.960D-03
         2   0.9032683439081598D+00      0.472E-13    -0.960D-03
         3   0.9075220637981604D+00      0.129E-11     0.210D-03
         4   0.9077316405366920D+00      0.297E-11    -0.210D-03
         5   0.9093026846956148D+00      0.219E-12     0.332D-03
         6   0.9117392539584653D+00      0.299E-14    -0.210D-02
         7   0.9139623171477664D+00      0.452E-13    -0.222D-02
         8   0.9215518596986422D+00      0.654E-14     0.371D-03
         9   0.9219228129327942D+00      0.356E-13    -0.371D-03
        10   0.9272414797172943D+00      0.198E-11    -0.301D-04
        11   0.9318128595792929D+00      0.386E-13     0.960D-03
        12   0.9327724989379482D+00      0.648E-13     0.850D-03
        13   0.9336220180228539D+00      0.488E-14    -0.850D-03
        14   0.9441957462784023D+00      0.164E-13     0.200D-03
        15   0.9443958907455297D+00      0.319E-13    -0.200D-03
        16   0.9538658202683405D+00      0.240E-09     0.394D-03
        17   0.9542594208233769D+00      0.232E-08    -0.394D-03
        18   0.9550432638722495D+00      0.201E-08    -0.784D-03
        19   0.9560950958355893D+00      0.189E-07     0.116D-03
        20   0.9562106619424611D+00      0.172E-07    -0.116D-03
        21   0.9577516552053169D+00      0.214E-08     0.112D-02
        22   0.9588724531626193D+00      0.474E-09     0.781D-03
        23   0.9596530964495994D+00      0.769E-09    -0.781D-03
        24   0.9618987632490549D+00      0.589E-10    -0.225D-02
        25   0.9644694963663571D+00      0.174E-10    -0.257D-02
        26   0.9673433632129476D+00      0.229E-10     0.103D-02
        27   0.9683690265546639D+00      0.727E-11    -0.103D-02
        28   0.9704958449702006D+00      0.514E-11    -0.213D-02
        29   0.9739000457753179D+00      0.301E-11     0.850D-03
        30   0.9747495648602239D+00      0.151E-11    -0.850D-03
        31   0.9775269221872765D+00      0.774E-12     0.163D-02
        32   0.9813455309753548D+00      0.765E-13    -0.219D-02
        33   0.9853232931157719D+00      0.137E-13    -0.398D-02
        34   0.9894262717444636D+00      0.436E-13    -0.410D-02
        35   0.9936194616944293D+00      0.114E-14    -0.311D-04
        36   0.9999999999998889D+00      0.155E-10    -0.638D-02
```

ABOVE ARE THE ERROR ESTIMATES OBTAINED FOR THE ISOLATED GOOD T-EIGENVALUES
OBTAINED VIA INVERSE ITERATION IN THE SUBROUTINE INVERR.
ALL OTHER GOOD T-EIGENVALUES HAVE CONVERGED.
ERROR ESTIMATE = BETAM*ABS(UM)
WHERE BETAM = BETA(MEV+1) AND UM = U(MEV).
U = UNIT EIGENVECTOR OF T WHERE T*U = EV*U AND EV = ISOLATED GOOD T-EIGENVALUE.
TMINGAP = GAP TO NEAREST DISTINCT EIGENVALUE OF T(1,MEV).
TMINGAP .LT. O. MEANS MINGAP IS DUE TO A LEFT NEIGHBOR.
ERROR ESTIMATE L.T. O MEANS INVERSE ITERATION DID NOT CONVERGE
------ END OF FILE 4 ERRINV -----------------------------

POISSONZ TRUE ERROR FOR GOOD EIGENVALUES
    N   NX   NY
  528   33   16

    A-DIAGONAL   X-CODIAGONAL   Y-CODIAGONAL
0.10000000D+01 0.25000000D+00 0.25000000D+00

```
  792    33   401 = MEV,NG,NEXACT
T-EV NO A-EV NO        GOOD EIGENVALUE       TRUEERROR         TMINGAP
      1     166   0.9023087045498001D+00   0.292E-12      0.960E-03
      2     167   0.9032683439082962D+00   0.145E-12      0.960E-03
      3     168   0.9075220662489024D+00   0.245E-08      0.210E-03
      4     169   0.9077316472575748D+00   0.672E-08      0.210E-03
      5     170   0.9093026847308840D+00   0.353E-10      0.157E-02
      6     171   0.9117392539653555D+00   0.688E-11      0.222E-02
      7     172   0.9139623171524736D+00   0.470E-11      0.222E-02
      8     173   0.9215518597033493D+00   0.471E-11      0.371E-03
      9     174   0.9219228129503947D+00   0.176E-10      0.371E-03
     10     175   0.9272414797174763D+00   0.187E-12      0.457E-02
     11     176   0.9318128595969843D+00   0.177E-10      0.960E-03
     12     177   0.9327725002595535D+00   0.132E-08      0.850E-03
     13     178   0.9336220180364069D+00   0.136E-10      0.850E-03
     14     179   0.9441957465130982D+00   0.235E-09      0.200E-03
     15     180   0.9443958908053349D+00   0.598E-10      0.200E-03
     16     181   0.9538733953927414D+00   0.758E-05      0.108E-02
     17     183   0.9549583603516555D+00   0.849E-04      0.108E-02
     18     185   0.9561549058563865D+00   0.558E-04      0.120E-02
     19     186   0.9582577195407838D+00   0.506E-03      0.750E-03
     20     187   0.9590079809845515D+00   0.136E-03      0.750E-03
     21     189   0.9618498555646015D+00   0.489E-04      0.259E-02
     22     190   0.9644445008375758D+00   0.250E-04      0.259E-02
     23     191   0.9672303320363761D+00   0.113E-03      0.113E-02
     24     192   0.9683565650437763D+00   0.125E-04      0.113E-02
     25     193   0.9704879822455046D+00   0.786E-05      0.213E-02
     26     194   0.9738976059208907D+00   0.244E-05      0.852E-03
     27     195   0.9747491588235540D+00   0.406E-06      0.852E-03
     28     196   0.9775269056442103D+00   0.165E-07      0.278E-02
     29     197   0.9813455306889490D+00   0.286E-09      0.382E-02
     30     198   0.9853232930778423D+00   0.379E-10      0.398E-02
     31     199   0.9894262717440089D+00   0.450E-12      0.410E-02
     32     200   0.9936194616944066D+00   0.268E-13      0.419E-02
     33     201   0.9999999999998889D+00   0.105E-13      0.638E-02
ABOVE ARE THE TRUE ERRORS FOR POISSON GOODEV
IF A-EV NO LT 0 THEN GOODEV HAS MULTIPLICITY GT 1
IF TRUE ERROR LT 0 THEN MORE THAN ONE GOODEV APPROXIMATES
THE SAME TRUE POISSON EIGENVALUE
IF TMINGAP LT 0 THE MINGAP IS DUE TO SPURIOUS EIGENVALUE


POISSONZ TRUE ERROR FOR GOOD EIGENVALUES
     N    NX    NY
   528    33    16

    A-DIAGONAL    X-CODIAGONAL    Y-CODIAGONAL
 0.10000000D+01 0.25000000D+00 0.25000000D+00



 1056    36   401 = MEV,NG,NEXACT
T-EV NO A-EV NO        GOOD EIGENVALUE       TRUEERROR         TMINGAP
      1     166   0.9023087045495045D+00   0.373E-14      0.960E-03
      2     167   0.9032683439081598D+00   0.840E-14      0.960E-03
      3     168   0.9075220637981604D+00   0.837E-14      0.210E-03
      4     169   0.9077316405366920D+00   0.568E-14      0.210E-03
      5     170   0.9093026846956148D+00   0.565E-14     -0.332E-03
      6     171   0.9117392539584653D+00   0.598E-14      0.210E-02
      7     172   0.9139623171477664D+00   0.543E-14      0.222E-02
      8     173   0.9215518596986422D+00   0.172E-14      0.371E-03
      9     174   0.9219228129327942D+00   0.393E-14      0.371E-03
     10     175   0.9272414797172943D+00   0.480E-14     -0.301E-04
     11     176   0.9318128595792929D+00   0.548E-14      0.960E-03
```

12	177	0.9327724989379482D+00	0.652E-14	0.850E-03
13	178	0.9336220180228539D+00	0.194E-14	0.850E-03
14	179	0.9441957462784023D+00	0.508E-14	0.200E-03
15	180	0.9443958907455297D+00	0.744E-14	0.200E-03
16	181	0.9538658202683405D+00	0.810E-14	0.394E-03
17	182	0.9542594208233769D+00	0.115E-13	0.394E-03
18	183	0.9550432638722495D+00	0.609E-14	0.784E-03
19	184	0.9560950958355893D+00	0.913E-14	0.116E-03
20	185	0.9562106619424611D+00	0.346E-14	0.116E-03
21	186	0.9577516552053169D+00	0.101E-13	0.112E-02
22	187	0.9588724531626193D+00	0.959E-14	0.781E-03
23	188	0.9596530964495994D+00	0.865E-14	0.781E-03
24	189	0.9618987632490549D+00	0.547E-14	0.225E-02
25	190	0.9644694963663571D+00	0.779E-14	0.257E-02
26	191	0.9673433632129476D+00	0.125E-14	0.103E-02
27	192	0.9683690265546639D+00	0.358E-14	0.103E-02
28	193	0.9704958449702006D+00	0.895E-14	0.213E-02
29	194	0.9739000457753179D+00	0.533E-14	0.850E-03
30	195	0.9747495648602239D+00	0.283E-14	0.850E-03
31	196	0.9775269221872765D+00	0.958E-15	-0.163E-02
32	197	0.9813455309753548D+00	0.171E-14	-0.219E-02
33	198	0.9853232931157719D+00	0.386E-14	0.398E-02
34	199	0.9894262717444636D+00	0.508E-14	0.410E-02
35	200	0.9936194616944293D+00	0.419E-14	-0.311E-04
36	201	0.9999999999998889D+00	0.105E-13	0.638E-02

ABOVE ARE THE TRUE ERRORS FOR POISSON GOODEV
IF A-EV NO LT O THEN GOODEV HAS MULTIPLICITY GT 1
IF TRUE ERROR LT O THEN MORE THAN ONE GOODEV APPROXIMATES
THE SAME TRUE POISSON EIGENVALUE
IF TMINGAP LT O THE MINGAP IS DUE TO SPURIOUS EIGENVALUE

-----SAMPLE OUTPUT FOR REAL SYMMETRIC EIGENVECTOR COMPUTATIONS-----

Below is a sample of the output from the LEVEC eigenvector program.
The output which is printed to file 6 is given first, followed by the
outputs which are generated and written to other files. Some of the
middle portion of the printout to file 6 has been condensed.

LANCZOS EIGENVECTOR PROCEDURE FOR REAL SYMMETRIC MATRICES

LANCZOS EIGENVECTOR COMPUTATIONS, NO REORTHOGONALIZATION
    POISSON INPUT

```
    N    KX    KY      DIAGONAL    X-CODIAGONAL   Y-CODIAGONAL
  528    33    16  0.100000000+01 0.250000000+00 0.250000000+00
```

MATRIX IDENTIFICATION NO. =   607840 ORDER OF A =   528

```
  MBOUND    NTVCON   SVTVEC   IREAD
      0        1        1       1

  TVSTOP    LVCONT   ERCONT   IWRITE
      0        1        0       1

  MDIMTV    MDIMRV    MBETA
   50000     20000     6000

      RELTOL    RHSEED
  0.1000D-09 45329517
```

T-MULTIPLICITY TOLERANCE USED IN THE EIGENVALUE COMPUTATIONS WAS   0.1398D-11
SCALED MACHINE EPSILON IS   0.6802D-15

EIGENVALUES SUPPLIED ARE READ IN FROM FILE 3
FILE 3 HEADER IS

```
  NG  NDIS   MEV  NOLD  MATOLD   SVSEED     MULTOL    IB      BTOL
  15    40  1056   528  607840  7892713  0.1400-11   616  0.10000-07
```

EIGENVALUES READ IN, T-MULTIPLICITIES, T-GAPS AND A-GAPS

```
   J       GOOD EIGENVALUE     MULT     TMINGAP      AMINGAP
   1    0.9023087045495045D+00    1     0.9600E-03   0.9600E-03
   2    0.9032683439081598D+00    1     0.9600E-03  -0.9600E-03
   3    0.9075220637981604D+00    1     0.2100E-03   0.2100E-03
   4    0.9077316405366920D+00    1     0.2100E-03  -0.2100E-03
   5    0.9093026846956148D+00    1    -0.3320E-03  -0.1570E-02
   6    0.9443958907455297D+00    1     0.2000E-03  -0.2000E-03
   7    0.9538658202683405D+00    1     0.3940E-03   0.3940E-03
   8    0.9542594208233769D+00    1     0.3940E-03  -0.3940E-03
   9    0.9550432638722495D+00    1     0.7840E-03  -0.7840E-03
  10    0.9560950958355893D+00    1     0.1160E-03   0.1160E-03
  11    0.9813455309753548D+00    1    -0.2190E-02  -0.3820E-02
  12    0.9853232931157719D+00    1     0.3980E-02  -0.3980E-02
  13    0.9894262717444636D+00    1     0.4100E-02  -0.4100E-02
  14    0.9936194616944293D+00    1    -0.3110E-04  -0.4190E-02
  15    0.9999999999999889D+00    1     0.6380E-02  -0.6380E-02
```

THESE EIGENVALUES WERE COMPUTED USING A T-MATRIX OF   ORDER 1056
AND SEED FOR RANDOM NUMBER GENERATOR =   7892713
ERROR ESTMATES =

```
   J      EIGENVALUE       ESTIMATE
   1   0.902308704550D+00   0.391E-13
   2   0.903268343908D+00   0.472E-13
   3   0.907522063798D+00   0.129E-11
   4   0.907731640537D+00   0.297E-11
```

```
 5   0.909302684696D+00        0.219E-12
 6   0.944395890746D+00        0.319E-13
 7   0.953865820268D+00        0.240E-09
 8   0.954259420823D+00        0.232E-08
 9   0.955043263872D+00        0.201E-08
10   0.956095095836D+00        0.189E-07
11   0.981345530975D+00        0.765E-13
12   0.985323293116D+00        0.137E-13
13   0.989426271744D+00        0.436E-13
14   0.993619461694D+00        0.114E-14
15   0.100000000000D+01        0.155E-10
```

READ IN THE T-MATRICES STORED ON FILE 2
FILE 2 HEADER IS
```
 KMAX   NOLD      SVSOLD   MATOLD
 1320    528      7892713  607840
```

ENLARGE KMAX FROM   1320 TO   1464

LANCZS SUBROUTINE GENERATES ALPHA(J), BETA(J+1), J =   1321 TO   1464

1ST GUESS AT APPROPRIATE SIZE T-MATRICES
ACTUAL VALUES WILL PROBABLY BE 1/4 AGAIN AS MUCH
```
  943    934    994    999    969    968   1112   1132   1129   1151    972    931    901
  863    837
```

     9TH EIGENVALUE
```
TSIZE              EIGENVALUE              U(M)              ERRORV
1197    0.9550432638722377D+00    0.18352D-09    0.34361D-12
```

CHANGE SIZE OF T-MATRIX TO   1231 RECOMPUTE T-EIGENVECTOR

     9TH EIGENVALUE
```
TSIZE              EIGENVALUE              U(M)              ERRORV
1231    0.9550432638722377D+00    0.30709D-10    0.34845D-12
```

     11TH EIGENVALUE
```
TSIZE              EIGENVALUE              U(M)              ERRORV
 972    0.9813455309753427D+00    0.38849D-11    0.13630D-10
```

THE CUMULATIVE LENGTH OF THE T-EIGENVECTORS IS   15314

     15 T-EIGENVECTORS OUT OF   15 REQUESTED WERE COMPUTED

  1343 = LARGEST SIZE T-MATRIX TO BE USED IN THE RITZ VECTOR COMPUTATIONS

     15 RITZ VECTORS WERE REQUESTED
     15 T-EIGENVECTORS WERE COMPUTED
     15 RITZ VECTORS WILL BE COMPUTED

   1 TH EIGENVALUE CONSIDERED =   0.9023087045500+00

NORM OF ERROR IN T-EIGENVECTOR =   0.300E-12
BETA(MA(J)+1)*U(MA(J)) =   0.617E-13
ABS(NORM(RITVEC) - 1.0) =   0.660D-12

RUNTIME = 1653 HNDRETHS OF A SECOND


```
   A-EIGENVALUE   MA(J)   A-MINGAP      AERROR    AERROR/GAP      TERROR
  0.902308070D+00   943   0.9600E-03  0.4410E-12  0.4594E-09  0.3001E-12
  0.903268340D+00   934  -0.9600E-03  0.2019E-12  0.2103E-09  0.1340E-12
  0.907522060D+00   994   0.2100E-03  0.2028E-11  0.9659E-08  0.6610E-13
```

```
0.9077316ЧD+00    999 -0.2100E-03   0.2570E-11   0.1224E-07   0.2588E-12
0.9093026BD+00    969 -0.1570E-02   0.2734E-11   0.1742E-08   0.1902E-11
0.94439589D+00    968 -0.2000E-03   0.2073E-10   0.1036E-06   0.2815E-12
0.95386582D+00   1112  0.3940E-03   0.2563E-10   0.6505E-07   0.2056E-12
0.95425942D+00   1268 -0.3940E-03   0.1024E-10   0.2600E-07   0.1103E-12
0.95504326D+00   1231 -0.7840E-03   0.1489E-10   0.1899E-07   0.3485E-12
0.95609510D+00   1343  0.1160E-03   0.1062E-10   0.9155E-07   0.2351E-11
0.98134553D+00    972 -0.3820E-02   0.1957E-10   0.5124E-08   0.1363E-10
0.98532329D+00    931 -0.3980E-02   0.2979E-11   0.7484E-09   0.2031E-11
0.98942627D+00    901 -0.4100E-02   0.7486E-12   0.1826E-09   0.4603E-12
0.99361946D+00    888 -0.4190E-02   0.3438E-11   0.8206E-09   0.3755E-12
0.10000000D+01    861 -0.6380E-02   0.2566E-10   0.4022E-08   0.5213E-13
```

ABOVE ARE ERROR ESTIMATES FOR THE A AND T EIGENVECTORS
ASSOCIATED WITH THE GOODEV LISTED IN COLUMN 1
AERROR = NORM(A*X - EV*X)   TERROR = NORM(T*Y - EV*Y)
WHERE T = T(1,MA(J))   X = RITZ VECTOR = V*Y   V = SUCCESSIVE
LANCZOS VECTORS. AMINGAP = GAP TO NEAREST A-EIGENVALUE

```
              GOODEV(J)       RITZNORM      AMINGAP       TBETA(J)      TLAST(J)
       0.90230870454952D+00  0.10000E+01   0.96000E-03   0.61736E-13  0.11215D-12
       0.90326834390815D+00  0.10000E+01  -0.96000E-03   0.58491E-13  0.16274D-12
       0.90752206379817D+00  0.10000E+01   0.21000E-03   0.20265E-11  0.37472D-11
       0.90773164053670D+00  0.10000E+01  -0.21000E-03   0.25446E-11  0.49470D-11
       0.90930268469560D+00  0.10000E+01  -0.15700E-02   0.12312E-12  0.32684D-12
       0.94439589074552D+00  0.10000E+01  -0.20000E-03   0.20734E-10  0.44109D-10
       0.95386582026835D+00  0.10000E+01   0.39400E-03   0.25604E-10  0.52969D-10
       0.95425942082337D+00  0.10000E+01  -0.39400E-03   0.10235E-10  0.30085D-10
       0.95504326387224D+00  0.10000E+01  -0.78400E-03   0.14891E-10  0.30709D-10
       0.95609509583558D+00  0.10000E+01   0.11600E-03   0.97828E-11  0.17603D-10
       0.98134553097534D+00  0.10000E+01  -0.38200E-02   0.20033E-11  0.38849D-11
       0.98532329311576D+00  0.10000E+01  -0.39800E-02   0.42732E-12  0.81168D-12
       0.98942627174445D+00  0.10000E+01  -0.41000E-02   0.31008E-12  0.11744D-11
       0.99361946169444D+00  0.10000E+01  -0.41900E-02   0.33970E-11  0.57069D-11
       0.10000000000000D+01  0.10000E+01  -0.63800E-02   0.25666E-10  0.76799D-10
```

ABOVE ARE ERROR ESTIMATES ASSOCIATED WITH THE GOODEV
RITZNORM = NORM(COMPUTED RITZ VECTOR)
TBETA(J) = BETA(MA(J)+1)*Y(MA(J)),   T*Y = EVAL*Y
TLAST(J) = Y(MA(J))
AMINGAP = GAP TO NEAREST A-EIGENVALUE

  528   1464 = ORDER OF USER MATRIX AND MAX ORDER OF T(1,MEV)

1ST GUESS AT APPROPRIATE SIZE T-MATRICES
ACTUAL VALUES WILL PROBABLY BE 1/4 AGAIN AS MUCH

```
  J    A-EIGENVALUE     M1(J) M2(J) MA(J)
  1  0.9023087045550D+00   754  1464   943
  2  0.9032683439080D+00   747  1464   934
  3  0.9075220637980D+00   795  1464   994
  4  0.9077316405370D+00   799  1464   999
  5  0.9093026846960D+00   775  1464   969
  6  0.9443958907460D+00   774  1464   968
  7  0.9538658202680D+00   889  1464  1112
  8  0.9542594208230D+00   905  1464  1132
  9  0.9550432638720D+00   903  1464  1129
 10  0.9560950958360D+00   920  1464  1151
 11  0.9813455309750D+00   777  1464   972
 12  0.9853232931160D+00   757  1452   931
 13  0.9894262717440D+00   725  1426   901
 14  0.9936194616940D+00   700  1350   863
 15  0.1000000000000D+01   679  1309   837
```

SIZES OF T-MATRICES THAT WILL BE USED IN THE RITZ COMPUTATIONS

J	GOODEV(J)	MA(J)
1	0.90230870454950D+00	943
2	0.90326834390816D+00	934
3	0.90752206379816D+00	994
4	0.90773164053669D+00	999
5	0.90930268469561D+00	969
6	0.94439589074553D+00	968
7	0.95386582026834D+00	1112
8	0.95425942082338D+00	1268
9	0.95504326387225D+00	1231
10	0.95609509583559D+00	1343
11	0.98134553097535D+00	972
12	0.98532329311577D+00	931
13	0.98942627174446D+00	901
14	0.99361946169443D+00	888
15	0.99999999999999D+00	861

```
EV = GOODEV(J) IS A GOOD EIGENVALUE OF T(1,MEV)
M1 = SMALLEST VALUE OF M SUCH THAT T(1,M) HAS AT LEAST
     ONE EIGENVALUE IN THE INTERVAL (EV-TOLN,EV+TOLN)
     TOLN(J) = DMAX1(GOODEV(J)*RELTOL, SCALE0*MULTOL)
M2 = SMALLEST M (IF ANY) SUCH THAT IN THE ABOVE INTERVAL
     T(1,M) HAS AT LEAST TWO EIGENVALUES
IABS(MA(J)) = APPROPRIATE SIZE T-MATRIX FOR GOODEV(J)
INITIAL VALUE OF MA(J) IS CHOSEN HEURISTICALLY
PROGRAM LOOPS ON SIZE OF T-MATRIX TO GET BETTER SIZE
END OF SIZES OF T-MATRICES FILE 10
```

CHAPTER 3

HERMITIAN MATRICES

SECTION 3.1   INTRODUCTION

The FORTRAN codes in this chapter address the question of computing distinct eigenvalues and corresponding eigenvectors of Hermitian matrices, using a single-vector Lanczos procedure. For a given Hermitian matrix A, these codes compute real scalars λ and corresponding complex vectors $x \neq 0$ such that

$$Ax = \lambda x . \qquad (3.1.1)$$

DEFINITION 3.1.1   A complex nxn matrix A, $A \equiv (a_{ij})$, $1 \leq i,j \leq n$, is a Hermitian matrix if and only if for every i and j, $a_{ij} = \bar{a}_{ji}$, where the overbar denotes the complex conjugate of the complex-valued entry a_{ij} .

It is straight-forward to demonstrate from Definition 3.1.1 that for any Hermitian matrix $A = B + Ci$, where B and C are real matrices and $i = \sqrt{-1}$, that B must be a real symmetric matrix and C must be a skew symmetric matrix. That is, $B^T = B$ and $C^T = -C$. Futhermore, it is not difficult to see that Hermitian matrices must have real diagonal entries and real eigenvalues. However, the eigenvectors are complex-valued. Any Hermitian matrix can be transformed into a real symmetric tridiagonal matrix for the purposes of computing the eigenvalues of the Hermitian matrix, Stewart [1973]. In fact, the Lanczos recursion which we use in the codes in this chapter transforms the given Hermitian matrix A into a family of real symmetric tridiagonal matrices rather than into a family of Hermitian tridiagonal matrices.

Hermitian matrices possess the 'same' properties as real symmetric matrices do, except that these properties are defined with respect to the complex or Hermitian norm, rather than with respect to the Euclidean norm, see Stewart [1973]. The Hermitian norm of a given complex-valued vector $x \equiv (x(i))$, $1 \leq i \leq n$, is defined as $\| x \|_{\mathscr{C}}^2 \equiv \sum_{i=1}^{n} \overline{x(i)}x(i)$. Three properties which we use are: (1) Hermitian matrices have complete eigensystems. That is, the dimension of the eigenspace corresponding to any given eigenvalue of a Hermitian matrix is the same as the multiplicity of that eigenvalue as a root of the characteristic polynomial of that matrix. (2) For any two distinct eigenvalues λ and μ, and corresponding eigenvectors x and y, $x^H y = 0$, where the superscript H denotes the complex conjugate transpose of the vector x. The complex conjugate transpose of a column vector x is the row vector whose ith component is $\overline{x(i)}$. Therefore there is a complete set of eigenvectors $X_n \equiv \{x_1,...,x_n\}$ such that X is a unitary matrix. (3) Small perturbations in a Hermitian matrix cause only small perturbations in the eigenvalues.

119

The single-vector Lanczos codes in this chapter can be used to compute either a very few or very many of the distinct eigenvalues of the given Hermitian matrix. The documentation for these codes is contained in Chapter 2, Section 2.2. As in the real symmetric case, the A-multiplicity of a given computed 'good' Lanczos eigenvalue can be obtained only with additional computation, and the modifications required to do this additional computation are not included in these versions of the codes. This implementation uses a Hermitian analog of the basic Lanczos recursion contained in Eqns(1.2.1) - (1.2.2) to generate a family of real symmetric tridiagonal matrices whose sizes are specified by the user. There is no reorthogonalization of the Lanczos vectors at any stage in any of the computations.

The Hermitian version of the Lanczos recursion which we use is given below. For $i=1,2,...,m$ and a randomly-generated complex starting vector v_1 with $\|v_1\|_{\mathscr{C}} = 1$, generate Lanczos vectors v_i using the following recursion.

$$\beta_{i+1} v_{i+1} = A v_i - \alpha_i v_i - \beta_i v_{i-1}, \qquad (3.1.2)$$

where

$$\alpha_i \equiv v_i^H A v_i \text{ and } \beta_{i+1} = \| A v_i - \alpha_i v_i - \beta_i v_{i-1} \|_{\mathscr{C}}. \qquad (3.1.3)$$

We see from Eqns(3.1.3), that the Hermitian inner product is used. This is the 'natural' inner product for Hermitian matrices. Gram-Schmidt orthogonalization is used, unlike the real symmetric case where a modified Gram-Schmidt orthogonalization was used. This change in the local orthogonalization procedure increases the storage requirements for the implementation of the Lanczos recursion by one additional complex vector of length equal to the order of the original A-matrix. Modified Gram-Schmidt orthogonalization cannot be used in the Hermitian case because corrections to the α_i defined by this modification are complex-valued not real, and it would not be legitimate to accept the real portions of these corrections and simply ignore the complex portions.

It is easy to demonstrate that as we stated earlier, each Lanczos matrix (T-matrix) generated by this Hermitian recursion is a real symmetric tridiagonal matrix. In particular, we see from the formulas in Eqn(3.1.3) that the diagonal entries of each of these matrices are Rayleigh quotients of the given Hermitian matrix A, and therefore must all be real-valued. Furthermore by construction, the nonzero off-diagonal entries β_{i+1} are all real-valued. This use of real-valued β_i requires some justification. This justification is given in Section 4.9 of Chapter 4 of Volume 1 of this book.

HLEVAL, the main program for the Hermitian eigenvalue computations, calls the subroutine BISEC to compute eigenvalues of the specified tridiagonal Lanczos matrices on the user-specified intervals. BISEC simultaneously computes these T-eigenvalues with their T-multiplicities and sorts the computed T-eigenvalues into two classes, the 'good' T-eigenvalues and the 'spurious' T-eigenvalues. The 'good' T-eigenvalues are accepted as approximations to eigenvalues of the user-specified matrix A. The accuracy of these 'good' T-eigenvalues as eigenvalues of A is then estimated using error estimates computed by subroutine INVERR. Error estimates are computed only for isolated 'good' T-eigenvalues. All other 'good' T-eigenvalues are assumed to have converged. Convergence is then checked. If convergence has not yet occurred and a larger T-matrix has been specified by the user, the program will continue on to the larger T-matrix, repeating the above procedure on this larger matrix.

Once the eigenvalues have been computed accurately enough, the user can select a subset of the 'converged' eigenvalues for which eigenvectors are to be computed. The main program HLEVEC, for computing eigenvectors of Hermitian matrices, is then used to compute these desired eigenvectors.

The computations in the Lanczos recursion are a mixture of double precision real arithmetic and of double precision complex arithmetic. Once the Lanczos matrices have been computed, the remaining computations are all done in double precision real arithmetic, using the same subroutines that are used in the real symmetric case. In addition to the programs and subroutines provided here, the user must supply a subroutine USPEC which defines and initializes the user-specified matrix A and a subroutine CMATV which computes matrix-vector multiplies Ax for any given vector x. These subroutines must be constructed in such a way as to take advantage of the sparsity (and/or structure) of the user-supplied A-matrix and such that these computations are done accurately.

SECTION 3.2   MAIN PROGRAM, EIGENVALUE COMPUTATIONS

```
C-----HLEVAL  (EIGENVALUES OF HERMITIAN MATRICES)----------------------HLE00010
C                                                                      HLE00020
C      CONTAINS MAIN PROGRAM FOR COMPUTING DISTINCT EIGENVALUES OF      HLE00030
C      A HERMITIAN MATRIX USING LANCZOS TRIDIAGONALIZATION WITHOUT      HLE00040
C      REORTHOGONALIZATION                                             HLE00050
C                                                                      HLE00060
C      PORTABILITY:                                                     HLE00070
C      THIS PROGRAM IS NOT PORTABLE DUE TO THE USE OF COMPLEX*16        HLE00080
C      VARIABLES.  MOREOVER, THE PFORT VERIFIER IDENTIFIED THE          HLE00090
C      FOLLOWING ADDITIONAL NONPORTABLE CONSTRUCTIONS:                  HLE00100
C                                                                      HLE00110
C      1.  DATA/MACHEP/ STATEMENT                                       HLE00120
C      2.  ALL READ(5,*) STATEMENTS (FREE FORMAT)                       HLE00130
C      3.  FORMAT(20A4) USED WITH EXPLANATORY HEADER EXPLAN.            HLE00140
C      4.  HEXADECIMAL FORMAT (4Z20) USED IN ALPHA/BETA FILES 1 AND 2.  HLE00150
C                                                                      HLE00160
C---------------------------------------------------------------------HLE00170
C                                                                      HLE00180
       DOUBLE PRECISION  ALPHA(5000),BETA(5001),VS(5000)               HLE00190
       COMPLEX*16  V1(5000),V2(5000)                                   HLE00200
       DOUBLE PRECISION  GR(1500),GC(1500),LB(20),UB(20)               HLE00210
       DOUBLE PRECISION  BTOL,GAPTOL,TTOL,MACHEP,EPSM,RELTOL           HLE00220
       DOUBLE PRECISION  SCALE1,SCALE2,SCALE3,SCALE4,BISTOL,CONTOL,MULTOLHLE00230
       DOUBLE PRECISION  ONE,ZERO,TEMP,TKMAX,BETAM,BKMIN,TO,T1         HLE00240
       REAL  G(5000),EXPLAN(20)                                        HLE00250
       INTEGER  MP(5000),NMEV(20)                                      HLE00260
       INTEGER  SVSEED,RHSEED,SVSOLD                                   HLE00270
       INTEGER IABS                                                    HLE00280
       REAL  ABS                                                       HLE00290
       DOUBLE PRECISION  DABS, DSQRT, DFLOAT                           HLE00300
       EXTERNAL CMATV                                                  HLE00310
C                                                                      HLE00320
C---------------------------------------------------------------------HLE00330
       DATA MACHEP/Z3410000000000000/                                  HLE00340
       EPSM = 2.0D0*MACHEP                                             HLE00350
C---------------------------------------------------------------------HLE00360
C      THE ARRAYS V1 AND V2 ARE DEFINED AS COMPLEX*16 IN THE MAIN PROGRAMHLE00370
C      AND IN THE SUBROUTINE LANCZS.  HOWEVER, IN THE OTHER SUBROUTINES HLE00380
C      THEY ARE DECLARED AS DOUBLE PRECISION ARRAYS. NOTE THAT THE      HLE00390
C      DIMENSION OF V2 ASSUMES THAT NO MORE THAN KMAX/2 EIGENVALUES OF  HLE00400
C      THE T-MATRICES ARE BEING COMPUTED IN ANY ONE OF THE SUB-INTERVALS HLE00410
C      BEING CONSIDERED.  V2 MUST CONTAIN UPPER AND LOWER BOUNDS        HLE00420
C      ON EACH T-EIGENVALUE COMPUTED BY BISEC IN ANY ONE GIVEN INTERVAL. HLE00430
C                                                                      HLE00440
C      ARRAYS MUST BE DIMENSIONED AS FOLLOWS:                          HLE00450
C         1. ALPHA:  >= KMAX.  BETA:  >= (KMAX+1)                       HLE00460
C         2. V1:  >= MAX(N,(KMAX+1)/2).   V2:  >= MAX(N,KMAX/2)         HLE00470
C         3. VS:  >= MAX(2*N,KMAX).                                     HLE00480
C         4. GR,GC:  >=  N                                              HLE00490
C         5. G:  >= MAX(2*KMAX,N)                                       HLE00500
C         6. MP:  >= KMAX                                               HLE00510
C         7. LB,UB:  >= NUMBER OF SUB-INTERVALS SPECIFIED               HLE00520
C         8. NMEV:  >= NUMBER OF T-MATRICES SPECIFIED                   HLE00530
C         9. EXPLAN: DIMENSION IS 20.                                   HLE00540
C                                                                      HLE00550
C                                                                      HLE00560
C      IMPORTANT TOLERANCES OR SCALES THAT ARE USED REPEATEDLY          HLE00570
C      THROUGHOUT THE PROGRAM ARE THE FOLLOWING:                       HLE00580
C      SCALED MACHINE EPSILON:  TTOL = TKMAX*EPSM WHERE                 HLE00590
C      EPSM = 2*MACHINE EPSILON AND                                     HLE00600
C      TKMAX = MAX(|ALPHA(J)|,BETA(J), J = 1,MEV)                       HLE00610
C      BISEC CONVERGENCE TOLERANCE:  BISTOL = DSQRT(1000+MEV)*TTOL      HLE00620
C      BISEC MULTIPLICITY TOLERANCE:  MULTOL = (1000+MEV)*TTOL          HLE00630
C      LANCZOS CONVERGENCE TOLERANCE:   CONTOL = BETA(MEV+1)*1.D-10     HLE00640
```

```
C--------------------------------------------------------------------------HLE00650
C     OUTPUT HEADER                                                        HLE00660
      WRITE(6,10)                                                          HLE00670
   10 FORMAT(/' LANCZOS PROCEDURE FOR HERMITIAN MATRICES'/)               HLE00680
C                                                                          HLE00690
C     SET PROGRAM PARAMETERS                                               HLE00700
C     SCALEK ARE USED IN TOLERANCES NEEDED IN SUBROUTINES LUMP,           HLE00710
C     ISOEV AND PRTEST.  USER MUST NOT MODIFY THESE SCALES.               HLE00720
      SCALE1 = 5.0D2                                                       HLE00730
      SCALE2 = 5.0D0                                                       HLE00740
      SCALE3 = 5.0D0                                                       HLE00750
      SCALE4 = 1.0D4                                                       HLE00760
      ONE  = 1.0D0                                                         HLE00770
      ZERO = 0.0D0                                                         HLE00780
C     BTOL = EPSM                                                          HLE00790
      BTOL = 1.0D-8                                                        HLE00800
      GAPTOL = 1.0D-8                                                      HLE00810
      ICONV = 0                                                            HLE00820
      MOLD = 0                                                             HLE00830
      MOLD1 = 1                                                            HLE00840
      ICT = 0                                                              HLE00850
      MMB = 0                                                              HLE00860
      IPROJ = 0                                                            HLE00870
C                                                                          HLE00880
C     READ USER-SPECIFIED PARAMETERS FROM INPUT FILE 5 (FREE FORMAT)      HLE00890
C                                                                          HLE00900
C     READ USER-PROVIDED HEADER FOR RUN                                    HLE00910
      READ(5,20) EXPLAN                                                    HLE00920
      WRITE(6,20) EXPLAN                                                   HLE00930
      READ(5,20) EXPLAN                                                    HLE00940
      WRITE(6,20) EXPLAN                                                   HLE00950
   20 FORMAT(20A4)                                                         HLE00960
C                                                                          HLE00970
C     READ ORDER OF MATRICES (N) , MAXIMUM ORDER OF T-MATRIX (KMAX),      HLE00980
C     NUMBER OF T-MATRICES ALLOWED (NMEVS), AND MATRIX IDENTIFICATION     HLE00990
C     NUMBERS (MATNO)                                                      HLE01000
      READ(5,20) EXPLAN                                                    HLE01010
      READ(5,*) N,KMAX,NMEVS,MATNO                                         HLE01020
C                                                                          HLE01030
C     READ SEEDS FOR LANCZS AND INVERR SUBROUTINES (SVSEED AND RHSEED)    HLE01040
C     READ MAXIMUM NUMBER OF ITERATIONS ALLOWED FOR EACH INVERSE          HLE01050
C     ITERATION (MXINIT) AND MAXIMUM NUMBER OF STURM SEQUENCES            HLE01060
C     ALLOWED (MXSTUR)                                                     HLE01070
      READ(5,20) EXPLAN                                                    HLE01080
      READ(5,*) SVSEED,RHSEED,MXINIT,MXSTUR                               HLE01090
C                                                                          HLE01100
C     ISTART = (0,1): ISTART = 0 MEANS ALPHA/BETA FILE IS NOT             HLE01110
C     AVAILABLE.  ISTART = 1 MEANS ALPHA/BETA FILE IS AVAILABLE ON        HLE01120
C     FILE 2.                                                              HLE01130
C     ISTOP = (0,1): ISTOP = 0 MEANS PROCEDURE GENERATES ALPHA/BETA       HLE01140
C     FILE AND THEN TERMINATES.  ISTOP = 1 MEANS PROCEDURE GENERATES      HLE01150
C     ALPHAS/BETAS IF NEEDED AND THEN COMPUTES EIGENVALUES AND ERROR      HLE01160
C     ESTIMATES AND THEN TERMINATES.                                      HLE01170
      READ(5,20) EXPLAN                                                    HLE01180
      READ(5,*) ISTART,ISTOP                                              HLE01190
C                                                                          HLE01200
C     IHIS = (0,1): IHIS = 0 MEANS ALPHA/BETA FILE IS NOT WRITTEN         HLE01210
C     TO FILE 1.  IHIS = 1 MEANS ALPHA/BETA FILE IS WRITTEN TO FILE 1.    HLE01220
C     IDIST = (0,1): IDIST = 0 MEANS DISTINCT T-EIGENVALUES               HLE01230
C     ARE NOT WRITTEN TO FILE 11.  IDIST = 1 MEANS DISTINCT               HLE01240
C     T-EIGENVALUES ARE WRITTEN TO FILE 11.                               HLE01250
C     IWRITE = (0,1): IWRITE = 0 MEANS NO INTERMEDIATE OUTPUT             HLE01260
C     FROM THE COMPUTATIONS IS WRITTEN TO FILE 6.  IWRITE = 1 MEANS       HLE01270
C     T-EIGENVALUES AND ERROR ESTIMATES ARE WRITTEN TO FILE 6             HLE01280
C     AS THEY ARE COMPUTED.                                                HLE01290
      READ(5,20) EXPLAN                                                    HLE01300
```

```
      READ(5,*) IHIS,IDIST,IWRITE                                      HLE01310
C                                                                      HLE01320
C     READ IN THE RELATIVE TOLERANCE (RELTOL) FOR USE IN THE           HLE01330
C     SPURIOUS, T-MULTIPLICITY, AND PRTESTS.                           HLE01340
      READ(5,20) EXPLAN                                                HLE01350
      READ(5,*) RELTOL                                                 HLE01360
C                                                                      HLE01370
C     READ IN THE SIZES OF THE T-MATRICES TO BE CONSIDERED.            HLE01380
      READ(5,20) EXPLAN                                                HLE01390
      READ(5,*) (NMEV(J), J=1,NMEVS)                                   HLE01400
C                                                                      HLE01410
C     READ IN THE NUMBER OF SUBINTERVALS TO BE CONSIDERED.             HLE01420
      READ(5,20) EXPLAN                                                HLE01430
      READ(5,*) NINT                                                   HLE01440
C                                                                      HLE01450
C     READ IN THE LEFT-END POINTS OF THE SUBINTERVALS TO BE CONSIDERED. HLE01460
C     THESE MUST BE IN ALGEBRAICALLY-INCREASING ORDER                 HLE01470
      READ(5,20) EXPLAN                                                HLE01480
      READ(5,*) (LB(J), J=1,NINT)                                      HLE01490
C                                                                      HLE01500
C     READ IN THE RIGHT-END POINTS OF THE SUBINTERVALS TO BE CONSIDERED.HLE01510
C     THESE MUST BE IN ALGEBRAICALLY-INCREASING ORDER                 HLE01520
      READ(5,20) EXPLAN                                                HLE01530
      READ(5,*) (UB(J), J=1,NINT)                                      HLE01540
C                                                                      HLE01550
C----------------------------------------------------------------------HLE01560
C                                                                      HLE01570
C     INITIALIZE THE ARRAYS FOR THE USER-SPECIFIED MATRIX             HLE01580
C     AND PASS THE STORAGE LOCATIONS OF THESE ARRAYS TO THE           HLE01590
C     MATRIX-VECTOR MULTIPLY SUBROUTINE CMATV.                         HLE01600
C                                                                      HLE01610
      CALL USPEC(N,MATNO)                                              HLE01620
C                                                                      HLE01630
C----------------------------------------------------------------------HLE01640
C                                                                      HLE01650
C     MASK UNDERFLOW AND OVERFLOW                                      HLE01660
C                                                                      HLE01670
      CALL MASK                                                        HLE01680
C                                                                      HLE01690
C----------------------------------------------------------------------HLE01700
C                                                                      HLE01710
C     WRITE TO FILE 6, A SUMMARY OF THE PARAMETERS FOR THIS RUN        HLE01720
C                                                                      HLE01730
      WRITE(6,30) MATNO,N,KMAX                                         HLE01740
   30 FORMAT(/3X,'MATRIX ID',4X,'ORDER OF A',4X,'MAX ORDER OF T'/      HLE01750
     1 I12,I14,I18/)                                                   HLE01760
C                                                                      HLE01770
      WRITE(6,40) ISTART,ISTOP                                         HLE01780
   40 FORMAT(/2X,'ISTART',3X,'ISTOP'/2I8/)                             HLE01790
C                                                                      HLE01800
      WRITE(6,50) IHIS,IDIST,IWRITE                                    HLE01810
   50 FORMAT(/4X,'IHIS',3X,'IDIST',2X,'IWRITE'/3I8/)                   HLE01820
C                                                                      HLE01830
      WRITE(6,60) SVSEED,RHSEED                                        HLE01840
   60 FORMAT(/' SEEDS FOR RANDOM NUMBER GENERATOR'//                   HLE01850
     1 4X,'LANCZS SEED',4X,'INVERR SEED'/2I15/)                        HLE01860
C                                                                      HLE01870
      WRITE(6,70) (NMEV(J), J=1,NMEVS)                                 HLE01880
   70 FORMAT(/' SIZES OF T-MATRICES TO BE CONSIDERED'/(6I12))          HLE01890
C                                                                      HLE01900
      WRITE(6,80) RELTOL,GAPTOL,BTOL                                   HLE01910
   80 FORMAT(/' RELATIVE TOLERANCE USED TO COMBINE COMPUTED T-EIGENVALUEHLE01920
     1S'/E15.3/' RELATIVE GAP TOLERANCES USED IN INVERSE ITERATION'/   HLE01930
     1E15.3/' RELATIVE TOLERANCE FOR CHECK ON SIZE OF BETAS'/E15.3/)   HLE01940
C                                                                      HLE01950
      WRITE(6,90) (J,LB(J),UB(J), J=1,NINT)                            HLE01960
   90 FORMAT(/' BISEC WILL BE USED ON THE FOLLOWING INTERVALS'/        HLE01970
```

```
    1 (I6,2E20.6))                                                   HLE01980
C                                                                    HLE01990
      IF (ISTART.EQ.0) GO TO 140                                     HLE02000
C                                                                    HLE02010
C     READ IN ALPHA BETA HISTORY                                     HLE02020
C                                                                    HLE02030
      READ(2,100)MOLD,NOLD,SVSOLD,MATOLD                             HLE02040
  100 FORMAT(2I6,I12,I8)                                             HLE02050
C                                                                    HLE02060
      IF (KMAX.LT.MOLD) KMAX = MOLD                                  HLE02070
      KMAX1 = KMAX + 1                                               HLE02080
C                                                                    HLE02090
C     CHECK THAT ORDER N, MATRIX ID MATNO, AND RANDOM SEED SVSEED    HLE02100
C     AGREE WITH THOSE IN THE HISTORY FILE.  IF NOT PROCEDURE STOPS. HLE02110
C                                                                    HLE02120
      ITEMP = (NOLD-N)**2+(MATNO-MATOLD)**2+(SVSEED-SVSOLD)**2        HLE02130
C                                                                    HLE02140
      IF (ITEMP.EQ.0) GO TO 120                                      HLE02150
C                                                                    HLE02160
      WRITE(6,110)                                                   HLE02170
  110 FORMAT(' PROGRAM TERMINATES'/    ' READ FROM FILE 2 CORRESPONDS TOHLE02180
     1 DIFFERENT MATRIX THAN MATRIX SPECIFIED'/)                     HLE02190
      GO TO 640                                                      HLE02200
C                                                                    HLE02210
  120 CONTINUE                                                       HLE02220
      MOLD1 = MOLD+1                                                 HLE02230
C                                                                    HLE02240
      READ(2,130)(ALPHA(J), J=1,MOLD)                                HLE02250
      READ(2,130)(BETA(J), J=1,MOLD1)                                HLE02260
  130 FORMAT(4Z20)                                                   HLE02270
C                                                                    HLE02280
      IF (KMAX.EQ.MOLD) GO TO 160                                    HLE02290
C                                                                    HLE02300
      READ(2,130)(V1(J), J=1,N)                                      HLE02310
      READ(2,130)(V2(J), J=1,N)                                      HLE02320
C                                                                    HLE02330
  140 CONTINUE                                                       HLE02340
      IIX = SVSEED                                                   HLE02350
C                                                                    HLE02360
C--------------------------------------------------------------------HLE02370
C                                                                    HLE02380
      CALL LANCZS(CMATV,V1,V2,VS,ALPHA,BETA,GR,GC,G,KMAX,MOLD1,N,IIX) HLE02390
C                                                                    HLE02400
C--------------------------------------------------------------------HLE02410
C                                                                    HLE02420
      KMAX1 = KMAX + 1                                               HLE02430
C                                                                    HLE02440
      IF (IHIS.EQ.0.AND.ISTOP.GT.0) GO TO 160                        HLE02450
C                                                                    HLE02460
      WRITE(1,150) KMAX,N,SVSEED,MATNO                               HLE02470
  150 FORMAT(2I6,I12,I8,' = KMAX,N,SVSEED,MATNO')                     HLE02480
C                                                                    HLE02490
C     TO AVOID PERTURBATIONS CAUSED BY HEX TO DECIMAL AND DECIMAL TO HEXHLE02500
C     CONVERSIONS, THE ALPHA AND BETA MUST BE WRITTEN OUT IN HEX.    HLE02510
      WRITE(1,130)(ALPHA(I), I=1,KMAX)                               HLE02520
      WRITE(1,130)(BETA(I), I=1,KMAX1)                               HLE02530
C                                                                    HLE02540
      WRITE(1,130)(V1(I), I=1,N)                                     HLE02550
      WRITE(1,130)(V2(I), I=1,N)                                     HLE02560
C                                                                    HLE02570
      IF (ISTOP.EQ.0) GO TO 540                                      HLE02580
C                                                                    HLE02590
  160 CONTINUE                                                       HLE02600
      BKMIN = BTOL                                                   HLE02610
      WRITE(6,170)                                                   HLE02620
  170 FORMAT(/' T-MATRICES (ALPHA AND BETA) ARE NOW AVAILABLE'/)     HLE02630
C                                                                    HLE02640
```

```
C----------------------------------------------------------------------HLE02650
C      SUBROUTINE TNORM CHECKS MIN(BETA)/(ESTIMATED NORM(A)) > BTOL .   HLE02660
C      IF THIS IS VIOLATED IB IS SET EQUAL TO THE NEGATIVE OF THE INDEX HLE02670
C      OF THE MINIMAL BETA.  IF(IB < 0) THEN SUBROUTINE TNORM IS        HLE02680
C      CALLED FOR EACH VALUE OF MEV TO DETERMINE WHETHER OR NOT THERE   HLE02690
C      IS A BETA IN THE T-MATRIX SPECIFIED THAT VIOLATES THIS TEST.     HLE02700
C      IF THERE IS SUCH A BETA THE PROGRAM TERMINATES FOR THE USER      HLE02710
C      TO DECIDE WHAT TO DO.  THIS TEST CAN BE OVER-RIDDEN BY           HLE02720
C      SIMPLY MAKING BTOL SMALLER, BUT THEN THERE IS THE POSSIBILITY    HLE02730
C      THAT LOSSES IN THE LOCAL ORTHOGONALITY MAY HURT THE COMPUTATIONS.HLE02740
C      BTOL = 1.D-8 IS HOWEVER A CONSERVATIVE CHOICE FOR BTOL.          HLE02750
C                                                                       HLE02760
C      TNORM ALSO COMPUTES TKMAX = MAX(|ALPHA(K)|,BETA(K), K=1,KMAX).   HLE02770
C      TKMAX IS USED TO SCALE THE TOLERANCES USED IN THE                HLE02780
C      T-MULTIPLICITY AND SPURIOUS TESTS IN BISEC. TKMAX IS ALSO USED INHLE02790
C      THE PROJECTION TEST FOR HIDDEN EIGENVALUES THAT HAD 'TOO SMALL'  HLE02800
C      A PROJECTION ON THE STARTING VECTOR.                             HLE02810
C                                                                       HLE02820
C      CALL TNORM(ALPHA,BETA,BKMIN,TKMAX,KMAX,IB)                       HLE02830
C                                                                       HLE02840
C----------------------------------------------------------------------HLE02850
C                                                                       HLE02860
C      TTOL = EPSM*TKMAX                                                HLE02870
C                                                                       HLE02880
C      LOOP ON THE SIZE OF THE T-MATRIX                                 HLE02890
C                                                                       HLE02900
  180 CONTINUE                                                          HLE02910
      MMB = MMB + 1                                                     HLE02920
      MEV = NMEV(MMB)                                                   HLE02930
C      IS MEV TOO LARGE ?                                               HLE02940
      IF(MEV.LE.KMAX) GO TO 200                                         HLE02950
      WRITE(6,190) MMB, MEV, KMAX                                       HLE02960
  190 FORMAT(/' TERMINATE PRIOR TO CONSIDERING THE',I6,'TH T-MATRIX'/   HLE02970
     1' BECAUSE THE SIZE REQUESTED',I6,' IS GREATER THAN THE MAXIMUM SIZHLE02980
     1E ALLOWED',I6/)                                                   HLE02990
      GO TO 540                                                         HLE03000
C                                                                       HLE03010
  200 MP1 = MEV + 1                                                     HLE03020
      BETAM = BETA(MP1)                                                 HLE03030
C                                                                       HLE03040
      IF (IB.GE.0) GO TO 210                                            HLE03050
C                                                                       HLE03060
      TO = BTOL                                                         HLE03070
C                                                                       HLE03080
C----------------------------------------------------------------------HLE03090
C                                                                       HLE03100
      CALL TNORM(ALPHA,BETA,TO,T1,MEV,IBMEV)                            HLE03110
C                                                                       HLE03120
C----------------------------------------------------------------------HLE03130
C                                                                       HLE03140
      TEMP = TO/TKMAX                                                   HLE03150
      IBMEV = IABS(IBMEV)                                               HLE03160
      IF (TEMP.GE.BTOL) GO TO 210                                       HLE03170
      IBMEV = -IBMEV                                                    HLE03180
      GO TO 600                                                         HLE03190
C                                                                       HLE03200
  210 CONTINUE                                                          HLE03210
      IC = MXSTUR-ICT                                                   HLE03220
C                                                                       HLE03230
C----------------------------------------------------------------------HLE03240
C      BISEC LOOP. THE SUBROUTINE BISEC INCORPORATES DIRECTLY THE       HLE03250
C      T-MULTIPLICITY AND SPURIOUS TESTS. T-EIGENVALUES WILL BE         HLE03260
C      CALCULATED BY BISEC SEQUENTIALLY ON INTERVALS                    HLE03270
C      (LB(J),UB(J)), J = 1,NINT).                                      HLE03280
C                                                                       HLE03290
C      ON RETURN FROM BISEC                                             HLE03300
C      NDIS = NUMBER OF DISTINCT EIGENVALUES OF T(1,MEV) ON UNION       HLE03310
```

```
C            OF THE (LB,UB) INTERVALS                              HLE03320
C    VS = DISTINCT T-EIGENVALUES IN ALGEBRAICALLY INCREASING ORDER HLE03330
C    MP = MULTIPLICITIES OF THE T-EIGENVALUES IN VS                HLE03340
C    MP(I) = (0,1,MI), MI>1, I=1,NDIS  MEANS:                      HLE03350
C       (0)  VS(I) IS SPURIOUS                                     HLE03360
C       (1)  VS(I) IS T-SIMPLE AND GOOD                            HLE03370
C       (MI) VS(I) IS MULTIPLE AND IS THEREFORE NOT ONLY GOOD BUT  HLE03380
C            ALSO A CONVERGED GOOD T-EIGENVALUE.                   HLE03390
C    WITHIN BISEC V1 AND V2 ARE DEFINED AS DOUBLE PRECISION ARRAYS HLE03400
C                                                                  HLE03410
C                                                                  HLE03420
      CALL BISEC(ALPHA,BETA,V1,V2,VS,LB,UB,EPSM,TTOL,MP,NINT,       HLE03430
     1 MEV,NDIS,IC,IWRITE)                                         HLE03440
C                                                                  HLE03450
C-----------------------------------------------------------------HLE03460
C                                                                  HLE03470
      IF (NDIS.EQ.0) GO TO 620                                     HLE03480
C                                                                  HLE03490
C    COMPUTE THE TOTAL NUMBER OF STURM SEQUENCES USED TO DATE      HLE03500
C    COMPUTE THE BISEC CONVERGENCE AND T-MULTIPLICITY TOLERANCES USED. HLE03510
C    COMPUTE THE CONVERGENCE TOLERANCE FOR EIGENVALUES OF A.       HLE03520
      ICT = ICT + IC                                               HLE03530
      TEMP = DFLOAT(MEV+1000)                                      HLE03540
      MULTOL = TEMP*TTOL                                           HLE03550
      TEMP = DSQRT(TEMP)                                           HLE03560
      BISTOL = TTOL*TEMP                                           HLE03570
      CONTOL = BETAM*1.D-10                                        HLE03580
C                                                                  HLE03590
C-----------------------------------------------------------------HLE03600
C    SUBROUTINE LUMP 'COMBINES' T-EIGENVALUES THAT ARE 'TOO CLOSE'. HLE03610
C    NOTE HOWEVER THAT CLOSE SPURIOUS T-EIGENVALUES ARE NOT AVERAGED HLE03620
C    WITH GOOD ONES. HOWEVER, THEY MAY BE USED TO INCREASE THE     HLE03630
C    MULTIPLICITY OF A GOOD T-EIGENVALUE.                          HLE03640
C                                                                  HLE03650
      LOOP = NDIS                                                  HLE03660
      CALL LUMP(VS,RELTOL,MULTOL,SCALE2,MP,LOOP)                   HLE03670
C                                                                  HLE03680
C-----------------------------------------------------------------HLE03690
C                                                                  HLE03700
      IF(NDIS.EQ.LOOP) GO TO 230                                   HLE03710
C                                                                  HLE03720
      WRITE(6,220) NDIS, MEV, LOOP                                 HLE03730
  220 FORMAT(/I6,' DISTINCT T-EIGENVALUES WERE COMPUTED IN BISEC AT MEV HLE03740
     1',I6/ 2X,' LUMP SUBROUTINE REDUCES NUMBER OF DISTINCT EIGENVALUES HLE03750
     1TO',I6)                                                      HLE03760
C                                                                  HLE03770
  230 CONTINUE                                                     HLE03780
      NDIS = LOOP                                                  HLE03790
      BETA(MP1) = BETAM                                            HLE03800
C                                                                  HLE03810
C-----------------------------------------------------------------HLE03820
C    THE SUBROUTINE ISOEV LABELS THOSE SIMPLE EIGENVALUES OF T(1,MEV) HLE03830
C    WITH VERY SMALL GAPS BETWEEN NEIGHBORING EIGENVALUES OF T(1,MEV) HLE03840
C    TO AVOID COMPUTING ERROR ESTIMATES FOR ANY SIMPLE GOOD        HLE03850
C    T-EIGENVALUE THAT IS TOO CLOSE TO A SPURIOUS T-EIGENVALUE.    HLE03860
C    ON RETURN FROM ISOEV, G CONTAINS CODED MINIMAL GAPS           HLE03870
C    BETWEEN THE DISTINCT EIGENVALUES OF T(1,MEV). (G IS REAL).    HLE03880
C    G(I) < 0 MEANS MINGAP IS DUE TO LEFT GAP G(I) > 0 MEANS DUE TO HLE03890
C    RIGHT GAP. MP(I) = -1 MEANS THAT THE GOOD T-EIGENVALUE IS SIMPLE HLE03900
C    AND HAS A VERY SMALL MINGAP IN T(1,MEV) DUE TO A SPURIOUS     HLE03910
C    T-EIGENVALUE.   NG = NUMBER OF GOOD EIGENVALUES.              HLE03920
C    NISO = NUMBER OF ISOLATED GOOD T-EIGENVALUES.                 HLE03930
C                                                                  HLE03940
      CALL ISOEV(VS,GAPTOL,MULTOL,SCALE1,G,MP,NDIS,NG,NISO)        HLE03950
C                                                                  HLE03960
C-----------------------------------------------------------------HLE03970
C                                                                  HLE03980
```

```
      WRITE(6,240)NG,NISO,NDIS                                       HLE03990
  240 FORMAT(/I6,' GOOD T-EIGENVALUES HAVE BEEN COMPUTED'/           HLE04000
     1 I6,' OF THESE ARE T-ISOLATED'/                                HLE04010
     2 I6,' = NUMBER OF DISTINCT T-EIGENVALUES COMPUTED'/)           HLE04020
C                                                                    HLE04030
C     DO WE WRITE DISTINCT EIGENVALUES OF T-MATRIX TO FILE 4?        HLE04040
      IF (IDIST.EQ.0) GO TO 280                                      HLE04050
C                                                                    HLE04060
      WRITE(11,250) NDIS,NISO,MEV,N,SVSEED,MATNO                     HLE04070
  250 FORMAT(/4I6,I12,I8,' = NDIS,NISO,MEV,N,SVSEED,MATNO'/)         HLE04080
C                                                                    HLE04090
      WRITE(11,260) (MP(I),VS(I),G(I), I=1,NDIS)                     HLE04100
  260 FORMAT(2(I3,E25.16,E12.3))                                     HLE04110
C                                                                    HLE04120
      WRITE(11,270) NDIS, (MP(I), I=1,NDIS)                          HLE04130
  270 FORMAT(/I6,' = NDIS, T-MULTIPLICITIES (0 MEANS  SPURIOUS)'/(20I4))HLE04140
C                                                                    HLE04150
  280 CONTINUE                                                       HLE04160
C                                                                    HLE04170
      IF (NISO.NE.0) GO TO 310                                       HLE04180
C                                                                    HLE04190
      WRITE(4,290) MEV                                               HLE04200
  290 FORMAT(/' AT MEV = ',I6,' THERE ARE NO ISOLATED T-EIGENVALUES'/ HLE04210
     1' SO NO ERROR ESTIMATES WERE COMPUTED/')                       HLE04220
C                                                                    HLE04230
      WRITE(6,300)                                                   HLE04240
  300 FORMAT(/' ALL COMPUTED GOOD T-EIGENVALUES ARE MULTIPLE'/       HLE04250
     1 ' THEREFORE ALL SUCH EIGENVALUES ARE ASSUMED TO HAVE CONVERGED') HLE04260
C                                                                    HLE04270
      ICONV = 1                                                      HLE04280
      GO TO 350                                                      HLE04290
C                                                                    HLE04300
  310 CONTINUE                                                       HLE04310
C                                                                    HLE04320
C-------------------------------------------------------------------HLE04330
C     SUBROUTINE INVERR COMPUTES ERROR ESTIMATES FOR ISOLATED GOOD  HLE04340
C     T-EIGENVALUES USING INVERSE ITERATION ON T(1,MEV). ON RETURN  HLE04350
C     G(J) = MINIMUM GAP IN T(1,MEV) FOR EACH VS(J), J=1,NDIS       HLE04360
C     G(MEV+1) = BETAM*|U(MEV)| = ERROR ESTIMATE FOR ISOLATED GOOD  HLE04370
C          T-EIGENVALUES, WHERE I = 1, NISO AND  BETAM = BETA(MEV+1)HLE04380
C          U(MEV) IS MEVTH COMPONENT OF THE UNIT EIGENVECTOR OF T   HLE04390
C          CORRESPONDING TO THE ITH ISOLATED GOOD T-EIGENVALUE.     HLE04400
C     A NEGATIVE ERROR ESTIMATE MEANS THAT FOR THAT PARTICULAR      HLE04410
C     EIGENVALUE THE INVERSE ITERATION DID NOT CONVERGE IN <= MXINIT HLE04420
C     STEPS AND THAT THE CORRESPONDING ERROR ESTIMATE IS QUESTIONABLE. HLE04430
C                                                                    HLE04440
C     V2 CONTAINS THE ISOLATED GOOD T-EIGENVALUES                   HLE04450
C     V1 CONTAINS THE MINGAPS TO THE NEAREST DISTINCT  EIGENVALUE   HLE04460
C        OF T(1,MEV) FOR EACH ISOLATED GOOD EIGENVALUE IN V2.       HLE04470
C     VS CONTAINS THE NDIS DISTINCT EIGENVALUES OF T(1,MEV)         HLE04480
C     MP CONTAINS THE CORRESPONDING CODED T-MULTIPLICITIES          HLE04490
C     WITHIN INVERR V1 AND V2 ARE DOUBLE PRECISION ARRAYS           HLE04500
C                                                                    HLE04510
      IT = MXINIT                                                    HLE04520
      CALL INVERR(ALPHA,BETA,V1,V2,VS,EPSM,G,MP,MEV,MMB,NDIS,NISO,N, HLE04530
     1 RHSEED,IT,IWRITE)                                            HLE04540
C                                                                    HLE04550
C-------------------------------------------------------------------HLE04560
C                                                                    HLE04570
C     SIMPLE CHECK FOR CONVERGENCE. CHECKS TO SEE IF ALL OF THE ERROR HLE04580
C     ESTIMATES ARE SMALLER THAN CONTOL.                            HLE04590
C     IF THIS TEST IS SATISFIED, THEN CONVERGENCE FLAG, ICONV IS SET HLE04600
C     TO 1.  TYPICALLY ERROR ESTIMATES ARE VERY CONSERVATIVE.       HLE04610
C                                                                    HLE04620
      WRITE(6,320) CONTOL                                            HLE04630
  320 FORMAT(/' CONVERGENCE IS TESTED USING THE CONVERGENCE TOLERANCE', HLE04640
```

```
      1E13.4/)                                                       HLE04650
C                                                                    HLE04660
      II = MEV +1                                                    HLE04670
      IF = MEV+NISO                                                  HLE04680
      DO 330 I = II,IF                                               HLE04690
      IF (ABS(G(I)).GT.CONTOL) GO TO 350                             HLE04700
  330 CONTINUE                                                       HLE04710
      ICONV = 1                                                      HLE04720
      MMB = NMEVS                                                    HLE04730
C                                                                    HLE04740
      WRITE(6,340) CONTOL                                            HLE04750
  340 FORMAT(' ALL COMPUTED ERROR ESTIMATES WERE LESS THAN',E15.4/   HLE04760
     1 ' THEREFORE PROCEDURE TERMINATES'/)                           HLE04770
C                                                                    HLE04780
  350 CONTINUE                                                       HLE04790
C                                                                    HLE04800
C     IF CONVERGENCE IS INDICATED, THAT IS ICONV = 1 ,THEN           HLE04810
C     THE SUBROUTINE PRTEST IS CALLED TO CHECK FOR ANY CONVERGED     HLE04820
C     EIGENVALUES THAT HAVE BEEN MISLABELLED AS SPURIOUS BECAUSE     HLE04830
C     THE PROJECTION OF THEIR EIGENVECTOR(S) ON THE STARTING         HLE04840
C     VECTOR WERE TOO SMALL.                                         HLE04850
C     NUMERICAL TESTS INDICATE THAT SUCH EIGENVALUES ARE RARE.       HLE04860
C     IF FOR SOME REASON MANY OF THESE HIDDEN EIGENVALUES APPEAR     HLE04870
C     ON SOME RUN, YOU CAN BE CERTAIN THAT SOMETHING IS FOULED UP.   HLE04880
C                                                                    HLE04890
      IF (ICONV.EQ.0) GO TO 480                                      HLE04900
C                                                                    HLE04910
C--------------------------------------------------------------------HLE04920
C                                                                    HLE04930
      CALL PRTEST (ALPHA,BETA,VS,TKMAX,EPSM,RELTOL,SCALE3,SCALE4,     HLE04940
     1 MP,NDIS,MEV,IPROJ)                                            HLE04950
C                                                                    HLE04960
C--------------------------------------------------------------------HLE04970
C                                                                    HLE04980
      IF(IPROJ.EQ.0) GO TO 470                                       HLE04990
C                                                                    HLE05000
      IF(IDIST.EQ.1)  WRITE(11,360) IPROJ                            HLE05010
  360 FORMAT(' SUBROUTINE PRTEST WANTS TO RELABEL',I6,' SPURIOUS EIGENVAHLE05020
     1LUES'/' WE ACCEPT RELABELLING ONLY IF LAST COMPONENT OF T-EIGENVECHLE05030
     1TOR IS L.T. 1.D-10'/)                                          HLE05040
C                                                                    HLE05050
      IIX = RHSEED                                                   HLE05060
C                                                                    HLE05070
C--------------------------------------------------------------------HLE05080
C                                                                    HLE05090
      CALL GENRAN(IIX,G,MEV)                                         HLE05100
C                                                                    HLE05110
C--------------------------------------------------------------------HLE05120
C                                                                    HLE05130
      ITEN = -10                                                     HLE05140
      NISOM = NISO + MEV                                             HLE05150
      IWRITO = IWRITE                                                HLE05160
      IWRITE = 0                                                     HLE05170
C                                                                    HLE05180
      DO 390 J = 1,NDIS                                              HLE05190
      IF(MP(J).NE.ITEN) GO TO 390                                    HLE05200
      TO = VS(J)                                                     HLE05210
C                                                                    HLE05220
C--------------------------------------------------------------------HLE05230
C                                                                    HLE05240
      IT = MXINIT                                                    HLE05250
      CALL INVERM(ALPHA,BETA,V1,V2,TO,TEMP,T1,EPSM,G,MEV,IT,IWRITE)   HLE05260
C                                                                    HLE05270
C--------------------------------------------------------------------HLE05280
C                                                                    HLE05290
      IF(TEMP.LE.1.D-10) GO TO 380                                   HLE05300
C     ERROR ESTIMATE WAS NOT SMALL REJECT RELABELLING OF THIS EIGENVALUEHLE05310
```

```
      IF(IDIST.EQ.1)  WRITE(11,370) J,TO,TEMP                        HLE05320
  370 FORMAT(/' LAST COMPONENT FOR',I6,'TH EIGENVALUE',E20.12/' IS TOO LHLE05330
     1ARGE = ',E15.6,' SO DO NOT ACCEPT PRTEST RELABELLING'/)        HLE05340
      MP(J) = 0                                                      HLE05350
      IPROJ = IPROJ - 1                                              HLE05360
      GO TO 390                                                      HLE05370
C     RELABELLING ACCEPTED                                          HLE05380
  380 NISOM = NISOM + 1                                              HLE05390
      G(NISOM) = BETAM*TEMP                                          HLE05400
  390 CONTINUE                                                       HLE05410
      IWRITE = IWRITO                                                HLE05420
C                                                                    HLE05430
      IF(IPROJ.EQ.0) GO TO 430                                       HLE05440
      WRITE(6,400) IPROJ                                             HLE05450
  400 FORMAT(/I6,' T-EIGENVALUES WERE RECLASSIFIED AS GOOD.'/        HLE05460
     1' THESE ARE IDENTIFIED IN FILE 3 BY A T-MULTIPLICITY OF -10'/' USEHLE05470
     2R SHOULD INSPECT EACH TO MAKE SURE NEIGHBORS HAVE CONVERGED'/) HLE05480
C                                                                    HLE05490
      IF(IDIST.EQ.1)  WRITE(11,410) IPROJ                            HLE05500
  410 FORMAT(/I6,' T-EIGENVALUES WERE RELABELLED AS GOOD'/           HLE05510
     1' BELOW IS CORRECTED T-MULTIPLICITY PATTERN'/)                 HLE05520
C                                                                    HLE05530
      WRITE(6,420) NDIS, (MP(I), I=1,NDIS)                           HLE05540
      IF(IDIST.EQ.1)  WRITE(11,420) NDIS, (MP(I), I=1,NDIS)          HLE05550
  420 FORMAT(/I6,' = NDIS, T-MULTIPLICITIES (0 MEANS   SPURIOUS)'/   HLE05560
     1 6X, ' (-10) MEANS SPURIOUS T-EIGENVALUE RELABELLED AS GOOD'/(2014HLE05570
     1))                                                             HLE05580
C                                                                    HLE05590
C     RECALCULATE MINGAPS FOR DISTINCT T(1,MEV) EIGENVALUES.        HLE05600
  430 NM1 = NDIS - 1                                                 HLE05610
      G(NDIS) = VS(NM1)-VS(NDIS)                                     HLE05620
      G(1) = VS(2)-VS(1)                                             HLE05630
C                                                                    HLE05640
      DO 440 J = 2,NM1                                               HLE05650
      TO = VS(J)-VS(J-1)                                             HLE05660
      T1 = VS(J+1)-VS(J)                                             HLE05670
      G(J) = T1                                                      HLE05680
      IF (TO.LT.T1) G(J) = -TO                                       HLE05690
  440 CONTINUE                                                       HLE05700
      IF(IPROJ.EQ.0) GO TO 470                                       HLE05710
C     WRITE TO FILE 4 ERROR ESTIMATES FOR THOSE T-EIGENVALUES RELABELLEDHLE05720
      NGOOD = 0                                                      HLE05730
      DO 450 J = 1,NDIS                                              HLE05740
      IF(MP(J).EQ.0) GO TO 450                                       HLE05750
      NGOOD = NGOOD + 1                                              HLE05760
      IF(MP(J).NE.ITEN) GO TO 450                                    HLE05770
      TO = VS(J)                                                     HLE05780
      NISO = NISO + 1                                                HLE05790
      NISOM = MEV + NISO                                             HLE05800
      WRITE(4,460) NGOOD,TO,G(NISOM),G(J)                            HLE05810
  450 CONTINUE                                                       HLE05820
  460 FORMAT(I10,E25.16,2E14.3)                                      HLE05830
C                                                                    HLE05840
  470 CONTINUE                                                       HLE05850
C                                                                    HLE05860
C     WRITE THE GOOD T-EIGENVALUES TO FILE 3.  FIRST TRANSFER THEM  HLE05870
C     TO V2 AND THEIR T-MULTIPLICITIES TO THE CORRESPONDING POSITIONS HLE05880
C     IN MP AND COMPUTE THE A-MINGAPS, THE MINIMAL GAPS BETWEEN THE HLE05890
C     GOOD T-EIGENVALUES.  THESE GAPS WILL BE PUT IN THE ARRAY G.   HLE05900
C     SINCE G CURRENTLY CONTAINS THE MINIMAL GAPS BETWEEN THE DISTINCT HLE05910
C     EIGENVALUES OF THE T-MATRIX, THESE GAPS WILL FIRST BE         HLE05920
C     TRANSFERRED TO GC.  NOTE THAT GC<0 MEANS THAT THAT MINIMAL GAP HLE05930
C     IN THE T-MATRIX IS DUE TO A SPURIOUS T-EIGENVALUE.           HLE05940
C     ALL THIS INFORMATION IS PRINTED TO FILE 3                     HLE05950
C                                                                    HLE05960
  480 CONTINUE                                                       HLE05970
C                                                                    HLE05980
```

```
      NG = 0                                                      HLE05990
      DO 490 I = 1,NDIS                                           HLE06000
      IF (MP(I).EQ.0) GO TO 490                                   HLE06010
      NG = NG+1                                                   HLE06020
      MP(NG) = MP(I)                                              HLE06030
      GR(NG) = VS(I)                                              HLE06040
      TEMP = G(I)                                                 HLE06050
      TEMP = DABS(TEMP)                                           HLE06060
      J = I+1                                                     HLE06070
      IF (G(I).LT.ZERO) J = I-1                                   HLE06080
      IF (MP(J).EQ.0) TEMP = -TEMP                                HLE06090
      GC(NG) = TEMP                                               HLE06100
  490 CONTINUE                                                    HLE06110
C                                                                 HLE06120
      WRITE(6,500)MEV                                             HLE06130
  500 FORMAT(//' T-EIGENVALUE CALCULATION AT MEV = ',I6,'    IS COMPLETEHLE06140
     1')                                                          HLE06150
C                                                                 HLE06160
C     NG = NUMBER OF COMPUTED DISTINCT GOOD T-EIGENVALUES.  NEXT  HLE06170
C     GENERATE GAPS BETWEEN GOOD T-EIGENVALUES (AMINGAPS) AND PUT THEM  HLE06180
C     IN G.  G(J) < 0 MEANS THE AMINGAP IS DUE TO THE LEFT-HAND GAP.    HLE06190
C                                                                 HLE06200
      NGM1 = NG - 1                                               HLE06210
      G(NG) = GR(NGM1)-GR(NG)                                     HLE06220
      G(1) = GR(2)-GR(1)                                          HLE06230
C                                                                 HLE06240
      DO 510 J = 2,NGM1                                           HLE06250
      T0 = GR(J)-GR(J-1)                                          HLE06260
      T1 = GR(J+1)-GR(J)                                          HLE06270
      G(J) = T1                                                   HLE06280
      IF (T0.LT.T1) G(J) = -T0                                    HLE06290
  510 CONTINUE                                                    HLE06300
C                                                                 HLE06310
C     WRITE GOOD T-EIGENVALUES OUT TO FILE 3.                     HLE06320
C                                                                 HLE06330
      WRITE(3,520)NG,NDIS,MEV,N,SVSEED,MATNO,MULTOL,IB,BTOL       HLE06340
  520 FORMAT(4I6,I12,I8,' = NG,NDIS,MEV,N,SVEED,MATNO'/           HLE06350
     1 E20.12,I6,E13.4,' = MUTOL,INDEX MINIMAL BETA,BTOL'/        HLE06360
     1' EV NO',2X,'TMULT',7X,'GOOD T-EIGENVALUE',7X,'TMINGAP',7X,'AMINGAHLE06370
     1P')                                                         HLE06380
C                                                                 HLE06390
      WRITE(3,530)(I,MP(I),GR(I),GC(I),G(I), I=1,NG)              HLE06400
  530 FORMAT(2I6,E25.16,2E14.3)                                   HLE06410
C                                                                 HLE06420
C     IF CONVERGENCE FLAG ICONV.NE.1 AND NUMBER OF T-MATRICES     HLE06430
C     CONSIDERED TO DATE IS LESS THAN NUMBER ALLOWED, INCREMENT MEV.    HLE06440
C     AND LOOP BACK TO 210 TO REPEAT COMPUTATIONS.  RESTORE BETA(MEV+1).HLE06450
C                                                                 HLE06460
      BETA(MP1) = BETAM                                           HLE06470
C                                                                 HLE06480
      IF (MMB.LT.NMEVS.AND.ICONV.NE.1) GO TO 180                  HLE06490
C                                                                 HLE06500
C     END OF LOOP ON DIFFERENT SIZE T-MATRICES ALLOWED.          HLE06510
C                                                                 HLE06520
  540 CONTINUE                                                    HLE06530
C                                                                 HLE06540
      IF(ISTOP.EQ.0)  WRITE(6,550)                                HLE06550
  550 FORMAT(/' T-MATRICES (ALPHA AND BETA) ARE NOW AVAILABLE, TERMINATEHLE06560
     1')                                                          HLE06570
      IF (IHIS.EQ.1.AND.KMAX.NE.MOLD) WRITE(1,560)                HLE06580
  560 FORMAT(/' ABOVE ARE THE FOLLOWING VECTORS '/                HLE06590
     1 '   ALPHA(I), I = 1,KMAX'/                                 HLE06600
     2 '   BETA(I), I = 1,KMAX+1'/                                HLE06610
     3 ' FINAL TWO LANCZOS VECTORS OF ORDER N FOR I = KMAX,KMAX+1'/HLE06620
     4 ' ALL ENTRIES IN THIS FILE HAVE FORMAT 4Z20 '/            HLE06630
     5 ' ----- END OF FILE 1 NEW ALPHA, BETA HISTORY--------------'///)HLE06640
C                                                                 HLE06650
```

```
      IF (ISTOP.EQ.0) GO TO 640                                     HLE06660
C                                                                   HLE06670
      WRITE(3,570)                                                  HLE06680
  570 FORMAT(/' ABOVE ARE COMPUTED GOOD T-EIGENVALUES'/             HLE06690
     1 ' NG = NUMBER OF GOOD T-EIGENVALUES COMPUTED'/               HLE06700
     2 ' NDIS = NUMBER OF COMPUTED DISTINCT EIGENVALUES OF T(1,MEV)'/ HLE06710
     3 ' N = ORDER OF A,  MATNO = MATRIX IDENT'/                    HLE06720
     4 ' MULTOL = MULTIPLICITY TOLERANCE FOR T-EIGENVALUES IN BISEC'/ HLE06730
     4 ' TMULT IS THE T-MULTIPLICITY OF GOOD T-EIGENVALUE'/         HLE06740
     5 ' TMULT = -1 MEANS SPURIOUS T-EIGENVALUE TOO CLOSE'/         HLE06750
     6 ' DO NOT COMPUTE ERROR ESTIMATES FOR SUCH EIGENVALUES'/      HLE06760
     7 ' AMINGAP = MINIMAL GAP BETWEEN THE COMPUTED A-EIGENVALUES'/ HLE06770
     8 ' AMINGAP .LT. 0. MEANS MINIMAL GAP IS DUE TO LEFT-HAND GAP'/ HLE06780
     9 ' TMINGAP= MINIMAL GAP W.R.T.  DISTINCT EIGENVALUES IN T(1,MEV)'/HLE06790
     1 ' TMINGAP .LT. 0. MEANS MINGAP IS DUE TO SPURIOUS EIGENVALUE'/ HLE06800
     2 ' ----- END OF FILE 3 GOODEIGENVALUES---------------------'///)HLE06810
C                                                                   HLE06820
      IF (IDIST.EQ.1) WRITE(11,580)                                 HLE06830
  580 FORMAT(/' ABOVE ARE THE DISTINCT EIGENVALUES OF T(1,MEV).'/   HLE06840
     2 ' THE FORMAT IS         T-MULTIPLICITY    EIGENVALUE    TMINGAP'/ HLE06850
     3 '             THIS FORMAT IS REPEATED TWICE ON EACH LINE.'/  HLE06860
     4 ' T-MULTIPLICITY = -1 MEANS THAT THE SUBROUTINE ISOEV HAS TAGGED'HLE06870
     5 /'   THIS SIMPLE T-EIGENVALUE AS HAVING A VERY CLOSE SPURIOUS'/ HLE06880
     6 '   T-EIGENVALUE SO THAT NO ERROR ESTIMATE WILL BE COMPUTED'/ HLE06890
     7 ' FOR THAT EIGENVALUE IN SUBROUTINE INVERR.'/                HLE06900
     8 ' TMINGAP .LT. 0, TMINGAP IS DUE TO LEFT GAP .GT. 0, RIGHT GAP.'/HLE06910
     9 ' EACH OF THE DISTINCT T-EIGENVALUE TABLES IS FOLLOWED'/     HLE06920
     9 ' BY THE T-MULTIPLICITY PATTERN.'/                           HLE06930
     1 ' NDIS = NUMBER OF COMPUTED DISTINCT EIGENVALUES OF T(1,MEV).'/ HLE06940
     2 ' NG = NUMBER OF GOOD T-EIGENVALUES. '/                      HLE06950
     3 ' NISO = NUMBER OF ISOLATED GOOD T-EIGENVALUES. '/           HLE06960
     4 ' NISO ALSO IS THE COUNT OF +1 ENTRIES IN T-MULTIPLICITY PATTERN.HLE06970
     5 '/' ----- END OF FILE 4 DISTINCT T-EIGENVALUES----------------'//HLE06980
     6 /)                                                           HLE06990
C                                                                   HLE07000
      IF(NISO.NE.0)  WRITE(4,590)                                   HLE07010
  590 FORMAT(/' ABOVE ARE THE ERROR ESTIMATES OBTAINED FOR THE ISOLATED HLE07020
     1GOOD EIGENVALUES'/                                            HLE07030
     1' OBTAINED VIA INVERSE ITERATION IN THE SUBROUTINE INVERR.'/  HLE07040
     1' ALL OTHER GOOD EIGENVALUES HAVE CONVERGED.'/                HLE07050
     2' ERROR ESTIMATE = BETAM*ABS(UM)'/                            HLE07060
     2' WHERE BETAM = BETA(MEV+1) AND UM = U(MEV).'/                HLE07070
     3' U = UNIT EIGENVECTOR OF T WHERE T*U = EV*U AND EV = ISOLATED GOOHLE07080
     3D EIGENVALUE.'/                                               HLE07090
     4' TMINGAP = GAP TO NEAREST DISTINCT EIGENVALUE OF T(1,MEV).'/ HLE07100
     5' TMINGAP .LT. 0. MEANS MINGAP IS DUE TO LEFT NEIGHBOR.'/     HLE07110
     6' ERROR ESTIMATE L.T. 0 MEANS INVERSE ITERATION DID NOT CONVERGE'/HLE07120
     7' ------ END OF FILE 7 ERRINV -------------------------------'//) HLE07130
      GO TO 640                                                     HLE07140
C                                                                   HLE07150
  600 CONTINUE                                                      HLE07160
C                                                                   HLE07170
      IBB = IABS(IBMEV)                                             HLE07180
      IF (IBMEV.LT.0) WRITE(6,610) MEV,IBB,BETA(IBB)                HLE07190
  610 FORMAT(/' PROGRAM TERMINATES BECAUSE MEV REQUESTED = ',I6,' IS .GTHLE07200
     1',I6/' AT WHICH AN ABNORMALLY SMALL BETA = ' , E13.4,' OCCURRED'/)HLE07210
      GO TO 640                                                     HLE07220
C                                                                   HLE07230
  620 IF (NDIS.EQ.0.AND.ISTOP.GT.0) WRITE(6,630)                    HLE07240
  630 FORMAT(/' INTERVALS SPECIFIED FOR BISECT DID NOT CONTAIN ANY EIGENHLE07250
     1VALUES'/' PROGRAM TERMINATES')                               HLE07260
C                                                                   HLE07270
  640 CONTINUE                                                      HLE07280
C                                                                   HLE07290
      STOP                                                          HLE07300
C-----END OF MAIN PROGRAM FOR LANCZOS HERMITIAN EIGENVALUE COMPUTATIONS-HLE07310
      END                                                           HLE07320
```

SECTION 3.3   MAIN PROGRAM, EIGENVECTOR COMPUTATIONS

```
C-----HLEVEC (EIGENVECTORS OF HERMITIAN MATRICES)----------------------HLE00010
C                                                                       HLE00020
C        CONTAINS MAIN PROGRAM FOR COMPUTING AN EIGENVECTOR CORRESPONDING HLE00030
C        TO EACH OF A SET OF EIGENVALUES THAT HAVE BEEN COMPUTED        HLE00040
C        ACCURATELY BY THE CORRESPONDING LANCZOS EIGENVALUE PROGRAM     HLE00050
C        (HLEVAL) FOR HERMITIAN MATRICES.  THIS PROGRAM COULD BE        HLE00060
C        MODIFIED TO COMPUTE ADDITIONAL EIGENVECTORS FOR ANY           HLE00070
C        MULTIPLE EIGENVALUE OF THE GIVEN A-MATRIX.  THE AMOUNT OF      HLE00080
C        ADDITIONAL COMPUTATION REQUIRED BY SUCH A MODIFICATION WOULD   HLE00090
C        DEPEND UPON THE GIVEN MATRIX AND UPON WHICH PART OF THE        HLE00100
C        SPECTRUM WAS INVOLVED.                                         HLE00110
C                                                                       HLE00120
C        THE LANCZOS EIGENVECTOR COMPUTATIONS ASSUME THAT EACH          HLE00130
C        EIGENVALUE THAT IS BEING CONSIDERED HAS CONVERGED AS AN        HLE00140
C        EIGENVALUE OF THE LANCZOS TRIDIAGONAL MATRICES.                HLE00150
C                                                                       HLE00160
C        PORTABILITY:                                                   HLE00170
C        THIS PROGRAM IS NOT PORTABLE BECAUSE OF THE USE OF COMPLEX*16  HLE00180
C        VARIABLES.  MOREOVER, THE PFORT VERIFIER IDENTIFIED THE        HLE00190
C        FOLLOWING ADDITIONAL NONPORTABLE CONSTRUCTIONS:                HLE00200
C                                                                       HLE00210
C        1.  DATA/MACHEP/ STATEMENT                                     HLE00220
C        2.  ALL READ(5,*) STATEMENTS (FREE FORMAT)                     HLE00230
C        3.  FORMAT(20A4) USED WITH THE EXPLANATORY HEADER, EXPLAN      HLE00240
C        4.  HEXADECIMAL FORMAT (4Z20) USED IN ALPHA/BETA FILES 1 AND 2. HLE00250
C                                                                       HLE00260
C        IMPORTANT NOTE:  PROGRAM ALLOWS ENLARGEMENT OF THE ALPHA, BETA HLE00270
C        ARRAYS.  IN PARTICULAR, IF ANY ONE OF THE EIGENVALUES SUPPLIED HLE00280
C        IS T-SIMPLE AND NOT CLOSE TO A SPURIOUS T-EIGENVALUE, THE PROGRAM HLE00290
C        REQUIRES THAT KMAX BE AT LEAST 11*MEV/8 + 12.  IF KMAX IS NOT  HLE00300
C        THIS LARGE, THEN THE PROGRAM WILL RESET KMAX TO THIS SIZE      HLE00310
C        AND EXTEND THE ALPHA, BETA HISTORY IF REQUIRED.                HLE00320
C        THUS, THE DIMENSIONS OF THE ALPHA AND BETA ARRAYS MUST BE      HLE00330
C        LARGE ENOUGH TO ALLOW FOR THIS POSSIBILITY.                    HLE00340
C        REMEMBER THAT THE BETA ARRAY, BETA(J), IS SUCH THAT            HLE00350
C        J = 1,..., KMAX+1.  SO IF THE KMAX USED BY THE PROGRAM         HLE00360
C        IS TO BE 3000, THEN BETA MUST BE OF LENGTH AT LEAST 3001.      HLE00370
C                                                                       HLE00380
C                                                                       HLE00390
C----------------------------------------------------------------------HLE00400
      COMPLEX*16  V1(1000),V2(1000),VS(1000),RITVEC(10000),ZEROC,TEMPC  HLE00410
      DOUBLE PRECISION  ALPHA(1000),BETA(1001),GR(1000),GC(1000)        HLE00420
      DOUBLE PRECISION  TVEC(20000),GOODEV(50),EVNEW(50)                HLE00430
      DOUBLE PRECISION  EVAL,EVALN,TOLN,TTOL,ERTOL,ALFA,BATA            HLE00440
      DOUBLE PRECISION  MULTOL,SCALEO,STUTOL,BTOL,LB,UB                 HLE00450
      DOUBLE PRECISION  ONE,ZERO,MACHEP,EPSM,TEMP,SUM,ERRMIN,BKMIN      HLE00460
      DOUBLE PRECISION  RELTOL,ERROR,TERROR,TLAST(50)                   HLE00470
      REAL  G(1000),AMINGP(50),TMINGP(50),EXPLAN(20)                    HLE00480
      REAL  TERR(50),ERR(50),ERRDGP(50),RNORM(50),TBETA(50)            HLE00490
      INTEGER  MP(50),M1(50),M2(50),MA(50),ML(50),MINT(50),MFIN(50)     HLE00500
      INTEGER  SVSEED,SVSOLD,RHSEED,IDELTA(50),MULEVA(50)               HLE00510
      INTEGER  MBOUND,NTVCON,SVTVEC,TVSTOP,LVCONT,ERCONT,TFLAG          HLE00520
      DOUBLE PRECISION  DABS, DMAX1, DSQRT, DFLOAT                      HLE00530
      REAL ABS                                                          HLE00540
      INTEGER  IABS                                                     HLE00550
C----------------------------------------------------------------------HLE00560
      EXTERNAL CMATV                                                    HLE00570
      DATA MACHEP/Z3410000000000000/                                    HLE00580
      EPSM = 2.D0*MACHEP                                                HLE00590
C----------------------------------------------------------------------HLE00600
C                                                                       HLE00610
C        ARRAYS MUST BE DIMENSIONED AS FOLLOWS:                         HLE00620
C        1.  ALPHA:  >= KMAXN, BETA:  >= (KMAXN+1) WHERE KMAXN, THE     HLE00630
C                    LARGEST SIZE T-MATRIX CONSIDERED BY THE PROGRAM,   HLE00640
```

```
C                IS THE LARGER OF THE SIZE OF THE ALPHA, BETA HISTORY   HLE00650
C                PROVIDED ON FILE 2 (IF ANY ) AND THE SIZE WHICH THE     HLE00660
C                PROGRAM SPECIFIES INTERNALLY, THIS LATTER IS ALWAYS     HLE00670
C                < = 11*MEV / 8  +  12, WHERE MEV IS THE SIZE           HLE00680
C                T-MATRIX THAT WAS USED IN THE CORRESPONDING EIGENVALUE HLE00690
C                COMPUTATIONS.                                          HLE00700
C       2.  V1:  >= MAX(N,KMAX/2)                                       HLE00710
C       3.  V2, VS:   >= N                                             HLE00720
C       4.  G:  >= MAX(N,KMAX).   GR, GC:  >= N                        HLE00730
C       5.  RITVEC: >= N*NGOOD, WHERE NGOOD IS NUMBER OF EIGENVALUES   HLE00740
C                SUPPLIED TO THIS PROGRAM.                             HLE00750
C       6.  TVEC: >= CUMULATIVE LENGTH OF ALL THE T-EIGENVECTORS       HLE00760
C                NEEDED TO GENERATE THE DESIRED RITZ VECTORS.  AN EDUCATED HLE00770
C                GUESS AT AN APPROPRIATE LENGTH CAN BE OBTAINED BY RUNNING THE HLE00780
C                PROGRAM WITH THE FLAG MBOUND = 1 AND MULTIPLYING THE   HLE00790
C                RESULTING SIZE BY 5/4.                                HLE00800
C       7.  GOODEV, EVNEW, AMINGP, TMINGP, TERR, ERR, ERRGDP, RNORM, TBETA HLE00810
C                TLAST, MP, MA, M1, M2, MINT, MFIN, MULEVA, AND IDELTA ALL HLE00820
C                MUST BE AT LEAST NGOOD.                               HLE00830
C                                                                       HLE00840
C------------------------------------------------------------------------HLE00850
C       OUTPUT HEADER                                                  HLE00860
        WRITE(6,10)                                                    HLE00870
     10 FORMAT(/' LANCZOS EIGENVECTOR PROCEDURE FOR HERMITIAN MATRICES'/) HLE00880
C                                                                       HLE00890
C       SET PROGRAM PARAMETERS                                         HLE00900
C       USER MUST NOT MODIFY SCALEO                                    HLE00910
        SCALEO = 5.0D0                                                 HLE00920
        ZERO = 0.0D0                                                   HLE00930
        ZEROC = DCMPLX(ZERO,ZERO)                                      HLE00940
        ONE = 1.0D0                                                    HLE00950
        MPMIN = -1000                                                  HLE00960
C       CONVERGENCE TOLERANCE FOR T-EIGENVECTORS FOR RITZ VECTORS     HLE00970
        ERTOL = 1.D-10                                                 HLE00980
        ISREAL = 0                                                     HLE00990
C                                                                       HLE01000
C       READ USER-SPECIFIED PARAMETER FROM INPUT FILE 5 (FREE FORMAT) HLE01010
C                                                                       HLE01020
C       READ USER-PROVIDED HEADER FOR RUN                             HLE01030
        READ(5,20) EXPLAN                                             HLE01040
        WRITE(6,20) EXPLAN                                            HLE01050
     20 FORMAT(20A4)                                                  HLE01060
C                                                                       HLE01070
C       READ IN THE MAXIMUM PERMISSIBLE DIMENSIONS FOR THE TVEC ARRAY HLE01080
C       (MDIMTV), FOR THE RITVEC ARRAY (MDIMRV), AND FOR THE BETA     HLE01090
C       ARRAY (MBETA).                                                HLE01100
C                                                                       HLE01110
        READ(5,20) EXPLAN                                             HLE01120
        READ(5,*) MDIMTV, MDIMRV, MBETA                               HLE01130
C                                                                       HLE01140
C       READ IN RELATIVE TOLERANCE (RELTOL) USED IN DETERMINING       HLE01150
C       APPROPRIATE SIZES FOR THE T-MATRICES USED IN THE EIGENVECTOR  HLE01160
C       COMPUTATIONS                                                  HLE01170
C                                                                       HLE01180
        READ(5,20) EXPLAN                                             HLE01190
        READ(5,*) RELTOL                                             HLE01200
C                                                                       HLE01210
C                                                                       HLE01220
C       SET FLAGS TO 0 OR 1:                                          HLE01230
C       MBOUND = 1:  PROGRAM TERMINATES AFTER COMPUTING 1ST GUESSES   HLE01240
C                    ON APPROPRIATE T-SIZES FOR USE IN THE RITZ VECTOR HLE01250
C                    COMPUTATIONS                                     HLE01260
C       NTVCON = 0:  PROGRAM TERMINATES IF THE TVEC ARRAY IS NOT      HLE01270
C                    LARGE ENOUGH TO HOLD ALL THE T-EIGENVECTORS REQUIRED.HLE01280
C       SVTVEC = 0:  THE T-EIGENVECTORS ARE NOT WRITTEN TO FILE 11    HLE01290
C                    UNLESS TVSTOP = 1                                HLE01300
C       SVTVEC = 1:  WRITE THE T-EIGENVECTORS TO FILE 11.             HLE01310
```

```
C     TVSTOP = 1:  PROGRAM TERMINATES AFTER COMPUTING THE            HLE01320
C                  T-EIGENVECTORS                                    HLE01330
C     LVCONT = 0:  PROGRAM TERMINATES IF THE NUMBER OF T-EIGENVECTORS HLE01340
C                  COMPUTED IS NOT EQUAL TO THE NUMBER OF RITZ       HLE01350
C                  VECTORS REQUESTED.                                HLE01360
C     ERCONT = 0:  MEANS FOR ANY GIVEN EIGENVALUE, A RITZ VECTOR     HLE01370
C                  WILL NOT BE COMPUTED FOR THAT EIGENVALUE UNLESS   HLE01380
C                  A T-EIGENVECTOR HAS BEEN IDENTIFIED WITH A LAST   HLE01390
C                  COMPONENT WHICH SATISFIES THE SPECIFIED           HLE01400
C                  CONVERGENCE CRITERION.                            HLE01410
C     ERCONT = 1:  MEANS FOR ANY GIVEN EIGENVALUE, A RITZ VECTOR     HLE01420
C                  WILL BE COMPUTED.  IF A T-EIGENVECTOR CANNOT      HLE01430
C                  BE IDENTIFIED WHICH SATISFIES THE LAST           HLE01440
C                  COMPONENT CRITERION, THEN THE PROGRAM WILL        HLE01450
C                  USE THE T-VECTOR THAT CAME CLOSEST TO            HLE01460
C                  SATISFYING THE CRITERION                          HLE01470
C     IWRITE = 1:  EXTENDED OUTPUT OF INTERMEDIATE COMPUTATIONS      HLE01480
C                  IS WRITTEN TO FILE 6                              HLE01490
C     IREAD = 0:   ALPHA/BETA FILE IS REGENERATED                   HLE01500
C     IREAD = 1:   ALPHA/BETA FILE USED IN EIGENVALUE COMPUTATIONS  HLE01510
C                  IS READ IN AND EXTENDED IF NECESSARY.  IN BOTH   HLE01520
C                  CASES IREAD = 0 OR 1, THE LANCZOS VECTORS ARE    HLE01530
C                  ALWAYS REGENERATED FOR THE RITZ VECTOR           HLE01540
C                  COMPUTATIONS                                      HLE01550
C                                                                    HLE01560
      READ(5,20) EXPLAN                                             HLE01570
      READ(5,*) MBOUND,NTVCON,SVTVEC,IREAD                          HLE01580
C                                                                    HLE01590
      READ(5,20) EXPLAN                                             HLE01600
      READ(5,*) TVSTOP,LVCONT,ERCONT,IWRITE                         HLE01610
      IF (TVSTOP.EQ.1) SVTVEC = 1                                   HLE01620
C                                                                    HLE01630
C     READ IN SEED FOR GENERATING RANDOM STARTING VECTOR FOR THE    HLE01640
C     INVERSE ITERATION ON THE T-MATRICES.                          HLE01650
C                                                                    HLE01660
      READ(5,20) EXPLAN                                             HLE01670
      READ(5,*) RHSEED                                             HLE01680
C                                                                    HLE01690
C     READ IN MATNO = MATRIX/RUN IDENTIFICATION NUMBER AND          HLE01700
C     N = ORDER OF A-MATRIX                                         HLE01710
C                                                                    HLE01720
      READ(5,20) EXPLAN                                             HLE01730
      READ(5,*) MATNO,N                                            HLE01740
C     IF MATNO < 0, THEN MATRIX SUPPLIED IS REAL AND USER WANTS TO  HLE01750
C     CHECK ON THE T-MULTIPLICITY OF THE EIGENVALUES OF GIVEN MATRIX HLE01760
      IF(MATNO.GT.0) GO TO 30                                       HLE01770
      ISREAL = 1                                                   HLE01780
      MATNO = - MATNO                                              HLE01790
   30 CONTINUE                                                     HLE01800
C                                                                    HLE01810
C---------------------------------------------------------------------HLE01820
C     INITIALIZE THE ARRAYS FOR THE USER-SPECIFIED MATRIX           HLE01830
C     AND PASS THE STORAGE LOCATIONS OF THESE ARRAYS TO THE         HLE01840
C     MATRIX-VECTOR MULTIPLY SUBROUTINE CMATV.                      HLE01850
C                                                                    HLE01860
      CALL USPEC(N,MATNO)                                          HLE01870
C                                                                    HLE01880
C---------------------------------------------------------------------HLE01890
C     MASK UNDERFLOW AND OVERFLOW                                   HLE01900
      CALL MASK                                                     HLE01910
C                                                                    HLE01920
C---------------------------------------------------------------------HLE01930
C                                                                    HLE01940
C     WRITE RUN PARAMETERS OUT TO FILE 6                            HLE01950
C                                                                    HLE01960
      WRITE(6,40) MATNO,N                                          HLE01970
```

```
   40 FORMAT(/' MATRIX IDENTIFICATION NO. = ',I10,' ORDER OF A = ',I5)   HLE01980
C                                                                        HLE01990
      WRITE(6,50) MBOUND,NTVCON,SVTVEC,IREAD                             HLE02000
   50 FORMAT(/3X,'MBOUND',3X,'NTVCON',3X,'SVTVEC',3X,'IREAD'/3I9,I8)     HLE02010
C                                                                        HLE02020
      WRITE(6,60) TVSTOP,LVCONT,ERCONT,IWRITE                           HLE02030
   60 FORMAT(/3X,'TVSTOP',3X,'LVCONT',3X,'ERCONT',3X,'IWRITE'/4I9)      HLE02040
C                                                                        HLE02050
      WRITE(6,70) MDIMTV,MDIMRV,MBETA                                    HLE02060
   70 FORMAT(/3X,'MDIMTV',3X,'MDIMRV',3X,'MBETA'/2I9,I8)                HLE02070
C                                                                        HLE02080
      WRITE(6,80) RELTOL,RHSEED                                          HLE02090
   80 FORMAT(/7X,'RELTOL',3X,'RHSEED'/E13.4,I9)                         HLE02100
C                                                                        HLE02110
C                                                                        HLE02120
C     FROM FILE 3 READ IN THE NUMBER OF EIGENVALUES (NGOOD) FOR WHICH    HLE02130
C     EIGENVECTORS ARE REQUESTED, THE ORDER (MEV) OF THE LANCZOS         HLE02140
C     TRIDIAGONAL MATRIX USED IN COMPUTING THESE EIGENVALUES, THE        HLE02150
C     ORDER (NOLD) OF THE USER-SPECIFIED MATRIX USED IN THE EIGENVALUE   HLE02160
C     COMPUTATIONS, THE SEED (SVSEED) USED FOR GENERATING THE STARTING   HLE02170
C     VECTOR THAT WAS USED IN THOSE LANCZOS EIGENVALUE COMPUTATIONS,     HLE02180
C     AND THE MATRIX/RUN IDENTIFICATION NUMBER (MATOLD) USED IN THOSE    HLE02190
C     COMPUTATIONS.  ALSO READ IN THE NUMBER (NDIS) OF DISTINCT          HLE02200
C     EIGENVALUES OF T(1,MEV) THAT WERE COMPUTED BUT THIS VALUE IS       HLE02210
C     NOT USED IN THE EIGENVECTOR COMPUTATIONS.                          HLE02220
C                                                                        HLE02230
      READ(3,90) NGOOD,NDIS,MEV,NOLD,SVSEED,MATOLD                       HLE02240
   90 FORMAT(4I6,I12,I8)                                                HLE02250
C                                                                        HLE02260
C     READ IN THE T-MULTIPLICITY TOLERANCE USED IN THE BISEC SUBROUTINE  HLE02270
C     DURING THE COMPUTATION OF THE GIVEN EIGENVALUES.                   HLE02280
C     ALSO READ IN THE FLAG IB.  IF IB < 0, THEN SOME BETA(I) IN THE     HLE02290
C     T-MATRIX FILE PROVIDED ON FILE 2 FAILED THE ORTHOGONALITY          HLE02300
C     TEST IN THE TNORM SUBROUTINE.  USER SHOULD NOTE THAT THIS          HLE02310
C     VECTOR PROGRAM PROCEEDS INDEPENDENTLY OF THE SIZE OF THE BETA.     HLE02320
C                                                                        HLE02330
      READ(3,100) MULTOL,IB,BTOL                                         HLE02340
  100 FORMAT(E20.12,I6,E13.4)                                           HLE02350
C                                                                        HLE02360
      TEMP = DFLOAT(MEV+1000)                                            HLE02370
      TTOL = MULTOL/TEMP                                                 HLE02380
      WRITE(6,110) MULTOL,TTOL                                           HLE02390
  110 FORMAT(/' T-MULTIPLICITY TOLERANCE USED IN THE EIGENVALUE COMPUTAT HLE02400
     1IONS WAS',E13.4/' SCALED MACHINE EPSILON IS',E13.4)              HLE02410
C                                                                        HLE02420
C     CONTINUE WRITE TO FILE 6 OF THE PARAMETERS FOR THIS RUN            HLE02430
C                                                                        HLE02440
      WRITE(6,120)NGOOD,NDIS,MEV,NOLD,MATOLD,SVSEED,MULTOL,IB,BTOL       HLE02450
  120 FORMAT(/' EIGENVALUES SUPPLIED ARE READ IN FROM FILE 3'/' FILE 3   HLE02460
     1HEADER IS'/4X,'NG',2X,'NDIS',3X,'MEV',2X,'NOLD',2X,'MATOLD',4X,    HLE02470
     1'SVSEED',6X,'MULTOL',6X,'IB',9X,'BTOL'/4I6,I8,I10,E12.3,I8,E13.4/)HLE02480
C                                                                        HLE02490
C     IS THE ARRAY RITVEC LONG ENOUGH TO HOLD ALL OF THE DESIRED         HLE02500
C     RITZ VECTORS (APPROXIMATE EIGENVECTORS)?                           HLE02510
      NMAX = NGOOD*N                                                     HLE02520
      IF(MBOUND.NE.0) GO TO 130                                          HLE02530
      IF(TVSTOP.NE.1.AND.NMAX.GT.MDIMRV) GO TO 1390                      HLE02540
C                                                                        HLE02550
C     CHECK THAT THE ORDER N AND THE MATRIX IDENTIFICATION NUMBER        HLE02560
C     MATNO SPECIFIED BY THE USER AGREE WITH THOSE READ IN FROM          HLE02570
C     FILE 3.                                                            HLE02580
  130 ITEMP = (NOLD-N)**2+(MATOLD-MATNO)**2                              HLE02590
      IF (ITEMP.NE.0) GO TO 1410                                         HLE02600
C                                                                        HLE02610
C                                                                        HLE02620
C     THE LANCZOS EIGENVECTOR COMPUTATIONS ASSUME THAT EACH              HLE02630
C     EIGENVALUE THAT IS BEING CONSIDERED HAS CONVERGED AS AN            HLE02640
```

```
C     EIGENVALUE OF THE LANCZOS TRIDIAGONAL MATRICES.              HLE02650
C                                                                   HLE02660
C     READ IN FROM FILE 3, THE T-MULTIPLICITIES OF THE EIGENVALUES  HLE02670
C     WHOSE EIGENVECTORS ARE TO BE COMPUTED, THE VALUES OF THESE    HLE02680
C     EIGENVALUES AND THEIR MINIMAL GAPS AS EIGENVALUES OF THE      HLE02690
C     USER-SPECIFIED MATRIX AND AS EIGENVALUES OF THE T-MATRIX.     HLE02700
C                                                                   HLE02710
      READ(3,20) EXPLAN                                            HLE02720
      READ(3,140) (MP(J),GOODEV(J),TMINGP(J),AMINGP(J), J=1,NGOOD) HLE02730
  140 FORMAT(6X,I6,E25.16,2E14.3)                                  HLE02740
C                                                                   HLE02750
      WRITE(6,150) (J,GOODEV(J),MP(J),TMINGP(J),AMINGP(J), J=1,NGOOD) HLE02760
  150 FORMAT(/' EIGENVALUES READ IN, T-MULTIPLICITIES, T-GAPS AND A-GAPSHLE02770
     1 '/4X,' J ',5X,'GOOD EIGENVALUE',5X,'MULT',4X,' TMINGAP ',4X, HLE02780
     1' AMINGAP '/(I6,E25.16,I4,2E15.4))                           HLE02790
C                                                                   HLE02800
C     READ IN ERROR ESTIMATES                                      HLE02810
      WRITE(6,180) MEV,SVSEED                                      HLE02820
C     CHECK WHETHER OR NOT THERE ARE ANY ISOLATED T-EIGENVALUES IN HLE02830
C     THE EIGENVALUES PROVIDED                                     HLE02840
      DO 160 J=1,NGOOD                                             HLE02850
      IF(MP(J).EQ.1) GO TO 170                                     HLE02860
  160 CONTINUE                                                     HLE02870
      GO TO 200                                                    HLE02880
  170 READ(4,20) EXPLAN                                            HLE02890
      READ(4,20) EXPLAN                                            HLE02900
      READ(4,20) EXPLAN                                            HLE02910
  180 FORMAT(/' THESE EIGENVALUES WERE COMPUTED USING A T-MATRIX OF HLE02920
     1ORDER ',I5/' AND SEED FOR RANDOM NUMBER GENERATOR =',I12)    HLE02930
      READ(4,190) NISO                                            HLE02940
  190 FORMAT(18X,I6)                                               HLE02950
      READ(4,20) EXPLAN                                            HLE02960
      READ(4,20) EXPLAN                                            HLE02970
      READ(4,20) EXPLAN                                            HLE02980
  200 DO 230 J=1,NGOOD                                             HLE02990
      ERR(J) = 0.D0                                               HLE03000
      IF(MP(J).NE.1) GO TO 230                                     HLE03010
      READ(4,210) EVAL, ERR(J)                                     HLE03020
  210 FORMAT(10X,E25.16,E14.3)                                     HLE03030
      IF(DABS(EVAL - GOODEV(J)).LT.1.D-10) GO TO 230               HLE03040
      WRITE(6,220) EVAL,GOODEV(J)                                  HLE03050
  220 FORMAT(' PROBLEM WITH READ IN OF ERROR ESTIMATES'/' EIGENVALUE REAHLE03060
     1D IN',E20.12,' DOES NOT MATCH GOODEV(J) ='/E20.12)           HLE03070
      GO TO 1630                                                   HLE03080
C                                                                   HLE03090
  230 CONTINUE                                                     HLE03100
C                                                                   HLE03110
      WRITE(6,240) (J,GOODEV(J),ERR(J), J=1,NGOOD)                 HLE03120
  240 FORMAT(' ERROR ESTIMATES ='/4X,' J',5X,'EIGENVALUE',10X,'ESTIMATE'HLE03130
     1 /(I6,E20.12,E14.3))                                         HLE03140
C                                                                   HLE03150
      IF(IREAD.EQ.0) GO TO 340                                     HLE03160
C                                                                   HLE03170
C     READ IN THE SIZE OF THE T-MATRIX PROVIDED ON FILE 2.  READ IN HLE03180
C     THE ORDER OF THE USER-SPECIFIED MATRIX , THE SEED FOR THE    HLE03190
C     RANDOM NUMBER GENERATOR, AND THE MATRIX/TEST IDENTIFICATION   HLE03200
C     NUMBER THAT WERE USED IN THE LANCZOS EIGENVALUE COMPUTATIONS. HLE03210
C     IF FLAG IREAD = 0, REGENERATE HISTORY.  HISTORY MUST BE       HLE03220
C     STORED IN HEXADECIMAL FORMAT TO AVOID ERRORS INCURRED IN      HLE03230
C     INPUT/OUTPUT CONVERSIONS.                                    HLE03240
C                                                                   HLE03250
      READ(2,250) KMAX,NOLD,SVSOLD,MATOLD                          HLE03260
  250 FORMAT(2I6,I12,I8)                                           HLE03270
C                                                                   HLE03280
      WRITE(6,260) KMAX,NOLD,SVSOLD,MATOLD                         HLE03290
  260 FORMAT(/' READ IN THE T-MATRICES STORED ON FILE 2'/' FILE 2 HEADERHLE03300
```

```
      1 IS'/2X,'KMAX',2X,'NOLD',6X,'SVSOLD',2X,'MATOLD'/2I6,I12,I8/)          HLE03310
C                                                                             HLE03320
C     CHECK THAT THE ORDER, THE MATRIX/TEST IDENTIFICATION NUMBER             HLE03330
C     AND THE SEED FOR THE RANDOM NUMBER GENERATOR USED IN THE                HLE03340
C     LANCZOS COMPUTATIONS THAT GENERATED THE HISTORY FILE                    HLE03350
C     BEING USED AGREE WITH WHAT THE USER HAS SPECIFIED.                      HLE03360
      IF (NOLD.NE.N.OR.MATOLD.NE.MATNO.OR.SVSOLD.NE.SVSEED) GO TO 1430        HLE03370
C                                                                             HLE03380
      KMAX1 = KMAX + 1                                                        HLE03390
C                                                                             HLE03400
C     READ IN THE T-MATRICES FROM FILE 2.  THESE ARE USED TO GENERATE         HLE03410
C     THE T-EIGENVECTORS THAT WILL BE USED IN THE RITZ VECTOR                 HLE03420
C     COMPUTATIONS.  ALPHA/BETA HISTORY MUST BE STORED IN                     HLE03430
C     MACHINE FORMAT, ((4Z20) FOR IBM/3081).                                  HLE03440
C                                                                             HLE03450
      READ(2,270) (ALPHA(J), J=1,KMAX)                                        HLE03460
      READ(2,270) (BETA(J), J=1,KMAX1)                                        HLE03470
  270 FORMAT(4Z20)                                                            HLE03480
C                                                                             HLE03490
      READ(2,270) (V1(J), J=1,N)                                              HLE03500
      READ(2,270) (V2(J), J=1,N)                                              HLE03510
C                                                                             HLE03520
C     ENLARGE KMAX IF THE SIZE AT WHICH THE EIGENVALUE                        HLE03530
C     COMPUTATIONS WERE PERFORMED IS ESSENTIALLY KMAX AND                     HLE03540
C     THERE IS AT LEAST ONE EIGENVALUE THAT IS T-SIMPLE AND                   HLE03550
C     T-ISOLATED, IN THE SENSE THAT IF ITS NEAREST T-NEIGHBOR IS TOO          HLE03560
C     CLOSE THAT NEIGHBOR IS A 'GOOD' T-EIGENVALUE.                           HLE03570
      DO 280 J = 1,NGOOD                                                      HLE03580
      IF(MP(J).EQ.1) GO TO 300                                                HLE03590
  280 CONTINUE                                                                HLE03600
      WRITE(6,290)                                                            HLE03610
  290 FORMAT(/' ALL EIGENVALUES USED ARE T-MULTIPLE OR CLOSE TO SPURIOUSHLE03620
      1 T-EIGENVALUES'/' SO DO NOT CHANGE KMAX')                             HLE03630
      IF(KMAX.LT.MEV) GO TO 1450                                              HLE03640
      GO TO 320                                                               HLE03650
C                                                                             HLE03660
  300 KMAXN= 11*MEV/8 + 12                                                    HLE03670
      IF(MBETA.LE.KMAXN) GO TO 1610                                           HLE03680
      IF(KMAX.GE.KMAXN ) GO TO 320                                            HLE03690
      WRITE(6,310) KMAX, KMAXN                                                HLE03700
  310 FORMAT(' ENLARGE KMAX FROM ',I6,' TO ',I6)                             HLE03710
      MOLD1 = KMAX + 1                                                        HLE03720
      KMAX = KMAXN                                                            HLE03730
      GO TO 390                                                               HLE03740
C                                                                             HLE03750
  320 WRITE(6,330) KMAX                                                       HLE03760
  330 FORMAT(/' T-MATRICES HAVE BEEN READ IN FROM FILE 2'/' THE LARGEST HLE03770
      1SIZE T-MATRIX ALLOWED IS',I6/)                                        HLE03780
C                                                                             HLE03790
      IF(IREAD.EQ.1) GO TO 410                                                HLE03800
C                                                                             HLE03810
C     REGENERATE THE ALPHA AND BETA                                          HLE03820
C                                                                             HLE03830
  340 MOLD1 = 1                                                               HLE03840
C                                                                             HLE03850
      DO 350 J = 1,NGOOD                                                      HLE03860
      IF(MP(J).EQ.1) GO TO 370                                                HLE03870
  350 CONTINUE                                                                HLE03880
      KMAX = MEV + 12                                                         HLE03890
      WRITE(6,360) KMAX                                                       HLE03900
  360 FORMAT(/' ALL EIGENVALUES FOR WHICH EIGENVECTORS ARE TO BE COMPUTEHLE03910
      1D ARE EITHER T-MULTIPLE OR CLOSE TO'/' A SPURIOUS EIGENVALUE. THERHLE03920
      1EFORE SET KMAX = MEV + 12 = ',I7)                                     HLE03930
      GO TO 390                                                               HLE03940
C                                                                             HLE03950
  370 KMAXN = 11*MEV/8 + 12                                                   HLE03960
      IF(MBETA.LE.KMAXN) GO TO 1610                                           HLE03970
```

```
      WRITE(6,380) KMAXN                                              HLE03980
  380 FORMAT(' SET KMAX EQUAL TO ',I6)                                HLE03990
      KMAX = KMAXN                                                    HLE04000
C                                                                     HLE04010
  390 WRITE(6,400) MOLD1,KMAX                                         HLE04020
  400 FORMAT(/' LANCZS SUBROUTINE GENERATES ALPHA(J), BETA(J+1), J =',HLE04030
     1 I6,' TO ', I6/)                                                HLE04040
C                                                                     HLE04050
C--------------------------------------------------------------------HLE04060
C                                                                     HLE04070
      CALL LANCZS(CMATV,V1,V2,VS,ALPHA,BETA,GR,GC,G,KMAX,MOLD1,N,SVSEED)HLE04080
C                                                                     HLE04090
C--------------------------------------------------------------------HLE04100
C                                                                     HLE04110
  410 CONTINUE                                                        HLE04120
C                                                                     HLE04130
C     THE SUBROUTINE STURMI DETERMINES THE SMALLEST SIZE T-MATRIX FOR HLE04140
C     WHICH THE EIGENVALUE IN QUESTION IS AN EIGENVALUE (TO WITHIN A  HLE04150
C     GIVEN TOLERANCE) AND IF POSSIBLE THE SMALLEST SIZE T-MATRIX     HLE04160
C     FOR WHICH IT IS A DOUBLE EIGENVALUE (TO WITHIN THE SAME         HLE04170
C     TOLERANCE).  THE SIZE T-MATRIX USED IN THE EIGENVECTOR          HLE04180
C     COMPUTATIONS IS THEN DETERMINED BY LOOPING ON THE SIZES OF THE  HLE04190
C     T-EIGENVECTORS, USING THE INFORMATION FROM STURMI TO OBTAIN     HLE04200
C     STARTING GUESSES AT THE T-SIZES.                               HLE04210
C                                                                     HLE04220
C                                                                     HLE04230
      STUTOL = SCALEO*MULTOL                                          HLE04240
      IF(IWRITE.EQ.1) WRITE(6,420)                                    HLE04250
  420 FORMAT(' FROM STURMI')                                          HLE04260
      DO 460 J = 1,NGOOD                                              HLE04270
      EVAL = GOODEV(J)                                                HLE04280
C     COMPUTE THE TOLERANCES USED BY STURMI TO DETERMINE AN INTERVAL  HLE04290
C     CONTAINING THE EIGENVALUE EVAL.                                HLE04300
      TEMP = DABS(EVAL)*RELTOL                                        HLE04310
      TOLN = DMAX1(TEMP,STUTOL)                                       HLE04320
C                                                                     HLE04330
C--------------------------------------------------------------------HLE04340
C                                                                     HLE04350
      CALL STURMI(ALPHA,BETA,EVAL,TOLN,EPSM,KMAX,MK1,MK2,IC,IWRITE)   HLE04360
C                                                                     HLE04370
C--------------------------------------------------------------------HLE04380
C                                                                     HLE04390
C     STORE THE COMPUTED ORDERS OF T-MATRICES FOR LATER PRINTOUT      HLE04400
      M1(J) = MK1                                                     HLE04410
      M2(J) = MK2                                                     HLE04420
      ML(J) = (MK1 + 3*MK2)/4                                         HLE04430
      IF(MK2.EQ.KMAX)  ML(J) = KMAX                                   HLE04440
C                                                                     HLE04450
      IF(IC.GT.0) GO TO 440                                           HLE04460
C     IC = 0 MEANS THERE WAS NO T-EIGENVALUE IN THE DESIGNATED INTERVALHLE04470
C     BY T-SIZE KMAX.  THIS MEANS THAT THE T-EIGENVALUE PROVIDED HAS  HLE04480
C     NOT YET CONVERGED AS AN EIGENVALUE OF THE TRIDIAGONAL MATRICES  HLE04490
C     SO PROGRAM SHOULD NOT COMPUTE ITS EIGENVECTOR.                 HLE04500
      WRITE(6,430) J,GOODEV(J),MK1,MK2                                HLE04510
  430 FORMAT(I6,'TH EIGENVALUE',E20.12,' HAS NOT CONVERGED '/         HLE04520
     1' SO DO NOT COMPUTE ANY T-EIGENVECTOR OR RITZ VECTOR FOR IT'    HLE04530
     1/' MK1 AND MK2 FOR THIS EIGENVALUE WERE',2I6)                   HLE04540
      MP(J) = MPMIN                                                   HLE04550
      MA(J) = -2*KMAX                                                 HLE04560
      GO TO 460                                                       HLE04570
C     COMPUTE AN APPROPRIATE SIZE T-MATRIX FOR THE GIVEN EIGENVALUE.  HLE04580
  440 IF(M2(J).EQ.KMAX) GO TO 450                                     HLE04590
C     M1 AND M2 WERE BOTH DETERMINED                                 HLE04600
      MA(J) = (3*M1(J) + M2(J))/4  + 1                                HLE04610
      GO TO 460                                                       HLE04620
C     M2 NOT DETERMINED                                              HLE04630
```

```
    450 MA(J) = 5*M1(J)/4  + 1                                        HLE04640
  C                                                                    HLE04650
    460 CONTINUE                                                       HLE04660
  C                                                                    HLE04670
        IF (IWRITE.EQ.1) WRITE(6,470) (MA(JJ), JJ=1,NGOOD)            HLE04680
    470 FORMAT(/' 1ST GUESS AT APPROPRIATE SIZE T-MATRICES'/          HLE04690
       1 ' ACTUAL VALUES WILL PROBABLY BE 1/4 AGAIN AS MUCH'/(13I6))  HLE04700
  C                                                                    HLE04710
  C     PRINT OUT TO FILE 10 1ST GUESSES AT SIZES OF T-MATRICES TO    HLE04720
  C     BE USED IN THE EIGENVECTOR COMPUTATIONS.                      HLE04730
  C     ACTUAL SIZES MAY BE 1/4 OR MORE LARGER THAN THESE SIZES.      HLE04740
        WRITE(10,480) N,KMAX                                          HLE04750
    480 FORMAT(2I8,' = ORDER OF USER MATRIX AND MAX ORDER OF T(1,MEV)')HLE04760
  C                                                                    HLE04770
        WRITE(10,490)                                                 HLE04780
    490 FORMAT(/' 1ST GUESS AT APPROPRIATE SIZE T-MATRICES'/          HLE04790
       1 ' ACTUAL VALUES WILL PROBABLY BE 1/4 AGAIN AS MUCH'/)        HLE04800
  C                                                                    HLE04810
        WRITE(10,500)                                                 HLE04820
    500 FORMAT(4X,'J',7X,'GOODEV(J)',4X,'M1(J)',1X,'M2(J)',1X,'MA(J)')HLE04830
  C                                                                    HLE04840
        WRITE(10,510) (J,GOODEV(J),M1(J),M2(J), MA(J), J=1,NGOOD)     HLE04850
    510 FORMAT(I5,E19.12,3I6)                                         HLE04860
  C                                                                    HLE04870
        IF(MBOUND.EQ.1) WRITE(10,520)                                 HLE04880
    520 FORMAT(/' GOODEV(J) IS A GOOD EIGENVALUE OF T(1,MEV)'/        HLE04890
       1 ' M1 = SMALLEST VALUE OF M SUCH THAT T(1,M) HAS AT LEAST'/   HLE04900
       1 '      ONE EIGENVALUE IN THE INTERVAL (EV-TOLN,EV+TOLN)'/    HLE04910
       1 '      TOLN(J) = DMAX1(GOODEV(J)*RELTOL, SCALE0*MULTOL)'/    HLE04920
       1 ' M2 = SMALLEST M (IF ANY) SUCH THAT IN THE ABOVE INTERVAL'/ HLE04930
       1 '      T(1,M) HAS AT LEAST TWO EIGENVALUES '/                HLE04940
       1 ' INITIAL VALUE OF MA(J) IS CHOSEN HEURISTICALLY'/           HLE04950
       1 ' PROGRAM LOOPS ON SIZE OF T-MATRIX TO GET BETTER SIZE'/     HLE04960
       1 ' END OF SIZES OF T-MATRICES FILE 10'///)                   HLE04970
  C                                                                    HLE04980
  C                                                                    HLE04990
  C     TERMINATE AFTER COMPUTING 1ST GUESSES AT  SIZES OF THE        HLE05000
  C     T-MATRICES REQUIRED FOR THE GIVEN EIGENVALUES?                HLE05010
        IF(MBOUND.EQ.1) GO TO 1470                                    HLE05020
  C                                                                    HLE05030
  C                                                                    HLE05040
  C     IS THERE ROOM FOR ALL OF THE REQUESTED T-EIGENVECTORS?        HLE05050
        MTOL = 0                                                      HLE05060
        DO 530 J = 1,NGOOD                                            HLE05070
        IF(MP(J).EQ.MPMIN) GO TO 530                                  HLE05080
        MTOL = MTOL + IABS(MA(J))                                     HLE05090
    530 CONTINUE                                                      HLE05100
        MTOL = (5*MTOL)/4                                             HLE05110
        IF(MTOL.GT.MDIMTV.AND.NTVCON.EQ.0) GO TO 1490                 HLE05120
  C                                                                    HLE05130
  C----------------------------------------------------------------HLE05140
  C     GENERATE A RANDOM VECTOR TO BE USED REPEATEDLY BY            HLE05150
  C     SUBROUTINE INVERM                                            HLE05160
  C                                                                    HLE05170
        IIL = RHSEED                                                  HLE05180
        CALL GENRAN(IIL,G,KMAX)                                       HLE05190
  C                                                                    HLE05200
  C---------------------------------------------------------------- HLE05210
  C                                                                    HLE05220
  C     LOOP ON GIVEN EIGENVALUES TO COMPUTE THE CORRESPONDING       HLE05230
  C     T-EIGENVECTOR.                                               HLE05240
  C                                                                    HLE05250
        MTOL = 0                                                      HLE05260
        NTVEC = 0                                                     HLE05270
        ILBIS = 0                                                     HLE05280
        DO 720 J = 1,NGOOD                                            HLE05290
        ICOUNT = 0                                                    HLE05300
```

```
      ERRMIN = 10.D0                                             HLE05310
      MABEST = MPMIN                                             HLE05320
      IF(MP(J).EQ.MPMIN) GO TO 720                               HLE05330
      TFLAG = 0                                                  HLE05340
      EVAL = GOODEV(J)                                           HLE05350
      TEMP = RELTOL* DABS(EVAL)                                  HLE05360
      UB = EVAL + DMAX1(STUTOL,TEMP)                             HLE05370
      LB = EVAL - DMAX1(STUTOL,TEMP)                             HLE05380
  540 KMAXU = IABS(MA(J))                                        HLE05390
C                                                                HLE05400
C     SELECT A SUITABLE INCREMENT FOR THE ORDERS OF THE T-MATRICES HLE05410
C     TO BE CONSIDERED IN DETERMINING APPROPRIATE SIZES FOR THE RITZ HLE05420
C     VECTOR COMPUTATIONS.                                       HLE05430
      IF(ICOUNT.GT.0) GO TO 560                                  HLE05440
C     SELECT IDELTA(J) BASED UPON THE T-MULTIPLICITY OBTAINED    HLE05450
      IF(M2(J).EQ.KMAX) GO TO 550                                HLE05460
C     M2 DETERMINED                                              HLE05470
      IDELTA(J) = ((3*M1(J) + 5*M2(J))/8  +  1 - IABS(MA(J)))/10 + 1  HLE05480
      GO TO 560                                                  HLE05490
C     M2 NOT DETERMINED                                          HLE05500
  550 MAMAX = MINO((11*MEV)/8 + 12, (13*M1(J))/8 + 1)            HLE05510
      IDELTA(J) = (MAMAX - IABS(MA(J)))/10  +  1                 HLE05520
  560 ICOUNT = ICOUNT + 1                                        HLE05530
C                                                                HLE05540
C----------------------------------------------------------------HLE05550
C     TO MIMIMIZE THE EFFECT OF THE ONE-SIDED ACCEPTANCE TEST FOR HLE05560
C     T-EIGENVALUES IN THE BISEC SUBROUTINE, RECOMPUTE THE GIVEN HLE05570
C     EIGENVALUE AT THE SPECIFIED KMAXU                          HLE05580
C                                                                HLE05590
      CALL LBISEC(ALPHA,BETA,EPSM,EVAL,EVALN,LB,UB,TTOL,KMAXU,NEVT) HLE05600
C                                                                HLE05610
C----------------------------------------------------------------HLE05620
C                                                                HLE05630
C     CHECK WHETHER OR NOT GIVEN T-MATRIX HAS AN EIGENVALUE IN THE HLE05640
C     SPECIFIED INTERVAL AND IF SO WHAT ITS T-MULTIPLICITY IS.   HLE05650
C                                                                HLE05660
      IF(NEVT.EQ.1) GO TO 600                                    HLE05670
      IF(NEVT.NE.0) GO TO 580                                    HLE05680
      ILBIS = 1                                                  HLE05690
      WRITE(6,570) EVAL,KMAXU                                    HLE05700
  570 FORMAT(/' PROBLEM ENCOUNTERED IN RECOMPUTATION OF USER-SUPPLIED EIHLE05710
     1GENVALUE',E20.12/' THE SIZE T-MATRIX SPECIFIED',I6,' DOES NOT   HLE05720
     1HAVE AN EIGENVALUE IN THE INTERVAL SPECIFIED'/' THEREFORE NO EIGENHLE05730
     1VECTOR WILL BE COMPUTED FOR THIS PARTICULAR EIGENVALUE'/)  HLE05740
      GO TO 620                                                  HLE05750
C                                                                HLE05760
  580 IF(NEVT.GT.1) WRITE(6,590) EVAL,KMAXU                      HLE05770
  590 FORMAT(/' PROBLEM ENCOUNTERED IN RECOMPUTATION OF USER-SUPPLIED  HLE05780
     1EIGENVALUE',E20.12/' FOR THE SIZE T-MATRIX SPECIFIED =',I6,' THE  HLE05790
     1GIVEN EIGENVALUE IS T-MULTIPLE IN THE INTERVAL SPECIFIED'/' SOMETHHLE05800
     1ING IS WRONG, THEREFORE NO EIGENVECTOR WILL BE COMPUTED FOR THIS EHLE05810
     1IGENVALUE'/)                                               HLE05820
C                                                                HLE05830
      MP(J) = MPMIN                                              HLE05840
      MA(J) = -2*KMAX                                            HLE05850
      GO TO 720                                                  HLE05860
C                                                                HLE05870
  600 CONTINUE                                                   HLE05880
      ILBIS = 0                                                  HLE05890
C                                                                HLE05900
      EVNEW(J) = EVALN                                           HLE05910
      EVAL = EVALN                                               HLE05920
      MTOL = MTOL+KMAXU                                          HLE05930
C                                                                HLE05940
C     IS THERE ROOM IN TVEC ARRAY FOR THE NEXT T-EIGENVECTOR?    HLE05950
C     IF NOT, SKIP TO RITZ VECTOR COMPUTATIONS.                  HLE05960
```

```
      IF (MTOL.GT.MDIMTV) GO TO 730                                 HLE05970
C                                                                   HLE05980
      IT = 3                                                        HLE05990
      KINT = MTOL - KMAXU +1                                        HLE06000
C                                                                   HLE06010
C     RECORD THE BEGINNING AND END OF THE T-EIGENVECTOR BEING COMPUTED HLE06020
      MINT(J) = KINT                                                HLE06030
      MFIN(J) = MTOL                                                HLE06040
C                                                                   HLE06050
C-------------------------------------------------------------------HLE06060
C     SUBROUTINE INVERM DOES INVERSE ITERATION, I.E. SOLVES         HLE06070
C     (T(1,KMAXU) - EVAL)*U = RHS  FOR EACH EIGENVALUE TO OBTAIN THE HLE06080
C     DESIRED T-EIGENVECTOR.                                        HLE06090
C                                                                   HLE06100
      IF(IWRITE.EQ.1)  WRITE(6,610) J                               HLE06110
  610 FORMAT(/I6,'TH EIGENVALUE')                                   HLE06120
C                                                                   HLE06130
      CALL INVERM(ALPHA,BETA,V1,TVEC(KINT),EVAL,ERROR,TERROR,EPSM,  HLE06140
     1 G,KMAXU,IT,IWRITE)                                           HLE06150
C                                                                   HLE06160
C-------------------------------------------------------------------HLE06170
C                                                                   HLE06180
      TERR(J) = TERROR                                              HLE06190
      TLAST(J) = ERROR                                              HLE06200
      KMAXU1 = KMAXU + 1                                            HLE06210
      TBETA(J) = BETA(KMAXU1)*ERROR                                 HLE06220
C                                                                   HLE06230
C     AFTER COMPUTING EACH OF THE T-EIGENVECTORS,                   HLE06240
C     CHECK THE SIZE OF THE ERROR ESTIMATE, ERROR.                  HLE06250
C     IF THIS ESTIMATE IS NOT AS SMALL AS DESIRED AND               HLE06260
C     |MA(J)| < ML(J), ATTEMPT TO INCREASE THE SIZE OF |MA(J)|      HLE06270
C     AND REPEAT THE T-EIGENVECTOR COMPUTATIONS.                    HLE06280
C                                                                   HLE06290
      IF(ERROR.LT.ERTOL.OR.TFLAG.EQ.1)  GO TO 710                   HLE06300
C                                                                   HLE06310
      IF(ERROR.GE.ERRMIN) GO TO 620                                 HLE06320
C     LAST COMPONENT IS LESS THAN MINIMAL TO DATE                   HLE06330
      ERRMIN = ERROR                                                HLE06340
      MABEST = MA(J)                                                HLE06350
  620 CONTINUE                                                      HLE06360
C                                                                   HLE06370
      IF(MA(J).GT.0)  ITEST = MA(J) + IDELTA(J)                     HLE06380
      IF(MA(J).LT.0)  ITEST = -(IABS(MA(J)) + IDELTA(J))            HLE06390
      IF(IABS(ITEST).LE.ML(J).AND.ICOUNT.LE.10) GO TO 640           HLE06400
C     NEW MA(J) IS GREATER THAN MAXIMUM ALLOWED.                    HLE06410
      IF(ERCONT.EQ.0.OR.MABEST.EQ.MPMIN) GO TO 660                  HLE06420
      TFLAG = 1                                                     HLE06430
      MA(J) = MABEST                                                HLE06440
      IF(ILBIS.EQ.0)  MTOL = MTOL - KMAXU                           HLE06450
      WRITE(6,630) MA(J)                                            HLE06460
  630 FORMAT(' 10 ORDERS WERE CONSIDERED.  NONE SATISFIED THE ERROR TESTHLE06470
     1'/' THEREFORE USE THE BEST ORDER OBTAINED FOR THE EIGENVECTORS' HLE06480
     1,16)                                                         HLE06490
      GO TO 540                                                     HLE06500
C                                                                   HLE06510
  640 MA(J) = ITEST                                                 HLE06520
C                                                                   HLE06530
      MT = IABS(MA(J))                                              HLE06540
      IF(IWRITE.EQ.1) WRITE(6,650)  MT                              HLE06550
  650 FORMAT(/' CHANGE SIZE OF T-MATRIX TO ',I6,' RECOMPUTE T-EIGENVECTOHLE06560
     1R')                                                          HLE06570
C                                                                   HLE06580
      IF(ILBIS.EQ.0)  MTOL = MTOL - KMAXU                           HLE06590
C                                                                   HLE06600
      GO TO 540                                                     HLE06610
C                                                                   HLE06620
C     APPROPRIATE SIZE T-MATRIX WAS NOT OBTAINED                    HLE06630
```

```
      660 CONTINUE                                                     HLE06640
          WRITE(10,670) J,EVAL,MP(J)                                   HLE06650
      670 FORMAT(/' ON 10 INCREMENTS NOT ABLE TO IDENTIFY APPROPRIATE SIZE HLE06660
         1T-MATRIX FOR'/                                               HLE06670
         1' EIGENVALUE(',14,') = ',E20.12,' T-MULTIPLICITY =',14/)     HLE06680
          IF(M2(J).EQ.KMAX) WRITE(10,680)                             HLE06690
          IF(M2(J).LT.KMAX) WRITE(10,690)                             HLE06700
      680 FORMAT(/' ORDERS TESTED RANGED FROM 5*M1(J)/4 TO APPROXIMATELY'/ HLE06710
         1'  MIN(11*MEV/8, 13*M1(J)/8)'/)                             HLE06720
      690 FORMAT(/' ORDERS TESTED RANGED FROM (3*M1(J)+M2(J)/4 TO APPROXIMATHLE06730
         1ELY'/'  (3*M1(J) + 5*M2(J))/8'/)                            HLE06740
          WRITE(10,700)                                               HLE06750
      700 FORMAT(' ALLOWING LARGER ORDERS FOR THE T-MATRICES MAY RESULT IN HLE06760
         1 SUCCESS'/' BUT PROBABLY WILL NOT.  PROBLEM IS PROBABLY DUE TO' HLE06770
         1 /' LACK OF CONVERGENCE OF GIVEN EIGENVALUE, CHECK THE ERROR ESTIMHLE06780
         1ATE')                                                        HLE06790
          MP(J) = MPMIN                                                HLE06800
          IF(ILBIS.EQ.0) MTOL = MTOL - KMAXU                          HLE06810
          GO TO 720                                                    HLE06820
      710 NTVEC = NTVEC + 1                                            HLE06830
    C                                                                  HLE06840
      720 CONTINUE                                                     HLE06850
          NGOODC = NGOOD                                               HLE06860
          GO TO 750                                                    HLE06870
    C                                                                  HLE06880
    C     COME HERE IF THERE IS NOT ENOUGH ROOM FOR ALL OF T-EIGENVECTORS HLE06890
      730 NGOODC = J-1                                                 HLE06900
          WRITE(6,740) J,MTOL,MDIMTV                                   HLE06910
      740 FORMAT(/' NOT ENOUGH ROOM IN TVEC ARRAY FOR ',14,'TH T-EIGENVECTORHLE06920
         1'/' TVEC DIMENSION REQUESTED = ',16,' BUT TVEC HAS DIMENSION ',16HLE06930
         1/)                                                           HLE06940
          IF(NGOODC.EQ.0) GO TO 1510                                  HLE06950
          MTOL = MTOL-KMAXU                                           HLE06960
    C                                                                  HLE06970
      750 CONTINUE                                                     HLE06980
    C                                                                  HLE06990
    C     THE LOOP ON T-EIGENVECTOR COMPUTATIONS IS COMPLETE.          HLE07000
    C     WRITE OUT THE SIZE T-MATRICES THAT WILL BE USED FOR          HLE07010
    C     THE RITZ VECTOR COMPUTATIONS.                                HLE07020
    C                                                                  HLE07030
          WRITE(10,760)                                               HLE07040
      760 FORMAT(/' SIZES OF T-MATRICES THAT WILL BE USED IN THE RITZ COMPUTHLE07050
         1ATIONS'/5X,'J',16X,'GOODEV(J)',1X,'MA(J)')                  HLE07060
    C                                                                  HLE07070
          WRITE(10,770)  (J,GOODEV(J),MA(J), J=1,NGOOD)               HLE07080
      770 FORMAT(16,E25.14,16)                                        HLE07090
          WRITE(10,520)                                               HLE07100
    C                                                                  HLE07110
          WRITE(6,780) MTOL                                           HLE07120
      780 FORMAT(/' THE CUMULATIVE LENGTH OF THE T-EIGENVECTORS IS',118) HLE07130
    C                                                                  HLE07140
          WRITE(6,790) NTVEC,NGOOD                                    HLE07150
      790 FORMAT(/16,' T-EIGENVECTORS OUT OF',16,' REQUESTED WERE COMPUTED')HLE07160
    C                                                                  HLE07170
    C     SAVE THE T-EIGENVECTORS ON FILE 11?                         HLE07180
          IF(TVSTOP.NE.1.AND.SVTVEC.EQ.0) GO TO 850                   HLE07190
    C                                                                  HLE07200
          WRITE(11,800) NTVEC,MTOL,MATNO,SVSEED                       HLE07210
      800 FORMAT(16,3112,' = NTVEC,MTOL,MATNO,SVSEED')                HLE07220
    C                                                                  HLE07230
          DO 830 J=1,NGOODC                                           HLE07240
    C     IF MP(J) = MPMIN THEN NO SUITABLE T-EIGENVECTOR IS AVAILABLE HLE07250
    C     FOR THAT EIGENVALUE.                                        HLE07260
          IF(MP(J).EQ.MPMIN) WRITE(11,810) J,MA(J),GOODEV(J),MP(J)    HLE07270
      810 FORMAT(216,E20.12,16/' TH EIGVAL,T-SIZE,EVALUE,FLAG,NO EIGVEC') HLE07280
          IF(MP(J).NE.MPMIN) WRITE(11,820) J,MA(J),GOODEV(J),MP(J)    HLE07290
      820 FORMAT(16,16,E20.12,16/' T-EIGVEC,SIZE T,EVALUE OF A,MP(J)') HLE07300
```

```
      IF(MP(J).EQ.MPMIN) GO TO 830                               HLE07310
      KI = MINT(J)                                               HLE07320
      KF = MFIN(J)                                               HLE07330
C                                                                HLE07340
      WRITE(11,270) (TVEC(K), K=KI,KF)                           HLE07350
C                                                                HLE07360
  830 CONTINUE                                                   HLE07370
C                                                                HLE07380
      IF(TVSTOP.NE.1) GO TO 850                                  HLE07390
C                                                                HLE07400
      WRITE(6,840) TVSTOP, NTVEC,NGOOD                           HLE07410
  840 FORMAT(/' USER SET TVSTOP = ',I1/                          HLE07420
     1' THEREFORE PROGRAM TERMINATES AFTER T-EIGENVECTOR COMPUTATIONS'/ HLE07430
     1' T-EIGENVECTORS THAT WERE COMPUTED ARE SAVED ON FILE 11'/ HLE07440
     1I8,' T-EIGENVECTORS WERE COMPUTED OUT OF',I7,' REQUESTED'/) HLE07450
C                                                                HLE07460
      GO TO 1630                                                 HLE07470
C                                                                HLE07480
  850 CONTINUE                                                   HLE07490
C     IF NOT ABLE TO COMPUTE ALL THE REQUESTED T-EIGENVECTORS    HLE07500
C     CONTINUE WITH THE LANCZOS VECTOR COMPUTATIONS ANYWAY?      HLE07510
      IF(NTVEC.NE.NGOOD.AND.LVCONT.EQ.0) GO TO 1530              HLE07520
C                                                                HLE07530
C     COMPUTE THE MAXIMUM SIZE OF THE T-MATRIX USED FOR THOSE    HLE07540
C     EIGENVALUES WITH GOOD ERROR ESTIMATES.                     HLE07550
C                                                                HLE07560
      KMAXU = 0                                                  HLE07570
      DO 86C J = 1,NGOODC                                        HLE07580
      MT = IABS(MA(J))                                           HLE07590
      IF(MT.LT.KMAXU.OR.MP(J).EQ.MPMIN) GO TO 860                HLE07600
      KMAXU = MT                                                 HLE07610
  860 CONTINUE                                                   HLE07620
C                                                                HLE07630
      IF(KMAXU.EQ.0) GO TO 1570                                  HLE07640
C                                                                HLE07650
      WRITE(6,870) KMAXU                                         HLE07660
  870 FORMAT(/I6,' = LARGEST SIZE T-MATRIX TO BE USED IN THE RITZ VECTORHLE07670
     1 COMPUTATIONS')                                            HLE07680
C                                                                HLE07690
C     COUNT THE NUMBER OF RITZ VECTORS NOT BEING COMPUTED        HLE07700
      MREJEC = 0                                                 HLE07710
      DO 880 J=1,NGOODC                                          HLE07720
  880 IF(MP(J).EQ.MPMIN)  MREJEC = MREJEC + 1                    HLE07730
      MREJET = MREJEC + (NGOOD-NGOODC)                           HLE07740
      IF(MREJET.NE.0) WRITE(6,890) MREJET                        HLE07750
  890 FORMAT(/' RITZ VECTORS ARE NOT COMPUTED FOR',I6,' OF THE EIGENVALUHLE07760
     1ES'/)                                                      HLE07770
      NACT = NGOODC - MREJEC                                     HLE07780
      WRITE(6,900) NGOOD,NTVEC,NACT                              HLE07790
  900 FORMAT(/I6,' RITZ VECTORS WERE REQUESTED'/I6,' T-EIGENVECTORS WEREHLE07800
     1 COMPUTED'/I6,' RITZ VECTORS WILL BE COMPUTED'/)           HLE07810
C     CHECK IF THERE ARE ANY RITZ VECTORS TO COMPUTE            HLE07820
      IF(MREJEC.EQ.NGOODC) GO TO 1550                            HLE07830
C                                                                HLE07840
C     CONTINUE WITH THE LANCZOS VECTOR COMPUTATIONS?             HLE07850
      IF(LVCONT.EQ.0.AND.MREJEC.NE.0) GO TO 1530                 HLE07860
C                                                                HLE07870
C     NOW COMPUTE THE RITZ VECTORS.  REGENERATE THE              HLE07880
C     LANCZOS VECTORS.                                           HLE07890
C                                                                HLE07900
      DO 910 I = 1,NMAX                                          HLE07910
  910 RITVEC(I) = ZEROC                                          HLE07920
C                                                                HLE07930
C     REGENERATE THE STARTING VECTOR. THIS MUST BE GENERATED AND HLE07940
C     NORMALIZED PRECISELY THE WAY IT WAS DONE IN THE EIGENVALUE HLE07950
C     COMPUTATIONS, OTHERWISE THERE WILL BE A MISMATCH BETWEEN   HLE07960
C     THE T-EIGENVECTORS THAT HAVE BEEN COMPUTED FROM THE T-MATRICES HLE07970
```

```
C      READ IN FROM FILE 2 AND THE LANCZOS VECTORS THAT ARE           HLE07980
C      BEING REGENERATED.                                             HLE07990
C                                                                     HLE08000
C---------------------------------------------------------------------HLE08010
C                                                                     HLE08020
       IIL = SVSEED                                                   HLE08030
       CALL GENRAN(IIL,G,N)                                           HLE08040
C                                                                     HLE08050
C---------------------------------------------------------------------HLE08060
C                                                                     HLE08070
       DO 920 I = 1,N                                                 HLE08080
   920 GR(I) = G(I)                                                   HLE08090
C                                                                     HLE08100
C---------------------------------------------------------------------HLE08110
C                                                                     HLE08120
       CALL GENRAN(IIL,G,N)                                           HLE08130
C                                                                     HLE08140
C---------------------------------------------------------------------HLE08150
C                                                                     HLE08160
       DO 930 I = 1,N                                                 HLE08170
   930 GC(I) = G(I)                                                   HLE08180
C                                                                     HLE08190
       DO 940 I = 1,N                                                 HLE08200
   940 V2(I) = DCMPLX(GR(I),GC(I))                                    HLE08210
C                                                                     HLE08220
C---------------------------------------------------------------------HLE08230
       CALL CINPRD(V2,V2,SUM,N)                                       HLE08240
C---------------------------------------------------------------------HLE08250
C                                                                     HLE08260
       SUM = ONE/DSQRT(SUM)                                           HLE08270
       DO 950 I = 1,N                                                 HLE08280
       V1(I) = ZEROC                                                  HLE08290
   950 V2(I) = V2(I)*SUM                                              HLE08300
C                                                                     HLE08310
C      LOOP FOR GENERATING REQUIRED RITZ VECTORS (IVEC = 1,KMAXU)     HLE08320
C      USES GRAM-SCHMIDT ORTHOGONALIZATION WITHOUT MODIFICATION       HLE08330
C                                                                     HLE08340
       IVEC = 1                                                       HLE08350
       BATA = ZERO                                                    HLE08360
C                                                                     HLE08370
       GO TO 1010                                                     HLE08380
C                                                                     HLE08390
   960 CONTINUE                                                       HLE08400
C                                                                     HLE08410
C---------------------------------------------------------------------HLE08420
C      CMATV(V2,VS,SUM) CALCULATES   VS = A*V2 - SUM*VS               HLE08430
       SUM = ZERO                                                     HLE08440
       CALL CMATV(V2,VS,SUM)                                          HLE08450
       CALL CINPRD(V2,VS,ALFA,N)                                      HLE08460
C                                                                     HLE08470
C---------------------------------------------------------------------HLE08480
C                                                                     HLE08490
       DO 970 J=1,N                                                   HLE08500
   970 V1(J) = (VS(J) - BATA*V1(J)) - ALFA*V2(J)                      HLE08510
C                                                                     HLE08520
C---------------------------------------------------------------------HLE08530
       CALL CINPRD(V1,V1,BATA,N)                                      HLE08540
C---------------------------------------------------------------------HLE08550
C                                                                     HLE08560
       BATA = DSQRT(BATA)                                             HLE08570
       SUM = ONE/BATA                                                 HLE08580
C                                                                     HLE08590
       TEMP = BETA(IVEC)                                              HLE08600
       TEMP = DABS(BATA - TEMP)/TEMP                                  HLE08610
       IF (TEMP.LT.1.0D-10)GO TO 990                                  HLE08620
C                                                                     HLE08630
C      THE BETA BEING REGENERATED DO NOT MATCH THE HISTORY FILE       HLE08640
```

```
C      SOMETHING IS WRONG IN THE LANCZOS VECTOR GENERATION            HLE08650
C      PROGRAM TERMINATES FOR USER TO CORRECT THE PROBLEM             HLE08660
C      WHICH MUST BE IN THE STARTING VECTOR GENERATION OR IN          HLE08670
C      THE MATRIX-VECTOR MULTIPLY SUBROUTINE CMATV SUPPLIED.          HLE08680
C      THIS SUBROUTINE MUST BE THE SAME ONE USED IN THE               HLE08690
C      EIGENVALUE COMPUTATIONS OR A MISMATCH WILL ENSUE.              HLE08700
C                                                                     HLE08710
       WRITE(6,980) IVEC,BATA,BETA(IVEC),TEMP                         HLE08720
   980 FORMAT(/2X,'IVEC',16X,'BATA',10X,'BETA(IVEC)',14X,'RELDIF'/16, HLE08730
      13E20.12/' IN LANCZOS VECTOR REGENERATION THE ENTRIES OF THE TRIDIAHLE08740
      1GONAL MATRICES BEING'/' GENERATED ARE NOT THE SAME AS THOSE IN THEHLE08750
      1 MATRIX SUPPLIED ON FILE 2.'/' THEREFORE SOMETHING IS BEING INITIAHLE08760
      1LIZED OR COMPUTED DIFFERENTLY FROM THE WAY'/' IT WAS COMPUTED IN THLE08770
      1HE EIGENVALUE COMPUTATIONS'/' THE PROGRAM TERMINATES FOR THE USER HLE08780
      1TO DETERMINE WHAT THE PROBLEM IS'/)                            HLE08790
       GO TO 1630                                                     HLE08800
C                                                                     HLE08810
C                                                                     HLE08820
   990 CONTINUE                                                       HLE08830
       DO 1000 J = 1,N                                                HLE08840
       TEMPC = SUM*V1(J)                                              HLE08850
       V1(J) = V2(J)                                                  HLE08860
  1000 V2(J) = TEMPC                                                  HLE08870
C                                                                     HLE08880
  1010 CONTINUE                                                       HLE08890
C                                                                     HLE08900
       LFIN = 0                                                       HLE08910
       DO 1030 J = 1,NGOODC                                           HLE08920
       LL = LFIN                                                      HLE08930
       LFIN = LFIN + N                                                HLE08940
C                                                                     HLE08950
       IF(IABS(MA(J)).LT.IVEC.OR.MP(J).EQ.MPMIN) GO TO 1030          HLE08960
       II = IVEC + MINT(J) - 1                                        HLE08970
       TEMP = TVEC(II)                                                HLE08980
C      II IS THE (IVEC)TH COMPONENT OF THE T-EIGENVECTOR CONTAINED    HLE08990
C      IN TVEC(MINT(J)).                                              HLE09000
C                                                                     HLE09010
       DO 1020 K = 1,N                                                HLE09020
       LL = LL + 1                                                    HLE09030
  1020 RITVEC(LL) = TEMP*V2(K) + RITVEC(LL)                           HLE09040
C                                                                     HLE09050
  1030 CONTINUE                                                       HLE09060
C                                                                     HLE09070
       IVEC = IVEC + 1                                                HLE09080
       IF (IVEC.LE.KMAXU) GO TO 960                                   HLE09090
C                                                                     HLE09100
C                                                                     HLE09110
C      RITZVECTOR GENERATION IS COMPLETE. NORMALIZE EACH RITZVECTOR.  HLE09120
C      NOTE THAT IF CERTAIN RITZ VECTORS WERE NOT COMPUTED THEN THAT  HLE09130
C      PORTION OF THE RITVEC ARRAY WAS NOT UTILIZED.                  HLE09140
C                                                                     HLE09150
       LFIN = 0                                                       HLE09160
       DO 1130 J = 1,NGOODC                                           HLE09170
C                                                                     HLE09180
       KK = LFIN                                                      HLE09190
       LFIN = LFIN + N                                                HLE09200
       IF(MP(J).EQ.MPMIN) GO TO 1130                                  HLE09210
C                                                                     HLE09220
       DO 1040 K = 1,N                                                HLE09230
       KK = KK + 1                                                    HLE09240
  1040 V2(K) = RITVEC(KK)                                             HLE09250
C                                                                     HLE09260
C---------------------------------------------------------------------HLE09270
       CALL CINPRD(V2,V2,SUM,N)                                       HLE09280
C---------------------------------------------------------------------HLE09300
C                                                                     HLE09300
       SUM = DSQRT(SUM)                                               HLE09310
```

```
            RNORM(J) = SUM                                          HLE09320
            TEMP = DABS(ONE-SUM)                                     HLE09330
            SUM = ONE/SUM                                           HLE09340
      C                                                             HLE09350
            KK = LFIN - N                                           HLE09360
            DO 1050 K = 1,N                                         HLE09370
            KK = KK + 1                                             HLE09380
            V2(K) = SUM*V2(K)                                       HLE09390
       1050 RITVEC(KK) = V2(K)                                      HLE09400
      C                                                             HLE09410
      C     ONLY ENTER NEXT PORTION IF GIVEN MATRIX IS REAL.        HLE09420
            IF(ISREAL.NE.1) GO TO 1100                              HLE09430
      C                                                             HLE09440
      C     AT THIS POINT RITZ VECTOR IS IN V2.                     HLE09450
      C     THIS PROGRAM CAN BE USED ON REAL MATRICES TO DETERMINE  HLE09460
      C     WHICH IF ANY EIGENVALUES ARE A-MULTIPLE AND IF SO TO COMPUTE  HLE09470
      C     TWO EIGENVECTORS FOR THOSE EIGENVALUES THAT ARE MULTIPLE AND ONE  HLE09480
      C     FOR THOSE THAT ARE NOT MULTIPLE. HERE ONLY IDENTIFIES WHETHER  HLE09490
      C     EIGENVALUE IS AT LEAST DOUBLE.  THIS IS DONE BY CHECKING THE  HLE09500
      C     RATIOS OF SUCCEEDING REAL AND IMAGINARY PARTS OF THE COMPUTED  HLE09510
      C     RITZ VECTORS.                                           HLE09520
      C                                                             HLE09530
            SUM = DIMAG(V2(1))/DREAL(V2(1))                         HLE09540
            DO 1060 K=2,N                                           HLE09550
            TEMP = DREAL(V2(K))                                     HLE09560
            IF(DABS(TEMP).LT.1.D-9) GO TO 1060                      HLE09570
            TEMP = DIMAG(V2(K))/DREAL(V2(K))                        HLE09580
            IF(DABS(TEMP - SUM).LE.1.D-6) GO TO 1060                HLE09590
            MULEVA(J) = 2                                           HLE09600
            GO TO 1070                                              HLE09610
       1060 CONTINUE                                                HLE09620
            MULEVA(J) = 1                                           HLE09630
       1070 IF(MULEVA(J).EQ.2) WRITE(6,1090) J,GOODEV(J)            HLE09640
            IF(MULEVA(J).EQ.1) WRITE(6,1080) J,GOODEV(J)            HLE09650
       1080 FORMAT(I6,'TH EIGENVALUE CONSIDERED =',E20.12,' IS SIMPLE')    HLE09660
       1090 FORMAT(I6,'TH EIGENVALUE CONSIDERED =',E20.12,' IS MULTIPLE')  HLE09670
      C                                                             HLE09680
       1100 CONTINUE                                                HLE09690
      C                                                             HLE09700
            IF (IWRITE.NE.0) WRITE(6,1110) J,GOODEV(J)              HLE09710
       1110 FORMAT(/I5,' TH EIGENVALUE CONSIDERED = ',E20.12/)      HLE09720
      C                                                             HLE09730
            IF (IWRITE.NE.0) WRITE(6,1120) TERR(J),TBETA(J),TEMP    HLE09740
       1120 FORMAT(' NORM OF ERROR IN T-EIGENVECTOR = ',E14.3/      HLE09750
          1 ' BETA(MA(J)+1)*U(MA(J))  = ',E14.3/                    HLE09760
          1 ' ABS(NORM(RITVEC) - 1.0)  = ',E14.3/)                  HLE09770
      C                                                             HLE09780
            LINT = LFIN - N + 1                                     HLE09790
            EVAL = EVNEW(J)                                         HLE09800
      C                                                             HLE09810
      C---------------------------------------------------------------HLE09820
      C                                                             HLE09830
            CALL CMATV(RITVEC(LINT),V2,EVAL)                        HLE09840
      C                                                             HLE09850
      C---------------------------------------------------------------HLE09860
      C                                                             HLE09870
      C     COMPUTE ERROR IN RITZ VECTOR CONSIDERED AS A EIGENVECTOR OF A.  HLE09880
      C     V2 = A*RITVEC - EVAL*RITVEC                             HLE09890
      C                                                             HLE09900
      C---------------------------------------------------------------HLE09910
            CALL CINPRD(V2,V2,SUM,N)                                HLE09920
      C---------------------------------------------------------------HLE09930
      C                                                             HLE09940
            SUM = DSQRT(SUM)                                        HLE09950
            ERR(J) = SUM                                            HLE09960
            GAP = ABS(AMINGP(J))                                    HLE09970
```

```
      ERRDGP(J) = SUM/GAP                                        HLE09980
C                                                                HLE09990
 1130 CONTINUE                                                   HLE10000
C                                                                HLE10010
C                                                                HLE10020
C     RITZVECTORS ARE NORMALIZED AND ERROR ESTIMATES ARE IN ERR ARRAY HLE10030
C     AND IN ERRDGP ARRAY. STORE EVERYTHING                     HLE10040
C                                                                HLE10050
C                                                                HLE10060
      WRITE(9,1140)                                              HLE10070
 1140 FORMAT(6X,'GOODEV(J)',1X,'MA(J)',4X,'A MINGAP',6X,'AERROR',2X, HLE10080
     1 'AERROR/GAP',6X,'TERROR')                                 HLE10090
C                                                                HLE10100
      WRITE(13,1150)                                             HLE10110
 1150 FORMAT(16X,'GOODEV(J)',5X,'RITZNORM',6X,'AMINGAP',5X,      HLE10120
     1 'TBETA(J)',5X,'TLAST(J)')                                 HLE10130
C                                                                HLE10140
      DO 1180 J=1,NGOODC                                         HLE10150
C                                                                HLE10160
      IF(MP(J).EQ.MPMIN) GO TO 1180                              HLE10170
C                                                                HLE10180
      WRITE(9,1160)EVNEW(J),MA(J),AMINGP(J),ERR(J),ERRDGP(J),TERR(J) HLE10190
 1160 FORMAT(E15.8,I6,4E12.4)                                    HLE10200
C                                                                HLE10210
      WRITE(13,1170) EVNEW(J),RNORM(J),AMINGP(J),TBETA(J),TLAST(J) HLE10220
 1170 FORMAT(E25.14,4E13.5)                                      HLE10230
C                                                                HLE10240
 1180 CONTINUE                                                   HLE10250
C                                                                HLE10260
      IF(MREJEC.EQ.0) GO TO 1260                                 HLE10270
      WRITE(9,1190)                                              HLE10280
 1190 FORMAT(/' RITZ VECTORS WERE NOT COMPUTED FOR THE FOLLOWING EIGENVAHLE10290
     1LUES'/' EITHER BECAUSE THEY HAD NOT CONVERGED OR BECAUSE THE ERRORHLE10300
     1 ESTIMATE'/'  WAS NOT AS SMALL AS DESIRED'/)               HLE10310
C                                                                HLE10320
      DO 1250 J = 1,NGOODC                                       HLE10330
      IF(MP(J).NE.MPMIN) GO TO 1250                              HLE10340
C     WRITE OUT MESSAGE FOR EACH EIGENVALUE FOR WHICH NO EIGENVECTOR HLE10350
C     WAS COMPUTED.                                              HLE10360
C                                                                HLE10370
      WRITE(9,1200)                                              HLE10380
 1200 FORMAT(6X,'GOODEV(J)',3X,'MA(J)',5X,'AMINGP(J)',6X,'TLAST(J)',3X, HLE10390
     1'MP(J)')                                                   HLE10400
      WRITE(9,1210) GOODEV(J),MA(J),AMINGP(J),TBETA(J),MP(J)     HLE10410
 1210 FORMAT(E15.8,I8,2E14.4,I8)                                 HLE10420
C                                                                HLE10430
      WRITE(13,1220)                                             HLE10440
 1220 FORMAT(/' RITZ VECTORS WERE NOT COMPUTED FOR THE FOLLOWING EIGENVAHLE10450
     1LUES'/' EITHER BECAUSE THEY HAD NOT CONVERGED OR BECAUSE'/' THE ERHLE10460
     1ROR ESTIMATE WAS NOT AS SMALL AS DESIRED'/)               HLE10470
C                                                                HLE10480
      WRITE(13,1230)                                             HLE10490
 1230 FORMAT(6X,'GOODEV(J)',3X,'MA(J)',3X,'M1(J)',3X,'M2(J)',3X,'MP(J)' HLE10500
     1/)                                                         HLE10510
      WRITE(13,1240) GOODEV(J),MA(J),M1(J),M2(J),MP(J)          HLE10520
 1240 FORMAT(E15.8,4I8)                                          HLE10530
C                                                                HLE10540
 1250 CONTINUE                                                   HLE10550
 1260 CONTINUE                                                   HLE10560
C                                                                HLE10570
      WRITE(9,1270)                                              HLE10580
 1270 FORMAT(/' ABOVE ARE ERROR ESTIMATES FOR THE A AND T EIGENVECTORS'/HLE10590
     1 ' ASSOCIATED WITH THE GOODEV LISTED IN COLUMN 1'/         HLE10600
     1 ' AERROR = NORM(A*X - EV*X)   TERROR = NORM(T*Y - EV*Y) '/ HLE10610
     1 ' WHERE T = T(1,MA(J))   X = RITZ VECTOR = V*Y  V = SUCCESSIVE'/HLE10620
     1 ' LANCZOS VECTORS. A MINGAP = GAP TO NEAREST A-EIGENVALUE'//) HLE10630
C                                                                HLE10640
```

```
      WRITE(13,1280)                                                    HLE10650
 1280 FORMAT(/' ABOVE ARE ERROR ESTIMATES ASSOCIATED WITH THE GOODEV'/  HLE10660
     1 ' RITZNORM = NORM(RITZ VECTOR)'/                                 HLE10670
     1 ' TBETA(J) = CDABS(BETA(MA(J)+1)*Y(MA(J))), T*Y = GOODEV*Y'/     HLE10680
     1 ' TLAST(J) = CDABS(Y(MA(J)))'/                                   HLE10690
     1 ' AMINGAP = DISTANCE TO CLOSEST COMPUTED GOOD T-EIGENVALUE'/)    HLE10700
C                                                                       HLE10710
C     NUMBER OF RITZ VECTORS COMPUTED                                   HLE10720
      NCOMPU = NGOODC - MREJEC                                          HLE10730
      WRITE(12,1290) N,NCOMPU,NGOODC,MATNO                              HLE10740
 1290 FORMAT(3I6,I12,' SIZE A, NO.RITZVECS, NO.EVALUES,MATNO')          HLE10750
C                                                                       HLE10760
      LFIN = 0                                                          HLE10770
      DO 1350 J = 1,NGOODC                                              HLE10780
      LINT = LFIN + 1                                                   HLE10790
      LFIN = LFIN + N                                                   HLE10800
C                                                                       HLE10810
      IF(MP(J).EQ.MPMIN) GO TO 1330                                     HLE10820
C     RITZ VECTOR WAS COMPUTED                                          HLE10830
      WRITE(12,1300) J, GOODEV(J), MP(J)                                HLE10840
 1300 FORMAT(I6,4X,E20.12,I6,' J, EIGENVAL, MP(J)')                     HLE10850
C                                                                       HLE10860
      WRITE(12,1310) ERR(J),ERRDGP(J)                                   HLE10870
 1310 FORMAT(2E15.5,' = NORM(A*Z-EVAL*Z) AND  NORM(A*Z-EVAL*Z)/MINGAP') HLE10880
C                                                                       HLE10890
      WRITE(12,1320) (RITVEC(LL), LL=LINT,LFIN)                         HLE10900
 1320 FORMAT(4Z20)                                                      HLE10910
      GO TO 1350                                                        HLE10920
C     NO RITZ VECTOR WAS COMPUTED FOR THIS EIGENVALUE                   HLE10930
 1330 WRITE(12,1340) J,GOODEV(J),MP(J)                                  HLE10940
 1340 FORMAT(I6,4X,E20.12,I6,' J,EIGVALUE,NO RITZ VECTOR COMPUTED')     HLE10950
C                                                                       HLE10960
 1350 CONTINUE                                                          HLE10970
C                                                                       HLE10980
C     DID ANY T-MATRICES INCLUDE OFF-DIAGONAL ENTRIES SMALLER THAN      HLE10990
C     DESIRED, AS SPECIFIED BY BTOL?                                    HLE11000
C                                                                       HLE11010
      IF(IB.GT.0) GO TO 1380                                            HLE11020
      WRITE(6,1360) KMAXU                                               HLE11030
 1360 FORMAT(/' FOR LARGEST T-MATRIX CONSIDERED',I7,' CHECK THE SIZE OF HLE11040
     1BETAS')                                                          HLE11050
C                                                                       HLE11060
C----------------------------------------------------------------------HLE11070
C                                                                       HLE11080
      CALL TNORM(ALPHA,BETA,BKMIN,TEMP,KMAXU,IBMT)                      HLE11090
C                                                                       HLE11100
C----------------------------------------------------------------------HLE11110
C                                                                       HLE11120
      IF(IBMT.LT.0) WRITE (6,1370)                                      HLE11130
 1370 FORMAT(/' WARNING THE T-MATRICES FOR ONE OR MORE OF THE EIGENVALUEHLE11140
     1S CONSIDERED'/' HAD AN OFF-DIAGONAL ENTRY THAT WAS SMALLER THAN THHLE11150
     1E BETA TOLERANCE THAT WAS SPECIFIED'/)                           HLE11160
 1380 CONTINUE                                                          HLE11170
C                                                                       HLE11180
      GO TO 1630                                                        HLE11190
C                                                                       HLE11200
 1390 WRITE(6,1400) NGOOD,NMAX,MDIMRV                                   HLE11210
 1400 FORMAT(/I4,' RITZ VECTORS WERE REQUESTED BUT THE REQUIRED DIMENSIOHLE11220
     1N',I6/' IS LARGER THAN THE USER-SPECIFIED DIMENSION OF RITVEC',I6 HLE11230
     1/' THEREFORE, THE EIGENVECTOR PROCEDURE TERMINATES FOR THE USER TOHLE11240
     1 INTERVENE')                                                     HLE11250
C                                                                       HLE11260
      GO TO 1630                                                        HLE11270
C                                                                       HLE11280
 1410 WRITE(6,1420) NOLD,N,MATOLD,MATNO                                 HLE11290
 1420 FORMAT(/' PARAMETERS READ FROM FILE 3 DO NOT AGREE WITH THOSE SPECHLE11300
     1 IFIED'/' BY THE USER.   NOLD,N,MATOLD,MATNO = '/2I6,2I12/        HLE11310
```

```
      1' THEREFORE, PROGRAM TERMINATES FOR USER TO RESOLVE THE DIFFERENCEHLE11320
      1S'/)                                                             HLE11330
C                                                                       HLE11340
      GO TO 1630                                                        HLE11350
C                                                                       HLE11360
 1430 WRITE(6,1440)                                                     HLE11370
 1440 FORMAT(/' PARAMETERS IN ALPHA,BETA FILE READ IN DO NOT AGREE WITH HLE11380
      1 THOSE'/' SPECIFIED BY THE USER.  THEREFORE, THE PROCEDURE TERMINAHLE11390
      1TES'/' FOR THE USER TO RESOLVE THE DIFFERENCES.'/)               HLE11400
C                                                                       HLE11410
      GO TO 1630                                                        HLE11420
C                                                                       HLE11430
 1450 WRITE(6,1460) KMAX,MEV                                            HLE11440
 1460 FORMAT(/' ON ALPHA,BETA HEADER KMAX = ',I6/                       HLE11450
      1' BUT EIGENVALUES WERE COMPUTED AT MEV = ',I6,' PROGRAM STOPS'/)  HLE11460
C                                                                       HLE11470
      GO TO 1630                                                        HLE11480
C                                                                       HLE11490
 1470 WRITE(6,1480)                                                     HLE11500
 1480 FORMAT(/' PROGRAM COMPUTED 1ST GUESSES ON T-MATRIX SIZES, READ THEHLE11510
      1M TO FILE 10'/' THEN TERMINATED AS REQUESTED.')                  HLE11520
      GO TO 1630                                                        HLE11530
C                                                                       HLE11540
 1490 WRITE(6,1500) MTOL, MDIMTV                                        HLE11550
 1500 FORMAT(/' PROGRAM TERMINATES BECAUSE THE TVEC DIMENSION ANTICIPATEHLE11560
      1D',I7/' IS LARGER THAN THE TVEC DIMENSION',I7,' SPECIFIED BY THE HLE11570
      1USER.'/' USER MAY RESET THE TVEC DIMENSION AND RESTART THE PROGRAHLE11580
      1M')                                                              HLE11590
      GO TO 1630                                                        HLE11600
C                                                                       HLE11610
 1510 WRITE(6,1520)                                                     HLE11620
 1520 FORMAT(/' PROGRAM TERMINATES BECAUSE NO SUITABLE T-EIGENVECTORS WEHLE11630
      1RE IDENTIFIED'/' FOR ANY OF THE EIGENVALUES SUPPLIED.  PROBLEM COHLE11640
      1ULD BE CAUSED'/' BY TOO SMALL A TVEC DIMENSION OR SIMPLY BE THAT HLE11650
      1IT WAS NOT POSSIBLE'/' TO IDENTIFY T-VECTORS.  USER SHOULD CHECK HLE11660
      1OUTPUT'/)                                                        HLE11670
      GO TO 1630                                                        HLE11680
C                                                                       HLE11690
 1530 WRITE(6,1540) LVCONT,NTVEC,NGOOD                                  HLE11700
 1540 FORMAT(/' LVCONT FLAG =',I2,' AND NUMBER ',I5,' OF T-EIGENVECTORS HLE11710
      1 COMPUTED N.E.'/' NUMBER',I5,' REQUESTED SO PROGRAM TERMINATES'/) HLE11720
      GO TO 1630                                                        HLE11730
 1550 WRITE(6,1560)                                                     HLE11740
 1560 FORMAT(/' PROGRAM TERMINATES WITHOUT COMPUTING ANY RITZ VECTORS'/ HLE11750
      1 ' BECAUSE ALL T-EIGENVECTORS WERE REJECTED AS NOT SUITABLE'/    HLE11760
      1 ' PROBABLE CAUSE IS LACK OF CONVERGENCE OF THE EIGENVALUES'/)   HLE11770
      GO TO 1630                                                        HLE11780
C                                                                       HLE11790
 1570 WRITE(6,1580)                                                     HLE11800
 1580 FORMAT(/' PROGRAM INDICATES THAT IT IS NOT POSSIBLE TO COMPUTE ANYHLE11810
      1 OF THE'/' REQUESTED EIGENVECTORS. THEREFORE PROGRAM TERMINATES') HLE11820
      DO 1590 J=1,NGOODC                                                HLE11830
 1590 WRITE(6,1600)  J,GOODEV(J),MP(J)                                  HLE11840
 1600 FORMAT(/4X,' J',11X,'GOODEV(J)',4X,'MP(J)'/I6,E20.12,I9)          HLE11850
      GO TO 1630                                                        HLE11860
C                                                                       HLE11870
 1610 WRITE(6,1620) MBETA,KMAXN                                         HLE11880
 1620 FORMAT(/' PROGRAM TERMINATES BECAUSE THE STORAGE ALLOTTED FOR THE HLE11890
      1BETA ARRAY',I8/' IS NOT SUFFICIENT FOR THE ENLARGED KMAX =',I8,' THLE11900
      1HAT THE PROGRAM WANTS'/' USER CAN ENLARGE THE ALPHA AND BETA ARRAYHLE11910
      1S AND RERUN THE PROGRAM.'/)                                      HLE11920
C                                                                       HLE11930
 1630 CONTINUE                                                          HLE11940
C                                                                       HLE11950
      STOP                                                              HLE11960
C-----END OF MAIN PROGRAM FOR LANCZOS HERMITIAN EIGENVECTOR COMPUTATIONSHLE11970
      END                                                               HLE11980
```

SECTION 3.4   LANCZS AND SAMPLE MATRIX-VECTOR MULTIPLY SUBROUTINES

```
C-----HLEMULT----HERMITIAN MATRICES----------------------------------------HLE00010
C                                                                          HLE00020
C     CONTAINS SUBROUTINE LANCZS AND SAMPLE USPEC, CMATV                   HLE00030
C     USED BY THE HERMITIAN VERSION OF THE LANCZOS ALGORITHMS              HLE00040
C                                                                          HLE00050
C     PORTABILITY:                                                         HLE00060
C     THESE PROGRAMS ARE NOT PORTABLE DUE TO THE USE OF COMPLEX*16         HLE00070
C     VARIABLES.  MOREOVER, THE PFORT VERIFIER IDENTIFIED THE              HLE00080
C     FOLLOWING ADDITIONAL NONPORTABLE CONSTRUCTIONS:                      HLE00090
C     1.  THE ENTRY MECHANISM USED TO PASS THE STORAGE                     HLE00100
C         LOCATIONS OF THE USER-SPECIFIED MATRIX FROM THE                  HLE00110
C         SUBROUTINE USPEC TO THE MATRIX-VECTOR SUBROUTINE CMATV.          HLE00120
C     2.  IN THE PROGRAMS PROVIDED FOR 'HERMITIAN POISSON' TEST MATRICESHLE00130
C         USPEC CONTAINS FREE FORMAT (8,*), AND FORMAT (20A4); AND         HLE00140
C         EXACT ERROR SUBROUTINE CONTAINS DATA/MACHEP DEFINITION.          HLE00150
C                                                                          HLE00160
C                                                                          HLE00170
C-----LANCZS-COMPUTE THE LANCZOS TRIDIAGONAL MATRICES-------------------HLE00180
C                                                                          HLE00190
C     GRAM-SCHMIDT ORTHOGONALIZATION WITHOUT MODIFICATION                  HLE00200
C     REQUIRES EXTRA VECTOR VS IN LANCZS.  MODIFICATION IS NOT             HLE00210
C     PERMISSIBLE IN THE HERMITIAN CASE BECAUSE CCMPLEX PORTION            HLE00220
C     OF THE MODIFICATION COULD NOT BE INCORPORATED.                       HLE00230
C                                                                          HLE00240
      SUBROUTINE LANCZS(MATVEC,V1,V2,VS,ALPHA,BETA,GR,GC,G,KMAX,MOLD1,N,HLE00250
     1 IIX)                                                                HLE00260
C                                                                          HLE00270
C----------------------------------------------------------------------HLE00280
      COMPLEX*16 V1(1), V2(1), VS(1), ZEROC, TEMP                          HLE00290
      DOUBLE PRECISION  ALPHA(1), BETA(1), BATA, SUM, ONE, ZERO           HLE00300
      DOUBLE PRECISION GR(1),GC(1)                                         HLE00310
      REAL  G(1)                                                           HLE00320
      EXTERNAL MATVEC                                                      HLE00330
      DOUBLE PRECISION  DSQRT                                              HLE00340
C----------------------------------------------------------------------HLE00350
C                                                                          HLE00360
      ZERO = 0.D0                                                          HLE00370
      ONE = 1.D0                                                           HLE00380
      ZEROC = DCMPLX(ZERO,ZERO)                                            HLE00390
C                                                                          HLE00400
      IF(MOLD1.GT.1)GO TO 50                                               HLE00410
C                                                                          HLE00420
C     ALPHA/BETA GENERATION STARTS AT I = 1                                HLE00430
C     MOLD1 = 1 SET V1 = 0. AND V2 = RANDOM UNIT VECTOR                    HLE00440
C     IIL=IIX                                                              HLE00450
C                                                                          HLE00460
C----------------------------------------------------------------------HLE00470
      CALL GENRAN(IIL,G,N)                                                 HLE00480
C----------------------------------------------------------------------HLE00490
C                                                                          HLE00500
      DO 10 I = 1,N                                                        HLE00510
   10 GR(I) = G(I)                                                         HLE00520
C                                                                          HLE00530
C----------------------------------------------------------------------HLE00540
      CALL GENRAN(IIL,G,N)                                                 HLE00550
C----------------------------------------------------------------------HLE00560
C                                                                          HLE00570
      DO 20 I = 1,N                                                        HLE00580
   20 GC(I) = G(I)                                                         HLE00590
C                                                                          HLE00600
      DO 30 I = 1,N                                                        HLE00610
   30 V2(I) = DCMPLX(GR(I),GC(I))                                          HLE00620
C                                                                          HLE00630
C----------------------------------------------------------------------HLE00640
```

```
      CALL CINPRD(V2,V2,SUM,N)                                      HLE00650
C--------------------------------------------------------------------------HLE00660
C                                                                   HLE00670
      SUM = ONE/DSQRT(SUM)                                          HLE00680
      DO 40 I = 1,N                                                 HLE00690
      V1(I) = ZEROC                                                 HLE00700
   40 V2(I) = V2(I)*SUM                                             HLE00710
      BETA(1) = ZERO                                                HLE00720
C                                                                   HLE00730
C     ALPHA BETA GENERATION LOOP                                    HLE00740
   50 CONTINUE                                                      HLE00750
C                                                                   HLE00760
      DO 80 I=MOLD1,KMAX                                            HLE00770
      SUM = ZERO                                                    HLE00780
C                                                                   HLE00790
C--------------------------------------------------------------------------HLE00800
C     MATVEC(V2,VS,SUM) CALCULATES   VS = A*V2 - SUM*VS             HLE00810
      CALL MATVEC(V2,VS,SUM)                                        HLE00820
      CALL CINPRD(V2,VS,SUM,N)                                      HLE00830
C--------------------------------------------------------------------------HLE00840
C                                                                   HLE00850
      ALPHA(I) = SUM                                                HLE00860
      BATA = BETA(I)                                                HLE00870
      DO 60 J=1,N                                                   HLE00880
   60 V1(J) = (VS(J)-BATA*V1(J)) - SUM*V2(J)                        HLE00890
C                                                                   HLE00900
C--------------------------------------------------------------------------HLE00910
      CALL CINPRD(V1,V1,SUM,N)                                      HLE00920
C--------------------------------------------------------------------------HLE00930
C                                                                   HLE00940
      IN = I+1                                                      HLE00950
      BETA(IN) = DSQRT(SUM)                                         HLE00960
      SUM = ONE/BETA(IN)                                            HLE00970
      DO 70 J=1,N                                                   HLE00980
      TEMP = SUM*V1(J)                                              HLE00990
      V1(J) = V2(J)                                                 HLE01000
   70 V2(J) = TEMP                                                  HLE01010
   80 CONTINUE                                                      HLE01020
C     END ALPHA, BETA GENERATION LOOP                              HLE01030
C                                                                   HLE01040
C-----END OF LANCZS---------------------------------------------------------HLE01050
C                                                                   HLE01060
      RETURN                                                        HLE01070
      END                                                           HLE01080
C                                                                   HLE01090
C-----USPEC-GENERAL SPARSE, HERMITIAN MATRIX-------------------------------HLE01100
C                                                                   HLE01110
C     SUBROUTINE USPEC(N,MATNO)                                     HLE01120
      SUBROUTINE GUSPEC(N,MATNO)                                    HLE01130
C                                                                   HLE01140
C--------------------------------------------------------------------------HLE01150
      COMPLEX*16 A(3000)                                            HLE01160
      DOUBLE PRECISION  AD(1000)                                    HLE01170
      INTEGER  IROW(3000),ICOL(1000)                                HLE01180
C--------------------------------------------------------------------------HLE01190
C     DIMENSION ARRAYS NEEDED TO DEFINE MATRIX, READ IN VALUES FOR  HLE01200
C     ARRAYS AND THEN PASS THE STORAGE LOCATIONS OF THESE ARRAYS TO HLE01210
C     THE MATRIX-VECTOR MULTIPLY SUBROUTINE CMATV.                  HLE01220
C                                                                   HLE01230
C     USER-SUPPLIED MATRIX IS STORED IN FOLLOWING SPARSE FORMAT:    HLE01240
C     N = ORDER OF A-MATRIX                                         HLE01250
C     NZS = NUMBER OF NONZERO SUBDIAGONAL ENTRIES IN A              HLE01260
C     NZL = INDEX OF LAST COLUMN CONTAINING NONZERO SUBDIAGONAL ENTRIES HLE01270
C     ICOL(J), J=1,NZL IS THE NUMBER OF NONZERO SUBDIAGONAL ELEMENTS HLE01280
C                 IN COLUMN J.                                      HLE01290
C     IROW(K), K = 1,NZS, IS THE ROW INDEX FOR CORRESPONDING A(K).  HLE01300
C     AD(I), I=1,N ARE DIAGONAL ENTRIES (INCLUDING ANY 0 DIAGONAL   HLE01310
```

```
C          ENTRIES)                                             HLE01320
C    A(K), K=1,NZS ARE NONZERO SUBDIAGONAL ENTRIES, LISTED BY COLUMN.  HLE01330
C    FOR J > NZL THERE ARE NO NONZERO SUBDIAGONAL ELEMENTS IN COLUMN J. HLE01340
C    ICOL(J) = 0 IS ALLOWED                                     HLE01350
C                                                               HLE01360
C-------------------------------------------------------------------HLE01370
C    IN THIS SAMPLE SUBROUTINE THE ARRAYS ARE READ IN FROM FILE 8  HLE01380
C                                                               HLE01390
     READ(8,10) NZS,NOLD,NZL,MATOLD                             HLE01400
  10 FORMAT(I10,2I6,I8)                                         HLE01410
C                                                               HLE01420
     WRITE(6,20) NZS,NOLD,NZL,MATOLD                            HLE01430
  20 FORMAT(I10,2I6,I8,' = NZS,NOLD,NZL,MATOLD'/)               HLE01440
C                                                               HLE01450
C    TEST OF PARAMETER CORRECTNESS                              HLE01460
     ITEMP = (NOLD-N)**2 + (MATNO-MATOLD)**2                    HLE01470
C                                                               HLE01480
     IF(ITEMP.EQ.0) GO TO 40                                    HLE01490
C                                                               HLE01500
     WRITE(6,30)                                                HLE01510
  30 FORMAT(/' PROGRAM TERMINATES BECAUSE EITHER ORDERS OF OR LABELS FOHLE01520
    1R MATRIX DISAGREE'/)                                       HLE01530
     GO TO 80                                                   HLE01540
C                                                               HLE01550
  40 CONTINUE                                                   HLE01560
C                                                               HLE01570
C    NUMBER OF NONZERO SUBDIAGONAL ENTRIES IN EACH COLUMN IS READ  HLE01580
C    THEN THE CORRESPONDING ROW INDEX FOR EACH SUCH ENTRY IS READ  HLE01590
     READ(8,50) (ICOL(K), K=1,NZL)                              HLE01600
     READ(8,50) (IROW(K), K=1,NZS)                              HLE01610
  50 FORMAT(13I6)                                               HLE01620
C    DIAGONAL IS READ FIRST, THEN NONZERO BELOW DIAGONAL ENTRIES  HLE01630
     READ(8,60) (AD(K), K=1,N)                                  HLE01640
  60 FORMAT(4E20.12)                                            HLE01650
     READ(8,70) (A(K), K=1,NZS)                                 HLE01660
C  50 FORMAT(4Z20)                                               HLE01670
  70 FORMAT(4E20.12)                                            HLE01680
C                                                               HLE01690
C-------------------------------------------------------------------HLE01700
C    PASS STORAGE LOCATIONS OF ARRAYS THAT DEFINE THE MATRIX TO  HLE01710
C    THE MATRIX-VECTOR MULTIPLY SUBROUTINE CMATV                HLE01720
     CALL CMATVE(A,AD,ICOL,IROW,N,NZL)                          HLE01730
C-------------------------------------------------------------------HLE01740
C                                                               HLE01750
     RETURN                                                     HLE01760
  80 STOP                                                       HLE01770
C                                                               HLE01780
C-----END OF USPEC FOR GENERAL, SPARSE HERMITIAN MATRICES-------------HLE01790
     END                                                        HLE01800
C                                                               HLE01810
C-----START OF MATRIX-VECTOR MULTIPLY-GENERAL SPARSE HERMITIAN--------- HLE01820
C                                                               HLE01830
C    SUBROUTINE CMATV(W,U,SUM)                                  HLE01840
     SUBROUTINE GCMATV(W,U,SUM)                                 HLE01850
C                                                               HLE01860
C-------------------------------------------------------------------HLE01870
     COMPLEX*16  U(1),W(1),A(1)                                 HLE01880
     DOUBLE PRECISION  AD(1),SUM                                HLE01890
     INTEGER  IROW(1),ICOL(1)                                   HLE01900
C-------------------------------------------------------------------HLE01910
C    SPARSE MATRIX-VECTOR MULTIPLY FOR LANCZS  U = A*W - SUM*U  HLE01920
C    SEE USPEC SUBROUTINE FOR DESCRIPTION OF THE ARRAYS THAT DEFINE  HLE01930
C    THE MATRIX                                                 HLE01940
C                                                               HLE01950
C    COMPUTE THE DIAGONAL TERMS                                 HLE01960
     DO 10 I = 1,N                                              HLE01970
```

```
      10 U(I) = AD(I)*W(I)-SUM*U(I)                                    HLE01980
C                                                                      HLE01990
C     COMPUTE BY COLUMN                                                HLE02000
      LLAST = 0                                                        HLE02010
      DO 30 J = 1,NZL                                                  HLE02020
C                                                                      HLE02030
      IF (ICOL(J).EQ.0) GO TO 30                                       HLE02040
      LFIRST = LLAST + 1                                               HLE02050
      LLAST = LLAST + ICOL(J)                                          HLE02060
C                                                                      HLE02070
      DO 20 L = LFIRST,LLAST                                           HLE02080
      I = IROW(L)                                                      HLE02090
C                                                                      HLE02100
      U(I) = U(I) + A(L)*W(J)                                          HLE02110
      U(J) = U(J) + DCONJG(A(L))*W(I)                                  HLE02120
C                                                                      HLE02130
   20 CONTINUE                                                         HLE02140
C                                                                      HLE02150
   30 CONTINUE                                                         HLE02160
C                                                                      HLE02170
      RETURN                                                           HLE02180
C                                                                      HLE02190
C---------------------------------------------------------------------HLE02200
C     STORAGE LOCATIONS OF ARRAYS ARE PASSED TO CMATV FROM USPEC       HLE02210
      ENTRY CMATVE(A,AD,ICOL,IROW,N,NZL)                               HLE02220
C---------------------------------------------------------------------HLE02230
C                                                                      HLE02240
C-----END OF CMATV-GENERAL, SPARSE, HERMITIAN MATRICES ---------------HLE02250
      RETURN                                                           HLE02260
      END                                                              HLE02270
C                                                                      HLE02280
C-----USPEC, CMATV, EXEVG, AND HEXVEC FOR HERMITIAN 'POISSON' MATRICES--HLE02290
C                                                                      HLE02300
C-----USPEC (HERMITIAN POISSON MATRICES)------------------------------HLE02310
C                                                                      HLE02320
C     SUBROUTINE HUSPEC(N,MATNO)                                       HLE02330
      SUBROUTINE USPEC(N,MATNO)                                        HLE02340
C                                                                      HLE02350
C---------------------------------------------------------------------HLE02360
      DOUBLE PRECISION  CO,C1,C2,HALF,ONE,SCR,SCI,ANGLE,TEMP           HLE02370
      COMPLEX*16 SC,TC,CL0,CL1,CL3,CL4                                 HLE02380
      REAL EXPLAN(20)                                                  HLE02390
      DOUBLE PRECISION EIGVAL(1000)                                    HLE02400
      REAL GAPS(1000)                                                  HLE02410
      INTEGER MULTS(1000)                                              HLE02420
C---------------------------------------------------------------------HLE02430
      HALF = 0.5D0                                                     HLE02440
      ONE  = 1.0D0                                                     HLE02450
C                                                                      HLE02460
C     READ IN PARAMETERS TO DEFINE MATRIX                             HLE02470
C     MATRIX IS COMPLEX DIAGONAL SIMILITARY TRANSFORM OF REAL SYMMETRIC HLE02480
C     POISSON MATRIX WHICH HAS SYMMETRIC TOEPLITZ BLOCKS ALONG         HLE02490
C     THE DIAGONAL, EACH ONE OF WHICH HAS THE PARAMETER C2 ALONG THE   HLE02500
C     DIAGONAL AND -CO ABOVE AND BELOW THE DIAGONAL, AND OFF-DIAGONAL  HLE02510
C     BLOCKS THAT ARE DIAGONAL WITH DIAGONAL ENTRIES -C1.  EACH BLOCK  HLE02520
C     IS KX*KX AND THERE ARE KY BLOCKS.  THE HERMITIAN VERSION IS      HLE02530
C     OBTAINED BY APPLYING A DIAGONAL SIMILARITY TRANSFORM TO THE      HLE02540
C     REAL MATRIX WHERE THIS TRANSFORMATION IS SUCH THAT ITS           HLE02550
C     DIAGONAL ENTRIES ARE (SC)**(K-1), K = 1,...,N, WHERE SC          HLE02560
C     HAS MODULUS 1.                                                   HLE02570
C                                                                      HLE02580
      READ(8,10) EXPLAN                                                HLE02590
      WRITE(6,10) EXPLAN                                               HLE02600
      READ(8,10) EXPLAN                                                HLE02610
   10 FORMAT(20A4)                                                     HLE02620
C     IF MTYPE = 0 WE HAVE ZERO BOUNDARY CONDITIONS                    HLE02630
C     IF MTYPE = 1 WE HAVE NORMAL DERIVATIVE BOUNDARY CONDITIONS       HLE02640
```

```
C     NOTE THAT SUBROUTINES EXEVG AND HEXVEC ARE VALID ONLY FOR      HLE02650
C     MTYPE = 0.                                                     HLE02660
      READ(8,*) NOLD,MATOLD,IVEC,MTYPE                               HLE02670
      WRITE(6,20) NOLD,MATOLD                                        HLE02680
   20 FORMAT(' ORDER OF MATRIX READ FROM FILE =',I6/' MATRIX NUMBER =', HLE02690
     118/)                                                           HLE02700
      IF(MTYPE.EQ.0) WRITE(6,30)                                     HLE02710
   30 FORMAT(/' HERMITIAN POISSON CORRESPONDING TO ZERO BOUNDARY CONDITIHLE02720
     10NS'/)                                                         HLE02730
      IF(MTYPE.EQ.1) WRITE(6,40)                                     HLE02740
   40 FORMAT(/' HERMITIAN POISSON CORRESPONDING TO NORMAL DERIVATIVE BOUHLE02750
     1NDARY CONDITIONS'/)                                            HLE02760
      IF(IVEC.NE.0.AND.MTYPE.EQ.0) WRITE(6,50)                       HLE02770
   50 FORMAT(' COMPUTE THE TRUE EIGENVALUES AND PUT IN FN TRUEEVAL'/) HLE02780
C                                                                    HLE02790
C     TEST OF PARAMETER CORRECTNESS                                  HLE02800
      ITEMP = (NOLD-N)**2 + (MATNO-MATOLD)**2                        HLE02810
C                                                                    HLE02820
      IF(ITEMP.EQ.0) GO TO 70                                        HLE02830
C                                                                    HLE02840
      WRITE(6,60)                                                    HLE02850
   60 FORMAT(' PROGRAM TERMINATES BECAUSE EITHER ORDERS OF OR LABELS FORHLE02860
     1 MATRIX DISAGREE')                                             HLE02870
      GO TO 150                                                      HLE02880
C                                                                    HLE02890
   70 CONTINUE                                                       HLE02900
C                                                                    HLE02910
      READ(8,10) EXPLAN                                              HLE02920
      READ(8,*)  CO,KX,KY                                            HLE02930
      IF (KX.GT.4.AND.KY.GT.4) GO TO 90                              HLE02940
      WRITE(6,80) KX,KY                                              HLE02950
   80 FORMAT(2I6,' = KX KY  ONE OR BOTH OF KX KY TOO SMALL SO STOP'/) HLE02960
      GO TO 150                                                      HLE02970
   90 CONTINUE                                                       HLE02980
      READ(8,10) EXPLAN                                              HLE02990
C     BELOW SC = COS(ANGLE) + I SIN(ANGLE)                          HLE03000
C     READ IN DESIRED COSINE, COMPUTE ANGLE, THEN SINE              HLE03010
      READ(8,*) SCR                                                 HLE03020
      ANGLE = DARCOS(SCR)                                           HLE03030
      SCI = DSIN(ANGLE)                                             HLE03040
      SC = DCMPLX(SCR,SCI)                                          HLE03050
      WRITE(6,100) SC                                               HLE03060
C     IF (IVEC.NE.0.AND.MTYPE.EQ.0) WRITE(9,7) SC                   HLE03070
  100 FORMAT(' GENERATOR OF DIAGONAL TRANSFORMATION ='/2E20.12)      HLE03080
C                                                                    HLE03090
      TC = SC                                                        HLE03100
      DO 110 J=2,KX                                                  HLE03110
  110 TC = SC*TC                                                     HLE03120
      WRITE(6,120) TC                                                HLE03130
  120 FORMAT(' TC = ',2E20.12)                                       HLE03140
C                                                                    HLE03150
      N = KX*KY                                                      HLE03160
      C2 = ONE                                                       HLE03170
      C1 = HALF-CO                                                   HLE03180
      TEMP = DSQRT(2.0D0)                                            HLE03190
      IF (MTYPE.EQ.0) TEMP = ONE                                     HLE03200
      CL0 = -SC*CO                                                   HLE03210
      CL1 = -TC*C1                                                   HLE03220
      CL3 = -SC*CO*TEMP                                              HLE03230
      CL4 = -TC*C1*TEMP                                              HLE03240
C                                                                    HLE03250
      WRITE(6,130) N,MTYPE,KX,KY,C2,CO,C1                            HLE03260
  130 FORMAT(/5X,'N',1X,'MTYPE',4X,'KX',4X,'KY',7X,'DIAGONAL',       HLE03270
     1 3X,'X-CODIAGONAL',3X,'Y-CODIAGONAL'/4I6,3E15.8/)             HLE03280
C                                                                    HLE03290
C------------------------------------------------------------------HLE03300
```

```
        CALL HMATVE(C2,CLO,CL1,CL3,CL4,KX,KY)                          HLE03310
C----------------------------------------------------------------------HLE03320
C                                                                       HLE03330
        IF(IVEC.EQ.0.OR.MTYPE.NE.0) GO TO 140                          HLE03340
C                                                                       HLE03350
C       COMPUTE THE EXACT EIGENVALUES                                   HLE03360
C                                                                       HLE03370
C----------------------------------------------------------------------HLE03380
        CALL EXEVG(EIGVAL,CO,C1,C2,GAPS,MULTS,KX,KY)                   HLE03390
C----------------------------------------------------------------------HLE03400
C                                                                       HLE03410
        IF(IVEC.LT.0) GO TO 150                                        HLE03420
C                                                                       HLE03430
  140 CONTINUE                                                          HLE03440
        RETURN                                                         HLE03450
C                                                                       HLE03460
C-----END OF USPEC-----------------------------------------------------HLE03470
  150 STOP                                                              HLE03480
        END                                                            HLE03490
C                                                                       HLE03500
C-----START OF CMATV FOR HERMITIAN POISSON MATRICES--------------------HLE03510
C                                                                       HLE03520
C       SUBROUTINE HMATV(W,U,SUM)                                       HLE03530
        SUBROUTINE CMATV(W,U,SUM)                                       HLE03540
C                                                                       HLE03550
C----------------------------------------------------------------------HLE03560
        DOUBLE PRECISION  C2,SUM                                        HLE03570
        COMPLEX*16 U(1),W(1)                                            HLE03580
        COMPLEX*16 CLO,CL1,CL3,CL4,CRO,CR1,CR3,CR4                      HLE03590
C----------------------------------------------------------------------HLE03600
C       CALCULATES U = A*W - SUM*U                                      HLE03610
C                                                                       HLE03620
        N = KK*LL                                                       HLE03630
        CRO = DCONJG(CLO)                                               HLE03640
        CR1 = DCONJG(CL1)                                               HLE03650
        CR3 = DCONJG(CL3)                                               HLE03660
        CR4 = DCONJG(CL4)                                               HLE03670
C                                                                       HLE03680
C----------------------------------------------------------------------HLE03690
C    FIRST AND LAST BLOCKS                                              HLE03700
        J = 1                                                           HLE03710
        U(J)=(C2*W(J)+CR3*W(J+1)+CR1*W(J+KK)) - SUM*U(J)               HLE03720
        J = 2                                                           HLE03730
        U(J)=(C2*W(J)+CL3*W(J-1)+CRO*W(J+1)+CR1*W(J+KK))-SUM*U(J)      HLE03740
        J = KK                                                          HLE03750
        U(J)=(C2*W(J)+CL3*W(J-1)+CR1*W(J+KK))-SUM*U(J)                 HLE03760
        J = KK - 1                                                      HLE03770
        U(J)=(C2*W(J)+CR3*W(J+1)+CLO*W(J-1)+CR1*W(J+KK))-SUM*U(J)      HLE03780
        J = N - KK + 1                                                  HLE03790
        U(J)=(C2*W(J)+CR3*W(J+1)+CL4*W(J-KK))-SUM*U(J)                 HLE03800
        J = N - KK + 2                                                  HLE03810
        U(J)=(C2*W(J)+CL3*W(J-1)+CRO*W(J+1)+CL4*W(J-KK))-SUM*U(J)      HLE03820
        J = N                                                           HLE03830
        U(J)=(C2*W(J)+CL3*W(J-1)+CL4*W(J-KK))-SUM*U(J)                 HLE03840
        J = N - 1                                                       HLE03850
        U(J)=(C2*W(J)+CLO*W(J-1)+CR3*W(J+1)+CL4*W(J-KK))-SUM*U(J)      HLE03860
C                                                                       HLE03870
        KK2 = KK - 2                                                    HLE03880
        DO 10 JJ = 3,KK2                                                HLE03890
        J = JJ                                                          HLE03900
        U(J)=(C2*W(J)+CLO*W(J-1)+CRO*W(J+1)+CR1*W(J+KK))-SUM*U(J)      HLE03910
        J = N - KK + JJ                                                 HLE03920
   10 U(J)=(C2*W(J)+CLO*W(J-1)+CRO*W(J+1)+CL4*W(J-KK))-SUM*U(J)        HLE03930
C                                                                       HLE03940
C     START BLOCKS 2 AND LL-1                                           HLE03950
        J = KK + 1                                                      HLE03960
        U(J)=(C2*W(J)+CR3*W(J+1)+CL1*W(J-KK)+CR1*W(J+KK))-SUM*U(J)     HLE03970
```

```
        J = KK + 2                                                      HLE03980
        U(J)=(C2*W(J)+CL3*W(J-1)+CRO*W(J+1)+CL1*W(J-KK)+CR1*W(J+KK))    HLE03990
       1 -SUM*U(J)                                                      HLE04000
        J = KK + KK                                                     HLE04010
        U(J)=(C2*W(J)+CL3*W(J-1)+CL1*W(J-KK)+CR1*W(J+KK))-SUM*U(J)      HLE04020
        J = KK + KK - 1                                                 HLE04030
        U(J)=(C2*W(J)+CR3*W(J+1)+CLO*W(J-1)+CL1*W(J-KK)+CR1*W(J+KK))    HLE04040
       1 -SUM*U(J)                                                      HLE04050
        J = N - 2*KK + 1                                                HLE04060
        U(J)=(C2*W(J)+CR3*W(J+1)+CR4*W(J+KK)+CL1*W(J-KK))               HLE04070
       1 -SUM*U(J)                                                      HLE04080
        J = N - 2*KK + 2                                                HLE04090
        U(J)=(C2*W(J)+CL3*W(J-1)+CRO*W(J+1)+CR4*W(J+KK)+CL1*W(J-KK))    HLE04100
       1 -SUM*U(J)                                                      HLE04110
        J = N - KK                                                      HLE04120
        U(J)=(C2*W(J)+CL3*W(J-1)+CR4*W(J+KK)+CL1*W(J-KK))-SUM*U(J)      HLE04130
        J = N - KK - 1                                                  HLE04140
        U(J)=(C2*W(J)+CR3*W(J+1)+CLO*W(J-1)+CR4*W(J+KK)+CL1*W(J-KK))    HLE04150
       1 -SUM*U(J)                                                      HLE04160
C                                                                       HLE04170
        DO 20 JJ = 3,KK2                                                HLE04180
        J = KK + JJ                                                     HLE04190
        U(J)=(C2*W(J)+CLO*W(J-1)+CRO*W(J+1)+CL1*W(J-KK)+CR1*W(J+KK))    HLE04200
       1 -SUM*U(J)                                                      HLE04210
        J = N - 2*KK + JJ                                               HLE04220
        U(J)=(C2*W(J)+CLO*W(J-1)+CRO*W(J+1)+CR4*W(J+KK)+CL1*W(J-KK))    HLE04230
       1 -SUM*U(J)                                                      HLE04240
     20 CONTINUE                                                        HLE04250
C                                                                       HLE04260
C       MIDDLE BLOCKS                                                   HLE04270
        LL2 = LL - 2                                                    HLE04280
        JP = KK                                                         HLE04290
        DO 40 JJ = 3,LL2                                                HLE04300
        JP = JP + KK                                                    HLE04310
C       JP = (JJ-1)*KK                                                  HLE04320
        J = JP + 1                                                      HLE04330
        U(J)=(C2*W(J)+CR3*W(J+1)+CL1*W(J-KK)+CR1*W(J+KK))-SUM*U(J)      HLE04340
        J = J + 1                                                       HLE04350
        U(J)=(C2*W(J)+CL3*W(J-1)+CRO*W(J+1)+CL1*W(J-KK)+              HLE04360
       1 CR1*W(J+KK))-SUM*U(J)                                          HLE04370
        J = J + KK - 2                                                  HLE04380
        U(J) = (C2*W(J)+CL3*W(J-1)+CL1*W(J-KK)+CR1*W(J+KK))-SUM*U(J)    HLE04390
        J = J - 1                                                       HLE04400
        U(J)=(C2*W(J)+CR3*W(J+1)+CLO*W(J-1)+CL1*W(J-KK)+              HLE04410
       1 CR1*W(J+KK))-SUM*U(J)                                          HLE04420
C                                                                       HLE04430
        DO 30 II = 3,KK2                                                HLE04440
        J = JP + II                                                     HLE04450
        U(J)=(C2*W(J)+CLO*W(J-1)+CRO*W(J+1)+CL1*W(J-KK)+CR1*W(J+KK))    HLE04460
       1 -SUM*U(J)                                                      HLE04470
     30 CONTINUE                                                        HLE04480
C                                                                       HLE04490
     40 CONTINUE                                                        HLE04500
C                                                                       HLE04510
        RETURN                                                          HLE04520
C                                                                       HLE04530
        ENTRY HMATVE(C2,CLO,CL1,CL3,CL4,KK,LL)                          HLE04540
C                                                                       HLE04550
C-----END OF HMATV--------------------------------------------------HLE04560
        RETURN                                                          HLE04570
        END                                                            HLE04580
C                                                                       HLE04590
C-----START OF EXEVG-----------------------------------------------HLE04600
C                                                                       HLE04610
C       FOR MTYPE = 0, ZERO BOUNDARY CONDITIONS:                        HLE04620
C       COMPUTES EXACT EIGENVALUES OF HERMITIAN POISSON MATRIX,         HLE04630
C       THEIR MULTIPLICITIES, AND THE GAPS BETWEEN THE EIGENVALUES AND  HLE04640
```

```
C      PUTS THEM RESPECTIVELY INTO VECTORS U, MP, AND G.  THESE         HLE04650
C      QUANTITIES ARE ALL WRITTEN TO FILE 9.                            HLE04660
C                                                                       HLE04670
       SUBROUTINE EXEVG(U,CO,C1,C2,G,MP,KX,KY)                          HLE04680
C                                                                       HLE04690
C-----------------------------------------------------------------------HLE04700
       DOUBLE PRECISION  U(1),MACHEP                                    HLE04710
       DOUBLE PRECISION  EPSM,CO,C1,C2,TO,T1,PIK,PIL,ONE,TWO,ATOLN,EE   HLE04720
       REAL G(1)                                                        HLE04730
       INTEGER MP(1)                                                    HLE04740
C-----------------------------------------------------------------------HLE04750
       DATA MACHEP/Z3410000000000000/                                  HLE04760
       EPSM = 2.0DO*MACHEP                                              HLE04770
C-----------------------------------------------------------------------HLE04780
       N = KX*KY                                                        HLE04790
       ONE  = 1.0DO                                                     HLE04800
       TWO  = 2.0DO                                                     HLE04810
       TO = DARCOS(-ONE)                                                HLE04820
       T1 = DFLOAT(KX+1)                                                HLE04830
       PIK = TO/T1                                                      HLE04840
       T1 = DFLOAT(KY+1)                                                HLE04850
       PIL = TO/T1                                                      HLE04860
C      GENERATE EXACT EIGENVALUES                                       HLE04870
       KP = 0                                                           HLE04880
       DO 20 J = 1,KY                                                   HLE04890
       T1 = PIL*DFLOAT(J)                                               HLE04900
       TO = C2 - TWO*C1*DCOS(T1)                                        HLE04910
       DO 10 I = 1,KX                                                   HLE04920
       KP = KP+1                                                        HLE04930
       T1 = PIK*DFLOAT(I)                                               HLE04940
   10  U(KP) = TO - TWO*CO*DCOS(T1)                                     HLE04950
   20  CONTINUE                                                         HLE04960
C                                                                       HLE04970
C      ORDER U VECTOR BY INCREASING ALGEBRAIC SIZE                      HLE04980
       DO 40 K = 2,N                                                    HLE04990
       KM1 = K-1                                                        HLE05000
       DO 30 L = 1,KM1                                                  HLE05010
       JJ = K-L                                                         HLE05020
       IF (U(JJ+1).GE.U(JJ)) GO TO 40                                   HLE05030
       TO = U(JJ)                                                       HLE05040
       U(JJ) = U(JJ+1)                                                  HLE05050
   30  U(JJ+1) = TO                                                     HLE05060
   40  CONTINUE                                                         HLE05070
       ATOLN = DMAX1(DABS(U(1)),DABS(U(N)))*EPSM                        HLE05080
C                                                                       HLE05090
       WRITE(9,50)                                                      HLE05100
   50  FORMAT(' TRUE EIGENVALUES FOR HERMITIAN POISSON')                HLE05110
C                                                                       HLE05120
       WRITE(9,60)N,KX,KY,C2,CO,C1,ATOLN                                HLE05130
       WRITE(6,60) N,KX,KY,C2,CO,C1,ATOLN                               HLE05140
   60  FORMAT(1X,'A-SIZE',2X,'X-DIM',2X,'Y-DIM'/3I7/                    HLE05150
      1 5X,'A-DIAGONAL',3X,'X-CODIAGONAL',3X,'Y-CODIAGONAL',10X,'ATOLN'/HLE05160
      2 4E15.8)                                                         HLE05170
C                                                                       HLE05180
C      DETERMINE TRUE MULTIPLICITIES FOR EXACT EIGENVALUES              HLE05190
       I = 1                                                            HLE05200
       IDEX = 1                                                         HLE05210
       J = 1                                                            HLE05220
       NEXACT = 0                                                       HLE05230
   70  J = J+1                                                          HLE05240
       IF (J.GT.N) GO TO 80                                             HLE05250
       EE = DABS(U(J)-U(I))                                             HLE05260
       IF (EE.GT.ATOLN) GO TO 80                                        HLE05270
       IDEX = IDEX+1                                                    HLE05280
       GO TO 70                                                         HLE05290
   80  NEXACT = NEXACT+1                                                HLE05300
       U(NEXACT) = U(I)                                                 HLE05310
```

```
      MP(NEXACT) = IDEX                                             HLE05320
C     MP(K) = MULTIPLICITY OF KTH EIGENVALUE CLUSTER FOR A          HLE05330
      IDEX = 1                                                      HLE05340
      I = J                                                         HLE05350
      IF (I.GT.N) GO TO 90                                          HLE05360
      GO TO 70                                                      HLE05370
   90 CONTINUE                                                      HLE05380
C                                                                   HLE05390
C     MULTIPLICITIES HAVE BEEN DETERMINED                          HLE05400
C     NEXACT = NUMBER OF DISTINCT A-EIGENVALUES                     HLE05410
C                                                                   HLE05420
      WRITE(9,100)NEXACT                                            HLE05430
      WRITE(6,100)NEXACT                                            HLE05440
  100 FORMAT(I6,' = NUMBER OF TRUE A-EIGENVALUES WHICH ARE DISTINCT'/) HLE05450
C                                                                   HLE05460
C     MINGAP CALCULATION FOR DISTINCT A-EIGENVALUES                 HLE05470
      NM1 = NEXACT - 1                                              HLE05480
      G(NEXACT) = U(NM1)-U(NEXACT)                                  HLE05490
      G(1) = U(2)-U(1)                                              HLE05500
C                                                                   HLE05510
      DO 110 J = 2,NM1                                              HLE05520
      TO = U(J)-U(J-1)                                              HLE05530
      T1 = U(J+1)-U(J)                                              HLE05540
      G(J) = T1                                                     HLE05550
      IF (TO.LT.T1) G(J) = -TO                                      HLE05560
  110 CONTINUE                                                      HLE05570
C                                                                   HLE05580
C     NEXACT DISTINCT A-EIGENVALUES ARE IN U IN ASCENDING ORDER     HLE05590
C     MP = MULTIPLICITIES OF THE DISTINCT EIGENVALUES OF A          HLE05600
C     G = TRUE MINIMUM GAP IN A FOR EACH OF THESE EIGENVALUES       HLE05610
C     G < 0 INDICATES THE LEFT-HAND GAP WAS MINIMAL.                HLE05620
C     OUTPUT MULTIPLICITIES, DISTINCT EVS, AND MINGAPS TO FILE 11   HLE05630
C                                                                   HLE05640
      WRITE(9,120)                                                  HLE05650
  120 FORMAT(5X,'I',1X,'AMULT',5X,'TRUE A-EIGENVALUE(I)',           HLE05660
     1 3X,'A-MINGAP(I)')                                            HLE05670
C                                                                   HLE05680
      WRITE(9,130)(J,MP(J),U(J),G(J), J=1,NEXACT)                   HLE05690
  130 FORMAT(2I6,E25.16,E14.3)                                      HLE05700
C                                                                   HLE05710
      WRITE(9,140)                                                  HLE05720
  140 FORMAT(' NEXACT DISTINCT A-EIGENVALUES ARE IN ASCENDING ORDER'/ HLE05730
     1 ' AMULT = MULTIPLICITIES OF THE DISTINCT EIGENVALUES OF A.'/ HLE05740
     2 ' A-MINGAP(I) = TRUE MINIMUM GAP IN A FOR EACH EIGENVALUE.'/ HLE05750
     3 ' A-MINGAP(I) LT 0 INDICATES THE LEFT-HAND GAP WAS MINIMAL.'//) HLE05760
C                                                                   HLE05770
C     WE ORDER U VECTOR BY INCREASING SIZE OF THE GAPS              HLE05780
C                                                                   HLE05790
      DO 150 K = 1,N                                                HLE05800
  150 MP(K) = K                                                     HLE05810
C                                                                   HLE05820
      DO 170 K = 2,N                                                HLE05830
      KM1 = K-1                                                     HLE05840
C                                                                   HLE05850
      DO 160 L = 1,KM1                                              HLE05860
      JJ = K - L                                                    HLE05870
      IF (ABS(G(JJ+1)).GE.ABS(G(JJ))) GO TO 170                     HLE05880
      EE = U(JJ)                                                    HLE05890
      U(JJ) = U(JJ+1)                                               HLE05900
      U(JJ+1) = EE                                                  HLE05910
      GG = G(JJ)                                                    HLE05920
      G(JJ) = G(JJ+1)                                               HLE05930
      G(JJ+1) = GG                                                  HLE05940
      IEE = MP(JJ)                                                  HLE05950
      MP(JJ) = MP(JJ+1)                                             HLE05960
  160 MP(JJ+1) = IEE                                                HLE05970
C                                                                   HLE05980
```

```
    170 CONTINUE                                                   HLE05990
C                                                                  HLE06000
       WRITE(9,180)                                                HLE06010
    180 FORMAT(5X,'K',6X,'A-MINGAP',5X,'TRUE A-EIGENVALUE(I)',2X,'A-EVNO')HLE06020
C                                                                  HLE06030
       WRITE(9,190)(J,G(J),U(J),MP(J), J=1,NEXACT)                 HLE06040
    190 FORMAT(I6,E14.3,E25.16,I8)                                 HLE06050
C                                                                  HLE06060
       WRITE(9,200)                                                HLE06070
    200 FORMAT(' NEXACT DISTINCT A-EIGENVALUES. GAPS IN ASCENDING ORDER'/ HLE06080
      2 ' A-MINGAP(I) = TRUE MINIMUM GAP IN A FOR EACH EIGENVALUE.'/  HLE06090
      3 ' A-MINGAP(I) LT O INDICATES THE LEFT-HAND GAP WAS MINIMAL.'/  HLE06100
      3 ' A-MATRIX IS BLOCK TRIDIAGONAL AND EACH DIAGONAL BLOCK IS OF ORDHLE06110
      3ER NX.'/                                                    HLE06120
      4 ' NX = NUMBER OF POINTS ON EACH X-LINE. THERE ARE NY DIAGONAL BLOHLE06130
      4CKS.'/                                                      HLE06140
      5 ' NY = NUMBER OF POINTS ON EACH Y-LINE.'/                  HLE06150
      5 ' A-DIAGONAL    = A(K,K)'/                                 HLE06160
      6 ' X-CODIAGONAL = A(I,I+1)'/                                HLE06170
      7 ' Y-CODIAGONAL = A(I,I+NX)'/                               HLE06180
      8 ' ----- END OF FILE 9 EXACTEV-------------------------'//) HLE06190
C                                                                  HLE06200
C-----END OF EXEVG------------------------------------------------HLE06210
C                                                                  HLE06220
       RETURN                                                      HLE06230
       END                                                         HLE06240
C                                                                  HLE06250
C-----START OF HEXVEC--------------------------------------------HLE06260
C                                                                  HLE06270
C      FOR THE HERMITIAN POISSON TEST CASES WITH MTYPE = 0 ONLY:   HLE06280
C      FOR A GIVEN RITZ VECTOR V AND EIGENVALUE X1, COMPUTES       HLE06290
C      THE CLOSEST TRUE EIGENVALUE Y1 AND CORRESPONDING TRUE       HLE06300
C      EIGENVECTOR Z, CALCULATES THE NORM OF V-Z AND THE MAXIMAL   HLE06310
C      DIFFERENCE OF THE COMPONENTS.  USER WOULD HAVE TO           HLE06320
C      INCORPORATE ENTRY AND CALL TO THIS SUBROUTINE INTO          HLE06330
C      HLEVEC PROGRAM IF THESE QUANTITIES ARE DESIRED.             HLE06340
C      U CONTAINS THE COMPUTED TRUE EIGENVALUES.                   HLE06350
C      W CONTAINS THE TRUE EIGENVECTOR FOR THE REAL POISSON MATRIX HLE06360
C                                                                  HLE06370
       SUBROUTINE HEXVEC(Z,V,U,W,X1,Y1,MP,JNUM)                    HLE06380
C                                                                  HLE06390
C---------------------------------------------------------------HLE06400
       DOUBLE PRECISION  U(1),W(1)                                 HLE06410
       DOUBLE PRECISION WI(110),WJ(110),WII(110)                   HLE06420
       DOUBLE PRECISION X1,Y1,EV,EE,WS,PIK,PIL,SUM,TEMP            HLE06430
       DOUBLE PRECISION ATOLN,EPSM,ZERO,HALF,ONE,TWO,MACHEP        HLE06440
       DOUBLE PRECISION CO,C1,C2,TO,T1,T2                          HLE06450
       COMPLEX*16 CONE,S,SB,STEMP,V(1),Z(1)                        HLE06460
       INTEGER MP(1)                                               HLE06470
C---------------------------------------------------------------HLE06480
       DATA MACHEP/Z3410000000000000/                             HLE06490
       EPSM = 2.0D0*MACHEP                                         HLE06500
C---------------------------------------------------------------HLE06510
C      THIS PROGRAM CALCULATES THE EXACT EIGENVALUES AND EIGENVECTORS HLE06520
C      OF THE HERMITIAN POISSON MATRIX A OF ORDER  N = KX BY KY    HLE06530
C      A CONSISTS OF KY TRIDIAGONAL BLOCKS OF ORDER KX            HLE06540
C      KX = X-DIMENSION    KY = Y-DIMENSION.                       HLE06550
C                                                                  HLE06560
C      C2 = DIAGONAL OF KX BY KX MATRIX                            HLE06570
C      -CO = CO-DIAGONAL OF THE KX BY KX MATRIX.                   HLE06580
C      -C1 = Y-CODIAGONAL.                                         HLE06590
C                                                                  HLE06600
C      NOTE THAT THE VECTORS WI,WJ,WII ARE DIMENSIONED INTERNALLY  HLE06610
C      THEY ARE USED JUST TO KEEP FROM REGENERATING INFORMATION.   HLE06620
C      WI,WII = REAL*8 ARRAYS OF DIMENSION AT LEAST KX             HLE06630
C      WJ     = REAL*8 ARRAY  OF DIMENSION AT LEAST KY.            HLE06640
C                                                                  HLE06650
```

```
C     NOTATION USED IN PROGRAM                                          HLE06660
C                                                                       HLE06670
C     PIK = ARCOS(-1)/(KX+1)      PIL = ARCOS(-1)/(KY+1)                HLE06680
C     WI(I) = PIK*I          WJ(J) = PIL*J                              HLE06690
C                                                                       HLE06700
C     TO = C2 - 2*C1*COS(PIL*J)   EV(I,J) = TO - 2*CO*COS(PIK*I)        HLE06710
C     I = 1,KX     J = 1,KY    KP = (J-1)*KX + I                        HLE06720
C                                                                       HLE06730
C     W(KV) = SIN(PIK*I*IK)*SIN(PIL*J*JK)                               HLE06740
C     IK = 1,KX    JK = 1,KY    KV = (JK-1)*KX + IK                     HLE06750
C     W IS UNSCALED EIGENVECTOR FOR EV(I,J)                             HLE06760
C     WS = 1/||W||: ||W|| = .5*DSQRT(T2*T3)  T2 = KX+1   T3 = KY+1      HLE06770
C     U(K) IS A-EV ORDERED BY INCREASING SIZE, K = 1,N                  HLE06780
C                                                                       HLE06790
C     GIVEN X1 FIND Y1 AND KVEC SUCH THAT                               HLE06800
C          Y1 = EV(KVEC) AND |X1-Y1| = MIN                             HLE06810
C          ALSO GIVEN UNIT RITZ VECTOR ASSOCIATED WITH X1              HLE06820
C          CALCULATE UNIT EIGENVECTOR W, A*W = Y1*W                    HLE06830
C          T2 = ||V-W||   T1 = MAX(|V(K)-W(K)|, K= 1,N)                HLE06840
C          MAX OCCURS FIRST AT K = KK                                  HLE06850
C                                                                       HLE06860
C-----------------------------------------------------------------------HLE06870
C     C2 = A(K,K)                                                       HLE06880
C     CO = A(K,K+1) = A(K+1,K)                                          HLE06890
C     C1 = A(K,K+KX) = A(K+KX,K)                                        HLE06900
C     CO + C1 = HALF                                                    HLE06910
C                                                                       HLE06920
C     SPECIFY PARAMETERS                                                HLE06930
      N = KX*KY                                                         HLE06940
      ZERO = 0.0D0                                                      HLE06950
      HALF = 0.5D0                                                      HLE06960
      ONE = 1.0D0                                                       HLE06970
      TWO = 2.0D0                                                       HLE06980
      TO = DARCOS(-ONE)                                                 HLE06990
      T1 = DFLOAT(KX+1)                                                 HLE07000
      PIK = TO/T1                                                       HLE07010
      T2 = DFLOAT(KY+1)                                                 HLE07020
      PIL = TO/T2                                                       HLE07030
      WS = TWO/DSQRT(T1*T2)                                             HLE07040
C                                                                       HLE07050
C     GENERATE WI WJ VECTORS                                            HLE07060
      KP = 0                                                            HLE07070
      DO 20 J = 1,KY                                                    HLE07080
      T1 = PIL*DFLOAT(J)                                                HLE07090
      WJ(J) = T1                                                        HLE07100
      TO = C2 - TWO*C1*DCOS(T1)                                         HLE07110
      DO 10 I = 1,KX                                                    HLE07120
      KP = KP+1                                                         HLE07130
      T1 = PIK*DFLOAT(I)                                                HLE07140
      WI(I) = T1                                                        HLE07150
   10 U(KP) = TO - TWO*CO*DCOS(T1)                                      HLE07160
   20 CONTINUE                                                          HLE07170
C     U(KP) = EV(I,J) = C2 - 2*C1*COS(PIL*J) - 2*CO*COS(PIK*I)          HLE07180
C                                                                       HLE07190
C     INITIALIZE MP VECTOR                                              HLE07200
      DO 30 K = 1,N                                                     HLE07210
   30 MP(K) = K                                                         HLE07220
C                                                                       HLE07230
C     WE ORDER U VECTOR BY INCREASING SIZE OF THE EVS                   HLE07240
      DO 50 K = 2,N                                                     HLE07250
      KM1 = K-1                                                         HLE07260
C                                                                       HLE07270
      DO 40 L = 1,KM1                                                   HLE07280
      JJ = K - L                                                        HLE07290
      IF (U(JJ+1).GE.U(JJ)) GO TO 50                                    HLE07300
      EE = U(JJ)                                                        HLE07310
      U(JJ) = U(JJ+1)                                                   HLE07320
```

```
      U(JJ+1) = EE                                             HLE07330
      IEE = MP(JJ)                                             HLE07340
      MP(JJ) = MP(JJ+1)                                        HLE07350
   40 MP(JJ+1) = IEE                                           HLE07360
C                                                              HLE07370
   50 CONTINUE                                                 HLE07380
C                                                              HLE07390
      ATOLN = DMAX1(DABS(U(1)),DABS(U(N)))*EPSM                HLE07400
C                                                              HLE07410
      WRITE(6,60) N,KX,KY,C2,CO,C1,ATOLN                       HLE07420
   60 FORMAT(/' EXACT ERRORS FOR CONVERGED GOODEV'/            HLE07430
     1 416,' = N KX KY'//                                      HLE07440
     1 4E12.5,' = C2 CO C1 ATOLN'//)                           HLE07450
C                                                              HLE07460
C     KP = MP(K) MEANS EIGENVALUE U(K) CORRESPONDS TO EIGENVECTOR W(KP) HLE07470
C     COMPUTE TOLERANCE USED IN COMPUTING TRUE MULTIPLICITIES  HLE07480
C                                                              HLE07490
C     X1 IS AN INPUT PARAMETER. WE CALCULATE EXACT             HLE07500
C     A-EIGENVALUE WHICH IS CLOSEST TO X1, LABEL IT Y1 AND CALCULATE    HLE07510
C     UNIT EIGENVECTOR OF A ASSOCIATED WITH Y1. A*W = Y1*W, ||W|| = 1.  HLE07520
C     Y1 = U(KEV). EIGENVALUES OF A ARE ORDERED BY INCREASING SIZE.     HLE07530
C     V = COMPLEX RITZ VECTOR ASSOCIATED WITH GOODEV X1        HLE07540
C     WE SHOULD HAVE V = D*W WHERE D = DIAG(D(1),D(2),..,D(N)) HLE07550
C     D(1) = ONE, D(K+1)/D(K) = SB, |SB| = ONE                HLE07560
C                                                              HLE07570
      KX1 = 0                                                  HLE07580
      IF (X1.LE.U(1)) KX1 = 1                                  HLE07590
      IF (X1.GE.U(N)) KX1 = N                                  HLE07600
      NM1 = N-1                                                HLE07610
      IF (KX1.NE.0) GO TO 80                                   HLE07620
C                                                              HLE07630
      DO 70 KVEC = 2,N                                         HLE07640
      IF (X1.GE.U(KVEC)) GO TO 70                              HLE07650
C     U(KVEC-1).LE.X1.LT.U(KVEC)                               HLE07660
      T1 = X1 - U(KVEC-1)                                      HLE07670
      T2 = U(KVEC) - X1                                        HLE07680
      KX1 = KVEC - 1                                           HLE07690
      IF (T1.GT.T2) KX1 = KVEC                                 HLE07700
      GO TO 80                                                 HLE07710
   70 CONTINUE                                                 HLE07720
C                                                              HLE07730
   80 Y1 = U(KX1)                                              HLE07740
C                                                              HLE07750
      IF (KX1.EQ.1) EE = U(2) - U(1)                           HLE07760
      IF (KX1.EQ.N) EE = U(N) - U(NM1)                         HLE07770
      IF (KX1.EQ.1.OR.KX1.EQ.N) GO TO 90                       HLE07780
      EE = DMIN1(U(KX1+1)-U(KX1),U(KX1)-U(KX1-1))              HLE07790
   90 CONTINUE                                                 HLE07800
C                                                              HLE07810
      TO = DABS(ONE - X1/Y1)                                   HLE07820
C                                                              HLE07830
      WRITE(6,100) N,KX1,JNUM,Y1,X1,TO,EE                      HLE07840
  100 FORMAT(3I8,' = N, A-EV NUMBER,GOODEV NO'//               HLE07850
     1 18X,'EXACTEV',19X,'GOODEV',4X,'RELERROR',4X,'A-MINGAP'/ HLE07860
     1 2E25.16,2E12.3/)                                        HLE07870
C                                                              HLE07880
      IF (EE.GT.ATOLN) GO TO 120                               HLE07890
C                                                              HLE07900
      WRITE(6,110)                                             HLE07910
  110 FORMAT(' Y1 IS A MULTIPLE EIGENVALUE OF A SO WE EXIT'/)  HLE07920
C                                                              HLE07930
      GO TO 200                                                HLE07940
C                                                              HLE07950
C     Y1 IS TOEPLITZ EIGENVALUE CLOSEST TO X1.                 HLE07960
C     CALCULATION OF EIGENVECTOR ASSOCIATED WITH EIGENVALUE Y1 HLE07970
C     A*W = Y1*W                                               HLE07980
```

```
C                                                                    HLE07990
C     DETERMINE I J FROM K: MP(K) = KP = (J-1)*KX+I                   HLE08000
  120 CONTINUE                                                        HLE08010
      K = KX1                                                         HLE08020
      KP = MP(K)                                                      HLE08030
      I = MOD(KP,KX)                                                  HLE08040
      IF (I.EQ.0) I = KX                                              HLE08050
      T1 = WI(I)                                                      HLE08060
      J = 1 + (KP-1)/KX                                               HLE08070
      T2 = WJ(J)                                                      HLE08080
C                                                                    HLE08090
      DO 130 II = 1,KX                                                HLE08100
      TO = T1*DFLOAT(II)                                              HLE08110
  130 WII(II) = WS*DSIN(TO)                                           HLE08120
C                                                                    HLE08130
      KV = 0                                                          HLE08140
      DO 150 JJ = 1,KY                                                HLE08150
      TO = T2*DFLOAT(JJ)                                              HLE08160
      TO = DSIN(TO)                                                   HLE08170
C                                                                    HLE08180
      DO 140 II = 1,KX                                                HLE08190
      KV = KV + 1                                                     HLE08200
  140 W(KV) = TO*WII(II)                                              HLE08210
C                                                                    HLE08220
  150 CONTINUE                                                        HLE08230
C                                                                    HLE08240
C     W IS UNIT EXACT EIGENVECTOR OF A ASSOCIATED WITH Y1             HLE08250
C     V IS UNIT COMPLEX RITZVECTOR OF B ASSOCIATED WITH X1            HLE08260
C                                                                    HLE08270
      CONE = DCMPLX(ONE,ZERO)                                         HLE08280
      STEMP = CONE                                                    HLE08290
      DO 160 K = 1,N                                                  HLE08300
      Z(K) = STEMP*W(K)                                               HLE08310
  160 STEMP = STEMP*SB                                                HLE08320
C                                                                    HLE08330
      T1 = ZERO                                                       HLE08340
      S = ONE                                                         HLE08350
      KK = 0                                                          HLE08360
      DO 170 K = 1,N                                                  HLE08370
      IF (CDABS(Z(K)).LE.T1) GO TO 170                                HLE08380
      T1 = CDABS(Z(K))                                                HLE08390
      KK = K                                                          HLE08400
  170 CONTINUE                                                        HLE08410
C                                                                    HLE08420
      S = V(KK)/Z(KK)                                                 HLE08430
C                                                                    HLE08440
      KK = 0                                                          HLE08450
      T1 = ZERO                                                       HLE08460
      T2 = ZERO                                                       HLE08470
      DO 180 K = 1,N                                                  HLE08480
      TEMP = CDABS(S*Z(K) - V(K))                                     HLE08490
      T2 = T2 + TEMP**2                                               HLE08500
      IF (TEMP.LE.T1) GO TO 180                                       HLE08510
      KK = K                                                          HLE08520
      T1 = TEMP                                                       HLE08530
  180 CONTINUE                                                        HLE08540
C                                                                    HLE08550
      T2 = DSQRT(T2)                                                  HLE08560
      WRITE(6,190) KK,T1,T2                                           HLE08570
  190 FORMAT(' EIGENVECTOR ERROR. MAX ERROR AT COMPONENT = ',I6/      HLE08580
     1 ' MAX CDABS(EXACTVEC(K)-RITZVEC(K)) = ',E12.5/                 HLE08590
     1 ' NORM(EXACTVEC-RITZVEC) = ',E12.5/)                           HLE08600
C                                                                    HLE08610
  200 CONTINUE                                                        HLE08620
C                                                                    HLE08630
```

```
      RETURN                                                            HLE08640
C                                                                       HLE08650
C----------------------------------------------------------------------HLE08660
      ENTRY EXVECP(SB,CO,C1,C2,KX,KY)                                   HLE08670
C----------------------------------------------------------------------HLE08680
C                                                                       HLE08690
C-----END OF HEXVEC----------------------------------------------------HLE08700
      RETURN                                                            HLE08710
      END                                                               HLE08720
C                                                                       HLE08730
C-----USPEC (TRIDIAGONAL HERMITIAN MATRICES)---------------------------HLE08740
C                                                                       HLE08750
C     SUBROUTINE USPEC(N,MATNO)                                         HLE08760
      SUBROUTINE TSPEC(N,MATNO)                                         HLE08770
C                                                                       HLE08780
C----------------------------------------------------------------------HLE08790
      DOUBLE PRECISION  D(100), DAR(100),DAI(100), PI, EIGVAL(100)      HLE08800
      DOUBLE PRECISION SPACE                                            HLE08810
      COMPLEX*16   DA(100),DB(100)                                      HLE08820
      REAL EXPLAN(20)                                                   HLE08830
C----------------------------------------------------------------------HLE08840
C     DIMENSION ARRAYS NEEDED TO DEFINE MATRIX.  THEN                   HLE08850
C     PASS THE STORAGE LOCATIONS OF THESE ARRAYS TO THE MATRIX-VECTOR   HLE08860
C     MULTIPLY SUBROUTINE CMATV.                                        HLE08870
C                                                                       HLE08880
C     DIAGONAL ENTRY = D, ABOVE DIAGONAL ENTRY = DA, BELOW DIAGONAL = DB.HLE08890
C                                                                       HLE08900
      READ(8,10) EXPLAN                                                 HLE08910
   10 FORMAT(20A4)                                                      HLE08920
      READ(8,*) NOLD,MATOLD                                             HLE08930
C                                                                       HLE08940
      WRITE(6,20) N,MATOLD                                              HLE08950
   20 FORMAT(I10,2I6,I8,' = N,MATOLD'/)                                 HLE08960
C                                                                       HLE08970
C     TEST OF PARAMETER CORRECTNESS                                     HLE08980
      ITEMP = (NOLD-N)**2 + (MATNO-MATOLD)**2                           HLE08990
C                                                                       HLE09000
      IF(ITEMP.EQ.0) GO TO 40                                           HLE09010
C                                                                       HLE09020
      WRITE(6,30)                                                       HLE09030
   30 FORMAT(' PROGRAM TERMINATES BECAUSE EITHER ORDERS OF OR LABELS FORHLE09040
     1 MATRIX DISAGREE')                                                HLE09050
      GO TO 250                                                         HLE09060
C                                                                       HLE09070
   40 CONTINUE                                                          HLE09080
C                                                                       HLE09090
C     IF ITOEP = 1 THEN MATRIX IS TOEPLITZ AND WE PRINT OUT TRUE        HLE09100
C     EIGENVALUES                                                       HLE09110
      READ(8,10) EXPLAN                                                 HLE09120
      READ(8,*) ITOEP                                                   HLE09130
      READ(8,10) EXPLAN                                                 HLE09140
C                                                                       HLE09150
      IF(ITOEP.EQ.1) WRITE(6,50)                                        HLE09160
   50 FORMAT(/' TEST MATRIX IS HERMITIAN TOEPLITZ'/)                    HLE09170
      IF(ITOEP.NE.1) GO TO 110                                          HLE09180
C                                                                       HLE09190
      READ(8,*) DAR(1),DAI(1),D(1)                                      HLE09200
      DA(1) = DCMPLX(DAR(1),DAI(1))                                     HLE09210
      DB(1) = DCONJG(DA(1))                                             HLE09220
      DO 60 J=2,N                                                       HLE09230
      D(J) = D(1)                                                       HLE09240
      DA(J) = DA(1)                                                     HLE09250
   60 DB(J) = DB(1)                                                     HLE09260
      WRITE(6,70) DB(1),D(1),DA(1)                                      HLE09270
      WRITE(9,70) DB(1),D(1),DA(1)                                      HLE09280
   70 FORMAT(' HERMITIAN TOEPLITZ MATRIX IS USED.'/' BELOW DIAGONAL ENTRHLE09290
     1Y = ',2E12.3/' DIAGONAL ENTRY = ',E12.3/' ABOVE DIAGONAL ENTRY =' HLE09300
```

```
      1,2E12.3)                                                      HLE09310
C                                                                    HLE09320
C     COMPUTE THE TRUE EIGENVALUES.  FORMULA IS CORRECT ONLY FOR THOSE HLE09330
C     MATRICES WHOSE DIAGONAL = 2., ABOVE DIAGONAL = A, BELOW DIAGONAL HLE09340
C     = A-CONJUGATE, AND A HAS NORM 1.                               HLE09350
C                                                                    HLE09360
      PI = DARCOS(-1.D0)                                             HLE09370
      DO 80 J=1,N                                                    HLE09380
   80 EIGVAL(J) = 2.D0 * (1.D0 -DCOS(PI*DFLOAT(J)/DFLOAT(N+1)))      HLE09390
      WRITE(9,90) N                                                  HLE09400
   90 FORMAT(I6, ' = ORDER OF MATRIX'/' TRUE EIGENVALUES ARE'/)      HLE09410
      WRITE(9,100) (J, EIGVAL(J), J=1,N)                            HLE09420
  100 FORMAT(I5,4X,E25.16,6X,I5,4X,E25.16)                          HLE09430
      GO TO 240                                                     HLE09440
C                                                                    HLE09450
C     NONTOEPLITZ HERMITIAN.  DIAGONAL ENTRIES ARE EQUALLY-SPACED.  HLE09460
C     ABOVE DIAGONAL ENTRIES ARE GENERATED BY GENERATING EQUALLY-SPACED HLE09470
C     REAL PARTS, AND EQUALLY-SPACED IMAGINARY PARTS.  THE BELOW    HLE09480
C     DIAGONAL ENTRIES ARE THEN OBTAINED BY TAKING THE COMPLEX CONJUGATE HLE09490
C     OF THE ABOVE DIAGONAL ENTRIES                                 HLE09500
C                                                                    HLE09510
  110 READ(8,*) D(1), SPACE                                         HLE09520
      WRITE(6,120) D(1),SPACE                                       HLE09530
  120 FORMAT(' 1ST DIAGONAL ENTRY =',E20.12,' SPACING =',E20.12)    HLE09540
      DO 130 J=2,N                                                  HLE09550
  130 D(J) = D(J-1) + SPACE                                         HLE09560
      WRITE(6,140) (D(J), J=1,3)                                    HLE09570
  140 FORMAT(' 1ST THREE DIAGONAL ENTRIES ='/(2E20.12))             HLE09580
      READ(8,10) EXPLAN                                             HLE09590
      READ(8,*) DAR(1), SPACE                                       HLE09600
      WRITE(6,150) DAR(1),SPACE                                     HLE09610
  150 FORMAT(' REAL PART OF 1ST ABOVE DIAGONAL ENTRY =',E20.12,/    HLE09620
     1' SPACING = ',E20.12)                                         HLE09630
      DO 160 J=2,N                                                  HLE09640
  160 DAR(J) = DAR(J-1) + SPACE                                     HLE09650
      WRITE(6,170) (DAR(J), J=1,3)                                  HLE09660
  170 FORMAT(' REAL PARTS OF 1ST THREE ABOVE DIAGONAL ENTRIES ='/   HLE09670
     1(2E20.12))                                                    HLE09680
      READ(8,10) EXPLAN                                             HLE09690
      READ(8,*) DAI(1), SPACE                                       HLE09700
      WRITE(6,180) DAI(1),SPACE                                     HLE09710
  180 FORMAT(' IMAGINARY PART OF 1ST ABOVE =',E20.12,/' SPACING =', HLE09720
     1 E20.12)                                                      HLE09730
      DO 190 J=2,N                                                  HLE09740
  190 DAI(J) = DAI(J-1) + SPACE                                     HLE09750
      WRITE(6,200) (DAI(J), J = 1,3)                                HLE09760
  200 FORMAT(' IMAGINARY PARTS OF 1ST THREE ABOVE DIAGONAL ENTRIES ='/ HLE09770
     1 (2E20.12))                                                   HLE09780
      DO 210 J=1,N                                                  HLE09790
      DA(J) = DCMPLX(DAR(J),DAI(J))                                 HLE09800
  210 DB(J) = DCONJG(DA(J))                                         HLE09810
C                                                                    HLE09820
      WRITE(9,220) (D(J), J=1,N)                                    HLE09830
  220 FORMAT(' DIAGONAL ENTRIES ='/(4E20.12))                       HLE09840
      WRITE(9,230) (DA(J), J=1,N)                                   HLE09850
  230 FORMAT(' ABOVE DIAGONAL ENTRIES'/(4E20.12))                   HLE09860
C                                                                    HLE09870
C     PASS STORAGE LOCATIONS OF ARRAYS THAT DEFINE THE MATRIX TO    HLE09880
C     THE MATRIX-VECTOR MULTIPLY SUBROUTINE CMATV                   HLE09890
C                                                                    HLE09900
  240 CONTINUE                                                      HLE09910
C                                                                    HLE09920
C--------------------------------------------------------------------HLE09930
      CALL TMATVE(DA,DB,D,N)                                        HLE09940
C--------------------------------------------------------------------HLE09950
C                                                                    HLE09960
      RETURN                                                        HLE09970
```

```
   250 STOP                                                             HLE09980
C                                                                       HLE09990
C-----END OF USPEC----------------------------------------------------HLE10000
      END                                                               HLE10010
C                                                                       HLE10020
C-----START OF MATRIX-VECTOR MULTIPLY (HERMITIAN TRIDIAGONAL)----------HLE10030
C                                                                       HLE10040
C     SUBROUTINE CMATV(W,U,SUM)                                         HLE10050
      SUBROUTINE TMATV(W,U,SUM)                                         HLE10060
C                                                                       HLE10070
C--------------------------------------------------------------------HLE10080
      COMPLEX*16   U(1),W(1),DA(1),DB(1)                                HLE10090
      DOUBLE PRECISION  D(1),SUM                                        HLE10100
C--------------------------------------------------------------------HLE10110
C   HERMITIAN MATRIX-VECTOR MULTIPLY FOR LANCZS   U = A*W - SUM*U       HLE10120
C   MATRIX IS TRIDIAGONAL HERMITIAN TOEPLITZ                           HLE10130
C--------------------------------------------------------------------HLE10140
C                                                                       HLE10150
C     COMPUTE A*W - SUM*U                                               HLE10160
      U(1) = D(1)*W(1) + DA(1)*W(2) - SUM*U(1)                          HLE10170
      N1 = N-1                                                          HLE10180
      DO 10 I = 2,N1                                                    HLE10190
   10 U(I) = DB(I-1)*W(I-1)+D(I)*W(I) + DA(I)*W(I+1) -SUM*U(I)          HLE10200
      U(N) = DB(N-1)*W(N-1) + D(N)*W(N) - SUM*U(N)                      HLE10210
C                                                                       HLE10220
      RETURN                                                            HLE10230
C                                                                       HLE10240
C    STORAGE LOCATIONS ARE PASSED TO CMATV FROM USPEC                   HLE10250
C                                                                       HLE10260
C--------------------------------------------------------------------HLE10270
      ENTRY TMATVE(DA,DB,D,N)                                           HLE10280
C--------------------------------------------------------------------HLE10290
C                                                                       HLE10300
C-----END OF CMATV----------------------------------------------------HLE10310
      RETURN                                                            HLE10320
      END                                                               HLE10330
```

SECTION 3.5   FILE DEFINITIONS AND SAMPLE INPUT FILE

Below is a listing of the input/output files which are accessed by the
Hermitian Lanczos eigenvalue program, HLEVAL. Included also is a
sample of the input file which HLEVAL requires on file 5. The
parameters supplied in this file are in free format. HLEVAL computes
eigenvalues of Hermitian matrices on user-supplied intervals.

```
SAMPLE DEFINITIONS OF THE INPUT/OUTPUT FILES FOR HLEVAL
-------------------------------------------------------------------
*HLEVAL EXEC HERMITIAN EIGENVALUE CALCULATION
FI 06 TERM
FILEDEF  1 DISK &1        NHISTORY  A (RECFM F LRECL 80 BLOCK 80
FILEDEF  2 DISK &1        HISTORY   A (RECFM F LRECL 80 BLOCK 80
FILEDEF  3 DISK &1        GOODEV    A (RECFM F LRECL 80 BLOCK 80
FILEDEF  4 DISK &1        ERRINV    A (RECFM F LRECL 80 BLOCK 80
FILEDEF  5 DISK HLEVAL    INPUT     A (RECFM F LRECL 80 BLOCK 80
FILEDEF  8 DISK &1        INPUT     A (RECFM F LRECL 80 BLOCK 80
FILEDEF 11 DISK &1        DISTINCT  A (RECFM F LRECL 80 BLOCK 80
LOAD  ·HLEVAL   LESUB    HLEMULT
-------------------------------------------------------------------
```

```
SAMPLE INPUT FILE FOR HLEVAL
-----------------------------------------------------------------------
 HLEVAL INPUT EIGENVALUE COMPUTATION, NO REORTHOGONALIZATION
 HERMITIAN TEST MATRIX
LINE 1     N      KMAX    NMEVS    MATNO
          528     2640      3     721830
LINE 2  SVSEED    RHSEED      MXINIT    MXSTUR
       49302312  5731029        5      100000
LINE 3  ISTART    ISTOP
          0         1
LINE 4   IHIS     IDIST  IWRITE
          1         0      1
LINE 5  RELTOL (RELATIVE TOLERANCE IN 'COMBINING' GOODEV)
       .0000000001
LINE 6   MB(1)    MB(2)   MB(3)   MB(4)   (ORDERS OF T(1,MEV) )
          528     1056    1584
LINE 7   NINT     (NUMBER OF SUB-INTERVALS FOR BISEC)
           1
LINE 8   LB(1)    LB(2)   LB(3)   LB(4)  (INTERVAL LOWER BOUNDS)
          1.3
LINE 9   UB(1)    UB(2)   UB(3)   UB(4)  (INTERVAL UPPER BOUNDS)
          1.34
-----------------------------------------------------------------------
```

Below is a listing of the input/output files which are accessed by the
Hermitian Lanczos eigenvector program, HLEVEC. Included also is a
sample of the input file which HLEVEC requires on file 5. The
parameters in this file are supplied in free format. HLEVEC computes
eigenvectors for each of a user-selected subset of the eigenvalues
computed by the eigenvalue program HLEVAL.

SAMPLE DEFINITIONS OF THE INPUT/OUTPUT FILES FOR HLEVEC

```
*HLEVEC EXEC TO RUN LANCZOS EIGENVECTOR PROGRAM, HERMITIAN MATRICES
FI 06 TERM
FILEDEF  2 DISK &1       HISTORY   A (RECFM F LRECL 80 BLOCK 80
FILEDEF  3 DISK &1       GOODEV    A (RECFM F LRECL 80 BLOCK 80
FILEDEF  4 DISK &1       ERRINV    A (RECFM F LRECL 80 BLOCK 80
FILEDEF  5 DISK HLEVEC   INPUT     A (RECFM F LRECL 80 BLOCK 80
FILEDEF  8 DISK &1       INPUT     A (RECFM F LRECL 80 BLOCK 80
FILEDEF  9 DISK &1       ERREST    A (RECFM F LRECL 80 BLOCK 80
FILEDEF 10 DISK &1       BOUNDS    A (RECFM F LRECL 80 BLOCK 80
FILEDEF 11 DISK &1       TEIGVECS  A (RECFM F LRECL 80 BLOCK 80
FILEDEF 12 DISK &1       RITZVECS  A (RECFM F LRECL 80 BLOCK 80
FILEDEF 13 DISK &1       PAIGE     A (RECFM F LRECL 80 BLOCK 80
LOAD  HLEVEC  LESUB  HLEMULT
```

SAMPLE INPUT FILE FOR HLEVEC

```
 HLEVEC EIGENVECTORS OF HERMITIAN MATRIX, NO REORTHOGONALIZATION
LINE 1  MDIMTV     MDIMRV  MBETA(MAX.DIMENSIONS, TVEC, RITVEC AND BETA
        10000      10000   2000
LINE 2      RELTOL
        .0000000001
LINE 3  MBOUND    NTVCON SVTVEC IREAD (FLAGS
        0           1      0     1
LINE 4  TVSTOP   LVCONT ERCONT  IWRITE (FLAGS
        0          1      1       1
LINE 5    RHSEED  (RANDOM GENERATOR SEED FOR STARTING VECTOR IN INVERM)
        45329517
LINE 6  MATNO       N
        100       100
```

CHAPTER 4

FACTORED INVERSES OF REAL SYMMETRIC MATRICES

SECTION 4.1   INTRODUCTION

The FORTRAN codes in this chapter address the question of computing distinct eigenvalues and corresponding eigenvectors of a real symmetric matrix by applying a single-vector Lanczos procedure to the inverse of an associated matrix $B \equiv PCP^T$, where $C = S0*A + SHIFT*I$. The scalars S0 and SHIFT are specified by the user, selected in such a way that the resulting matrix C (or B) has a reasonable numerical condition. The permutation matrix P is chosen so that for a sparse matrix A, the resulting factorization of B is also sparse.

For a given real symmetric matrix A, these codes compute real scalars λ and corresponding real-valued vectors $x \neq 0$ such that

$$B^{-1}x = \lambda x, \tag{4.1.1}$$

where B is as defined above. Note that the eigenvectors of B^{-1} are simple permutations of the eigenvectors of A. The eigenvalues of A are obtained from those of B by a simple scalar modification, which is incorporated in the codes. These codes do not require the matrix A. The Lanczos computations use only the user-supplied factorization of the associated matrix B, the scalars S0 and SHIFT, and the permutation P (if any).

Real symmetric matrices and factorizations of such matrices are discussed in Stewart [1973]. See also Bunch and Kaufman [1977] and George and Liu [1981]. Chapter 2, Section 2.1 contains a brief summary of the properties of real symmetric matrices which we use in these codes.

Given a real symmetric matrix A, the user may decide to use the codes in this chapter rather than those in Chapter 2 if the eigenvalues to be computed are 'small' with 'small' gaps between them and the required factorization can be obtained with a reasonable amount of computation and storage. The user should note however that this type of transformation of the given matrix may not yield an eigenvalue distribution which is better for these Lanczos codes. Such a transformation will accelerate the Lanczos computations only if the desired eigenvalues either become larger in size relative to the other eigenvalues and/or the gaps between the desired eigenvalues become larger relative to the gaps between the other eigenvalues. This type of transformation can be very effective in compressing the big end of the spectrum of a given matrix and enhancing the small end of the spectrum. The Lanczos procedure, however, does not require large gaps between the desired eigenvalues, all it really requires is a reasonable overall gap ratio. That is, the ratio of the largest gap between two neighboring eigenvalues to the smallest such gap must be a reasonable size.

The single-vector Lanczos codes in this chapter can be used to compute either a very few or very many of the distinct eigenvalues of the given real symmetric matrix. The documentation for these codes is contained in Chapter 2, Section 2.2. As in the direct real symmetric case (Chapter 2, Section 2.1), the A-multiplicity of a given computed eigenvalue can be obtained only with additional computation, and the modifications required to do this additional computation are not included in these versions of the codes. This implementation uses the basic Lanczos recursion contained in Eqns(1.2.1) - (1.2.2) to generate a family of real symmetric tridiagonal matrices (T-matrices) for the matrix B^{-1}, whose sizes are specified by the user. Specifically, for $i = 1,2,...,m$ and a randomly-generated starting vector v_1 with $\|v_1\| = 1$, generate Lanczos vectors v_i using the following recursion and Eqn(1.2.2) applied to the matrix B^{-1}.

$$\beta_{i+1}v_{i+1} = B^{-1}v_i - \alpha_i v_i - \beta_i v_{i-1}. \tag{4.1.2}$$

B is the matrix defined above in terms of the scalars S0 and SHIFT and the permutation P, and each $B^{-1}v_i$ is evaluated by solving the system of equations $Bz = v_i$.

LIVAL, the main program for the factored inverse computations, calls the subroutine BISEC to compute eigenvalues of the specified Lanczos tridiagonal matrices on the user-specified intervals. BISEC simultaneously computes these T-eigenvalues with their T-multiplicities and sorts the computed T-eigenvalues into two classes, the 'good' T-eigenvalues and the 'spurious' T-eigenvalues. The 'good' T-eigenvalues are accepted as approximations to eigenvalues of the B^{-1} matrix associated with the user-specified matrix A, scalars S0 and SHIFT, and the permutation matrix P (if any). The accuracy of these 'good' T-eigenvalues as eigenvalues of B^{-1} is then estimated using error estimates computed by subroutine INVERR. Error estimates are computed only for isolated 'good' T-eigenvalues. All other 'good' T-eigenvalues are assumed to have converged. Convergence is then checked. If convergence has not yet occurred and a larger T-matrix has been specified by the user, the program will continue on to the larger T-matrix, repeating the above procedure on this larger matrix. After each T-matrix eigenvalue computation, the corresponding approximations to the eigenvalues of the user-specified matrix A are computed and included in the output.

Once the eigenvalues of B^{-1} have been computed accurately enough, the user can select a subset of the 'converged' eigenvalues for which eigenvectors are to be computed. The main program LIVEC, for computing eigenvectors of the inverse of a real symmetric matrix given a factorization, is used to compute the desired eigenvectors. If the matrix B is a permutation of the matrix C, then LIVEC unwinds the permutation to obtain the corresponding eigenvectors of the user-supplied A-matrix.

All of the computations are done in double precision real arithmetic. Once the Lanczos T-matrices have been computed, the remaining computations use the same subroutines that are used in the real symmetric case discussed in Chapter 2. In addition to the programs and subroutines provided here, the user must supply a subroutine USPEC which defines and initializes the factorization of the scaled, shifted, and permuted version B of the original matrix A, and a subroutine BSOLV which computes matrix-vector multiplies $B^{-1}x$ for any given vector x. These subroutines must be constructed in such a way as to take advantage of the sparsity (and/or structure) of the user-supplied A-matrix and such that these computations are done accurately.

The sample subroutines USPEC and BSOLV provided assume that the associated matrix B is positive definite and that its Cholesky factorization

$$B = LL^T, \quad L \text{ a lower triangular matrix,} \tag{4.1.3}$$

is used to compute $B^{-1}y$, for any given y. Thus, the sample USPEC subroutine provided for this chapter defines and initializes arrays which define the Cholesky factor L of the associated matrix B. The sample BSOLV subroutine provided computes the required matrix-vector multiplies $u = B^{-1}y$ by solving sequentially the two equations $Lz = y$ and $L^T u = z$. These two equations are very easy to solve since L is a triangular matrix. The main portions of these Lanczos codes do not however require that the B-matrix be positive definite, only that a factorization be available. Therefore, the user could replace the sample USPEC and BSOLV subroutines by subroutines which use a more general factorization of B, for example $B = LDL^T$, where D is a diagonal matrix. All that is necessary is that the BSOLV subroutine provide the matrix-vector products $B^{-1}x$, rapidly and accurately. The information supplied to the Lanczos procedures about the matrix being processed must be consistent.

Several optional preprocessing programs are provided, PERMUT, LORDER, LFACT, and LTEST. PERMUT calls the SPARSPAK Library [1979] to attempt to identify a reordering or permutation P of the given matrix A for which sparseness will be preserved under factorization of the permuted matrix. LORDER takes a given matrix C and permutation P and computes the sparse matrix format for the permuted matrix, $B \equiv PCP^T$. LFACT computes the Cholesky factors of a given positive definite matrix. LTEST performs a very crude check on the numerical condition of the matrix supplied to it, by solving a system of equations with and without iterative refinement LINPACK [1979]. More details about the single-vector Lanczos procedures are given in Chapter 4 of Volume 1 of this book.

SECTION 4.2   MAIN PROGRAM, EIGENVALUE COMPUTATIONS

```
C-----LIVAL---(EIGENVALUES OF INVERSES OF REAL SYMMETRIC MATRICES)------LIV00010
C                                                                       LIV00020
C       CONTAINS MAIN PROGRAM FOR COMPUTING DISTINCT EIGENVALUES OF      LIV00030
C       INVERSES OF REAL SYMMETRIC MATRICES USING REORDERING            LIV00040
C       AND SPARSE FACTORIZATION.  THE LANCZOS RECURSION IS APPLIED      LIV00050
C       TO A SCALED, SHIFTED, AND REORDERED VERSION B OF THE            LIV00060
C       ORIGINAL A-MATRIX.  THE PROCEDURE USES LANCZOS                   LIV00070
C       TRIDIAGONALIZATION WITHOUT REORTHOGONALIZATION                   LIV00080
C                                                                       LIV00090
C       PFORT VERIFIER IDENITIFIED THE FOLLOWING NONPORTABLE            LIV00100
C       CONSTRUCTIONS                                                    LIV00110
C                                                                       LIV00120
C       1.  DATA/MACHEP/ STATEMENT                                       LIV00130
C       2.  ALL READ(5,*) STATEMENTS (FREE FORMAT)                       LIV00140
C       3.  FORMAT(20A4) USED WITH EXPLANATORY HEADER EXPLAN.            LIV00150
C       4.  HEXADECIMAL FORMAT (4Z20) USED IN ALPHA/BETA FILES 1 AND 2.  LIV00160
C                                                                       LIV00170
C-----------------------------------------------------------------------LIV00180
C                                                                       LIV00190
        DOUBLE PRECISION  ALPHA(3000),BETA(3001)                        LIV00200
        DOUBLE PRECISION  V1(3001),V2(3000),VS(3000)                    LIV00210
        DOUBLE PRECISION  LB(20),UB(20)                                 LIV00220
        DOUBLE PRECISION  BTOL,GAPTOL,TTOL,MACHEP,EPSM,SHIFT,SHIFTO,RELTOLLIV00230
        DOUBLE PRECISION  SCALE1,SCALE2,SCALE3,SCALE4,BISTOL,CONTOL,MULTOLLIV00240
        DOUBLE PRECISION  ONE,ZERO,TEMP,TKMAX,BETAM,BKMIN,TO,T1,SO       LIV00250
        REAL  G(3000),GG(3000),EXPLAN(20)                               LIV00260
        INTEGER  MP(3000),NMEV(20)                                       LIV00270
        INTEGER  SVSEED,RHSEED,SVSOLD                                    LIV00280
        INTEGER IABS                                                     LIV00290
        REAL  ABS                                                        LIV00300
        DOUBLE PRECISION  DABS, DSQRT, DFLOAT                            LIV00310
        EXTERNAL BSOLV                                                   LIV00320
C                                                                       LIV00330
C-----------------------------------------------------------------------LIV00340
        DATA MACHEP/Z3410000000000000/                                  LIV00350
        EPSM = 2.0D0*MACHEP                                              LIV00360
C-----------------------------------------------------------------------LIV00370
C                                                                       LIV00380
C       ARRAYS MUST BE DIMENSIONED AS FOLLOWS:                           LIV00390
C       1.  ALPHA: >= KMAX,   BETA: >= (KMAX+1) WHERE KMAX MAY           LIV00400
C           IS THE LARGEST SIZE T-MATRIX TO BE CONSIDERED.               LIV00410
C       2.  V1:  >= MAX(N,KMAX+1)                                        LIV00420
C       3.  V2,VS:  >= MAX(N,KMAX)                                       LIV00430
C       4.  GG:  >= KMAX                                                 LIV00440
C       5.  G:  >= MAX(N,2*KMAX)                                         LIV00450
C       6.  MP:  >= KMAX                                                 LIV00460
C       7.  LB,UB: >= NUMBER OF SUBINTERVALS SUPPLIED TO BISEC.          LIV00470
C       8.  NMEV:  >= NUMBER OF T-MATRICES ALLOWED.                      LIV00480
C       9.  EXPLAN:  DIMENSION IS 20.                                    LIV00490
C                                                                       LIV00500
C                                                                       LIV00510
C       IMPORTANT TOLERANCES OR SCALES THAT ARE USED REPEATEDLY          LIV00520
C       THROUGHOUT THE PROGRAM ARE THE FOLLOWING:                        LIV00530
C       SCALED MACHINE EPSILON:  TTOL = TKMAX*EPSM WHERE                 LIV00540
C       EPSM = 2*MACHINE EPSILON AND                                     LIV00550
C       TKMAX = MAX(|ALPHA(J)|,BETA(J), J = I,MEV)                       LIV00560
C       BISEC CONVERGENCE TOLERANCE:  BISTOL  = DSQRT(1000+MEV)*TTOL     LIV00570
C       BISEC MULTIPLICITY TOLERANCE:  MULTOL = (1000+MEV)*TTOL          LIV00580
C       LANCZOS CONVERGENCE TOLERANCE: CONTOL = BETA(MEV+1)*1.D-10       LIV00590
C                                                                       LIV00600
C-----------------------------------------------------------------------LIV00610
C       OUTPUT HEADER                              .                     LIV00620
        WRITE(6,10)                                                     LIV00630
     10 FORMAT(/' LANCZOS PROCEDURE FOR FACTORED INVERSES OF REAL SYMMETRILIV00640
```

```
      1C MATRICES')                                              LIV00650
C                                                               LIV00660
C     SET PROGRAM PARAMETERS                                    LIV00670
C     SCALEK ARE USED IN TOLERANCES NEEDED IN SUBROUTINES LUMP, LIV00680
C     ISOEV AND PRTEST. USER MUST NOT MODIFY THESE SCALES.      LIV00690
      SCALE1 = 5.0D2                                            LIV00700
      SCALE2 = 5.0D0                                            LIV00710
      SCALE3 = 5.0D0                                            LIV00720
      SCALE4 = 1.0D4                                            LIV00730
      ONE   = 1.0D0                                             LIV00740
      ZERO  = 0.0D0                                             LIV00750
C     BTOL = 1.0D-8                                             LIV00760
      BTOL = EPSM                                               LIV00770
      GAPTOL = 1.0D-8                                           LIV00780
      ICONV = 0                                                 LIV00790
      MOLD = 0                                                  LIV00800
      MOLD1 = 1                                                 LIV00810
      ICT = 0                                                   LIV00820
      MMB = 0                                                   LIV00830
      IPROJ = 0                                                 LIV00840
C----------------------------------------------------------------LIV00850
C     READ USER-SPECIFIED PARAMETERS FROM INPUT FILE 5 (FREE FORMAT)  LIV00860
C                                                               LIV00870
C     READ USER-PROVIDED HEADER FOR RUN                         LIV00880
      READ(5,20) EXPLAN                                         LIV00890
      WRITE(6,20) EXPLAN                                        LIV00900
      READ(5,20) EXPLAN                                         LIV00910
      WRITE(6,20) EXPLAN                                        LIV00920
   20 FORMAT(20A4)                                              LIV00930
C                                                               LIV00940
C     READ ORDER OF MATRICES (N) , MAXIMUM ORDER OF T-MATRIX (KMAX),  LIV00950
C     NUMBER OF T-MATRICES ALLOWED (NMEVS), AND MATRIX IDENTIFICATION LIV00960
C     NUMBERS (MATNO), SHIFT APPLIED TO MATRIX (SHIFT) AND      LIV00970
C     SCALE (SO).                                               LIV00980
      READ(5,20) EXPLAN                                         LIV00990
      READ(5,*) N,KMAX,NMEVS,MATNO,SO,SHIFT                     LIV01000
C                                                               LIV01010
C     READ SEEDS FOR LANCZS AND INVERR SUBROUTINES (SVSEED AND RHSEED) LIV01020
C     READ MAXIMUM NUMBER OF ITERATIONS ALLOWED FOR EACH INVERSE LIV01030
C     ITERATION (MXINIT) AND MAXIMUM NUMBER OF STURM SEQUENCES  LIV01040
C     ALLOWED (MXSTUR)                                          LIV01050
      READ(5,20) EXPLAN                                         LIV01060
      READ(5,*) SVSEED,RHSEED,MXINIT,MXSTUR                     LIV01070
C                                                               LIV01080
C     ISTART = (0,1): ISTART = 0 MEANS ALPHA/BETA FILE IS NOT   LIV01090
C     AVAILABLE. ISTART = 1 MEANS ALPHA/BETA FILE IS AVAILABLE ON LIV01100
C     FILE 2.                                                   LIV01110
C     ISTOP = (0,1): ISTOP = 0 MEANS PROCEDURE GENERATES ALPHA/BETA LIV01120
C     FILE AND THEN TERMINATES. ISTOP = 1 MEANS PROCEDURE GENERATES LIV01130
C     ALPHAS/BETAS IF NEEDED AND THEN COMPUTES EIGENVALUES AND ERROR LIV01140
C     ESTIMATES AND THEN TERMINATES.                            LIV01150
      READ(5,20) EXPLAN                                         LIV01160
      READ(5,*) ISTART,ISTOP                                    LIV01170
C                                                               LIV01180
C     IHIS = (0,1): IHIS = 0 MEANS ALPHA/BETA FILE IS NOT WRITTEN LIV01190
C     TO FILE 1. IHIS = 1 MEANS ALPHA/BETA FILE IS WRITTEN TO FILE 1. LIV01200
C     IDIST = (0,1): IDIST = 0 MEANS DISTINCT T-EIGENVALUES     LIV01210
C     ARE NOT WRITTEN TO FILE 11. IDIST = 1 MEANS DISTINCT      LIV01220
C     T-EIGENVALUES ARE WRITTEN TO FILE 11.                     LIV01230
C     IWRITE = (0,1): IWRITE = 0 MEANS NO INTERMEDIATE OUTPUT   LIV01240
C     FROM THE COMPUTATIONS IS WRITTEN TO FILE 6. IWRITE = 1 MEANS LIV01250
C     T-EIGENVALUES AND ERROR ESTIMATES ARE WRITTEN TO FILE 6   LIV01260
C     AS THEY ARE COMPUTED.                                     LIV01270
      READ(5,20) EXPLAN                                         LIV01280
      READ(5,*) IHIS,IDIST,IWRITE                               LIV01290
C                                                               LIV01300
```

```
C      READ IN THE RELATIVE TOLERANCE (RELTOL) FOR USE IN THE            LIVO1310
C      SPURIOUS, T-MULTIPLICITY, AND PRTESTS.                            LIVO1320
       READ(5,20) EXPLAN                                                 LIVO1330
       READ(5,*) RELTOL                                                  LIVO1340
C                                                                        LIVO1350
C      READ IN THE SIZES OF THE T-MATRICES TO BE CONSIDERED.             LIVO1360
       READ(5,20) EXPLAN                                                 LIVO1370
       READ(5,*) (NMEV(J), J=1,NMEVS)                                    LIVO1380
C                                                                        LIVO1390
C      READ IN THE NUMBER OF SUBINTERVALS TO BE CONSIDERED.              LIVO1400
       READ(5,20) EXPLAN                                                 LIVO1410
       READ(5,*) NINT                                                    LIVO1420
C                                                                        LIVO1430
C      READ IN THE LEFT-END POINTS OF THE SUBINTERVALS TO BE CONSIDERED. LIVO1440
C      THESE MUST BE IN ALGEBRAICALLY-INCREASING ORDER                   LIVO1450
       READ(5,20) EXPLAN                                                 LIVO1460
       READ(5,*) (LB(J), J=1,NINT)                                       LIVO1470
C                                                                        LIVO1480
C      READ IN THE RIGHT-END POINTS OF THE SUBINTERVALS TO BE CONSIDERED.LIVO1490
C      THESE MUST BE IN ALGEBRAICALLY-INCREASING ORDER                   LIVO1500
       READ(5,20) EXPLAN                                                 LIVO1510
       READ(5,*) (UB(J), J=1,NINT)                                       LIVO1520
C                                                                        LIVO1530
C------------------------------------------------------------------------LIVO1540
C      INITIALIZE THE ARRAYS FOR THE FACTORIZATION OF THE ASSOCIATED     LIVO1550
C      SCALED, SHIFTED AND PERMUTED VERSION OF THE A-MATRIX.             LIVO1560
C      THE STORAGE LOCATIONS OF THESE ARRAYS ARE PASSED TO THE BSOLV     LIVO1570
C      SUBROUTINE WHICH WILL BE CALLED FROM LANCZS FOR THE T-MATRIX      LIVO1580
C      GENERATION.                                                       LIVO1590
C                                                                        LIVO1600
       CALL USPEC(N,MATNO)                                               LIVO1610
C                                                                        LIVO1620
C------------------------------------------------------------------------LIVO1630
C                                                                        LIVO1640
C      MASKS UNDERFLOW AND OVERFLOW, USER MUST SUPPLY OR COMMENT OUT.     LIVO1650
       CALL MASK                                                         LIVO1660
C                                                                        LIVO1670
C------------------------------------------------------------------------LIVO1680
C                                                                        LIVO1690
C      WRITE TO FILE 6, A SUMMARY OF THE PARAMETERS FOR THIS RUN         LIVO1700
C                                                                        LIVO1710
       WRITE(6,30) MATNO,N,KMAX,SHIFT,SO                                 LIVO1720
 30 FORMAT(/3X,'MATRIX ID',4X,'ORDER OF A',4X,'MAX ORDER OF T'//         LIVO1730
    1 I12,I14,I18//8X,' SHIFT',8X,'SCALE'/2E15.6//                       LIVO1740
    1 ' C = SCALE*A + SHIFT*I '/                                         LIVO1750
    1 ' B = P*C*P-TRANSPOSE WHERE P IS A REORDERING OF C'/               LIVO1760
    1 ' LANCZOS PROCEDURE USES THE FACTORIZATION OF B'/)                 LIVO1770
C                                                                        LIVO1780
       WRITE(6,40) ISTART,ISTOP                                          LIVO1790
 40 FORMAT(/2X,'ISTART',3X,'ISTOP'/2I8/)                                 LIVO1800
C                                                                        LIVO1810
       WRITE(6,50) IHIS,IDIST,IWRITE                                     LIVO1820
 50 FORMAT(/4X,'IHIS',3X,'IDIST',2X,'IWRITE'/3I8/)                       LIVO1830
C                                                                        LIVO1840
       WRITE(6,60) SVSEED,RHSEED                                         LIVO1850
 60 FORMAT(/' SEEDS FOR RANDOM NUMBER GENERATOR'//                       LIVO1860
    1 4X,'LANCZS SEED',4X,'INVERR SEED'/2I15/)                           LIVO1870
C                                                                        LIVO1880
       WRITE(6,70) (NMEV(J), J=1,NMEVS)                                  LIVO1890
 70 FORMAT(/' SIZES OF T-MATRICES TO BE CONSIDERED'/(6I12))              LIVO1900
C                                                                        LIVO1910
       WRITE(6,80) RELTOL,GAPTOL,BTOL                                    LIVO1920
 80 FORMAT(/' RELATIVE TOLERANCE USED TO COMBINE COMPUTED T-EIGENVALUE   LIVO1930
    1S'/E15.3/' RELATIVE GAP TOLERANCES USED IN INVERSE ITERATION'/      LIVO1940
    1E15.3/' RELATIVE TOLERANCE FOR CHECK ON SIZE OF BETAS'/E15.3/)      LIVO1950
C                                                                        LIVO1960
       WRITE(6,90) (J,LB(J),UB(J), J=1,NINT)                             LIVO1970
```

```
      90 FORMAT(/' BISEC WILL BE USED ON THE FOLLOWING INTERVALS'/      LIV01980
     1 (I6,2E20.6))                                                     LIV01990
C                                                                       LIV02000
         IF (ISTART.EQ.0) GO TO 140                                     LIV02010
C                                                                       LIV02020
C     READ IN ALPHA BETA HISTORY                                        LIV02030
C                                                                       LIV02040
         READ(2,100)MOLD,NOLD,SVSOLD,MATOLD,SHIFTO                      LIV02050
     100 FORMAT(2I6,I12,I8,E13.4)                                       LIV02060
C                                                                       LIV02070
         IF (KMAX.LT.MOLD) KMAX = MOLD                                  LIV02080
         KMAX1 = KMAX + 1                                               LIV02090
C                                                                       LIV02100
C     CHECK THAT ORDER N, MATRIX ID MATNO, AND RANDOM SEED SVSEED       LIV02110
C     AGREE WITH THOSE IN THE HISTORY FILE.  IF NOT PROCEDURE STOPS.    LIV02120
C                                                                       LIV02130
         ITEMP = (NOLD-N)**2+(MATNO-MATOLD)**2+(SVSEED-SVSOLD)**2       LIV02140
C                                                                       LIV02150
         IF (ITEMP.EQ.0.AND.SHIFT.EQ.SHIFTO) GO TO 120                  LIV02160
C                                                                       LIV02170
         WRITE(6,110)                                                   LIV02180
     110 FORMAT(' PROGRAM TERMINATES'/     ' READ FROM FILE 2 CORRESPONDS TOLIV02190
     1 DIFFERENT MATRIX THAN MATRIX SPECIFIED'/)                        LIV02200
         GO TO 700                                                      LIV02210
C                                                                       LIV02220
     120 CONTINUE                                                       LIV02230
         MOLD1 = MOLD+1                                                 LIV02240
C                                                                       LIV02250
         READ(2,130)(ALPHA(J), J=1,MOLD)                               LIV02260
         READ(2,130)(BETA(J), J=1,MOLD1)                               LIV02270
     130 FORMAT(4Z20)                                                   LIV02280
C                                                                       LIV02290
         IF (KMAX.EQ.MOLD) GO TO 170                                    LIV02300
C                                                                       LIV02310
         READ(2,130)(V1(J), J=1,N)                                      LIV02320
         READ(2,130)(V2(J), J=1,N)                                      LIV02330
C                                                                       LIV02340
     140 CONTINUE                                                       LIV02350
         IIX = SVSEED                                                   LIV02360
C                                                                       LIV02370
         WRITE(6,150)                                                   LIV02380
     150 FORMAT(' ENTERING LANCZS'/)                                    LIV02390
C                                                                       LIV02400
C----------------------------------------------------------------------LIV02410
C                                                                       LIV02420
         CALL LANCZS(BSOLV,ALPHA,BETA,V1,V2,VS,G,KMAX,MOLD1,N,IIX)      LIV02430
C                                                                       LIV02440
C----------------------------------------------------------------------LIV02450
C                                                                       LIV02460
C     ALPHA BETA WRITE                                                  LIV02470
         KMAX1 = KMAX + 1                                               LIV02480
C                                                                       LIV02490
         IF(IHIS.EQ.0.AND.ISTOP.GT.0) GO TO 170                         LIV02500
C                                                                       LIV02510
         WRITE(1,160) KMAX,N,SVSEED,MATNO,SHIFT                         LIV02520
     160 FORMAT(2I6,I12,I8,E13.4,' = KMAX,N,SVSEED,MATNO,SHIFT')        LIV02530
C                                                                       LIV02540
         WRITE(1,130)(ALPHA(I), I=1,KMAX)                               LIV02550
         WRITE(1,130)(BETA(I), I=1,KMAX1)                               LIV02560
C                                                                       LIV02570
         WRITE(1,130)(V1(I), I=1,N)                                     LIV02580
         WRITE(1,130)(V2(I), I=1,N)                                     LIV02590
C                                                                       LIV02600
         IF (ISTOP.EQ.0) GO TO 600                                      LIV02610
C                                                                       LIV02620
     170 CONTINUE                                                       LIV02630
         KMAX1 = KMAX + 1                                               LIV02640
```

```
      BKMIN = BTOL                                                   LIV02650
C                                                                    LIV02660
      WRITE(6,180)                                                   LIV02670
  180 FORMAT(/' T-MATRICES (ALPHA AND BETA) ARE NOW AVAILABLE'/)     LIV02680
C                                                                    LIV02690
C---------------------------------------------------------------------LIV02700
C     SUBROUTINE TNORM CHECKS MIN(BETA)/(ESTIMATED NORM(A)) > BTOL .  LIV02710
C     IF THIS IS VIOLATED IB IS SET EQUAL TO THE NEGATIVE OF THE INDEX LIV02720
C     OF THE MINIMAL BETA.  IF(IB < 0) THEN SUBROUTINE TNORM IS       LIV02730
C     CALLED FOR EACH VALUE OF MEV TO DETERMINE WHETHER OR NOT THERE  LIV02740
C     IS A BETA IN THE T-MATRIX SPECIFIED THAT VIOLATES THIS TEST.    LIV02750
C     IF THERE IS SUCH A BETA THE PROGRAM TERMINATES FOR THE USER     LIV02760
C     TO DECIDE WHAT TO DO.  THIS TEST CAN BE OVER-RIDDEN BY          LIV02770
C     SIMPLY MAKING BTOL SMALLER, BUT THEN THERE IS THE POSSIBILITY   LIV02780
C     THAT LOSSES IN THE LOCAL ORTHOGONALITY MAY HURT THE COMPUTATIONS. LIV02790
C     BTOL = 1.D-8 IS HOWEVER A CONSERVATIVE CHOICE FOR BTOL.         LIV02800
C                                                                    LIV02810
C     TNORM ALSO COMPUTES TKMAX = MAX(|ALPHA(K)|,BETA(K), K=1,KMAX).  LIV02820
C     TKMAX IS USED TO SCALE THE TOLERANCES USED IN THE              LIV02830
C     T-MULTIPLICITY AND SPURIOUS TESTS IN BISEC. TKMAX IS ALSO USED IN LIV02840
C     THE PROJECTION TEST FOR HIDDEN EIGENVALUES THAT HAD 'TOO SMALL' LIV02850
C     A PROJECTION ON THE STARTING VECTOR.                           LIV02860
C                                                                    LIV02870
      CALL TNORM(ALPHA,BETA,BKMIN,TKMAX,KMAX,IB)                      LIV02880
C                                                                    LIV02890
C---------------------------------------------------------------------LIV02900
      TTOL = EPSM*TKMAX                                               LIV02910
C                                                                    LIV02920
C     LOOP ON THE SIZE OF THE T-MATRIX                               LIV02930
  190 CONTINUE                                                       LIV02940
      MMB = MMB + 1                                                   LIV02950
      MEV = NMEV(MMB)                                                 LIV02960
C     IS MEV TOO LARGE ?                                             LIV02970
      IF(MEV.LE.KMAX) GO TO 210                                       LIV02980
C                                                                    LIV02990
      WRITE(6,200) MMB, MEV, KMAX                                     LIV03000
  200 FORMAT(/' TERMINATE PRIOR TO CONSIDERING THE',I6,'TH T-MATRIX'/ LIV03010
     1' BECAUSE THE SIZE REQUESTED',I6,' IS GREATER THAN THE MAXIMUM SIZLIV03020
     1E ALLOWED',I6/)                                                 LIV03030
      GO TO 600                                                       LIV03040
C                                                                    LIV03050
  210 MP1 = MEV + 1                                                   LIV03060
      BETAM = BETA(MP1)                                               LIV03070
      WRITE(6,220) MEV,MEV,BETA(MEV),MEV,BETAM                        LIV03080
  220 FORMAT(/' AT T-SIZE = ',I6,' BETA(',I4,') = ',E13.4/' BETA(',I4,'+LIV03090
     11) =',E13.4)                                                    LIV03100
      IF (IB.GE.0) GO TO 230                                          LIV03110
      TO = BTOL  .                                                    LIV03120
C---------------------------------------------------------------------LIV03130
C                                                                    LIV03140
      CALL TNORM(ALPHA,BETA,TO,T1,MEV,IBMEV)                          LIV03150
C                                                                    LIV03160
C---------------------------------------------------------------------LIV03170
      TEMP = TO/TKMAX                                                 LIV03180
      IBMEV = IABS(IBMEV)                                             LIV03190
      IF (TEMP.GE.BTOL) GO TO 230                                     LIV03200
      IBMEV = -IBMEV                                                  LIV03210
      GO TO 660                                                       LIV03220
  230 CONTINUE                                                        LIV03230
      IC = MXSTUR-ICT                                                 LIV03240
C                                                                    LIV03250
C---------------------------------------------------------------------LIV03260
C     BISEC LOOP. THE SUBROUTINE BISEC INCORPORATES DIRECTLY THE      LIV03270
C     T-MULTIPLICITY AND SPURIOUS TESTS. T-EIGENVALUES WILL BE        LIV03280
C     CALCULATED BY BISEC SEQUENTIALLY ON INTERVALS                   LIV03290
C     (LB(J),UB(J)), J = 1,NINT).                                     LIV03300
C                                                                    LIV03310
```

```
C     ON RETURN FROM BISEC                                          LIV03320
C     NDIS = NUMBER OF DISTINCT EIGENVALUES OF T(1,MEV) ON UNION     LIV03330
C       OF THE (LB,UB) INTERVALS                                    LIV03340
C     VS = DISTINCT T-EIGENVALUES IN ALGEBRAICALLY INCREASING ORDER  LIV03350
C     MP = T-MULTIPLICITIES OF THE T-EIGENVALUES IN VS              LIV03360
C     MP(I) = (0,1,MI), MI>1, I=1,NDIS  MEANS:                      LIV03370
C       (0)  VS(I) IS SPURIOUS                                      LIV03380
C       (1)  VS(I) IS T-SIMPLE AND GOOD                             LIV03390
C       (MI) VS(I) IS MULTIPLE AND IS THEREFORE NOT ONLY GOOD BUT   LIV03400
C            ALSO A CONVERGED GOOD T-EIGENVALUE.                    LIV03410
C                                                                   LIV03420
      CALL BISEC(ALPHA,BETA,V1,V2,VS,LB,UB,EPSM,TTOL,MP,NINT,       LIV03430
     1 MEV,NDIS,IC,IWRITE)                                          LIV03440
C                                                                   LIV03450
C-----------------------------------------------------------------LIV03460
      IF (NDIS.EQ.0) GO TO 680                                      LIV03470
C                                                                   LIV03480
C     COMPUTE THE TOTAL NUMBER OF STURM SEQUENCES USED TO DATE      LIV03490
C     COMPUTE THE BISEC CONVERGENCE AND T-MULTIPLICITY TOLERANCES USED. LIV03500
C     COMPUTE THE CONVERGENCE TOLERANCE FOR EIGENVALUES OF A.       LIV03510
      ICT = ICT + IC                                                LIV03520
      TEMP = DFLOAT(MEV+1000)                                       LIV03530
      MULTOL = TEMP*TTOL                                            LIV03540
      TEMP = DSQRT(TEMP)                                            LIV03550
      BISTOL = TTOL*TEMP                                            LIV03560
      CONTOL = BETAM*1.D-10                                         LIV03570
C                                                                   LIV03580
C-----------------------------------------------------------------LIV03590
C     SUBROUTINE LUMP 'COMBINES' T-EIGENVALUES THAT ARE 'TOO CLOSE'. LIV03600
C     NOTE HOWEVER THAT CLOSE SPURIOUS T-EIGENVALUES ARE NOT AVERAGED LIV03610
C     WITH GOOD ONES. HOWEVER, THEY MAY BE USED TO INCREASE THE     LIV03620
C     T-MULTIPLICITY OF A GOOD T-EIGENVALUE.                        LIV03630
C                                                                   LIV03640
      LOOP = NDIS                                                   LIV03650
      CALL LUMP(VS,RELTOL,MULTOL,SCALE2,MP,LOOP)                    LIV03660
C                                                                   LIV03670
C-----------------------------------------------------------------LIV03680
      IF(NDIS.EQ.LOOP) GO TO 250                                    LIV03690
C                                                                   LIV03700
      WRITE(6,240) NDIS, MEV, LOOP                                  LIV03710
  240 FORMAT(/16,' DISTINCT T-EIGENVALUES WERE COMPUTED IN BISEC AT MEV LIV03720
     1',16/ 2X,' LUMP SUBROUTINE REDUCES NUMBER OF DISTINCT T-EIGENVALUELIV03730
     IS TO',16)                                                    LIV03740
C                                                                   LIV03750
  250 CONTINUE                                                      LIV03760
      NDIS = LOOP                                                   LIV03770
      BETA(MP1) = BETAM                                             LIV03780
C-----------------------------------------------------------------LIV03790
C     THE SUBROUTINE ISOEV LABELS THOSE SIMPLE EIGENVALUES OF T(1,MEV) LIV03800
C     WITH VERY SMALL GAPS BETWEEN NEIGHBORING EIGENVALUES OF T(1,MEV) LIV03810
C     TO AVOID COMPUTING ERROR ESTIMATES FOR ANY SIMPLE GOOD        LIV03820
C     T-EIGENVALUE THAT IS TOO CLOSE TO A SPURIOUS EIGENVALUE.      LIV03830
C     ON RETURN FROM ISOEV, G CONTAINS CODED MINIMAL GAPS           LIV03840
C     BETWEEN THE DISTINCT EIGENVALUES OF T(1,MEV). (G IS REAL).    LIV03850
C     G(I) < 0 MEANS MINGAP IS DUE TO LEFT GAP G(I) > 0 MEANS DUE TO LIV03860
C     RIGHT GAP. MP(I) = -1 MEANS THAT THE GOOD T-EIGENVALUE IS SIMPLE LIV03870
C     AND HAS A VERY SMALL MINGAP IN T(1,MEV) DUE TO A SPURIOUS     LIV03880
C     T-EIGENVALUE.  NG = NUMBER OF GOOD T-EIGENVALUES.             LIV03890
C     NISO = NUMBER OF ISOLATED GOOD T-EIGENVALUES.                 LIV03900
C                                                                   LIV03910
      CALL ISOEV(VS,GAPTOL,MULTOL,SCALE1,G,MP,NDIS,NG,NISO)         LIV03920
C                                                                   LIV03930
C-----------------------------------------------------------------LIV03940
C                                                                   LIV03950
      WRITE(6,260)NG,NISO,NDIS                                      LIV03960
  260 FORMAT(/16,' GOOD T-EIGENVALUES HAVE BEEN COMPUTED'/          LIV03970
     1 16,' OF THESE ARE T-ISOLATED'/                              LIV03980
```

```
      2 16,' = NUMBER OF DISTINCT T-EIGENVALUES COMPUTED'/)              LIVO3990
C                                                                        LIVO4000
C     DO WE WRITE DISTINCT EIGENVALUES OF T-MATRIX TO FILE 11?           LIVO4010
      IF (IDIST.EQ.0) GO TO 310                                          LIVO4020
C                                                                        LIVO4030
      WRITE(11,270) NDIS,NISO,MEV,N,SVSEED,MATNO                         LIVO4040
  270 FORMAT(/4I6,I12,I8,' = NDIS,NISO,MEV,N,SVSEED,MATNO'/)             LIVO4050
C                                                                        LIVO4060
      WRITE(11,280)                                                      LIVO4070
  280 FORMAT(/1X,'MP',21X,'EVBI',5X,'TMINGAP',1X,'MP',21X,'EVBI',5X,     LIVO4080
     1'TMINGAP'/)                                                        LIVO4090
C                                                                        LIVO4100
      WRITE(11,290) (MP(I),VS(I),G(I), I=1,NDIS)                         LIVO4110
  290 FORMAT(2(I3,E25.16,E12.3))                                         LIVO4120
C                                                                        LIVO4130
      WRITE(11,300) NDIS, (MP(I), I=1,NDIS)                              LIVO4140
  300 FORMAT(/I6,' = NDIS, T-MULTIPLICITIES (0 MEANS  SPURIOUS)'/(20I4))LIVO4150
C                                                                        LIVO4160
  310 CONTINUE                                                           LIVO4170
      IF (NISO.NE.0) GO TO 340                                           LIVO4180
C                                                                        LIVO4190
      WRITE(4,320) MEV                                                   LIVO4200
  320 FORMAT(/' AT MEV = ',I6,' THERE ARE NO ISOLATED T-EIGENVALUES'/    LIVO4210
     1' SO NO ERROR ESTIMATES WERE COMPUTED/')                          LIVO4220
C                                                                        LIVO4230
      WRITE(6,330)                                                       LIVO4240
  330 FORMAT(/' ALL COMPUTED GOOD T-EIGENVALUES ARE MULTIPLE'/           LIVO4250
     1 ' THEREFORE ALL SUCH EIGENVALUES ARE ASSUMED TO HAVE CONVERGED') LIVO4260
C                                                                        LIVO4270
      ICONV = 1                                                          LIVO4280
      GO TO 380                                                          LIVO4290
  340 CONTINUE                                                           LIVO4300
C------------------------------------------------------------------------LIVO4310
C     SUBROUTINE INVERR COMPUTES ERROR ESTIMATES FOR ISOLATED GOOD       LIVO4320
C     T-EIGENVALUES USING INVERSE ITERATION ON T(1,MEV). ON RETURN       LIVO4330
C     G(J) = MINIMUM GAP IN T(1,MEV) FOR EACH VS(J), J=1,NDIS            LIVO4340
C     G(MEV+I) = BETAM*|U(MEV)| = ERROR ESTIMATE FOR ISOLATED GOOD       LIVO4350
C              T-EIGENVALUES, WHERE I = 1, NISO AND  BETAM = BETA(MEV+1)LIVO4360
C              U(MEV) IS MEVTH COMPONENT OF THE UNIT EIGENVECTOR OF T    LIVO4370
C              CORRESPONDING TO THE ITH ISOLATED GOOD T-EIGENVALUE.      LIVO4380
C     A NEGATIVE ERROR ESTIMATE MEANS THAT FOR THAT PARTICULAR           LIVO4390
C     EIGENVALUE THE INVERSE ITERATION DID NOT CONVERGE IN <= MXINIT     LIVO4400
C     STEPS AND THAT THE CORRESPONDING ERROR ESTIMATE IS QUESTIONABLE.   LIVO4410
C                                                                        LIVO4420
C     V2 CONTAINS THE ISOLATED GOOD T-EIGENVALUES                        LIVO4430
C     V1 CONTAINS THE MINGAPS TO THE NEAREST DISTINCT  EIGENVALUE        LIVO4440
C        OF T(1,MEV) FOR EACH ISOLATED GOOD T-EIGENVALUE IN V2.          LIVO4450
C     VS CONTAINS THE NDIS DISTINCT EIGENVALUES OF T(1,MEV)              LIVO4460
C     MP CONTAINS THE CORRESPONDING CODED T-MULTIPLICITIES               LIVO4470
C                                                                        LIVO4480
      IT = MXINIT                                                        LIVO4490
      CALL INVERR(ALPHA,BETA,V1,V2,VS,EPSM,G,MP,MEV,MMB,NDIS,NISO,N,     LIVO4500
     1 RHSEED,IT,IWRITE)                                                 LIVO4510
C                                                                        LIVO4520
C------------------------------------------------------------------------LIVO4530
C     SIMPLE CHECK FOR CONVERGENCE. CHECKS TO SEE IF ALL OF THE          LIVO4540
C     LAST COMPONENTS OF EIGENVECTORS ARE L.T. CONTOL.                   LIVO4550
C     IF THIS TEST IS SATISFIED, THEN CONVERGENCE FLAG, ICONV IS SET     LIVO4560
C     TO 1.  TYPICALLY ERROR ESTIMATES ARE VERY CONSERVATIVE.           LIVO4570
C                                                                        LIVO4580
      WRITE(6,350) CONTOL                                                LIVO4590
  350 FORMAT(/' CONVERGENCE IS TESTED USING THE CONVERGENCE TOLERANCE', LIVO4600
     1E13.4/)                                                            LIVO4610
C                                                                        LIVO4620
      II = MEV +1                                                        LIVO4630
      IF = MEV+NISO                                                      LIVO4640
      DO 360 I = II,IF                                                   LIVO4650
```

```
         IF (ABS(G(I)).GT.CONTOL) GO TO 380              LIV04660
  360 CONTINUE                                           LIV04670
         ICONV = 1                                       LIV04680
         MMB = NMEVS                                     LIV04690
C                                                        LIV04700
         WRITE(6,370) CONTOL                             LIV04710
  370 FORMAT(' ALL COMPUTED ERROR ESTIMATES WERE LESS THAN',E15.4/  LIV04720
     1 ' THEREFORE PROCEDURE TERMINATES'/)               LIV04730
C                                                        LIV04740
  380 CONTINUE                                           LIV04750
C                                                        LIV04760
         IF (ICONV.EQ.0) GO TO 510                       LIV04770
C                                                        LIV04780
C------------------------------------------------------------- LIV04790
C      IF CONVERGENCE IS INDICATED, THAT IS ICONV = 1 ,THEN    LIV04800
C      THE SUBROUTINE PRTEST IS CALLED TO CHECK FOR ANY CONVERGED  LIV04810
C      T-EIGENVALUES THAT HAVE BEEN MISLABELLED AS SPURIOUS BECAUSE  LIV04820
C      THE PROJECTION OF THEIR EIGENVECTOR(S) ON THE STARTING  LIV04830
C      VECTOR WAS(WERE) TOO SMALL.                       LIV04840
C      NUMERICAL TESTS INDICATE THAT SUCH EIGENVALUES ARE RARE.  LIV04850
C      IF FOR SOME REASON MANY OF THESE HIDDEN EIGENVALUES APPEAR  LIV04860
C      ON SOME RUN, YOU CAN BE CERTAIN THAT SOMETHING IS FOULED UP.  LIV04870
C                                                        LIV04880
         CALL PRTEST(ALPHA,BETA,VS,TKMAX,EPSM,RELTOL,SCALE3,SCALE4,  LIV04890
     1 MP,NDIS,MEV,IPROJ)                                LIV04900
C                                                        LIV04910
C-------------------------------------------------------------LIV04920
C                                                        LIV04930
         IF(IPROJ.EQ.0) GO TO 500                        LIV04940
C                                                        LIV04950
         IF(IDIST.EQ.1)  WRITE(11,390) IPROJ             LIV04960
  390 FORMAT(' SUBROUTINE PRTEST WANTS TO RELABEL',I6,' SPURIOUS T-EIGENLIV04970
     1VALUES'/' WE ACCEPT RELABELLING ONLY IF LAST COMPONENT OF T-EIGENVLIV04980
     1ECTOR IS L.T. 1.D-10'/)                            LIV04990
C                                                        LIV05000
         IIX = RHSEED                                    LIV05010
C                                                        LIV05020
C-------------------------------------------------------------LIV05030
C                                                        LIV05040
         CALL GENRAN(IIX,G,MEV)                          LIV05050
C                                                        LIV05060
C-------------------------------------------------------------LIV05070
C                                                        LIV05080
         ITEN = -10                                      LIV05090
         NISOM = NISO + MEV                              LIV05100
         IWRITO = IWRITE                                 LIV05110
         IWRITE = 0                                      LIV05120
C                                                        LIV05130
         DO 420 J = 1,NDIS                               LIV05140
         IF(MP(J).NE.ITEN) GO TO 420                     LIV05150
         TO = VS(J)                                      LIV05160
C                                                        LIV05170
C-------------------------------------------------------------LIV05180
C                                                        LIV05190
         IT = MXINIT                                     LIV05200
         CALL INVERM(ALPHA,BETA,V1,V2,TO,TEMP,T1,EPSM,G,MEV,IT,IWRITE)  LIV05210
C                                                        LIV05220
C-------------------------------------------------------------LIV05230
C                                                        LIV05240
         IF(TEMP.LE.1.D-10) GO TO 410                    LIV05250
C      ERROR ESTIMATE WAS NOT SMALL REJECT RELABELLING OF THIS  LIV05260
C      T-EIGENVALUE                                      LIV05270
         IF(IDIST.EQ.1)  WRITE(11,400) J,TO,TEMP         LIV05280
  400 FORMAT(/' LAST COMPONENT FOR',I6,'TH T-EIGENVALUE',E20.12/' IS TOOLIV05290
     1 LARGE = ',E15.6,' SO DO NOT ACCEPT PRTEST RELABELLING'/)  LIV05300
         MP(J) = 0                                       LIV05310
         IPROJ = IPROJ - 1                               LIV05320
```

```
      GO TO 420                                                        LIV05330
C     RELABELLING ACCEPTED                                            LIV05340
  410 NISOM = NISOM + 1                                               LIV05350
      G(NISOM) = BETAM*TEMP                                           LIV05360
  420 CONTINUE                                                        LIV05370
      IWRITE = IWRITO                                                 LIV05380
C                                                                     LIV05390
      IF(IPROJ.EQ.0) GO TO 460                                        LIV05400
      WRITE(6,430) IPROJ                                              LIV05410
  430 FORMAT(/I6,' T-EIGENVALUES WERE RECLASSIFIED AS GOOD.'/         LIV05420
     1' THESE ARE IDENTIFIED IN FILE 3 BY A T-MULTIPLICITY OF -10'/' USELIV05430
     2R SHOULD INSPECT EACH TO MAKE SURE NEIGHBORS HAVE CONVERGED'/)  LIV05440
C                                                                     LIV05450
      IF(IDIST.EQ.1)  WRITE(11,440) IPROJ                             LIV05460
  440 FORMAT(/I6,' T-EIGENVALUES WERE RELABELLED AS GOOD'/            LIV05470
     1' BELOW IS CORRECTED T-MULTIPLICITY PATTERN'/)                  LIV05480
C                                                                     LIV05490
      WRITE(6,450) NDIS, (MP(I), I=1,NDIS)                            LIV05500
      IF(IDIST.EQ.1)  WRITE(11,450) NDIS, (MP(I), I=1,NDIS)           LIV05510
  450 FORMAT(/I6,' = NDIS, T-MULTIPLICITIES (0 MEANS  SPURIOUS)'/     LIV05520
     1 6X, ' (-10) MEANS SPURIOUS T-EIGENVALUE RELABELLED AS GOOD'/(20I4LIV05530
     1))                                                              LIV05540
C                                                                     LIV05550
C     RECALCULATE MINGAPS FOR DISTINCT T(1,MEV) EIGENVALUES.          LIV05560
  460 NM1 = NDIS - 1                                                  LIV05570
      G(NDIS) = VS(NM1)-VS(NDIS)                                      LIV05580
      G(1) = VS(2)-VS(1)                                              LIV05590
C                                                                     LIV05600
      DO 470 J = 2,NM1                                                LIV05610
      TO = VS(J)-VS(J-1)                                              LIV05620
      T1 = VS(J+1)-VS(J)                                              LIV05630
      G(J) = T1                                                       LIV05640
      IF (TO.LT.T1) G(J) = -TO                                        LIV05650
  470 CONTINUE                                                        LIV05660
      IF(IPROJ.EQ.0) GO TO 500                                        LIV05670
C     WRITE TO FILE 4 ERROR ESTIMATES FOR THOSE T-EIGENVALUES RELABELLEDLIV05680
      NGOOD = 0                                                       LIV05690
      DO 480 J = 1,NDIS                                               LIV05700
      IF(MP(J).EQ.0) GO TO 480                                        LIV05710
      NGOOD = NGOOD + 1                                               LIV05720
      IF(MP(J).NE.ITEN) GO TO 480                                     LIV05730
      TO = VS(J)                                                      LIV05740
      NISO = NISO + 1                                                 LIV05750
      NISOM = MEV + NISO                                              LIV05760
      WRITE(4,490) NGOOD,TO,G(NISOM),G(J)                            LIV05770
  480 CONTINUE                                                        LIV05780
  490 FORMAT(I10,E25.16,2E14.3)                                       LIV05790
C                                                                     LIV05800
  500 CONTINUE                                                        LIV05810
C                                                                     LIV05820
C     WRITE THE GOOD T-EIGENVALUES TO FILE 3.  FIRST TRANSFER THEM    LIV05830
C     TO V2 AND THEIR T-MULTIPLICITIES TO THE CORRESPONDING POSITIONS LIV05840
C     IN MP AND COMPUTE THE A-MINGAPS, THE MINIMAL GAPS BETWEEN THE   LIV05850
C     GOOD T-EIGENVALUES.  THESE GAPS WILL BE PUT IN THE ARRAY G.     LIV05860
C     SINCE G CURRENTLY CONTAINS THE MINIMAL GAPS BETWEEN THE DISTINCT LIV05870
C     EIGENVALUES OF THE T-MATRIX, THESE GAPS WILL FIRST BE           LIV05880
C     TRANSFERRED TO V1.  NOTE THAT V1<0 MEANS THAT THAT MINIMAL GAP  LIV05890
C     IN THE T-MATRIX IS DUE TO A SPURIOUS T-EIGENVALUE.             LIV05900
C     ALL THIS INFORMATION IS PRINTED TO FILE 3                       LIV05910
C                                                                     LIV05920
  510 CONTINUE                                                        LIV05930
      NG = 0                                                          LIV05940
      DO 520 I = 1,NDIS                                               LIV05950
      IF (MP(I).EQ.0) GO TO 520                                       LIV05960
      NG = NG+1                                                       LIV05970
      MP(NG) = MP(I)                                                  LIV05980
      V2(NG) = VS(I)                                                  LIV05990
```

```
      TEMP = G(I)                                                LIV06000
      TEMP = DABS(TEMP)                                          LIV06010
      J = I+1                                                    LIV06020
      IF (G(I).LT.ZERO) J = I-1                                  LIV06030
      IF (MP(J).EQ.0) TEMP = -TEMP                               LIV06040
      VI(NG) = TEMP                                              LIV06050
  520 CONTINUE                                                   LIV06060
C                                                                LIV06070
      WRITE(6,530)MEV                                            LIV06080
  530 FORMAT(//' T-EIGENVALUE CALCULATION AT MEV = ',I6,' IS COMPLETE'/LIV06090
     1)                                                          LIV06100
C                                                                LIV06110
C     NG = NUMBER OF COMPUTED DISTINCT GOOD T-EIGENVALUES.  NEXT LIV06120
C     GENERATE GAPS BETWEEN GOOD T-EIGENVALUES (BIMINGAPS) AND PUT THEM LIV06130
C     G.  G(J) < 0 MEANS THE MINIMAL GAP IS DUE TO THE LEFT-HAND GAP.   LIV06140
C                                                                LIV06150
C     GG(J) = BIMINGAP FOR EIGENVALUES OF B-INVERSE MATRIX.      LIV06160
      NGM1 = NG - 1                                              LIV06170
      GG(NG) = V2(NGM1)-V2(NG)                                   LIV06180
      GG(1) = V2(2)-V2(1)                                        LIV06190
C                                                                LIV06200
      DO 540 J = 2,NGM1                                          LIV06210
      TO = V2(J)-V2(J-1)                                         LIV06220
      T1 = V2(J+1)-V2(J)                                         LIV06230
      GG(J) = T1                                                 LIV06240
      IF (TO.LT.T1) GG(J) = -TO                                  LIV06250
  540 CONTINUE                                                   LIV06260
C                                                                LIV06270
C     WRITE GOOD BI EIGENVALUES TO FILE 3.                       LIV06280
      WRITE(3,550)NG,NDIS,MEV,N,SVSEED,MATNO,MULTOL,IB,BTOL,SHIFT LIV06290
  550 FORMAT(4I6,I12,I8,' = NG,NDIS,MEV,N,SVSEED,MATNO'/         LIV06300
     1 E20.12,I6,2E10.3,' = MULTOL,I(MINBETA),BTOL,SHIFT')       LIV06310
C                                                                LIV06320
C     CALCULATE EIGENVALUES OF ORIGINAL INPUT MATRIX CORRESPONDING LIV06330
C     TO COMPUTED GOOD T-EIGENVALUES.                            LIV06340
      TEMP = -ONE/SO                                             LIV06350
      DO 560 K = 1,NG                                            LIV06360
      VS(K) = (SHIFT - (ONE/V2(K)))*TEMP                         LIV06370
  560 CONTINUE                                                   LIV06380
C                                                                LIV06390
      NGM1 = NG - 1                                              LIV06400
      G(NG) = DABS(VS(NGM1)-VS(NG))                              LIV06410
      G(1) = DABS(VS(2)-VS(1))                                   LIV06420
C                                                                LIV06430
      DO 570 J = 2,NGM1                                          LIV06440
      TO = DABS(VS(J)-VS(J-1))                                   LIV06450
      T1 = DABS(VS(J+1)-VS(J))                                   LIV06460
      G(J) = T1                                                  LIV06470
      IF (TO.LT.T1) G(J)=-TO                                     LIV06480
  570 CONTINUE                                                   LIV06490
C                                                                LIV06500
      WRITE(3,580)                                               LIV06510
  580 FORMAT(' EVNO',1X,'TMULT',20X,'EVBI',5X,'BIGAP',6X,'AGAP',6X, LIV06520
     1'TGAP',12X,'EVA')                                          LIV06530
C                                                                LIV06540
      WRITE(3,590)(I,MP(I),V2(I),GG(I),G(I),VI(I),VS(I),  I=1,NG) LIV06550
  590 FORMAT(2I5,E25.16,3E10.3,E15.8)                            LIV06560
C                                                                LIV06570
C     IF CONVERGENCE FLAG ICONV.NE.1 AND NUMBER OF T-MATRICES    LIV06580
C     CONSIDERED TO DATE IS LESS THAN NUMBER ALLOWED, INCREMENT MEV. LIV06590
C     AND LOOP BACK TO 210 TO REPEAT COMPUTATIONS.  RESTORE BETA(MEV+1).LIV06600
C                                                                LIV06610
      BETA(MP1) = BETAM                                          LIV06620
      IF (MMB.LT.NMEVS.AND.ICONV.NE.1) GO TO 190                 LIV06630
C     END OF LOOP ON DIFFERENT SIZE T-MATRICES ALLOWED.          LIV06640
  600 CONTINUE                                                   LIV06650
C                                                                LIV06660
```

```
      IF(ISTOP.EQ.0)  WRITE(6,610)                                     LIV06670
  610 FORMAT(/' T-MATRICES (ALPHA AND BETA) ARE NOW AVAILABLE, TERMINATELIV06680
     1')                                                               LIV06690
      IF (IHIS.EQ.1.AND.KMAX.NE.MOLD) WRITE(1,620)                     LIV06700
  620 FORMAT(/' ABOVE ARE THE FOLLOWING VECTORS '/                     LIV06710
     1  '  ALPHA(I), I = 1,KMAX'/                                      LIV06720
     2  '  BETA(I), I = 1,KMAX+1'/                                     LIV06730
     3  ' FINAL TWO LANCZOS VECTORS OF ORDER N FOR I = KMAX,KMAX+1'/   LIV06740
     4  ' ALL VECTORS IN THIS FILE HAVE HEX FORMAT 4Z20'/             LIV06750
     5  ' ----- END OF FILE 1 NEW ALPHA, BETA HISTORY--------------'///)LIV06760
C                                                                      LIV06770
      IF (ISTOP.EQ.0) GO TO 700                                        LIV06780
C                                                                      LIV06790
      WRITE(3,630)                                                     LIV06800
  630 FORMAT(/' ABOVE ARE COMPUTED GOOD T-EIGENVALUES'/                LIV06810
     1  ' NG = NUMBER OF GOOD T-EIGENVALUES COMPUTED'/                 LIV06820
     2  ' NDIS = NUMBER OF COMPUTED DISTINCT EIGENVALUES OF T(1,MEV)'/ LIV06830
     3  ' N = ORDER OF A,  MATNO = MATRIX IDENT'/                      LIV06840
     3  ' THERE ARE TWO SETS OF EIGENVALUES, THOSE FOR A AND THOSE FOR'/ LIV06850
     3  ' B-INVERSE WHERE C=SO*A + SHIFT*I, B = P*C*P-TRANS = L*L-TRANS'/LIV06860
     3  ' THE LANCZOS RECURSIONS ARE APPLIED TO B-INVERSE, USING L'/   LIV06870
     3  ' IF EVBI IS A GOOD EIGENVALUE OF B-INVERSE, THEN EVA IS A'/   LIV06880
     3  ' GOOD EIGENVALUE OF A WHERE EVA = (SHIFT-ONE/EVBI)(-ONE/SO)'/ LIV06890
     4  ' MULTOL = T-MULTIPLICITY TOLERANCE FOR T-EIGENVALUES IN BISEC'/ LIV06900
     4  ' TMULT IS THE T-MULTIPLICITY OF GOOD T-EIGENVALUE'/           LIV06910
     5  ' TMULT = -1 MEANS SPURIOUS T-EIGENVALUE TOO CLOSE'/           LIV06920
     6  ' DO NOT COMPUTE ERROR ESTIMATES FOR SUCH EIGENVALUES'/        LIV06930
     7  ' AMINGAP = MINIMAL GAP BETWEEN THE COMPUTED A-EIGENVALUES'/   LIV06940
     8  ' AMINGAP .LT. 0. MEANS MINIMAL GAP IS DUE TO LEFT-HAND GAP'/  LIV06950
     9  ' TMINGAP= MINIMAL GAP W.R.T.  DISTINCT EIGENVALUES IN T(1,MEV)'/LIV06960
     1  ' TMINGAP .LT. 0. MEANS MINGAP IS DUE TO SPURIOUS T-EIGENVALUE'/ LIV06970
     2  ' ----- END OF FILE 3 GOODEIGENVALUES----------------------'///)LIV06980
C                                                                      LIV06990
      IF (IDIST.EQ.1) WRITE(11,640)                                    LIV07000
  640 FORMAT(/' ABOVE ARE THE DISTINCT EIGENVALUES OF T(1,MEV).'/      LIV07010
     2  ' THE FORMAT IS       T-MULTIPLICITY     T-EIGENVALUE   TMINGAP'/LIV07020
     3  '        THIS FORMAT IS REPEATED TWICE ON EACH LINE.'/         LIV07030
     4  ' T-MULTIPLICITY = -1 MEANS THAT THE SUBROUTINE ISOEV HAS TAGGED'LIV07040
     5/'   THIS SIMPLE T-EIGENVALUE AS HAVING A VERY CLOSE SPURIOUS'/  LIV07050
     6  '   T-EIGENVALUE SO THAT NO ERROR ESTIMATE WILL BE COMPUTED'/  LIV07060
     7  '   FOR THAT EIGENVALUE IN SUBROUTINE INVERR.'/                LIV07070
     8  ' TMINGAP .LT. 0, TMINGAP IS DUE TO LEFT GAP .GT. 0, RIGHT GAP.'/LIV07080
     9  ' EACH OF THE DISTINCT T-EIGENVALUE TABLES IS FOLLOWED'/       LIV07090
     9  ' BY THE T-MULTIPLICITY PATTERN.'/                             LIV07100
     1  ' NDIS = NUMBER OF COMPUTED DISTINCT EIGENVALUES OF T(1,MEV).'/ LIV07110
     2  ' NG = NUMBER OF GOOD T-EIGENVALUES. '/                        LIV07120
     3  ' NISO = NUMBER OF ISOLATED GOOD T-EIGENVALUES. '/             LIV07130
     4  ' NISO ALSO IS THE COUNT OF +1 ENTRIES IN T-MULTIPLICITY PATTERN.LIV07140
     5'/ ' ----- END OF FILE 11 DISTINCT T-EIGENVALUES--------------'///LIV07150
     6)                                                                LIV07160
C                                                                      LIV07170
      IF(NIOS.NE.0)  WRITE(4,650)                                      LIV07180
  650 FORMAT(/' ABOVE ARE THE ERROR ESTIMATES OBTAINED FOR THE ISOLATED LIV07190
     1GOOD T-EIGENVALUES'/                                             LIV07200
     1' OBTAINED VIA INVERSE ITERATION IN THE SUBROUTINE INVERR.'/     LIV07210
     1' ALL OTHER GOOD T-EIGENVALUES HAVE CONVERGED.'/                 LIV07220
     2' ERROR ESTIMATE = BETAM*ABS(UM)'/                              LIV07230
     2' WHERE BETAM = BETA(MEV+1) AND UM = U(MEV).'/                   LIV07240
     3' U = UNIT EIGENVECTOR OF T WHERE T*U = EV*U AND EV = ISOLATED GOOLIV07250
     3D T-EIGENVALUE.'/                                                LIV07260
     4' TMINGAP = GAP TO NEAREST DISTINCT EIGENVALUE OF T(1,MEV).'/    LIV07270
     5' TMINGAP .LT. 0. MEANS MINGAP IS DUE TO LEFT NEIGHBOR'/         LIV07280
     6' ERROR ESTIMATE L.T. 0 MEANS INVERSE ITERATION DID NOT CONVERGE'/LIV07290
     7' ------ END OF FILE 4 ERRINV ----------------------------'//)   LIV07300
      GO TO 700                                                        LIV07310
C                                                                      LIV07320
```

```
  660 CONTINUE                                                        LIV07330
C                                                                     LIV07340
      IBB = IABS(IBMEV)                                               LIV07350
      IF (IBMEV.LT.0) WRITE(6,670) MEV,IBB,BETA(IBB)                  LIV07360
  670 FORMAT(/' PROGRAM TERMINATES BECAUSE MEV REQUESTED = ',I6,' IS .GTLIV07370
     1',I6/' AT WHICH AN ABNORMALLY SMALL BETA = ' , E13.4,' OCCURRED'/)LIV07380
      GO TO 700                                                       LIV07390
C                                                                     LIV07400
  680 IF (NDIS.EQ.0.AND.ISTOP.GT.0) WRITE(6,690)                      LIV07410
  690 FORMAT(/' INTERVALS SPECIFIED FOR BISECT DID NOT CONTAIN ANY T-EIGLIV07420
     1ENVALUES'/' PROGRAM TERMINATES')                               LIV07430
C                                                                     LIV07440
  700 CONTINUE                                                        LIV07450
C                                                                     LIV07460
      STOP                                                            LIV07470
C-----END OF LIVAL (INVERSES OF REAL SYMMETRIC MATRICES)-------------LIV07480
      END                                                             LIV07490
```

SECTION 4.3   MAIN PROGRAM, EIGENVECTOR COMPUTATIONS

```
C-----LIVEC (EIGENVECTORS OF INVERSES OF REAL SYMMETRIC MATRICES)-------LIV00010
C                                                                       LIV00020
C       CONTAINS MAIN PROGRAM FOR COMPUTING AN EIGENVECTOR CORRESPONDING LIV00030
C       TO EACH OF A SET OF EIGENVALUES WHICH HAVE BEEN COMPUTED         LIV00040
C       ACCURATELY BY THE CORRESPONDING LANCZOS EIGENVALUE PROGRAM       LIV00050
C       (LIVAL) FOR FACTORED INVERSES OF REAL, SYMMETRIC MATRICES.       LIV00060
C       THIS PROGRAM COULD BE MODIFIED TO COMPUTE ADDITIONAL EIGENVECTORS LIV00070
C       FOR ANY EIGENVALUES WHICH ARE MULTIPLE EIGENVALUES OF THE        LIV00080
C       A-MATRIX.  THE AMOUNT OF ADDITIONAL COMPUTATION REQUIRED BY      LIV00090
C       SUCH A MODIFICATION DEPENDS UPON THE GIVEN A-MATRIX AND UPON     LIV00100
C       WHICH PORTION OF THE SPECTRUM IS INVOLVED.                       LIV00110
C                                                                       LIV00120
C       THESE LANCZOS EIGENVECTOR COMPUTATIONS ASSUME THAT EACH         LIV00130
C       EIGENVALUE THAT IS BEING CONSIDERED HAS CONVERGED AS AN         LIV00140
C       EIGENVALUE OF THE LANCZOS TRIDIAGONAL MATRICES.                 LIV00150
C                                                                       LIV00160
C       PFORT VERIFIER IDENTIFIED THE FOLLOWING NONPORTABLE             LIV00170
C       CONSTRUCTIONS                                                   LIV00180
C                                                                       LIV00190
C       1.  DATA/MACHEP/ STATEMENT                                      LIV00200
C       2.  ALL READ(5,*) STATEMENTS (FREE FORMAT)                      LIV00210
C       3.  FORMAT(20A4) USED WITH THE EXPLANATORY HEADER, EXPLAN       LIV00220
C       4.  HEXADECIMAL FORMAT (4Z20) USED FOR ALPHA/BETA FILES 1 AND 2. LIV00230
C                                                                       LIV00240
C       IMPORTANT NOTE:  PROGRAM ALLOWS ENLARGEMENT OF THE ALPHA, BETA  LIV00250
C       ARRAYS.  IN PARTICULAR, IF ANY ONE OF THE EIGENVALUES SUPPLIED  LIV00260
C       IS T-SIMPLE AND NOT CLOSE TO A SPURIOUS EIGENVALUE, THE PROGRAM LIV00270
C       REQUIRES THAT KMAX BE AT LEAST 11*MEV/8 + 12.  IF KMAX IS NOT   LIV00280
C       THIS LARGE, THEN THE PROGRAM RESETS KMAX TO THIS SIZE           LIV00290
C       AND EXTENDS THE ALPHA, BETA HISTORY IF REQUIRED.                LIV00300
C       THUS, THE DIMENSIONS OF THE ALPHA AND BETA ARRAYS MUST BE       LIV00310
C       LARGE ENOUGH TO ALLOW FOR THIS POSSIBILITY.                     LIV00320
C       REMEMBER THAT THE BETA ARRAY, BETA(J), IS SUCH THAT             LIV00330
C       J = 1,..., KMAX+1,  SO IF THE KMAX USED BY THE PROGRAM          LIV00340
C       IS TO BE 3000, THEN BETA MUST BE OF LENGTH AT LEAST 3001.       LIV00350
C                                                                       LIV00360
C----------------------------------------------------------------------LIV00370
        DOUBLE PRECISION    ALPHA(1000),BETA(1001)                       LIV00380
        DOUBLE PRECISION    V1(2200),V2(2200),VS(2200)                   LIV00390
        DOUBLE PRECISION    RITVEC(40000),TVEC(5000)                     LIV00400
        DOUBLE PRECISION    GOODA(50),GOODBI(50),EVNEW(50),TLAST(50)     LIV00410
        DOUBLE PRECISION    EVAL,EVALN,TOLN,TTOL,ERTOL,ALFA,BATA         LIV00420
        DOUBLE PRECISION    MULTOL,SCALEO,STUTOL,BTOL,LB,UB,SO,RNORME    LIV00430
        DOUBLE PRECISION    ONE,ZERO,MACHEP,EPSM,TEMP,SUM,SHIFT,SHIFTO   LIV00440
        DOUBLE PRECISION    RELTOL,ERROR,TERROR,BKMIN,ERRMIN            LIV00450
        REAL G(5000),AMINGP(50),TMINGP(50),BIERR(50),BIEVER(50),BIERRG(50)LIV00460
        REAL TERR(50),RNORM(50),TBETA(50),BIMING(50)                    LIV00470
        REAL EXPLAN(20)                                                  LIV00480
        INTEGER  MP(50),IDELTA(50)                                       LIV00490
        INTEGER  M1(50),M2(50),MA(50),ML(50),MINT(50),MFIN(50)           LIV00500
        INTEGER  SVSEED,SVSOLD,RHSEED                                    LIV00510
        INTEGER  MBOUND,NTVCON,SVTVEC,TVSTOP,LVCONT,ERCONT,TFLAG         LIV00520
        DOUBLE PRECISION  FINPRO                                         LIV00530
        DOUBLE PRECISION  DABS, DMAX1, DSQRT, DFLOAT                     LIV00540
        REAL ABS                                                         LIV00550
        INTEGER  IABS                                                    LIV00560
        EXTERNAL BSOLV                                                   LIV00570
C----------------------------------------------------------------------LIV00580
        DATA MACHEP/Z3410000000000000/                                  LIV00590
        EPSM = 2.D0*MACHEP                                              LIV00600
C----------------------------------------------------------------------LIV00610
C       ARRAYS MUST BE DIMENSIONED AS FOLLOWS:                          LIV00620
C       1.  ALPHA: >= KMAXN, BETA: >= (KMAXN+1) WHERE KMAXN, THE        LIV00630
C               LARGEST SIZE T-MATRIX CONSIDERED BY THE PROGRAM,        LIV00640
```

```
C                    IS THE LARGER OF THE SIZE OF THE ALPHA, BETA HISTORY  LIV00650
C                    PROVIDED ON FILE 2 (IF ANY ) AND THE SIZE WHICH THE   LIV00660
C                    PROGRAM SPECIFIES INTERNALLY, THIS LATTER IS ALWAYS   LIV00670
C                    < = 11*MEV / 8 + 12, WHERE MEV IS THE SIZE           LIV00680
C                    T-MATRIX THAT WAS USED IN THE CORRESPONDING EIGENVALUE LIV00690
C                    COMPUTATIONS.                                         LIV00700
C       2.  V1:   >= MAX(N,KMAX)                                          LIV00710
C       3.  V2, VS:   >= N                                                LIV00720
C       4.  G:    >= MAX(N,KMAX)                                          LIV00730
C       5.  RITVEC: >= N*NGOOD, WHERE NGOOD IS NUMBER OF EIGENVALUES      LIV00740
C                    SUPPLIED TO THIS PROGRAM.                            LIV00750
C       6.  TVEC:  >= CUMULATIVE LENGTH OF ALL THE T-EIGENVECTORS        LIV00760
C                    NEEDED TO GENERATE THE DESIRED RITZ VECTORS.  AN EDUCATED LIV00770
C                    GUESS AT AN APPROPRIATE LENGTH CAN BE OBTAINED BY RUNNING THE LIV00780
C                    PROGRAM WITH THE FLAG MBOUND = 1 AND MULTIPLYING THE LIV00790
C                    RESULTING SIZE BY 5/4.                               LIV00800
C       7.  GOODA, GOODBI, EVNEW, AMINGP, TMINGP, TERR, RNORM,           LIV00810
C           TBETA, TLAST, BIERR, BIERRG, MP, MA, M1, M2, MINT,           LIV00820
C           MFIN AND IDELTA MUST BE OF DIMENSION AT LEAST NGOOD.         LIV00830
C                                                                         LIV00840
C----------------------------------------------------------------------  LIV00850
C       OUTPUT HEADER                                                     LIV00860
        WRITE(6,10)                                                       LIV00870
     10 FORMAT(/' LANCZOS PROCEDURE FOR FACTORED INVERSES OF REAL SYMMETRILIV00880
     1C MATRICES'/'       COMPUTE EIGENVECTORS'/)                         LIV00890
C                                                                         LIV00900
C       SET PROGRAM PARAMETERS                                            LIV00910
C       USER MUST NOT MODIFY SCALEO                                       LIV00920
        SCALEO = 5.0D0                                                    LIV00930
        ZERO = 0.0D0                                                      LIV00940
        ONE = 1.0D0                                                       LIV00950
        MPMIN = -1000                                                     LIV00960
C       CONVERGENCE TOLERANCE FOR T-EIGENVECTORS FOR RITZ COMPUTATIONS    LIV00970
        ERTOL = 1.D-10                                                    LIV00980
C                                                                         LIV00990
C       READ USER-SPECIFIED PARAMETER FROM INPUT FILE 5 (FREE FORMAT)     LIV01000
C                                                                         LIV01010
C       READ USER-PROVIDED HEADER FOR RUN                                 LIV01020
        READ(5,20) EXPLAN                                                 LIV01030
        WRITE(6,20) EXPLAN                                                LIV01040
     20 FORMAT(20A4)                                                      LIV01050
C                                                                         LIV01060
C       READ IN MATNO = MATRIX/RUN IDENTIFICATION NUMBER AND             LIV01070
C       N = ORDER OF A-MATRIX                                             LIV01080
C       READ IN SCALE (SO) AND SHIFT (SHIFT) APPLIED TO GIVEN            LIV01090
C       MATRIX AND FLAG JPERM.   JPERM = (0,1):                          LIV01100
C       JPERM = 1 MEANS THAT A-MATRIX HAS BEEN PERMUTED.                 LIV01110
C                                                                         LIV01120
        READ(5,20) EXPLAN                                                 LIV01130
        READ(5,*) MATNO,N,SO,SHIFT,JPERM                                  LIV01140
C                                                                         LIV01150
C       READ IN THE MAXIMUM PERMISSIBLE DIMENSIONS FOR THE TVEC ARRAY    LIV01160
C       (MDIMTV), FOR THE RITVEC ARRAY (MDIMRV), AND FOR THE BETA        LIV01170
C       ARRAY (MBETA).                                                    LIV01180
C                                                                         LIV01190
        READ(5,20) EXPLAN                                                 LIV01200
        READ(5,*) MDIMTV, MDIMRV, MBETA                                   LIV01210
C                                                                         LIV01220
C       READ IN RELATIVE TOLERANCE (RELTOL) USED IN DETERMINING          LIV01230
C       APPROPRIATE SIZES FOR THE T-MATRICES USED IN THE EIGENVECTOR     LIV01240
C       COMPUTATIONS                                                      LIV01250
C                                                                         LIV01260
        READ(5,20) EXPLAN                                                 LIV01270
        READ(5,*) RELTOL                                                  LIV01280
C                                                                         LIV01290
C       SET FLAGS TO 0 OR 1:                                             LIV01300
C       MBOUND = 1:  PROGRAM TERMINATES AFTER COMPUTING 1ST GUESSES      LIV01310
```

```
C                      ON APPROPRIATE T-SIZES FOR USE IN THE RITZ VECTOR     LIV01320
C                      COMPUTATIONS                                          LIV01330
C         NTVCON = 0:  PROGRAM TERMINATES IF THE TVEC ARRAY IS NOT           LIV01340
C                      LARGE ENOUGH TO HOLD ALL THE T-EIGENVECTORS REQUIRED. LIV01350
C         SVTVEC = 0:  THE T-EIGENVECTORS ARE NOT WRITTEN TO FILE 11         LIV01360
C                      UNLESS TVSTOP = 1                                     LIV01370
C         SVTVEC = 1:  WRITE THE T-EIGENVECTORS TO FILE 11.                  LIV01380
C         TVSTOP = 1:  PROGRAM TERMINATES AFTER COMPUTING THE                LIV01390
C                      T-EIGENVECTORS                                        LIV01400
C         LVCONT = 0:  PROGRAM TERMINATES IF THE NUMBER OF T-EIGENVECTORS    LIV01410
C                      COMPUTED IS NOT EQUAL TO THE NUMBER OF RITZ           LIV01420
C                      VECTORS REQUESTED.                                    LIV01430
C         ERCONT = 0:  MEANS FOR ANY GIVEN EIGENVALUE, A RITZ VECTOR         LIV01440
C                      WILL NOT BE COMPUTED FOR THAT EIGENVALUE UNLESS       LIV01450
C                      A T-EIGENVECTOR HAS BEEN IDENTIFIED WITH A LAST       LIV01460
C                      COMPONENT WHICH SATISFIES THE SPECIFIED               LIV01470
C                      CONVERGENCE CRITERION.                                LIV01480
C         ERCONT = 1:  MEANS FOR ANY GIVEN EIGENVALUE, A RITZ VECTOR         LIV01490
C                      WILL BE COMPUTED.  IF A T-EIGENVECTOR CANNOT          LIV01500
C                      BE IDENTIFIED WHICH SATISFIES THE LAST               LIV01510
C                      COMPONENT CRITERION, THEN THE PROGRAM WILL            LIV01520
C                      USE THE T-VECTOR THAT CAME CLOSEST TO                 LIV01530
C                      SATISFYING THE CRITERION                              LIV01540
C         IWRITE = 1:  EXTENDED OUTPUT OF INTERMEDIATE COMPUTATIONS          LIV01550
C                      IS WRITTEN TO FILE 6                                  LIV01560
C         IREAD = 0:   ALPHA/BETA FILE IS REGENERATED.                       LIV01570
C         IREAD = 1:   ALPHA/BETA FILE USED IN EIGENVALUE COMPUTATIONS       LIV01580
C                      IS READ IN AND EXTENDED IF NECESSARY.  IN BOTH        LIV01590
C                      CASES IREAD = 0 OR 1, THE LANCZOS VECTORS ARE         LIV01600
C                      ALWAYS REGENERATED FOR THE RITZ VECTOR                LIV01610
C                      COMPUTATIONS                                          LIV01620
C                                                                            LIV01630
      READ(5,20) EXPLAN                                                      LIV01640
      READ(5,*) MBOUND,NTVCON,SVTVEC,IREAD                                   LIV01650
C                                                                            LIV01660
      READ(5,20) EXPLAN                                                      LIV01670
      READ(5,*) TVSTOP,LVCONT,ERCONT,IWRITE                                  LIV01680
      IF (TVSTOP.EQ.1) SVTVEC = 1                                            LIV01690
C                                                                            LIV01700
C     READ IN SEED (RHSEED) FOR GENERATING RANDOM STARTING VECTOR            LIV01710
C     FOR THE INVERSE ITERATION ON THE T-MATRICES.                          LIV01720
C                                                                            LIV01730
      READ(5,20) EXPLAN                                                      LIV01740
      READ(5,*) RHSEED                                                       LIV01750
C                                                                            LIV01760
C-----------------------------------------------------------------------LIV01770
C     INITIALIZE THE ARRAYS THAT DEFINE THE FACTORIZATION OF                 LIV01780
C     THE B-MATRIX AND PASS THE STORAGE LOCATIONS OF THESE ARRAYS            LIV01790
C     TO THE SUBROUTINE BSOLV.                                               LIV01800
C                                                                            LIV01810
      CALL USPEC(N,MATNO)                                                    LIV01820
C-----------------------------------------------------------------------LIV01830
C     MASK UNDERFLOW AND OVERFLOW                                            LIV01840
C                                                                            LIV01850
      CALL MASK                                                              LIV01860
C-----------------------------------------------------------------------LIV01870
C     WRITE RUN PARAMETERS OUT TO FILE 6                                     LIV01880
C                                                                            LIV01890
      WRITE(6,30) MATNO,N,JPERM                                              LIV01900
   30 FORMAT(/4X,'MATRIX IDENTIFICATION NO.',4X,'SIZE OF A-MATRIX',4X,  LIV01910
     1'JPERM'/I29,I21,I9)                                                    LIV01920
C                                                                            LIV01930
      WRITE(6,40) SO,SHIFT                                                   LIV01940
   40 FORMAT(/4X,'SCALE APPLIED TO MATRIX',4X,'SHIFT APPLIED TO MATRIX'/LIV01950
     1E27.4,E27.4)                                                          LIV01960
C                                                                            LIV01970
      WRITE(6,50) MBOUND,NTVCON,SVTVEC,IREAD                                 LIV01980
```

```
      50 FORMAT(/3X,'MBOUND',3X,'NTVCON',3X,'SVTVEC',3X,'IREAD'/3I9,I8/)      LIVO1990
   C                                                                          LIVO2000
         WRITE(6,60) TVSTOP,LVCONT,ERCONT,IWRITE                             LIVO2010
      60 FORMAT(3X,'TVSTOP',3X,'LVCONT',3X,'ERCONT',3X,'IWRITE'/4I9)         LIVO2020
   C                                                                          LIVO2030
         WRITE(6,70) MDIMTV,MDIMRV,MBETA                                     LIVO2040
      70 FORMAT(/3X,'MDIMTV',3X,'MDIMRV',3X,'MBETA'/2I9,I8)                  LIVO2050
   C                                                                          LIVO2060
         WRITE(6,80) RELTOL,RHSEED                                           LIVO2070
      80 FORMAT(/7X,'RELTOL',3X,'RHSEED'/E13.4,I9)                           LIVO2080
   C                                                                          LIVO2090
   C     FROM FILE 3 READ IN THE NUMBER OF EIGENVALUES (NGOOD) FOR WHICH      LIVO2100
   C     EIGENVECTORS ARE REQUESTED, THE ORDER (MEV) OF THE LANCZOS          LIVO2110
   C     TRIDIAGONAL MATRIX USED IN COMPUTING THESE EIGENVALUES, THE         LIVO2120
   C     ORDER (NOLD) OF THE USER-SPECIFIED MATRIX USED IN THE EIGENVALUE    LIVO2130
   C     COMPUTATIONS, THE SEED (SVSEED) USED FOR GENERATING THE STARTING    LIVO2140
   C     VECTOR THAT WAS USED IN THOSE LANCZOS EIGENVALUE COMPUTATIONS,      LIVO2150
   C     AND THE MATRIX/RUN IDENTIFICATION NUMBER (MATOLD) USED IN THOSE     LIVO2160
   C     COMPUTATIONS.  ALSO READ IN THE NUMBER (NDIS) OF DISTINCT           LIVO2170
   C     EIGENVALUES OF T(1,MEV) THAT WERE COMPUTED BUT THIS VALUE IS        LIVO2180
   C     NOT USED IN THE EIGENVECTOR COMPUTATIONS.                           LIVO2190
   C                                                                          LIVO2200
         READ(3,90) NGOOD,NDIS,MEV,NOLD,SVSEED,MATOLD                       LIVO2210
      90 FORMAT(4I6,I12,I8)                                                  LIVO2220
   C                                                                          LIVO2230
   C     READ IN THE MULTIPLICITY TOLERANCE USED IN THE BISEC SUBROUTINE     LIVO2240
   C     DURING THE COMPUTATION OF THE GIVEN EIGENVALUES.                    LIVO2250
   C     ALSO READ IN THE FLAG IB.  IF IB < 0, THEN SOME BETA(I) IN THE      LIVO2260
   C     T-MATRIX FILE PROVIDED ON FILE 2 FAILED THE ORTHOGONALITY          LIVO2270
   C     TEST IN THE TNORM SUBROUTINE.  USER SHOULD NOTE THAT THIS VECTOR    LIVO2280
   C     PROGRAM PROCEEDS INDEPENDENTLY OF THE SIZE OF THE BETA USED.        LIVO2290
   C                                                                          LIVO2300
         READ(3,100) MULTOL,IB,BTOL,SHIFTO                                   LIVO2310
     100 FORMAT(E20.12,I6,2E10.3)                                            LIVO2320
   C                                                                          LIVO2330
         TEMP = DFLOAT(MEV+1000)                                             LIVO2340
         TTOL = MULTOL/TEMP                                                  LIVO2350
   C                                                                          LIVO2360
         WRITE(6,110) MULTOL,TTOL                                            LIVO2370
     110 FORMAT(/' T-MULTIPLICITY TOLERANCE USED IN THE EIGENVALUE COMPUTATLIVO2380
        110NS WAS',E13.4/' SCALED MACHINE EPSILON TTOL IS',E13.4)           LIVO2390
   C                                                                          LIVO2400
   C     CONTINUE WRITE TO FILE 6 OF THE PARAMETERS FOR THIS RUN             LIVO2410
   C                                                                          LIVO2420
         NG = NGOOD                                                          LIVO2430
         WRITE(6,120)NG,NDIS,MEV,NOLD,MATOLD,SVSEED,IB,MULTOL,BTOL,SHIFTO    LIVO2440
     120 FORMAT(/' EIGENVALUES ARE READ IN FROM FILE 3.  THE HEADER IS'/     LIVO2450
        1 4X,'NG',2X,'NDIS',3X,'MEV',2X,'NOLD',2X,'MATOLD',6X,'SVSEED'/     LIVO2460
        1 4I6,I8,I12/                                                       LIVO2470
        1 6X,'IB',6X,'MULTOL',8X,'BTOL',6X,'SHIFTO'/                        LIVO2480
        1 I8,E12.3,E12.3,E12.3/)                                            LIVO2490
   C                                                                          LIVO2500
   C     IS THE ARRAY RITVEC LONG ENOUGH TO HOLD ALL OF THE DESIRED          LIVO2510
   C     RITZ VECTORS (APPROXIMATE EIGENVECTORS)?                            LIVO2520
   C                                                                          LIVO2530
         NMAX = NGOOD*N                                                      LIVO2540
         IF(MBOUND.EQ.1) GO TO 130                                          LIVO2550
         IF(TVSTOP.NE.1.AND.NMAX.GT.MDIMRV) GO TO 1430                      LIVO2560
   C                                                                          LIVO2570
   C     CHECK THAT THE ORDER N AND THE MATRIX IDENTIFICATION NUMBER         LIVO2580
   C     MATNO SPECIFIED BY THE USER AGREE WITH THOSE READ IN FROM FILE 3.   LIVO2590
   C                                                                          LIVO2600
     130 ITEMP = (NOLD-N)**2+(MATOLD-MATNO)**2                              LIVO2610
         IF (ITEMP.NE.0.OR.SHIFTO.NE.SHIFT) GO TO 1450                      LIVO2620
   C                                                                          LIVO2630
   C     READ IN FROM FILE 3, THE T-MULTIPLICITIES OF THE EIGENVALUES        LIVO2640
   C     WHOSE EIGENVECTORS ARE TO BE COMPUTED, THE VALUES OF THESE          LIVO2650
```

```
C     EIGENVALUES AND THEIR MINIMAL GAPS AS EIGENVALUES OF THE              LIV02660
C     USER-SPECIFIED MATRIX AND AS EIGENVALUES OF THE T-MATRIX.             LIV02670
C                                                                           LIV02680
      READ(3,20) EXPLAN                                                     LIV02690
      READ(3,140) (MP(J),GOODBI(J),BIMING(J),AMINGP(J),TMINGP(J),          LIV02700
     1 J = 1,NGOOD)                                                         LIV02710
  140 FORMAT(5X,I5,E25.16,3E10.3)                                          LIV02720
C                                                                           LIV02730
C                                                                           LIV02740
      DO 150 J=1,NGOOD                                                      LIV02750
  150 GOODA(J) = (ONE/GOODBI(J) - SHIFT)/SO                                LIV02760
C                                                                           LIV02770
      WRITE(6,160) (J,GOODA(J),MP(J),GOODBI(J), J=1,NGOOD)                LIV02780
  160 FORMAT(/' EIGENVALUES READ IN, T-MULTIPLICITIES'/                    LIV02790
     1 4X,' J ',5X,' A-EIGENVALUE',6X,'TMULT',3X,'B-INVERSE EIGENVALUE'/LIV02800
     1(I6,E25.16,I4,E25.16))                                               LIV02810
      WRITE(6,170) (J,GOODBI(J),TMINGP(J),BIMING(J), J=1,NGOOD)           LIV02820
  170 FORMAT(/' B(INVERSE) EIGENVALUES READ IN, T-GAPS AND B(INVERSE)-GALIV02830
     1PS'/4X,' J ',3X,'B-INVERSE EIGENVALUE',6X,' TMINGAP ',6X,           LIV02840
     1' BIMINGAP '/(I6,E25.16,2E15.4))                                    LIV02850
      WRITE(6,180) (J,GOODA(J),AMINGP(J), J=1,NGOOD)                      LIV02860
  180 FORMAT(/' A-EIGENVALUES READ IN AND A-GAPS'/                         LIV02870
     1 4X,' J ',5X,'A-EIGENVALUE',10X,' AMINGAP '                         LIV02880
     1/(I6,E25.16,E15.4))                                                  LIV02890
C                                                                           LIV02900
C     READ IN ERROR ESTIMATES                                              LIV02910
      WRITE(6,210) MEV,SVSEED                                              LIV02920
C     CHECK WHETHER OR NOT THERE ARE ANY T-ISOLATED EIGENVALUES IN         LIV02930
C     THE EIGENVALUES PROVIDED                                             LIV02940
      DO 190 J=1,NGOOD                                                      LIV02950
      IF(MP(J).EQ.1) GO TO 200                                            LIV02960
  190 CONTINUE                                                             LIV02970
      GO TO 230                                                            LIV02980
  200 READ(4,20) EXPLAN                                                    LIV02990
      READ(4,20) EXPLAN                                                    LIV03000
      READ(4,20) EXPLAN                                                    LIV03010
  210 FORMAT(/' THESE EIGENVALUES WERE COMPUTED USING A T-MATRIX OF       LIV03020
     1ORDER ',I5/' AND SEED FOR RANDOM NUMBER GENERATOR =',I12)           LIV03030
      READ(4,220) NISO                                                    LIV03040
  220 FORMAT(18X,I6)                                                       LIV03050
      READ(4,20) EXPLAN                                                    LIV03060
      READ(4,20) EXPLAN                                                    LIV03070
      READ(4,20) EXPLAN                                                    LIV03080
  230 DO 260 J=1,NGOOD                                                     LIV03090
      BIERR(J) = 0.D0                                                      LIV03100
      IF(MP(J).NE.1) GO TO 260                                            LIV03110
      READ(4,240) EVAL, BIERR(J)                                          LIV03120
  240 FORMAT(10X,E25.16,E14.3)                                            LIV03130
      IF(DABS(EVAL - GOODBI(J)).LT.1.D-10) GO TO 260                      LIV03140
      WRITE(6,250) EVAL,GOODBI(J)                                         LIV03150
  250 FORMAT(' PROBLEM WITH READ IN OF ERROR ESTIMATES'/' EIGENVALUE REALIV03160
     1D IN',E20.12,' DOES NOT MATCH GOODBI(J) ='/E20.12)                  LIV03170
      GO TO 1670                                                          LIV03180
C                                                                           LIV03190
  260 CONTINUE                                                             LIV03200
C                                                                           LIV03210
      WRITE(6,270) (J,GOODBI(J),BIERR(J), J=1,NGOOD)                      LIV03220
  270 FORMAT(' B(INVERSE) ERROR ESTIMATES '/4X,' J',5X,'EIGENVALUE',10X LIV03230
     1,'ESTIMATE'/(I6,E20.12,E14.3))                                      LIV03240
C                                                                           LIV03250
      IF(IREAD.EQ.0)  GO TO 370                                           LIV03260
C                                                                           LIV03270
C     READ IN THE SIZE OF THE T-MATRIX PROVIDED ON FILE 2.  READ IN       LIV03280
C     THE ORDER OF THE USER-SPECIFIED MATRIX , THE SEED FOR THE           LIV03290
C     RANDOM NUMBER GENERATOR, AND THE MATRIX/TEST IDENTIFICATION         LIV03300
C     NUMBER THAT WERE USED IN THE LANCZOS EIGENVALUE COMPUTATIONS.       LIV03310
```

```
C       IF FLAG IREAD = 0, REGENERATE ALPHA,BETA ARRAYS            LIVO3320
C                                                                  LIVO3330
        READ(2,280) KMAX,NOLD,SVSOLD,MATOLD,SHIFTO                 LIVO3340
    280 FORMAT(2I6,I12,I8,E13.4)                                   LIVO3350
C                                                                  LIVO3360
        WRITE(6,290) KMAX,NOLD,SVSOLD,MATOLD,SHIFTO                LIVO3370
    290 FORMAT(/' READ IN THE T-MATRICES STORED ON FILE 2'/' FILE 2 HEADERLIVO3380
     1 IS'/2X,'KMAX',2X,'NOLD',6X,'SVSOLD',2X,'MATOLD',4X,'SHIFTO'/  LIVO3390
     1 2I6,I12,I8,E10.3/)                                         LIVO3400
C                                                                  LIVO3410
C       CHECK THAT THE ORDER, THE MATRIX/TEST IDENTIFICATION NUMBER LIVO3420
C       AND THE SEED FOR THE RANDOM NUMBER GENERATOR USED IN THE   LIVO3430
C       LANCZOS COMPUTATIONS THAT GENERATED THE ALPHA,BETA FILE    LIVO3440
C       BEING USED AGREE WITH WHAT THE USER HAS SPECIFIED.         LIVO3450
        IF (NOLD.NE.N.OR.MATOLD.NE.MATNO.OR.SVSOLD.NE.SVSEED) GO TO 1470 LIVO3460
C                                                                  LIVO3470
        KMAX1 = KMAX + 1                                           LIVO3480
C                                                                  LIVO3490
C       READ IN THE T-MATRICES FROM FILE 2.  THESE ARE USED TO GENERATE LIVO3500
C       THE T-EIGENVECTORS THAT WILL BE USED IN THE RITZ VECTOR    LIVO3510
C       COMPUTATIONS.  ALPHA,BETA MUST BE STORED IN MACHINE FORMAT LIVO3520
C       ((4Z20) ON IBM/3081)                                       LIVO3530
C                                                                  LIVO3540
        READ(2,300) (ALPHA(J), J=1,KMAX)                           LIVO3550
        READ(2,300) (BETA(J), J=1,KMAX1)                           LIVO3560
    300 FORMAT(4Z20)                                               LIVO3570
C                                                                  LIVO3580
        READ(2,300) (V1(J), J=1,N)                                 LIVO3590
        READ(2,300) (V2(J), J=1,N)                                 LIVO3600
C                                                                  LIVO3610
C       ENLARGE KMAX IF THE SIZE AT WHICH THE EIGENVALUE           LIVO3620
C       COMPUTATIONS WERE PERFORMED IS ESSENTIALLY KMAX AND        LIVO3630
C       THERE IS AT LEAST ONE EIGENVALUE THAT IS T-SIMPLE AND      LIVO3640
C       T-ISOLATED IN THE SENSE THAT IF ITS NEAREST NEIGHBOR IS    LIVO3650
C       TOO CLOSE THEN THAT NEIGHBOR IS A GOOD T-EIGENVALUE.       LIVO3660
        DO 310 J = 1,NGOOD                                         LIVO3670
        IF(MP(J).EQ.1) GO TO 330                                   LIVO3680
    310 CONTINUE                                                   LIVO3690
        WRITE(6,320)                                               LIVO3700
    320 FORMAT(/' ALL EIGENVALUES USED ARE T-MULTIPLE OR CLOSE TO SPURIOUSLIVO3710
     1 T-EIGENVALUES'/' SO DO NOT CHANGE KMAX')                    LIVO3720
        IF(KMAX.LT.MEV) GO TO 1490                                 LIVO3730
        GO TO 350                                                  LIVO3740
C                                                                  LIVO3750
    330 KMAXN= 11*MEV/8 + 12                                       LIVO3760
        IF(MBETA.LE.KMAXN) GO TO 1650                              LIVO3770
        IF(KMAX.GE.KMAXN ) GO TO 350                               LIVO3780
        WRITE(6,340) KMAX, KMAXN                                   LIVO3790
    340 FORMAT(' ENLARGE KMAX FROM ',I6,' TO ',I6)                 LIVO3800
        MOLD1 = KMAX + 1                                           LIVO3810
        KMAX = KMAXN                                               LIVO3820
        GO TO 420                                                  LIVO3830
C                                                                  LIVO3840
    350 WRITE(6,360) KMAX                                          LIVO3850
    360 FORMAT(/' T-MATRICES HAVE BEEN READ IN FROM FILE 2'/' THE LARGEST LIVO3860
     1SIZE T-MATRIX ALLOWED IS',I6/)                              LIVO3870
C                                                                  LIVO3880
        IF(IREAD.EQ.1) GO TO 440                                   LIVO3890
C                                                                  LIVO3900
C       REGENERATE THE ALPHA AND BETA                             LIVO3910
C                                                                  LIVO3920
    370 MOLD1 = 1                                                  LIVO3930
C                                                                  LIVO3940
        DO 380 J = 1,NGOOD                                         LIVO3950
        IF(MP(J).EQ.1) GO TO 400                                   LIVO3960
    380 CONTINUE                                                   LIVO3970
        KMAX = MEV + 12                                            LIVO3980
```

```
      WRITE(6,390) KMAX                                             LIV03990
  390 FORMAT(/' ALL EIGENVALUES FOR WHICH EIGENVECTORS ARE TO BE COMPUTELIV04000
     1D ARE EITHER T-MULTIPLE OR CLOSE TO'/' A SPURIOUS T-EIGENVALUE. THLIV04010
     1EREFORE SET KMAX = MEV + 12 = ',I7)                           LIV04020
      GO TO 420                                                     LIV04030
C                                                                   LIV04040
  400 KMAXN = 11*MEV/8 + 12                                         LIV04050
      IF(MBETA.LE.KMAXN) GO TO 1650                                 LIV04060
      WRITE(6,410) KMAXN                                            LIV04070
  410 FORMAT(' SET KMAX EQUAL TO ',I6)                              LIV04080
      KMAX = KMAXN                                                  LIV04090
C                                                                   LIV04100
  420 WRITE(6,430) MOLD1,KMAX                                       LIV04110
  430 FORMAT(/' LANCZS SUBROUTINE GENERATES ALPHA(J), BETA(J+1), J =',LIV04120
     1 I6,' TO ', I6/)                                              LIV04130
C                                                                   LIV04140
C------------------------------------------------------------------LIV04150
C                                                                   LIV04160
      CALL LANCZS(BSOLV,ALPHA,BETA,V1,V2,VS,G,KMAX,MOLD1,N,SVSEED)  LIV04170
C                                                                   LIV04180
C------------------------------------------------------------------LIV04190
C                                                                   LIV04200
  440 CONTINUE                                                      LIV04210
C                                                                   LIV04220
C     THE SUBROUTINE STURMI DETERMINES THE SMALLEST SIZE T-MATRIX FOR LIV04230
C     WHICH THE EIGENVALUE IN QUESTION IS AN EIGENVALUE (TO WITHIN A LIV04240
C     GIVEN TOLERANCE) AND IF POSSIBLE THE SMALLEST SIZE T-MATRIX  LIV04250
C     FOR WHICH IT IS A DOUBLE EIGENVALUE (TO WITHIN THE SAME      LIV04260
C     TOLERANCE).  THE SIZE T-MATRIX USED IN THE EIGENVECTOR       LIV04270
C     COMPUTATIONS IS THEN DETERMINED BY LOOPING ON SIZE OF THE    LIV04280
C     T-EIGENVECTORS, USING THE VALUES FROM STURMI TO DETERMINE    LIV04290
C     FIRST GUESSES AT THE APPROPRIATE T-SIZES.                    LIV04300
C                                                                   LIV04310
C                                                                   LIV04320
      STUTOL = SCALE0*MULTOL                                        LIV04330
      IF(IWRITE.EQ.1) WRITE(6,450)                                 LIV04340
  450 FORMAT(' FROM STURMI')                                       LIV04350
      DO 490 J = 1,NGOOD                                           LIV04360
      EVAL = GOODBI(J)                                             LIV04370
C     COMPUTE THE TOLERANCES USED BY STURMI TO DETERMINE AN INTERVAL LIV04380
C     CONTAINING THE EIGENVALUE EVAL.                             L'V04390
      TEMP = DABS(EVAL)*RELTOL                                     LIV04400
      TOLN = DMAX1(TEMP,STUTOL)                                    LIV04410
C                                                                   LIV04420
C------------------------------------------------------------------LIV04430
C                                                                   LIV04440
      CALL STURMI(ALPHA,BETA,EVAL,TOLN,EPSM,KMAX,MK1,MK2,IC,IWRITE) LIV04450
C                                                                   LIV04460
C------------------------------------------------------------------LIV04470
C                                                                   LIV04480
C     STORE THE COMPUTED ORDERS OF T-MATRICES FOR LATER PRINTOUT   LIV04490
      M1(J) = MK1                                                  LIV04500
      M2(J) = MK2                                                  LIV04510
      ML(J) = (MK1 + 3*MK2)/4                                      LIV04520
      IF(MK2.EQ.KMAX)  ML(J) = KMAX                                LIV04530
C                                                                   LIV04540
      IF(IC.GT.0) GO TO 470                                        LIV04550
C     IC = 0 MEANS THERE WAS NO T-EIGENVALUE IN THE DESIGNATED INTERVAL LIV04560
C     BY T-SIZE KMAX.  THIS MEANS THAT THE EIGENVALUE PROVIDED HAS LIV04570
C     NOT YET CONVERGED SO ITS EIGENVECTOR SHOULD NOT BE COMPUTED. LIV04580
      WRITE(6,460) J,GOODBI(J),MK1,MK2                            LIV04590
  460 FORMAT(I6,'TH EIGENVALUE',E20.12,' HAS NOT CONVERGED '/      LIV04600
     1' SO DO NOT COMPUTE ANY T-EIGENVECTOR OR RITZ VECTOR FOR IT' LIV04610
     1/' MK1 AND MK2 FOR THIS EIGENVALUE WERE',2I6)               LIV04620
      MP(J) = MPMIN                                                LIV04630
      MA(J) = -2*KMAX                                              LIV04640
```

```
      GO TO 490                                                   LIV04650
C     COMPUTE AN APPROPRIATE SIZE T-MATRIX FOR THE GIVEN EIGENVALUE.  LIV04660
  470 IF(M2(J).EQ.KMAX) GO TO 480                                 LIV04670
C     M1 AND M2 WERE BOTH DETERMINED                              LIV04680
      MA(J) = (3*M1(J) + M2(J))/4 + 1                             LIV04690
      GO TO 490                                                   LIV04700
C     M2 NOT DETERMINED                                           LIV04710
  480 MA(J) = (5*M1(J))/4 + 1                                     LIV04720
C                                                                 LIV04730
  490 CONTINUE                                                    LIV04740
C                                                                 LIV04750
      IF (IWRITE.EQ.1) WRITE(6,500) (MA(JJ), JJ=1,NGOOD)          LIV04760
  500 FORMAT(/' 1ST GUESS AT APPROPRIATE SIZE T-MATRICES'/        LIV04770
     1 ' ACTUAL VALUES WILL PROBABLY BE 1/4 AGAIN AS MUCH'/(1316))  LIV04780
C                                                                 LIV04790
C     PRINT OUT TO FILE 10 1ST GUESSES AT SIZES OF THE T-MATRICES TO  LIV04800
C     BE USED IN THE EIGENVECTOR COMPUTATIONS.                    LIV04810
C     ACTUAL VALUES USED MAY BE 1/4 OR MORE LARGER THAN THESE VALUES.  LIV04820
      WRITE(10,510) N,KMAX                                        LIV04830
  510 FORMAT(2I8,' = ORDER OF USER MATRIX AND MAX ORDER OF T(1,MEV)')  LIV04840
C                                                                 LIV04850
      WRITE(10,520)                                               LIV04860
  520 FORMAT(/' 1ST GUESS AT APPROPRIATE SIZE T-MATRICES'/        LIV04870
     1 ' ACTUAL VALUES WILL PROBABLY BE 1/4 AGAIN AS MUCH'/)      LIV04880
C                                                                 LIV04890
      WRITE(10,530)                                               LIV04900
  530 FORMAT(4X,'J',7X,'GOODBI(J)',4X,'M1(J)',1X,'M2(J)',1X,'MA(J)')  LIV04910
C                                                                 LIV04920
      WRITE(10,540) (J,GOODBI(J),M1(J),M2(J), MA(J), J=1,NGOOD)   LIV04930
  540 FORMAT(I5,E19.12,3I6)                                       LIV04940
C                                                                 LIV04950
      IF(MBOUND.EQ.1) WRITE(10,550)                               LIV04960
  550 FORMAT(/' EV = GOODBI(J) IS A GOOD EIGENVALUE OF T(1,MEV)'/  LIV04970
     1 ' M1 = SMALLEST VALUE OF M SUCH THAT T(1,M) HAS AT LEAST'/  LIV04980
     1 '      ONE EIGENVALUE IN THE INTERVAL (EV-TOLN,EV+TOLN)'/  LIV04990
     1 ' TOLN(J) = DMAX1(GOODBI(J)*RELTOL, SCALEO*MULTOL)'/       LIV05000
     1 ' M2 = SMALLEST M (IF ANY) SUCH THAT IN THE ABOVE INTERVAL'/  LIV05010
     1 '      T(1,M) HAS AT LEAST TWO EIGENVALUES '/              LIV05020
     1 ' INITIAL VALUE OF MA(J) IS CHOSEN HEURISTICALLY'/         LIV05030
     1 ' PROGRAM LOOPS ON SIZE OF T-MATRIX TO GET APPROPRIATE SIZE'/  LIV05040
     1 ' END OF SIZES OF T-MATRICES FILE 10'///)                 LIV05050
C                                                                 LIV05060
C                                                                 LIV05070
C     TERMINATE AFTER COMPUTING 1ST GUESSES AT  SIZES OF THE      LIV05080
C     T-MATRICES REQUIRED FOR THE GIVEN EIGENVALUES?             LIV05090
      IF(MBOUND.EQ.1) GO TO 1510                                  LIV05100
C                                                                 LIV05110
C                                                                 LIV05120
C     WILL THERE BE ROOM FOR ALL OF THE REQUESTED T-EIGENVECTORS?  LIV05130
      MTOL = 0                                                    LIV05140
      DO 560 J = 1,NGOOD                                          LIV05150
      IF(MP(J).EQ.MPMIN) GO TO 560                                LIV05160
      MTOL = MTOL + IABS(MA(J))                                   LIV05170
  560 CONTINUE                                                    LIV05180
      MTOL = (5*MTOL)/4                                           LIV05190
      IF(MTOL.GT.MDIMTV.AND.NTVCON.EQ.0) GO TO 1530               LIV05200
C                                                                 LIV05210
C----------------------------------------------------------------LIV05220
C     GENERATE A RANDOM VECTOR TO BE USED REPEATEDLY BY           LIV05230
C     SUBROUTINE INVERM                                           LIV05240
C                                                                 LIV05250
      IIL = RHSEED                                                LIV05260
      CALL GENRAN(IIL,G,KMAX)                                     LIV05270
C                                                                 LIV05280
C---------------------------------------------------------------- LIV05290
C                                                                 LIV05300
C     FOR EACH EIGENVALUE LOOP ON T-EIGENVECTOR COMPUTATIONS TO   LIV05310
```

```
C     COMPUTE AN APPROPRIATE T-EIGENVECTOR TO USE IN THE RITZ          LIV05320
C     VECTOR COMPUTATIONS.                                             LIV05330
C                                                                      LIV05340
      MTOL = 0                                                         LIV05350
      NTVEC = 0                                                        LIV05360
      ILBIS = 0                                                        LIV05370
      DO 750 J = 1,NGOOD                                               LIV05380
      ICOUNT = 0                                                       LIV05390
      ERRMIN = 10.DO                                                   LIV05400
      MABEST = MPMIN                                                   LIV05410
      IF(MP(J).EQ.MPMIN) GO TO 750                                     LIV05420
      TFLAG = 0                                                        LIV05430
      EVAL = GOODBI(J)                                                 LIV05440
      TEMP = RELTOL*DABS(EVAL)                                         LIV05450
      UB = EVAL + DMAX1(STUTOL,TEMP)                                   LIV05460
      LB = EVAL - DMAX1(STUTOL,TEMP)                                   LIV05470
  570 KMAXU = IABS(MA(J))                                              LIV05480
C                                                                      LIV05490
C     SELECT A SUITABLE INCREMENT FOR THE ORDERS OF THE T-MATRICES     LIV05500
C     TO BE CONSIDERED IN DETERMINING APPROPRIATE SIZES FOR THE RITZ   LIV05510
C     VECTOR COMPUTATIONS.                                             LIV05520
      IF(ICOUNT.GT.0) GO TO 590                                        LIV05530
C     SELECT IDELTA(J) BASED UPON THE T-MULTIPLICITY OBTAINED          LIV05540
      IF(M2(J).EQ.KMAX) GO TO 580                                      LIV05550
C     M2 DETERMINED                                                    LIV05560
      IDELTA(J) = ((3*M1(J) + 5*M2(J))/8  +  1 - IABS(MA(J)))/10 +  1  LIV05570
      GO TO 590                                                        LIV05580
C     M2 NOT DETERMINED                                                LIV05590
  580 MAMAX = MINO((11*MEV)/8 + 12, (13*M1(J))/8 + 1)                  LIV05600
      IDELTA(J) = (MAMAX - IABS(MA(J)))/10 +  1                        LIV05610
  590 ICOUNT = ICOUNT + 1                                              LIV05620
C                                                                      LIV05630
C----------------------------------------------------------------------LIV05640
C     TO MIMIMIZE THE EFFECT OF THE ONE-SIDED ACCEPTANCE TEST FOR      LIV05650
C     EIGENVALUES IN THE BISEC SUBROUTINE, RECOMPUTE THE GIVEN         LIV05660
C     EIGENVALUE AT THE SPECIFIED KMAXU                                LIV05670
C                                                                      LIV05680
      CALL LBISEC(ALPHA,BETA,EPSM,EVAL,EVALN,LB,UB,TTOL,KMAXU,NEVT)    LIV05690
C                                                                      LIV05700
C----------------------------------------------------------------------LIV05710
C                                                                      LIV05720
C     CHECK WHETHER OR NOT GIVEN T-MATRIX HAS AN EIGENVALUE IN THE     LIV05730
C     SPECIFIED INTERVAL AND IF SO WHAT ITS T-MULTIPLICITY IS.         LIV05740
C                                                                      LIV05750
      IF(NEVT.EQ.1) GO TO 630                                          LIV05760
      IF(NEVT.NE.0) GO TO 610                                          LIV05770
      ILBIS = 1                                                        LIV05780
      WRITE(6,600) EVAL,KMAXU                                          LIV05790
  600 FORMAT(/' PROBLEM ENCOUNTERED IN RECOMPUTATION OF USER-SUPPLIED EILIV05800
     1GENVALUE',E20.12/' THE SIZE T-MATRIX SPECIFIED',I6,' DOES NOT     LIV05810
     1HAVE AN EIGENVALUE IN THE INTERVAL SPECIFIED'/' THEREFORE NO EIGENLIV05820
     1VECTOR WILL BE COMPUTED FOR THIS PARTICULAR EIGENVALUE'/)        LIV05830
      GO TO 650                                                        LIV05840
C                                                                      LIV05850
  610 IF(NEVT.GT.1)  WRITE(6,620) EVAL,KMAXU                           LIV05860
  620 FORMAT(/' PROBLEM ENCOUNTERED IN RECOMPUTATION OF USER-SUPPLIED   LIV05870
     1EIGENVALUE',E20.12/' FOR THE SIZE T-MATRIX SPECIFIED =',I6,' THE  LIV05880
     1GIVEN EIGENVALUE IS MULTIPLE IN THE INTERVAL SPECIFIED'/' SOMETHINLIV05890
     1G IS WRONG, THEREFORE NO EIGENVECTOR WILL BE COMPUTED FOR THIS EIGLIV05900
     1NVALUE'/)                                                        LIV05910
C                                                                      LIV05920
      MP(J) = MPMIN                                                    LIV05930
      MA(J) = -2*KMAX                                                  LIV05940
      GO TO 750                                                        LIV05950
C                                                                      LIV05960
  630 CONTINUE                                                         LIV05970
```

```
      ILBIS = 0                                                      LIV05980
C                                                                    LIV05990
      EVNEW(J) = EVALN                                               LIV06000
      EVAL = EVALN                                                   LIV06010
      MTOL = MTOL+KMAXU                                              LIV06020
C                                                                    LIV06030
C     IS THERE ROOM IN TVEC ARRAY FOR THE NEXT T-EIGENVECTOR?        LIV06040
C     IF NOT, SKIP TO RITZ VECTOR COMPUTATIONS.                      LIV06050
      IF (MTOL.GT.MDIMTV) GO TO 760                                  LIV06060
C                                                                    LIV06070
      IT = 3                                                         LIV06080
      KINT = MTOL - KMAXU +1                                         LIV06090
C                                                                    LIV06100
C     RECORD THE BEGINNING AND END OF THE T-EIGENVECTOR BEING COMPUTED LIV06110
      MINT(J) = KINT                                                 LIV06120
      MFIN(J) = MTOL                                                 LIV06130
C                                                                    LIV06140
C----------------------------------------------------------------------LIV06150
C     SUBROUTINE INVERM DOES INVERSE ITERATION, I.E. SOLVES          LIV06160
C     (T(1,KMAXU) - EVAL)*U = RHS  FOR EACH EIGENVALUE TO OBTAIN     LIV06170
C     THE DESIRED T-EIGENVECTOR.                                     LIV06180
C                                                                    LIV06190
      IF(IWRITE.EQ.1)   WRITE(6,640) J                              LIV06200
  640 FORMAT(/I6,'TH EIGENVALUE')                                    LIV06210
C                                                                    LIV06220
      CALL INVERM(ALPHA,BETA,V1,TVEC(KINT),EVAL,ERROR,TERROR,EPSM,   LIV06230
     1 G,KMAXU,IT,IWRITE)                                            LIV06240
C                                                                    LIV06250
C----------------------------------------------------------------------LIV06260
C                                                                    LIV06270
      TERR(J) = TERROR                                               LIV06280
      TLAST(J) = ERROR                                               LIV06290
      KMAXU1 = KMAXU + 1                                             LIV06300
      TBETA(J) = BETA(KMAXU1)*ERROR                                  LIV06310
C                                                                    LIV06320
C     AFTER COMPUTING EACH OF THE T-EIGENVECTORS,                    LIV06330
C     CHECK THE SIZE OF THE ERROR ESTIMATE, ERROR.                   LIV06340
C     IF THIS ESTIMATE IS NOT AS SMALL AS DESIRED AND                LIV06350
C     |MA(J)| < ML(J), ATTEMPT TO INCREASE THE SIZE OF |MA(J)|       LIV06360
C     AND REPEAT THE T-EIGENVECTOR COMPUTATIONS.                     LIV06370
C                                                                    LIV06380
      IF(ERROR.LT.ERTOL.OR.TFLAG.EQ.1)  GO TO 740                    LIV06390
C                                                                    LIV06400
      IF(ERROR.GE.ERRMIN) GO TO 650                                  LIV06410
C     LAST COMPONENT IS LESS THAN MINIMAL TO DATE                    LIV06420
      ERRMIN = ERROR                                                 LIV06430
      MABEST = MA(J)                                                 LIV06440
  650 CONTINUE                                                       LIV06450
C                                                                    LIV06460
      IF(MA(J).GT.0)  ITEST = MA(J) + IDELTA(J)                      LIV06470
      IF(MA(J).LT.0)  ITEST = -(IABS(MA(J)) + IDELTA(J))             LIV06480
      IF(IABS(ITEST).LE.ML(J).AND.ICOUNT.LE.10) GO TO 670            LIV06490
C     NEW MA(J) IS GREATER THAN MAXIMUM ALLOWED.                     LIV06500
      IF(ERCONT.EQ.0.OR.MABEST.EQ.MPMIN) GO TO 690                   LIV06510
      TFLAG = 1                                                      LIV06520
      MA(J) = MABEST                                                 LIV06530
      IF(ILBIS.EQ.0) MTOL = MTOL - KMAXU                             LIV06540
      WRITE(6,660) MA(J)                                             LIV06550
  660 FORMAT(' 10 ORDERS WERE CONSIDERED.  NONE SATISFIED THE ERROR TESTLIV06560
     1'/' THEREFORE USE THE BEST ORDER OBTAINED FOR THE EIGENVECTORS' LIV06570
     1,I6)                                                           LIV06580
      GO TO 570                                                      LIV06590
C                                                                    LIV06600
  670 MA(J) = ITEST                                                  LIV06610
C                                                                    LIV06620
      MT = IABS(MA(J))                                               LIV06630
      IF(IWRITE.EQ.1) WRITE(6,680)  MT                              LIV06640
```

```
  680 FORMAT(/' CHANGE SIZE OF T-MATRIX TO ',I6,' RECOMPUTE T-EIGENVECTOLIV06650
     1R')                                                            LIV06660
C                                                                    LIV06670
      IF(ILBIS.EQ.0)  MTOL = MTOL - KMAXU                            LIV06680
C                                                                    LIV06690
      GO TO 570                                                      LIV06700
C                                                                    LIV06710
C     APPROPRIATE SIZE T-MATRIX WAS NOT OBTAINED                     LIV06720
  690 CONTINUE                                                       LIV06730
      WRITE(10,700) J,EVAL,MP(J)                                     LIV06740
  700 FORMAT(/' ON 10 INCREMENTS NOT ABLE TO IDENTIFY APPROPRIATE SIZE LIV06750
     1T-MATRIX FOR'/                                                 LIV06760
     1' EIGENVALUE(',I4,') = ',E20.12,' T-MULTIPLICITY =',I4/)       LIV06770
      IF(M2(J).EQ.KMAX) WRITE(10,710)                                LIV06780
      IF(M2(J).LT.KMAX) WRITE(10,720)                                LIV06790
  710 FORMAT(' ORDERS TESTED RANGED FROM 5*M1(J)/4 TO APPROXIMATELY   LIV06800
     1 MIN(11*MEV/8,13*M1(J)/8)')                                    LIV06810
  720 FORMAT(' ORDERS TESTED RANGED FROM APPROX. (3*M1(J)+M2(J))/4 TO (3LIV06820
     1*M1(J)+5*M2(J))/8')                                            LIV06830
      WRITE(10,730)                                                  LIV06840
  730 FORMAT(' ALLOWING LARGER ORDERS FOR THE T-MATRICES MAY RESULT IN  LIV06850
     1 SUCCESS'/' BUT PROBABLY WILL NOT.  PROBLEM IS PROBABLY DUE TO' LIV06860
     1 /' LACK OF CONVERGENCE OF GIVEN EIGENVALUE, CHECK THE ERROR ESTIMLIV06870
     1ATE')                                                          LIV06880
      MP(J) = MPMIN                                                  LIV06890
      IF(ILBIS.EQ.0)  MTOL = MTOL - KMAXU                            LIV06900
      GO TO 750                                                      LIV06910
  740 NTVEC = NTVEC + 1                                              LIV06920
C                                                                    LIV06930
  750 CONTINUE                                                       LIV06940
      NGOODC = NGOOD                                                 LIV06950
      GO TO 780                                                      LIV06960
C                                                                    LIV06970
C     COME HERE IF THERE IS NOT ENOUGH ROOM FOR ALL OF T-EIGENVECTORS LIV06980
  760 NGOODC = J-1                                                   LIV06990
      WRITE(6,770)  J,MTOL,MDIMTV                                    LIV07000
  770 FORMAT(/' NOT ENOUGH ROOM IN TVEC ARRAY FOR ',I4,'TH T-EIGENVECTORLIV07010
     1'/'  TVEC DIMENSION REQUESTED = ',I6,' BUT TVEC HAS DIMENSION ',I6LIV07020
     1/)                                                             LIV07030
      IF(NGOODC.EQ.0)  GO TO 1550                                    LIV07040
      MTOL = MTOL-KMAXU                                              LIV07050
C                                                                    LIV07060
  780 CONTINUE                                                       LIV07070
C                                                                    LIV07080
C     THE LOOP ON T-EIGENVECTOR COMPUTATIONS IS COMPLETE.            LIV07090
C     WRITE OUT THE SIZE T-MATRICES THAT WILL BE USED FOR            LIV07100
C     THE RITZ VECTOR COMPUTATIONS.                                  LIV07110
C                                                                    LIV07120
      WRITE(10,790)                                                  LIV07130
  790 FORMAT(/' SIZES OF T-MATRICES THAT WILL BE USED IN THE RITZ COMPUTLIV07140
     1ATIONS'/5X,'J',8X,'  GOODBI(J) ',13X,'  GOODA(J) ',7X,'MA(J)')  LIV07150
C                                                                    LIV07160
      WRITE(10,800)   (J,GOODBI(J),GOODA(J),MA(J), J=1,NGOOD)        LIV07170
  800 FORMAT(I6,2E25.14,I6)                                          LIV07180
      WRITE(10,550)                                                  LIV07190
C  .                                                                 LIV07200
      WRITE(6,810) MTOL                                              LIV07210
  810 FORMAT(/' THE CUMULATIVE LENGTH OF THE T-EIGENVECTORS IS',I18)  LIV07220
C                                                                    LIV07230
      WRITE(6,820) NTVEC,NGOOD                                       LIV07240
  820 FORMAT(/I6,' T-EIGENVECTORS OUT OF',I6,' REQUESTED WERE COMPUTED')LIV07250
C                                                                    LIV07260
C     SAVE THE T-EIGENVECTORS ON FILE 11?                            LIV07270
      IF(TVSTOP.NE.1.AND.SVTVEC.EQ.0) GO TO 880                      LIV07280
C                                                                    LIV07290
      WRITE(11,830) NTVEC,MTOL,MATNO,SVSEED                          LIV07300
```

```
      830 FORMAT(I6,3I12,' = NTVEC,MTOL,MATNO,SVSEED')                   LIV07310
    C                                                                    LIV07320
          DO 860 J=1,NGOODC                                             LIV07330
    C     IF MP(J) = MPMIN THEN NO SUITABLE T-EIGENVECTOR IS AVAILABLE   LIV07340
    C     FOR THAT EIGENVALUE.                                          LIV07350
          IF(MP(J).EQ.MPMIN) WRITE(11,840) J,MA(J),GOODBI(J),MP(J)      LIVO7360
      840 FORMAT(2I6,E20.12,I6/' TH EIGVAL,T-SIZE,EVALUE,FLAG,NO EIGVEC')LIVO7370
          IF(MP(J).NE.MPMIN) WRITE(11,850) J,MA(J),GOODBI(J),MP(J)      LIV07380
      850 FORMAT(I6,I6,E20.12,I6/' T-EIGENVECTOR, T-SIZE , BI-EIGENVALUE, TLIV07390
         1-MULTIPLICITY')                                               LIV07400
          IF(MP(J).EQ.MPMIN) GO TO 860                                  LIV07410
          KI = MINT(J)                                                  LIV07420
          KF = MFIN(J)                                                  LIV07430
    C                                                                    LIV07440
          WRITE(11,300) (TVEC(K), K=KI,KF)                             LIV07450
    C                                                                    LIV07460
      860 CONTINUE                                                       LIV07470
    C                                                                    LIV07480
          IF(TVSTOP.NE.1) GO TO 880                                     LIV07490
    C                                                                    LIV07500
          WRITE(6,870) TVSTOP, NTVEC,NGOOD                              LIV07510
      870 FORMAT(/' USER SET TVSTOP = ',I1/                             LIV07520
         1' THEREFORE PROGRAM TERMINATES AFTER T-EIGENVECTOR COMPUTATIONS'/ LIV07530
         1' T-EIGENVECTORS THAT WERE COMPUTED ARE SAVED ON FILE 11'/    LIV07540
         1I8,' T-EIGENVECTORS WERE COMPUTED OUT OF',I7,' REQUESTED'/)   LIV07550
    C                                                                    LIV07560
          GO TO 1670                                                    LIV07570
    C                                                                    LIV07580
      880 CONTINUE                                                       LIV07590
    C     IF NOT ABLE TO COMPUTE ALL THE REQUESTED T-EIGENVECTORS        LIV07600
    C     CONTINUE WITH THE LANCZOS VECTOR COMPUTATIONS ANYWAY?          LIV07610
          IF(NTVEC.NE.NGOOD.AND.LVCONT.EQ.0) GO TO 1570                 LIV07620
    C                                                                    LIV07630
    C     COMPUTE THE MAXIMUM SIZE OF THE T-MATRIX USED FOR THOSE        LIV07640
    C     EIGENVALUES WITH GOOD ERROR ESTIMATES.                        LIV07650
    C                                                                    LIV07660
          KMAXU = 0                                                     LIV07670
          DO 890 J = 1,NGOODC                                           LIV07680
          MT = IABS(MA(J))                                              LIV07690
          IF(MT.LT.KMAXU.OR.MP(J).EQ.MPMIN) GO TO 890                   LIV07700
          KMAXU = MT                                                    LIV07710
      890 CONTINUE                                                       LIV07720
    C                                                                    LIV07730
          IF(KMAXU.EQ.0) GO TO 1610                                     LIV07740
    C                                                                    LIV07750
          WRITE(6,900) KMAXU                                            LIV07760
      900 FORMAT(/I6,' = LARGEST SIZE T-MATRIX TO BE USED IN THE RITZ VECTORLIV07770
         1 COMPUTATIONS')                                               LIV07780
    C                                                                    LIV07790
    C     COUNT THE NUMBER OF RITZ VECTORS NOT BEING COMPUTED           LIV07800
          MREJEC = 0                                                    LIV07810
          DO 910 J=1,NGOODC                                             LIV07820
      910 IF(MP(J).EQ.MPMIN)  MREJEC = MREJEC + 1                       LIV07830
          MREJET = MREJEC + (NGOOD-NGOODC)                             LIV07840
          IF(MREJET.NE.0) WRITE(6,920) MREJET                          LIV07850
      920 FORMAT(/' RITZ VECTORS ARE NOT COMPUTED FOR',I6,' OF THE EIGNEVALULIV07860
         1ES'/)                                                         LIV07870
          NACT = NGOODC - MREJEC                                        LIV07880
          WRITE(6,930) NGOOD,NTVEC,NACT                                LIV07890
      930 FORMAT(/I6,' RITZ VECTORS WERE REQUESTED'/I6,' T-EIGENVECTORS WERELIV07900
         1 COMPUTED'/I6,' RITZ VECTORS WILL BE COMPUTED'/)              LIV07910
    C     CHECK IF THERE ARE ANY RITZ VECTORS TO COMPUTE                LIV07920
          IF(MREJEC.EQ.NGOODC) GO TO 1590                               LIV07930
    C                                                                    LIV07940
    C     CONTINUE WITH THE LANCZOS VECTOR COMPUTATIONS?                 LIV07950
          IF(LVCONT.EQ.0.AND.MREJEC.NE.0) GO TO 1570                    LIV07960
    C                                                                    LIV07970
```

```
C      NOW COMPUTE THE RITZ VECTORS.   REGENERATE THE                    LIV07980
C      LANCZOS VECTORS.                                                  LIV07990
C                                                                        LIV08000
       DO 940 I = 1,NMAX                                                 LIV08010
  940 RITVEC(I) = ZERO                                                   LIV08020
C                                                                        LIV08030
C------------------------------------------------------------------------LIV08040
C      REGENERATE THE STARTING VECTOR. THIS MUST BE GENERATED AND        LIV08050
C      NORMALIZED PRECISELY THE WAY IT WAS DONE IN THE EIGENVALUE        LIV08060
C      COMPUTATIONS, OTHERWISE THERE WILL BE A MISMATCH BETWEEN          LIV08070
C      THE T-EIGENVECTORS THAT HAVE BEEN COMPUTED FROM THE T-MATRICES    LIV08080
C      READ IN FROM FILE 2 AND THE LANCZOS VECTORS THAT ARE             LIV08090
C      BEING REGENERATED.                                                LIV08100
C                                                                        LIV08110
       CALL GENRAN(SVSEED,G,N)                                           LIV08120
C                                                                        LIV08130
C------------------------------------------------------------------------LIV08140
C                                                                        LIV08150
       DO 950 J = 1,N                                                    LIV08160
  950 V2(J) = G(J)                                                       LIV08170
C                                                                        LIV08180
C------------------------------------------------------------------------LIV08190
       SUM = FINPRO(N,V2(1),1,V2(1),1)                                   LIV08200
C------------------------------------------------------------------------LIV08210
C                                                                        LIV08220
       SUM = ONE/DSQRT(SUM)                                              LIV08230
C                                                                        LIV08240
       DO 960 I = 1,N                                                    LIV08250
       V1(I) = ZERO                                                      LIV08260
  960 V2(I) = V2(I)*SUM                                                  LIV08270
C                                                                        LIV08280
       WRITE(6,970)                                                      LIV08290
  970 FORMAT(' STARTING LANCZOS VECTOR HAS BEEN CALCULATED'/)            LIV08300
C                                                                        LIV08310
C      LOOP FOR GENERATING RITZ VECTORS  (IVEC = 1,KMAXU)               LIV08320
       IVEC = 1                                                          LIV08330
       BATA = ZERO                                                       LIV08340
C                                                                        LIV08350
       GO TO 1050                                                        LIV08360
C                                                                        LIV08370
  980 CONTINUE                                                           LIV08380
C                                                                        LIV08390
C      SOLVE B*VS = V2 FOR VS                                            LIV08400
       DO 990 K = 1,N                                                    LIV08410
  990 VS(K) = V2(K)                                                      LIV08420
C                                                                        LIV08430
C------------------------------------------------------------------------LIV08440
       JBSOLV = 2                                                        LIV08450
       CALL BSOLV(VS,VS,JBSOLV)                                          LIV08460
C------------------------------------------------------------------------LIV08470
C                                                                        LIV08480
C      VS = BI*V2    BI = B(INVERSE)                                     LIV08490
C      COMPUTE V1 = BI*V2 - BATA*V1                                      LIV08500
       DO 1000 K = 1,N                                                   LIV08510
 1000 V1(K) = VS(K) - BATA*V1(K)                                         LIV08520
C                                                                        LIV08530
C------------------------------------------------------------------------LIV08540
       ALFA = FINPRO(N,V1(1),1,V2(1),1)                                  LIV08550
C------------------------------------------------------------------------LIV08560
C                                                                        LIV08570
       DO 1010 J = 1,N                                                   LIV08580
 1010 V1(J) = V1(J)-ALFA*V2(J)                                           LIV08590
C                                                                        LIV08600
C------------------------------------------------------------------------LIV08610
       BATA = FINPRO(N,V1(1),1,V1(1),1)                                  LIV08620
C------------------------------------------------------------------------LIV08630
C                                                                        LIV08640
```

```
         BATA = DSQRT(BATA)                                            LIV08650
         SUM = ONE/BATA                                               LIV08660
C                                                                      LIV08670
         TEMP = BETA(IVEC)                                            LIV08680
         TEMP = DABS(BATA - TEMP)/TEMP                                LIV08690
         IF (TEMP.LT.1.0D-10)GO TO 1030                               LIV08700
C                                                                      LIV08710
C        THE BETA BEING REGENERATED DO NOT MATCH THE HISTORY FILE     LIV08720
C        SOMETHING IS WRONG IN THE LANCZOS VECTOR GENERATION          LIV08730
C        PROGRAM TERMINATES FOR USER TO CORRECT THE PROBLEM           LIV08740
C        WHICH MUST BE IN THE STARTING VECTOR GENERATION OR IN        LIV08750
C        THE MATRIX-VECTOR MULTIPLY SUBROUTINE CMATV SUPPLIED.        LIV08760
C        THIS SUBROUTINE MUST BE THE SAME ONE USED IN THE             LIV08770
C        EIGENVALUE COMPUTATIONS OR AGAIN A MISMATCH WILL ENSUE.      LIV08780
C                                                                      LIV08790
         WRITE(6,1020) IVEC,BATA,BETA(IVEC),TEMP                      LIV08800
    1020 FORMAT(/2X,'IVEC',16X,'BATA',10X,'BETA(IVEC)',14X,'RELDIF'/16,LIV08810
        13E20.12/' IN LANCZOS VECTOR REGENERATION THE ENTRIES OF THE TRIDIALIV08820
        1GONAL MATRICES BEING'/' GENERATED ARE NOT THE SAME AS THOSE IN THELIV08830
        1 MATRIX SUPPLIED ON FILE 2.'/' THEREFORE SOMETHING IS BEING INITIALIV08840
        1LIZED OR COMPUTED DIFFERENTLY FROM THE WAY'/' IT WAS COMPUTED IN TLIV08850
        1HE EIGENVALUE COMPUTATIONS'/' THE PROGRAM TERMINATES FOR THE USER LIV08860
        1TO DETERMINE WHAT THE PROBLEM IS'/)                          LIV08870
         GO TO 1670                                                   LIV08880
C                                                                      LIV08890
    1030 CONTINUE                                                     LIV08900
         DO 1040 J = 1,N                                              LIV08910
         TEMP = SUM*V1(J)                                             LIV08920
         V1(J) = V2(J)                                                LIV08930
    1040 V2(J) = TEMP                                                 LIV08940
C                                                                      LIV08950
    1050 CONTINUE                                                     LIV08960
C                                                                      LIV08970
         LFIN = 0                                                     LIV08980
         DO 1070 J = 1,NGOODC                                         LIV08990
         LL = LFIN                                                    LIV09000
         LFIN = LFIN + N                                              LIV09010
C                                                                      LIV09020
         IF(IABS(MA(J)).LT.IVEC.OR.MP(J).EQ.MPMIN) GO TO 1070        LIV09030
         II = IVEC + MINT(J) - 1                                      LIV09040
         TEMP = TVEC(II)                                              LIV09050
C        II IS THE (IVEC)TH COMPONENT OF THE T-EIGENVECTOR CONTAINED  LIV09060
C        IN TVEC(MINT(J)).                                            LIV09070
C                                                                      LIV09080
         DO 1060 K = 1,N                                              LIV09090
         LL = LL + 1                                                  LIV09100
    1060 RITVEC(LL) = TEMP*V2(K) + RITVEC(LL)                         LIV09110
C                                                                      LIV09120
    1070 CONTINUE                                                     LIV09130
C                                                                      LIV09140
         IVEC = IVEC + 1                                              LIV09150
         IF (IVEC.LE.KMAXU) GO TO 980                                 LIV09160
C                                                                      LIV09170
C        RITZVECTOR GENERATION IS COMPLETE. NORMALIZE EACH RITZVECTOR. LIV09180
C        NOTE THAT IF CERTAIN RITZ VECTORS WERE NOT COMPUTED THEN THAT LIV09190
C        PORTION OF THE RITVEC ARRAY WAS NOT UTILIZED.               LIV09200
C                                                                      LIV09210
         LFIN = 0                                                     LIV09220
         DO 1140 J = 1,NGOODC                                         LIV09230
C                                                                      LIV09240
         KK = LFIN                                                    LIV09250
         LFIN = LFIN + N                                              LIV09260
         IF(MP(J).EQ.MPMIN) GO TO 1140                                LIV09270
C                                                                      LIV09280
         DO 1080 K = 1,N                                              LIV09290
         KK = KK + 1                                                  LIV09300
         V1(K) = RITVEC(KK)                                           LIV09310
```

```
 1080 VS(K) = V1(K)                                              LIV09320
C                                                                LIV09330
      IF(JPERM.EQ.0)  GO TO 1090                                 LIV09340
C                                                                LIV09350
C----------------------------------------------------------------LIV09360
C     V2 = V1 = (L-TRANSPOSE)*V1                                 LIV09370
      IPERM = 2                                                  LIV09380
      CALL LPERM(V1,V2,IPERM)                                    LIV09390
C----------------------------------------------------------------LIV09400
C                                                                LIV09410
C     V2 CONTAINS RITZ VECTOR FOR A, VS CONTAINS THE RITZ VECTOR FOR B  LIV09420
C                                                                LIV09430
 1090 CONTINUE                                                   LIV09440
C                                                                LIV09450
C----------------------------------------------------------------LIV09460
      SUM = FINPRO(N,V1(1),1,V1(1),1)                            LIV09470
C----------------------------------------------------------------LIV09480
C                                                                LIV09490
      SUM = DSQRT(SUM)                                           LIV09500
      RNORM(J) = SUM                                             LIV09510
      RNORME = DABS(ONE-SUM)                                     LIV09520
      SUM = ONE/SUM                                              LIV09530
C                                                                LIV09540
      KK = LFIN - N                                              LIV09550
      DO 1100 K = 1,N                                            LIV09560
      KK = KK + 1                                                LIV09570
      VS(K) = SUM*VS(K)                                          LIV09580
 1100 RITVEC(KK) = SUM*V1(K)                                     LIV09590
C                                                                LIV09600
C     VS IS RITZ VECTOR FOR B1:  RITVEC IS RITZ VECTOR FOR A-MATRIX  LIV09610
C     B = SO*P*A*P' + SHIFT*I                                    LIV09620
C     BIERR = ||B1*VS - GOODBI(J)*VS||                           LIV09630
C     BIEVER = |(VS-TRANS)*B1*VS - GOODBI(J)|                    LIV09640
C                                                                LIV09650
C----------------------------------------------------------------LIV09660
C     V1 = (B-INVERSE)*VS                                        LIV09670
      JBSOLV = 2                                                 LIV09680
      CALL BSOLV(VS,V1,JBSOLV)                                   LIV09690
C----------------------------------------------------------------LIV09700
C                                                                LIV09710
      EVALN = EVNEW(J)                                           LIV09720
C                                                                LIV09730
C----------------------------------------------------------------LIV09740
      TEMP = FINPRO(N,V1(1),1,VS(1),1)                           LIV09750
C----------------------------------------------------------------LIV09760
C                                                                LIV09770
      TEMP = DABS(TEMP - EVALN)                                  LIV09780
      BIEVER(J) = TEMP                                           LIV09790
      DO 1110 K = 1,N                                            LIV09800
 1110 V1(K) = V1(K) - EVALN*VS(K)                                LIV09810
C                                                                LIV09820
C----------------------------------------------------------------LIV09830
      SUM = FINPRO(N,V1(1),1,V1(1),1)                            LIV09840
C----------------------------------------------------------------LIV09850
C                                                                LIV09860
      SUM = DSQRT(SUM)                                           LIV09870
      BIERR(J) = SUM                                             LIV09880
      BIERRG(J) = SUM/ABS(BIMING(J))                             LIV09890
C                                                                LIV09900
      LINT = LFIN - N + 1                                        LIV09910
      EVAL = (ONE/EVALN - SHIFT)/SO                              LIV09920
      GOODA(J) = EVAL                                            LIV09930
      TEMP = BIEVER(J)                                           LIV09940
C                                                                LIV09950
      IF(IWRITE.EQ.0)  GO TO 1140                                LIV09960
      WRITE(6,1120) J,GOODBI(J)                                  LIV09970
```

```
 1120 FORMAT(/I5,' TH B-INVERSE EIGENVALUE COMPUTED = ',E20.12/)        LIV09980
C                                                                       LIV09990
      WRITE(6,1130)  TERR(J),TBETA(J),RNORME                            LIV10000
 1130 FORMAT(' NORM OF ERROR IN T-EIGENVECTOR = ',E14.3/               LIV10010
     1' BETA(MA(J)+I)*U(MA(J)) = ',E14.3/                               LIV10020
     1' ABS(NORM(RITVEC) - 1.0) = ',E14.3/)                             LIV10030
C                                                                       LIV10040
 1140 CONTINUE                                                          LIV10050
C                                                                       LIV10060
C     RITZVECTORS ARE NORMALIZED AND ERROR ESTIMATES ARE IN BIERR       LIV10070
C     AND BIERRG ARRAYS. STORE EVERYTHING                               LIV10080
C                                                                       LIV10090
      WRITE(13,1150)                                                    LIV10100
 1150 FORMAT(6X,'BIEIGENVALUE',6X,'RITZNORM',7X,'TBETA',7X,'TLAST',5X,  LIV10110
     1 'BIERROR',6X,'BIEVER')                                           LIV10120
C                                                                       LIV10130
      WRITE(9,1160)                                                     LIV10140
 1160 FORMAT(5X,'BIEIGENVALUE',4X,'MA(J)',4X,'BIMINGAP',5X,'BIERROR',3X LIV10150
     1 ,'BIERR/GAP',6X,'TERROR')                                        LIV10160
C                                                                       LIV10170
      DO 1190 J=1,NGOODC                                                LIV10180
C                                                                       LIV10190
      IF(MP(J).EQ.MPMIN) GO TO 1190                                     LIV10200
C                                                                       LIV10210
      WRITE(9,1170) GOODBI(J),MA(J),BIMING(J),BIERR(J),BIERRG(J),TERR(J)LIV10220
 1170 FORMAT(E20.12,I6,4E12.4)                                          LIV10230
C                                                                       LIV10240
      WRITE(13,1180) EVNEW(J),RNORM(J),TBETA(J),TLAST(J),BIERR(J),      LIV10250
     1 BIEVER(J)                                                        LIV10260
 1180 FORMAT(E20.12,5E12.4)                                             LIV10270
C                                                                       LIV10280
 1190 CONTINUE                                                          LIV10290
C                                                                       LIV10300
      WRITE(9,1200)                                                     LIV10310
 1200 FORMAT(/5X, 'J',7X,'AEIGENVALUE',3X,'MA(J)',5X,'AMINGAP')         LIV10320
C                                                                       LIV10330
      DO 1210 J = 1,NGOOD                                               LIV10340
      IF(MP(J).EQ.MPMIN) GO TO 1210                                     LIV10350
      WRITE(9,1220) J,GOODA(J),MA(J),AMINGP(J)                          LIV10360
 1210 CONTINUE                                                          LIV10370
 1220 FORMAT(I6,E20.12,I6,E12.4)                                        LIV10380
C                                                                       LIV10390
      IF (MREJEC.EQ.0) GO TO 1300                                       LIV10400
C                                                                       LIV10410
      WRITE(9,1230)                                                     LIV10420
 1230 FORMAT(/' RITZ VECTORS WERE NOT COMPUTED FOR THE FOLLOWING EIGENVALLIV10430
     1LUES'/' EITHER BECAUSE THEY HAD NOT CONVERGED OR BECAUSE THE ERRORLIV10440
     1 ESTIMATE'/'  WAS NOT AS SMALL AS DESIRED'/)                      LIV10450
C                                                                       LIV10460
      WRITE(9,1240)                                                     LIV10470
 1240 FORMAT(6X,'GOODBI(J)',3X,'MA(J)',5X,'BIMING(J)',6X,'TBETA(J)',3X, LIV10480
     1'MP(J)')                                                          LIV10490
C                                                                       LIV10500
      WRITE(13,1250)                                                    LIV10510
 1250 FORMAT(/' RITZ VECTORS WERE NOT COMPUTED FOR THE FOLLOWING EIGENVALLIV10520
     1LUES'/' EITHER BECAUSE THEY HAD NOT CONVERGED OR BECAUSE'/' THE ERLIV10530
     1ROR ESTIMATE WAS NOT AS SMALL AS DESIRED'/)                       LIV10540
C                                                                       LIV10550
      WRITE(13,1260)                                                    LIV10560
 1260 FORMAT(3X,'BIEIGENVALUE',3X,'MA(J)',3X,'M1(J)',3X,'M2(J)',3X,'MP(JLIV10570
     1)')                                                               LIV10580
C                                                                       LIV10590
      DO 1290 J = 1,NGOODC                                              LIV10600
C                                                                       LIV10610
      IF(MP(J).NE.MPMIN) GO TO 1290                                     LIV10620
C                                                                       LIV10630
C     WRITE OUT MESSAGE FOR EACH EIGENVALUE FOR WHICH NO EIGENVECTOR     LIV10640
```

```
C     WAS COMPUTED.                                                    LIV10650
C                                                                      LIV10660
      WRITE(9,1270) GOODBI(J),MA(J),BIMING(J),TBETA(J),MP(J)           LIV10670
 1270 FORMAT(E15.8,I8,2E14.4,I8)                                       LIV10680
C                                                                      LIV10690
      WRITE(13,1280) GOODBI(J),MA(J),M1(J),M2(J),MP(J)                 LIV10700
 1280 FORMAT(E15.8,4I8)                                                LIV10710
C                                                                      LIV10720
 1290 CONTINUE                                                         LIV10730
C                                                                      LIV10740
 1300 CONTINUE                                                         LIV10750
C                                                                      LIV10760
      WRITE(9,1310)                                                    LIV10770
 1310 FORMAT(/' ABOVE ARE ERROR ESTIMATES FOR THE BI AND T EIGENVECTORS'LIV10780
     1/ ' ASSOCIATED WITH THE GOODBI LISTED, DENOTED BY EV '/          LIV10790
     1 ' BIERROR = NORM(BI*X-EV*X),  TERROR = NORM(T*Y - EV*Y)'/       LIV10800
     1 ' WHERE T = T(1,MA(J)), P*X = RITZVEC = V*Y, T*Y = GOODBI*Y'/   LIV10810
     1 ' BIMINGAP = GAP TO NEAREST BI-EIGENVALUE'/)                    LIV10820
C                                                                      LIV10830
      WRITE(13,1320)                                                   LIV10840
 1320 FORMAT(/' ABOVE ARE ERROR ESTIMATES FOR THE EIGENVECTORS'/       LIV10850
     1 ' ASSOCIATED WITH THE BI-EIGENVALUES'/                          LIV10860
     1 ' RITZNORM = NORM(COMPUTED RITZ VECTOR FOR B-INVERSE)'/         LIV10870
     1 ' TBETA(J) = BETA(MA(J)+1)*Y(MA(J)),  T*Y = BIEVAL*Y'/          LIV10880
     1 ' TLAST(J) = DABS(Y(MA(J)))'/                                   LIV10890
     1 ' BIERROR = NORM(BI*X - BIEVAL*X) WHERE X = V*Y'/               LIV10900
     1 ' BIEVER = DABS(BIEIGENVALUE - (X-TRANSPOSE*BINVERSE*X))'/)     LIV10910
C                                                                      LIV10920
C     NUMBER OF RITZ VECTORS COMPUTED                                  LIV10930
      NCOMPU = NGOODC - MREJEC                                         LIV10940
      WRITE(12,1330) N,NCOMPU,NGOODC,MATNO                             LIV10950
 1330 FORMAT(3I6,I8,' = SIZE A, NO.RITZVECS, NO.GOODEVALUES,MATNO')    LIV10960
C                                                                      LIV10970
      LFIN = 0                                                         LIV10980
      DO 1390 J = 1,NGOODC                                             LIV10990
      LINT = LFIN + 1                                                  LIV11000
      LFIN = LFIN + N                                                  LIV11010
C                                                                      LIV11020
      IF(MP(J).EQ.MPMIN) GO TO 1370                                    LIV11030
C     RITZ VECTOR WAS COMPUTED                                         LIV11040
      WRITE(12,1340) J, EVNEW(J), GOODA(J),MP(J)                       LIV11050
 1340 FORMAT(I6,4X,2E20.12,I6,' J,GOODBI,GOODA,MP(J)')                 LIV11060
C                                                                      LIV11070
      WRITE(12,1350) BIERR(J), BIERRG(J), BIMING(J),AMINGP(J)          LIV11080
 1350 FORMAT(4X,' BIRESIDUAL  ',2X,'BIRESIDUAL/GAP',                   LIV11090
     12X,'BIMINGAP',3X,' AMINGAP'/                                     LIV11100
     1 E15.5,E16.5,2E11.3)                                             LIV11110
C                                                                      LIV11120
      WRITE(12,1360) (RITVEC(LL), LL=LINT,LFIN)                        LIV11130
 1360 FORMAT(4E20.12)                                                  LIV11140
      GO TO 1390                                                       LIV11150
C     NO RITZ VECTOR WAS COMPUTED FOR THIS EIGENVALUE                  LIV11160
 1370 CONTINUE                                                         LIV11170
      WRITE(12,1380) J,GOODBI(J),GOODA(J),MP(J)                        LIV11180
 1380 FORMAT(/I5,E20.12,E20.12,I6,' = J,GOODBI,GOODA,MP'/' NO RITZ VECTOLIV11190
     1R WAS COMPUTED FOR THIS EIGENVALUE'/)                            LIV11200
C                                                                      LIV11210
 1390 CONTINUE                                                         LIV11220
C                                                                      LIV11230
C     DID ANY T-MATRICES INCLUDE OFF-DIAGONAL ENTRIES SMALLER THAN     LIV11240
C     DESIRED, AS SPECIFIED BY BTOL?                                   LIV11250
C                                                                      LIV11260
      IF(IB.GT.0) GO TO 1420                                           LIV11270
      WRITE(6,1400) KMAXU                                              LIV11280
 1400 FORMAT(/' FOR LARGEST T-MATRIX CONSIDERED',I7,' CHECK THE SIZE OF LIV11290
     1BETAS')                                                          LIV11300
C                                                                      LIV11310
```

```
C-----------------------------------------------------------------------LIV11320
C                                                                        LIV11330
      CALL TNORM(ALPHA,BETA,BKMIN,TEMP,KMAXU,IBMT)                       LIV11340
C                                                                        LIV11350
C-----------------------------------------------------------------------LIV11360
C                                                                        LIV11370
      IF(IBMT.LT.0) WRITE(6,1410)                                        LIV11380
 1410 FORMAT(/' WARNING THE T-MATRICES FOR ONE OR MORE OF THE EIGENVALUELIV11390
     1S CONSIDERED'/' HAD AN OFF DIAGONAL ENTRY THAT WAS SMALLER THAN THLIV11400
     1E BETA TOLERANCE THAT WAS SPECIFIED'/)                             LIV11410
 1420 CONTINUE                                                           LIV11420
C                                                                        LIV11430
      GO TO 1670                                                         LIV11440
C                                                                        LIV11450
 1430 WRITE(6,1440) NGOOD,NMAX,MDIMRV                                    LIV11460
 1440 FORMAT(/I4,' RITZ VECTORS WERE REQUESTED BUT THE REQUIRED DIMENSIOLIV11470
     1N',I6/' IS LARGER THAN USER-SPECIFIED DIMENSION OF RITVEC',I6/     LIV11480
     1' THEREFORE, THE EIGENVECTOR PROCEDURE TERMINATES FOR THE USER TO LIV11490
     1 INTERVENE'/)                                                      LIV11500
C                                                                        LIV11510
      GO TO 1670                                                         LIV11520
C                                                                        LIV11530
 1450 WRITE(6,1460) NOLD,N,MATOLD,MATNO,SHIFTO,SHiFT                     LIV11540
 1460 FORMAT(/' PARAMETERS READ FROM FILE 3 DO NOT AGREE WITH WHAT USER LIV11550
     1SPECIFIED'/ ' NOLD,N,MATOLD,MATNO,SHIFTO,SHIFT = '/2I6,2I8,2E10.3 LIV11560
     1/' THEREFORE PROGRAM TERMINATES FOR USER TO RESOLVE THE DIFFERENCELIV11570
     1S'/)                                                               LIV11580
C                                                                        LIV11590
      GO TO 1670                                                         LIV11600
C                                                                        LIV11610
 1470 WRITE(6,1480)                                                      LIV11620
 1480 FORMAT(/' PARAMETERS READ FROM ALPHA,BETA FILE DO NOT AGREE WITH WLIV11630
     1HAT USER SPECIFIED'/' PROGRAM TERMINATES FOR USER TO RESOLVE THE DLIV11640
     1IFFERENCES'/)                                                      LIV11650
C                                                                        LIV11660
      GO TO 1670                                                         LIV11670
C                                                                        LIV11680
 1490 WRITE(6,1500) KMAX,MEV                                             LIV11690
 1500 FORMAT(/' IN ALPHA, BETA FILE KMAX = ',I6/                         LIV11700
     1' BUT EIGENVALUES WERE COMPUTED AT MEV =',I6,' PROGRAM STOPS'/)    LIV11710
C                                                                        LIV11720
      GO TO 1670                                                         LIV11730
C                                                                        LIV11740
 1510 WRITE(6,1520)                                                      LIV11750
 1520 FORMAT(/' PROGRAM COMPUTED 1ST GUESSES ON T-MATRIX SIZES AND READ LIV11760
     1THEM TO FILE 10'/' THEN TERMINATED AS REQUESTED.'/)                LIV11770
      GO TO 1670                                                         LIV11780
C                                                                        LIV11790
 1530 WRITE(6,1540) MTOL, MDIMTV                                         LIV11800
 1540 FORMAT(/' PROGRAM TERMINATES BECAUSE THE TVEC DIMENSION ANTICIPATELIV11810
     1D',I7/' IS LARGER THAN THE TVEC DIMENSION',I7,' SPECIFIED BY THE LIV11820
     1USER.'/' USER MAY RESET THE TVEC DIMENSION AND RESTART THE PROGRALIV11830
     1M')                                                                LIV11840
      GO TO 1670                                                         LIV11850
C                                                                        LIV11860
 1550 WRITE(6,1560)                                                      LIV11870
 1560 FORMAT(/' PROGRAM TERMINATES BECAUSE NO SUITABLE T-EIGENVECTORS WELIV11880
     1RE IDENTIFIED'/' FOR ANY OF THE EIGENVALUES SUPPLIED. PROBLEM COLIV11890
     1ULD BE CAUSED'/' BY TOO SMALL A TVEC DIMENSION OR SIMPLY THAT SUILIV11900
     1TABLE T-VECTORS COULD'/' NOT BE IDENTIFIED. USER SHOULD EXAMINE OLIV11910
     1UTPUT'/)                                                           LIV11920
      GO TO 1670                                                         LIV11930
C                                                                        LIV11940
 1570 WRITE(6,1580) LVCONT,NTVEC,NGOOD                                   LIV11950
 1580 FORMAT(/' LVCONT FLAG =',I2,' AND NUMBER ',I5,' OF T-EIGENVECTORS LIV11960
     1 COMPUTED N.E.'/' NUMBER',I5,' REQUESTED SO PROGRAM TERMINATES'/) LIV11970
```

```
      GO TO 1670                                                  LIV11980
C                                                                 LIV11990
 1590 WRITE(6,1600)                                               LIV12000
 1600 FORMAT(/' PROGRAM TERMINATES WITHOUT COMPUTING RITZ VECTORS'/   LIV12010
     1' BECAUSE ALL T-EIGENVECTORS WERE REJECTED AS NOT SUITABLE FOR THELIV12020
     1RITZ VECTOR'/' COMPUTATIONS.  PROBABLE CAUSE IS LACK OF CONVERGENCLIV12030
     1E OF EIGENVALUES SUPPLIED'/)                                LIV12040
      GO TO 1670                                                  LIV12050
C                                                                 LIV12060
 1610 WRITE(6,1620)                                               LIV12070
 1620 FORMAT(/' PROGRAM INDICATES THAT IT IS NOT POSSIBLE TO COMPUTE ANYLIV12080
     1 OF THE REQUESTED EIGENVECTORS.'/' THEREFORE PROGRAM TERMINATES') LIV12090
      DO 1630 J=1,NGOODC                                          LIV12100
 1630 WRITE(6,1640)  J,GOODBI(J),MP(J)                            LIV12110
 1640 FORMAT(/4X,' J',11X,'GOODBI(J)',4X,'MP(J)'/I6,E20.12,I9/)   LIV12120
      GO TO 1670                                                  LIV12130
C                                                                 LIV12140
 1650 WRITE(6,1660) MBETA,KMAXN                                   LIV12150
 1660 FORMAT(/' PROGRAM TERMINATES BECAUSE THE STORAGE ALLOTTED FOR THE LIV12160
     1BETA ARRAY',I8,/' IS NOT SUFFICIENT FOR THE ENLARGED KMAX =',I8,' LIV12170
     1THAT THE PROGRAM WANTS.'/' USER CAN ENLARGE THE ALPHA,BETA ARRAYS LIV12180
     1 AND RERUN THE PROGRAM'/)                                   LIV12190
C                                                                 LIV12200
 1670 CONTINUE                                                    LIV12210
C                                                                 LIV12220
      STOP                                                        LIV12230
C-----END EIGENVECTOR COMPUTATIONS FOR INVERSES OF REAL SYMMETRIC-------LIV12240
      END                                                         LIV12250
```

SECTION 4.4   LANCZS AND SAMPLE MATRIX-VECTOR SOLVE SUBROUTINES

```
C---LIMULT-(INVERSES OF REAL SYMMETRIC MATRICES)-----------------------LIM00010
C                                                                       LIM00020
C     CONTAINS SUBROUTINE LANCZS AND SAMPLE USPEC AND BSOLV             LIM00030
C     USED BY THE VERSION OF THE LANCZOS ALGORITHMS FOR                 LIM00040
C     FACTORED INVERSES OF REAL SYMMETRIC MATRICES, LIVAL AND LIVEC.    LIM00050
C                                                                       LIM00060
C     NONPORTABLE CONSTRUCTIONS:                                        LIM00070
C     1.  THE ENTRY MECHANISM USED TO PASS THE STORAGE LOCATIONS        LIM00080
C         OF THE FACTORIZATION OF THE MATRIX TO BE USED BY              LIM00090
C         LANCZS TO THE SOLVE SUBROUTINE BSOLV.                         LIM00100
C     2.  IN THE SAMPLE USPEC SUBROUTINES PROVIDED:                     LIM00110
C         THE FREE FORMAT (7,*) AND FORMATS (20A4) AND   (4Z20)         LIM00120
C         USED IN DEFINING THE MATRICES.                                LIM00130
C                                                                       LIM00140
C-----LANCZS-COMPUTE LANCZOS TRIDIAGONAL MATRICES----------------------LIM00150
C                                                                       LIM00160
      SUBROUTINE LANCZS(MATVEC,ALPHA,BETA,V1,V2,VS,G,KMAX,MOLD1,N,IIX)  LIM00170
C                                                                       LIM00180
C---------------------------------------------------------------------LIM00190
      DOUBLE PRECISION  ALPHA(1), BETA(1), V1(1), V2(1), VS(1)          LIM00200
      DOUBLE PRECISION SUM, ONE, ZERO, TEMP                             LIM00210
      REAL  G(1)                                                        LIM00220
      EXTERNAL MATVEC                                                   LIM00230
      DOUBLE PRECISION  FINPRO, DSQRT                                   LIM00240
C---------------------------------------------------------------------LIM00250
C     ALPHA, BETA, LANCZOS VECTOR GENERATION                           LIM00260
C     ALPHA BETA GENERATION STARTS WITH IVEC = 1, BETA(1) = ZERO       LIM00270
C     V2 = RANDOM UNIT VECTOR AND V1 = ZERO, OR EXTENDS                 LIM00280
C     AN EXISTING ALPHA/BETA FILE.                                      LIM00290
C                                                                       LIM00300
      ZERO = 0.0D0                                                      LIM00310
      ONE  = 1.0D0                                                      LIM00320
      IF (MOLD1.GT.1) GO TO 30                                          LIM00330
      BETA(1) = ZERO                                                    LIM00340
      IIL = IIX                                                         LIM00350
C                                                                       LIM00360
C---------------------------------------------------------------------LIM00370
      CALL GENRAN(IIL,G,N)                                              LIM00380
C---------------------------------------------------------------------LIM00390
C                                                                       LIM00400
      DO 10 K = 1,N                                                     LIM00410
   10 V2(K) = G(K)                                                      LIM00420
C                                                                       LIM00430
C---------------------------------------------------------------------LIM00440
      SUM = FINPRO(N,V2(1),1,V2(1),1)                                   LIM00450
C---------------------------------------------------------------------LIM00460
C                                                                       LIM00470
      SUM = ONE/DSQRT(SUM)                                              LIM00480
C                                                                       LIM00490
      DO 20 K = 1,N                                                     LIM00500
      V1(K) = ZERO                                                      LIM00510
   20 V2(K) = SUM*V2(K)                                                 LIM00520
C                                                                       LIM00530
   30 CONTINUE                                                          LIM00540
C                                                                       LIM00550
      DO 80 IVEC = MOLD1,KMAX                                           LIM00560
C                                                                       LIM00570
      DO 40 K = 1,N                                                     LIM00580
   40 VS(K) = V2(K)                                                     LIM00590
C                                                                       LIM00600
C---------------------------------------------------------------------LIM00610
      JBSOLV = 2                                                        LIM00620
      CALL MATVEC(VS,VS,JBSOLV)                                         LIM00630
C---------------------------------------------------------------------LIM00640
```

```
C                                                                  LIM00650
C        VS = B(INVERSE)*V2                                        LIM00660
C                                                                  LIM00670
         SUM = BETA(IVEC)                                          LIM00680
C                                                                  LIM00690
         DO 50 K = 1,N                                             LIM00700
   50    V1(K) = VS(K)-SUM*V1(K)                                   LIM00710
C                                                                  LIM00720
C-----------------------------------------------------------------LIM00730
         SUM = FINPRO(N,V1(1),1,V2(1),1)                           LIM00740
C-----------------------------------------------------------------LIM00750
C                                                                  LIM00760
         ALPHA(IVEC) = SUM                                         LIM00770
C                                                                  LIM00780
         DO 60 K = 1,N                                             LIM00790
   60    V1(K) = V1(K)-SUM*V2(K)                                   LIM00800
C                                                                  LIM00810
C-----------------------------------------------------------------LIM00820
         SUM = FINPRO(N,V1(1),1,V1(1),1)                           LIM00830
C-----------------------------------------------------------------LIM00840
C                                                                  LIM00850
         IN = IVEC+1                                               LIM00860
C                                                                  LIM00870
         BETA(IN) = DSQRT(SUM)                                     LIM00880
         SUM = ONE/BETA(IN)                                        LIM00890
C                                                                  LIM00900
         DO 70 K = 1,N                                             LIM00910
         TEMP = SUM*V1(K)                                          LIM00920
         V1(K) = V2(K)                                             LIM00930
   70    V2(K) = TEMP                                              LIM00940
C                                                                  LIM00950
   80    CONTINUE                                                  LIM00960
C                                                                  LIM00970
         RETURN                                                    LIM00980
C-----END LANCZS-------------------------------------------------- LIM00990
         END                                                       LIM01000
C                                                                  LIM01010
C-----USPEC FOR FACTORED INVERSES OF REAL SYMMETRIC MATRICES-------LIM01020
C                                                                  LIM01030
         SUBROUTINE CUSPEC(N,MATNO)                                LIM01040
C        SUBROUTINE USPEC(N,MATNO)                                 LIM01050
C                                                                  LIM01060
C-----------------------------------------------------------------LIM01070
         DOUBLE PRECISION  BD(2200),BSD(10000)                     LIM01080
         INTEGER  KCOL(2200),KROW(10000),IPR(2200),IPT(2200)       LIM01090
C-----------------------------------------------------------------LIM01100
C     NOTE THAT THIS SUBROUTINE ASSUMES THAT B IS POSITIVE DEFINITE. LIM01110
C     USER COULD REPLACE THIS SUBROUTINE AND CORRESPONDING SAMPLE  LIM01120
C     USPEC SUBROUTINE BY ONE THAT WORKS WITH GENERAL FACTORIZATION. LIM01130
C                                                                  LIM01140
C     DIMENSIONS ARRAYS NEEDED TO DEFINE CHOLESKY FACTOR OF B-MATRIX, LIM01150
C     READS CHOLESKY FACTOR FROM FILE 7, AND THEN PASSES STORAGE   LIM01160
C     LOCATIONS OF THESE ARRAYS TO THE B-MATRIX SOLVE SUBROUTINE BSOLV. LIM01170
C                                                                  LIM01180
C     HERE WE HAVE B = P*C*P' = L*L'  WHERE  C = SO*A + SHIFT*I.   LIM01190
C     P IS A PERMUTATION MATRIX DEFINED BY THE VECTOR MAPS IPR AND IPT. LIM01200
C     THE ITH ROW OF B CORRESPONDS TO THE JTH ROW OF C (A) WHERE   LIM01210
C     J = IPR(I) AND I = IPT(J).   A IS THE ORIGINAL MATRIX.       LIM01220
C                                                                  LIM01230
C     THE B-CHOLESKY FACTOR IS STORED IN THE FOLLOWING SPARSE FORMAT: LIM01240
C     N = ORDER OF THE B-MATRIX.                                   LIM01250
C     NZT = NUMBER OF NONZERO SUBDIAGONAL ENTRIES IN THE CHOLESKY  LIM01260
C           FACTOR, L.                                             LIM01270
C     KCOL(J), J=1,N IS THE NUMBER OF NONZERO SUBDIAGONAL ELEMENTS IN LIM01280
C             COLUMN J OF L.                                       LIM01290
C     KROW(K), K=1,NZT IS THE ROW INDEX FOR CORRESPONDING ENTRY BSD(K). LIM01300
C     BD(J), J = 1,N CONTAINS THE DIAGONAL ENTRIES OF L.           LIM01310
```

```
C     BSD(K), K =1,NZT CONTAINS THE NONZERO SUBDIAGONAL ENTRIES OF L     LIM01320
C           BY COLUMN.                                                   LIM01330
C     JPERM = (0,1):  1 MEANS CHOLEKSY FACTOR CORRESPONDS TO             LIM01340
C                     PERMUTED C.  O MEANS NO PERMUTATION WAS USED.      LIM01350
C---------------------------------------------------------------------- LIM01360
C     READ CHOLESKY FACTOR FROM FILE 7.  MUST BE STORED                 LIM01370
C     IN SPARSE MATRIX FORMAT.                                          LIM01380
      READ(7,10) NZT,NOLD,NZL,MATOLD,JPERM                              LIM01390
   10 FORMAT(I10,2I6,I8,I6)                                             LIM01400
C                                                                       LIM01410
      WRITE(6,20) NZT,NZL,N,NOLD,MATOLD,JPERM                           LIM01420
   20 FORMAT(' HEADER, CHOLESKY FACTOR FILE'/                           LIM01430
     1 3X,'NZT',3X,'NZL',5X,'N',2X,'NOLD',2X,'MATOLD',1X,'JPERM'/       LIM01440
     1 4I6,I3,I6/)                                                      LIM01450
C                                                                       LIM01460
      IF (N.NE.NOLD.OR.MATNO.NE.MATOLD) GO TO 70                        LIM01470
C                                                                       LIM01480
      READ(7,30) (KCOL(K), K = 1,NZL)                                   LIM01490
      READ(7,30) (KROW(K), K = 1,NZT)                                   LIM01500
   30 FORMAT(13I6)                                                      LIM01510
      READ(7,40) (BD(K), K = 1,N)                                       LIM01520
      READ(7,40) (BSD(K), K = 1,NZT)                                    LIM01530
   40 FORMAT(4Z20)                                                      LIM01540
C  20 FORMAT(3E25.16)                                                   LIM01550
C                                                                       LIM01560
C     DOES CHOLESKY FACTOR CORRESPOND TO PERMUTED B?                    LIM01570
      IF(JPERM.EQ.0)  GO TO 60                                          LIM01580
      READ(7,30) (IPR(K), K = 1,N)                                      LIM01590
C                                                                       LIM01600
      DO 50 K = 1,N                                                     LIM01610
      J = IPR(K)                                                        LIM01620
   50 IPT(J) = K                                                        LIM01630
C                                                                       LIM01640
C---------------------------------------------------------------------- LIM01650
      CALL LPERME(IPR,IPT,N)                                            LIM01660
C---------------------------------------------------------------------- LIM01670
C                                                                       LIM01680
   60 CONTINUE                                                          LIM01690
C                                                                       LIM01700
C---------------------------------------------------------------------- LIM01710
C     PASS STORAGE LOCATIONS OF FACTORS TO INVERSION SUBROUTINE BSOLV   LIM01720
      CALL BSOLVE(BSD,BD,KCOL,KROW,N,NZT,NZL)                           LIM01730
C---------------------------------------------------------------------- LIM01740
C                                                                       LIM01750
      GO TO 90                                                          LIM01760
C                                                                       LIM01770
   70 CONTINUE                                                          LIM01780
C     DEFAULT EXIT                                                      LIM01790
      WRITE(6,80)                                                       LIM01800
   80 FORMAT(' TERMINATE.  PARAMETERS IN CHOLESKY FACTOR FILE'/         LIM01810
     1' DO NOT AGREE WITH THOSE SPECIFIED BY THE USER'/)                LIM01820
      STOP                                                              LIM01830
C                                                                       LIM01840
   90 CONTINUE                                                          LIM01850
C-----END OF USPEC----------------------------------------------------- LIM01860
      RETURN                                                            LIM01870
      END                                                              LIM01880
C                                                                       LIM01890
C-----BSOLV-(FACTORED INVERSE OR L*L-TRANS MULTIPLY)------------------- LIM01900
C          (FOR POSITIVE DEFINITE SYMMETRIC SPARSE MATRICES)            LIM01910
C                                                                       LIM01920
C     SUBROUTINE BSOLV(V,U,JBSOLV)                                      LIM01930
      SUBROUTINE CBSOLV(V,U,JBSOLV)                                     LIM01940
C                                                                       LIM01950
C---------------------------------------------------------------------- LIM01960
      DOUBLE PRECISION  BD(1),BSD(1),U(1),V(1),TEMP,ZERO,ONE            LIM01970
```

```
      INTEGER  KCOL(1),KROW(1)                                          LIM01980
C----------------------------------------------------------------------LIM01990
C     JBSOLV = 2 MEANS SOLVE B*U = V                                    LIM02000
C     JBSOLV = 1 MEANS COMPUTE U = B*V:  NOTE THAT IN THIS CASE V IS    LIM02010
C     DESTROYED.  LANCZOS PROGRAMS AS WRITTEN DO NOT USE JBSOLV = 1     LIM02020
C     PATH.                                                             LIM02030
      ZERO = 0.0D0                                                      LIM02040
      ONE  = 1.0D0                                                      LIM02050
      IF (JBSOLV .EQ.2) GO TO 60                                        LIM02060
C     U = B*V WHERE  B = L*L'                                           LIM02070
      KL = 0                                                            LIM02080
      DO 20 J = 1,N                                                     LIM02090
      TEMP = V(J)*BD(J)                                                 LIM02100
      IF (KCOL(J).EQ.0.OR.J.EQ.N) GO TO 20                             LIM02110
      KF = KL + 1                                                       LIM02120
      KL  = KL + KCOL(J)                                                LIM02130
      DO 10 K = KF,KL                                                   LIM02140
      IK = KROW(K)                                                      LIM02150
   10 TEMP = BSD(K)*V(IK) + TEMP                                        LIM02160
   20 V(J) = TEMP                                                       LIM02170
C     V = L'*V                                                          LIM02180
      DO 30 K = 1,N                                                     LIM02190
   30 U(K) = V(K)*BD(K)                                                 LIM02200
      KL = 0                                                            LIM02210
      DO 50 K = 1,N                                                     LIM02220
      TEMP = V(K)                                                       LIM02230
      IF (KCOL(K).EQ.0.OR.K.EQ.N) GO TO 50                             LIM02240
      KF = KL + 1                                                       LIM02250
      KL = KL + KCOL(K)                                                 LIM02260
      DO 40 KK = KF,KL                                                  LIM02270
      KR = KROW(KK)                                                     LIM02280
   40 U(KR) = U(KR) + TEMP*BSD(KK)                                      LIM02290
   50 CONTINUE                                                          LIM02300
      GO TO 120                                                         LIM02310
C     U = B*V                                                           LIM02320
C--------------------------------------------------------------------- LIM02330
   60 CONTINUE                                                          LIM02340
C     SOLVE B*U = V FOR U WHERE  B = L*L'                               LIM02350
C     SET U = V. FIRST SOLVE L*U = U FOR U, THEN SOLVE L'*U = U FOR U   LIM02360
      KL = 0                                                            LIM02370
      DO 70 K = 1,N                                                     LIM02380
   70 U(K) = V(K)                                                       LIM02390
      DO 90 K = 1,N                                                     LIM02400
      TEMP = U(K)/BD(K)                                                 LIM02410
      U(K) = TEMP                                                       LIM02420
      IF (KCOL(K).EQ.0.OR.K.EQ.N) GO TO 90                             LIM02430
      KF = KL + 1                                                       LIM02440
      KL = KL + KCOL(K)                                                 LIM02450
      DO 80 KK = KF,KL                                                  LIM02460
      KR = KROW(KK)                                                     LIM02470
   80 U(KR) = U(KR) - TEMP*BSD(KK)                                      LIM02480
   90 CONTINUE                                                          LIM02490
      NP1 = N+1                                                         LIM02500
      KF = NZT + 1                                                      LIM02510
      DO 110 K = 1,N                                                    LIM02520
      L = NP1 - K                                                       LIM02530
      TEMP = U(L)                                                       LIM02540
      IF (KCOL(L).EQ.0.OR.L.EQ.N) GO TO 110                            LIM02550
      KL = KF - 1                                                       LIM02560
      KF = KF - KCOL(L)                                                 LIM02570
      DO 100 LL = KF,KL                                                 LIM02580
      LR = KROW(LL)                                                     LIM02590
  100 TEMP = TEMP - BSD(LL)*U(LR)                                       LIM02600
  110 U(L) = TEMP/BD(L)                                                 LIM02610
  120 CONTINUE                                                          LIM02620
C                                                                       LIM02630
```

```
      RETURN                                                         LIM02640
C                                                                    LIM02650
C---------------------------------------------------------------------LIM02660
      ENTRY BSOLVE(BSD,BD,KCOL,KROW,N,NZT,NZL)                       LIM02670
C---------------------------------------------------------------------LIM02680
C                                                                    LIM02690
      RETURN                                                         LIM02700
C-----END OF BSOLV--------------------------------------------------- LIM02710
      END                                                            LIM02720
C                                                                    LIM02730
C-----SUBROUTINES FOR DIAGONAL TEST MATRICES------------------------- LIM02740
C                                                                    LIM02750
C     BSOLV AND USPEC SUBROUTINES FOR DIAGONAL TEST MATRICES         LIM02760
C                                                                    LIM02770
C-----BSOLV DIAGONAL TEST MATRIX------------------------------------- LIM02780
C                                                                    LIM02790
C     SUBROUTINE DBSOLV(V,U,JBSOLV)                                  LIM02800
      SUBROUTINE BSOLV(V,U,JBSOLV)                                   LIM02810
C                                                                    LIM02820
C---------------------------------------------------------------------LIM02830
      DOUBLE PRECISION  V(1),U(1),D(1)                               LIM02840
C---------------------------------------------------------------------LIM02850
C     JBSOLV = 1, COMPUTE U = D*V. (NOTE THIS IS NOT USED)           LIM02860
C     JBSOLV = 2, COMPUTE U = (D-INVERSE)*V                          LIM02870
      IF(JBSOLV.EQ.2) GO TO 20                                       LIM02880
      DO 10 I=1,N                                                    LIM02890
   10 U(I) = D(I)*V(I)                                               LIM02900
      GO TO 40                                                       LIM02910
C                                                                    LIM02920
   20 DO 30 I=1,N                                                    LIM02930
   30 U(I)= V(I)/D(I)                                                LIM02940
C                                                                    LIM02950
   40 CONTINUE                                                       LIM02960
      RETURN                                                         LIM02970
C                                                                    LIM02980
C---------------------------------------------------------------------LIM02990
C     BELOW ENTRY IS FOR A DIAGONAL TEST MATRIX                      LIM03000
      ENTRY DSOLVE(D,N)                                              LIM03010
C---------------------------------------------------------------------LIM03020
C                                                                    LIM03030
      RETURN                                                         LIM03040
C-----END OF BSOLV FOR DIAGONAL TEST MATRIX ------------------------- LIM03050
      END                                                            LIM03060
C                                                                    LIM03070
C-----START OF USPEC FOR DIAGONAL TEST MATRIX----------------------- LIM03080
C                                                                    LIM03090
      SUBROUTINE USPEC(N,MATNO)                                      LIM03100
C     SUBROUTINE DUSPEC(N,MATNO)                                     LIM03110
C                                                                    LIM03120
C---------------------------------------------------------------------LIM03130
      DOUBLE PRECISION  D(1000), DI(1000), SHIFT, SPACE              LIM03140
      DOUBLE PRECISION  DABS, DFLOAT                                 LIM03150
      REAL  EXPLAN(20)                                               LIM03160
C---------------------------------------------------------------------LIM03170
C                                                                    LIM03180
      READ(7,10) EXPLAN                                              LIM03190
   10 FORMAT(20A4)                                                   LIM03200
      READ(7,*) NOLD,NUNIF,SPACE,D(1),SHIFT                          LIM03210
      NNUNIF = NOLD - NUNIF                                          LIM03220
      WRITE(6,20) NOLD,SPACE,NNUNIF,D(1),SHIFT                       LIM03230
   20 FORMAT(/' DIAGONAL TEST MATRIX, SIZE = ',I4/' IS THE INVERSE OF MALIM03240
     1TRIX WITH MOST ENTRIES',E10.3/' UNITS APART AND WITH ',I3,' ENTRIELIM03250
     IS IRREGULARLY SPACED'/' FIRST ENTRY WAS ',E13.4,' SHIFT = ',E10.3 LIM03260
     1/)                                                             LIM03270
C                                                                    LIM03280
      IF(N.NE.NOLD) GO TO 100                                        LIM03290
C     COMPUTE THE UNIFORM PORTION OF THE SPECTRUM                    LIM03300
```

```
      DO 30 J=2,NUNIF                                              LIM03310
   30 D(J) = D(1) - DFLOAT(J-1)*SPACE                              LIM03320
      NUNIF1=NUNIF + 1                                             LIM03330
      READ(7,10) EXPLAN                                            LIM03340
      DO 40 J=NUNIF1,N                                             LIM03350
   40 READ(7,*) D(J)                                               LIM03360
      NB = NUNIF - 2                                               LIM03370
C                                                                  LIM03380
      IF(SHIFT.EQ.0.) GO TO 60                                     LIM03390
      DO 50 J=1,N                                                  LIM03400
   50 D(J) = D(J) + SHIFT                                          LIM03410
C                                                                  LIM03420
C     COMPUTE EIGENVALUES OF INVERSE FOR PRINTOUT ONLY             LIM03430
   60 DO 70 J = 1,N                                                LIM03440
   70 DI(J) = 1.D0/D(J)                                            LIM03450
      WRITE(6,80) (DI(I), I=1,10 )                                 LIM03460
      WRITE(6,90) (DI(I), I = NB,N)                                LIM03470
   80 FORMAT(/' INVERSE LANCZOS TEST, LANCZS USES INVERSE OF GIVEN MATRILIM03480
     1X'/' 1ST 10 ENTRIES OF INVERSE OF DIAGONAL TEST MATRIX = '/(3E22.1LIM03490
     14))                                                          LIM03500
   90 FORMAT(/' MIDDLE (ORIGINALLY UNIFORM) PORTION OF MATRIX IS NOT PRILIM03510
     1NTED OUT'/' END OF (UNIFORM) PLUS NONUNIFORM SECTION = '/(3E25.16)LIM03520
     1)                                                            LIM03530
C                                                                  LIM03540
C     DIAGONAL GENERATION COMPLETE                                 LIM03550
C                                                                  LIM03560
C------------------------------------------------------------------LIM03570
C     PASS STORAGE LOCATIONS OF D AND N TO DSOLV SUBROUTINE        LIM03580
      CALL DSOLVE(D,N)                                             LIM0359C
C------------------------------------------------------------------LIM03600
C                                                                  LIM03610
      RETURN                                                       LIM03620
  100 WRITE(6,110) NOLD,N                                          LIM03630
  110 FORMAT(' PROGRAM TERMINATES BECAUSE NOLD = ',I5,'DOES NOT EQUAL N LIM03640
     1 =',I5)                                                      LIM03650
C-----END OF USPEC SUBROUTINE FOR DIAGONAL TEST MATRICES-----------LIM03660
      STOP                                                         LIM03670
      END                                                          LIM03680
```

SECTION 4.5   OPTIONAL PREPROCESSING PROGRAMS FOR CHAPTERS 4,5,9

```
C-----PERMUT-(USES SPARSPAK PACKAGE)-------------------------------------PER00010
C                                                                        PER00020
C     OPTIONAL PREPROCESSING PROGRAM FOR USE WITH LANCZOS CODES.         PER00030
C     GIVEN A REAL SYMMETRIC A-MATRIX IN SPARSE MATRIX FORMAT, PERMUT    PER00040
C     CALLS THE SPARSPAK PACKAGE (A. GEORGE, J. LIU, E. NG, U. WATERLOO)PER00050
C     TO DETERMINE A REORDERING OF A, THAT IS A PERMUTATION MATRIX       PER00060
C     P, SUCH THAT SPARSITY IS PRESERVED IN THE FACTORIZATION OF         PER00070
C     THE PERMUTED MATRIX.  PERMUT ALSO MODIFIES THE GIVEN A-MATRIX      PER00080
C     TO FORM THE MATRIX C = SO*A + SHIFT*I, WHERE SO AND SHIFT          PER00090
C     ARE SCALARS PROVIDED BY THE USER, AND THEN WRITES THIS            PER00100
C     C-MATRIX OUT TO FILE 9 ALONG WITH THE PERMUTATION P WHICH          PER00110
C     IS DEFINED BY THE VECTOR IPR.  IPR IS ALSO WRITTEN SEPARATELY      PER00120
C     TO FILE 14.                                                        PER00130
C                                                                        PER00140
C     NONPORTABLE CONSTRUCTIONS:                                         PER00150
C     1.   INTEGER*2 VARIABLE NPERM.  NOTE THAT THIS VARIABLE CANNOT     PER00160
C          BE CHANGED TO INTEGER*4.                                      PER00170
C     2.   FREE FORMAT (5,*) AND THE FORMAT (20A4).                      PER00180
C     3.   TO AVOID COMPOUNDING FORMAT CONVERSION ERRORS, THE MATRIX     PER00190
C          ENTRIES SHOULD BE STORED IN MACHINE FORMAT, ((4Z20) FOR       PER00200
C          IBM/3081)                                                     PER00210
C                                                                        PER00220
C----------------------------------------------------------------------PER00230
C     SYMMETRIC A-MATRIX IS READ FROM FILE 8.  MATRIX IS STORED          PER00240
C     IN FOLLOWING SPARSE FORMAT:                                        PER00250
C                                                                        PER00260
C     NZL = INDEX OF LAST COLUMN CONTAINING NONZEROS BELOW THE DIAGONAL.PER00270
C     NZS = NUMBER OF NONZERO SUBDIAGONAL ENTRIES                        PER00280
C     ICOL(K), K=1,NZL CONTAINS THE NUMBER OF NONZERO SUBDIAGONAL        PER00290
C              ENTRIES IN COLUMN K.                                      PER00300
C     IROW(K), K=1,NZS CONTAINS ROW INDEX OF KTH NONZERO SUBDIAGONAL     PER00310
C              ENTRY, ENTRIES STORED COLUMN BY COLUMN.                   PER00320
C     AD(K), K=1,N CONTAINS THE DIAGONAL ENTRIES OF A, INCLUDING ANY     PER00330
C            ZERO ENTRIES.                                               PER00340
C     ASD(K), K=1,NZS CONTAINS THE NONZERO SUBDIAGONAL ENTRIES OF A,     PER00350
C             COLUMN BY COLUMN.                                          PER00360
C                                                                        PER00370
C-----INPUT/OUTPUT FILES ------------------------------------------------PER00380
C                                                                        PER00390
C     INPUT FILES:                                                       PER00400
C     FILE 5      CONTAINS THE PROGRAM PARAMETERS SET BY USER            PER00410
C     FILE 8      CONTAINS THE SPARSE A-MATRIX                           PER00420
C                                                                        PER00430
C     OUTPUT FILES:                                                      PER00440
C     FILE 6      INTERACTIVE TERMINAL FILE                              PER00450
C     FILE 9      CONTAINS THE SPARSE DATA FOR C = SO*A + SHIFT*I.       PER00460
C     FILE 14     CONTAINS PERMUTATION IPR DEFINING THE REORDERING.      PER00470
C                 IN PARTICULAR J = IPR(I) MEANS ROW(COL) I OF           PER00480
C                 B = P*C*(P-TRANSPOSE) CORRESPONDS TO ROW(COL) J        PER00490
C                 OF THE A-MATRIX.                                       PER00500
C                                                                        PER00510
C-----SPARSPAK----------------------------------------------------------PER00520
C     ARRAYS AND PARAMETERS THAT ARE REQUIRED BY SPARSPAK.               PER00530
C     NOTE THAT THE CALL FOR SPARSPAK IS SPRSPK.  SUBROUTINES            PER00540
C     IJBEGN, INIJ, IJEND, ORDRB5, AND PSTATS ARE SPARSPAK              PER00550
C     SUBROUTINES.                                                       PER00560
C                                                                        PER00570
C     S = VECTOR WHOSE ACTUAL DIMENSION IS DETERMINED BY SPARSPAK        PER00580
C         WHEN THE REORDERING IS OBTAINED.  USER SPECIFIES MAXIMUM       PER00590
C         DIMENSION MAXS ALLOWED; SPARSPAK DEFAULTS IF THIS MAXIMUM      PER00600
C         IS EXCEEDED.  SPARSPAK IS DESIGNED FOR SOLVING SYSTEMS         PER00610
C         OF EQUATIONS, THUS THE VECTOR S IS DESIGNED TO CONTAIN         PER00620
C         THE SOLUTION VECTOR IF THERE IS ONE, FOLLOWED BY THE           PER00630
C         PERMUTATION VECTOR IPR, FOLLOWED BY OTHER INFORMATION          PER00640
```

```
C           GENERATED BY SPARSPAK.  A CORRECT SIZE FOR MAXS CAN BE        PER00650
C           DETERMINED ONLY AFTER THE FACT.  AS A FIRST GUESS ONE         PER00660
C           CAN SET MAXS = K*N WHERE K >= 10.                             PER00670
C                                                                         PER00680
C       MSGLVL = CONTROL FOR WRITES TO FILE 6                             PER00690
C       NEQNS = ORDER OF A, THIS IS COMPUTED BY SPARSPAK                  PER00700
C       IERR = CONTROLS WRITING OF ERROR MESSAGES BY SPARSPAK.           PER00710
C       MAXS = USER-SPECIFIED MAXIMUM ALLOWED DIMENSION OF S-ARRAY.       PER00720
C                                                                         PER00730
C                                                                         PER00740
C------------------------------------------------------------------------PER00750
      DOUBLE PRECISION AD(3000),ASD(10000),SO,SHIFT                       PER00760
      DOUBLE PRECISION S(30000),STEMP                                     PER00770
      REAL EXPLAN(20)                                                     PER00780
      INTEGER ICOL(3000),IROW(10000),IPR(3000)                            PER00790
      INTEGER*2 NPERM(4)                                                  PER00800
      COMMON /SPKUSR/ MSGLVL,IERR,MAXS,NEQNS                              PER00810
      EQUIVALENCE (STEMP,NPERM(1))                                        PER00820
C------------------------------------------------------------------------ PER00830
C                                                                         PER00840
C       ARRAYS MUST BE DIMENSIONED AS FOLLOWS:                            PER00850
C       1.  AD:  >= N, THE ORDER OF A-MATRIX.                             PER00860
C       2.  ASD: >= NZS, THE NUMBER OF NONZERO SUBDIAGONAL ENTRIES IN A.  PER00870
C       4.  ICOL:   >= N                                                  PER00880
C       5.  IROW:   >= NZS                                                PER00890
C       6.  IPR:    >= N + 4                                              PER00900
C       7.  S:   >= MAXS                                                  PER00910
C                                                                         PER00920
C------------------------------------------------------------------------ PER00930
C                                                                         PER00940
      WRITE(6,10)                                                         PER00950
   10 FORMAT(/' CALL SPARSPAK TO FIND REORDERING OF THE GIVEN MATRIX'/    PER00960
     1' THAT PRESERVES SPARSITY IN THE FACTORIZATION'/)                   PER00970
C                                                                         PER00980
C       READ IN USER-SPECIFIED PARAMETERS                                 PER00990
C       SPECIFY THE MAXIMUM DIMENSION (MAXS) ALLOWED FOR ARRAY S,         PER01000
C       AND WHETHER OR NOT A-MATRIX IS BEING SCALED BY SO OR SHIFTED      PER01010
C       ISCALE = 0, THEN NO SCALING OR SHIFTING                           PER01020
      READ (5,20) EXPLAN                                                  PER01030
      WRITE (6,20) EXPLAN                                                 PER01040
      READ (5,20) EXPLAN                                                  PER01050
   20 FORMAT(20A4)                                                        PER01060
      READ (5,*) MAXS,ISCALE,SO,SHIFT                                     PER01070
C                                                                         PER01080
C       READ IN INDICES FOR A-MATRIX STORED IN SPARSE FORMAT ON FILE 8.   PER01090
C       ONLY THE NONZERO STRUCTURE IS NEEDED TO OBTAIN THE PERMUTATION.   PER01100
C                                                                         PER01110
      READ(8,30) NZS,N,NZL,MATNO                                          PER01120
   30 FORMAT(I10,2I6,I8)                                                  PER01130
C                                                                         PER01140
      WRITE(6,40) NZS,N,NZL,MATNO,MAXS,ISCALE,SHIFT,SO                    PER01150
   40 FORMAT(/I10,2I6,I10,I10,' = NZS,N,NZL,MATNO,MAXS'/                  PER01160
     1 I6,2E12.5,' = ISCALE,SHIFT,SO'/)                                   PER01170
C                                                                         PER01180
      READ(8,50) (ICOL(K), K=1,NZL)                                       PER01190
      READ(8,50) (IROW(K), K=1,NZS)                                       PER01200
   50 FORMAT(13I6)                                                        PER01210
C                                                                         PER01220
C       DIAGONAL IS READ (INCLUDING ANY ZERO ENTRIES), THEN NONZERO       PER01230
C       BELOW DIAGONAL ENTRIES ARE READ IN                               PER01240
      READ(8,60) (AD(K), K=1,N)                                           PER01250
      READ(8,60) (ASD(K), K=1,NZS)                                        PER01260
   60 FORMAT(4E19.10)                                                     PER01270
C                                                                         PER01280
      IF (ISCALE.EQ.0) GO TO 90                                           PER01290
C                                                                         PER01300
C       CALCULATE C = SO*A + SHIFT*I AND PUT IN A-ARRAY                   PER01310
```

```
      DO 70 K = 1,N                                               PER01320
   70 AD(K) = SO*AD(K) + SHIFT                                     PER01330
      DO 80 K = 1,NZS                                             PER01340
   80 ASD(K) = SO*ASD(K)                                           PER01350
   90 CONTINUE                                                     PER01360
C                                                                 PER01370
C----------------------------------------------------------------PER01380
C     INPUT THE SPARSENESS STRUCTURE OF GIVEN A-MATRIX TO SPARSPAK PER01390
      CALL SPRSPK                                                 PER01400
C----------------------------------------------------------------PER01410
C                                                                 PER01420
      MSGLVL = 4                                                  PER01430
C                                                                 PER01440
C----------------------------------------------------------------PER01450
      CALL IJBEGN                                                 PER01460
C----------------------------------------------------------------PER01470
C                                                                 PER01480
      LLAST = 0                                                   PER01490
      DO 110 J = 1,NZL                                            PER01500
      IF (ICOL(J).EQ.0) GO TO 110                                 PER01510
      JJ = J                                                      PER01520
      LFIRST = LLAST + 1                                          PER01530
      LLAST = LLAST + ICOL(J)                                     PER01540
      DO 100 L = LFIRST,LLAST                                     PER01550
      II = IROW(L)                                                PER01560
C                                                                 PER01570
C----------------------------------------------------------------PER01580
      CALL INIJ(II,JJ,S)                                          PER01590
C----------------------------------------------------------------PER01600
C                                                                 PER01610
  100 CONTINUE                                                    PER01620
C                                                                 PER01630
  110 CONTINUE                                                    PER01640
C                                                                 PER01650
C     SPARSENESS STRUCTURE HAS BEEN INPUTED TO SPARSPAK.          PER01660
C                                                                 PER01670
C----------------------------------------------------------------PER01680
      CALL IJEND(S)                                               PER01690
C----------------------------------------------------------------PER01700
C                                                                 PER01710
      WRITE(6,120) N,NEQNS                                        PER01720
  120 FORMAT(/2I6,' = N,NEQNS'/)                                  PER01730
      IF (N.NE.NEQNS) GO TO 230                                   PER01740
C                                                                 PER01750
C----------------------------------------------------------------PER01760
C     USE SPARSPAK TO GENERATE REORDERING OF A THAT PRESERVES     PER01770
C     SPARSITY.  CORRESPONDING FACTORIZATION CAN BE COMPUTED BY   PER01780
C     PREPROCESSING PROGRAM LFACT WHEN C = SO*A + SHIFT*I IS POSITIVE PER01790
C     DEFINITE. BELOW CALLS THE MINIMUM DEGREE ALGORITHM PROVIDED PER01800
C     IN SPARSPAK.                                                PER01810
      CALL ORDRB5(S)                                              PER01820
      CALL PSTATS                                                 PER01830
C----------------------------------------------------------------PER01840
C                                                                 PER01850
C     EXTRACT THE REORDERING FROM SPARSPAK S VECTOR AND STORE IN FILE 14PER01860
      L = 1                                                       PER01870
      KNUM = N                                                    PER01880
      DO 130 K = 1,N                                              PER01890
      KNUM = KNUM + 1                                             PER01900
      STEMP = S(KNUM)                                             PER01910
      IPR(L) = NPERM(1)                                           PER01920
      IPR(L+1) = NPERM(2)                                         PER01930
      IPR(L+2) = NPERM(3)                                         PER01940
      IPR(L+3) = NPERM(4)                                         PER01950
      L = L+4                                                     PER01960
      IF (L.GT.N) GO TO 140                                       PER01970
  130 CONTINUE                                                    PER01980
```

```
     140 CONTINUE                                                        PER01990
C                                                                        PER02000
         WRITE(14,150) N,MATNO                                           PER02010
     150 FORMAT(I6,I8,' = N MATNO  K IPR(K)    A-MATRIX PERMUTATION')    PER02020
         WRITE(14,160) (K,IPR(K), K = 1,N)                              PER02030
     160 FORMAT(6(1X,2I6))                                              PER02040
C                                                                        PER02050
C                                                                        PER02060
C        WRITE C = SO*A + SHIFT*I WITH THE PERMUTATION IPR TO FILE 9.    PER02070
C                                                                        PER02080
         JPERM = 1                                                       PER02090
         WRITE(9,170) NZS,N,NZL,MATNO,JPERM                            PER02100
     170 FORMAT(I10,2I6,I8,I6,' = NZS,N,NZL,MATNO,JPERM. ACOMPAC')      PER02110
C                                                                        PER02120
C        NUMBER OF NONZERO SUBDIAGONAL ENTRIES IN EACH COLUMN IS WRITTEN PER02130
C        THEN THE CORRESPONDING ROW INDEX FOR EACH SUCH ENTRY IS WRITTEN PER02140
         WRITE(9,180) (ICOL(K), K=1,NZL)                               PER02150
         WRITE(9,180) (IROW(K), K=1,NZS)                               PER02160
     180 FORMAT(13I6)                                                   PER02170
C        DIAGONAL IS WRITTEN FIRST, THEN NONZERO BELOW DIAGONAL ENTRIES  PER02180
         WRITE(9,190) (AD(K), K=1,N)                                    PER02190
         WRITE(9,190) (ASD(K), K=1,NZS)                                 PER02200
     190 FORMAT(4E19.10)                                                PER02210
         WRITE(9,180) (IPR(K), K=1,N)                                   PER02220
C                                                                        PER02230
         IF(ISCALE.NE.0) GO TO 200                                      PER02240
C        ISCALE = 0, SET DEFAULT VALUES OF SO AND SHIFT                 PER02250
         SO = 1.D0                                                      PER02260
         SHIFT = 0.D0                                                   PER02270
     200 WRITE(9,210) SO,SHIFT                                          PER02280
     210 FORMAT(2E12.5,' = SO SHIFT'/                                   PER02290
        1 ' ABOVE IS SPARSE DATA FOLLOWED BY PERMUTATION IPR'/          PER02300
        1 ' FOR THE MATRIX C = SO*A+SHIFT*I '/                          PER02310
        1 ' B = P*C*PTRANS CAN BE GENERATED IN SUBROUTINE LORDER'/      PER02320
        1 ' ROW(COL) I OF B CORRESPONDS TO ROW(COL) J OF C, J = IPR(I)'/ PER02330
        1 ' NZS = TOTAL NUMBER OF SUBDIAGONAL NONZEROS IN C'/           PER02340
        1 ' KCOL(K) = NUMBER OF SUBDIAGONAL NONZEROS IN COL K OF C'/    PER02350
        1 ' KROW(K) = ROW INDEX OF SUBDIAGONAL NONZERO'/               PER02360
        1 ' SUBDIAGONAL NONZEROS IN C ARE STORED COLUMN BY COLUMN'/     PER02370
        1 ' AD(K) = THE KTH DIAGONAL ELEMENT OF C'/                     PER02380
        1 ' ASD(K) =  KTH SUBDIAGONAL NONZERO IN C'/)                   PER02390
C                                                                        PER02400
         WRITE(6,220)                                                    PER02410
     220 FORMAT(/' PERMUT IS FINISHED MATRIX IS ON FILE 9'/)            PER02420
C                                                                        PER02430
     230 CONTINUE                                                        PER02440
         STOP                                                            PER02450
C-----END PERMUT---------------------------------------------------------PER02460
         END                                                            PER02470
```

```
C-----LORDER-(STAND ALONE PROGRAM)-------------------------------------LOR00010
C                                                                       LOR00020
C      ACCORDING TO THE PFORT VERIFIER THIS PROGRAM IS PORTABLE.        LOR00030
C      HOWEVER TO AVOID COMPOUNDING FORMAT CONVERSION ERRORS,           LOR00040
C      MATRIX ENTRIES SHOULD BE STORED IN MACHINE FORMAT, ((4Z20)       LOR00050
C      FOR IBM/3081).                                                   LOR00060
C                                                                       LOR00070
C      LORDER TAKES A SPARSE MATRIX C AND A PERMUTATION P GIVEN BY       LOR00080
C      THE VECTOR IPR AND COMPUTES THE PERMUTED MATRIX B = P*C*P' ,     LOR00090
C      AND THEN WRITES B TO FILE 9 ALONG WITH IPR AND ANY SCALE SO      LOR00100
C      AND SHIFT THAT WERE USED TO OBTAIN THE INPUT MATRIX C. (HERE      LOR00110
C      ROW(COL) I OF B CORRESPONDS TO ROW(COL) J OF A WHERE J = IPR(I), LOR00120
C      AND INPUT MATRIX C = SO*A + SHIFT*I.                             LOR00130
C                                                                       LOR00140
C---------------------------------------------------------------------LOR00150
       DOUBLE PRECISION ASD(10000),AD(3000),BSD(10000),BD(3000)         LOR00160
       DOUBLE PRECISION SHIFT,SO                                        LOR00170
       INTEGER  IPR(3000),IPT(3000)                                     LOR00180
       INTEGER  IROW(10000),INUM(10000),ICOL(3000)                      LOR00190
       INTEGER  KROW(10000),KNUM(10000),KCOL(3000)                      LOR00200
C---------------------------------------------------------------------LOR00210
C                                                                       LOR00220
C      ARRAYS MUST BE DIMENSIONED AS FOLLOWS:                           LOR00230
C      1. AD, BD: >= N, THE ORDER OF C-MATRIX.                          LOR00240
C      2. ASD: >= NZS, THE NUMBER OF NONZERO SUBDIAGONAL ENTRIES IN C.  LOR00250
C      3. BSD:   >= NZS, THE NUMBER OF NONZERO SUBDIAGONAL ENTRIES IN   LOR00260
C                 B = P*C*P-TRANSPOSE                                   LOR00270
C      4. IPR, IPT:  >= N                                               LOR00280
C      5. ICOL, KCOL:  >= N                                             LOR00290
C      6. IROW, KROW, INUM, KNUM:  >= NZ = 2*NZS + N                    LOR00300
C                                                                       LOR00310
C---------------------------------------------------------------------LOR00320
C      OUTPUT HEADER                                                    LOR00330
       WRITE(6,10)                                                      LOR00340
   10 FORMAT(/' LORDER PROGRAM, COMPUTE B = P*C*(P-TRANSPOSE), STORE ON LOR00350
      1FILE 9'/)                                                        LOR00360
C                                                                       LOR00370
C      READ NUMBER OF NONZERO SUBDIAGONAL ENTRIES (NZS), ORDER OF MATRIXLOR00380
C      (N),INDEX OF LAST COLUMN CONTAINING NONZERO ENTRIES BELOW THE    LOR00390
C      DIAGONAL (NZL), MATRIX IDENTIFICATION NUMBER (MATNO), PERMUTATIONLOR00400
C      FLAG (JPERM).                                                    LOR00410
       READ(8,20) NZS,N,NZL,MATNO,JPERM                                 LOR00420
   20 FORMAT(I10,2I6,I8,I6)                                             LOR00430
C                                                                       LOR00440
       WRITE(6,30) NZS,N,NZL,MATNO,JPERM                                LOR00450
   30 FORMAT(/I10,2I6,I8,I3,' = NZS,N,NZL,MATNO,JPERM'/)                LOR00460
C                                                                       LOR00470
C      NUMBER OF NONZERO SUBDIAGONAL ENTRIES IN EACH COLUMN IS READ     LOR00480
C      THEN THE CORRESPONDING ROW INDEX FOR EACH SUCH ENTRY IS READ     LOR00490
       READ(8,40) (ICOL(K), K=1,NZL)                                    LOR00500
       READ(8,40) (IROW(K), K=1,NZS)                                    LOR00510
   40 FORMAT(13I6)                                                      LOR00520
C                                                                       LOR00530
       NZL1 = NZL + 1                                                   LOR00540
       DO 50 K = NZL1,N                                                 LOR00550
   50 ICOL(K) = 0                                                       LOR00560
C                                                                       LOR00570
C      DIAGONAL OF C-MATRIX IS READ (INCLUDING ANY ZERO ENTRIES), THEN  LOR00580
C      NONZERO SUBDIAGONAL ENTRIES ARE READ IN                          LOR00590
       READ(8,60) (AD(K), K=1,N)                                        LOR00600
       READ(8,60) (ASD(K), K=1,NZS)                                     LOR00610
   60 FORMAT(4E19.10)                                                   LOR00620
C                                                                       LOR00630
       IF(JPERM.EQ.0) GO TO 390                                         LOR00640
C      READ PERMUTATION                                                 LOR00650
```

```
      READ(8,40) (IPR(K), K = 1,N)                               LOR00660
C                                                                LOR00670
      DO 70 K = 1,N                                              LOR00680
      J = IPR(K)                                                 LOR00690
   70 IPT(J) = K                                                 LOR00700
C                                                                LOR00710
      READ(8,80) SO,SHIFT                                        LOR00720
   80 FORMAT(2E12.5)                                             LOR00730
C                                                                LOR00740
      WRITE(6,90)                                                LOR00750
   90 FORMAT(/' MATRIX HAS BEEN READ IN FROM FILE 8'/            LOR00760
     1 ' PERMUTATION IPR HAS BEEN READ IN'/)                     LOR00770
C                                                                LOR00780
C     EXPAND IROW AND ICOL TO INCLUDE DIAGONAL AND SUPER DIAGONAL LOR00790
      KCOL(1) = 1 + ICOL(1)                                      LOR00800
      KNUM(1) = -1                                               LOR00810
      KROW(1) = 1                                                LOR00820
      IF (ICOL(1).EQ.0) GO TO 110                                LOR00830
      KL = ICOL(1)                                               LOR00840
      DO 100 K = 1,KL                                            LOR00850
      KP1 = K+1                                                  LOR00860
      KROW(KP1) = IROW(K)                                        LOR00870
  100 KNUM(KP1) = K                                              LOR00880
  110 KCOUNT = KCOL(1)                                           LOR00890
C                                                                LOR00900
      DO 160 K = 2,N                                             LOR00910
      K1 = MIN(K-1,NZL)                                          LOR00920
      JL = 0                                                     LOR00930
      JCOUNT = 0                                                 LOR00940
      DO 140 J = 1,K1                                            LOR00950
      IF (ICOL(J).EQ.0) GO TO 140                                LOR00960
      JF = JL + 1                                                LOR00970
      JL = JL + ICOL(J)                                          LOR00980
      DO 130 JJ = JF,JL                                          LOR00990
      IF (IROW(JJ)-K) 130,120,140                                LOR01000
  120 KCOUNT = KCOUNT + 1                                        LOR01010
      JCOUNT = JCOUNT + 1                                        LOR01020
      KROW(KCOUNT) = J                                           LOR01030
      KNUM(KCOUNT) = JJ                                          LOR01040
      GO TO 140                                                  LOR01050
  130 CONTINUE                                                   LOR01060
  140 CONTINUE                                                   LOR01070
      KCOUNT = KCOUNT + 1                                        LOR01080
      KROW(KCOUNT) = K                                           LOR01090
      KNUM(KCOUNT) = -K                                          LOR01100
      ITEMP = 0                                                  LOR01110
      IF (K.LE.NZL) ITEMP = ICOL(K)                              LOR01120
      KCOL(K) = JCOUNT + 1 + ITEMP                               LOR01130
      IF (K.GT.NZL.OR.ICOL(K).EQ.0) GO TO 160                    LOR01140
      KF = 1 + KL                                                LOR01150
      KL = KL + ICOL(K)                                          LOR01160
      DO 150 J = KF,KL                                           LOR01170
      KCOUNT = KCOUNT + 1                                        LOR01180
      KROW(KCOUNT) = IROW(J)                                     LOR01190
  150 KNUM(KCOUNT) = J                                           LOR01200
  160 CONTINUE                                                   LOR01210
C     NTOTAL = N + 2*NZS                                         LOR01220
C     A-MATRIX INDEX LISTS HAVE BEEN EXPANDED                    LOR01230
C                                                                LOR01240
      WRITE(6,170)                                               LOR01250
  170 FORMAT(/' EXPANSION OF INDEX LISTS FOR C-MATRIX IS COMPLETED'/) LOR01260
C                                                                LOR01270
C     DETERMINE STRUCTURE OF B = P*C*P-TRANSPOSE                 LOR01280
      IL = 0                                                     LOR01290
      KCOUNT = 0                                                 LOR01300
      DO 180 K = 1,N                                             LOR01310
  180 ICOL(K) = 0                                                LOR01320
```

```
      DO 270 K = 1,N                                            LOR01330
      J = IPR(K)                                                LOR01340
      JL = 0                                                    LOR01350
      IF (J.EQ.1) GO TO 200                                     LOR01360
      JM1 = J - 1                                               LOR01370
      DO 190 JJ = 1,JM1                                         LOR01380
  190 JL = JL + KCOL(JJ)                                        LOR01390
  200 CONTINUE                                                  LOR01400
      JF = JL + 1                                               LOR01410
      JL = JL + KCOL(J)                                         LOR01420
      ICOL(K) = KCOL(J)                                         LOR01430
      IF = IL + 1                                               LOR01440
      IL = IL + ICOL(K)                                         LOR01450
C                                                               LOR01460
      DO 210 JJ = JF,JL                                         LOR01470
      KCOUNT = KCOUNT + 1                                       LOR01480
      JR = KROW(JJ)                                             LOR01490
      JK = IPT(JR)                                              LOR01500
      INUM(KCOUNT) = KNUM(JJ)                                   LOR01510
  210 IROW(KCOUNT) = JK                                         LOR01520
C                                                               LOR01530
C     ORDER IROW VECTOR BY INCREASING SIZE                     LOR01540
      IF (IF.EQ.IL) GO TO 240                                   LOR01550
      IF1 = IF + 1                                              LOR01560
      DO 230 I = IF1,IL                                         LOR01570
      IM1 = I-1                                                 LOR01580
      IMF = IM1 + IF                                            LOR01590
      DO 220 L = IF,IM1                                         LOR01600
      II = IMF - L                                              LOR01610
      IF (IROW(II+1).GE.IROW(II)) GO TO 230                     LOR01620
      IO = IROW(II)                                             LOR01630
      IROW(II) = IROW(II+1)                                     LOR01640
      IROW(II+1) = IO                                           LOR01650
      IO = INUM(II)                                             LOR01660
      INUM(II) = INUM(II+1)                                     LOR01670
      INUM(II+1) = IO                                           LOR01680
  220 CONTINUE                                                  LOR01690
  230 CONTINUE                                                  LOR01700
  240 CONTINUE                                                  LOR01710
C                                                               LOR01720
      DO 250 I = IF,IL                                          LOR01730
      IF (INUM(I).LT.0) GO TO 260                               LOR01740
  250 CONTINUE                                                  LOR01750
  260 INUM(I) = -J                                              LOR01760
  270 CONTINUE                                                  LOR01770
C                                                               LOR01780
C     GENERATE SPARSE MATRIX REPRESENTATION OF B-MATRIX        LOR01790
      KCOUNT = 0                                                LOR01800
      DO 280 K = 1,N                                            LOR01810
  280 KCOL(K) = 0                                               LOR01820
      DO 320 K = 1,N                                            LOR01830
      KL = 0                                                    LOR01840
      DO 290 KK = 1,K                                           LOR01850
  290 KL = KL + ICOL(KK)                                        LOR01860
      KK = KL+ 1                                                LOR01870
  300 KK = KK - 1                                               LOR01880
      IF (INUM(KK).GE.0) GO TO 300                              LOR01890
      KCOL(K) = KL - KK                                         LOR01900
      J = IPR(K)                                                LOR01910
      BD(K) = AD(J)                                             LOR01920
      KF = KK + 1                                               LOR01930
      IF (KCOL(K).EQ.0) GO TO 320                               LOR01940
      DO 310 JJ = KF,KL                                         LOR01950
      KCOUNT = KCOUNT + 1                                       LOR01960
      KROW(KCOUNT) = IROW(JJ)                                   LOR01970
      KK = INUM(JJ)                                             LOR01980
  310 BSD(KCOUNT) = ASD(KK)                                     LOR01990
```

```
  320 CONTINUE                                                       LORO2000
      NZL = 0                                                        LORO2010
      DO 330 K = 1,N                                                 LORO2020
      IF (KCOL(K).NE.0) NZL = K                                      LORO2030
  330 CONTINUE                                                       LORO2040
C     WE NOW HAVE B = P*A*P-TRANSPOSE IN SPARSE MATRIX FORMAT, WRITE TO LORO2050
C     FILE 9                                                         LORO2060
C                                                                    LORO2070
      JPERM = 1                                                      LORO2080
      WRITE(9,340) NZS,N,NZL,MATNO,JPERM                             LORO2090
  340 FORMAT(I10,2I6,I8,I6,' = NZS,N,NZL,MATNO,JPERM. BCOMPAC')      LORO2100
C                                                                    LORO2110
C     NUMBER OF NONZERO SUBDIAGONAL ENTRIES IN EACH COLUMN IS WRITTEN LORO2120
C     THEN THE CORRESPONDING ROW INDEX FOR EACH SUCH ENTRY IS WRITTEN LORO2130
      WRITE(9,350) (KCOL(K), K=1,NZL)                                LORO2140
      WRITE(9,350) (KROW(K), K=1,NZS)                                LORO2150
  350 FORMAT(13I6)                                                   LORO2160
C     DIAGONAL IS WRITTEN FIRST, THEN NONZERO BELOW DIAGONAL ENTRIES  LORO2170
      WRITE(9,360) (BD(K), K=1,N)                                    LORO2180
      WRITE(9,360) (BSD(K), K=1,NZS)                                 LORO2190
  360 FORMAT(4E19.10)                                                LORO2200
C                                                                    LORO2210
C     WRITE PERMUTATION                                              LORO2220
      WRITE(9,350) (IPR(K), K=1,N)                                   LORO2230
C                                                                    LORO2240
      WRITE(9,370) SO,SHIFT                                          LORO2250
  370 FORMAT(2E12.5,' = SO SHIFT'/                                   LORO2260
     1 ' ABOVE IS REORDERED MATRIX, B'/                             LORO2270
     1 ' INPUT MATRIX SUPPLIED WAS C = SO*A + SHIFT*I'/             LORO2280
     1 ' B = P*C*(P-TRANSPOSE),  B IS STORED IN SPARSE MATRIX FORMAT'/ LORO2290
     1 ' ROW(COL) I OF B CORRESPONDS TO ROW(COL) J OF C, J = IPR(I)'/ LORO2300
     1 ' NZS = TOTAL NUMBER OF SUBDIAGONAL NONZEROS IN B-MATRIX'/   LORO2310
     1 ' KCOL(K) = NUMBER OF SUBDIAGONAL NONZEROS IN COL K OF B'/   LORO2320
     1 ' KROW(K) = ROW INDEX OF SUBDIAGONAL NONZERO'/               LORO2330
     1 ' SUBDIAGONAL NONZEROS IN B ARE STORED COLUMN BY COLUMN'/    LORO2340
     1 ' BD(K) = THE KTH DIAGONAL ELEMENT OF B'/                    LORO2350
     1 ' BSD(K) = NUMERICAL VALUE OF KTH SUBDIAGONAL NONZERO IN B'/ LORO2360
     1 ' IPR(K) = J MEANS THAT ROW J OF C CORRESPONDS TO ROW K OF B'/) LORO2370
C                                                                    LORO2380
      WRITE(6,380)                                                   LORO2390
  380 FORMAT(' SPARSE FORMAT FOR B-MATRIX HAS BEEN WRITTEN TO FILE 9'/) LORO2400
      GO TO 410                                                      LORO2410
C                                                                    LORO2420
  390 WRITE(6,400)                                                   LORO2430
  400 FORMAT(/' LORDER PROGRAM TERMINATES BECAUSE MATRIX FILE SUPPLIED DLORO2440
     1ID NOT'/' CONTAIN A PERMUTATION'/)                            LORO2450
C                                                                    LORO2460
  410 CONTINUE                                                       LORO2470
C                                                                    LORO2480
      STOP                                                           LORO2490
C-----END OF LORDER---------------------------------------------------- LORO2500
      END                                                            LORO2510
```

```
C-----LFACT--------------------------------------------------------------- LFA00010
C                                                                           LFA00020
C     NONPORTABLE CONSTRUCTIONS:                                            LFA00030
C     1.  FORMAT (4Z20).  TO AVOID COMPOUNDING FORMAT CONVERSION            LFA00040
C         ERRORS, THE MATRIX ENTRIES SHOULD BE IN MACHINE FORMAT,           LFA00050
C         (4Z20) FOR IBM/3081.                                              LFA00060
C                                                                           LFA00070
C     LFACT COMPUTES THE CHOLESKY FACTOR L FOR THE MATRIX B AND STORES      LFA00080
C     THIS FACTOR ON FILE 7.  B MUST BE A POSITIVE DEFINITE MATRIX.         LFA00090
C     THE PERMUTATION P (IN IPR), THE SCALE SO AND THE SHIFT (IF ANY)       LFA00100
C     USED TO OBTAIN B FROM THE ORIGINAL MATRIX A ARE STORED AT THE END     LFA00110
C     OF FILE 7.   THAT IS,  B = SO*P*A*P' + SHIFT*I.  THE PROGRAM          LFA00120
C     ASSUMES THAT THE DATA READ FROM FILE 9 IS FOR THE B-MATRIX.           LFA00130
C                                                                           LFA00140
C------------------------------------------------------------------------LFA00150
C                                                                           LFA00160
C     ARRAYS MUST BE DIMENSIONED AS FOLLOWS:                                LFA00170
C     1.  AD:   >= N, THE ORDER OF A-MATRIX.                                LFA00180
C     3.  ASD:  >= NZT, THE NUMBER OF NONZERO SUBDIAGONAL ENTRIES           LFA00190
C               IN THE CHOLESKY FACTOR OF B.                                LFA00200
C     4.  ICOL,IPR:  >= N                                                   LFA00210
C     5.  IROW:  >=  NZT                                                    LFA00220
C                                                                           LFA00230
C------------------------------------------------------------------------LFA00240
      DOUBLE PRECISION ASD(10000),AD(3000)                                 LFA00250
      DOUBLE PRECISION ZERO,ONE,TEMP,SO,SHIFT                              LFA00260
      INTEGER IROW(10000),ICOL(3000),IPR(3000)                            LFA00270
      DOUBLE PRECISION DSQRT                                               LFA00280
C------------------------------------------------------------------------LFA0C290
C     OUTPUT HEADER                                                         LFA00300
      WRITE(6,5)                                                            LFA00310
    5 FORMAT(/' LFACT PROGRAM, COMPUTE CHOLESKY FACTOR FOR POSITIVED DEFLFA00320
     1INITE B-MATRIX'/' AND STORE THE FACTOR ON FILE 7'/)                  LFA00330
C                                                                           LFA00340
C     SET PROGRAM PARAMETERS                                                LFA00350
      ONE = 1.0D0                                                          LFA00360
      ZERO = 0.0D0                                                         LFA00370
C                                                                           LFA00380
C     READ NUMBER OF NONZERO BELOW DIAGONAL ENTRIES, ORDER OF MATRIX,      LFA00390
C     INDEX OF LAST COLUMN CONTAINING NONZERO ENTRIES BELOW THE            LFA00400
C     DIAGONAL, MATRIX IDENTIFICATION NUMBER                                LFA00410
      READ(9,15) NZS,N,NZL,MATNO,JPERM                                     LFA00420
   15 FORMAT(I10,2I6,I8,I6)                                                LFA00430
C                                                                           LFA00440
      WRITE(6,20) NZS,N,NZL,JPERM,MATNO                                    LFA00450
   20 FORMAT(I10,3I6,I8,' = NZS,N,NZL,JPERM,MATNO'/)                       LFA00460
C                                                                           LFA00470
C     NUMBER OF NONZERO SUBDIAGONAL ENTRIES IN EACH COLUMN IS READ         LFA00480
C     THEN THE CORRESPONDING ROW INDEX FOR EACH SUCH ENTRY IS READ         LFA00490
      READ(9,30) (ICOL(K), K=1,NZL)                                        LFA00500
      READ(9,30) (IROW(K), K=1,NZS)                                        LFA00510
   30 FORMAT(13I6)                                                         LFA00520
C                                                                           LFA00530
C                                                                           LFA00540
      NZL1 = NZL + 1                                                       LFA00550
      DO 40 K = NZL1,N                                                     LFA00560
   40 ICOL(K) = 0                                                          LFA00570
C                                                                           LFA00580
C     DIAGONAL IS READ (INCLUDING ANY ZERO ENTRIES), THEN NONZERO          LFA00590
C     BELOW DIAGONAL ENTRIES ARE READ IN                                   LFA00600
      READ(9,50) (AD(K), K=1,N)                                            LFA00610
      READ(9,50) (ASD(K), K=1,NZS)                                         LFA00620
   50 FORMAT(4E19.10)                                                      LFA00630
C  50 FORMAT(4Z20)                                                         LFA00640
C                                                                           LFA00650
```

```
      IF (JPERM.NE.0) READ(9,30) (IPR(K), K = 1,N)                LFA00660
C                                                                  LFA00670
      READ(9,55) SO,SHIFT                                          LFA00680
   55 FORMAT(2E12.5)                                               LFA00690
C                                                                  LFA00700
      WRITE(6,60)                                                  LFA00710
   60 FORMAT(/' B-MATRIX HAS BEEN READ IN FROM FILE 9'/)           LFA00720
C                                                                  LFA00730
      IF (JPERM.NE.0) WRITE(6,65)                                  LFA00740
   65 FORMAT( ' PERMUTATION IPR HAS BEEN READ IN'/)                LFA00750
C                                                                  LFA00760
C     CALCULATE CHOLESKY FACTOR,  B = BL*(BL-TRANSPOSE)            LFA00770
      NZT = NZS                                                    LFA00780
      NZL = N-1                                                    LFA00790
      KL = 0                                                       LFA00800
      DO 70 K = 1,N                                                LFA00810
C     CALCULATE KTH PIVOT FOR BL                                   LFA00820
      TEMP = AD(K)                                                 LFA00830
C                                                                  LFA00840
      IF (AD(K).GT.ZERO) GO TO 80                                  LFA00850
C                                                                  LFA00860
      WRITE(6,90) K,AD(K)                                          LFA00870
   90 FORMAT(/I6,E15.8,' = K,AD(K)'/                               LFA00880
     1' PIVOT IS NEGATIVE SO B-MATRIX IS NOT POSITIVE DEFINITE'/   LFA00890
     1' THEREFORE COMPUTATION OF CHOLESKY FACTOR TERMINATES'/)     LFA00900
      GO TO 240                                                    LFA00910
C                                                                  LFA00920
   80 CONTINUE                                                     LFA00930
      TEMP = DSQRT(TEMP)                                           LFA00940
      AD(K) = TEMP                                                 LFA00950
      TEMP = ONE/TEMP                                              LFA00960
      IF(K.EQ.N.OR.ICOL(K).EQ.0) GO TO 70                          LFA00970
      KF = KL + 1                                                  LFA00980
      KL = KL + ICOL(K)                                            LFA00990
      DO 100 KK = KF,KL                                            LFA01000
      KR = IROW(KK)                                                LFA01010
      ASD(KK) = TEMP*ASD(KK)                                       LFA01020
  100 AD(KR) = AD(KR) - ASD(KK)**2                                 LFA01030
      IF (KF.EQ.KL) GO TO 70                                       LFA01040
      K1 = K+1                                                     LFA01050
      DO 110 KK = KF,KL                                            LFA01060
      KR = IROW(KK)                                                LFA01070
      IF (KK.EQ.KL) GO TO 110                                      LFA01080
      KE = KL                                                      LFA01090
      DO 120 KC = K1,KR                                            LFA01100
  120 KE= KE + ICOL(KC)                                            LFA01110
      KB = KE - ICOL(KR) + 1                                       LFA01120
      KK1 = KK + 1                                                 LFA01130
      L = KB                                                       LFA01140
      DO 130 LL = KK1,KL                                           LFA01150
      LR = IROW(LL)                                                LFA01160
      IF (ICOL(KR).EQ.0.OR.L.GT.KE) GO TO 140                      LFA01170
  150 LC = IROW(L)                                                 LFA01180
      IF (LC - LR) 160,170,140                                     LFA01190
  160 L = L + 1                                                    LFA01200
      IF (L.LE.KE) GO TO 150                                       LFA01210
C     NEW NONZERO IN CHOLESKY FACTOR L                             LFA01220
  140 NZT = NZT + 1                                                LFA01230
      L1 = L + 1                                                   LFA01240
      NT = NZT + L1                                                LFA01250
      DO 180 KM = L1,NZT                                           LFA01260
      MK = NT - KM                                                 LFA01270
      ASD(MK) = ASD(MK-1)                                          LFA01280
  180 IROW(MK) = IROW(MK-1)                                        LFA01290
      ICOL(KR) = ICOL(KR) + 1                                      LFA01300
      KE = KE + 1                                                  LFA01310
      ASD(L) = -ASD(KK)*ASD(LL)                                    LFA01320
```

```
      IROW(L) = LR                                              LFA01330
      GO TO 130                                                 LFA01340
C     UPDATE EXISTING ELEMENT                                   LFA01350
  170 ASD(L) = ASD(L) - ASD(KK)*ASD(LL)                         LFA01360
  130 L = L + 1                                                 LFA01370
  110 CONTINUE                                                  LFA01380
   70 CONTINUE                                                  LFA01390
C                                                               LFA01400
C                                                               LFA01410
C     FACTOR L HAS BEEN COMPUTED, STORE IN SPARSE FORMAT ON FILE 7   LFA01420
C                                                               LFA01430
      WRITE(7,190) NZT,N,NZL,MATNO,JPERM                        LFA01440
  190 FORMAT(I10,2I6,I8,I6,' = NZT,N,NZL,MATNO,JPERM. LCOMPAC') LFA01450
C                                                               LFA01460
C     NUMBER OF NONZERO SUBDIAGONAL ENTRIES IN EACH COLUMN IS WRITTEN   LFA01470
C     THEN THE CORRESPONDING ROW INDEX FOR EACH SUCH ENTRY IS WRITTEN   LFA01480
      WRITE(7,200) (ICOL(K), K=1,NZL)                           LFA01490
      WRITE(7,200) (IROW(K), K=1,NZT)                           LFA01500
  200 FORMAT(13I6)                                              LFA01510
C     DIAGONAL IS WRITTEN FIRST, THEN NONZERO BELOW DIAGONAL ENTRIES   LFA01520
      WRITE(7,210) (AD(K), K=1,N)                               LFA01530
      WRITE(7,210) (ASD(K), K=1,NZT)                            LFA01540
  210 FORMAT(4Z20)                                              LFA01550
C 210 FORMAT(3E25.16)                                           LFA01560
      IF (JPERM.NE.0) WRITE(7,200) (IPR(K), K=1,N)              LFA01570
C                                                               LFA01580
      WRITE(7,220) SO,SHIFT                                     LFA01590
  220 FORMAT(2E12.5,' = SO SHIFT'/                              LFA01600
     1 ' ABOVE IS CHOLESKY FACTOR FOR B-MATRIX'/                LFA01610
     1 ' IF JPERM = 0, THEN P = I.  C = SO*A * SHIFT*I'/        LFA01620
     1 ' B = P*C*P-TRANS = L*L-TRANS, L IS STORED IN SPARSE FORMAT'/   LFA01630
     1 ' ROW(COL) I OF B CORRESPONDS TO ROW(COL) J OF C, J = IPR(I)'/  LFA01640
     1 ' NZT = TOTAL NUMBER OF SUBDIAGONAL NONZEROS IN L'/      LFA01650
     1 ' ICOL(K) = NUMBER OF SUBDIAGONAL NONZEROS IN COL K OF L'/   LFA01660
     1 ' IROW(K) = ROW INDEX OF SUBDIAGONAL NONZERO'/           LFA01670
     1 ' SUBDIAGONAL NONZEROS IN L ARE STORED COLUMN BY COLUMN'/   LFA01680
     1 ' AD(K) = KTH DIAGONAL ELEMENT OF L'/                    LFA01690
     1 ' ASD(K) = KTH SUBDIAGONAL NONZERO IN L'/)               LFA01700
C                                                               LFA01710
      WRITE(6,230)                                              LFA01720
  230 FORMAT(' CHOLESKY FACTOR HAS BEEN WRITTEN TO FILE 7 '/)   LFA01730
C                                                               LFA01740
  240 CONTINUE                                                  LFA01750
      STOP                                                      LFA01760
C-----END OF IFACT---------------------------------------------   LFA01770
      END                                                       LFA01780
```

```
C-----LTEST---------------------------------------------------------- LTE00010
C                                                                       LTE00020
C      CONTAINS MAIN PROGRAM LTEST AND SAMPLE CMATS, CMATV, BSOLV       LTE00030
C      LTEST ALSO REQUIRES A RANDOM NUMBER GENERATOR.                   LTE00040
C                                                                       LTE00050
C      LTEST GIVES A ROUGH CHECK ON THE CONDITION OF A MATRIX B BY      LTE00060
C      SOLVING B*X = B*V1 FOR X WHERE V1 IS A KNOWN, RANDOMLY-GENERATED LTE00070
C      VECTOR.  SOLVING IS DONE, WITH AND WITHOUT ITERATIVE REFINEMENT. LTE00080
C      IN BOTH CASES, X IS COMPARED WITH V1 AND THE ERRORS ARE          LTE00090
C      WRITTEN TO FILE 6.                                               LTE00100
C                                                                       LTE00110
C      VECTORS V0, V1, V2, VS, AND G ARE USED IN THE COMPUTATIONS.      LTE00120
C      NOTE THAT THE SUBROUTINE CMATS USED TO COMPUTE THE RESIDUAL      LTE00130
C      IN EXTENDED PRECISION FOR THE ITERATIVE REFINEMENT CALCULATION   LTE00140
C      REQUIRES AN EXTRA LONG V0 VECTOR OF LENGTH TWICE THE SIZE OF B.  LTE00150
C                                                                       LTE00160
C      NONPORTABLE CONSTRUCTIONS:                                       LTE00170
C      1.  THE ENTRY MECHANISM WHICH PASSES THE STORAGE LOCATIONS OF    LTE00180
C          ARRAYS AND PARAMETERS THAT DEFINE THE B-MATRIX TO THE        LTE00190
C          SUBROUTINES CMATV, CMATS, AND BSOLV.                         LTE00200
C      2.  FORMATS (20A4) AND (4Z20).  TO AVOID COMPOUNDING FORMAT      LTE00210
C          CONVERSION ERRORS, MATRIX ENTRIES SHOULD BE STORED IN        LTE00220
C          MACHINE FORMAT, ((4Z20) FOR IBM/3081). ALSO FREE FORMAT      LTE00230
C          (5,*).                                                       LTE00240
C      3.  REAL*16 VARIABLES IN CMATS SUBROUTINE.                       LTE00250
C                                                                       LTE00260
C                                                                       LTE00270
C--------------------------------------------------------------------LTE00280
C                                                                       LTE00290
      DOUBLE PRECISION ASD(10000),AD(3000),BSD(20000),BD(3000)          LTE00290
      DOUBLE PRECISION V0(6000),V1(3000),V2(3000),VS(3000)              LTE00300
      DOUBLE PRECISION ZERO,ONE,TEMP,SUM                                LTE00310
      DOUBLE PRECISION ERROR0,ERROR1,ENORM0,ENORM1                      LTE00320
      REAL EXPLAN(20),G(3000)                                           LTE00330
      INTEGER IROW(20000),ICOL(3000),KROW(30000),KCOL(3000),SVSEED      LTE00340
      DOUBLE PRECISION FINPRO                                           LTE00350
      DOUBLE PRECISION DABS, DMAX1, DSQRT                               LTE00360
C--------------------------------------------------------------------LTE00370
C                                                                       LTE00380
C      ARRAYS MUST BE DIMENSIONED AS FOLLOWS:                           LTE00390
C      1.  AD, BD: >= N, THE ORDER OF A-MATRIX.                         LTE00400
C      2.  ASD:  >= NZS, THE NUMBER OF NONZERO SUBDIAGONAL ENTRIES IN B.LTE00410
C      3.  BSD:  >= NZT, THE NUMBER OF NONZERO SUBDIAGONAL ENTRIES      LTE00420
C          IN THE CHOLESKY FACTOR OF B.                                 LTE00430
C      5.  ICOL, KCOL:  >= N                                            LTE00440
C      6.  KROW:  >= NZS                                                LTE00450
C      7.  IROW:  >= NZT                                                LTE00460
C      8.  V1,V2,VS:  >=  N                                             LTE00470
C      9.  V0:  >=  2*N                                                 LTE00480
C                                                                       LTE00490
C--------------------------------------------------------------------LTE00500
C      OUTPUT HEADER                                                    LTE00510
       WRITE(6,10)                                                      LTE00520
   10 FORMAT(/' LTEST PROGRAM, ROUGH CHECK ON NUMERICAL CONDITION OF GIVLTE00530
     1EN MATRIX'/)                                                      LTE00540
C                                                                       LTE00550
C      SET PROGRAM PARAMETERS                                           LTE00560
       ONE = 1.0D0                                                      LTE00570
       ZERO = 0.0D0                                                     LTE00580
C                                                                       LTE00590
C      READ INPUT HEADER                                                LTE00600
       READ(5,20) EXPLAN                                                LTE00610
       WRITE(6,20) EXPLAN                                               LTE00620
   20 FORMAT(20A4)                                                      LTE00630
C                                                                       LTE00640
C      READ IN IN FREE FORMAT USER-SPECIFIED PARAMETERS FROM FILE 5     LTE00650
       READ(5,20) EXPLAN                                                LTE00660
```

```
       READ(5,*) SVSEED                                                 LTE00670
C                                                                       LTE00680
C      READ NUMBER OF NONZERO BELOW DIAGONAL ENTRIES, ORDER OF MATRIX,  LTE00690
C      INDEX OF LAST COLUMN CONTAINING NONZERO ENTRIES BELOW THE        LTE00700
C      DIAGONAL, MATRIX IDENTIFICATION NUMBER                           LTE00710
       READ(9,30) NZS,N,NZL,MATNO,JPERM                                 LTE00720
    30 FORMAT(110,216,18,16)                                            LTE00730
C                                                                       LTE00740
       WRITE(6,40) NZS,N,NZL,JPERM,MATNO,SVSEED                         LTE00750
    40 FORMAT(110,316,' = NZS,N,NZL,JPERM'/                             LTE00760
      1 18,112,' = MATNO,SVSEED'/)                                      LTE00770
C                                                                       LTE00780
C      NUMBER OF NONZERO SUBDIAGONAL ENTRIES IN EACH COLUMN IS READ     LTE00790
C      THEN THE CORRESPONDING ROW INDEX FOR EACH SUCH ENTRY IS READ     LTE00800
       READ(9,50) (KCOL(K), K=1,NZL)                                    LTE00810
       READ(9,50) (KROW(K), K=1,NZS)                                    LTE00820
    50 FORMAT(1316)                                                     LTE00830
C                                                                       LTE00840
C                                                                       LTE00850
       NZL1 = NZL + 1                                                   LTE00860
       DO 60 K = NZL1,N                                                 LTE00870
    60 KCOL(K) = 0                                                      LTE00880
C                                                                       LTE00890
C      DIAGONAL IS READ (INCLUDING ANY ZERO ENTRIES), THEN NONZERO      LTE00900
C      BELOW DIAGONAL ENTRIES ARE READ IN                              LTE00910
       READ(9,70) (AD(K), K=1,N)                                        LTE00920
       READ(9,70) (ASD(K), K=1,NZS)                                     LTE00930
    70 FORMAT(4E19.10)                                                  LTE00940
C                                                                       LTE00950
       WRITE(6,80)                                                      LTE00960
    80 FORMAT(/' B-MATRIX HAS BEEN READ IN FROM FILE 9'/)               LTE00970
C                                                                       LTE00980
C----------------------------------------------------------------------LTE00990
C      ENTRIES TO CMATS AND CMATV SUBROUTINES                           LTE01000
       CALL CMATSE(ASD,AD,KCOL,KROW,N,NZL)                              LTE01010
       CALL CMATVE(ASD,AD,KCOL,KROW,N,NZL)                              LTE01020
C----------------------------------------------------------------------LTE01030
C                                                                       LTE01040
C      READ CHOLESKY FACTOR FROM FILE 7                                 LTE01050
C                                                                       LTE01060
       READ(7,90) NZT,N,NZL,MATNO,JPERM                                 LTE01070
    90 FORMAT(110,216,18,16)                                            LTE01080
C                                                                       LTE01090
C      NUMBER OF NONZERO SUBDIAGONAL ENTRIES IN EACH COLUMN IS READ     LTE01100
C      THEN THE CORRESPONDING ROW INDEX FOR EACH SUCH ENTRY IS READ     LTE01110
       READ(7,100) (ICOL(K), K=1,NZL)                                   LTE01120
       READ(7,100) (IROW(K), K=1,NZT)                                   LTE01130
   100 FORMAT(1316)                                                     LTE01140
C      DIAGONAL IS READ FIRST, THEN NONZERO BELOW DIAGONAL ENTRIES      LTE01150
       READ(7,110) (BD(K), K=1,N)                                       LTE01160
       READ(7,110) (BSD(K), K=1,NZT)                                    LTE01170
   110 FORMAT(4Z20)                                                     LTE01180
C   90 FORMAT(3E25.16)                                                  LTE01190
C                                                                       LTE01200
C----------------------------------------------------------------------LTE01210
C      ENTRY TO BSOLV SUBROUTINE, PASS FACTOR OF B                      LTE01220
       CALL BSOLVE(BSD,BD,ICOL,IROW,N,NZT,NZL)                          LTE01230
C----------------------------------------------------------------------LTE01240
C                                                                       LTE01250
C      SOLVE B*X = B*V1 WITH AND WITHOUT ITERATIVE REFINEMENT, COMPARE  LTE01260
C      ERRORS IN SOLVING AS A ROUGH CHECK ON THE CONDITION OF THE       LTE01270
C      MATRIX B.                                                        LTE01280
C                                                                       LTE01290
       IIX = SVSEED                                                     LTE01300
C                                                                       LTE01310
C----------------------------------------------------------------------LTE01320
C      COMPUTES RANDOM VECTOR FOR USE IN RIGHT-HAND SIDE                LTE01330
```

```
      CALL GENRAN(IIX,G,N)                                           LTE01340
C-----------------------------------------------------------------LTE01350
C                                                                   LTE01360
      DO 120 K = 1,N                                                LTE01370
  120 V1(K) = G(K)                                                  LTE01380
C                                                                   LTE01390
C-----------------------------------------------------------------LTE01400
      SUM = FINPRO(N,V1(1),1,V1(1),1)                              LTE01410
C-----------------------------------------------------------------LTE01420
      SUM = ONE/DSQRT(SUM)                                          LTE01430
C                                                                   LTE01440
      DO 130 K = 1,N                                                LTE01450
  130 V1(K) = V1(K)*SUM                                             LTE01460
C                                                                   LTE01470
      SUM = ZERO                                                    LTE01480
C                                                                   LTE01490
C-----------------------------------------------------------------  LTE01500
C     COMPUTE V2 = RHS = B*V1   C = SO*A + SHIFT*I   B = P*C*P'     LTE01510
C     VS = B(INVERSE)*V2                                            LTE01520
      CALL CMATV(V1,V2,SUM)                                         LTE01530
      CALL BSOLV(VS,V2)                                             LTE01540
C-----------------------------------------------------------------  LTE01550
C                                                                   LTE01560
      SUM = ZERO                                                    LTE01570
      ERRORO = ZERO                                                 LTE01580
      DO 140 K = 1,N                                                LTE01590
      TEMP = DABS(V1(K) - VS(K))                                    LTE01600
      SUM = SUM + TEMP*TEMP                                         LTE01610
  140 ERRORO = DMAX1(ERRORO,TEMP)                                   LTE01620
      ENORMO = DSQRT(SUM)                                           LTE01630
C                                                                   LTE01640
      WRITE(6,150) ENORMO,ERRORO                                    LTE01650
  150 FORMAT(6X,'ENORMO',6X,'ERRORO'/2E12.4/                        LTE01660
     1 ' ENORMO  = NORM (V1 - VS),      VS = BI*(B*V1)'/            LTE01670
     1 ' ERRORO = MAX DABS(V1(K) - VS(K)),  K = 1,N'/)             LTE01680
C                                                                   LTE01690
      SUM = ONE                                                     LTE01700
C                                                                   LTE01710
C-----------------------------------------------------------------  LTE01720
C     CALCULATE RESIDUAL IN EXTENDED PRECISION V2 = B*VS - V2       LTE01730
C     THEN DO ITERATIVE REFINEMENT                                  LTE01740
      CALL CMATS(VS,V2,VO,SUM)                                      L-E01750
      CALL BSOLV(V2,V2)                                             LTE01760
C-----------------------------------------------------------------  LTE01770
C                                                                   LTE01780
      DO 160 K = 1,N                                                LTE01790
  160 VS(K) = VS(K) - V2(K)                                         LTE01800
C                                                                   LTE01810
      SUM = ZERO                                                    LTE01820
      ERROR1 = ZERO                                                 LTE01830
      DO 170 K = 1,N                                                LTE01840
      TEMP = DABS(V1(K) - VS(K))                                    LTE01850
      SUM = SUM + TEMP*TEMP                                         LTE01860
  170 ERROR1 = DMAX1(ERROR1,TEMP)                                   LTE01870
      ENORM1 = DSQRT(SUM)                                           LTE01880
C                                                                   LTE01890
      WRITE(6,180) ENORM1,ERROR1                                    LTE01900
  180 FORMAT(6X,'ENORM1',6X,'ERROR1'/2E12.4/                        LTE01910
     1 ' ERROR AFTER ITERATIVE REFINEMENT'/                         LTE01920
     1 ' ENORM1  = NORM (V1 - VS),      VS = BI*(B*V1)'/            LTE01930
     1 ' ERROR1 = MAX DABS(V1(K) - VS(K)),  K = 1,N'/)             LTE01940
C                                                                   LTE01950
      STOP                                                          LTE01960
C-----END OF LTEST-----------------------------------------------  LTE01970
      END                                                           LTE01980
C                                                                   LTE01990
C----CMATS-------------------------------------------------------LTE02000
```

```
C                                                                      LTE02010
C     REAL, SYMMETRIC, SPARSE MATRIX-VECTOR MULTIPLY USING EXTENDED    LTE02020
C     PRECISION.  CALCULATES U = B*W - SUM*U FOR USE IN ITERATIVE      LTE02030
C     REFINEMENT.  MATRIX B STORED IN SPARSE FORMAT.                   LTE02040
C                                                                      LTE02050
      SUBROUTINE CMATS(W,U,Z,SUM)                                      LTE02060
C                                                                      LTE02070
C------------------------------------------------------------------------LTE02080
      DOUBLE PRECISION  U(1),W(1),BSD(1),BD(1),SUM                     LTE02090
      REAL*16 Z(1),T0,T1,T2,S0                                         LTE02100
      INTEGER  IROW(1),ICOL(1)                                         LTE02110
C------------------------------------------------------------------------LTE02120
      S0 = SUM                                                         LTE02130
C                                                                      LTE02140
      DO 10 I = 1,N                                                    LTE02150
      T0 = BD(I)                                                       LTE02160
      T1 = W(I)                                                        LTE02170
      T2 = U(I)                                                        LTE02180
   10 Z(I) = T0*T1-S0*T2                                               LTE02190
C                                                                      LTE02200
      LLAST = 0                                                        LTE02210
C                                                                      LTE02220
      DO 30 J = 1,NZL                                                  LTE02230
C                                                                      LTE02240
      IF (ICOL(J).EQ.0) GO TO 30                                       LTE02250
C                                                                      LTE02260
      LFIRST = LLAST + 1                                               LTE02270
      LLAST = LLAST + ICOL(J)                                          LTE02280
C                                                                      LTE02290
      DO 20 L = LFIRST,LLAST                                           LTE02300
      I = IROW(L)                                                      LTE02310
      T0 = BSD(L)                                                      LTE02320
      T1 = W(J)                                                        LTE02330
      T2 = W(I)                                                        LTE02340
C                                                                      LTE02350
      Z(I) = Z(I) + T0*T1                                              LTE02360
      Z(J) = Z(J) + T0*T2                                              LTE02370
C                                                                      LTE02380
   20 CONTINUE                                                         LTE02390
C                                                                      LTE02400
   30 CONTINUE                                                         LTE02410
C                                                                      LTE02420
      DO 40 I =1,N                                                     LTE02430
   40 U(I) = Z(I)                                                      LTE02440
C                                                                      LTE02450
      RETURN                                                           LTE02460
C                                                                      LTE02470
C------------------------------------------------------------------------LTE02480
      ENTRY CMATSE(BSD,BD,ICOL,IROW,N,NZL)                            LTE02490
C------------------------------------------------------------------------LTE02500
C                                                                      LTE02510
      RETURN                                                           LTE02520
C-----END OF CMATS-------------------------------------------------------LTE02530
      END                                                              LTE02540
C                                                                      LTE02550
C-----CMATV-------------------------------------------------------------LTE02560
C                                                                      LTE02570
C     SYMMETRIC, SPARSE MATRIX-VECTOR MULTIPLY, B MATRIX STORED        LTE02580
C     IN SPARSE FORMAT.  CMATV CALCULATES U = B*W - SUM*U              LTE02590
C                                                                      LTE02600
      SUBROUTINE CMATV(W,U,SUM)                                        LTE02610
C                                                                      LTE02620
C------------------------------------------------------------------------LTE02630
      DOUBLE PRECISION  U(1),W(1),BSD(i),BD(1),SUM                     LTE02640
      INTEGER  KROW(1),KCOL(1)                                         LTE02650
C------------------------------------------------------------------------LTE02660
C                                                                      LTE02670
```

```
      DO 10 I = 1,N                                              LTE02680
   10 U(I) = BD(I)*W(I) - SUM*U(I)                               LTE02690
C                                                                LTE02700
      LLAST = 0                                                  LTE02710
C                                                                LTE02720
      DO 30 J = 1,NZL                                            LTE02730
C                                                                LTE02740
      IF (KCOL(J).EQ.0) GO TO 30                                 LTE02750
C                                                                LTE02760
      LFIRST = LLAST + 1                                         LTE02770
      LLAST = LLAST + KCOL(J)                                    LTE02780
C                                                                LTE02790
      DO 20 L = LFIRST,LLAST                                     LTE02800
      I = KROW(L)                                                LTE02810
C                                                                LTE02820
      U(I) = U(I) + BSD(L)*W(J)                                  LTE02830
      U(J) = U(J) + BSD(L)*W(I)                                  LTE02840
C                                                                LTE02850
   20 CONTINUE                                                   LTE02860
C                                                                LTE02870
   30 CONTINUE                                                   LTE02880
C                                                                LTE02890
      RETURN                                                     LTE02900
C                                                                LTE02910
C-----------------------------------------------------------------LTE02920
      ENTRY CMATVE(BSD,BD,KCOL,KROW,N,NZL)                       LTE02930
C-----------------------------------------------------------------LTE02940
C                                                                LTE02950
      RETURN                                                     LTE02960
C-----END OF CMATV-----------------------------------------------LTE02970
      END                                                        LTE02980
C                                                                LTE02990
C-----BSOLV------------------------------------------------- LTE03000
C                                                                LTE03010
C     SOLVES B*U = V WHERE  B = L*L'.                            LTE03020
C     FIRST SOLVES L*U = V FOR U, THEN SOLVES L'*U = U FOR U     LTE03030
C                                                                LTE03040
      SUBROUTINE BSOLV(U,V)                                      LTE03050
C                                                                LTE03060
C-----------------------------------------------------------------LTE03070
      DOUBLE PRECISION   AD(1),ASD(1),U(1),V(1),TEMP             LTE03080
      INTEGER  ICOL(1),IROW(1)                                   LTE03090
C-----------------------------------------------------------------LTE03100
      KL = 0                                                     LTE03110
      DO 10 K = 1,N                                              LTE03120
   10 U(K) = V(K)                                                LTE03130
      DO 30 K = 1,N                                              LTE03140
      TEMP = U(K)/AD(K)                                          LTE03150
      U(K) = TEMP                                                LTE03160
      IF (ICOL(K).EQ.0.OR.K.EQ.N) GO TO 30                       LTE03170
      KF = KL + 1                                                LTE03180
      KL = KL + ICOL(K)                                          LTE03190
      DO 20 KK = KF,KL                                           LTE03200
      KR = IROW(KK)                                              LTE03210
   20 U(KR) = U(KR) - TEMP*ASD(KK)                               LTE03220
   30 CONTINUE                                                   LTE03230
C                                                                LTE03240
      NP1 = N+1                                                  LTE03250
      KF = NZT + 1                                               LTE03260
      DO 50 K = 1,N                                              LTE03270
      L = NP1 - K                                                LTE03280
      TEMP = U(L)                                                LTE03290
      IF (ICOL(L).EQ.0.OR.L.EQ.N) GO TO 50                       LTE03300
      KL = KF - 1                                                LTE03310
      KF = KF - ICOL(L)                                          LTE03320
      DO 40 LL = KF,KL                                           LTE03330
      LR = IROW(LL)                                              LTE03340
```

```
   40 TEMP = TEMP - ASD(LL)*U(LR)                                    LTE03350
   50 U(L) = TEMP/AD(L)                                              LTE03360
C                                                                    LTE03370
      RETURN                                                         LTE03380
C                                                                    LTE03390
C-------------------------------------------------------------------LTE03400
      ENTRY BSOLVE(ASD,AD,ICOL,IROW,N,NZT,NZL)                       LTE03410
C-------------------------------------------------------------------LTE03420
C                                                                    LTE03430
      RETURN                                                         LTE03440
C-----END OF BSOLV--------------------------------------------------LTE03450
      END                                                            LTE03460
```

SECTION 4.6   FILE DEFINITIONS AND SAMPLE INPUT FILE

Below is a listing of the input/output files which are accessed by the
real symmetric Lanczos eigenvalue program, LIVAL. Included also is a
sample of the input file which LIVAL requires on file 5. The parameters
in this file are supplied in free format. LIVAL computes eigenvalues of
real symmetric matrices BI on user-specified intervals where BI is
the inverse of the matrix $B = P*C*P'$ with $C = SO*A + SHIFT*I$ and
SO is a scale and SHIFT is a shift. The sample codes assume that
C is positive definite and has a reasonable condition number. The
permutation matrix P Is used to preserve the sparseness of the
given matrix in the Cholesky factorization, $B = L*L'$. The user
could replace tne BSOLVE subroutine provided here by another more
general factorization subroutine.

SAMPLE DEFINITIONS OF THE INPUT/OUTPUT FILES FOR LIVAL

```
*LIVAL EXEC LANCZOS EIGENVALUE CALCULATION USING FACTORIZATION
FI 06 TERM
FILEDEF  1 DISK &1        NHISTORY  A (RECFM F LRECL 80 BLOCK 80
FILEDEF  2 DISK &1        HISTORY   A (RECFM F LRECL 80 BLOCK 80
FILEDEF  3 DISK &1        GOODEV    A (RECFM F LRECL 80 BLOCK 80
FILEDEF  4 DISK &1        ERRINV    A (RECFM F LRECL 80 BLOCK 80
FILEDEF  5 DISK LIVAL     INPUT     A (RECFM F LRECL 80 BLOCK 80
FILEDEF  7 DISK &1        LCOMPAC   A (RECFM F LRECL 80 BLOCK 80
FILEDEF 11 DISK &1        DISTINCT  A (RECFM F LRECL 80 BLOCK 80
LOAD   LIVAL   LESUB   LIMULT
```

SAMPLE INPUT FILE FOR LIVAL

```
 LIVAL EIGENVALUE COMPUTATION, NO REORTHOGONALIZATION
 USING INVERSE OF REAL SYMMETRIC MATRIX VIA FACTORIZATION
LINE 1     N      KMAX    NMEVS    MATNO     SO      SHIFT
          528    2640       2     721830    1.0      0.
LINE 2   SVSEED    RHSEED      MXINIT    MXSTUR
        49302312  5731029         5     100000
LINE 3   ISTART     ISTOP
           0          1
LINE 4    IHIS     IDIST   IWRITE
           1         0        1
LINE 5   RELTOL (RELATIVE TOLERANCE IN 'COMBINING' GOODEV)
        .0000000001
LINE 6   MB(1)   MB(2)   MB(3)   MB(4)    (ORDERS OF T(1,MEV) )
          100     125
LINE 7    NINT      (NUMBER OF SUB-INTERVALS FOR BISEC)
           1
LINE 8   LB(1)   LB(2)   LB(3)   LB(4)  (INTERVAL LOWER BOUNDS)
          1.0
LINE 9   UB(1)   UB(2)   UB(3)   UB(4)  (INTERVAL UPPER BOUNDS)
          100.0
```

Below is a listing of the input/output files which are accessed by the
real symmetric Lanczos eigenvector program, LIVEC. Included also is a
sample of the input file which LIVEC requires on file 5. The parameters
in this file are supplied in free format. LIVEC computes eigenvectors
for each of a user-specified subset of the eigenvalues computed by the
companion program LIVAL. The matrix used in the eigenvector computation
is a scaled, shifted and inverted version of a given matrix. Inversion
is accomplished via matrix factorization.

SAMPLE DEFINITIONS OF THE INPUT/OUTPUT FILES FOR LIVEC
----
*LIVEC EXEC, CALCULATE EIGENVECTORS FOR INVERSE OF REAL SYMMETRIC MATRIX
FI 06 TERM
FILEDEF 2 DISK &1   HISTORY A (RECFM F LRECL 80 BLOCK 80
FILEDEF 3 DISK &1   GOODEV   A (RECFM F LRECL 80 BLOCK 80
FILEDEF 4 DISK &1   ERRINV   A (RECFM F LRECL 80 BLOCK 80
FILEDEF 5 DISK LIVEC   INPUT   A (RECFM F LRECL 80 BLOCK 80
FILEDEF 7 DISK &1   LCOMPAC A (RECFM F LRECL 80 BLOCK 80
FILEDEF 9 DISK &1   ERREST   A (RECFM F LRECL 80 BLOCK 80
FILEDEF 10 DISK &1   BOUNDS   A (RECFM F LRECL 80 BLOCK 80
FILEDEF 11 DISK &1   TEIGVECS A (RECFM F LRECL 80 BLOCK 80
FILEDEF 12 DISK &1   RITZVECS A (RECFM F LRECL 80 BLOCK 80
FILEDEF 13 DISK &1   PAIGE   A (RECFM F LRECL 80 BLOCK 80
LOAD LIVEC LESUB LIMULT

SAMPLE INPUT FILE FOR LIVEC

 LIVEC INPUT LANCZOS EIGENVECTOR COMPUTATIONS, NO REORTHOGONALIZATION
 LINE 1 MATNO   N   SO   SHIFT JPERM (ID,SIZE,SCALE,SHIFT,PERMUT?
          20   2161   -1.0   0.01   0
 LINE 2 MDIMTV   MDIMRV MBETA (MAX.DIMENSIONS,TVEC,RITVEC AND BETA
          10000   10000   2000
 LINE 3   RELTOL
         .0000000001
 LINE 4 MBOUND   NTVCON SVTVEC IREAD (FLAGS
          0   1   0   1
 LINE 5 TVSTOP   LVCONT ERCONT IWRITE (FLAGS
          0   1   1   1
 LINE 6   RHSEED (RANDOM GENERATOR SEED FOR STARTING VECTOR IN INVERM)
         45329517

CHAPTER 5

REAL SYMMETRIC GENERALIZED PROBLEMS

SECTION 5.1   INTRODUCTION

The FORTRAN codes in this Chapter address the question of computing distinct eigenvalues and corresponding eigenvectors of a real symmetric generalized eigenvalue problem. Given two real symmetric matrices A and B, where B is positive definite and its Cholesky factors are available, these codes compute real scalars λ and corresponding real-valued vectors $x \neq 0$ such that

$$Ax = \lambda Bx . \tag{5.1.1}$$

Given a real symmetric positive definite matrix B, the Cholesky decomposition of B has the form

$$B = LL^T \text{ where L is a lower triangular matrix.} \tag{5.1.2}$$

Real symmetric matrices and Cholesky factorizations are discussed in detail in Stewart [1973]. See Section 2.1 for a brief summary of the properties of real symmetric matrices which we use.

Theoretically, this type of real symmetric generalized problem is equivalent to the following real symmetric problem:

$$L^{-1}AL^{-T}y = \lambda y, \text{ where } y = L^T x . \tag{5.1.3}$$

Therefore, we could solve this type of generalized problem by applying the real symmetric Lanczos procedure given in Chapter 2 directly to the composite matrix $C \equiv L^{-1}AL^{-T}$ given in Eqn(5.1.3). However, we prefer to work directly with the generalized problem. In this setting the role of the B-matrix in the single-vector Lanczos computations is clearly displayed.

The single-vector Lanczos codes in this chapter can be used to compute either a very few or very many of the distinct eigenvalues of the given real symmetric generalized problem. The documentation for these codes is contained in Section 2.2. As in the real symmetric case, the AB-multiplicity of a given computed 'good' Lanczos eigenvalue can be obtained only with additional computation, and the modifications required to do this additional computation are not included in the enclosed versions of these codes.

We use the following 'generalized' Lanczos recursion. For $i = 1,2,...,m$ and a randomly-generated starting vector v_1 with $\| v_1 \|_B = 1$, generate Lanczos vectors v_i using the following recursion.

$$\beta_{i+1}Bv_{i+1} = Av_i - \alpha_i Bv_i - \beta_i Bv_{i-1} \tag{5.1.4}$$

228

where

$$\alpha_i \equiv v_i^T(Av_i - \beta_i Bv_{i-1}) \text{ and } \beta_{i+1} \equiv \| L^{-1}(Av_i - \alpha_i Bv_i - \beta_i Bv_{i-1}) \| \qquad (5.1.5)$$

By construction, the B-norm of each Lanczos vector is one. That is, for all i, $\| v_i \|_B \equiv [v_i^T Bv_i]^{1/2} = 1$.

The B-norm is used because it is the 'natural' norm for real symmetric generalized problems when the B-matrix is positive definite. Given any two distinct eigenvalues λ and μ of Eqn(5.1.1), and corresponding eigenvectors x and y, we have that $x^T By = 0$. That is, the eigenvectors are orthogonal w.r.t. the B-norm, and the eigenvectors form a complete set of vectors. The positive definiteness of B is essential. The closer B is to being singular or indefinite, the less stable these computations will be. The generalized Lanczos recursion in Eqns(5.1.4) and (5.1.5) generates a family of real symmetric tridiagonal matrices (T-matrices) whose sizes are specified by the user.

LGVAL, the main program for the real symmetric generalized computations, calls the subroutine BISEC to compute eigenvalues of the specified tridiagonal T-matrices on the user-specified intervals. BISEC simultaneously computes these T-eigenvalues with their T-multiplicities and sorts the computed T-eigenvalues into two classes, the 'good' T-eigenvalues and the 'spurious' T-eigenvalues. The 'good' T-eigenvalues are accepted as approximations to eigenvalues of the generalized problem. The accuracy of these 'good' T-eigenvalues as eigenvalues of the generalized problem is then estimated using error estimates computed by the subroutine INVERR. Error estimates are computed only for isolated 'good' T-eigenvalues. All other 'good' T-eigenvalues are assumed to have converged. Convergence is then checked. If convergence has not yet occurred and a larger T-matrix has been specified by the user, the program will continue on to the larger T-matrix, repeating the above procedure on this larger matrix. After each T-matrix eigenvalue computation the corresponding approximations to the eigenvalues of the user-specified matrix A are computed and included in the output.

Once the eigenvalues have been computed accurately enough, the user can select a subset of the 'converged' eigenvalues for which eigenvectors are to be computed. The main program LGVEC, for computing eigenvectors of the real symmetric generalized problem using a factorization of B, is used to compute the desired eigenvectors.

All of the computations are done in double precision arithmetic. Once the Lanczos matrices have been computed, the remaining computations use the same subroutines which are used in the real symmetric case discussed in Chapter 2. In addition to the programs and subroutines provided here, the user must supply a subroutine USPECA which defines and initializes the A-matrix and a subroutine USPECB which defines and initializes the factors of the B-matrix. A subroutine AMATV which computes matrix-vector multiplies Ax for the A-matrix, and a subroutine BSOLV which solves the system of equations Bz=v must also be supplied. These subroutines must be constructed in such a way as to take advantage of the sparsity (and/or structure) of the two user-supplied matrices A and B and such that they are accurate.

The optional preprocessing programs PERMUT, LORDER, LFACT, and LTEST listed in Chapter 4 can also be used with the codes in this chapter. PERMUT calls the SPARSPAK Library [1979] to attempt to identify a reordering or permutation P of the given matrix B for which the sparseness of B is preserved under the factorization of the permuted matrix. LORDER takes a given matrix C and permutation P and computes the sparse format for the permuted matrix, PCP^T. LFACT computes the Cholesky factors of a given positive definite matrix. LTEST performs a very crude check on the numerical condition of the matrix supplied to it, by solving a system of equations with and without iterative refinement, LINPACK [1979]. Obviously, if the B-matrix is permuted then the A-matrix must be subjected to the same permutation. These codes assume that the Cholesky factor supplied in the subroutine USPECB corresponds to the permuted B-matrix and that the AMATV subroutine supplied corresponds to the corresponding permuted A-matrix. Thus, the Lanczos codes compute the eigenvalues and eigenvectors of the permuted problem. The permutation (if any) is then unwrapped in the eigenvector program LGVEC.

SECTION 5.2   MAIN PROGRAM, EIGENVALUE CALCULATIONS

```
C-----LGVAL  (EIGENVALUES, GENERALIZED SYMMETRIC PROBLEM)---------------LGV00010
C                                                                        LGV00020
C       CONTAINS MAIN PROGRAM FOR COMPUTING DISTINCT EIGENVALUES OF      LGV00030
C       A*X = EVAL*B*X WHERE A AND B ARE REAL SYMMETRIC MATRICES,        LGV00040
C       B IS POSITIVE DEFINITE, AND THE CHOLESKY FACTORS OF B            LGV00050
C       ARE AVAILABLE FOR USE IN THE PROCEDURE.  PROCEDURE USES          LGV00060
C       GENERALIZATION OF LANCZOS TRIDIAGONALIZATION WITHOUT ANY         LGV00070
C       REORTHGONALIZATION.                                              LGV00080
C                                                                        LGV00090
C       PFORT VERIFIER IDENTIFIED THE FOLLOWING NONPORTABLE              LGV00100
C       CONSTRUCTIONS                                                    LGV00110
C                                                                        LGV00120
C       1.  DATA/MACHEP/ STATEMENT                                       LGV00130
C       2.  ALL READ(5,*) STATEMENTS (FREE FORMAT)                       LGV00140
C       3.  FORMAT(20A4) USED WITH EXPLANATORY HEADER EXPLAN.            LGV00150
C       4.  HEXADECIMAL FORMAT (4Z20) USED IN ALPHA/BETA FILES 1 AND 2.  LGV00160
C                                                                        LGV00170
C----------------------------------------------------------------------LGV00180
C                                                                        LGV00190
        DOUBLE PRECISION  ALPHA(5000),BETA(5001)                         LGV00200
        DOUBLE PRECISION  V1(5000),V2(5000),VS(5000)                     LGV00210
        DOUBLE PRECISION  LB(20),UB(20)                                  LGV00220
        DOUBLE PRECISION  BTOL,GAPTOL,TTOL,MACHEP,EPSM,RELTOL            LGV00230
        DOUBLE PRECISION  SCALE1,SCALE2,SCALE3,SCALE4,BISTOL,CONTOL,MULTOLLGV00240
        DOUBLE PRECISION  ONE,ZERO,TEMP,TKMAX,BETAM,BKMIN,TO,T1          LGV00250
        REAL  G(5000),EXPLAN(20)                                         LGV00260
        INTEGER  MP(5000),NMEV(20)                                       LGV00270
        INTEGER  SVSEED,RHSEED,SVSOLD                                    LGV00280
        INTEGER  IABS                                                    LGV00290
        REAL  ABS                                                        LGV00300
        DOUBLE PRECISION  DABS, DSQRT, DFLOAT                            LGV00310
        EXTERNAL LSOLV, AMATV                                            LGV00320
C                                                                        LGV00330
C----------------------------------------------------------------------LGV00340
        DATA MACHEP/Z3410000000000000/                                   LGV00350
        EPSM = 2.0D0*MACHEP                                              LGV00360
C----------------------------------------------------------------------LGV00370
C                                                                        LGV00380
C       ARRAYS MUST BE DIMENSIONED AS FOLLOWS:                           LGV00390
C       DIMENSION OF V2 ASSUMES THAT NO MORE THAN KMAX/2 EIGENVALUES     LGV00400
C       OF THE LANCZOS T-MATRICES ARE BEING COMPUTED IN ANY ONE OF THE   LGV00410
C       SUB-INTERVALS BEING CONSIDERED.  V2 CONTAINS THE UPPER AND LOWER LGV00420
C       BOUNDS FOR EACH T-EIGENVALUE BEING COMPUTED BY BISEC IN ANY ONE  LGV00430
C       GIVEN INTERVAL.                                                  LGV00440
C                                                                        LGV00450
C       1.  ALPHA: >= KMAX,    BETA: >= (KMAX+1)                         LGV00460
C       2.  V1:    >= MAX(N,KMAX+1)                                      LGV00470
C       3.  V2,VS:   >= MAX(N,KMAX)                                      LGV00480
C       4.  G:   >= MAX(N,2*KMAX)                                        LGV00490
C       5.  MP:    >= KMAX                                               LGV00500
C       6.  LB,UB:  >= NUMBER OF SUBINTERVALS SUPPLIED TO BISEC.         LGV00510
C       7.  NMEV:  >= NUMBER OF T-MATRICES ALLOWED.                      LGV00520
C       8.  EXPLAN:  DIMENSION IS 20.                                    LGV00530
C                                                                        LGV00540
C                                                                        LGV00550
C       IMPORTANT TOLERANCES OR SCALES THAT ARE USED REPEATEDLY          LGV00560
C       THROUGHOUT THE PROGRAM ARE THE FOLLOWING:                        LGV00570
C       SCALED MACHINE EPSILON:  TTOL = TKMAX*EPSM WHERE                 LGV00580
C       EPSM = 2*MACHINE EPSILON AND                                     LGV00590
C       TKMAX = MAX(|ALPHA(J)|,BETA(J), J = 1,MEV)                       LGV00600
C       BISEC CONVERGENCE TOLERANCE:  BISTOL = DSQRT(1000+MEV)*TTOL      LGV00610
C       BISEC T-MULTIPLICITY TOLERANCE:  MULTOL = (1000+MEV)*TTOL        LGV00620
C       LANCZOS CONVERGENCE TOLERANCE:  CONTOL = BETA(MEV+1)*1.D-10      LGV00630
```

```
C-------------------------------------------------------------------LGV00640
C     OUTPUT HEADER                                                  LGV00650
      WRiTE(6,10)                                                    LGV00660
   10 FORMAT(/' LANCZOS EIGENVALUE PROCEDURE FOR REAL SYMMETRIC GENERALILGV00670
     1ZED PROBLEMS,'/' A*X = EVAL*B*X, B POSITIVE DEFINITE WITH CHOLESKYLGV00680
     1 FACTORS AVAILABLE'/)                                          LGV00690
C                                                                    LGV00700
C     SET PROGRAM PARAMETERS                                         LGV00710
C     SCALEK ARE USED IN TOLERANCES NEEDED IN SUBROUTINES LUMP,      LGV00720
C     ISOEV AND PRTEST.  USER MUST NOT MODIFY THEM.                  LGV00730
      SCALE1 = 5.0D2                                                 LGV00740
      SCALE2 = 5.0D0                                                 LGV00750
      SCALE3 = 5.0D0                                                 LGV00760
      SCALE4 = 1.0D4                                                 LGV00770
      ONE  = 1.0D0                                                   LGV00780
      ZERO = 0.0D0                                                   LGV00790
      BTOL = 1.0D-8                                                  LGV00800
C     BTOL = EPSM                                                    LGV00810
      GAPTOL = 1.0D-8                                                LGV00820
      ICONV = 0                                                      LGV00830
      MOLD = 0                                                       LGV00840
      MOLD1 = 1                                                      LGV00850
      ICT = 0                                                        LGV00860
      MMB = 0                                                        LGV00870
      IPROJ = 0                                                      LGV00880
C-------------------------------------------------------------------LGV00890
C     READ USER-SPECIFIED PARAMETERS FROM INPUT FILE 5 (FREE FORMAT) LGV00900
C                                                                    LGV00910
C     READ USER-PROVIDED HEADER FOR RUN                              LGV00920
      READ(5,20) EXPLAN                                              LGV00930
      WRITE(6,20) EXPLAN                                             LGV00940
      READ(5,20) EXPLAN                                              LGV00950
      WRITE(6,20) EXPLAN                                             LGV00960
   20 FORMAT(20A4)                                                   LGV00970
C                                                                    LGV00980
C     READ ORDER OF MATRICES (N) , MAXIMUM ORDER OF T-MATRIX (KMAX), LGV00990
C     NUMBER OF T-MATRICES ALLOWED (NMEVS), AND MATRIX IDENTIFICATION LGV01000
C     NUMBERS (MATNOA AND MATNOB)                                    LGV01010
      READ(5,20) EXPLAN                                              LGV01020
      READ(5,*) N,KMAX,NMEVS,MATNOA,MATNOB                           LGV01030
C                                                                    LGV01040
C     READ SEEDS FOR LANCZS AND INVERR SUBROUTINES (SVSEED AND RHSEED) LGV01050
C     READ MAXIMUM NUMBER OF ITERATIONS ALLOWED FOR EACH INVERSE     LGV01060
C     ITERATION (MXINIT) AND MAXIMUM NUMBER OF STURM SEQUENCES       LGV01070
C     ALLOWED (MXSTUR)                                               LGV01080
      READ(5,20) EXPLAN                                              LGV01090
      READ(5,*) SVSEED,RHSEED,MXINIT,MXSTUR                          LGV01100
C                                                                    LGV01110
C     ISTART = (0,1):  ISTART = 0 MEANS ALPHA/BETA FILE IS NOT       LGV01120
C     AVAILABLE.  ISTART = 1 MEANS ALPHA/BETA FILE IS AVAILABLE ON   LGV01130
C     FILE 2.                                                        LGV01140
C     ISTOP = (0,1):  ISTOP = 0 MEANS PROCEDURE GENERATES ALPHA/BETA LGV01150
C     FILE AND THEN TERMINATES.  ISTOP = 1 MEANS PROCEDURE GENERATES LGV01160
C     ALPHAS/BETAS IF NEEDED AND THEN COMPUTES EIGENVALUES AND ERROR LGV01170
C     ESTIMATES AND THEN TERMINATES.                                 LGV01180
      READ(5,20) EXPLAN                                              LGV01190
      READ(5,*) ISTART,ISTOP                                         LGV01200
C                                                                    LGV01210
C     IHIS = (0,1):  IHIS = 0 MEANS ALPHA/BETA FILE IS NOT WRITTEN   LGV01220
C     TO FILE 1.  IHIS = 1 MEANS ALPHA/BETA FILE IS WRITTEN TO FILE 1. LGV01230
C     IDIST = (0,1):  IDIST = 0 MEANS DISTINCT EIGENVALUES OF        LGV01240
C     ARE NOT WRITTEN TO FILE 11.  IDIST = 1 MEANS DISTINCT          LGV01250
C     EIGENVALUES ARE WRITTEN TO FILE 11.                            LGV01260
C     IWRITE = (0,1):  IWRITE = 0 MEANS NO INTERMEDIATE OUTPUT       LGV01270
C     FROM THE COMPUTATIONS IS WRITTEN TO FILE 6.  IWRITE = 1 MEANS  LGV01280
C     EIGENVALUES AND ERROR ESTIMATES ARE WRITTEN TO FILE 6          LGV01290
C     AS THEY ARE COMPUTED.                                          LGV01300
```

```
      READ(5,20) EXPLAN                                              LGV01310
      READ(5,*) IHIS,IDIST,IWRITE                                    LGV01320
C                                                                    LGV01330
C     READ IN THE RELATIVE TOLERANCE (RELTOL) FOR USE IN THE         LGV01340
C     SPURIOUS, T-MULTIPLICITY, AND PRTESTS.                         LGV01350
      READ(5,20) EXPLAN                                              LGV01360
      READ(5,*) RELTOL                                               LGV01370
C                                                                    LGV01380
C     READ IN THE SIZES OF THE T-MATRICES TO BE CONSIDERED.          LGV01390
      READ(5,20) EXPLAN                                              LGV01400
      READ(5,*) (NMEV(J), J=1,NMEVS)                                 LGV01410
C                                                                    LGV01420
C     READ IN THE NUMBER OF SUBINTERVALS TO BE CONSIDERED.           LGV01430
      READ(5,20) EXPLAN                                              LGV01440
      READ(5,*) NINT                                                 LGV01450
C                                                                    LGV01460
C     READ IN THE LEFT-END POINTS OF THE SUBINTERVALS TO BE CONSIDERED. LGV01470
C     THESE MUST BE IN ALGEBRAICALLY-INCREASING ORDER               LGV01480
      READ(5,20) EXPLAN                                              LGV01490
      READ(5,*) (LB(J), J=1,NINT)                                    LGV01500
C                                                                    LGV01510
C     READ IN THE RIGHT-END POINTS OF THE SUBINTERVALS TO BE CONSIDERED.LGV01520
C     THESE MUST BE IN ALGEBRAICALLY-INCREASING ORDER               LGV01530
      READ(5,20) EXPLAN                                              LGV01540
      READ(5,*) (UB(J), J=1,NINT)                                    LGV01550
C                                                                    LGV01560
C-------------------------------------------------------------------LGV01570
C     INITIALIZE THE ARRAYS FOR THE USER-SPECIFIED MATRICES          LGV01580
C     AND PASS THE STORAGE LOCATIONS OF THESE ARRAYS TO THE          LGV01590
C     MATRIX-VECTOR MULTIPLY SUBROUTINE AMATV AND THE SOLVE          LGV01600
C     SUBROUTINE LSOLV.                                              LGV01610
C                                                                    LGV01620
      CALL USPECA(N,MATNOA)                                          LGV01630
      CALL USPECB(N,MATNOB)                                          LGV01640
C                                                                    LGV01650
C-------------------------------------------------------------------LGV01660
C                                                                    LGV01670
C     MASK UNDERFLOW AND OVERFLOW                                    LGV01680
      CALL MASK                                                      LGV01690
C                                                                    LGV01700
C-------------------------------------------------------------------LGV01710
C                                                                    LGV01720
C     WRITE TO FILE 6, A SUMMARY OF THE PARAMETERS FOR THIS RUN      LGV01730
C                                                                    LGV01740
      WRITE(6,30) MATNOA,MATNOB,N,KMAX                               LGV01750
   30 FORMAT(/3X,'A-MATRIX ID',3X,'B-MATRIX ID',4X,'ORDER OF A',4X,  LGV01760
     1'MAX ORDER OF T'/I14,I14,I14,I18/)                            LGV01770
C                                                                    LGV01780
      WRITE(6,40) ISTART,ISTOP                                       LGV01790
   40 FORMAT(/2X,'ISTART',3X,'ISTOP'/2I8/)                          LGV01800
C                                                                    LGV01810
      WRITE(6,50) IHIS,IDIST,IWRITE                                  LGV01820
   50 FORMAT(/4X,'IHIS',3X,'IDIST',2X,'IWRITE'/3I8/)                LGV01830
C                                                                    LGV01840
      WRITE(6,60) SVSEED,RHSEED                                      LGV01850
   60 FORMAT(/' SEEDS FOR RANDOM NUMBER GENERATOR'//                LGV01860
     1 4X,'LANCZS SEED',4X,'INVERR SEED'/2I15/)                     LGV01870
C                                                                    LGV01880
      WRITE(6,70) (NMEV(J), J=1,NMEVS)                               LGV01890
   70 FORMAT(/' SIZES OF T-MATRICES TO BE CONSIDERED'/(6I12))       LGV01900
C                                                                    LGV01910
      WRITE(6,80) RELTOL,GAPTOL,BTOL                                 LGV01920
   80 FORMAT(/' RELATIVE TOLERANCE USED TO COMBINE COMPUTED T-EIGENVALUELGV01930
     1S'/E15.3/' RELATIVE GAP TOLERANCES USED IN INVERSE ITERATION'/ LGV01940
     1E15.3/' RELATIVE TOLERANCE FOR CHECK ON SIZE OF BETAS'/E15.3/) LGV01950
C                                                                    LGV01960
      WRITE(6,90) (J,LB(J),UB(J), J=1,NINT)                          LGV01970
```

```
   90 FORMAT(/' BISEC WILL BE USED ON THE FOLLOWING INTERVALS'/      LGV01980
     1 (I6,2E20.6))                                                   LGV01990
C                                                                     LGV02000
      IF (ISTART.EQ.0) GO TO 140                                      LGV02010
C                                                                     LGV02020
C     READ IN ALPHA BETA HISTORY                                      LGV02030
C                                                                     LGV02040
      READ(2,100)MOLD,NOLD,SVSOLD,MATAO,MATBO                         LGV02050
  100 FORMAT(2I6,I12,2I8)                                             LGV02060
C                                                                     LGV02070
      IF (KMAX.LT.MOLD) KMAX = MOLD                                   LGV02080
      KMAX1 = KMAX + 1                                                LGV02090
C                                                                     LGV02100
C     CHECK THAT ORDER N, MATRIX IDS (MATNOA AND MATNOB), AND RANDOM  LGV02110
C     SEED (SVSEED) AGREE WITH THOSE IN THE HISTORY FILE.  IF NOT     LGV02120
C     PROCEDURE STOPS.                                                LGV02130
C                                                                     LGV02140
      ITEMP = (NOLD-N)**2 + (MATNOA-MATAO)**2 + (SVSEED-SVSOLD)**2    LGV02150
     1 + (MATNOB-MATBO)**2                                            LGV02160
C                                                                     LGV02170
      IF (ITEMP.EQ.0) GO TO 120                                       LGV02180
C                                                                     LGV02190
      WRITE(6,110)                                                    LGV02200
  110 FORMAT(' PROGRAM TERMINATES'/    ' READ FROM FILE 2 CORRESPONDS TOLGV02210
     1 DIFFERENT MATRIX THAN MATRIX SPECIFIED'/)                      LGV02220
      GO TO 640                                                       LGV02230
C                                                                     LGV02240
  120 CONTINUE                                                        LGV02250
      MOLD1 = MOLD+1                                                  LGV02260
C                                                                     LGV02270
      READ(2,130)(ALPHA(J), J=1,MOLD)                                 LGV02280
      READ(2,130)(BETA(J), J=1,MOLD1)                                 LGV02290
  130 FORMAT(4Z20)                                                    LGV02300
C                                                                     LGV02310
      IF (KMAX.EQ.MOLD) GO TO 160                                     LGV02320
C                                                                     LGV02330
C     SAVE V1 = B*V(KMAX), VS = B*V(KMAX+1), V2 = V(KMAX+1)           LGV02340
      READ(2,130) (V1(J), J=1,N)                                      LGV02350
      READ(2,130) (VS(J), J=1,N)                                      LGV02360
      READ(2,130) (V2(J), J=1,N)                                      LGV02370
C                                                                     LGV02380
  140 CONTINUE                                                        LGV02390
      IIX = SVSEED                                                    LGV02400
C                                                                     LGV02410
C--------------------------------------------------------------------LGV02420
C                                                                     LGV02430
      CALL LANCZS(LSOLV,AMATV,ALPHA,BETA,V1,V2,VS,G,KMAX,MOLD1,N,IIX) LGV02440
C                                                                     LGV02450
C--------------------------------------------------------------------LGV02460
C                                                                     LGV02470
      KMAX1 = KMAX + 1                                                LGV02480
C                                                                     LGV02490
      IF (IHIS.EQ.0.AND.ISTOP.GT.0) GO TO 160                         LGV02500
C                                                                     LGV02510
      WRITE(1,150) KMAX,N,SVSEED,MATNOA,MATNOB                        LGV02520
  150 FORMAT(2I6,I12,2I8,' = KMAX,N,SVSEED,MATNOA,MATNOB')            LGV02530
C                                                                     LGV02540
      WRITE(1,130)(ALPHA(I), I=1,KMAX)                                LGV02550
      WRITE(1,130)(BETA(I), I=1,KMAX1)                                LGV02560
C                                                                     LGV02570
C     SAVE V1 = B*V(KMAX), VS = B*V(KMAX+1), V2 = V(KMAX+1)           LGV02580
      WRITE(1,130) (V1(I), I=1,N)                                     LGV02590
      WRITE(1,130) (VS(I), I=1,N)                                     LGV02600
      WRITE(1,130) (V2(I), I=1,N)                                     LGV02610
C                                                                     LGV02620
      IF (ISTOP.EQ.0) GO TO 540                                       LGV02630
C                                                                     LGV02640
```

```
      160 CONTINUE                                                   LGV02650
          BKMIN = BTOL                                               LGV02660
          WRITE(6,170)                                               LGV02670
      170 FORMAT(/' T-MATRICES (ALPHA AND BETA) ARE NOW AVAILABLE'/) LGV02680
    C                                                                LGV02690
    C---------------------------------------------------------------LGV02700
    C        SUBROUTINE TNORM CHECKS MIN(BETA)/(ESTIMATED NORM(A)) > BTOL . LGV02710
    C        IF THIS IS VIOLATED IB IS SET EQUAL TO THE NEGATIVE OF THE INDEX LGV02720
    C        OF THE MINIMAL BETA.  IF(IB < 0) THEN SUBROUTINE TNORM IS LGV02730
    C        CALLED FOR EACH VALUE OF MEV TO DETERMINE WHETHER OR NOT THERE LGV02740
    C        IS A BETA IN THE T-MATRIX SPECIFIED THAT VIOLATES THIS TEST. LGV02750
    C        IF THERE IS SUCH A BETA THE PROGRAM TERMINATES FOR THE USER LGV02760
    C        TO DECIDE WHAT TO DO.  THIS TEST CAN BE OVER-RIDDEN BY   LGV02770
    C        SIMPLY MAKING BTOL SMALLER, BUT THEN THERE IS THE POSSIBILITY LGV02780
    C        THAT LOSSES IN THE LOCAL ORTHOGONALITY MAY HURT THE COMPUTATIONS. LGV02790
    C        BTOL = 1.D-8 IS HOWEVER A CONSERVATIVE CHOICE FOR BTOL.  LGV02800
    C                                                                LGV02810
    C        TNORM ALSO COMPUTES TKMAX = MAX(|ALPHA(K)|,BETA(K), K=1,KMAX). LGV02820
    C        TKMAX IS USED TO SCALE THE TOLERANCES USED IN THE       LGV02830
    C        T-MULTIPLICITY AND SPURIOUS TESTS IN BISEC. TKMAX IS ALSO USED IN LGV02840
    C        THE PROJECTION TEST FOR HIDDEN EIGENVALUES THAT HAD 'TOO SMALL' LGV02850
    C        A PROJECTION ON THE STARTING VECTOR.                    LGV02860
    C                                                                LGV02870
          CALL TNORM(ALPHA,BETA,BKMIN,TKMAX,KMAX,IB)                 LGV02880
    C                                                                LGV02890
    C---------------------------------------------------------------LGV02900
    C                                                                LGV02910
          TTOL = EPSM*TKMAX                                          LGV02920
    C                                                                LGV02930
    C        LOOP ON THE SIZE OF THE T-MATRIX                        LGV02940
    C                                                                LGV02950
      180 CONTINUE                                                   LGV02960
          MMB = MMB + 1                                              LGV02970
          MEV = NMEV(MMB)                                            LGV02980
    C        IS MEV TOO LARGE ?                                      LGV02990
          IF(MEV.LE.KMAX) GO TO 200                                 LGV03000
          WRITE(6,190) MMB, MEV, KMAX                               LGV03010
      190 FORMAT(/' TERMINATE PRIOR TO CONSIDERING THE',I6,'TH T-MATRIX'/ LGV03020
         1' BECAUSE THE SIZE REQUESTED',I6,' IS GREATER THAN THE MAXIMUM SIZLGV03030
         1E ALLOWED',I6/)                                           LGV03040
          GO TO 540                                                 LGV03050
    C                                                                LGV03060
      200 MP1 = MEV + 1                                              LGV03070
          BETAM = BETA(MP1)                                         LGV03080
    C                                                                LGV03090
          IF (IB.GE.0) GO TO 210                                    LGV03100
    C                                                                LGV03110
          TO = BTOL                                                 LGV03120
    C                                                                LGV03130
    C---------------------------------------------------------------LGV03140
    C                                                                LGV03150
          CALL TNORM(ALPHA,BETA,TO,T1,MEV,IBMEV)                    LGV03160
    C                                                                LGV03170
    C---------------------------------------------------------------LGV03180
    C                                                                LGV03190
          TEMP = TO/TKMAX                                           LGV03200
          IBMEV = IABS(IBMEV)                                       LGV03210
          IF (TEMP.GE.BTOL) GO TO 210                               LGV03220
          IBMEV = -IBMEV                                            LGV03230
          GO TO 600                                                 LGV03240
    C                                                                LGV03250
      210 CONTINUE                                                   LGV03260
          IC = MXSTUR-ICT                                           LGV03270
    C                                                                LGV03280
    C---------------------------------------------------------------LGV03290
    C        BISEC LOOP. THE SUBROUTINE BISEC INCORPORATES DIRECTLY THE LGV03300
    C        T-MULTIPLICITY AND SPURIOUS TESTS. T-EIGENVALUES WILL BE LGV03310
```

```
C       CALCULATED BY BISEC SEQUENTIALLY ON INTERVALS               LGV03320
C       (LB(J),UB(J)), J = 1,NINT).                                 LGV03330
C                                                                   LGV03340
C       ON RETURN FROM BISEC                                        LGV03350
C       NDIS = NUMBER OF DISTINCT EIGENVALUES OF T(1,MEV) ON UNION  LGV03360
C              OF THE (LB,UB) INTERVALS                             LGV03370
C       VS = DISTINCT T-EIGENVALUES IN ALGEBRAICALLY INCREASING ORDER LGV03380
C       MP = MULTIPLICITIES OF THE T-EIGENVALUES IN VS              LGV03390
C       MP(I) = (0,1,MI), MI>1, I=1,NDIS  MEANS:                    LGV03400
C          (0)  VS(I) IS SPURIOUS                                   LGV03410
C          (1)  VS(I) IS T-SIMPLE AND GOOD                          LGV03420
C          (MI) VS(I) IS MULTIPLE AND IS THEREFORE NOT ONLY GOOD BUT LGV03430
C               ALSO A CONVERGED GOOD T-EIGENVALUE.                 LGV03440
C                                                                   LGV03450
C                                                                   LGV03460
        CALL BISEC(ALPHA,BETA,V1,V2,VS,LB,UB,EPSM,TTOL,MP,NINT,     LGV03470
      1 MEV,NDIS,IC,IWRITE)                                         LGV03480
C                                                                   LGV03490
C-------------------------------------------------------------------LGV03500
C                                                                   LGV03510
        IF (NDIS.EQ.0) GO TO 620                                    LGV03520
C                                                                   LGV03530
C       COMPUTE THE TOTAL NUMBER OF STURM SEQUENCES USED TO DATE    LGV03540
C       COMPUTE THE BISEC CONVERGENCE AND T-MULTIPLICITY TOLERANCES USED. LGV03550
C       COMPUTE THE CONVERGENCE TOLERANCE FOR EIGENVALUES OF A.     LGV03560
        ICT = ICT + IC                                              LGV03570
        TEMP = DFLOAT(MEV+1000)                                     LGV03580
        MULTOL = TEMP*TTOL                                          LGV03590
        TEMP = DSQRT(TEMP)                                          LGV03600
        BISTOL = TTOL*TEMP                                          LGV03610
        CONTOL = BETAM*1.D-10                                       LGV03620
C                                                                   LGV03630
C-------------------------------------------------------------------LGV03640
C       SUBROUTINE LUMP 'COMBINES' T-EIGENVALUES THAT ARE 'TOO CLOSE'. LGV03650
C       NOTE HOWEVER THAT CLOSE SPURIOUS T-EIGENVALUES ARE NOT AVERAGED LGV03660
C       WITH GOOD ONES. HOWEVER, THEY MAY BE USED TO INCREASE THE   LGV03670
C       MULTIPLICITY OF A GOOD T-EIGENVALUE.                        LGV03680
C                                                                   LGV03690
        LOOP = NDIS                                                 LGV03700
        CALL LUMP(VS,RELTOL,MULTOL,SCALE2,MP,LOOP)                  LGV03710
C                                                                   LGV03720
C-------------------------------------------------------------------LGV03730
C                                                                   LGV03740
        IF(NDIS.EQ.LOOP) GO TO 230                                  LGV03750
C                                                                   LGV03760
        WRITE(6,220) NDIS, MEV, LOOP                                LGV03770
  220 FORMAT(/I6,' DISTINCT T-EIGENVALUES WERE COMPUTED IN BISEC AT MEV LGV03780
     1',I6/ 2X,' LUMP SUBROUTINE REDUCES NUMBER OF DISTINCT EIGENVALUES LGV03790
     10',I6)                                                        LGV03800
C                                                                   LGV03810
  230 CONTINUE                                                      LGV03820
        NDIS = LOOP                                                 LGV03830
        BETA(MP1) = BETAM                                           LGV03840
C                                                                   LGV03850
C-------------------------------------------------------------------LGV03860
C       THE SUBROUTINE ISOEV LABELS THOSE SIMPLE EIGENVALUES OF T(1,MEV) LGV03870
C       WITH VERY SMALL GAPS BETWEEN NEIGHBORING EIGENVALUES OF T(1,MEV) LGV03880
C       TO AVOID COMPUTING ERROR ESTIMATES FOR ANY SIMPLE GOOD      LGV03890
C       T-EIGENVALUE THAT IS TOO CLOSE TO A SPURIOUS EIGENVALUE.    LGV03900
C       ON RETURN FROM ISOEV, G CONTAINS CODED MINIMAL GAPS         LGV03910
C       BETWEEN THE DISTINCT EIGENVALUES OF T(1,MEV). (G IS REAL).  LGV03920
C       G(I) < 0 MEANS MINGAP IS DUE TO LEFT GAP G(I) > 0 MEANS DUE TO LGV03930
C       RIGHT GAP. MP(I) = -1 MEANS THAT THE GOOD T-EIGENVALUE IS SIMPLE LGV03940
C       AND HAS A VERY SMALL MINGAP IN T(1,MEV) DUE TO A SPURIOUS   LGV03950
C       EIGENVALUE.  NG = NUMBER OF GOOD T-EIGENVALUES.             LGV03960
C       NISO = NUMBER OF ISOLATED GOOD T-EIGENVALUES.               LGV03970
C                                                                   LGV03980
```

```
      CALL ISOEV(VS,GAPTOL,MULTOL,SCALE1,G,MP,NDIS,NG,NISO)              LGV03990
C                                                                        LGV04000
C-----------------------------------------------------------------------LGV04010
C                                                                        LGV04020
      WRITE(6,240)NG,NISO,NDIS                                           LGV04030
  240 FORMAT(/I6,' GOOD T-EIGENVALUES HAVE BEEN COMPUTED'/               LGV04040
     1 I6,' OF THESE ARE T-ISOLATED'/                                    LGV04050
     2 I6,' = NUMBER OF DISTINCT T-EIGENVALUES COMPUTED'/)               LGV04060
C                                                                        LGV04070
C     DO WE WRITE DISTINCT EIGENVALUES OF T-MATRIX TO FILE 11?           LGV04080
      IF (IDIST.EQ.0) GO TO 280                                          LGV04090
C                                                                        LGV04100
      WRITE(11,250) NDIS,NISO,MEV,N,SVSEED,MATNOA,MATNOB                 LGV04110
  250 FORMAT(/4I6,I12,I8,'=ND,NIS,MEV,N,SEED,MNA,MNB'/)                  LGV04120
C                                                                        LGV04130
      WRITE(11,260) (MP(I),VS(I),G(I), I=1,NDIS)                         LGV04140
  260 FORMAT(2(I3,E25.16,E12.3))                                         LGV04150
C                                                                        LGV04160
      WRITE(11,270) NDIS, (MP(I), I=1,NDIS)                              LGV04170
  270 FORMAT(/I6,' = NDIS, T-MULTIPLICITIES (0 MEANS  SPURIOUS)'/(20I4)) LGV04180
C                                                                        LGV04190
  280 CONTINUE                                                           LGV04200
C                                                                        LGV04210
      IF (NISO.NE.0) GO TO 310                                           LGV04220
C                                                                        LGV04230
      WRITE(4,290) MEV                                                   LGV04240
  290 FORMAT(/' AT MEV = ',I6,' THERE ARE NO ISOLATED T-EIGENVALUES'/    LGV04250
     1' SO NO ERROR ESTIMATES WERE COMPUTED/')                          LGV04260
C                                                                        LGV04270
      WRITE(6,300)                                                       LGV04280
  300 FORMAT(/' ALL COMPUTED GOOD T-EIGENVALUES ARE MULTIPLE'/           LGV04290
     1 ' THEREFORE ALL SUCH EIGENVALUES ARE ASSUMED TO HAVE CONVERGED')  LGV04300
C                                                                        LGV04310
      ICONV = 1                                                          LGV04320
      GO TO 350                                                          LGV04330
C                                                                        LGV04340
  310 CONTINUE                                                           LGV04350
C                                                                        LGV04360
C-----------------------------------------------------------------------LGV04370
C     SUBROUTINE INVERR COMPUTES ERROR ESTIMATES FOR ISOLATED GOOD       LGV04380
C     T-EIGENVALUES USING INVERSE ITERATION ON T(1,MEV). ON RETURN       LGV04390
C     G(J) = MINIMUM GAP IN T(1,MEV) FOR EACH VS(J), J=1,NDIS            LGV04400
C     G(MEV+I) = BETAM*|U(MEV)| = ERROR ESTIMATE FOR ISOLATED GOOD       LGV04410
C                T-EIGENVALUES, WHERE I = 1, NISO AND  BETAM = BETA(MEV+1)LGV04420
C                U(MEV) IS MEVTH COMPONENT OF THE UNIT EIGENVECTOR OF T   LGV04430
C                CORRESPONDING TO THE ITH ISOLATED GOOD T-EIGENVALUE.     LGV04440
C     A NEGATIVE ERROR ESTIMATE MEANS THAT FOR THAT PARTICULAR           LGV04450
C     EIGENVALUE THE INVERSE ITERATION DID NOT CONVERGE IN <= MXINIT     LGV04460
C     STEPS AND THAT THE CORRESPONDING ERROR ESTIMATE IS QUESTIONABLE.   LGV04470
C                                                                        LGV04480
C     V2 CONTAINS THE ISOLATED GOOD T-EIGENVALUES                        LGV04490
C     V1 CONTAINS THE MINGAPS TO THE NEAREST DISTINCT  EIGENVALUE        LGV04500
C        OF T(1,MEV) FOR EACH ISOLATED GOOD T-EIGENVALUE IN V2.          LGV04510
C     VS CONTAINS THE NDIS DISTINCT EIGENVALUES OF T(1,MEV)              LGV04520
C     MP CONTAINS THE CORRESPONDING CODED T-MULTIPLICITIES               LGV04530
C                                                                        LGV04540
      IT = MXINIT                                                        LGV04550
      CALL INVERR(ALPHA,BETA,V1,V2,VS,EPSM,G,MP,MEV,MMB,NDIS,NISO,N,     LGV04560
     1 RHSEED,IT,IWRITE)                                                 LGV04570
C                                                                        LGV04580
C-----------------------------------------------------------------------LGV04590
C                                                                        LGV04600
C     SIMPLE CHECK FOR CONVERGENCE. CHECKS TO SEE IF ALL OF THE ERROR    LGV04610
C     ESTIMATES ARE SMALLER THAN CONTOL = BETAM*1.D-10.                  LGV04620
C     IF THIS TEST IS SATISFIED, THEN CONVERGENCE FLAG, ICONV IS SET     LGV04630
C     TO 1.  TYPICALLY ERROR ESTIMATES ARE VERY CONSERVATIVE.           LGV04640
C                                                                        LGV04650
```

```
      WRITE(6,320) CONTOL                                         LGV04660
  320 FORMAT(/' CONVERGENCE IS TESTED USING THE CONVERGENCE TOLERANCE', LGV04670
     1E13.4/)                                                     LGV04680
C                                                                 LGV04690
      II = MEV +1                                                 LGV04700
      IF = MEV+NISO                                               LGV04710
      DO 330 I = II,IF                                            LGV04720
      IF (ABS(G(I)).GT.CONTOL) GO TO 350                          LGV04730
  330 CONTINUE                                                    LGV04740
      ICONV = 1                                                   LGV04750
      MMB = NMEVS                                                 LGV04760
C                                                                 LGV04770
      WRITE(6,340) CONTOL                                         LGV04780
  340 FORMAT(' ALL COMPUTED ERROR ESTIMATES WERE LESS THAN',E15.4/ LGV04790
     1 ' THEREFORE PROCEDURE TERMINATES'/)                        LGV04800
C                                                                 LGV04810
  350 CONTINUE                                                    LGV04820
C                                                                 LGV04830
C     IF CONVERGENCE IS INDICATED, THAT IS ICONV = 1 ,THEN        LGV04840
C     THE SUBROUTINE PRTEST IS CALLED TO CHECK FOR ANY CONVERGED  LGV04850
C     T-EIGENVALUES THAT HAVE BEEN MISLABELLED AS SPURIOUS BECAUSE LGV04860
C     THE PROJECTION OF THEIR EIGENVECTOR(S) ON THE STARTING      LGV04870
C     VECTOR WERE TOO SMALL.                                      LGV04880
C     NUMERICAL TESTS INDICATE THAT SUCH EIGENVALUES ARE RARE.    LGV04890
C     IF FOR SOME REASON MANY OF THESE HIDDEN EIGENVALUES APPEAR   LGV04900
C     ON SOME RUN, YOU CAN BE CERTAIN THAT SOMETHING IS FOULED UP. LGV04910
C                                                                 LGV04920
      IF (ICONV.EQ.0) GO TO 480                                   LGV04930
C                                                                 LGV04940
C-----------------------------------------------------------------LGV04950
C                                                                 LGV04960
      CALL PRTEST (ALPHA,BETA,VS,TKMAX,EPSM,RELTOL,SCALE3,SCALE4, LGV04970
     1 MP,NDIS,MEV,IPROJ)                                         LGV04980
C                                                                 LGV04990
C-----------------------------------------------------------------LGV05000
C                                                                 LGV05010
      IF(IPROJ.EQ.0) GO TO 470                                    LGV05020
C                                                                 LGV05030
      IF(IDIST.EQ.1)  WRITE(11,360) IPROJ                         LGV05040
  360 FORMAT(' SUBROUTINE PRTEST WANTS TO RELABEL',I6,' SPURIOUS T-EIGENLGV05050
     1VALUES'/' WE ACCEPT RELABELLING ONLY IF LAST COMPONENT OF T-EIGENVLGV05060
     1ECTOR IS L.T. 1.D-10'/)                                     LGV05070
C                                                                 LGV05080
      IIX = RHSEED                                                LGV05090
C                                                                 LGV05100
C-----------------------------------------------------------------LGV05110
C                                                                 LGV05120
      CALL GENRAN(IIX,G,MEV)                                      LGV05130
C                                                                 LGV05140
C-----------------------------------------------------------------LGV05150
C                                                                 LGV05160
      ITEN = -10                                                  LGV05170
      NISOM = NISO + MEV                                          LGV05180
      IWRITO = IWRITE                                             LGV05190
      IWRITE = 0                                                  LGV05200
C                                                                 LGV05210
      DO 390 J = 1,NDIS                                           LGV05220
      IF(MP(J).NE.ITEN) GO TO 390                                 LGV05230
      TO = VS(J)                                                  LGV05240
C                                                                 LGV05250
C-----------------------------------------------------------------LGV05260
C                                                                 LGV05270
      IT = MXINIT                                                 LGV05280
      CALL INVERM(ALPHA,BETA,V1,V2,TO,TEMP,T1,EPSM,G,MEV,IT,IWRITE) LGV05290
C                                                                 LGV05300
C-----------------------------------------------------------------LGV05310
C                                                                 LGV05320
```

```
      IF(TEMP.LE.1.D-10) GO TO 380                                      LGV05330
C     ERROR ESTIMATE WAS NOT SMALL REJECT RELABELLING OF THIS EIGENVALUELGV05340
      IF(IDIST.EQ.1)  WRITE(11,370) J,TO,TEMP                           LGV05350
  370 FORMAT(/' LAST COMPONENT FOR',I6,'TH EIGENVALUE',E20.12/' IS TOO LLGV05360
     1ARGE = ',E15.6,' SO DO NOT ACCEPT PRTEST RELABELLING'/)           LGV05370
      MP(J) = 0                                                         LGV05380
      IPROJ = IPROJ - 1                                                 LGV05390
      GO TO 390                                                         LGV05400
C     RELABELLING ACCEPTED                                              LGV05410
  380 NISOM = NISOM + 1                                                 LGV05420
      G(NISOM) = BETAM*TEMP                                             LGV05430
  390 CONTINUE                                                          LGV05440
      IWRITE = IWRITO                                                   LGV05450
C                                                                       LGV05460
      IF(IPROJ.EQ.0) GO TO 430                                          LGV05470
      WRITE(6,400) IPROJ                                                LGV05480
  400 FORMAT(/I6,' T-EIGENVALUES WERE RECLASSIFIED AS GOOD.'/           LGV05490
     1' THESE ARE IDENTIFIED IN FILE 3 BY A T-MULTIPLICITY OF -10'/' USELGV05500
     2R SHOULD INSPECT EACH TO MAKE SURE NEIGHBORS HAVE CONVERGED'/)    LGV05510
C                                                                       LGV05520
      IF(IDIST.EQ.1)  WRITE(11,410) IPROJ                               LGV05530
  410 FORMAT(/I6,' T-EIGENVALUES WERE RELABELLED AS GOOD'/              LGV05540
     1' BELOW IS CORRECTED T-MULTIPLICITY PATTERN'/)                    LGV05550
C                                                                       LGV05560
      WRITE(6,420) NDIS, (MP(I), I=1,NDIS)                              LGV05570
      IF(IDIST.EQ.1)  WRITE(11,420) NDIS, (MP(I), I=1,NDIS)             LGV05580
  420 FORMAT(/I6,' = NDIS, T-MULTIPLICITIES (0 MEANS  SPURIOUS)'/       LGV05590
     1 6X, ' (-10) MEANS SPURIOUS T-EIGENVALUE RELABELLED AS GOOD'/(2014LGV05600
     1))                                                                LGV05610
C                                                                       LGV05620
C     RECALCULATE MINGAPS FOR DISTINCT T(1,MEV) EIGENVALUES.            LGV05630
  430 NM1 = NDIS - 1                                                    LGV05640
      G(NDIS) = VS(NM1)-VS(NDIS)                                        LGV05650
      G(1) = VS(2)-VS(1)                                                LGV05660
C                                                                       LGV05670
      DO 440 J = 2,NM1                                                  LGV05680
      TO = VS(J)-VS(J-1)                                                LGV05690
      T1 = VS(J+1)-VS(J)                                                LGV05700
      G(J) = T1                                                         LGV05710
      IF (TO.LT.T1) G(J) = -TO                                          LGV05720
  440 CONTINUE                                                          LGV05730
      IF(IPROJ.EQ.0) GO TO 470                                          LGV05740
C     WRITE TO FILE 4 ERROR ESTIMATES FOR THOSE T-EIGENVALUES RELABELLEDLGV05750
      NGOOD = 0                                                         LGV05760
      DO 450 J = 1,NDIS                                                 LGV05770
      IF(MP(J).EQ.0) GO TO 450                                          LGV05780
      NGOOD = NGOOD + 1                                                 LGV05790
      IF(MP(J).NE.ITEN) GO TO 450                                       LGV05800
      TO = VS(J)                                                        LGV05810
      NISO = NISO + 1                                                   LGV05820
      NISOM = MEV + NISO                                                LGV05830
      WRITE(4,460) NGOOD,TO,G(NISOM),G(J)                               LGV05840
  450 CONTINUE                                                          LGV05850
  460 FORMAT(I10,E25.16,2E14.3)                                         LGV05860
C                                                                       LGV05870
  470 CONTINUE                                                          LGV05880
C                                                                       LGV05890
C     WRITE THE GOOD T-EIGENVALUES TO FILE 3.  FIRST TRANSFER THEM      LGV05900
C     TO V2 AND THEIR T-MULTIPLICITIES TO THE CORRESPONDING POSITIONS   LGV05910
C     IN MP AND COMPUTE THE AB-MINGAPS, THE MINIMAL GAPS BETWEEN THE    LGV05920
C     GOOD T-EIGENVALUES.  THESE GAPS WILL BE PUT IN THE ARRAY G.       LGV05930
C     SINCE G CURRENTLY CONTAINS THE MINIMAL GAPS BETWEEN THE DISTINCT  LGV05940
C     EIGENVALUES OF THE T-MATRIX, THESE GAPS WILL FIRST BE             LGV05950
C     TRANSFERRED TO V1.  NOTE THAT V1<0 MEANS THAT THAT MINIMAL GAP    LGV05960
C     IN THE T-MATRIX IS DUE TO A SPURIOUS T-EIGENVALUE.                LGV05970
C     ALL THIS INFORMATION IS PRINTED TO FILE 3                         LGV05980
C                                                                       LGV05990
```

```
   480 CONTINUE                                                      LGV06000
C                                                                    LGV06010
      NG = 0                                                         LGV06020
      DO 490 I = 1,NDIS                                              LGV06030
      IF (MP(I).EQ.0) GO TO 490                                      LGV06040
      NG = NG+1                                                      LGV06050
      MP(NG) = MP(I)                                                 LGV06060
      V2(NG) = VS(I)                                                 LGV06070
      TEMP = G(I)                                                    LGV06080
      TEMP = DABS(TEMP)                                              LGV06090
      J = I+1                                                        LGV06100
      IF (G(I).LT.ZERO) J = I-1                                      LGV06110
      IF (MP(J).EQ.0) TEMP = -TEMP                                   LGV06120
      V1(NG) = TEMP                                                  LGV06130
   490 CONTINUE                                                      LGV06140
C                                                                    LGV06150
      WRITE(6,500)MEV                                                LGV06160
   500 FORMAT(//' T-EIGENVALUE CALCULATION AT MEV = ',I6,'   IS COMPLETELGV06170
     1')                                                             LGV06180
C                                                                    LGV06190
C     NG = NUMBER OF COMPUTED DISTINCT GOOD T-EIGENVALUES.  NEXT     LGV06200
C     GENERATE GAPS BETWEEN GOOD T-EIGENVALUES (ABMINGAPS) AND PUT THEM LGV06210
C     IN G.  G(J) < 0 MEANS THE ABMINGAP IS DUE TO THE LEFT-HAND GAP. LGV06220
C                                                                    LGV06230
      NGM1 = NG - 1                                                  LGV06240
      G(NG) = V2(NGM1)-V2(NG)                                        LGV06250
      G(1) = V2(2)-V2(1)                                             LGV06260
C                                                                    LGV06270
      DO 510 J = 2,NGM1                                              LGV06280
      T0 = V2(J)-V2(J-1)                                             LGV06290
      T1 = V2(J+1)-V2(J)                                             LGV06300
      G(J) = T1                                                      LGV06310
      IF (T0.LT.T1) G(J) = -T0                                       LGV06320
   510 CONTINUE                                                      LGV06330
C                                                                    LGV06340
C     WRITE GOOD T-EIGENVALUES OUT TO FILE 3.                        LGV06350
C                                                                    LGV06360
      WRITE(3,520)NG,NDIS,MEV,N,SVSEED,MATNOA,MATNOB,MULTOL,IB,BTOL  LGV06370
   520 FORMAT(4I6,I12,2I8,'=NG,ND,MEV,N,SEED,MNA,MNB'/               LGV06380
     1 E20.12,I6,E13.4,' = MUTOL,INDEX MINIMAL BETA,BTOL'/           LGV06390
     1' EV NO',1X,'TMULT',10X,'GOOD EIGENVALUE',7X,'TMINGAP',6X,'ABMINGALGV06400
     1P')                                                            LGV06410
C                                                                    LGV06420
      WRITE(3,530)(I,MP(I),V2(I),V1(I),G(I), I=1,NG)                 LGV06430
   530 FORMAT(2I6,E25.16,2E14.3)                                     LGV06440
C                                                                    LGV06450
C     IF CONVERGENCE FLAG ICONV.NE.1 AND NUMBER OF T-MATRICES        LGV06460
C     CONSIDERED TO DATE IS LESS THAN NUMBER ALLOWED, INCREMENT MEV. LGV06470
C     AND LOOP BACK TO 210 TO REPEAT COMPUTATIONS.  RESTORE BETA(MEV+1).LGV06480
C                                                                    LGV06490
      BETA(MP1) = BETAM                                              LGV06500
C                                                                    LGV06510
      IF (MMB.LT.NMEVS.AND.ICONV.NE.1) GO TO 180                     LGV06520
C                                                                    LGV06530
C     END OF LOOP ON DIFFERENT SIZE T-MATRICES ALLOWED.             LGV06540
C                                                                    LGV06550
   540 CONTINUE                                                      LGV06560
C                                                                    LGV06570
      IF(ISTOP.EQ.0)  WRITE(6,550)                                   LGV06580
   550 FORMAT(/' T-MATRICES (ALPHA AND BETA) ARE NOW AVAILABLE, TERMINATELGV06590
     1')                                                             LGV06600
      IF (IHIS.EQ.1.AND.KMAX.NE.MOLD) WRITE(1,560)                   LGV06610
   560 FORMAT(/' ABOVE ARE THE FOLLOWING VECTORS '/                  LGV06620
     1 '  ALPHA(I), I = 1,KMAX'/                                     LGV06630
     2 '  BETA(I), I = 1,KMAX+1'/                                    LGV06640
     3 ' FINAL THREE VECTORS USED IN LANCZS SUBROUTINE'/            LGV06650
     3 ' V1 = B*V(KMAX), VS = B*V(KMAX+1), V2 = V(KMAX+1)'/          LGV06660
```

```
      4 ' ALL VECTORS IN THIS FILE HAVE HEX FORMAT 4Z20'/            LGV06670
      5 ' ----- END OF FILE 1 NEW ALPHA, BETA HISTORY---------------'///)LGV06680
C                                                                    LGV06690
      IF (ISTOP.EQ.0) GO TO 640                                      LGV06700
C                                                                    LGV06710
      WRITE(3,570)                                                   LGV06720
  570 FORMAT(/' ABOVE ARE COMPUTED GOOD T-EIGENVALUES'/              LGV06730
      1 ' NG = NUMBER OF GOOD T-EIGENVALUES COMPUTED'/               LGV06740
      2 ' ND = NUMBER OF COMPUTED DISTINCT EIGENVALUES OF T(1,MEV)'/ LGV06750
      3 ' N = ORDER OF A AND B-MATRIX, MNA, MNB = MATRIX IDENTS'/    LGV06760
      4 ' MULTOL = T-MULTIPLICITY TOLERANCE FOR T-EIGENVALUES IN BISEC'/LGV06770
      4 ' TMULT IS THE T-MULTIPLICITY OF GOOD T-EIGENVALUE'/         LGV06780
      5 ' TMULT = -1 MEANS SPURIOUS T-EIGENVALUE TOO CLOSE'/         LGV06790
      6 ' DO NOT COMPUTE ERROR ESTIMATES FOR SUCH EIGENVALUES'/      LGV06800
      7 ' ABMINGAP = MINIMAL GAP BETWEEN THE COMPUTED EIGENVALUES'/  LGV06810
      8 ' ABMINGAP .LT. 0. MEANS MINIMAL GAP IS DUE TO LEFT-HAND GAP'/ LGV06820
      9 ' TMINGAP= MINIMAL GAP W.R.T. DISTINCT EIGENVALUES IN T(1,MEV)'/LGV06830
      1 ' TMINGAP .LT. 0. MEANS MINGAP IS DUE TO SPURIOUS T-EIGENVALUE'/ LGV06840
      2 ' ----- END OF FILE 3 GOODEIGENVALUES----------------------'///)LGV06850
C                                                                    LGV06860
      IF (IDIST.EQ.1) WRITE(11,580)                                  LGV06870
  580 FORMAT(/' ABOVE ARE THE DISTINCT EIGENVALUES OF T(1,MEV).'/    LGV06880
      2 ' THE FORMAT IS      T-MULTIPLICITY    T-EIGENVALUE   TMINGAP'/ LGV06890
      3 '          THIS FORMAT IS REPEATED TWICE ON EACH LINE.'/     LGV06900
      4 ' T-MULTIPLICITY = -1 MEANS THAT THE SUBROUTINE ISOEV HAS TAGGED'LGV06910
      5/'   THIS SIMPLE T-EIGENVALUE AS HAVING A VERY CLOSE SPURIOUS'/ LGV06920
      6 '   T-EIGENVALUE SO THAT NO ERROR ESTIMATE WILL BE COMPUTED'/ LGV06930
      7 '   FOR THAT EIGENVALUE IN SUBROUTINE INVERR.'/              LGV06940
      8 ' TMINGAP .LT. 0, TMINGAP IS DUE TO LEFT GAP .GT. 0, RIGHT GAP.'/LGV06950
      9 ' EACH OF THE DISTINCT T-EIGENVALUE TABLES IS FOLLOWED'/     LGV06960
      9 ' BY THE T-MULTIPLICITY PATTERN.'/                           LGV06970
      1 ' NDIS = NUMBER OF COMPUTED DISTINCT EIGENVALUES OF T(1,MEV).'/ LGV06980
      2 ' NG = NUMBER OF GOOD T-EIGENVALUES. '/                      LGV06990
      3 ' NISO = NUMBER OF ISOLATED GOOD T-EIGENVALUES. '/           LGV07000
      4 ' NISO ALSO IS THE COUNT OF +1 ENTRIES IN T-MULTIPLICITY PATTERN.LGV07010
      5 '/' ----- END OF FILE 11 DISTINCT T-EIGENVALUES--------------'///LGV07020
      6 )                                                            LGV07030
C                                                                    LGV07040
      IF(NISO.NE.0)  WRITE(4,590)                                    LGV07050
  590 FORMAT(/' ABOVE ARE THE ERROR ESTIMATES OBTAINED FOR THE ISOLATED LGV07060
      1GOOD T-EIGENVALUES'/                                          LGV07070
      1' OBTAINED VIA INVERSE ITERATION IN THE SUBROUTINE INVERR.'/  LGV07080
      1' ALL OTHER GOOD T-EIGENVALUES HAVE CONVERGED.'/              LGV07090
      2' ERROR ESTIMATE = BETAM*ABS(UM)'/                            LGV07100
      2' WHERE BETAM = BETA(MEV+1) AND UM = U(MEV).'/                LGV07110
      3' U = UNIT EIGENVECTOR OF T WHERE T*U = EV*U AND EV = ISOLATED GOOLGV07120
      3D T-EIGENVALUE.'/                                             LGV07130
      4' TMINGAP = GAP TO NEAREST DISTINCT EIGENVALUE OF T(1,MEV).'/ LGV07140
      5' TMINGAP .LT. 0. MEANS MINGAP IS DUE TO A LEFT NEIGHBOR.'/   LGV07150
      6' ERROR ESTIMATE L.T. 0 MEANS INVERSE ITERATION DID NOT CONVERGE'/LGV07160
      7' ------ END OF FILE 4 ERRINV ------------------------------'//) LGV07170
      GO TO 640                                                      LGV07180
C                                                                    LGV07190
  600 CONTINUE                                                       LGV07200
C                                                                    LGV07210
      IBB = IABS(IBMEV)                                              LGV07220
      IF (IBMEV.LT.0) WRITE(6,610) MEV,IBB,BETA(IBB)                 LGV07230
  610 FORMAT(/' PROGRAM TERMINATES BECAUSE MEV REQUESTED = ',I6,' IS .GTLGV07240
      1',I6/' AT WHICH AN ABNORMALLY SMALL BETA = ' , E13.4,' OCCURRED'/)LGV07250
      GO TO 640                                                      LGV07260
C                                                                    LGV07270
  620 IF (NDIS.EQ.0.AND.ISTOP.GT.0) WRITE(6,630)                     LGV07280
  630 FORMAT(/' INTERVALS SPECIFIED FOR BISECT DID NOT CONTAIN ANY T-EIGLGV07290
      1ENVALUES'/' PROGRAM TERMINATES')                             LGV07300
C                                                                    LGV07310
  640 CONTINUE                                                       LGV07320
C                                                                    LGV07330
```

```
      STOP                                                  LGV07340
C-----END OF MAIN PROGRAM FOR LANCZOS EIGENVALUE COMPUTATIONS----------LGV07350
      END                                                   LGV07360
```

SECTION 5.3   MAIN PROGRAM, EIGENVECTOR CALCULATIONS

```
C-----LGVEC (EIGENVECTORS OF A*X = EVAL*B*X)--------------------------LGV00010
C                                                                     LGV00020
C      CONTAINS MAIN PROGRAM FOR COMPUTING AN EIGENVECTOR CORRESPONDING LGV00030
C      TO EACH OF A SET OF EIGENVALUES WHICH HAVE BEEN COMPUTED        LGV00040
C      ACCURATELY BY THE CORRESPONDING LANCZOS EIGENVALUE PROGRAM      LGV00050
C      (LGVAL) FOR THE SYMMETRIC, GENERALIZED PROBLEM A*X = EVAL*B*X.  LGV00060
C      LGVAL AND LGVEC ASSUME THAT B IS POSITIVE DEFINITE AND THAT THE LGV00070
C      CHOLESKY FACTORS OF B (OR OF A PERMUTATION OF B) ARE AVAILABLE  LGV00080
C      FOR USE IN THE LANCZOS PROCEDURES.  IF B HAS BEEN PERMUTED,     LGV00090
C      THEN THESE PROCEDURES ASSUME THAT THE DATA PRESENTED FOR THE    LGV00100
C      A-MATRIX HAS BEEN SUBJECTED TO THE SAME PERMUTATION.  THAT      LGV00110
C      PERMUTATION WILL THEN BE USED AFTER THE RITZ VECTORS FOR THE    LGV00120
C      PERMUTED VERSION OF THE ORIGINAL PROBLEM HAVE BEEN COMPUTED,    LGV00130
C      TO OBTAIN THE ASSOCIATED RITZ VECTORS FOR THE ORIGINAL PROBLEM. LGV00140
C      NOTE THAT THIS PROGRAM COULD BE MODIFIED TO COMPUTE ADDITIONAL  LGV00150
C      EIGENVECTORS FOR ANY COMPUTED EIGENVALUE WHICH IS A MULTIPLE    LGV00160
C      EIGENVALUE OF THE GIVEN A-MATRIX.  THE AMOUNT OF ADDITIONAL     LGV00170
C      COMPUTATION REQUIRED WOULD DEPEND UPON THE PARTICULAR           LGV00180
C      A-MATRIX AND B-MATRIX USED AND UPON WHAT PART OF THE            LGV00190
C      SPECTRUM OF EIGENVALUES IS BEING CONSIDERED.                    LGV00200
C                                                                     LGV00210
C      THE LANCZOS EIGENVECTOR COMPUTATIONS ASSUME THAT EACH          LGV00220
C      EIGENVALUE THAT IS BEING CONSIDERED HAS CONVERGED AS AN        LGV00230
C      EIGENVALUE OF THE ASSOCIATED LANCZOS TRIDIAGONAL MATRICES.     LGV00240
C                                                                     LGV00250
C      PFORT VERIFIER IDENTIFIED THE FOLLOWING NONPORTABLE            LGV00260
C      CONSTRUCTIONS                                                  LGV00270
C                                                                     LGV00280
C      1.  DATA/MACHEP/ STATEMENT                                     LGV00290
C      2.  ALL READ(5,*) STATEMENTS (FREE FORMAT)                     LGV00300
C      3.  FORMAT(20A4) USED WITH THE EXPLANATORY HEADER, EXPLAN      LGV00310
C      4.  HEXADECIMAL FORMAT (4Z20) USED IN ALPHA/BETA FILES 1 AND 2. LGV00320
C                                                                     LGV00330
C      IMPORTANT NOTE:  PROGRAM ALLOWS ENLARGEMENT OF THE ALPHA, BETA LGV00340
C      ARRAYS.  IN PARTICULAR, IF ANY ONE OF THE EIGENVALUES SUPPLIED LGV00350
C      IS T-SIMPLE AND NOT CLOSE TO A SPURIOUS EIGENVALUE, THE PROGRAM LGV00360
C      REQUIRES THAT KMAX BE AT LEAST 11*MEV/8 + 12.  IF KMAX IS NOT  LGV00370
C      THIS LARGE, THEN THE PROGRAM RESETS KMAX TO THIS SIZE          LGV00380
C      AND EXTENDS THE ALPHA, BETA HISTORY IF REQUIRED.               LGV00390
C      THUS, THE DIMENSIONS OF THE ALPHA AND BETA ARRAYS MUST BE      LGV00400
C      LARGE ENOUGH TO ALLOW FOR THIS POSSIBILITY.                    LGV00410
C      REMEMBER THAT THE BETA ARRAY, BETA(J), IS SUCH THAT            LGV00420
C      J = 1,..., KMAX+1.  SO IF THE KMAX USED BY THE PROGRAM         LGV00430
C      IS TO BE 3000, THEN BETA MUST BE OF LENGTH AT LEAST 3001.      LGV00440
C                                                                     LGV00450
C--------------------------------------------------------------------LGV00460
       DOUBLE PRECISION  ALPHA(5000),BETA(5001)                       LGV00470
       DOUBLE PRECISION  V1(5000),V2(5000),VS(5000)                   LGV00480
       DOUBLE PRECISION  RITVEC(30000),TVEC(30000),GOODEV(50),EVNEW(50) LGV00490
       DOUBLE PRECISION  EVAL,EVALN,TOLN,TTOL,ERTOL,ALFA,BATA          LGV00500
       DOUBLE PRECISION  MULTOL,SCALEO,STUTOL,BTOL,LB,UB               LGV00510
       DOUBLE PRECISION  ONE,ZERO,MACHEP,EPSM,TEMP,SUM,ERRMIN,BKMIN    LGV00520
       DOUBLE PRECISION  RELTOL,ERROR,TERROR,TLAST(50)                 LGV00530
       REAL  G(10000),AMINGP(50),TMINGP(50),EXPLAN(20)                 LGV00540
       REAL  TERR(50),ERR(50),ERRDGP(50),RNORM(50),TBETA(50)           LGV00550
       INTEGER  MP(50),M1(50),M2(50),MA(50),ML(50),MINT(50),MFIN(50)   LGV00560
       INTEGER  SVSEED,SVSOLD,RHSEED,IDELTA(50)                        LGV00570
       INTEGER  MBOUND,NTVCON,SVTVEC,TVSTOP,LVCONT,ERCONT,TFLAG        LGV00580
       DOUBLE PRECISION  FINPRO                                        LGV00590
       DOUBLE PRECISION  DABS, DMAX1, DSQRT, DFLOAT                    LGV00600
       REAL ABS                                                        LGV00610
       INTEGER  IABS                                                   LGV00620
       EXTERNAL LSOLV, AMATV                                           LGV00630
C--------------------------------------------------------------------LGV00640
```

```
      DATA MACHEP/Z3410000000000000/                                 LGV00650
      EPSM = 2.DO*MACHEP                                             LGV00660
C-----------------------------------------------------------------------LGV00670
C                                                                   LGV00680
C     ARRAYS MUST BE DIMENSIONED AS FOLLOWS:                        LGV00690
C     1.  ALPHA:  >= KMAXN,  BETA: >= (KMAXN+1) WHERE KMAXN, THE    LGV00700
C                 LARGEST SIZE T-MATRIX CONSIDERED BY THE PROGRAM,  LGV00710
C                 IS THE LARGER OF THE SIZE OF THE ALPHA, BETA HISTORY LGV00720
C                 PROVIDED ON FILE 2 (IF ANY ) AND THE SIZE WHICH THE LGV00730
C                 PROGRAM SPECIFIES INTERNALLY, THIS LATTER IS ALWAYS LGV00740
C                 < = 11*MEV / 8  +  12, WHERE MEV IS THE SIZE      LGV00750
C                 T-MATRIX THAT WAS USED IN THE CORRESPONDING EIGENVALUE LGV00760
C                 COMPUTATIONS.                                     LGV00770
C     2.  V1:  >= MAX(N,KMAX)                                       LGV00780
C     3.  V2,VS:   >= N                                             LGV00790
C     4.  G:  >= MAX(N,KMAX)                                        LGV00800
C     5.  RITVEC:  >= N*NGOOD, WHERE NGOOD IS NUMBER OF EIGENVALUES LGV00810
C                  SUPPLIED TO THIS PROGRAM.                        LGV00820
C     6.  TVEC:  >= CUMULATIVE LENGTH OF ALL THE T-EIGENVECTORS     LGV00830
C                 NEEDED TO GENERATE THE DESIRED RITZ VECTORS.  AN EDUCATED LGV00840
C                 GUESS AT AN APPROPRIATE LENGTH CAN BE OBTAINED BY RUNNING THE LGV00850
C                 PROGRAM WITH THE FLAG MBOUND = 1 AND MULTIPLYING THE LGV00860
C                 RESULTING SIZE BY 5/4.                            LGV00870
C     7.  GOODEV, AMINGP, TMINGP, TERR, ERR, ERRGDP, RNORM, TBETA,  LGV00880
C                 TLAST, EVNEW, MP, MA, M1, M2, MINT, MFIN AND IDELTA ALL MUST LGV00890
C                 BE >= NGOOD.                                      LGV00900
C                                                                   LGV00910
C-----------------------------------------------------------------------LGV00920
C     OUTPUT HEADER                                                 LGV00930
      WRITE(6,10)                                                   LGV00940
   10 FORMAT(/' LANCZOS EIGENVECTOR PROCEDURE FOR REAL SYMMETRIC MATRICELGV00950
     1S'/)                                                         LGV00960
C                                                                   LGV00970
C     SET PROGRAM PARAMETERS                                        LGV00980
C     USER MUST NOT MODIFY SCALE0                                   LGV00990
      SCALE0 = 5.0D0                                                LGV01000
      ZERO = 0.0D0                                                  LGV01010
      ONE = 1.0D0                                                   LGV01020
      MPMIN = -1000                                                 LGV01030
C     SET CONVERGENCE CRITERION FOR T-EIGENVECTORS.                 LGV01040
      ERTOL = 1.D-10                                                LGV01050
C                                                                   LGV01060
C     READ USER-SPECIFIED PARAMETER FROM INPUT FILE 5 (FREE FORMAT) LGV01070
C                                                                   LGV01080
C     READ USER-PROVIDED HEADER FOR RUN                             LGV01090
      READ(5,20) EXPLAN                                             LGV01100
      WRITE(6,20) EXPLAN                                            LGV01110
   20 FORMAT(20A4)                                                  LGV01120
C                                                                   LGV01130
C     READ IN THE MAXIMUM PERMISSIBLE DIMENSIONS FOR THE TVEC ARRAY LGV01140
C     (MDIMTV), FOR THE RITVEC ARRAY (MDIMRV), AND FOR THE BETA     LGV01150
C     ARRAY (MBETA).                                                LGV01160
      READ(5,20) EXPLAN                                             LGV01170
      READ(5,*) MDIMTV, MDIMRV, MBETA                               LGV01180
C                                                                   LGV01190
C     READ IN RELATIVE TOLERANCE (RELTOL) USED IN DETERMINING       LGV01200
C     APPROPRIATE SIZES FOR THE T-MATRICES TO BE USED IN THE RITZ   LGV01210
C     VECTOR COMPUTATIONS.                                          LGV01220
      READ(5,20) EXPLAN                                             LGV01230
      READ(5,*) RELTOL                                              LGV01240
C                                                                   LGV01250
C     SET FLAGS TO 0 OR 1:                                          LGV01260
C     MBOUND = 1:  PROGRAM TERMINATES AFTER COMPUTING 1ST GUESSES   LGV01270
C                  ON APPROPRIATE T-SIZES FOR USE IN THE RITZ VECTOR LGV01280
C                  COMPUTATIONS                                     LGV01290
C     NTVCON = 0:  PROGRAM TERMINATES IF THE TVEC ARRAY IS NOT      LGV01300
C                  LARGE ENOUGH TO HOLD ALL THE T-EIGENVECTORS REQUIRED.LGV01310
```

```
C          SVTVEC = 0:   THE T-EIGENVECTORS ARE NOT WRITTEN TO FILE 11      LGVO1320
C                        UNLESS TVSTOP = 1                                  LGVO1330
C          SVTVEC = 1:   WRITE THE T-EIGENVECTORS TO FILE 11.               LGVO1340
C          TVSTOP = 1:   PROGRAM TERMINATES AFTER COMPUTING THE             LGVO1350
C                        T-EIGENVECTORS                                     LGVO1360
C          LVCONT = 0:   PROGRAM TERMINATES IF THE NUMBER OF T-EIGENVECTORS LGVO1370
C                        COMPUTED IS NOT EQUAL TO THE NUMBER OF RITZ        LGVO1380
C                        VECTORS REQUESTED.                                 LGVO1390
C          ERCONT = 0:   MEANS FOR ANY GIVEN EIGENVALUE, A RITZ VECTOR      LGVO1400
C                        WILL NOT BE COMPUTED FOR THAT EIGENVALUE UNLESS    LGVO1410
C                        A T-EIGENVECTOR HAS BEEN IDENTIFIED WITH A LAST    LGVO1420
C                        COMPONENT WHICH SATISFIES THE SPECIFIED           LGVO1430
C                        CONVERGENCE CRITERION.                             LGVO1440
C          ERCONT = 1:   MEANS FOR ANY GIVEN EIGENVALUE, A RITZ VECTOR      LGVO1450
C                        WILL BE COMPUTED.  IF A T-EIGENVECTOR CANNOT       LGVO1460
C                        BE IDENTIFIED WHICH SATISFIES THE LAST            LGVO1470
C                        COMPONENT CRITERION, THEN THE PROGRAM WILL        LGVO1480
C                        USE THE T-VECTOR THAT CAME CLOSEST TO            LGVO1490
C                        SATISFYING THE CRITERION                          LGVO1500
C          IWRITE = 1:   EXTENDED OUTPUT OF INTERMEDIATE COMPUTATIONS       LGVO1510
C                        IS WRITTEN TO FILE 6                               LGVO1520
C          IREAD = 0:    ALPHA/BETA FILE IS REGENERATED.                    LGVO1530
C          IREAD = 1:    ALPHA/BETA FILE USED IN EIGENVALUE COMPUTATIONS    LGVO1540
C                        IS READ IN AND EXTENDED IF NECESSARY.  IN BOTH     LGVO1550
C                        CASES IREAD = 0 OR 1, THE LANCZOS VECTORS ARE      LGVO1560
C                        ALWAYS REGENERATED FOR THE RITZ VECTOR             LGVO1570
C                        COMPUTATIONS                                       LGVO1580
C                                                                           LGVO1590
           READ(5,20) EXPLAN                                                LGVO1600
           READ(5,*) MBOUND,NTVCON,SVTVEC,IREAD                             LGVO1610
C                                                                           LGVO1620
           READ(5,20) EXPLAN                                                LGVO1630
           READ(5,*) TVSTOP,LVCONT,ERCONT,IWRITE                            LGVO1640
           IF (TVSTOP.EQ.1) SVTVEC = 1                                      LGVO1650
C                                                                           LGVO1660
C     READ IN SEED (RHSEED) FOR GENERATING RANDOM STARTING VECTOR           LGVO1670
C     FOR INVERSE ITERATION ON THE T-MATRICES.                             LGVO1680
           READ(5,20) EXPLAN                                                LGVO1690
           READ(5,*) RHSEED                                                 LGVO1700
C                                                                           LGVO1710
C     READ IN MATNOA, MATNOB = MATRIX/RUN IDENTIFICATION NUMBERS,           LGVO1720
C     N = ORDER OF A-MATRIX AND B-MATRIX AND FLAG, JPERM.                   LGVO1730
C     JPERM = (0,1):  1 MEANS PERMUTED A AND B ARE BEING USED, 0            LGVO1740
C     MEANS A AND B HAVE NOT BEEN PERMUTED.                                LGVO1750
           READ(5,20) EXPLAN                                                LGVO1760
           READ(5,*) MATNOA,MATNOB,N,JPERM                                  LGVO1770
C                                                                           LGVO1780
C-----------------------------------------------------------------------LGVO1790
C     INITIALIZE THE ARRAYS FOR THE USER-SPECIFIED MATRICES                 LGVO1800
C     AND PASS THE STORAGE LOCATIONS OF THESE ARRAYS TO THE                 LGVO1810
C     MATRIX-VECTOR MULTIPLY SUBROUTINE AMATV AND THE SOLVE                 LGVO1820
C     SUBROUTINE LSOLV.                                                     LGVO1830
           CALL USPECA(N,MATNOA)                                           LGVO1840
           CALL USPECB(N,MATNOB)                                           LGVO1850
C                                                                           LGVO1860
C-----------------------------------------------------------------------LGVO1870
C                                                                           LGVO1880
C     MASK UNDERFLOW AND OVERFLOW                                           LGVO1890
           CALL MASK                                                        LGVO1900
C                                                                           LGVO1910
C-----------------------------------------------------------------------LGVO1920
C     WRITE RUN PARAMETERS OUT TO FILE 6                                    LGVO1930
C                                                                           LGVO1940
           WRITE(6,30) MATNOA,MATNOB,N,JPERM                                LGVO1950
 30    FORMAT(/4X,'A-MATRIX ID',4X,'B-MATRIX ID',4X,'SIZES OF MATRICES', LGVO1960
      14X,'JPERM'/I15,I15,I21,I9)                                          LGVO1970
C                                                                           LGVO1980
```

```
      WRITE(6,40) MBOUND,NTVCON,SVTVEC,IREAD                            LGV01990
   40 FORMAT(/3X,'MBOUND',3X,'NTVCON',3X,'SVTVEC',3X,'IREAD'/3I9,I8)    LGV02000
C                                                                       LGV02010
      WRITE(6,50) TVSTOP,LVCONT,ERCONT,IWRITE                           LGV02020
   50 FORMAT(/3X,'TVSTOP',3X,'LVCONT',3X,'ERCONT',3X,'IWRITE'/4I9)      LGV02030
C                                                                       LGV02040
      WRITE(6,60) MDIMTV,MDIMRV,MBETA                                   LGV02050
   60 FORMAT(/3X,'MDIMTV',3X,'MDIMRV',3X,'MBETA'/2I9,I8)               LGV02060
C                                                                       LGV02070
      WRITE(6,70) RELTOL,RHSEED                                         LGV02080
   70 FORMAT(/7X,'RELTOL',3X,'RHSEED'/E13.4,I9)                        LGV02090
C                                                                       LGV02100
C                                                                       LGV02110
C     FROM FILE 3 READ IN THE NUMBER OF EIGENVALUES (NGOOD) FOR WHICH   LGV02120
C     EIGENVECTORS ARE REQUESTED, THE ORDER (MEV) OF THE LANCZOS        LGV02130
C     TRIDIAGONAL MATRIX USED IN COMPUTING THESE EIGENVALUES, THE       LGV02140
C     ORDER (NOLD) OF THE USER-SPECIFIED MATRIX USED IN THE EIGENVALUE  LGV02150
C     COMPUTATIONS, THE SEED (SVSEED) USED FOR GENERATING THE STARTING  LGV02160
C     VECTOR THAT WAS USED IN THOSE LANCZOS EIGENVALUE COMPUTATIONS,    LGV02170
C     AND THE MATRIX/RUN IDENTIFICATION NUMBERS (MATA, MATB) USED IN    LGV02180
C     THOSE COMPUTATIONS.  ALSO READ IN THE NUMBER (NDIS) OF DISTINCT   LGV02190
C     EIGENVALUES OF T(1,MEV) THAT WERE COMPUTED BUT THIS VALUE IS      LGV02200
C     NOT USED IN THE EIGENVECTOR COMPUTATIONS.                        LGV02210
C                                                                       LGV02220
      READ(3,80) NGOOD,NDIS,MEV,NOLD,SVSEED,MATA,MATB                   LGV02230
   80 FORMAT(4I6,I12,2I8)                                               LGV02240
C                                                                       LGV02250
C     READ IN THE T-MULTIPLICITY TOLERANCE USED IN THE BISEC SUBROUTINE LGV02260
C     DURING THE COMPUTATION OF THE GIVEN EIGENVALUES.                  LGV02270
C     ALSO READ IN THE FLAG IB.  IF IB < 0, THEN SOME BETA(I) IN THE    LGV02280
C     T-MATRIX FILE PROVIDED ON FILE 2 FAILED THE ORTHOGONALITY         LGV02290
C     TEST IN THE TNORM SUBROUTINE.  USER SHOULD NOTE THAT THIS VECTOR  LGV02300
C     PROGRAM PROCEEDS INDEPENDENTLY OF THE SIZE OF THE BETA USED.      LGV02310
C                                                                       LGV02320
      READ(3,90) MULTOL,IB,BTOL                                         LGV02330
   90 FORMAT(E20.12,I6,E13.4)                                           LGV02340
C                                                                       LGV02350
      TEMP = DFLOAT(MEV+1000)                                           LGV02360
      TTOL = MULTOL/TEMP                                                LGV02370
      WRITE(6,100) MULTOL,TTOL                                          LGV02380
  100 FORMAT(/' T-MULTIPLICITY TOLERANCE USED IN THE EIGENVALUE COMPUTATLGV02390
     1IONS WAS',E13.4/' SCALED MACHINE EPSILON IS',E13.4)              LGV02400
C                                                                       LGV02410
C     CONTINUE WRITE TO FILE 6 OF THE PARAMETERS FOR THIS RUN           LGV02420
C                                                                       LGV02430
      WRITE(6,110)NGOOD,NDIS,MEV,NOLD,MATA,MATB,SVSEED,MULTOL,IB,BTOL   LGV02440
  110 FORMAT(/' EIGENVALUES SUPPLIED ARE READ IN FROM FILE 3'/' FILE 3  LGV02450
     1HEADER IS'/4X,'NG',2X,'NDIS',3X,'MEV',2X,'NOLD',2X,'MATNOA',2X,   LGV02460
     1'MATNOB'/(4I6,2I8)/4X,'SVSEED',6X,'MULTOL',6X,'IB',9X,'BTOL'/     LGV02470
     1I12,E12.3,I8,E13.4/)                                              LGV02480
C                                                                       LGV02490
C     IS THE ARRAY RITVEC LONG ENOUGH TO HOLD ALL OF THE DESIRED        LGV02500
C     RITZ VECTORS (APPROXIMATE EIGENVECTORS)?                          LGV02510
      NMAX = NGOOD*N                                                    LGV02520
      IF(MBOUND.NE.0) GO TO 120                                         LGV02530
      IF(TVSTOP.NE.1.AND.NMAX.GT.MDIMRV) GO TO 1350                     LGV02540
C                                                                       LGV02550
C     CHECK THAT THE ORDER N AND THE MATRIX IDENTIFICATION NUMBERS      LGV02560
C     MATNOA AND MATNOB SPECIFIED BY THE USER AGREE WITH THOSE READ     LGV02570
C     IN FROM FILE 3.                                                   LGV02580
  120 ITEMP = (NOLD-N)**2 + (MATA-MATNOA)**2 + (MATB-MATNOB)**2         LGV02590
      IF (ITEMP.NE.0) GO TO 1370                                        LGV02600
C                                                                       LGV02610
C     READ IN FROM FILE 3, THE T-MULTIPLICITIES OF THE EIGENVALUES      LGV02620
C     WHOSE EIGENVECTORS ARE TO BE COMPUTED, THE VALUES OF THESE        LGV02630
C     EIGENVALUES AND THEIR MINIMAL GAPS AS EIGENVALUES OF THE          LGV02640
```

```
C      USER-SPECIFIED MATRIX AND AS EIGENVALUES OF THE T-MATRIX.         LGV02650
C                                                                        LGV02660
       READ(3,20) EXPLAN                                                 LGV02670
       READ(3,130) (MP(J),GOODEV(J),TMINGP(J),AMINGP(J), J=1,NGOOD)      LGV02680
 130   FORMAT(6X,I6,E25.16,2E14.3)                                       LGV02690
C                                                                        LGV02700
       WRITE(6,140) (J,GOODEV(J),MP(J),TMINGP(J),AMINGP(J), J=1,NGOOD)   LGV02710
 140   FORMAT(/' EIGENVALUES READ IN, T-MULTIPLICITIES, T-GAPS AND A-GAPSLGV02720
      1 '/4X,' J ',5X,'GOOD EIGENVALUE',5X,'MULT',4X,' TMINGAP ',4X,     LGV02730
      1' ABMINGAP '/(I6,E25.16,I4,2E15.4))                              LGV02740
C                                                                        LGV02750
C      READ IN ERROR ESTIMATES                                           LGV02760
       WRITE(6,150) MEV,SVSEED                                           LGV02770
 150   FORMAT(/' THESE EIGENVALUES WERE COMPUTED USING A T-MATRIX OF     LGV02780
      1ORDER ',I5/' AND SEED FOR RANDOM NUMBER GENERATOR =',I12)         LGV02790
C      CHECK WHETHER OR NOT THERE ARE ANY T-ISOLATED EIGENVALUES IN      LGV02800
C      THE EIGENVALUES PROVIDED                                          LGV02810
       DO 160 J=1,NGOOD                                                  LGV02820
       IF(MP(J).EQ.1) GO TO 170                                          LGV02830
 160   CONTINUE                                                          LGV02840
       GO TO 190                                                         LGV02850
 170   READ(4,20) EXPLAN                                                 LGV02860
       READ(4,20) EXPLAN                                                 LGV02870
       READ(4,20) EXPLAN                                                 LGV02880
       READ(4,180) NISO                                                  LGV02890
 180   FORMAT(18X,I6)                                                    LGV02900
       READ(4,20) EXPLAN                                                 LGV02910
       READ(4,20) EXPLAN                                                 LGV02920
       READ(4,20) EXPLAN                                                 LGV02930
 190   DO 220 J=1,NGOOD                                                  LGV02940
       ERR(J) = 0.D0                                                     LGV02950
       IF(MP(J).NE.1) GO TO 220                                          LGV02960
       READ(4,200) EVAL, ERR(J)                                          LGV02970
 200   FORMAT(10X,E25.16,E14.3)                                          LGV02980
       IF(DABS(EVAL - GOODEV(J)).LT.1.D-10) GO TO 220                    LGV02990
       WRITE(6,210) EVAL,GOODEV(J)                                       LGV03000
 210   FORMAT(' PROBLEM WITH READ IN OF ERROR ESTIMATES'/' EIGENVALUE REALGV03010
      1D IN',E20.12,' DOES NOT MATCH GOODEV(J) ='/E20.12)               LGV03020
       GO TO 1590                                                        LGV03030
C                                                                        LGV03040
 220   CONTINUE                                                          LGV03050
C                                                                        LGV03060
       WRITE(6,230) (J,GOODEV(J),ERR(J), J=1,NGOOD)                      LGV03070
 230   FORMAT(' ERROR ESTMATES ='/4X,' J',5X,'EIGENVALUE',10X,' ESTIMATE LGV03080
      1'/(I6,E20.12,E14.3))                                             LGV03090
C                                                                        LGV03100
       IF(IREAD.EQ.0)  GO TO 330                                         LGV03110
C                                                                        LGV03120
C      READ IN THE SIZE OF THE T-MATRIX PROVIDED ON FILE 2.  READ IN     LGV03130
C      THE ORDER OF THE USER-SPECIFIED MATRIX , THE SEED FOR THE         LGV03140
C      RANDOM NUMBER GENERATOR, AND THE MATRIX/TEST IDENTIFICATION       LGV03150
C      NUMBERS THAT WERE USED IN THE LANCZOS EIGENVALUE COMPUTATIONS.    LGV03160
C      THESE ARE USED IN A CONSISTENCY CHECK                            LGV03170
C      IF FLAG IREAD = 0 REGENERATE ALPHA, BETA                          LGV03180
C                                                                        LGV03190
       READ(2,240) KMAX,NOLD,SVSOLD,MATA,MATB                           LGV03200
 240   FORMAT(2I6,I12,2I8)                                               LGV03210
C                                                                        LGV03220
       WRITE(6,250) KMAX,NOLD,SVSOLD,MATA,MATB                          LGV03230
 250   FORMAT(/' READ IN THE T-MATRICES STORED ON FILE 2'/' FILE 2 HEADERLGV03240
      1 IS'/2X,'KMAX',2X,'NOLD',6X,'SVSOLD',2X,'MATNOA',2X,'MATNOB'/     LGV03250
      1 2I6,I12,2I8/)                                                    LGV03260
C                                                                        LGV03270
C      CHECK THAT THE ORDER, THE MATRIX/TEST IDENTIFICATION NUMBERS      LGV03280
C      AND THE SEED FOR THE RANDOM NUMBER GENERATOR USED IN THE          LGV03290
C      LANCZOS COMPUTATIONS THAT GENERATED THE HISTORY FILE              LGV03300
C      BEING USED AGREE WITH WHAT THE USER HAS SPECIFIED.                LGV03310
```

```
         IF (NOLD.NE.N.OR.MATA.NE.MATNOA.OR.MATNOB.NE.MATB.OR.SVSOLD.NE.   LGV03320
        1 SVSEED) GO TO 1390                                               LGV03330
C                                                                          LGV03340
         KMAX1 = KMAX + 1                                                  LGV03350
C                                                                          LGV03360
C        READ IN THE T-MATRICES FROM FILE 2.  THESE ARE USED TO GENERATE   LGV03370
C        THE T-EIGENVECTORS THAT WILL BE USED IN THE RITZ VECTOR           LGV03380
C        COMPUTATIONS.  HISTORY MUST BE STORED IN MACHINE FORMAT           LGV03390
C        ((4Z20) FOR IBM/3081).                                           LGV03400
C                                                                          LGV03410
         READ(2,260) (ALPHA(J), J=1,KMAX)                                  LGV03420
         READ(2,260) (BETA(J), J=1,KMAX1)                                  LGV03430
     260 FORMAT(4Z20)                                                      LGV03440
C                                                                          LGV03450
         READ(2,260) (V1(J), J=1,N)                                        LGV03460
         READ(2,260) (VS(J), J=1,N)                                        LGV03470
         READ(2,260) (V2(J), J=1,N)                                        LGV03480
C                                                                          LGV03490
C        KMAX MAY BE ENLARGED IF THE SIZE AT WHICH THE EIGENVALUE          LGV03500
C        COMPUTATIONS WERE PERFORMED IS ESSENTIALLY KMAX AND               LGV03510
C        THERE IS AT LEAST ONE EIGENVALUE THAT IS T-SIMPLE AND             LGV03520
C        T-ISOLATED, IN THE SENSE THAT IF ITS NEAREST NEIGHBOR IS TOO      LGV03530
C        CLOSE THAT NEIGHBOR IS A 'GOOD' T-EIGENVALUE.                     LGV03540
         DO 270 J = 1,NGOOD                                                LGV03550
         IF(MP(J).EQ.1) GO TO 290                                          LGV03560
     270 CONTINUE                                                          LGV03570
         WRITE(6,280)                                                      LGV03580
     280 FORMAT(/' ALL EIGENVALUES USED ARE T-MULTIPLE OR CLOSE TO SPURIOUSLGV03590
        1 T-EIGENVALUES'/' SO DO NOT CHANGE KMAX')                         LGV03600
         IF(KMAX.LT.MEV) GO TO 1410                                        LGV03610
         GO TO 310                                                         LGV03620
C                                                                          LGV03630
     290 KMAXN= 11*MEV/8 + 12                                              LGV03640
         IF(MBETA.LE.KMAXN) GO TO 1570                                     LGV03650
         IF(KMAX.GE.KMAXN ) GO TO 310                                      LGV03660
         WRITE(6,300) KMAX, KMAXN                                          LGV03670
     300 FORMAT(' ENLARGE KMAX FROM ',I6,' TO ',I6)                        LGV03680
         MOLD1 = KMAX + 1                                                  LGV03690
         KMAX = KMAXN                                                      LGV03700
         GO TO 380                                                         LGV03710
C                                                                          LGV03720
     310 WRITE(6,320) KMAX                                                 LGV03730
     320 FORMAT(/' T-MATRICES HAVE BEEN READ IN FROM FILE 2'/' THE LARGEST LGV03740
        1SIZE T-MATRIX ALLOWED IS',I6/)                                    LGV03750
C                                                                          LGV03760
         IF(IREAD.EQ.1) GO TO 400                                          LGV03770
C                                                                          LGV03780
C        REGENERATE THE ALPHA AND BETA                                     LGV03790
C                                                                          LGV03800
     330 MOLD1 = 1                                                         LGV03810
C                                                                          LGV03820
         DO 340 J = 1,NGOOD                                                LGV03830
         IF(MP(J).EQ.1) GO TO 360                                          LGV03840
     340 CONTINUE                                                          LGV03850
         KMAX = MEV + 12                                                   LGV03860
         WRITE(6,350) KMAX                                                 LGV03870
     350 FORMAT(/' ALL EIGENVALUES FOR WHICH EIGENVECTORS ARE TO BE COMPUTELGV03880
        1D ARE EITHER T-MULTIPLE OR CLOSE TO'/' A SPURIOUS T-EIGENVALUE. THLGV03890
        1EREFORE SET KMAX = MEV + 12 = ',I7)                              LGV03900
         GO TO 380                                                         LGV03910
C                                                                          LGV03920
     360 KMAXN = 11*MEV/8 + 12                                             LGV03930
         IF(MBETA.LE.KMAXN) GO TO 1570                                     LGV03940
         WRITE(6,370) KMAXN                                                LGV03950
     370 FORMAT(' SET KMAX EQUAL TO ',I6)                                  LGV03960
         KMAX = KMAXN                                                      LGV03970
C                                                                          LGV03980
```

```
    380 WRITE(6,390) MOLD1,KMAX                                          LGV03990
    390 FORMAT(/' LANCZS SUBROUTINE GENERATES ALPHA(J), BETA(J+1), J =',  LGV04000
       1 16,' TO ', 16/)                                                 LGV04010
C                                                                        LGV04020
C------------------------------------------------------------------------LGV04030
C                                                                        LGV04040
      IIX = SVSEED                                                       LGV04050
      CALL LANCZS(LSOLV,AMATV,ALPHA,BETA,V1,V2,VS,G,KMAX,MOLD1,N,IIX)    LGV04060
C                                                                        LGV04070
C------------------------------------------------------------------------LGV04080
C                                                                        LGV04090
  400 CONTINUE                                                           LGV04100
C                                                                        LGV04110
C     THE SUBROUTINE STURMI DETERMINES THE SMALLEST SIZE T-MATRIX FOR    LGV04120
C     WHICH THE EIGENVALUE IN QUESTION IS A T-EIGENVALUE (TO WITHIN A    LGV04130
C     GIVEN TOLERANCE) AND IF POSSIBLE THE SMALLEST SIZE T-MATRIX        LGV04140
C     FOR WHICH IT IS A DOUBLE T-EIGENVALUE (TO WITHIN THE SAME          LGV04150
C     TOLERANCE).  THE SIZE T-MATRIX USED IN THE RITZ VECTOR            LGV04160
C     COMPUTATIONS IS THEN DETERMINED BY LOOPING ON SIZE OF THE          LGV04170
C     T-EIGENVECTORS, STARTING WITH A T-SIZE DETERMINED BY STURMI.       LGV04180
C                                                                        LGV04190
C                                                                        LGV04200
      STUTOL = SCALEO*MULTOL                                             LGV04210
      IF(IWRITE.EQ.1) WRITE(6,410)                                       LGV04220
  410 FORMAT(' FROM STURMI')                                             LGV04230
      DO 450 J = 1,NGOOD                                                 LGV04240
      EVAL = GOODEV(J)                                                   LGV04250
C     COMPUTE THE TOLERANCES USED BY STURMI TO DETERMINE AN INTERVAL     LGV04260
C     CONTAINING THE EIGENVALUE EVAL.                                    LGV04270
      TEMP = DABS(EVAL)*RELTOL                                           LGV04280
      TOLN = DMAX1(TEMP,STUTOL)                                          LGV04290
C                                                                        LGV04300
C------------------------------------------------------------------------LGV04310
C                                                                        LGV04320
      CALL STURMI(ALPHA,BETA,EVAL,TOLN,EPSM,KMAX,MK1,MK2,IC,IWRITE)      LGV04330
C                                                                        LGV04340
C------------------------------------------------------------------------LGV04350
C                                                                        LGV04360
C     STORE THE COMPUTED ORDERS OF T-MATRICES FOR LATER PRINTOUT         LGV04370
      M1(J) = MK1                                                        LGV04380
      M2(J) = MK2                                                        LGV04390
      ML(J) = (MK1 + 3*MK2)/4                                            LGV04400
      IF(MK2.EQ.KMAX)  ML(J) = KMAX                                      LGV04410
C                                                                        LGV04420
      IF(IC.GT.0) GO TO 430                                              LGV04430
C     IC = 0 MEANS THERE WAS NO EIGENVALUE IN THE DESIGNATED INTERVAL    LGV04440
C     BY T-SIZE KMAX.  THIS MEANS THAT THE EIGENVALUE PROVIDED HAS       LGV04450
C     NOT YET CONVERGED SO ITS EIGENVECTOR IS NOT COMPUTED.             LGV04460
      WRITE(6,420) J,GOODEV(J),MK1,MK2                                   LGV04470
  420 FORMAT(I6,'TH EIGENVALUE',E20.12,' HAS NOT CONVERGED '/           LGV04480
       1' SO DO NOT COMPUTE ANY T-EIGENVECTOR OR RITZ VECTOR FOR IT'     LGV04490
       1/' MK1 AND MK2 FOR THIS EIGENVALUE WERE',2I6)                    LGV04500
      MP(J) = MPMIN                                                      LGV04510
      MA(J) = -2*KMAX                                                    LGV04520
      GO TO 450                                                          LGV04530
C     COMPUTE AN APPROPRIATE SIZE T-MATRIX FOR THE GIVEN EIGENVALUE.     LGV04540
  430 IF(M2(J).EQ.KMAX) GO TO 440                                        LGV04550
C     M1 AND M2 WERE BOTH DETERMINED                                     LGV04560
      MA(J) = (3*M1(J) + M2(J))/4  + 1                                   LGV04570
      GO TO 450                                                          LGV04580
C     M2 NOT DETERMINED                                                  LGV04590
  440 MA(J) = (5*M1(J))/4  + 1                                           LGV04600
C                                                                        LGV04610
  450 CONTINUE                                                           LGV04620
C                                                                        LGV04630
      IF (IWRITE.EQ.1) WRITE(6,460) (MA(JJ), JJ=1,NGOOD)                 LGV04640
  460 FORMAT(/' 1ST GUESS AT APPROPRIATE SIZE T-MATRICES'/               LGV04650
```

```
      1 ' ACTUAL VALUES WILL PROBABLY BE 1/4 AGAIN AS MUCH'/(13I6))      LGV04660
C                                                                        LGV04670
C     PRINT OUT TO FILE 10 1ST GUESSES AT SIZES OF THE T-MATRICES TO     LGV04680
C     BE USED IN THE EIGENVECTOR COMPUTATIONS.                           LGV04690
C     PROGRAM LOOPS ON T-SIZE TO DETERMINE APPROPRIATE SIZE T-MATRIX.    LGV04700
      WRITE(10,470) N,KMAX                                               LGV04710
  470 FORMAT(2I8,' = ORDER OF USER MATRIX AND MAX ORDER OF T(1,MEV)')    LGV04720
C                                                                        LGV04730
      WRITE(10,480)                                                      LGV04740
  480 FORMAT(/' 1ST GUESS AT APPROPRIATE SIZE T-MATRICES'/               LGV04750
      1 ' ACTUAL VALUES WILL PROBABLY BE 1/4 AGAIN AS MUCH'/)            LGV04760
C                                                                        LGV04770
      WRITE(10,490)                                                      LGV04780
  490 FORMAT(4X,'J',3X,'AB-EIGENVALUE',4X,'M1(J)',1X,'M2(J)',1X,'MA(J)')LGV04790
C                                                                        LGV04800
      WRITE(10,500) (J,GOODEV(J),M1(J),M2(J), MA(J), J=1,NGOOD)          LGV04810
  500 FORMAT(I5,E19.12,3I6)                                              LGV04820
C                                                                        LGV04830
      IF(MBOUND.EQ.1) WRITE(10,510)                                      LGV04840
  510 FORMAT(/'  EV = AB-EIGENVALUE IS A GOOD EIGENVALUE OF T(1,MEV)'/   LGV04850
      1' M1 = SMALLEST VALUE OF M SUCH THAT T(1,M) HAS AT LEAST'/        LGV04860
      1 '    ONE EIGENVALUE IN THE INTERVAL (EV-TOLN,EV+TOLN)'/          LGV04870
      1 '    TOLN(J) = DMAX1(EV(J)*RELTOL, SCALEO*MULTOL)'/              LGV04880
      1 ' M2 = SMALLEST M (IF ANY) SUCH THAT IN THE ABOVE INTERVAL'/     LGV04890
      1 '    T(1,M) HAS AT LEAST TWO EIGENVALUES '/                      LGV04900
      1 ' IABS(MA(J)) = APPROPRIATE SIZE T-MATRIX FOR EV(J)'/            LGV04910
      1 ' INITIAL VALUE OF MA(J) IS CHOSEN HEURISTICALLY'/               LGV04920
      1 ' PROGRAM LOOPS ON SIZE OF T-MATRIX TO GET BETTER SIZE'/         LGV04930
      1 ' END OF SIZES OF T-MATRICES FILE 10'///)                       LGV04940
C                                                                        LGV04950
C                                                                        LGV04960
C     TERMINATE AFTER COMPUTING 1ST GUESSES AT  SIZES OF THE            LGV04970
C     T-MATRICES REQUIRED FOR THE GIVEN EIGENVALUES?                     LGV04980
      IF(MBOUND.EQ.1) GO TO 1430                                         LGV04990
C                                                                        LGV05000
C                                                                        LGV05010
C     IS THERE ROOM FOR ALL OF THE REQUESTED T-EIGENVECTORS?             LGV05020
      MTOL = 0                                                           LGV05030
      DO 520 J = 1,NGOOD                                                 LGV05040
      IF(MP(J).EQ.MPMIN) GO TO 520                                       LGV05050
      MTOL = MTOL + IABS(MA(J))                                          LGV05060
  520 CONTINUE                                                           LGV05070
      MTOL = (5*MTOL)/4                                                  LGV05080
      IF(MTOL.GT.MDIMTV.AND.NTVCON.EQ.0) GO TO 1450                      LGV05090
C                                                                        LGV05100
C------------------------------------------------------------------------LGV05110
C     GENERATE A RANDOM VECTOR TO BE USED REPEATEDLY BY                  LGV05120
C     SUBROUTINE INVERM                                                  LGV05130
C                                                                        LGV05140
      IIL = RHSEED                                                       LGV05150
      CALL GENRAN(IIL,G,KMAX)                                            LGV05160
C                                                                        LGV05170
C------------------------------------------------------------------------ LGV05180
C                                                                        LGV05190
C     LOOP ON GIVEN EIGENVALUES TO COMPUTE THE CORRESPONDING             LGV05200
C     T-EIGENVECTOR.                                                     LGV05210
C                                                                        LGV05220
      MTOL = 0                                                           LGV05230
      NTVEC = 0                                                          LGV05240
      ILBIS = 0                                                          LGV05250
      DO 710 J = 1,NGOOD                                                 LGV05260
      ICOUNT = 0                                                         LGV05270
      ERRMIN = 10.D0                                                     LGV05280
      MABEST = MPMIN                                                     LGV05290
      IF(MP(J).EQ.MPMIN) GO TO 710                                       LGV05300
      TFLAG = 0                                                          LGV05310
      EVAL = GOODEV(J)                                                   LGV05320
```

```
      TEMP = DABS(EVAL)*RELTOL                                         LGV05330
      UB = EVAL + DMAX1(STUTOL,TEMP)                                   LGV05340
      LB = EVAL - DMAX1(STUTOL,TEMP)                                   LGV05350
  530 KMAXU = IABS(MA(J))                                              LGV05360
C                                                                      LGV05370
C     SELECT A SUITABLE INCREMENT FOR THE ORDERS OF THE T-MATRICES     LGV05380
C     TO BE CONSIDERED IN DETERMINING APPROPRIATE SIZES FOR THE RITZ   LGV05390
C     VECTOR COMPUTATIONS.                                             LGV05400
      IF(ICOUNT.GT.0) GO TO 550                                        LGV05410
C     SELECT IDELTA(J) BASED UPON THE T-MULTIPLICITY OBTAINED          LGV05420
      IF(M2(J).EQ.KMAX) GO TO 540                                      LGV05430
C     M2 DETERMINED                                                    LGV05440
      IDELTA(J) = ((3*M1(J) + 5*M2(J))/8  +  1 - IABS(MA(J)))/10  +  1 LGV05450
      GO TO 550                                                        LGV05460
C     M2 NOT DETERMINED                                                LGV05470
  540 MAMAX = MINO((11*MEV)/8 + 12, (13*M1(J))/8 + 1)                  LGV05480
      IDELTA(J) = (MAMAX - IABS(MA(J)))/10  +  1                       LGV05490
  550 ICOUNT = ICOUNT + 1                                              LGV05500
C                                                                      LGV05510
C---------------------------------------------------------------------LGV05520
C     TO MIMIMIZE THE EFFECT OF THE ONE-SIDED ACCEPTANCE TEST FOR      LGV05530
C     EIGENVALUES IN THE BISEC SUBROUTINE, RECOMPUTE THE GIVEN         LGV05540
C     EIGENVALUE AT THE SPECIFIED KMAXU                                LGV05550
C                                                                      LGV05560
      CALL LBISEC(ALPHA,BETA,EPSM,EVAL,EVALN,LB,UB,TTOL,KMAXU,NEVT)    LGV05570
C                                                                      LGV05580
C---------------------------------------------------------------------LGV05590
C                                                                      LGV05600
C     CHECK WHETHER OR NOT GIVEN T-MATRIX HAS AN EIGENVALUE IN THE     LGV05610
C     SPECIFIED INTERVAL AND IF SO WHAT ITS T-MULTIPLICITY IS.         LGV05620
C                                                                      LGV05630
      IF(NEVT.EQ.1) GO TO 590                                          LGV05640
      IF(NEVT.NE.0) GO TO 570                                          LGV05650
      ILBIS = 1                                                        LGV05660
      WRITE(6,560) EVAL,KMAXU                                          LGV05670
  560 FORMAT(/' PROBLEM ENCOUNTERED IN RECOMPUTATION OF USER-SUPPLIED EILGV05680
     1GENVALUE',E20.12/' THE SIZE T-MATRIX SPECIFIED',I6,' DOES NOT    LGV05690
     1HAVE AN EIGENVALUE IN THE INTERVAL SPECIFIED'/' THEREFORE NO EIGENLGV05700
     1VECTOR WILL BE COMPUTED FOR THIS PARTICULAR EIGENVALUE'/)        LGV05710
      GO TO 610                                                        LGV05720
C                                                                      LGV05730
  570 IF(NEVT.GT.1)  WRITE(6,580) EVAL,KMAXU                           LGV05740
  580 FORMAT(/' PROBLEM ENCOUNTERED IN RECOMPUTATION OF USER-SUPPLIED  LGV05750
     1EIGENVALUE',E20.12/' FOR THE SIZE T-MATRIX SPECIFIED =',I6,' THE LGV05760
     1GIVEN EIGENVALUE IS T-MULTIPLE IN THE INTERVAL SPECIFIED'/' SOMETHLGV05770
     1ING IS WRONG, THEREFORE NO EIGENVECTOR WILL BE COMPUTED FOR THIS ELGV05780
     1IGENVALUE'/)                                                     LGV05790
C                                                                      LGV05800
      MP(J) = MPMIN                                                    LGV05810
      MA(J) = -2*KMAX                                                  LGV05820
      GO TO 710                                                        LGV05830
C                                                                      LGV05840
  590 CONTINUE                                                         LGV05850
      ILBIS = 0                                                        LGV05860
C                                                                      LGV05870
      EVNEW(J) = EVALN                                                 LGV05880
      EVAL = EVALN                                                     LGV05890
      MTOL = MTOL+KMAXU                                                LGV05900
C                                                                      LGV05910
C     IS THERE ROOM IN TVEC ARRAY FOR THE NEXT T-EIGENVECTOR?          LGV05920
C     IF NOT, SKIP TO RITZ VECTOR COMPUTATIONS.                        LGV05930
      IF (MTOL.GT.MDIMTV) GO TO 720                                    LGV05940
C                                                                      LGV05950
      IT = 3                                                           LGV05960
      KINT = MTOL - KMAXU +1                                           LGV05970
C                                                                      LGV05980
C     RECORD THE BEGINNING AND END OF THE T-EIGENVECTOR BEING COMPUTED LGV05990
```

```
      MINT(J) = KINT                                                  LGV06000
      MFIN(J) = MTOL                                                  LGV06010
C                                                                     LGV06020
C---------------------------------------------------------------------LGV06030
C     SUBROUTINE INVERM DOES INVERSE ITERATION, I.E. SOLVES           LGV06040
C     (T(1,KMAXU) - EVAL)*U = RHS  FOR EACH EIGENVALUE TO OBTAIN THE   LGV06050
C     DESIRED T-EIGENVECTOR.                                          LGV06060
C                                                                     LGV06070
      IF(IWRITE.EQ.1)  WRITE(6,600) J                                 LGV06080
  600 FORMAT(/I6,'TH EIGENVALUE')                                     LGV06090
C                                                                     LGV06100
      CALL INVERM(ALPHA,BETA,V1,TVEC(KINT),EVAL,ERROR,TERROR,EPSM,     LGV06110
     1 G,KMAXU,IT,IWRITE)                                             LGV06120
C                                                                     LGV06130
C---------------------------------------------------------------------LGV06140
C                                                                     LGV06150
      TERR(J) = TERROR                                                LGV06160
      TLAST(J) = ERROR                                                LGV06170
      KMAXU1 = KMAXU + 1                                              LGV06180
      TBETA(J) = BETA(KMAXU1)*ERROR                                   LGV06190
C                                                                     LGV06200
C     AFTER COMPUTING EACH OF THE T-EIGENVECTORS,                     LGV06210
C     CHECK THE SIZE OF THE ERROR ESTIMATE, ERROR.                    LGV06220
C     IF THIS ESTIMATE IS NOT AS SMALL AS DESIRED AND                 LGV06230
C     |MA(J)| < ML(J), ATTEMPT TO INCREASE THE SIZE OF |MA(J)|        LGV06240
C     AND REPEAT THE T-EIGENVECTOR COMPUTATIONS.                      LGV06250
C                                                                     LGV06260
      IF(ERROR.LT.ERTOL.OR.TFLAG.EQ.1)  GO TO 700                     LGV06270
C                                                                     LGV06280
      IF(ERROR.GE.ERRMIN) GO TO 610                                   LGV06290
C     LAST COMPONENT IS LESS THAN MINIMAL TO DATE                     LGV06300
      ERRMIN = ERROR                                                  LGV06310
      MABEST = MA(J)                                                  LGV06320
  610 CONTINUE                                                        LGV06330
C                                                                     LGV06340
      IF(MA(J).GT.0)  ITEST = MA(J) + IDELTA(J)                       LGV06350
      IF(MA(J).LT.0)  ITEST = -(IABS(MA(J)) + IDELTA(J))              LGV06360
      IF(IABS(ITEST).LE.ML(J).AND.ICOUNT.LE.10) GO TO 630             LGV06370
C     NEW MA(J) IS GREATER THAN MAXIMUM ALLOWED.                      LGV06380
      IF(ERCONT.EQ.0.OR.MABEST.EQ.MPMIN) GO TO 650                    LGV06390
      TFLAG = 1                                                       LGV06400
      MA(J) = MABEST                                                  LGV06410
      IF(ILBIS.EQ.0)  MTOL = MTOL - KMAXU                             LGV06420
      WRITE(6,620) MA(J)                                              LGV06430
  620 FORMAT(' 10 ORDERS WERE CONSIDERED.  NONE SATISFIED THE ERROR TESTLGV06440
     1'/' THEREFORE USE THE BEST ORDER OBTAINED FOR THE EIGENVECTORS'  LGV06450
     1,I6)                                                            LGV06460
      GO TO 530                                                       LGV06470
C                                                                     LGV06480
  630 MA(J) = ITEST                                                   LGV06490
C                                                                     LGV06500
      MT = IABS(MA(J))                                                LGV06510
      IF(IWRITE.EQ.1) WRITE(6,640)  MT                                LGV06520
  640 FORMAT(/' CHANGE SIZE OF T-MATRIX TO ',I6,' RECOMPUTE T-EIGENVECTOLGV06530
     1R')                                                             LGV06540
C                                                                     LGV06550
      IF(ILBIS.EQ.0)  MTOL = MTOL - KMAXU                             LGV06560
C                                                                     LGV06570
      GO TO 530                                                       LGV06580
C                                                                     LGV06590
C     APPROPRIATE SIZE T-MATRIX WAS NOT OBTAINED                      LGV06600
  650 CONTINUE                                                        LGV06610
      WRITE(10,660) J,EVAL,MP(J)                                      LGV06620
  660 FORMAT(/' ON 10 INCREMENTS NOT ABLE TO IDENTIFY APPROPRIATE SIZE  LGV06630
     1T-MATRIX FOR'/                                                  LGV06640
     1' EIGENVALUE(',I4,') = ',E20.12,' T-MULTIPLICITY =',I4/)        LGV06650
      IF(M2(J).EQ.KMAX) WRITE(10,670)                                 LGV06660
```

```
      IF(M2(J).LT.KMAX) WRITE(10,680)                               LGV06670
  670 FORMAT(/' ORDERS TESTED RANGED FROM 5*M1(J)/4 TO APPROXIMATELY  LGV06680
     1 '/'  MIN(11*MEV/8,13*M1(J)/8)'/)                              LGV06690
  680 FORMAT(/' ORDERS TESTED RANGED FROM (3*M1(J)+M2(J))/4 TO APPROXIMALGV06700
     1TELY'/'  (3*M1(J) + 5*M2(J))/8.'/)                            LGV06710
      WRITE(10,690)                                                 LGV06720
  690 FORMAT(' ALLOWING LARGER ORDERS FOR THE T-MATRICES MAY RESULT IN LGV06730
     1 SUCCESS'/' BUT PROBABLY WILL NOT.  PROBLEM IS PROBABLY DUE TO'  LGV06740
     1 /' LACK OF CONVERGENCE OF GIVEN EIGENVALUE, CHECK THE ERROR ESTIMLGV06750
     1ATE'/)                                                        LGV06760
      MP(J) = MPMIN                                                 LGV06770
      IF(ILBIS.EQ.0)  MTOL = MTOL - KMAXU                          LGV06780
      GO TO 710                                                    LGV06790
  700 NTVEC = NTVEC + 1                                            LGV06800
C                                                                  LGV06810
  710 CONTINUE                                                     LGV06820
      NGOODC = NGOOD                                               LGV06830
      GO TO 740                                                    LGV06840
C                                                                  LGV06850
C     COME HERE IF THERE IS NOT ENOUGH ROOM FOR ALL OF T-EIGENVECTORS LGV06860
  720 NGOODC = J-1                                                 LGV06870
      WRITE(6,730)  J, MTOL, MDIMTV                                LGV06880
  730 FORMAT(/' NOT ENOUGH ROOM IN TVEC FOR ',I4,'TH T-VECTOR'/'  T-DIMLGV06890
     1ENSION REQUESTED = ',I6,' BUT TVEC HAS DIMENSION = ',I6/)     LGV06900
      IF(NGOODC.EQ.0)  GO TO 1470                                  LGV06910
      MTOL = MTOL-KMAXU                                            LGV06920
C                                                                  LGV06930
  740 CONTINUE                                                     LGV06940
C                                                                  LGV06950
C     THE LOOP ON T-EIGENVECTOR COMPUTATIONS IS COMPLETE.          LGV06960
C     WRITE OUT THE SIZE T-MATRICES THAT WILL BE USED FOR          LGV06970
C     THE RITZ VECTOR COMPUTATIONS.                                LGV06980
C                                                                  LGV06990
      WRITE(10,750)                                                LGV07000
  750 FORMAT(/' SIZES OF T-MATRICES THAT WILL BE USED IN THE RITZ COMPUTLGV07010
     1ATIONS'/5X,'J',16X,'GOODEV(J)',1X,'MA(J)')                   LGV07020
C                                                                  LGV07030
      WRITE(10,760)  (J,GOODEV(J),MA(J), J=1,NGOOD)                LGV07040
  760 FORMAT(I6,E25.14,I6)                                         LGV07050
      WRITE(10,510)                                                LGV07060
C                                                                  LGV07070
      WRITE(6,770) MTOL                                            LGV07080
  770 FORMAT(/' THE CUMULATIVE LENGTH OF THE T-EIGENVECTORS IS',I18) LGV07090
C                                                                  LGV07100
      WRITE(6,780) NTVEC,NGOOD                                     LGV07110
  780 FORMAT(/I6,' T-EIGENVECTORS OUT OF',I6,' REQUESTED WERE COMPUTED')LGV07120
C                                                                  LGV07130
C     SAVE THE T-EIGENVECTORS ON FILE 11?                          LGV07140
      IF(TVSTOP.NE.1.AND.SVTVEC.EQ.0) GO TO 840                    LGV07150
C                                                                  LGV07160
      WRITE(11,790) NTVEC,MTOL,MATNOA,MATNOB,SVSEED                LGV07170
  790 FORMAT(I6,3I8,I12,' = NTVEC,MTOL,MATNOA,MATNOB,SVSEED')       LGV07180
C                                                                  LGV07190
      DO 820 J=1,NGOODC                                            LGV07200
C     IF MP(J) = MPMIN THEN NO SUITABLE T-EIGENVECTOR IS AVAILABLE LGV07210
C     FOR THAT EIGENVALUE.                                         LGV07220
      IF(MP(J).EQ.MPMIN) WRITE(11,800) J,MA(J),GOODEV(J),MP(J)     LGV07230
  800 FORMAT(2I6,E20.12,I6/' TH EIGVAL,T-SIZE,EVALUE,FLAG,NO EIGVEC') LGV07240
      IF(MP(J).NE.MPMIN) WRITE(11,810) J,MA(J),GOODEV(J),MP(J)     LGV07250
  810 FORMAT(I6,I6,E20.12,I6/' T-EIGVEC,SIZE T,EVALUE OF A,MP(J)')  LGV07260
      IF(MP(J).EQ.MPMIN) GO TO 820                                 LGV07270
      KI = MINT(J)                                                 LGV07280
      KF = MFIN(J)                                                 LGV07290
C                                                                  LGV07300
      WRITE(11,260) (TVEC(K), K=KI,KF)                             LGV07310
C                                                                  LGV07320
```

```
    820 CONTINUE                                                          LGV07330
C                                                                         LGV07340
        IF(TVSTOP.NE.1) GO TO 840                                         LGV07350
C                                                                         LGV07360
        WRITE(6,830) TVSTOP, NTVEC,NGOOD                                  LGV07370
    830 FORMAT(/' USER SET TVSTOP = ',I1/                                 LGV07380
       1' THEREFORE PROGRAM TERMINATES AFTER T-EIGENVECTOR COMPUTATIONS'/ LGV07390
       1' T-EIGENVECTORS THAT WERE COMPUTED ARE SAVED ON FILE 11'/        LGV07400
       118,' T-EIGENVECTORS WERE COMPUTED OUT OF',I7,' REQUESTED'/)       LGV07410
C                                                                         LGV07420
        GO TO 1590                                                        LGV07430
C                                                                         LGV07440
    840 CONTINUE                                                          LGV07450
C       IF NOT ABLE TO COMPUTE ALL THE REQUESTED T-EIGENVECTORS,          LGV07460
C       CONTINUE WITH THE LANCZOS VECTOR COMPUTATIONS ANYWAY?             LGV07470
        IF(NTVEC.NE.NGOOD.AND.LVCONT.EQ.0) GO TO 1490                     LGV07480
C                                                                         LGV07490
C       COMPUTE THE MAXIMUM SIZE OF THE T-MATRIX USED FOR THOSE           LGV07500
C       EIGENVALUES WITH GOOD ERROR ESTIMATES.                           LGV07510
C                                                                         LGV07520
        KMAXU = 0                                                        LGV07530
        DO 850 J = 1,NGOODC                                               LGV07540
        MT = IABS(MA(J))                                                  LGV07550
        IF(MT.LT.KMAXU.OR.MP(J).EQ.MPMIN) GO TO 850                       LGV07560
        KMAXU = MT                                                        LGV07570
    850 CONTINUE                                                          LGV07580
C                                                                         LGV07590
        IF(KMAXU.EQ.0) GO TO 1530                                         LGV07600
C                                                                         LGV07610
        WRITE(6,860) KMAXU                                                LGV07620
    860 FORMAT(/I6,' = LARGEST SIZE T-MATRIX TO BE USED IN THE RITZ VECTORLGV07630
       1 COMPUTATIONS')                                                   LGV07640
C                                                                         LGV07650
C       COUNT THE NUMBER OF RITZ VECTORS NOT BEING COMPUTED               LGV07660
        MREJEC = 0                                                        LGV07670
        DO 870 J=1,NGOODC                                                 LGV07680
    870 IF(MP(J).EQ.MPMIN)  MREJEC = MREJEC + 1                           LGV07690
        MREJET = MREJEC + (NGOOD-NGOODC)                                  LGV07700
        IF(MREJET.NE.0) WRITE(6,880) MREJET                              LGV07710
    880 FORMAT(/' RITZ VECTORS ARE NOT COMPUTED FOR',I6,' OF THE EIGENVALULGV07720
       1ES'/)                                                            LGV07730
        NACT = NGOODC - MREJEC                                            LGV07740
        WRITE(6,890) NGOOD,NTVEC,NACT                                     LGV07750
    890 FORMAT(/I6,' RITZ VECTORS WERE REQUESTED'/I6,' T-EIGENVECTORS WERELGV07760
       1 COMPUTED'/I6,' RITZ VECTORS WILL BE COMPUTED'/)                 LGV07770
C       CHECK IF THERE ARE ANY RITZ VECTORS TO COMPUTE                    LGV07780
        IF(MREJEC.EQ.NGOODC) GO TO 1510                                   LGV07790
C                                                                         LGV07800
C       CONTINUE WITH THE LANCZOS VECTOR COMPUTATIONS?                    LGV07810
        IF(LVCONT.EQ.0.AND.MREJEC.NE.0) GO TO 1490                        LGV07820
C                                                                         LGV07830
C       NOW COMPUTE THE RITZ VECTORS.  REGENERATE THE                     LGV07840
C       LANCZOS VECTORS.                                                  LGV07850
C                                                                         LGV07860
        DO 900 I = 1,NMAX                                                 LGV07870
    900 RITVEC(I) = ZERO                                                  LGV07880
C                                                                         LGV07890
C------------------------------------------------------------------------LGV07900
C       REGENERATE THE STARTING VECTOR. THIS MUST BE GENERATED AND        LGV07910
C       NORMALIZED PRECISELY THE WAY IT WAS DONE IN THE EIGENVALUE        LGV07920
C       COMPUTATIONS, OTHERWISE THERE WILL BE A MISMATCH BETWEEN          LGV07930
C       THE T-EIGENVECTORS THAT HAVE BEEN COMPUTED FROM THE T-MATRICES    LGV07940
C       READ IN FROM FILE 2 AND THE LANCZOS VECTORS THAT ARE             LGV07950
C       BEING REGENERATED.                                               LGV07960
C                                                                         LGV07970
        IIL = SVSEED                                                      LGV07980
```

```
      CALL GENRAN(IIL,G,N)                                        LGV07990
C                                                                 LGV08000
C----------------------------------------------------------------LGV08010
C                                                                 LGV08020
      DO 910 K = 1,N                                              LGV08030
  910 V2(K) = G(K)                                                LGV08040
C                                                                 LGV08050
C----------------------------------------------------------------LGV08060
C     COMPUTE L-TRANSPOSE*V2 AND ITS NORM                         LGV08070
      ISOLV = 2                                                   LGV08080
      CALL LSOLV(V2,VS,ISOLV)                                     LGV08090
      SUM = FINPRO(N,VS(1),1,VS(1),1)                             LGV08100
C----------------------------------------------------------------LGV08110
C                                                                 LGV08120
C     NORMALIZE STARTING VECTORS:  (V2-TRANSPOSE*B*V2) = 1        LGV08130
      SUM = ONE/DSQRT(SUM)                                        LGV08140
      DO 920 K = 1,N                                              LGV08150
      VS(K) = SUM*VS(K)                                           LGV08160
  920 V2(K) = SUM*V2(K)                                           LGV08170
C                                                                 LGV08180
C----------------------------------------------------------------LGV08190
C     INITIALIZE V1 = B*V2 = L*VS                                 LGV08200
      ISOLV = 1                                                   LGV08210
      CALL LSOLV(VS,V1,ISOLV)                                     LGV08220
C----------------------------------------------------------------LGV08230
C                                                                 LGV08240
      DO 930 K = 1,N                                              LGV08250
      VS(K) = V1(K)                                               LGV08260
  930 V1(K) = ZERO                                                LGV08270
C                                                                 LGV08280
      IVEC = 1                                                    LGV08290
      BATA = ZERO                                                 LGV08300
C                                                                 LGV08310
      GO TO 1000                                                  LGV08320
C                                                                 LGV08330
C     VS = B*V(I),  V1 = B*V(I-1),  V2 = V(I)                     LGV08340
  940 CONTINUE                                                    LGV08350
      SUM = BATA                                                  LGV08360
C                                                                 LGV08370
C----------------------------------------------------------------LGV08380
C     COMPUTE V1 = A*V2 - SUM*V1                                  LGV08390
      CALL MATVEC(V2,V1,SUM)                                      LGV08400
C     COMPUTE ALFA                                                LGV08410
      ALFA = FINPRO(N,V1(1),1,V2(1),1)                            LGV08420
C----------------------------------------------------------------LGV08430
C                                                                 LGV08440
      DO 950 K = 1,N                                              LGV08450
  950 V1(K) = V1(K)-ALFA*VS(K)                                    LGV08460
C                                                                 LGV08470
C     SET V1 = B*V(IVEC) AND VS = (NEW BATA)*B*V(IVEC+1)          LGV08480
      DO 960 K = 1,N                                              LGV08490
      TEMP = V1(K)                                                LGV08500
      V1(K) = VS(K)                                               LGV08510
  960 VS(K) = TEMP                                                LGV08520
C                                                                 LGV08530
C----------------------------------------------------------------LGV08540
C     COMPUTE V2 = (L-INVERSE)*VS                                 LGV08550
      ISOLV = 3                                                   LGV08560
      CALL LSOLV(VS,V2,ISOLV)                                     LGV08570
C     COMPUTE NEXT BATA                                           LGV08580
      SUM = FINPRO(N,V2(1),1,V2(1),1)                             LGV08590
C----------------------------------------------------------------LGV08600
C                                                                 LGV08610
      BATA = DSQRT(SUM)                                           LGV08620
      TEMP = BETA(IVEC)                                           LGV08630
      TEMP = DABS(BATA - TEMP)/TEMP                               LGV08640
```

```
      IF (TEMP.LT.1.0D-10)GO TO 980                                  LGV08650
C                                                                    LGV08660
C     THE BETA BEING REGENERATED DO NOT MATCH THE BETA IN FILE 2.    LGV08670
C     SOMETHING IS WRONG IN THE LANCZOS VECTOR GENERATION.           LGV08680
C     PROGRAM TERMINATES FOR USER TO CORRECT THE PROBLEM             LGV08690
C     WHICH MUST BE IN THE STARTING VECTOR GENERATION OR IN          LGV08700
C     THE SUBROUTINES AMATV AND LSOLV SUPPLIED.                      LGV08710
C     THESE SUBROUTINES MUST BE THE SAME ONES USED IN THE            LGV08720
C     EIGENVALUE COMPUTATIONS OR A MISMATCH WILL ENSUE.              LGVC8730
C                                                                    LGV08740
      WRITE(6,970) IVEC,BATA,BETA(IVEC),TEMP                         LGV08750
  970 FORMAT(/2X,'IVEC',16X,'BATA',10X,'BETA(IVEC)',14X,'RELDIF'/16, LGV08760
     13E20.12/' IN LANCZOS VECTOR REGENERATION THE ENTRIES OF THE TRIDIALGV08770
     1GONAL MATRICES BEING'/' GENERATED ARE NOT THE SAME AS THOSE IN THELGV08780
     1 MATRIX SUPPLIED ON FILE 2.'/' THEREFORE SOMETHING IS BEING INITIALGV08790
     1LIZED OR COMPUTED DIFFERENTLY FROM THE WAY'/' IT WAS COMPUTED IN TLGV08800
     1HE EIGENVALUE COMPUTATIONS'/' THE PROGRAM TERMINATES FOR THE USER LGV08810
     1TO DETERMINE WHAT THE PROBLEM IS'/)                            LGV08820
      GO TO 1590                                                     LGV08830
  980 CONTINUE                                                       LGV08840
C                                                                    LGV08850
C--------------------------------------------------------------------LGV08860
      ISOLV = 4                                                      LGV08870
      CALL LSOLV(V2,V2,ISOLV)                                        LGV08880
C--------------------------------------------------------------------LGV08890
C                                                                    LGV08900
      SUM = ONE/BATA                                                 LGV08910
      DO 990 K = 1,N                                                 LGV08920
      V2(K) = SUM*V2(K)                                              LGV0893C
  990 VS(K) = SUM*VS(K)                                              LGV08940
C                                                                    LGV08950
 1000 CONTINUE                                                       LGV08960
C                                                                    LGV08970
      LFIN = 0                                                       LGV08980
      DO 1020 J = 1,NGOODC                                           LGV08990
      LL = LFIN                                                      LGV09000
      LFIN = LFIN + N                                                LGV09010
C                                                                    LGV09020
      IF(IABS(MA(J)).LT.IVEC.OR.MP(J).EQ.MPMIN) GO TO 1020           LGV09030
      II = IVEC + MINT(J) - 1                                        LGV09040
      TEMP = TVEC(II)                                                LGV09050
C     II IS THE (IVEC)TH COMPONENT OF THE T-EIGENVECTOR CONTAINED    LGV09060
C     IN TVEC(MINT(J)).                                              LGV09070
C                                                                    LGV09080
      DO 1010 K = 1,N                                                LGV09090
      LL = LL + 1                                                    LGV09100
 1010 RITVEC(LL) = TEMP*V2(K) + RITVEC(LL)                           LGV09110
C                                                                    LGV09120
 1020 CONTINUE                                                       LGV09130
C                                                                    LGV09140
      IVEC = IVEC + 1                                                LGV09150
      IF (IVEC.LE.KMAXU) GO TO 940                                   LGV09160
C                                                                    LGV09170
C     RITZVECTOR GENERATION IS COMPLETE. B-NORMALIZE EACH RITZVECTOR.LGV09180
C     NOTE THAT IF CERTAIN RITZ VECTORS WERE NOT COMPUTED THEN THAT  LGV09190
C     PORTION OF THE RITVEC ARRAY WAS NOT UTILIZED.                  LGV09200
C                                                                    LGV09210
      LFIN = 0                                                       LGV09220
      DO 1090 J = 1,NGOODC                                           LGV09230
C                                                                    LGV09240
      KK = LFIN                                                      LGV09250
      LFIN = LFIN + N                                                LGV09260
      IF(MP(J).EQ.MPMIN) GO TO 1090                                  LGV09270
C                                                                    LGV09280
      DO 1030 K = 1,N                                                LGV09290
      KK = KK + 1                                                    LGV09300
```

```
 1030 V2(K) = RITVEC(KK)                                              LGV09310
C                                                                     LGV09320
C--------------------------------------------------------------------LGV09330
      ISOLV = 2                                                       LGV09340
      CALL LSOLV(V2,VS,ISOLV)                                         LGV09350
      SUM = FINPRO(N,VS(1),1,VS(1),1)                                 LGV09360
C--------------------------------------------------------------------LGV09370
C                                                                     LGV09380
      SUM = DSQRT(SUM)                                                LGV09390
      RNORM(J) = SUM                                                  LGV09400
      TEMP = DABS(ONE-SUM)                                            LGV09410
      SUM = ONE/SUM                                                   LGV09420
C                                                                     LGV09430
      DO 1040 K = 1,N                                                 LGV09440
      VS(K) = SUM*VS(K)                                               LGV09450
      V2(K) = SUM*V2(K)                                               LGV09460
 1040 CONTINUE                                                        LGV09470
C                                                                     LGV09480
C--------------------------------------------------------------------LGV09490
      ISOLV = 1                                                       LGV09500
      CALL LSOLV(VS,V1,ISOLV)                                         LGV09510
C--------------------------------------------------------------------LGV09520
C                                                                     LGV09530
C     V1 = B*V2                                                       LGV09540
      EVAL = EVNEW(J)                                                 LGV09550
C                                                                     LGV09560
C     COMPUTE ERROR IN RITZ VECTOR CONSIDERED AS A EIGENVECTOR OF A.  LGV09570
C     V1 = A*RITVEC - EVAL*B*RITVEC                                   LGV09580
C                                                                     LGV09590
C--------------------------------------------------------------------LGV09600
      CALL AMATV(V2,V1,EVAL)                                          LGV09610
      SUM = FINPRO(N,V1(1),1,V1(1),1)                                 LGV09620
C--------------------------------------------------------------------LGV09630
C                                                                     LGV09640
      SUM = DSQRT(SUM)                                                LGV09650
      ERR(J) = SUM                                                    LGV09660
      GAP = ABS(AMINGP(J))                                            LGV09670
      ERRDGP(J) = SUM/GAP                                             LGV09680
C                                                                     LGV09690
C                                                                     LGV09700
      IF (JPERM.EQ.0) GO TO 1050                                      LGV09710
C                                                                     LGV09720
C--------------------------------------------------------------------LGV09730
C     ON RETURN V2 = P(TRANSPOSE)*V2                                  LGV09740
      IPERM = 2                                                       LGV09750
      CALL LPERM(V2,V1,IPERM)                                         LGV09760
C--------------------------------------------------------------------LGV09770
C                                                                     LGV09780
 1050 CONTINUE                                                        LGV09790
      KK = LFIN - N                                                   LGV09800
      DO 1060 K = 1,N                                                 LGV09810
      KK = KK + 1                                                     LGV09820
 1060 RITVEC(KK) = V2(K)                                              LGV09830
C                                                                     LGV09840
      IF (IWRITE.NE.0) WRITE(6,1070) J,GOODEV(J)                      LGV09850
 1070 FORMAT(/I5,' TH EIGENVALUE CONSIDERED = ',E20.12/)             LGV09860
C                                                                     LGV09870
      IF (IWRITE.NE.0) WRITE(6,1080) TERR(J),TBETA(J),TEMP            LGV09880
 1080 FORMAT(' NORM OF ERROR IN T-EIGENVECTOR = ',E14.3/             LGV09890
     1 ' BETA(MA(J)+1)*U(MA(J)) = ',E14.3/                           LGV09900
     1 ' ABS(NORM(RITVEC) - 1.0)  = ',E14.3/)                        LGV09910
C                                                                     LGV09920
 1090 CONTINUE                                                        LGV09930
C                                                                     LGV09940
C     RITZVECTORS ARE NORMALIZED AND ERROR ESTIMATES ARE IN ERR ARRAY LGV09950
C     AND IN ERRDGP ARRAY. STORE EVERYTHING                          LGV09960
```

```
C                                                                    LGV09970
C                                                                    LGV09980
      WRITE(9,1100)                                                  LGV09990
 1100 FORMAT(2X,'AB-EIGENVALUE',2X,'MA(J)',2X,'AB-MINGAP',5X,'ABERROR',1LGV10000
     1X, 'ABERROR/GAP',6X,'TERROR')                                  LGV10010
C                                                                    LGV10020
      WRITE(13,1110)                                                 LGV10030
 1110 FORMAT(12X,'AB-EIGENVALUE',5X,'RITZNORM',5X,'ABMINGAP',5X,     LGV10040
     1 'TBETA(J)',5X,'TLAST(J)')                                     LGV10050
C                                                                    LGV10060
      DO 1140 J=1,NGOODC                                             LGV10070
C                                                                    LGV10080
      IF(MP(J).EQ.MPMIN) GO TO 1140                                 LGV10090
C                                                                    LGV10100
      WRITE(9,1120)EVNEW(J),MA(J),AMINGP(J),ERR(J),ERRDGP(J),TERR(J) LGV10110
 1120 FORMAT(E15.8,I6,4E12.4)                                        LGV10120
C                                                                    LGV10130
      WRITE(13,1130) EVNEW(J),RNORM(J),AMINGP(J),TBETA(J),TLAST(J)   LGV10140
 1130 FORMAT(E25.14,4E13.5)                                          LGV10150
C                                                                    LGV10160
 1140 CONTINUE                                                       LGV10170
C                                                                    LGV10180
      IF(MREJEC.EQ.0) GO TO 1220                                     LGV10190
      WRITE(9,1150)                                                  LGV10200
 1150 FORMAT(/' RITZ VECTORS WERE NOT COMPUTED FOR THE FOLLOWING EIGENVALGV10210
     1LUES'/' EITHER BECAUSE THEY HAD NOT CONVERGED OR BECAUSE THE ERRORLGV10220
     1 ESTIMATE'/' WAS NOT AS SMALL AS DESIRED'/)                    LGV10230
C                                                                    LGV10240
      DO 1210 J = 1,NGOODC                                           LGV10250
      IF(MP(J).NE.MPMIN) GO TO 1210                                 LGV10260
C     WRITE OUT MESSAGE FOR EACH EIGENVALUE FOR WHICH NO EIGENVECTOR LGV10270
C     WAS COMPUTED.                                                  LGV10280
C                                                                    LGV10290
      WRITE(9,1160)                                                  LGV10300
 1160 FORMAT(2X,'AB-EIGENVALUE',3X,'MA(J)',5X,'AMINGP(J)',6X,'TLAST(J)',LGV10310
     13X,'MP(J)')                                                    LGV10320
      WRITE(9,1170) GOODEV(J),MA(J),AMINGP(J),TBETA(J),MP(J)         LGV10330
 1170 FORMAT(E15.8,I8,2E14.4,I8)                                     LGV10340
C                                                                    LGV10350
      WRITE(13,1180)                                                 LGV10360
 1180 FORMAT(/' RITZ VECTORS WERE NOT COMPUTED FOR THE FOLLOWING EIGENVALGV10370
     1LUES'/' BECAUSE THEY HAD NOT CONVERGED'/)                      LGV10380
C                                                                    LGV10390
      WRITE(13,1190)                                                 LGV10400
 1190 FORMAT(2X,'AB-EIGENVALUE',3X,'MA(J)',3X,'M1(J)',3X,'M2(J)',3X,'MP(LGV10410
     1J)'/)                                                          LGV10420
      WRITE(13,1200) GOODEV(J),MA(J),M1(J),M2(J),MP(J)               LGV10430
 1200 FORMAT(E15.8,4I8)                                              LGV10440
C                                                                    LGV10450
 1210 CONTINUE                                                       LGV10460
 1220 CONTINUE                                                       LGV10470
C                                                                    LGV10480
      WRITE(9,1230)                                                  LGV10490
 1230 FORMAT(/' ABOVE ARE ERROR ESTIMATES FOR THE AB AND T EIGENVECTORS'LGV10500
     1 /' ASSOCIATED WITH THE AB-EIGENVALUES LISTED IN COLUMN 1'/    LGV10510
     1 ' ABERROR = NORM(A*X - EV*B*X)   TERROR = NORM(T*Y - EV*Y)    LGV10520
     1 '/' WHERE T = T(1,MA(J))   X = RITZ VECTOR = V*Y   V = SUCCESSIVELGV10530
     1 '/' LANCZOS VECTORS. ABMINGAP = GAP TO NEAREST AB-EIGENVALUE'//) LGV10540
C                                                                    LGV10550
      WRITE(13,1240)                                                 LGV10560
 1240 FORMAT(/' ABOVE ARE ERROR ESTIMATES ASSOCIATED WITH THE AB-EIGVALSLGV10570
     1 '/' RITZNORM = NORM(COMPUTED RITZ VECTOR)'/                   LGV10580
     1 ' TBETA(J) = BETA(MA(J)+1)*Y(MA(J)),   T*Y = EVAL*Y'/         LGV10590
     1 ' TLAST(J) = Y(MA(J))'/                                       LGV10600
     1 ' ABMINGAP = GAP TO NEAREST AB-EIGENVALUE'/)                  LGV10610
C                                                                    LGV10620
C     NUMBER OF RITZ VECTORS COMPUTED                                LGV10630
```

```
         NCOMPU = NGOODC - MREJEC                                       LGV10640
         WRITE(12,1250) N,NCOMPU,NGOODC,MATNOA,MATNOB                   LGV10650
 1250 FORMAT(3I6,2I8,' SIZE A, NO.RITZVECS, NO.EVALS,MATNOA,MATNOB')    LGV10660
C                                                                       LGV10670
         LFIN = 0                                                       LGV10680
         DO 1310 J = 1,NGOODC                                           LGV10690
         LINT = LFIN + 1                                                LGV10700
         LFIN = LFIN + N                                                LGV10710
C                                                                       LGV10720
         IF(MP(J).EQ.MPMIN) GO TO 1290                                  LGV10730
C     RITZ VECTOR WAS COMPUTED                                          LGV10740
         WRITE(12,1260) J, GOODEV(J), MP(J)                             LGV10750
 1260 FORMAT(I6,4X,E20.12,I6,' J, AB-EIGENVAL, MP(J)')                  LGV10760
C                                                                       LGV10770
         WRITE(12,1270) ERR(J),ERRDGP(J)                               LGV10780
 1270 FORMAT(2E15.5,'= NORM(A*Z-EVAL*B*Z), NORM(A*Z-EVAL*B*Z)/ABMINGAP')LGV10790
C                                                                       LGV10800
         WRITE(12,1280) (RITVEC(LL), LL=LINT,LFIN)                      LGV10810
 1280 FORMAT(4E20.12)                                                   LGV10820
         GO TO 1310                                                     LGV10830
C     NO RITZ VECTOR WAS COMPUTED FOR THIS EIGENVALUE                   LGV10840
 1290 WRITE(12,1300) J,GOODEV(J),MP(J)                                  LGV10850
 1300 FORMAT(I6,4X,E20.12,I6,' J,AB-EIGVALUE,NO RITZ VECTOR COMPUTED')  LGV10860
C                                                                       LGV10870
 1310 CONTINUE                                                          LGV10880
C                                                                       LGV10890
C     DID ANY T-MATRICES INCLUDE OFF-DIAGONAL ENTRIES SMALLER THAN      LGV10900
C     DESIRED, AS SPECIFIED BY BTOL?                                    LGV10910
C                                                                       LGV10920
         IF(IB.GT.0) GO TO 1340                                         LGV10930
C                                                                       LGV10940
         WRITE(6,1320) KMAXU                                            LGV10950
 1320 FORMAT(/' FOR LARGEST T-MATRIX CONSIDERED',I7,' CHECK THE SIZE OF LGV10960
      1BETAS')                                                          LGV10970
C                                                                       LGV10980
C----------------------------------------------------------------------LGV10990
C                                                                       LGV11000
         CALL TNORM(ALPHA,BETA,BKMIN,TEMP,KMAXU,IBMT)                   LGV11010
C                                                                       LGV11020
C----------------------------------------------------------------------LGV11030
C                                                                       LGV11040
         IF(IBMT.LT.0) WRITE (6,1330)                                   LGV11050
 1330 FORMAT(/' WARNING THE T-MATRICES FOR ONE OR MORE OF THE EIGENVALUELGV11060
      1S CONSIDERED'/' HAD AN OFF-DIAGONAL ENTRY THAT WAS SMALLER THAN THLGV11070
      1E BETA TOLERANCE THAT WAS SPECIFIED'/)                           LGV11080
 1340 CONTINUE                                                          LGV11090
C                                                                       LGV11100
         GO TO 1590                                                     LGV11110
C                                                                       LGV11120
 1350 WRITE(6,1360) NGOOD,NMAX,MDIMRV                                   LGV11130
 1360 FORMAT(/I4,' RITZ VECTORS WERE REQUESTED BUT THE REQUIRED DIMENSIOLGV11140
      1N',I6/' IS LARGER THAN THE USER-SPECIFIED DIMENSION OF RITVEC',I6 LGV11150
      1/' THEREFORE, THE EIGENVECTOR PROCEDURE TERMINATES FOR THE USER TOLGV11160
      1 INTERVENE')                                                     LGV11170
C                                                                       LGV11180
         GO TO 1590                                                     LGV11190
C                                                                       LGV11200
 1370 WRITE(6,1380) NOLD,N,MATA,MATNOA,MATB,MATNOB                      LGV11210
 1380 FORMAT(/' PARAMETERS READ FROM FILE 3 DO NOT AGREE WITH USER-SPECILGV11220
      1FIED VALUES'/' NOLD,N,MATA,MATNOA,MATB,MATNOB = '/2I6,4I12/       LGV11230
      1' THEREFORE PROGRAM TERMINATES FOR USER TO RESOLVE DIFFERENCES'/) LGV11240
C                                                                       LGV11250
         GO TO 1590                                                     LGV11260
C                                                                       LGV11270
 1390 WRITE(6,1400)                                                     LGV11280
 1400 FORMAT(/' PARAMETERS IN ALPHA,BETA FILE READ IN DO NOT AGREE WITH LGV11290
      1THOSE'/' SPECIFIED BY THE USER.  THEREFORE PROGRAM TERMINATES FOR'LGV11300
```

```
      1/' USER TO RESOLVE DIFFERENCES'/)                           LGV11310
C                                                                  LGV11320
      GO TO 1590                                                   LGV11330
C                                                                  LGV11340
 1410 WRITE(6,1420) KMAX,MEV                                       LGV11350
 1420 FORMAT(/' ALPHA,BETA HEADER HAS KMAX = ',16/                 LGV11360
     1' BUT EIGENVALUES WERE COMPUTED AT MEV = ',16,' PROGRAM STOPS'/) LGV11370
C                                                                  LGV11380
      GO TO 1590                                                   LGV11390
C                                                                  LGV11400
 1430 WRITE(6,1440)                                                LGV11410
 1440 FORMAT(/' PROGRAM COMPUTED 1ST GUESSES AT T-MATRIX SIZES AND READ LGV11420
     1THEM TO FILE 10'/' THEN TERMINATED AS REQUESTED.')           LGV11430
      GO TO 1590                                                   LGV11440
C                                                                  LGV11450
 1450 WRITE(6,1460) MTOL, MDIMTV                                   LGV11460
 1460 FORMAT(/' PROGRAM TERMINATES BECAUSE THE TVEC DIMENSION ANTICIPATELGV11470
     1D',17/' IS LARGER THAN THE TVEC DIMENSION',17,' SPECIFIED BY THE LGV11480
     1USER.'/' USER MAY RESET THE TVEC DIMENSION AND RESTART THE PROGRALGV11490
     1M')                                                          LGV11500
      GO TO 1590                                                   LGV11510
C                                                                  LGV11520
 1470 WRITE(6,1480)                                                LGV11530
 1480 FORMAT(/' PROGRAM TERMINATES BECAUSE NO SUITABLE T-EIGENVECTORS WELGV11540
     1RE IDENTIFIED'/' FOR ANY OF THE EIGENVALUES SUPPLIED.  PROBLEM COLGV11550
     1ULD BE CAUSED'/' BY TOO SMALL A TVEC DIMENSION OR SIMPLY THAT SUILGV11560
     1TABLE T-VECTORS COULD'/' NOT BE IDENTIFIED.  USER SHOULD CHECK OULGV11570
     1TPUT'/)                                                      LGV11580
      GO TO 1590                                                   LGV11590
C                                                                  LGV11600
 1490 WRITE(6,1500) LVCONT,NTVEC,NGOOD                             LGV11610
 1500 FORMAT(/' LVCONT FLAG =',12,' AND NUMBER ',15,' OF T-EIGENVECTORS LGV11620
     1 COMPUTED N.E.'/' NUMBER',15,' REQUESTED SO PROGRAM TERMINATES'/) LGV11630
      GO TO 1590                                                   LGV11640
C                                                                  LGV11650
 1510 WRITE(6,1520)                                                LGV11660
 1520 FORMAT(/' PROGRAM TERMINATES WITHOUT COMPUTING RITZ VECTORS'/ LGV11670
     1' BECAUSE ALL T-EIGENVECTORS WERE REJECTED AS NOT SUITABLE FOR THELGV11680
     1 RITZ VECTOR'/' COMPUTATIONS.  PROBABLE CAUSE IS LACK OF CONVERGENLGV11690
     1CE OF THE EIGENVALUES SUPPLIED'/)                            LGV11700
      GO TO 1590                                                   LGV11710
C                                                                  LGV11720
 1530 WRITE(6,1540)                                                LGV11730
 1540 FORMAT(/' PROGRAM INDICATES THAT IT IS NOT POSSIBLE TO COMPUTE ANYLGV11740
     1 OF THE'/' REQUESTED EIGENVECTORS. THEREFORE PROGRAM TERMINATES') LGV11750
      DO 1550 J=1,NGOODC                                          LGV11760
 1550 WRITE(6,1560)  J,GOODEV(J),MP(J)                            LGV11770
 1560 FORMAT(/4X,' J',9X,'AB-EIGENVALUE',4X,'MP(J)'/16,E20.12,19) LGV11780
      GO TO 1590                                                   LGV11790
C                                                                  LGV11800
 1570 WRITE(6,1580) MBETA,KMAXN                                    LGV11810
 1580 FORMAT(/' PROGRAM TERMINATES BECAUSE THE STORAGE ALLOTTED FOR THE LGV11820
     1BETA ARRAY',18/' IS NOT SUFFICIENT FOR THE ENLARGED KMAX =',18,' TLGV11830
     1HAT THE PROGRAM WANTS'/' USER CAN ENLARGE THE ALPHA AND BETA ARRAYLGV11840
     1IS AND RERUN THE PROGRAM'/)                                 LGV11850
C                                                                  LGV11860
 1590 CONTINUE                                                     LGV11870
C                                                                  LGV11880
      STOP                                                         LGV11890
C-----END OF MAIN PROGRAM FOR LANCZOS EIGENVECTOR COMPUTATIONS---------LGV11900
      END                                                          LGV11910
```

SECTION 5.4   LANCZS AND SAMPLE MATRIX-VECTOR MULTIPLY AND SOLVE SUBROUTINES

```
C-----LGMULT---------------------------------------------------------LGM00010
C                                                                     LGM00020
C     CONTAINS SUBROUTINES  LANCZS, USPECA, USPECB, AMATV, AND LSOLV.  LGM00030
C     TO BE USED WITH THE LANCZOS CODES FOR THE GENERALIZED EIGENVALUE LGM00040
C     PROBLEM, A*X = EVAL*B*X, WHERE A AND B ARE REAL SYMMETRIC, AND   LGM00050
C     B IS POSITIVE DEFINITE WITH ITS CHOLESKY FACTORS AVAILABLE.      LGM00060
C                                                                     LGM00070
C     NONPORTABLE CONSTRUCTIONS:                                       LGM00080
C     1.  THE ENTRY MECHANISM USED TO PASS THE STORAGE                 LGM00090
C         LOCATIONS OF THE USER-SPECIFIED MATRICES FROM THE            LGM00100
C         SUBROUTINES USPECA AND USPECB TO THE MATRIX-VECTOR           LGM00110
C         SUBROUTINE, AMATV AND TO THE SOLVE SUBROUTINE, LSOLV.        LGM00120
C     2.  IN SAMPLE USPECA AND USPECB:  FREE FORMAT (8,*); FORMAT      LGM00130
C         (20A4), AND FORMAT (4Z20).                                   LGM00140
C                                                                     LGM00150
C-----LANCZS-COMPUTE LANCZOS TRIDIAGONAL MATRICES---------------------LGM00160
C                                                                     LGM00170
      SUBROUTINE LANCZS(LSOLV,MATVEC,ALPHA,BETA,V1,V2,VS,G,KMAX,MOLD1,N,LGM00180
     1 IIX)                                                            LGM00190
C                                                                     LGM00200
C-------------------------------------------------------------------LGM00210
      DOUBLE PRECISION  ALPHA(1), BETA(1), V1(1), V2(1), VS(1)         LGM00220
      DOUBLE PRECISION SUM, ONE, ZERO, TEMP                           LGM00230
      REAL  G(1)                                                       LGM00240
      DOUBLE PRECISION  FINPRO,DSQRT                                   LGM00250
      EXTERNAL MATVEC, LSOLV                                           LGM00260
C-------------------------------------------------------------------LGM00270
C     ALPHA, BETA, AND LANCZOS VECTOR GENERATION                       LGM00280
C     ALPHA BETA GENERATION STARTS WITH IVEC = 1, BETA(1) = ZERO,      LGM00290
C     V2 =  RANDOM VECTOR WITH UNIT B-NORM, VS = B*V2, AND V1 = 0.;    LGM00300
C     OR STARTS WITH AN EXISTING ALPHA/BETA FILE AND THE MOST          LGM00310
C     RECENTLY GENERATED V2, VS, AND V1.                               LGM00320
C                                                                     LGM00330
      ZERO = 0.0D0                                                     LGM00340
      ONE  = 1.0D0                                                     LGM00350
      IF (MOLD1.GT.1) GO TO 40                                         LGM00360
      BETA(1) = ZERO                                                   LGM00370
      IIL = IIX                                                        LGM00380
C                                                                     LGM00390
C-----------------------------------------------------------------LGM00400
      CALL GENRAN(IIL,G,N)                                             LGM00410
C-----------------------------------------------------------------LGM00420
C                                                                     LGM00430
      DO 10 K = 1,N                                                    LGM00440
   10 V2(K) = G(K)                                                     LGM00450
C                                                                     LGM00460
C-----------------------------------------------------------------LGM00470
C     COMPUTE L-TRANSPOSE*V2 AND ITS NORM                             LGM00480
      ISOLV = 2                                                        LGM00490
      CALL LSOLV(V2,VS,ISOLV)                                          LGM00500
      SUM = FINPRO(N,VS(1),1,VS(1),1)                                  LGM00510
C-----------------------------------------------------------------LGM00520
C                                                                     LGM00530
C     NORMALIZE STARTING VECTORS:  (V2-TRANSPOSE*B*V2) = 1            LGM00540
      SUM = ONE/DSQRT(SUM)                                             LGM00550
      DO 20 K = 1,N                                                    LGM00560
      VS(K) = SUM*VS(K)                                                LGM00570
   20 V2(K) = SUM*V2(K)                                                LGM00580
C                                                                     LGM00590
C-----------------------------------------------------------------LGM00600
C     INITIALIZE V1 = B*V2 = L*VS                                     LGM00610
      ISOLV = 1                                                        LGM00620
```

```
      CALL LSOLV(VS,V1,ISOLV)                                      LGM00630
C-----------------------------------------------------------------LGM00640
C                                                                  LGM00650
      DO 30 K = 1,N                                                LGM00660
      VS(K) = V1(K)                                                LGM00670
   30 V1(K) = 0.D0                                                 LGM00680
   40 CONTINUE                                                     LGM00690
C                                                                  LGM00700
C     INITIALIZATIONS ARE:  VS = B*V(I),  V1 = B*V(I-1),  V2 = V(I) LGM00710
C                                                                  LGM00720
      DO 80 IVEC = MOLD1,KMAX                                      LGM00730
      SUM = BETA(IVEC)                                             LGM00740
C                                                                  LGM00750
C-----------------------------------------------------------------LGM00760
C     COMPUTE V1 = A*V2 - SUM*V1                                   LGM00770
      CALL MATVEC(V2,V1,SUM)                                       LGM00780
C     COMPUTE ALPHA(I)                                             LGM00790
      SUM = FINPRO(N,V1(1),1,V2(1),1)                              LGM00800
C-----------------------------------------------------------------LGM00810
C                                                                  LGM00820
      ALPHA(IVEC) = SUM                                            LGM00830
      DO 50 K = 1,N                                                LGM00840
   50 V1(K) = V1(K)-SUM*VS(K)                                      LGM00850
C                                                                  LGM00860
C     SET V1 = B*V(IVEC) AND VS = BETA(IVEC+1)*B*V(IVEC+1)         LGM00870
      DO 60 K = 1,N                                                LGM00880
      TEMP = V1(K)                                                 LGM00890
      V1(K) = VS(K)                                                LGM00900
   60 VS(K) = TEMP                                                 LGM00910
C                                                                  LGM00920
C-----------------------------------------------------------------LGM00930
C     CCMPUTE V2 = (L-INVERSE)*VS                                  LGM00940
      ISOLV = 3                                                    LGM00950
      CALL LSOLV(VS,V2,ISOLV)                                      LGM00960
C     COMPUTE BETA(IVEC+1)                                         LGM00970
      SUM = FINPRO(N,V2(1),1,V2(1),1)                              LGM00980
C-----------------------------------------------------------------LGM00990
C                                                                  LGM01000
      IN = IVEC+1                                                  LGM01010
      BETA(IN) = DSQRT(SUM)                                        LGM01020
C                                                                  LGM01030
C-----------------------------------------------------------------LGM01040
      ISOLV = 4                                                    LGM01050
      CALL LSOLV(V2,V2,ISOLV)                                      LGM01060
C-----------------------------------------------------------------LGM01070
C                                                                  LGM01080
      SUM = ONE/BETA(IN)                                           LGM01090
      DO 70 K = 1,N                                                LGM01100
      V2(K) = SUM*V2(K)                                            LGM01110
   70 VS(K) = SUM*VS(K)                                            LGM01120
C                                                                  LGM01130
   80 CONTINUE                                                     LGM01140
C                                                                  LGM01150
      RETURN                                                       LGM01160
C-----END LANCZS--------------------------------------------------LGM01170
      END                                                          LGM01180
C                                                                  LGM01190
C-----USPEC (GENERAL SYMMETRIC SPARSE MATRICES)-------------------LGM01200
C                                                                  LGM01210
C     SUBROUTINE USPECA(N,MATNOA)                                  LGM01220
      SUBROUTINE GUSPEC(N,MATNOA)                                  LGM01230
C                                                                  LGM01240
C-----------------------------------------------------------------LGM01250
      DOUBLE PRECISION  ASD(10000),AD(5010)                        LGM01260
      INTEGER  IROW(10000),ICOL(5010)                              LGM01270
C-----------------------------------------------------------------LGM01280
C     USPEC DIMENSIONS AND INITIALIZES THE ARRAYS NEEDED TO DEFINE LGM01290
```

```
C     THE USER-SPECIFIED A-MATRIX AND THEN PASSES THE STORAGE LOCATIONS   LGM01300
C     OF THESE ARRAYS TO THE MULTIPLY SUBROUTINE AMATV.                    LGM01310
C                                                                         LGM01320
C     MATRIX IS STORED IN FOLLOWING SPARSE MATRIX FORMAT:                 LGM01330
C     N = ORDER OF A-MATRIX,                                              LGM01340
C     NZS = NUMBER OF NONZERO SUBDIAGONAL ENTRIES,                        LGM01350
C     NZL = INDEX OF LAST COLUMN CONTAINING NONZERO SUBDIAGONAL ENTRIES,  LGM01360
C     ICOL(J), J=1,NZL IS THE NUMBER OF NONZERO SUBDIAGONAL ELEMENTS      LGM01370
C              IN COLUMN J.                                               LGM01380
C     IROW(K), K = 1,NZS IS THE CORRESPONDING ROW INDEX FOR ASD(K).       LGM01390
C     AD(I), I=1,N CONTAINS DIAGONAL ENTRIES (INCLUDING ANY 0             LGM01400
C              DIAGONAL ENTRIES).                                         LGM01410
C     ASD(K), K=1,NZS CONTAINS NONZERO SUBDIAGONAL ENTRIES, BY COLUMN     LGM01420
C     FOR J > NZL THERE ARE NO NONZERO SUBDIAGONAL ELEMENTS IN COLUMN J.  LGM01430
C     ICOL(J) = 0 IS ALLOWED                                             LGM01440
C                                                                         LGM01450
C-------------------------------------------------------------------------LGM01460
C     ARRAYS THAT DEFINE THE A-MATRIX ARE READ IN FROM FILE 8.  NOTE      LGM01470
C     THAT IF THE B-MATRIX IS PERMUTED, THEN LANCZOS PROGRAM ASSUMES      LGM01480
C     THAT THE DATA ON FILE 8 CORRESPONDS TO THE CORRESPONDING           LGM01490
C     PERMUTED A-MATRIX.  LANCZOS PROCEDURE WORKS DIRECTLY WITH THE       LGM01500
C     PERMUTED MATRICES.  EIGENVECTOR CODE, LGVEC, THEN PERMUTES THE      LGM01510
C     COMPUTED EIGENVECTORS TO GET THOSE CORRESPONDING TO THE ORIGINAL    LGM01520
C     MATRICES.                                                          LGM01530
C                                                                         LGM01540
      READ(8,10) NZS,NOLD,NZL,MATOLD                                      LGM01550
   10 FORMAT(I10,2I6,I8)                                                  LGM01560
C                                                                         LGM01570
      WRITE(6,20) NZS,NOLD,NZL,MATOLD                                     LGM01580
   20 FORMAT(I10,2I6,I8,' = NZS,NOLD,NZL,MATOLD'/)                        LGM01590
C                                                                         LGM01600
C     TEST OF PARAMETER CORRECTNESS                                       LGM01610
      ITEMP = (NOLD-N)**2 + (MATNOA-MATOLD)**2                            LGM01620
C                                                                         LGM01630
      IF(ITEMP.EQ.0) GO TO 40                                            LGM01640
C                                                                         LGM01650
      WRITE(6,30)                                                         LGM01660
   30 FORMAT(' PROGRAM TERMINATES BECAUSE EITHER ORDERS OF OR LABELS FORLGM01670
     1 MATRIX DISAGREE')                                                 LGM01680
      GO TO 70                                                           LGM01690
C                                                                         LGM01700
   40 CONTINUE                                                           LGM01710
C                                                                         LGM01720
C     NUMBER OF NONZERO SUBDIAGONAL ENTRIES IN EACH COLUMN IS READ        LGM01730
C     THEN THE CORRESPONDING ROW INDEX FOR EACH SUCH ENTRY IS READ        LGM01740
      READ(8,50) (ICOL(K), K=1,NZL)                                      LGM01750
      READ(8,50) (IROW(K), K=1,NZS)                                      LGM01760
   50 FORMAT(13I6)                                                       LGM01770
C                                                                         LGM01780
C     DIAGONAL IS READ FIRST, THEN NONZERO BELOW DIAGONAL ENTRIES         LGM01790
      READ(8,60) (AD(K), K=1,N)                                          LGM01800
      READ(8,60) (ASD(K), K=1,NZS)                                       LGM01810
   60 FORMAT(4E19.10)                                                    LGM01820
C                                                                         LGM01830
C-------------------------------------------------------------------------LGM01840
C     PASS STORAGE LOCATIONS OF ARRAYS THAT DEFINE THE A-MATRIX TO        LGM01850
C     THE MATRIX-VECTOR MULTIPLY SUBROUTINE AMATV                        LGM01860
      CALL AMATVE(ASD,AD,ICOL,IROW,N,NZL)                                LGM01870
C-------------------------------------------------------------------------LGM01880
C                                                                         LGM01890
      RETURN                                                             LGM01900
   70 STOP                                                               LGM01910
C-----END OF USPECA-------------------------------------------------------LGM01920
      END                                                                LGM01930
C                                                                         LGM01940
C-----USPECB FOR CHOLESKY FACTORS OF GENERAL SPARSE SYMMETRIC MATRIX----LGM01950
```

```
C                                                                      LGM01960
C     SUBROUTINE USPECB(N,MATNOB)                                      LGM01970
      SUBROUTINE CUSPEC(N,MATNOB)                                      LGM01980
C                                                                      LGM01990
C----------------------------------------------------------------------LGM02000
      DOUBLE PRECISION  BD(2200),BSD(10000)                            LGM02010
      INTEGER  KCOL(2200),KROW(10000),IPR(2200),IPT(2200)             LGM02020
C----------------------------------------------------------------------LGM02030
C     DIMENSIONS ARRAYS NEEDED TO DEFINE CHOLESKY FACTOR OF B-MATRIX,  LGM02040
C     READS CHOLESKY FACTOR FROM FILE 7, AND THEN PASSES STORAGE       LGM02050
C     LOCATIONS OF THESE ARRAYS TO THE MATRIX SOLVE SUBROUTINE LSOLV   LGM02060
C                                                                      LGM02070
C     THE LANCZOS PROCEDURE LGVAL WILL USE THE CHOLESKY FACTORS ON     LGM02080
C     FILE 7.  THESE FACTORS MAY CORRESPOND TO A PERMUTED VERSION OF   LGM02090
C     THE GIVEN B-MATRIX IN WHICH CASE THIS PERMUTATION WILL BE STORED LGM02100
C     IN IPR. THE ITH ROW OF THE PERMUTED B WILL CORRESPOND TO THE     LGM02110
C     JTH ROW OF B WHERE J = IPR(I) AND I = IPT(J).  IF B IS           LGM02120
C     PERMUTED, THE LANCZOS PROCEDURE ASSUMES THAT THE USER-PROVIDED   LGM02130
C     A-MATRIX IS IN FACT, THE CORRESPONDING PERMUTED VERSION OF THE   LGM02140
C     ORIGINAL A-MATRIX.                                              LGM02150
C                                                                      LGM02160
C     THE CHOLESKY FACTOR IS STORED IN THE FOLLOWING SPARSE FORMAT:    LGM02170
C     N = ORDER OF THE B-MATRIX.                                       LGM02180
C     NZT = NUMBER OF NONZERO SUBDIAGONAL ENTRIES IN THE CHOLESKY      LGM02190
C           FACTOR, L.                                                 LGM02200
C     KCOL(J), J=1,N IS THE NUMBER OF NONZERO SUBDIAGONAL ELEMENTS IN  LGM02210
C           COLUMN J OF L.                                             LGM02220
C     KROW(K), K=1,NZT IS THE ROW INDEX FOR CORRESPONDING ENTRY BSD(K).LGM02230
C     BD(J), J = 1,N CONTAINS THE DIAGONAL ENTRIES OF L.               LGM0224C
C     BSD(K), K =1,NZT CONTAINS THE NONZERO SUBDIAGONAL ENTRIES OF L   LGM02250
C           BY COLUMN.                                                 LGM02260
C----------------------------------------------------------------------LGM02270
C                                                                      LGM02280
C     READ CHOLESKY FACTOR FROM FILE 7.  MUST BE STORED                LGM02290
C     IN SPARSE MATRIX FORMAT.                                         LGM02300
      READ(7,10) NZT,NOLD,NZL,MATOLD,JPERM                             LGM02310
   10 FORMAT(I10,2I6,I8,I6)                                            LGM02320
C                                                                      LGM02330
      WRITE(6,20) NZT,NZL,N,NOLD,MATOLD,JPERM                          LGM02340
   20 FORMAT(' HEADER, CHOLESKY FACTOR FILE'/                          LGM02350
     1 3X,'NZT',3X,'NZL',5X,'N',2X,'NOLD',2X,'MATOLD',1X,'JPERM'/      LGM02360
     1 4I6,I8,I6/)                                                     LGM02370
C                                                                      LGM02380
      IF (N.NE.NOLD.OR.MATNOB.NE.MATOLD) GO TO 70                      LGM02390
C                                                                      LGM02400
      READ(7,30) (KCOL(K), K = 1,NZL)                                  LGM02410
      READ(7,30) (KROW(K), K = 1,NZT)                                  LGM02420
   30 FORMAT(13I6)                                                     LGM02430
      READ(7,40) (BD(K), K = 1,N)                                      LGM02440
      READ(7,40) (BSD(K), K = 1,NZT)                                   LGM02450
   40 FORMAT(4Z20)                                                     LGM02460
C  20 FORMAT(3E25.16)                                                  LGM02470
C                                                                      LGM02480
      IF(JPERM.EQ.0)  GO TO 60                                         LGM02490
C                                                                      LGM02500
      READ(7,30) (IPR(K), K = 1,N)                                     LGM02510
      DO 50 K = 1,N                                                    LGM02520
      J = IPR(K)                                                       LGM02530
   50 IPT(J) = K                                                       LGM02540
C                                                                      LGM02550
C----------------------------------------------------------------------LGM02560
      CALL LPERME(IPR,IPT,N)                                           LGM02570
C----------------------------------------------------------------------LGM02580
C                                                                      LGM02590
   60 CONTINUE                                                         LGM02600
C                                                                      LGM02610
```

```
C-----------------------------------------------------------------------LGM02620
C     PASS STORAGE LOCATIONS OF FACTORS TO SUBROUTINE LSOLV          LGM02630
      CALL LSOLVE(BSD,BD,KCOL,KROW,N,NZT,NZL)                        LGM02640
C-----------------------------------------------------------------------LGM02650
C                                                                   LGM02660
      GO TO 90                                                      LGM02670
C                                                                   LGM02680
   70 CONTINUE                                                      LGM02690
C     DEFAULT EXIT                                                  LGM02700
      WRITE(6,80) MATNOB,MATOLD                                     LGM02710
   80 FORMAT(' TERMINATE.  PARAMETERS IN CHOLESKY FACTOR FILE'/     LGM02720
     1' DO NOT AGREE WITH THOSE SPECIFIED BY THE USER'/             LGM02730
     1' MATNOB = ',I8,' MATOLD = ',I8/)                            LGM02740
      STOP                                                          LGM02750
C                                                                   LGM02760
   90 CONTINUE                                                      LGM02770
C-----END OF USPECB-----------------------------------------------------LGM02780
      RETURN                                                        LGM02790
      END                                                           LGM02800
C                                                                   LGM02810
C-----MATRIX-VECTOR MULTIPLY FOR REAL SPARSE SYMMETRIC MATRICES---------LGM02820
C                                                                   LGM02830
C     SUBROUTINE AMATV(W,U,SUM)                                     LGM02840
      SUBROUTINE GCMATV(W,U,SUM)                                    LGM02850
C                                                                   LGM02860
C-----------------------------------------------------------------------LGM02870
      DOUBLE PRECISION   U(1),W(1),ASD(1),AD(1),SUM                 LGM02880
      INTEGER  IROW(1),ICOL(1)                                      LGM02890
C-----------------------------------------------------------------------LGM02900
C     SPARSE MATRIX-VECTOR MULTIPLY FOR LANCZS   U = A*W - SUM*U    LGM02910
C     SEE USPECA SUBROUTINE FOR DESCRIPTION OF THE ARRAYS           LGM02920
C     THAT DEFINE THE A-MATRIX                                      LGM02930
C-----------------------------------------------------------------------LGM02940
C                                                                   LGM02950
C     COMPUTE THE DIAGONAL TERMS                                    LGM02960
      DO 10 I = 1,N                                                 LGM02970
   10 U(I) = AD(I)*W(I)-SUM*U(I)                                    LGM02980
C                                                                   LGM02990
C     COMPUTE BY COLUMN                                             LGM03000
      LLAST = 0                                                     LGM03010
      DO 30 J = 1,NZL                                               LGM03020
C                                                                   LGM03030
      IF (ICOL(J).EQ.0) GO TO 30                                    LGM03040
      LFIRST = LLAST + 1                                            LGM03050
      LLAST = LLAST + ICOL(J)                                       LGM03060
C                                                                   LGM03070
      DO 20 L = LFIRST,LLAST                                        LGM03080
      I = IROW(L)                                                   LGM03090
C                                                                   LGM03100
      U(I) = U(I) + ASD(L)*W(J)                                     LGM03110
      U(J) = U(J) + ASD(L)*W(I)                                     LGM03120
C                                                                   LGM03130
   20 CONTINUE                                                      LGM03140
C                                                                   LGM03150
   30 CONTINUE                                                      LGM03160
C                                                                   LGM03170
      RETURN                                                        LGM03180
C                                                                   LGM03190
C-----------------------------------------------------------------------LGM03200
C     STORAGE LOCATIONS OF ARRAYS ARE PASSED TO AMATV FROM USPECA   LGM03210
      ENTRY AMATVE(ASD,AD,ICOL,IROW,N,NZL)                          LGM03220
C-----------------------------------------------------------------------LGM03230
C                                                                   LGM03240
      RETURN                                                        LGM03250
C-----END OF AMATV----------------------------------------------------- LGM03260
      END                                                           LGM03270
C                                                                   LGM03280
```

```
C-----LSOLV-GENERAL SPARSE, POSITIVE DEFINITE B-MATRIX------------------LGM03290
C     (USES THE CHOLESKY FACTORS OF B, B = L*(L-TRANSPOSE))              LGM03300
C                                                                       LGM03310
      SUBROUTINE TLSOLV(W,U,ISOLV)                                      LGM03320
C     SUBROUTINE LSOLV(W,U,ISOLV)                                       LGM03330
C                                                                       LGM03340
C----------------------------------------------------------------------LGM03350
      DOUBLE PRECISION  U(1),W(1),BD(1),BSD(1), TEMP                    LGM03360
      INTEGER  KCOL(1),KROW(1)                                          LGM03370
C----------------------------------------------------------------------LGM03380
C     SUBROUTINE HAS 4 BRANCHES:  ISOLV = (1,2,3,4)  CALCULATES         LGM03390
C     ISOLV = 1    U = L*W                                              LGM03400
C     ISOLV = 2    U = L'*W                                             LGM03410
C     ISOLV = 3    SOLVE FOR U IN  L*U = W                              LGM03420
C     ISOLV = 4    SOLVE FOR U IN  L'*U = W                             LGM03430
C----------------------------------------------------------------------LGM03440
      GO TO (10,50,80,120), ISOLV                                       LGM03450
C                                                                       LGM03460
C     ISOLV = 1,   U=L*W                                                LGM03470
   10 CONTINUE                                                          LGM03480
      KL = 0                                                            LGM03490
      DO 20 K = 1,N                                                     LGM03500
   20 U(K) = W(K)*BD(K)                                                 LGM03510
      DO 40 K = 1,N                                                     LGM03520
      TEMP = W(K)                                                       LGM03530
      IF (KCOL(K).EQ.0.OR.K.EQ.N) GO TO 40                              LGM03540
      KF = KL + 1                                                       LGM03550
      KL = KL + KCOL(K)                                                 LGM03560
      DO 30 KK = KF,KL                                                  LGM03570
      KR = KROW(KK)                                                     LGM03580
   30 U(KR) = U(KR) + TEMP*BSD(KK)                                      LGM03590
   40 CONTINUE                                                          LGM03600
      GO TO 150                                                         LGM03610
C                                                                       LGM03620
C     ISOLV = 2,  U = (L-TRANSPOSE)*W                                   LGM03630
   50 CONTINUE                                                          LGM03640
      KL = 0                                                            LGM03650
      DO 70 J = 1,N                                                     LGM03660
      TEMP = W(J)*BD(J)                                                 LGM03670
      IF (KCOL(J).EQ.0.OR.J.EQ.N) GO TO 70                              LGM03680
      KF = KL + 1                                                       LGM03690
      KL  = KL + KCOL(J)                                                LGM03700
      DO 60 K = KF,KL                                                   LGM03710
      IK = KROW(K)                                                      LGM03720
   60 TEMP = BSD(K)*W(IK) + TEMP                                        LGM03730
   70 U(J) = TEMP                                                       LGM03740
      GO TO 150                                                         LGM03750
C                                                                       LGM03760
C     ISOLV = 3,  U = (L-INVERSE)*W                                     LGM03770
   80 CONTINUE                                                          LGM03780
      DO 90 K = 1,N                                                     LGM03790
   90 U(K) = W(K)                                                       LGM03800
      KL = 0                                                            LGM03810
      DO 110 K = 1,N                                                    LGM03820
      TEMP = U(K)/BD(K)                                                 LGM03830
      U(K) = TEMP                                                       LGM03840
      IF (KCOL(K).EQ.0.OR.K.EQ.N) GO TO 110                             LGM03850
      KF = KL + 1                                                       LGM03860
      KL = KL + KCOL(K)                                                 LGM03870
      DO 100 KK = KF,KL                                                 LGM03880
      KR = KROW(KK)                                                     LGM03890
  100 U(KR) = U(KR) - TEMP*BSD(KK)                                      LGM03900
  110 CONTINUE                                                          LGM03910
      GO TO 150                                                         LGM03920
C                                                                       LGM03930
C     ISOLV = 4,  U = (L-TRANSPOSE)-INVERSE*W                           LGM03940
  120 CONTINUE                                                          LGM03950
```

```
      NP1 = N+1                                                   LGM03960
      KF = NZT + 1                                                LGM03970
      DO 140 K = 1,N                                              LGM03980
      L = NP1 - K                                                 LGM03990
      TEMP = W(L)                                                 LGM04000
      IF (KCOL(L).EQ.0.OR.L.EQ.N) GO TO 140                       LGM04010
      KL = KF - 1                                                 LGM04020
      KF = KF - KCOL(L)                                           LGM04030
      DO 130 LL = KF,KL                                           LGM04040
      LR = KROW(LL)                                               LGM04050
  130 TEMP = TEMP - BSD(LL)*U(LR)                                 LGM04060
  140 U(L) = TEMP/BD(L)                                           LGM04070
      GO TO 150                                                   LGM04080
  150 CONTINUE                                                    LGM04090
C                                                                 LGM04100
      RETURN                                                      LGM04110
C                                                                 LGM04120
C-----------------------------------------------------------------LGM04130
      ENTRY LSOLVE(BSD,BD,KCOL,KROW,N,NZT,NZL)                    LGM04140
C-----------------------------------------------------------------LGM04150
C                                                                 LGM04160
C-----END OF LSOLV-----------------------------------             LGM04170
      RETURN                                                      LGM04180
      END                                                         LGM04190
C                                                                 LGM04200
C-----START OF USPEC FOR DIAGONAL TEST A-MATRIX-------------------LGM04210
C                                                                 LGM04220
      SUBROUTINE USPECA(N,MATNO)                                  LGM04230
C     SUBROUTINE DUSPEC(N,MATNO)                                  LGM04240
C                                                                 LGM04250
C-----------------------------------------------------------------LGM04260
      DOUBLE PRECISION  D(1000), SPACE, SHIFT                     LGM04270
      DOUBLE PRECISION  DABS, DFLOAT                              LGM04280
      REAL  EXPLAN(20)                                            LGM04290
C-----------------------------------------------------------------LGM04300
C                                                                 LGM04310
      READ(8,10) EXPLAN                                           LGM04320
   10 FORMAT(20A4)                                                LGM04330
      READ(8,*) NOLD,NUNIF,SPACE,D(1),SHIFT                       LGM04340
      NNUNIF = NOLD - NUNIF                                       LGM04350
      WRITE(6,20) NOLD,SPACE,NNUNIF,D(1),SHIFT                    LGM04360
   20 FORMAT(/' DIAGONAL TEST A-MATRIX, SIZE = ',I4/' MOST ENTRIES ARE 'LGM04370
     1,E10.3,' UNITS APART.',I3,' ENTRIES'/' ARE IRREGULARLY SPACED. FIRLGM04380
     1ST ENTRY IS ',E10.3,' SHIFT = ',E10.3/)                    LGM04390
C                                                                 LGM04400
      IF(N.NE.NOLD) GO TO 90                                      LGM04410
C     COMPUTE THE UNIFORM PORTION OF THE SPECTRUM                 LGM04420
      DO 30 J=2,NUNIF                                             LGM04430
   30 D(J) = D(1) - DFLOAT(J-1)*SPACE                             LGM04440
      NUNIF1=NUNIF + 1                                            LGM04450
      READ(8,10)  EXPLAN                                          LGM04460
      DO 40 J=NUNIF1,N                                            LGM04470
   40 READ(8,*) D(J)                                              LGM04480
      NB = NUNIF - 2                                              LGM04490
C                                                                 LGM04500
      IF(SHIFT.EQ.0.) GO TO 60                                    LGM04510
      DO 50 J=1,N                                                 LGM04520
   50 D(J) = D(J) + SHIFT                                         LGM04530
C                                                                 LGM04540
C     PRINT OUT A-MATRIX                                          LGM04550
   60 WRITE(6,70) (D(I), I=1,10 )                                 LGM04560
      WRITE(6,80) (D(I), I = NB,N)                                LGM04570
   70 FORMAT(/' GENERALIZED LANCZOS TEST, 1ST 10 ENTRIES OF DIAGONAL A-MLGM04580
     1ATRIX = '/(3E22.14))                                       LGM04590
   80 FORMAT(/' MIDDLE UNIFORM PORTION OF MATRIX IS NOT PRINTED OUT'/    LGM04600
     1' END OF UNIFORM PLUS NONUNIFORM SECTION = '/(3E25.16))    LGM04610
C                                                                 LGM04620
```

```
C    DIAGONAL GENERATION COMPLETE                                       LGM04630
C                                                                       LGM04640
C-----------------------------------------------------------------------LGM04650
C    CALL ENTRY TO MATRIX-VECTOR MULTIPLY SUBROUTINE  TO PASS           LGM04660
C    STORAGE LOCATION OF D-ARRAY AND ORDER OF A-MATRIX.                 LGM04670
     CALL MVDIAE(D,N)                                                   LGM04680
C-----------------------------------------------------------------------LGM04690
C                                                                       LGM04700
     RETURN                                                             LGM04710
  90 WRITE(6,100) NOLD,N                                                LGM04720
 100 FORMAT(' PROGRAM TERMINATES BECAUSE NOLD = ',I5,'DOES NOT EQUAL N LGM04730
    1 =',I5)                                                            LGM04740
C-----END OF USPECA SUBROUTINE FOR DIAGONAL TEST MATRICES---------------LGM04750
     STOP                                                               LGM04760
     END                                                                LGM04770
C                                                                       LGM04780
C-----USPECB--DIAGONAL TEST B-MATRIX------------------------------------LGM04790
C                                                                       LGM04800
     SUBROUTINE USPECB(N,MATNO)                                         LGM04810
C    SUBROUTINE USPECD(N,MATNO)                                         LGM04820
C                                                                       LGM04830
C-----------------------------------------------------------------------LGM04840
     DOUBLE PRECISION  D(1000), DS(1000), SPACE, SHIFT                  LGM04850
     DOUBLE PRECISION  DFLOAT, DSQRT                                    LGM04860
     REAL  EXPLAN(20)                                                   LGM04870
C-----------------------------------------------------------------------LGM04880
C                                                                       LGM04890
     READ(7,10) EXPLAN                                                  LGM04900
  10 FORMAT(20A4)                                                       LGM04910
     READ(7,*) NOLD,NUNIF,SPACE,D(1),SHIFT                              LGM04920
     NNUNIF = NOLD - NUNIF                                              LGM04930
     WRITE(6,20) NOLD,SPACE,NNUNIF,D(1),SHIFT                           LGM04940
  20 FORMAT(/' DIAGONAL TEST B-MATRIX, SIZE = ',I4/' MOST ENTRIES ARE 'LGM04950
    1,E10.3,' UNITS APART.',I3,' ENTRIES'/' ARE IRREGULARLY SPACED. FIRLGM04960
    1ST ENTRY IS ',E10.3,' SHIFT = ',E10.3/)                           LGM04970
C                                                                       LGM04980
     IF(N.NE.NOLD) GO TO 100                                           LGM04990
C    COMPUTE THE UNIFORM PORTION OF THE SPECTRUM                        LGM05000
     DO 30 J=2,NUNIF                                                    LGM05010
  30 D(J) = D(1) - DFLOAT(J-1)*SPACE                                    LGM05020
     NUNIF1=NUNIF + 1                                                   LGM05030
     READ(7,10)  EXPLAN                                                 LGM05040
     DO 40 J=NUNIF1,N                                                   LGM05050
  40 READ(7,*) D(J)                                                     LGM05060
     NB = NUNIF - 2                                                     LGM05070
C                                                                       LGM05080
     IF(SHIFT.EQ.0.) GO TO 60                                          LGM05090
     DO 50 J=1,N                                                        LGM05100
  50 D(J) = D(J) + SHIFT                                                LGM05110
C                                                                       LGM05120
C    PRINT OUT B-MATRIX                                                 LGM05130
  60 WRITE(6,70) (D(I), I=1,10 )                                        LGM05140
     WRITE(6,80) (D(I), I = NB,N)                                       LGM05150
  70 FORMAT(/' GENERALIZED LANCZOS TEST, 1ST 10 ENTRIES OF DIAGONAL B-MLGM05160
    1ATRIX = '/(3E22.14))                                              LGM05170
  80 FORMAT(/' MIDDLE UNIFORM PORTION OF MATRIX IS NOT PRINTED OUT'/    LGM05180
    1' END OF UNIFORM PLUS NONUNIFORM SECTION = '/(3E25.16))           LGM05190
C                                                                       LGM05200
C    DIAGONAL GENERATION COMPLETE                                       LGM05210
C                                                                       LGM05220
     DO 90 K = 1,N                                                      LGM05230
  90 DS(K) = DSQRT(D(K))                                                LGM05240
C                                                                       LGM05250
C-----------------------------------------------------------------------LGM05260
C    PASS STORAGE LOCATION OF THE L-FACTOR (THE DS-ARRAY) AND ORDER OF  LGM05270
C    B-MATRIX TO LSOLV SUBROUTINE.                                      LGM05280
```

```
      CALL DSOLVE(DS,N)                                             LGM05290
C----------------------------------------------------------------------LGM05300
C                                                                      LGM05310
      RETURN                                                           LGM05320
  100 WRITE(6,110) NOLD,N                                              LGM05330
  110 FORMAT(' PROGRAM TERMINATES BECAUSE NOLD = ',I5,'DOES NOT EQUAL N LGM05340
     1 =',I5)                                                          LGM05350
C-----END OF USPECB SUBROUTINE FOR DIAGONAL TEST MATRICES-------------LGM05360
      STOP                                                             LGM05370
      END                                                              LGM05380
C                                                                      LGM05390
C-----MATRIX-VECTOR MULTIPLY FOR DIAGONAL TEST MATRICES--------------LGM05400
C                                                                      LGM05410
      SUBROUTINE AMATV(W,U,SUM)                                        LGM05420
C     SUBROUTINE DCMATV(W,U,SUM)                                       LGM05430
C                                                                      LGM05440
C     AMATV COMPUTES  U = (DIAGONAL MATRIX) * W - SUM * U              LGM05450
C----------------------------------------------------------------------LGM05460
      DOUBLE PRECISION  W(1),U(1),D(1),SUM                             LGM05470
C----------------------------------------------------------------------LGM05480
C                                                                      LGM05490
      DO 10 I=1,N                                                      LGM05500
   10 U(I)= D(I)*W(I) - SUM*U(I)                                       LGM05510
C                                                                      LGM05520
      RETURN                                                           LGM05530
C                                                                      LGM05540
C----------------------------------------------------------------------LGM05550
      ENTRY MVDIAE(D,N)                                                LGM05560
C----------------------------------------------------------------------LGM05570
C                                                                      LGM05580
      RETURN                                                           LGM05590
C-----END OF DIAGONAL TEST MATRIX  MULTIPLY-------------------------LGM05600
      END                                                              LGM05610
C                                                                      LGM05620
C-----LSOLV FOR DIAGONAL MATRIX-------------------------------------LGM05630
C                                                                      LGM05640
      SUBROUTINE LSOLV(W,U,ISOLV)                                      LGM05650
C     SUBROUTINE DSOLV(W,U,ISOLV)                                      LGM05660
C                                                                      LGM05670
C----------------------------------------------------------------------LGM05680
      DOUBLE PRECISION U(1), W(1),  DS(1)                              LGM05690
C----------------------------------------------------------------------LGM05700
      GO TO (10,30,50,70), ISOLV                                       LGM05710
C                                                                      LGM05720
C     ISOLV = 1                                                        LGM05730
   10 CONTINUE                                                         LGM05740
      DO 20 K = 1,N                                                    LGM05750
   20 U(K) = DS(K)*W(K)                                                LGM05760
      GO TO 90                                                         LGM05770
C                                                                      LGM05780
C     ISOLV = 2                                                        LGM05790
   30 CONTINUE                                                         LGM05800
      DO 40 K = 1,N                                                    LGM05810
   40 U(K) = DS(K)*W(K)                                                LGM05820
      GO TO 90                                                         LGM05830
C                                                                      LGM05840
C     ISOLV = 3                                                        LGM05850
   50 CONTINUE                                                         LGM05860
      DO 60 K = 1,N                                                    LGM05870
   60 U(K) = W(K)/DS(K)                                                LGM05880
      GO TO 90                                                         LGM05890
C                                                                      LGM05900
C     ISOLV = 4                                                        LGM05910
   70 CONTINUE                                                         LGM05920
      DO 80 K = 1,N                                                    LGM05930
   80 U(K) = W(K)/DS(K)                                                LGM05940
C                                                                      LGM05950
```

```
   90 CONTINUE                                                        LGM05960
C                                                                     LGM05970
      RETURN                                                          LGM05980
C                                                                     LGM05990
C------------------------------------------------------------------LGM06000
      ENTRY DSOLVE(DS,N)                                              LGM06010
C------------------------------------------------------------------LGM06020
C                                                                     LGM06030
      RETURN                                                          LGM06040
C-----END OF DSOLV-------------------------------------------------LGM06050
      END                                                             LGM06060
```

SECTION 5.5   FILE DEFINITIONS AND SAMPLE INPUT FILES

Below is a listing of the input/output files which are accessed by the
Lanczos eigenvalue program LGVAL for real symmetric generalized
problems where one of the two matrices is positive definite. Included
also is a sample of the input file which LGVAL requires on
file 5. The parameters in this file are supplied in free format.
LGVAL computes eigenvalues of the matrix eigenvalue problem
AX = EV*BX on user-specified intervals where A and B are both
real symmetric matrices and B is positive definite. The program uses
the Cholesky factor L OF B = L*L'.


```
SAMPLE DEFINITIONS OF THE INPUT/OUTPUT FILES FOR LGVAL
-----------------------------------------------------------------
*LGVAL EXEC LANCZOS EIGENVALUE CALCULATION AX = EV*BX CASE
FI 06 TERM
FILEDEF  1 DISK &1       NHISTORY  A (RECFM F LRECL 80 BLOCK 80
FILEDEF  2 DISK &1       HISTORY   A (RECFM F LRECL 80 BLOCK 80
FILEDEF  3 DISK &1       GOODEV    A (RECFM F LRECL 80 BLOCK 80
FILEDEF  4 DISK &1       ERRINV    A (RECFM F LRECL 80 BLOCK 80
FILEDEF  5 DISK LGVAL    INPUT     A (RECFM F LRECL 80 BLOCK 80
FILEDEF  7 DISK &1       LCOMPAC   A (RECFM F LRECL 80 BLOCK 80
FILEDEF  8 DISK &1       ACOMPAC   A (RECFM F LRECL 80 BLOCK 80
FILEDEF 11 DISK &1       DISTINCT  A (RECFM F LRECL 80 BLOCK 80
LOAD  LGVAL   LESUB  LGMULT
-----------------------------------------------------------------
```


```
SAMPLE INPUT FOR LGVAL
-----------------------------------------------------------------
 LGVAL INPUT LANCZOS EIGENVALUE COMPUTATION, NO REORTHOGONALIZATION
 AX = EV*BX GENERALIZED EIGENVALUE PROBLEM
LINE 1    N      KMAX     NMEVS     MATNOA    MATNOB
          100      300       1        100       100
LINE 2  SVSEED   RHSEED       MXINIT   MXSTUR
       49302312  5731029         5     100000
LINE 3  ISTART    ISTOP
           0        1
LINE 4   IHIS     IDIST  IWRITE
           1        0       1
LINE 5  RELTOL (RELATIVE TOLERANCE IN 'COMBINING' GOODEV)
     .0000000001
LINE 6   MB(1)   MB(2)   MB(3)   MB(4)    (ORDERS OF T(1,MEV) )
          300
LINE 7   NINT      (NUMBER OF SUB-INTERVALS FOR BISEC)
           1
LINE 8   LB(1)   LB(2)   LB(3)   LB(4)   (INTERVAL LOWER BOUNDS)
          1.5
LINE 9   UB(1)   UB(2)   UB(3)   UB(4)   (INTERVAL UPPER BOUNDS)
         2100.
-----------------------------------------------------------------
```

Below is a listing of the input/output files which are accessed by the
Lanczos eigenvector program for real symmetric generalized problems,
LGVEC. Also included below is a sample of the input file which LGVEC
requires on file 5. The parameters in this file are supplied in free
format. LGVEC computes eigenvectors for each of a user-specified
subset of the eigenvalues computed by the companion program LGVAL.

SAMPLE DEFINITIONS OF THE INPUT/OUTPUT FILES FOR LGVEC
----
*LGVEC EXEC TO RUN LANCZOS EIGENVECTOR PROGRAM, REAL SYMMETRIC MATRICES
FI 06 TERM
```
FILEDEF  2 DISK &1        HISTORY    A (RECFM F LRECL 80 BLOCK 80
FILEDEF  3 DISK &1        GOODEV     A (RECFM F LRECL 80 BLOCK 80
FILEDEF  4 DISK &1        ERRINV     A (RECFM F LRECL 80 BLOCK 80
FILEDEF  5 DISK LGVEC     INPUT      A (RECFM F LRECL 80 BLOCK 80
FILEDEF  7 DISK &1        LCOMPAC    A (RECFM F LRECL 80 BLOCK 80
FILEDEF  8 DISK &1        ACOMPAC    A (RECFM F LRECL 80 BLOCK 80
FILEDEF  9 DISK &1        ERREST     A (RECFM F LRECL 80 BLOCK 80
FILEDEF 10 DISK &1        BOUNDS     A (RECFM F LRECL 80 BLOCK 80
FILEDEF 11 DISK &1        TEIGVECS   A (RECFM F LRECL 80 BLOCK 80
FILEDEF 12 DISK &1        RITZVECS   A (RECFM F LRECL 80 BLOCK 80
FILEDEF 13 DISK &1        PAIGE      A (RECFM F LRECL 80 BLOCK 80
LOAD  LGVEC  LESUB  LGMULT
```

SAMPLE INPUT FILE FOR LGVEC

```
 LGVEC EIGENVECTOR COMPUTATIONS AX = EV*BX NO REORTHOGONALIZATION
LINE 1  MDIMTV     MDIMRV  MBETA (MAX.DIMENSIONS, TVEC, RITVEC AND BETA
        10000      10000   2000
LINE 2      RELTOL
        .0000000001
LINE 3  MBOUND    NTVCON SVTVEC IREAD (FLAGS
        0          1      0       1
LINE 4  TVSTOP   LVCONT ERCONT  IWRITE (FLAGS
        0          1      1       1
LINE 5   RHSEED  (RANDOM GENERATOR SEED FOR STARTING VECTOR IN INVERM)
        45329517
LINE 6  MATNOA     MATNOB    N   JPERM
        100        100     100   0
```

CHAPTER 6

REAL RECTANGULAR MATRICES

SECTION 6.1   INTRODUCTION

The FORTRAN codes in this Chapter address the question of computing distinct singular values and corresponding left and right singular vectors of real rectangular matrices, using a single-vector Lanczos procedure. For a given real rectangular ℓxn matrix A, these codes compute nonnegative scalars σ and corresponding real vectors $x \neq 0$ and $y \neq 0$ such that

$$Ax = \sigma y \text{ and } A^T y = \sigma x. \tag{6.1.1}$$

Every real rectangular ℓxn matrix, where $\ell \geq n$, has a singular value decomposition,

$$A = Y\Sigma X^T \text{ with } X^T X = I, \ Y^T Y = I, \text{ and } \Sigma = \begin{bmatrix} \Sigma_1 \\ 0 \end{bmatrix} \tag{6.1.2}$$

where Σ is ℓxn and $\Sigma_1 = \text{diag}\{\sigma_1,...,\sigma_n\}$ with σ_i, $1 \leq i \leq n$, the singular values of A. X is a nxn orthogonal matrix, Y is a ℓxℓ orthogonal matrix, and the columns of X and of Y are respectively, right and left singular vectors of A. There are many applications for this type of decomposition. Singular values and vectors are discussed in detail for example in Stewart [1973].

Using Eqn(6.1.1), it is not difficult to demonstrate that the singular values of a given real matrix A are just the nonnegative square roots of the eigenvalues of the associated real symmetric matrix $A^T A$ (or AA^T). Thus from the perturbation theorems for real symmetric matrices, we have that a small perturbation in the given matrix A causes small perturbations in the singular values. The same arguments demonstrate that the right singular vectors of a matrix A are eigenvectors of the matrix $A^T A$, and the left singular vectors are eigenvectors of the matrix AA^T. Therefore, we also have that the perturbation theorems for eigenvectors of real symmetric matrices apply to the singular vectors.

The Lanczos recursion as presented in Eqns(1.2.1) - (1.2.2) is only applicable to real symmetric matrices. Therefore we ask the question: How do we construct a real symmetric matrix which will give us the desired singular values? Obviously, we could just apply the real symmetric Lanczos recursion to either $A^T A$ or to AA^T. However in general, these matrices are not suitable because of the effects that squaring a matrix can have on the eigenvalues. Small singular values of A which are close together correspond to eigenvalues of $A^T A$ which are smaller and even closer together. Large singular values of A which are far apart correspond to eigenvalues of $A^T A$ which are larger and further apart. When a matrix A has both small and large singular values, dealing numerically with the square of that matrix is difficult. Lanczos [1961] suggested the use of an alternative real symmetric matrix. He proposed that the

273

following larger but real symmetric $[\ell + n] \times [\ell + n]$ matrix be used.

$$B \equiv \begin{bmatrix} 0 & A \\ A^T & 0 \end{bmatrix}. \qquad (6.1.3)$$

The relationships between the eigenvalues and the eigenvectors of B and the singular values and singular vectors of A are discussed in detail in Section 5.4 of Chapter 5 in Volume 1.

We could apply the real symmetric version of the Lanczos recursion directly to the matrix B in Eqn(6.1.3). However, because this matrix is considerably larger than the A-matrix, we use a modification of the real symmetric Lanczos recursion which incorporates the following choice of starting vector suggested by Golub and Kahan [1965]. We choose a starting vector either of the form $(0, u^T)^T$ or of the form $(v^T, 0)^T$ where u is of length n, the column order of the A-matrix, and v is of length ℓ, the row order of the A-matrix. If we use such a starting vector in the basic Lanczos recursion in Eqns(1.2.1) - (1.2.2), we obtain a version of the Lanczos recursion designed specifically for the B-matrix in Eqn(6.1.3). The Lanczos vectors generated by this recursion alternate in form from either $(0, u^T)^T$ to $(v^T, 0)^T$ or vice-versa, as the iterations proceed. Furthermore, on each iteration of this recursion it is only necessary to either compute Au_i or $A^T v_i$. Therefore, the amount of work per iteration of this recursion is no more than applying the real symmetric Lanczos recursion to a real symmetric matrix of order $\max(\ell, n)$. For details on the corresponding Lanczos recursion see Section 5.4 of Chapter 5 in Volume 1.

These codes can compute either a very few or very many of the distinct singular values of a given real rectangular matrix. As the documentation in Section 6.2 indicates, the A-multiplicity of a computed singular value can be obtained only with additional computation, and the modifications required to do this additional computation are not included in these versions of the codes.

The Lanczos recursions which we use generate a family of real symmetric, tridiagonal matrices (T-matrices). The diagonal entries of each of these T-matrices are all 0. The eigenvalues of any even-ordered T-matrix occur in ± pairs. This latter property is inherited from the B-matrix whose eigenvalues are just $\pm\sigma_i$, the ± pairs of singular values plus $\ell - n$ additional zero eigenvalues if $\ell \neq n$. Only even-ordered T-matrices may be used in the Lanczos computations. There is no reorthogonalization of the Lanczos vectors at any stage in any of the computations.

LSVAL, the main program for the single-vector, Lanczos singular value computations, calls the subroutine BISEC to compute eigenvalues of those Lanczos tridiagonal matrices specified by the user and on those sub-intervals specified by the user. The BISEC subroutine used in this Chapter is a modification of the BISEC subroutine given in Chapter 2 which assumes that that the diagonal entries of the T-matrices supplied to it are all 0. BISEC simultaneously computes the T-eigenvalues and T-multiplicities and then sorts the computed T-eigenvalues into two categories, the 'good' T-eigenvalues and the 'spurious' T-eigenvalues. The 'good' T-eigenvalues are accepted as approximations to singular values of the user-specified matrix A. The accuracy of these 'good' T-eigenvalues as singular values of A is then estimated using error estimates computed by subroutine INVERR. The subroutine INVERR in this chapter is a modification of the INVERR subroutine in Chapter 2 which assumes the diagonal entries of the tridiagonal matrices supplied to it are all 0. Error estimates are computed only for isolated 'good' T-eigenvalues. All other 'good' T-eigenvalues are assumed to have converged. If convergence has not yet occurred and a larger Lanczos matrix has been specified by the user, these programs will continue on repeating the above procedure on a larger Lanczos matrix.

Once the singular values been computed accurately enough, the user can select a subset of the 'converged' singular values for which singular vectors are to be computed. The main program LSVEC, for computing singular vectors of real rectangular matrices, is then used to compute these desired singular vectors. These singular vectors are obtained by computing Ritz vectors for the B-matrix and then splitting each of these $(\ell + n)$ – dimensional Ritz vectors into approximate left and right singular vectors of A. The user should note that if the singular value being considered is very small, then LSVEC is not able to accurately compute both a left and a right singular vector approximation simultaneously. In this situation one of the two singular vectors will be more accurate than the other one is. If the starting vector is of the form $(0,u^T)^T$, then the right singular vector will be more accurate than the corresponding left vector. Similarly, if we use a starting vector of the form $(v^T,0)^T$, then the left vector will be more accurate than the right vector will be. This loss in accuracy in one of the two vectors increases as the size of the singular value is decreased, and in the limit for a zero singular value, one of the two computed singular vectors will have no accuracy at all. See Section 5.4 of Chapter 5 in Volume 1.

All computations are in double precision real arithmetic. The user must supply a subroutine USPEC which defines and initializes the user-specified matrix A, and subroutines SVMAT and STRAN which compute respectively, matrix-vector multiplies Ax and A^Ty for any given vectors x and y. These subroutines must be constructed in such a way as to take advantage of the sparsity (and/or structure) of the user-supplied A-matrix and such that these computations are done accurately. More details about these real rectangular, single-vector Lanczos procedures are given in Section 5.4 of Chapter 5 in Volume 1.

SECTION 6.2   DOCUMENTATION FOR PROGRAMS IN CHAPTER 6

```
C-----DOCUMENTATION FOR THE SINGLE-VECTOR-------------------------------LSV00010
C     LANCZOS SINGULAR VALUE/VECTOR PROGRAMS                             LSV00020
C     FOR REAL, RECTANGULAR MATRICES                                     LSV00030
C                                                                        LSV00040
C                                                                        LSV00050
C     AUTHORS:   JANE CULLUM AND RALPH A. WILLOUGHBY, IBM RESEARCH,      LSV00060
C                YORKTOWN HEIGHTS, NY 10598.  PHONE: 914-945-2227        LSV00070
C                                                                        LSV00080
C                                                                        LSV00090
C     GIVEN A REAL RECTANGULAR MATRIX A OF ORDER M X N THE THREE         LSV00100
C     SETS OF FORTRAN FILES LABELLED LSVAL, LSSUB, AND LSMULT            LSV00110
C     CAN BE USED TO COMPUTE DISTINCT SINGULAR VALUES OF A IN            LSV00120
C     USER-SPECIFIED INTERVALS.                                          LSV00130
C                                                                        LSV00140
C     CORRESPONDING SINGULAR VECTORS FOR SELECTED, COMPUTED             LSV00150
C     SINGULAR VALUES CAN BE COMPUTED USING THE SETS OF FILES            LSV00160
C     LABELLED LSVEC, LSSUB AND LSMULT.                                  LSV00170
C                                                                        LSV00180
C     THESE PROGRAMS USE LANCZOS TRIDIAGONALIZATION WITHOUT              LSV00190
C     REORTHOGONALIZATION ON THE ASSOCIATED REAL SYMMETRIC MATRIX        LSV00200
C                                                                        LSV00210
C                         ----                    ----                   LSV00220
C                        |  0                 A  |                       LSV00230
C                 B  =   |                       |                       LSV00240
C                        |  A-TRANSPOSE       0  |                       LSV00250
C                         ----                    ----                   LSV00260
C                                                                        LSV00270
C     OF ORDER M + N  TO GENERATE REAL SYMMETRIC TRIDIAGONAL             LSV00280
C     MATRICES, T(1,MEV), OF ORDER MEV.  SUBSETS OF THE EIGENVALUES OF   LSV00290
C     THESE T-MATRICES, LABELLED AS THE 'GOOD EIGENVALUES' OF T(1,MEV),  LSV00300
C     ARE APPROXIMATIONS TO THE DESIRED SINGULAR VALUES OF A.            LSV00310
C     CORRESPONDING RITZ VECTORS FOR B ARE APPROXIMATIONS TO            LSV00320
C     EIGENVECTORS OF B WHICH IN TURN CONTAIN APPROXIMATIONS TO          LSV00330
C     THE DESIRED LEFT AND RIGHT SINGULAR VECTORS OF A.  THIS            LSV00340
C     PROCEDURE USES A SPECIAL STARTING VECTOR SUGGESTED BY GOLUB        LSV00350
C     AND KAHAN.  THUS, THE STARTING LANCZOS VECTOR IS EITHER OF         LSV00360
C     THE FORM (V1,0) OR (0,V2) WHERE V1 IS MX1 AND V2 IS NX1 AND        LSV00370
C     ALL SUCCEEDING LANCZOS VECTORS GENERATED ALTERNATE BETWEEN         LSV00380
C     THESE 2 FORMS.  THIS SPECIAL CHOICE OF STARTING VECTOR RESULTS     LSV00390
C     IN SIGNIFICANT GAINS IN STORAGE AND OPERATION COUNTS AND           LSV00400
C     ALSO IN CONVERGENCE RELATIVE TO A 'BRUTE FORCE' APPLICATION        LSV00410
C     OF THE REAL SYMMETRIC LANCZOS PROCEDURE DIRECTLY TO THE            LSV00420
C     MATRIX B ABOVE.  FOR MORE DETAILS SEE REFERENCE 1 BELOW.           LSV00430
C     IN THE DISCUSSIONS T(1,MEV) DENOTES THE LANCZOS T-MATRIX           LSV00440
C     OF SIZE MEV.                                                       LSV00450
C                                                                        LSV00460
C     THE IDEAS USED IN THESE PROGRAMS ARE DISCUSSED IN THE FOLLOWING    LSV00470
C     REFERENCES.                                                        LSV00480
C                                                                        LSV00490
C     1.  JANE CULLUM, RALPH A. WILLOUGHBY AND MARK LAKE, A LANCZOS      LSV00500
C         ALGORITHM FOR COMPUTING SINGULAR VALUES AND VECTORS OF LARGE   LSV00510
C         MATRICES, SIAM J. SCIENTIFIC AND STATISTICAL COMPUTING,        LSV00520
C         VOL. 4, JUNE 1983, PP. 197-215.                               LSV00530
C                                                                        LSV00540
C     2.  JANE CULLUM AND RALPH A. WILLOUGHBY, LANCZOS ALGORITHMS        LSV00550
C         FOR LARGE SYMMETRIC MATRICES,  PROGRESS IN                     LSV00560
C         SCIENTIFIC COMPUTING, EDITORS, G. GOLUB, H.O. KREISS,          LSV00570
C         S. ARBARBANEL, AND R. GLOWINSKI,  BIRKHAUSER BOSTON INC.,      LSV00580
C         CAMBRIDGE, MASSACHUSETTS, 1984.                                LSV00590
C                                                                        LSV00600
C     3.  JANE CULLUM AND RALPH A. WILLOUGHBY, COMPUTING EIGENVECTORS    LSV00610
C         (AND EIGENVALUES) OF LARGE, SYMMETRIC MATRICES USING           LSV00620
C         LANCZOS TRIDIAGONALIZATION, LECTURE NOTES IN MATHEMATICS,      LSV00630
C         773, NUMERICAL ANALYSIS PROCEEDINGS, DUNDEE 1979, EDITED BY    LSV00640
```

```
C         G. A. WATSON, SPRINGER-VERLAG, (1980), BERLIN, PP.46-63.      LSV00650
C                                                                       LSV00660
C      4. IBID, LANCZOS AND THE COMPUTATION IN SPECIFIED INTERVALS OF    LSV00670
C         THE SPECTRUM OF LARGE SPARSE, REAL SYMMETRIC MATRICES, SPARSE  LSV00680
C         MATRIX PROCEEDINGS 1978, ED. I.S. DUFF AND G. W. STEWART,      LSV00690
C         SIAM, PHILADELPHIA, PP.220-255, 1979.                         LSV00700
C                                                                       LSV00710
C      5. IBID, COMPUTING EIGENVALUES OF VERY LARGE SYMMETRIC MATRICES-  LSV00720
C         AN IMPLEMENTATION OF A LANCZOS ALGORITHM WITHOUT              LSV00730
C         REORTHOGONALIZATION, J. COMPUT. PHYS. 44(1981), 329-358.      LSV00740
C                                                                       LSV00750
C                                                                       LSV00760
C-----PORTABILITY-------------------------------------------------------LSV00770
C                                                                       LSV00780
C                                                                       LSV00790
C      PROGRAMS WERE TESTED FOR PORTABILITY USING THE PFORT VERIFIER.   LSV00800
C      FOR DETAILS OF THE VERIFIER SEE FOR EXAMPLE, B. G. RYDER AND     LSV00810
C      A. D. HALL, "THE PFORT VERIFIER", COMPUTING SCIENCE TECHNICAL    LSV00820
C      REPORT 12, BELL LABORATORIES, MURRAY HILL, NEW JERSEY 07974,     LSV00830
C      (REVISED), JANUARY 1981.                                        LSV00840
C                                                                       LSV00850
C      EXCEPT FOR THE FOLLOWING CONSTRUCTIONS WHICH CAN BE EASILY       LSV00860
C      MODIFIED BY THE USER TO MATCH THE PARTICULAR COMPUTER BEING      LSV00870
C      USED, THE PROGRAM STATEMENTS ARE PORTABLE.                      LSV00880
C                                                                       LSV00890
C      NONPORTABLE CONSTRUCTIONS.                                      LSV00900
C                                                                       LSV00910
C      IN LSVAL AND IN LSVEC                                          LSV00920
C         1.  DATA/MACHEP STATEMENT                                   LSV00930
C         2.  ALL READ(5,*) STATEMENTS (FREE FORMAT)                  LSV00940
C         3.  FORMAT(20A4) USED FOR THE EXPLANATORY HEADER ARRAY, EXPLANLSV00950
C         4.  FORMAT(4Z20) USED TO READ AND WRITE BETA FILES 1 AND 2.  LSV00960
C      IN LSMULT                                                      LSV00970
C         1.  IN SVMAT, STRAN, AND  USPEC THE ENTRY THAT PASSES THE    LSV00980
C             STORAGE LOCATIONS OF THE ARRAYS DEFINING THE            LSV00990
C             USER-SPECIFIED MATRIX.                                  LSV01000
C         2.  IN SAMPLE USPEC FOR 'DIAGONAL' MATRICES:  THE FREE       LSV01010
C             FORMAT (8,*) AND THE FORMAT (20A4).                     LSV01020
C      IN LSSUB                                                       LSV01030
C         1.  ALL STATEMENTS ARE PORTABLE.                           LSV01040
C                                                                       LSV01050
C                                                                       LSV01060
C      IN THE COMMENTS BELOW:                                         LSV01070
C      COMPLEX*16 = COMPLEX VARIABLE, 16 BYTES OF STORAGE             LSV01080
C      REAL*8 = REAL VARIABLE, 8 BYTES OF STORAGE                     LSV01090
C      REAL*4 = REAL VARIABLE, 4 BYTES OF STORAGE                     LSV01100
C      INTEGER*4 = INTEGER VARIABLE, 4 BYTES                          LSV01110
C                                                                       LSV01120
C                                                                       LSV01130
C-----A-MATRIX SPECIFICATION--------------------------------------------LSV01140
C                                                                       LSV01150
C                                                                       LSV01160
C      SUBROUTINE USPEC IS USED TO SPECIFY THE USER-SUPPLIED A-MATRIX.  LSV01170
C      SUBROUTINES SVMAT AND STRAN ARE, RESPECTIVELY, CORRESPONDING    LSV01180
C      MATRIX-VECTOR MULTIPLE SUBROUTINES FOR A AND FOR A-TRANSPOSE.   LSV01190
C      THESE SUBROUTINES SHOULD BE DESIGNED TO TAKE ADVANTAGE OF       LSV01200
C      ANY SPECIAL PROPERTIES OF THE USER-SUPPLIED MATRIX.   THE       LSV01210
C      MATRIX-VECTOR MULTIPLIES REQUIRED BY THE LANCZOS PROCEDURES     LSV01220
C      MUST BE COMPUTED RAPIDLY AND ACCURATELY.                       LSV01230
C                                                                       LSV01240
C      SUBROUTINE USPEC HAS THE CALLING SEQUENCE                       LSV01250
C                                                                       LSV01260
C           CALL USPEC(M,N,MATNO)                                     LSV01270
C                                                                       LSV01280
C      WHERE M IS THE NUMBER OF ROWS IN THE USER-SPECIFIED             LSV01290
C      A-MATRIX AND N IS THE NUMBER OF COLUMNS. MATNO IS A             LSV01300
C      <= 8 DIGIT INTEGER USED AS A MATRIX AND TEST IDENTIFICATION     LSV01310
```

```
C      NUMBER.  THIS SUBROUTINE DEFINES (DIMENSIONS) THE ARRAYS          LSV01320
C      REQUIRED TO SPECIFY THE A-MATRIX.  THIS SUBROUTINE ALSO           LSV01330
C      INITIALIZES THESE ARRAYS AND ANY OTHER PARAMETERS NEEDED TO       LSV01340
C      DEFINE THE MATRIX.  THE STORAGE LOCATIONS OF THESE PARAMETERS     LSV01350
C      AND ARRAYS ARE THEN PASSED TO THE MATRIX-VECTOR MULTIPLY          LSV01360
C      SUBROUTINES SVMAT AND STRAN VIA ENTRIES.  SAMPLE SUBROUTINES      LSV01370
C      ARE INCLUDED IN THE FORTRAN FILE LSMULT.                          LSV01380
C                                                                        LSV01390
C      IMPORTANT NOTE:                                                   LSV01400
C      THE SAMPLE MATRIX-VECTOR MULTIPLY SUBROUTINES IN LSMULT           LSV01410
C      ASSUME THAT M >= N.  THEY ALSO ASSUME THAT THE USER-SUPPLIED      LSV01420
C      INFORMATION ABOUT THE GIVEN MATRIX IS STORED ON FILE 8.           LSV01430
C      THE USER SHOULD SEE THE LSMULT PROGRAMS FOR MORE DETAILS.         LSV01440
C                                                                        LSV01450
C      SUBROUTINE SVMAT HAS THE CALLING SEQUENCE                         LSV01460
C                                                                        LSV01470
C           CALL SVMAT(W,U,SUM)                                          LSV01480
C                                                                        LSV01490
C      WHERE U AND W ARE REAL*8 VECTORS AND SUM IS A REAL*8              LSV01500
C      SCALAR.  SVMAT CALCULATES U = A*W - SUM*U FOR THE                 LSV01510
C      USER-SPECIFIED A-MATRIX.  SUBROUTINE STRAN HAS THE                LSV01520
C      CALLING SEQUENCE                                                  LSV01530
C                                                                        LSV01540
C           CALL STRAN(W,U,SUM)                                          LSV01550
C                                                                        LSV01560
C      STRAN CALCULATES U = (A-TRANSPOSE)*W - SUM*U FOR THE              LSV01570
C      TRANSPOSE OF THE USER-SUPPLIED A-MATRIX.  THE ARRAY AND PARAMETER LSV01580
C      INFORMATION NEEDED TO PERFORM THE MATRIX-VECTOR MULTIPLIES        LSV01590
C      IS PASSED TO THE SVMAT AND THE STRAN SUBROUTINES FROM THE         LSV01600
C      USPEC SUBROUTINE VIA ENTRIES.  ONE SET OF THE SAMPLE SVMAT        LSV01610
C      AND STRAN SUBROUTINES INCLUDED IN LSMULT COMPUTES                 LSV01620
C      MATRIX-VECTOR MULTIPLIES FOR AN ARBITRARY SPARSE,                 LSV01630
C      RECTANGULAR MATRIX STORED IN THE SPARSE FORMAT SPECIFIED          LSV01640
C      IN THE CORRESPONDING SAMPLE USPEC SUBROUTINE.  THE LANCZS         LSV01650
C      SUBROUTINE CALLS SVMAT AND STRAN IN THE GENERATION OF             LSV01660
C      THE LANCZOS T-MATRICES FOR THE B MATRIX.                          LSV01670
C                                                                        LSV01680
C      THE DATA FOR THE A-MATRIX IS ASSUMED TO BE ON FILE 8 AND          LSV01690
C      IN THE FOLLOWING SPARSE FORMAT:                                   LSV01700
C      NZ = NUMBER OF NONZERO ELEMENTS OF A                              LSV01710
C      ICOL(K), K = 1,N, NUMBER OF NONZEROS OF A IN COLUMN K.            LSV01720
C      IROW(K), K = 1,NZ, ROW INDEX OF A(K).                             LSV01730
C      A(K), K=1,NZ  CONTAINS THE ELEMENTS OF A BY COLUMN.               LSV01740
C                                                                        LSV01750
C                                                                        LSV01760
C      SVMATV AND STRAN ARE CALLED FROM THE SUBROUTINE LANCZS            LSV01770
C      WHICH GENERATES THE LANCZOS TRIDIAGONAL MATRICES, THE             LSV01780
C      BETA HISTORY.  SIMILARLY, THSE SUBROUTINES ARE CALLED FROM        LSV01790
C      THE CORRESPONDING SINGULAR VECTOR PROGRAM, LSVEC.                 LSV01800
C      SVMAT AND STRAN ARE DECLARED AS EXTERNAL VARIABLES.               LSV01810
C      EACH IS AN ARGUMENT FOR THE LANCZS SUBROUTINE.                    LSV01820
C                                                                        LSV01830
C      USPEC, SVMAT, AND STRAN SUBROUTINES SUITABLE FOR THE              LSV01840
C      USER-SPECIFIED MATRIX MUST BE SUPPLIED BY THE USER.               LSV01850
C                                                                        LSV01860
C      THE MAIN PROGRAMS FOR THE SINGULAR VALUE AND SINGULAR VECTOR      LSV01870
C      CALCULATIONS ASSUME THAT INPUT FILE 5 CONTAINS THE ROW ORDER      LSV01880
C      M AND THE COLUMN ORDER N OF THE GIVEN A-MATRIX AND MATNO,         LSV01890
C      AN IDENTIFICATION NUMBER OF <= 8 DIGITS FOR THE GIVEN MATRIX.     LSV01900
C                                                                        LSV01910
C                                                                        LSV01920
C-----MACHEP----------------------------------------------------------LSV01930
C                                                                        LSV01940
C                                                                        LSV01950
C      MACHEP IS A MACHINE DEPENDENT PARAMETER SPECIFYING THE RELATIVE   LSV01960
C      PRECISION OF THE FLOATING POINT ARITHMETIC USED.                  LSV01970
C      MACHEP = 2.2 * 10**-16 FOR DOUBLE PRECISION ARITHMETIC ON         LSV01980
```

```
C        IBM 370-3081.                                               LSV01990
C                                                                    LSV02000
C        THE USER WILL HAVE TO RESET THIS PARAMETER TO               LSV02010
C        THE CORRESPONDING VALUE FOR THE MACHINE BEING USED.  NOTE THAT  LSV02020
C        IF A MACHINE WITH A MACHINE EPSILON THAT IS MUCH LARGER THAN THE  LSV02030
C        VALUE GIVEN HERE IS BEING USED, THEN THERE COULD BE         LSV02040
C        PROBLEMS WITH THE TOLERANCES.                               LSV02050
C                                                                    LSV02060
C                                                                    LSV02070
C-----SUBROUTINES AND FUNCTIONS USER MUST SUPPLY----------------------LSV02080
C                                                                    LSV02090
C                                                                    LSV02100
C        GENRAN,  FINPRO,  MASK,  USPEC,  SVMAT AND  STRAN           LSV02110
C                                                                    LSV02120
C        GENRAN = COMPUTES K PSEUDO-RANDOM NUMBERS AND STORES THEM IN  LSV02130
C                 THE REAL ARRAY, G.  THIS SUBROUTINE IS USED TO     LSV02140
C                 GENERATE A STARTING VECTOR FOR THE LANCZOS PROCEDURE  LSV02150
C                 IN THE SUBROUTINE LANCZS AND A STARTING RIGHT-HAND SIDE  LSV02160
C                 FOR INVERSE ITERATION IN THE SUBROUTINE INVERR.    LSV02170
C                                                                    LSV02180
C                 TESTS REPORTED IN THE REFERENCES USED EITHER GGL1 OR  LSV02190
C                 GGL2 FROM THE IBM LIBRARY  SLMATH.                 LSV02200
C                 THE EXISTING CALLING SEQUENCE IS:                  LSV02210
C                                                                    LSV02220
C                      CALL GENRAN(IIX,G,K).                         LSV02230
C                                                                    LSV02240
C                 WHERE IIX =INTEGER SEED, G = REAL*4 ARRAY WHOSE    LSV02250
C                 DIMENSION MUST BE >= K.  K RANDOM NUMBERS ARE GENERATED  LSV02260
C                 AND PLACED IN G.                                   LSV02270
C                                                                    LSV02280
C        FINPRO = DOUBLE PRECISION FUNCTION WHICH COMPUTES THE INNER  LSV02290
C                 PRODUCT OF 2 DOUBLE PRECISION VECTORS OF DIMENSION  K.  LSV02300
C                 TESTS REPORTED IN THE REFERENCES USED THE HARWELL  LSV02310
C                 LIBRARY SUBROUTINE FM02AD.                         LSV02320
C                 EXISTING CALLING SEQUENCE IS                       LSV02330
C                                                                    LSV02340
C                      CALL FINPRO(N,V,J,W,K).                       LSV02350
C                                                                    LSV02360
C                 COMPUTES THE INNER PRODUCT OF DIMENSION N OF THE VECTORS  LSV02370
C                 V AND W.  SUCCESSIVE COMPONENTS OF V AND OF W ARE STORED  LSV02330
C                 AT LOCATIONS THAT ARE ,RESPECTIVELY, J AND K UNITS APART.LSV02390
C                                                                    LSV02400
C        MASK = MASKS OVERFLOW AND UNDERFLOW.                        LSV02410
C               USER MUST SUPPLY OR COMMENT OUT CALL.                LSV02420
C                                                                    LSV02430
C        USPEC = DIMENSIONS AND INITIALIZES ARRAYS NEEDED TO SPECIFY  LSV02440
C                USER-SUPPLIED A-MATRIX.  SEE A-MATRIX SPECIFICATION SECTIONLSV02450
C                                                                    LSV02460
C        SVMAT = MATRIX-VECTOR MULTIPLY FOR USER-SUPPLIED A-MATRIX.  LSV02470
C                SEE A-MATRIX SPECIFICATION SECTION.                 LSV02480
C                                                                    LSV02490
C        STRAN = MATRIX-VECTOR MULTIPLY FOR TRANSPOSE OF USER-SUPPLIED  LSV02500
C                A-MATRIX.  SEE A-MATRIX SPECIFICATION SECTION.      LSV02510
C                                                                    LSV02520
C                                                                    LSV02530
C--------------------------------------------------------------------LSV02540
C                                                                    LSV02550
C        COMMENTS FOR SINGULAR VALUE COMPUTATIONS                    LSV02560
C                                                                    LSV02570
C--------------------------------------------------------------------LSV02580
C                                                                    LSV02590
C                                                                    LSV02600
C-----PARAMETER CONTROLS FOR SINGULAR VALUE PROGRAMS-----------------LSV02610
C                                                                    LSV02620
C                                                                    LSV02630
C        PARAMETER CONTROLS ARE INTRODUCED TO ALLOW SEGMENTATION OF THE  LSV02640
C        SINGULAR VALUE COMPUTATIONS AND TO ALLOW VARIOUS COMBINATIONS  LSV02650
```

```
C     OF READ/WRITES.                                                  LSV02660
C                                                                      LSV02670
C     THE FLAG ISTART CONTROLS THE T-MATRIX (BETA HISTORY)             LSV02680
C     GENERATION.                                                      LSV02690
C                                                                      LSV02700
C     ISTART = (0,1)  MEANS                                            LSV02710
C                                                                      LSV02720
C              (0) THERE IS NO EXISTING BETA HISTORY AND ONE           LSV02730
C                  MUST BE GENERATED.                                  LSV02740
C                                                                      LSV02750
C              (1) THERE IS AN EXISTING BETA HISTORY AND IT IS         LSV02760
C                  TO BE READ IN FROM FILE 2 AND EXTENDED IF NECESSARY. LSV02770
C                                                                      LSV02780
C     THE FLAG ISTOP CAN BE USED IN CONJUNCTION WITH THE FLAG ISTART TO LSV02790
C     ALLOW SEGMENTATION OF THE SINGULAR VALUE COMPUTATIONS.          LSV02800
C                                                                      LSV02810
C     ISTOP  = (0,1)  MEANS                                           LSV02820
C                                                                      LSV02830
C              (0) PROGRAM COMPUTES ONLY THE REQUESTED BETAS,          LSV02840
C                  STORES THEM AND THE LAST 2 LANCZOS VECTORS GENERATED LSV02850
C                  IN FILE 1 AND THEN TERMINATES.                      LSV02860
C                                                                      LSV02870
C              (1) PROGRAM COMPUTES REQUESTED BETAS AND THEN           LSV02880
C                  USES THE BISEC SUBROUTINE TO CALCULATE EIGENVALUES  LSV02890
C                  OF THE TRIDIAGONAL MATRICES GENERATED FOR THE ORDERS LSV02900
C                  SPECIFIED BY THE USER AND ON THE USER-SPECIFIED     LSV02910
C                  INTERVALS.  PROGRAM THEN USES THE SUBROUTINE INVERR LSV02920
C                  TO COMPUTE ERROR ESTIMATES FOR THE ISOLATED GOOD    LSV02930
C                  T-EIGENVALUES WHICH ARE USED TO CHECK THE          LSV02940
C                  CONVERGENCE OF THESE T-EIGENVALUES.                LSV02950
C                                                                      LSV02960
C     CONTROL PARAMETERS FOR WRITES                                    LSV02970
C                                                                      LSV02980
C     IHIS  = (0,1)  MEANS                                            LSV02990
C                                                                      LSV03000
C              (0) IF ISTOP .GT. 0 THEN BETAS ARE NOT SAVED ON FILE 1. LSV03010
C                                                                      LSV03020
C              (1) PROGRAM WRITES BETAS AND LAST 2 LANCZOS            LSV03030
C                  VECTORS TO FILE 1 SO THAT THE T-MATRIX GENERATION   LSV03040
C                  MAY BE REUSED OR CONTINUED LATER IF NECESSARY.      LSV03050
C                  TYPICALLY ONE WOULD ALWAYS DO THIS ON ANY RUN WHERE LSV03060
C                  A HISTORY FILE IS BEING GENERATED.  HISTORY MUST BE LSV03070
C                  SAVED IN MACHINE FORMAT ((4Z20) FOR IBM/3081) SO    LSV03080
C                  THAT NO ERRORS DUE TO FORMAT CONVERSIONS OCCUR.     LSV03090
C                                                                      LSV03100
C     IDIST  = (0,1)  MEANS                                           LSV03110
C                                                                      LSV03120
C              (0) DISTINCT EIGENVALUES OF T-MATRICES ARE NOT SAVED.   LSV03130
C                                                                      LSV03140
C              (1) PROGRAM WRITES COMPUTED DISTINCT EIGENVALUES OF     LSV03150
C                  T-MATRICES ALONG WITH THEIR T-MULTIPLICITIES        LSV03160
C                  TO FILE 11.                                         LSV03170
C                                                                      LSV03180
C     IWRITE = (0,1)  MEANS                                           LSV03190
C                                                                      LSV03200
C              (0) NO EXTENDED OUTPUT FROM SUBROUTINES BISEC AND INVERR LSV03210
C                  IS SENT TO FILE 6.                                  LSV03220
C                                                                      LSV03230
C              (1) INDIVIDUAL COMPUTED T-EIGENVALUES AND CORRESPONDING LSV03240
C                  ERROR ESTIMATES FROM THE SUBROUTINES BISEC AND INVERR LSV03250
C                  ARE PRINTED OUT TO FILE 6 AS THEY ARE COMPUTED.     LSV03260
C                                                                      LSV03270
C     THE PROGRAM ALWAYS MAKES A SEPARATE LIST OF THE COMPUTED GOOD    LSV03280
C     EIGENVALUES OF THE LANCZOS MATRICES T(1,MEV) CONSIDERED,         LSV03290
C     THESE ARE THE APPROXIMATIONS TO THE DESIRED SINGULAR VALUES,     LSV03300
C     ALONG WITH THEIR MINIMAL GAPS AS SINGULAR VALUES OF A AND        LSV03310
C     WRITES THEM TO FILE 3.  CORRESPONDING ERROR ESTIMATES FOR ANY    LSV03320
```

```
C      ISOLATED COMPUTED GOOD T-EIGENVALUES (SINGULAR VALUES OF A)      LSV03330
C      ARE ALWAYS WRITTEN TO FILE 4.                                    LSV03340
C                                                                       LSV03350
C                                                                       LSV03360
C-----INPUT/OUTPUT FILES FOR SINGULAR VALUE PROGRAMS-----------------LSV03370
C                                                                       LSV03380
C      ANY INPUT DATA OTHER THAN THE BETA HISTORY SHOULD BE STORED      LSV03390
C      ON FILE 5. SEE SAMPLE INPUT/OUTPUT FROM TYPICAL RUN.            LSV03400
C      THE READ STATEMENTS IN THE GIVEN FORTRAN PROGRAM ASSUME THAT     LSV03410
C      THE DATA STORED ON FILE 5 IS IN FREE FORMAT.  USER SHOULD NOTE   LSV03420
C      THAT 'FREE FORMAT' IS NOT CLASSIFIED AS PORTABLE BY PFORT SO THAT LSV03430
C      THE USER MAY HAVE TO MODIFY THE READ STATEMENTS FROM FILE 5 TO   LSV03440
C      CONFORM TO WHAT IS PERMISSIBLE ON THE MACHINE BEING USED.        LSV03450
C                                                                       LSV03460
C      FILE 6 WAS USED AS THE INTERACTIVE TERMINAL OUTPUT FILE.         LSV03470
C      THIS FILE PROVIDES A RUNNING ACCOUNT OF THE PROGRESS OF THE      LSV03480
C      COMPUTATIONS.  THE AMOUNT OF INFORMATION PRINTED OUT IS          LSV03490
C      CONTROLLED BY THE PARAMETER IWRITE.                             LSV03500
C                                                                       LSV03510
C DESCRIPTION OF OTHER I/O FILES                                        LSV03520
C                                                                       LSV03530
C FILE (K)   CONTAINS:                                                  LSV03540
C                                                                       LSV03550
C      (1)    OUTPUT FILE:                                              LSV03560
C             HISTORY FILE OF NEWLY-GENERATED T-MATRIX                  LSV03570
C             (BETA VECTOR) AND LAST 2 LANCZOS VECTORS USED             LSV03580
C             IN THE T-MATRIX GENERATION.                              LSV03590
C             IF IHIS = 0 AND ISTOP = 1, FILE 1 IS NOT WRITTEN.         LSV03600
C                                                                       LSV03610
C      (2)    INPUT FILE:                                               LSV03620
C             SAME AS FILE 1 EXCEPT THAT IT CONTAINS A                  LSV03630
C             PREVIOUSLY-GENERATED T-MATRIX (IF ANY). IF ISTART = 1,    LSV03640
C             PROGRAM ASSUMES THAT THERE IS A HISTORY FILE OF           LSV03650
C             BETAS ON FILE 2.  THESE BETAS AND THE LAST TWO LANCZOS    LSV03660
C             VECTORS USED IN THE T-MATRIX GENERATION ARE READ IN.      LSV03670
C                                                                       LSV03680
C      (3)    OUTPUT FILE:                                              LSV03690
C             COMPUTED GOOD EIGENVALUES OF THE T-MATRICES CONSIDERED.   LSV03700
C             ALSO CONTAINS T-MULTIPLICITIES OF THESE T-EIGENVALUES AS  LSV03710
C             EIGENVALUES OF THE T-MATRIX, AND THEIR GAPS AS            LSV03720
C             EIGENVALUES IN THE B MATRIX AND IN THE T-MATRIX.          LSV03730
C             NOTE THAT THESE GOOD T-EIGENVALUES ARE THE COMPUTED       LSV03740
C             SINGULAR VALUES OF THE A-MATRIX AND THAT THE GAPS         LSV03750
C             OF THESE EIGENVALUES AS EIGENVALUES OF THE B-MATRIX       LSV03760
C             ARE EQUAL TO THEIR GAPS AS SINGULAR VALUES OF A.  FILE    LSV03770
C             3 IS ALWAYS WRITTEN.                                      LSV03780
C                                                                       LSV03790
C      (4)    OUTPUT FILE:                                              LSV03800
C             ERROR ESTIMATES FOR THE ISOLATED COMPUTED SINGULAR        LSV03810
C             SINGULAR VALUES (ISOLATED GOOD EIGENVALUES OF T(1,MEV))   LSV03820
C             THESE ARE OBTAINED USING THE SUBROUTINE INVERR. THESE     LSV03830
C             ESTIMATES USE THE LAST COMPONENTS OF THE ASSOCIATED       LSV03840
C             T-EIGENVECTORS WHICH ARE COMPUTED USING INVERSE          LSV03850
C             ITERATION.  FILE 4 IS ALWAYS WRITTEN.                    LSV03860
C                                                                       LSV03870
C                                                                       LSV03880
C      (8)    INPUT FILE:                                               LSV03890
C             SAMPLE USPEC SUBROUTINE ASSUMES THAT THE ARRAYS           LSV03900
C             REQUIRED TO SPECIFY THE USER'S MATRIX ARE STORED ON       LSV03910
C             FILE 8.  USERS MUST MAKE WHATEVER DEFINITIONS ARE         LSV03920
C             APPROPRIATE FOR THEIR MATRICES.                          LSV03930
C                                                                       LSV03940
C      (9)    OUTPUT FILE: OPTIONAL                                     LSV03950
C             CAN BE USED TO STORE THE TRUE SINGULAR VALUES OF          LSV03960
C             A GIVEN TEST MATRIX, WHEN THE SINGULAR VALUE PROCEDURE    LSV03970
C             IS BEING EXERCISED ON A TEST MATRIX.                      LSV03980
C                                                                       LSV03990
```

```
C     (11)     OUTPUT FILE:                                          LSV04000
C               COMPUTED DISTINCT EIGENVALUES OF T-MATRICES USED.    LSV04010
C               ALSO CONTAINS THEIR T-MULTIPLICITIES AND T-GAPS TO   LSV04020
C               NEAREST DISTINCT T-EIGENVALUES, AND THE T-MULTIPLICITY LSV04030
C               PATTERN OF THE GOOD AND THE SPURIOUS T-EIGENVALUES.  LSV04040
C               FILE 11 IS WRITTEN ONLY IF IDIST = 1.                LSV04050
C                                                                    LSV04060
C                                                                    LSV04070
C-----PARAMETERS SET BY THE SINGULAR VALUE PROGRAMS-----------------LSV04080
C                                                                    LSV04090
C                                                                    LSV04100
C     THESE PARAMETERS ARE SET INTERNALLY IN THE PROGRAM            LSV04110
C                                                                    LSV04120
C     SCALEK     K = 1,2,3,4                                         LSV04130
C                                                                    LSV04140
C               THE SCALING FACTORS SCALEK HAVE BEEN INTRODUCED IN AN LSV04150
C               ATTEMPT TO MAKE THE TOLERANCES USED IN THE          LSV04160
C               T-MULTIPLICITY, SPURIOUS, ISOLATION AND PRTESTS ADJUST LSV04170
C               TO THE SCALE OF THE GIVEN MATRIX.  THESE FACTORS MUST LSV04180
C               NOT BE MODIFIED.                                     LSV04190
C                                                                    LSV04200
C     NOTE:    THE USER SHOULD NOTE THAT IF THE MATRIX BEING        LSV04210
C     PROCESSED IS VERY STIFF, THAT IS THE RATIO OF THE LARGEST     LSV04220
C     SINGULAR VALUE TO THE SMALLEST SINGULAR VALUE IS VERY         LSV04230
C     LARGE, THEN THE TOLERANCES BEING USED IN BISEC, LUMP, ISOEV   LSV04240
C     AND PRTEST MAY NOT TREAT THE SMALLEST SINGULAR VALUES         LSV04250
C     VERY WELL.  IN SOME SUCH CASES A USER-INTRODUCED REDUCTION    LSV04260
C     IN THE SIZE OF TKMAX AND THE SUBSEQUENT RECOMPUTATION OF      LSV04270
C     THE T-MATRIX EIGENVALUES CORRESPONDING TO THE SMALLEST        LSV04280
C     SINGULAR VALUES USING THIS TKMAX MAY RESULT IN IMPROVED       LSV04290
C     COMPUTATIONS AT THE LOW END.                                  LSV04300
C                                                                    LSV04310
C     THE LUMP, ISOEV, AND PRTEST TOLERANCES THAT WERE USED         LSV04320
C     MOST IN  THE TESTING OF THIS ALGORITHM WERE NOT               LSV04330
C     SCALE INVARIANT BUT SEEMED TO WORK WELL ON MATRICES THAT      LSV04340
C     HAD SINGULAR VALUES BOTH GREATER THAN AND LESS                LSV04350
C     THAN 1.  THESE TOLERANCES ARE ALSO INCLUDED IN THESE THREE    LSV04360
C     SUBROUTINES BUT AS COMMENTED OUT STATEMENTS.  THEY CAN BE     LSV04370
C     REVIVED BY COMMENTING OUT THE CORRESPONDING TOLERANCES        LSV04380
C     SPECIFIED IN THE STATEMENT ABOVE EACH OF THESE.               LSV04390
C                                                                    LSV04400
C     IMPORTANT TOLERANCES OR SCALES THAT ARE USED REPEATEDLY       LSV04410
C     THROUGHOUT THIS PROGRAM ARE THE FOLLOWING:                    LSV04420
C     SCALED MACHINE EPSILON:  TTOL = TKMAX*EPSM WHERE              LSV04430
C     EPSM = 2*MACHINE EPSILON AND                                  LSV04440
C     TKMAX = MAX(BETA(J), J = 1,MEV)                               LSV04450
C     BISEC CONVERGENCE TOLERANCE:  BISTOL = DSQRT(1000+MEV)*TTOL   LSV04460
C     BISEC T-MULTIPLICITY TOLERANCE:  MULTOL = (1000+MEV)*TTOL     LSV04470
C     LANCZOS CONVERGENCE TOLERANCE:   CONTOL = BETA(MEV+1)*1.D-10  LSV04480
C                                                                    LSV04490
C                                                                    LSV04500
C     BTOL = RELATIVE TOLERANCE USED TO ESTIMATE ANY LOSS OF LOCAL  LSV04510
C            ORTHOGONALITY OF THE LANCZOS VECTORS AFTER THE T-MATRIX LSV04520
C            HAS BEEN GENERATED.  THE LANCZOS PROCEDURE WORKS WELL   LSV04530
C            ONLY IF LOCAL ORTHOGONALITY BETWEEN SUCCESSIVE LANCZOS  LSV04540
C            VECTORS IS MAINTAINED.  THE TNORM SUBROUTINE TESTS      LSV04550
C            WHETHER OR NOT                                          LSV04560
C                                                                    LSV04570
C               MINIMUM  |BETA(I)|/||A|| > BTOL.                     LSV04580
C               I=2,KMAX                                             LSV04590
C                                                                    LSV04600
C            IF THIS TEST IS VIOLATED BY SOME BETA AND A T-MATRIX THAT LSV04610
C            WOULD INCLUDE SUCH A BETA IS REQUESTED, THEN THE LANCZOS LSV04620
C            PROCEDURE WILL TERMINATE FOR THE USER TO DECIDE WHAT TO LSV04630
C            DO.  THE USER CAN OVER-RIDE THIS TEST BY SIMPLY DECREASING LSV04640
C            THE SIZE OF BTOL, BUT THEN CONVERGENCE IS NOT AS CERTAIN. LSV04650
C            THE PROGRAM SETS BTOL = 1.D-8 WHICH IS A VERY CONSERVATIVE LSV04660
```

```
C                  CHOICE. THE || A || IS ESTIMATED BY USING AN ESTIMATE     LSV04670
C                  OF THE NORM OF THE T-MATRIX, T(1,KMAX).                    LSV04680
C                                                                            LSV04690
C        GAPTOL = RELATIVE TOLERANCE USED IN THE SUBROUTINE ISOEV            LSV04700
C                  TO DETERMINE FOR WHICH OF THE GOOD T-EIGENVALUES,          LSV04710
C                  THE COMPUTED SINGULAR VALUES, ERROR ESTIMATES SHOULD       LSV04720
C                  BE COMPUTED.  THE PROGRAM SETS GAPTOL = 1.D-8.             LSV04730
C                  IF FOR A GIVEN 'GOOD' T-EIGENVALUE OF THE GIVEN            LSV04740
C                  T-MATRIX  THE COMPUTED GAP IN THE T-MATRIX IS TOO          LSV04750
C                  SMALL AND IS DUE TO A 'SPURIOUS' EIGENVALUE OF             LSV04760
C                  THE T-MATRIX, THEN THE 'GOOD' T-EIGENVALUE IS ASSUMED      LSV04770
C                  TO HAVE CONVERGED AND AN ERROR ESTIMATE IS NOT             LSV04780
C                  COMPUTED.                                                  LSV04790
C                                                                            LSV04800
C                                                                            LSV04810
C-----USER-SPECIFIED PARAMETERS FOR SINGULAR VALUE PROGRAMS-------------LSV04820
C                                                                            LSV04830
C                                                                            LSV04840
C        RELTOL = RELATIVE TOLERANCE USED IN 'COMBINING' COMPUTED            LSV04850
C                  EIGENVALUES OF T(1,MEV) PRIOR TO COMPUTING ERROR           LSV04860
C                  ESTIMATES.                                                 LSV04870
C                                                                            LSV04880
C        THE LUMPING OF T-EIGENVALUES OCCURS IN SUBROUTINE LUMP.             LSV04890
C        LUMPING IS NECESSARY BECAUSE IT IS IMPOSSIBLE TO ACCURATELY         LSV04900
C        PREDICT THE ACCURACY OF THE BISEC SUBROUTINE.  LUMP 'COMBINES'      LSV04910
C        T-EIGENVALUES THAT HAVE SLIPPED BY THE TOLERANCE THAT WAS USED      LSV04920
C        IN THE T-MULTIPLICITY TESTS.  IN PARTICULAR IF FOR SOME J,          LSV04930
C                                                                            LSV04940
C        |EVALUE(J)-EVALUE(J-1)| < DMAX1(RELTOL*|EVALUE(J)|,SCALE2*MULTOL)   LSV04950
C                                                                            LSV04960
C        THEN THESE T-EIGENVALUES ARE 'COMBINED'.  MULTOL IS THE TOLERANCE  LSV04970
C        THAT WAS USED IN THE T-MULTIPLICITY TEST IN BISEC.  SEE THE HEADER LSV04980
C        ON THE LUMP SUBROUTINE FOR MORE DETAILS.                           LSV04990
C                                                                            LSV05000
C        RELTOL IS SET TO 1.D-10.                                           LSV05010
C                                                                            LSV05020
C        MXINIT = MAXIMUM NUMBER OF INVERSE ITERATIONS ALLOWED IN           LSV05030
C                  SUBROUTINE INVERR FOR EACH ISOLATED GOOD T-EIGENVALUE      LSV05040
C                  CONSIDERED.  TYPICALLY ONLY ONE IS REQUIRED.              LSV05050
C                                                                            LSV05060
C        SEEDS FOR RANDOM NUMBER GENERATORS = INTEGER*4 SCALARS.            LSV05070
C                                                                            LSV05080
C                  (1) SVSEED = SEED FOR STARTING VECTOR USED IN            LSV05090
C                      T-MATRIX GENERATION IN LANCZS SUBROUTINE             LSV05100
C                                                                            LSV05110
C                  (2) RHSEED = SEED FOR RIGHT-HAND SIDE USED IN            LSV05120
C                      INVERSE ITERATION COMPUTATIONS IN INVERR.            LSV05130
C                                                                            LSV05140
C        BISEC DATA                                                         LSV05150
C                                                                            LSV05160
C        (1) NINT  =  NUMBER OF SUBINTERVALS ON WHICH SINGULAR VALUES       LSV05170
C                      ARE TO BE COMPUTED.                                   LSV05180
C                                                                            LSV05190
C        (2) LB(J) = (J = 1,NINT) = LEFT END POINTS OF THESE INTERVALS.     LSV05200
C                      MUST BE PROVIDED IN INCREASING ORDER.  THAT IS,       LSV05210
C                      LB(J) < LB(J+1) FOR J = 1,NINT.                       LSV05220
C                                                                            LSV05230
C        (3) UB(J) = (J = 1,NINT) = RIGHT END POINTS OF THESE INTERVALS.    LSV05240
C                      MUST BE PROVIDED IN INCREASING ORDER.  THAT IS,       LSV05250
C                      UB(J) < UB(J+1) FOR J = 1,NINT.                       LSV05260
C                                                                            LSV05270
C        (4) MXSTUR  =  MAXIMUM NUMBER OF STURM ITERATIONS ALLOWED FOR      LSV05280
C                      ENTIRE SET OF SINGULAR VALUE CALCULATIONS OVER        LSV05290
C                      ALL SPECIFIED SIZE T-MATRICES.  PROGRAM WILL          LSV05300
C                      TERMINATE IF THIS LIMIT IS EXCEEDED.                  LSV05310
C                                                                            LSV05320
C        T-MATRICES                                                         LSV05330
```

284

```
C                                                                      LSV05340
C       SIZES OF T-MATRICES                                            LSV05350
C                                                                      LSV05360
C           (1) KMAX= MAXIMUM ORDER FOR T-MATRIX THAT USER IS WILLING  LSV05370
C                   TO CONSIDER.                                       LSV05380
C                                                                      LSV05390
C           (2) NMEVS = MAXIMUM NUMBER OF T-MATRICES THAT WILL BE      LSV05400
C                   CONSIDERED.                                        LSV05410
C                                                                      LSV05420
C           (3) NMEV(J)  (J=1,NMEVS)  = SIZES OF T-MATRIX TO BE        LSV05430
C                               CONSIDERED SEQUENTIALLY.               LSV05440
C                                                                      LSV05450
C       T-MATRIX-GENERATION                                            LSV05460
C                                                                      LSV05470
C       IPAR  = (1,2)   MEANS                                          LSV05480
C                                                                      LSV05490
C           (1) STARTING VECTOR IS OF FORM (0,V2) WHERE V2 IS          LSV05500
C               NX1.  USE WHEN M > N .                                 LSV05510
C                                                                      LSV05520
C           (2) STARTING VECTOR IF OF FORM (V1,0) WHERE V1 IS          LSV05530
C               MX1.  USE WHEN M < N .                                 LSV05540
C                                                                      LSV05550
C       USER SHOULD NOTE THAT THIS PROGRAM FIRST COMPUTES A T-MATRIX   LSV05560
C       OF ORDER KMAX AND THEN CYCLES THROUGH THE T-MATRICES SPECIFIED LSV05570
C       A PRIORI BY THE USER, USING THE SUBROUTINE BISEC TO COMPUTE    LSV05580
C       EIGENVALUES OF THE T-MATRICES ON THE INTERVALS SPECIFIED BY    LSV05590
C       THE USER.  SUBSETS OF THESE T-EIGENVALUES ARE THEN SELECTED    LSV05600
C       AS APPROXIMATIONS TO THE DESIRED SINGULAR VALUES.              LSV05610
C                                                                      LSV05620
C       IDEALLY, ONE WOULD COMPUTE THE SINGULAR VALUE APPROXIMATIONS   LSV05630
C       AT A REASONABLE SIZE T-MATRIX, LOOK AT THE ACCURACY OF THE     LSV05640
C       COMPUTED RESULTS AND USE THAT TO DETERMINE AN APPROPRIATE      LSV05650
C       INCREMENT FOR THE SIZE OF THE T-MATRIX BASED UPON WHAT         LSV05660
C       HAS ALREADY CONVERGED AND UPON THE SIZES OF THE ERROR ESTIMATES LSV05670
C       ON THOSE SINGULAR VALUES THAT ARE DESIRED BUT THAT HAVE NOT    LSV05680
C       YET CONVERGED.  HOWEVER, IN THE INTERESTS OF GENERALITY AND    LSV05690
C       SIMPLICITY WE CHOSE NOT TO DO THAT HERE.                       LSV05700
C                                                                      LSV05710
C                                                                      LSV05720
C-----CONVERGENCE TESTS FOR THE SINGULAR VALUE PROGRAMS---------------LSV05730
C                                                                      LSV05740
C                                                                      LSV05750
C       THE CONVERGENCE TEST INCORPORATED IN THIS PROGRAM IS           LSV05760
C       BASED UPON THE ASSUMPTION THAT THOSE T-EIGENVALUES AND         LSV05770
C       THEIR ASSOCIATED T-EIGENVECTORS THAT CORRESPOND TO             LSV05780
C       THE SINGULAR VALUES AND VECTORS WHICH WE WISH TO COMPUTE       LSV05790
C       CONVERGE AS THE T-SIZE IS INCREASED.                           LSV05800
C                                                                      LSV05810
C       AS CURRENTLY PROGRAMMED, CONVERGENCE IS CHECKED BY EXAMINING   LSV05820
C       THE SIZES OF ALL OF THE COMPUTED ERROR ESTIMATES ON ALL OF THE LSV05830
C       INTERVALS SPECIFIED BY THE USER.  IDEALLY CONVERGENCE SHOULD   LSV05840
C       BE CHECKED ONLY ON THOSE SINGULAR VALUES OF INTEREST AND       LSV05850
C       ONCE THE SINGULAR VALUES ON SUB-INTERVALS OF THESE INTERVALS   LSV05860
C       HAVE CONVERGED, ANY SUBSEQUENT SINGULAR VALUE COMPUTATIONS     LSV05870
C       SHOULD BE MADE ONLY ON THE UNCONVERGED PORTIONS.  OBVIOUSLY,   LSV05880
C       IT WOULD BE DIFFICULT TO INCORPORATE CODE TO DO THE ABOVE      LSV05890
C       WITHOUT KNOWING A PRIORI PRECISELY WHAT THE USER IS TRYING     LSV05900
C       TO COMPUTE.  THEREFORE, WE DID NOT ATTEMPT TO DO THIS.  IF     LSV05910
C       ONE WISHES TO MAKE SUCH A MODIFICATION THEN ONE MUST ALSO      LSV05920
C       MODIFY THE PROGRAM SO THAT IT CREATES AN OVERALL LIST OF THE   LSV05930
C       CONVERGED SINGULAR VALUES AS THEY ARE COMPUTED, SINCE          LSV05940
C       CONVERGED SINGULAR VALUES OBTAINED AT A PARTICULAR VALUE OF    LSV05950
C       MEV WOULD NO LONGER BE RECOMPUTED AT LARGER VALUES OF MEV.     LSV05960
C                                                                      LSV05970
C       IF ONLY A FEW SINGULAR VALUES ARE TO BE COMPUTED THEN SUCH     LSV05980
C       CHANGES WOULD NOT MAKE MUCH DIFFERENCE IN THE RUNNING TIME.    LSV05990
C                                                                      LSV06000
```

```
C                                                                    LSV06010
C-----ARRAYS REQUIRED BY THE SINGULAR VALUE PROGRAMS------------------LSV06020
C                                                                    LSV06030
C                                                                    LSV06040
C     BETA(J) = REAL*8 ARRAY.  ITS DIMENSION MUST BE AT LEAST KMAX+1. LSV06050
C               THE LENGTH OF THE LARGEST T-MATRIX ALLOWED.  THIS     LSV06060
C               ARRAY CONTAINS THE SUBDIAGONAL ENTRIES OF THE         LSV06070
C               T-MATRICES.  THE DIAGONAL ENTRIES ARE ALL ZERO.       LSV06080
C                                                                    LSV06090
C               THE BETA VECTOR IS NOT ALTERED DURING THE            LSV06100
C               CALCULATIONS.  IMPORTANT NOTE:  ONLY EVEN ORDER       LSV06110
C               T-MATRICES ARE PERMISSIBLE.                          LSV06120
C                                                                    LSV06130
C     V1(J),V2(J),VS(J) = REAL*8 ARRAYS.   VS MUST BE OF             LSV06140
C                         DIMENSION AT LEAST KMAX.  V1 MUST BE       LSV06150
C                         OF DIMENSION AT LEAST MAX(M,KMAX+1).        LSV06160
C                         V2 MUST BE OF DIMENSION AT LEAST           LSV06170
C                         MAX(N,KMAX).  M IS THE ROW DIMENSION OF    LSV06180
C                         A, AND N IS THE COLUMN DIMENSION.          LSV06190
C                         HOWEVER, THE DIMENSION                     LSV06200
C                         FOR V2 IS VALID ONLY IF NO MORE            LSV06210
C                         THAN KMAX/2 EIGENVALUES OF THE GIVEN       LSV06220
C                         T-MATRICES ARE TO BE COMPUTED IN ANY GIVEN LSV06230
C                         SUBINTERVAL. V2 IS USED IN THE SUBROUTINE  LSV06240
C                         BISEC TO HOLD THE UPPER AND LOWER          LSV06250
C                         ENDPOINTS OF THE SUBINTERVALS GENERATED    LSV06260
C                         DURING THE BISECTIONS.  THEREFORE, ITS     LSV06270
C                         DIMENSION MUST ALWAYS BE AT LEAST 2*Q      LSV06280
C                         WHERE Q IS THE MAXIMUM NUMBER OF           LSV06290
C                         EIGENVALUES OF THE SPECIFIED T-MATRIX IN ANY LSV06300
C                         ONE OF THE SPECIFIED INTERVALS.            LSV06310
C                                                                    LSV06320
C     LB(J),UB(J) = REAL*8 ARRAYS. EACH MUST BE OF DIMENSION AT LEAST LSV06330
C                   NINT, THE NUMBER OF SUBINTERVALS TO BE CONSIDERED. LSV06340
C                   LB CONTAINS THE LEFT-END POINTS OF THE INTERVALS  LSV06350
C                   ON WHICH SINGULAR VALUES ARE TO BE COMPUTED.      LSV06360
C                   UB CONTAINS THE RIGHT-END POINTS.                 LSV06370
C                                                                    LSV06380
C     EXPLAN(J) = REAL*4 ARRAY.  ITS DIMENSION IS 20.  THIS ARRAY IS  LSV06390
C                 USED TO ALLOW EXPLANATORY COMMENTS IN THE INPUT FILES.LSV06400
C                                                                    LSV06410
C     G(J) = REAL*4 ARRAY.  ITS DIMENSION MUST BE >= MAX(2*KMAX,M,N)  LSV06420
C            IT IS USED FOR HOLDING THE RANDOM VECTORS GENERATED,     LSV06430
C            HOLDING THE COMPUTED ERROR ESTIMATES AND THE COMPUTED    LSV06440
C            MINIMAL GAPS FOR THE SINGULAR VALUES.                    LSV06450
C                                                                    LSV06460
C     MP(J) = INTEGER*4 ARRAY.  ITS DIMENSION MUST BE AT LEAST KMAX,  LSV06470
C             THE MAXIMUM SIZE OF THE T-MATRICES ALLOWED.  IT CONTAINS LSV06480
C             THE T-MULTIPLICITIES OF THE COMPUTED T-EIGENVALUES OF   LSV06490
C             THE T-MATRICES.  NOTE THAT 'SPURIOUS' EIGENVALUES       LSV06500
C             OF THE T-MATRICES ARE DENOTED BY A T-MULTIPLICITY OF    LSV06510
C             0.  T-EIGENVALUES THAT THE SUBROUTINE PRTEST HAS        LSV06520
C             IDENTIFIED AS 'GOOD' BUT HIDDEN ARE IDENTIFIED BY A     LSV06530
C             T-MULTIPLICITY OF -10 AND SUBSEQUENTLY ADDED TO THE LIST LSV06540
C             OF COMPUTED SINGULAR VALUES.                            LSV06550
C                                                                    LSV06560
C     NMEV(J) = INTEGER*4 ARRAY.  ITS DIMENSION MUST BE AT LEAST THE  LSV06570
C               NUMBER OF T-MATRICES ALLOWED.  IT CONTAINS THE ORDERS LSV06580
C               OF THE T-MATRICES TO BE CONSIDERED.                   LSV06590
C                                                                    LSV06600
C                                                                    LSV06610
C     OTHER ARRAYS                                                    LSV06620
C                                                                    LSV06630
C     THE USER MUST SPECIFY IN THE SUBROUTINE USPEC WHATEVER ARRAYS   LSV06640
C     ARE REQUIRED TO DEFINE THE MATRIX BEING USED.                   LSV06650
C                                                                    LSV06660
C                                                                    LSV06670
```

```
C-----SUBROUTINES INCLUDED-------------------------------------------LSV06680
C                                                                    LSV06690
C                                                                    LSV06700
C     LANCZS = COMPUTES THE BETA HISTORY. USES SUBROUTINES           LSV06710
C              FINPRO, GENRAN, SVMAT AND STRAN.                      LSV06720
C                                                                    LSV06730
C     BISEC = COMPUTES EIGENVALUES OF THE SPECIFIED T-MATRIX USING   LSV06740
C             STURM SEQUENCING, ON SEQUENCE OF INTERVALS SPECIFIED   LSV06750
C             BY THE USER.  EACH SUBINTERVAL IS TREATED AS OPEN      LSV06760
C             ON THE LEFT AND CLOSED ON THE RIGHT.  EIGENVALUES      LSV06770
C             ARE COMPUTED WITH SIMULTANEOUS DETERMINATION OF THE    LSV06780
C             T-MULTIPLICITIES AND OF WHICH T-EIGENVALUES ARE SPURIOUS. LSV06790
C                                                                    LSV06800
C     INVERR = USES INVERSE ITERATION ON T-MATRICES TO COMPUTE ERROR LSV06810
C              ESTIMATES ON COMPUTED SINGULAR VALUES. (USES GENRAN)  LSV06820
C                                                                    LSV06830
C     LUMP = 'COMBINES' EIGENVALUES OF T-MATRIX USING THE RELATIVE   LSV06840
C            TOLERANCE RELTOL.                                       LSV06850
C                                                                    LSV06860
C     ISOEV = CALCULATES GAPS BETWEEN DISTINCT EIGENVALUES OF T-MATRIX LSV06870
C             AND THEN USES THESE GAPS TO LABEL THOSE 'GOOD'         LSV06880
C             T-EIGENVALUES FOR WHICH ERROR ESTIMATES ARE NOT COMPUTED. LSV06890
C                                                                    LSV06900
C     TNORM = COMPUTES THE SCALE TKMAX USED IN DETERMINING THE       LSV06910
C             TOLERANCES FOR THE SPURIOUS, T-MULTIPLICITY AND PRTESTS. LSV06920
C             IT ALSO CHECKS FOR LOCAL ORTHOGONALITY OF THE LANCZOS  LSV06930
C             VECTORS BY TESTING THE RELATIVE SIZE OF THE BETAS USING LSV06940
C             THE RELATIVE TOLERANCE BTOL.                           LSV06950
C                                                                    LSV06960
C     PRTEST = LOOKS FOR 'GOOD' T-EIGENVALUES THAT HAVE BEEN MISLABELLEDLSV06970
C              BY THE SPURIOUS TEST BECAUSE THEY HAD 'TOO SMALL' A   LSV06980
C              PROJECTION ON THE STARTING LANCZOS VECTOR.            LSV06990
C              (LESS THAN SINGLE PRECISION)                          LSV07000
C              TESTS INDICATE THAT SUCH T-EIGENVALUES ARE RARE.      LSV07010
C              PRTEST SHOULD BE CALLED ONLY AFTER CONVERGENCE        LSV07020
C              HAS BEEN ESTABLISHED.                                 LSV07030
C                                                                    LSV07040
C     INVERM = USED TO COMPUTE ERROR ESTIMATES FOR ANY T-EIGENVALUES LSV07050
C              WHICH PRTEST INDICATES MAY HAVE BEEN MISLABELLED.     LSV07060
C              SUCH T-EIGENVALUES ARE RELABELLED ONLY IF THEIR ERROR LSV07070
C              ESTIMATES ARE SUFFICIENTLY SMALL.  PRIMARY USE OF     LSV07080
C              INVERM IS IN THE CORRESPONDING SINGULAR VECTOR PROGRAM. LSV07090
C                                                                    LSV07100
C     SAMPLE USPEC, SVMAT AND STRAN SUBROUTINES ARE INCLUDED.        LSV07110
C                                                                    LSV07120
C     ALSO INCLUDED IS A STAND-ALONE PROGRAM, LSCOMPAC, THAT         LSV07130
C     TRANSLATES A MATRIX GIVEN IN THE I,J, A(I,J) FORMAT INTO       LSV07140
C     THE PARTICULAR SPARSE MATRIX FORMAT USED IN THE SAMPLE USPEC,  LSV07150
C     SVMAT AND STRAN SUBROUTINES PROVIDED.                          LSV07160
C                                                                    LSV07170
C                                                                    LSV07180
C-----OTHER PROGRAMS PROVIDED----------------------------------------LSV07190
C                                                                    LSV07200
C                                                                    LSV07210
C     LSCOMPAC = STAND-ALONE PROGRAM THAT TRANSLATES A SPARSE        LSV07220
C                RECTANGULAR M X N MATRIX A, GIVEN AS I, J, A(I,J),  LSV07230
C                INTO THE SPARSE MATRIX FORMAT REQUIRED BY THE SAMPLE LSV07240
C                USPEC, STRAN AND SVMAT SUBROUTINES PROVIDED FOR USE LSV07250
C                IN THE SINGULAR VALUE/VECTOR PROGRAMS.              LSV07260
C                THIS PROGRAM ASSUMES THAT THE MATRIX ENTRIES ARE    LSV07270
C                GIVEN EITHER COLUMN BY COLUMN OR ROW BY ROW.  IT    LSV07280
C                CANNOT HANDLE ANY OTHER ORDERINGS.  IN FACT IF      LSV07290
C                THE ENTRIES ARE GIVEN ROW BY ROW, THE DATA SET      LSV07300
C                CREATED ON FILE 8 CORRESPONDS TO A-TRANSPOSE AND    LSV07310
C                NOT TO A.  THUS, IN THIS SITUATION, IN ANY          LSV07320
C                SUBSEQUENT USE OF THE LANCZOS SINGULAR VALUE/VECTOR LSV07330
C                PROGRAMS THE USER WILL HAVE TO INTERCHANGE THE      LSV07340
```

```
C               ROLES OF M AND OF N.                              LSV07350
C                                                                 LSV07360
C                                                                 LSV07370
C-----COMMENTS ON THE STORAGE REQUIRED FOR SINGULAR VALUE PROGRAMS------LSV07380
C                                                                 LSV07390
C                                                                 LSV07400
C      THE ARRAYS IN THE REAL SINGULAR VALUE PROGRAM REQUIRE      LSV07410
C      APPROXIMATELY THE EQUIVALENT OF ONE REAL*8 ARRAY OF DIMENSION  LSV07420
C                                                                 LSV07430
C      2.5*KMAX + MAX(KMAX,M) + MAX(KMAX,N) + .5* MAX(2*KMAX,M,N) LSV07440
C                                                                 LSV07450
C      PLUS WHATEVER IS NEEDED TO GENERATE A*X FOR THE GIVEN MATRIX A.  LSV07460
C      THE ARRAYS BETA, VS AND MP CONSUME 2.5*KMAX*8 BYTES.       LSV07470
C      THE ARRAY V1 CONSUMES MAXIMUM(KMAX+1,M)*8 BYTES,  THE      LSV07480
C      ARRAY V2 CONSUMES MAXIMUM(KMAX,N)*8 BYTES, WITH THE        LSV07490
C      QUALIFICATION STATED ABOVE WHERE V2 IS DEFINED.  THE G-ARRAY  LSV07500
C      CONSUMES .5*MAX(2*KMAX,M,N)*8 BYTES.                       LSV07510
C                                                                 LSV07520
C                                                                 LSV07530
C------------------------------------------------------------------LSV07540
C                                                                 LSV07550
C      COMMENTS FOR SINGULAR VECTOR COMPUTATIONS                  LSV07560
C                                                                 LSV07570
C------------------------------------------------------------------LSV07580
C                                                                 LSV07590
C                                                                 LSV07600
C      THE SINGULAR VALUES WHOSE SINGULAR VECTORS ARE TO BE COMPUTED  LSV07610
C      MUST HAVE BEEN COMPUTED USING THE CORRESPONDING LANCZOS    LSV07620
C      SINGULAR VALUE PROGRAMS FOR REAL RECTANGULAR MATRICES BECAUSE  LSV07630
C      THESE SINGULAR VECTOR PROGRAMS USE THE SAME FAMILY OF LANCZOS  LSV07640
C      TRIDIAGONAL MATRICES THAT WAS USED IN THE CORRESPONDING    LSV07650
C      SINGULAR VALUE COMPUTATIONS.                               LSV07660
C                                                                 LSV07670
C      THESE PROGRAMS ASSUME THAT THE SINGULAR VALUES SUPPLIED TO IT  LSV07680
C      HAVE BEEN COMPUTED ACCURATELY, AS MEASURED BY THE          LSV07690
C      ERROR ESTIMATES COMPUTED IN THE CORRESPONDING LANCZOS      LSV07700
C      SINGULAR VALUE COMPUTATIONS, ALTHOUGH THESE ESTIMATES      LSV07710
C      ARE TYPICALLY CONSERVATIVE.   THE SINGULAR VALUES SUPPLIED LSV07720
C      ARE STORED IN THE ARRAY GOODSV(J), J=1,NGOOD.              LSV07730
C                                                                 LSV07740
C      FOR EACH GOODSV(J), THE SUBROUTINE STURMI COMPUTES THE     LSV07750
C      SMALLEST SIZE LANCZOS TRIDIAGONAL MATRIX, T(1,M1(J)), FOR  LSV07760
C      WHICH  GOODSV(J) IS A T-EIGENVALUE TO WITHIN A SPECIFIED   LSV07770
C      TOLERANCE.  IT ALSO ATTEMPTS TO COMPUTE THE SIZE, M2(J),   LSV07780
C      BY WHICH THE GIVEN SINGULAR VALUE BECOMES A DOUBLE         LSV07790
C      T-EIGENVALUE TO WITHIN THE GIVEN TOLERANCE.  THESE SIZES ARE  LSV07800
C      USED TO DETERMINE 1ST GUESSES AT SIZES FOR THE T-EIGENVECTORS  LSV07810
C      THAT WILL BE USED IN THE SINGULAR VECTOR COMPUTATIONS.     LSV07820
C      SUBROUTINE INVERM SUCCESSIVELY COMPUTES CORRESPONDING      LSV07830
C      T-EIGENVECTORS OF ENLARGED T-MATRICES UNTIL A SUITABLE     LSV07840
C      SIZE T-MATRIX IS DETERMINED FOR EACH J.  UP TO 10 SUCH     LSV07850
C      T-EIGENVECTOR COMPUTATIONS ARE ALLOWED FOR EACH SINGULAR   LSV07860
C      VALUE SUPPLIED.                                            LSV07870
C                                                                 LSV07880
C      AFTER APPROPRIATE T-EIGENVECTORS HAVE BEEN COMPUTED,       LSV07890
C      RITZ VECTORS FOR THE MATRIX B CORRESPONDING TO THESE       LSV07900
C      T-EIGENVECTORS ARE THEN COMPUTED.  SECTIONS OF THESE       LSV07910
C      RITZ VECTORS ARE THEN TAKEN AS APPROXIMATE LEFT AND        LSV07920
C      RIGHT SINGULAR VECTORS CORRESPONDING TO THE GIVEN          LSV07930
C      SINGULAR VALUES GOODSV(J),  J = 1,...,NGOOD.               LSV07940
C                                                                 LSV07950
C      THIS IMPLEMENTATION FIRST COMPUTES ALL OF THE RELEVANT     LSV07960
C      T-EIGENVECTORS OF THE SYMMETRIC TRIDIAGONAL MATRICES       LSV07970
C      IN THE VECTOR, TVEC.                                       LSV07980
C                                                                 LSV07990
C      THEN, AS EACH OF THE LANCZOS VECTORS IS REGENERATED, ALL   LSV08000
C      OF THE B-MATRIX RITZ VECTORS CORRESPONDING TO THESE        LSV08010
```

```
C        T-EIGENVECTORS ARE UPDATED USING THE CURRENTLY-GENERATED        LSV08020
C        LANCZOS VECTOR.  LANCZOS VECTORS ARE GENERATED (NOTE            LSV08030
C        THAT THEY ARE NOT BEING KEPT), UNTIL ENOUGH HAVE               LSV08040
C        BEEN GENERATED TO MAP THE LONGEST T-EIGENVECTOR INTO ITS       LSV08050
C        CORRESPONDING B-MATRIX RITZ VECTOR.  THE ARRAY RITVEC          LSV08060
C        CONTAINS THE SUCCESSIVE RITZ VECTORS WHICH ARE THEN            LSV08070
C        SPLIT INTO APPROXIMATIONS TO THE LEFT AND RIGHT SINGULAR       LSV08080
C        VECTORS OF THE USER-SUPPLIED MATRIX A.                         LSV08090
C                                                                       LSV08100
C                                                                       LSV08110
C-----PARAMETER CONTROLS FOR SINGULAR VECTOR PROGRAMS------------------LSV08120
C                                                                       LSV08130
C                                                                       LSV08140
C        PARAMETER CONTROLS ARE INTRODUCED TO ALLOW SEGMENTATION OF THE LSV08150
C        SINGULAR VECTOR COMPUTATIONS AND TO ALLOW VARIOUS COMBINATIONS LSV08160
C        OF READ/WRITES.                                                LSV08170
C                                                                       LSV08180
C        THE FLAG MBOUND ALLOWS THE USER TO DETERMINE A FIRST GUESS ON THE LSV08190
C        STORAGE THAT WILL BE REQUIRED BY THE T-EIGENVECTORS FOR THE    LSV08200
C        SINGULAR VALUES WHOSE SINGULAR VECTORS ARE TO BE COMPUTED.     LSV08210
C        THIS CAN BE USED TO ESTIMATE THE REQUIRED SIZE OF THE TVEC ARRAY. LSV08220
C                                                                       LSV08230
C        MBOUND = (0,1) MEANS                                           LSV08240
C                                                                       LSV08250
C              (0)  PROGRAM COMPUTES FIRST GUESSES AT THE SIZES         LSV08260
C                   OF THE T-MATRICES REQUIRED BY EACH OF THE           LSV08270
C                   SINGULAR VALUES SUPPLIED AND THEN CONTINUES         LSV08280
C                   WITH THE CORRESPONDING T-EIGENVECTOR                LSV08290
C                   COMPUTATIONS.                                       LSV08300
C                                                                       LSV08310
C              (1)  PROGRAM COMPUTES FIRST GUESSES AT THE SIZES         LSV08320
C                   OF THE T-MATRICES REQUIRED BY EACH OF THE           LSV08330
C                   SINGULAR VALUES SUPPLIED, STORES THESE IN FILE      LSV08340
C                   10 AND THEN TERMINATES.  THE USER CAN USE THESE     LSV08350
C                   SIZES TO ESTIMATE THE SIZE TVEC ARRAY NEEDED        LSV08360
C                   FOR THE DESIRED T-EIGENVECTOR COMPUTATIONS.         LSV08370
C                                                                       LSV08380
C        THE FLAGS NTVCON, TVSTOP, LVCONT, AND ERCONT CONTROL THE STOPPING LSV08390
C        CRITERIA FOR INTERMEDIATE POINTS IN THE LANCZOS PROCEDURE.  THEY LSV08400
C        TERMINATE THE PROCEDURE IF VARIOUS SPECIFIED QUANTITIES COULD  LSV08410
C        NOT BE COMPUTED AS DESIRED.                                    LSV08420
C                                                                       LSV08430
C        NTVCON = (0,1) MEANS                                           LSV08440
C                                                                       LSV08450
C              (0)  IF THE ESTIMATED STORAGE FOR THE T-EIGENVECTORS     LSV08460
C                   EXCEEDS THE USER-SPECIFIED DIMENSION OF THE         LSV08470
C                   TVEC ARRAY PROGRAM DOES NOT CONTINUE WITH THE       LSV08480
C                   T-EIGENVECTOR COMPUTATIONS. TERMINATION OCCURS.     LSV08490
C                                                                       LSV08500
C              (1)  CONTINUE WITH THE T-EIGENVECTOR COMPUTATIONS        LSV08510
C                   EVEN IF THE ESTIMATED STORAGE FOR TVEC EXCEEDS      LSV08520
C                   THE USER-SPECIFIED DIMENSION OF THE TVEC ARRAY.     LSV08530
C                   IN THIS SITUATION THE PROGRAM COMPUTES AS MANY      LSV08540
C                   T-EIGENVECTORS AS IT HAS ROOM FOR, IN THE SAME      LSV08550
C                   ORDER IN WHICH THE SINGULAR VALUES ARE SUPPLIED.    LSV08560
C                                                                       LSV08570
C        SVTVEC = (0,1) MEANS                                           LSV08580
C                                                                       LSV08590
C              (0)  DO NOT STORE THE COMPUTED T-EIGENVECTORS ON         LSV08600
C                   FILE 11 UNLESS ALSO HAVE THE FLAG TVSTOP = 1,       LSV08610
C                   IN WHICH CASE THE T-EIGENVECTORS ARE ALWAYS         LSV08620
C                   WRITTEN TO FILE 11.                                 LSV08630
C                                                                       LSV08640
C              (1)  STORE THE COMPUTED T-EIGENVECTORS ON FILE 11.       LSV08650
C                                                                       LSV08660
C        TVSTOP = (0,1) MEANS                                           LSV08670
C                                                                       LSV08680
```

```
C                        (0)   ATTEMPT TO CONTINUE ON TO THE COMPUTATION       LSV08690
C                              OF THE B-MATRIX RITZVECTORS AFTER COMPLETING THE LSV08700
C                              COMPUTATION OF THE T-EIGENVECTORS.               LSV08710
C                                                                              LSV08720
C                        (1)   TERMINATE AFTER COMPUTING THE                    LSV08730
C                              T-EIGENVECTORS AND STORING THEM ON FILE 11.      LSV08740
C                                                                              LSV08750
C          LVCONT = (0,1) MEANS                                                 LSV08760
C                                                                              LSV08770
C                        (0)   IF SOME OF THE T-EIGENVECTORS THAT WERE          LSV08780
C                              REQUESTED WERE NOT COMPUTED, EXIT                LSV08790
C                              FROM THE PROGRAM WITHOUT COMPUTING THE           LSV08800
C                              CORRESPONDING RITZ VECTORS.                      LSV08810
C                                                                              LSV08820
C                        (1)   CONTINUE ON TO THE RITZ VECTOR COMPUTATIONS      LSV08830
C                              EVEN IF NOT ALL OF THE T-EIGENVECTORS THAT       LSV08840
C                              WERE REQUESTED WERE COMPUTED.                    LSV08850
C                                                                              LSV08860
C          ERCONT = (0,1) MEANS                                                 LSV08870
C                                                                              LSV08880
C                        (0)   PROGRAM WILL NOT COMPUTE THE RITZ                LSV08890
C                              VECTOR FOR ANY SINGULAR VALUE FOR WHICH NO       LSV08900
C                              T-EIGENVECTOR WHICH SATISFIES THE ERROR          LSV08910
C                              ESTIMATE TEST (ERTOL) HAS BEEN IDENTIFIED.       LSV08920
C                                                                              LSV08930
C                        (1)   A RITZ VECTOR WILL BE COMPUTED FOR EVERY         LSV08940
C                              SINGULAR VALUE FOR WHICH A T-EIGENVECTOR HAS BEEN LSV08950
C                              COMPUTED REGARDLESS OF WHETHER OR NOT THAT       LSV08960
C                              T-EIGENVECTOR SATISFIES THE ERROR ESTIMATE TEST. LSV08970
C                                                                              LSV08980
C                                                                              LSV08990
C-----INPUT/OUTPUT FILES FOR THE SINGULAR VECTOR COMPUTATIONS-----------LSV09000
C                                                                              LSV09010
C                                                                              LSV09020
C     ANY INPUT DATA OTHER THAN THE T-MATRIX HISTORY FILE AND THE              LSV09030
C     PREVIOUSLY COMPUTED SINGULAR VALUES AND ERROR ESTIMATES                  LSV09040
C     SHOULD BE STORED ON FILE 5 IN FREE FORMAT.  SEE SAMPLE                   LSV09050
C     INPUT/OUTPUT FOR TYPICAL INPUT FILE.                                     LSV09060
C                                                                              LSV09070
C     FILE 6 WAS USED AS THE INTERACTIVE TERMINAL OUTPUT FILE.                 LSV09080
C     THIS FILE PROVIDES A RUNNING ACCOUNT OF THE PROGRESS OF THE              LSV09090
C     COMPUTATIONS.  ADDITIONAL PRINTOUT IS GENERATED WHEN                     LSV09100
C     THE FLAG IWRITE = 1.                                                     LSV09110
C                                                                              LSV09120
C                                                                              LSV09130
C DESCRIPTION OF OTHER I/O FILES                                               LSV09140
C                                                                              LSV09150
C FILE (K)    CONTAINS:                                                        LSV09160
C                                                                              LSV09170
C     (2)     INPUT FILE:                                                      LSV09180
C             PREVIOUSLY-GENERATED T-MATRICES (BETA ARRAY)                     LSV09190
C             AND THE FINAL TWO LANCZOS VECTORS USED ON THAT                   LSV09200
C             COMPUTATION.  THIS PROGRAM ALLOWS ENLARGEMENT                    LSV09210
C             OF ANY T-MATRICES PROVIDED ON FILE 2.                            LSV09220
C                                                                              LSV09230
C     (3)     INPUT FILE:                                                      LSV09240
C             THE SINGULAR VALUES FOR WHICH CORRESPONDING                      LSV09250
C             SINGULAR VECTORS ARE REQUESTED.  FILE 3 ALSO                     LSV09260
C             CONTAINS THE T-MULTIPLICITIES OF THESE SINGULAR                  LSV09270
C             VALUES (AS T-EIGENVALUES) AND THEIR COMPUTED GAPS                LSV09280
C             BOTH THE T-MATRICES AND IN THE USER-SUPPLIED MATRIX.             LSV09290
C             THIS FILE IS CREATED IN THE LANCZOS SINGULAR                     LSV09300
C             VALUE COMPUTATIONS.                                              LSV09310
C                                                                              LSV09320
C     (4)     INPUT FILE:                                                      LSV09330
C             ERROR ESTIMATES FOR THE ISOLATED SINGULAR VALUES                 LSV09340
C             OF FILE 3.  THIS FILE IS CREATED DURING THE LANCZOS              LSV09350
```

```
C               SINGULAR VALUE COMPUTATIONS.                         LSV09360
C                                                                    LSV09370
C       (8)     INPUT FILE:                                          LSV09380
C               USPEC SUBROUTINE ASSUMES THAT THE USER-              LSV09390
C               SUPPLIED MATRIX IS ON FILE 8.                        LSV09400
C                                                                    LSV09410
C       (9)     OUTPUT FILE:                                         LSV09420
C               ERROR ESTIMATES FOR THE COMPUTED RITZ VECTORS CONSIDERED LSV09430
C               AS EIGENVECTORS OF THE B-MATRIX. THESE ESTIMATES     LSV09440
C               ARE OF THE FORM                                      LSV09450
C                   BERROR = || B*RITVEC - SVAL*RITVEC ||            LSV09460
C               WHERE B DENOTES THE M+N ORDER SYMMETRIC MATRIX       LSV09470
C               ASSOCIATED WITH THE USER-SUPPLIED MATRIX A, SVAL     LSV09480
C               DENOTES THE SINGULAR VALUE BEING CONSIDERED AND      LSV09490
C               RITVEC DENOTES THE ASSOCIATED COMPUTED RITZ VECTOR.  LSV09500
C                                                                    LSV09510
C       (10)    OUTPUT FILE:                                         LSV09520
C               GUESSES AT APPROPRIATE SIZE T-MATRICES FOR THE       LSV09530
C               T-EIGENVECTORS FOR EACH SUPPLIED SINGULAR VALUE      LSV09540
C               IN THE ARRAY GOODSV(J),  J = 1,...,NGOOD.            LSV09550
C                                                                    LSV09560
C       (11)    OUTPUT FILE:                                         LSV09570
C               COMPUTED T-EIGENVECTORS CORRESPONDING TO SINGULAR    LSV09580
C               VALUES IN THE GOODSV ARRAY.  NOTE THAT IT IS POSSIBLE LSV09590
C               IN CERTAIN SITUATIONS THAT FOR SOME SINGULAR VALUES  LSV09600
C               SUPPLIED IN THE GOODSV ARRAY A T-EIGENVECTOR WILL    LSV09610
C               NOT BE COMPUTED.                                     LSV09620
C                                                                    LSV09630
C       (12)    OUTPUT FILE:                                         LSV09640
C               CONTAINS COMPUTED RITZ VECTORS CORRESPONDING TO      LSV09650
C               THE T-EIGENVECTORS ON FILE 11. NOTE THAT IN          LSV09660
C               SOME SITUATIONS THAT FOR SOME SINGULAR VALUES IN     LSV09670
C               THE GOODSV ARRAY FOR WHICH T-EIGENVECTORS HAVE       LSV09680
C               BEEN COMPUTED NO CORRESPONDING RITZ VECTOR WILL      LSV09690
C               HAVE BEEN COMPUTED.                                  LSV09700
C                                                                    LSV09710
C       (13)    OUTPUT FILE:                                         LSV09720
C               ADDITIONAL INFORMATION ABOUT THE BOUNDS AND ERROR    LSV09730
C               ESTIMATES OBTAINED.                                  LSV09740
C                                                                    LSV09750
C                                                                    LSV09760
C-----SEEDS FOR SINGULAR VECTOR PROGRAMS-------------------------------LSV09770
C                                                                    LSV09780
C       SEEDS FOR RANDOM NUMBER GENERATOR GENRAN                     LSV09790
C               (1) SVSEED = INTEGER*4 SCALAR USED IN THE SUBROUTINE LSV09800
C                            GENRAN TO GENERATE THE STARTING VECTOR FORLSV09810
C                            THE REGENERATION OF THE LANCZOS VECTORS. LSV09820
C                                                                    LSV09830
C               (2) RHSEED = INTEGER*4 SCALAR USED IN THE SUBROUTINE LSV09840
C                            GENRAN TO GENERATE A RANDOM VECTOR FOR   LSV09850
C                            USE IN SUBROUTINE INVERM.               LSV09860
C                                                                    LSV09870
C       USER SHOULD NOTE THAT SVSEED MUST BE THE SAME SEED THAT      LSV09880
C       WAS USED TO GENERATE THE T-MATRICES THAT WERE USED TO        LSV09890
C       COMPUTE THE SINGULAR VALUES WHOSE SINGULAR VECTORS ARE TO BE LSV09900
C       COMPUTED.  SVSEED IS READ IN FROM FILE 3.                    LSV09910
C                                                                    LSV09920
C                                                                    LSV09930
C-----USER-SPECIFIED PARAMETERS FOR THE SINGULAR VECTOR PROGRAMS--------LSV09940
C                                                                    LSV09950
C                                                                    LSV09960
C       NGOOD   = NUMBER OF SINGULAR VALUES READ INTO THE GOODSV ARRAY LSV09970
C                 READ FROM FILE 3.                                  LSV09980
C                                                                    LSV09990
C       M       = ROW ORDER OF THE USER-SUPPLIED MATRIX.             LSV10000
C                                                                    LSV10010
C       N       = COLUMN ORDER OF THE USER-SUPPLIED MATRIX.          LSV10020
```

```
C                                                             LSV10030
C     MEV       = SIZE OF THE T-MATRIX THAT WAS USED TO COMPUTE   LSV10040
C                 THE SINGULAR VALUES WHOSE SINGULAR VECTORS ARE  LSV10050
C                 REQUESTED.  MEV IS READ IN FROM FILE 3.         LSV10060
C                                                             LSV10070
C     KMAX      = SIZE OF THE T-MATRIX PROVIDED ON FILE 2.        LSV10080
C                                                             LSV10090
C     MDIMTV    = MAXIMUM CUMULATIVE SIZE OF THE TVEC ARRAY ALLOWED  LSV10100
C                 FOR ALL OF THE T-EIGENVECTORS REQUIRED.  MDIMTV   LSV10110
C                 MUST NOT EXCEED THE USER-SPECIFIED DIMENSION OF   LSV10120
C                 THE TVEC ARRAY.  PROGRAM CAN BE RUN WITH THE FLAG  LSV10130
C                 MBOUND = 1 TO DETERMINE AN EDUCATED GUESS ON AN   LSV10140
C                 APPROPRIATE DIMENSION FOR THE TVEC ARRAY.         LSV10150
C                                                             LSV10160
C     MDIMRV    = MAXIMUM CUMULATIVE SIZE OF THE RITVEC ARRAY ALLOWED  LSV10170
C                 FOR ALL OF THE RITZ VECTORS TO BE COMPUTED. MDIMRV  LSV10180
C                 MUST NOT EXCEED THE USER-SPECIFIED DIMENSION OF   LSV10190
C                 THE RITVEC ARRAY.  MUST BE SELECTED SO THAT       LSV10200
C                 THERE IS ENOUGH ROOM FOR A RITZ VECTOR FOR EVERY  LSV10210
C                 GOODEV(J) READ INTO PROGRAM.  (>= NGOOD*(M+N))    LSV10220
C                                                             LSV10230
C                                                             LSV10240
C-----ARRAYS REQUIRED BY THE SINGULAR VECTOR PROGRAMS------------------LSV10250
C                                                             LSV10260
C                                                             LSV10270
C     BETA(J) = REAL*8 ARRAY.  ITS DIMENSION MUST BE AT LEAST   LSV10280
C                 KMAXN+1, WHERE KMAXN IS THE LARGEST SIZE T-MATRIX  LSV10290
C                 CONSIDERED BY THE PROGRAM.  NOTE THAT KMAXN IS THE  LSV10300
C                 LARGER OF THE SIZE OF THE BETA HISTORY PROVIDED    LSV10310
C                 ON FILE 2 (IF ANY ) AND THE SIZE WHICH THE PROGRAM  LSV10320
C                 SPECIFIES INTERNALLY, THIS LATTER IS ALWAYS        LSV10330
C                 < = 11*MEV / 8  +  12, WHERE MEV IS THE SIZE       LSV10340
C                 T-MATRIX THAT WAS USED IN THE CORRESPONDING        LSV10350
C                 SINGULAR VALUE COMPUTATIONS.  BETA CONTAINS THE    LSV10360
C                 NONZERO ENTRIES OF THE LANCZOS T-MATRICES.         LSV10370
C                 BETA IS NOT DESTROYED IN THE COMPUTATIONS.         LSV10380
C                 THE DIAGONAL ENTRIES OF THE T-MATRICES ARE ALL ZERO.  LSV10390
C                                                             LSV10400
C     RITVEC(J) = REAL*8 ARRAY. IT DIMENSION MUST BE > = NGOOD*(M+N)  LSV10410
C                 WHERE THE USER-SUPPLIED MATRIX IS MXN            LSV10420
C                 AND NGOOD IS THE NUMBER OF SINGULAR VALUES WHOSE   LSV10430
C                 SINGULAR VECTORS ARE TO BE COMPUTED.  IT CONTAINS  LSV10440
C                 THE COMPUTED APPROXIMATE SINGULAR VECTORS OF A.    LSV10450
C                 THESE COMPUTED RITZ VECTORS ARE STORED ON FILE 12. LSV10460
C                                                             LSV10470
C     TVEC(J)   = REAL*8 ARRAY.  ITS DIMENSION MUST BE AT LEAST    LSV10480
C                 MTOL = |MA(1)| + |MA(2)| + ... + |MA(NGOOD)|     LSV10490
C                 WHERE NGOOD IS THE NUMBER OF SINGULAR VALUES BEING  LSV10500
C                 CONSIDERED AND |MA(J)| IS THE SIZE OF THE         LSV10510
C                 T-MATRIX BEING USED FOR THE B-MATRIX RITZ VECTOR  LSV10520
C                 COMPUTATION FOR GOODSV(J).  THESE SIZES          LSV10530
C                 ARE COMPUTED BY THE PROGRAM.  AN ESTIMATE OF      LSV10540
C                 MTOL CAN BE OBTAINED BY SETTING MBOUND = 1,       LSV10550
C                 RUNNING THE PROGRAM, AND THEN MULTIPLYING THE     LSV10560
C                 RESULTING TOTAL T-SIZE SPECIFIED BY 5/4.  THE TVEC  LSV10570
C                 ARRAY CONTAINS THE COMPUTED T-EIGENVECTORS.  IF    LSV10580
C                 THE FLAG SVTVEC = 1 OR THE FLAG TVSTOP = 1, THEN   LSV10590
C                 THESE VECTORS ARE SAVED ON FILE 11.               LSV10600
C                                                             LSV10610
C     V1(J)     = REAL*8 ARRAY.  ITS DIMENSION MUST BE GREATER     LSV10620
C                 THAN THE MAXIMUM OF KMAX AND M, WHERE M IS        LSV10630
C                 THE ROW ORDER OF THE GIVEN MATRIX.  V1 IS USED    LSV10640
C                 IN THE SUBROUTINE INVERM AND IN THE REGENERATION  LSV10650
C                 OF THE LANCZOS VECTORS.                          LSV10660
C                                                             LSV10670
C     V2(J)     = REAL*8 ARRAY.  ITS DIMENSION MUST BE GREATER     LSV10680
C                 THAN MAX(KMAX,N), WHERE N IS THE COLUMN ORDER OF  LSV10690
```

```
C                    THE GIVEN MATRIX.  IT IS USED IN THE REGENERATION      LSV10700
C                    OF THE LANCZOS VECTORS AND IN SUBROUTINE INVERM.        LSV10710
C                                                                           LSV10720
C     GOODSV(J),   = REAL*8 ARRAYS EACH OF DIMENSION AT LEAST NGOOD.         LSV10730
C     SVNEW(J)       CONTAIN THE SINGULAR VALUES FOR WHICH                   LSV10740
C                    SINGULAR VECTORS ARE REQUESTED.  SINGULAR VALUES        LSV10750
C                    IN GOODSV ARE READ IN FROM FILE 3.                      LSV10760
C                                                                           LSV10770
C     BMINGP(J),   = REAL*4 ARRAYS OF DIMENSION AT LEAST NGOOD.              LSV10780
C     TMINGP(J)      CONTAIN, RESPECTIVELY, THE MINIMAL GAPS FOR             LSV10790
C                    CORRESPONDING SINGULAR VALUES IN GOODSV ARRAY IN        LSV10800
C                    B-MATRIX AND IN T-MATRIX.                              LSV10810
C                                                                           LSV10820
C     TERR(J), ERR(J),     = REAL*4 ARRAYS (EXCEPT TLAST WHICH IS           LSV10830
C     ERRDGP(J), TLAST(J)    REAL*8).  EACH MUST BE OF DIMENSION            LSV10840
C     RNORM(J), TBETA(J)     AT LEAST NGOOD.  USED TO STORE QUANTITIES      LSV10850
C                            GENERATED DURING THE COMPUTATIONS FOR          LSV10860
C                            LATER PRINTOUT.                                LSV10870
C                                                                           LSV10880
C     G(J)         = REAL*4 ARRAY WHOSE DIMENSION MUST BE AT LEAST          LSV10890
C                    MAX(KMAX,M,N).  USED IN SUBROUTINE GENRAN TO HOLD      LSV10900
C                    RANDOM NUMBERS NEEDED FOR THE LANCZOS VECTOR           LSV10910
C                    REGENERATION AND FOR THE INVERSE ITERATION            LSV10920
C                    COMPUTATIONS IN THE SUBROUTINE INVERM.                LSV10930
C                                                                           LSV10940
C     MP(J)  = INTEGER*4 ARRAY WHOSE DIMENSION IS AT LEAST NGOOD.           LSV10950
C              INITIALLY CONTAINS THE T-MULTIPLICITY OF THE SINGULAR        LSV10960
C              VALUE GOODSV(J) AS AN EIGENVALUE OF THE T-MATRIX.           LSV10970
C              USED TO FLAG SINGULAR VALUES FOR WHICH NO T-EIGENVECTOR     LSV10980
C              OR NO RITZ VECTOR IS TO BE COMPUTED.                        LSV10990
C                                                                           LSV11000
C     MA(J)   = INTEGER*4 ARRAYS EACH OF WHOSE DIMENSIONS                   LSV11010
C               IS AT LEAST NGOOD.  USED IN DETERMINING                     LSV11020
C               AN APPROPRIATE T-MATRIX FOR EACH SINGULAR VALUE             LSV11030
C               IN GOODSV ARRAY.                                           LSV11040
C                                                                           LSV11050
C     MINT(J),MFIN(J) = INTEGER*4 ARRAYS WHOSE DIMENSIONS MUST BE AT        LSV11060
C                       LEAST NGOOD.  USED TO POINT TO THE BEGINNINGS       LSV11070
C                       AND THE ENDS OF THE COMPUTED EIGENVECTOR            LSV11080
C                       OF THE T-MATRIX, T(1,|MA(J)|).                      LSV11090
C                                                                           LSV11100
C     IDELTA(J)  = INTEGER*4 ARRAY WHOSE DIMENSION MUST BE AT               LSV11110
C                  LEAST NGOOD.  CONTAINS INCREMENTS USED IN LOOPS          LSV11120
C                  ON APPROPRIATE SIZE T-MATRIX FOR THE T-EIGENVECTOR       LSV11130
C                  COMPUTATIONS.                                           LSV11140
C                                                                           LSV11150
C                                                                           LSV11160
C-----SUBROUTINES INCLUDED FOR THE SINGULAR VECTOR COMPUTATIONS---------LSV11170
C                                                                           LSV11180
C                                                                           LSV11190
C     STURMI = FOR EACH GIVEN SINGULAR VALUE GOODSV(J) DETERMINES           LSV11200
C              THE SMALLEST SIZE T-MATRIX FOR WHICH GOODSV(J) IS            LSV11210
C              A T-EIGENVALUE (TO WITHIN A GIVEN TOLERANCE) AND IF          LSV11220
C              POSSIBLE THE SMALLEST SIZE T-MATRIX FOR WHICH               LSV11230
C              IT IS A DOUBLE T-EIGENVALUE (TO WITHIN THE SAME             LSV11240
C              TOLERANCE).  THE SIZE T-MATRIX USED IN THE                  LSV11250
C              T-EIGENVECTOR COMPUTATIONS IS THEN DETERMINED BY            LSV11260
C              STARTING WITH AN INITIAL GUESS BASED ON THE                 LSV11270
C              INFORMATION FROM STURMI, AND THEN LOOPING ON THE            LSV11280
C              SIZE OF THE T-EIGENVECTOR COMPUTATIONS.                     LSV11290
C                                                                           LSV11300
C     LBISEC = RECOMPUTES THE VALUE OF THE GIVEN SINGULAR VALUE             LSV11310
C              AT THE T-SIZE SPECIFIED FOR THE T-EIGENVECTOR                LSV11320
C              COMPUTATION.  LBISEC IS A SIMPLIFICATION OF THE             LSV11330
C              BISEC SUBROUTINE USED IN THE LANCZOS SINGULAR               LSV11340
C              VALUE COMPUTATIONS.                                         LSV11350
C                                                                           LSV11360
```

```
C      INVERM = FOR THE T-SIZES CONSIDERED BY THE PROGRAM COMPUTES        LSV11370
C               THE CORRESPONDING EIGENVECTORS OF THESE T-MATRICES        LSV11380
C               CORRESPONDING TO THE USER-SUPPLIED SINGULAR VALUES        LSV11390
C               IN THE GOODSV ARRAY.                                      LSV11400
C                                                                         LSV11410
C      LANCZS AND TNORM  SUBROUTINES ARE ALSO USED HERE AS WELL AS        LSV11420
C      IN THE CORRESPONDING SINGULAR VALUE COMPUTATIONS.                  LSV11430
C                                                                         LSV11440
C                                                                         LSV11450
C-------------------------------------------------------------------------LSV11460
```

SECTION 6.3   MAIN PROGRAM, SINGULAR VALUE COMPUTATIONS

```
C-----LSVAL  (SINGULAR VALUES OF REAL, RECTANGULAR MATRICES------------LSV00010
C                                                                      LSV00020
C     CONTAINS MAIN PROGRAM FOR COMPUTING DISTINCT SINGULAR VALUES OF   LSV00030
C     A REAL M X N MATRIX USING LANCZOS TRIDIAGONALIZATION WITHOUT      LSV00040
C     REORTHOGONALIZATION AND WITH SPECIAL STARTING VECTORS.            LSV00050
C                                                                      LSV00060
C     FOR A GIVEN REAL MATRIX A OF ORDER M X N THE LANCZOS RECURSION    LSV00070
C     IS APPLIED TO THE ASSOCIATED REAL SYMMETRIC MATRIX B OF ORDER     LSV00080
C     MN = M + N                                                        LSV00090
C                                                                      LSV00100
C                            ----              ----                     LSV00110
C                           |  0            A    |                      LSV00120
C                  B  =     |                    |                      LSV00130
C                           |  A-TRANSPOSE     0 |                      LSV00140
C                            ----              ----                     LSV00150
C                                                                      LSV00160
C     USING SPECIAL STARTING VECTORS.  PLEASE NOTE: ONLY EVEN ORDER     LSV00170
C     LANCZOS TRIDIAGONAL MATRICES AND ONLY NONNEGATIVE SUBINTERVALS    LSV00180
C     ARE PERMISSIBLE.                                                  LSV00190
C                                                                      LSV00200
C     PFORT VERIFIER IDENTIFIED THE FOLLOWING NONPORTABLE               LSV00210
C     CONSTRUCTIONS                                                     LSV00220
C                                                                      LSV00230
C     1.  DATA/MACHEP/ STATEMENT                                        LSV00240
C     2.  ALL READ(5,*) STATEMENTS (FREE FORMAT)                        LSV00250
C     3.  FORMAT(20A4) USED WITH EXPLANATORY HEADER EXPLAN.             LSV00260
C     4.  HEXADECIMAL FORMAT (4Z20) USED IN BETA FILES.                 LSV00270
C                                                                      LSV00280
C---------------------------------------------------------------------LSV00290
      DOUBLE PRECISION  BETA(5001),V1(5000),V2(5000),VS(5000)           LSV00300
      DOUBLE PRECISION  LB(20),UB(20)                                   LSV00310
      DOUBLE PRECISION  BTOL,GAPTOL,TTOL,MACHEP,EPSM,RELTOL             LSV00320
      DOUBLE PRECISION  SCALE1,SCALE2,SCALE3,SCALE4,BISTOL,CONTOL,MULTOLLSV00330
      DOUBLE PRECISION  ONE,ZERO,TEMP,TKMAX,BETAM,BKMIN,T0,T1           LSV00340
      REAL  G(5000),EXPLAN(20)                                          LSV00350
      INTEGER  MP(5000),NMEV(20)                                        LSV00360
      INTEGER  SVSEED,RHSEED,SVSOLD                                     LSV00370
      INTEGER IABS                                                      LSV00380
      REAL  ABS                                                         LSV00390
      DOUBLE PRECISION  DABS, DSQRT, DFLOAT                             LSV00400
      EXTERNAL SVMAT,STRAN                                              LSV00410
C---------------------------------------------------------------------LSV00420
      DATA MACHEP/Z3410000000000000/                                    LSV00430
      EPSM = 2.0D0*MACHEP                                               LSV00440
C---------------------------------------------------------------------LSV00450
C                                                                      LSV00460
C     ARRAYS MUST BE DIMENSIONED AS FOLLOWS:                            LSV00470
C     1.  BETA: >= (KMAX+1) WHERE KMAX IS READ IN AND IS                LSV00480
C         THE SIZE OF THE LARGEST T-MATRIX THAT CAN BE CONSIDERED.      LSV00490
C     2.  V1:   >= MAX(M,KMAX+1)                                        LSV00500
C     3.  V2:   >= MAX(N,KMAX)                                          LSV00510
C     4.  VS:   >= KMAX                                                 LSV00520
C     5.  G:    >= MAX(2*KMAX,M,N)                                      LSV00530
C     6.  MP:   >= KMAX                                                 LSV00540
C     7.  LB,UB: >= NUMBER OF SUBINTERVALS SUPPLIED TO BISEC.           LSV00550
C     8.  NMEV: >= NUMBER OF T-MATRICES ALLOWED.                        LSV00560
C     9.  EXPLAN:  DIMENSION IS 20.                                     LSV00570
C                                                                      LSV00580
C                                                                      LSV00590
C     IMPORTANT TOLERANCES OR SCALES THAT ARE USED REPEATEDLY           LSV00600
C     THROUGHOUT THIS PROGRAM ARE THE FOLLOWING:                        LSV00610
C     SCALED MACHINE EPSILON:  TTOL = TKMAX*EPSM WHERE                  LSV00620
C     EPSM = 2*MACHINE EPSILON AND                                      LSV00630
C     TKMAX = MAX(BETA(J), J = 1,MEV)                                   LSV00640
```

```
C       BISEC CONVERGENCE TOLERANCE:  BISTOL = DSQRT(1000+MEV)*TTOL         LSV00650
C       BISEC MULTIPLICITY TOLERANCE:  MULTOL = (1000+MEV)*TTOL             LSV00660
C       LANCZOS CONVERGENCE TOLERANCE:   CONTOL = BETA(MEV+1)*1.D-10        LSV00670
C------------------------------------------------------------------------LSV00680
C       OUTPUT HEADER                                                      LSV00690
        WRITE(6,10)                                                        LSV00700
    10 FORMAT(/' LANCZOS PROCEDURE FOR REAL, RECTANGULAR MATRICES'/)        LSV00710
C                                                                          LSV00720
C       SET PROGRAM PARAMETERS                                             LSV00730
C       SCALEK ARE USED IN TOLERANCES NEEDED IN SUBROUTINES LUMP,          LSV00740
C       ISOEV AND PRTEST.  USER MUST NOT MODIFY THESE SCALES.              LSV00750
        SCALE1 = 5.0D2                                                     LSV00760
        SCALE2 = 5.0D0                                                     LSV00770
        SCALE3 = 5.0D0                                                     LSV00780
        SCALE4 = 1.0D4                                                     LSV00790
        ONE  = 1.0D0                                                       LSV00800
        ZERO = 0.0D0                                                       LSV00810
        BTOL = 1.0D-8                                                      LSV00820
C       BTOL = EPSM                                                        LSV00830
        GAPTOL = 1.0D-8                                                    LSV00840
        ICONV = 0                                                          LSV00850
        MOLD = 0                                                           LSV00860
        MOLD1 = 1                                                          LSV00870
        ICT = 0                                                            LSV00880
        MMB = 0                                                            LSV00890
        IPROJ = 0                                                          LSV00900
C                                                                          LSV00910
C       READ USER-SPECIFIED PARAMETERS FROM INPUT FILE 5 (FREE FORMAT)     LSV00920
C                                                                          LSV00930
C       READ USER-PROVIDED HEADERS FOR RUN                                 LSV00940
        READ(5,20) EXPLAN                                                  LSV00950
        WRITE(6,20) EXPLAN                                                 LSV00960
        READ(5,20) EXPLAN                                                  LSV00970
        WRITE(6,20) EXPLAN                                                 LSV00980
    20 FORMAT(20A4)                                                        LSV00990
C                                                                          LSV01000
C       READ THE ROW ORDER M OF THE MATRIX AND THE COLUMN ORDER N.         LSV01010
C       READ THE MAXIMUM ORDER OF THE T-MATRICES ALLOWED (KMAX),           LSV01020
C       THE NUMBER OF T-MATRICES ALLOWED (NMEVS), AND A                    LSV01030
C       MATRIX IDENTIFICATION NUMBER (MATNO).                              LSV01040
        READ(5,20) EXPLAN                                                  LSV01050
        READ(5,*) M,N,KMAX,NMEVS,MATNO                                     LSV01060
        NM = M + N                                                         LSV01070
C                                                                          LSV01080
C       READ SEEDS FOR LANCZS AND INVERR SUBROUTINES (SVSEED AND RHSEED)   LSV01090
C       READ MAXIMUM NUMBER OF ITERATIONS ALLOWED FOR EACH INVERSE         LSV01100
C       ITERATION (MXINIT) AND MAXIMUM NUMBER OF STURM SEQUENCES           LSV01110
C       ALLOWED (MXSTUR)                                                   LSV01120
        READ(5,20) EXPLAN                                                  LSV01130
        READ(5,*) SVSEED,RHSEED,MXINIT,MXSTUR                              LSV01140
C                                                                          LSV01150
C       ISTART = (0,1):  ISTART = 0 MEANS BETA FILE IS NOT                 LSV01160
C       AVAILABLE.  ISTART = 1 MEANS BETA FILE IS AVAILABLE ON             LSV01170
C       FILE 2.                                                            LSV01180
C       ISTOP = (0,1):  ISTOP = 0 MEANS PROCEDURE GENERATES BETA           LSV01190
C       FILE AND THEN TERMINATES.  ISTOP = 1 MEANS PROCEDURE GENERATES     LSV01200
C       BETAS IF NEEDED AND THEN COMPUTES SINGULAR VALUES AND              LSV01210
C       ERROR ESTIMATES AND THEN TERMINATES.                              LSV01220
        READ(5,20) EXPLAN                                                  LSV01230
        READ(5,*) ISTART,ISTOP                                            LSV01240
C                                                                          LSV01250
C       IHIS = (0,1):  IHIS = 0 MEANS BETA FILE IS NOT WRITTEN             LSV01260
C       TO FILE 1.  IHIS = 1 MEANS BETA FILE IS WRITTEN TO FILE 1.         LSV01270
C       IDIST = (0,1):  IDIST = 0 MEANS DISTINCT T-EIGENVALUES             LSV01280
C       ARE NOT WRITTEN TO FILE 11.  IDIST = 1 MEANS DISTINCT              LSV01290
C       T-EIGENVALUES ARE WRITTEN TO FILE 11.                             LSV01300
C       IWRITE = (0,1):  IWRITE = 0 MEANS NO INTERMEDIATE OUTPUT           LSV01310
```

```
C       FROM THE COMPUTATIONS IS WRITTEN TO FILE 6.   IWRITE = 1 MEANS     LSV01320
C       T-EIGENVALUES AND ERROR ESTIMATES ARE WRITTEN TO FILE 6           LSV01330
C       AS THEY ARE COMPUTED.   SPECIFY THE PARITY (IPAR) OF THE          LSV01340
C       LANCZOS STARTING VECTOR.  IF M > N, THEN IPAR = 1,                LSV01350
C       IF M < N, THEN IPAR = 2.                                          LSV01360
        READ(5,20) EXPLAN                                                 LSV01370
        READ(5,*) IHIS,IDIST,IWRITE,IPAR                                  LSV01380
        IF(M.GT.N) IPAR = 1                                               LSV01390
        IF(M.LT.N) IPAR = 2                                               LSV01400
        IPARO = IPAR                                                      LSV01410
C                                                                         LSV01420
C       READ IN THE RELATIVE TOLERANCE (RELTOL) FOR USE IN THE            LSV01430
C       SPURIOUS, T-MULTIPLICITY, AND PRTEST TESTS.                       LSV01440
        READ(5,20) EXPLAN                                                 LSV01450
        READ(5,*) RELTOL                                                  LSV01460
C                                                                         LSV01470
C       READ IN THE SIZES OF THE T-MATRICES TO BE CONSIDERED.             LSV01480
C       NOTE THAT ONLY EVEN ORDER T-SIZES ARE PERMISSIBLE.                LSV01490
        READ(5,20) EXPLAN                                                 LSV01500
        READ(5,*) (NMEV(J), J=1,NMEVS)                                    LSV01510
C                                                                         LSV01520
C       CHECK TO SEE THAT ALL T-SIZES PROVIDED ARE EVEN ORDERED.          LSV01530
C       TERMINATE IF THAT IS NOT THE CASE.                               LSV01540
        DO 30 I = 1,NMEVS                                                 LSV01550
        NMEV2 = NMEV(I)/2                                                 LSV01560
        IF(2*NMEV2.NE.NMEV(I)) GO TO 670                                  LSV01570
   30 CONTINUE                                                            LSV01580
C                                                                         LSV01590
C       READ IN THE NUMBER OF SUBINTERVALS TO BE CONSIDERED.              LSV01600
        READ(5,20) EXPLAN                                                 LSV01610
        READ(5,*) NINT                                                    LSV01620
C                                                                         LSV01630
C       READ IN THE LEFT-END POINTS OF THE SUBINTERVALS TO BE CONSIDERED. LSV01640
C       THESE MUST BE IN ALGEBRAICALLY INCREASING ORDER                  LSV01650
        READ(5,20) EXPLAN                                                 LSV01660
        READ(5,*) (LB(J), J=1,NINT)                                       LSV01670
C                                                                         LSV01680
C       READ IN THE RIGHT-END POINTS OF THE SUBINTERVALS TO BE CONSIDERED.LSV01690
C       THESE MUST BE IN ALGEBRAICALLY INCREASING ORDER                  LSV01700
        READ(5,20) EXPLAN                                                 LSV01710
        READ(5,*) (UB(J), J=1,NINT)                                       LSV01720
C                                                                         LSV01730
C------------------------------------------------------------------------LSV01740
C       INITIALIZE THE ARRAYS FOR THE USER-SPECIFIED MATRIX              LSV01750
C       AND PASS THE STORAGE LOCATIONS OF THESE ARRAYS TO THE            LSV01760
C       MATRIX-VECTOR MULTIPLY SUBROUTINES SVMAT AND STRAN.              LSV01770
C                                                                         LSV01780
        CALL USPEC(M,N,MATNO)                                             LSV01790
C                                                                         LSV01800
C------------------------------------------------------------------------LSV01810
C       MASK UNDERFLOW AND OVERFLOW                                       LSV01820
C                                                                         LSV01830
        CALL MASK                                                         LSV01840
C                                                                         LSV01850
C------------------------------------------------------------------------LSV01860
C                                                                         LSV01870
C       WRITE TO FILE 6, A SUMMARY OF THE PARAMETERS FOR THIS RUN         LSV01880
C                                                                         LSV01890
        WRITE(6,40) MATNO,M,N,KMAX                                        LSV01900
   40 FORMAT(/3X,'MATRIX ID',5X,'M',5X,'N',4X,'MAX ORDER OF T'/           LSV01910
     1 I12,2I6,I18/)                                                      LSV01920
C                                                                         LSV01930
        WRITE(6,50) ISTART,ISTOP                                          LSV01940
   50 FORMAT(/2X,'ISTART',3X,'ISTOP'/2I8/)                                LSV01950
C                                                                         LSV01960
        WRITE(6,60) IHIS,IDIST,IWRITE,IPAR                                LSV01970
```

```
   60 FORMAT(/4X,'IHIS',3X,'IDIST',2X,'IWRITE',4X,'IPAR'/4I8/)        LSV01980
C                                                                     LSV01990
      WRITE(6,70) SVSEED,RHSEED                                       LSV02000
   70 FORMAT(/' SEEDS FOR RANDOM NUMBER GENERATOR'//                  LSV02010
     1 4X,'LANCZS SEED',4X,'INVERR SEED'/2I15/)                       LSV02020
C                                                                     LSV02030
      WRITE(6,80) (NMEV(J), J=1,NMEVS)                                LSV02040
   80 FORMAT(/' SIZES OF T-MATRICES TO BE CONSIDERED'/(6I12))         LSV02050
C                                                                     LSV02060
      WRITE(6,90) RELTOL,GAPTOL,BTOL                                  LSV02070
   90 FORMAT(/' RELATIVE TOLERANCE USED TO COMBINE COMPUTED T-EIGENVALUELSV02080
     1S'/E15.3/' RELATIVE GAP TOLERANCES USED IN INVERSE ITERATION'/  LSV02090
     1E15.3/' RELATIVE TOLERANCE FOR CHECK ON SIZE OF BETAS'/E15.3/)  LSV02100
C                                                                     LSV02110
      WRITE(6,100) (J,LB(J),UB(J), J=1,NINT)                          LSV02120
  100 FORMAT(/' BISEC WILL BE USED ON THE FOLLOWING INTERVALS'/       LSV02130
     1 (I6,2E20.6)/)                                                  LSV02140
C                                                                     LSV02150
      IF (ISTART.EQ.0.AND.IPAR.EQ.1) WRITE(6,110)                     LSV02160
      IF (ISTART.EQ.0.AND.IPAR.EQ.2) WRITE(6,120)                     LSV02170
  110 FORMAT(/' STARTING VECTOR IS OF FORM (0,V2)'/)                  LSV02180
  120 FORMAT(/' STARTING VECTOR IS OF FORM (V1,0)'/)                  LSV02190
C                                                                     LSV02200
      IF (ISTART.EQ.0) GO TO 170                                      LSV02210
C                                                                     LSV02220
C     READ IN BETA HISTORY FROM FILE 2                               LSV02230
C                                                                     LSV02240
      READ(2,130)MOLD,MO,NO,IPARO,IPAR,SVSOLD,MATOLD                  LSV02250
  130 FORMAT(3I6,2I3,I12,I8)                                          LSV02260
C                                                                     LSV02270
      IF (KMAX.LT.MOLD) KMAX = MOLD                                   LSV02280
      KMAX1 = KMAX + 1                                                LSV02290
C                                                                     LSV02300
C     CHECK THAT M, N, MATRIX ID MATNO, AND RANDOM SEED SVSEED        LSV02310
C     AGREE WITH THOSE IN THE HISTORY FILE.  IF NOT PROCEDURE STOPS.  LSV02320
C                                                                     LSV02330
      ITEMP = (MO-M)**2+(NO-N)**2+(MATNO-MATOLD)**2+(SVSEED-SVSOLD)**2 LSV02340
C                                                                     LSV02350
      IF (ITEMP.EQ.0) GO TO 150                                       LSV02360
C                                                                     LSV02370
      WRITE(6,140)                                                    LSV02380
  140 FORMAT(' PROGRAM TERMINATES'/    ' READ FROM FILE 2 CORRESPONDS TOLSV02390
     1 DIFFERENT MATRIX THAN MATRIX SPECIFIED'/)                      LSV02400
      GO TO 690                                                       LSV02410
C                                                                     LSV02420
  150 CONTINUE                                                        LSV02430
      MOLD1 = MOLD+1                                                  LSV02440
C                                                                     LSV02450
      READ(2,160)(BETA(J), J=1,MOLD1)                                 LSV02460
  160 FORMAT(4Z20)                                                    LSV02470
C                                                                     LSV02480
      IF (KMAX.EQ.MOLD) GO TO 190                                     LSV02490
C                                                                     LSV02500
      READ(2,160)(V1(J), J=1,M)                                       LSV02510
      READ(2,160)(V2(J), J=1,N)                                       LSV02520
C                                                                     LSV02530
  170 CONTINUE                                                        LSV02540
      IIX = SVSEED                                                    LSV02550
C                                                                     LSV02560
C---------------------------------------------------------------------LSV02570
C                                                                     LSV02580
      CALL LANCZS(SVMAT,STRAN,BETA,V1,V2,G,KMAX,MOLD1,M,N,IPAR,IIX)   LSV02590
C                                                                     LSV02600
C---------------------------------------------------------------------LSV02610
C                                                                     LSV02620
      KMAX1 = KMAX + 1                                                LSV02630
C                                                                     LSV02640
```

```
      IF (IHIS.EQ.0.AND.ISTOP.GT.0) GO TO 190                           LSV02650
C                                                                       LSV02660
      WRITE(1,180) KMAX,M,N,IPARO,IPAR,SVSEED,MATNO                      LSV02670
  180 FORMAT(3I6,2I3,I12,I8,' = KMAX,M,N,IPARO,IPAR,SVSEED,MATNO')       LSV02680
C                                                                       LSV02690
      WRITE(1,160)(BETA(I), I=1,KMAX1)                                  LSV02700
C                                                                       LSV02710
      WRITE(1,160)(V1(I), I=1,M)                                        LSV02720
      WRITE(1,160)(V2(I), I=1,N)                                        LSV02730
C                                                                       LSV02740
      IF (ISTOP.EQ.0) GO TO 570                                         LSV02750
C                                                                       LSV02760
  190 CONTINUE                                                          LSV02770
      BKMIN = BTOL                                                      LSV02780
      WRITE(6,200)                                                      LSV02790
  200 FORMAT(/' T-MATRICES (BETA) ARE NOW AVAILABLE'/)                  LSV02800
C                                                                       LSV02810
C----------------------------------------------------------------------LSV02820
C     SUBROUTINE TNORM CHECKS MIN(BETA)/(ESTIMATED NORM(A)) > BTOL .    LSV02830
C     IF THIS IS VIOLATED IB IS SET EQUAL TO THE NEGATIVE OF THE INDEX  LSV02840
C     OF THE MINIMAL BETA.  IF(IB < 0) THEN SUBROUTINE TNORM IS         LSV02850
C     CALLED FOR EACH VALUE OF MEV TO DETERMINE WHETHER OR NOT THERE    LSV02860
C     IS A BETA IN THE T-MATRIX SPECIFIED THAT VIOLATES THIS TEST.      LSV02870
C     IF THERE IS SUCH A BETA THE PROGRAM TERMINATES FOR THE USER       LSV02880
C     TO DECIDE WHAT TO DO.  THIS TEST CAN BE OVER-RIDDEN BY            LSV02890
C     SIMPLY MAKING BTOL SMALLER, BUT THEN THERE IS THE POSSIBILITY     LSV02900
C     THAT LOSSES IN THE LOCAL ORTHOGONALITY MAY HURT THE COMPUTATIONS. LSV02910
C     BTOL = 1.D-8 IS HOWEVER A CONSERVATIVE CHOICE FOR BTOL.           LSV02920
C                                                                       LSV02930
C     TNORM ALSO COMPUTES TKMAX = MAX(BETA(K), K=1,KMAX).               LSV02940
C     TKMAX IS USED TO SCALE THE TOLERANCES USED IN THE                 LSV02950
C     T-MULTIPLICITY AND SPURIOUS TESTS IN BISEC. TKMAX IS ALSO USED IN LSV02960
C     THE PROJECTION TEST FOR HIDDEN T-EIGENVALUES THAT HAD 'TOO SMALL' LSV02970
C     A PROJECTION ON THE STARTING VECTOR.                             LSV02980
C                                                                       LSV02990
      CALL TNORM(BETA,BKMIN,TKMAX,KMAX,IB)                              LSV03000
C                                                                       LSV03010
C----------------------------------------------------------------------LSV03020
C                                                                       LSV03030
      TTOL = EPSM*TKMAX                                                 LSV03040
C                                                                       LSV03050
C     LOOP ON THE SIZE OF THE T-MATRIX                                 LSV03060
C                                                                       LSV03070
  210 CONTINUE                                                          LSV03080
      MMB = MMB + 1                                                     LSV03090
C     NOTE THAT ONLY EVEN ORDER T-SIZES ARE PERMISSIBLE.               LSV03100
      MEV = NMEV(MMB)                                                   LSV03110
C     IS MEV TOO LARGE ?                                               LSV03120
      IF(MEV.LE.KMAX) GO TO 230                                         LSV03130
      WRITE(6,220) MMB, MEV, KMAX                                       LSV03140
  220 FORMAT(/' TERMINATE PRIOR TO CONSIDERING THE',I6,'TH T-MATRIX'/   LSV03150
     1' BECAUSE THE SIZE REQUESTED',I6,' IS GREATER THAN THE MAXIMUM SIZLSV03160
     1E ALLOWED',I6/)                                                   LSV03170
      GO TO 570                                                         LSV03180
C                                                                       LSV03190
  230 MP1 = MEV + 1                                                     LSV03200
      BETAM = BETA(MP1)                                                 LSV03210
C                                                                       LSV03220
      IF (IB.GE.0) GO TO 240                                            LSV03230
C                                                                       LSV03240
      TO = BTOL                                                         LSV03250
C                                                                       LSV03260
C----------------------------------------------------------------------LSV03270
C                                                                       LSV03280
      CALL TNORM(BETA,TO,T1,MEV,IBMEV)                                  LSV03290
C                                                                       LSV03300
```

```
C-------------------------------------------------------------------LSV03310
C                                                                    LSV03320
      TEMP = TO/TKMAX                                                LSV03330
      IBMEV = IABS(IBMEV)                                            LSV03340
      IF (TEMP.GE.BTOL) GO TO 240                                    LSV03350
      IBMEV = -IBMEV                                                 LSV03360
      GO TO 630                                                      LSV03370
C                                                                    LSV03380
  240 CONTINUE                                                       LSV03390
      IC = MXSTUR-ICT                                                LSV03400
C                                                                    LSV03410
C-------------------------------------------------------------------LSV03420
C     BISEC LOOP. THE SUBROUTINE BISEC INCORPORATES DIRECTLY THE     LSV03430
C     T-MULTIPLICITY AND SPURIOUS TESTS. T-EIGENVALUES WILL BE       LSV03440
C     CALCULATED BY BISEC SEQUENTIALLY ON INTERVALS                  LSV03450
C     (LB(J),UB(J)), J = 1,NINT).                                    LSV03460
C                                                                    LSV03470
C     ON RETURN FROM BISEC                                           LSV03480
C     NDIS = NUMBER OF DISTINCT EIGENVALUES OF T(1,MEV) ON UNION     LSV03490
C            OF THE (LB,UB) INTERVALS                                LSV03500
C     VS = DISTINCT T-EIGENVALUES IN ALGEBRAICALLY INCREASING ORDER  LSV03510
C     MP = T-MULTIPLICITIES OF THE T-EIGENVALUES STORED IN VS        LSV03520
C     MP(I) = (0,1,MI), MI>1, I=1,NDIS  MEANS:                       LSV03530
C        (0) VS(I) IS SPURIOUS                                       LSV03540
C        (1) VS(I) IS T-SIMPLE AND GOOD                              LSV03550
C        (MI) VS(I) IS T-MULTIPLE AND IS THEREFORE NOT ONLY GOOD BUT LSV03560
C             ALSO A CONVERGED GOOD T-EIGENVALUE.                    LSV03570
C                                                                    LSV03580
C                                                                    LSV03590
      CALL BISEC(BETA,V1,V2,VS,LB,UB,EPSM,TTOL,MP,NINT,              LSV03600
     1 MEV,NDIS,IC,IWRITE)                                           LSV03610
C                                                                    LSV03620
C-------------------------------------------------------------------LSV03630
C                                                                    LSV03640
      IF (NDIS.EQ.0) GO TO 650                                       LSV03650
C                                                                    LSV03660
C     COMPUTE THE TOTAL NUMBER OF STURM SEQUENCES USED TO DATE       LSV03670
C     COMPUTE THE BISEC CONVERGENCE AND T-MULTIPLICITY TOLERANCES USED. LSV03680
C     COMPUTE THE CONVERGENCE TOLERANCE FOR T-EIGENVALUES.          LSV03690
      ICT = ICT + IC                                                 LSV03700
      TEMP = DFLOAT(MEV+1000)                                        LSV03710
      MULTOL = TEMP*TTOL                                             LSV03720
      TEMP = DSQRT(TEMP)                                             LSV03730
      BISTOL = TTOL*TEMP                                             LSV03740
      CONTOL = BETAM*1.D-10                                          LSV03750
C                                                                    LSV03760
C-------------------------------------------------------------------LSV03770
C     SUBROUTINE LUMP 'COMBINES' T-EIGENVALUES THAT ARE 'TOO CLOSE'. LSV03780
C     NOTE HOWEVER THAT CLOSE SPURIOUS T-EIGENVALUES ARE NOT AVERAGED LSV03790
C     WITH GOOD ONES. HOWEVER, THEY MAY BE USED TO INCREASE THE      LSV03800
C     T-MULTIPLICITY OF A GOOD T-EIGENVALUE.                         LSV03810
C                                                                    LSV03820
      LOOP = NDIS                                                    LSV03830
      CALL LUMP(VS,RELTOL,MULTOL,SCALE2,MP,LOOP)                     LSV03840
C                                                                    LSV03850
C-------------------------------------------------------------------LSV03860
C                                                                    LSV03870
      IF(NDIS.EQ.LOOP) GO TO 260                                     LSV03880
C                                                                    LSV03890
      WRITE(6,250) NDIS, MEV, LOOP                                   LSV03900
  250 FORMAT(/I6,' DISTINCT T-EIGENVALUES WERE COMPUTED IN BISEC AT MEV LSV03910
     1=',I6/ 2X,' LUMP SUBROUTINE REDUCES NUMBER OF DISTINCT T-EIGENVALULSV03920
     1ES TO',I6)                                                     LSV03930
C                                                                    LSV03940
  260 CONTINUE                                                       LSV03950
      NDIS = LOOP                                                    LSV03960
```

```
      BETA(MP1) = BETAM                                              LSV03970
C                                                                    LSV03980
C--------------------------------------------------------------------LSV03990
C     THE SUBROUTINE ISOEV LABELS THOSE SIMPLE T-EIGENVALUES OF T(1,MEV)LSV04000
C     WITH VERY SMALL GAPS BETWEEN NEIGHBORING T-EIGENVALUES OF T(1,MEV)LSV04010
C     TO AVOID COMPUTING ERROR ESTIMATES FOR ANY SIMPLE GOOD          LSV04020
C     T-EIGENVALUE THAT IS TOO CLOSE TO A SPURIOUS T-EIGENVALUE.      LSV04030
C     ON RETURN FROM ISOEV, G CONTAINS CODED MINIMAL GAPS            LSV04040
C     BETWEEN THE DISTINCT EIGENVALUES OF T(1,MEV). (G IS REAL).     LSV04050
C     G(I) < 0 MEANS MINGAP IS DUE TO LEFT GAP G(I) > 0 MEANS DUE TO LSV04060
C     RIGHT GAP. MP(I) = -1 MEANS THAT THE GOOD T-EIGENVALUE IS SIMPLE LSV04070
C     AND HAS A VERY SMALL MINGAP IN T(1,MEV) DUE TO A SPURIOUS      LSV04080
C     T-EIGENVALUE.                                                  LSV04090
C     NG = NUMBER OF GOOD T-EIGENVALUES.                             LSV04100
C     NISO = NUMBER OF ISOLATED, GOOD T-EIGENVALUES.                 LSV04110
C                                                                    LSV04120
      CALL ISOEV(VS,GAPTOL,MULTOL,SCALE1,G,MP,NDIS,NG,NISO)          LSV04130
C                                                                    LSV04140
C--------------------------------------------------------------------LSV04150
C                                                                    LSV04160
      WRITE(6,270)NG,NISO,NDIS                                       LSV04170
  270 FORMAT(/I6,' SINGULAR VALUES HAVE BEEN COMPUTED'/              LSV04180
     1 I6,' OF THESE ARE ISOLATED'/                                  LSV04190
     2 I6,' = NUMBER OF DISTINCT T-EIGENVALUES COMPUTED'/)           LSV04200
C                                                                    LSV04210
C     DO WE WRITE DISTINCT T-EIGENVALUES TO FILE 11?                 LSV04220
      IF (IDIST.EQ.0) GO TO 310                                      LSV04230
C                                                                    LSV04240
      WRITE(11,280) NDIS,NISO,MEV,M,N,SVSEED,MATNO                   LSV04250
  280 FORMAT(5I5,I12,I8,' = NDIS,NISO,MEV,M,N,SVSEED,MATNO'/)        LSV04260
C                                                                    LSV04270
      WRITE(11,290) (MP(I),VS(I),G(I), I=1,NDIS)                     LSV04280
  290 FORMAT(2(I3,E25.16,E12.3))                                     LSV04290
C                                                                    LSV04300
      WRITE(11,300) NDIS, (MP(I), I=1,NDIS)                          LSV04310
  300 FORMAT(/I6,' = NDIS, T-MULTIPLICITIES (0 MEANS  SPURIOUS)'/(20I4))LSV04320
C                                                                    LSV04330
  310 CONTINUE                                                       LSV04340
C                                                                    LSV04350
      IF (NISO.NE.0) GO TO 340                                       LSV04360
C                                                                    LSV04370
      WRITE(4,320) MEV                                               LSV04380
  320 FORMAT(/' AT MEV = ',I6,' THERE ARE NO ISOLATED T-EIGENVALUES'/LSV04390
     1' SO NO ERROR ESTIMATES WERE COMPUTED/')                       LSV04400
C                                                                    LSV04410
      WRITE(6,330)                                                   LSV04420
  330 FORMAT(/' ALL COMPUTED SINGULAR VALUES ARE T-MULTIPLE'/        LSV04430
     1 ' THEREFORE ALL COMPUTED SINGULAR VALUES ARE ASSUMED TO HAVE CONVLSV04440
     1ERGED'/)                                                       LSV04450
C                                                                    LSV04460
      ICONV = 1                                                      LSV04470
      GO TO 380                                                      LSV04480
C                                                                    LSV04490
  340 CONTINUE                                                       LSV04500
C                                                                    LSV04510
C--------------------------------------------------------------------LSV04520
C     SUBROUTINE INVERR COMPUTES ERROR ESTIMATES FOR ISOLATED GOOD   LSV04530
C     T-EIGENVALUES USING INVERSE ITERATION ON T(1,MEV). ON RETURN   LSV04540
C     G(J) = MINIMUM GAP IN T(1,MEV) FOR EACH VS(J), J=1,NDIS        LSV04550
C     G(MEV+I) = BETAM*|U(MEV)| = ERROR ESTIMATE FOR ISOLATED GOOD   LSV04560
C          T-EIGENVALUES, WHERE I = 1, NISO AND  BETAM = BETA(MEV+1) LSV04570
C          U(MEV) IS MEVTH COMPONENT OF THE UNIT EIGENVECTOR OF T    LSV04580
C          CORRESPONDING TO THE ITH ISOLATED GOOD T-EIGENVALUE.      LSV04590
C     A NEGATIVE ERROR ESTIMATE MEANS THAT FOR THAT PARTICULAR       LSV04600
C     T-EIGENVALUE THE INVERSE ITERATION DID NOT CONVERGE IN <= MXINIT LSV04610
C     STEPS AND THAT THE CORRESPONDING ERROR ESTIMATE IS QUESTIONABLE.LSV04620
C                                                                    LSV04630
```

```
C     V2 CONTAINS THE ISOLATED GOOD T-EIGENVALUES                      LSV04640
C     V1 CONTAINS THE MINGAPS TO THE NEAREST DISTINCT  EIGENVALUE      LSV04650
C       OF T(1,MEV) FOR EACH ISOLATED GOOD T-EIGENVALUE IN V2.         LSV04660
C     VS CONTAINS THE NDIS DISTINCT EIGENVALUES OF T(1,MEV)            LSV04670
C     MP CONTAINS THE CORRESPONDING CODED T-MULTIPLICITIES             LSV04680
C                                                                      LSV04690
      IT = MXINIT                                                      LSV04700
      CALL INVERR(BETA,V1,V2,VS,EPSM,G,MP,MEV,MMB,NDIS,NISO,NM,        LSV04710
     1 RHSEED,IT,IWRITE)                                               LSV04720
C                                                                      LSV04730
C--------------------------------------------------------------------LSV04740
C                                                                      LSV04750
C     SIMPLE CHECK FOR CONVERGENCE. CHECKS TO SEE IF ALL OF THE ERROR  LSV04760
C     ESTIMATES ARE SMALLER THAN CONTOL.                              LSV04770
C     IF THIS TEST IS SATISFIED, THEN CONVERGENCE FLAG, ICONV IS SET   LSV04780
C     TO 1.  TYPICALLY ERROR ESTIMATES ARE VERY CONSERVATIVE.          LSV04790
C                                                                      LSV04800
      WRITE(6,350) CONTOL                                              LSV04810
  350 FORMAT(/' CONVERGENCE IS TESTED USING THE CONVERGENCE TOLERANCE', LSV04820
     1E13.4/)                                                          LSV04830
C                                                                      LSV04840
      II = MEV +1                                                      LSV04850
      IF = MEV+NISO                                                    LSV04860
      DO 360 I = II,IF                                                 LSV04870
      IF (ABS(G(I)).GT.CONTOL) GO TO 380                               LSV04880
  360 CONTINUE                                                         LSV04890
      ICONV = 1                                                        LSV04900
      MMB = NMEVS                                                      LSV04910
C                                                                      LSV04920
      WRITE(6,370) CONTOL                                              LSV04930
  370 FORMAT(' ALL COMPUTED ERROR ESTIMATES WERE LESS THAN',E15.4/     LSV04940
     1 ' THEREFORE PROCEDURE TERMINATES'/)                             LSV04950
C                                                                      LSV04960
  380 CONTINUE                                                         LSV04970
C                                                                      LSV04980
C     IF CONVERGENCE IS INDICATED, THAT IS ICONV = 1 ,THEN             LSV04990
C     THE SUBROUTINE PRTEST IS CALLED TO CHECK FOR ANY CONVERGED       LSV05000
C     T-EIGENVALUES THAT HAVE BEEN MISLABELLED AS SPURIOUS BECAUSE     LSV05010
C     THE PROJECTION OF THEIR SINGULAR VECTOR ON THE STARTING          LSV05020
C     VECTOR WAS TOO SMALL.  NUMERICAL TESTS INDICATE THAT             LSV05030
C     SUCH SINGULAR VALUES ARE RARE.  THEREFORE, IF MANY OF            LSV05040
C     THESE HIDDEN SINGULAR VALUES APPEAR ON SOME RUN, THE USER        LSV05050
C     CAN BE CERTAIN THAT SOMETHING IS FOULED UP.                      LSV05060
C                                                                      LSV05070
      IF (ICONV.EQ.0) GO TO 510                                        LSV05080
C                                                                      LSV05090
C--------------------------------------------------------------------LSV05100
C                                                                      LSV05110
      CALL PRTEST (BETA,VS,TKMAX,EPSM,RELTOL,SCALE3,SCALE4,            LSV05120
     1 MP,NDIS,MEV,IPROJ)                                              LSV05130
C                                                                      LSV05140
C--------------------------------------------------------------------LSV05150
C                                                                      LSV05160
      IF(IPROJ.EQ.0) GO TO 500                                         LSV05170
C                                                                      LSV05180
      IF(IDIST.EQ.1) WRITE(11,390) IPROJ                               LSV05190
  390 FORMAT(' SUBROUTINE PRTEST WANTS TO RELABEL',I6,' SPURIOUS T-EIGENLSV05200
     1VALUES'/' WE ACCEPT RELABELLING ONLY IF LAST COMPONENT OF T-EIGENVLSV05210
     1ECTOR IS L.T. 1.D-10'/)                                          LSV05220
C                                                                      LSV05230
      IIX = RHSEED                                                     LSV05240
C                                                                      LSV05250
C--------------------------------------------------------------------LSV05260
C                                                                      LSV05270
      CALL GENRAN(IIX,G,MEV)                                           LSV05280
C                                                                      LSV05290
```

```
C-------------------------------------------------------------------LSV05300
C                                                                    LSV05310
      ITEN = -10                                                     LSV05320
      NISOM = NISO + MEV                                             LSV05330
      IWRITO = IWRITE                                                LSV05340
      IWRITE = 0                                                     LSV05350
C                                                                    LSV05360
      DO 420 J = 1,NDIS                                              LSV05370
      IF(MP(J).NE.ITEN) GO TO 420                                   LSV05380
      TO = VS(J)                                                     LSV05390
C                                                                    LSV05400
C-------------------------------------------------------------------LSV05410
C                                                                    LSV05420
      IT = MXINIT                                                    LSV05430
      CALL INVERM(BETA,V1,V2,TO,TEMP,T1,EPSM,G,MEV,IT,IWRITE)       LSV05440
C                                                                    LSV05450
C-------------------------------------------------------------------LSV05460
C                                                                    LSV05470
      IF(TEMP.LE.1.D-10) GO TO 410                                  LSV05480
C     ERROR ESTIMATE WAS NOT SMALL REJECT RELABELLING OF THIS        LSV05490
C     T-EIGENVALUE.                                                  LSV05500
      IF(IDIST.EQ.1) WRITE(11,400) J,TO,TEMP                        LSV05510
  400 FORMAT(/' LAST COMPONENT FOR',I6,'TH T-EIGENVALUE',E20.12/' IS TOOLSV05520
     1 LARGE = ',E15.6,' SO DO NOT ACCEPT PRTEST RELABELLING'/)      LSV05530
      MP(J) = 0                                                      LSV05540
      IPROJ = IPROJ - 1                                              LSV05550
      GO TO 420                                                      LSV05560
C     RELABELLING ACCEPTED                                           LSV05570
  410 NISOM = NISOM + 1                                              LSV05530
      G(NISOM) = BETAM*TEMP                                          LSV05590
  420 CONTINUE                                                       LSV05600
      WRITE = IWRITO                                                 LSV05610
C                                                                    LSV05620
      IF(IPROJ.EQ.0) GO TO 460                                      LSV05630
      WRITE(6,430) IPROJ                                             LSV05640
  430 FORMAT(/I6,' T-EIGENVALUES WERE RECLASSIFIED AS GOOD.'/        LSV05650
     1' THESE ARE IDENTIFIED IN FILE 3 BY A T-MULTIPLICITY OF -10'/' USELSV05660
     2R SHOULD INSPECT EACH TO MAKE SURE NEIGHBORS HAVE CONVERGED'/) LSV05670
C                                                                    LSV05680
      IF(IDIST.EQ.1)  WRITE(11,440) IPROJ                           LSV05690
  440 FORMAT(/I6,' T-EIGENVALUES WERE RELABELLED AS GOOD'/           LSV05700
     1' BELOW IS CORRECTED T-MULTIPLICITY PATTERN'/)                 LSV05710
C                                                                    LSV05720
      WRITE(6,450) NDIS, (MP(I), I=1,NDIS)                          LSV05730
      IF(IDIST.EQ.1) WRITE(11,450) NDIS, (MP(I), I=1,NDIS)         LSV05740
  450 FORMAT(/I6,' = NDIS, T-MULTIPLICITIES (0 MEANS  SPURIOUS)'/    LSV05750
     1 6X, ' (-10) MEANS SPURIOUS T-EIGENVALUE RELABELLED AS GOOD'/(20I4LSV05760
     1))                                                             LSV05770
C                                                                    LSV05780
C     RECALCULATE MINGAPS FOR DISTINCT T(1,MEV) EIGENVALUES.         LSV05790
  460 NDIS1 = NDIS - 1                                               LSV05800
      G(NDIS) = VS(NDIS1)-VS(NDIS)                                   LSV05810
      G(1) = VS(2)-VS(1)                                             LSV05820
C                                                                    LSV05830
      DO 470 J = 2,NDIS1                                             LSV05840
      TO = VS(J)-VS(J-1)                                             LSV05850
      T1 = VS(J+1)-VS(J)                                             LSV05860
      G(J) = T1                                                      LSV05870
      IF (TO.LT.T1) G(J) = -TO                                       LSV05880
  470 CONTINUE                                                       LSV05890
      IF(IPROJ.EQ.0) GO TO 500                                      LSV05900
C     WRITE TO FILE 4 ERROR ESTIMATES FOR THOSE T-EIGENVALUES RELABELLEDLSV05910
      NGOOD = 0                                                      LSV05920
      DO 480 J = 1,NDIS                                              LSV05930
      IF(MP(J).EQ.0) GO TO 480                                      LSV05940
      NGOOD = NGOOD + 1                                              LSV05950
      IF(MP(J).NE.ITEN) GO TO 480                                   LSV05960
```

```
      TO = VS(J)                                             LSV05970
      NISO = NISO + 1                                        LSV05980
      NISOM = MEV + NISO                                     LSV05990
      WRITE(4,490) NGOOD,TO,G(NISOM),G(J)                    LSV06000
  480 CONTINUE                                               LSV06010
  490 FORMAT(I10,E25.16,2E14.3)                              LSV06020
C                                                            LSV06030
  500 CONTINUE                                               LSV06040
C                                                            LSV06050
C     WRITE THE COMPUTED SINGULAR VALUES TO FILE 3.  FIRST TRANSFER THEMLSV06060
C     TO V2 AND THEIR T-MULTIPLICITIES TO THE CORRESPONDING POSITIONS   LSV06070
C     IN MP AND COMPUTE THE B-MINGAPS, THE MINIMAL GAPS BETWEEN THE      LSV06080
C     SINGULAR VALUES CONSIDERED AS EIGENVALUES OF THE B-MATRIX.         LSV06090
C     THESE GAPS WILL BE PUT IN THE ARRAY G.                 LSV06100
C     SINCE G CURRENTLY CONTAINS THE MINIMAL GAPS BETWEEN THE DISTINCT   LSV06110
C     EIGENVALUES OF THE T-MATRIX, THESE GAPS WILL FIRST BE  LSV06120
C     TRANSFERRED TO V1.  NOTE THAT V1<0 MEANS THAT THAT MINIMAL GAP     LSV06130
C     IN THE T-MATRIX IS DUE TO A SPURIOUS T-EIGENVALUE.     LSV06140
C     ALL THIS INFORMATION IS PRINTED TO FILE 3              LSV06150
C                                                            LSV06160
  510 CONTINUE                                               LSV06170
C                                                            LSV06180
      NG = 0                                                 LSV06190
      DO 520 I = 1,NDIS                                      LSV06200
      IF (MP(I).EQ.0) GO TO 520                              LSV06210
      NG = NG+1                                              LSV06220
      MP(NG) = MP(I)                                         LSV06230
      V2(NG) = VS(I)                                         LSV06240
      TEMP = G(I)                                            LSV06250
      TEMP = DABS(TEMP)                                      LSV06260
      J = I+1                                                LSV06270
      IF (G(I).LT.ZERO) J = I-1                              LSV06280
      IF (MP(J).EQ.0) TEMP = -TEMP                           LSV06290
      V1(NG) = TEMP                                          LSV06300
  520 CONTINUE                                               LSV06310
C                                                            LSV06320
      WRITE(6,530)MEV                                        LSV06330
  530 FORMAT(//' SINGULAR VALUE CALCULATION AT MEV = ',I6,'   IS COMPLELSV06340
     1TE'//)                                                 LSV06350
C                                                            LSV06360
C     NG = NUMBER OF COMPUTED DISTINCT GOOD T-EIGENVALUES.  NEXT         LSV06370
C     GENERATE GAPS BETWEEN GOOD T-EIGENVALUES (BMINGAPS) AND PUT THEM   LSV06380
C     IN G.  G(J) < 0 MEANS THE BMINGAP IS DUE TO THE LEFT-HAND GAP.     LSV06390
C                                                            LSV06400
      NGM1 = NG - 1                                          LSV06410
      G(NG) = V2(NGM1)-V2(NG)                                LSV06420
      G(1) = V2(2)-V2(1)                                     LSV06430
C                                                            LSV06440
      DO 540 J = 2,NGM1                                      LSV06450
      TO = V2(J)-V2(J-1)                                     LSV06460
      T1 = V2(J+1)-V2(J)                                     LSV06470
      G(J) = T1                                              LSV06480
      IF (TO.LT.T1) G(J) = -TO                               LSV06490
  540 CONTINUE                                               LSV06500
C                                                            LSV06510
C     WRITE GOOD T-EIGENVALUES (COMPUTED SINGULAR VALUES) OUT TO FILE 3.LSV06520
C                                                            LSV06530
      WRITE(3,550)NG,NDIS,MEV,M,N,SVSEED,MATNO,IPARO,MULTOL,IB,BTOL      LSV06540
  550 FORMAT(5I6,,I12,I8,I2,'=NG,ND,MEV,M,N,SEED,MN,IPARO'/  LSV06550
     1 E20.12,I6,E13.4,' = MUTOL,INDEX MINIMAL BETA,BTOL'/   LSV06560
     1' SV NO',2X,'T-MULT',10X,'SINGULAR VALUE',7X,'BMINGAP',7X,'TMINGAPLSV06570
     1')                                                     LSV06580
C                                                            LSV06590
      WRITE(3,560)(I,MP(I),V2(I),G(I),V1(I),  I=1,NG)        LSV06600
  560 FORMAT(I6,I8,E25.16,2E14.3)                            LSV06610
C                                                            LSV06620
C     IF CONVERGENCE FLAG ICONV.NE.1 AND NUMBER OF T-MATRICES LSV06630
```

```
C     CONSIDERED TO DATE IS LESS THAN NUMBER ALLOWED, INCREMENT MEV.     LSV06640
C     AND LOOP BACK TO 210 TO REPEAT COMPUTATIONS.  RESTORE BETA(MEV+1).LSV06650
C                                                                       LSV06660
      BETA(MP1) = BETAM                                                 LSV06670
C                                                                       LSV06680
      IF (MMB.LT.NMEVS.AND.ICONV.NE.1) GO TO 210                        LSV06690
C                                                                       LSV06700
C     END OF LOOP ON DIFFERENT SIZE T-MATRICES ALLOWED.                 LSV06710
C                                                                       LSV06720
  570 CONTINUE                                                          LSV06730
C                                                                       LSV06740
      IF(ISTOP.EQ.0)  WRITE(6,580)                                      LSV06750
  580 FORMAT(/' T-MATRICES (BETA) ARE NOW AVAILABLE, TERMINATE'/)       LSV06760
      IF (IHIS.EQ.1.AND.KMAX.NE.MOLD) WRITE(1,590)                      LSV06770
  590 FORMAT(/' ABOVE ARE THE FOLLOWING VECTORS '/                      LSV06780
     2 '   BETA(I), I = 1,KMAX+1'/                                      LSV06790
     3 ' FINAL TWO LANCZOS VECTORS OF ORDERS M,N FOR I = KMAX,KMAX+1'/  LSV06800
     4 ' ALL VECTORS IN THIS FILE HAVE FORMAT 4Z20'/                    LSV06810
     5 ' ----- END OF FILE 1 NEW BETA HISTORY---------------'////)      LSV06820
C                                                                       LSV06830
      IF (ISTOP.EQ.0) GO TO 690                                         LSV06840
C                                                                       LSV06850
      WRITE(3,600)                                                      LSV06860
  600 FORMAT(/' ABOVE ARE COMPUTED SINGULAR VALUES'/                    LSV06870
     1 ' NG = NUMBER OF SINGULAR VALUES COMPUTED'/                      LSV06880
     2 ' NDIS = NUMBER OF COMPUTED DISTINCT EIGENVALUES OF T(1,MEV)'/   LSV06890
     3 ' M = ROW ORDER OF A N = COLUMN ORDER,  MATNO = MATRIX IDENT'/   LSV06900
     4 ' MULTOL = T-MULTIPLICITY TOLERANCE FOR T-EIGENVALUES IN BISEC'/ LSV06910
     4 ' T-MULT IS THE T-MULTIPLICITY OF SINGULAR VALUE'/               LSV06920
     5 ' T-MULT = -1 MEANS SPURIOUS T-EIGENVALUE TOO CLOSE'/            LSV06930
     6 ' DO NOT COMPUTE ERROR ESTIMATES FOR SUCH T-EIGENVALUES'/        LSV06940
     7 ' BMINGAP = MINIMAL GAP BETWEEN THE COMPUTED SINGULAR VALUES'/   LSV06950
     8 ' BMINGAP .LT. 0. MEANS MINIMAL GAP IS DUE TO LEFT-HAND GAP'/    LSV06960
     9 ' TMINGAP= MINIMAL GAP W.R.T.  DISTINCT EIGENVALUES IN T(1,MEV)'/LSV06970
     1 ' TMINGAP .LT. 0. MEANS MINGAP IS DUE TO SPURIOUS T-EIGENVALUE'/ LSV06980
     2 ' ----- END OF FILE 3 SINGULAR VALUES-----------------------'//)LSV06990
C                                                                       LSV07000
      IF (IDIST.EQ.1) WRITE(11,610)                                     LSV07010
  610 FORMAT(/' ABOVE ARE THE DISTINCT EIGENVALUES OF T(1,MEV).'/       LSV07020
     2 ' THE FORMAT IS      T-MULTIPLICITY    T-EIGENVALUE   TMINGAP'/  LSV07030
     3 '          THIS FORMAT IS REPEATED TWICE ON EACH LINE.'/         LSV07040
     4 ' T-MULTIPLICITY = -1 MEANS THAT THE SUBROUTINE ISOEV HAS TAGGED'LSV07050
     5 /'   THIS COMPUTED SINGULAR VALUE AS HAVING A VERY CLOSE SPURIOUSLSV07060
     6 '/'   T-EIGENVALUE SO THAT NO ERROR ESTIMATE WILL BE COMPUTED'/  LSV07070
     7 '   FOR THAT SINGULAR VALUE IN SUBROUTINE INVERR.'/              LSV07080
     8 ' TMINGAP .LT. 0, TMINGAP IS DUE TO LEFT GAP .GT. 0, RIGHT GAP.'/LSV07090
     9 ' EACH OF THE DISTINCT T-EIGENVALUE TABLES IS FOLLOWED'/         LSV07100
     9 ' BY THE T-MULTIPLICITY PATTERN.'/                               LSV07110
     1 ' NDIS = NUMBER OF COMPUTED DISTINCT EIGENVALUES OF T(1,MEV).'/  LSV07120
     2 ' NG = NUMBER OF COMPUTED SINGULAR VALUES. '/                    LSV07130
     3 ' NISO = NUMBER OF ISOLATED (IN T-MATRIX) SINGULAR VALUES. '/    LSV07140
     4 ' NISO ALSO IS THE COUNT OF +1 ENTRIES IN T-MULTIPLICITY PATTERN.LSV07150
     5 '/' ----- END OF FILE 11 DISTINCT T-EIGENVALUES--------------'//)LSV07160
C                                                                       LSV07170
      IF(NISO.NE.0)  WRITE(4,620)                                       LSV07180
  620 FORMAT(/' ABOVE ARE THE ERROR ESTIMATES OBTAINED FOR THE ISOLATED LSV07190
     1GOOD T-EIGENVALUES'/                                              LSV07200
     1' OBTAINED VIA INVERSE ITERATION IN THE SUBROUTINE INVERR.'/      LSV07210
     1' ALL OTHER GOOD T-EIGENVALUES HAVE CONVERGED.'/                  LSV07220
     2' ERROR ESTIMATE = BETAM*ABS(UM)'/                                LSV07230
     2' WHERE BETAM = BETA(MEV+1) AND UM = U(MEV).'/                    LSV07240
     3' U = UNIT EIGENVECTOR OF T WHERE T*U = SV*U AND SV = ISOLATED GOOLSV07250
     3D T-EIGENVALUE.'/                                                 LSV07260
     4' TMINGAP = GAP TO NEAREST DISTINCT EIGENVALUE OF T(1,MEV).'/     LSV07270
     5' TMINGAP .LT. 0. MEANS MINGAP IS DUE TO A SPURIOUS T-EIGENVALUE.'LSV07280
     6/' ------ END OF FILE 4 ERRINV -------------------------'//)LSV07290
```

```
      GO TO 690                                                        LSV07300
C                                                                      LSV07310
  630 CONTINUE                                                         LSV07320
C                                                                      LSV07330
      IBB = IABS(IBMEV)                                                LSV07340
      IF (IBMEV.LT.0) WRITE(6,640) MEV,IBB,BETA(IBB)                   LSV07350
  640 FORMAT(/' PROGRAM TERMINATES BECAUSE MEV REQUESTED = ',I6,' IS .GTLSV07360
     1',I6/' AT WHICH AN ABNORMALLY SMALL BETA = ' , E13.4,' OCCURRED'/)LSV07370
      GO TO 690                                                        LSV07380
C                                                                      LSV07390
  650 IF (NDIS.EQ.0.AND.ISTOP.GT.0) WRITE(6,660)                       LSV07400
  660 FORMAT(/' INTERVALS SPECIFIED FOR BISECT DID NOT CONTAIN ANY T-EIGLSV07410
     1ENVALUES'/' PROGRAM TERMINATES')                                 LSV07420
      GO TO 690                                                        LSV07430
C                                                                      LSV07440
  670 WRITE(6,680) I, NMEV(I)                                          LSV07450
  680 FORMAT(//I6,'TH T-SIZE REQUESTED ',I6,' IS ODD'/                 LSV07460
     1' BUT ONLY EVEN T-SIZES ARE PERMISSIBLE.  PROGRAM TERMINATES FOR ULSV07470
     1SER TO FIX'//)                                                   LSV07480
      GO TO 690                                                        LSV07490
C                                                                      LSV07500
  690 CONTINUE                                                         LSV07510
C                                                                      LSV07520
      STOP                                                             LSV07530
C-----END OF MAIN PROGRAM FOR LANCZOS SINGULAR VALUE COMPUTATIONS-------LSV07540
      END                                                              LSV07550
```

SECTION 6.4   MAIN PROGRAM, SINGULAR VECTOR CALCULATIONS

```
C-----LSVEC (SINGULAR VECTORS OF REAL RECTANGULAR MATRICES)-------------LSV00010
C                                                                       LSV00020
C       CONTAINS MAIN PROGRAM FOR COMPUTING A LEFT AND A                LSV00030
C       RIGHT SINGULAR VECTOR CORRESPONDING TO EACH OF A SET            LSV00040
C       OF SINGULAR VALUES WHICH HAVE BEEN COMPUTED ACCURATELY BY THE   LSV00050
C       CORRESPONDING LANCZOS SINGULAR VALUE PROGRAM (LSVAL)            LSV00060
C       FOR REAL RECTANGULAR MATRICES.  THIS PROGRAM COULD BE           LSV00070
C       MODIFIED TO COMPUTE ADDITIONAL SINGULAR VECTORS FOR ANY         LSV00080
C       SINGULAR VALUE THAT IS A MULTIPLE SINGULAR VALUE OF A.          LSV00090
C       THE AMOUNT OF ADDITIONAL COMPUTATION REQUIRED BY SUCH A         LSV00100
C       MODIFICATION DEPENDS UPON THE GIVEN A-MATRIX AND UPON           LSV00110
C       THE PART OF THE SPECTRUM INVOLVED.                              LSV00120
C                                                                       LSV00130
C       FOR A GIVEN REAL MATRIX A OF ORDER M X N THE LANCZOS RECURSION  LSV00140
C       IS APPLIED TO THE ASSOCIATED REAL SYMMETRIC MATRIX B OF ORDER   LSV00150
C       MN = M+N                                                        LSV00160
C                                                                       LSV00170
C                                 ----            ----                  LSV00180
C                                |  0            A   |                  LSV00190
C                     B   =      |                   |                  LSV00200
C                                |  A-TRANSPOSE   0  |                  LSV00210
C                                 ----            ----                  LSV00220
C       USING SPECIAL STARTING VECTORS.                                 LSV00230
C                                                                       LSV00240
C       THESE SINGULAR VECTOR COMPUTATIONS ASSUME THAT EACH             LSV00250
C       SINGULAR VALUE THAT IS BEING CONSIDERED HAS CONVERGED AS        LSV00260
C       AN EIGENVALUE OF THE LANCZOS TRIDIAGONAL MATRICES GENERATED.    LSV00270
C                                                                       LSV00280
C       THE EIGENVALUES OF EACH EVEN-ORDERED LANCZOS MATRIX OCCUR       LSV00290
C       IN + AND - PAIRS, AND THE RITZ VECTOR COMPUTATION RESTS ON      LSV00300
C       AN INVERSE ITERATION COMPUTATION FOR A LANCZOS MATRIX.          LSV00310
C       THIS CAUSES AN ANOMALY IN THE SINGULAR VECTOR COMPUTATIONS      LSV00320
C       FOR VERY SMALL SINGULAR VALUES.  IN PRACTICE WE SEE THAT        LSV00330
C       FOR ANY SUCH SINGULAR VALUE THAT ONE MEMBER OF EACH PAIR OF     LSV00340
C       APPROXIMATE SINGULAR VECTORS WILL BE MORE ACCURATE THAN THE     LSV00350
C       OTHER MEMBER OF THAT PAIR IS.  IF IPAR = 1 (STARTING LANCZOS    LSV00360
C       VECTOR IS OF FORM (0,V2) WHERE V2 IS NX1) THEN THE RIGHT        LSV00370
C       SINGULAR VECTOR WILL BE OBTAINED MORE ACCURATELY THAN THE       LSV00380
C       LEFT SINGULAR VECTOR.  IF IPAR = 2 (STARTING LANCZOS VECTOR     LSV00390
C       IS OF FORM (V1,0) WHERE V1 IS MX1) THEN THE LEFT SINGULAR       LSV00400
C       VECTOR WILL BE MORE ACCURATE THAN THE RIGHT SINGULAR VECTOR.    LSV00410
C       PRIOR TO NORMALIZATION THE SIZES OF THESE INACCURATE VECTORS    LSV00420
C       WILL BE THE SAME AS THE SIZE OF THE ASSOCIATED VERY SMALL       LSV00430
C       SINGULAR VALUE.  IN FACT IN THE LIMIT, FOR A ZERO SINGULAR VALUE LSV00440
C       AND IPAR = 1, THE VECTOR COMPUTED AS THE APPROXIMATION TO THE   LSV00450
C       LEFT SINGULAR VECTOR WILL BE THE O VECTOR.  (IF IPAR = 2 THEN   LSV00460
C       THIS WOULD BE THE RIGHT SINGULAR VECTOR).  THE CORRESPONDING    LSV00470
C       ERROR ESTIMATES WILL REFLECT THE INACCURACY OF THE ONE MEMBER   LSV00480
C       OF EACH SUCH PAIR, SINCE THESE ESTIMATES ARE A SUM OF ESTIMATES LSV00490
C       FOR THE INDIVIDUAL MEMBERS OF THE PAIR.  THEREFORE, FOR ANY VERY LSV00500
C       SMALL SINGULAR VALUE A CORRESPONDING SINGULAR VECTOR WILL BE    LSV00510
C       COMPUTED ONLY IF THE USER HAS SET THE FLAG ERCONT TO 1.         LSV00520
C                                                                       LSV00530
C-----------------------------------------------------------------------LSV00540
C                                                                       LSV00550
C       PFORT VERIFIER IDENTIFIED THE FOLLOWING NONPORTABLE             LSV00560
C       CONSTRUCTIONS                                                   LSV00570
C                                                                       LSV00580
C       1.  DATA/MACHEP/ STATEMENT                                      LSV00590
C       2.  ALL READ(5,*) STATEMENTS (FREE FORMAT)                      LSV00600
C       3.  FORMAT(20A4) USED WITH THE EXPLANATORY HEADER, EXPLAN       LSV00610
C       4.  HEXADECIMAL FORMAT (4Z20) USED FOR BETA HISTORY.            LSV00620
C                                                                       LSV00630
C       IMPORTANT NOTE:  THIS PROGRAM ALLOWS ENLARGEMENT OF THE         LSV00640
```

```
C      BETA ARRAY.  IN PARTICULAR, IF ANY ONE OF THE SINGULAR VALUES      LSV00650
C      SUPPLIED IS T-SIMPLE AND AS AN EIGENVALUE OF THE ASSOCIATED        LSV00660
C      LANCZOS TRIDIAGONAL MATRIX IS NOT CLOSE TO A SPURIOUS             LSV00670
C      EIGENVALUE OF THAT MATRIX, THIS PROGRAM WILL REQUIRE              LSV00680
C      THAT KMAX BE AT LEAST THE LARGEST EVEN NUMBER LESS                LSV00690
C      THAN OR EQUAL TO  (11*MEV)/8 + 13.  IF KMAX IS NOT THAT           LSV00700
C      LARGE, THEN THIS PROGRAM WILL RESET KMAX TO THIS SIZE             LSV00710
C      AND EXTEND THE BETA HISTORY IF REQUIRED.                          LSV00720
C      THUS, THE DIMENSION OF THE BETA ARRAY MUST BE                     LSV00730
C      LARGE ENOUGH TO ALLOW FOR THIS POSSIBILITY.                       LSV00740
C      REMEMBER THAT THE BETA ARRAY, BETA(J), IS SUCH THAT               LSV00750
C      J = 1,..., KMAX+1.  SO IF THE KMAX USED BY THE PROGRAM            LSV00760
C      IS TO BE 3000, THEN BETA MUST BE OF LENGTH AT LEAST 3001.         LSV00770
C                                                                        LSV00780
C------------------------------------------------------------------------LSV00790
       DOUBLE PRECISION   BETA(5001),V1(5000),V2(5000),RITVEC(30000)     LSV00800
       DOUBLE PRECISION   TVEC(30000),GOODSV(50),SVNEW(50),TLAST(50)     LSV00810
       DOUBLE PRECISION   SVAL,SVALN,TOLN,TTOL,ERTOL,BATA               LSV00820
       DOUBLE PRECISION   MULTOL,SCALEO,STUTOL,BTOL,LB,UB                LSV00830
       DOUBLE PRECISION   ONE,ZERO,MACHEP,EPSM,TEMP,SUM                  LSV00840
       DOUBLE PRECISION   RELTOL,ERROR,TERROR,ERRMIN,BKMIN               LSV00850
       REAL       G(10000),BMINGP(50),TMINGP(50),EXPLAN(20)             LSV00860
       REAL       TERR(50),BERR(50),BERRGP(50),RNORM(50),TBETA(50)      LSV00870
       INTEGER    MP(50),M1(50),M2(50),MA(50),ML(50),MINT(50),MFIN(50)   LSV00880
       INTEGER    SVSEED,SVSOLD,RHSEED,IDELTA(50)                        LSV00890
       INTEGER    MBOUND,NTVCON,SVTVEC,TVSTOP,LVCONT,ERCONT,TFLAG        LSV00900
       DOUBLE PRECISION   FINPRO                                         LSV00910
       DOUBLE PRECISION   DABS, DMAX1, DSQRT, DFLOAT                     LSV00920
       REAL       ABS                                                    LSV00930
       INTEGER    IABS                                                   LSV00940
C------------------------------------------------------------------------LSV00950
       EXTERNAL SVMAT, STRAN                                             LSV00960
       DATA MACHEP/Z3410000000000000/                                   LSV00970
       EPSM = 2.D0*MACHEP                                               LSV00980
C------------------------------------------------------------------------LSV00990
C                                                                        LSV01000
C      ARRAYS MUST BE DIMENSIONED AS FOLLOWS:                            LSV01010
C      1.  BETA: >= (KMAX+1) WHERE KMAX, THE LARGEST SIZE                LSV01020
C                T-MATRIX CONSIDERED BY THE PROGRAM, IS THE              LSV01030
C                LARGER OF THE SIZE OF THE BETA HISTORY PROVIDED         LSV01040
C                ON FILE 2 (IF ANY ) AND THE SIZE WHICH THE PROGRAM      LSV01050
C                SPECIFIES INTERNALLY, THIS LATTER IS ALWAYS             LSV01060
C                < = (11*MEV)/8 + 13, WHERE MEV IS THE SIZE             LSV01070
C                T-MATRIX THAT WAS USED IN THE CORRESPONDING             LSV01080
C                SINGULAR VALUE COMPUTATIONS.  NOTE THAT ALL             LSV01090
C                T-MATRICES CONSIDERED MUST HAVE EVEN ORDER.             LSV01100
C      2.  V1:   >= MAX(M,KMAX)                                          LSV01110
C      3.  V2:   >= N                                                    LSV01120
C      4.  G:    >= MAX(M,N,KMAX)                                        LSV01130
C      5.  RITVEC: >= (N+M)*NGOOD, WHERE NGOOD IS THE NUMBER OF          LSV01140
C                SINGULAR VALUES SUPPLIED TO THIS PROGRAM.               LSV01150
C      6.  TVEC: >= CUMULATIVE LENGTH OF ALL THE T-EIGENVECTORS          LSV01160
C                NEEDED TO GENERATE THE DESIRED RITZ VECTORS.  AN        LSV01170
C                EDUCATED GUESS AT AN APPROPRIATE LENGTH CAN BE          LSV01180
C                OBTAINED BY RUNNING THE PROGRAM WITH THE FLAG           LSV01190
C                MBOUND = 1 AND MULTIPLYING THE RESULTING SIZE BY 5/4.   LSV01200
C      7.  GOODSV, TMINGP, BMINGP,TERR, BERR, BERRGP, RNORM,             LSV01210
C          TBETA, TLAST, SVNEW, MP, MA, M1, M2, MINT, MFIN AND           LSV01220
C          IDELTA MUST ALL BE >= NGOOD.                                  LSV01230
C                                                                        LSV01240
C------------------------------------------------------------------------LSV01250
C      OUTPUT HEADER                                                     LSV01260
       WRITE(6,10)                                                       LSV01270
   10 FORMAT(/' LANCZOS PROCEDURE FOR REAL, RECTANGULAR MATRICES'/       LSV01280
      1'      COMPUTE SINGULAR VECTORS'/)                                LSV01290
C                                                                        LSV01300
```

```
C     SET PROGRAM PARAMETERS                                          LSV01310
C     USER MUST NOT MODIFY SCALE0                                     LSV01320
      SCALE0 = 5.0D0                                                  LSV01330
      ZERO = 0.0D0                                                    LSV01340
      ONE = 1.0D0                                                     LSV01350
      MPMIN = -1000                                                   LSV01360
C     CONVERGENCE TOLERANCE FOR T-EIGENVECTORS FOR RITZ COMPUTATIONS  LSV01370
      ERTOL = 1.D-10                                                  LSV01380
C                                                                     LSV01390
C     READ USER-SPECIFIED PARAMETER FROM INPUT FILE 5 (FREE FORMAT)   LSV01400
C                                                                     LSV01410
C     READ USER-PROVIDED HEADER FOR RUN                              LSV01420
      READ(5,20) EXPLAN                                               LSV01430
      WRITE(6,20) EXPLAN                                              LSV01440
   20 FORMAT(20A4)                                                    LSV01450
C                                                                     LSV01460
C     READ IN MATNO = MATRIX/RUN IDENTIFICATION NUMBER, 8 DIGITS OR LESSLSV01470
C     AND THE ORDER OF THE MATRIX M X N .                            LSV01480
C                                                                     LSV01490
      READ(5,20) EXPLAN                                               LSV01500
      READ(5,*) MATNO, M, N                                           LSV01510
      MN = M + N                                                      LSV01520
C                                                                     LSV01530
C     READ IN THE MAXIMUM PERMISSIBLE DIMENSIONS FOR THE TVEC ARRAY  LSV01540
C     (MDIMTV), FOR THE RITVEC ARRAY (MDIMRV), AND FOR THE BETA      LSV01550
C     ARRAY (MBETA).                                                 LSV01560
C                                                                     LSV01570
      READ(5,20) EXPLAN                                               LSV01580
      READ(5,*) MDIMTV, MDIMRV, MBETA                                 LSV01590
C                                                                     LSV01600
C     READ IN RELATIVE TOLERANCE USED IN DETERMINING APPROPRIATE     LSV01610
C     SIZES FOR THE T-MATRICES USED IN THE SINGULAR VECTOR COMPUTATIONS.LSV01620
C                                                                     LSV01630
      READ(5,20) EXPLAN                                               LSV01640
      READ(5,*) RELTOL                                                LSV01650
C                                                                     LSV01660
C     SET FLAGS TO 0 OR 1:                                           LSV01670
C     MBOUND = 1:  PROGRAM TERMINATES AFTER COMPUTING 1ST GUESSES    LSV01680
C                  ON APPROPRIATE T-SIZES FOR USE IN THE RITZ VECTOR LSV01690
C                  COMPUTATIONS                                      LSV01700
C     NTVCON = 0:  PROGRAM TERMINATES IF THE TVEC ARRAY IS NOT       LSV01710
C                  LARGE ENOUGH TO HOLD ALL THE T-EIGENVECTORS REQUIRED.LSV01720
C     SVTVEC = 0:  THE T-EIGENVECTORS ARE NOT WRITTEN TO FILE 11     LSV01730
C                  UNLESS TVSTOP = 1                                 LSV01740
C     SVTVEC = 1:  WRITE THE T-EIGENVECTORS TO FILE 11.              LSV01750
C     TVSTOP = 1:  PROGRAM TERMINATES AFTER COMPUTING THE            LSV01760
C                  T-EIGENVECTORS                                    LSV01770
C     LVCONT = 0:  PROGRAM TERMINATES IF THE NUMBER OF T-EIGENVECTORS LSV01780
C                  COMPUTED IS NOT EQUAL TO THE NUMBER OF RITZ       LSV01790
C                  VECTORS (SINGULAR VECTORS) REQUESTED.             LSV01800
C     ERCONT = 0:  MEANS FOR ANY GIVEN SINGULAR VALUE, A RITZ VECTOR LSV01810
C                  WILL NOT BE COMPUTED FOR THAT SINGULAR VALUE UNLESS LSV01820
C                  A T-EIGENVECTOR HAS BEEN IDENTIFIED WITH A LAST   LSV01830
C                  COMPONENT WHICH SATISFIES THE SPECIFIED           LSV01840
C                  CONVERGENCE CRITERION.                            LSV01850
C     ERCONT = 1:  MEANS FOR ANY GIVEN SINGULAR VALUE, A RITZ VECTOR LSV01860
C                  WILL BE COMPUTED.  IF A T-EIGENVECTOR CANNOT      LSV01870
C                  BE IDENTIFIED WHICH SATISFIES THE LAST           LSV01880
C                  COMPONENT CRITERION, THEN THE PROGRAM WILL        LSV01890
C                  USE THE T-VECTOR THAT CAME CLOSEST TO             LSV01900
C                  SATISFYING THE CRITERION                          LSV01910
C     IWRITE = 1:  EXTENDED OUTPUT OF INTERMEDIATE COMPUTATIONS      LSV01920
C                  IS WRITTEN TO FILE 6                              LSV01930
C     IREAD = 0:   BETA FILE IS REGENERATED.                         LSV01940
C     IREAD = 1:   BETA FILE USED IN SINGULAR VALUE COMPUTATIONS     LSV01950
C                  IS READ IN AND EXTENDED IF NECESSARY.  IN BOTH    LSV01960
C                  CASES IREAD = 0 OR 1, THE LANCZOS VECTORS ARE     LSV01970
```

```
C                     ALWAYS REGENERATED FOR THE RITZ VECTOR           LSV01980
C                     COMPUTATIONS                                     LSV01990
C                                                                      LSV02000
      READ(5,20) EXPLAN                                                LSV02010
      READ(5,*) MBOUND,NTVCON,SVTVEC,IREAD                             LSV02020
C                                                                      LSV02030
      READ(5,20) EXPLAN                                                LSV02040
      READ(5,*) TVSTOP,LVCONT,ERCONT,IWRITE                           LSV02050
      IF (TVSTOP.EQ.1) SVTVEC = 1                                      LSV02060
C                                                                      LSV02070
C     READ IN SEED (RHSEED) FOR GENERATING RANDOM STARTING VECTOR      LSV02080
C     FOR THE INVERSE ITERATION ON THE T-MATRICES.                     LSV02090
C                                                                      LSV02100
      READ(5,20) EXPLAN                                                LSV02110
      READ(5,*) RHSEED                                                 LSV02120
C                                                                      LSV02130
C----------------------------------------------------------------------LSV02140
C     INITIALIZE THE ARRAYS FOR THE USER-SPECIFIED MATRIX AND          LSV02150
C     PASS THE STORAGE LOCATIONS OF THESE ARRAYS TO THE MATRIX-VECTOR  LSV02160
C     MULTIPLY SUBROUTINES SVMAT AND STRAN.                            LSV02170
C                                                                      LSV02180
      CALL USPEC(M,N,MATNO)                                            LSV02190
C                                                                      LSV02200
C----------------------------------------------------------------------LSV02210
C     MASK UNDERFLOW AND OVERFLOW                                      LSV02220
      CALL MASK                                                        LSV02230
C                                                                      LSV02240
C----------------------------------------------------------------------LSV02250
C     WRITE RUN PARAMETERS OUT TO FILE 6                               LSV02260
C                                                                      LSV02270
      WRITE(6,30) M,N,MATNO                                            LSV02280
   30 FORMAT(/' MATRIX ORDER =',I5,' BY ',I5/                          LSV02290
     1 ' A-MATRIX AND CASE IDENTIFIER = ',I10/)                       LSV02300
C                                                                      LSV02310
      WRITE(6,40) MBOUND,NTVCON,SVTVEC,IREAD                           LSV02320
   40 FORMAT(/3X,'MBOUND',3X,'NTVCON',3X,'SVTVEC',3X,'IREAD'/3I9,I8/)  LSV02330
C                                                                      LSV02340
      WRITE(6,50) TVSTOP,LVCONT,ERCONT,IWRITE                         LSV02350
   50 FORMAT(/3X,'TVSTOP',3X,'LVCONT',3X,'ERCONT',3X,'IWRITE'/4I9)    LSV02360
C                                                                      LSV02370
      WRITE(6,60) MDIMTV,MDIMRV,MBETA                                  LSV02380
   60 FORMAT(/3X,'MDIMTV',3X,'MDIMRV',3X,'MBETA'/2I9,I8)              LSV02390
C                                                                      LSV02400
      WRITE(6,70) RELTOL,RHSEED                                        LSV02410
   70 FORMAT(/7X,'RELTOL',3X,'RHSEED'/E13.4,I9)                       LSV02420
C                                                                      LSV02430
C     FROM FILE 3 READ IN THE NUMBER OF SINGULAR VALUES (NGOOD)        LSV02440
C     FOR WHICH SINGULAR VECTORS ARE REQUESTED, THE ORDER (MEV) OF     LSV02450
C     THE LANCZOS TRIDIAGONAL MATRIX USED IN COMPUTING THESE           LSV02460
C     SINGULAR VALUES, THE ORDER MOLD X NOLD OF THE USER-SPECIFIED     LSV02470
C     MATRIX USED IN THOSE COMPUTATIONS, THE SEED (SVSEED) USED FOR    LSV02480
C     GENERATING THE STARTING VECTOR THAT WAS USED IN THOSE            LSV02490
C     COMPUTATIONS, AND THE MATRIX/RUN IDENTIFICATION NUMBER (MATOLD)  LSV02500
C     USED IN THOSE COMPUTATIONS.  ALSO READ IN THE NUMBER (NDIS) OF   LSV02510
C     DISTINCT EIGENVALUES OF THE MATRIX T(1,MEV) THAT WERE COMPUTED   LSV02520
C     BUT THIS VALUE IS NOT USED IN THE SINGULAR VECTOR                LSV02530
C     COMPUTATIONS.                                                    LSV02540
C                                                                      LSV02550
      READ(3,80) NGOOD,NDIS,MEV,MOLD,NOLD,SVSEED,MATOLD,IPARO         LSV02560
   80 FORMAT(5I6,I12,I8,I2)                                           LSV02570
C                                                                      LSV02580
C     READ IN THE T-MULTIPLICITY TOLERANCE USED IN THE BISEC SUBROUTINE LSV02590
C     DURING THE COMPUTATION OF THE GIVEN SINGULAR VALUES.             LSV02600
C     ALSO READ IN THE FLAG IB.  IF IB < 0, THEN SOME BETA(I) IN THE   LSV02610
C     T-MATRIX FILE PROVIDED ON FILE 2 FAILED THE ORTHOGONALITY        LSV02620
C     TEST IN THE TNORM SUBROUTINE.  USER SHOULD NOTE THAT THIS        LSV02630
```

```
C     PROGRAM PROCEEDS INDEPENDENTLY OF THE SIZE OF THE BETA USED.        LSV02640
C                                                                         LSV02650
      READ(3,90) MULTOL,IB,BTOL                                           LSV02660
   90 FORMAT(E20.12,I6,E13.4)                                             LSV02670
C                                                                         LSV02680
      TEMP = DFLOAT(MEV+1000)                                             LSV02690
      TTOL = MULTOL/TEMP                                                  LSV02700
      WRITE(6,100) MULTOL,TTOL                                            LSV02710
  100 FORMAT(/' T-MULTIPLICITY TOLERANCE USED IN THE SINGULAR VALUE COMPLSV02720
     IUTATIONS WAS',E13.4/' SCALED MACHINE EPSILON IS',E13.4)            LSV02730
C                                                                         LSV02740
C     CONTINUE WRITE TO FILE 6 OF THE PARAMETERS FOR THIS RUN             LSV02750
C                                                                         LSV02760
      WRITE(6,110)NGOOD,NDIS,MEV,MOLD,NOLD,MATOLD,SVSEED,MULTOL,IB,       LSV02770
     IBTOL,IPARO                                                          LSV02780
  110 FORMAT(/' SINGULAR VALUES SUPPLIED ARE READ IN FROM FILE 3'/        LSV02790
     1 6X,'NG',2X,'NDIS',3X,'MEV',2X,'MOLD',2X,'NOLD',2X,'MATOLD',4X/     LSV02800
     118,4I6,I8//6X,'SVSEED',6X,'MULTOL',9X,'IB',8X,'BTOL',4X,'IPARO'/    LSV02810
     1I12,E12.3,I11,E12.4,I9/)                                           LSV02820
C                                                                         LSV02830
C     IS THE ARRAY RITVEC LONG ENOUGH TO HOLD ALL OF THE DESIRED          LSV02840
C     RITZ VECTORS (APPROXIMATE EIGENVECTORS OF B)?                       LSV02850
      MNMAX = NGOOD*MN                                                    LSV02860
      IF(MBOUND.EQ.1) GO TO 120                                           LSV02870
      IF(TVSTOP.NE.1.AND.MNMAX.GT.MDIMRV) GO TO 1600                      LSV02880
C                                                                         LSV02890
C     CHECK THAT THE ORDERS M,N AND THE MATRIX IDENTIFICATION NUMBER      LSV02900
C     MATNO SPECIFIED BY THE USER AGREE WITH THOSE READ IN FROM           LSV02910
C     FILE 3.                                                             LSV02920
  120 ITEMP = (MOLD-M)**2+(NOLD-N)**2+(MATOLD-MATNO)**2                   LSV02930
      IF (ITEMP.NE.0) GO TO 1620                                          LSV02940
C                                                                         LSV02950
C     READ IN FROM FILE 3, THE T-MULTIPLICITIES OF THE SINGULAR VALUES    LSV02960
C     WHOSE SINGULAR VECTORS ARE TO BE COMPUTED, THE VALUES OF THESE      LSV02970
C     SINGULAR VALUES AND THEIR MINIMAL GAPS AS SINGULAR VALUES OF THE    LSV02980
C     USER-SPECIFIED MATRIX AND OF THE RELATED T-MATRIX.                  LSV02990
C                                                                         LSV03000
      READ(3,20) EXPLAN                                                   LSV03010
      READ(3,130) (MP(J),GOODSV(J),BMINGP(J),TMINGP(J), J=1,NGOOD)        LSV03020
  130 FORMAT(6X,I8,E25.16,2E14.3)                                         LSV03030
C                                                                         LSV03040
      WRITE(6,140) (J,GOODSV(J),MP(J),BMINGP(J),   J=1,NGOOD)             LSV03050
  140 FORMAT(/' SINGULAR VALUES READ IN FROM FILE 3 AND THEIR T-MULTIPLILSV03060
     1CITIES'/4X,' J ',4X,' SINGULAR VALUE',5X,'TMULT',4X,'BMINGP'/       LSV03070
     1(I6,E20.12,I6,E13.4))                                               LSV03080
C                                                                         LSV03090
      WRITE(6,150) MEV,SVSEED                                             LSV03100
  150 FORMAT(/' THESE SINGULAR VALUES WERE COMPUTED USING A T-MATRIX OF LSV03110
     IORDER ',I5/' AND SEED FOR RANDOM NUMBER GENERATOR =',I12)           LSV03120
C                                                                         LSV03130
C     READ IN THE ERROR ESTIMATES                                        LSV03140
C                                                                         LSV03150
C     CHECK WHETHER OR NOT THERE ARE ANY ISOLATED T-EIGENVALUES IN        LSV03160
C     THE T-EIGENVALUES PROVIDED (HERE THE SINGULAR VALUES ARE            LSV03170
C     CONSIDERED AS EIGENVALUES OF THE ASSOCIATED LANCZOS TRIDIAGONAL     LSV03180
C     MATRICES.)                                                          LSV03190
      DO 160 J=1,NGOOD                                                    LSV03200
      IF(MP(J).EQ.1) GO TO 170                                            LSV03210
  160 CONTINUE                                                            LSV03220
      GO TO 190                                                           LSV03230
  170 READ(4,20) EXPLAN                                                   LSV03240
      READ(4,20) EXPLAN                                                   LSV03250
      READ(4,20) EXPLAN                                                   LSV03260
      READ(4,180) NISO                                                    LSV03270
  180 FORMAT(18X,I6)                                                      LSV03280
      READ(4,20) EXPLAN                                                   LSV03290
      READ(4,20) EXPLAN                                                   LSV03300
```

```
      READ(4,20) EXPLAN                                                LSV03310
  190 DO 220 J=1,NGOOD                                                 LSV03320
      BERR(J) = 0.D0                                                   LSV03330
      IF(MP(J).NE.1) GO TO 220                                         LSV03340
      READ(4,200) SVAL, BERR(J)                                        LSV03350
  200 FORMAT(10X,E25.16,E14.3)                                         LSV03360
      IF(DABS(SVAL - GOODSV(J)).LT.1.D-10) GO TO 220                   LSV03370
      WRITE(6,210) SVAL,GOODSV(J)                                      LSV03380
  210 FORMAT(' PROBLEM WITH READ IN OF ERROR ESTIMATES'/' SINGULAR VALUELSV03390
     1READ IN',E20.12,' DOES NOT MATCH GOODSV(J) ='/E20.12)            LSV03400
      GO TO 1860                                                       LSV03410
C                                                                      LSV03420
  220 CONTINUE                                                         LSV03430
C                                                                      LSV03440
      WRITE(6,230) (J,GOODSV(J),BERR(J), J=1,NGOOD)                    LSV03450
  230 FORMAT(' ERROR ESTIMATES ='/4X,' J',3X,'SINGULAR VALUE',8X,      LSV03460
     1'ESTIMATE'/(I6,E20.12,E14.3))                                    LSV03470
C                                                                      LSV03480
      IF(IREAD.EQ.0) IPAR = IPARO                                      LSV03490
      IF(IREAD.EQ.0)  GO TO 350                                        LSV03500
C                                                                      LSV03510
C     READ IN THE SIZE OF THE T-MATRIX PROVIDED ON FILE 2.  READ IN    LSV03520
C     THE ORDER OF THE USER-SPECIFIED MATRIX , THE FLAGS IPARO         LSV03530
C     AND IPAR WHICH INDICATE RESPECTIVELY THE PARITY OF THE           LSV03540
C     STARTING VECTOR USED IN THE GENERATION OF THE EXISTING           LSV03550
C     BETA AND THE PARITY OF THE NEXT LANCZOS VECTOR THAT              LSV03560
C     HAS TO BE GENERATED IF THE BETA HISTORY IS EXTENDED,             LSV03570
C     THE SEED USED BY THE RANDOM NUMBER GENERATOR WHEN                LSV03580
C     GENERATING THE STARTING VECTOR THAT WAS USED, AND THE            LSV03590
C     MATRIX/TEST IDENTIFICATION NUMBER THAT WERE USED IN              LSV03600
C     THE LANCZOS SINGULAR VALUE COMPUTATIONS.  IF THE FLAG            LSV03610
C     IREAD = 0, REGENERATE HISTORY AND DO NOT READ ANYTHING           LSV03620
C     FROM FILE 2.  HISTORY MUST BE STORED IN MACHINE FORMAT,          LSV03630
C     ((4Z20) FOR IBM 3081).                                          LSV03640
C                                                                      LSV03650
      READ(2,240) KMAX,MOLD,NOLD,IPARO,IPAR,SVSOLD,MATOLD              LSV03660
  240 FORMAT(3I6,2I3,I12,I8)                                           LSV03670
C                                                                      LSV03680
      WRITE(6,250) KMAX,MOLD,NOLD,IPARO,IPAR,SVSOLD,MATOLD             LSV03690
  250 FORMAT(/' READ IN HEADER FROM BETA FILE 2'/                      LSV03700
     1 2X,'KMAX',2X,'MOLD',2X,'NOLD',2X,'IPARO',2X,'IPAR',6X,'SVSOLD   LSV03710
     1 ',2X,'MATOLD'/3I6,I7,I6,I12,I12)                                LSV03720
C                                                                      LSV03730
C     CHECK THAT THE PARAMETERS READ IN AGREE WITH WHAT THE USER       LSV03740
C     HAS SPECIFIED                                                    LSV03750
      IF(MOLD.NE.M.OR.NOLD.NE.N.OR.MATOLD.NE.MATNO.OR.SVSOLD.NE.SVSEED) LSV03760
     1 GO TO 1640                                                      LSV03770
C                                                                      LSV03780
      IF(IPARO.EQ.1) WRITE(6,260)                                      LSV03790
      IF(IPARO.EQ.2) WRITE(6,270)                                      LSV03800
  260 FORMAT(/' STARTING VECTOR USED IN EXISTING SINGULAR VALUE HISTORY LSV03810
     1WAS'/' OF THE FORM (0,V2)')                                      LSV03820
  270 FORMAT(/' STARTING VECTOR USED IN EXISTING SINGULAR VALUE HISTORY LSV03830
     1WAS'/' OF THE FORM (V1,0)')                                      LSV03840
C                                                                      LSV03850
      KMAX1 = KMAX + 1                                                 LSV03860
C                                                                      LSV03870
C     READ IN THE T-MATRICES FROM FILE 2.  THESE ARE USED TO GENERATE  LSV03880
C     THE T-EIGENVECTORS THAT WILL BE USED IN THE RITZ VECTOR          LSV03890
C     COMPUTATIONS.  HISTORY MUST BE STORED IN 4Z20 FORMAT.            LSV03900
C                                                                      LSV03910
      READ(2,280) (BETA(J), J=1,KMAX1)                                 LSV03920
  280 FORMAT(4Z20)                                                     LSV03930
C                                                                      LSV03940
      READ(2,280) (V1(J), J=1,M)                                       LSV03950
      READ(2,280) (V2(J), J=1,N)                                       LSV03960
C                                                                      LSV03970
```

```
C       KMAX MAY BE ENLARGED IF THE SIZE AT WHICH THE SINGULAR VALUE      LSV03980
C       COMPUTATIONS WERE PERFORMED IS ESSENTIALLY KMAX AND              LSV03990
C       THERE IS AT LEAST ONE SINGULAR VALUE THAT IS SIMPLE AS AN        LSV04000
C       EIGENVALUE OF T(1,MEV), AND IF ITS NEAREST NEIGHBOR IN THE       LSV04010
C       T-MATRIX IS TOO CLOSE, THAT NEIGHBOR IS A 'GOOD' T-EIGENVALUE.   LSV04020
        DO 290 J = 1,NGOOD                                               LSV04030
        IF(MP(J).EQ.1) GO TO 310                                         LSV04040
  290 CONTINUE                                                           LSV04050
      WRITE(6,300)                                                       LSVC4060
  300 FORMAT(/' ALL SINGULAR VALUES USED ARE T-MULTIPLE OR CLOSE TO SPURLSV04070
     1IOUS EIGENVALUES'/' (AS EIGENVALUES OF T(1,MEV)) SO KMAX IS NOT CHLSV04080
     1ANGED'/)                                                          LSV04090
      IF(KMAX.LT.MEV) GO TO 1660                                         LSV04100
      GO TO 330                                                          LSV04110
C                                                                        LSV04120
  310 KMAXN= (11*MEV)/8 + 12                                             LSV04130
      IF((KMAXN/2)*2.NE.KMAXN) KMAXN = KMAXN + 1                         LSV04140
      IF(MBETA.LE.KMAXN) GO TO 1840                                      LSV04150
      IF(KMAX.GE.KMAXN )  GO TO 330                                      LSV04160
      WRITE(6,320) KMAX, KMAXN                                           LSV04170
  320 FORMAT(' ENLARGE KMAX FROM ',I6,' TO ',I6)                         LSV04180
      MOLD1 = KMAX + 1                                                   LSV04190
      KMAX = KMAXN                                                       LSV04200
      GO TO 420                                                          LSV04210
C                                                                        LSV04220
  330 WRITE(6,340) KMAX                                                  LSV04230
  340 FORMAT(/' T-MATRICES HAVE BEEN READ IN FROM FILE 2'/' THE LARGEST LSV04240
     1SIZE T-MATRIX ALLOWED IS',I6/)                                     LSV04250
C                                                                        LSV04260
      IF(IREAD.EQ.1) GO TO 460                                           LSV04270
C                                                                        LSV04280
C     REGENERATE THE BETA                                               LSV04290
C                                                                        LSV04300
  350 MOLD1 = 1                                                          LSV04310
C                                                                        LSV04320
      IF(IPAR.EQ.1) WRITE(6,360)                                         LSV04330
      IF(IPAR.EQ.2) WRITE(6,370)                                         LSV04340
  360 FORMAT(/' STARTING VECTOR USED IN HISTORY REGENERATION IS OF THE   LSV04350
     1FORM (0,V2)')                                                     LSV04360
  370 FORMAT(/' STARTING VECTOR USED IN HISTORY REGENERATION IS OF THE   LSV04370
     1FORM (V1,0)')                                                     LSV04380
C                                                                        LSV04390
      DO 380 J = 1,NGOOD                                                 LSV04400
      IF(MP(J).EQ.1) GO TO 400                                           LSV04410
  380 CONTINUE                                                           LSV04420
      KMAX = MEV + 12                                                    LSV04430
      IF((KMAX/2)*2.NE.KMAX) GO TO 1680                                  LSV04440
      WRITE(6,390) KMAX                                                  LSV04450
  390 FORMAT(/' ALL SINGULAR VALUES FOR WHICH SINGULAR VECTORS ARE TO BELSV04460
     1COMPUTED ARE EITHER T-MULTIPLE OR CLOSE TO'/' A SPURIOUS T-EIGENVALSV04470
     1LUE THEREFORE SET KMAX = MEV + 12 = ',I7)                         LSV04480
      GO TO 420                                                          LSV04490
C                                                                        LSV04500
  400 KMAXN = (11*MEV)/8 + 12                                            LSV04510
      IF((KMAXN/2)*2.NE.KMAXN) KMAXN = KMAXN + 1                         LSV04520
      IF(MBETA.LE.KMAXN) GO TO 1840                                      LSV04530
      WRITE(6,410) KMAXN                                                 LSV04540
  410 FORMAT(' SET KMAX EQUAL TO ',I6)                                   LSV04550
      KMAX = KMAXN                                                       LSV04560
C                                                                        LSV04570
  420 KMAX1 = KMAX + 1                                                   LSV04580
      WRITE(6,430) MOLD1,KMAX1                                           LSV04590
  430 FORMAT(/' LANCZS SUBROUTINE GENERATES BETA(J+1), J =',             LSV04600
     1 I6,' TO ', I6/)                                                  LSV04610
      IF(IREAD.EQ.1.AND.IPAR.EQ.1) WRITE(6,440)                          LSV04620
      IF(IREAD.EQ.1.AND.IPAR.EQ.2) WRITE(6,450)                          LSV04630
  440 FORMAT(/' FIRST LANCZOS VECTOR IN HISTORY EXTENSION IF OF THE FORMLSV04640
```

```
      1 (0,V2)')                                                    LSV04650
   450 FORMAT(/' FIRST LANCZOS VECTOR IN HISTORY EXTENSION IF OF THE FORMLSV04660
      1 (V1,0)')                                                    LSV04670
C                                                                   LSV04680
C-------------------------------------------------------------------LSV04690
C                                                                   LSV04700
      CALL LANCZS(SVMAT,STRAN,BETA,V1,V2,G,KMAX,MOLD1,M,N,IPAR,SVSEED) LSV04710
C                                                                   LSV04720
C-------------------------------------------------------------------LSV04730
C                                                                   LSV04740
   460 CONTINUE                                                     LSV04750
C                                                                   LSV04760
C     THE SUBROUTINE STURMI DETERMINES THE SMALLEST SIZE T-MATRIX FOR LSV04770
C     WHICH THE SINGULAR VALUE IN QUESTION IS AN EIGENVALUE (TO      LSV04780
C     WITHIN A SPECIFIED TOLERANCE) AND IF POSSIBLE THE SMALLEST     LSV04790
C     SIZE T-MATRIX FOR WHICH THE SINGULAR VALUE IS A DOUBLE         LSV04800
C     EIGENVALUE (TO WITHIN THE SAME TOLERANCE).  THE SIZE           LSV04810
C     T-MATRIX THAT WILL BE USED IN EACH OF THE RITZ VECTOR COMPUTATIONSLSV04820
C     IS THEN DETERMINED BY LOOPING ON THE SIZE OF THE T-EIGENVECTOR LSV04830
C     COMPUTATIONS, STARTING WITH A SIZE DETERMINED FROM THE         LSV04840
C     INFORMATION OBTAINED FROM STURMI.                             LSV04850
C                                                                   LSV04860
      STUTOL = SCALEO*MULTOL                                        LSV04870
      IF(IWRITE.EQ.1) WRITE(6,470)                                  LSV04880
   470 FORMAT(' FROM STURMI')                                       LSV04890
      DO 510 J = 1,NGOOD                                            LSV04900
      SVAL = GOODSV(J)                                              LSV04910
C     COMPUTE THE TOLERANCES USED BY STURMI TO DETERMINE AN INTERVAL LSV04920
C     CONTAINING THE SINGULAR VALUE SVAL.                           LSV04930
      TEMP = DABS(SVAL)*RELTOL                                      LSV04940
      TOLN = DMAX1(TEMP,STUTOL)                                     LSV04950
C                                                                   LSV04960
C-------------------------------------------------------------------LSV04970
C                                                                   LSV04980
      CALL STURMI(BETA,SVAL,TOLN,EPSM,KMAX,MK1,MK2,IC,IWRITE)       LSV04990
C                                                                   LSV05000
C-------------------------------------------------------------------LSV05010
C                                                                   LSV05020
C     STORE THE COMPUTED ORDERS OF T-MATRICES FOR LATER PRINTOUT    LSV05030
      IF(MK1.GT.1) GO TO 475                                        LSV05040
C     SVAL IS VERY SMALL SINGULAR VALUE, RESET MK1 TO CORRECT VALUE LSV05050
      MK1 = MK2                                                     LSV05060
      MK2 = MINO(2*MK1,KMAX)                                        LSV05070
      M1(J) = MK1                                                   LSV05080
      M2(J) = MK2                                                   LSV05090
      ML(J) = MK2                                                   LSV05100
      GO TO 476                                                     LSV05110
   475 M1(J) = MK1                                                  LSV05120
      M2(J) = MK2                                                   LSV05130
      ML(J) = (MK1 + 3*MK2)/4                                       LSV05140
      IF(MK2.EQ.KMAX)  ML(J) = KMAX                                 LSV05150
C                                                                   LSV05160
   476 IF(IC.GT.0) GO TO 490                                        LSV05170
C     IC = 0 MEANS THERE WAS NO T-EIGENVALUE IN THE DESIGNATED INTERVAL LSV05180
C     EVEN BY T-SIZE KMAX.  THIS MEANS THAT THE SINGULAR VALUE       LSV05190
C     PROVIDED HAS NOT YET CONVERGED SO PROGRAM DOES NOT COMPUTE     LSV05200
C     A SINGULAR VECTOR FOR IT.                                     LSV05210
      WRITE(6,480) J,GOODSV(J),MK1,MK2                              LSV05220
   480 FORMAT(I6,'TH SINGULAR VALUE',E20.12,' HAS NOT CONVERGED '/  LSV05230
      1' SO DO NOT COMPUTE ANY T-EIGENVECTOR OR RITZ VECTOR FOR IT'  LSV05240
      1/' MK1 AND MK2 FOR THIS SINGULAR VALUE WERE',2I6)            LSV05250
      MP(J) = MPMIN                                                 LSV05260
      MA(J) = -2*KMAX                                               LSV05270
      GO TO 510                                                     LSV05280
C     COMPUTE AN APPROPRIATE SIZE T-MATRIX FOR THE GIVEN SINGULAR    LSV05290
C     VALUE.                                                        LSV05300
```

```
      490 IF(M2(J).EQ.KMAX) GO TO 500                                LSV05310
    C      M1 AND M2 WERE BOTH DETERMINED                            LSV05320
          MAJ = (3*M1(J) + M2(J))/4  + 1                             LSV05330
          IF((MAJ/2)*2.NE.MAJ)  MAJ = MAJ + 1                        LSV05340
          MA(J) = MAJ                                                LSV05350
          GO TO 510                                                  LSV05360
    C      M2 NOT DETERMINED                                         LSV05370
      500 MAJ = (5*M1(J))/4  + 1                                     LSV05380
          IF((MAJ/2)*2.NE.MAJ)  MAJ = MAJ + 1                        LSV05390
          MA(J) = MAJ                                                LSV05400
    C                                                                LSV05410
      510 CONTINUE                                                   LSV05420
    C                                                                LSV05430
          IF (IWRITE.EQ.1) WRITE(6,520) (MA(JJ), JJ=1,NGOOD)         LSV05440
      520 FORMAT(/' 1ST GUESS AT APPROPRIATE SIZE T-MATRICES'/       LSV05450
         1 ' ACTUAL VALUES WILL PROBABLY BE 1/4 AGAIN AS MUCH'/(13I6)) LSV05460
    C                                                                LSV05470
    C      PRINT OUT TO FILE 10 1ST GUESSES AT SIZES OF THE T-MATRICES TO LSV05480
    C      BE USED IN THE SINGULAR VECTOR COMPUTATIONS.              LSV05490
    C      PROGRAM LOOPS ON T-SIZE TO DETERMINE APPROPRIATE SIZE T-MATRIX. LSV05500
          WRITE(10,530) N,KMAX                                       LSV05510
      530 FORMAT(2I8,' = ORDER OF USER MATRIX AND MAX ORDER OF T(1,MEV)') LSV05520
    C                                                                LSV05530
          WRITE(10,540)                                              LSV05540
      540 FORMAT(/' 1ST GUESS AT APPROPRIATE SIZE T-MATRICES'/       LSV05550
         1 ' ACTUAL VALUES WILL PROBABLY BE 1/4 AGAIN AS MUCH'/)     LSV05560
    C                                                                LSV05570
          WRITE(10,550)                                              LSV05580
      550 FORMAT(4X,'J',7X,'GOODSV(J)',4X,'M1(J)',1X,'M2(J)',1X,'MA(J)') LSV05590
    C                                                                LSV05600
          WRITE(10,560) (J,GOODSV(J),M1(J),M2(J), MA(J), J=1,NGOOD)  LSV05610
      560 FORMAT(I5,E19.12,3I6)                                      LSV05620
    C                                                                LSV05630
          IF(MBOUND.EQ.1) WRITE(10,570)                              LSV05640
      570 FORMAT(/' GOODSV(J) IS A GOOD EIGENVALUE OF T(1,MEV)'/      LSV05650
         1 ' M1 = SMALLEST VALUE OF M SUCH THAT T(1,M) HAS AT LEAST'/ LSV05660
         1 '      ONE EIGENVALUE IN THE INTERVAL (SV-TOLN,SV+TOLN)'/  LSV05670
         1 '      TOLN(J) = DMAX1(GOODSV(J)*RELTOL, SCALEO*MULTOL)'/  LSV05680
         1 ' M2 = SMALLEST M (IF ANY) SUCH THAT IN THE ABOVE INTERVAL'/ LSV05690
         1 '      T(1,M) HAS AT LEAST TWO EIGENVALUES '/             LSV05700
         1 ' INITIAL VALUE OF MA(J) IS CHOSEN HEURISTICALLY'/        LSV05710
         1 ' PROGRAM LOOPS ON SIZE OF T-MATRIX TO GET APPROPRIATE SIZE'/ LSV05720
         1 ' END OF SIZES OF T-MATRICES FILE 10'///)                 LSV05730
    C                                                                LSV05740
    C                                                                LSV05750
    C      TERMINATE AFTER COMPUTING 1ST GUESSES AT  SIZES OF THE    LSV05760
    C      T-MATRICES REQUIRED FOR THE GIVEN SINGULAR VALUES?        LSV05770
          IF(MBOUND.EQ.1) GO TO 1700                                 LSV05780
    C                                                                LSV05790
    C                                                                LSV05800
    C      IS THERE ROOM FOR ALL OF THE REQUESTED T-EIGENVECTORS?    LSV05810
          MTOL = 0                                                   LSV05820
          DO 580 J = 1,NGOOD                                         LSV05830
          IF(MP(J).EQ.MPMIN) GO TO 580                               LSV05840
          MTOL = MTOL + IABS(MA(J))                                  LSV05850
      580 CONTINUE                                                   LSV05860
          MTOL = (5*MTOL)/4                                          LSV05870
          IF(MTOL.GT.MDIMTV.AND.NTVCON.EQ.0) GO TO 1720              LSV05880
    C                                                                LSV05890
    C----------------------------------------------------------------LSV05900
    C      GENERATE A RANDOM VECTOR TO BE USED REPEATEDLY BY         LSV05910
    C      SUBROUTINE INVERM                                         LSV05920
    C                                                                LSV05930
          IIL = RHSEED                                               LSV05940
          CALL GENRAN(IIL,G,KMAX)                                    LSV05950
    C                                                                LSV05960
    C----------------------------------------------------------------LSV05970
```

```
C                                                                      LSV05980
C       FOR EACH SINGULAR VALUE LOOP ON T-EIGENVECTOR COMPUTATIONS      LSV05990
C       TO COMPUTE AN APPROPRIATE T-EIGENVECTOR TO USE IN THE           LSV06000
C       RITZ VECTOR COMPUTATIONS.                                       LSV06010
C                                                                       LSV06020
        MTOL = 0                                                        LSV06030
        NTVEC = 0                                                       LSV06040
        ILBIS = 0                                                       LSV06050
        DO 770 J = 1,NGOOD                                              LSV06060
        ICOUNT = 0                                                      LSV06070
        ERRMIN = 10.DO                                                  LSV06080
        MABEST = MPMIN                                                  LSV06090
        IF(MP(J).EQ.MPMIN) GO TO 770                                    LSV06100
        TFLAG = 0                                                       LSV06110
        SVAL = GOODSV(J)                                                LSV06120
        TEMP = RELTOL*DABS(SVAL)                                        LSV06130
        UB = SVAL + DMAX1(STUTOL,TEMP)                                  LSV06140
        LB = SVAL - DMAX1(STUTOL,TEMP)                                  LSV06150
        LB = DMAX1(LB,ZERO)                                             LSV06160
   590 KMAXU = IABS(MA(J))                                              LSV06170
C                                                                       LSV06180
C       SELECT A SUITABLE INCREMENT FOR THE ORDERS OF THE T-MATRICES    LSV06190
C       TO BE CONSIDERED IN DETERMINING APPROPRIATE SIZES FOR THE RITZ  LSV06200
C       VECTOR COMPUTATIONS.  ALL ORDERS CONSIDERED MUST BE EVEN.       LSV06210
        IF(ICOUNT.GT.0) GO TO 610                                       LSV06220
C       SELECT IDELTA(J) BASED UPON THE T-MULTIPLICITY OBTAINED         LSV06230
        IF(M2(J).EQ.KMAX) GO TO 600                                     LSV06240
C       M2 DETERMINED                                                   LSV06250
        IDEL = ((3*M1(J) + 5*M2(J))/8  + 1 - IABS(MA(J)))/10  + 1       LSV06260
        IF((IDEL/2)*2.NE.IDEL) IDEL = IDEL + 1                          LSV06270
        IDELTA(J) = IDEL                                                LSV06280
        GO TO 610                                                       LSV06290
C       M2 NOT DETERMINED                                               LSV06300
   600 MAMAX = MINO((11*MEV)/8 + 12, (13*M1(J))/8 + 1)                  LSV06310
        IDEL = (MAMAX - IABS(MA(J)))/10  + 1                            LSV06320
        IF((IDEL/2)*2.NE.IDEL)  IDEL = IDEL + 1                         LSV06330
        IDELTA(J) = IDEL                                                LSV06340
   610 ICOUNT = ICOUNT + 1                                              LSV06350
C                                                                       LSV06360
C----------------------------------------------------------------------LSV06370
C       TO MIMIMIZE THE EFFECT OF THE ONE-SIDED ACCEPTANCE TEST FOR     LSV06380
C       EIGENVALUES IN THE BISEC SUBROUTINE, RECOMPUTE THE GIVEN        LSV06390
C       SINGULAR VALUE AT THE SPECIFIED KMAXU                           LSV06400
C                                                                       LSV06410
        CALL LBISEC(BETA,EPSM,SVAL,SVALN,LB,UB,TTOL,KMAXU,NEVT)         LSV06420
C                                                                       LSV06430
C----------------------------------------------------------------------LSV06440
C                                                                       LSV06450
C       CHECK WHETHER OR NOT GIVEN T-MATRIX HAS AN EIGENVALUE IN THE    LSV06460
C       SPECIFIED INTERVAL AND IF SO WHAT ITS T-MULTIPLICITY IS.        LSV06470
C                                                                       LSV06480
        IF(NEVT.EQ.1) GO TO 650                                         LSV06490
        IF(NEVT.NE.0)  GO TO 630                                        LSV06500
        ILBIS = 1                                                       LSV06510
        WRITE(6,620) SVAL,KMAXU                                         LSV06520
   620 FORMAT(/' PROBLEM ENCOUNTERED IN RECOMPUTATION OF USER-SUPPLIED SILSV06530
       1NGULAR VALUE',E20.12/' THE SIZE T-MATRIX SPECIFIED',I6,' DOES NOT LSV06540
       1HAVE A SINGULAR VALUE IN THE INTERVAL SPECIFIED'/' INCREASE SIZE ALSV06550
       1ND TRY AGAIN'/)                                                 LSV06560
        GO TO 670                                                       LSV06570
C                                                                       LSV06580
   630 IF(NEVT.GT.1)  WRITE(6,640) SVAL,KMAXU                           LSV06590
   640 FORMAT(/' PROBLEM ENCOUNTERED IN RECOMPUTATION OF USER-SUPPLIED   LSV06600
       1SINGULAR VALUE',E20.12/' FOR THE SIZE T-MATRIX SPECIFIED =',I6,' TLSV06610
       1HE GIVEN SINGULAR VALUE IS T-MULTIPLE IN THE INTERVAL SPECIFIED'/'LSV06620
       1SOMETHING IS WRONG, THEREFORE NO SINGULAR VECTORS WILL BE COMPUTEDLSV06630
```

```
      1 FOR THIS SINGULAR VALUE'/)                                LSV06640
C                                                                 LSV06650
         MP(J) = MPMIN                                            LSV06660
         MA(J) = -2*KMAX                                          LSV06670
         GO TO 770                                                LSV06680
C                                                                 LSV06690
  650 CONTINUE                                                    LSV06700
         ILBIS = 0                                                LSV06710
C                                                                 LSVC6720
C                                                                 LSV06730
         SVNEW(J) = SVALN                                         LSV06740
         SVAL = SVALN                                             LSV06750
         MTOL = MTOL+KMAXU                                        LSV06760
C                                                                 LSV06770
C     IS THERE ROOM IN TVEC ARRAY FOR THE NEXT T-EIGENVECTOR?     LSV06780
C     IF NOT, SKIP TO RITZ VECTOR COMPUTATIONS.                   LSV06790
      IF (MTOL.GT.MDIMTV) GO TO 780                               LSV06800
C                                                                 LSV06810
         IT = 3                                                   LSV06820
         KINT = MTOL - KMAXU +1                                   LSV06830
C                                                                 LSV06840
C     RECORD THE BEGINNING AND END OF THE T-EIGENVECTOR BEING COMPUTED LSV06850
         MINT(J) = KINT                                           LSV06860
         MFIN(J) = MTOL                                           LSV06870
C                                                                 LSV06880
C----------------------------------------------------------------LSV06890
C     SUBROUTINE INVERM DOES INVERSE ITERATION, I.E. SOLVES       LSV06900
C     (T(1,KMAXU) - SVAL)*U = RHS  FOR EACH SINGULAR VALUE TO     LSV06910
C     OBTAIN THE DESIRED T-EIGENVECTOR.                           LSV06920
C                                                                 LSV06930
      IF(IWRITE.EQ.1)  WRITE(6,660) J                             LSV06940
  660 FORMAT(/I6,'TH SINGULAR VALUE ')                            LSV06950
C                                                                 LSV06960
         CALL INVERM(BETA,V1,TVEC(KINT),SVAL,ERROR,TERROR,EPSM,G,KMAXU, LSV06970
      1 IT,IWRITE)                                                LSV06980
C                                                                 LSV06990
C----------------------------------------------------------------LSV07000
C                                                                 LSV07010
         TERR(J) = TERROR                                         LSV07020
         TLAST(J) = ERROR                                         LSV07030
         KMAXU1 = KMAXU + 1                                       LSV07040
         TBETA(J) = BETA(KMAXU1)*ERROR                            LSV07050
C                                                                 LSV07060
C     AFTER COMPUTING EACH OF THE T-EIGENVECTORS,                 LSV07070
C     CHECK THE SIZE OF THE ERROR ESTIMATE, ERROR.                LSV07080
C     IF THIS ESTIMATE IS NOT AS SMALL AS DESIRED AND             LSV07090
C     |MA(J)| < ML(J), ATTEMPT. TO INCREASE THE SIZE OF |MA(J)|   LSV07100
C     AND REPEAT THE T-EIGENVECTOR COMPUTATIONS.                  LSV07110
C                                                                 LSV07120
      IF(ERROR.LT.ERTOL.OR.TFLAG.EQ.1)  GO TO 760                 LSV07130
C                                                                 LSV07140
      IF(ERROR.GE.ERRMIN) GO TO 670                               LSV07150
C     LAST COMPONENT IS LESS THAN MINIMAL TO DATE                 LSV07160
         ERRMIN = ERROR                                           LSV07170
         MABEST = MA(J)                                           LSV07180
  670 CONTINUE                                                    LSV07190
C                                                                 LSV07200
      IF(MA(J).GT.0)  ITEST = MA(J) + IDELTA(J)                   LSV07210
      IF(MA(J).LT.0)  ITEST = -(IABS(MA(J)) + IDELTA(J))          LSV07220
      IF(IABS(ITEST).LE.ML(J).AND.ICOUNT.LE.10) GO TO 690         LSV07230
C     NEW MA(J) IS GREATER THAN MAXIMUM ALLOWED.                  LSV07240
      IF(ERCONT.EQ.0.OR.MABEST.EQ.MPMIN) GO TO 710                LSV07250
         TFLAG = 1                                                LSV07260
         MA(J) = MABEST                                           LSV07270
         IF(ILBIS.EQ.0)   MTOL = MTOL - KMAXU                     LSV07280
         WRITE(6,680) MA(J)                                       LSV07290
  680 FORMAT(' 10 ORDERS WERE CONSIDERED.  NONE SATISFIED THE ERROR TESTLSV07300
```

```
         1'/' THEREFORE USE THE BEST ORDER OBTAINED FOR THE T-EIGENVECTORS' LSV07310
         1,16)                                                              LSV07320
            GO TO 590                                                       LSV07330
C                                                                           LSV07340
      690 MA(J) = ITEST                                                     LSV07350
C                                                                           LSV07360
            MT = IABS(MA(J))                                                LSV07370
            IF(IWRITE.EQ.1.AND.ILBIS.EQ.0) WRITE(6,700) MT                  LSV07380
      700 FORMAT(/' CHANGE SIZE OF T-MATRIX TO ',I6,' RECOMPUTE T-EIGENVECTOLSV07390
         1R')                                                               LSV07400
C                                                                           LSV07410
            IF(ILBIS.EQ.0)  MTOL = MTOL - KMAXU                             LSV07420
C                                                                           LSV07430
            GO TO 590                                                       LSV07440
C                                                                           LSV07450
C         APPROPRIATE SIZE T-MATRIX WAS NOT OBTAINED                        LSV07460
      710 CONTINUE                                                          LSV07470
            WRITE(10,720) J,SVAL,MP(J)                                      LSV07480
      720 FORMAT(/' ON 10 INCREMENTS NOT ABLE TO IDENTIFY APPROPRIATE SIZE  LSV07490
         1T-MATRIX FOR'/                                                    LSV07500
         114,' TH SINGULAR VALUE  = ',E20.12,' T-MULTIPLICITY =',I4/)       LSV07510
            IF(M2(J).EQ.KMAX) WRITE(10,730)                                 LSV07520
            IF(M2(J).LT.KMAX) WRITE(10,740)                                 LSV07530
      730 FORMAT(/' ORDERS TESTED RANGED FROM 5*M1(J)/4 TO APPROXIMATELY'/   LSV07540
         1 '   MIN(11*MEV/8, 13*M1(J)/8)'/)                                 LSV07550
      740 FORMAT(/' ORDERS TESTED RANGED FROM (3*M1(J)+M2(J))/4  TO APPROXIMLSV07560
         1ATELY'/'  (3*M1(J)+5*M2(J))/8'/)                                  LSV07570
            WR:TE(10,750)                                                   LSV07580
      750 FORMAT(' ALLOWING LARGER ORDERS FOR THE T-MATRICES MAY RESULT IN  LSV07590
         1 SUCCESS'/' BUT PROBABLY WILL NOT.  PROBLEM IS PROBABLY DUE TO'   LSV07600
         1 /' LACK OF CONVERGENCE OF GIVEN SINGULAR VALUE, CHECK THE ERROR ELSV07610
         1STIMATE')                                                         LSV07620
            MP(J) = MPMIN                                                   LSV07630
            IF(ILBIS.EQ.0)  MTOL = MTOL - KMAXU                             LSV07640
            GO TO 770                                                       LSV07650
      760 NTVEC = NTVEC + 1                                                 LSV07660
C                                                                           LSV07670
      770 CONTINUE                                                          LSV07680
            NGOODC = NGOOD                                                  LSV07690
            GO TO 800                                                       LSV07700
C                                                                           LSV07710
C         COME HERE IF THERE IS NOT ENOUGH ROOM FOR ALL OF T-EIGENVECTORS   LSV07720
      780 NGOODC = J-1                                                      LSV07730
            WRITE(6,790) J,MTOL,MDIMTV                                      LSV07740
      790 FORMAT(/' NOT ENOUGH ROOM IN TVEC ARRAY FOR ',I4,'TH T-EIGENVECTORLSV07750
         1'/' TVEC DIMENSION REQUESTED = ',I6,' BUT TVEC HAS DIMENSION ',I6LSV07760
         1/)                                                               LSV07770
            IF(NGOODC.EQ.0) GO TO 1740                                      LSV07780
            MTOL = MTOL-KMAXU                                               LSV07790
C                                                                           LSV07800
      800 CONTINUE                                                          LSV07810
C                                                                           LSV07820
C         THE LOOP ON T-EIGENVECTOR COMPUTATIONS IS COMPLETE.               LSV07830
C         WRITE OUT THE SIZE T-MATRICES THAT WILL BE USED FOR               LSV07840
C         THE RITZ VECTOR COMPUTATIONS.                                     LSV07850
C                                                                           LSV07860
            WRITE(10,810)                                                   LSV07870
      810 FORMAT(/' SIZES OF T-MATRICES THAT WILL BE USED IN THE RITZ COMPUTLSV07880
         1ATIONS'/5X,'J',8X,' SINGULAR VALUE  ',1X,'MA(J)')                 LSV07890
C                                                                           LSV07900
            WRITE(10,820)   (J,GOODSV(J),MA(J),  J=1,NGOOD)                 LSV07910
      820 FORMAT(I6,E25.14,I6)                                              LSV07920
            WR:TE(10,570)                                                   LSV07930
C                                                                           LSV07940
            WRITE(6,830) MTOL                                               LSV07950
      830 FORMAT(/' THE CUMULATIVE LENGTH OF THE T-EIGENVECTORS IS',I18)     LSV07960
C                                                                           LSV07970
```

```
      WRITE(6,840) NTVEC,NGOOD                                      LSV07980
  840 FORMAT(/I6,' T-EIGENVECTORS OUT OF',I6,' REQUESTED WERE COMPUTED')LSV07990
C                                                                   LSV08000
C     SAVE THE T-EIGENVECTORS ON FILE 11?                           LSV08010
      IF(TVSTOP.NE.1.AND.SVTVEC.EQ.0) GO TO 900                     LSV08020
C                                                                   LSV08030
      WRITE(11,850) NTVEC,MTOL,MATNO,SVSEED                         LSV08040
  850 FORMAT(I6,3I12,' = NTVEC,MTOL,MATNO,SVSEED')                  LSV08050
C                                                                   LSV08060
      DO 880 J=1,NGOODC                                             LSV08070
C     IF MP(J) = MPMIN THEN NO SUITABLE T-EIGENVECTOR IS AVAILABLE  LSV08080
C     FOR THAT SINGULAR VALUE.                                      LSV08090
      IF(MP(J).EQ.MPMIN) WRITE(11,860) J,MA(J),GOODSV(J),MP(J)      LSV08100
  860 FORMAT(2I6,E20.12,I6/' TH SINGVAL,T-SIZE,SVALUE,FLAG,NO EIGVEC') LSV08110
      IF(MP(J).NE.MPMIN) WRITE(11,870) J,MA(J),GOODSV(J),MP(J)      LSV08120
  870 FORMAT(I6,I6,E20.12,I6/' T-EIGVEC,SIZE T,SVALUE OF A,MP(J)')  LSV08130
      IF(MP(J).EQ.MPMIN) GO TO 880                                 LSV08140
      KI = MINT(J)                                                  LSV08150
      KF = MFIN(J)                                                  LSV08160
C                                                                   LSV08170
      WRITE(11,280) (TVEC(K), K=KI,KF)                             LSV08180
C                                                                   LSV08190
  880 CONTINUE                                                      LSV08200
C                                                                   LSV08210
      IF(TVSTOP.NE.1) GO TO 900                                     LSV08220
C                                                                   LSV08230
      WRITE(6,890) TVSTOP, NTVEC,NGOOD                              LSV08240
  890 FORMAT(/' USER SET TVSTOP = ',I1/                             LSV08250
     1' THEREFORE PROGRAM TERMINATES AFTER T-EIGENVECTOR COMPUTATIONS'/ LSV08260
     1' T-EIGENVECTORS THAT WERE COMPUTED ARE SAVED ON FILE 11'/    LSV08270
     1I8,' T-EIGENVECTORS WERE COMPUTED OUT OF',I7,' REQUESTED'/)   LSV08280
C                                                                   LSV08290
      GO TO 1860                                                    LSV08300
C                                                                   LSV08310
  900 CONTINUE                                                      LSV08320
C     IF NOT ALL OF THE REQUESTED T-EIGENVECTORS WERE COMPUTED,     LSV08330
C     ARE THE LANCZOS SINGULAR VECTOR COMPUTATIONS CONTINUED?       LSV08340
C                                                                   LSV08350
      IF(NTVEC.NE.NGOOD.AND.LVCONT.EQ.0) GO TO 1760                 LSV08360
C                                                                   LSV08370
C     COMPUTE THE MAXIMUM SIZE OF THE T-MATRIX USED FOR THOSE       LSV08380
C     SINGULAR VALUES WITH GOOD ERROR ESTIMATES.                    LSV08390
C                                                                   LSV08400
      KMAXU = 0                                                     LSV08410
      DO 910 J = 1,NGOODC                                           LSV08420
      MT = IABS(MA(J))                                              LSV08430
      IF(MT.LT.KMAXU.OR.MP(J).EQ.MPMIN) GO TO 910                   LSV08440
      KMAXU = MT                                                    LSV08450
  910 CONTINUE                                                      LSV08460
C                                                                   LSV08470
      IF(KMAXU.EQ.0) GO TO 1800                                     LSV08480
C                                                                   LSV08490
      WRITE(6,920) KMAXU                                            LSV08500
  920 FORMAT(/I6,' = LARGEST SIZE T-MATRIX TO BE USED IN THE RITZ VECTORLSV08510
     1 COMPUTATIONS')                                               LSV08520
C                                                                   LSV08530
C     COUNT THE NUMBER OF RITZ VECTORS NOT BEING COMPUTED           LSV08540
      MREJEC = 0                                                    LSV08550
      DO 930 J=1,NGOODC                                             LSV08560
  930 IF(MP(J).EQ.MPMIN)  MREJEC = MREJEC + 1                       LSV08570
      MREJET = MREJEC + (NGOOD-NGOODC)                              LSV08580
      IF(MREJET.NE.0) WRITE(6,940) MREJET                          LSV08590
  940 FORMAT(/' RITZ VECTORS ARE NOT COMPUTED FOR',I6,' OF THE SINGULAR LSV08600
     1VALUES'/)                                                     LSV08610
      NACT = NGOODC - MREJEC                                        LSV08620
      WRITE(6,950) NGOOD,NTVEC,NACT                                 LSV08630
  950 FORMAT(/I6,' RITZ VECTORS WERE REQUESTED'/I6,' T-EIGENVECTORS WERELSV08640
```

```
      1 COMPUTED'/I6,' RITZ VECTORS WILL BE COMPUTED'/)          LSV08650
C       CHECK IF THERE ARE ANY RITZ VECTORS TO COMPUTE             LSV08660
        IF(MREJEC.EQ.NGOODC) GO TO 1780                            LSV08670
C                                                                  LSV08680
C       CONTINUE WITH THE LANCZOS VECTOR COMPUTATIONS?             LSV08690
        IF(LVCONT.EQ.0.AND.MREJEC.NE.0) GO TO 1760                 LSV08700
C                                                                  LSV08710
C       NOW COMPUTE THE RITZ VECTORS. "REGENERATE THE              LSV08720
C       LANCZOS VECTORS.                                           LSV08730
C                                                                  LSV08740
        DO 960 I = 1,MNMAX                                         LSV08750
    960 RITVEC(I) = ZERO                                           LSV08760
C                                                                  LSV08770
C       REGENERATE THE STARTING VECTOR. THIS MUST BE GENERATED AND LSV08780
C       NORMALIZED PRECISELY THE WAY IT WAS DONE IN THE CORRESPONDINGLSV08790
C       SINGULAR VALUE COMPUTATIONS, OTHERWISE THERE WILL BE A     LSV08800
C       MISMATCH BETWEEN THE T-EIGENVECTORS THAT HAVE BEEN COMPUTED LSV08810
C       FROM THE T-MATRICES READ IN FROM FILE 2 (IF THEY WERE READ IN)LSV08820
C       AND THE LANCZOS TRIDIAGONAL MATRICES  THAT ARE BEING REGENERATED.LSV08830
C                                                                  LSV08840
C       STARTING VECTORS ARE OF THE FORM (V1,0) OR (0,V2) WHERE V1 IS LSV08850
C       OF LENGTH M AND V2 IS OF LENGTH N.  SUCCEEDING LANCZOS VECTORS LSV08860
C       ALTERNATE BETWEEN THESE TWO FORMS AND THE DIAGONAL ENTRIES OF THE LSV08870
C       T-MATRICES ALL VANISH.   THE PARAMETER IPARO DETERMINES THE SHAPE LSV08880
C       OF THE STARTING VECTOR.  IF IPARO=1, THEN STARTING VECTOR WAS LSV08890
C       OF THE FORM (0,V2).   IF IPARO=2, THEN STARTING VECTOR WAS OF LSV08900
C       THE FORM (V1,0).                                           LSV08910
C       REGENERATE STARTING VECTOR                                 LSV08920
        BATA = ZERO                                                LSV08930
        IPAR = IPARO                                               LSV08940
        ITNUM = 1                                                  LSV08950
        IF (IPAR.EQ.2) GO TO 1020                                  LSV08960
C                                                                  LSV08970
C------------------------------------------------------------------LSV08980
C       IPAR = 1 SO SET V2 TO RANDOM UNIT VECTOR AND SET V1 = 0.   LSV08990
C                                                                  LSV09000
        IIL = SVSEED                                               LSV09010
        CALL GENRAN(IIL,G,N)                                       LSV09020
C                                                                  LSV09030
C------------------------------------------------------------------LSV09040
C                                                                  LSV09050
        DO 970 J = 1,N                                             LSV09060
    970 V2(J) = G(J)                                               LSV09070
C------------------------------------------------------------------LSV09080
        SUM = ONE/DSQRT(FINPRO(N,V2,1,V2,1))                       LSV09090
C------------------------------------------------------------------LSV09100
C                                                                  LSV09110
        DO 980 J = 1,M                                             LSV09120
    980 V1(J) = ZERO                                               LSV09130
C                                                                  LSV09140
        DO 990 J = 1,N                                             LSV09150
    990 V2(J) = V2(J)*SUM                                          LSV09160
C                                                                  LSV09170
C       INITIALIZE RITZ VECTORS                                    LSV09180
        DO 1010 J = 1,NGOODC                                       LSV09190
        IF (MP(J).EQ.MPMIN) GO TO 1010                             LSV09200
        LL = MN*J - N                                              LSV09210
        II = MINT(J)                                               LSV09220
        TEMP = TVEC(II)                                            LSV09230
C                                                                  LSV09240
        DO 1000 K = 1,N                                            LSV09250
        LL = LL + 1                                                LSV09260
   1000 RITVEC(LL) = TEMP*V2(K)                                    LSV09270
C                                                                  LSV09280
   1010 CONTINUE                                                   LSV09290
C                                                                  LSV09300
```

```
      GO TO 1150                                                  LSV09310
C                                                                 LSV09320
 1020 CONTiNUE                                                    LSV09330
C                                                                 LSV09340
C-----------------------------------------------------------------LSV09350
C    IPAR = 2 SO SET V1 TO RANDOM UNIT VECTOR AND SET V2 = 0.     LSV09360
C                                                                 LSV09370
      CALL GENRAN(SVSEED,G,M)                                     LSV09380
C                                                                 LSV09390
C-----------------------------------------------------------------LSV09400
C                                                                 LSV09410
      DO 1030 J = 1,M                                             LSV09420
 1030 V1(J) = G(J)                                                LSV09430
C-----------------------------------------------------------------LSV09440
      SUM = ONE/DSQRT(FINPRO(M,V1,1,V1,1))                        LSV09450
C-----------------------------------------------------------------LSV09460
C                                                                 LSV09470
      DO 1040 J = 1,N                                             LSV09480
 1040 V2(J) = ZERO                                                LSV09490
C                                                                 LSV09500
      DO 1050 J = 1,M                                             LSV09510
 1050 V1(J) = V1(J)*SUM                                           LSV09520
C                                                                 LSV09530
C    INITIALIZE RITZ VECTORS                                      LSV09540
      DO 1070 J = 1,NGOODC                                        LSV09550
      IF (MP(J).EQ.MPMIN) GO TO 1070                              LSV09560
      LL = MN*(J-1)                                               LSV09570
      II = MINT(J)                                                LSV09580
      TEMP = TVEC(II)                                             LSV09590
C                                                                 LSV09600
      DO 1060 K = 1,M                                             LSV09610
      LL = LL + 1                                                 LSV09620
 1060 RITVEC(LL) = TEMP*V1(K)                                     LSV09630
C                                                                 LSV09640
 1070 CONTINUE                                                    LSV09650
C                                                                 LSV09660
 1080 CONTINUE                                                    LSV09670
C                                                                 LSV09680
C DO ONE ITERATION OF LANCZOS WHERE NEW LANCZOS VECTOR WILL HAVE THE LSV09690
C FORM  (0,V2).                                                  LSV09700
C                                                                 LSV09710
C-----------------------------------------------------------------LSVC9720
C                                                                 LSV09730
      CALL STRAN(V1,V2,BATA)                                      LSV09740
C                                                                 LSV09750
C-----------------------------------------------------------------LSV09760
C                                                                 LSV09770
C-----------------------------------------------------------------LSV09780
      BATA = DSQRT(FINPRO(N,V2,1,V2,1))                           LSV09790
C-----------------------------------------------------------------LSV09800
      SUM = ONE/BATA                                              LSV09810
      ITNUM = ITNUM + 1                                           LSV09820
      IPAR = 2                                                    LSV09830
C                                                                 LSV09840
      TEMP = BETA(ITNUM)                                          LSV09850
      TEMP = DABS(BATA - TEMP)/TEMP                               LSV09860
      IF (TEMP.LT.1.0D-10) GO TO 1110                             LSV09870
C                                                                 LSV09880
C    HISTORY MISMATCH ON REGENERATION THUS DEFAULT               LSV09890
 1090 WRITE(6,1100) ITNUM,IPAR,BATA,BETA(ITNUM),TEMP             LSV09900
 1100 FORMAT(1X,'ITNUM',2X,'IPAR',16X,'BATA',16X,'BETA',14X,'RELERR'/ LSV09910
     1 216,3E20.12/' BATA AND BETA DO NOT AGREE SO PROGRAM STOPS'/) LSV09920
      GO TO 1860                                                  LSV09930
C                                                                 LSV09940
 1110 CONTINUE                                                    LSV09950
C    NORMALIZE LANCZOS VECTOR                                     LSV09960
      DO 1120 J = 1,N                                             LSV09970
```

```
 1120 V2(J) = V2(J)*SUM                                              LSV09980
C                                                                    LSV09990
C  UPDATE RITZ VECTORS                                               LSV10000
      DO 1140 J = 1,NGOODC                                           LSV10010
      IF (IABS(MA(J)).LT.ITNUM.OR.MP(J).EQ.MPMIN) GO TO 1140         LSV10020
      LL = MN*J - N                                                  LSV10030
      II = MINT(J) + ITNUM - 1                                       LSV10040
      TEMP = TVEC(II)                                                LSV10050
C                                                                    LSV10060
      DO 1130 K = 1,N                                                LSV10070
      LL = LL + 1                                                    LSV10080
 1130 RITVEC(LL) = TEMP*V2(K) + RITVEC(LL)                           LSV10090
C                                                                    LSV10100
 1140 CONTINUE                                                       LSV10110
C      HAVE ALL REQUIRED LANCZOS VECTORS BEEN REGENERATED ?          LSV10120
C                                                                    LSV10130
      IF(ITNUM.EQ.KMAXU) GO TO 1190                                  LSV10140
C                                                                    LSV10150
 1150 CONTINUE                                                       LSV10160
C                                                                    LSV10170
C      DO ONE ITERATION OF LANCZOS WHERE NEW LANCZOS VECTOR WILL HAVE LSV10180
C      THE FORM  (V1,0).                                             LSV10190
C                                                                    LSV10200
C-------------------------------------------------------------------LSV10210
C                                                                    LSV10220
      CALL SVMAT(V2,V1,BATA)                                         LSV10230
C                                                                    LSV10240
C-------------------------------------------------------------------LSV10250
C                                                                    LSV10260
C-------------------------------------------------------------------LSV10270
      BATA = DSQRT(FINPRO(M,V1,1,V1,1))                              LSV10280
C-------------------------------------------------------------------LSV10290
      SUM = ONE/BATA                                                 LSV10300
      ITNUM = ITNUM + 1                                             LSV10310
      IPAR = 1                                                       LSV10320
C                                                                    LSV10330
      TEMP = BETA(ITNUM)                                             LSV10340
      TEMP = DABS(BATA - TEMP)/TEMP                                  LSV10350
      IF (TEMP.GE.1.0D-10) GO TO 1090                                LSV10360
C                                                                    LSV10370
C      NORMALIZE LANCZOS VECTOR                                      LSV10380
      DO 1160 J = 1,M                                                LSV10390
 1160 V1(J) = V1(J)*SUM                                              LSV10400
C                                                                    LSV10410
C      UPDATE RITZ VECTORS                                           LSV10420
      DO 1180 J = 1,NGOODC                                           LSV10430
      IF (IABS(MA(J)).LT.ITNUM.OR.MP(J).EQ.MPMIN) GO TO 1180         LSV10440
      LL = MN*(J-1)                                                  LSV10450
      II = MINT(J) + ITNUM - 1                                       LSV10460
      TEMP = TVEC(II)                                                LSV10470
C                                                                    LSV10480
      DO 1170 K = 1,M                                                LSV10490
      LL = LL + 1                                                    LSV10500
 1170 RITVEC(LL) = TEMP*V1(K) + RITVEC(LL)                           LSV10510
C                                                                    LSV10520
 1180 CONTINUE                                                       LSV10530
C  HAVE ALL REQUIRED LANCZOS VECTORS BEEN COMPUTED ?                 LSV10540
      IF (ITNUM.LT.KMAXU) GO TO 1080                                 LSV10550
C                                                                    LSV10560
 1190 CONTINUE                                                       LSV10570
C                                                                    LSV10580
C      RITZVECTOR GENERATION IS COMPLETE. NORMALIZE EACH RITZVECTOR  LSV10590
C      AS AN EIGENVECTOR OF THE ASSOCIATED SYMMETRIC MATRIX B.       LSV10600
C      THEN COMPUTE THE ERRORS IN THESE VECTORS AS EIGENVECTORS      LSV10610
C      OF B AND WRITE THESE OUT TO FILE 9.  THEN INDIVIDUALLY        LSV10620
C      NORMALIZE THE FIRST M AND THE LAST N COMPONENTS OF EACH OF    LSV10630
C      THESE RITZ VECTORS AND TAKE THESE NORMALIZED VECTORS AS       LSV10640
```

```
C      RESPECTIVELY APPROXIMATIONS TO THE LEFT AND TO THE RIGHT         LSV10650
C      SINGULAR VECTORS OF THE CORRESPONDING SINGULAR VALUE OF          LSV10660
C      THE ORIGINAL MATRIX.                                             LSV10670
C                                                                       LSV10680
C                                                                       LSV10690
C      NORMALIZE THE RITZ VECTORS AS EIGENVECTORS OF B                  LSV10700
       DO 1280 J = 1,NGOODC                                             LSV10710
       IF (MP(J).EQ.MPMIN) GO TO 1280                                   LSV10720
       LINT = MN*(J-1) + 1                                              LSV10730
       LFIN = MN*J                                                      LSV10740
       SUM = ZERO                                                       LSV10750
       SVAL = SVNEW(J)                                                  LSV10760
C                                                                       LSV10770
       DO 1200 K = LINT,LFIN                                            LSV10780
 1200  SUM = SUM + RITVEC(K)*RITVEC(K)                                  LSV10790
C                                                                       LSV10800
       SUM = DSQRT(SUM)                                                 LSV10810
       RNORM(J) = SUM                                                   LSV10820
       TEMP = ONE - SUM                                                 LSV10830
       SUM = ONE/SUM                                                    LSV10840
C                                                                       LSV10850
       DO 1210 K = LINT,LFIN                                            LSV10860
 1210  RITVEC(K) = RITVEC(K)*SUM                                        LSV10870
C                                                                       LSV10880
C      COMPUTE ERROR IN RITZ VECTOR CONSIDERED AS AN EIGENVECTOR OF B.  LSV10890
       LINTM = LINT + M                                                 LSV10900
       L = LINT - 1                                                     LSV10910
       DO 1220 K = 1,M                                                  LSV10920
       L = L + 1                                                        LSV10930
 1220  V1(K) = RITVEC(L)                                                LSV10940
       DO 1230 K = 1,N                                                  LSV10950
       L = L + 1                                                        LSV10960
 1230  V2(K) = RITVEC(L)                                                LSV10970
C                                                                       LSV10980
C-----------------------------------------------------------------------LSV10990
C                                                                       LSV11000
       CALL SVMAT(RITVEC(LINTM),V1,SVAL)                                LSV11010
       CALL STRAN(RITVEC(LINT),V2,SVAL)                                 LSV11020
C                                                                       LSV11030
C-----------------------------------------------------------------------LSV11040
C                                                                       LSV11050
       SUM = ZERO                                                       LSV11060
       DO 1240 JJ = 1,M                                                 LSV11070
 1240  SUM = SUM + V1(JJ)*V1(JJ)                                        LSV11080
C                                                                       LSV11090
       DO 1250 JJ = 1,N                                                 LSV11100
 1250  SUM = SUM + V2(JJ)*V2(JJ)                                        LSV11110
C                                                                       LSV11120
       IF(IWRITE.NE.0)  WRITE(6,1260) J,GOODSV(J)                       LSV11130
 1260  FORMAT(/I5,'TH SINGULAR VALUE CONSIDERED =',E20.12/)             LSV11140
C                                                                       LSV11150
       IF(IWRITE.NE.0) WRITE(6,1270) TERR(J), TBETA(J), RNORM(J)        LSV11160
 1270  FORMAT(' RESIDUAL FOR T-EIGENVECTOR = ',E14.3/                   LSV11170
      1' DABS(BETA(MA(J)+1)*U(MA(J))) = ' ,E14.3/                       LSV11180
      1' NORM(RITZVEC) = ', E14.3/)                                     LSV11190
C                                                                       LSV11200
       SUM = DSQRT(SUM)                                                 LSV11210
       BERR(J) = SUM                                                    LSV11220
       BERRGP(J) = SUM/ABS(BMINGP(J))                                   LSV11230
 1280  CONTINUE                                                         LSV11240
C                                                                       LSV11250
C      RITZVECTORS ARE NORMALIZED AND B-MATRIX ESTIMATES ARE IN BERR    LSV11260
C      AND IN BERRGP ARRAYS.   STORE THESE ESTIMATES BUT NOT THE        LSV11270
C      VECTORS.                                                         LSV11280
C                                                                       LSV11290
       WRITE(9,1290)                                                    LSV11300
 1290  FORMAT(11X,'GOODSV(J)',3X,'MA(J)',2X,' BMINGAP',6X,' BERROR',2X, LSV11310
```

```
      1 'BERROR/BGAP',4X,' TERROR')                               LSV11320
C                                                                  LSV11330
      WRITE(13,1300)                                               LSV11340
 1300 FORMAT(11X,'GOODSV(J)',5X,'RITZNORM',5X,' BMINGAP',5X,'TBETA(J)', LSV11350
     1 5X,'TLAST(J)')                                              LSV11360
C                                                                  LSV11370
      DO 1330 J=1,NGOODC                                           LSV11380
C                                                                  LSV11390
      IF(MP(J).EQ.MPMIN) GO TO 1330                                LSV11400
C                                                                  LSV11410
      WRITE(9,1310)SVNEW(J),MA(J),BMINGP(J),BERR(J),BERRGP(J),TERR(J) LSV11420
 1310 FORMAT(E20.12,I6,4E13.5)                                     LSV11430
C                                                                  LSV11440
      WRITE(13,1320) SVNEW(J),RNORM(J),BMINGP(J),TBETA(J),TLAST(J) LSV11450
 1320 FORMAT(E20.12,4E13.5)                                        LSV11460
C                                                                  LSV11470
 1330 CONTINUE                                                     LSV11480
C                                                                  LSV11490
      IF (MREJEC.EQ.0) GO TO 1410                                  LSV11500
C                                                                  LSV11510
      WRITE(9,1340)                                                LSV11520
 1340 FORMAT(/' RITZ VECTORS WERE NOT COMPUTED FOR THE FOLLOWING SINGULALSV11530
     1R VALUES'/' EITHER BECAUSE THEY HAD NOT CONVERGED OR BECAUSE THE ELSV11540
     1RROR ESTIMATE'/' WAS NOT AS SMALL AS DESIRED'/)              LSV11550
C                                                                  LSV11560
      WRITE(13,1350)                                               LSV11570
 1350 FORMAT(/' RITZ VECTORS WERE NOT COMPUTED FOR THE FOLLOWING SINGULALSV11580
     1R VALUES'/' EITHER BECAUSE THEY HAD NOT CONVERGED OR BECAUSE THE ELSV11590
     1ERROR ESTIMATE'/' WAS NOT AS SMALL AS DESIRED'/)             LSV11600
C                                                                  LSV11610
      DO 1400 J = 1,NGOODC                                         LSV11620
      IF(MP(J).NE.MPMIN) GO TO 1400                                LSV11630
C     EACH SINGULAR VALUE FOR WHICH NO SINGULAR VECTOR WAS CALCULATED  LSV11640
C     HAS INFORMATION OUTPUTTED TO FILES 4 AND 13                  LSV11650
C                                                                  LSV11660
      WRITE(9,1360)                                                LSV11670
 1360 FORMAT(6X,'GOODSV(J)',3X,'MA(J)',5X,'BMINGP(J)',6X,'TLAST(J)', LSV11680
     1 6X,'TBETA(J)',3X,'MP(J)')                                   LSV11690
      WRITE(9,1370) GOODSV(J),MA(J),BMINGP(J),TLAST(J),TBETA(J),MP(J) LSV11700
 1370 FORMAT(E15.8,I8,3E14.4,I8)                                   LSV11710
C                                                                  LSV11720
      WRITE(13,1380)                                               LSV11730
 1380 FORMAT(6X,'GOODSV(J)',3X,'MA(J)',3X,'M1(J)',3X,'M2(J)',3X,'MP(J)' LSV11740
     1/)                                                           LSV11750
      WRITE(13,1390) GOODSV(J),MA(J),M1(J),M2(J),MP(J)             LSV11760
 1390 FORMAT(E15.8,4I8)                                            LSV11770
C                                                                  LSV11780
 1400 CONTINUE                                                     LSV11790
C                                                                  LSV11800
 1410 CONTINUE                                                     LSV11810
C                                                                  LSV11820
      WRITE(9,1420)                                                LSV11830
 1420 FORMAT(/' ABOVE ARE ERROR ESTIMATES FOR THE B AND T EIGENVECTORS'/LSV11840
     1 ' ASSOCIATED WITH THE GOODSV LISTED IN COLUMN 1'/           LSV11850
     1 ' BERROR = NORM(B*X - SV*X)   TERROR = NORM(T*Y - SV*Y) '/  LSV11860
     1 ' WHERE T = T(1,MA(J))    X = RITZ VECTOR = V*Y  V = SUCCESSIVE'/LSV11870
     1 ' LANCZOS VECTORS. BMINGAP = GAP TO NEAREST B-EIGENVALUE'//) LSV11880
C                                                                  LSV11890
      WRITE(13,1430)                                               LSV11900
 1430 FORMAT(/' ABOVE ARE ERROR ESTIMATES FOR THE GOODSVS'/        LSV11910
     1 ' RITZNORM = NORM(COMPUTED RITZ VECTOR FOR B-MATRIX'/       LSV11920
     1 ' TBETA(J) = BETA(MA(J)+1)*Y(MA(J)),  T*Y = SV*Y '/         LSV11930
     1 ' TLAST(J) = DABS(Y(MA(J)) '/)                              LSV11940
C                                                                  LSV11950
C     NUMBER OF RITZ VECTORS COMPUTED                              LSV11960
      NCOMPU = NGOODC - MREJEC                                     LSV11970
      WRITE(12,1440) N,NCOMPU,NGOODC,MATNO                         LSV11980
```

```
 1440 FORMAT(3I6,I12,' SIZE A, NO.RITZVECS, NO.SVALUES,MATNO')          LSV11990
C                                                                       LSV12000
C     INDIVIDUALLY NORMALIZE THE FIRST M AND THE LAST N COMPONENTS OF    LSV12010
C     EACH RITZ VECTOR.                                                 LSV12020
C                                                                       LSV12030
      LFIN = 0                                                          LSV12040
      DO 1560 J = 1,NGOODC                                              LSV12050
C                                                                       LSV12060
      IF(MP(J).EQ.MPMIN) GO TO 1540                                     LSV12070
C                                                                       LSV12080
C     RITZ VECTOR WAS COMPUTED                                          LSV12090
      LINT = MN*(J-1) + 1                                               LSV12100
      LFIN = MN*J                                                       LSV12110
      LFIN1 = LINT + M - 1                                              LSV12120
      LINT1 = LFIN1 + 1                                                 LSV12130
C                                                                       LSV12140
      SUM = 0.D0                                                        LSV12150
      TEMP = 0.D0                                                       LSV12160
      DO 1450 I = LINT,LFIN1                                            LSV12170
 1450 SUM = SUM + RITVEC(I)*RITVEC(I)                                   LSV12180
      SUM = ONE/DSQRT(SUM)                                              LSV12190
      DO 1460 I = LINT,LFIN1                                            LSV12200
 1460 RITVEC(I) = SUM*RITVEC(I)                                         LSV12210
      DO 1470 I = LINT1,LFIN                                            LSV12220
 1470 TEMP = TEMP + RITVEC(I)*RITVEC(I)                                 LSV12230
      TEMP = ONE/DSQRT(TEMP)                                            LSV12240
      DO 1480 I = LINT1,LFIN                                            LSV12250
 1480 RITVEC(I) = TEMP*RITVEC(I)                                        LSV12260
C                                                                       LSV12270
      WRITE(12,1490) J, GOODSV(J), MP(J)                                LSV12280
 1490 FORMAT(/I6,4X,E20.12,I6,' J, SINGULAR VALUE, MP(J)')              LSV12290
C                                                                       LSV12300
      WRITE(12,1500) BERR(J),BERRGP(J)                                  LSV12310
 1500 FORMAT(2E15.5,' = NORM(B*Z-SVAL*Z) AND   NORM(B*Z-SVAL*Z)/BMINGAP')LSV12320
C                                                                       LSV12330
      WRITE(12,1510) J                                                  LSV12340
 1510 FORMAT(/I6,'TH LEFT SINGULAR VECTOR'/)                            LSV12350
C     WRITE(12,170) (RITVEC(LL), LL=LINT,LFIN1)                         LSV12360
      WRITE(12,1520) (RITVEC(LL), LL=LINT,LFIN1)                        LSV12370
 1520 FORMAT(4E20.12)                                                   LSV12380
C                                                                       LSV12390
      WRITE(12,1530) J                                                  LSV12400
 1530 FORMAT(/I6,'TH RIGHT SINGULAR VECTOR'/)                           LSV12410
C     WRITE(12,170) (RITVEC(LL), LL=LINT1,LFIN)                         LSV12420
      WRITE(12,1520) (RITVEC(LL), LL=LINT1,LFIN)                        LSV12430
C                                                                       LSV12440
      GO TO 1560                                                        LSV12450
C                                                                       LSV12460
C     NO RITZ VECTOR WAS COMPUTED FOR THIS SINGULAR VALUE               LSV12470
 1540 WRITE(12,1550) J,GOODSV(J),MP(J)                                  LSV12480
 1550 FORMAT(I6,4X,E20.12,I6,' J,SINGVALUE,MP(J),NO RITZ VECTOR COMPUTEDLSV12490
     1')                                                                LSV12500
C                                                                       LSV12510
 1560 CONTINUE                                                          LSV12520
C                                                                       LSV12530
C     DID ANY T-MATRICES INCLUDE OFF-DIAGONAL ENTRIES SMALLER THAN      LSV12540
C     DESIRED, AS SPECIFIED BY BTOL?                                    LSV12550
C                                                                       LSV12560
      IF(IB.GT.0) GO TO 1590                                            LSV12570
      WRITE(6,1570) KMAXU                                               LSV12580
 1570 FORMAT(/' FOR LARGEST T-MATRIX CONSIDERED',I7,' CHECK THE SIZE OF LSV12590
     1BETAS')                                                           LSV12600
C                                                                       LSV12610
C--------------------------------------------------------------------LSV12620
C                                                                       LSV12630
      CALL TNORM(BETA,BKMIN,TEMP,KMAXU,IBMT)                            LSV12640
C                                                                       LSV12650
```

```
C-------------------------------------------------------------------LSV12660
C                                                                    LSV12670
      IF(IBMT.LT.0) WRITE (6,1580)                                   LSV12680
 1580 FORMAT(/' WARNING THE T-MATRICES FOR ONE OR MORE OF THE SINGULAR VLSV12690
     1ALUES CONSIDERED'/' HAD AN OFF-DIAGONAL ENTRY THAT WAS SMALLER THALSV12700
     1N THE BETA TOLERANCE THAT WAS SPECIFIED'/)                     LSV12710
 1590 CONTINUE                                                       LSV12720
C                                                                    LSV12730
      GO TO 1860                                                     LSV12740
C                                                                    LSV12750
 1600 WRITE(6,1610) NGOOD,MNMAX,MDIMRV                               LSV12760
 1610 FORMAT(/I4,' RITZ VECTORS WERE REQUESTED BUT THE REQUIRED DIMENSIOLSV12770
     1N',I6/' IS LARGER THAN THE USER-SPECIFIED DIMENSION OF RITVEC',I6 LSV12780
     1/' THEREFORE, THE SINGULAR VECTOR PROCEDURE TERMINATES FOR THE USELSV12790
     1R TO INTERVENE')                                               LSV12800
C                                                                    LSV12810
      GO TO 1860                                                     LSV12820
C                                                                    LSV12830
 1620 WRITE(6,1630)  MOLD,M,NOLD,N,MATOLD,MATNO                      LSV12840
 1630 FORMAT(/' GOODSV PARAMETERS READ FROM FILE 3 DO NOT AGREE WITH THOLSV12850
     1SE'/' SPECIFIED BY THE USER. MOLD,M,NOLD,N,MATOLD,MATNO ='/    LSV12860
     1416, 2I12/' THEREFORE PROGRAM TERMINATES FOR USER TO RESOLVE DIFFELSV12870
     1RENCES'/)                                                      LSV12880
C                                                                    LSV12890
      GO TO 1860                                                     LSV12900
C                                                                    LSV12910
 1640 WRITE(6,1650)                                                  LSV12920
 1650 FORMAT(/' PARAMETERS IN BETA FILE DO NOT AGREE WITH THOSE SPECIFIELSV12930
     1D BY THE USER.'/' THEREFORE, THE PROGRAM TERMINATES FOR THE USER TLSV12940
     1O RESOLVE THE DIFFERENCES'/)                                   LSV12950
C                                                                    LSV12960
      GO TO 1860                                                     LSV12970
C                                                                    LSV12980
 1660 WRITE(6,1670) KMAX,MEV                                         LSV12990
 1670 FORMAT(/' IN BETA HISTORY HEADER KMAX =',I6/                   LSV13000
     1' BUT SINGULAR VALUES WERE COMPUTED AT MEV = ',I6,' PROGRAM STOPS'LSV13010
     1)                                                              LSV13020
C                                                                    LSV13030
      GO TO 1860                                                     LSV13040
C                                                                    LSV13050
 1680 WRITE(6,1690) MEV                                              LSV13060
 1690 FORMAT(/' SOMETHING IS WRONG.'/' HEADER SAYS THAT SIZE T-MATRIX USLSV13070
     1ED IN THE SINGULAR VALUE COMPUTATIONS WAS = ',I6/' BUT THIS IS AN LSV13080
     1ODD ORDER AND THAT IS NOT ALLOWED.  PROGRAM STOPS'/)           LSV13090
C                                                                    LSV13100
      GO TO 1860                                                     LSV13110
C                                                                    LSV13120
 1700 WRITE(6,1710)                                                  LSV13130
 1710 FORMAT(/' PROGRAM COMPUTED 1ST GUESSES AT T-MATRIX SIZES, READ THELSV13140
     1M TO FILE 10'/' THEN TERMINATED AS REQUESTED.')               LSV13150
      GO TO 1860                                                     LSV13160
C                                                                    LSV13170
 1720 WRITE(6,1730) MTOL, MDIMTV                                     LSV13180
 1730 FORMAT(/' PROGRAM TERMINATES BECAUSE THE TVEC DIMENSION ANTICIPATELSV13190
     1D',I7/' IS LARGER THAN THE TVEC DIMENSION',I7,' SPECIFIED BY THE LSV13200
     1USER.'/' USER MAY RESET THE TVEC DIMENSION AND RESTART THE PROGRALSV13210
     1M'/)                                                           LSV13220
      GO TO 1860                                                     LSV13230
C                                                                    LSV13240
 1740 WRITE(6,1750)                                                  LSV13250
 1750 FORMAT(/' PROGRAM TERMINATES BECAUSE NO SUITABLE T-EIGENVECTORS WELSV13260
     1RE IDENTIFIED'/' FOR ANY OF THE SINGULAR VALUES SUPPLIED. PROBLEMLSV13270
     1 COULD BE CAUSED BY'/' TOO SMALL A TVEC DIMENSION OR SIMPLY BE THALSV13280
     1T NO SUITABLE T-VECTORS'/' WERE IDENTIFIED.  USER SHOULD CHECK OUTLSV13290
     1PUT'/)                                                         LSV13300
      GO TO 1860                                                     LSV13310
C                                                                    LSV13320
```

326

```
 1760 WRITE(6,1770) LVCONT,NTVEC,NGOOD                                    LSV13330
 1770 FORMAT(/' LVCONT FLAG =',I2,' AND NUMBER ',I5,' OF T-EIGENVECTORS LSV13340
     1 COMPUTED N.E.'/' NUMBER',I5,' REQUESTED SO PROGRAM TERMINATES'/) LSV13350
      GO TO 1860                                                         LSV13360
 1780 WRITE(6,1790)                                                      LSV13370
 1790 FORMAT(/' PROGRAM TERMINATES WITHOUT COMPUTING ANY RITZ VECTORS'/ LSV13380
     1/' BECAUSE ALL OF THE T-EIGENVECTORS WERE REJECTED AS NOT SUITABLELSV13390
     1 FOR'/' THE RITZ VECTOR COMPUTATIONS.  PROBABLE CAUSE WAS LACK OF LSV13400
     1CONVERGENCE'/' OF THE SINGULAR VALUES'/)                          LSV13410
      GO TO 1860                                                         LSV13420
C                                                                        LSV13430
 1800 WRITE(6,1810)                                                      LSV13440
 1810 FORMAT(/' PROGRAM INDICATES THAT IT IS NOT POSSIBLE TO COMPUTE ANYLSV13450
     1 OF THE'/' REQUESTED EIGENVECTORS. THEREFORE PROGRAM TERMINATES') LSV13460
      DO 1820 J=1,NGOODC                                                 LSV13470
 1820 WRITE(6,1830)  J,GOODSV(J),MP(J)                                   LSV13480
 1830 FORMAT(/4X,' J',11X,'GOODSV(J)',4X,'MP(J)'/I6,E20.12,I9)          LSV13490
      GO TO 1860                                                         LSV13500
C                                                                        LSV13510
 1840 WRITE(6,1850) MBETA,KMAXN                                          LSV13520
 1850 FORMAT(/' PROGRAM TERMINATES BECAUSE THE STORAGE ALLOTTED FOR THE LSV13530
     1BETA ARRAY',I8/' IS NOT SUFFICIENT FOR THE ENLARGED KMAX =',I8,' TLSV13540
     1HAT THE PROGRAM WANTS'/' USER CAN ENLARGE THE BETA ARRAY AND RERUNLSV13550
     1 THE PROGRAM'/)                                                    LSV13560
C                                                                        LSV13570
 1860 CONTINUE                                                           LSV13580
C                                                                        LSV13590
      STOP                                                               LSV13600
C-----END OF MAIN PROGRAM FOR LANCZOS SINGULAR VECTOR COMPUTATIONS------LSV13610
      END                                                                LSV13620
```

SECTION 6.5   LANCZS AND SAMPLE MATRIX-VECTOR MULTIPLY SUBROUTINES

```
C-----LSMULT--------------------------------------------------------------LSM00010
C                                                                          LSM00020
C     CONTAINS SUBROUTINES LANCZS, USPECS, STRAN, AND SVMAT                LSM00030
C     FOR USE WITH THE LANCZOS SINGULAR VALUE/VECTOR PROGRAMS              LSM00040
C                                                                          LSM00050
C     NONPORTABLE CONSTRUCTIONS:                                           LSM00060
C     1.   THE ENTRY MECHANISM USED TO PASS THE STORAGE LOCATIONS          LSM00070
C          OF THE USER-SPECIFIED MATRIX FROM THE SUBROUTINE USPEC          LSM00080
C          TO THE MATRIX-VECTOR MULTIPLY SUBROUTINES SVMAT AND             LSM00090
C          STRAN.                                                          LSM00100
C     2.   IN THE SAMPLE USPEC PROVIDED:  THE FREE FORMAT (8,*),           LSM00110
C          AND THE FORMAT (20A4).                                          LSM00120
C                                                                          LSM00130
C-----START OF LANCZS-----------------------------------------------------LSM00140
C                                                                          LSM00150
      SUBROUTINE LANCZS(MATVEC,MTRAN,BETA,V1,V2,G,KMAX,MOLD1,             LSM00160
     1                  M,N,IPAR,IIX)                                      LSM00170
C                                                                          LSM00180
C------------------------------------------------------------------------LSM00190
      DOUBLE PRECISION BETA(1),V1(1),V2(1),SUM,TEMP,ONE,ZERO              LSM00200
      REAL G(1)                                                            LSM00210
      DOUBLE PRECISION FINPRO                                             LSM00220
      INTEGER IPAR                                                        LSM00230
      EXTERNAL MATVEC,MTRAN                                               LSM00240
C------------------------------------------------------------------------LSM00250
C     COMPUTE T(1,MEV) FOR SYMMETRIZED VERSION OF GIVEN A-MATRIX.          LSM00260
C                                                                          LSM00270
C                      ----                ----                           LSM00280
C                     |  0                A  |                             LSM00290
C             B  =    |                      |                             LSM00300
C                     | A-TRANSPOSE       0  |                             LSM00310
C                      ----                ----                           LSM00320
C                                                                          LSM00330
C     WHERE A IS AN M BY N REAL SPARSE MATRIX, USING STARTING              LSM00340
C     VECTORS OF THE FORM (V1,0) WHEN THE FLAG IPAR = 2 AND                LSM00350
C     OF THE FORM (0,V2) WHEN THE FLAG IPAR = 1.  V1 IS OF                 LSM00360
C     DIMENSION M, THE ROW DIMENSION OF A, AND V2 IS OF DIMENSION          LSM00370
C     N, THE COLUMN DIMENSION OF A.                                        LSM00380
C                                                                          LSM00390
C     WITH STARTING VECTORS OF THESE FORMS, THE LANCZOS VECTORS            LSM00400
C     GENERATED ALTERNATE BETWEEN THESE 2 FORMS AND ALL OF THE             LSM00410
C     DIAGONAL ENTRIES OF THE LANCZOS TRIDIAGONAL MATRICES T(1,MEV)        LSM00420
C     GENERATED ARE 0.                                                     LSM00430
C                                                                          LSM00440
C     LANCZS USES 2 USER-SUPPLIED SUBROUTINES MATVEC AND MTRAN.            LSM00450
C     MAIN PROGRAM CALLS THESE SVMAT AND STRAN, RESPECTIVELY.              LSM00460
C     CALLING SEQUENCES ARE                                                LSM00470
C                                                                          LSM00480
C          CALL MATVEC(V2,V1,SUM)                                          LSM00490
C          CALL MTRAN(V1,V2,SUM)                                           LSM00500
C                                                                          LSM00510
C     MATVEC  COMPUTES V1 = A*V2 - SUM*V1.                                 LSM00520
C     MTRAN COMPUTES V2 = (A-TRANSPOSE)*V1 - SUM*V2.                       LSM00530
C                                                                          LSM00540
C     ON EXIT V1 AND V2 CONTAIN THE NONZERO PARTS OF THE                   LSM00550
C     LAST TWO LANCZOS VECTORS.                                            LSM00560
C                                                                          LSM00570
C     IF MOLD1 = 1 THEN T(1,KMAX) IS GENERATED FROM SCRATCH.              LSM00580
C     IF MOLD1 > 1 THEN A PREVIOUSLY-GENERATED T-MATRIX OF SIZE            LSM00590
C     (MOLD1-1) IS EXTENDED TO ONE OF SIZE KMAX. SINGULAR VALUE            LSM00600
C     PRGORAMS CAN ONLY UTILIZE T-MATRICES OF EVEN ORDER.                  LSM00610
C     BETA(KMAX+1) IS ALSO COMPUTED FOR USE IN THE ERROR ESTIMATES.        LSM00620
C                                                                          LSM00630
C------------------------------------------------------------------------LSM00640
```

```
      ONE  = 1.0D0                                                LSM00650
      ZERO = 0.0D0                                                LSM00660
      ITNUM = MOLD1                                               LSM00670
C                                                                 LSM00680
      IF (ITNUM .GT. 1) GO TO (80,100), IPAR                      LSM00690
C                                                                 LSM00700
C     NO PREVIOUS BETA HISTORY                                    LSM00710
      BETA(1) = ZERO                                              LSM00720
      IIL = IIX                                                   LSM00730
      IF (IPAR .EQ. 2) GO TO 40                                   LSM00740
C                                                                 LSM00750
C-----------------------------------------------------------------LSM00760
C     IPAR = 1 SO SET V2 EQUAL TO A UNIT RANDOM VECTOR AND SET V1 = 0.  LSM00770
      CALL GENRAN(IIL,G,N)                                        LSM00780
C-----------------------------------------------------------------LSM00790
C                                                                 LSM00800
      DO 10 J = 1,N                                               LSM00810
   10 V2(J) = G(J)                                                LSM00820
C                                                                 LSM00830
C-----------------------------------------------------------------LSM00840
      TEMP = FINPRO(N,V2(1),1,V2(1),1)                            LSM00850
C-----------------------------------------------------------------LSM00860
C                                                                 LSM00870
      SUM = ONE/DSQRT(TEMP)                                       LSM00880
      DO 20 J = 1,M                                               LSM00890
   20 V1(J) = ZERO                                                LSM00900
C                                                                 LSM00910
      DO 30 J = 1,N                                               LSM00920
   30 V2(J) = V2(J)*SUM                                           LSM00930
      GO TO 100                                                   LSM00940
C                                                                 LSM00950
   40 CONTINUE                                                    LSM00960
C                                                                 LSM00970
C-----------------------------------------------------------------LSM00980
C     IPAR = 2 SO SET V1 EQUAL TO A UNIT RANDOM VECTOR AND SET V2 = 0.  LSM00990
      CALL GENRAN(IIL,G,M)                                        LSM01000
C-----------------------------------------------------------------LSM01010
C                                                                 LSM01020
      DO 50 J=1,M                                                 LSM01030
   50 V1(J) = G(J)                                                LSM01040
C                                                                 LSM01050
C-----------------------------------------------------------------LSM01060
      TEMP = FINPRO(M,V1(1),1,V1(1),1)                            LSM01070
C-----------------------------------------------------------------LSM01080
C                                                                 LSM01090
      SUM = ONE/DSQRT(TEMP)                                       LSM01100
      DO 60 J = 1,N                                               LSM01110
   60 V2(J) = ZERO                                                LSM01120
      DO 70 J = 1,M                                               LSM01130
   70 V1(J) = V1(J)*SUM                                           LSM01140
C                                                                 LSM01150
C     BELOW IS START FOR MOLD1 > 1 AND IPAR = 1                   LSM01160
C     DO ONE ITERATION OF LANCZOS TO OBTAIN (0,V2)                LSM01170
C                                                                 LSM01180
   80 CONTINUE                                                    LSM01190
      SUM = BETA(ITNUM)                                           LSM01200
C                                                                 LSM01210
C-----------------------------------------------------------------LSM01220
      CALL MTRAN(V1,V2,SUM)                                       LSM01230
C-----------------------------------------------------------------LSM01240
C                                                                 LSM01250
C-----------------------------------------------------------------LSM01260
      SUM = FINPRO(N,V2(1),1,V2(1),1)                             LSM01270
C-----------------------------------------------------------------LSM01280
C                                                                 LSM01290
      ITNUM = ITNUM + 1                                           LSM01300
      BETA(ITNUM) = DSQRT(SUM)                                    LSM01310
```

```
      SUM = ONE/BETA(ITNUM)                                      LSM01320
C                                                                LSM01330
      DO 90 J = 1,N                                              LSM01340
   90 V2(J) = V2(J)*SUM                                          LSM01350
C                                                                LSM01360
      IPAR = 2                                                   LSM01370
      IF (ITNUM .GT. KMAX) GO TO 120                             LSM01380
C                                                                LSM01390
C     BELOW IS START FOR MOLD1 > 1 AND IPAR = 2                  LSM01400
C     DO ONE ITERATION OF LANCZOS TO OBTAIN (V1,0)               LSM01410
C                                                                LSM01420
  100 CONTINUE                                                   LSM01430
      SUM = BETA(ITNUM)                                          LSM01440
C                                                                LSM01450
C----------------------------------------------------------------LSM01460
      CALL MATVEC(V2,V1,SUM)                                     LSM01470
C----------------------------------------------------------------LSM01480
C                                                                LSM01490
C----------------------------------------------------------------LSM01500
      SUM = FINPRO(M,V1(1),1,V1(1),1)                            LSM01510
C----------------------------------------------------------------LSM01520
C                                                                LSM01530
      ITNUM = ITNUM + 1                                          LSM01540
      BETA(ITNUM) = DSQRT(SUM)                                   LSM01550
      SUM = ONE/BETA(ITNUM)                                      LSM01560
C                                                                LSM01570
      DO 110 J = 1,M                                             LSM01580
  110 V1(J)= V1(J) * SUM                                         LSM01590
C                                                                LSM01600
      IPAR = 1                                                   LSM01610
      IF (ITNUM .GT. KMAX) GO TO 120                             LSM01620
      GO TO 80                                                   LSM01630
C                                                                LSM01640
  120 CONTINUE                                                   LSM01650
C                                                                LSM01660
      RETURN                                                     LSM01670
C-----END OF LANCZS----------------------------------------------LSM01680
      END                                                        LSM01690
C                                                                LSM01700
C-----START OF USPEC (GENERAL SPARSE, RECTANGULAR MATRIX)--------LSM01710
C                                                                LSM01720
C     SUBROUTINE USPEC(M,N,MATNO)                                LSM01730
      SUBROUTINE SUSPEC(M,N,MATNO)                               LSM01740
C                                                                LSM01750
C-------------------------------------------------------------   LSM01760
      DOUBLE PRECISION A(10000)                                  LSM01770
      INTEGER IROW(10000),ICOL(3010)                             LSM01780
C-------------------------------------------------------------   LSM01790
C     DIMENSIONS ARRAYS NEEDED TO DEFINE THE USER-SUPPLIED       LSM01800
C     M X N RECTANGULAR A-MATRIX, READS IN VALUES OF THESE       LSM01810
C     ARRAYS AND THEN PASSES THE STORAGE LOCATIONS OF THESE      LSM01820
C     ARRAYS TO THE CORRESPONDING MATRIX-VECTOR MULTIPLY         LSM01830
C     SUBROUTINES SVMAT AND STRAN.                               LSM01840
C                                                                LSM01850
C     THE A-MATRIX IS STORED IN THE FOLLOWING SPARSE FORMAT:     LSM01860
C     M = NUMBER OF ROWS IN A.                                   LSM01870
C     N = NUMBER OF COLUMNS IN A.                                LSM01880
C     NZ = NUMBER OF NONZERO ENTRIES IN A-MATRIX.                LSM01890
C     ICOL(J), J=1,N IS NUMBER OF NONZERO ENTRIES IN COLUMN J.   LSM01900
C     IROW(K), K = 1,NZ IS THE ROW INDEX FOR CORRESPONDING A(K). LSM01910
C     A(K), K=1,NZ IS NONZERO ENTRIES IN A, COLUMN BY COLUMN.    LSM01920
C     IT IS ASSUMED THAT ICOL(J) > 0 FOR ALL J                   LSM01930
C                                                                LSM01940
C     NOTE:  ASSOCIATED SUBROUTINES SVMAT AND STRAN ASSUME THAT  LSM01950
C            M >= N.                                             LSM01960
C                                                                LSM01970
C-------------------------------------------------------------   LSM01980
```

```
C     READ IN  MATRIX FROM  FILE 8                                    LSM01990
C                                                                     LSM02000
      READ(8,10) NZ,MOLD,NOLD,MATOLD                                  LSM02010
   10 FORMAT(I10,2I6,I8)                                              LSM02020
C                                                                     LSM02030
      WRITE(6,20) NZ,MOLD,NOLD,MATOLD                                 LSM02040
   20 FORMAT(6X,'NZ',4X,'MOLD',4X,'NOLD',4X,'MATOLD'/I10,2I6,I10/)    LSM02050
C                                                                     LSM02060
C     TEST OF PARAMETER CORRECTNESS                                   LSM02070
      ITEMP = (MOLD-M)**2 + (NOLD-N)**2 + (MATOLD-MATNO)**2           LSM02080
C                                                                     LSM02090
      IF (ITEMP.EQ.0) GO TO 40                                        LSM02100
C                                                                     LSM02110
      WRITE(6,30)                                                     LSM02120
   30 FORMAT(' PROGRAM TERMINATES BECAUSE EITHER ORDERS OF OR LABELS FORLSM02130
     1 MATRIX DISAGREE')                                              LSM02140
      GO TO 70                                                        LSM02150
C                                                                     LSM02160
   40 CONTINUE                                                        LSM02170
C                                                                     LSM02180
C     NUMBER OF NONZERO ENTRIES IN EACH COLUMN IS READ IN             LSM02190
C     THEN THE CORRESPONDING ROW INDEX FOR EACH SUCH ENTRY IS READ    LSM02200
      READ(8,50) (ICOL(K), K=1,N)                                     LSM02210
      READ(8,50) (IROW(K), K=1,NZ)                                    LSM02220
   50 FORMAT(13I6)                                                    LSM02230
C                                                                     LSM02240
C     READ IN THE NONZERO ENTRIES IN THE MATRIX                       LSM02250
      READ(9,60) (A(K), K=1,NZ)                                       LSM02260
   60 FORMAT(3E25.16)                                                 LSM02270
C  50 FORMAT(4E19.10)                                                 LSM02280
C                                                                     LSM02290
C----------------------------------------------------------------------LSM02300
C     PASS STORAGE LOCATIONS OF ARRAYS THAT DEFINE THE MATRIX TO       LSM02310
C     THE MATRIX-VECTOR MULTIPLY SUBROUTINES SVMAT AND STRAN          LSM02320
      CALL SMATVE(A,ICOL,IROW,M,N)                                    LSM02330
      CALL STRANE(A,ICOL,IROW,M,N)                                    LSM02340
C----------------------------------------------------------------------LSM02350
C                                                                     LSM02360
C-----END OF USPEC-----------------------------------------------------LSM02370
      RETURN                                                          LSM02380
   70 STOP                                                            LSM02390
      END                                                             LSM02400
C                                                                     LSM02410
C-----STRAN (GENERAL SPARSE MATRIX)------------------------------------LSM02420
C                                                                     LSM02430
C     SUBROUTINE STRAN(W,U,SUM)                                       LSM02440
      SUBROUTINE SSTRAN(W,U,SUM)                                      LSM02450
C                                                                     LSM02460
C----------------------------------------------------------------------LSM02470
      DOUBLE PRECISION W(1),U(1),A(1),SUM,TEMP                        LSM02480
      INTEGER IROW(1),ICOL(1)                                         LSM02490
C----------------------------------------------------------------------LSM02500
C     SUBROUTINE TO COMPUTE U = (A-TRANSPOSE)*W - SUM*U WHERE A IS     LSM02510
C     A GENERAL, SPARSE M X N MATRIX WITH M >= N.                     LSM02520
C                                                                     LSM02530
C     ASSUMES MATRIX IS STORED IN SPARSE FORMAT GIVEN IN              LSM02540
C     CORRESPONDING USPEC SUBROUTINE.                                 LSM02550
C----------------------------------------------------------------------LSM02560
      JLAST = 0                                                       LSM02570
      DO 20 J = 1,N                                                   LSM02580
      JFIRST = JLAST + 1                                              LSM02590
      JLAST  = JLAST + ICOL(J)                                        LSM02600
      TEMP   = -SUM*U(J)                                              LSM02610
C                                                                     LSM02620
      DO 10 K = JFIRST,JLAST                                          LSM02630
      IK = IROW(K)                                                    LSM02640
```

```
   10 TEMP = A(K)*W(IK) + TEMP                                      LSM02650
C                                                                   LSM02660
   20 U(J) = TEMP                                                   LSM02670
C                                                                   LSM02680
      RETURN                                                        LSM02690
C                                                                   LSM02700
C------------------------------------------------------------------LSM02710
      ENTRY STRANE(A,ICOL,IROW,M,N)                                 LSM02720
C------------------------------------------------------------------LSM02730
C                                                                   LSM02740
C-----END OF STRAN FOR GENERAL SPARSE MATRIX-----------------------LSM02750
      RETURN                                                        LSM02760
      END                                                           LSM02770
C                                                                   LSM02780
C-----SVMAT (GENERAL SPARSE MATRIX)--------------------------------LSM02790
C                                                                   LSM02800
C     SUBROUTINE SVMAT(W,U,SUM)                                     LSM02810
      SUBROUTINE SSVMAT(W,U,SUM)                                    LSM02820
C                                                                   LSM02830
C------------------------------------------------------------------LSM02840
      DOUBLE PRECISION  W(1),U(1),A(1),SUM,TEMP                     LSM02850
      INTEGER IROW(1),ICOL(1)                                       LSM02860
C------------------------------------------------------------------LSM02870
C     SUBROUTINE TO COMPUTE U = A*W - SUM*U WHERE A IS A            LSM02880
C     GENERAL, SPARSE M X N MATRIX WITH M >= N.                     LSM02890
C                                                                   LSM02900
C     ASSUMES THAT THE MATRIX IS STORED IN THE SPARSE FORMAT        LSM02910
C     GIVEN IN THE CORRESPONDING USPEC SUBROUTINE.                  LSM02920
C------------------------------------------------------------------LSM02930
      DO 10 I = 1,M                                                 LSM02940
   10 U(I) = -SUM*U(I)                                              LSM02950
C                                                                   LSM02960
C MAIN LOOP.  PROCESSING PROCEEDS COL BY COL.  JFIRST AND JLAST ARE LSM02970
C POINTERS TO THE FIRST AND LAST NONZEROS IN COLUMN J.             LSM02980
C                                                                   LSM02990
      JLAST = 0                                                     LSM03000
      DO 30 J = 1,N                                                 LSM03010
      JFIRST = JLAST + 1                                            LSM03020
      JLAST = JLAST + ICOL(J)                                       LSM03030
      TEMP = W(J)                                                   LSM03040
C                                                                   LSM03050
      DO 20 K = JFIRST,JLAST                                        LSM03060
      IK = IROW(K)                                                  LSM03070
   20 U(IK) = U(IK) + A(K)*TEMP                                     LSM03080
C                                                                   LSM03090
   30 CONTINUE                                                      LSM03100
C                                                                   LSM03110
      RETURN                                                        LSM03120
C                                                                   LSM03130
C------------------------------------------------------------------LSM03140
      ENTRY SMATVE(A,ICOL,IROW,M,N)                                 LSM03150
C------------------------------------------------------------------LSM03160
C                                                                   LSM03170
C----END OF SVMAT FOR GENERAL SPARSE MATRICES----------------------LSM03180
      RETURN                                                        LSM03190
      END                                                           LSM03200
C                                                                   LSM03210
C-----ROUTINES FOR 'DIAGONAL' TEST MATRICES------------------------LSM03220
C     DMATV,DMTRAN,DIAGSP SUBROUTINES ARE FOR RECTANGULAR DIAGONAL  LSM03230
C     TEST MATRICES.                                                LSM03240
C                                                                   LSM03250
C-----START OF USPEC FOR 'DIAGONAL' TEST MATRIX--------------------LSM03260
C                                                                   LSM03270
      SUBROUTINE USPEC(M,N,MATNO)                                   LSM03280
C     SUBROUTINE DIAGSP(M,N,MATNO)                                  LSM03290
C                                                                   LSM03300
C     DEFINES 'DIAGONAL' MATRIX OF FOLLOWING FORM                   LSM03310
```

```
C                                                                LSM03320
C                     -----       -----                          LSM03330
C                    | 0      0       D |                        LSM03340
C            A =     | 0      0       0 |                        LSM03350
C                    |D-TRANS 0       0 |                        LSM03360
C                     -----       -----                          LSM03370
C                                                                LSM03380
C     WHERE D IS DIAGONAL MATRIX OF ORDER N, AND IN THE          LSM03390
C     MIDDLE THERE ARE (M-N) ROWS OF ZEROES.                     LSM03400
C     CALLS ENTRY TO MATRIX-VECTOR MULTIPLY SUBROUTINE TO PASS   LSM03410
C     STORAGE LOCATION OF THE D-ARRAY AND THE ORDERS M AND N.    LSM03420
C                                                                LSM03430
C     NOTE:  ASSOCIATED MATRIX-VECTOR SUBROUTINES ASSUME THAT    LSM03440
C            M >= N.                                             LSM03450
C---------------------------------------------------------------LSM03460
      DOUBLE PRECISION  D(1000), SPACE                           LSM03470
      REAL  EXPLAN(20)                                           LSM03480
C---------------------------------------------------------------LSM03490
C                                                                LSM03500
      READ(8,10) EXPLAN                                          LSM03510
   10 FORMAT(20A4)                                               LSM03520
      READ(8,*) MOLD,NOLD,NUNIF,SPACE,D(1)                       LSM03530
C                                                                LSM03540
      IF(N.NE.NOLD.OR.M.NE.MOLD) GO TO 80                        LSM03550
C     COMPUTE THE UNIFORM PORTION OF THE SPECTRUM                LSM03560
      DO 20 J=2,NUNIF                                            LSM03570
   20 D(J) = D(1) - DFLOAT(J-1)*SPACE                            LSM03580
      NUNIF1=NUNIF + 1                                           LSM03590
      READ(8,10)  EXPLAN                                         LSM03600
      DO 30 J=NUNIF1,N                                           LSM03610
   30 READ(8,*) D(J)                                             LSM03620
      NNUNIF = NOLD - NUNIF                                      LSM03630
      WRITE(6,40) NOLD,SPACE,NNUNIF,D(1)                         LSM03640
   40 FORMAT(/' DIAGONAL TEST MATRIX, SIZE = ',I4/' MOST ENTRIES ARE ',LSM03650
     1E10.3,' UNITS APART.',I3,' ENTRIES'/' ARE IRREGULARLY SPACED. FIRSLSM03660
     1T ENTRY IS ',E10.3/)                                      LSM03670
      NB = NUNIF - 2                                             LSM03680
C                                                                LSM03690
C                                                                LSM03700
C     PRINT OUT DIAGONAL PORTION OF A-MATRIX                     LSM03710
      WRITE(6,50) (D(I), I=1,10 )                                LSM03720
      WRITE(6,60) (D(I), I = NB,N)                               LSM03730
      MNDIF = MOLD - NOLD                                        LSM03740
      IF(MNDIF.NE.0) WRITE(6,70) MNDIF                           LSM03750
   50 FORMAT(/' SINGULAR VALUE LANCZOS TEST, 1ST 10 ENTRIES OF DIAGONAL LSM03760
     1A-MATRIX = '/(3E22.14))                                   LSM03770
   60 FORMAT(/' MIDDLE UNIFORM PORTION OF MATRIX IS NOT PRINTED OUT'/    LSM03780
     1' END OF UNIFORM PLUS NONUNIFORM SECTION = '/(3E22.14))   LSM03790
   70 FORMAT(I4,' ZERO ROWS ARE ADDED TO THE DIAGONAL TO MAKE IT RECTANGLSM03800
     1ULAR'/)                                                   LSM03810
C                                                                LSM03820
C     DIAGONAL GENERATION COMPLETE                              LSM03830
C                                                                LSM03840
C---------------------------------------------------------------LSM03850
C     CALL ENTRY TO MATRIX-VECTOR MULTIPLY SUBROUTINES   TO PASS LSM03860
C     STORAGE LOCATION OF D-ARRAY AND ORDER OF A-MATRIX.        LSM03870
      CALL DMATVE(D,M,N)                                         LSM03880
      CALL DMTRAE(D,M,N)                                         LSM03890
C---------------------------------------------------------------LSM03900
C                                                                LSM03910
      RETURN                                                     LSM03920
   80 WRITE(6,90) MOLD,NOLD,M,N                                  LSM03930
   90 FORMAT(' PROGRAM TERMINATES MOLD=',I5,' N.E. M=',I5,' OR NOLD=',   LSM03940
     1I5,' N.E. N=',I5)                                          LSM03950
C-----END OF USPEC SUBROUTINE FOR 'DIAGONAL' TEST MATRICES-------LSM03960
      STOP                                                       LSM03970
```

```
      END                                                       LSM03980
C                                                               LSM03990
C-----DSVMAT ('DIAGONAL' TEST MATRICES)-------------------------------LSM04000
C                                                               LSM04010
C      SUBROUTINE DSVMAT(Z,W,SUM)                               LSM04020
       SUBROUTINE SVMAT(Z,W,SUM)                                LSM04030
C                                                               LSM04040
C-----------------------------------------------------------------------LSM04050
       DOUBLE PRECISION  A(1),Z(1),W(1),SUM                     LSM04060
C-----------------------------------------------------------------------LSM04070
C                                                               LSM04080
C      COMPUTES W = A*Z - SUM*W .  ASSUMES THAT M >= N.         LSM04090
       DO 10 I = 1,N                                            LSM04100
   10  W(I) = A(I)*Z(I) - SUM *W(I)                             LSM04110
       IF(M.EQ.N) RETURN                                        LSM04120
       N1 = N+1                                                 LSM04130
       DO 20 I = N1,M                                           LSM04140
   20  W(I) = -SUM*W(I)                                         LSM04150
       RETURN                                                   LSM04160
C                                                               LSM04170
C-----------------------------------------------------------------------LSM04180
C      STORAGE LOCATIONS OF THE A-ARRAY                         LSM04190
C      AND THE ORDER OF THE A-MATRIX ARE PASSED TO THE MATVEC SUBROUTINE.LSM04200
C      ENTRY MATVE(A,M,N)                                       LSM04210
       ENTRY DMATVE(A,M,N)                                      LSM04220
C-----------------------------------------------------------------------LSM04230
C                                                               LSM04240
C-----END OF MATRIX -VECTOR MULTIPLY 'DIAGONAL' TEST PROBLEMS----------LSM04250
       RETURN                                                   LSM04260
       END                                                      LSM04270
C                                                               LSM04280
C-----MATRIX-VECTOR MULTIPLY FOR 'DIAGONAL' TEST MATRICES--------------LSM04290
C                                                               LSM04300
       SUBROUTINE STRAN(Z,W,SUM)                                LSM04310
C      SUBROUTINE DSTRAN(Z,W,SUM)                               LSM04320
C                                                               LSM04330
C-----------------------------------------------------------------------LSM04340
       DOUBLE PRECISION  A(1),Z(1),W(1),SUM                     LSM04350
C-----------------------------------------------------------------------LSM04360
C                                                               LSM04370
C      COMPUTES W = A-TRANSPOSE*Z - SUM*W .  ASSUMES M >= N.    LSM04380
       DO 10 I = 1,N                                            LSM04390
   10  W(I) = A(I)*Z(I)- SUM*W(I)                               LSM04400
       RETURN                                                   LSM04410
C                                                               LSM04420
C-----------------------------------------------------------------------LSM04430
C      STORAGE LOCATIONS OF THE A-ARRAY AND THE ORDER           LSM04440
C      OF THE A-MATRIX ARE OBTAINED FROM USPEC SUBROUTINE.      LSM04450
C      ENTRY MTRANE(A,M,N)                                      LSM04460
       ENTRY DMTRAE(A,M,N)                                      LSM04470
C-----------------------------------------------------------------------LSM04480
C                                                               LSM04490
C-----END OF SPARSE SYMMETRIC MATRIX-VECTOR MULTIPLY------------------LSM04500
       RETURN                                                   LSM04510
       END                                                      LSM04520
```

SECTION 6.6   OTHER SUBROUTINES USED BY THE PROGRAMS IN CHAPTER 6

```
C-----LSSUB-------(SINGULAR VALUES AND VECTORS)------------------------LSS00010
C     ACCORDING TO PFORT THESE SUBROUTINES ARE PORTABLE                 LSS00020
C                                                                       LSS00030
C                                                                       LSS00040
C     SUBROUTINES    BISEC, INVERR, TNORM, LUMP, ISOEV, PRTEST, AND     LSS00050
C                    INVERM ARE USED WITH LANCZOS SINGULAR VALUE        LSS00060
C                    PROGRAM LSVAL.  STURMI, INVERM, LBISEC, TNORM      LSS00070
C                    ARE USED WITH THE LANCZOS SINGULAR VECTOR          LSS00080
C                    PROGRAM LSVEC.                                     LSS00090
C                                                                       LSS00100
C                                                                       LSS00110
C-----COMPUTE T-EIGENVALUES BY BISECTION------------------------------LSS00120
C                                                                       LSS00130
      SUBROUTINE BISEC(BETA,BETA2,VB,VS,LBD,UBD,EPS,TTOL,MP,            LSS00140
     1 NINT,MEV,NDIS,IC,IWRITE)                                         LSS00150
C                                                                       LSS00160
C--------------------------------------------------------------------LSS00170
      DOUBLE PRECISION  BETA(1),BETA2(1),VB(1),VS(1)                   LSS00180
      DOUBLE PRECISION  LBD(1),UBD(1),EPS,EPT,EPO,EP1,TEMP,TTOL        LSS00190
      DOUBLE PRECISION  ZERO,ONE,HALF,YU,YV,LB,UB,XL,XU,X1,X0,XS,BETAM LSS00200
      INTEGER MP(1),IDEF(10)                                           LSS00210
      DOUBLE PRECISION  DABS, DSQRT, DMAX1, DMIN1, DFLOAT              LSS00220
C--------------------------------------------------------------------LSS00230
C     COMPUTES EIGENVALUES OF T(1,MEV) BY LOOPING INTERNALLY ON THE    LSS00240
C     USER-SPECIFIED INTERVALS,  (LB(J),UB(J)), J = 1,NINT.  INTERVALS LSS00250
C     ARE TREATED AS OPEN ON THE LEFT AND CLOSED ON THE RIGHT.         LSS00260
C     THE BISEC SUBROUTINE SIMULTANEOUSLY LABELS SPURIOUS T-EIGENVALUES LSS00270
C     AND DETERMINES THE T-MULTIPLICITIES OF EACH GOOD T-EIGENVALUE.   LSS00280
C     SPURIOUS T-EIGENVALUES ARE LABELLED BY A T-MULTIPLICITY = 0.     LSS00290
C     ANY T-EIGENVALUE WITH A T-MULTIPLICITY >= 1 IS 'GOOD'.           LSS00300
C                                                                       LSS00310
C     IF IWRITE = 0 THEN MOST OF THE WRITES TO FILE 6 ARE NOT          LSS00320
C     ACTIVATED.                                                       LSS00330
C                                                                       LSS00340
C     NOTE THAT PROGRAM ASSUMES THAT NO MORE THAN MMAX/2 T-EIGENVALUES LSS00350
C     OF T(1,MEV) ARE TO BE COMPUTED IN ANY ONE OF THE SUBINTERVALS    LSS00360
C     CONSIDERED, WHERE MMAX = DIMENSION OF VB SPECIFIED BY THE USER   LSS00370
C     IN THE MAIN PROGRAM LEVAL.                                       LSS00380
C                                                                       LSS00390
C     ON ENTRY                                                         LSS00400
C     BETA2(J) IS SET = BETA(J)*BETA(J).  THE STORAGE FOR BETA2 COULD  LSS00410
C     BE ELIMINATED BY RECOMPUTING THE BETA(J)**2 FOR EACH STURM       LSS00420
C     SEQUENCE.                                                        LSS00430
C                                                                       LSS00440
C     EPS = 2*MACHEP =  4.4 * 10**-16 ON IBM 3081.                     LSS00450
C     TTOL = EPS*TKMAX WHERE                                           LSS00460
C     TKMAX = MAX(BETA(K), K=1,KMAX)                                   LSS00470
C                                                                       LSS00480
C     ON EXIT                                                          LSS00490
C     NDIS = TOTAL NUMBER OF COMPUTED DISTINCT T-EIGENVALUES OF        LSS00500
C          T(1,MEV) ON THE UNION OF THE (LB,UB) INTERVALS.            LSS00510
C     VS = COMPUTED DISTINCT T-EIGENVALUES OF T(1,MEV) IN ALGEBRAICALLY-LSS00520
C          INCREASING ORDER                                           LSS00530
C     MP = CORRESPONDING T-MULTIPLICITIES OF THESE T-EIGENVALUES       LSS00540
C     MP(I) = (0,1,MI), MI>1, I=1,NDIS  MEANS:                         LSS00550
C        (0) V(I) IS SPURIOUS                                          LSS00560
C        (1) V(I) IS T-ISOLATED AND GOOD                               LSS00570
C       (MI) V(I) IS T-MULTIPLE AND HENCE A CONVERGED GOOD T-EIGENVALUELSS00580
C     IC = TOTAL NUMBER OF STURMS USED                                 LSS00590
C                                                                       LSS00600
C     DEFAULTS                                                         LSS00610
C     ISKIP = 0 INITIALLY. IF DEFAULT OCCURS ON J-TH SUB-INTERVAL, SET LSS00620
C          ISKIP=ISKIP+1 AND IDEF(ISKIP) = J                          LSS00630
C          DEFAULTS OCCUR IF THERE ARE NO T-EIGENVALUES  IN THE       LSS00640
```

```
C              SUBINTERVAL SPECIFIED OR IF THE NUMBER              LSS00650
C              OF STURMS SEQUENCES REQUIRED EXCEEDS MXSTUR.        LSS00660
C              WHEN A DEFAULT OCCURS THE PROGRAM                   LSS00670
C              SKIPS THE INTERVAL INVOLVED AND GOES ON TO THE NEXT LSS00680
C              INTERVAL.                                           LSS00690
C                                                                  LSS00700
C------------------------------------------------------------------LSS00710
C     SPECIFY PARAMETERS                                           LSS00720
      ZERO = 0.0D0                                                 LSS00730
      ONE  = 1.0D0                                                 LSS00740
      HALF = 0.5D0                                                 LSS00750
      MXSTUR = IC                                                  LSS00760
      NDIS = 0                                                     LSS00770
      IC = 0                                                       LSS00780
      ISKIP = 0                                                    LSS00790
      MP1 = MEV+1                                                  LSS00800
C     SAVE THEN SET BETA(MEV+1) = 0. GENERATE BETA**2              LSS00810
      BETAM = BETA(MP1)                                            LSS00820
      BETA(MP1) = ZERO                                             LSS00830
C                                                                  LSS00840
      DO 10 I = 1,MP1                                              LSS00850
   10 BETA2(I) = BETA(I)*BETA(I)                                   LSS00860
C                                                                  LSS00870
C     EPO IS USED IN T-MULTIPLICITY AND SPURIOUS TESTS             LSS00880
C     EP1 AND EPS ARE USED IN THE BISEC CONVERGENCE TEST           LSS00890
C                                                                  LSS00900
      TEMP = DFLOAT(MEV+1000)                                      LSS00910
      EPO  = TEMP*TTOL                                             LSS00920
      EP1  = DSQRT(TEMP)*TTOL                                      LSS00930
C                                                                  LSS00940
      WRITE(6,20)MEV,NINT                                          LSS00950
   20 FORMAT(/' BISEC CALCULATION'/' ORDER OF T IS',I6/            LSS00960
     1' NUMBER OF INTERVALS IS',I6/)                               LSS00970
C                                                                  LSS00980
      WRITE(6,30) EPO,EP1                                          LSS00990
   30 FORMAT(/' MULTOL, TOLERANCE USED IN T-MULTIPLICITY AND SPURIOUS TELSS01000
     1STS = ',E10.3/' BISTOL, TOLERANCE USED IN BISEC CONVERGENCE TEST =LSS01010
     1',E10.3/)                                                    LSS01020
C                                                                  LSS01030
C     LOOP ON THE NINT INTERVALS  (LB(J),UB(J)), J=1,NINT          LSS01040
      DO 430 JIND = 1,NINT                                         LSS01050
      LB = LBD(JIND)                                               LSS01060
      UB = UBD(JIND)                                               LSS01070
C                                                                  LSS01080
      WRITE(6,40)JIND,LB,UB                                        LSS01090
   40 FORMAT(//1X,'BISEC INTERVAL NO',2X,'LOWER BOUND',2X,'UPPER BOUND'/LSS01100
     1I18,2E13.5/)                                                 LSS01110
C                                                                  LSS01120
C     INITIALIZATION AND PARAMETER SPECIFICATION                   LSS01130
C     ICT IS TOTAL STURM COUNT ON (LB,UB)                          LSS01140
C                                                                  LSS01150
      NA = 0                                                       LSS01160
      MD = 0                                                       LSS01170
      NG = 0                                                       LSS01180
      ICT = 0                                                      LSS01190
C                                                                  LSS01200
C     START OF T-EIGENVALUE CALCULATIONS                           LSS01210
      X1 = UB                                                      LSS01220
      ISTURM = 1                                                   LSS01230
      GO TO 330                                                    LSS01240
C     FORWARD STURM CALCULATION TO DETERMINE NA = NO. T-EIGENVALUES > UBLSS01250
   50 NA = NEV                                                     LSS01260
C                                                                  LSS01270
      X1 = LB                                                      LSS01280
      ISTURM = 2                                                   LSS01290
      GO TO 330                                                    LSS01300
C     FORWARD STURM CALC TO DETERMINE MT = NO. T-EIGENVALUES ON (LB,UB) LSS01310
```

```
      60 CONTINUE                                                        LSS01320
         MT=NEV                                                         LSS01330
         ICT = ICT +2                                                   LSS01340
C                                                                       LSS01350
         WRITE(6,70)MT,NA                                               LSS01360
      70 FORMAT(/2I6,' = NO. TMEV ON (LB,UB) AND NO. .GT. UB'/)         LSS01370
C                                                                       LSS01380
C        DEFAULT TEST: IS ESTIMATED NUMBER OF STURMS > MXSTUR?          LSS01390
         IEST = 30*MT                                                   LSS01400
         IF (IEST.LT.MXSTUR) GO TO 90                                   LSS01410
C                                                                       LSS01420
         WRITE(6,80)                                                    LSS01430
      80 FORMAT(//' ESTIMATED NUMBER OF STURMS REQUIRED EXCEEDS USER LIMIT'LSS01440
        1/' SKIP THIS SUBINTERVAL')                                     LSS01450
         GO TO 110                                                      LSS01460
C                                                                       LSS01470
      90 CONTINUE                                                       LSS01480
C                                                                       LSS01490
         IF (MT.GE.1) GO TO 120                                         LSS01500
C                                                                       LSS01510
         WRITE(6,100)                                                   LSS01520
     100 FORMAT(//' THERE ARE NO T-EIGENVALUES ON THIS INTERVAL)'/)     LSS01530
C                                                                       LSS01540
     110 ISKIP = ISKIP+1                                                LSS01550
         IDEF(ISKIP) = JIND                                             LSS01560
         GO TO 430                                                      LSS01570
C                                                                       LSS01580
C        REGULAR CASE.                                                  LSS01590
     120 CONTINUE                                                       LSS01600
C                                                                       LSS01610
         IF (IWRITE.NE.0) WRITE(6,130)                                  LSS01620
     130 FORMAT(/' DISTINCT T-EIGENVALUES COMPUTED USING BISEC'/        LSS01630
        1 13X,'T-EIGENVALUE',2X,'TMULT',3X,'MD',4X,'NG')                LSS01640
C                                                                       LSS01650
C        SET UP INITIAL UPPER AND LOWER BOUNDS FOR T-EIGENVALUES        LSS01660
         DO 140 I=1,MT                                                  LSS01670
         VB(I) = LB                                                     LSS01680
         MTI = MT + I                                                   LSS01690
     140 VB(MTI) = UB                                                   LSS01700
C                                                                       LSS01710
C        CALCULATE T-EIGENVALUES FROM LB UP TO UB  K = MT,...,1         LSS01720
C        MAIN LOOP FOR FINDING KTH T-EIGENVALUE                         LSS01730
C                                                                       LSS01740
         K = MT                                                         LSS01750
     150 CONTINUE                                                       LSS01760
         ICO = 0                                                        LSS01770
         XL = VB(K)                                                     LSS01780
         MTK = MT+K                                                     LSS01790
         XU = VB(MTK)                                                   LSS01800
C                                                                       LSS01810
         ISTURM = 3                                                     LSS01820
         X1 = XU                                                        LSS01830
         ICO = ICO + 1                                                  LSS01840
         GO TO 330                                                      LSS01850
C        FORWARD STURM CALCULATION AT XU                                LSS01860
     160 NU=NEV                                                         LSS01870
C                                                                       LSS01880
C        BISECTION LOOP FOR KTH T-EIGENVALUE. TEST X1=MIDPOINT OF (XL,XU)LSS01890
         ISTURM = 4                                                     LSS01900
     170 CONTINUE                                                       LSS01910
         X1 = (XL+XU)*HALF                                              LSS01920
         XS = DABS(XL)+DABS(XU)                                         LSS01930
         XO = XU-XL                                                     LSS01940
         EPT = EPS*XS+EP1                                               LSS01950
C                                                                       LSS01960
C        EPT IS CONVERGENCE TOLERANCE FOR KTH T-EIGENVALUE             LSS01970
C                                                                       LSS01980
```

```
       IF (XO.LE.EPT) GO TO 230                                   LSS01990
C                                                                 LSS02000
C      T-EIGENVALUE HAS NOT YET CONVERGED                         LSS02010
C                                                                 LSS02020
       ICO = ICO + 1                                              LSS02030
       GO TO 330                                                  LSS02040
C      FORWARD STURM CALCULATION AT CURRENT T-EIGENVALUE APPROXIMATION. LSS02050
  180 CONTINUE                                                    LSS02060
C                                                                 LSS02070
C      UPDATE T-EIGENVALUE INTERVAL (XL,XU)                       LSS02080
C                                                                 LSS02090
       IF (NEV.LT.K) GO TO 190                                    LSS02100
C                                                                 LSS02110
C      NUMBER OF T-EIGENVALUES NEV = K                            LSS02120
       XL = X1                                                    LSS02130
       GO TO !70                                                  LSS02140
  190 CONTINUE                                                    LSS02150
C      NUMBER OF T-EIGENVALUES NEV<K                              LSS02160
       XU = X1                                                    LSS02170
       NU = NEV                                                   LSS02180
C                                                                 LSS02190
C      UPDATE OF T-EIGENVALUE BOUNDS                              LSS02200
C                                                                 LSS02210
       IF (NEV.EQ.0) GO TO 210                                    LSS02220
C                                                                 LSS02230
       DO 200 I = 1,NEV                                           LSS02240
  200 VB(I) = DMAX1(X1,VB(I))                                     LSS02250
C                                                                 LSS02260
  210 NEV1 = NEV+1                                                LSS02270
C                                                                 LSS02280
       DO 220 II = NEV1,K                                         LSS02290
       I = MT+II                                                  LSS02300
  220 VB(I) = DMIN1(X1,VB(I))                                     LSS02310
C                                                                 LSS02320
       GO TO 170                                                  LSS02330
C                                                                 LSS02340
C      END (XL,XU) BISECTION LOOP FOR KTH T-EIGENVALUE ON (LB,UB) LSS02350
C      TEST FOR T-MULTIPLICITY AND IF SIMPLE THEN TEST FOR SPURIOUSNESS LSS02360
C                                                                 LSS02370
  230 CONTINUE                                                    LSS02380
       NDIS = NDIS+1                                              LSS02390
       MD = MD+1                                                  LSS02400
       VS(NDIS) = X1                                              LSS02410
C                                                                 LSS02420
       JSTURM = 1                                                 LSS02430
       X1 = XL-EPO                                                LSS02440
       GO TO 370                                                  LSS02450
C      BACKWARD STURM CALCULATION                                 LSS02460
  240 KL = KEV                                                    LSS02470
       JL = JEV                                                   LSS02480
C                                                                 LSS02490
       JSTURM = 2                                                 LSS02500
       ICO = ICO + 2                                              LSS02510
       X1 = XU+EPO                                                LSS02520
       GO TO 370                                                  LSS02530
C      BACKWARD STURM CALCULATION                                 LSS02540
  250 JU = JEV                                                    LSS02550
       KU = KEV                                                   LSS02560
C                                                                 LSS02570
C      FOR T(1,MEV)                                               LSS02580
C      NU - KU = NO. T-EIGENVALUES ON (XU, XU + EPO)              LSS02590
C      KL - KU = NO. T-EIGENVALUES ON (XL - EPO, XU + EPO)        LSS02600
C                                                                 LSS02610
C      FOR T(2,MEV)                                               LSS02620
C      JL -JU = NO. T-EIGENVALUES ON (XL - EPO, XU + EPO)         LSS02630
C                                                                 LSS02640
```

```
C       IS THIS A SIMPLE T-EIGENVALUE?                              LSS02650
C                                                                   LSS02660
        IF (KL-KU-1.EQ.0) GO TO 290                                LSS02670
C                                                                   LSS02680
C       VS(NDIS) = KTH-T-EIGENVALUE OF (LB,UB) IS T-MULTIPLE AND HENCE  LSS02690
C       GOOD                                                       LSS02700
        IF (KU.EQ.NU) GO TO 280                                    LSS02710
C       CONTINUE TO CHECK FOR T-MULTIPLICITY                       LSS02720
  260 CONTINUE                                                     LSS02730
        ISTURM = 5                                                 LSS02740
        X1 = X1+EPO                                                LSS02750
        ICO = ICO + 1                                              LSS02760
        GO TO 330                                                  LSS02770
C       FORWARD STURM CALCULATION                                  LSS02780
  270 KNE = KU-NEV                                                 LSS02790
        KU = NEV                                                   LSS02800
        IF (KNE.NE.0) GO TO 260                                    LSS02810
C       SPECIFY T-MULTIPLICITY = MP(NDIS)                          LSS02820
  280 MPEV = KL-KU                                                 LSS02830
        KNEW = KU                                                  LSS02840
        GO TO 300                                                  LSS02850
C       END T-MULTIPLE CASE                                        LSS02860
C                                                                   LSS02870
C       T-EIGENVALUE IS SIMPLE   CHECK IF IT IS SPURIOUS           LSS02880
  290 CONTINUE                                                     LSS02890
        MPEV = 1                                                   LSS02900
        IF (JU.LT.JL) MPEV=0                                       LSS02910
        KNEW = K-1                                                 LSS02920
C                                                                   LSS02930
C       X1 >= XU+EPO                                               LSS02940
C       SPURIOUS TEST AND SIMPLE CASE COMPLETED                    LSS02950
C       START OF NEXT T-EIGENVALUE COMPUTATION                     LSS02960
C                                                                   LSS02970
  300 K = KNEW                                                     LSS02980
        MP(NDIS) = MPEV                                            LSS02990
        IF (MPEV.GE.1) NG = NG + 1                                 LSS03000
C                                                                   LSS03010
        IF (IWRITE.NE.0) WRITE(6,310) VS(NDIS),MPEV,MD,NG          LSS03020
  310 FORMAT(E25.16,3I6)                                           LSS03030
C                                                                   LSS03040
C       UPDATE STURM COUNT. ICO = STURM COUNT FOR KTH T-EIGENVALUE LSS03050
        ICT = ICT + ICO                                            LSS03060
C                                                                   LSS03070
C       EXIT TEST FOR K DO LOOP                                    LSS03080
C                                                                   LSS03090
        IF (K.LE.0) GO TO 410                                      LSS03100
C                                                                   LSS03110
C       UPDATE LOWER BOUNDS                                        LSS03120
        DO 320 I=1,KNEW                                            LSS03130
  320 VB(I) = DMAX1(X1,VB(I))                                      LSS03140
C                                                                   LSS03150
        GO TO 150                                                  LSS03160
C       END OF BISECTION LOOP FOR KTH EIGENVALUE                   LSS03170
C                                                                   LSS03180
C       FORWARD STURM CALCULATION                                  LSS03190
  330 NEV = -NA                                                    LSS03200
        YU = ONE                                                   LSS03210
C                                                                   LSS03220
        DO 360 I = 1,MEV                                           LSS03230
        IF (YU.NE.ZERO) GO TO 340                                  LSS03240
        YV = BETA(I)/EPS                                           LSS03250
        GO TO 350                                                  LSS03260
  340 YV = BETA2(I)/YU                                             LSS03270
  350 YU = X1 - YV                                                 LSS03280
        IF (YU.GE.ZERO) GO TO 360                                  LSS03290
        NEV = NEV + 1                                              LSS03300
```

```
  360 CONTINUE                                                       LSS03310
C     NEV = NUMBER OF T-EIGENVALUES ON (X1,UB)                       LSS03320
C                                                                    LSS03330
      GO TO (50,60,160,180,270), ISTURM                             LSS03340
C                                                                    LSS03350
C     BACKWARD STURM CALCULATION FOR T(1,MEV) AND T(2,MEV)           LSS03360
  370 KEV = -NA                                                      LSS03370
      YU  = ONE                                                      LSS03380
C                                                                    LSS03390
      DO 400 II = 1,MEV                                              LSS03400
      I = MP1-II                                                     LSS03410
      IF (YU.NE.ZERO) GO TO 380                                      LSS03420
      YV = BETA(I+1)/EPS                                             LSS03430
      GO TO 390                                                      LSS03440
  380 YV = BETA2(I+1)/YU                                             LSS03450
  390 YU = X1-YV                                                     LSS03460
      JEV = 0                                                        LSS03470
      IF (YU.GE.ZERO) GO TO 400                                      LSS03480
      KEV = KEV+1                                                    LSS03490
      JEV = 1                                                        LSS03500
  400 CONTINUE                                                       LSS03510
      JEV = KEV-JEV                                                  LSS03520
C                                                                    LSS03530
      GO TO (240,250), JSTURM                                       LSS03540
C                                                                    LSS03550
C     KEV = -NA + (NUMBER OF T(1,MEV) T-EIGENVALUES) > X1            LSS03560
C     JEV = -NA + (NUMBER OF T(2,MEV) T-EIGENVALUES) > X1            LSS03570
C                                                                    LSS03580
C     SET PARAMETERS FOR NEXT INTERVAL                               LSS03590
  410 CONTINUE                                                       LSS03600
      IC = ICT+IC                                                    LSS03610
      MXSTUR = MXSTUR-ICT                                            LSS03620
C                                                                    LSS03630
      WRITE(6,420) JIND,NG,MD,ICT                                    LSS03640
  420 FORMAT(/' T-EIGENVALUE CALCULATION ON INTERVAL',I6,'  IS COMPLETE'LSS03650
     1 /3X,'NO. GOOD',3X,'NO. DISTINCT',4X,'STURMS'/I10,I13,I10)     LSS03660
C                                                                    LSS03670
  430 CONTINUE                                                       LSS03680
C                                                                    LSS03690
C     END LOOP ON THE SUBINTERVALS (LB(J),UB(J)), J=1,NINT           LSS03700
C     ISKIP OUTPUT                                                   LSS03710
C                                                                    LSS03720
      IF (ISKIP.GT.0) WRITE(6,440)ISKIP                              LSS03730
  440 FORMAT(' BISEC DEFAULTED ON',I3,3X,'INTERVALS'/                LSS03740
     1 ' DEFAULTS OCCUR IF AN INTERVAL HAS NO T-EIGENVALUES'/        LSS03750
     2 ' OR THE STURM ESTIMATE EXCEEDS THE USER-SPECIFIED LIMIT'/)   LSS03760
C                                                                    LSS03770
      IF (ISKIP.GT.0) WRITE(6,450)(IDEF(I), I=1,ISKIP)               LSS03780
  450 FORMAT(' BISEC DEFAULTED ON INTERVALS'/(10I8))                 LSS03790
C                                                                    LSS03800
C     RESET BETA AT I = MP1                                          LSS03810
      BETA(MP1) = BETAM                                              LSS03820
C-----END OF BISEC--------------------------------------------------LSS03830
      RETURN                                                         LSS03840
      END                                                            LSS03850
C                                                                    LSS03860
C-----INVERSE ITERATION ON T(1,MEV)--------------------------------LSS03870
C                                                                    LSS03880
      SUBROUTINE INVERR(BETA,V1,V2,VS,EPS,G,MP,MEV,MMB,NDIS,NISO,    LSS03890
     1 NM,IKL,IT,IWRITE)                                             LSS03900
C                                                                    LSS03910
C-----------------------------------------------------------------LSS03920
      DOUBLE PRECISION BETA(1),V1(1),V2(1),VS(1)                     LSS03930
      DOUBLE PRECISION X1,U,Z,EST,TEMP,TO,T1,RATIO,SUM,XU,NORM,TSUM  LSS03940
      DOUBLE PRECISION  BETAM,EPS,EPS3,EPS4,ZERO,ONE                 LSS03950
      REAL G(1)                                                      LSS03960
```

```
      INTEGER  MP(1)                                                    LSS03970
C-----------------------------------------------------------------------LSS03980
      DOUBLE PRECISION FINPRO                                           LSS03990
      REAL  ABS                                                         LSS04000
      DOUBLE PRECISION  DABS, DMIN1, DSQRT, DFLOAT                      LSS04010
C-----------------------------------------------------------------------LSS04020
C     COMPUTES ERROR ESTIMATES FOR COMPUTED ISOLATED GOOD T-EIGENVALUES LSS04030
C     IN VS AND WRITES THESE T-EIGENVALUES AND ESTIMATES TO FILE 4.     LSS04040
C     BY DEFINITION A GOOD T-EIGENVALUE IS ISOLATED IF ITS             LSS04050
C     CLOSEST NEIGHBOR IS ALSO GOOD, OR IF ONE OF ITS NEIGHBORS IS     LSS04060
C     SPURIOUS BUT THAT NEIGHBOR IS FAR ENOUGH AWAY.  SO              LSS04070
C     IN PARTICULAR, WE COMPUTE ESTIMATES FOR GOOD T-EIGENVALUES       LSS04080
C     THAT ARE IN CLUSTERS OF GOOD T-EIGENVALUES.                      LSS04090
C                                                                       LSS04100
C     USES INVERSE ITERATION ON T(1,MEV) SOLVING THE EQUATION          LSS04110
C     (T - X1*I)V2 = RIGHT-HAND SIDE (RANDOMLY-GENERATED)              LSS04120
C     FOR EACH SUCH GOOD T-EIGENVALUE X1.                              LSS04130
C                                                                       LSS04140
C     PROGRAM REFACTORS T-X1*I ON EACH ITERATION OF INVERSE ITERATION. LSS04150
C     TYPICALLY ONLY ONE ITERATION IS NEEDED PER T-EIGENVALUE X1.      LSS04160
C                                                                       LSS04170
C     POSSIBLE STORAGE COMPRESSION                                     LSS04180
C     G STORAGE COULD BE ELIMINATED BY REGENERATING THE RANDOM         LSS04190
C     RIGHT-HAND SIDE ON EACH ITERATION AND PRINTING OUT THE           LSS04200
C     ERROR ESTIMATES AS THEY ARE GENERATED.                           LSS04210
C                                                                       LSS04220
C     ON ENTRY AND EXIT                                                LSS04230
C     MEV = ORDER OF T                                                 LSS04240
C     BETA CONTAINS THE NONZERO ENTRIES OF THE T-MATRIX               LSS04250
C     VS = COMPUTED DISTINCT EIGENVALUES OF T(1,MEV)                   LSS04260
C     MP = T-MULTIPLICITY OF EACH EIGENVALUE IN VS. MP(I) = -1 MEANS   LSS04270
C          VS(I) IS A GOOD T-EIGENVALUE BUT THAT IT IS SITTING CLOSE TO LSS04280
C          A SPURIOUS T-EIGENVALUE.  MP(I) = 0 MEANS VS(I) IS SPURIOUS. LSS04290
C          ESTIMATES ARE COMPUTED ONLY FOR THOSE T-EIGENVALUES         LSS04300
C          WITH MP(I) = 1. FLAGGING WAS DONE IN SUBROUTINE ISOEV       LSS04310
C          PRIOR TO ENTERING INVERR.                                   LSS04320
C     NISO = NUMBER OF ISOLATED GOOD T-EIGENVALUES CONTAINED IN VS     LSS04330
C     NDIS =  NUMBER OF DISTINCT T-EIGENVALUES IN VS                   LSS04340
C     IKL = SEED FOR RANDOM NUMBER GENERATOR                           LSS04350
C     EPS = 2. * MACHINE EPSILON                                       LSS04360
C                                                                       LSS04370
C     IN PROGRAM:                                                      LSS04380
C     ITER = MAXIMUM NUMBER OF INVERSE ITERATION STEPS ALLOWED FOR EACH LSS04390
C            X1.  ITER = IT ON ENTRY.                                  LSS04400
C     G = ARRAY OF DIMENSION AT LEAST MEV + NISO.  USED TO STORE       LSS04410
C         RANDOMLY-GENERATED RIGHT-HAND SIDE.  THIS IS NOT             LSS04420
C         REGENERATED FOR EACH X1. G IS ALSO USED TO STORE ERROR       LSS04430
C         ESTIMATES AS THEY ARE COMPUTED FOR LATER PRINTOUT.           LSS04440
C     V1,V2 = WORK SPACES USED IN THE FACTORIZATION OF T(1,MEV).       LSS04450
C     AT THE END OF THE INVERSE ITERATION COMPUTATION FOR X1, V2       LSS04460
C     CONTAINS THE UNIT EIGENVECTOR OF T(1,MEV) CORRESPONDING TO X1.   LSS04470
C     V1 AND V2 MUST BE OF DIMENSION AT LEAST MEV.                     LSS04480
C                                                                       LSS04490
C     ON EXIT                                                          LSS04500
C     G(J) = MINIMUM GAP IN T(1,MEV) FOR EACH VS(J), J=1,NDIS          LSS04510
C     G(MEV+I) = BETAM*|V2(MEV)| = ERROR ESTIMATE FOR ISOLATED GOOD    LSS04520
C          T-EIGENVALUES, WHERE I = 1,NISO  AND  BETAM = BETA(MEV+1)   LSS04530
C          V2(MEV) IS LAST COMPONENT OF THE UNIT T-EIGENVECTOR OF      LSS04540
C          T(1,MEV) CORRESPONDING TO ITH ISOLATED GOOD T-EIGENVALUE.   LSS04550
C                                                                       LSS04560
C     IF FOR SOME X1 IT.GT.ITER THEN THE ERROR ESTIMATE IN G IS MARKED LSS04570
C     WITH A - SIGN.                                                   LSS04580
C                                                                       LSS04590
C     V2 = ISOLATED GOOD T-EIGENVALUES                                 LSS04600
C     V1 = MINIMAL T-GAPS FOR THE EIGENVALUES IN V2.                   LSS04610
C     THESE ARE CONSTRUCTED FOR WRITE-OUT PURPOSES ONLY AND NOT        LSS04620
C     NEEDED ELSEWHERE IN THE PROGRAM.                                 LSS04630
```

```
C------------------------------------------------------------------LSS04640
C                                                                   LSS04650
C     LABEL OUTPUT FILE  4                                          LSS04660
      IF (MMB.EQ.1) WRITE(4,10)                                     LSS04670
   10 FORMAT(' INVERSE ITERATION ERROR ESTIMATES'/)                 LSS04680
C                                                                   LSS04690
C     FILE 6 (TERMINAL) OUTPUT OF ERROR ESTIMATES                  LSS04700
      IF (IWRITE.NE.0.AND.NISO.NE.0) WRITE(6,20)                    LSS04710
   20 FORMAT(/' INVERSE ITERATION ERROR ESTIMATES'/'  JISO',' JDIST',8X LSS04720
     1,'GOOD T-EIGENVALUE',4X,'BETAM*UM',5X,'TMINGAP')              LSS04730
C                                                                   LSS04740
C     INITIALIZATION AND PARAMETER SPECIFICATION                   LSS04750
      ZERO = 0.0D0                                                  LSS04760
      ONE = 1.0D0                                                   LSS04770
      NG = 0                                                        LSS04780
      NISO = 0                                                      LSS04790
      ITER = IT                                                     LSS04800
      MP1 = MEV+1                                                   LSS04810
      MM1 = MEV-1                                                   LSS04820
      BETAM = BETA(MP1)                                             LSS04830
      BETA(MP1) = ZERO                                              LSS04840
C                                                                   LSS04850
C     CALCULATE SCALE AND TOLERANCES                               LSS04860
      TSUM = ZERO                                                   LSS04870
      DO 30 I = 2,MEV                                               LSS04880
   30 TSUM = TSUM + BETA(I)                                         LSS04890
C                                                                   LSS04900
      EPS3 = EPS*TSUM                                               LSS04910
      EPS4 = DFLOAT(MEV)*EPS3                                       LSS04920
C                                                                   LSS04930
C     GENERATE SCALED RANDOM RIGHT-HAND SIDE                       LSS04940
      ILL = IKL                                                     LSS04950
C                                                                   LSS04960
C------------------------------------------------------------------LSS04970
      CALL GENRAN(ILL,G,MEV)                                        LSS04980
C------------------------------------------------------------------LSS04990
C                                                                   LSS05000
      GSUM = ZERO                                                   LSS05010
      DO 40 I = 1,MEV                                               LSS05020
   40 GSUM = GSUM+ABS(G(I))                                         LSS05030
      GSUM = EPS4/GSUM                                              LSS05040
C                                                                   LSS05050
      DO 50 I = 1,MEV                                               LSS05060
   50 G(I) = GSUM*G(I)                                              LSS05070
C                                                                   LSS05080
C     LOOP ON ISOLATED GOOD T-EIGENVALUES IN VS (MP(I) = 1) TO     LSS05090
C     CALCULATE CORRESPONDING UNIT T-EIGENVECTOR OF T(1,MEV)       LSS05100
C                                                                   LSS05110
      DO 180 JEV = 1,NDIS                                           LSS05120
      IF (MP(JEV).EQ.0) GO TO 180                                   LSS05130
      NG = NG + 1                                                   LSS05140
      IF (MP(JEV).NE.1) GO TO 180                                   LSS05150
      IT = 1                                                        LSS05160
      NISO = NISO + 1                                               LSS05170
      X1 = VS(JEV)                                                  LSS05180
C                                                                   LSS05190
C     INITIALIZE RIGHT HAND SIDE FOR INVERSE ITERATION             LSS05200
      DO 60 I = 1,MEV                                               LSS05210
   60 V2(I) = G(I)                                                  LSS05220
C                                                                   LSS05230
C     TRIANGULAR FACTORIZATION WITH NEAREST NEIGHBOR PIVOT         LSS05240
C     STRATEGY. INTERCHANGES ARE LABELLED BY SETTING BETA < 0.     LSS05250
C                                                                   LSS05260
   70 CONTINUE                                                      LSS05270
      U = -X1                                                       LSS05280
      Z = BETA(2)                                                   LSS05290
C                                                                   LSS05300
```

```
      DO 90 I = 2,MEV                                          LSS05310
      IF (BETA(I).GT.DABS(U)) GO TO 80                         LSS05320
C     NO INTERCHANGE                                           LSS05330
      V1(I-1) = Z/U                                            LSS05340
      V2(I-1) = V2(I-1)/U                                      LSS05350
      V2(I) = V2(I)-BETA(I)*V2(I-1)                            LSS05360
      RATIO = BETA(I)/U                                        LSS05370
      U = -X1-Z*RATIO                                          LSS05380
      Z = BETA(I+1)                                            LSS05390
      GO TO 90                                                 LSS05400
   80 CONTINUE                                                 LSS05410
C     INTERCHANGE CASE                                         LSS05420
      RATIO = U/BETA(I)                                        LSS05430
      BETA(I) = -BETA(I)                                       LSS05440
      V1(I-1) = -X1                                            LSS05450
      U = Z-RATIO*V1(I-1)                                      LSS05460
      Z = -RATIO*BETA(I+1)                                     LSS05470
      TEMP = V2(I-1)                                           LSS05480
      V2(I-1) = V2(I)                                          LSS05490
      V2(I) = TEMP-RATIO*V2(I)                                 LSS05500
   90 CONTINUE                                                 LSS05510
      IF (U.EQ.ZERO) U = EPS3                                  LSS05520
C                                                              LSS05530
C     SMALLNESS TEST AND DEFAULT VALUE FOR LAST COMPONENT      LSS05540
C     PIVOT(I-1) = |BETA(I)| FOR INTERCHANGE CASE              LSS05550
C     (I-1,I+1) ELEMENT IN RIGHT FACTOR = BETA(I+1)            LSS05560
C     END OF FACTORIZATION AND FORWARD SUBSTITUTION            LSS05570
C                                                              LSS05580
C     BACK SUBSTITUTION                                        LSS05590
      V2(MEV) = V2(MEV)/U                                      LSS05600
      DO 110 II = 1,MM1                                        LSS05610
      I = MEV-II                                               LSS05620
      IF (BETA(I+1).LT.ZERO) GO TO 100                         LSS05630
C     NO INTERCHANGE                                           LSS05640
      V2(I) = V2(I)-V1(I)*V2(I+1)                              LSS05650
      GO TO 110                                                LSS05660
C     INTERCHANGE CASE                                         LSS05670
  100 BETA(I+1) = -BETA(I+1)                                   LSS05680
      V2(I) = (V2(I)-V1(I)*V2(I+1)-BETA(I+2)*V2(I+2))/BETA(I+1) LSS05690
  110 CONTINUE                                                 LSS05700
C                                                              LSS05710
C     TESTS FOR CONVERGENCE OF INVERSE ITERATION               LSS05720
C     IF SUM |V2| COMPS. LE. 1 AND IT. LE. ITER DO ANOTHER INVIT STEP LSS05730
C                                                              LSS05740
      NORM = DABS(V2(MEV))                                     LSS05750
      DO 120 II = 1,MM1                                        LSS05760
      I = MEV-II                                               LSS05770
  120 NORM = NORM+DABS(V2(I))                                  LSS05780
C                                                              LSS05790
      IF (NORM.GE.ONE) GO TO 140                               LSS05800
      IT = IT+1                                                LSS05810
      IF (IT.GT.ITER) GO TO 140                                LSS05820
      XU = EPS4/NORM                                           LSS05830
C                                                              LSS05840
      DO 130 I = 1,MEV                                         LSS05850
  130 V2(I) = V2(I)*XU                                         LSS05860
C                                                              LSS05870
      GO TO 70                                                 LSS05880
C     ANOTHER INVERSE ITERATION STEP                           LSS05890
C                                                              LSS05900
C     INVERSE ITERATION FINISHED                               LSS05910
C     NORMALIZE COMPUTED T-EIGENVECTOR : V2 = V2/||V2||        LSS05920
  140 CONTINUE                                                 LSS05930
      SUM = FINPRO(MEV,V2(1),1,V2(1),1)                        LSS05940
      SUM = ONE/DSQRT(SUM)                                     LSS05950
C                                                              LSS05960
      DO 150 II = 1,MEV                                        LSS05970
```

```
    150 V2(II) = SUM*V2(II)                                           LSS05980
C                                                                     LSS05990
C       SAVE ERROR ESTIMATE FOR LATER OUTPUT                          LSS06000
        EST = BETAM*DABS(V2(MEV))                                     LSS06010
        IF (IT.GT.ITER) EST = -EST                                    LSS06020
        MEVPNI = MEV + NISO                                           LSS06030
        G(MEVPNI) = EST                                               LSS06040
        IF (IWRITE.EQ.0) GO TO 180                                    LSS06050
C                                                                     LSS06060
C       FILE 6 (TERMINAL) OUTPUT OF ERROR ESTIMATES.                  LSS06070
        IF (JEV.EQ.1) GAP = VS(2) - VS(1)                             LSS06080
        IF (JEV.EQ.MEV) GAP = VS(MEV) - VS(MEV-1)                     LSS06090
        IF (JEV.EQ.MEV.OR.JEV.EQ.1) GO TO 160                         LSS06100
        TEMP = DMIN1(VS(JEV+1)-VS(JEV),VS(JEV)-VS(JEV-1))             LSS06110
        GAP = TEMP                                                    LSS06120
    160 CONTINUE                                                      LSS06130
C                                                                     LSS06140
        WRITE(6,170) NISO,JEV,X1,EST,GAP                              LSS06150
    170 FORMAT(2I6,E25.16,2E12.3)                                     LSS06160
C                                                                     LSS06170
    180 CONTINUE                                                      LSS06180
C                                                                     LSS06190
C       END ERROR ESTIMATE LOOP ON ISOLATED GOOD T-EIGENVALUES.       LSS06200
C       GENERATE DISTINCT MINGAPS FOR T(1,MEV).  THIS IS USEFUL AS AN LSS06210
C       INDICATOR OF THE GOODNESS OF THE INVERSE ITERATION ESTIMATES. LSS06220
C       TRANSFER ISOLATED GOOD T-EIGENVALUES AND CORRESPONDING TMINGAPS LSS06230
C       TO V2 AND V1 FOR OUTPUT PURPOSES ONLY.                        LSS06240
C                                                                     LSS06250
        NM1 = NDIS - 1                                                LSS06260
        G(NDIS) = VS(NM1)-VS(NDIS)                                    LSS06270
        G(1) = VS(2)-VS(1)                                            LSS06280
C                                                                     LSS06290
        DO 190 J = 2,NM1                                              LSS06300
        T0 = VS(J)-VS(J-1)                                            LSS06310
        T1 = VS(J+1)-VS(J)                                            LSS06320
        G(J) = T1                                                     LSS06330
        IF (T0.LT.T1) G(J)=-T0                                        LSS06340
    190 CONTINUE                                                      LSS06350
        ISO = 0                                                       LSS06360
        DO 200 J = 1,NDIS                                             LSS06370
        IF (MP(J).NE.1) GO TO 200                                     LSS06380
        ISO = ISO+1                                                   LSS06390
        V1(ISO) = G(J)                                                LSS06400
        V2(ISO) = VS(J)                                               LSS06410
    200 CONTINUE                                                      LSS06420
C                                                                     LSS06430
        IF(NISO.EQ.0) GO TO 250                                       LSS06440
C                                                                     LSS06450
C       ERROR ESTIMATES ARE WRITTEN TO FILE 4                         LSS06460
        WRITE(4,210)MEV,NDIS,NG,NISO,NM,IKL,ITER,BETAM                LSS06470
    210 FORMAT(1X,'TSIZE',2X,'NDIS',1X,'NGOOD',2X,'NISO',3X,'M+N'/5I6/ LSS06480
     1 4X,'RHSEED',2X,'MXINIT',5X,'BETAM'/I10,I8,E10.3/               LSS06490
     2 2X,'GOODEVNO',8X,'GOOD T-EIGENVALUE',6X,'BETAM*UM',7X,'TMINGAP') LSS06500
C                                                                     LSS06510
        ISPUR = 0                                                     LSS06520
        I = 0                                                         LSS06530
        DO 240 J = 1,NDIS                                             LSS06540
        IF(MP(J).NE.0) GO TO 220                                      LSS06550
        ISPUR = ISPUR + 1                                             LSS06560
        GO TO 240                                                     LSS06570
    220 IF(MP(J).NE.1) GO TO 240                                      LSS06580
        I = I + 1                                                     LSS06590
        MEVI = MEV + I                                                LSS06600
        IGOOD = J - ISPUR                                             LSS06610
        WRITE(4,230) IGOOD,V2(I),G(MEVI),V1(I)                        LSS06620
    230 FORMAT(I10,E25.16,2E14.3)                                     LSS06630
    240 CONTINUE                                                      LSS06640
```

```
      GO TO 270                                                      LSS06650
C                                                                    LSS06660
  250 WRITE(4,260)                                                   LSS06670
  260 FORMAT(/' THERE ARE NO ISOLATED T-EIGENVALUES SO NO ERROR ESTIMATELSS06680
     1S WERE COMPUTED')                                              LSS06690
C     RESTORE BETA(MEV+1) = BETAM                                    LSS06700
  270 BETA(MP1) = BETAM                                              LSS06710
C-----END OF INVERR--------------------------------------------------LSS06720
      RETURN                                                         LSS06730
      END                                                            LSS06740
C                                                                    LSS06750
C-----START OF TNORM-------------------------------------------------LSS06760
C                                                                    LSS06770
      SUBROUTINE TNORM(BETA,BMIN,TMAX,MEV,IB)                        LSS06780
C                                                                    LSS06790
C------------------------------------------------------------------- LSS06800
      DOUBLE PRECISION  BETA(1)                                      LSS06810
      DOUBLE PRECISION  TMAX,BMIN,BSIZE,BTOL                         LSS06820
      DOUBLE PRECISION  DABS, DMAX1                                  LSS06830
C------------------------------------------------------------------- LSS06840
C     COMPUTE SCALING FACTOR USED IN THE T-MULTIPLICITY, SPURIOUS AND LSS06850
C     PRTESTS.   CHECK RELATIVE SIZE OF THE BETA(K), K=1,MEV         LSS06860
C     AS A TEST ON THE LOCAL ORTHOGONALITY OF THE LANCZOS VECTORS.   LSS06870
C                                                                    LSS06880
C             TMAX = MAX (BETA(I), I=1,MEV)                          LSS06890
C             BMIN = MIN (BETA(I) I=2,MEV)                           LSS06900
C             BSIZE = BMIN/TMAX                                      LSS06910
C             |IB| = INDEX OF MINIMAL(BETA)                          LSS06920
C             IB < 0 IF BMIN/TMAX < BTOL                             LSS06930
C------------------------------------------------------------------- LSS06940
C     SPECIFY PARAMETERS                                             LSS06950
      IB = 2                                                         LSS06960
      BTOL = BMIN                                                    LSS06970
      BMIN = BETA(2)                                                 LSS06980
      TMAX = BETA(2)                                                 LSS06990
C                                                                    LSS07000
      DO 20 I = 2,MEV                                                LSS07010
      IF (BETA(I).GE.BMIN) GO TO 10                                  LSS07020
      IB = I                                                         LSS07030
      BMIN = BETA(I)                                                 LSS07040
   10 TMAX = DMAX1(TMAX,BETA(I))                                     LSS07050
   20 CONTINUE                                                       LSS07060
C                                                                    LSS07070
C     TEST OF LOCAL ORTHOGONALITY USING SCALED BETAS                 LSS07080
      BSIZE = BMIN/TMAX                                              LSS07090
      IF (BSIZE.GE.BTOL) GO TO 40                                    LSS07100
C                                                                    LSS07110
C     DEFAULT.  BSIZE IS SMALLER THAN TOLERANCE BTOL SPECIFIED IN MAIN LSS07120
C     PROGRAM.  PROGRAM TERMINATES FOR USER TO DECIDE WHAT TO DO     LSS07130
C     BECAUSE LOCAL ORTHOGONALITY OF THE LANCZOS VECTORS COULD BE    LSS07140
C     LOST.                                                          LSS07150
C                                                                    LSS07160
      IB = -IB                                                       LSS07170
      WRITE(6,30) MEV                                                LSS07180
   30 FORMAT(/' BETA TEST INDICATES POSSIBLE LOSS OF LOCAL ORTHOGONALITYLSS07190
     1OVER 1ST',I6,' LANCZOS VECTORS'/)                             LSS07200
C                                                                    LSS07210
   40 CONTINUE                                                       LSS07220
C                                                                    LSS07230
      WRITE(6,50) IB                                                 LSS07240
   50 FORMAT(/' MINIMUM BETA RATIO OCCURS AT',I6,' TH BETA'/)        LSS07250
C                                                                    LSS07260
      WRITE(6,60) MEV,BMIN,TMAX,BSIZE                                LSS07270
   60 FORMAT(/1X,'TSIZE',6X,'MIN BETA',5X,'TKMAX',6X,'MIN RATIO'/    LSS07280
     1 I6,E14.3,E10.3,E15.3/)                                        LSS07290
C                                                                    LSS07300
C-----END OF TNORM--------------------------------------------------LSS07310
```

```
      RETURN                                                          LSS07320
      END                                                             LSS07330
C                                                                     LSS07340
C                                                                     LSS07350
C-----START OF LUMP-------------------------------------------------------LSS07360
C                                                                     LSS07370
      SUBROUTINE LUMP(V1,RELTOL,MULTOL,SCALE2,LINDEX,LOOP)            LSS07380
C                                                                     LSS07390
C------------------------------------------------------------------------LSS07400
      DOUBLE PRECISION  V1(1),SUM,RELTOL,MULTOL,THOLD,ZERO,SCALE2     LSS07410
      INTEGER  LINDEX(1)                                              LSS07420
      DOUBLE PRECISION  DABS, DFLOAT, DMAX1                           LSS07430
C------------------------------------------------------------------------LSS07440
C     LINDEX(J) = T-MULTIPLICITY OF JTH DISTINCT T-EIGENVALUE         LSS07450
C     LOOP = NUMBER OF DISTINCT T-EIGENVALUES                         LSS07460
C     LUMP 'COMBINES' COMPUTED 'GOOD' T-EIGENVALUES THAT ARE          LSS07470
C     'TOO CLOSE'.                                                    LSS07480
C     VALUE FOR RELTOL IS 1.D-10.                                     LSS07490
C                                                                     LSS07500
C     IF IN A SET OF T-EIGENVALUES TO BE COMBINED THERE IS AN EIGENVALUELSS07510
C     WITH LINDEX=1, THEN THE VALUE OF THE COMBINED T-EIGENVALUES IS SETLSS07520
C     EQUAL TO THE VALUE OF THAT T-EIGENVALUE.  NOTE THAT IF A SPURIOUS LSS07530
C     T-EIGENVALUE IS TO BE 'COMBINED' WITH A GOOD T-EIGENVALUE, THEN  LSS07540
C     THIS IS DONE ONLY BY INCREASING THE INDEX, LINDEX, FOR THAT     LSS07550
C     T-EIGENVALUE. NUMERICAL VALUES OF SPURIOUS T-EIGENVALUES ARE    LSS07560
C     NEVER COMBINED WITH THOSE OF GOOD T-EIGENVALUES.                LSS07570
C------------------------------------------------------------------------LSS07580
      ZERO = 0.0D0                                                    LSS07590
      NLOOP = 0                                                       LSS07600
      J = 0                                                           LSS07610
      ICOUNT = 1                                                      LSS07620
      JI = 1                                                          LSS07630
      THOLD = DMAX1(RELTOL*DABS(V1(1)),SCALE2*MULTOL)                 LSS07640
C     THOLD = DMAX1(RELTOL*DABS(V1(1)),RELTOL)                        LSS07650
C                                                                     LSS07660
   10 J = J+1                                                         LSS07670
      IF (J.EQ.LOOP) GO TO 20                                         LSS07680
      SUM = DABS(V1(J)-V1(J+1))                                       LSS07690
      IF (SUM.LT.THOLD) GO TO 60                                      LSS07700
   20 JF = JI + ICOUNT - 1                                            LSS07710
      INDSUM = 0                                                      LSS07720
      ISPUR = 0                                                       LSS07730
C                                                                     LSS07740
      DO 30 KK = JI,JF                                                LSS07750
      IF (LINDEX(KK).NE.0) GO TO 30                                   LSS07760
      ISPUR = ISPUR + 1                                               LSS07770
      INDSUM = INDSUM + 1                                             LSS07780
   30 INDSUM = INDSUM + LINDEX(KK)                                    LSS07790
C                                                                     LSS07800
C     IF (JF-JI.GE.1) WRITE(6,40) (V1(KKK), KKK=JI,JF)                LSS07810
   40 FORMAT(/' LUMP LUMPS THE T-EIGENVALUES'/(4E20.13))              LSS07820
C                                                                     LSS07830
C     COMPUTE THE 'COMBINED' T-EIGENVALUE AND THE RESULTING          LSS07840
C     T-MULTIPLICITY                                                  LSS07850
      K = JI - 1                                                      LSS07860
   50 K = K+1                                                         LSS07870
      IF (K.GT.JF) GO TO 70                                           LSS07880
      IF (LINDEX(K) .NE.1) GO TO 50                                   LSS07890
      NLOOP = NLOOP + 1                                               LSS07900
      V1(NLOOP) = V1(K)                                               LSS07910
      GO TO 100                                                       LSS07920
   60 ICOUNT = ICOUNT + 1                                             LSS07930
      GO TO 10                                                        LSS07940
C                                                                     LSS07950
C     ALL INDICES WERE 0 OR >1                                        LSS07960
   70 NLOOP = NLOOP + 1                                               LSS07970
      IDIF = INDSUM - ISPUR                                           LSS07980
```

```
      IF (IDIF.EQ.0) GO TO 90                                          LSS07990
C                                                                      LSS08000
      SUM = ZERO                                                       LSS08010
      DO 80 KK = JI,JF                                                 LSS08020
   80 SUM = SUM + V1(KK) * DFLOAT(LINDEX(KK))                          LSS08030
C                                                                      LSS08040
      V1(NLOOP) = SUM/DFLOAT(IDIF)                                     LSS08050
      GO TO 100                                                        LSS08060
   90 V1(NLOOP) = V1(JI)                                               LSS08070
  100 LINDEX(NLOOP) = INDSUM                                           LSS08080
      IDIF = INDSUM - ISPUR                                            LSS08090
      IF (IDIF.EQ.0.AND.ISPUR.EQ.1) LINDEX(NLOOP) = 0                  LSS08100
      IF (J.EQ.LOOP) GO TO 110                                         LSS08110
      ICOUNT = 1                                                       LSS08120
      JI= J+1                                                          LSS08130
      THOLD = DMAX1(RELTOL*DABS(V1(JI)),SCALE2*MULTOL)                 LSS08140
C     THOLD = DMAX1(RELTOL*DABS(V1(JI)),RELTOL)                        LSS08150
      IF (JI.LT.LOOP) GO TO 10                                         LSS08160
      NLOOP = NLOOP + 1                                                LSS08170
      V1(NLOOP)= V1(JI)                                                LSS08180
      LINDEX(NLOOP) = LINDEX(JI)                                       LSS08190
  110 CONTINUE                                                         LSS08200
C                                                                      LSS08210
C     ON RETURN V1 CONTAINS THE DISTINCT T-EIGENVALUES                 LSS08220
C     LINDEX CONTAINS THE CORRESPONDING T-MULTIPLICITIES               LSS08230
C                                                                      LSS08240
      LOOP = NLOOP                                                     LSS08250
      RETURN                                                           LSS08260
C-----END OF LUMP-----------------------------------------------------LSS08270
      END                                                             LSS08280
C                                                                      LSS08290
C                                                                      LSS08300
C-----START OF ISOEV--------------------------------------------------LSS08310
C                                                                      LSS08320
      SUBROUTINE ISOEV(VS,GAPTOL,MULTOL,SCALE1,G,MP,NDIS,NG,NISO)      LSS08330
C                                                                      LSS08340
C---------------------------------------------------------------------LSS08350
      DOUBLE PRECISION  VS(1),T0,T1,MULTOL,GAPTOL,SCALE1,TEMP          LSS08360
      REAL   G(1),GAP                                                  LSS08370
      INTEGER  MP(1)                                                   LSS08380
      REAL  ABS                                                        LSS08390
      DOUBLE PRECISION  DABS, DMAX1                                    LSS08400
C---------------------------------------------------------------------LSS08410
C     GENERATE DISTINCT TMINGAPS AND USE THEM TO LABEL THE ISOLATED    LSS08420
C     GOOD T-EIGENVALUES THAT ARE VERY CLOSE TO SPURIOUS ONES.         LSS08430
C     ERROR ESTIMATES WILL NOT BE COMPUTED FOR THESE T-EIGENVALUES.    LSS08440
C                                                                      LSS08450
C     ON ENTRY AND EXIT                                                LSS08460
C     VS CONTAINS THE COMPUTED DISTINCT T-EIGENVALUES OF T(1,MEV)      LSS08470
C     MP CONTAINS THE CORRESPONDING T-MULTIPLICITIES                   LSS08480
C     NDIS = NUMBER OF DISTINCT T-EIGENVALUES                          LSS08490
C     GAPTOL = RELATIVE GAP TOLERANCE SET IN MAIN                      LSS08500
C                                                                      LSS08510
C     ON EXIT                                                          LSS08520
C     G CONTAINS THE TMINGAPS.                                         LSS08530
C     G(I) < 0 MEANS MINGAP IS DUE TO LEFT GAP                         LSS08540
C     MP(I) IS NOT CHANGED EXCEPT THAT   MP(I)=-1, IF MP(I)=1,         LSS08550
C     TMINGAP WAS TOO SMALL AND DUE TO A SPURIOUS T-EIGENVALUE.        LSS08560
C                                                                      LSS08570
C     IF MP(I)=-1 THAT SIMPLE GOOD T-EIGENVALUE WILL BE SKIPPED        LSS08580
C     IN THE SUBSEQUENT ERROR ESTIMATE COMPUTATIONS IN INVERR          LSS08590
C     THAT IS, WE COMPUTE ERROR ESTIMATES ONLY FOR THOSE GOOD          LSS08600
C     T-EIGENVALUES WITH MP(I)=1.                                      LSS08610
C---------------------------------------------------------------------LSS08620
C     CALCULATE MINGAPS FOR DISTINCT T(1,MEV) EIGENVALUES.             LSS08630
      NM1 = NDIS - 1                                                   LSS08640
      G(NDIS) = VS(NM1)-VS(NDIS)                                       LSS08650
```

```
      G(1) = VS(2)-VS(1)                                              LSS08660
C                                                                     LSS08670
      DO 10 J = 2,NM1                                                 LSS08680
      TO = VS(J)-VS(J-1)                                              LSS08690
      T1 = VS(J+1)-VS(J)                                              LSS08700
      G(J) = T1                                                       LSS08710
      IF (TO.LT.T1) G(J) = -TO                                        LSS08720
   10 CONTINUE                                                        LSS08730
C                                                                     LSS08740
C     SET MP(I)=-1 FOR SIMPLE GOOD T-EIGENVALUES WHOSE MINGAPS  ARE   LSS08750
C     'TOO SMALL' AND DUE TO SPURIOUS T-EIGENVALUES.                  LSS08760
C                                                                     LSS08770
      NISO = 0                                                        LSS08780
      NG = 0                                                          LSS08790
      DO 20 J = 1,NDIS                                                LSS08800
      IF (MP(J).EQ.0) GO TO 20                                        LSS08810
      NG = NG+1                                                       LSS08820
      IF (MP(J).NE.1) GO TO 20                                        LSS08830
C     VS(J) IS NEXT TO SIMPLE GOOD T-EIGENVALUE                       LSS08840
      NISO = NISO + 1                                                 LSS08850
      I = J+1                                                         LSS08860
      IF (G(J).LT.0.0) I = J-1                                        LSS08870
      IF (MP(I).NE.0) GO TO 20                                        LSS08880
      GAP = ABS(G(J))                                                 LSS08890
      TO = DMAX1(SCALE1*MULTOL,GAPTOL*DABS(VS(J)))                    LSS08900
C     TO = DMAX1(GAPTOL,GAPTOL*DABS(VS(J)))                           LSS08910
      TEMP = TO                                                       LSS08920
      IF (GAP.GT.TEMP) GO TO 20                                       LSS08930
      MP(J) = -MP(J)                                                  LSS08940
      NISO = NISO-1                                                   LSS08950
   20 CONTINUE                                                        LSS08960
C                                                                     LSS08970
C-----END OF ISOEV--------------------------------------------------------LSS08980
      RETURN                                                          LSS08990
      END                                                             LSS09000
C                                                                     LSS09010
C-----START OF PRTEST-----------------------------------------------------LSS09020
C                                                                     LSS09030
      SUBROUTINE PRTEST(BETA,TEIG,TKMAX,EPSM,RELTOL,SCALE3,SCALE4,     LSS09040
     1 TMULT,NDIST,MEV,IPROJ)                                         LSS09050
C                                                                     LSS09060
C------------------------------------------------------------------------LSS09070
      DOUBLE PRECISION   BETA(1),TEIG(1),SIGMA(4)                     LSS09080
      DOUBLE PRECISION   EPSM,RELTOL,PRTOL,TKMAX,LRATIO,URATIO        LSS09090
      DOUBLE PRECISION   EPS,EPS1,BETAM,LBD,UBD,SIG,YU,YV,LRATS,URATS LSS09100
      DOUBLE PRECISION   ZERO,ONE,TEN,BISTOL,SCALE3,SCALE4,AEV,TEMP   LSS09110
      INTEGER TMULT(1),ISIGMA(4)                                      LSS09120
      DOUBLE PRECISION DABS, DMAX1, DSQRT, DFLOAT                     LSS09130
C------------------------------------------------------------------------LSS09140
C     AFTER CONVERGENCE HAS BEEN ESTABLISHED, SUBROUTINE PRTEST      LSS09150
C     TESTS COMPUTED EIGENVALUES OF T(1,MEV) THAT HAVE BEEN LABELLED LSS09160
C     SPURIOUS TO DETERMINE IF ANY SINGULAR VALUES OF A HAVE BEEN    LSS09170
C     MISSED BY LANCZOS PROCEDURE.  A SINGULAR VALUE WHOSE           LSS09180
C     SINGULAR VECTOR(S) HAS A VERY SMALL PROJECTION ON THE          LSS09190
C     STARTING VECTOR (< SINGLE PRECISION) CAN BE MISSED BECAUSE     LSS09200
C     IT WILL THEN ALSO BE AN EIGENVALUE OF T(2,MEV) TO WITHIN       LSS09210
C     THE SQUARE OF THIS ORIGINAL PROJECTION. HOWEVER,              LSS09220
C     OUR EXPERIENCE IS THAT SUCH SMALL PROJECTIONS OCCUR ONLY       LSS09230
C     VERY INFREQUENTLY.                                             LSS09240
C                                                                     LSS09250
C     THIS SUBROUTINE IS CALLED ONLY AFTER CONVERGENCE HAS BEEN      LSS09260
C     ESTABLISHED. ONCE CONVERGENCE HAS BEEN OBSERVED ON THE         LSS09270
C     OTHER SINGULAR VALUES, THEN ONE CAN EXPECT TO ALSO HAVE        LSS09280
C     CONVERGENCE ON ANY SUCH 'HIDDEN' SINGULAR VALUES. (IF THERE    LSS09290
C     ARE ANY).  PROCEDURE CONSIDERS ONLY SPURIOUS T-EIGENVALUES AND LSS09300
C     ONLY THOSE SPURIOUS T-EIGENVALUES THAT ARE ISOLATED FROM GOOD  LSS09310
C     T-EIGENVALUES.  FOR EACH SUCH T-EIGENVALUE IT DOES 2 STURM     LSS09320
```

```
C     SEQUENCES AND A FEW SCALAR MULTIPLICATIONS.  UPON RETURN TO MAIN   LSS09330
C     PROGRAM ERROR ESTIMATES WILL BE COMPUTED FOR ANY T-EIGENVALUES      LSS09340
C     THAT HAVE BEEN LABELLED AS 'HIDDEN'.  SUCH T-EIGENVALUES            LSS09350
C     WILL BE RELABELLED AS 'GOOD' ONLY IF THESE ERROR ESTIMATES          LSS09360
C     ARE SUFFICIENTLY SMALL.                                             LSS09370
C---------------------------------------------------------------------LSS09380
      ZERO = 0.0D0                                                        LSS09390
      ONE  = 1.0D0                                                        LSS09400
      TEN  = 10.0D0                                                       LSS09410
      PRTOL = 1.D-6                                                       LSS09420
      TEMP = DFLOAT(MEV+1000)                                             LSS09430
      TEMP = DSQRT(TEMP)                                                  LSS09440
      BISTOL = TKMAX*EPSM*TEMP                                            LSS09450
      NSIGMA = 4                                                          LSS09460
      SIGMA(1) = TEN*TKMAX                                                LSS09470
C                                                                         LSS09480
      DO 10 J = 2,NSIGMA                                                  LSS09490
   10 SIGMA(J) = TEN*SIGMA(J-1)                                           LSS09500
C                                                                         LSS09510
      IFIN = 0                                                            LSS09520
      MF = 1                                                              LSS09530
      ML = MEV                                                            LSS09540
      BETAM = BETA(MF)                                                    LSS09550
      BETA(MF) = ZERO                                                     LSS09560
      IPROJ = 0                                                           LSS09570
      J = 1                                                               LSS09580
C                                                                         LSS09590
      IF (TMULT(1).NE.0) GO TO 110                                        LSS09600
C                                                                         LSS09610
      AEV = DABS(TEIG(1))                                                 LSS09620
      TEMP = PRTOL*AEV                                                    LSS09630
      EPS1 = DMAX1(TEMP,SCALE4*BISTOL)                                    LSS09640
C     EPS1 = DMAX1(TEMP,PRTOL)                                            LSS09650
      TEMP = RELTOL*AEV                                                   LSS09660
      EPS  = DMAX1(TEMP,SCALE3*BISTOL)                                    LSS09670
C     EPS  = DMAX1(TEMP,RELTOL)                                           LSS09680
C                                                                         LSS09690
      IF (TEIG(2)-TEIG(1).LT.EPS1.AND.TMULT(2).NE.0) GO TO 110            LSS09700
C                                                                         LSS09710
   20 LBD = TEIG(J) - EPS                                                 LSS09720
      UBD = TEIG(J) + EPS                                                 LSS09730
      MEVL = 0                                                            LSS09740
      IL = 0                                                              LSS09750
      YU = ONE                                                            LSS09760
C                                                                         LSS09770
      DO 50 I=MF,ML                                                       LSS09780
      IF (YU.NE.ZERO) GO TO 30                                            LSS09790
      YV = BETA(I)/EPSM                                                   LSS09800
      GO TO 40                                                            LSS09810
   30 YV = BETA(I)*BETA(I)/YU                                             LSS09820
   40 YU = -LBD-YV                                                        LSS09830
      IF (YU.GE.ZERO) GO TO 50                                            LSS09840
C     MEVL INCREMENTED                                                    LSS09850
      MEVL = MEVL + 1                                                     LSS09860
      IL = I                                                              LSS09870
   50 CONTINUE                                                            LSS09880
C                                                                         LSS09890
      LRATIO = YU                                                         LSS09900
      MEV1L = MEVL                                                        LSS09910
      IF (IL.EQ.ML) MEV1L=MEVL-1                                          LSS09920
C                                                                         LSS09930
C     MEVL = NUMBER OF EVS OF T(1,MEV) WHICH ARE < LBD                    LSS09940
C     MEV1L = NUMBER OF EVS OF T(1,MEV-1) WHICH ARE < LBD                 LSS09950
C     LRATIO = DET(T(1,MEV)-LBD)/DET(T(1,MEV-1)-LBD):                     LSS09960
C                                                                         LSS09970
      MEVU = 0                                                            LSS09980
      IL = 0                                                              LSS09990
```

```
      YU = ONE                                                      LSS10000
C                                                                   LSS10010
      DO 80 I=MF,ML                                                 LSS10020
      IF (YU.NE.ZERO) GO TO 60                                      LSS10030
      YV = BETA(I)/EPSM                                             LSS10040
      GO TO 70                                                      LSS10050
   60 YV = BETA(I)*BETA(I)/YU                                       LSS10060
   70 YU = -UBD-YV                                                  LSS10070
      IF (YU.GE.ZERO) GO TO 80                                      LSS10080
C     MEVU INCREMENTED                                              LSS10090
      MEVU = MEVU + 1                                               LSS10100
      IL = I                                                        LSS10110
   80 CONTINUE                                                      LSS10120
C                                                                   LSS10130
      URATIO = YU                                                   LSS10140
      MEV1U = MEVU                                                  LSS10150
      IF (IL.EQ.ML) MEV1U=MEVU-1                                    LSS10160
C                                                                   LSS10170
C     MEVU = NUMBER OF EVS OF T(MEV) WHICH ARE < UBD               LSS10180
C     MEV1U = NUMBER OF EVS OF T(MEV-1) WHICH ARE < UBD            LSS10190
C     URATIO = DET(TM-UBD)/DET(T(M-1)-UBD): TM=T(MF,ML)            LSS10200
C                                                                   LSS10210
      NEV1 = MEV1U-MEV1L                                            LSS10220
C                                                                   LSS10230
      DO 90 K=1,NSIGMA                                              LSS10240
      SIG = SIGMA(K)                                                LSS10250
      LRATS = LRATIO-SIG                                            LSS10260
      URATS = URATIO-SIG                                            LSS10270
C     NOTE THE INCREMENT IS ON NUMBER OF EVALUES OF T(M-1)         LSS10280
      MEVLS = MEV1L                                                 LSS10290
      IF (LRATS.LT.0.) MEVLS=MEV1L+1                                LSS10300
      MEVUS = MEV1U                                                 LSS10310
      IF (URATS.LT.0.) MEVUS=MEV1U+1                                LSS10320
      ISIGMA(K) = MEVUS - MEVLS                                     LSS10330
   90 CONTINUE                                                      LSS10340
C                                                                   LSS10350
      ICOUNT = 0                                                    LSS10360
      DO 100 K=1,NSIGMA                                             LSS10370
  100 IF (ISIGMA(K).EQ.1) ICOUNT=ICOUNT + 1                         LSS10380
C                                                                   LSS10390
      IF (ICOUNT.LT.2.OR.NEV1.EQ.0) GO TO 110                       LSS10400
      TMULT(J) = -10                                                LSS10410
      IPROJ=IPROJ+1                                                 LSS10420
C                                                                   LSS10430
  110 J=J+1                                                         LSS10440
C                                                                   LSS10450
      IF (J.GE.NDIST) GO TO 120                                     LSS10460
      IF (TMULT(J).NE.0) GO TO 110                                  LSS10470
C                                                                   LSS10480
      AEV = DABS(TEIG(J))                                           LSS10490
      TEMP = PRTOL*AEV                                              LSS10500
      EPS1 = DMAX1(TEMP,SCALE4*BISTOL)                              LSS10510
C     EPS1 = DMAX1(TEMP,PRTOL)                                      LSS10520
      TEMP = RELTOL*AEV                                             LSS10530
      EPS = DMAX1(TEMP,SCALE3*BISTOL)                               LSS10540
C     EPS = DMAX1(TEMP,RELTOL)                                      LSS10550
C                                                                   LSS10560
      IF (TEIG(J)-TEIG(J-1).LT.EPS1.AND.TMULT(J-1).NE.0) GO TO 110 LSS10570
      IF (TEIG(J+1)-TEIG(J).LT.EPS1.AND.TMULT(J+1).NE.0) GO TO 110 LSS10580
C                                                                   LSS10590
      GO TO 20                                                      LSS10600
C                                                                   LSS10610
  120 IF (IFIN.EQ.1) GO TO 130                                      LSS10620
      IF (TMULT(NDIST).NE.0) GO TO 130                              LSS10630
C                                                                   LSS10640
      AEV = DABS(TEIG(NDIST))                                       LSS10650
      TEMP = PRTOL*AEV                                              LSS10660
```

```
      EPS1 = DMAX1(TEMP,SCALE4*BISTOL)                            LSS10670
C     EPS1 = DMAX1(TEMP,PRTOL)                                    LSS10680
      TEMP = RELTOL*AEV                                           LSS10690
      EPS  = DMAX1(TEMP,SCALE3*BISTOL)                            LSS10700
C     EPS  = DMAX1(TEMP,RELTOL)                                   LSS10710
C                                                                 LSS10720
      NDIST1=NDIST -1                                             LSS10730
      TEMP = TEIG(NDIST)-TEIG(NDIST1)                             LSS10740
      IF (TEMP.LT.EPS1.AND.TMULT(NDIST1).NE.0) GO TO 130          LSS10750
      IFIN = 1                                                    LSS10760
C                                                                 LSS10770
      GO TO 20                                                    LSS10780
C                                                                 LSS10790
  130 BETA(MF) = BETAM                                            LSS10800
C                                                                 LSS10810
C-----END OF PRTEST----------------------------------------------LSS10820
      RETURN                                                      LSS10830
      END                                                         LSS10840
C                                                                 LSS10850
C------START OF STURMI-------------------------------------------LSS10860
C                                                                 LSS10870
      SUBROUTINE STURMI(BETA,X1,TOLN,EPSM,MMAX,MK1,MK2,IC,IWRITE) LSS10880
C                                                                 LSS10890
C---------------------------------------------------------------LSS10900
      DOUBLE PRECISION  BETA(1)                                   LSS10910
      DOUBLE PRECISION  EPSM,X1,TOLN,EVL,EVU,BETA2                LSS10920
      DOUBLE PRECISION  U1,U2,V1,V2,ZERO,ONE                      LSS10930
      INTEGER I,IC,ICD,ICO,IC1,IC2,MK1,MK2,MMAX                   LSS10940
C---------------------------------------------------------------LSS10950
C                                                                 LSS10960
C     FOR ANY GOOD T-EIGENVALUE THAT HAS CONVERGED AS AN EIGENVALUE LSS10970
C     OF THE T-MATRICES THIS SUBROUTINE CALCULATES               LSS10980
C     THE SMALLEST SIZE OF THE T-MATRIX, T(1,MK1) DEFINED        LSS10990
C     BY THE BETA ARRAY SUCH THAT MK1.LE.MMAX                    LSS11000
C     AND THE INTERVAL (X1-TOLN,X1+TOLN) CONTAINS AT LEAST ONE   LSS11010
C     EIGENVALUE OF T(1,MK1). IT ALSO CALCULATES MK2 <= MMAX     LSS11020
C     AS THE SMALLEST SIZE T-MATRIX (IF ANY) SUCH THAT THIS INTERVAL LSS11030
C     CONTAINS AT LEAST TWO EIGENVALUES OF T(1,MK2).             LSS11040
C     IF NO T-MATRIX OF ORDER < MMAX SATISFIES THIS REQUIREMENT  LSS11050
C     THEN MK2 IS SET EQUAL TO MMAX.  THE SINGULAR VECTOR PROGRAM LSS11060
C     USES THESE VALUES TO DETERMINE A 1ST GUESS AT AN APPROPRIATE LSS11070
C     SIZE T-MATRIX FOR THE SINGULAR VALUE  X1.                  LSS11080
C                                                                 LSS11090
C     ON EXIT IC = NUMBER OF EIGENVALUES OF T(1,MK2) IN THIS INTERVAL LSS11100
C                                                                 LSS11110
C     STURMI REGENERATES THE QUANTITIES BETA(I)**2 EACH TIME IT IS LSS11120
C     CALLED, OBVIOUSLY FOR THE PRICE OF ANOTHER VECTOR OF LENGTH LSS11130
C     MMAX THIS GENERATION COULD BE DONE ONCE IN THE MAIN        LSS11140
C     PROGRAM BEFORE THE LOOP ON THE CALLS TO SUBROUTINE STURMI. LSS11150
C                                                                 LSS11160
C     IF ANY OF THE GOOD T-EIGENVALUES BEING CONSIDERED WERE MULTIPLE LSS11170
C     AS SINGULAR VALUES OF THE USER-SPECIFIED MATRIX, THEN      LSS11180
C     THIS SUBROUTINE COULD BE MODIFIED TO COMPUTE ADDITIONAL    LSS11190
C     SIZES MKJ, J = 3, ...  WHICH COULD THEN BE USED IN THE     LSS11200
C     MAIN LANCZOS SINGULAR VECTOR PROGRAM TO COMPUTE ADDITIONAL LSS11210
C     SINGULAR VECTORS CORRESPONDING TO THESE MULTIPLE SINGULAR  LSS11220
C     VALUES. THE MAIN PROGRAM LSVEC PROVIDED DOES NOT INCLUDE   LSS11230
C     THIS OPTION.                                               LSS11240
C                                                                 LSS11250
C---------------------------------------------------------------LSS11260
C     INITIALIZATION OF PARAMETERS                               LSS11270
      MK1 = 0                                                     LSS11280
      MK2 = 0                                                     LSS11290
      ZERO = 0.0D0                                                LSS11300
      ONE  = 1.0D0                                                LSS11310
      BETA(1) = ZERO                                              LSS11320
      EVL = X1-TOLN                                               LSS11330
```

```
      EVU = X1+TOLN                                               LSS11340
      U1 = ONE                                                    LSS11350
      U2 = ONE                                                    LSS11360
      ICO = 0                                                     LSS11370
      IC1 = 0                                                     LSS11380
      IC2 = 0                                                     LSS11390
C                                                                 LSS11400
C     MAIN LOOP FOR CALCULATING THE SIZES MK1,MK2                 LSS11410
      DO 60 I = 1,MMAX                                            LSS11420
      BETA2 = BETA(I)*BETA(I)                                     LSS11430
      IF (U1.NE.ZERO) GO TO 10                                    LSS11440
      V1 = BETA(I)/EPSM                                           LSS11450
      GO TO 20                                                    LSS11460
   10 V1 = BETA2/U1                                               LSS11470
   20 U1 = EVL - V1                                               LSS11480
      IF (U1.LT.ZERO) IC1 = IC1+1                                 LSS11490
      IF (U2.NE.ZERO) GO TO 30                                    LSS11500
      V2 = BETA(I)/EPSM                                           LSS11510
      GO TO 40                                                    LSS11520
   30 V2 = BETA2/U2                                               LSS11530
   40 U2 = EVU - V2                                               LSS11540
      IF (U2.LT.ZERO) IC2 = IC2+1                                 LSS11550
C     TEST FOR CHANGE IN NUMBER OF T-EIGENVALUES ON (EVL,EVU)     LSS11560
      ICD = IC1-IC2                                               LSS11570
      IC = ICD-ICO                                                LSS11580
      IF (IC.GE.1) GO TO 50                                       LSS11590
      GO TO 60                                                    LSS11600
   50 CONTINUE                                                    LSS11610
      IF (ICO.EQ.0) MK1 = I                                       LSS11620
      ICO = ICO+1                                                 LSS11630
      IF (ICO.GT.1) GO TO 70                                      LSS11640
   6C CONTINUE                                                    LSS11650
C                                                                 LSS11660
      I = I-1                                                     LSS11670
      IF (ICO.EQ.0) MK1 = MMAX                                    LSS11680
   70 MK2 = I                                                     LSS11690
      IC = ICD                                                    LSS11700
C                                                                 LSS11710
      IF (IWRITE.EQ.1) WRITE(6,80) X1,MK1,MK2,IC                  LSS11720
   80 FORMAT(' EVAL =',E20.12,' MK1 =',I6,' MK2 =',I6,' IC =',I3/)LSS11730
C                                                                 LSS11740
      RETURN                                                      LSS11750
C-----END OF STURMI----------------------------------------------LSS11760
      END                                                         LSS11770
C                                                                 LSS11780
C                                                                 LSS11790
C-----START OF INVERM--------------------------------------------LSS11800
C                                                                 LSS11810
      SUBROUTINE INVERM(BETA,V1,V2,X1,ERROR,ERRORV,EPS,G,MEV,IT,  LSS11820
     1 IWRITE)                                                    LSS11830
C                                                                 LSS11840
C---------------------------------------------------------------LSS11850
      DOUBLE PRECISION  BETA(1),V1(1),V2(1)                       LSS11860
      DOUBLE PRECISION  X1,U,Z,TEMP,RATIO,SUM,XU,NORM,TSUM,BETAM  LSS11870
      DOUBLE PRECISION  EPS,EPS3,EPS4,ERROR,ERRORV,ZERO,ONE       LSS11880
      REAL  G(1)                                                  LSS11890
      DOUBLE PRECISION  DABS, DSQRT, DFLOAT                       LSS11900
      DOUBLE PRECISION  FINPRO                                    LSS11910
      REAL  ABS                                                   LSS11920
C---------------------------------------------------------------LSS11930
C                                                                 LSS11940
C     COMPUTES T-EIGENVECTORS FOR ISOLATED GOOD T-EIGENVALUES X1  LSS11950
C     USING INVERSE ITERATION ON T(1,MEV(X1)) SOLVING EQUATION    LSS11960
C     (T - X1*I)V2 = RIGHT-HAND SIDE (RANDOMLY-GENERATED) .       LSS11970
C     PROGRAM REFACTORS T- X1*I ON EACH ITERATION OF INVERSE ITERATION. LSS11980
C     TYPICALLY ONLY ONE ITERATION IS NEEDED PER T-EIGENVALUE X1. LSS11990
C                                                                 LSS12000
```

```
C      IF IWRITE = 1 THEN THERE ARE EXTENDED WRITES TO FILE 6 (TERMINAL) LSS12010
C                                                                        LSS12020
C      ON ENTRY G CONTAINS A REAL*4 RANDOM VECTOR WHICH WAS GENERATED     LSS12030
C      IN MAIN PROGRAM.                                                   LSS12040
C                                                                         LSS12050
C      ON ENTRY AND EXIT                                                  LSS12060
C      MEV = ORDER OF T                                                   LSS12070
C      BETA CONTAINS THE OFFDIAGONAL ENTRIES OF T.                        LSS12080
C      EPS = 2. * MACHINE EPSILON                                         LSS12090
C                                                                         LSS12100
C      IN PROGRAM:                                                        LSS12110
C      ITER = MAXIMUM NUMBER STEPS ALLOWED FOR INVERSE ITERATION          LSS12120
C      ITER = IT ON ENTRY.                                                LSS12130
C      V1,V2 = WORK SPACES USED IN THE FACTORIZATION OF T(1,MEV).         LSS12140
C      V1 AND V2 MUST BE OF DIMENSION AT LEAST MEV.                       LSS12150
C                                                                         LSS12160
C      ON EXIT                                                            LSS12170
C      V2 = THE UNIT EIGENVECTOR OF T(1,MEV) CORRESPONDING TO X1.         LSS12180
C      ERROR = |V2(MEV)| = ERROR ESTIMATE FOR CORRESPONDING              LSS12190
C              RITZ VECTOR FOR X1.                                        LSS12200
C                                                                         LSS12210
C      ERRORV = || T*V2 - X1*V2 || = ERROR ESTIMATE ON T-EIGENVECTOR.     LSS12220
C      IF IT.GT.ITER THEN ERRORV = -ERRORV                                LSS12230
C      IT = NUMBER OF ITERATIONS ACTUALLY REQUIRED                        LSS12240
C-----------------------------------------------------------------------LSS12250
C      INITIALIZATION AND PARAMETER SPECIFICATION                         LSS12260
       ONE  = 1.0D0                                                       LSS12270
       ZERO = 0.0D0                                                       LSS12280
       ITER = IT                                                          LSS12290
       MP1 = MEV+1                                                        LSS12300
       MM1 = MEV-1                                                        LSS12310
       BETAM = BETA(MP1)                                                  LSS12320
       BETA(MP1) = ZERO                                                   LSS12330
C                                                                         LSS12340
C      CALCULATE SCALE AND TOLERANCES                                     LSS12350
       TSUM = ZERO                                                        LSS12360
       DO 10 I = 2,MEV                                                    LSS12370
   10 TSUM = TSUM + BETA(I)                                               LSS12380
C                                                                         LSS12390
       EPS3 = EPS*TSUM                                                    LSS12400
       EPS4 = DFLOAT(MEV)*EPS3                                            LSS12410
C                                                                         LSS12420
C      GENERATE SCALED RANDOM RIGHT-HAND SIDE                             LSS12430
       GSUM = ZERO                                                        LSS12440
       DO 20 I = 1,MEV                                                    LSS12450
   20 GSUM = GSUM+ABS(G(I))                                               LSS12460
       GSUM = EPS4/GSUM                                                   LSS12470
C                                                                         LSS12480
C      INITIALIZE RIGHT HAND SIDE FOR INVERSE ITERATION                   LSS12490
       DO 30 I = 1,MEV                                                    LSS12500
   30 V2(I) = GSUM*G(I)                                                   LSS12510
       IT = 1                                                             LSS12520
C                                                                         LSS12530
C      CALCULATE UNIT EIGENVECTOR OF T(1,MEV) FOR ISOLATED GOOD           LSS12540
C      T-EIGENVALUE X1.                                                   LSS12550
C                                                                         LSS12560
C      TRIANGULAR FACTORIZATION WITH NEAREST NEIGHBOR PIVOT               LSS12570
C      STRATEGY. INTERCHANGES ARE LABELLED BY SETTING BETA < 0.           LSS12580
C                                                                         LSS12590
   40 CONTINUE                                                            LSS12600
       U = -X1                                                            LSS12610
       Z = BETA(2)                                                        LSS12620
C                                                                         LSS12630
       DO 60 I=2,MEV                                                      LSS12640
       IF (BETA(I).GT.DABS(U)) GO TO 50                                   LSS12650
C      NO PIVOT INTERCHANGE                                               LSS12660
       V1(I-1) = Z/U                                                      LSS12670
```

```
      V2(I-1) = V2(I-1)/U                                           LSS12680
      V2(I) = V2(I)-BETA(I)*V2(I-1)                                 LSS12690
      RATIO = BETA(I)/U                                             LSS12700
      U = -X1-Z*RATIO                                               LSS12710
      Z = BETA(I+1)                                                 LSS12720
      GO TO 60                                                      LSS12730
C     PIVOT INTERCHANGE                                             LSS12740
   50 CONTINUE                                                      LSS12750
      RATIO = U/BETA(I)                                             LSS12760
      BETA(I) = -BETA(I)                                            LSS12770
      V1(I-1) = -X1                                                 LSS12780
      U = Z-RATIO*V1(I-1)                                           LSS12790
      Z = -RATIO*BETA(I+1)                                          LSS12800
      TEMP = V2(I-1)                                                LSS12810
      V2(I-1) = V2(I)                                               LSS12820
      V2(I) = TEMP-RATIO*V2(I)                                      LSS12830
   60 CONTINUE                                                      LSS12840
C                                                                   LSS12850
      IF (U.EQ.ZERO) U=EPS3                                         LSS12860
C                                                                   LSS12870
C     SMALLNESS TEST AND DEFAULT VALUE FOR LAST COMPONENT           LSS12880
C     PIVOT(I-1) = |BETA(I)| FOR INTERCHANGE CASE                   LSS12890
C     (I-1,I+1) ELEMENT IN RIGHT FACTOR = BETA(I+1)                 LSS12900
C     END OF FACTORIZATION AND FORWARD SUBSTITUTION                 LSS12910
C                                                                   LSS12920
C     BACK SUBSTITUTION                                             LSS12930
      V2(MEV) = V2(MEV)/U                                           LSS12940
      DO 80 II = 1,MM1                                              LSS12950
      I = MEV-II                                                    LSS12960
      IF (BETA(I+1).LT.ZERO) GO TO 70                              LSS12970
C     NO PIVOT INTERCHANGE                                          LSS12980
      V2(I) = V2(I)-V1(I)*V2(I+1)                                   LSS12990
      GO TO 80                                                      LSS13000
C     PIVOT INTERCHANGE                                             LSS13010
   70 BETA(I+1) = -BETA(I+1)                                        LSS13020
      V2(I) = (V2(I)-V1(I)*V2(I+1)-BETA(I+2)*V2(I+2))/BETA(I+1)     LSS13030
   80 CONTINUE                                                      LSS13040
C                                                                   LSS13050
C                                                                   LSS13060
C     TESTS FOR CONVERGENCE OF INVERSE ITERATION                    LSS13070
C     IF SUM |V2| COMPS. LE. 1 AND IT. LE. ITER DO ANOTHER INVIT STEP LSS13080
C                                                                   LSS13090
      NORM = DABS(V2(MEV))                                          LSS13100
      DO 90 II = 1,MM1                                              LSS13110
      I = MEV-II                                                    LSS13120
   90 NORM = NORM+DABS(V2(I))                                       LSS13130
C                                                                   LSS13140
C     IS DESIRED GROWTH IN VECTOR ACHIEVED ?                        LSS13150
C     IF NOT, DO ANOTHER INVERSE ITERATION STEP UNLESS NUMBER ALLOWED ISLSS13160
C     EXCEEDED.                                                     LSS13170
      IF (NORM.GE.ONE) GO TO 110                                    LSS13180
C                                                                   LSS13190
      IT=IT+1                                                       LSS13200
      IF (IT.GT.ITER) GO TO 110                                     LSS13210
C                                                                   LSS13220
      XU = EPS4/NORM                                                LSS13230
      DO 100 I=1,MEV                                                LSS13240
  100 V2(I) = V2(I)*XU                                              LSS13250
C                                                                   LSS13260
      GO TO 40                                                      LSS13270
C                                                                   LSS13280
C     NORMALIZE COMPUTED T-EIGENVECTOR : V2 = V2/||V2||             LSS13290
C                                                                   LSS13300
  110 CONTINUE                                                      LSS13310
C                                                                   LSS13320
      SUM = FINPRO(MEV,V2(1),1,V2(1),1)                             LSS13330
      SUM = ONE/DSQRT(SUM)                                          LSS13340
```

```
         DO 120 II = 1,MEV                                          LSS13350
     120 V2(II) = SUM*V2(II)                                        LSS13360
C                                                                   LSS13370
C       SAVE ERROR ESTIMATE FOR LATER OUTPUT                        LSS13380
         ERROR = DABS(V2(MEV))                                      LSS13390
C                                                                   LSS13400
C       GENERATE ERRCRV = ||T*V2 - X1*V2||.                         LSS13410
         V1(MEV) = BETA(MEV)*V2(MEV-1)-X1*V2(MEV)                   LSS13420
         DO 130 J = 2,MM1                                           LSS13430
         JM = MP1 - J                                               LSS13440
         V1(JM) = BETA(JM)*V2(JM-1) + BETA(JM+1)*V2(JM+1            LSS13450
     1) - X1*V2(JM)                                                 LSS13460
     130 CONTINUE                                                   LSS13470
C                                                                   LSS13480
         V1(1) = BETA(2)*V2(2) - X1*V2(1)                           LSS13490
         ERRORV = FINPRO(MEV,V1(1),1,V1(1),1)                       LSS13500
         ERRORV = DSQRT(ERRORV)                                     LSS13510
         IF (IT.GT.ITER) ERRORV = -ERRORV                           LSS13520
         IF (IWRITE.EQ.0) GO TO 150                                 LSS13530
C                                                                   LSS13540
C       FILE 6 (TERMINAL) OUTPUT OF ERROR ESTIMATES.                LSS13550
         WRITE(6,140) MEV,X1,ERROR,ERRORV                           LSS13560
     140 FORMAT(' INVERSE ITERATION OUTPUT'/                        LSS13570
     1 2X,'TSIZE',13X,'T-EIGENVALUE',11X,'U(M)',9X,'ERRORV'/        LSS13580
     1 I6,E25.16,2E15.5)                                            LSS13590
C                                                                   LSS13600
C       RESTORE BETA(MEV+1) = BETAM                                 LSS13610
     150 CONTINUE                                                   LSS13620
         BETA(MP1) = BETAM                                          LSS13630
C-----END OF INVERM-------------------------------------------------LSS13640
         RETURN                                                     LSS13650
         END                                                        LSS13660
C                                                                   LSS13670
C-----START OF LBISEC-----------------------------------------------LSS13680
C                                                                   LSS13690
         SUBROUTINE LBISEC(BETA,EPSM,EVAL,EVALN,LB,UB,TTOL,M,NEVT)  LSS13700
C                                                                   LSS13710
C-----------------------------------------------------------------LSS13720
         DOUBLE PRECISION  BETA(1),XO,X1,XL,XU,YU,YV,LB,UB          LSS13730
         DOUBLE PRECISION  EPSM,EP1,EVAL,EVALN,EVD,EPT              LSS13740
         DOUBLE PRECISION  ZERO,ONE,HALF,TTOL,TEMP                  LSS13750
         DOUBLE PRECISION  DABS,DSQRT,DFLOAT                        LSS13760
C-----------------------------------------------------------------LSS13770
C       SPECIFY PARAMETERS                                          LSS13780
         ZERO = 0.0D0                                               LSS13790
         HALF = 0.5D0                                               LSS13800
         ONE  = 1.0D0                                               LSS13810
         XL = LB                                                    LSS13820
         XU = UB                                                    LSS13830
C                                                                   LSS13840
C       EP1 = DSQRT(1000+M)*TTOL      TTOL = EPSM*TKMAX             LSS13850
C       TKMAX = MAX(BETA(K), K= 1,KMAX)                             LSS13860
C                                                                   LSS13870
         TEMP = DFLOAT(1000+M)                                      LSS13880
         EP1 = DSQRT(TEMP)*TTOL                                     LSS13890
C                                                                   LSS13900
         NA = 0                                                     LSS13910
         X1 = XU                                                    LSS13920
         JSTURM = 1                                                 LSS13930
         GO TO 60                                                   LSS13940
C       FORWARD STURM CALCULATION                                   LSS13950
      10 NA = NEV                                                   LSS13960
         X1 = XL                                                    LSS13970
         JSTURM = 2                                                 LSS13980
         GO TO 60                                                   LSS13990
C       FORWARD STURM CALCULATION                                   LSS14000
```

```
   20 NEVT = NEV                                                    LSS14010
C                                                                   LSS14020
C     WRITE(6,30) M,EVAL,NEVT,EP1                                   LSS14030
   30 FORMAT(/3X,'TSIZE',23X,'EV',9X/18,E25.16/                     LSS14040
     1 16,' = NUMBER OF T(1,M) EIGENVALUES ON TEST INTERVAL'/       LSS14050
     1 E12.3,' = CONVERGENCE TOLERANCE'/)                           LSS14060
C                                                                   LSS14070
      IF (NEVT.NE.1) GO TO 120                                      LSS14080
C                                                                   LSS14090
C     BISECTION LOOP                                                LSS14100
      JSTURM = 3                                                    LSS14110
   40 X1 = HALF*(XL+XU)                                             LSS14120
      X0 = XU-XL                                                    LSS14130
      EPT = EPSM*(DABS(XL) + DABS(XU)) + EP1                        LSS14140
C     CONVERGENCE TEST                                              LSS14150
      IF (X0.LE.EPT) GO TO 100                                      LSS14160
      GO TO 60                                                      LSS14170
C     FORWARD STURM CALCULATION                                     LSS14180
   50 CONTINUE                                                      LSS14190
      IF(NEV.EQ.0) XU = X1                                          LSS14200
      IF(NEV.EQ.1) XL = X1                                          LSS14210
      GO TO 40                                                      LSS14220
C     NEV = NUMBER OF EIGENVALUES OF T(1,M) ON (X1,XU)              LSS14230
C     THERE IS EXACTLY ONE EIGENVALUE OF T(1,M) ON (XL,XU)          LSS14240
C                                                                   LSS14250
C     FORWARD STURM CALCULATION                                     LSS14260
   60 NEV = -NA                                                     LSS14270
      YU = ONE                                                      LSS14280
      DO 90 I = 1,M                                                 LSS14290
      IF (YU.NE.ZERO) GO TO 70                                      LSS14300
      YV = BETA(I)/EPSM                                             LSS14310
      GO TO 80                                                      LSS14320
   70 YV = BETA(I)*BETA(I)/YU                                       LSS14330
   80 YU = X1 - YV                                                  LSS14340
      IF (YU.GE.ZERO) GO TO 90                                      LSS14350
      NEV = NEV+1                                                   LSS14360
   90 CONTINUE                                                      LSS14370
      GO TO (10,20,50), JSTURM                                      LSS14380
C                                                                   LSS14390
  100 CONTINUE                                                      LSS14400
C                                                                   LSS14410
      EVALN = X1                                                    LSS14420
      EVD = DABS(EVALN-EVAL)                                        LSS14430
C     WRITE(6,110) EVALN,EVAL,EVD                                   LSS14440
  110 FORMAT(/20X,'EVALN',21X,'EVAL',6X,'CHANGE'/2E25.16,E12.3/)    LSS14450
C                                                                   LSS14460
  120 CONTINUE                                                      LSS14470
      RETURN                                                        LSS14480
C-----END OF LBISEC-----------------------------------------------LSS14490
      END                                                           LSS14500
```

SECTION 6.7   OPTIONAL PREPROCESSING PROGRAM, FILE DEFINITIONS, AND SAMPLE INPUT FILES

```
C-----LSCOMPAC-(STAND ALONE PROGRAM)-----------------------------------LSC00010
C                                                                       LSC00020
C     THIS PROGRAM TRANSLATES A SPARSE RECTANGULAR M X N MATRIX A,       LSC00030
C     GIVEN AS I, J, A(I,J), INTO THE SPARSE MATRIX FORMAT              LSC00040
C     REQUIRED BY THE SAMPLE USPEC AND CMATV PROGRAMS PROVIDED          LSC00050
C     FOR USE WITH THE LANCZOS SINGULAR VALUE PROCEDURES.               LSC00060
C     THIS PROGRAM ASSUMES THAT THE MATRIX ENTRIES ARE GIVEN EITHER     LSC00070
C     COLUMN BY COLUMN OR ROW BY ROW.  MOREOVER, IF THE ENTRIES         LSC00080
C     ARE GIVEN ROW BY ROW, THEN THE FILE 8 CREATED WILL CORRESPOND     LSC00090
C     TO A-TRANSPOSE NOT A.   THUS, IN ANY SUBSEQUENT USE OF THIS FILE  LSC00100
C     IN THE LANCZOS SINGULAR VALUE/VECTOR PROGRAMS, THE USER WILL HAVE LSC00110
C     TO INTERCHANGE THE ROLES OF M AND N IN THE INPUT FILES.           LSC00120
C                                                                       LSC00130
C     NONPORTABLE STATEMENTS:  PFORT VERIFIER INDICATES THIS PROGRAM    LSC00140
C                              IS PORTABLE                              LSC00150
C                                                                       LSC00160
C----------------------------------------------------------------------LSC00170
      DOUBLE PRECISION A(15000)                                         LSC00180
      INTEGER IROW(15000),ICOL(15000)                                   LSC00190
C----------------------------------------------------------------------LSC00200
C     INPUT FILE 7 CONTAINS THE SPARSE REAL MXN MATRIX STORED AS:       LSC00210
C                                                                       LSC00220
C                 NZ,M,N,MATNO                                          LSC00230
C                 I(K)  J(K)  A(K)   K = 1,NZ                           LSC00240
C                                                                       LSC00250
C     WHERE NZ IS THE TOTAL NUMBER OF NONZEROS IN THE MATRIX A,         LSC00260
C     M IS THE ROW DIMENSION OF A, N IS THE COLUMN DIMENSION,           LSC00270
C     AND A(K) ARE THE NONZERO ENTRIES STORED ROW BY ROW OR            LSC00280
C     COLUMN BY COLUMN.  PROGRAM READS THIS IN AS IROW(K) = I(K),       LSC00290
C     ICOL(K) = J(K), AND A(K) = A(K).                                  LSC00300
C                                                                       LSC00310
C     OUTPUT FILE = 8 CONTAINS THE A-MATRIX IN SPARSE FORMAT            LSC00320
C                                                                       LSC00330
C                 NZ,M,N,MATNO                                          LSC00340
C                 ICOL(K)    K = 1,N                                    LSC00350
C                 IROW(K)    K = 1,NZ                                   LSC00360
C                  A(K)      K = 1,NZ                                    LSC00370
C                                                                       LSC00380
C     ON OUTPUT A(K), K=1,NZ CONTAINS THE KTH NONZERO ELEMENT           LSC00390
C     OF THE GIVEN MATRIX, ORDERED COLUMN BY COLUMN.  IROW(K),          LSC00400
C     K=1,NZ, EQUALS THE ROW INDEX OF THE NONZERO ENTRY A(K).           LSC00410
C     ICOL(K), K=1,N, EQUALS THE NUMBER OF NONZEROS OF A IN            LSC00420
C     COLUMN K.                                                         LSC00430
C                                                                       LSC00440
C----------------------------------------------------------------------LSC00450
C                                                                       LSC00460
      READ(7,10) NZ,M,N,MATNO,IIROW                                     LSC00470
   10 FORMAT(I10,2I6,I8,I6)                                             LSC00480
C                                                                       LSC00490
      WRITE(6,20) NZ,M,N,MATNO,IIROW                                    LSC00500
   20 FORMAT(I10,2I6,I8,' = NO. NONZERO AIJ,M,N,MATNO'/                 LSC00510
     1 I6,' = IIROW IF IIROW=0 ORDERING IS BY COLS IIROW=1 BY ROWS'/)   LSC00520
C                                                                       LSC00530
      IF (IIROW.EQ.0) READ(7,30) (IROW(K),ICOL(K),A(K), K=1,NZ)         LSC00540
      IF (IIROW.EQ.0) GO TO 40                                          LSC00550
C                                                                       LSC00560
      READ(7,30) (ICOL(K),IROW(K),A(K), K=1,NZ)                         LSC00570
   30 FORMAT(2I5,E14.7)                                                 LSC00580
      ITEMP = M                                                         LSC00590
      M = N                                                             LSC00600
      N = ITEMP                                                         LSC00610
```

```
      40 CONTINUE                                                LSC00620
C                                                                LSC00630
         LCOUNT = 0                                              LSC00640
         K = 1                                                   LSC00650
C                                                                LSC00660
C        START OF A NEW COLUMN                                   LSC00670
      50 CONTINUE                                                LSC00680
         J = ICOL(K)                                             LSC00690
         ICOL(J) = 0                                             LSC00700
      60 CONTINUE                                                LSC00710
         LCOUNT = LCOUNT + 1                                     LSC00720
         A(LCOUNT) = A(K)                                        LSC00730
         IROW(LCOUNT) = IROW(K)                                  LSC00740
         ICOL(J) = ICOL(J) + 1                                   LSC00750
C                                                                LSC00760
      70 CONTINUE                                                LSC00770
         K = K+1                                                 LSC00780
C                                                                LSC00790
         IF(K.GT.NZ) GO TO 80                                    LSC00800
C                                                                LSC00810
         IF(ICOL(K).GT.J) GO TO 50                               LSC00820
C                                                                LSC00830
         GO TO 60                                                LSC00840
C                                                                LSC00850
      80 CONTINUE                                                LSC00860
C        NZ = LCOUNT                                             LSC00870
C                                                                LSC00880
         WRITE(8,90) NZ,M,N,MATNO                                LSC00890
         WRITE(6,90) NZ,M,N,MATNO                                LSC00900
      90 FORMAT(I10,2I6,I8,' = NZ M N MATNO')                    LSC00910
C                                                                LSC00920
         WRITE(8,100) (ICOL(I), I=1,N)                           LSC00930
         WRITE(8,100) (IROW(K), K=1,NZ)                          LSC00940
     100 FORMAT(13I6)                                            LSC00950
C                                                                LSC00960
         WRITE(8,110) (A(K), K=1,NZ)                             LSC00970
     110 FORMAT(4E19.10)                                         LSC00980
C                                                                LSC00990
C-----END LSCOMPAC----------------------------------------------LSC01000
         STOP                                                    LSC01010
         END                                                     LSC01020
```

-----INPUT/OUTPUT FILE DEFINITIONS FOR LSVAL AND LSVEC-----

Below is a listing of the input/output files which are accessed by the
Lanczos program LSVAL for computing singular values of real
rectangular matrices on user-specified intervals. Included also is a
sample of the input file which LSVAL requires on file 5. The parameters
in this file are supplied in free format.

SAMPLE DEFINITIONS OF THE INPUT/OUTPUT FILES FOR LSVAL
----
*LSVAL EXEC FOR LANCZOS SINGULAR VALUE CALCULATIONS
FI 06 TERM
FILEDEF 1 DISK &1   NSHISTORY A (RECFM F LRECL 80 BLOCK 80
FILEDEF 2 DISK &1   SVHISTORY A (RECFM F LRECL 80 BLOCK 80
FILEDEF 3 DISK &1   GOODEV   A (RECFM F LRECL 80 BLOCK 80
FILEDEF 4 DISK &1   ERRINV   A (RECFM F LRECL 80 BLOCK 80
FILEDEF 5 DISK LSVAL   INPUT   A (RECFM F LRECL 80 BLOCK 80
FILEDEF 8 DISK &1   ACOMPAC   A (RECFM F LRECL 80 BLOCK 80
FILEDEF 11 DISK &1   DISTINCT A (RECFM F LRECL 80 BLOCK 80
LOAD LSVAL LSSUB LSMULT

SAMPLE INPUT FILE FOR LSVAL

 LANCZOS SINGULAR VALUE PROCEDURE,
 WITHOUT REORTHOGONALIZATION BUT WITH BIDIAGONALIZATION.
 LINE 1   M   N   KMAX   NMEVS   MATNO
          100   100   300   1   2220
 LINE 2   SVSEED   RHSEED   MXINIT   MXSTUR
         49302312 7549309   5   100000
 LINE 3   ISTART   ISTOP
                 0   1
 LINE 4   IHIS   IDIST   IWRITE   IPAR
               1   0   1   2
 LINE 5 RELTOL(RELATIVE TOLERANCE USED IN 'COMBINING' COMPUTED GOOD EIGENVALUES)
    .0000000001
 LINE 6 MB(1)   MB(2)   MB(3)   MB(4) (ORDERS OF T(1,MEV) MUST BE EVEN
           280
 LINE 7   NINT   (NUMBER OF BISEC INTERVALS)
            1
 LINE 8   LB(1)   LB(2)   LB(3)   LB(4) (LOWER BOUNDS OF THESE INTERVALS)
            0.0
 LINE 9   UB(1)   UB(2)   UB(3)   UB(4) (UPPER BOUNDS OF THESE INTERVALS)
            1.0

Below is a listing of the input/output files which are accessed by the
Lanczos program for computing singular vectors, LSVEC. Included also
is a sample of the input file which LSVEC requires on file 5. The
parameters in this file are supplied in free format. LSVEC computes
singular vectors for each of a user-specified subset of the
singular values computed by the companion program LSVAL.

SAMPLE DEFINITIONS OF THE INPUT/OUTPUT FILES FOR LSVEC
----
*LSVEC EXEC TO RUN LANCZOS SINGULAR VECTOR PROGRAM
FI 06 TERM
FILEDEF 2 DISK &1   SVHISTORY A (RECFM F LRECL 80 BLOCK 80
FILEDEF 3 DISK &1   GOODSV   A (RECFM F LRECL 80 BLOCK 80
FILEDEF 4 DISK &1   ERRINV   A (RECFM F LRECL 80 BLOCK 80
FILEDEF 5 DISK LSVEC   INPUT   A (RECFM F LRECL 80 BLOCK 80
FILEDEF 8 DISK &1   ACOMPAC   A (RECFM F LRECL 80 BLOCK 80
FILEDEF 9 DISK &1   ERREST   A (RECFM F LREC'_ 80 BLOCK 80
FILEDEF 10 DISK &1   BOUNDS   A (RECFM F LRECL 80 BLOCK 80
FILEDEF 11 DISK &1   TEIGVECS A (RECFM F LRECL 80 BLOCK 80
FILEDEF 12 DISK &1   RITZVECS A (RECFM F LRECL 80 BLOCK 80
FILEDEF 13 DISK &1   PAIGE   A (RECFM F LRECL 80 BLOCK 80
LOAD   LSVEC   LSSUB   LSMULT

SAMPLE INPUT FILE FOR LSVEC

LSVEC SINGULAR VECTORS, NO REORTHOGONALIZATION BUT BIDIAGONALIZATION
LINE 1 MATNO   M   N
        100   100   80
LINE 2 MDIMTV   MDIMRV MBETA (MAX.DIMENSIONS,TVEC,RITVEC AND BETA
        10000   10000   2000
LINE 3   RELTOL
        .0000000001
LINE 4 MBOUND   NTVCON SVTVEC IREAD (FLAGS
        0   1   0   1
LINE 5 TVSTOP   LVCONT ERCONT IWRITE (FLAGS
        0   1   1   1
LINE 6   RHSEED (RANDOM GENERATOR SEED FOR STARTING VECTOR IN INVERM)
        45329517

CHAPTER 7

NONDEFECTIVE COMPLEX SYMMETRIC MATRICES

SECTION 7.1   INTRODUCTION

The FORTRAN codes in this chapter address the question of computing distinct eigenvalues and eigenvectors of a nondefective, complex symmetric matrix, using a single-vector Lanczos procedure. For a given nondefective, complex symmetric matrix A, these codes compute complex scalars λ and corresponding complex vectors $x \neq 0$ such that

$$Ax = \lambda x. \qquad (7.1.1)$$

DEFINITION 7.1.1   A complex nxn matrix $A \equiv (a_{ij})$, $1 \leq i, j \leq n$, is complex symmetric if and only if for every i and j, $a_{ij} = a_{ji}$. A complex symmetric matrix is nondefective if and only if it has a complete set of eigenvectors.

It is straight-forward to show from Definition 7.1.1 that if $A = B + iC$, where A and B are real matrices and $i = \sqrt{-1}$, is a complex symmetric matrix then B and C are real symmetric matrices. It is also easy to prove that if λ and μ are two distinct eigenvalues of A and x and y are corresponding eigenvectors of A, then the Euclidean inner product applied to the complex vectors x and y satisfies

$$x^T y = 0. \qquad (7.1.2)$$

In Eqn(7.1.2) the superscript T denotes transpose. Thus, although the eigenvectors of a complex symmetric matrix are not orthogonal with respect to the complex norm, $\|x\|_{\mathscr{C}}^2 = \sum_{i=1}^{n} \overline{x(i)} x(i)$, they are real orthogonal in the sense specified in Eqn(7.1.2). Therefore, when we consider generalizing the Lanczos recursion to the complex symmetric case we are led to consider an 'inner product' which is a mixture of real and complex quantities. In fact the Euclidean inner product, which of course is not an inner product for complex vectors, is the natural 'inner product' to use in the complex symmetric case.

Complex symmetric matrices are not 'easy' like real symmetric matrices. They bear little resemblance to real symmetric matrices. Complex symmetric matrices need not have complete sets of eigenvectors. Even if a complete set of eigenvectors exists, eigenvectors corresponding to different eigenvalues are only real orthogonal in the sense of Eqn(7.1.2). If a small perturbation is applied to a complex symmetric matrix, then large perturbations in the eigenvalues may result. See Wilkinson [1965] for a discussion of the properties of complex symmetric matrices.

The Lanczos recursion as presented in Eqns(1.2.1) - (1.2.2) is only applicable to real symmetric matrices so we ask the question: How do we construct a complex symmetric version of the basic Lanczos recursion which will give us the desired eigenvalues? We have used what has been suggested elsewhere, Moro and Freed [1981]. In particular, we use the recursion in Eqn(1.2.1) with the formulas for the scalars α_i and β_{i+1} given in Eqn(1.2.2), except that the quantities involved are now complex-valued, but the real Euclidean inner product is used. See Section 6.3 of Chapter 6 in Volume 1.

There are some fundamental differences between the amount of computation required by the complex symmetric codes versus that required by the real symmetric codes. First, all of the complex symmetric computations are done in double precision complex arithmetic. All the vectors used are complex vectors. Each of the Lanczos matrices generated is a complex symmetric tridiagonal matrix. Unfortunately, there is no simple analog of the bisection procedure used in the real symmetric case which would allow us to compute the eigenvalues of a given complex symmetric tridiagonal matrix on only some small portion of the spectrum. We are therefore forced to do a complete eigenvalue computation on each complex symmetric tridiagonal matrix which we consider. Actually in the complex symmetric case we are forced to do two complete eigenvalue computations for each Lanczos tridiagonal matrix which we consider. Two are required because the identification test for categorizing the eigenvalues of the Lanczos T-matrices into 'good' and 'spurious' ones uses the eigenvalues of the corresponding tridiagonal matrix obtained from the Lanczos T-matrix by crossing out the first row and column of that matrix. This is the same identification test as that used in the procedures for real symmetric problems. However, in the real symmetric cases this test is directly incorporated into the BISEC subroutine which is used to compute the eigenvalues of the Lanczos matrices, and the resulting cost of this test is negligible for those types of problems.

These codes can be used to compute either a very few or very many of the distinct eigenvalues of a nondefective, complex symmetric matrix. As the documentation in the next section indicates, the A-multiplicity of a given computed eigenvalue can be obtained only with additional computation, and the modifications required to do this additional computation are not included in these versions of the codes.

The Lanczos recursions used generate a family of complex symmetric, tridiagonal matrices. A real orthogonal analog of the EISPACK [1976,1977] subroutine IMTQL1 which we call CMTQL1 was developed to compute the eigenvalues of the complex symmetric, tridiagonal Lanczos matrices generated. There is no reorthogonalization of the Lanczos vectors at any stage in any of the computations.

CSLEVAL, the main program for the complex symmetric eigenvalue computations, calls the subroutines COMPEV and CMTQL1 to compute the eigenvalues of the Lanczos T-matrices specified by the user. The eigenvalues of the related complex symmetric tridiagonal matrices obtained by deleting the first row and first column from the given Lanczos T-matrix are also computed. COMPEV then determines the T-multiplicities of the T-eigenvalues and sorts the computed T-eigenvalues into two classes, the 'good' T-eigenvalues and the 'spurious' T-eigenvalues. The 'good' T-eigenvalues are accepted as approximations to eigenvalues of the user-specified matrix A. The accuracy of these 'good' T-eigenvalues as eigenvalues of A is then estimated using error estimates computed by a complex version of the subroutine INVERR. Error estimates are computed only for isolated 'good' T-eigenvalues. All other 'good' T-eigenvalues are assumed to have converged. Convergence is then checked. If convergence has not yet occurred and a larger Lanczos matrix has been specified by the user, the program will continue on to the larger T-matrix, repeating the above procedure on this larger matrix.

Once the eigenvalues been computed accurately enough, the user can select a subset of the 'converged' eigenvalues for which eigenvectors are to be computed. The main program CSLEVEC, for computing eigenvectors of complex symmetric matrices, is then used to compute these desired eigenvectors.

As stated earlier, all computations are in double precision complex arithmetic. The user must supply a subroutine USPEC which defines and initializes the user-specified matrix A and a subroutine CMATV which computes matrix-vector multiplies Ax for any given vector x. These subroutines must be constructed in such a way as to take advantage of the sparsity (and/or structure) of the user-supplied A-matrix and such that these computations are done accurately.

The user should note that the complex symmetric computations are considerably more expensive than the corresponding real symmetric ones. Two complete T-matrix eigenvalue computations must be done for each T-size. Moreover, the accuracy of these computations is noticeably less than that achievable in the real symmetric case. This is to be expected from the perturbation analysis for the complex symmetric case. Therefore we reduced the anticipated accuracy of the computed eigenvalues and used larger tolerances in our multiplicity and spuriousness tests. These larger tolerances decrease the resolution capabilities of these codes. However, these tolerances are realistic. Moreover, these complex symmetric codes cannot be expected to handle stiff problems effectively. More details about these complex symmetric, single-vector Lanczos procedures are included in Chapter 6 of Volume 1.

SECTION 7.2   DOCUMENTATION FOR PROGRAMS IN CHAPTER 7

```
C-----DOCUMENTATION FOR SINGLE-VECTOR---------------------------------CSL00010
C     LANCZOS EIGENVALUE/EIGENVECTOR PROGRAMS FOR                      CSL00020
C     NONDEFECTIVE COMPLEX SYMMETRIC MATRICES                          CSL00030
C                                                                      CSL00040
C     AUTHORS:  JANE CULLUM AND RALPH A. WILLOUGHBY, IBM RESEARCH,     CSL00050
C               YORKTOWN HEIGHTS, NY 10598.  PHONE: 914-945-2227       CSL00060
C                                                                      CSL00070
C     GIVEN A NONDEFECTIVE COMPLEX SYMMETRIC MATRIX A OF ORDER N       CSL00080
C     THE THREE SETS OF FORTRAN FILES LABELLED CSLEVAL, CSLESUB,       CSL00090
C     AND CSLEMULT CAN BE USED TO COMPUTE DISTINCT EIGENVALUES OF      CSL00100
C     A.  NOTE THAT THESE PROGRAMS DIFFER FROM THE REAL SYMMETRIC      CSL00110
C     AND HERMITIAN PROGRAMS IN THAT IT IS NOT POSSIBLE TO             CSL00120
C     COMPUTE THE EIGENVALUES OF THE LANCZOS TRIDIAGONAL MATRICES      CSL00130
C     ONLY IN SPECIFIED INTERVALS.  THUS, ON ANY GIVEN                 CSL00140
C     ITERATION ALL OF THE EIGENVALUES OF THESE TRIDIAGONAL MATRICES   CSL00150
C     MUST BE COMPUTED.  IN FACT TWO COMPLETE TRIDIAGONAL EIGENVALUE   CSL00160
C     COMPUTATIONS ARE USED.                                           CSL00170
C                                                                      CSL00180
C     CORRESPONDING EIGENVECTORS FOR SELECTED, COMPUTED EIGENVALUES CAN CSL00190
C     BE COMPUTED USING THE CORRESPONDING SETS OF FILES LABELLED       CSL00200
C     CSLEVEC, CSLESUB AND CSLEMULT.                                   CSL00210
C                                                                      CSL00220
C     THESE PROGRAMS ALL USE A GENERALIZATION OF LANCZOS              CSL00230
C     TRIDIAGONALIZATION TO COMPLEX SYMMETRIC MATRICES TO              CSL00240
C     GENERATE COMPLEX SYMMETRIC TRIDIAGONAL MATRICES, T(1,MEV)        CSL00250
C     OF ORDER MEV.  NO REORTHOGONALIZATION IS USED.  SUBSETS OF       CSL00260
C     THE EIGENVALUES OF THESE T-MATRICES, LABELLED AS THE             CSL00270
C     'GOOD EIGENVALUES', YIELD APPROXIMATIONS TO THE DESIRED          CSL00280
C     EIGENVALUES OF A.  CORRESPONDING RITZ VECTORS ARE APPROXIMATIONS CSL00290
C     TO THE DESIRED EIGENVECTORS OF A.  NOTE THAT IN THE DISCUSSION   CSL00300
C     T(1,MEV) DENOTES THE LANCZOS MATRIX OF ORDER MEV AND T(2,MEV)    CSL00310
C     DENOTES THE MATRIX OF SIZE MEV-1 OBTAINED FROM T(1,MEV) BY       CSL00320
C     DELETING THE FIRST ROW AND COLUMN OF T(1,MEV).                   CSL00330
C                                                                      CSL00340
C     THE IDEAS USED IN THESE PROGRAMS ARE DISCUSSED IN THE FOLLOWING  CSL00350
C     REFERENCES.                                                      CSL00360
C                                                                      CSL00370
C     1.  JANE CULLUM AND RALPH A. WILLOUGHBY, LANCZOS ALGORITHMS      CSL00380
C         FOR LARGE SYMMETRIC MATRICES,  PROGRESS IN                   CSL00390
C         SCIENTIFIC COMPUTING, EDITORS, G. GOLUB, H.C. KREISS,        CSL00400
C         S. ARBARBANEL, AND R. GLOWINSKI,  BIRKHAUSER BOSTON INC.,    CSL00410
C         CAMBRIDGE, MASSACHUSETTS, 1984.                              CSL00420
C                                                                      CSL00430
C     2.  JANE CULLUM AND RALPH A. WILLOUGHBY, COMPUTING EIGENVECTORS  CSL00440
C         (AND EIGENVALUES) OF LARGE, SYMMETRIC MATRICES USING         CSL00450
C         LANCZOS TRIDIAGONALIZATION, LECTURE NOTES IN MATHEMATICS,    CSL00460
C         773, NUMERICAL ANALYSIS PROCEEDINGS, DUNDEE 1979, EDITED BY  CSL00470
C         G. A. WATSON, SPRINGER-VERLAG, (1980), BERLIN, PP.46-63.     CSL00480
C                                                                      CSL00490
C     3. IBID, LANCZOS AND THE COMPUTATION IN SPECIFIED INTERVALS OF   CSL00500
C         THE SPECTRUM OF LARGE SPARSE, REAL SYMMETRIC MATRICES, SPARSE CSL00510
C         MATRIX PROCEEDINGS 1978, ED. I.S. DUFF AND G. W. STEWART,    CSL00520
C         SIAM, PHILADELPHIA, PP.220-255, 1979.                        CSL00530
C                                                                      CSL00540
C     4. IBID, COMPUTING EIGENVALUES OF VERY LARGE SYMMETRIC MATRICES- CSL00550
C         AN IMPLEMENTATION OF A LANCZOS ALGORITHM WITHOUT             CSL00560
C         REORTHOGONALIZATION, J. COMPUT. PHYS. 44(1981), 329-358.     CSL00570
C                                                                      CSL00580
C     5. IBID, A LANCZOS ALGORITHM FOR NONDEFECTIVE COMPLEX SYMMETRIC  CSL00590
C         MATRICES, IBM RESEARCH REPORT, 1984.                         CSL00600
C                                                                      CSL00610
C-----PORTABILITY-----------------------------------------------------CSL00620
C                                                                      CSL00630
C     PROGRAMS WERE TESTED FOR PORTABILITY USING THE PFORT VERIFIER.   CSL00640
```

```
C       FOR DETAILS OF THE VERIFIER SEE FOR EXAMPLE, B. G. RYDER AND        CSL00650
C       A. D. HALL, 'THE PFORT VERIFIER', COMPUTING SCIENCE TECHNICAL       CSL00660
C       REPORT 12, BELL LABORATORIES, MURRAY HILL, NEW JERSEY 07974,        CSL00670
C       (REVISED), JANUARY 1981.                                           CSL00680
C                                                                          CSL00690
C       PORTABILITY:                                                       CSL00700
C       THESE PROGRAMS ARE NOT PORTABLE DUE TO THE USE OF COMPLEX*16       CSL00710
C       VARIABLES AND CORRESPONDING COMPLEX FUNCTIONS.  IN ADDITION, THE   CSL00720
C       PFORT VERIFIER IDENTIFIED THE FOLLOWING NONPORTABLE                CSL00730
C       CONSTRUCTIONS.                                                     CSL00740
C       IN CSLEVAL AND IN CSLEVEC                                          CSL00750
C           1.  DATA/MACHEP STATEMENT                                      CSL00760
C           2.  ALL READ(5,*) STATEMENTS (FREE FORMAT)                     CSL00770
C           3.  FORMAT(20A4) USED FOR THE EXPLANATORY HEADER ARRAY, EXPLANCSL00780
C           4.  HEXADECIMAL FORMAT (4Z20) FOR ALPHA/BETA FILES 1 AND 2.    CSL00790
C       IN CSLEMULT                                                        CSL00800
C           1.  IN CMATV AND  USPEC THE ENTRY THAT PASSES THE STORAGE      CSL00810
C               LOCATIONS OF THE ARRAYS DEFINING THE USER-SPECIFIED        CSL00820
C               MATRIX.                                                    CSL00830
C           2.  IN SAMPLE USPEC PROVIDED : FREE FORMAT (8,*), THE          CSL00840
C               FORMAT (20A4), AND THE DATA/MACHEP STATEMENT.              CSL00850
C                                                                          CSL00860
C       IN THE COMMENTS BELOW :                                            CSL00870
C       REAL*16 = COMPLEX VARIABLE, 16 BYTES OF STORAGE                    CSL00880
C       REAL*8 = REAL VARIABLE, 8 BYTES OF STORAGE                         CSL00890
C       REAL*4 = REAL VARIABLE, 4 BYTES OF STORAGE                         CSL00900
C       INTEGER*4 = INTEGER VARIABLE, 4 BYTES OF STORAGE                   CSL00910
C                                                                          CSL00920
C-----A-MATRIX SPECIFICATION----------------------------------------------CSL00930
C                                                                          CSL00940
C       SUBROUTINE USPEC IS USED TO SPECIFY THE USER-SUPPLIED MATRIX.      CSL00950
C       SUBROUTINE CMATV IS A CORRESPONDING MATRIX-VECTOR MULTIPLY         CSL00960
C       SUBROUTINE WHICH SHOULD BE DESIGNED TO TAKE ADVANTAGE OF           CSL00970
C       ANY SPECIAL PROPERTIES OF THE USER-SUPPLIED MATRIX.  THE           CSL00980
C       MATRIX-VECTOR MULTIPLIES REQUIRED BY THE LANCZOS PROCEDURES        CSL00990
C       MUST BE COMPUTED RAPIDLY AND ACCURATELY.                          CSL01000
C                                                                          CSL01010
C       SUBROUTINE USPEC HAS THE CALLING SEQUENCE                          CSL01020
C                                                                          CSL01030
C             CALL USPEC(N,MATNO)                                          CSL01040
C                                                                          CSL01050
C       WHERE N IS THE ORDER OF THE USER-SUPPLIED MATRIX A AND             CSL01060
C       MATNO IS A <= 8 DIGIT INTEGER USED AS A MATRIX AND                 CSL01070
C       TEST IDENTIFICATION NUMBER.  THIS SUBROUTINE DEFINES (DIMENSIONS)  CSL01080
C       THE ARRAYS REQUIRED TO SPECIFY THE USER-SUPPLIED MATRIX AND        CSL01090
C       INITIALIZES THESE ARRAYS AND ANY OTHER PARAMETERS NEEDED TO        CSL01100
C       DEFINE THE MATRIX.  THE STORAGE LOCATIONS OF THESE PARAMETERS      CSL01110
C       AND ARRAYS ARE THEN PASSED TO THE MATRIX-VECTOR MULTIPLY           CSL01120
C       SUBROUTINE CMATV VIA AN ENTRY.  A SAMPLE USPEC SUBROUTINE          CSL01130
C       IS INCLUDED.  THIS SAMPLE SUBROUTINE ASSUMES THAT THE MATRIX       CSL01140
C       IS STORED ON FILE 8 IN A TYPICAL SPARSE MATRIX FORMAT.             CSL01150
C       SEE THE HEADER ON THE SUBROUTINE USPEC FOR DETAILS ON THIS         CSL01160
C       PARTICULAR STORAGE FORMAT.                                         CSL01170
C                                                                          CSL01180
C       SUBROUTINE CMATV HAS THE CALLING SEQUENCE                          CSL01190
C                                                                          CSL01200
C             CALL CMATV(W,U,SUM)                                          CSL01210
C                                                                          CSL01220
C       IN THE COMPLEX SYMMETRIC CASE, U AND W ARE                         CSL01230
C       COMPLEX*16 VECTORS AND SUM IS A COMPLEX*16                         CSL01240
C       SCALAR.   CMATV CALCULATES U = A*W - SUM*U FOR THE                 CSL01250
C       USER-SPECIFIED MATRIX A.  THE ARRAY AND PARAMETER INFORMATION      CSL01260
C       NEEDED TO PERFORM THE MATRIX-VECTOR MULTIPLIES IS PASSED TO        CSL01270
C       THE CMATV SUBROUTINE FROM THE USPEC SUBROUTINE VIA THE CMATVE      CSL01280
C       ENTRY IN CMATV.  A SAMPLE CMATV SUBROUTINE IS INCLUDED WHICH       CSL01290
C       COMPUTES MATRIX-VECTOR MULTIPLIES FOR AN ARBITRARY SPARSE,         CSL01300
C       COMPLEX SYMMETRIC MATRIX STORED IN THE SPARSE FORMAT               CSL01310
```

```
C       SPECIFIED IN THE SAMPLE USPEC SUBROUTINE.                    CSL01320
C                                                                    CSL01330
C       CMATV IS CALLED FROM THE SUBROUTINE LANCZS WHICH GENERATES   CSL01340
C       THE T-MATRICES IN THE ALPHA, BETA ARRAYS.  IT IS ALSO CALLED CSL01350
C       FROM THE MAIN PROGRAM CSLEVEC FOR THE EIGENVECTOR COMPUTATIONS. CSL01360
C       CMATV IS DECLARED AS AN EXTERNAL VARIABLE AND IS AN ARGUMENT CSL01370
C       FOR THE SUBROUTINE LANCZS.                                   CSL01380
C                                                                    CSL01390
C       THE USPEC AND CMATV SUBROUTINES MUST BE MODIFIED BY THE USER CSL01400
C       TO ACCOMODATE THE USER'S SPECIFIED MATRIX.                   CSL01410
C                                                                    CSL01420
C       THE MAIN PROGRAMS FOR THE EIGENVALUE AND EIGENVECTOR         CSL01430
C       CALCULATIONS ASSUME THAT INPUT FILE 5 CONTAINS N = ORDER OF  CSL01440
C       THE MATRIX AND MATNO = AN IDENTIFICATION NUMBER OF <= 8 DIGITS CSL01450
C       FOR THE MATRIX AND THE RUN.                                  CSL01460
C                                                                    CSL01470
C                                                                    CSL01480
C-----MACHEP-------------------------------------------------------CSL01490
C                                                                    CSL01500
C                                                                    CSL01510
C       MACHEP IS A MACHINE DEPENDENT PARAMETER SPECIFYING THE RELATIVE CSL01520
C       PRECISION OF THE FLOATING POINT ARITHMETIC USED.             CSL01530
C       MACHEP = 2.2 * 10**-16 FOR DOUBLE PRECISION ARITHMETIC ON    CSL01540
C       IBM 370-3081.                                                CSL01550
C                                                                    CSL01560
C       THE USER WILL HAVE TO RESET THIS PARAMETER TO                CSL01570
C       THE CORRESPONDING VALUE FOR THE MACHINE BEING USED.  NOTE THAT CSL01580
C       IF A MACHINE WITH A MACHINE EPSILON THAT IS MUCH LARGER THAN THE CSL01590
C       VALUE GIVEN HERE IS BEING USED, THEN THERE COULD BE          CSL01600
C       PROBLEMS WITH THE TOLERANCES.                                CSL01610
C                                                                    CSL01620
C                                                                    CSL01630
C-----SUBROUTINES AND FUNCTIONS USER MUST SUPPLY----------------------CSL01640
C                                                                    CSL01650
C       GENRAN,  MASK,  USPEC,  AND  CMATV                           CSL01660
C                                                                    CSL01670
C       GENRAN = COMPUTES K PSEUDO-RANDOM NUMBERS AND STORES THEM IN CSL01680
C                THE REAL ARRAY, G.  THIS SUBROUTINE IS USED TO      CSL01690
C                GENERATE A STARTING VECTOR FOR THE LANCZOS PROCEDURE CSL01700
C                IN THE SUBROUTINE LANCZS AND A STARTING RIGHT-HAND SIDE CSL01710
C                FOR INVERSE ITERATION IN THE SUBROUTINE INVERR.     CSL01720
C                                                                    CSL01730
C                TESTS REPORTED IN THE REFERENCES USED EITHER GGL1 OR CSL01740
C                GGL2 FROM THE IBM LIBRARY SLMATH.                   CSL01750
C                THE EXISTING CALLING SEQUENCE IS:                   CSL01760
C                                                                    CSL01770
C                     CALL GENRAN(IIX,G,K).                          CSL01780
C                                                                    CSL01790
C                WHERE IIX =INTEGER SEED, G = REAL*4 ARRAY WHOSE      CSL01800
C                DIMENSION MUST BE >= K.  K RANDOM NUMBERS ARE GENERATED CSL01810
C                AND PLACED IN G.                                    CSL01820
C                                                                    CSL01830
C       MASK = MASKS OVERFLOW AND UNDERFLOW.                         CSL01840
C              USER MUST SUPPLY OR COMMENT OUT CALL.                 CSL01850
C                                                                    CSL01860
C       USPEC = DIMENSIONS AND INITIALIZES ARRAYS NEEDED TO SPECIFY  CSL01870
C               USER-SUPPLIED MATRIX.  SEE A-MATRIX SPECIFICATION SECTION.CSL01880
C                                                                    CSL01890
C       CMATV = MATRIX-VECTOR MULTIPLY FOR USER-SUPPLIED MATRIX.     CSL01900
C               SEE A-MATRIX SPECIFICATION SECTION.                  CSL01910
C                                                                    CSL01920
C                                                                    CSL01930
C-------------------------------------------------------------------CSL01940
C                                                                    CSL01950
C       COMMENTS FOR EIGENVALUE COMPUTATIONS                         CSL01960
C                                                                    CSL01970
C-------------------------------------------------------------------CSL01980
```

```
C                                                                        CSLO1990
C                                                                        CSLO2000
C-----PARAMETER CONTROLS FOR EIGENVALUE PROGRAMS------------------------CSLO2010
C                                                                        CSLO2020
C        PARAMETER CONTROLS ARE INTRODUCED TO ALLOW SEGMENTATION OF THE  CSLO2030
C        EIGENVALUE COMPUTATIONS AND TO ALLOW VARIOUS COMBINATIONS OF    CSLO2040
C        READ/WRITES.                                                    CSLO2050
C                                                                        CSLO2060
C        THE FLAG ISTART CONTROLS THE T-MATRIX (ALPHA/BETA HISTORY)      CSLO2070
C        GENERATION.                                                     CSLO2080
C                                                                        CSLO2090
C        ISTART = (0,1)  MEANS                                           CSLO2100
C                                                                        CSLO2110
C                (0) THERE IS NO EXISTING ALPHA/BETA HISTORY AND ONE     CSLO2120
C                    MUST BE GENERATED.                                  CSLO2130
C                                                                        CSLO2140
C                (1) THERE IS AN EXISTING ALPHA/BETA HISTORY AND IT IS   CSLO2150
C                    TO BE READ IN FROM FILE 2 AND EXTENDED IF NECESSARY. CSLO2160
C                                                                        CSLO2170
C        THE FLAG ISTOP CAN BE USED IN CONJUNCTION WITH THE FLAG ISTART TO CSLO2180
C        ALLOW SEGMENTATION OF THE EIGENVALUE COMPUTATIONS.             CSLO2190
C                                                                        CSLO2200
C        ISTOP  = (0,1)  MEANS                                           CSLO2210
C                                                                        CSLO2220
C                (0) PROGRAM COMPUTES ONLY THE REQUESTED ALPHAS/BETAS,   CSLO2230
C                    STORES THEM AND THE LAST 2 LANCZOS VECTORS GENERATED CSLO2240
C                    IN FILE 1 AND THEN TERMINATES.                      CSLO2250
C                                                                        CSLO2260
C                (1) PROGRAM COMPUTES REQUESTED ALPHAS/BETAS AND THEN    CSLO2270
C                    USES THE CMTQL1 SUBROUTINE TO CALCULATE EIGENVALUES CSLO2280
C                    OF THE TRIDIAGONAL MATRICES GENERATED FOR THE ORDERS CSLO2290
C                    SPECIFIED BY THE USER.  PROGRAM THEN USES THE       CSLO2300
C                    SUBROUTINE INVERR TO COMPUTE ERROR ESTIMATES FOR    CSLO2310
C                    THE ISOLATED GOOD T-EIGENVALUES WHICH ARE USED TO   CSLO2320
C                    CHECK THE CONVERGENCE OF THESE GOOD T-EIGENVALUES.  CSLO2330
C                                                                        CSLO2340
C        CONTROL PARAMETERS FOR WRITES                                   CSLO2350
C                                                                        CSLO2360
C        IHIS  = (0,1)  MEANS                                            CSLO2370
C                                                                        CSLO2380
C                (0) IF ISTOP .GT. 0 THEN ALPHAS/BETAS ARE NOT SAVED     CSLO2390
C                    ON FILE 1.                                          CSLO2400
C                                                                        CSLO2410
C                (1) PROGRAM WRITES ALPHAS/BETAS AND LAST 2 LANCZOS      CSLO2420
C                    VECTORS TO FILE 1 SO THAT THE T-MATRIX GENERATION   CSLO2430
C                    MAY BE REUSED OR CONTINUED LATER IF NECESSARY.      CSLO2440
C                    TYPICALLY ONE WOULD ALWAYS DO THIS ON ANY RUN WHERE CSLO2450
C                    A HISTORY FILE IS BEING GENERATED.  HISTORY MUST    CSLO2460
C                    BE SAVED IN MACHINE FORMAT ((4Z20) FOR IBM/3081)    CSLO2470
C                    SO THAT NO ERRORS ARE INTRODUCED DUE TO FORMAT      CSLO2480
C                    CONVERSIONS.                                        CSLO2490
C                                                                        CSLO2500
C        IDIST = (0,1)  MEANS                                            CSLO2510
C                                                                        CSLO2520
C                (0) DISTINCT EIGENVALUES OF T-MATRICES ARE NOT SAVED.   CSLO2530
C                                                                        CSLO2540
C                (1) PROGRAM WRITES COMPUTED DISTINCT EIGENVALUES OF     CSLO2550
C                    T-MATRICES ALONG WITH THEIR T-MULTIPLICITIES        CSLO2560
C                    TO FILE 11.                                         CSLO2570
C                                                                        CSLO2580
C        IWRITE = (0,1)   MEANS                                          CSLO2590
C                                                                        CSLO2600
C                (0) NO EXTENDED OUTPUT FROM SUBROUTINES COMPEV AND INVERRCSLO2610
C                    IS SENT TO FILE 6.                                  CSLO2620
C                                                                        CSLO2630
C                (1) INDIVIDUAL COMPUTED EIGENVALUES AND CORRESPONDING   CSLO2640
C                    ERROR ESTIMATES FROM THE SUBROUTINES COMPEV AND     CSLO2650
```

```
C                    INVERR ARE PRINTED OUT TO FILE 6 AS THEY ARE COMPUTEDCSL02660
C                                                                    CSL02670
C      SAVTEV = (-1,0,1)    MEANS                                     CSL02680
C                                                                    CSL02690
C               (-1) NO T-EIGENVALUE COMPUTATIONS.  PREVIOUSLY-COMPUTED CSL02700
C                    EIGENVALUES OF T(1,MEV) AND T(2,MEV) ARE TO     CSL02710
C                    BE READ IN FROM FILE 10.                        CSL02720
C                                                                    CSL02730
C               (0) COMPUTED EIGENVALUES OF T(1,MEV) AND OF T(2,MEV) CSL02740
C                    ARE NOT TO BE SAVED ON FILE 10.  THIS IS NOT    CSL02750
C                    RECOMMENDED IF THE T-MATRICES BEING USED ARE VERY CSL02760
C                    LARGE BECAUSE IN THAT CASE THE TRIDIAGONAL      CSL02770
C                    EIGENVALUE COMPUTATIONS ARE VERY EXPENSIVE.     CSL02780
C                                                                    CSL02790
C               (1) COMPUTED EIGENVALUES OF T(1,MEV) AND OF T(2,MEV) CSL02800
C                    WILL BE SAVED ON FILE 10.  THIS IS RECOMMENDED  CSL02810
C                    BECAUSE ONCE THESE T-EIGENVALUES ARE COMPUTED THE CSL02820
C                    LATTER PORTION OF THE EIGENVALUE PROGRAM IS EASILY CSL02830
C                    RESTARTED FROM THE POINT OF THESE EIGENVALUE    CSL02840
C                    COMPUTATIONS.                                   CSL02850
C                                                                    CSL02860
C      THE PROGRAM ALWAYS MAKES A SEPARATE LIST OF THE COMPUTED GOOD CSL02870
C      T-EIGENVALUES ALONG WITH THEIR MINIMAL GAPS AND WRITES THEM OUT CSL02880
C      TO FILE 3.  CORRESPONDING ERROR ESTIMATES FOR ANY ISOLATED   CSL02890
C      GOOD T-EIGENVALUES ARE ALWAYS WRITTEN TO FILE 4.             CSL02900
C                                                                    CSL02910
C                                                                    CSL02920
C-----INPUT/OUTPUT FILES FOR EIGENVALUE PROGRAMS----------------------CSL02930
C                                                                    CSL02940
C      ANY INPUT DATA OTHER THAN THE ALPHA/BETA HISTORY OR PREVIOUSLY- CSL02950
C      COMPUTED EIGENVALUES OF T(1,MEV) AND T(2,MEV) SHOULD BE STORED CSL02960
C      ON FILE 5. SEE SAMPLE INPUT/OUTPUT FROM TYPICAL RUN.         CSL02970
C      THE READ STATEMENTS IN THE GIVEN FORTRAN PROGRAM ASSUME THAT CSL02980
C      THE DATA STORED ON FILE 5 IS IN FREE FORMAT.  USER SHOULD NOTE CSL02990
C      THAT 'FREE FORMAT' IS NOT CLASSIFIED AS PORTABLE BY PFORT SO THAT CSL03000
C      THE USER MAY HAVE TO MODIFY THE READ STATEMENTS FROM FILE 5 TO CSL03010
C      CONFORM TO WHAT IS PERMISSIBLE ON THE MACHINE BEING USED.    CSL03020
C                                                                    CSL03030
C      FILE 6 WAS USED AS THE INTERACTIVE TERMINAL OUTPUT FILE.     CSL03040
C      THIS FILE PROVIDES A RUNNING ACCOUNT OF THE PROGRESS OF THE  CSL03050
C      COMPUTATIONS.  THE AMOUNT OF INFORMATION PRINTED OUT IS      CSL03060
C      CONTROLLED BY THE PARAMETER IWRITE.                          CSL03070
C                                                                    CSL03080
C DESCRIPTION OF OTHER I/O FILES                                     CSL03090
C                                                                    CSL03100
C FILE (K)    CONTAINS:                                              CSL03110
C                                                                    CSL03120
C      (1)    OUTPUT FILE:                                           CSL03130
C             HISTORY FILE OF NEWLY-GENERATED T-MATRIX (ALPHA AND    CSL03140
C             BETA VECTORS) AND LAST 2 LANCZOS VECTORS USED          CSL03150
C             IN THE T-MATRIX GENERATION.                            CSL03160
C             IF IHIS = 0 AND ISTOP = 1, FILE 1 IS NOT WRITTEN.      CSL03170
C                                                                    CSL03180
C      (2)    INPUT FILE:                                            CSL03190
C             SAME AS FILE 1 EXCEPT THAT IT CONTAINS A               CSL03200
C             PREVIOUSLY-GENERATED T-MATRIX (IF ANY).  IF ISTART = 1, CSL03210
C             PROGRAM ASSUMES THAT THERE IS A HISTORY FILE OF ALPHAS CSL03220
C             AND BETAS ON FILE 2.  THESE ALPHAS AND BETAS ARE       CSL03230
C             READ IN ALONG WITH THE LAST TWO LANCZOS VECTORS        CSL03240
C             USED IN THE T-MATRIX GENERATION.                       CSL03250
C                                                                    CSL03260
C      (3)    OUTPUT FILE:                                           CSL03270
C             COMPUTED GOOD EIGENVALUES OF THE T-MATRICES USED. ALSO CSL03280
C             CONTAINS T-MULTIPLICITIES OF THESE EIGENVALUES AS      CSL03290
C             EIGENVALUES OF THE T-MATRIX, AND THEIR GAPS AS         CSL03300
C             EIGENVALUES IN THE A MATRIX AND IN THE T-MATRIX.       CSL03310
C             FILE 3 IS ALWAYS WRITTEN.                              CSL03320
```

```
C                                                                  CSL03330
C      (4)      OUTPUT FILE:                                        CSL03340
C               ERROR ESTIMATES FOR THE ISOLATED GOOD T-EIGENVALUES WHICHCSL03350
C               ARE OBTAINED USING THE SUBROUTINE INVERR. THESE    CSL03360
C               ESITMATES USE THE LAST COMPONENTS OF THE ASSOCIATED CSL03370
C               T-EIGENVECTORS WHICH ARE COMPUTED USING INVERSE    CSL03380
C               ITERATION.  FILE 4 IS ALWAYS WRITTEN.              CSL03390
C                                                                  CSL03400
C      (8)      INPUT FILE:                                        CSL03410
C               SAMPLE USPEC SUBROUTINE ASSUMES THAT THE ARRAYS    CSL03420
C               REQUIRED TO SPECIFY THE USER'S-MATRIX ARE STORED ON CSL03430
C               FILE 8.  USERS MUST MAKE WHATEVER DEFINITIONS ARE  CSL03440
C               APPROPRIATE FOR THEIR MATRICES.                    CSL03450
C                                                                  CSL03460
C      (10)     OUTPUT OR INPUT FILE DEPENDING UPON VALUE OF SAVTEV: CSL03470
C               COMPUTED EIGENVALUES OF EACH T(1,MEV) FOLLOWED     CSL03480
C               BY THE COMPUTED EIGENVALUES OF THE CORRESPONDING   CSL03490
C               T(2,MEV).                                          CSL03500
C                                                                  CSL03510
C      (11)     OUTPUT FILE:                                       CSL03520
C               COMPUTED DISTINCT EIGENVALUES OF T-MATRICES USED.  CSL03530
C               ALSO CONTAINS THEIR T-MULTIPLICITIES AND T-GAPS TO CSL03540
C               NEAREST DISTINCT EIGENVALUES, AND THE T-MULTIPLICITY CSL03550
C               PATTERN OF THE GOOD AND THE SPURIOUS T-EIGENVALUES. CSL03560
C               FILE 11 IS WRITTEN ONLY IF IDIST = 1.              CSL03570
C                                                                  CSL03580
C                                                                  CSL03590
C-----PARAMETERS SET BY THE EIGENVALUE PROGRAM--------------------------CSL03600
C                                                                  CSL03610
C      THESE PARAMETERS ARE SET INTERNALLY IN THE PROGRAM          CSL03620
C                                                                  CSL03630
C      SCALEK    K = 1,2,3,4                                       CSL03640
C                                                                  CSL03650
C                THE SCALING FACTORS SCALEK HAVE BEEN INTRODUCED IN AN CSL03660
C                ATTEMPT TO MAKE THE TOLERANCES USED IN THE        CSL03670
C                T-MULTIPLICITY, SPURIOUS, AND ISOLATION TESTS ADJUST CSL03680
C                TO THE SCALE OF THE GIVEN MATRIX.  THESE FACTORS MUST CSL03690
C                NOT BE MODIFIED.                                  CSL03700
C                                                                  CSL03710
C      BTOL = RELATIVE TOLERANCE USED TO ESTIMATE ANY LOSS OF LOCAL CSL03720
C             ORTHOGONALITY OF THE LANCZOS VECTORS AFTER THE T-MATRIX CSL03730
C             HAS BEEN GENERATED.  THE LANCZOS PROCEDURE WORKS WELL CSL03740
C             ONLY IF LOCAL ORTHOGONALITY BETWEEN SUCCESSIVE LANCZOS CSL03750
C             VECTORS IS MAINTAINED.  THE TNORM SUBROUTINE TESTS   CSL03760
C             WHETHER OR NOT                                       CSL03770
C                                                                  CSL03780
C                 MINIMUM  |BETA(I)|/||A|| > BTOL.                 CSL03790
C                 I=2,KMAX                                         CSL03800
C                                                                  CSL03810
C             IF THIS TEST IS VIOLATED BY SOME BETA AND A T-MATRIX THAT CSL03820
C             WOULD INCLUDE SUCH A BETA IS REQUESTED, THEN THE LANCZOS CSL03830
C             PROCEDURE WILL TERMINATE FOR THE USER TO DECIDE WHAT TO CSL03840
C             DO.  THE USER CAN OVER-RIDE THIS TEST BY SIMPLY DECREASING CSL03850
C             THE SIZE OF BTOL, BUT THEN CONVERGENCE IS NOT AS CERTAIN. CSL03860
C             THE PROGRAM SETS BTOL = 1.D-8 WHICH IS A VERY CONSERVATIVE CSL03870
C             CHOICE. THE || A || IS ESTIMATED BY USING            CSL03880
C             AN ESTIMATE OF THE NORM OF THE T-MATRIX, T(1,KMAX).  CSL03890
C                                                                  CSL03900
C      GAPTOL = RELATIVE TOLERANCE USED IN THE SUBROUTINE ISOEV    CSL03910
C               TO DETERMINE WHICH OF THE GOOD T-EIGENVALUES NEED  CSL03920
C               ERROR ESTIMATES.  THE PROGRAM SETS GAPTOL = 1.D-7. CSL03930
C               IF FOR A GIVEN 'GOOD' T-EIGENVALUE THE COMPUTED GAP CSL03940
C               IS TOO SMALL AND IS DUE TO A 'SPURIOUS' T-EIGENVALUE CSL03950
C               THEN THE 'GOOD' T-EIGENVALUE IS ASSUMED TO HAVE CONVERGEDCSL03960
C               AND NO ERROR ESTIMATES ARE COMPUTED.               CSL03970
C                                                                  CSL03980
C-----USER-SPECIFIED PARAMETERS FOR EIGENVALUE PROGRAMS----------------CSL03990
```

```
C                                                                        CSL04000
C     RELTOL = RELATIVE TOLERANCE USED IN 'COMBINING' COMPUTED           CSL04010
C              EIGENVALUES OF T(1,MEV) PRIOR TO COMPUTING ERROR          CSL04020
C              ESTIMATES.                                                CSL04030
C                                                                        CSL04040
C     THE LUMPING OF T-EIGENVALUES OCCURS IN SUBROUTINE LUMP.            CSL04050
C     LUMPING IS NECESSARY BECAUSE IT IS IMPOSSIBLE TO ACCURATELY        CSL04060
C     PREDICT THE ACCURACY OF THE CMTQL1 SUBROUTINE.  LUMP 'COMBINES'    CSL04070
C     T-EIGENVALUES THAT HAVE SLIPPED BY THE TOLERANCE THAT WAS USED     CSL04080
C     IN THE T-MULTIPLICITY TESTS.  IN PARTICULAR IF FOR SOME J,         CSL04090
C                                                                        CSL04100
C     |EVALUE(J)-EVALUE(J-1)| < DMAX1(RELTOL*|EVALUE(J)|,SCALE2*MULTOL)  CSL04110
C                                                                        CSL04120
C     THEN THESE T-EIGENVALUES ARE 'COMBINED'.  MULTOL IS THE TOLERANCE  CSL04130
C     THAT WAS USED IN THE T-MULTIPLICITY TEST IN COMPEV.  SEE THE       CSL04140
C     HEADER ON THE LUMP SUBROUTINE FOR MORE DETAILS.                    CSL04150
C                                                                        CSL04160
C     THE RECOMMENDED VALUE OF RELTOL (ONLY IN THE COMPLEX SYMMETRIC     CSL04170
C     CASE) IS 1.D-8 BECAUSE THE OBSERVED ACCURACY OF THE                CSL04180
C     COMPUTED EIGENVALUES OF THE T-MATRICES IS SEVERAL DIGITS           CSL04190
C     LESS THAN THAT OBSERVED IN THE REAL SYMMETRIC CASE.                CSL04200
C     THUS, THE OBSERVED RESOLUTION OF THE COMPLEX SYMMETRIC             CSL04210
C     VERSION IS LESS THAN THAT OBTAINABLE IN THE REAL SYMMETRIC CASE.   CSL04220
C                                                                        CSL04230
C     MXINIT = MAXIMUM NUMBER OF INVERSE ITERATIONS ALLOWED IN           CSL04240
C              SUBROUTINE INVERR FOR EACH ISOLATED GOOD T-EIGENVALUE.    CSL04250
C              TYPICALLY ONLY ONE ITERATION IS REQUIRED.                 CSL04260
C                                                                        CSL04270
C     SEEDS FOR RANDOM NUMBER GENERATORS = INTEGER*4 SCALARS.            CSL04280
C                                                                        CSL04290
C              (1) SVSEED = SEED FOR STARTING VECTOR USED IN             CSL04300
C                  T-MATRIX GENERATION IN LANCZS SUBROUTINE              CSL04310
C                                                                        CSL04320
C              (2) RHSEED = SEED FOR RIGHT-HAND SIDE USED IN             CSL04330
C                  INVERSE ITERATION COMPUTATIONS IN INVERR.             CSL04340
C                                                                        CSL04350
C                                                                        CSL04360
C     T-MATRICES                                                         CSL04370
C                                                                        CSL04380
C     SIZES OF T-MATRICES                                                CSL04390
C                                                                        CSL04400
C              (1) KMAX= MAXIMUM ORDER FOR T-MATRIX THAT USER IS WILLING CSL04410
C                  TO CONSIDER.                                          CSL04420
C                                                                        CSL04430
C              (2) NMEVS = MAXIMUM NUMBER OF T-MATRICES THAT WILL BE      CSL04440
C                  CONSIDERED.                                           CSL04450
C                                                                        CSL04460
C              (3) NMEV(J)  (J=1,NMEVS)  = SIZES OF T-MATRIX TO BE        CSL04470
C                  CONSIDERED SEQUENTIALLY.                              CSL04480
C                                                                        CSL04490
C     T-MATRIX-GENERATION                                                CSL04500
C                                                                        CSL04510
C     USER SHOULD NOTE THAT THIS PROGRAM FIRST COMPUTES A T-MATRIX       CSL04520
C     OF ORDER KMAX AND THEN CYCLES THROUGH THE T-MATRICES SPECIFIED     CSL04530
C     A PRIORI BY THE USER, USING THE SUBROUTINE CMTQL1 TO COMPUTE THE   CSL04540
C     EIGENVALUES OF THE T-MATRICES.  THE EIGENVALUE COMPUTATION         CSL04550
C     FOR THE COMPLEX SYMMETRIC CASE WILL BE                            CSL04560
C     CONSIDERABLY MORE EXPENSIVE THAN FOR THE REAL SYMMETRIC OR         CSL04570
C     HERMITIAN CASES BECAUSE WE DO NOT HAVE AN ANALOG OF                CSL04580
C     THE BISECTION SUBROUTINE FOR THE COMPLEX SYMMETRIC CASE.           CSL04590
C     THUS, ANY RECYCLING AND SUBSEQUENT ENLARGEMENT OF THE T-MATRIX     CSL04600
C     REQUIRES THE RECOMPUTATION OF ALL OF THE EIGENVALUES OF            CSL04610
C     THE RESULTING T-MATRIX.  WE CANNOT GO IN AND COMPUTE ONLY THOSE    CSL04620
C     T-EIGENVALUES ON SOME SUBINTERVAL OF THE SPECTRUM OF THE           CSL04630
C     T-MATRIX AS WE DID IN THE REAL SYMMETRIC AND HERMITIAN CASES.      CSL04640
C     OF COURSE, IF THE T-MATRICES BEING CONSIDERED ARE NOT              CSL04650
C     VERY LARGE, THEN THIS IS NOT REALLY A PROBLEM.  HOWEVER, IF THEY   CSL04660
```

```
C      ARE VERY LARGE, THEN THE USER SHOULD PROBABLY DO ONE EIGENVALUE    CSL04670
C      COMPUTATION OF A LARGE T-MATRIX RATHER THAN START WITH             CSL04680
C      A SMALLER T-MATRIX AND WORK UP TO A BIG ONE.                       CSL04690
C                                                                         CSL04700
C-----CONVERGENCE TESTS FOR THE EIGENVALUE PROGRAMS--------------------CSL04710
C                                                                         CSL04720
C      THE CONVERGENCE TEST INCORPORATED IN THIS PROGRAM IS              CSL04730
C      BASED UPON THE ASSUMPTION THAT THOSE T-EIGENVALUES AND             CSL04740
C      THEIR ASSOCIATED T-EIGENVECTORS WHICH CORRESPOND TO THE            CSL04750
C      EIGENVALUES AND RITZVECTORS WHICH ARE TO BE COMPUTED               CSL04760
C      CONVERGE AS THE T-SIZE IS INCREASED.                               CSL04770
C                                                                         CSL04780
C-----ARRAYS REQUIRED BY THE EIGENVALUE PROGRAM------------------------CSL04790
C                                                                         CSL04800
C      ALPHA(J) = COMPLEX*16  ARRAY. ITS DIMENSION MUST BE AT LEAST       CSL04810
C                 KMAX, THE LENGTH OF THE LARGEST T-MATRIX ALLOWED.       CSL04820
C                 THIS ARRAY CONTAINS THE DIAGONAL ENTRIES OF THE         CSL04830
C                 T-MATRICES GENERATED.                                   CSL04840
C                                                                         CSL04850
C      BETA(J) = COMPLEX*16 ARRAY.  ITS DIMENSION MUST BE AT LEAST        CSL04860
C                KMAX+1. THIS ARRAY CONTAINS THE SUBDIAGONAL ENTRIES OF   CSL04870
C                THE T-MATRICES.                                          CSL04880
C                                                                         CSL04890
C                THE ALPHA AND BETA VECTORS ARE NOT ALTERED               CSL04900
C                DURING THE CALCULATIONS.                                 CSL04910
C                                                                         CSL04920
C      V1(J),V2(J),VS(J) = COMPLEX*16 ARRAYS.  V1 AND V2                  CSL04930
C                     MUST BE OF DIMENSION AT LEAST MAX(KMAX,N).          CSL04940
C                     VS MUST BE OF DIMENSION AT LEAST KMAX.              CSL04950
C                                                                         CSL04960
C      GR(J),GC(J) = REAL*8 ARRAYS. USED FOR RANDOM VECTOR GENERATION.    CSL04970
C                  EACH MUST BE OF DIMENSION AT LEAST MAX(KMAX,N).        CSL04980
C                                                                         CSL04990
C      EXPLAN(J) = REAL*4 ARRAY.  ITS DIMENSION IS 20.  THIS ARRAY IS     CSL05000
C                  USED TO ALLOW EXPLANATORY COMMENTS IN THE INPUT FILES.CSL05010
C                                                                         CSL05020
C      G(J),GG(J) = REAL*4 ARRAYS.  G MUST BE OF DIMENSION AT LEAST       CSL05030
C                   MAX(N,KMAX).  GG MUST BE OF DIMENSION AT LEAST        CSL05040
C                   KMAX.  G AND GG ARE USED IN RANDOM VECTOR GENERATIONSCSL05050
C                   AND TO STORE GAPS IN T-MATRIX, GAPS IN A-MATRIX,      CSL05060
C                   AND ERROR ESTIMATES.                                  CSL05070
C                                                                         CSL05080
C      MP(J),MP2(J) = INTEGER*4 ARRAYS.  EACH MUST HAVE DIMENSION         CSL05090
C                   AT LEAST KMAX, THE MAXIMUM SIZE OF THE T-MATRICES.    CSL05100
C                   MP CONTAINS THE T-MULTIPLICITIES OF THE COMPUTED      CSL05110
C                   T-EIGENVALUES. 'SPURIOUS' T-EIGENVALUES ARE DENOTEDCSL05120
C                   BY A T-MULTIPLICITY OF 0.  NOTE THAT WE DO NOT HAVECSL05130
C                   AN ANALOG OF THE SUBROUTINE PRTEST FOR THE           CSL05140
C                   COMPLEX SYMMETRIC CASE, SO NO RELABELLING OF          CSL05150
C                   MP OCCURS.  MP2 IS USED TO KEEP TRACK OF WHICH        CSL05160
C                   EIGENVALUES OF T(1,MEV) HAVE BEEN USED IN THE         CSL05170
C                   T-MULTIPLICITY TEST AND WHICH EIGENVALUES OF          CSL05180
C                   T(2,MEV) HAVE BEEN USED IN THE SPURIOUS TEST.         CSL05190
C                                                                         CSL05200
C      NMEV(J) = INTEGER*4 ARRAY.  ITS DIMENSION MUST BE AT LEAST THE     CSL05210
C                NUMBER OF T-MATRICES ALLOWED.  IT CONTAINS THE ORDERS    CSL05220
C                OF THE T-MATRICES TO BE CONSIDERED.                      CSL05230
C                                                                         CSL05240
C      OTHER ARRAYS                                                       CSL05250
C                                                                         CSL05260
C      THE USER MUST SPECIFY IN THE SUBROUTINE USPEC WHATEVER ARRAYS      CSL05270
C      ARE REQUIRED TO DEFINE THE MATRIX BEING USED.                      CSL05280
C                                                                         CSL05290
C                                                                         CSL05300
C-----SUBROUTINES INCLUDED FOR EIGENVALUE COMPUTATIONS-----------------CSL05310
C                                                                         CSL05320
C      LANCZS = COMPUTES THE ALPHA/BETA HISTORY. USES SUBROUTINES         CSL05330
```

```
C               CINPRD, INPRDC, GENRAN, AND CMATV.                    CSL05340
C                                                                     CSL05350
C      COMPEV = CALLS CMTQL1 TO COMPUTE THE EIGENVALUES OF T(1,MEV)   CSL05360
C               AND OF T(2,MEV), THEN DETERMINES T-MULTIPLE AND       CSL05370
C               SPURIOUS T-EIGENVALUES.                               CSL05380
C                                                                     CSL05390
C      COMGAP = COMPUTES MINIMAL GAPS BETWEEN T-EIGENVALUES           CSL05400
C               SUPPLIED.                                             CSL05410
C                                                                     CSL05420
C      CMTQL1 = COMPUTES EIGENVALUES OF THE SPECIFIED T-MATRIX USING  CSL05430
C               A REAL ORTHOGONAL ANALOG OF THE QL ALGORITHM IMTQL1   CSL05440
C               IN EISPACK.                                           CSL05450
C                                                                     CSL05460
C      INVERR = USES INVERSE ITERATION ON T-MATRICES TO COMPUTE ERROR CSL05470
C               ESTIMATES ON COMPUTED T-EIGENVALUES. (USES GENRAN)    CSL05480
C                                                                     CSL05490
C      LUMP = 'COMBINES' EIGENVALUES OF T-MATRIX USING THE RELATIVE   CSL05500
C               TOLERANCE RELTOL.                                     CSL05510
C                                                                     CSL05520
C      ISOEV = CALCULATES GAPS BETWEEN DISTINCT EIGENVALUES OF T-MATRIX CSL05530
C               AND THEN USES THESE GAPS TO LABEL THOSE 'GOOD'        CSL05540
C               T-EIGENVALUES FOR WHICH ERROR ESTIMATES ARE NOT COMPUTED. CSL05550
C                                                                     CSL05560
C      TNORM = COMPUTES THE SCALE TKMAX USED IN CHECKING              CSL05570
C               FOR LOCAL ORTHOGONALITY OF THE LANCZOS VECTORS        CSL05580
C               BY TESTING THE RELATIVE SIZE OF THE BETAS USING       CSL05590
C               THE RELATIVE TOLERANCE BTOL.                          CSL05600
C                                                                     CSL05610
C      CINPRD = COMPUTES THE HERMITIAN INNER PRODUCT OF TWO           CSL05620
C               COMPLEX*16 VECTORS, USED IN SUBROUTINE INVERR         CSL05630
C               AND IN THE MAIN PROGRAM.                              CSL05640
C                                                                     CSL05650
C      INPRDC = COMPUTES THE EUCLIDEAN INNER PRODUCT OF TWO           CSL05660
C               COMPLEX*16 VECTORS. USED IN SUBROUTINE LANCZS.        CSL05670
C                                                                     CSL05680
C                                                                     CSL05690
C-----OTHER PROGRAMS SUPPLIED----------------------------------------CSL05700
C                                                                     CSL05710
C                                                                     CSL05720
C      LCCOMPAC = PROGRAM TO TRANSLATE A SPARSE, COMPLEX SYMMETRIC    CSL05730
C               MATRIX GIVEN AS I, J, A(I,J), INTO THE SPARSE MATRIX  CSL05740
C               FORMAT USED IN THE SAMPLE USPEC AND CMATV SUBROUTINES CSL05750
C               PROVIDED.  PROGRAM ASSUMES THAT THE MATRIX ENTRIES    CSL05760
C               ARE GIVEN EITHER COLUMN BY COLUMN OR ROW BY ROW.      CSL05770
C                                                                     CSL05780
C                                                                     CSL05790
C-----COMMENTS ON THE STORAGE REQUIRED FOR EIGENVALUE COMPUTATIONS------CSL05800
C                                                                     CSL05810
C      THE ARRAYS USED IN THIS EIGENVALUE PROGRAM USE THE EQUIVALENT OF CSL05820
C      ONE REAL*8 ARRAY OF DIMENSION                                  CSL05830
C                                                                     CSL05840
C         8*KMAX + 4*MAX(KMAX,N)                                      CSL05850
C                                                                     CSL05860
C      PLUS WHATEVER IS NEEDED TO GENERATE A*X FOR THE GIVEN MATRIX A. CSL05870
C      THE ARRAYS ALPHA, BETA, VS, G, GG, MP, AND MP2 CONSUME         CSL05880
C      8*KMAX*8 BYTES.  THE ARRAYS V1 AND V2 CONSUME                  CSL05890
C      4*MAXIMUM(KMAX,N)*8 BYTES.                                     CSL05900
C                                                                     CSL05910
C                                                                     CSL05920
C-------------------------------------------------------------------CSL05930
C                                                                     CSL05940
C      COMMENTS FOR EIGENVECTOR COMPUTATIONS                          CSL05950
C                                                                     CSL05960
C-------------------------------------------------------------------CSL05970
C                                                                     CSL05980
C                                                                     CSL05990
C      THE EIGENVALUES WHOSE EIGENVECTORS ARE TO BE COMPUTED MUST     CSL06000
```

```
C     HAVE BEEN COMPUTED USING THE CORRESPONDING LANCZOS EIGENVALUE      CSL06010
C     FILES:  CSLEVAL + CSLESUB + CSLEMULT, FOR COMPLEX SYMMETRIC        CSL06020
C     MATRICES BECAUSE THE EIGENVECTOR PROGRAMS WILL USE THE SAME        CSL06030
C     FAMILY OF LANCZOS TRIDIAGONAL MATRICES AND LANCZOS VECTORS         CSL06040
C     THAT WAS USED IN THE EIGENVALUE COMPUTATIONS.                      CSL06050
C                                                                        CSL06060
C     THESE PROGRAMS ASSUME THAT THE EIGENVALUES SUPPLIED TO IT          CSL06070
C     HAVE BEEN COMPUTED ACCURATELY, AS MEASURED BY THE                  CSL06080
C     ERROR ESTIMATES COMPUTED IN THE CORRESPONDING LANCZOS              CSL06090
C     EIGENVALUE COMPUTATIONS, ALTHOUGH THESE ESTIMATES ARE              CSL06100
C     TYPICALLY CONSERVATIVE.  THE EIGENVALUES OF INTEREST                CSL06110
C     ARE IN THE ARRAY GOODEV(J), J=1,NGOOD.                             CSL06120
C                                                                        CSL06130
C     FOR EACH GOODEV(J), AN INITIAL ESTIMATE IS MADE OF AN              CSL06140
C     APPROPRIATE ORDER, MA(J), J=1,NGOOD, FOR A LANCZOS TRIDIAGONAL     CSL06150
C     FOR THE JTH EIGENVECTOR COMPUTATION.  THEN FOR EACH J,             CSL06160
C     SUBROUTINE INVERM SUCCESSIVELY COMPUTES CORRESPONDING              CSL06170
C     EIGENVECTORS OF ENLARGED T-MATRICES UNTIL A SUITABLE               CSL06180
C     SIZE T-MATRIX IS DETERMINED FOR EACH J.  UP TO 10 SUCH             CSL06190
C     EIGENVECTOR COMPUTATIONS ARE ALLOWED FOR EACH EIGENVALUE.          CSL06200
C                                                                        CSL06210
C     ONCE SUITABLE T-EIGENVECTORS HAVE BEEN OBTAINED THEN THE           CSL06220
C     RITZ VECTOR CORRESPONDING TO THESE T-EIGENVECTORS ARE              CSL06230
C     COMPUTED AND TAKEN AS APPROXIMATE EIGENVECTORS OF A FOR THE        CSL06240
C     GIVEN EIGENVALUES, GOODEV(J), J = 1, ..., NGOOD.                   CSL06250
C                                                                        CSL06260
C     THIS IMPLEMENTATION FIRST COMPUTES ALL OF THE RELEVANT             CSL06270
C     EIGENVECTORS OF THE COMPLEX SYMMETRIC TRIDIAGONAL MATRICES         CSL06280
C     IN THE VECTOR, TVEC.                                               CSL06290
C                                                                        CSL06300
C     THEN, AS EACH OF THE LANCZOS VECTORS IS REGENERATED, ALL           CSL06310
C     OF THE RITZ VECTORS CORRESPONDING TO THESE                         CSL06320
C     T-EIGENVECTORS ARE UPDATED USING THE CURRENTLY-GENERATED           CSL06330
C     LANCZOS VECTOR.  LANCZOS VECTORS ARE GENERATED (NOTE               CSL06340
C     THAT THEY ARE NOT BEING KEPT), UNTIL ENOUGH HAVE                   CSL06350
C     BEEN GENERATED TO MAP THE LONGEST T-EIGENVECTOR INTO ITS           CSL06360
C     CORRESPONDING RITZ VECTOR.  THE ARRAY RITVEC CONTAINS THE          CSL06370
C     SUCCESSIVE RITZ VECTORS WHICH ARE THE APPROXIMATE                  CSL06380
C     EIGENVECTORS OF A.                                                 CSL06390
C                                                                        CSL06400
C                                                                        CSL06410
C-----PARAMETER CONTROLS FOR EIGENVECTOR PROGRAMS----------------------CSL06420
C                                                                        CSL06430
C                                                                        CSL06440
C     PARAMETER CONTROLS ARE INTRODUCED TO ALLOW SEGMENTATION OF THE     CSL06450
C     EIGENVECTOR COMPUTATIONS AND TO ALLOW VARIOUS COMBINATIONS OF      CSL06460
C     READ/WRITES.                                                       CSL06470
C                                                                        CSL06480
C     THE FLAG MBOUND ALLOWS THE USER TO DETERMINE A FIRST GUESS ON THE CSL06490
C     STORAGE THAT WILL BE REQUIRED BY THE T-EIGENVECTORS FOR THE        CSL06500
C     EIGENVALUES WHOSE EIGENVECTORS ARE TO BE COMPUTED.                 CSL06510
C     THIS CAN BE USED TO ESTIMATE THE REQUIRED SIZE OF THE TVEC ARRAY. CSL06520
C                                                                        CSL06530
C     MBOUND = (0,1) MEANS                                               CSL06540
C                                                                        CSL06550
C          (0)  PROGRAM COMPUTES FIRST GUESSES AT THE SIZES             CSL06560
C               OF THE T-MATRICES REQUIRED BY EACH OF THE                CSL06570
C               EIGENVALUES SUPPLIED AND THEN CONTINUES WITH             CSL06580
C               THE CORRESPONDING T-EIGENVECTOR COMPUTATIONS.            CSL06590
C                                                                        CSL06600
C          (1)  PROGRAM COMPUTES FIRST GUESSES AT THE SIZES             CSL06610
C               OF THE T-MATRICES REQUIRED BY EACH OF THE                CSL06620
C               EIGENVALUES SUPPLIED, STORES THESE IN FILE 10            CSL06630
C               AND THEN TERMINATES.  THE USER CAN USE THESE             CSL06640
C               SIZES TO ESTIMATE THE SIZE TVEC ARRAY NEEDED             CSL06650
C               FOR THE DESIRED T-EIGENVECTOR COMPUTATIONS.              CSL06660
C                                                                        CSL06670
```

```
C       THE FLAGS NTVCON, TVSTOP, LVCONT, AND ERCONT CONTROL THE STOPPING CSL06680
C       CRITIERIA FOR INTERMEDIATE POINTS IN THE LANCZOS PROCEDURE.       CSL06690
C       THEY CAUSE TERMINATION OF THE LANCZOS PROCEDURE IF VARIOUS        CSL06700
C       QUANTITIES CANNOT BE COMPUTED AS DESIRED.                         CSL06710
C                                                                         CSL06720
C       NTVCON = (0,1) MEANS                                              CSL06730
C                                                                         CSL06740
C              (0)  IF THE ESTIMATED STORAGE FOR THE T-EIGENVECTORS       CSL06750
C                   EXCEEDS THE USER-SPECIFIED DIMENSION OF THE           CSL06760
C                   TVEC ARRAY PROGRAM DOES NOT CONTINUE WITH THE         CSL06770
C                   T-EIGENVECTOR COMPUTATIONS. TERMINATION OCCURS.       CSL06780
C                                                                         CSL06790
C              (1)  CONTINUE WITH THE T-EIGENVECTOR COMPUTATIONS          CSL06800
C                   EVEN IF THE ESTIMATED STORAGE FOR TVEC EXCEEDS        CSL06810
C                   THE USER-SPECIFIED DIMENSION OF THE TVEC ARRAY.       CSL06820
C                   IN THIS SITUATION THE PROGRAM COMPUTES AS MANY        CSL06830
C                   T-EIGENVECTORS AS IT HAS ROOM FOR, IN THE SAME        CSL06840
C                   ORDER IN WHICH THE EIGENVALUES ARE PROVIDED.          CSL06850
C                                                                         CSL06860
C       SVTVEC = (0,1) MEANS                                              CSL06870
C                                                                         CSL06880
C              (0)  DO NOT STORE THE COMPUTED T-EIGENVECTORS ON           CSL06890
C                   FILE 11 UNLESS ALSO HAVE THE FLAG TVSTOP = 1,         CSL06900
C                   IN WHICH CASE THE T-EIGENVECTORS ARE ALWAYS           CSL06910
C                   WRITTEN TO FILE 11.                                   CSL06920
C                                                                         CSL06930
C              (1)  STORE THE COMPUTED T-EIGENVECTORS ON FILE 11.         CSL06940
C                                                                         CSL06950
C       TVSTOP = (0,1) MEANS                                              CSL06960
C                                                                         CSL06970
C              (0)  ATTEMPT TO CONTINUE ON TO THE COMPUTATION             CSL06980
C                   OF THE RITZVECTORS AFTER COMPLETING THE               CSL06990
C                   COMPUTATION OF THE T-EIGENVECTORS.                    CSL07000
C                                                                         CSL07010
C              (1)  TERMINATE AFTER COMPUTING THE                         CSL07020
C                   T-EIGENVECTORS AND STORING THEM ON FILE 11.           CSL07030
C                                                                         CSL07040
C       LVCONT = (0,1) MEANS                                              CSL07050
C                                                                         CSL07060
C              (0)  IF SOME OF THE T-EIGENVECTORS THAT WERE               CSL07070
C                   REQUIRED WERE NOT COMPUTED, EXIT                      CSL07080
C                   FROM THE PROGRAM WITHOUT COMPUTING THE                CSL07090
C                   CORRESPONDING RITZ VECTORS.                           CSL07100
C                                                                         CSL07110
C              (1)  CONTINUE ON TO THE RITZ VECTOR COMPUTATIONS           CSL07120
C                   EVEN IF NOT ALL OF THE T-EIGENVECTORS THAT            CSL07130
C                   WERE REQUESTED WERE COMPUTED.                         CSL07140
C                                                                         CSL07150
C       ERCONT = (0,1) MEANS                                              CSL07160
C                                                                         CSL07170
C              (0)  PROGRAM WILL NOT COMPUTE THE RITZ                     CSL07180
C                   VECTOR FOR ANY EIGENVALUE FOR WHICH NO                CSL07190
C                   T-EIGENVECTOR WHICH SATISFIES THE ERROR ESTIMATE      CSL07200
C                   TEST (ERTOL) HAS BEEN IDENTIFIED.                     CSL07210
C                                                                         CSL07220
C              (1)  A RITZ VECTOR WILL BE COMPUTED FOR EVERY              CSL07230
C                   EIGENVALUE FOR WHICH A T-EIGENVECTOR HAS BEEN         CSL07240
C                   COMPUTED REGARDLESS OF WHETHER OR NOT THAT            CSL07250
C                   T-EIGENVECTOR SATISFIED THE ERROR ESTIMATE TEST.      CSL07260
C                                                                         CSL07270
C                                                                         CSL07280
C-----INPUT/OUTPUT FILES FOR THE EIGENVECTOR COMPUTATIONS---------------CSL07290
C                                                                         CSL07300
C                                                                         CSL07310
C       INPUT DATA OTHER THAN THE T-MATRIX HISTORY FILE AND THE           CSL07320
C       EIGENVALUES AND ERROR ESTIMATES SUPPLIED SHOULD BE STORED ON      CSL07330
C       FILE 5 IN FREE FORMAT.  SEE SAMPLE INPUT/OUTPUT FOR TYPICAL       CSL07340
```

```
C     INPUT/OUTPUT FILE.                                        CSL07350
C                                                               CSL07360
C     FILE 6 WAS USED AS THE INTERACTIVE TERMINAL OUTPUT FILE.  CSL07370
C     THIS FILE PROVIDES A RUNNING ACCOUNT OF THE PROGRESS OF THE CSL07380
C     COMPUTATIONS.  ADDITIONAL PRINTOUT IS GENERATED WHEN      CSL07390
C     THE FLAG IWRITE = 1.                                      CSL07400
C                                                               CSL07410
C                                                               CSL07420
C DESCRIPTION OF OTHER I/O FILES                                CSL07430
C                                                               CSL07440
C FILE (K)     CONTAINS:                                        CSL07450
C                                                               CSL07460
C     (2)      INPUT FILE:                                      CSL07470
C              PREVIOUSLY-GENERATED T-MATRICES (ALPHA/BETA ARRAYS) CSL07480
C              AND THE FINAL TWO LANCZOS VECTORS USED ON THAT   CSL07490
C              COMPUTATION.  THIS PROGRAM ALLOWS ENLARGEMENT    CSL07500
C              OF ANY T-MATRICES PROVIDED ON FILE 2.            CSL07510
C                                                               CSL07520
C     (3)      INPUT FILE:                                      CSL07530
C              THE GOOD EIGENVALUES OF THE T-MATRIX  T(1,MEV)   CSL07540
C              FOR WHICH EIGENVECTORS ARE REQUESTED.           CSL07550
C              FILE 3 ALSO CONTAINS THE T-MULTIPLICITIES OF THESE CSL07560
C              EIGENVALUES AND THEIR COMPUTED GAPS IN THE       CSL07570
C              T-MATRICES AND IN THE USER-SUPPLIED MATRIX. THIS CSL07580
C              FILE IS CREATED IN THE LANCZOS EIGENVALUE COMPUTATIONS. CSL07590
C                                                               CSL07600
C     (4)      INPUT FILE:                                      CSL07610
C              ERROR ESTIMATES FOR THE ISOLATED GOOD T-EIGENVALUES CSL07620
C              IN FILE 3.  THIS FILE IS CREATED DURING THE LANCZOS CSL07630
C              EIGENVALUE COMPUTATIONS.                         CSL07640
C                                                               CSL07650
C     (8)      INPUT FILE:                                      CSL07660
C              SAMPLE USPEC SUBROUTINE ASSUMES THAT THE ARRAYS  CSL07670
C              REQUIRED TO SPECIFY THE USER'S-MATRIX ARE STORED ON CSL07680
C              FILE 8.  USERS MUST MAKE WHATEVER DEFINITIONS ARE CSL07690
C              APPROPRIATE FOR THEIR MATRICES.                  CSL07700
C                                                               CSL07710
C     (9)      OUTPUT FILE:                                     CSL07720
C              ERROR ESTIMATES FOR THE COMPUTED RITZ VECTORS CONSIDERED CSL07730
C              AS EIGENVECTORS OF THE ORIGINAL MATRIX. THESE ESTIMATES CSL07740
C              ARE OF THE FORM                                  CSL07750
C                  AERROR = || A*RITVEC - EVAL*RITVEC ||        CSL07760
C              WHERE A DENOTES THE USER-SUPPLIED MATRIX, EVAL DENOTES CSL07770
C              THE EIGENVALUE BEING CONSIDERED AND RITVEC DENOTES CSL07780
C              THE COMPUTED RITZ VECTOR.                        CSL07790
C                                                               CSL07800
C     (10)     OUTPUT FILE:                                     CSL07810
C              GUESSES AT APPROPRIATE SIZE T-MATRICES FOR THE   CSL07820
C              T-EIGENVECTORS FOR EACH SUPPLIED EIGENVALUE GOODEV(J). CSL07830
C                                                               CSL07840
C     (11)     OUTPUT FILE:                                     CSL07850
C              COMPUTED T-EIGENVECTORS CORRESPONDING TO EIGENVALUES CSL07860
C              IN THE GOODEV ARRAY.  NOTE THAT IT IS POSSIBLE IN CSL07870
C              CERTAIN SITUATIONS THAT FOR SOME EIGENVALUES IN THE CSL07880
C              GOODEV ARRAY A T-EIGENVECTOR WILL NOT BE COMPUTED. CSL07890
C              (WRITTEN ONLY IF FLAG SVTVEC = 1).               CSL07900
C                                                               CSL07910
C     (12)     OUTPUT FILE:                                     CSL07920
C              CONTAINS COMPUTED RITZ VECTORS CORRESPONDING TO  CSL07930
C              THE T-EIGENVECTORS ON FILE 11. NOTE THAT IN      CSL07940
C              SOME SITUATIONS THAT FOR SOME EIGENVALUES IN     CSL07950
C              THE GOODEV ARRAY FOR WHICH T-EIGENVECTORS HAVE   CSL07960
C              BEEN COMPUTED NO RITZ VECTOR WILL HAVE BEEN      CSL07970
C              COMPUTED.                                        CSL07980
C                                                               CSL07990
C     (13)     OUTPUT FILE:                                     CSL08000
C              ADDITIONAL INFORMATION ABOUT THE BOUNDS AND ERROR CSL08010
```

```
C              ESTIMATES OBTAINED.                                    CSL08020
C                                                                     CSL08030
C                                                                     CSL08040
C-----SEEDS FOR EIGENVECTOR PROGRAMS----------------------------------CSL08050
C                                                                     CSL08060
C      SEEDS FOR RANDOM NUMBER GENERATOR GENRAN                       CSL08070
C            (1) SVSEED = INTEGER*4 SCALAR USED IN THE SUBROUTINE     CSL08080
C                         GENRAN TO GENERATE THE STARTING VECTOR FORCSL08090
C                         THE REGENERATION OF THE LANCZOS VECTORS.    CSL08100
C                                                                     CSL08110
C            (2) RHSEED = INTEGER*4 SCALAR USED IN THE SUBROUTINE     CSL08120
C                         GENRAN TO GENERATE A RANDOM VECTOR FOR      CSL08130
C                         USE IN SUBROUTINE INVERM.                   CSL08140
C                                                                     CSL08150
C      USER SHOULD NOTE THAT SVSEED MUST BE THE SAME SEED THAT        CSL08160
C      WAS USED TO GENERATE THE T-MATRICES THAT WERE USED TO          CSL08170
C      COMPUTE THE EIGENVALUES WHOSE EIGENVECTORS ARE TO BE COMPUTED. CSL08180
C      SVSEED IS READ IN FROM FILE 3.                                 CSL08190
C                                                                     CSL08200
C                                                                     CSL08210
C-----USER-SPECIFIED PARAMETERS FOR THE EIGENVECTOR PROGRAMS----------CSL08220
C                                                                     CSL08230
C      NGOOD   = NUMBER OF EIGENVALUES READ INTO THE GOODEV ARRAY     CSL08240
C                READ FROM FILE 3.                                    CSL08250
C                                                                     CSL08260
C      N       = SIZE OF THE USER-SUPPLIED MATRIX.                    CSL08270
C                                                                     CSL08280
C      MEV     = SIZE OF THE T-MATRIX THAT WAS USED TO COMPUTE        CSL08290
C                THE EIGENVALUES WHOSE EIGENVECTORS ARE REQUESTED.    CSL08300
C                MEV IS READ IN FROM FILE 3.                          CSL08310
C                                                                     CSL08320
C      KMAX    = SIZE OF THE T-MATRIX PROVIDED ON FILE 2.             CSL08330
C                                                                     CSL08340
C      MDIMTV  = MAXIMUM CUMULATIVE SIZE OF THE TVEC ARRAY ALLOWED    CSL08350
C                FOR ALL OF THE T-EIGENVECTORS REQUIRED.  MDIMTV      CSL08360
C                MUST NOT EXCEED THE USER-SPECIFIED DIMENSION OF      CSL08370
C                THE TVEC ARRAY.  PROGRAM CAN BE RUN WITH THE FLAG    CSL08380
C                MBOUND = 1 TO DETERMINE AN EDUCATED GUESS ON AN      CSL08390
C                APPROPRIATE DIMENSION FOR THE TVEC ARRAY.            CSL08400
C                                                                     CSL08410
C      MDIMRV = MAXIMUM CUMULATIVE SIZE OF THE RITVEC ARRAY ALLOWED   CSL08420
C                FOR ALL OF THE RITZ VECTORS TO BE COMPUTED. MDIMRV   CSL08430
C                MUST NOT EXCEED THE USER-SPECIFIED DIMENSION OF      CSL08440
C                THE RITVEC ARRAY.  MUST BE SELECTED SO THAT          CSL08450
C                THERE IS ENOUGH ROOM FOR A RITZ VECTOR FOR EVERY     CSL08460
C                GOODEV(J) READ INTO PROGRAM.  (>= NGOOD*N)           CSL08470
C                                                                     CSL08480
C                                                                     CSL08490
C-----ARRAYS REQUIRED BY THE EIGENVECTOR PROGRAMS--------------------CSL08500
C                                                                     CSL08510
C                                                                     CSL08520
C      ALPHA(J) = COMPLEX*16 ARRAY WHOSE DIMENSION MUST BE AT LEAST   CSL08530
C                 KMAXN, THE LARGEST SIZE T-MATRIX CONSIDERED BY      CSL08540
C                 THE PROGRAM.  NOTE THAT KMAXN IS THE LARGER OF      CSL08550
C                 THE SIZE OF THE ALPHA, BETA HISTORY PROVIDED        CSL08560
C                 ON FILE 2 (IF ANY ) AND THE SIZE WHICH THE PROGRAM  CSL08570
C                 SPECIFIES INTERNALLY, THIS LATTER IS ALWAYS         CSL08580
C                 < = 11*MEV / 8  +  12, WHERE MEV IS THE SIZE        CSL08590
C                 T-MATRIX THAT WAS USED IN THE CORRESPONDING EIGENVALUE CSL08600
C                 COMPUTATIONS.  ALPHA CONTAINS THE DIAGONAL ENTRIES  CSL08610
C                 OF THE LANCZOS T-MATRICES.  ALPHA IS NOT DESTROYED  CSL08620
C                 IN THE COMPUTATIONS.                                CSL08630
C                                                                     CSL08640
C      BETA(J) = COMPLEX*16 ARRAY WHOSE DIMENSION MUST BE AT LEAST 1  CSL08650
C                 MORE THAN THAT OF ALPHA.  DIMENSION COMMENTS ABOVE  CSL08660
C                 ABOUT ALPHA APPLY ALSO TO THE BETA ARRAY.  BETA     CSL08670
C                 CONTAINS THE SUBDIAGONAL ENTRIES OF THE T-MATRICES. CSL08680
```

```
C                 BETA IS NOT DESTROYED IN THE COMPUTATIONS.            CSL08690
C                                                                       CSL08700
C     RITVEC(J) = COMPLEX*16 ARRAY WHOSE DIMENSION MUST BE AT LEAST     CSL08710
C                 NGOOD*N WHERE N IS THE ORDER OF THE USER-SUPPLIED     CSL08720
C                 MATRIX AND NGOOD IS THE NUMBER OF EIGENVALUES         CSL08730
C                 WHOSE EIGENVECTORS ARE TO BE COMPUTED.  IT CONTAINS   CSL08740
C                 THE COMPUTED RITZ VECTORS  (THE APPROXIMATE           CSL08750
C                 EIGENVECTORS OF A).  THESE VECTORS ARE STORED         CSL08760
C                 ON FILE 12.                                           CSL08770
C                                                                       CSL08780
C     TVEC(J)   = COMPLEX*16 ARRAY WHOSE DIMENSION MUST BE AT LEAST     CSL08790
C                 MTOL = |MA(1)| + |MA(2)| + ... + |MA(NGOOD)|          CSL08800
C                 WHERE NGOOD IS THE NUMBER OF EIGENVALUES BEING        CSL08810
C                 CONSIDERED AND |MA(J)| IS THE SIZE OF THE             CSL08820
C                 T-MATRIX BEING USED IN THE RITZ VECTOR COMPUTATIONS   CSL08830
C                 FOR GOODEV(J).  THESE SIZES ARE DETERMINED BY THE     CSL08840
C                 PROGRAM.  AN ESTIMATE OF MTOL CAN BE OBTAINED BY      CSL08850
C                 SETTING MBOUND = 1, RUNNING THE PROGRAM, AND          CSL08860
C                 MULTIPLYING THE RESULTING TOTAL T-SIZES BY 5/4.       CSL08870
C                 THE ARRAY TVEC IS USED TO HOLD THE COMPUTED           CSL08880
C                 T-EIGENVECTORS.  IF THE FLAG SVTVEC = 1 OR THE        CSL08890
C                 FLAG TVSTOP = 1, THESE VECTORS ARE SAVED ON FILE 11.  CSL08900
C                                                                       CSL08910
C     V1(J)     = COMPLEX*16 ARRAY WHOSE DIMENSION MUST BE AT LEAST     CSL08920
C                 MAX(KMAX,N)  WHERE KMAX IS THE                        CSL08930
C                 LARGEST SIZE T-MATRIX THAT CAN BE CONSIDERED          CSL08940
C                 IN THE T-EIGENVECTOR COMPUTATIONS.  V1 IS USED        CSL08950
C                 IN THE SUBROUTINE INVERM AND IN THE REGENERATION      CSL08960
C                 OF THE LANCZOS VECTORS.                               CSL08970
C                                                                       CSL08980
C     V2(J)     = COMPLEX*16 ARRAY WHOSE DIMENSION MUST BE AT LEAST     CSL08990
C                 MAX(KMAX,N). IT IS USED IN THE REGENERATION OF        CSL09000
C                 THE LANCZOS VECTORS AND IN THE SUBROUTINE INVERM.     CSL09010
C                                                                       CSL09020
C     GOODEV(J) = COMPLEX*16 ARRAY OF DIMENSION AT LEAST NGOOD.         CSL09030
C                 CONTAINS THE EIGENVALUES FOR WHICH EIGENVECTORS       CSL09040
C                 ARE REQUESTED. THESE EIGENVALUES ARE READ IN          CSL09050
C                 FROM FILE 3.                                          CSL09060
C                                                                       CSL09070
C     GR(J),GC(J)   = REAL*8 ARRAYS WHOSE DIMENSION MUST BE AT          CSL09080
C                     LEAST MAX(N,KMAX).  USED TO HOLD RANDOMLY-        CSL09090
C                     GENERATED STARTING VECTORS FOR LANCZS             CSL09100
C                     COMPUTATIONS AND FOR THE INVERM SUBROUTINE.       CSL09110
C                                                                       CSL09120
C     AMINGP(J), = REAL*4 ARRAYS OF DIMENSION AT LEAST NGOOD.           CSL09130
C     TMINGP(J)    CONTAIN, RESPECTIVELY, THE MINIMAL GAPS FOR          CSL09140
C                  CORRESPONDING EIGENVALUES IN GOODEV ARRAY IN         CSL09150
C                  A-MATRIX AND IN T-MATRIX.                            CSL09160
C                                                                       CSL09170
C     TERR(J), ERR(J),    = REAL*4 ARRAYS (EXCEPT TLAST WHICH IS        CSL09180
C     ERRDGP(J), TLAST(J)   REAL*8) EACH OF WHOSE DIMENSIONS MUST BE    CSL09190
C     RNORM(J), TBETA(J)    AT LEAST NGOOD.  USED TO STORE QUANTITIES   CSL09200
C                           GENERATED DURING THE COMPUTATIONS FOR       CSL09210
C                           LATER PRINTOUT.                             CSL09220
C                                                                       CSL09230
C     G(J)      = REAL*4 ARRAY WHOSE DIMENSION MUST BE AT LEAST         CSL09240
C                 MAX(KMAX,N).  USED IN SUBROUTINE GENRAN TO HOLD       CSL09250
C                 RANDOM NUMBERS NEEDED FOR THE LANCZOS VECTORS         CSL09260
C                 REGENERATION AND FOR THE INVERSE ITERATION            CSL09270
C                 COMPUTATIONS IN THE SUBROUTINE INVERM.                CSL09280
C                                                                       CSL09290
C     MP(J) = INTEGER*4 ARRAY WHOSE DIMENSION IS AT LEAST NGOOD.        CSL09300
C             INITIALLY CONTAINS THE T-MULTIPLICITY OF THE EIGENVALUE   CSL09310
C             GOODEV(J) AS AN EIGENVALUE OF THE T-MATRIX T(1,MEV).      CSL09320
C             USED TO FLAG EIGENVALUES FOR WHICH NO T-EIGENVECTOR       CSL09330
C             OR NO RITZ VECTOR IS TO BE COMPUTED.                      CSL09340
C                                                                       CSL09350
```

```
C     MA(J)   = INTEGER*4 ARRAYS EACH OF WHOSE DIMENSIONS           CSL09360
C               IS AT LEAST NGOOD.  USED IN DETERMINING             CSL09370
C               AN APPROPRIATE T-MATRIX FOR EACH EIGENVALUE         CSL09380
C               IN GOODEV ARRAY.                                    CSL09390
C                                                                   CSL09400
C     MINT(J),MFIN(J) = INTEGER*4 ARRAYS WHOSE DIMENSIONS MUST BE AT CSL09410
C               LEAST NGOOD.  USED TO POINT TO THE BEGINNINGS       CSL09420
C               AND THE ENDS OF THE COMPUTED EIGENVECTOR            CSL09430
C               OF THE T-MATRIX, T(1,|MA(J)|).                      CSL09440
C                                                                   CSL09450
C     IDELTA(J) = INTEGER*4 ARRAY WHOSE DIMENSION MUST BE AT        CSL09460
C               LEAST NGOOD.  CONTAINS INCREMENTS USED IN LOOPS     CSL09470
C               ON APPROPRIATE SIZE T-MATRIX FOR THE T-EIGENVECTOR  CSL09480
C               COMPUTATIONS.                                       CSL09490
C                                                                   CSL09500
C                                                                   CSL09510
C     INTERC(J) = INTEGER*4 ARRAY WHOSE DIMENSION MUST BE AT        CSL09520
C               LEAST KMAX.  WORK SPACE USED IN INVERM.             CSL09530
C                                                                   CSL09540
C-----SUBROUTINES INCLUDED FOR THE EIGENVECTOR COMPUTATIONS------------CSL09550
C                                                                   CSL09560
C                                                                   CSL09570
C     INVERM = FOR THE T-SIZES CONSIDERED BY THE PROGRAM COMPUTES   CSL09580
C               THE CORRESPONDING EIGENVECTORS OF THESE T-MATRICES  CSL09590
C               CORRESPONDING TO THE USER-SUPPLIED EIGENVALUES IN   CSL09600
C               THE GOODEV ARRAY.                                   CSL09610
C                                                                   CSL09620
C     LANCZS, TNORM , CINPRD, INPRDC, CMATV AND GENRAN ARE USED     CSL09630
C     HERE AS WELL AS IN THE EIGENVALUE COMPUTATIONS.               CSL09640
C                                                                   CSL09650
C                                                                   CSL09660
C-------------------------------------------------------------------CSL09670
```

SECTION 7.3   MAIN PROGRAM, EIGENVALUE CALCULATIONS

```
C-----CSLEVAL (EIGENVALUES OF COMPLEX SYMMETRIC MATRICES)--------------CSL00010
C                                                                      CSL00020
C      CONTAINS MAIN PROGRAM FOR COMPUTING DISTINCT EIGENVALUES OF      CSL00030
C      A NONDEFECTIVE COMPLEX SYMMETRIC MATRIX USING LANCZOS            CSL00040
C      TRIDIAGONALIZATION WITHOUT REORTHOGONALIZATION                   CSL00050
C                                                                      CSL00060
C      PORTABILITY:                                                     CSL00070
C      THESE PROGRAMS ARE NOT PORTABLE DUE TO THE USE OF COMPLEX*16     CSL00080
C      VARIABLES AND CORRESPONDING COMPLEX FUNCTIONS SUCH AS DCMPLX     CSL00090
C      AND CDABS. FURTHERMORE, OTHER NONPORTABLE CONSTRUCTIONS          CSL00100
C      IDENTIFIED BY THE PFORT VERIFIER ARE THE FOLLOWING:              CSL00110
C                                                                  .-   CSL00120
C      1.  DATA/MACHEP/ STATEMENT THAT DEFINES MACHINE EPSILON          CSL00130
C      2.  ALL READ(5,*) INPUT STATEMENTS IN FREE FORMAT                CSL00140
C      3.  FORMAT(20A4) USED WITH EXPLANATORY HEADER EXPLAN.            CSL00150
C      4.  HEXADECIMAL FORMAT (4Z20) USED WITH ALPHA/BETA FILES 1 AND 2. CSL00160
C                                                                      CSL00170
C----------------------------------------------------------------------CSL00180
C                                                                      CSL00190
       COMPLEX*16  ALPHA(3000),BETA(3000),VS(3000)                      CSL00200
       COMPLEX*16  V1(3000),V2(3000),ZEROC,BETAM,Z                      CSL00210
       DOUBLE PRECISION  GR(3000),GC(3000)                              CSL00220
       DOUBLE PRECISION  BTOL,GAPTOL,TTOL,MACHEP,EPSM,RELTOL            CSL00230
       DOUBLE PRECISION  SCALE1,SCALE2,SPUTOL,CONTOL,MULTOL,EVMAX       CSL00240
       DOUBLE PRECISION  ONE,ZERO,TEMP,TKMAX,EVMAX,BKMIN,TO,T1          CSL00250
       REAL  G(3000),GG(3000),EXPLAN(20),GTEMP                          CSL00260
       INTEGER  MP(3000),MP2(3000),NMEV(20)                             CSL00270
       INTEGER  SVSEED,RHSEED,SVSOLD,SAVTEV                             CSL00280
       INTEGER  IABS                                                    CSL00290
       REAL  ABS                                                        CSL00300
       DOUBLE PRECISION  DABS, DFLOAT                                   CSL00310
       EXTERNAL CMATV                                                   CSL00320
C                                                                      CSL00330
C----------------------------------------------------------------------CSL00340
       DATA MACHEP/Z3410000000000000/                                   CSL00350
       EPSM = 2.0D0*MACHEP                                              CSL00360
C----------------------------------------------------------------------CSL00370
C                                                                      CSL00380
C      ARRAYS MUST BE DIMENSIONED AS FOLLOWS:                           CSL00390
C          1. ALPHA AND VS: >= KMAX.   BETA: >= (KMAX+1)                CSL00400
C          2. V1, V2, GR, GC: >= MAX(N,KMAX)                            CSL00410
C          3. G: >= MAX(N,KMAX).   GG: >= KMAX.                         CSL00420
C          4. MP, MP2:  >= KMAX                                         CSL00430
C          5. NMEV: >= NUMBER OF T-MATRICES ALLOWED                     CSL00440
C          6. EXPLAN:  DIMENSION IS 20.                                 CSL00450
C                                                                      CSL00460
C      NOTE:  THE OBSERVED ACHIEVABLE ACCURACY FOR THE COMPLEX          CSL00470
C      SYMMETRIC MATRICES TESTED WAS SIGNIFICANTLY LESS THAN THAT       CSL00480
C      OBTAINED WITH THE REAL SYMMETRIC AND HERMITIAN VERSIONS          CSL00490
C      OF THESE LANCZOS CODES AND IT IS DOUBTFUL THAT THIS CODE         CSL00500
C      CAN HANDLE VERY STIFF COMPLEX SYMMETRIC MATRICES.                CSL00510
C                                                                      CSL00520
C      IMPORTANT TOLERANCES OR SCALES THAT ARE USED REPEATEDLY          CSL00530
C      THROUGHOUT THE PROGRAM ARE THE FOLLOWING:                        CSL00540
C      SCALED MACHINE EPSILON: TTOL = EVMAX*EPSM WHERE                  CSL00550
C      EPSM = 2*MACHINE EPSILON AND                                     CSL00560
C      EVMAX = MAX(|LAMBDA(J)|), J =1,MEV OF EIGENVALUES OF T(1,MEV).   CSL00570
C      TOLERANCE: T-MULTIPLICITY TESTS:  MULTOL = 500*(1000+MEV)*TTOL   CSL00580
C      TOLERANCE: SPURIOUS TESTS   SPUTOL = MULTOL                      CSL00590
C      NOTE THAT IN THE MAIN PROGRAM THESE TOLERANCES ARE INITIALIZED   CSL00600
C      TO QUANTITIES THAT ARE NOT A FUNCTION OF THE SIZE OF THE         CSL00610
C      T-EIGENVALUES AND THEN THE SIZES OF THE T-EIGENVALUES ARE        CSL00620
C      INTRODUCED IN THE SUBROUTINE COMPEV.                             CSL00630
C                                                                      CSL00640
```

```
C      LANCZOS CONVERGENCE TOLERANCE: CONTOL = CDABS(BETA(MEV+1)*1.D-10  CSL00650
C------------------------------------------------------------------------CSL00660
C      OUTPUT HEADER                                                      CSL00670
       WRITE(6,10)                                                        CSL00680
    10 FORMAT(/' LANCZOS EIGENVALUE PROCEDURE FOR COMPLEX SYMMETRIC MATRICSL00690
      1CES'/)                                                             CSL00700
C                                                                         CSL00710
C      SET PROGRAM PARAMETERS                                             CSL00720
C      SCALEK ARE USED IN TOLERANCES NEEDED IN SUBROUTINES LUMP           CSL00730
C      AND ISOEV.  USER MUST NOT MODIFY THESE SCALES.                     CSL00740
       SCALE1 = 5.0D2                                                     CSL00750
       SCALE2 = 5.0D0                                                     CSL00760
       ONE  = 1.0D0                                                       CSL00770
       ZERO = 0.0D0                                                       CSL00780
       ZEROC = DCMPLX(ZERO,ZERO)                                          CSL00790
       BTOL = 1.0D-8                                                      CSL00800
C      BTOL = MACHEP                                                      CSL00810
       GAPTOL = 1.0D-7                                                    CSL00820
       ICONV = 0                                                          CSL00830
       MOLD = 0                                                           CSL00840
       MOLD1 = 1                                                          CSL00850
       MMB = 0                                                            CSL00860
C                                                                         CSL00870
C      READ USER-SPECIFIED PARAMETERS FROM INPUT FILE 5 (FREE FORMAT)     CSL00880
C                                                                         CSL00890
C      READ USER-PROVIDED HEADER FOR RUN                                  CSL00900
       READ(5,20) EXPLAN                                                  CSL00910
       WRITE(6,20) EXPLAN                                                 CSL00920
       READ(5,20) EXPLAN                                                  CSL00930
       WRITE(6,20) EXPLAN                                                 CSL00940
    20 FORMAT(20A4)                                                       CSL00950
C                                                                         CSL00960
C      READ ORDER OF MATRICES (N) , MAXIMUM ORDER OF T-MATRIX (KMAX),      CSL00970
C      NUMBER OF T-MATRICES ALLOWED (NMEVS), AND MATRIX IDENTIFICATION      CSL00980
C      NUMBERS (MATNO)                                                    CSL00990
       READ(5,20) EXPLAN                                                  CSL01000
       READ(5,*) N,KMAX,NMEVS,MATNO                                       CSL01010
C                                                                         CSL01020
C      READ SEEDS FOR LANCZS AND INVERR SUBROUTINES (SVSEED AND RHSEED)    CSL01030
C      READ MAXIMUM NUMBER OF ITERATIONS ALLOWED FOR EACH INVERSE          CSL01040
C      ITERATION (MXINIT).                                                CSL01050
       READ(5,20) EXPLAN                                                  CSL01060
       READ(5,*) SVSEED,RHSEED,MXINIT                                     CSL01070
C                                                                         CSL01080
C      ISTART = (0,1):  ISTART = 0 MEANS ALPHA/BETA FILE IS NOT            CSL01090
C      AVAILABLE.  ISTART = 1 MEANS ALPHA/BETA FILE IS AVAILABLE ON        CSL01100
C      FILE 2.  COMPLEX SYMMETRIC HISTORIES MUST BE STORED                 CSL01110
C      IN HEX FORMAT (4Z20).                                              CSL01120
C      ISTOP = (0,1):  ISTOP = 0 MEANS PROCEDURE GENERATES ALPHA/BETA      CSL01130
C      FILE AND THEN TERMINATES.  ISTOP = 1 MEANS PROCEDURE GENERATES      CSL01140
C      ALPHAS/BETAS IF NEEDED AND THEN COMPUTES EIGENVALUES AND ERROR      CSL01150
C      ESTIMATES AND THEN TERMINATES.                                     CSL01160
       READ(5,20) EXPLAN                                                  CSL01170
       READ(5,*) ISTART,ISTOP                                            CSL01180
C                                                                         CSL01190
C      IHIS = (0,1):  IHIS = 0 MEANS ALPHA/BETA FILE IS NOT WRITTEN        CSL01200
C      TO FILE 1.  IHIS = 1 MEANS ALPHA/BETA FILE IS WRITTEN TO FILE 1.    CSL01210
C      IDIST = (0,1):  IDIST = 0 MEANS DISTINCT T(1,MEV)-EIGENVALUES       CSL01220
C      ARE NOT WRITTEN TO FILE 11.  IDIST = 1 MEANS DISTINCT              CSL01230
C      T(1,MEV)-EIGENVALUES ARE WRITTEN TO FILE 11.                       CSL01240
C      SAVTEV = (-1,0,1):  SAVTEV = - 1 MEANS T(1,MEV) AND T(2,MEV)        CSL01250
C      EIGENVALUES ARE AVAILABLE ON FILE 10 FROM AN EARLIER RUN.          CSL01260
C      IN THIS CASE, ALPHA/BETA FILE FROM THAT RUN MUST BE                 CSL01270
C      AVAILABLE ON FILE 2.                                               CSL01280
C      SAVTEV = 0 MEANS WE WILL NOT SAVE THE T(1,MEV) AND T(2,MEV)         CSL01290
C      EIGENVALUES.  SAVTEV = 1 MEANS WE WRITE THE T(1,MEV) AND            CSL01300
C      T(2,MEV) EIGENVALUES TO FILE 10.                                   CSL01310
```

```
C      IWRITE = (0,1):  IWRITE = 0 MEANS NO INTERMEDIATE OUTPUT           CSLO1320
C      FROM THE COMPUTATIONS IS WRITTEN TO FILE 6.  IWRITE = 1 MEANS      CSLO1330
C      EIGENVALUES AND ERROR ESTIMATES ARE WRITTEN TO FILE 6             CSLO1340
C      AS THEY ARE COMPUTED.                                             CSLO1350
       READ(5,20) EXPLAN                                                 CSLO1360
       READ(5,*) IHIS,IDIST,SAVTEV,IWRITE                                CSLO1370
C                                                                        CSLO1380
       IF(SAVTEV.GE.0) GO TO 30                                          CSLO1390
       NMEVS = 1                                                         CSLO1400
       IF(ISTART.EQ.0) GO TO 610                                         CSLO1410
C                                                                        CSLO1420
   30 CONTINUE                                                           CSLO1430
C      READ IN THE RELATIVE TOLERANCE (RELTOL) FOR USE IN THE LUMP       CSLO1440
C      SUBROUTINE                                                        CSLO1450
       READ(5,20) EXPLAN                                                 CSLO1460
       READ(5,*) RELTOL                                                  CSLO1470
C                                                                        CSLO1480
C      READ IN THE SIZES OF THE T(1,MEV) MATRICES TO BE CONSIDERED.      CSLO1490
       READ(5,20) EXPLAN                                                 CSLO1500
       READ(5,*) (NMEV(J), J=1,NMEVS)                                    CSLO1510
C                                                                        CSLO1520
C-----------------------------------------------------------------------CSLO1530
C      INITIALIZE THE ARRAYS FOR THE USER-SPECIFIED MATRIX              CSLO1540
C      AND PASS THE STORAGE LOCATIONS OF THESE ARRAYS TO THE           CSLO1550
C      MATRIX-VECTOR MULTIPLY SUBROUTINE CMATV.                         CSLO1560
C                                                                        CSLO1570
       CALL USPEC(N,MATNO)                                               CSLO1580
C                                                                        CSLO1590
C-----------------------------------------------------------------------CSLO1600
C      MASK UNDERFLOW AND OVERFLOW                                       CSLO1610
C                                                                        CSLO1620
       CALL MASK                                                         CSLO1630
C                                                                        CSLO1640
C-----------------------------------------------------------------------CSLO1650
C                                                                        CSLO1660
C      WRITE TO FILE 6, A SUMMARY OF THE PARAMETERS FOR THIS RUN         CSLO1670
C                                                                        CSLO1680
       WRITE(6,40) MATNO,N,KMAX                                          CSLO1690
   40 FORMAT(/3X,'MATRIX ID',4X,'ORDER OF A',4X,'MAX ORDER OF T'/        CSLO1700
     1  I12,I14,I18/)                                                    CSLO1710
C                                                                        CSLO1720
       WRITE(6,50) ISTART,ISTOP                                          CSLO1730
   50 FORMAT(/2X,'ISTART',3X,'ISTOP'/2I8/)                               CSLO1740
C                                                                        CSLO1750
       WRITE(6,60) IHIS,IDIST,SAVTEV,IWRITE                              CSLO1760
   60 FORMAT(/4X,'IHIS',3X,'IDIST',3X,'SAVTEV',2X,'IWRITE'/2I8,I9,I8/)   CSLO1770
C                                                                        CSLO1780
       WRITE(6,70) SVSEED,RHSEED                                         CSLO1790
   70 FORMAT(/' SEEDS FOR RANDOM NUMBER GENERATOR'//                     CSLO1800
     1  4X,'LANCZS SEED',4X,'INVERR SEED'/2I15/)                         CSLO1810
C                                                                        CSLO1820
       WRITE(6,80) (NMEV(J), J=1,NMEVS)                                  CSLO1830
   80 FORMAT(/' SIZES OF T-MATRICES TO BE CONSIDERED'/(6I12))            CSLO1840
C                                                                        CSLO1850
       WRITE(6,90) RELTOL,GAPTOL,BTOL                                    CSLO1860
   90 FORMAT(/' RELATIVE TOLERANCE USED TO COMBINE COMPUTED T-EIGENVALUECSLO1870
     1S'/E15.3/' RELATIVE GAP TOLERANCES USED IN INVERSE ITERATION'/     CSLO1880
     1E15.3/' RELATIVE TOLERANCE FOR CHECK ON SIZE OF BETAS'/E15.3/)     CSLO1890
C                                                                        CSLO1900
       IF (ISTART.EQ.0) GO TO 140                                        CSLO1910
C                                                                        CSLO1920
C      READ IN ALPHA BETA HISTORY                                        CSLO1930
C      HISTORY MUST BE STORED IN MACHINE FORMAT TO PREVENT              CSLO1940
C      ERRORS CAUSED BY INPUT/OUTPUT CONVERSIONS.                       CSLO1950
C                                                                        CSLO1960
       READ(2,100)MOLD,NOLD,SVSOLD,MATOLD                                CSLO1970
```

```
    100 FORMAT(2I6,I12,I8)                                                  CSL01980
  C                                                                         CSL01990
        IF (KMAX.LT.MOLD) KMAX = MOLD                                       CSL02000
        KMAX1 = KMAX + 1                                                    CSL02010
  C                                                                         CSL02020
  C     CHECK THAT ORDER N, MATRIX ID MATNO, AND RANDOM SEED SVSEED         CSL02030
  C     AGREE WITH THOSE IN THE HISTORY FILE.  IF NOT PROCEDURE STOPS.      CSL02040
  C                                                                         CSL02050
        ITEMP = (NOLD-N)**2+(MATNO-MATOLD)**2+(SVSEED-SVSOLD)**2            CSL02060
  C                                                                         CSL02070
        IF (ITEMP.EQ.0) GO TO 120                                          CSL02080
  C                                                                         CSL02090
        WRITE(6,110)                                                        CSL02100
    110 FORMAT(' PROGRAM TERMINATES'/    ' READ FROM FILE 2 CORRESPONDS TOCSL02110
       1 DIFFERENT MATRIX THAN MATRIX SPECIFIED'/)                          CSL02120
        GO TO 650                                                           CSL02130
  C                                                                         CSL02140
    120 CONTINUE                                                            CSL02150
        MOLD1 = MOLD+1                                                      CSL02160
  C                                                                         CSL02170
        READ(2,130)(ALPHA(J), J=1,MOLD)                                     CSL02180
        READ(2,130)(BETA(J), J=1,MOLD1)                                     CSL02190
    130 FORMAT(4Z20)                                                        CSL02200
  C                                                                         CSL02210
        IF (KMAX.EQ.MOLD) GO TO 160                                         CSL02220
  C                                                                         CSL02230
        READ(2,130)(V1(J), J=1,N)                                           CSL02240
        READ(2,130)(V2(J), J=1,N)                                           CSL02250
  C                                                                         CSL02260
    140 CONTINUE                                                            CSL02270
        IIX = SVSEED                                                        CSL02280
  C                                                                         CSL02290
  C-----------------------------------------------------------------------CSL02300
  C                                                                         CSL02310
        CALL LANCZS(CMATV,V1,V2,ALPHA,BETA,GR,GC,G,KMAX,MOLD1,N,IIX)        CSL02320
  C                                                                         CSL02330
  C-----------------------------------------------------------------------CSL02340
  C                                                                         CSL02350
        KMAX1 = KMAX + 1                                                    CSL02360
  C                                                                         CSL02370
        IF (IHIS.EQ.0.AND.ISTOP.GT.0) GO TO 160                            CSL02380
  C                                                                         CSL02390
        WRITE(1,150) KMAX,N,SVSEED,MATNO                                    CSL02400
    150 FORMAT(2I6,I12,I8,' = KMAX,N,SVSEED,MATNO')                         CSL02410
  C                                                                         CSL02420
        WRITE(1,130)(ALPHA(I), I=1,KMAX)                                    CSL02430
        WRITE(1,130)(BETA(I), I=1,KMAX1)                                    CSL02440
  C                                                                         CSL02450
        WRITE(1,130)(V1(I), I=1,N)                                          CSL02460
        WRITE(1,130)(V2(I), I=1,N)                                          CSL02470
  C                                                                         CSL02480
        IF (ISTOP.EQ.0) GO TO 520                                           CSL02490
  C                                                                         CSL02500
    160 CONTINUE                                                            CSL02510
        BKMIN = BTOL                                                        CSL02520
        WRITE(6,170)                                                        CSL02530
    170 FORMAT(/' T-MATRICES (ALPHA AND BETA) ARE NOW AVAILABLE'/)          CSL02540
  C                                                                         CSL02550
  C-----------------------------------------------------------------------CSL02560
  C     SUBROUTINE TNORM CHECKS MIN|BETA|/(ESTIMATED NORM(A)) > BTOL .      CSL02570
  C     IF THIS IS VIOLATED IB IS SET EQUAL TO THE NEGATIVE OF THE INDEX    CSL02580
  C     OF THE MINIMAL BETA.  IF(IB < 0) THEN SUBROUTINE TNORM IS           CSL02590
  C     CALLED FOR EACH VALUE OF MEV TO DETERMINE WHETHER OR NOT THERE      CSL02600
  C     IS A BETA IN THE T-MATRIX SPECIFIED THAT VIOLATES THIS TEST.        CSL02610
  C     IF THERE IS SUCH A BETA THE PROGRAM TERMINATES FOR THE USER         CSL02620
  C     TO DECIDE WHAT TO DO.  THIS TEST CAN BE OVER-RIDDEN BY              CSL02630
  C     SIMPLY MAKING BTOL SMALLER, BUT THEN THERE IS THE POSSIBILITY       CSL02640
```

```
C      THAT LOSSES IN THE LOCAL ORTHOGONALITY MAY HURT THE COMPUTATIONS.  CSL02650
C                                                                         CSL02660
C      TNORM ALSO COMPUTES TKMAX = MAX(|ALPHA(K)|,|BETA(K)|, K=1,KMAX).    CSL02670
C      HOWEVER, IN THE COMPLEX SYMMETRIC CASE SINCE ALL OF THE             CSL02680
C      EIGENVALUES OF T(1,MEV) ARE COMPUTED, TKMAX IS NOT USED TO SCALE    CSL02690
C      THE T-MULTIPLICITY AND SPURIOUS TOLERANCES.  THE COMPUTED           CSL02700
C      T-EIGENVALUE LARGEST IN MAGNITUDE IS USED INSTEAD.                  CSL02710
C                                                                         CSL02720
       CALL TNORM(ALPHA,BETA,BKMIN,TKMAX,KMAX,IB)                          CSL02730
C                                                                         CSL02740
C------------------------------------------------------------------------CSL02750
C                                                                         CSL02760
C      LOOP ON THE SIZE OF THE T-MATRIX                                    CSL02770
C                                                                         CSL02780
  180 CONTINUE                                                             CSL02790
       MMB = MMB + 1                                                       CSL02800
       MEV = NMEV(MMB)                                                     CSL02810
C      IS MEV TOO LARGE ?                                                  CSL02820
       IF(MEV.LE.KMAX) GO TO 200                                          CSL02830
       WRITE(6,190) MMB, MEV, KMAX                                         CSL02840
  190 FORMAT(/' TERMINATE PRIOR TO CONSIDERING THE',I6,'TH T-MATRIX'/     CSL02850
      1' BECAUSE THE SIZE REQUESTED',I6,' IS GREATER THAN THE MAXIMUM SIZCSL02860
      1E ALLOWED',I6/)                                                    CSL02870
       GO TO 520                                                          CSL02880
C                                                                         CSL02890
  200 MP1 = MEV + 1                                                       CSL02900
       BETAM = BETA(MP1)                                                  CSL02910
C                                                                         CSL02920
       IF (IB.GE.0) GO TO 220                                             CSL02930
C                                                                         CSL02940
       TO = BTOL                                                          CSL02950
C                                                                         CSL02960
C------------------------------------------------------------------------CSL02970
C                                                                         CSL02980
       CALL TNORM(ALPHA,BETA,TO,T1,MEV,IBMEV)                             CSL02990
C                                                                         CSL03000
C------------------------------------------------------------------------CSL03010
C                                                                         CSL03020
  210 TEMP = TO/TKMAX                                                     CSL03030
       IBMEV = IABS(IBMEV)                                                CSL03040
       IF (TEMP.GE.BTOL) GO TO 220                                        CSL03050
       IBMEV = -IBMEV                                                     CSL03060
       GO TO 590                                                          CSL03070
  220 CONTINUE                                                            CSL03080
C                                                                         CSL03090
C------------------------------------------------------------------------CSL03100
C      SUBROUTINE COMPEV CALLS SUBROUTINE CMTQL1 TO COMPUTE THE           CSL03110
C      T-EIGENVALUES.  COMPEV THEN APPLIES THE T-MULTIPLICITY AND         CSL03120
C      SPURIOUS TESTS TO THE COMPUTED T-EIGENVALUES.  HERE INITIALIZE     CSL03130
C      THE TOLERANCES USED IN THE T-MULTIPLICITY AND THE SPURIOUS         CSL03140
C      TESTS.  THE MAX(|LAMBDA(T(1,MEV)|) WILL BE INCORPORATED            CSL03150
C      INSIDE THE SUBROUTINE COMPEV.  NOTE THAT THE OBSERVED ACCURACY     CSL03160
C      OF THE COMPUTED T-EIGENVALUES FOR THE COMPLEX SYMMETRIC CASE       CSL03170
C      IS APPROXIMATELY 3 DIGITS LESS THAN THAT ACHIEVED IN THE REAL      CSL03180
C      CASE. THUS, A FACTOR OF 500 HAS BEEN INTRODUCED.  THIS HOWEVER     CSL03190
C      MEANS THAT THIS TEST IS NOT AS SHARP AS IT WAS IN THE              CSL03200
C      REAL SYMMETRIC AND HERMITIAN CASES.  THUS, IT HAS LOWER            CSL03210
C      RESOLUTION AND CAN OCCASIONALLY MAKE A MISTAKE.                    CSL03220
C                                                                         CSL03230
       MULTOL = 500.D0 * DFLOAT(MEV+1000) * EPSM                          CSL03240
       SPUTOL =  MULTOL                                                   CSL03250
C                                                                         CSL03260
C      ON RETURN FROM COMPEV                                              CSL03270
C      NDIS = NUMBER OF DISTINCT EIGENVALUES OF T(1,MEV)                  CSL03280
C      VS = DISTINCT T-EIGENVALUES IN INCREASING ORDER OF MAGNITUDE       CSL03290
C      GR(K) = |VS(K)|, K = 1,NDIS,  GR(K).LE.GR(K+1)                     CSL03300
C      MP = T-MULTIPLICITIES OF THE T-EIGENVALUES IN VS                   CSL03310
```

```
C      MP(I) = (0,1,MI), MI>1, I=1,NDIS  MEANS:                        CSL03320
C         (0)  VS(I) IS SPURIOUS                                       CSL03330
C         (1)  VS(I) IS SIMPLE AND GOOD                                CSL03340
C         (MI) VS(I) IS T-MULTIPLE AND IS THEREFORE NOT ONLY GOOD BUT  CSL03350
C              ALSO A CONVERGED GOOD T-EIGENVALUE.                     CSL03360
C                                                                      CSL03370
C                                                                      CSL03380
       CALL COMPEV(ALPHA,BETA,V1,V2,VS,GR,MULTOL,SPUTOL,MP,MP2,        CSL03390
      1MEV,NDIS,SAVTEV)                                                CSL03400
C                                                                      CSL03410
C----------------------------------------------------------------------CSL03420
C                                                                      CSL03430
       IF (NDIS.EQ.0) GO TO 630                                        CSL03440
C                                                                      CSL03450
C      ON EXIT FROM COMPEV MULTOL AND SPUTOL SHOULD BE SCALED          CSL03460
C      BY THE SIZES OF THE T-EIGENVALUES                              CSL03470
       EVMAX = GR(NDIS)                                                CSL03480
       LOOP = NDIS                                                     CSL03490
C                                                                      CSL03500
C----------------------------------------------------------------------CSL03510
C                                                                      CSL03520
       CALL LUMP(VS,V1,GR,RELTOL,SPUTOL,SCALE2,MP,MP2,LOOP)            CSL03530
C                                                                      CSL03540
C----------------------------------------------------------------------CSL03550
C                                                                      CSL03560
       IF (LOOP.LT.0) GO TO 650                                        CSL03570
C                                                                      CSL03580
       IF (NDIS.EQ.LOOP) GO TO 240                                     CSL03590
C                                                                      CSL03600
       WRITE(6,230) NDIS,LOOP,MEV                                      CSL03610
  230 FORMAT(/' AFTER LUMP NDIS,LOOP,MEV = ',3I6/)                     CSL03620
C                                                                      CSL03630
  240 CONTINUE                                                         CSL03640
       NDIS = LOOP                                                     CSL03650
C                                                                      CSL03660
C-------------CALCULATE MINGAPS FOR DISTINCT T(1,MEV) EIGENVALUES.-----CSL03670
C      CALCULATE MINGAPS FOR DISTINCT T(1,MEV) EIGENVALUES.           CSL03680
C      ON EXIT |GG(K)| = MIN(J.NE.K,|VS(K)-VS(J)|), MP2(K)=J INDEX     CSL03690
C      FOR MINIMUM.  GG(K)< 0 MEANS NEAREST NEIGHBOR IS SPURIOUS.      CSL03700
       IGAP = 0                                                        CSL03710
       ITAG = 1                                                        CSL03720
C                                                                      CSL03730
       CALL COMGAP(VS,GR,GG,MP,MP2,NDIS,IGAP,ITAG)                     CSL03740
C                                                                      CSL03750
C----------------------------------------------------------------------CSL03760
C                                                                      CSL03770
C      SET CONVERGENCE CRITIERION                                      CSL03780
       TTOL = EPSM * EVMAX                                             CSL03790
       CONTOL = CDABS(BETAM)*1.D-10                                    CSL03800
C                                                                      CSL03810
  250 CONTINUE                                                         CSL03820
       BETA(MP1) = BETAM                                               CSL03830
C                                                                      CSL03840
C----------------------------------------------------------------------CSL03850
C      THE SUBROUTINE ISOEV LABELS THOSE SIMPLE EIGENVALUES OF T(1,MEV)CSL03860
C      WITH VERY SMALL GAPS BETWEEN NEIGHBORING EIGENVALUES OF T(1,MEV)CSL03870
C      TO AVOID COMPUTING ERROR ESTIMATES FOR ANY SIMPLE GOOD          CSL03880
C      T-EIGENVALUE THAT IS TOO CLOSE TO A SPURIOUS T-EIGENVALUE.      CSL03890
C      MP(I) = -1 MEANS THAT THE GOOD T-EIGENVALUE IS SIMPLE AND       CSL03900
C      IS TOO CLOSE TO A SPURIOUS T-EIGENVALUE.                        CSL03910
C                                                                      CSL03920
C      NG = NUMBER OF GOOD T-EIGENVALUES.                              CSL03930
C      NISO = NUMBER OF ISOLATED GOOD T-EIGENVALUES.                   CSL03940
C      GG = MINIMAL GAPS IN T(1,MEV)                                   CSL03950
C      GR(K) = |VS(K)|, K=1,NDIS                                       CSL03960
C                                                                      CSL03970
```

```
      CALL ISOEV(VS,GR,GG,GAPTOL,SPUTOL,SCALE1,MP,NDIS,NG,NISO)       CSL03980
C                                                                     CSL03990
C--------------------------------------------------------------------CSL04000
C                                                                     CSL04010
      WRITE(6,260)NG,NISO,NDIS                                        CSL04020
  260 FORMAT(/I6,' GOOD T-EIGENVALUES HAVE BEEN COMPUTED'/            CSL04030
     1 I6,' OF THESE ARE ISOLATED'/                                   CSL04040
     2 I6,' = NUMBER OF DISTINCT T-EIGENVALUES COMPUTED'/)            CSL04050
C                                                                     CSL04060
C     DO WE WRITE DISTINCT EIGENVALUES OF T-MATRIX TO FILE 11?        CSL04070
      IF (IDIST.EQ.0) GO TO 300                                       CSL04080
C                                                                     CSL04090
      WRITE(11,270) NDIS,NISO,MEV,N,SVSEED,MATNO                      CSL04100
  270 FORMAT(/4I6,I12,I8,' = NDIS,NISO,MEV,N,SVSEED,MATNO'/)          CSL04110
C                                                                     CSL04120
      WRITE(11,280) (I,MP(I),VS(I),GG(I),MP2(I), I=1,NDIS)            CSL04130
  280 FORMAT(I4,I4,2E20.12,E12.3,I6)                                  CSL04140
C                                                                     CSL04150
      WRITE(11,290) NDIS, (MP(I), I=1,NDIS)                           CSL04160
  290 FORMAT(/I6,' = NDIS, T-MULTIPLICITIES (0 MEANS  SPURIOUS)'/(20I4))CSL04170
C                                                                     CSL04180
  300 CONTINUE                                                        CSL04190
C                                                                     CSL04200
      IF (NISO.NE.0) GO TO 330                                        CSL04210
C                                                                     CSL04220
      WRITE(4,310) MEV                                                CSL04230
  310 FORMAT(/' AT MEV = ',I6,' THERE ARE NO ISOLATED T-EIGENVALUES'/ CSL04240
     1' SO NO ERROR ESTIMATES WERE COMPUTED/')                       CSL04250
C                                                                     CSL04260
      WRITE(6,320)                                                    CSL04270
  320 FORMAT(/' ALL COMPUTED GOOD T-EIGENVALUES ARE T-MULTIPLE'/      CSL04280
     1 ' THEREFORE THESE EIGENVALUES ARE ASSUMED TO HAVE CONVERGED')  CSL04290
C                                                                     CSL04300
      ICONV = 1                                                       CSL04310
      GO TO 370                                                       CSL04320
C                                                                     CSL04330
  330 CONTINUE                                                        CSL04340
C                                                                     CSL04350
C--------------------------------------------------------------------CSL04360
C     SUBROUTINE INVERR COMPUTES ERROR ESTIMATES FOR ISOLATED GOOD    CSL04370
C     T-EIGENVALUES USING INVERSE ITERATION ON T(1,MEV). ON RETURN    CSL04380
C     GG(J) = MINIMUM GAP IN T(1,MEV) FOR EACH VS(J), J=1,NDIS        CSL04390
C     G(I) = |BETAM|*|U(MEV)| = ERROR ESTIMATE FOR ISOLATED GOOD      CSL04400
C             T-EIGENVALUES, WHERE I = 1, NISO AND  BETAM = BETA(MEV+1)CSL04410
C             U(MEV) IS MEVTH COMPONENT OF THE UNIT EIGENVECTOR OF T   CSL04420
C             CORRESPONDING TO THE ITH ISOLATED GOOD T-EIGENVALUE.    CSL04430
C     A NEGATIVE ERROR ESTIMATE MEANS THAT FOR THAT PARTICULAR        CSL04440
C     T-EIGENVALUE THE INVERSE ITERATION DID NOT CONVERGE IN <= MXINIT CSL04450
C     STEPS AND THAT THE CORRESPONDING ERROR ESTIMATE IS QUESTIONABLE. CSL04460
C                                                                     CSL04470
C     ON EXIT                                                         CSL04480
C     V2 CONTAINS THE ISOLATED GOOD T-EIGENVALUES                     CSL04490
C     GR CONTAINS THE MINGAPS TO THE NEAREST DISTINCT EIGENVALUE      CSL04500
C        OF T(1,MEV) FOR EACH ISOLATED GOOD T-EIGENVALUE IN V2.       CSL04510
C     VS CONTAINS THE NDIS DISTINCT EIGENVALUES OF T(1,MEV)           CSL04520
C     MP CONTAINS THE CORRESPONDING CODED T-MULTIPLICITIES            CSL04530
C                                                                     CSL04540
C     IT = MXINIT                                                     CSL04550
C                                                                     CSL04560
      CALL INVERR(ALPHA,BETA,V1,V2,VS,EPSM,GR,GC,G,GG,MP,MP2,MEV,MMB,  CSL04570
     1NDIS,NISO,N,RHSEED,IT,IWRITE)                                   CSL04580
C                                                                     CSL04590
C--------------------------------------------------------------------CSL04600
C                                                                     CSL04610
C     SIMPLE CHECK FOR CONVERGENCE. CHECKS TO SEE IF ALL OF THE ERROR CSL04620
C     ESTIMATES ARE SMALLER THAN CONTOL = CDABS(BETA(MEV+1)*1.D-10    CSL04630
C     IF THIS TEST IS SATISFIED, THEN CONVERGENCE FLAG, ICONV IS SET  CSL04640
```

```
C      TO 1.  TYPICALLY ERROR ESTIMATES ARE VERY CONSERVATIVE.          CSLO4650
C                                                                        CSLO4660
       WRITE(6,340) CONTOL                                               CSLO4670
  340 FORMAT(/' CONVERGENCE IS TESTED USING THE CONVERGENCE TOLERANCE',  CSLO4680
     1E13.4/)                                                            CSLO4690
C                                                                        CSLO4700
       DO 350 I = 1,NISO                                                 CSLO4710
       IF (ABS(G(I)).GT.CONTOL) GO TO 370                               CSLO4720
  350 CONTINUE                                                           CSLO4730
       ICONV = 1                                                         CSLO4740
       MMB = NMEVS                                                       CSLO4750
C                                                                        CSLO4760
       WRITE(6,360) CONTOL                                               CSLO4770
  360 FORMAT(' ALL COMPUTED ERROR ESTIMATES WERE LESS THAN',E15.4/       CSLO4780
     1 ' THEREFORE PROCEDURE TERMINATES'/)                              CSLO4790
C                                                                        CSLO4800
  370 CONTINUE                                                           CSLO4810
C                                                                        CSLO4820
C      IN REAL SYMMETRIC AND HERMITIAN LANCZOS PROGRAMS                  CSLO4830
C      AT THIS CORRESPONDING POINT THE SUBROUTINE PRTEST IS CALLED       CSLO4840
C      TO IDENTIFY ANY T-EIGENVALUES THAT MAY HAVE BEEN MISLABELLED      CSLO4850
C      AS SPURIOUS BECAUSE THEIR PROJECTIONS ON THE STARTING VECTOR      CSLO4860
C      WERE TOO SMALL.  THIS CHECK WAS MADE ONLY AFTER CONVERGENCE       CSLO4870
C      HAD OCCURRED.  HOWEVER, THE PRTEST SUBROUTINE IS BASED UPON       CSLO4880
C      STURM SEQUENCING AND THAT IS NOT VALID FOR COMPLEX SYMMETRIC      CSLO4890
C      MATRICES.  PERHAPS THERE IS SOME RECTANGLE ANALOG OF THE          CSLO4900
C      PRTEST BUT WE HAVE NOT ATTEMPTED TO IDENTIFY AND INCLUDE          CSLO4910
C      SUCH A TEST BECAUSE WE EXPECT, AS IN THE REAL SYMMETRIC AND       CSLO4920
C      HERMITIAN CASES THAT HIDDEN EIGENVALUES WILL BE RARE.            CSLO4930
C                                                                        CSLO4940
C      WRITE THE GOOD T-EIGENVALUES TO FILE 3.  FIRST TRANSFER THEM      CSLO4950
C      TO V2 AND THEIR T-MULTIPLICITIES TO THE CORRESPONDING POSITIONS   CSLO4960
C      IN MP AND COMPUTE THE A-MINGAPS, THE MINIMAL GAPS BETWEEN THE     CSLO4970
C      GOOD T-EIGENVALUES.  THESE GAPS WILL BE PUT IN THE ARRAY GG.      CSLO4980
C      NOTE THAT AFTER THE SECOND CALL TO COMGAP THE ARRAY GC            CSLO4990
C      WILL CONTAIN THE CORRESPONDING MINIMAL GAPS IN THE                CSLO5000
C      T-MATRIX, T(1,MEV).                                               CSLO5010
C                                                                        CSLO5020
  380 CONTINUE                                                           CSLO5030
C                                                                        CSLO5040
       NG = 0                                                            CSLO5050
       DO 390 I = 1,NDIS                                                 CSLO5060
       IF (MP(I).EQ.0) GO TO 390                                         CSLO5070
       NG = NG+1                                                         CSLO5080
       MP(NG) = MP(I)                                                    CSLO5090
       V2(NG) = VS(I)                                                    CSLO5100
       GC(NG) = GG(I)                                                    CSLO5110
  390 CONTINUE                                                           CSLO5120
C                                                                        CSLO5130
       DO 400 I = 1,NG                                                   CSLO5140
  400 GR(I) = CDABS(V2(I))                                               CSLO5150
C                                                                        CSLO5160
C-------------------------------------------------------------------CSLO5170
C      CALCULATE MINGAPS FOR GOODEV                                      CSLO5180
C      ON EXIT GG(K) = MIN(J.NE.K,|V2(K)-V2(J)|), MP2(K)=J INDEX FOR MIN CSLO5190
C      NG = NUMBER OF COMPUTED DISTINCT GOOD T-EIGENVALUES.              CSLO5200
       IGAP = 0                                                          CSLO5210
       ITAG = 0                                                          CSLO5220
C                                                                        CSLO5230
       CALL COMGAP(V2,GR,GG,MP,MP2,NG,IGAP,ITAG)                         CSLO5240
C                                                                        CSLO5250
C-------------------------------------------------------------------CSLO5260
C                                                                        CSLO5270
C      WRITE GOOD T-EIGENVALUES OUT TO FILE 3.                           CSLO5280
C                                                                        CSLO5290
       WRITE(6,410)MEV                                                   CSLO5300
```

```
  410 FORMAT(//' EIGENVALUE CALCULATION AT MEV = ',I6,' IS COMPLETE'/) CSL05310
C                                                                       CSL05320
      WRITE(3,420)NG,NDIS,MEV,N,SVSEED,MATNO,MULTOL,SPUTOL,IB,BTOL       CSL05330
  420 FORMAT(4I6,I12,I8,' = NG,NDIS,MEV,N,SVEED,MATNO'/                  CSL05340
     1 2E15.5,I6,E13.4,' = MULTOL,SPUTOL,IB,BTOL'/                       CSL05350
     1' EVNO',1X,'MULT',13X,'R(GOODEV)',13X,'I(GOODEV)',                 CSL05360
     1 3X,'TMINGAP',3X,'AMINGAP',1X,'NEIGH')                             CSL05370
C                                                                       CSL05380
      WRITE(3,430)(I,MP(I),V2(I),GC(I),GG(I),MP2(I), I=1,NG)            CSL05390
  430 FORMAT(2I5,2E22.14,2E10.3,I6)                                      CSL05400
C                                                                       CSL05410
C     ORDER GOODEV BY INCREASING GAP SIZE                               CSL05420
      DO 440 I = 1,NG                                                    CSL05430
      MP(I) = I                                                          CSL05440
      V1(I) = V2(I)                                                      CSL05450
      G(I) = GG(I)                                                       CSL05460
  440 CONTINUE                                                           CSL05470
C                                                                       CSL05480
C     WRITE(12,436)                                                      CSL05490
  450 FORMAT(' MINGAPS FOR GOOD T-EIGENVALUES'/                         CSL05500
     1 1X,'EVNUM',1X,'NEIGH',15X,'R(EV)',15X,'I(EV)',4X,'MINGAP')       CSL05510
C                                                                       CSL05520
C     WRITE(12,439) (K,MP2(K),V2(K),G(K), K = 1,NG)                     CSL05530
  460 FORMAT(2I6,2E20.12,E10.3)                                          CSL05540
C                                                                       CSL05550
      DO 480 K = 2,NG                                                    CSL05560
      KM1 = K-1                                                          CSL05570
      DO 470 L = 1,KM1                                                   CSL05580
      KK = K-L                                                           CSL05590
      KP1 = KK+1                                                         CSL05600
      IF (G(KP1).GE.G(KK)) GO TO 480                                     CSL05610
      Z = V1(KK)                                                         CSL05620
      V1(KK) = V1(KP1)                                                   CSL05630
      V1(KP1) = Z                                                        CSL05640
      GTEMP = G(KK)                                                      CSL05650
      G(KK) = G(KP1)                                                     CSL05660
      G(KP1) = GTEMP                                                     CSL05670
      ITEMP = MP(KK)                                                     CSL05680
      MP(KK) = MP(KP1)                                                   CSL05690
      MP(KP1) = ITEMP                                                    CSL05700
  470 CONTINUE                                                           CSL05710
  480 CONTINUE                                                           CSL05720
C                                                                       CSL05730
C     WRITE(12,441)                                                      CSL05740
      WRITE(3,490)                                                       CSL05750
  490 FORMAT(' T-EIGENVALUES ORDERED BY INCREASING MINGAP'/             CSL05760
     1 1X,'GAPNUM',1X,'EVNUM',15X,'R(EV)',15X,'I(EV)',4X,'MINGAP')      CSL05770
C                                                                       CSL05780
C     WRITE(12,442) (K,MP(K),V1(K),G(K), K = 1,NG)                      CSL05790
      WRITE(3,500) (K,MP(K),V1(K),G(K), K = 1,NG)                       CSL05800
  500 FORMAT(I7,I6,2E20.12,E10.3)                                        CSL05810
C                                                                       CSL05820
  510 CONTINUE                                                           CSL05830
C                                                                       CSL05840
C     IF CONVERGENCE FLAG ICONV.NE.1 AND NUMBER OF T-MATRICES           CSL05850
C     CONSIDERED TO DATE IS LESS THAN NUMBER ALLOWED, INCREMENT MEV.    CSL05860
C     AND LOOP BACK TO 210 TO REPEAT COMPUTATIONS.  RESTORE BETA(MEV+1).CSL05870
C                                                                       CSL05880
      BETA(MP1) = BETAM                                                  CSL05890
C                                                                       CSL05900
      IF (MMB.LT.NMEVS.AND.ICONV.NE.1) GO TO 180                        CSL05910
C                                                                       CSL05920
C     END OF LOOP ON DIFFERENT SIZE T-MATRICES ALLOWED.                 CSL05930
C                                                                       CSL05940
  520 CONTINUE                                                           CSL05950
C                                                                       CSL05960
      IF(ISTOP.EQ.0)  WRITE(6,530)                                      CSL05970
```

```
     530 FORMAT(/' T-MATRICES (ALPHA AND BETA) ARE NOW AVAILABLE, TERMINATECSL05980
        1')                                                           CSL05990
          IF (ISTOP.EQ.0.AND.KMAX.NE.MOLD) WRITE(1,540)               CSL06000
          IF (IHIS.EQ.1.AND.KMAX.NE.MOLD) WRITE(1,540)                CSL06010
     540 FORMAT(/' ABOVE ARE THE FOLLOWING VECTORS '/                 CSL06020
        1 '  ALPHA(I), I = 1,KMAX'/                                   CSL06030
        2 '  BETA(I), I = 1,KMAX+1'/                                  CSL06040
        3 ' FINAL TWO LANCZOS VECTORS OF ORDER N FOR I = KMAX,KMAX+1'/ CSL06050
        4 ' ALPHA BETA ARE IN HEX FORMAT 4Z20 '/                      CSL06060
        4 ' LANCZOS VECTORS ARE IN HEX FORMAT 4Z20 '/                 CSL06070
        5 ' ----- END OF FILE 1 NEW ALPHA, BETA HISTORY--------------'///)CSL06080
    C                                                                 CSL06090
          IF (ISTOP.EQ.0) GO TO 650                                   CSL06100
    C                                                                 CSL06110
          WRITE(3,550)                                                CSL06120
     550 FORMAT(/' ABOVE ARE COMPUTED GOOD T-EIGENVALUES'/            CSL06130
        1 ' NG = NUMBER OF GOOD T-EIGENVALUES COMPUTED'/              CSL06140
        2 ' NDIS = NUMBER OF COMPUTED DISTINCT EIGENVALUES OF T(1,MEV)'/ CSL06150
        3 ' N = ORDER OF A, MATNO = MATRIX IDENT'/                    CSL06160
        4 ' MULTOL = T-MULTIPLICITY TOLERANCE FOR T-EIGENVALUES'/     CSL06170
        4 ' SPUTOL = SPURIOUS TOLERANCE FOR T-EIGENVALUES'/           CSL06180
        4 ' MULT IS THE T-MULTIPLICITY OF GOOD T-EIGENVALUE'/         CSL06190
        5 ' MULT = -1 MEANS SPURIOUS T-EIGENVALUE TOO CLOSE'/         CSL06200
        6 ' DO NOT COMPUTE ERROR ESTIMATES FOR SUCH T-EIGENVALUES'/   CSL06210
        7 ' AMINGAP = MINIMAL GAP BETWEEN THE COMPUTED A-EIGENVALUES'/ CSL06220
        9 ' TMINGAP= MINIMAL GAP W.R.T. DISTINCT EIGENVALUES IN T(1,MEV)'/CSL06230
        2 ' ----- END OF FILE 3 GOOD T-EIGENVALUES--------------------'///CSL06240
        3 )                                                           CSL06250
    C                                                                 CSL06260
          IF (IDIST.NE.0) WRITE(11,560)                               CSL06270
     560 FORMAT(/' ABOVE ARE THE DISTINCT EIGENVALUES OF T(1,MEV).'/  CSL06280
        2 ' THE FORMAT IS     T-MULTIPLICITY    T-EIGENVALUE    TMINGAP'/ CSL06290
        4 ' T-MULTIPLICITY = -1 MEANS THAT THE SUBROUTINE ISOEV HAS TAGGED'CSL06300
        5 /'   THIS SIMPLE T-EIGENVALUE AS HAVING A VERY CLOSE SPURIOUS'/ CSL06310
        6 '   T-EIGENVALUE SO THAT NO ERROR ESTIMATE WILL BE COMPUTED'/ CSL06320
        7 '   FOR THAT EIGENVALUE IN SUBROUTINE INVERR.'/             CSL06330
        9 ' EACH OF THE DISTINCT T-EIGENVALUE TABLES IS FOLLOWED'/    CSL06340
        9 ' BY THE T-MULTIPLICITY PATTERN.'/                          CSL06350
        1 ' NDIS = NUMBER OF COMPUTED DISTINCT EIGENVALUES OF T(1,MEV).'/ CSL06360
        2 ' NG = NUMBER OF GOOD T-EIGENVALUES. '/                     CSL06370
        3 ' NISO = NUMBER OF ISOLATED GOOD T-EIGENVALUES. '/          CSL06380
        4 ' NISO ALSO IS THE COUNT OF +1 ENTRIES IN MULTIPLICITY PATTERN.'/CSL06390
        5 ' -----END OF FILE 11 DISTINCT T-EIGENVALUES----------------'///)CSL06400
    C                                                                 CSL06410
          WRITE(4,570)                                                CSL06420
     570 FORMAT(/' ABOVE ARE THE ERROR ESTIMATES OBTAINED FOR THE ISOLATED CSL06430
        1GOOD T-EIGENVALUES'/                                         CSL06440
        1' OBTAINED VIA INVERSE ITERATION IN THE SUBROUTINE INVERR.'/ CSL06450
        1' ALL OTHER GOOD T-EIGENVALUES HAVE CONVERGED.'/             CSL06460
        2' ERROR ESTIMATE = CDABS(BETAM*(UM))'/                       CSL06470
        2' WHERE BETAM = BETA(MEV+1) AND UM = U(MEV).'/               CSL06480
        3' U = UNIT EIGENVECTOR OF T WHERE T*U = EV*U AND EV = ISOLATED GOOCSL06490
        3D T-EIGENVALUE.'/                                            CSL06500
        4' TMINGAP = GAP TO NEAREST DISTINCT EIGENVALUE OF T(1,MEV).'/ CSL06510
        6' ------ END OF FILE 4 ERRINV --------------------------------'///)CSL06520
    C                                                                 CSL06530
          IF(SAVTEV.LT.0) GO TO 650                                   CSL06540
          WRITE(10,580)                                               CSL06550
     580 FORMAT(//' ABOVE ARE THE T(1,MEV) EIGENVALUES FOLLOWED BY THE'/ CSL06560
        1 ' T(2,MEV) EIGENVALUES FOR MEV = NMEV(J), J = 1,NMEVS'/     CSL06570
        1 ' ------END OF FILE 10 T-T2EVAL--------------------'///)     CSL06580
    C                                                                 CSL06590
          GO TO 650                                                   CSL06600
    C                                                                 CSL06610
     590 CONTINUE                                                     CSL06620
    C                                                                 CSL06630
          IBB = IABS(IBMEV)                                           CSL06640
```

```
      TEMP = CDABS(BETA(IBB))                                      CSL06650
      IF (IBMEV.LT.0) WRITE(6,600) MEV,IBB,TEMP                    CSL06660
  600 FORMAT(/' PROGRAM TERMINATES BECAUSE MEV REQUESTED = ',I6,' IS .GTCSL06670
     1',I6/' AT WHICH AN ABNORMALLY SMALL BETA = ' ,E13.4,' OCCURRED'/) CSL06680
      GO TO 650                                                   CSL06690
C                                                                  CSL06700
  610 WRITE(6,620) SAVTEV,ISTART                                  CSL06710
  620 FORMAT(2I6,' = SAVTEV,ISTART'/' WHEN SAVTEV = -1, WE MUST HAVE ISTCSL06720
     1ART = 1'/)                                                  CSLC6730
      GO TO 650                                                   CSL06740
C                                                                  CSL06750
  630 IF (NDIS.EQ.0.AND.ISTOP.GT.0) WRITE(6,640)                  CSL06760
  640 FORMAT(/' INTERVALS SPECIFIED FOR BISECT DID NOT CONTAIN ANY T-EIGCSL06770
     1ENVALUES'/' PROGRAM TERMINATES')                           CSL06780
C                                                                  CSL06790
  650 CONTINUE                                                    CSL06800
C                                                                  CSL06810
      STOP                                                        CSL06820
C-----END OF MAIN PROGRAM FOR COMPLEX SYMMETRIC EIGENVALUE COMPUTATIONS-CSL06830
      END                                                         CSL06840
```

SECTION 7.4 MAIN PROGRAM, EIGENVECTOR CALCULATIONS

```
C-----CSLEVEC (EIGENVECTORS OF COMPLEX SYMMETRIC MATRICES)-------------CSL00010
C                                                                      CSL00020
C     CONTAINS MAIN PROGRAM FOR COMPUTING AN EIGENVECTOR CORRESPONDING  CSL00030
C     TO EACH OF A SET OF EIGENVALUES THAT HAVE BEEN COMPUTED           CSL00040
C     ACCURATELY BY THE CORRESPONDING LANCZOS EIGENVALUE PROGRAM        CSL00050
C     (CSLEVAL) FOR NONDEFECTIVE COMPLEX SYMMETRIC MATRICES.            CSL00060
C     THIS PROGRAM COULD BE MODIFIED TO COMPUTE ADDITIONAL             CSL00070
C     EIGENVECTORS FOR THOSE EIGENVALUES WHICH ARE MULTIPLE EIGENVALUES CSL00080
C     OF THE GIVEN A-MATRIX.  THE AMOUNT OF ADDITIONAL COMPUTATION      CSL00090
C     REQUIRED WOULD DEPEND UPON THE GIVEN A-MATRIX AND UPON WHAT       CSL00100
C     PART OF THE SPECTRUM OF A IS INVOLVED.                            CSL00110
C                                                                      CSL00120
C     THESE LANCZOS EIGENVECTOR COMPUTATIONS ASSUME THAT EACH          CSL00130
C     EIGENVALUE THAT IS BEING CONSIDERED HAS CONVERGED AS AN          CSL00140
C     EIGENVALUE OF THE CORRESPONDING LANCZOS TRIDIAGONAL MATRICES.    CSL00150
C                                                                      CSL00160
C     PORTABILITY:                                                     CSL00170
C     THIS PROGRAM IS NOT PORTABLE DUE TO THE USE OF THE COMPLEX*16    CSL00180
C     VARIABLES AND CORRESPONDING COMPLEX FUNCTIONS.  MOREOVER, PFORT  CSL00190
C     IDENTIFIED THE FOLLOWING ADDITIONAL NONPORTABLE CONSTRUCTIONS:   CSL00200
C                                                                      CSL00210
C     1.  DATA/MACHEP/ STATEMENT                                       CSL00220
C     2.  ALL READ(5,*) STATEMENTS (FREE FORMAT)                       CSL00230
C     3.  FORMAT(20A4) USED WITH THE EXPLANATORY HEADER, EXPLAN        CSL00240
C     4.  FORMAT (4Z20) USED FOR ALPHA/ BETA FILE 2.                   CSL00250
C                                                                      CSL00260
C     IMPORTANT NOTE:  PROGRAM ALLOWS ENLARGEMENT OF THE ALPHA,BETA    CSL00270
C     ARRAYS.  IN PARTICULAR, IF ANY ONE OF THE EIGENVALUES SUPPLIED   CSL00280
C     IS T-SIMPLE AND NOT CLOSE TO A SPURIOUS T-EIGENVALUE, THE PROGRAM CSL00290
C     REQUIRES THAT KMAX BE AT LEAST 11*MEV/8 + 12.  IF KMAX IS NOT    CSL00300
C     THIS LARGE, THEN THE PROGRAM WILL RESET KMAX TO THIS SIZE        CSL00310
C     AND EXTEND THE ALPHA, BETA HISTORY IF REQUIRED.                  CSL00320
C     THUS, THE DIMENSIONS OF THE ALPHA AND BETA ARRAYS MUST BE        CSL00330
C     LARGE ENOUGH TO ALLOW FOR THIS POSSIBILITY.                      CSL00340
C     REMEMBER THAT THE BETA ARRAY, BETA(J), IS SUCH THAT             CSL00350
C     J = 1,..., KMAX+1.  SO IF THE KMAX USED BY THE PROGRAM          CSL00360
C     IS TO BE 3000, THEN BETA MUST BE OF LENGTH AT LEAST 3001.       CSL00370
C                                                                      CSL00380
C----------------------------------------------------------------------CSL00390
      COMPLEX*16  V1(1600),V2(1600),RITVEC(10000),ZEROC,TEMPC          CSL00400
      COMPLEX*16  ALPHA(1600),BETA(1601),GOODEV(50),TVEC(20000)        CSL00410
      COMPLEX*16  EVAL,ALFA,BATA,SUMC                                  CSL00420
      DOUBLE PRECISION  GR(1600),GC(1600)                              CSL00430
      DOUBLE PRECISION  ERTOL,SUM,TEMP,BKMIN                           CSL00440
      DOUBLE PRECISION  MULTOL,SPUTOL,SCALEO,BTOL                      CSL00450
      DOUBLE PRECISION  ONE,ZERO,MACHEP,EPSM                           CSL00460
      DOUBLE PRECISION  RELTOL,ERROR,ERRMIN,TERROR,TLAST(50)           CSL00470
      REAL  G(1600),AMINGP(50),TMINGP(50),EXPLAN(20)                   CSL00480
      REAL  TERR(50),ERR(50),ERRDGP(50),RNORM(50),TBETA(50)           CSL00490
      INTEGER  MP(50),MA(50),ML(50),MINT(50),MFIN(50),IDELTA(50)      CSL00500
      INTEGER  SVSEED,SVSOLD,RHSEED                                    CSL00510
      INTEGER INTERC(1600)                                             CSL00520
      INTEGER  MBOUND,NTVCON,SVTVEC,TVSTOP,LVCONT,ERCONT,TFLAG        CSL00530
      DOUBLE PRECISION  DABS, DMAX1, DSQRT                             CSL00540
      REAL ABS                                                         CSL00550
      INTEGER  IABS                                                    CSL00560
C----------------------------------------------------------------------CSL00570
      EXTERNAL CMATV                                                   CSL00580
      DATA MACHEP/Z3410000000000000/                                   CSL00590
      EPSM = 2.D0*MACHEP                                               CSL00600
C----------------------------------------------------------------------CSL00610
C                                                                      CSL00620
C     ARRAYS MUST BE DIMENSIONED AS FOLLOWS:                           CSL00630
C     1.  ALPHA: >= KMAXN,  BETA: >= (KMAXN+1) WHERE KMAXN, THE        CSL00640
```

```
C                   LARGEST SIZE T-MATRIX CONSIDERED BY THE PROGRAM,      CSL00650
C                   IS THE LARGER OF THE SIZE OF THE ALPHA, BETA HISTORY  CSL00660
C                   PROVIDED ON FILE 2 (IF ANY ) AND THE SIZE WHICH THE   CSL00670
C                   PROGRAM SPECIFIES INTERNALLY, THIS LATTER IS ALWAYS   CSL00680
C                   < = 11*MEV / 8  +  12, WHERE MEV IS THE SIZE          CSL00690
C                   T-MATRIX THAT WAS USED IN THE CORRESPONDING EIGENVALUE CSL00700
C                   COMPUTATIONS.                                         CSL00710
C       2.  V1:  >= MAX(N,KMAX)                                           CSL00720
C       3.  V2:  >= N                                                     CSL00730
C       4.  G, GR, GC:  >= MAX(N,KMAX)                                    CSL00740
C       5.  RITVEC:  >= N*NGOOD, WHERE NGOOD IS THE NUMBER OF EIGENVALUES CSL00750
C                   SUPPLIED TO THIS PROGRAM.                             CSL00760
C       6.  TVEC:  >= CUMULATIVE LENGTH OF ALL THE T-EIGENVECTORS NEEDED  CSL00770
C                   TO GENERATE THE DESIRED RITZ VECTORS.  AN EDUCATED    CSL00780
C                   GUESS AT AN APPROPRIATE LENGTH CAN BE OBTAINED        CSL00790
C                   BY RUNNING THE PROGRAM WITH THE FLAG MBOUND = 1       CSL00800
C                   AND MULTIPLYING THE RESULTING SIZE BY 5/4.            CSL00810
C       7.  INTERC:  >= KMAX                                             CSL00820
C       8.  GOODEV, AMINGP, TMINGP, TERR, ERR, ERRDGP, RNORM, TBETA,     CSL00830
C           TLAST, MP, MA, MINT, MFIN, AND IDELTA :    >= NUMBER OF       CSL00840
C           EIGENVALUES SUPPLIED.                                         CSL00850
C                                                                         CSL00860
C       OUTPUT HEADER                                                     CSL00870
        WRITE(6,10)                                                       CSL00880
   10 FORMAT(/' LANCZOS EIGENVECTOR PROCEDURE FOR COMPLEX SYMMETRIC MATRCSL00890
     1ICES'/)                                                             CSL00900
C                                                                         CSL00910
C       SET PROGRAM PARAMETERS                                            CSL00920
C       USER MUST NOT MODIFY SCALE0                                       CSL00930
        SCALE0 = 5.0D0                                                    CSL00940
        ZERO = 0.0D0                                                      CSL00950
        ZEROC = DCMPLX(ZERO,ZERO)                                         CSL00960
        ONE = 1.0D0                                                       CSL00970
        MPMIN = -1000                                                     CSL00980
        MONE = -1                                                         CSL00990
C       CONVERGENCE TOLERANCE FOR T-EIGENVECTORS FOR RITZ COMPUTATIONS    CSL01000
        ERTOL = 1.D-10                                                    CSL01010
C------------------------------------------------------------------------CSL01020
C       READ USER-SPECIFIED PARAMETERS FROM INPUT FILE 5 (FREE FORMAT)    CSL01030
C                                                                         CSL01040
C       READ USER-PROVIDED HEADER FOR RUN                                 CSL01050
        READ(5,20) EXPLAN                                                 CSL01060
        WRITE(6,20) EXPLAN                                                CSL01070
   20 FORMAT(20A4)                                                        CSL01080
C                                                                         CSL01090
C       READ IN THE MAXIMUM PERMISSIBLE DIMENSIONS FOR THE TVEC ARRAY     CSL01100
C       (MDIMTV), FOR THE RITVEC ARRAY (MDIMRV), AND FOR THE BETA         CSL01110
C       ARRAY (MBETA).                                                    CSL01120
C                                                                         CSL01130
        READ(5,20) EXPLAN                                                 CSL01140
        READ(5,*) MDIMTV, MDIMRV, MBETA                                   CSL01150
C                                                                         CSL01160
C       READ IN RELATIVE TOLERANCE (RELTOL) USED IN DETERMINING           CSL01170
C       APPROPRIATE SIZES FOR THE T-MATRICES USED IN THE RITZ            CSL01180
C       VECTOR COMPUTATIONS                                               CSL01190
C                                                                         CSL01200
        READ(5,20) EXPLAN                                                 CSL01210
        READ(5,*) RELTOL                                                  CSL01220
C                                                                         CSL01230
C       SET FLAGS TO 0 OR 1:                                              CSL01240
C       MBOUND = 1:  PROGRAM TERMINATES AFTER COMPUTING 1ST GUESSES       CSL01250
C                    ON APPROPRIATE T-SIZES FOR USE IN THE RITZ VECTOR    CSL01260
C                    COMPUTATIONS                                         CSL01270
C       NTVCON = 0:  PROGRAM TERMINATES IF THE TVEC ARRAY IS NOT          CSL01280
C                    LARGE ENOUGH TO HOLD ALL THE T-EIGENVECTORS REQUIRED.CSL01290
C       SVTVEC = 0:  THE T-EIGENVECTORS ARE NOT WRITTEN TO FILE 11        CSL01300
C                    UNLESS TVSTOP = 1                                    CSL01310
```

```
C     SVTVEC = 1:  WRITE THE T-EIGENVECTORS TO FILE 11.              CSL01320
C     TVSTOP = 1:  PROGRAM TERMINATES AFTER COMPUTING THE            CSL01330
C                  T-EIGENVECTORS                                    CSL01340
C     LVCONT = 0:  PROGRAM TERMINATES IF THE NUMBER OF T-EIGENVECTORS CSL01350
C                  COMPUTED IS NOT EQUAL TO THE NUMBER OF RITZ       CSL01360
C                  VECTORS REQUESTED.                                CSL01370
C     ERCONT = 0:  MEANS FOR ANY GIVEN EIGENVALUE, A RITZ VECTOR     CSL01380
C                  WILL NOT BE COMPUTED FOR THAT EIGENVALUE UNLESS   CSL01390
C                  A T-EIGENVECTOR HAS BEEN IDENTIFIED WITH A LAST   CSL01400
C                  COMPONENT WHICH SATISFIES THE SPECIFIED           CSL01410
C                  CONVERGENCE CRITERION.                            CSL01420
C     ERCONT = 1:  MEANS FOR ANY GIVEN EIGENVALUE, A RITZ VECTOR     CSL01430
C                  WILL BE COMPUTED.  IF A T-EIGENVECTOR CANNOT      CSL01440
C                  BE IDENTIFIED WHICH SATISFIES THE LAST           CSL01450
C                  COMPONENT CRITERION, THEN THE PROGRAM WILL        CSL01460
C                  USE THE T-VECTOR THAT CAME CLOSEST TO            CSL01470
C                  SATISFYING THE CRITERION                          CSL01480
C     IWRITE = 1:  EXTENDED OUTPUT OF INTERMEDIATE COMPUTATIONS      CSL01490
C                  IS WRITTEN TO FILE 6                              CSL01500
C     IREAD = 0:   ALPHA/BETA FILE IS REGENERATED.                   CSL01510
C     IREAD = 1:   ALPHA/BETA FILE USED IN EIGENVALUE COMPUTATIONS   CSL01520
C                  IS READ IN AND EXTENDED IF NECESSARY.  IN BOTH    CSL01530
C                  CASES IREAD = 0 OR 1, THE LANCZOS VECTORS ARE     CSL01540
C                  ALWAYS REGENERATED FOR THE RITZ VECTOR            CSL01550
C                  COMPUTATIONS                                      CSL01560
C                                                                    CSL01570
      READ(5,20) EXPLAN                                              CSL01580
      READ(5,*) MBOUND,NTVCON,SVTVEC,IREAD                           CSL01590
C                                                                    CSL01600
      READ(5,20) EXPLAN                                              CSL01610
      READ(5,*) TVSTOP,LVCONT,ERCONT,IWRITE                          CSL01620
      IF (TVSTOP.EQ.1) SVTVEC = 1                                    CSL01630
C                                                                    CSL01640
C     READ IN SEED (RHSEED) FOR GENERATING RANDOM STARTING VECTOR    CSL01650
C     FOR INVERSE ITERATION ON THE T-MATRICES.                      CSL01660
C                                                                    CSL01670
      READ(5,20) EXPLAN                                              CSL01680
      READ(5,*) RHSEED                                               CSL01690
C                                                                    CSL01700
C     READ IN MATNO = MATRIX/RUN IDENTIFICATION NUMBER AND           CSL01710
C     N = ORDER OF A-MATRIX                                          CSL01720
C                                                                    CSL01730
      READ(5,20) EXPLAN                                              CSL01740
      READ(5,*) MATNO,N                                             CSL01750
C                                                                    CSL01760
C-------------------------------------------------------------------CSL01770
C     INITIALIZE THE ARRAYS FOR THE USER-SPECIFIED MATRIX            CSL01780
C     AND PASS THE STORAGE LOCATIONS OF THESE ARRAYS TO THE          CSL01790
C     MATRIX-VECTOR MULTIPLY SUBROUTINE CMATV.                       CSL01800
C                                                                    CSL01810
      CALL USPEC(N,MATNO)                                            CSL01820
C                                                                    CSL01830
C-------------------------------------------------------------------CSL01840
C                                                                    CSL01850
C     MASK UNDERFLOW AND OVERFLOW                                    CSL01860
      CALL MASK                                                      CSL01870
C                                                                    CSL01880
C-------------------------------------------------------------------CSL01890
C     WRITE RUN PARAMETERS OUT TO FILE 6                             CSL01900
C                                                                    CSL01910
      WRITE(6,30) MATNO,N                                            CSL01920
   30 FORMAT(/' MATRIX IDENTIFICATION NO. = ',I10,' ORDER OF A = ',I5) CSL01930
C                                                                    CSL01940
      WRITE(6,40) MBOUND,NTVCON,SVTVEC,IREAD                         CSL01950
   40 FORMAT(/3X,'MBOUND',3X,'NTVCON',3X,'SVTVEC',3X,'IREAD'/3I9,I8) CSL01960
C                                                                    CSL01970
      WRITE(6,50) TVSTOP,LVCONT,ERCONT,IWRITE                        CSL01980
```

```
   50 FORMAT(/3X,'TVSTOP',3X,'LVCONT',3X,'ERCONT',3X,'IWRITE'/4I9)    CSL01990
C                                                                      CSL02000
      WRITE(6,60) MDIMTV,MDIMRV,MBETA                                  CSL02010
   60 FORMAT(/3X,'MDIMTV',3X,'MDIMRV',3X,'MBETA'/2I9,I8)              CSL02020
C                                                                      CSL02030
      WRITE(6,70) RELTOL,RHSEED                                        CSL02040
   70 FORMAT(/7X,'RELTOL',3X,'RHSEED'/E13.4,I9)                        CSL02050
C                                                                      CSL02060
C                                                                      CSL02070
C     FROM FILE 3 READ IN THE NUMBER OF EIGENVALUES (NGOOD) FOR WHICH   CSL02080
C     EIGENVECTORS ARE REQUESTED, THE ORDER (MEV) OF THE LANCZOS        CSL02090
C     TRIDIAGONAL MATRIX USED IN COMPUTING THESE EIGENVALUES, THE       CSL02100
C     ORDER (NOLD) OF THE USER-SPECIFIED MATRIX USED IN THE EIGENVALUE  CSL02110
C     COMPUTATIONS, THE SEED (SVSEED) USED FOR GENERATING THE STARTING  CSL02120
C     VECTOR THAT WAS USED IN THOSE LANCZOS EIGENVALUE COMPUTATIONS,    CSL02130
C     AND THE MATRIX/RUN IDENTIFICATION NUMBER (MATOLD) USED IN THOSE   CSL02140
C     COMPUTATIONS.  ALSO READ IN THE NUMBER (NDIS) OF DISTINCT         CSL02150
C     EIGENVALUES OF T(1,MEV) THAT WERE COMPUTED BUT THIS VALUE IS      CSL02160
C     NOT USED IN THE EIGENVECTOR COMPUTATIONS.                        CSL02170
C                                                                      CSL02180
      READ(3,80) NGOOD,NDIS,MEV,NOLD,SVSEED,MATOLD                     CSL02190
   80 FORMAT(4I6,I12,I8)                                               CSL02200
C                                                                      CSL02210
C     READ IN THE TOLERANCES USED IN THE T-MULTIPLICITY AND SPURIOUS    CSL02220
C     TESTS DURING THE EIGENVALUE COMPUTATIONS.                         CSL02230
C     ALSO READ IN THE FLAG IB.  IF IB < 0, THEN SOME BETA(I) IN THE    CSL02240
C     T-MATRIX FILE PROVIDED ON FILE 2 FAILED THE ORTHOGONALITY         CSL02250
C     TEST IN THE TNORM SUBROUTINE.  USER SHOULD NOTE THAT THIS         CSL02260
C     PROGRAM PROCEEDS INDEPENDENTLY OF THE SIZES OF THE BETA USED.     CSL02270
C                                                                      CSL02280
      READ(3,90) MULTOL,SPUTOL,IB,BTOL                                 CSL02290
   90 FORMAT(2E15.5,I6,E13.4)                                          CSL02300
C                                                                      CSL02310
      WRITE(6,100) MULTOL,SPUTOL                                       CSL02320
  100 FORMAT(/' MULTIPLICITY TOLERANCE USED IN THE T-EIGENVALUE COMPUTATCSL02330
     1IONS WAS',E13.4/' TOLERANCE USED IN SPURIOUS CHECK',E13.4)       CSL02340
C                                                                      CSL02350
C     CONTINUE WRITE TO FILE 6 OF THE PARAMETERS FOR THIS RUN          CSL02360
C                                                                      CSL02370
      WRITE(6,110)NGOOD,NDIS,MEV,NOLD,MATOLD,SVSEED,MULTOL,SPUTOL,IB,   CSL02380
     1BTOL                                                             CSL02390
  110 FORMAT(/' EIGENVALUES SUPPLIED ARE READ IN FROM FILE 3'/' FILE 3  CSL02400
     1HEADER IS'/4X,'NG',2X,'NDIS',3X,'MEV',2X,'NOLD',2X,'MATOLD',4X,   CSL02410
     1'SVSEED'/4I6,I8,I10/7X,'MULTOL',7X,'SPUTOL',6X,'IB',9X,'BTOL'/    CSL02420
     12E13.4,I8,E13.4)                                                 CSL02430
C                                                                      CSL02440
C     IS THE ARRAY RITVEC LONG ENOUGH TO HOLD ALL OF THE DESIRED       CSL02450
C     RITZ VECTORS (APPROXIMATE EIGENVECTORS)?                         CSL02460
      NMAX = NGOOD*N                                                   CSL02470
      IF(MBOUND.EQ.1) GO TO 120                                        CSL02480
      IF(TVSTOP.NE.1.AND.NMAX.GT.MDIMRV) GO TO 1310                    CSL02490
C                                                                      CSL02500
C     CHECK THAT THE ORDER N AND THE MATRIX IDENTIFICATION NUMBER       CSL02510
C     MATNO SPECIFIED BY THE USER AGREE WITH THOSE READ IN FROM         CSL02520
C     FILE 3.                                                         CSL02530
  120 ITEMP = (NOLD-N)**2+(MATOLD-MATNO)**2                            CSL02540
      IF (ITEMP.NE.0) GO TO 1330                                       CSL02550
C                                                                      CSL02560
C     READ IN FROM FILE 3, THE T(1,MEV)-MULTIPLICITIES OF THE          CSL02570
C     EIGENVALUES WHOSE EIGENVECTORS ARE TO BE COMPUTED, THE VALUES     CSL02580
C     OF THESE EIGENVALUES AND THEIR MINIMAL GAPS AS EIGENVALUES        CSL02590
C     OF THE USER-SPECIFIED MATRIX AND AS EIGENVALUES OF THE T-MATRIX.  CSL02600
C                                                                      CSL02610
      READ(3,20) EXPLAN                                                CSL02620
      READ(3,130) (MP(J),GOODEV(J),TMINGP(J),AMINGP(J), J=1,NGOOD)     CSL02630
  130 FORMAT(5X,I5,2E22.14,2E10.3)                                     CSL02640
C                                                                      CSL02650
```

```
      WRITE(6,140) (J,GOODEV(J),MP(J),TMINGP(J),AMINGP(J), J=1,NGOOD)   CSL02660
  140 FORMAT(/' EIGENVALUES READ IN, T-MULTIPLICITIES, T-GAPS AND A-GAPSCSL02670
     1 '/4X,' J ',15X,' EIGENVALUE',14X,'TMULT',4X,' TMINGAP ',4X,       CSL02680
     1' AMINGAP '/(I6,2E20.12,I4,2E15.4))                               CSL02690
C                                                                        CSL02700
C     READ IN ERROR ESTIMATES                                           CSL02710
      WRITE(6,170) MEV,SVSEED                                           CSL02720
C     CHECK WHETHER OR NOT THERE ARE ANY T-ISOLATED EIGENVALUES IN      CSL02730
C     THE EIGENVALUES PROVIDED                                          CSL02740
      DO 150 J=1,NGOOD                                                  CSL02750
      IF(MP(J).EQ.1) GO TO 160                                          CSL02760
  150 CONTINUE                                                          CSL02770
      GO TO 190                                                         CSL02780
  160 READ(4,20) EXPLAN                                                 CSL02790
      READ(4,20) EXPLAN                                                 CSL02800
      READ(4,20) EXPLAN                                                 CSL02810
  170 FORMAT(/' THESE EIGENVALUES WERE COMPUTED USING A T-MATRIX OF     CSL02820
     1ORDER ',I5/' AND SEED FOR RANDOM NUMBER GENERATOR =',I12)         CSL02830
      READ(4,180) NISO                                                  CSL02840
  180 FORMAT(18X,I6)                                                    CSL02850
      READ(4,20) EXPLAN                                                 CSL02860
      READ(4,20) EXPLAN                                                 CSL02870
      READ(4,20) EXPLAN                                                 CSL02880
  190 DO 220 J=1,NGOOD                                                  CSL02890
      ERR(J) = 0.D0                                                     CSL02900
      IF(MP(J).NE.1) GO TO 220                                          CSL02910
      READ(4,200) EVAL, ERR(J)                                          CSL02920
  200 FORMAT(10X,2E20.12,E14.3)                                         CSL02930
      IF(CDABS(EVAL - GOODEV(J)).LT.1.D-10) GO TO 220                   CSL02940
      WRITE(6,210) EVAL,GOODEV(J)                                       CSL02950
  210 FORMAT(' PROBLEM WITH READ IN OF ERROR ESTIMATES'/' EIGENVALUE REACSL02960
     1D IN',2E20.12,' DOES NOT MATCH GOODEV(J) ='/2E20.12)             CSL02970
      GO TO 1550                                                        CSL02980
C                                                                        CSL02990
  220 CONTINUE                                                          CSL03000
C                                                                        CSL03010
      WRITE(6,230) (J,GOODEV(J),ERR(J), J=1,NGOOD)                      CSL03020
  230 FORMAT(' ERROR ESTIMATES ='/4X,' J',15X,'EIGENVALUE',20X,'ESTIMATECSL03030
     1'/(I6,2E20.12,E14.3))                                            CSL03040
C                                                                        CSL03050
C     READ IN THE SIZE OF THE T-MATRIX PROVIDED ON FILE 2.  READ IN     CSL03060
C     THE ORDER OF THE USER-SPECIFIED MATRIX , THE SEED FOR THE         CSL03070
C     RANDOM NUMBER GENERATOR, AND THE MATRIX/TEST IDENTIFICATION       CSL03080
C     NUMBER THAT WERE USED IN THE LANCZOS EIGENVALUE COMPUTATIONS.     CSL03090
C     IF FLAG IREAD = 0, REGENERATE HISTORY FROM SCRATCH                CSL03100
C     HISTORY MUST BE STORED IN MACHINE FORMAT, ((4Z20) FOR            CSL03110
C     IBM/3081)                                                         CSL03120
C                                                                        CSL03130
      IF(IREAD.EQ.0)  GO TO 330                                         CSL03140
C                                                                        CSL03150
      READ(2,240) KMAX,NOLD,SVSOLD,MATOLD                              CSL03160
  240 FORMAT(2I6,I12,I8)                                               CSL03170
C                                                                        CSL03180
      WRITE(6,250) KMAX,NOLD,SVSOLD,MATOLD                             CSL03190
  250 FORMAT(/' READ IN HEADER FOR T-MATRICES'/' FILE 2 HEADER IS'/    CSL03200
     1 2X,'KMAX',2X,'NOLD',6X,'SVSOLD',2X,'MATOLD'/2I6,I12,I8/)        CSL03210
C                                                                        CSL03220
C     CHECK THAT THE ORDER, THE MATRIX/TEST IDENTIFICATION NUMBER       CSL03230
C     AND THE SEED FOR THE RANDOM NUMBER GENERATOR USED IN THE          CSL03240
C     LANCZOS COMPUTATIONS THAT GENERATED THE HISTORY FILE              CSL03250
C     BEING USED AGREE WITH WHAT THE USER HAS SPECIFIED.                CSL03260
      IF (NOLD.NE.N.OR.MATOLD.NE.MATNO.OR.SVSOLD.NE.SVSEED) GO TO 1350  CSL03270
C                                                                        CSL03280
      KMAX1 = KMAX + 1                                                  CSL03290
C                                                                        CSL03300
C     READ IN THE T-MATRICES FROM FILE 2.  THESE ARE USED TO GENERATE   CSL03310
C     THE T-EIGENVECTORS THAT WILL BE USED IN THE RITZ VECTOR           CSL03320
```

```
C     COMPUTATIONS.  HISTORY MUST BE IN MACHINE FORMAT.              CSL03330
C                                                                    CSL03340
      READ(2,260) (ALPHA(J), J=1,KMAX)                               CSL03350
      READ(2,260) (BETA(J), J=1,KMAX1)                               CSL03360
  260 FORMAT(4Z20)                                                   CSL03370
C                                                                    CSL03380
      READ(2,260) (V1(J), J=1,N)                                     CSL03390
      READ(2,260) (V2(J), J=1,N)                                     CSL03400
C                                                                    CSL03410
C     KMAX MAY BE ENLARGED IF THE SIZE AT WHICH THE EIGENVALUE       CSL03420
C     COMPUTATIONS WERE PERFORMED IS ESSENTIALLY KMAX AND            CSL03430
C     THERE IS AT LEAST ONE EIGENVALUE THAT IS T-SIMPLE AND          CSL03440
C     T-ISOLATED, IN THE SENSE THAT IF ITS CLOSEST NEIGHBOR IS TOO   CSL03450
C     CLOSE THAT NEIGHBOR IS A 'GOOD' T-EIGENVALUE.                  CSL03460
      DO 270 J = 1,NGOOD                                             CSL03470
      IF(MP(J).EQ.1) GO TO 290                                       CSL03480
  270 CONTINUE                                                       CSL03490
      WRITE(6,280)                                                   CSL03500
  280 FORMAT(/' ALL EIGENVALUES USED ARE T-MULTIPLE OR CLOSE TO SPURIOUSCSL03510
     1 T-EIGENVALUES'/' SO DO NOT CHANGE KMAX')                      CSL03520
      IF(KMAX.LT.MEV) GO TO 1370                                     CSL03530
      GO TO 310                                                      CSL03540
C                                                                    CSL03550
  290 KMAXN= 11*MEV/8 + 12                                           CSL03560
      IF(MBETA.LE.KMAXN) GO TO 1530                                  CSL03570
      IF(KMAX.GE.KMAXN ) GO TO 310                                   CSL03580
      WRITE(6,300) KMAX, KMAXN                                       CSL03590
  300 FORMAT(' ENLARGE KMAX FROM ',I6,' TO ',I6)                     CSL03600
      MOLD1 = KMAX + 1                                               CSL03610
      KMAX = KMAXN                                                   CSL03620
      GO TO 380                                                      CSL03630
C                                                                    CSL03640
  310 WRITE(6,320) KMAX                                              CSL03650
  320 FORMAT(/' T-MATRICES HAVE BEEN READ IN FROM FILE 2'/' THE LARGEST CSL03660
     1SIZE T-MATRIX ALLOWED IS',I6/)                                 CSL03670
C                                                                    CSL03680
      IF(IREAD.EQ.1) GO TO 400                                       CSL03690
C                                                                    CSL03700
C     REGENERATE THE ALPHA AND BETA                                  CSL03710
C                                                                    CSL03720
  330 MOLD1 = 1                                                      CSL03730
C                                                                    CSL03740
C     SET KMAX                                                       CSL03750
      DO 340 J = 1,NGOOD                                             CSL03760
      IF(MP(J).EQ.1) GO TO 360                                       CSL03770
  340 CONTINUE                                                       CSL03780
      KMAX = MEV + 12                                                CSL03790
      WRITE(6,350) KMAX                                              CSL03800
  350 FORMAT(/' ALL EIGENVALUES FOR WHICH EIGENVECTORS ARE TO BE COMPUTECSL03810
     1D ARE EITHER T-MULTIPLE OR CLOSE TO'/' A SPURIOUS EIGENVALUE. THERCSL03820
     1EFORE SET KMAX = MEV + 12 = ',I7)                              CSL03830
      GO TO 380                                                      CSL03840
C                                                                    CSL03850
  360 KMAXN = 11*MEV/8 + 12                                          CSL03860
      IF(MBETA.LE.KMAXN) GO TO 1530                                  CSL03870
      WRITE(6,370) KMAXN                                             CSL03880
  370 FORMAT(' SET KMAX EQUAL TO ',I6)                               CSL03890
      KMAX = KMAXN                                                   CSL03900
C                                                                    CSL03910
  380 WRITE(6,390) MOLD1,KMAX                                        CSL03920
  390 FORMAT(/' LANCZS SUBROUTINE GENRATES ALPHA(J), BETA(J+1), J =', CSL03930
     1 I6,' TO ', I6/)                                               CSL03940
C                                                                    CSL03950
C--------------------------------------------------------------------CSL03960
C                                                                    CSL03970
      CALL LANCZS(CMATV,V1,V2,ALPHA,BETA,GR,GC,G,KMAX,MOLD1,N,SVSEED) CSL03980
C                                                                    CSL03990
```

```
C-------------------------------------------------------------------CSL04000
C                                                                    CSL04010
  400 CONTINUE                                                        CSL04020
C                                                                    CSL04030
C     SIMPLE STURM SEQUENCING IS NOT VALID FOR COMPLEX SYMMETRIC     CSL04040
C     MATRICES.  THUS, THE STRATEGY USED HERE FOR SELECTING          CSL04050
C     APPROPRIATE SIZE T-MATRICES FOR THE EIGENVECTOR COMPUTATIONS   CSL04060
C     MUST BE DIFFERENT FROM THAT USED IN THE REAL SYMMETRIC,        CSL04070
C     HERMITIAN, AND SINGULAR VALUE CASES.  AS IN THOSE CASES,       CSL04080
C     FOR EACH EIGENVALUE, A FIRST GUESS IS SELECTED AND THEN        CSL04090
C     LOOPING ON THE SIZE OF THE T-EIGENVECTOR COMPUTATIONS          CSL04100
C     DETERMINES APPROPRIATE SIZES FOR THE EIGENVECTOR COMPUTATIONS. CSL04110
C     FIRST GUESSES AT APPROPRIATE SIZES ARE SPECIFIED BELOW.        CSL04120
C                                                                    CSL04130
      DO 430 J = 1,NGOOD                                             CSL04140
      EVAL = GOODEV(J)                                               CSL04150
C     COMPUTE A FIRST GUESS ON AN APPROPRIATE SIZE T-MATRIX EACH     CSL04160
C     EIGENVALUE.                                                    CSL04170
      IF(MP(J).GT.1) GO TO 410                                       CSL04180
C     EIGENVALUE IS T-SIMPLE                                         CSL04190
      IF(MP(J).EQ.MONE) GO TO 420                                    CSL04200
C     EIGENVALUE IS T-SIMPLE AND T-ISOLATED                         CSL04210
      MA(J) = (8*MEV)/9  +  1                                        CSL04220
      ML(J) = ((11*MEV)/8 + 12)                                      CSL04230
      GO TO 430                                                      CSL04240
C     EIGENVALUE IS T-MULTIPLE                                       CSL04250
  410 MA(J) = (5*MEV)/(4*MP(J))  +  1                                CSL04260
      ML(J) = (7*MEV)/(4*MP(J)) + 1                                  CSL04270
      GO TO 430                                                      CSL04280
C     EIGENVALUE IS T-SIMPLE AND NOT T-ISOLATED                     CSL04290
  420 MA(J) = (5*MEV)/8 + 1                                          CSL04300
      ML(J) = MEV                                                    CSL04310
  430 CONTINUE                                                       CSL04320
C                                                                    CSL04330
      IF (IWRITE.EQ.1) WRITE(6,440) (MA(JJ), JJ=1,NGOOD)             CSL04340
  440 FORMAT(/' 1ST GUESS AT APPROPRIATE SIZES FOR T-MATRICES '/     CSL04350
     1' ACTUAL VALUES WILL PROBABLY BE 1/4 AGAIN AS MUCH'/(1316))    CSL04360
C                                                                    CSL04370
      WRITE(10,450) N,KMAX                                           CSL04380
  450 FORMAT(2I8,' = ORDER OF USER MATRIX AND MAX ORDER OF T(1,MEV)') CSL04390
C                                                                    CSL04400
      WRITE(10,460)                                                  CSL04410
  460 FORMAT(/' 1ST GUESS AT APPROPRIATE SIZES FOR T-MATRICES '/     CSL04420
     1' ACTUAL VALUES WILL PROBABLY BE 1/4 AGAIN AS MUCH'/)          CSL04430
      WRITE(10,470)                                                  CSL04440
  470 FORMAT(5X,'J',8X,'REAL(GOODEV)',8X,'IMAG(GOODEV)',7X,'MA(J)',  CSL04450
     17X,'MP(J)')                                                    CSL04460
C                                                                    CSL04470
      WRITE(10,480) (J,GOODEV(J), MA(J), MP(J), J=1,NGOOD)           CSL04480
  480 FORMAT(I6,2E20.12,I12,I12)                                     CSL04490
C                                                                    CSL04500
      IF(MBOUND.EQ.1) WRITE(10,490)                                  CSL04510
  490 FORMAT(/' GOODEV(J) IS A GOOD EIGENVALUE OF T(1,MEV)'/         CSL04520
     1 ' IABS(MA(J)) = APPROPRIATE SIZE T-MATRIX FOR GOODEV(J)'/     CSL04530
     1 ' INITIAL VALUE OF MA(J) IS CHOSEN HEURISTICALLY'/            CSL04540
     1 ' PROGRAM LOOPS ON SIZE OF T-MATRIX TO GET BETTER SIZE'/      CSL04550
     1 ' END OF SIZES OF T-MATRICES FILE 10'///)                     CSL04560
C                                                                    CSL04570
C                                                                    CSL04580
C     TERMINATE AFTER COMPUTING 1ST GUESSES ON SIZES OF T-MATRICES   CSL04590
C     REQUIRED FOR THE GIVEN EIGENVALUES?                            CSL04600
      IF(MBOUND.EQ.1) GO TO 1390                                     CSL04610
C                                                                    CSL04620
C                                                                    CSL04630
C     IS THERE ROOM FOR ALL OF THE REQUESTED T-EIGENVECTORS?         CSL04640
      MTOL = 0                                                       CSL04650
      DO 500 J = 1,NGOOD                                             CSL04660
```

```
      MTOL = MTOL + IABS(MA(J))                                  CSL04670
  500 CONTINUE                                                   CSL04680
      MTOL = (5*MTOL)/4                                          CSL04690
      IF(MTOL.GT.MDIMTV.AND.NTVCON.EQ.0) GO TO 1410             CSL04700
C                                                                CSL04710
C------------------------------------------------------------------CSL04720
C     GENERATE A RANDOM VECTOR TO BE USED REPEATEDLY BY          CSL04730
C     SUBROUTINE INVERM                                          CSL04740
C                                                                CSL04750
      ILL = RHSEED                                               CSL04760
      CALL GENRAN(ILL,G,KMAX)                                    CSL04770
C                                                                CSL04780
C------------------------------------------------------------------CSL04790
C                                                                CSL04800
      DO 510 I = 1,KMAX                                          CSL04810
  510 GR(I) = G(I)                                               CSL04820
C                                                                CSL04830
C------------------------------------------------------------------CSL04840
C                                                                CSL04850
      CALL GENRAN(ILL,G,KMAX)                                    CSL04860
C                                                                CSL04870
C------------------------------------------------------------------CSL04880
C                                                                CSL04890
      DO 520 I = 1,KMAX                                          CSL04900
  520 GC(I) = G(I)                                               CSL04910
C                                                                CSL04920
C     FOR EACH EIGENVALUE LOOP ON T-EIGENVECTOR COMPUTATIONS TO  CSL04930
C     COMPUTE AN APPROPRIATE T-EIGENVECTOR TO USE IN THE RITZ    CSL04940
C     VECTOR COMPUTATIONS.                                       CSL04950
C                                                                CSL04960
      MTOL = 0                                                   CSL04970
      NTVEC = 0                                                  CSL04980
      DO 690 J = 1,NGOOD                                         CSL04990
      ICOUNT = 0                                                 CSL05000
      TFLAG = 0                                                  CSL05010
      ERRMIN = 10.D0                                             CSL05020
      MABEST = MPMIN                                             CSL05030
      IF(MP(J).EQ.MPMIN) GO TO 690                               CSL05040
      EVAL = GOODEV(J)                                           CSL05050
  530 KMAXU = IABS(MA(J))                                        CSL05060
C     SELECT A SUITABLE INCREMENT FOR THE ORDERS OF T-MATRICES   CSL05070
C     TO BE CONSIDERED IN DETERMINING APPROPRIATE SIZES FOR THE RITZ CSL05080
C     VECTOR COMPUTATIONS                                        CSL05090
      IF(ICOUNT.GT.0) GO TO 560                                  CSL05100
C     SELECT IDELTA(J) BASED UPON THE MULTIPLICITY IN T(1,MEV)   CSL05110
      IF(MP(J).GT.1) GO TO 540                                   CSL05120
      IF(MP(J).LT.0) GO TO 550                                   CSL05130
C     MP(J) = 1, INITIAL MA(J) = 8*MEV/9 + 1                     CSL05140
      IDELTA(J) = (ML(J) - IABS(MA(J)))/10  + 1                  CSL05150
      GO TO 560                                                  CSL05160
C     MULTIPLE T-EIGENVALUE: INITIAL MA(J) = 5*MEV/4*MP + 1      CSL05170
  540 IDELTA(J) = (ML(J) - IABS(MA(J)))/10  + 1                  CSL05180
      GO TO 560                                                  CSL05190
C     T-SIMPLE EVALUE, NEAR SPURIOUS ONE, INITIAL MA(J) = 5*MEV/8 + 1 CSL05200
  550 IDELTA(J) = (ML(J) - IABS(MA(J)))/10  + 1                  CSL05210
  560 ICOUNT = ICOUNT + 1                                        CSL05220
      MTOL = MTOL+KMAXU                                          CSL05230
C                                                                CSL05240
C     IS THERE ROOM IN TVEC ARRAY FOR THE NEXT T-EIGENVECTOR?    CSL05250
C     IF NOT, SKIP TO RITZ VECTOR COMPUTATIONS.                  CSL05260
      IF (MTOL.GT.MDIMTV) GO TO 700                              CSL05270
C                                                                CSL05280
      IT = 3                                                     CSL05290
      KINT = MTOL - KMAXU +1                                     CSL05300
C                                                                CSL05310
C     RECORD THE BEGINNING AND END OF THE T-EIGENVECTOR BEING COMPUTED CSL05320
      MINT(J) = KINT                                             CSL05330
```

```
      MFIN(J) = MTOL                                              CSL05340
C                                                                 CSL05350
C----------------------------------------------------------------CSL05360
C     SUBROUTINE INVERM DOES INVERSE ITERATION, I.E. SOLVES       CSL05370
C     (T(1,KMAXU) - EVAL)*U = RHS  FOR EACH EIGENVALUE TO OBTAIN THE CSL05380
C     DESIRED T-EIGENVECTOR.                                      CSL05390
C                                                                 CSL05400
      IF(IWRITE.EQ.1)  WRITE(6,570) J                            CSL05410
  570 FORMAT(/16,'TH EIGENVALUE')                                 CSL05420
C                                                                 CSL05430
      CALL INVERM(ALPHA,BETA,V1,TVEC(KINT),EVAL,ERROR,TERROR,EPSM, CSL05440
     1    GR,GC,INTERC,KMAXU,IT,IWRITE)                          CSL05450
C                                                                 CSL05460
C----------------------------------------------------------------CSL05470
C                                                                 CSL05480
      TERR(J) = TERROR                                           CSL05490
      TLAST(J) = ERROR                                           CSL05500
      KMAXU1 = KMAXU + 1                                         CSL05510
      TBETA(J) = CDABS(BETA(KMAXU1))*ERROR                       CSL05520
C                                                                 CSL05530
C     AFTER COMPUTING EACH OF THE T-EIGENVECTORS,                CSL05540
C     CHECK THE SIZE OF THE ERROR ESTIMATE, ERROR.               CSL05550
C     IF THIS ESTIMATE IS NOT AS SMALL AS DESIRED AND            CSL05560
C     |MA(J)| < ML(J), ATTEMPT TO INCREASE THE SIZE OF |MA(J)|   CSL05570
C     AND REPEAT THE T-EIGENVECTOR COMPUTATIONS.                 CSL05580
C                                                                 CSL05590
      IF(ERROR.LT.ERTOL.OR.TFLAG.EQ.1)  GO TO 680                CSL05600
C                                                                 CSL05610
      IF(ERROR.GE.ERRMIN) GO TO 580                              CSL05620
C     LAST COMPONENT IS LESS THAN MINIMAL TO DATE                CSL05630
      ERRMIN = ERROR                                            CSL05640
      MABEST = MA(J)                                            CSL05650
  580 CONTINUE                                                   CSL05660
C                                                                 CSL05670
      IF(MA(J).GT.0)  ITEST = MA(J) + IDELTA(J)                  CSL05680
      IF(MA(J).LT.0)  ITEST = -(IABS(MA(J)) + IDELTA(J))         CSL05690
      IF(IABS(ITEST).LE.ML(J).AND.ICOUNT.LE.10) GO TO 600        CSL05700
C     NEW MA(J) IS GREATER THAN MAXIMUM ALLOWED.                 CSL05710
      IF(ERCONT.EQ.0.OR.MABEST.EQ.MPMIN) GO TO 620               CSL05720
      TFLAG = 1                                                 CSL05730
      MA(J) = MABEST                                            CSL05740
      MTOL = MTOL - KMAXU                                        CSL05750
      WRITE(6,590) MA(J)                                        CSL05760
  590 FORMAT(' 10 ORDERS WERE CONSIDERED.  NONE SATISFIED THE ERROR TESTCSL05770
     1'/' THEREFORE USE THE BEST ORDER OBTAINED FOR THE EIGENVECTORS' CSL05780
     1,16)                                                      CSL05790
      GO TO 530                                                 CSL05800
C                                                                 CSL05810
  600 MA(J) = ITEST                                              CSL05820
C                                                                 CSL05830
      MT = IABS(MA(J))                                          CSL05840
      IF(IWRITE.EQ.1) WRITE(6,610) MT                           CSL05850
  610 FORMAT(/' CHANGE SIZE OF T-MATRIX TO',16,' RECOMPUTE T-EIGENVECTORCSL05860
     1')                                                        CSL05870
C                                                                 CSL05880
      MTOL = MTOL - KMAXU                                        CSL05890
C                                                                 CSL05900
      GO TO 530                                                 CSL05910
C                                                                 CSL05920
C        APPROPRIATE SIZE T-MATRIX WAS NOT OBTAINED              CSL05930
  620 CONTINUE                                                   CSL05940
      WRITE(10,630) J,EVAL,MP(J)                                 CSL05950
  630 FORMAT(/' ON 10 INCREMENTS NOT ABLE TO IDENTIFY APPROPRIATE SIZE  CSL05960
     1T-MATRIX FOR'/                                            CSL05970
     1' EIGENVALUE(',14,') =',2E20.12,' T-MULTIPLICITY =',14/)  CSL05980
      IF(MP(J).GT.1) WRITE(10,640)                              CSL05990
      IF(MP(J).LT.0)  WRITE(10,650)                             CSL06000
```

```
      IF(MP(J).EQ.1) WRITE(10,660)                                 CSL06010
 640 FORMAT(/' ORDERS TESTED RANGED FROM (5*MEV/4*MP(J)) TO APPROXIMATECSL06020
    1LY'/'  (7*MEV)/(4*MP(J)'/)                                    CSL06030
 650 FORMAT(/' ORDERS TESTED RANGED FROM (5*MEV/8)  TO MEV'/)      CSL06040
 660 FORMAT(/' ORDERS TESTED RANGED FROM 8*MEV/9 TO APPROXIMATELY 11*MECSL06050
    1V/8'/)                                                        CSL06060
      WRITE(10,670)                                                CSL06070
 670 FORMAT(' ALLOWING LARGER ORDERS FOR THE T-MATRICES MAY RESULT IN CSL06080
    1 SUCCESS'/' BUT PROBABLY WILL NOT.  PROBLEM IS PROBABLY DUE TO' CSL06090
    1 /' LACK OF CONVERGENCE OF GIVEN EIGENVALUE, CHECK THE ERROR ESTIMCSL06100
    1ATE')                                                         CSL06110
      MP(J) = MPMIN                                                CSL06120
      MTOL = MTOL - KMAXU                                          CSL06130
      GO TO 690                                                    CSL06140
 680 NTVEC = NTVEC + 1                                             CSL06150
C                                                                  CSL06160
 690 CONTINUE                                                      CSL06170
      NGOODC = NGOOD                                               CSL06180
      GO TO 720                                                    CSL06190
C                                                                  CSL06200
C     COME HERE IF THERE IS NOT ENOUGH ROOM FOR ALL OF T-EIGENVECTORS CSL06210
 700 NGOODC = J-1                                                  CSL06220
      WRITE(6,710) J,MTOL,MDIMTV                                   CSL06230
 710 FORMAT(/' NOT ENOUGH ROOM IN TVEC ARRAY FOR ',I4,'TH T-EIGENVECTORCSL06240
    1'/' TVEC-DIMENSION REQUESTED = ',I6,' BUT TVEC HAS DIMENSION ',I6/CSL06250
    1)                                                             CSL06260
      IF(NGOODC.EQ.0) GO TO 1430                                   CSL06270
      MTOL = MTOL-KMAXU                                            CSL06280
C                                                                  CSL06290
 720 CONTINUE                                                      CSL06300
C                                                                  CSL06310
C     THE LOOP ON T-EIGENVECTOR COMPUTATIONS IS COMPLETE.          CSL06320
C     WRITE OUT THE SIZE T-MATRICES THAT WILL BE USED FOR          CSL06330
C     THE RITZ VECTOR COMPUTATIONS.                                CSL06340
C                                                                  CSL06350
      WRITE(10,730)                                                CSL06360
 730 FORMAT(/' SIZES OF T-MATRICES THAT WILL BE USED IN THE RITZ COMPUTCSL06370
    1ATIONS'/5X,'J',13X,'REAL(GOODEV)',13X,'IMAG(GOODEV)',1X,'MA(J)') CSL06380
C                                                                  CSL06390
      WRITE(10,740)   (J,GOODEV(J),MA(J), J=1,NGOOD)               CSL06400
 740 FORMAT(I6,2E25.14,I6)                                         CSL06410
      WRITE(10,490)                                                CSL06420
C                                                                  CSL06430
      WRITE(6,750) MTOL                                            CSL06440
 750 FORMAT(/' THE CUMULATIVE LENGTH OF THE T-EIGENVECTORS IS',I18) CSL06450
C                                                                  CSL06460
      WRITE(6,760) NTVEC,NGOOD                                     CSL06470
 760 FORMAT(/I6,' T-EIGENVECTORS OUT OF',I6,' REQUESTED WERE COMPUTED')CSL06480
C                                                                  CSL06490
C     SAVE THE T-EIGENVECTORS ON FILE 11?                          CSL06500
      IF(TVSTOP.NE.1.AND.SVTVEC.EQ.0) GO TO 820                    CSL06510
C                                                                  CSL06520
      WRITE(11,770) NTVEC,MTOL,MATNO,SVSEED                        CSL06530
 770 FORMAT(I6,3I12,' = NTVEC,MTOL,MATNO,SVSEED')                  CSL06540
C                                                                  CSL06550
      DO 800 J=1,NGOODC                                            CSL06560
C     IF MP(J) = MPMIN THEN NO SUITABLE T-EIGENVECTOR IS AVAILABLE CSL06570
C     FOR THAT EIGENVALUE.                                         CSL06580
      IF(MP(J).EQ.MPMIN) WRITE(11,780) J,MA(J),GOODEV(J),MP(J)     CSL06590
 780 FORMAT(2I6,2E20.12,I6/' TH EIGVAL,T-SIZE,EVALUE,FLAG,NO EIGVEC') CSL06600
      IF(MP(J).NE.MPMIN) WRITE(11,790) J,MA(J),GOODEV(J),MP(J)     CSL06610
 790 FORMAT(I6,I6,2E20.12,I6/' T-EIGVEC,SIZE T,EVALUE OF A,MP(J)') CSL06620
      IF(MP(J).EQ.MPMIN) GO TO 800                                 CSL06630
      KI = MINT(J)                                                 CSL06640
      KF = MFIN(J)                                                 CSL06650
C                                                                  CSL06660
```

```
      WRITE(11,260) (TVEC(K), K=KI,KF)                           CSL06670
C                                                                CSL06680
  800 CONTINUE                                                   CSL06690
C                                                                CSL06700
      IF(TVSTOP.NE.1) GO TO 820                                  CSL06710
C                                                                CSL06720
      WRITE(6,810) TVSTOP, NTVEC,NGOOD                           CSL06730
  810 FORMAT(/' USER SET TVSTOP = ',I1/                          CSL06740
     1' THEREFORE PROGRAM TERMINATES AFTER T-EIGENVECTOR COMPUTATIONS'/ CSL06750
     1' T-EIGENVECTORS THAT WERE COMPUTED ARE SAVED ON FILE 11'/ CSL06760
     118,' T-EIGENVECTORS WERE COMPUTED OUT OF',I7,' REQUESTED'/) CSL06770
C                                                                CSL06780
      GO TO 1550                                                 CSL06790
C                                                                CSL06800
  820 CONTINUE                                                   CSL06810
C     IF NOT ABLE TO COMPUTE ALL THE REQUESTED T-EIGENVECTORS    CSL06820
C     CONTINUE WITH THE LANCZOS VECTOR COMPUTATIONS ANYWAY?      CSL06830
C                                                                CSL06840
      IF(NTVEC.NE.NGOOD.AND.LVCONT.EQ.0) GO TO 1450              CSL06850
C                                                                CSL06860
C     COMPUTE THE MAXIMUM SIZE OF THE T-MATRIX USED FOR THOSE    CSL06870
C     EIGENVALUES WITH GOOD ERROR ESTIMATES.                     CSL06880
C                                                                CSL06890
      KMAXU = 0                                                  CSL06900
      DO 830 J = 1,NGOODC                                        CSL06910
      MT = IABS(MA(J))                                           CSL06920
      IF(MT.LT.KMAXU.OR.MP(J).EQ.MPMIN) GO TO 830                CSL06930
      KMAXU = MT                                                 CSL06940
  830 CONTINUE                                                   CSL06950
C                                                                CSL06960
      IF(KMAXU.EQ.0) GO TO 1490                                  CSL06970
C                                                                CSL06980
      WRITE(6,840) KMAXU                                         CSL06990
  840 FORMAT(/I6,' = LARGEST SIZE T-MATRIX TO BE USED IN THE RITZ VECTORCSL07000
     1 COMPUTATIONS')                                            CSL07010
C                                                                CSL07020
C     COUNT THE NUMBER OF RITZ VECTORS NOT BEING COMPUTED        CSL07030
      MREJEC = 0                                                 CSL07040
      DO 850 J=1,NGOODC                                          CSL07050
  850 IF(MP(J).EQ.MPMIN)  MREJEC = MREJEC + 1                    CSL07060
      MREJET = MREJEC + (NGOOD-NGOODC)                           CSL07070
      IF(MREJET.NE.0) WRITE(6,860) MREJET                        CSL07080
  860 FORMAT(/' RITZ VECTORS ARE NOT COMPUTED FOR',I6,' OF THE EIGENVALUCSL07090
     1ES'/)                                                      CSL07100
      NACT = NGOODC - MREJEC                                     CSL07110
      WRITE(6,870) NGOOD,NTVEC,NACT                              CSL07120
  870 FORMAT(/I6,' RITZ VECTORS WERE REQUESTED'/I6,' T-EIGENVECTORS WERECSL07130
     1COMPUTED'/I6,' RITZ VECTORS WILL BE COMPUTED'/)            CSL07140
C     CHECK IF THERE ARE ANY RITZ VECTORS TO COMPUTE             CSL07150
      IF(MREJEC.EQ.NGOODC) GO TO 1470                            CSL07160
C                                                                CSL07170
C     CONTINUE WITH THE LANCZOS VECTOR COMPUTATIONS?             CSL07180
      IF(LVCONT.EQ.0.AND.MREJEC.NE.0) GO TO 1450                 CSL07190
C                                                                CSL07200
C     NOW COMPUTE THE RITZ VECTORS.  REGENERATE THE              CSL07210
C     LANCZOS VECTORS.                                           CSL07220
C                                                                CSL07230
      DO 880 I = 1,NMAX                                          CSL07240
  880 RITVEC(I) = ZEROC                                          CSL07250
C                                                                CSL07260
C----------------------------------------------------------------CSL07270
C     REGENERATE THE STARTING VECTOR. THIS MUST BE GENERATED AND CSL07280
C     NORMALIZED PRECISELY THE WAY IT WAS DONE IN THE EIGENVALUE CSL07290
C     COMPUTATIONS, OTHERWISE THERE WILL BE A MISMATCH BETWEEN   CSL07300
C     THE T-EIGENVECTORS THAT HAVE BEEN COMPUTED FROM THE T-MATRICES CSL07310
C     READ IN FROM FILE 2 AND THE LANCZOS VECTORS THAT ARE       CSL07320
```

```
C     BEING REGENERATED.                                      CSL07330
C                                                             CSL07340
      IIL = SVSEED                                            CSL07350
      CALL GENRAN(IIL,G,N)                                    CSL07360
C                                                             CSL07370
C------------------------------------------------------------CSL07380
C                                                             CSL07390
      DO 890 I = 1,N                                          CSL07400
  890 GR(I) = G(I)                                            CSL07410
C                                                             CSL07420
C------------------------------------------------------------CSL07430
C                                                             CSL07440
      CALL GENRAN(IIL,G,N)                                    CSL07450
C                                                             CSL07460
C------------------------------------------------------------CSL07470
C                                                             CSL07480
      DO 900 I = 1,N                                          CSL07490
  900 GC(I) = G(I)                                            CSL07500
C                                                             CSL07510
      DO 910 I = 1,N                                          CSL07520
  910 V2(I) = DCMPLX(GR(I),GC(I))                             CSL07530
C                                                             CSL07540
C------------------------------------------------------------CSL07550
      CALL INPRDC(V2,V2,SUMC,N)                               CSL07560
C------------------------------------------------------------CSL07570
C                                                             CSL07580
      SUMC = ONE/CDSQRT(SUMC)                                 CSL07590
      DO 920 I = 1,N                                          CSL07600
      V1(I) = ZEROC                                           CSL07610
  920 V2(I) = V2(I)*SUMC                                      CSL07620
C                                                             CSL07630
C     LOOP FOR GENERATING REQUIRED RITZ VECTORS (IVEC = 1,KMAXU) CSL07640
C                                                             CSL07650
      IVEC = 1                                                CSL07660
      BATA = ZEROC                                            CSL07670
C                                                             CSL07680
      GO TO 980                                               CSL07690
C                                                             CSL07700
  930 CONTINUE                                                CSL07710
C                                                             CSL07720
C------------------------------------------------------------CSL07730
C                                                             CSL07740
C     CMATV(V2,V1,BATA) CALCULATES   V1 = A*V2 - BATA*V1      CSL07750
      CALL CMATV(V2,V1,BATA)                                  CSL07760
      CALL INPRDC(V2,V1,ALFA,N)                               CSL07770
C                                                             CSL07780
C------------------------------------------------------------CSL07790
C                                                             CSL07800
      DO 940 J=1,N                                            CSL07810
  940 V1(J) = V1(J)-ALFA*V2(J)                                CSL07820
C                                                             CSL07830
C------------------------------------------------------------CSL07840
      CALL INPRDC(V1,V1,BATA,N)                               CSL07850
C------------------------------------------------------------CSL07860
C                                                             CSL07870
      BATA = CDSQRT(BATA)                                     CSL07880
      SUMC = ONE/BATA                                         CSL07890
C                                                             CSL07900
      TEMPC = BETA(IVEC)                                      CSL07910
      TEMP = CDABS(BATA - TEMPC)/CDABS(TEMPC)                 CSL07920
      IF (TEMP.LT.1.0D-10)GO TO 960                           CSL07930
C                                                             CSL07940
C     IF THE BETA BEING REGENERATED DO NOT MATCH THE HISTORY FILE CSL07950
C     THEN SOMETHING IS WRONG IN THE LANCZOS VECTOR GENERATION CSL07960
C     AND PROGRAM TERMINATES FOR USER TO CORRECT THE PROBLEM  CSL07970
C     WHICH MUST BE IN THE STARTING VECTOR GENERATION OR IN   CSL07980
C     THE MATRIX-VECTOR MULTIPLY SUBROUTINE CMATV SUPPLIED.   CSL07990
```

```
C       THIS PART OF THE COMPUTATIONS MUST BE IDENTICAL TO THE          CSL08000
C       CORRESPONDING PART IN THE EIGENVALUE COMPUTATIONS.              CSL08010
C                                                                       CSL08020
        WRITE(6,950) IVEC,BATA,BETA(IVEC),TEMP                          CSL08030
  950 FORMAT(/2X,'IVEC',16X,'BATA',10X,'BETA(IVEC)',14X,'RELDIF'/16,    CSL08040
     13E20.12/' IN LANCZOS VECTOR REGENERATION THE ENTRIES OF THE TRIDIACSL08050
     1GONAL MATRICES BEING'/' GENERATED ARE NOT THE SAME AS THOSE IN THECSL08060
     1 MATRIX SUPPLIED ON FILE 2.'/' THEREFORE SOMETHING IS BEING INITIACSL08070
     1LIZED OR COMPUTED DIFFERENTLY FROM THE WAY'/' IT WAS COMPUTED IN TCSL08080
     1HE EIGENVALUE COMPUTATIONS'/' THE PROGRAM TERMINATES FOR THE USER CSL08090
     1TO DETERMINE WHAT THE PROBLEM IS'/)                               CSL08100
        GO TO 1550                                                      CSL08110
C                                                                       CSL08120
C                                                                       CSL08130
  960 CONTINUE                                                          CSL08140
        DO 970 J = 1,N                                                  CSL08150
        TEMPC = SUMC*V1(J)                                              CSL08160
        V1(J) = V2(J)                                                   CSL08170
  970 V2(J) = TEMPC                                                     CSL08180
C                                                                       CSL08190
  980 CONTINUE                                                          CSL08200
C                                                                       CSL08210
        LFIN = 0                                                        CSL08220
        DO 1000 J = 1,NGOODC                                            CSL08230
        LL = LFIN                                                       CSL08240
        LFIN = LFIN + N                                                 CSL08250
C                                                                       CSL08260
        IF(IABS(MA(J)).LT.IVEC.OR.MP(J).EQ.MPMIN) GO TO 1000           CSL08270
        II = IVEC + MINT(J) - 1                                         CSL08280
        TEMPC = TVEC(II)                                                CSL08290
C       II IS THE (IVEC)TH COMPONENT OF THE T-EIGENVECTOR CONTAINED     CSL08300
C       IN TVEC(MINT(J)).                                               CSL08310
C                                                                       CSL08320
        DO 990 K = 1,N                                                  CSL08330
        LL = LL + 1                                                     CSL08340
  990 RITVEC(LL) = TEMPC*V2(K) + RITVEC(LL)                            CSL08350
C                                                                       CSL08360
 1000 CONTINUE                                                          CSL08370
C                                                                       CSL08380
        IVEC = IVEC + 1                                                 CSL08390
        IF (IVEC.LE.KMAXU) GO TO 930                                    CSL08400
C                                                                       CSL08410
C                                                                       CSL08420
C       RITZVECTOR GENERATION IS COMPLETE. NORMALIZE EACH RITZVECTOR.   CSL08430
C       NOTE THAT IF CERTAIN RITZ VECTORS WERE NOT COMPUTED THEN THE    CSL08440
C       CORRESPONDING PORTION OF THE RITVEC ARRAY WAS NOT UTILIZED.     CSL08450
C                                                                       CSL08460
        LFIN = 0                                                        CSL08470
        DO 1050 J = 1,NGOODC                                            CSL08480
C                                                                       CSL08490
        KK = LFIN                                                       CSL08500
        LFIN = LFIN + N                                                 CSL08510
        IF(MP(J).EQ.MPMIN) GO TO 1050                                   CSL08520
C                                                                       CSL08530
        DO 1010 K = 1,N                                                 CSL08540
        KK = KK + 1                                                     CSL08550
 1010 V2(K) = RITVEC(KK)                                               CSL08560
C                                                                       CSL08570
C-----------------------------------------------------------------------CSL08580
        CALL INPRDC(V2,V2,SUMC,N)                                       CSL08590
C-----------------------------------------------------------------------CSL08600
C                                                                       CSL08610
        SUMC = CDSQRT(SUMC)                                             CSL08620
        RNORM(J) = CDABS(SUMC)                                          CSL08630
        TEMP = DABS(ONE-RNORM(J))                                       CSL08640
        SUMC = DCMPLX(ONE,ZERO)/SUMC                                    CSL08650
C                                                                       CSL08660
```

```
      KK = LFIN - N                                             CSL08670
      DO 1020 K = 1,N                                           CSL08680
      KK = KK + 1                                               CSL08690
      V2(K) = SUMC*V2(K)                                        CSL08700
 1020 RITVEC(KK) = V2(K)                                        CSL08710
C                                                               CSL08720
C     COMPUTE THE 'REAL' NORM                                   CSL08730
C                                                               CSL08740
C-------------------------------------------------------------CSL08750
      CALL CINPRD(V2,V2,SUM,N)                                  CSL08760
C-------------------------------------------------------------CSL08770
C                                                               CSL08780
      IF (IWRITE.NE.0) WRITE(6,1030) J,GOODEV(J)                CSL08790
 1030 FORMAT(/I5,' TH EIGENVALUE CONSIDERED = ',2E20.12/)       CSL08800
C                                                               CSL08810
      IF (IWRITE.NE.0) WRITE(6,1040) TERR(J),TBETA(J),RNORM(J),SUM CSL08820
 1040 FORMAT(' NORM OF ERROR IN T-EIGENVECTOR = ',E14.3/        CSL08830
     1 ' CDABS(BETA(MA(J)+1)*U(MA(J)))  ',E14.3/               CSL08840
     1 ' CDABS(EUCLIDEAN-NORM(RITVEC))  = ',E14.3/             CSL08850
     1 ' HERMITIAN-NORM(RITVEC)**2  = ',E14.3/)                CSL08860
C                                                               CSL08870
      LINT = LFIN - N + 1                                       CSL08880
      EVAL = GOODEV(J)                                          CSL08890
C                                                               CSL08900
C-------------------------------------------------------------CSL08910
C                                                               CSL08920
      CALL CMATV(RITVEC(LINT),V2,EVAL)                          CSL08930
C                                                               CSL08940
C-------------------------------------------------------------CSL08950
C                                                               CSL08960
C     COMPUTE ERROR IN RITZ VECTOR CONSIDERED AS A EIGENVECTOR OF A. CSL08970
C     V2 = A*RITVEC - EVAL*RITVEC                               CSL08980
C                                                               CSL08990
C-------------------------------------------------------------CSL09000
      CALL CINPRD(V2,V2,SUM,N)                                  CSL09010
C-------------------------------------------------------------CSL09020
C                                                               CSL09030
      SUM = DSQRT(SUM)                                          CSL09040
      ERR(J) = SUM                                              CSL09050
      GAP = ABS(AMINGP(J))                                      CSL09060
      ERRDGP(J) = SUM/GAP                                       CSL09070
C                                                               CSL09080
 1050 CONTINUE                                                  CSL09090
C                                                               CSL09100
C                                                               CSL09110
C     RITZVECTORS ARE NORMALIZED AND ERROR ESTIMATES ARE IN ERR ARRAY CSL09120
C     AND IN ERRDGP ARRAY. STORE EVERYTHING                    CSL09130
C                                                               CSL09140
C                                                               CSL09150
      WRITE(9,1060)                                             CSL09160
 1060 FORMAT(3X,'REAL(GOODEV)',3X,'IMAG(GOODEV)',1X,'MA(J)',7X,'AMINGAP'CSL09170
     1 ,4X,'AERROR',2X,'AERR/GAP',4X,'TERROR')                 CSL09180
C                                                               CSL09190
      WRITE(13,1070)                                            CSL09200
 1070 FORMAT(8X,'REAL(GOODEV)',8X,'IMAG(GOODEV)',2X,'RITZNORM',3X,'AMINGCSL09210
     1AP',2X,'TBETA(J)',2X,'TLAST(J)')                         CSL09220
C                                                               CSL09230
      DO 1100 J=1,NGOODC                                        CSL09240
C                                                               CSL09250
      IF(MP(J).EQ.MPMIN) GO TO 1100                             CSL09260
C                                                               CSL09270
      WRITE(9,1080)GOODEV(J),MA(J),AMINGP(J),ERR(J),ERRDGP(J),TERR(J) CSL09280
 1080 FORMAT(2E15.8,I6,E14.6,3E10.3)                           CSL09290
C                                                               CSL09300
      WRITE(13,1090) GOODEV(J),RNORM(J),AMINGP(J),TBETA(J),TLAST(J) CSL09310
 1090 FORMAT(2E20.12,4E10.3)                                    CSL09320
C                                                               CSL09330
```

```
      1100 CONTINUE                                                  CSL09340
    C                                                               CSL09350
          IF(MREJEC.EQ.0) GO TO 1180                                CSL09360
          WRITE(9,1110)                                             CSL09370
      1110 FORMAT(/' RITZ VECTORS WERE NOT COMPUTED FOR THE FOLLOWING EIGENVACSL09380
         1LUES'/' EITHER BECAUSE THEY HAD NOT CONVERGED OR BECAUSE THE ERRORCSL09390
         1 ESTIMATE'/' WAS NOT AS SMALL AS DESIRED'/)               CSL09400
    C                                                               CSL09410
          WRITE(13,1120)                                            CSL09420
      1120 FORMAT(/' RITZ VECTORS WERE NOT COMPUTED FOR THE FOLLOWING EIGENVACSL09430
         1LUES'/' EITHER BECAUSE THEY HAD NOT CONVERGED OR BECAUSE'/' THE ERCSL09440
         1ROR ESTIMATE WAS NOT AS SMALL AS DESIRED'/)               CSL09450
    C                                                               CSL09460
          DO 1170 J = 1,NGOODC                                      CSL09470
          IF(MP(J).NE.MPMIN) GO TO 1170                             CSL09480
    C     WRITE OUT MESSAGE FOR EACH EIGENVALUE FOR WHICH NO EIGENVECTOR CSL09490
    C     WAS COMPUTED.                                             CSL09500
    C                                                               CSL09510
          WRITE(9,1130)                                             CSL09520
      1130 FORMAT(6X,'GOODEV(J)',3X,'MA(J)',5X,'AMINGP(J)',6X,'TLAST(J)',3X, CSL09530
         1'MP(J)')                                                  CSL09540
          WRITE(9,1140) GOODEV(J),MA(J),AMINGP(J),TBETA(J),MP(J)    CSL09550
      1140 FORMAT(2E15.8,I8,2E14.4,I8)                              CSL09560
    C                                                               CSL09570
          WRITE(13,1150)                                            CSL09580
      1150 FORMAT(6X,'REAL(GOODEV(J))',6X,'IMAG(GOODEV(J))',4X,'MA(J)',3X, CSL09590
         1'MP(J)')                                                  CSL09600
          WRITE(13,1160) GOODEV(J),MA(J),MP(J)                      CSL09610
      1160 FORMAT(2E15.8,2I8)                                       CSL09620
    C                                                               CSL09630
      1170 CONTINUE                                                  CSL09640
      1180 CONTINUE                                                  CSL09650
    C                                                               CSL09660
          WRITE(9,1190)                                             CSL09670
      1190 FORMAT(/' ABOVE ARE ERROR ESTIMATES FOR THE A AND T EIGENVECTORS'/CSL09680
         1 ' ASSOCIATED WITH THE GOODEV LISTED IN COLUMN 1'/        CSL09690
         1 ' AERROR = NORM(A*X - EV*X)  TERROR = NORM(T*Y - EV*Y) '/ CSL09700
         1 ' WHERE T = T(1,MA(J))   X = RITZ VECTOR = V*Y  V = SUCCESSIVE'/CSL09710
         1 ' LANCZOS VECTORS. A MINGAP = GAP TO NEAREST A-EIGENVALUE'//) CSL09720
    C                                                               CSL09730
          WRITE(13,1200)                                            CSL09740
      1200 FORMAT(/' ABOVE ARE ERROR ESTIMATES FOR THE A AND T EIGENVECTORS'/CSL09750
         1 ' ASSOCIATED WITH THE GOODEV LISTED IN COLUMN 1'/        CSL09760
         1 ' AERROR = NORM(A*X-EV*X)  TERROR = NORM(T*Y-EV*Y) WHERE' CSL09770
         1/' T = T(1,MA(J))   X = RITZ VECTOR = V*Y  V = SUCCESSIVE '/ CSL09780
         1 ' LANCZOS VECTORS. A MINGAP = GAP TO NEAREST A-EIGENVALUE'/ CSL09790
         1 ' AERROR AND TERROR ARE GIVEN IN FILE 9. RNORM = NORM(X)'/ CSL09800
         1 ' BETA(M+1)*ABS(Y(M)) IS AN ESTIMATOR OF NORM(A*X-EV*X)'//) CSL09810
    C                                                               CSL09820
    C     NUMBER OF RITZ VECTORS COMPUTED                           CSL09830
          NCOMPU = NGOODC - MREJEC                                  CSL09840
          WRITE(12,1210) N,NCOMPU,NGOODC,MATNO                      CSL09850
      1210 FORMAT(3I6,I12,' SIZE A, NO.RITZVECS, NO.EVALUES,MATNO') CSL09860
    C                                                               CSL09870
          LFIN = 0                                                  CSL09880
          DO 1270 J = 1,NGOODC                                      CSL09890
          LINT = LFIN + 1                                           CSL09900
          LFIN = LFIN + N                                           CSL09910
    C                                                               CSL09920
          IF(MP(J).EQ.MPMIN) GO TO 1250                             CSL09930
    C     RITZ VECTOR WAS COMPUTED                                  CSL09940
          WRITE(12,1220) J, GOODEV(J), MP(J)                        CSL09950
      1220 FORMAT(I6,4X,2E20.12,I6,' J, EIGENVAL, MP(J)')           CSL09960
    C                                                               CSL09970
          WRITE(12,1230) ERR(J),ERRDGP(J)                          CSL09980
      1230 FORMAT(2E15.5,' = NORM(A*Z-EVAL*Z) AND  NORM(A*Z-EVAL*Z)/MINGAP') CSL09990
    C                                                               CSL10000
```

```
       WRITE(12,1240) (RITVEC(LL), LL=LINT,LFIN)                       CSL10010
C1240 FORMAT(4Z20)                                                     CSL10020
 1240 FORMAT(2(2E20.12))                                               CSL10030
      GO TO 1270                                                       CSL10040
C     NO RITZ VECTOR WAS COMPUTED FOR THIS EIGENVALUE                  CSL10050
 1250 WRITE(12,1260) J,GOODEV(J),MP(J)                                 CSL10060
 1260 FORMAT(I6,4X,E20.12,I6,' J,EIGVALUE,NO RITZ VECTOR COMPUTED')    CSL10070
C                                                                      CSL10080
 1270 CONTINUE                                                         CSL10090
C                                                                      CSL10100
C     DID ANY T-MATRICES INCLUDE OFF-DIAGONAL ENTRIES SMALLER THAN     CSL10110
C     DESIRED, AS SPECIFIED BY BTOL?                                   CSL10120
C                                                                      CSL10130
      IF(IB.GT.0) GO TO 1300                                           CSL10140
      WRITE(6,1280) KMAXU                                              CSL10150
 1280 FORMAT(/' FOR LARGEST T-MATRIX CONSIDERED',I7,' CHECK THE SIZE OF CSL10160
     1BETAS')                                                          CSL10170
C                                                                      CSL10180
C----------------------------------------------------------------------CSL10190
C                                                                      CSL10200
      CALL TNORM(ALPHA,BETA,BKMIN,TEMP,KMAXU,IBMT)                     CSL10210
C                                                                      CSL10220
C----------------------------------------------------------------------CSL10230
C                                                                      CSL10240
      IF(IBMT.LT.0) WRITE (6,1290)                                     CSL10250
 1290 FORMAT(/' WARNING THE T-MATRICES FOR ONE OR MORE OF THE EIGENVALUECSL10260
     1S CONSIDERED'/' HAD AN OFF-DIAGONAL ENTRY THAT WAS SMALLER THAN THCSL10270
     1E BETA TOLERANCE THAT WAS SPECIFIED'/)                           CSL10280
 1300 CONTINUE                                                         CSL10290
C                                                                      CSL10300
      GO TO 1550                                                       CSL10310
C                                                                      CSL10320
 1310 WRITE(6,1320) NGOOD,NMAX,MDIMRV                                  CSL10330
 1320 FORMAT(/I4,' RITZ VECTORS WERE REQUESTED BUT THE REQUIRED DIMENSIOCSL10340
     1N',I6/' IS LARGER THAN THE USER-SPECIFIED DIMENSION OF RITVEC',I6 CSL10350
     1/' THEREFORE, THE EIGENVECTOR PROCEDURE TERMINATES FOR THE USER TOCSL10360
     1 INTERVENE')                                                     CSL10370
C                                                                      CSL10380
      GO TO 1550                                                       CSL10390
C                                                                      CSL10400
 1330 WRITE(6,1340) NOLD,N,MATOLD,MATNO                                CSL10410
 1340 FORMAT(/' PARAMETERS READ FROM FILE 3 DO NOT AGREE WITH USER-SPECICSL10420
     1FIED'/' PARAMETERS, NOLD,N,MATOLD,MATNO = '/2I6,2I12/            CSL10430
     1' THEREFORE PROGRAM TERMINATES FOR USER TO RESOLVE DIFFERENCES'/) CSL10440
C                                                                      CSL10450
      GO TO 1550                                                       CSL10460
C                                                                      CSL10470
 1350 WRITE(6,1360)                                                    CSL10480
 1360 FORMAT(/' PARAMETERS IN ALPHA,BETA FILE READ IN DO NOT AGREE WITH CSL10490
     1 THOSE'/' SPECIFIED BY THE USER.  THEREFORE, THE PROGRAM TERMINATECSL10500
     1S FOR'/' THE USER TO RESOLVE THE DIFFERENCES'/)                  CSL10510
C                                                                      CSL10520
      GO TO 1550                                                       CSL10530
C                                                                      CSL10540
 1370 WRITE(6,1380) KMAX,MEV                                           CSL10550
 1380 FORMAT(/' IN ALPHA, BETA HISTORY HEADER KMAX =',I6/              CSL10560
     1' BUT EIGENVALUES WERE COMPUTED AT MEV = ',I6,' PROGRAM STOPS'/)  CSL10570
C                                                                      CSL10580
      GO TO 1550                                                       CSL10590
C                                                                      CSL10600
 1390 WRITE(6,1400)                                                    CSL10610
 1400 FORMAT(/' PROGRAM COMPUTED 1ST GUESSES AT T-MATRIX SIZES'/' READ TCSL10620
     1HEM TO FILE 10, THEN TERMINATED AS REQUESTED.')                  CSL10630
      GO TO 1550                                                       CSL10640
C                                                                      CSL10650
 1410 WRITE(6,1420) MTOL, MDIMTV                                       CSL10660
 1420 FORMAT(/' PROGRAM TERMINATES BECAUSE THE MINIMAL TVEC DIMENSION ANCSL10670
```

```
      1TICIPATED',17/'  IS LARGER THAN THE TVEC DIMENSION',17,' SPECIFIEDCSL10680
      1 BY THE USER.'/'  USER MAY RESET THE TVEC DIMENSION AND RESTART THCSL10690
      1E PROGRAM')                                                      CSL10700
         GO TO 1550                                                     CSL10710
C                                                                       CSL10720
 1430 WRITE(6,1440)                                                     CSL10730
 1440 FORMAT(/' PROGRAM TERMINATES BECAUSE NO SUITABLE T-EIGENVECTORS WECSL10740
      1RE IDENTIFIED'/' FOR ANY OF THE EIGENVALUES SUPPLIED.  PROBLEM COUCSL10750
      1LD BE CAUSED BY'/'  TOO SMALL A TVEC DIMENSION OR SIMPLY BE THAT TCSL10760
      1-EIGENVECTORS COULD'/'  NOT BE IDENTIFIED.  USER SHOULD CHECK OUTPCSL10770
      1UT'/)                                                            CSL10780
         GO TO 1550                                                     CSL10790
C                                                                       CSL10800
 1450 WRITE(6,1460) LVCONT,NTVEC,NGOOD                                  CSL10810
 1460 FORMAT(/' LVCONT FLAG =',12,' AND NUMBER ',15,' OF T-EIGENVECTORS CSL10820
      1 COMPUTED N.E.'/' NUMBER',15,' REQUESTED SO PROGRAM TERMINATES'/) CSL10830
         GO TO 1550                                                     CSL10840
 1470 WRITE(6,1480)                                                     CSL10850
 1480 FORMAT(/' PROGRAM TERMINATES WITHOUT COMPUTING ANY RITZ VECTORS'/ CSL10860
      1' BECAUSE ALL T-EIGENVECTORS COMPUTED WERE REJECTED AS NOT SUITABLCSL10870
      1E'/' FOR THE RITZ VECTOR COMPUTATIONS.  PROBABLE CAUSE IS LACK OF CSL10880
      1'/' CONVERGENCE OF THE EIGENVALUES'/)                            CSL10890
         GO TO 1550                                                     CSL10900
C                                                                       CSL10910
 1490 WRITE(6,1500)                                                     CSL10920
 1500 FORMAT(/' PROGRAM INDICATES THAT IT IS NOT POSSIBLE TO COMPUTE ANYCSL10930
      1 OF THE'/' REQUESTED EIGENVECTORS. THEREFORE PROGRAM TERMINATES') CSL10940
         DO 1510 J=1,NGOODC                                             CSL10950
 1510 WRITE(6,1520)  J,GOODEV(J),MP(J)                                  CSL10960
 1520 FORMAT(/4X,' J',11X,'GOODEV(J)',4X,'MP(J)'/16,2E20.12,19)         CSL10970
         GO TO 1550                                                     CSL10980
C                                                                       CSL10990
 1530 WRITE(6,1540) MBETA,KMAXN                                         CSL11000
 1540 FORMAT(/' PROGRAM TERMINATES BECAUSE THE STORAGE ALLOTTED FOR THE CSL11010
      1BETA ARRAY',18/' IS NOT SUFFICIENT FOR THE ENLARGED KMAX =',18,' TCSL11020
      1HAT THE PROGRAM WANTS'/' USER CAN ENLARGE THE ALPHA AND BETA ARRAYCSL11030
      1S AND RERUN THE PROGRAM'/)                                       CSL11040
C                                                                       CSL11050
 1550 CONTINUE                                                          CSL11060
C                                                                       CSL11070
      STOP                                                              CSL11080
C-----END OF MAIN PROGRAM FOR COMPLEX SYMMETRIC EIGENVECTORS-----------CSL11090
      END                                                               CSL11100
```

SECTION 7.5   LANCZS AND SAMPLE MATRIX-VECTOR MULTIPLY SUBROUTINES

```
C-----CSLEMULT--(COMPLEX SYMMETRIC MATRICES)---------------------------CSL00010
C                                                                      CSL00020
C       CONTAINS SUBROUTINE LANCZS USED IN THE COMPLEX SYMMETRIC       CSL00030
C       VERSION OF THE LANCZOS PROCEDURES PLUS SAMPLE USPEC AND        CSL00040
C       CMATV SUBROUTINES.                                             CSL00050
C                                                                      CSL00060
C       PORTABILITY:                                                   CSL00070
C       THESE PROGRAMS ARE NOT PORTABLE DUE TO THE USE OF COMPLEX*16   CSL00080
C       VARIABLES AND CORRESPONDING FUNCTIONS.  MOREOVER, THE PFORT    CSL00090
C       VERIFIER IDENTIFIED THE FOLLOWING ADDITIONAL NONPORTABLE       CSL00100
C       CONSTRUCTIONS:                                                 CSL00110
C                                                                      CSL00120
C       1.  ENTRIES USED TO PASS THE STORAGE LOCATIONS OF THE          CSL00130
C           ARRAYS AND PARAMETERS NEEDED TO SPECIFY THE GIVEN MATRIX   CSL00140
C           FROM THE USPEC SUBROUTINE TO THE MATRIX-VECTOR MULTIPLY    CSL00150
C           SUBROUTINE CMATV.                                          CSL00160
C       2.  IN THE SAMPLE USPEC SUBROUTINES PROVIDED: THE FREE FORMAT  CSL00170
C           READ(8,*) AND THE FORMAT (20A4).  IN THE SAMPLE CMATV:     CSL00180
C           THE COMPUTATION OF INDICES:  IN THE AUXILIARY SUBROUTINE   CSL00190
C           USED FOR COMPUTING THE KNOWN EIGENVALUES OF TEST CLASS 2   CSL00200
C           MATRICES, THE DATA/MACHEP DEFINITION.                      CSL00210
C                                                                      CSL00220
C-----LANCZS-COMPUTE THE LANCZOS TRIDIAGONAL MATRICES------------------CSL00230
C                                                                      CSL00240
      SUBROUTINE LANCZS(MATVEC,V1,V2,ALPHA,BETA,                        CSL00250
     1GR,GC,G,KMAX,MOLD1,N,IIX)                                        CSL00260
C                                                                      CSL00270
C---------------------------------------------------------------------CSL00280
      COMPLEX*16 V1(1), V2(1), BATA, ZEROC, TEMP, SUMC                 CSL00290
      COMPLEX*16  ALPHA(1), BETA(1)                                    CSL00300
      DOUBLE PRECISION SUM, ONE, ZERO, GR(1), GC(1)                    CSL00310
      REAL  G(1)                                                       CSL00320
      EXTERNAL MATVEC                                                  CSL00330
C       COMPLEX*16 CDSQRT, DCMPLX                                      CSL00340
C---------------------------------------------------------------------CSL00350
C                                                                      CSL00360
      ZERO = 0.D0                                                      CSL00370
      ONE = 1.D0                                                       CSL00380
      ZEROC = DCMPLX(ZERO,ZERO)                                        CSL00390
C                                                                      CSL00400
      IF(MOLD1.GT.1)GO TO 50                                           CSL00410
C                                                                      CSL00420
C     ALPHA/BETA GENERATION STARTS AT I = 1                            CSL00430
C     MOLD1 = 1 SET V1 = 0. AND V2 = RANDOM UNIT VECTOR                CSL00440
      IIL=IIX                                                          CSL00450
C                                                                      CSL00460
C---------------------------------------------------------------------CSL00470
      CALL GENRAN(IIL,G,N)                                             CSL00480
C---------------------------------------------------------------------CSL00490
C                                                                      CSL00500
      DO 10 I = 1,N                                                    CSL00510
   10 GR(I) = G(I)                                                     CSL00520
C                                                                      CSL00530
C---------------------------------------------------------------------CSL00540
      CALL GENRAN(IIL,G,N)                                             CSL00550
C---------------------------------------------------------------------CSL00560
C                                                                      CSL00570
      DO 20 I = 1,N                                                    CSL00580
   20 GC(I) = G(I)                                                     CSL00590
C                                                                      CSL00600
      DO 30 I = 1,N                                                    CSL00610
   30 V2(I) = DCMPLX(GR(I),GC(I))                                      CSL00620
C                                                                      CSL00630
C---------------------------------------------------------------------CSL00640
```

```
      CALL INPRDC(V2,V2,SUMC,N)                                   CSL00650
C------------------------------------------------------------------CSL00660
C                                                                  CSL00670
      SUMC = ONE/CDSQRT(SUMC)                                      CSL00680
      DO 40 I = 1,N                                                CSL00690
      V1(I) = ZEROC                                                CSL00700
   40 V2(I) = V2(I)*SUMC                                           CSL00710
      BETA(1) = ZEROC                                              CSL00720
C                                                                  CSL00730
C    ALPHA BETA GENERATION LOOP                                    CSL00740
   50 CONTINUE                                                     CSL00750
C                                                                  CSL00760
      DO 80 I=MOLD1,KMAX                                           CSL00770
      SUMC = BETA(I)                                               CSL00780
C                                                                  CSL00790
C------------------------------------------------------------------CSL00800
C    MATVEC(V2,V1,SUMC) CALCULATES   V1 = A*V2 - SUMC*V1           CSL00810
      CALL MATVEC(V2,V1,SUMC)                                      CSL00820
      CALL INPRDC(V2,V1,SUMC,N)                                    CSL00830
C------------------------------------------------------------------CSL00840
C                                                                  CSL00850
      ALPHA(I) = SUMC                                              CSL00860
      DO 60 J=1,N                                                  CSL00870
   60 V1(J) = V1(J)-SUMC*V2(J)                                     CSL00880
C                                                                  CSL00890
C------------------------------------------------------------------CSL00900
      CALL INPRDC(V1,V1,SUMC,N)                                    CSL00910
C------------------------------------------------------------------CSL00920
C                                                                  CSL00930
      IN = I+1                                                     CSL00940
      BATA = CDSQRT(SUMC)                                          CSL00950
      BETA(IN) = BATA                                              CSL00960
      SUMC = ONE/BATA                                              CSL00970
      DO 70 J=1,N                                                  CSL00980
      TEMP = SUMC*V1(J)                                            CSL00990
      V1(J) = V2(J)                                                CSL01000
   70 V2(J) = TEMP                                                 CSL01010
   80 CONTINUE                                                     CSL01020
C    END ALPHA, BETA GENERATION LOOP                              CSL01030
C                                                                  CSL01040
C-----END OF LANCZS------------------------------------------------CSL01050
C                                                                  CSL01060
      RETURN                                                       CSL01070
      END                                                          CSL01080
C                                                                  CSL01090
C-----USPEC, AND CMATV FOR COMPLEX SYMMETRIC TEST MATRICES 1-------CSL01100
C                                                                  CSL01110
C-----START OF USPEC-(COMPLEX SYMMETRIC TEST MATRICES 1)-----------CSL01120
C                                                                  CSL01130
C    SUBROUTINE CSPEC(N,MATNO)                                     CSL01140
      SUBROUTINE USPEC(N,MATNO)                                    CSL01150
C                                                                  CSL01160
C------------------------------------------------------------------CSL01170
      DOUBLE PRECISION  CO,C1,C2,HALF,ONE,SCR,SCI,ANGLE            CSL01180
      COMPLEX*16 SC,TC,CLO,CL1                                     CSL01190
      REAL EXPLAN(20)                                              CSL01200
      DOUBLE PRECISION DARCOS                                      CSL01210
C    COMPLEX*16 DCMPLX                                             CSL01220
C------------------------------------------------------------------CSL01230
      HALF = 0.5D0                                                 CSL01240
      ONE  = 1.0D0                                                 CSL01250
C                                                                  CSL01260
C    READ IN PARAMETERS TO DEFINE MATRIX                          CSL01270
C    MATRIX IS COMPLEX DIAGONAL SIMILIARITY TRANSFORM OF THE BLOCK CSL01280
C    TOEPLITZ POISSON MATRICES USED TO TEST REAL SYMMETRIC MATRICES. CSL01290
C    THE REAL POISSON MATRIX HAS SYMMETRIC TOEPLITZ BLOCKS ALONG THE CSL01300
C    DIAGONAL.  EACH ONE OF THESE HAS THE PARAMETER C2 ALONG THE   CSL01310
```

```
C        DIAGONAL AND -CO ABOVE AND BELOW THE DIAGONAL.  THE OFF-DIAGONAL   CSLO1320
C        BLOCKS ARE DIAGONAL WITH DIAGONAL ENTRIES -C1.   EACH BLOCK IS     CSLO1330
C        KX*KX AND THERE ARE KY BLOCKS.  A HERMITIAN VERSION IS OBTAINED    CSLO1340
C        BY APPLYING A DIAGONAL SIMILARITY TRANSFORM TO THE ABOVE           CSLO1350
C        MATRIX WHERE THE DIAGONAL MATRIX IS SUCH THAT ITS                  CSLO1360
C        DIAGONAL ENTRIES ARE (SC)**(K-1), K=1,...,N-1.                     CSLO1370
C        THIS HERMITIAN VERSION IS TURNED INTO A COMPLEX SYMMETRIC ONE      CSLO1380
C        IN THE MATRIX VECTOR MULTIPLY BY TREATING THE BELOW DIAGONAL       CSLO1390
C        ENTRIES AS BEING EQUAL TO THE ABOVE DIAGONAL ENTRIES RATHER        CSLO1400
C        THAN THEIR COMPLEX CONJUGATES.                                     CSLO1410
C                                                                           CSLO1420
         READ(8,10) EXPLAN                                                  CSLO1430
      10 FORMAT(20A4)                                                       CSLO1440
         READ(8,*) NOLD,MATOLD                                             CSLO1450
         WRITE(6,20) NOLD,MATOLD                                            CSLO1460
      20 FORMAT(' ORDER OF MATRIX READ FROM FILE =',I6/' MATRIX NUMBER =', CSLO1470
        1I8)                                                                CSLO1480
C                                                                           CSLO1490
C        TEST OF PARAMETER CORRECTNESS                                      CSLO1500
         ITEMP = (NOLD-N)**2 + (MATNO-MATOLD)**2                            CSLO1510
C                                                                           CSLO1520
         IF(ITEMP.EQ.0) GO TO 40                                           CSLO1530
C                                                                           CSLO1540
         WRITE(6,30)                                                        CSLO1550
      30 FORMAT(' PROGRAM TERMINATES BECAUSE EITHER ORDERS OF OR LABELS FORCSLO1560
        1 MATRIX DISAGREE')                                                 CSLO1570
         GO TO 100                                                         CSLO1580
C                                                                           CSLO1590
      40 CONTINUE                                                           CSLO1600
C                                                                           CSLO1610
         READ(8,10) EXPLAN                                                  CSLO1620
         READ(8,*)  CO,KX,KY                                               CSLO1630
         READ(8,10) EXPLAN                                                  CSLO1640
         READ(8,*) SCR                                                     CSLO1650
         ANGLE = DARCOS(SCR)                                                CSLO1660
         SCI = DSIN(ANGLE)                                                  CSLO1670
         SC = DCMPLX(SCR,SCI)                                               CSLO1680
         WRITE(6,50) SC                                                     CSLO1690
         WRITE(9,50) SC                                                     CSLO1700
      50 FORMAT(' GENERATOR OF DIAGONAL TRANSFORMATION ='/2E20.12)          CSLO1710
C                                                                           CSLO1720
         TC = SC                                                            CSLO1730
         DO 60 J=2,KX                                                       CSLO1740
      60 TC = SC*TC                                                         CSLO1750
         WRITE(6,70) TC                                                     CSLO1760
      70 FORMAT(' TC = ',2E20.12)                                           CSLO1770
C                                                                           CSLO1780
         N = KX*KY                                                          CSLO1790
         C2 = ONE                                                           CSLO1800
         C1 = HALF-CO                                                       CSLO1810
         CLO = -SC*CO                                                       CSLO1820
         CL1 = -TC*C1                                                       CSLO1830
C                                                                           CSLO1840
         WRITE(6,80) N,KX,KY,C2,CO,C1                                       CSLO1850
      80 FORMAT(/5X,'N',4X,'KX',4X,'KY',7X,'DIAGONAL',3X,'X-CODIAGONAL',    CSLO1860
        1 3X,'Y-CODIAGONAL'/3I6,3E15.8/)                                    CSLO1870
C                                                                           CSLO1880
C------------------------------------------------------------------------CSLO1890
         CALL HMATVE(C2,CLO,CL1,KX,KY)                                      CSLO1900
C------------------------------------------------------------------------CSLO1910
C                                                                           CSLO1920
      90 CONTINUE                                                           CSLO1930
         RETURN                                                             CSLO1940
C                                                                           CSLO1950
C-----END OF USPEC-------------------------------------------------------CSLO1960
     100 STOP                                                               CSLO1970
```

```
      END                                                          CSL01980
C                                                                  CSL01990
C-----START OF CSMATV (FOR TEST MATRICES 1)------------------------CSL02000
C     CALCULATE U = A*W - SUMC*U FOR COMPLEX SYMMETRIC MATRICES     CSL02010
C     HERE WE HAVE TAKEN A HERMITIAN VERSION OF POISSON MATRICES    CSL02020
C     AND TURNED IT INTO A COMPLEX SYMMETRIC TEST PROBLEM (WHOSE    CSL02030
C     EIGENVALUES WE DO NOT KNOW)                                   CSL02040
C                                                                  CSL02050
C     SUBROUTINE CSMATV(W,U,CSUM)                                   CSL02060
      SUBROUTINE CMATV(W,U,CSUM)                                    CSL02070
C                                                                  CSL02080
C-----------------------------------------------------------------CSL02090
      DOUBLE PRECISION  C2                                          CSL02100
      COMPLEX*16 U(1),W(1)                                          CSL02110
      COMPLEX*16 CLO,CL1,CRO,CR1,CSUM                               CSL02120
C-----------------------------------------------------------------CSL02130
C                                                                  CSL02140
      N = KX*KY                                                     CSL02150
      KX1 = KX-1                                                    CSL02160
      KY1 = KY-1                                                    CSL02170
      CRO = CLO                                                     CSL02180
      CR1 = CL1                                                     CSL02190
C                                                                  CSL02200
      KK = 1                                                        CSL02210
      U(KK)=(C2*W(KK)+CRO*W(KK+1)+CR1*W(KK+KX)) - CSUM*U(KK)        CSL02220
      KK = KX                                                       CSL02230
      U(KK)=(C2*W(KK)+CLO*W(KK-1)+CR1*W(KK+KX)) - CSUM*U(KK)        CSL02240
      KK = N - KX + 1                                               CSL02250
      U(KK)=(C2*W(KK)+CRO*W(KK+1)+CL1*W(KK-KX)) - CSUM*U(KK)        CSL02260
      KK = N                                                        CSL02270
      U(KK)=(C2*W(KK)+CLO*W(KK-1)+CL1*W(KK-KX)) - CSUM*U(KK)        CSL02280
C                                                                  CSL02290
      DO 10 J = 2,KX1                                               CSL02300
      KK = J                                                        CSL02310
      U(KK)=(C2*W(KK)+CLO*W(KK-1)+CRO*W(KK+1)+CR1*W(KK+KX))-CSUM*U(KK) CSL02320
      KK = J+N-KX                                                   CSL02330
      U(KK)=(C2*W(KK)+CLO*W(KK-1)+CRO*W(KK+1)+CL1*W(KK-KX))-CSUM*U(KK) CSL02340
   10 CONTINUE                                                      CSL02350
C                                                                  CSL02360
      DO 30 J = 2,KY1                                               CSL02370
      KK = (J-1)*KX + 1                                             CSL02380
      U(KK)=(C2*W(KK)+CRO*W(KK+1)+CL1*W(KK-KX)+CR1*W(KK+KX))-CSUM*U(KK) CSL02390
      KK = J*KX                                                     CSL02400
      U(KK)=(C2*W(KK)+CLO*W(KK-1)+CL1*W(KK-KX)+CR1*W(KK+KX))-CSUM*U(KK) CSL02410
      DO 20 I = 2,KX1                                               CSL02420
      KK = (J-1)*KX + I                                             CSL02430
      U(KK)=(C2*W(KK)+CLO*W(KK-1)+CRO*W(KK+1)+CL1*W(KK-KX)          CSL02440
     1 +CR1*W(KK+KX)) - CSUM*U(KK)                                  CSL02450
   20 CONTINUE                                                      CSL02460
   30 CONTINUE                                                      CSL02470
C                                                                  CSL02480
      RETURN                                                        CSL02490
C                                                                  CSL02500
C-----------------------------------------------------------------CSL02510
      ENTRY HMATVE(C2,CLO,CL1,KX,KY)                                CSL02520
C-----------------------------------------------------------------CSL02530
C                                                                  CSL02540
C-----END OF CSMATV-----------------------------------------------CSL02550
      RETURN                                                        CSL02560
      END                                                           CSL02570
C                                                                  CSL02580
C     BELOW IS USPEC AND CMATV FOR TEST MATRICES 2.  IN THIS CASE   CSL02590
C     THE EIGENVALUES ARE KNOWN AND WE COMPUTE THEM TO CHECK        CSL02600
C     VALUES OBTAINED FROM THE LANCZOS PROGRAMS.                    CSL02610
C                                                                  CSL02620
C     USES 3 SUBROUTINES BELOW, USPEC CMATV EXEVG                   CSL02630
C                                                                  CSL02640
```

```
C-----START OF USPEC (TEST MATRICES 2)--------------------------------CSL02650
C                                                                    CSL02660
C     SUBROUTINE USPEC(N,MATNO)                                      CSL02670
      SUBROUTINE CSPEC(N,MATNO)                                      CSL02680
C                                                                    CSL02690
C--------------------------------------------------------------------CSL02700
      COMPLEX*16 CPAR,CCO,CC1,CC2                                    CSL02710
      DOUBLE PRECISION   CO,C1,C2,HALF,ONE                           CSL02720
      REAL  EXPLAN(20)                                               CSL02730
C     COMPLEX*16  DCMPLX                                             CSL02740
C--------------------------------------------------------------------CSL02750
C     IVEC  = (0,-1,1) MEANS                                         CSL02760
C               (0) ONLY SET ENTRY FOR CMATV                         CSL02770
C               (-1) CALCULATE EXACTEV AND MINGAPS AND STOP.         CSL02780
C               (1) CALCULATE EXACTEV AND MINGAPS AND THEN CONTINUE. CSL02790
C--------------------------------------------------------------------CSL02800
      HALF = 0.5D0                                                   CSL02810
      ONE  = 1.0D0                                                   CSL02820
      CPAR = DCMPLX(ONE,ONE)                                         CSL02830
C--------------------------------------------------------------------CSL02840
C     READ USER-SPECIFIED PARAMETERS FROM INPUT FILE 8 (FREE FORMAT) CSL02850
C                                                                    CSL02860
      READ(8,10) EXPLAN                                              CSL02870
      WRITE(6,10) EXPLAN                                             CSL02880
   10 FORMAT(20A4)                                                   CSL02890
C                                                                    CSL02900
      READ(8,10) EXPLAN                                              CSL02910
      READ(8,*) KX,KY,IVEC,CO                                        CSL02920
      N = KX*KY                                                      CSL02930
      C1 = HALF-CO                                                   CSL02940
      C2 = ONE                                                       CSL02950
      CCO = CPAR*CO                                                  CSL02960
      CC1 = CPAR*C1                                                  CSL02970
      CC2 = CPAR*C2                                                  CSL02980
C                                                                    CSL02990
      WRITE(6,20) N,KX,KY,C2,CO,C1,CPAR                             CSL03000
   20 FORMAT(/5X,'N',4X,'KX',4X,'KY',7X,'DIAGONAL',3X,'X-CODIAGONAL',CSL03010
     1 3X,'Y-CODIAGONAL'/3I6,3E15.8/7X,' COMPLEX SCALAR MULTIPLIER'/ CSL03020
     1 13X,2E15.4)                                                   CSL03030
C                                                                    CSL03040
C--------------------------------------------------------------------CSL03050
      CALL CMATVE(CCO,CC1,CC2,KX,KY)                                 CSL03060
C--------------------------------------------------------------------CSL03070
C                                                                    CSL03080
      IF (IVEC.EQ.0) GO TO 30                                        CSL03090
C                                                                    CSL03100
C--------------------------------------------------------------------CSL03110
C     COMPUTE TRUE EIGENVALUES FOR CORRESPONDING REAL POISSON MATRIX CSL03120
      CALL EXEVG(CO,C1,C2,KX,KY)                                     CSL03130
C--------------------------------------------------------------------CSL03140
C                                                                    CSL03150
      IF (IVEC.LT.0) STOP                                            CSL03160
C                                                                    CSL03170
   30 CONTINUE                                                       CSL03180
C                                                                    CSL03190
C-----END OF USPEC---------------------------------------------------CSL03200
      RETURN                                                         CSL03210
      END                                                            CSL03220
C                                                                    CSL03230
C-----START OF CMATV (USES TEST MATRICES 2)-------------------------CSL03240
C     CALCULATE U = A*W - SUM*U                                      CSL03250
C                                                                    CSL03260
C     SUBROUTINE CMATV(W,U,CSUM)                                     CSL03270
      SUBROUTINE CSRMAT(W,U,CSUM)                                    CSL03280
C                                                                    CSL03290
C--------------------------------------------------------------------CSL03300
      COMPLEX*16 U(1),W(1)                                           CSL03310
```

```
      COMPLEX*16 CCO,CC1,CC2,CLO,CL1,CRO,CR1,CSUM                      CSL03320
C-----------------------------------------------------------------------CSL03330
C                                                                      CSL03340
      N = KX*KY                                                        CSL03350
      KX1 = KX-1                                                       CSL03360
      KY1 = KY-1                                                       CSL03370
      CRO = CCO                                                        CSL03380
      CR1 = CC1                                                        CSL03390
      CLO = CCO                                                        CSL03400
      CL1 = CC1                                                        CSL03410
C                                                                      CSL03420
      KK = 1                                                           CSL03430
      U(KK)=(CC2*W(KK)+CRO*W(KK+1)+CR1*W(KK+KX)) - CSUM*U(KK)          CSL03440
      KK = KX                                                          CSL03450
      U(KK)=(CC2*W(KK)+CLO*W(KK-1)+CR1*W(KK+KX)) - CSUM*U(KK)          CSL03460
      KK = N - KX + 1                                                  CSL03470
      U(KK)=(CC2*W(KK)+CRO*W(KK+1)+CL1*W(KK-KX)) - CSUM*U(KK)          CSL03480
      KK = N                                                           CSL03490
      U(KK)=(CC2*W(KK)+CLO*W(KK-1)+CL1*W(KK-KX)) - CSUM*U(KK)          CSL03500
C                                                                      CSL03510
      DO 10 J = 2,KX1                                                  CSL03520
      KK = J                                                           CSL03530
      U(KK)=(CC2*W(KK)+CLO*W(KK-1)+CRO*W(KK+1)+CR1*W(KK+KX))-CSUM*U(KK) CSL03540
      KK = J+N-KX                                                      CSL03550
      U(KK)=(CC2*W(KK)+CLO*W(KK-1)+CRO*W(KK+1)+CL1*W(KK-KX))-CSUM*U(KK) CSL03560
   10 CONTINUE                                                         CSL03570
C                                                                      CSL03580
      DO 30 J = 2,KY1                                                  CSL03590
      KK = (J-1)*KX + 1                                                CSL03600
      U(KK)=(CC2*W(KK)+CRO*W(KK+1)+CL1*W(KK-KX)+CR1*W(KK+KX))-CSUM*U(KK)CSL03610
      DO 20 I = 2,KX1                                                  CSL03620
      KK = KK + 1                                                      CSL03630
      U(KK)=(CC2*W(KK)+CLO*W(KK-1)+CRO*W(KK+1)+CL1*W(KK-KX)            CSL03640
     1 +CR1*W(KK+KX)) - CSUM*U(KK)                                     CSL03650
   20 CONTINUE                                                         CSL03660
      KK = KK + 1                                                      CSL03670
      U(KK)=(CC2*W(KK)+CLO*W(KK-1)+CL1*W(KK-KX)+CR1*W(KK+KX))-CSUM*U(KK)CSL03680
   30 CONTINUE                                                         CSL03690
C                                                                      CSL03700
      RETURN                                                           CSL03710
C                                                                      CSL03720
C-----------------------------------------------------------------------CSL03730
      ENTRY CMATVE(CCO,CC1,CC2,KX,KY)                                  CSL03740
C-----------------------------------------------------------------------CSL03750
C                                                                      CSL03760
C-----END OF CMATV------------------------------------------------------CSL03770
      RETURN                                                           CSL03780
      END                                                              CSL03790
C                                                                      CSL03800
C-----START OF EXEVG (COMPUTES EXACT EIGENVALUES FOR TEST MATRICES 2)---CSL03810
C                                                                      CSL03820
      SUBROUTINE EXEVG(CO,C1,C2,KX,KY)                                 CSL03830
C                                                                      CSL03840
C-----------------------------------------------------------------------CSL03850
      DOUBLE PRECISION U(2000),MACHEP                                  CSL03860
      DOUBLE PRECISION EPSM,CO,C1,C2,TO,T1,PIK,PIL,ONE,TWO,ATOLN,EE    CSL03870
      REAL G(2000)                                                     CSL03880
      INTEGER MP(2000)                                                 CSL03890
      REAL ABS                                                         CSL03900
      DOUBLE PRECISION DABS, DARCOS, DFLOAT, DCOS, DMAX1               CSL03910
C-----------------------------------------------------------------------CSL03920
      DATA MACHEP/Z3410000000000000/                                  CSL03930
      EPSM = 2.0D0*MACHEP                                              CSL03940
C-----------------------------------------------------------------------CSL03950
      N = KX*KY                                                        CSL03960
      ONE  = 1.0D0                                                     CSL03970
      TWO  = 2.0D0                                                     CSL03980
```

```
        TO = DARCOS(-ONE)                                        CSL03990
        T1 = DFLOAT(KX+1)                                        CSL04000
        PIK = TO/T1                                              CSL04010
        T1 = DFLOAT(KY+1)                                        CSL04020
        PIL = TO/T1                                              CSL04030
C       GENERATE EXACT EIGENVALUES                               CSL04040
        KP = 0                                                   CSL04050
        DO 20 J = 1,KY                                           CSL04060
        T1 = PIL*DFLOAT(J)                                       CSL04070
        TO = C2 - TWO*C1*DCOS(T1)                                CSL04080
        DO 10 I = 1,KX                                           CSL04090
        KP = KP+1                                                CSL04100
        T1 = PIK*DFLOAT(I)                                       CSL04110
     10 U(KP) = TO - TWO*CO*DCOS(T1)                             CSL04120
     20 CONTINUE                                                 CSL04130
C                                                                CSL04140
C       ORDER U VECTOR BY INCREASING ALGEBRAIC SIZE              CSL04150
        DO 40 K = 2,N                                            CSL04160
        KM1 = K-1                                                CSL04170
        DO 30 L = 1,KM1                                          CSL04180
        JJ = K-L                                                 CSL04190
        IF (U(JJ+1).GE.U(JJ)) GO TO 40                           CSL04200
        TO = U(JJ)                                               CSL04210
        U(JJ) = U(JJ+1)                                          CSL04220
     30 U(JJ+1) = TO                                             CSL04230
     40 CONTINUE                                                 CSL04240
        ATOLN = DMAX1(DABS(U(1)),DABS(U(N)))*EPSM                CSL04250
C                                                                CSL04260
        WRITE(9,50)                                              CSL04270
     50 FORMAT(' TRUE EIGENVALUES FOR POISSON'/)                 CSL04280
C                                                                CSL04290
        WRITE(9,60)N,KX,KY,C2,CO,C1,ATOLN                        CSL04300
        WRITE(6,60) N,KX,KY,C2,CO,C1,ATOLN                       CSL04310
     60 FORMAT(1X,'A-SIZE',2X,'X-DIM',2X,'Y-DIM'/3I7/            CSL04320
      1 5X,'A-DIAGONAL',3X,'X-CODIAGONAL',3X,'Y-CODIAGONAL',10X,'ATOLN'/ CSL04330
      2 4E15.8)                                                  CSL04340
C                                                                CSL04350
C       DETERMINE MULTIPLICITIES FOR EXACT EIGENVALUES           CSL04360
        I = 1                                                    CSL04370
        IDEX = 1                                                 CSL04380
        J = 1                                                    CSL04390
        NEXACT = 0                                               CSL04400
     70 J = J+1                                                  CSL04410
        IF (J.GT.N) GO TO 80                                     CSL04420
        EE = DABS(U(J)-U(I))                                     CSL04430
        IF (EE.GT.ATOLN) GO TO 80                                CSL04440
        IDEX = IDEX+1                                            CSL04450
        GO TO 70                                                 CSL04460
     80 NEXACT = NEXACT+1                                        CSL04470
        U(NEXACT) = U(I)                                         CSL04480
        MP(NEXACT) = IDEX                                        CSL04490
C       MP(K) = MULTIPLICITY OF KTH EIGENVALUE CLUSTER FOR A     CSL04500
        IDEX = 1                                                 CSL04510
        I = J                                                    CSL04520
        IF (I.GT.N) GO TO 90                                     CSL04530
        GO TO 70                                                 CSL04540
     90 CONTINUE                                                 CSL04550
C                                                                CSL04560
C       MULTIPLICITIES HAVE BEEN DETERMINED                      CSL04570
C       NEXACT = NUMBER OF DISTINCT A-EIGENVALUES                CSL04580
C                                                                CSL04590
C                                                                CSL04600
        WRITE(9,100)NEXACT                                       CSL04610
        WRITE(6,100)NEXACT                                       CSL04620
    100 FORMAT(I6,' = NUMBER OF TRUE A-EIGENVALUES WHICH ARE DISTINCT'/) CSL04630
C                                                                CSL04640
C       MINGAP CALCULATION FOR DISTINCT A-EIGENVALUES            CSL04650
```

```
      NM1 = NEXACT - 1                                          CSL04660
      G(NEXACT) = U(NM1)-U(NEXACT)                              CSL04670
      G(1) = U(2)-U(1)                                          CSL04680
C                                                               CSL04690
      DO 110 J = 2,NM1                                          CSL04700
      TO = U(J)-U(J-1)                                          CSL04710
      T1 = U(J+1)-U(J)                                          CSL04720
      G(J) = T1                                                 CSL04730
      IF (TO.LT.T1) G(J) = -TO                                  CSL04740
  110 CONTINUE                                                  CSL04750
C                                                               CSL04760
C     NEXACT DISTINCT A-EIGENVALUES ARE IN U IN ASCENDING ORDER CSL04770
C     MP = MULTIPLICITIES OF THE DISTINCT EIGENVALUES OF A      CSL04780
C     G = TRUE MINIMUM GAP IN A FOR EACH OF THESE EIGENVALUES   CSL04790
C     G < 0 INDICATES THE LEFT-HAND GAP WAS MINIMAL.            CSL04800
C     OUTPUT MULTIPLICITIES, DISTINCT EVS, AND MINGAPS TO FILE 9 CSL04810
C                                                               CSL04820
      WRITE(9,120)                                              CSL04830
  120 FORMAT(5X,'I',1X,'AMULT',5X,'TRUE A-EIGENVALUE(I)',       CSL04840
     1 3X,'A-MINGAP(I)')                                        CSL04850
C                                                               CSL04860
      WRITE(9,130)(J,MP(J),U(J),G(J), J=1,NEXACT)               CSL04870
  130 FORMAT(2I6,E25.16,E14.3)                                  CSL04880
C                                                               CSL04890
      WRITE(9,140)                                              CSL04900
  140 FORMAT(' NEXACT DISTINCT A-EIGENVALUES ARE IN ASCENDING ORDER'/ CSL04910
     1 ' AMULT = MULTIPLICITIES OF THE DISTINCT EIGENVALUES OF A.'/ CSL04920
     2 ' A-MINGAP(I) = TRUE MINIMUM GAP IN A FOR EACH EIGENVALUE.'/ CSL04930
     3 ' A-MINGAP(I).LT.0 INDICATES THE LEFT-HAND GAP WAS MINIMAL.'//) CSL04940
C                                                               CSL04950
C     WE ORDER U VECTOR BY INCREASING SIZE OF THE GAPS          CSL04960
C                                                               CSL04970
      DO 150 K = 1,N                                            CSL04980
  150 MP(K) = K                                                 CSL04990
C                                                               CSL05000
      DO 170 K = 2,N                                            CSL05010
      KM1 = K-1                                                 CSL05020
C                                                               CSL05030
      DO 160 L = 1,KM1                                          CSL05040
      JJ = K - L                                                CSL05050
      IF (ABS(G(JJ+1)).GE.ABS(G(JJ))) GO TO 170                 CSL05060
      EE = U(JJ)                                                CSL05070
      U(JJ) = U(JJ+1)                                           CSL05080
      U(JJ+1) = EE                                              CSL05090
      GG = G(JJ)                                                CSL05100
      G(JJ) = G(JJ+1)                                           CSL05110
      G(JJ+1) = GG                                              CSL05120
      IEE = MP(JJ)                                              CSL05130
      MP(JJ) = MP(JJ+1)                                         CSL05140
  160 MP(JJ+1) = IEE                                            CSL05150
C                                                               CSL05160
  170 CONTINUE                                                  CSL05170
C                                                               CSL05180
      WRITE(9,180)                                              CSL05190
  180 FORMAT(5X,'K',6X,'A-MINGAP',5X,'TRUE A-EIGENVALUE(I)',2X,'A-EVNO')CSL05200
C                                                               CSL05210
      WRITE(9,190)(J,G(J),U(J),MP(J), J=1,NEXACT)               CSL05220
  190 FORMAT(I6,E14.3,E25.16,I8)                                CSL05230
C                                                               CSL05240
      WRITE(9,200)                                              CSL05250
  200 FORMAT(' NEXACT DISTINCT A-EIGENVALUES. GAPS IN ASCENDING ORDER'/ CSL05260
     2 ' A-MINGAP(I) = TRUE MINIMUM GAP IN A FOR EACH EIGENVALUE.'/ CSL05270
     3 ' A-MINGAP(I).LT.0 INDICATES THE LEFT-HAND GAP WAS MINIMAL.'/ CSL05280
     3 ' A-MATRIX IS BLOCK TRIDIAGONAL AND EACH DIAGONAL BLOCK IS OF ORD CSL05290
     3ER NX.'/                                                  CSL05300
     4 ' NX = NUMBER OF POINTS ON EACH X-LINE. THERE ARE NY DIAGONAL BLOC CSL05310
     4CKS.'/                                                    CSL05320
```

```
      5 ' NY = NUMBER OF POINTS ON EACH Y-LINE.'/                    CSL05330
      5 ' A-DIAGONAL    = A(K,K)'/                                   CSL05340
      6 ' X-CODIAGONAL = A(I,I+1)'/                                  CSL05350
      7 ' Y-CODIAGONAL = A(I,I+NX)'/                                 CSL05360
      8 ' ----- END OF FILE 9 EXACTEV----------------------------'//) CSL05370
C                                                                   CSL05380
C-----END OF EXEVG--------------------------------------------------CSL05390
C                                                                   CSL05400
      RETURN                                                        CSL05410
      END                                                           CSL05420
```

SECTION 7.6   OTHER SUBROUTINES USED BY THE PROGRAMS IN CHAPTER 7

```
C-----CSLESUB-(NONDEFECTIVE COMPLEX SYMMETRIC MATRICES)-----------------CSL00010
C                                                                       CSL00020
C     NONPORTABLE CONSTRUCTIONS:                                        CSL00030
C     THESE SUBROUTINES ARE NOT PORTABLE DUE TO THE USE OF THE          CSL00040
C     COMPLEX*16 VARIABLES AND THE CORRESPONDING COMPLEX FUNCTIONS,     CSL00050
C     CDABS, DCMPLX, DREAL, DIMAG.  MOREOVER, IN SUBROUTINE             CSL00060
C     COMPEV THE NONPORTABLE FORMATS (4Z20) AND (20A4) ARE USED,        CSL00070
C     AND IN SUBROUTINE CMTQL1 THE MACHINE EPSILON IS INTRODUCED        CSL00080
C     VIA A NONPORTABLE DATA DEFINITION.                                CSL00090
C                                                                       CSL00100
C     CONTAINS SUBROUTINES USED BY THE COMPLEX SYMMETRIC VERSION OF     CSL00110
C     THE LANCZOS EIGENVALUE/EIGENVECTOR CODES.                         CSL00120
C                                                                       CSL00130
C     SUBROUTINES    COMPEV, CMTQL1, INVERR, TNORM, LUMP, ISOEV AND     CSL00140
C                    COMGAP ARE USED WITH THE LANCZOS EIGENVALUE        CSL00150
C                    PROGRAM CSLEVAL.  INVERM IS USED                   CSL00160
C                    IN THE EIGENVECTOR PROGRAM CSLEVEC.  THE INNER     CSL00170
C                    PRODUCT SUBROUTINES CINPRD AND INPRDC ARE USED     CSL00180
C                    BY BOTH PROGRAMS.                                  CSL00190
C                                                                       CSL00200
C-----INVERSE ITERATION ON COMPLEX SYMMETRIC T(1,MEV)------------------CSL00210
C                                                                       CSL00220
      SUBROUTINE INVERR(ALPHA,BETA,V1,V2,VS,EPS,GR,GC,G,GG,MP,INTERC,   CSL00230
     1MEV,MMB,NDIS,NISO,N,IKL,IT,IWRITE)                               CSL00240
C                                                                       CSL00250
C--------------------------------------------------------------------CSL00260
      COMPLEX*16  ALPHA(1),BETA(1),V1(1),V2(1),VS(1)                    CSL00270
      COMPLEX*16  U,Z,X1,RATIO,BETAM,TEMP,ZEROC                         CSL00280
      DOUBLE PRECISION EST,ESTR,ESTC,SUM,XU,NORM,TSUM,GSUM              CSL00290
      DOUBLE PRECISION  EPS,EPS3,EPS4,ZERO,ONE,GR(1),GC(1),GAP          CSL00300
      REAL G(1),GG(1)                                                   CSL00310
      INTEGER  MP(1), INTERC(1)                                         CSL00320
      REAL  ABS                                                         CSL00330
      DOUBLE PRECISION  DABS, DMIN1, DSQRT, DFLOAT, CDABS, DIMAG, DREAL CSL00340
C     COMPLEX*16 DCMPLX                                                 CSL00350
C--------------------------------------------------------------------CSL00360
C                                                                       CSL00370
C     COMPUTES ERROR ESTIMATES FOR COMPUTED ISOLATED GOOD T-EIGENVALUES CSL00380
C     IN VS AND WRITES THESE EIGENVALUES AND ESTIMATES TO FILE 4.       CSL00390
C     BY DEFINITION A GOOD T-EIGENVALUE IS ISOLATED IF ITS CLOSEST      CSL00400
C     NEIGHBOR IS ALSO GOOD, OR IF ITS CLOSEST NEIGHBOR IS              CSL00410
C     SPURIOUS BUT THAT NEIGHBOR IS FAR ENOUGH AWAY.  SO                CSL00420
C     IN PARTICULAR, WE WILL COMPUTE ESTIMATES FOR ANY GOOD             CSL00430
C     T-EIGENVALUE THAT IS IN A CLUSTER OF GOOD T-EIGENVALUES.          CSL00440
C                                                                       CSL00450
C     USES INVERSE ITERATION ON T(1,MEV) SOLVING THE EQUATION           CSL00460
C     (T - X1*I)V2 = RIGHT-HAND SIDE (RANDOMLY-GENERATED)               CSL00470
C     FOR EACH SUCH GOOD T-EIGENVALUE X1.                               CSL00480
C                                                                       CSL00490
C     PROGRAM REFACTORS T-X1*I ON EACH ITERATION OF INVERSE ITERATION.  CSL00500
C     TYPICALLY ONLY ONE ITERATION IS NEEDED PER T-EIGENVALUE X1.       CSL00510
C                                                                       CSL00520
C     ON ENTRY AND EXIT                                                 CSL00530
C     MEV = ORDER OF T :  N = ORDER OF ORIGINAL MATRIX A                CSL00540
C     ALPHA, BETA CONTAIN THE NONZERO ENTRIES OF THE T-MATRIX           CSL00550
C     VS = COMPUTED DISTINCT EIGENVALUES OF T(1,MEV)                    CSL00560
C     MP = T-MULTIPLICITY OF EACH T-EIGENVALUE IN VS. MP(I) = -1 MEANS  CSL00570
C          VS(I) IS A GOOD T-EIGENVALUE BUT THAT IT IS SITTING CLOSE TO CSL00580
C          A SPURIOUS T-EIGENVALUE.  MP(I) = 0 MEANS VS(I) IS SPURIOUS. CSL00590
C          ESTIMATES ARE COMPUTED ONLY FOR THOSE T-EIGENVALUES          CSL00600
C          WITH MP(I) = 1. FLAGGING WAS DONE IN SUBROUTINE ISOEV        CSL00610
C          PRIOR TO ENTERING INVERR.                                    CSL00620
C     NISO = NUMBER OF ISOLATED GOOD T-EIGENVALUES CONTAINED IN VS      CSL00630
C     NDIS =  NUMBER OF DISTINCT T-EIGENVALUES IN VS                    CSL00640
```

```
C     IKL = SEED FOR RANDOM NUMBER GENERATOR                       CSL00650
C     EPS = 2. * MACHINE EPSILON                                   CSL00660
C                                                                  CSL00670
C     IN PROGRAM:                                                  CSL00680
C     ITER = MAXIMUM NUMBER OF INVERSE ITERATION STEPS ALLOWED FOR EACH CSL00690
C           X1.  ITER = IT ON ENTRY.                               CSL00700
C     GR,GC = ARRAYS OF DIMENSION AT LEAST MEV + NISO.  USED TO STORE   CSL00710
C           RANDOMLY-GENERATED RIGHT-HAND SIDE.  THIS IS NOT       CSL00720
C           REGENERATED FOR EACH X1.  G IS ALSO USED TO STORE ERROR CSL00730
C           ESTIMATES AS THEY ARE COMPUTED FOR LATER PRINTOUT.     CSL00740
C     V1,V2 = WORK SPACES USED IN THE FACTORIZATION OF T(1,MEV).   CSL00750
C     AT THE END OF THE INVERSE ITERATION COMPUTATION FOR X1, V2   CSL00760
C     CONTAINS THE UNIT EIGENVECTOR OF T(1,MEV) CORRESPONDING TO X1. CSL00770
C     V1 AND V2 MUST BE OF DIMENSION AT LEAST MEV.                 CSL00780
C                                                                  CSL00790
C     ON EXIT                                                      CSL00800
C     GG(J) = MINIMUM GAP IN T(1,MEV) FOR EACH VS(J), J=1,NDIS     CSL00810
C     G(I) = |BETAM|*|V2(MEV)| = ERROR ESTIMATE FOR ISOLATED GOOD  CSL00820
C             T-EIGENVALUES, WHERE I = 1,NISO  AND  BETAM = BETA(MEV+1)CSL00830
C             T(1,MEV) CORRESPONDING TO ITH ISOLATED GOOD T-EIGENVALUE.CSL00840
C                                                                  CSL00850
C     IF FOR SOME X1 IT.GT.ITER THEN THE ERROR ESTIMATE IN G IS MARKED CSL00860
C     WITH A - SIGN.                                               CSL00870
C                                                                  CSL00880
C     V2 = ISOLATED GOOD T-EIGENVALUES                            CSL00890
C     V1 = MINIMAL T-GAPS FOR THE T-EIGENVALUES IN V2.            CSL00900
C     THESE ARE CONSTRUCTED FOR WRITE-OUT PURPOSES ONLY AND NOT   CSL00910
C     NEEDED ELSEWHERE IN THE PROGRAM.                            CSL00920
C------------------------------------------------------------------CSL00930
C                                                                  CSL00940
C     LABEL OUTPUT FILE  4                                         CSL00950
      IF (MMB.EQ.1) WRITE(4,10)                                   CSL00960
   10 FORMAT(' INVERSE ITERATION ERROR ESTIMATES'/)               CSL00970
C                                                                  CSL00980
C     FILE 6 (TERMINAL) OUTPUT OF ERROR ESTIMATES                 CSL00990
      IF (IWRITE.NE.0.AND.NISO.NE.0) WRITE(6,20)                  CSL01000
   20 FORMAT(/' INVERSE ITERATION ERROR ESTIMATES'/'  JISO',' JDIST',8X CSL01010
     1,'GOOD T-EIGENVALUE',4X,'BETAM*UM',5X,'TMINGAP')             CSL01020
C                                                                  CSL01030
C     INITIALIZATION AND PARAMETER SPECIFICATION                  CSL01040
      ZERO = 0.0D0                                                CSL01050
      ONE = 1.0D0                                                 CSL01060
      ZEROC = DCMPLX(ZERO,ZERO)                                   CSL01070
      NG = 0                                                      CSL01080
      NISO = 0                                                    CSL01090
      ITER = IT                                                   CSL01100
      MP1 = MEV+1                                                 CSL01110
      MM1 = MEV-1                                                 CSL01120
      BETAM = BETA(MP1)                                           CSL01130
      BETA(MP1) = ZEROC                                           CSL01140
C                                                                  CSL01150
C     CALCULATE SCALE AND TOLERANCES                              CSL01160
      TSUM = CDABS(ALPHA(1))                                      CSL01170
      DO 30 I = 2,MEV                                             CSL01180
   30 TSUM = TSUM + CDABS(ALPHA(I)) + CDABS(BETA(I))              CSL01190
C                                                                  CSL01200
      EPS3 = EPS*TSUM                                             CSL01210
      EPS4 = DFLOAT(MEV)*EPS3                                     CSL01220
C                                                                  CSL01230
C     GENERATE SCALED RANDOM RIGHT-HAND SIDE                      CSL01240
      ILL = IKL                                                   CSL01250
C                                                                  CSL01260
C------------------------------------------------------------------CSL01270
      CALL GENRAN(ILL,G,MEV)                                      CSL01280
C------------------------------------------------------------------CSL01290
C                                                                  CSL01300
      DO 40 I = 1,MEV                                             CSL01310
```

```
   40 GR(I) = G(I)                                                    CSL01320
C                                                                     CSL01330
C---------------------------------------------------------------------CSL01340
      CALL GENRAN(ILL,G,MEV)                                          CSL01350
C---------------------------------------------------------------------CSL01360
C                                                                     CSL01370
      DO 50 I = 1,MEV                                                 CSL01380
   50 GC(I) = G(I)                                                    CSL01390
C                                                                     CSL01400
      GSUM = ZERO                                                     CSL01410
      DO 60 I = 1,MEV                                                 CSL01420
   60 GSUM = GSUM + DABS(GR(I)) + DABS(GC(I))                         CSL01430
      GSUM = EPS4/GSUM                                                CSL01440
C                                                                     CSL01450
      DO 70 I = 1,MEV                                                 CSL01460
      GR(I) = GSUM*GR(I)                                              CSL01470
   70 GC(I) = GSUM*GC(I)                                              CSL01480
C                                                                     CSL01490
C     LOOP ON ISOLATED GOOD T-EIGENVALUES IN VS (MP(I) = 1) TO        CSL01500
C     CALCULATE CORRESPONDING UNIT EIGENVECTOR OF T(1,MEV)            CSL01510
C                                                                     CSL01520
      DO 200 JEV = 1,NDIS                                             CSL01530
      IF (MP(JEV).EQ.0) GO TO 200                                     CSL01540
      NG = NG + 1                                                     CSL01550
      IF (MP(JEV).NE.1) GO TO 200                                     CSL01560
      IT = 1                                                          CSL01570
      NISO = NISO + 1                                                 CSL01580
      X1 = VS(JEV)                                                    CSL01590
C                                                                     CSL01600
C     INITIALIZE RIGHT HAND SIDE FOR INVERSE ITERATION               CSL01610
C     AND THE FLAG ON WHICH ROWS ARE INTERCHANGED                    CSL01620
      DO 80 I = 1,MEV                                                 CSL01630
      INTERC(I) = 0                                                   CSL01640
   80 V2(I) = DCMPLX(GR(I),GC(I))                                     CSL01650
C                                                                     CSL01660
C     TRIANGULAR FACTORIZATION WITH NEAREST NEIGHBOR PIVOT           CSL01670
C     STRATEGY. INTERCHANGES ARE LABELLED BY SETTING INTERC = 1.     CSL01680
C                                                                     CSL01690
   90 CONTINUE                                                        CSL01700
      U = ALPHA(1)-X1                                                 CSL01710
      Z = BETA(2)                                                     CSL01720
C                                                                     CSL01730
      DO 110 I = 2,MEV                                                CSL01740
      IF (CDABS(BETA(I)).GT.CDABS(U)) GO TO 100                       CSL01750
C     NO INTERCHANGE                                                  CSL01760
      V1(I-1) = Z/U                                                   CSL01770
      V2(I-1) = V2(I-1)/U                                             CSL01780
      V2(I) = V2(I)-BETA(I)*V2(I-1)                                   CSL01790
      RATIO = BETA(I)/U                                               CSL01800
      U = ALPHA(I)-X1-Z*RATIO                                         CSL01810
      Z = BETA(I+1)                                                   CSL01820
      GO TO 110                                                       CSL01830
  100 CONTINUE                                                        CSL01840
C     INTERCHANGE CASE                                                CSL01850
      RATIO = U/BETA(I)                                               CSL01860
      INTERC(I) = 1                                                   CSL01870
      V1(I-1) = ALPHA(I)-X1                                           CSL01880
      U = Z-RATIO*V1(I-1)                                             CSL01890
      Z = -RATIO*BETA(I+1)                                           CSL01900
      TEMP = V2(I-1)                                                  CSL01910
      V2(I-1) = V2(I)                                                 CSL01920
      V2(I) = TEMP-RATIO*V2(I)                                        CSL01930
  110 CONTINUE                                                        CSL01940
      IF (CDABS(U).EQ.ZERO) U = DCMPLX(EPS3,EPS3)                     CSL01950
C                                                                     CSL01960
C     SMALLNESS TEST AND DEFAULT VALUE FOR LAST COMPONENT            CSL01970
C     PIVOT(I-1) = BETA(I) FOR INTERCHANGE CASE                      CSL01980
```

```
C     (I-1,I+1) ELEMENT IN RIGHT FACTOR = BETA(I+1)                    CSL01990
C     END OF FACTORIZATION AND FORWARD SUBSTITUTION                    CSL02000
C                                                                      CSL02010
C     BACK SUBSTITUTION                                                CSL02020
      V2(MEV) = V2(MEV)/U                                              CSL02030
      DO 130 II = 1,MM1                                                CSL02040
      I = MEV-II                                                       CSL02050
      IF (INTERC(I+1).EQ.1) GO TO 120                                  CSL02060
C     NO INTERCHANGE                                                   CSL02070
      V2(I) = V2(I)-V1(I)*V2(I+1)                                      CSL02080
      GO TO 130                                                        CSL02090
C     INTERCHANGE CASE                                                 CSL02100
  120 CONTINUE                                                         CSL02110
      V2(I) = (V2(I)-V1(I)*V2(I+1)-BETA(I+2)*V2(I+2))/BETA(I+1)        CSL02120
  130 CONTINUE                                                         CSL02130
C                                                                      CSL02140
C     TESTS FOR CONVERGENCE OF INVERSE ITERATION                       CSL02150
C     IF SUM |V2| COMPS. LE. 1 AND IT. LE. ITER DO ANOTHER INVIT STEP  CSL02160
C                                                                      CSL02170
      NORM = CDABS(V2(MEV))                                            CSL02180
      DO 140 II = 1,MM1                                                CSL02190
      I = MEV-II                                                       CSL02200
  140 NORM = NORM + CDABS(V2(I))                                       CSL02210
C                                                                      CSL02220
      IF (NORM.GE.ONE) GO TO 160                                       CSL02230
      IT = IT+1                                                        CSL02240
      IF (IT.GT.ITER) GO TO 160                                        CSL02250
      XU = EPS4/NORM                                                   CSL02260
C                                                                      CSL02270
      DO 150 I = 1,MEV                                                 CSL02280
  150 V2(I) = V2(I)*XU                                                 CSL02290
C                                                                      CSL02300
      GO TO 90                                                         CSL02310
C     ANOTHER INVERSE ITERATION STEP                                   CSL02320
C                                                                      CSL02330
C     INVERSE ITERATION FINISHED                                       CSL02340
C     NORMALIZE COMPUTED T-EIGENVECTOR : V2 = V2/||V2||                CSL02350
  160 CONTINUE                                                         CSL02360
C                                                                      CSL02370
C----------------------------------------------------------------------CSL02380
      CALL CINPRD(V2,V2,SUM,MEV)                                       CSL02390
C----------------------------------------------------------------------CSL02400
C                                                                      CSL02410
      SUM = ONE/DSQRT(SUM)                                             CSL02420
C                                                                      CSL02430
      DO 170 II = 1,MEV                                                CSL02440
  170 V2(II) = SUM*V2(II)                                              CSL02450
C                                                                      CSL02460
C     SAVE ERROR ESTIMATE FOR LATER OUTPUT                             CSL02470
      EST = CDABS(BETAM)*CDABS(V2(MEV))                                CSL02480
      ESTR = DABS(DREAL(V2(MEV)))                                      CSL02490
      ESTC = DABS(DIMAG(V2(MEV)))                                      CSL02500
      GSUM = CDABS(BETAM)                                              CSL02510
      IF (IT.GT.ITER) EST = -EST                                       CSL02520
      G(NISO) = EST                                                    CSL02530
      IF (IWRITE.EQ.0) GO TO 200                                       CSL02540
C                                                                      CSL02550
C     FILE 6 (TERMINAL) OUTPUT OF ERROR ESTIMATES.                     CSL02560
      GAP = GG(JEV)                                                    CSL02570
      WRITE(6,180) NISO,JEV,X1,EST,GAP                                 CSL02580
  180 FORMAT(2I6,2E20.12,2E12.3)                                       CSL02590
      WRITE(6,190) JEV, X1, EST,ESTR,ESTC                              CSL02600
  190 FORMAT(I6,2E20.12,3E11.3)                                        CSL02610
C                                                                      CSL02620
  200 CONTINUE                                                         CSL02630
C                                                                      CSL02640
C     END ERROR ESTIMATE LOOP ON ISOLATED GOOD T-EIGENVALUES.          CSL02650
```

```
C      GENERATE DISTINCT MINGAPS FOR T(1,MEV). THIS IS USEFUL AS AN      CSL02660
C      INDICATOR OF THE GOODNESS OF THE INVERSE ITERATION ESTIMATES.     CSL02670
C      TRANSFER ISOLATED GOOD T-EIGENVALUES AND CORRESPONDING TMINGAPS   CSL02680
C      TO V2 AND V1 FOR OUTPUT PURPOSES ONLY.                            CSL02690
C                                                                        CSL02700
       ISO = 0                                                           CSL02710
       DO 210 J = 1,NDIS                                                 CSL02720
       IF (MP(J).NE.1) GO TO 210                                         CSL02730
       ISO = ISO+1                                                       CSL02740
       GR(ISO) = GG(J)                                                   CSL02750
       V2(ISO) = VS(J)                                                   CSL02760
   210 CONTINUE                                                          CSL02770
       IF(NISO.EQ.0) GO TO 270                                          CSL02780
C                                                                        CSL02790
C      ERROR ESTIMATES ARE WRITTEN TO FILE 4                             CSL02800
       WRITE(4,220)MEV,NDIS,NG,NISO,N,IKL,ITER,GSUM                      CSL02810
   220 FORMAT(1X,'TSIZE',2X,'NDIS',1X,'NGOOD',2X,'NISO',1X,'ASIZE'/5I6/  CSL02820
      1 4X,'RHSEED',2X,'MXINIT',5X,'BETAM'/I10,I8,E10.3)                 CSL02830
C                                                                        CSL02840
       WRITE(4,230)                                                      CSL02850
   230 FORMAT(2X,'GOODEVNO',11X,'R(GOODEV)',11X,'I(GOODEV)',             CSL02860
      1 6X,'BETAM*UM',7X,'TMINGAP')                                     CSL02870
C                                                                        CSL02880
       ISPUR = 0                                                         CSL02890
       I = 0                                                             CSL02900
       DO 260 J = 1,NDIS                                                 CSL02910
       IF(MP(J).NE.0) GO TO 240                                          CSL02920
       ISPUR = ISPUR + 1                                                 CSL02930
       GO TO 260                                                         CSL02940
   240 IF(MP(J).NE.1) GO TO 260                                          CSL02950
       I = I + 1                                                         CSL02960
       IGOOD = J - ISPUR                                                 CSL02970
       WRITE(4,250) IGOOD,V2(I),G(I),GR(I)                              CSL02980
   250 FORMAT(I10,2E20.12,2E14.3)                                        CSL02990
   260 CONTINUE                                                          CSL03000
       GO TO 290                                                         CSL03010
C                                                                        CSL03020
   270 WRITE(4,280)                                                      CSL03030
   280 FORMAT(/' THERE ARE NO ISOLATED T-EIGENVALUES SO NO ERROR ESTIMATECSL03040
      1S WERE COMPUTED')                                                 CSL03050
C      RESTORE BETA(MEV+1) = BETAM                                       CSL03060
   290 BETA(MP1) = BETAM                                                 CSL03070
C-----END OF INVERR---------------------------------------------------CSL03080
       RETURN                                                           CSL03090
       END                                                              CSL03100
C-----START OF TNORM--------------------------------------------------CSL03110
C                                                                        CSL03120
       SUBROUTINE TNORM(ALPHA,BETA,BMIN,TMAX,MEV,IB)                    CSL03130
C                                                                        CSL03140
C--------------------------------------------------------------------CSL03150
       COMPLEX*16  ALPHA(1),BETA(1)                                     CSL03160
       DOUBLE PRECISION  TMAX,BMIN,BMAX,BSIZE,BTOL,ABATA,AALFA          CSL03170
       DOUBLE PRECISION  DMAX1, CDABS                                   CSL03180
C      COMPLEX*16 DCMPLX                                                CSL03190
C--------------------------------------------------------------------CSL03200
C      IN REAL SYMMETRIC AND HERMITIAN VERSIONS TMAX IS USED            CSL03210
C      TO DETERMINE THE TOLERANCES USED IN THE T-MULTIPLICITY AND IN     CSL03220
C      THE SPURIOUS TESTS. FOR THE COMPLEX SYMMETRIC CASE WE            CSL03230
C      HAVE TO COMPUTE ALL OF THE T-EIGENVALUES SO WE USE THEM INSTEAD   CSL03240
C      OF TMAX TO DETERMINE THESE TOLERANCES. WE USE TMAX TO            CSL03250
C      CHECK THE RELATIVE SIZES OF THE BETA(K), K=1,...,MEV AS A         CSL03260
C      TEST ON THE LOCAL ORTHOGONALITY OF THE LANCZOS VECTORS.          CSL03270
C                                                                        CSL03280
C              TMAX = MAX (|ALPHA(I)|, |BETA(I)|,  I=1,MEV)             CSL03290
C              BMIN = MIN (|BETA(I)|, I=2,MEV)                          CSL03300
C              BSIZE = BMIN/TMAX                                        CSL03310
C              |IB| = INDEX OF MINIMAL(BETA)                           CSL03320
```

```
C            IB < 0 IF BMIN/TMAX < BTOL                              CSL03330
C-----------------------------------------------------------------------CSL03340
C     SPECIFY PARAMETERS                                             CSL03350
      IB = 2                                                         CSL03360
      BTOL = BMIN                                                    CSL03370
      BMIN = CDABS(BETA(2))                                          CSL03380
      BMAX = BMIN                                                    CSL03390
      TMAX = CDABS(ALPHA(1))                                         CSL03400
C                                                                    CSL03410
      DO 20 I = 2,MEV                                                CSL03420
      ABATA = CDABS(BETA(I))                                         CSL03430
      IF (ABATA.GE.BMIN) GO TO 10                                    CSL03440
      IB = I                                                         CSL03450
      BMIN = ABATA                                                   CSL03460
   10 AALFA = CDABS(ALPHA(I))                                        CSL03470
      TMAX = DMAX1(TMAX,AALFA)                                       CSL03480
      BMAX = DMAX1(ABATA,BMAX)                                       CSL03490
   20 CONTINUE                                                       CSL03500
      TMAX = DMAX1(BMAX,TMAX)                                        CSL03510
C                                                                    CSL03520
C     TEST OF LOCAL ORTHOGONALITY USING SCALED BETAS                 CSL03530
      BSIZE = BMIN/TMAX                                              CSL03540
      IF (BSIZE.GE.BTOL) GO TO 40                                    CSL03550
C                                                                    CSL03560
C     DEFAULT.  BSIZE IS SMALLER THAN TOLERANCE BTOL SPECIFIED IN MAIN CSL03570
C     PROGRAM.  PROGRAM TERMINATES FOR USER TO DECIDE WHAT TO DO     CSL03580
C     BECAUSE LOCAL ORTHOGONALITY OF THE LANCZOS VECTORS COULD BE    CSL03590
C     LOST.                                                          CSL03600
C                                                                    CSL0361C
      IB = -IB                                                       CSL03620
      WRITE(6,30) MEV                                                CSL03630
   30 FORMAT(/' BETA TEST INDICATES POSSIBLE LOSS OF LOCAL ORTHOGONALITYCSL03640
     1 OVER 1ST',I6,' LANCZOS VECTORS'/)                            CSL03650
C                                                                    CSL03660
   40 CONTINUE                                                       CSL03670
C                                                                    CSL03680
      WRITE(6,50) IB                                                 CSL03690
   50 FORMAT(/' MINIMUM BETA RATIO OCCURS AT',I6,' TH BETA'/)        CSL03700
C                                                                    CSL03710
      WRITE(6,60) MEV,BMIN,TMAX,BSIZE                                CSL03720
   60 FORMAT(/1X,'TSIZE',6X,'MIN BETA',5X,'TKMAX',6X,'MIN RATIO'/    CSL03730
     1 I6,E14.3,E10.3,E15.3/)                                        CSL03740
C                                                                    CSL03750
C-----END OF TNORM------------------------------------------------------CSL03760
      RETURN                                                         CSL03770
      END                                                            CSL03780
C                                                                    CSL03790
C-----START OF LUMP-----------------------------------------------------CSL03800
C                                                                    CSL03810
      SUBROUTINE LUMP(VC,V1,VA,RELTOL,SPUTOL,SCALE2,LINDEX,TFLAG,LOOP) CSL03820
C                                                                    CSL03830
C-----------------------------------------------------------------------CSL03840
      COMPLEX*16 VC(1),V1(1),ZEROC,SUMC                              CSL03850
      DOUBLE PRECISION  VA(1),RELTOL,SPUTOL,SCALE2                   CSL03860
      DOUBLE PRECISION  THOLD,TH1,TH2,DGAP,ZERO,ONE                  CSL03870
      INTEGER  LINDEX(1),TFLAG(1)                                    CSL03880
      DOUBLE PRECISION  DFLOAT, DMAX1, CDABS                         CSL03890
C     COMPLEX*16 DCMPLX                                              CSL03900
C-----------------------------------------------------------------------CSL03910
C     VC(J) = JTH DISTINCT T-EIGENVALUE,  VA(J) = |VC(J)|, IN ORDER  CSL03920
C     OF INCREASING MAGNITUDE.                                       CSL03930
C     LINDEX(J) = T-MULTIPLICITY OF JTH DISTINCT T-EIGENVALUE        CSL03940
C     LOOP = NUMBER OF DISTINCT T-EIGENVALUES                        CSL03950
C     LUMP 'COMBINES' COMPUTED 'GOOD' T-EIGENVALUES THAT ARE 'TOO CLOSE'CSL03960
C     VALUE OF RELTOL IS 1.D-8.                                      CSL03970
C                                                                    CSL03980
C     IF IN A SET OF T-EIGENVALUES TO BE COMBINED THERE IS AN EIGENVALUECSL03990
```

```
C       WITH LINDEX=1, THEN THE VALUE OF THE COMBINED T-EIGENVALUES IS SETCSL04000
C       EQUAL TO THE VALUE OF THAT EIGENVALUE.  NOTE THAT IF A SPURIOUS  CSL04010
C       T-EIGENVALUE IS TO BE 'COMBINED' WITH A GOOD EIGENVALUE, THEN THISCSL04020
C       IS DONE ONLY BY INCREASING THE INDEX, LINDEX, FOR THAT EIGENVALUE CSL04030
C       NUMERICAL VALUES OF SPURIOUS T-EIGENVALUES ARE NEVER COMBINED WITHCSL04040
C       THOSE OF GOOD T-EIGENVALUES.                                      CSL04050
C-------------------------------------------------------------------------CSL04060
        ZERO = 0.0D0                                                     CSL04070
        ONE = 1.D0                                                       CSL04080
        ZEROC = DCMPLX(ZERO,ZERO)                                        CSL04090
        TH2 = SCALE2*SPUTOL                                              CSL04100
        DO 10 K = 1,LOOP                                                 CSL04110
     10 TFLAG(K) = 0                                                     CSL04120
        NLOOP = 0                                                        CSL04130
        J = 0                                                            CSL04140
     20 J = J+1                                                          CSL04150
        IF (J.GT.LOOP) GO TO 130                                         CSL04160
        IF (TFLAG(J).EQ.1) GO TO 20                                      CSL04170
        NLOOP = NLOOP + 1                                                CSL04180
        TFLAG(J) = 1                                                     CSL04190
        V1(1) = VC(J)                                                    CSL04200
        ICOUNT = 1                                                       CSL04210
        JN = LINDEX(J)                                                   CSL04220
        TH1 = RELTOL*VA(J)                                               CSL04230
        THOLD = DMAX1(TH1,TH2)                                           CSL04240
C       THOLD = RELTOL*DMAX1(ONE,VA(J))                                  CSL04250
        IF (JN.EQ.0) GO TO 30                                            CSL04260
        INDSUM = JN                                                      CSL04270
        ISPUR = 0                                                        CSL04280
        SUMC = DFLOAT(JN)*VC(J)                                          CSL04290
        GO TO 40                                                         CSL04300
     30 INDSUM = 1                                                       CSL04310
        ISPUR = 1                                                        CSL04320
        SUMC = ZEROC                                                     CSL04330
     40 IF (J.EQ.LOOP) GO TO 70                                          CSL04340
        I = J                                                            CSL04350
     50 I = I + 1                                                        CSL04360
        IF (I.GT.LOOP) GO TO 70                                          CSL04370
        IF (TFLAG(I).EQ.1) GO TO 50                                      CSL04380
        DGAP = VA(I) - VA(J)                                             CSL04390
        IF (DGAP.GE.THOLD) GO TO 70                                      CSL04400
        DGAP = CDABS(VC(I)-VC(J))                                        CSL04410
        IF (DGAP.GE.THOLD) GO TO 50                                      CSL04420
C       LUMP VC(I) WITH VC(J)                                            CSL04430
        ICOUNT = ICOUNT + 1                                              CSL04440
        TFLAG(I) = 1                                                     CSL04450
        V1(ICOUNT) = VC(I)                                               CSL04460
        IN = LINDEX(I)                                                   CSL04470
        IF (IN.NE.0) GO TO 60                                            CSL04480
        ISPUR = ISPUR + 1                                               CSL04490
        INDSUM = INDSUM + 1                                              CSL04500
        GO TO 50                                                         CSL04510
     60 INDSUM = INDSUM + IN                                             CSL04520
        SUMC = SUMC + DFLOAT(IN)*VC(I)                                   CSL04530
        GO TO 50                                                         CSL04540
C       COMPUTE THE 'COMBINED' T-EIGENVALUE AND THE RESULTING            CSL04550
C       T-MULTIPLICITY                                                   CSL04560
     70 CONTINUE                                                         CSL04570
C                                                                        CSL04580
C       IF (ICOUNT.GT.1) WRITE(6,80) (K,V1(K), K = 1,ICOUNT)             CSL04590
     80 FORMAT(/' T-EIGENVALUES ARE LUMPED '/                            CSL04600
      1 5X,'J',12X,'REAL(EV)',12X,'IMAG(EV)'/(I6,2E20.12))               CSL04610
C                                                                        CSL04620
        IF (ICOUNT.EQ.1) INDSUM = JN                                     CSL04630
        IDIF = INDSUM - ISPUR                                            CSL04640
        IF (IDIF.EQ.0.AND.ICOUNT.GT.1) GO TO 110                         CSL04650
```

```
      IF (ICOUNT.EQ.1) GO TO 90                                    CSL04660
C     ICOUNT.GT.1 AND IDIF.GT.0                                    CSL04670
      SUMC = SUMC/DFLOAT(IDIF)                                     CSL04680
      VC(NLOOP) = SUMC                                             CSL04690
      VA(NLOOP) = CDABS(SUMC)                                      CSL04700
      GO TO 100                                                    CSL04710
   90 VC(NLOOP) = VC(J)                                            CSL04720
      VA(NLOOP) = VA(J)                                            CSL04730
  100 LINDEX(NLOOP) = INDSUM                                       CSL04740
      GO TO 20                                                     CSL04750
C     INDEX J IS FINISHED                                          CSL04760
C                                                                  CSL04770
C     ON RETURN VC CONTAINS THE DISTINCT T-EIGENVALUES  VA = |VC|  CSL04780
C     LINDEX CONTAINS THE CORRESPONDING T-MULTIPLICITIES           CSL04790
C                                                                  CSL04800
  110 WRITE(6,120) (K,V1(K), K = 1,ICOUNT)                         CSL04810
  120 FORMAT(/' DEFAULT IN LUMP LUMPING INDICATED FOR TWO SPURIOUS EV'/ CSL04820
     1 ' K,V1(K) = '/(16,2E20.12))                                CSL04830
      NLOOP = -NLOOP                                               CSL04840
  130 CONTINUE                                                     CSL04850
      LOOP = NLOOP                                                 CSL04860
      RETURN                                                       CSL04870
C-----END OF LUMP----------------------------------------------------CSL04880
      END                                                          CSL04890
C                                                                  CSL04900
C-----START OF ISOEV------------------------------------------------CSL04910
C                                                                  CSL04920
      SUBROUTINE ISOEV(VS,GR,GG,GAPTOL,SPUTOL,SCALE1,MP,NDIS,NG,NISO) CSL04930
C                                                                  CSL04940
C------------------------------------------------------------------CSL04950
      COMPLEX*16 VS(1),TO                                          CSL04960
      DOUBLE PRECISION  GR(1),SPUTOL,GAPTOL,SCALE1,TEMP,TOL,TJ,DGAP,ONE CSL04970
      REAL GG(1)                                                   CSL04980
      INTEGER  MP(1)                                               CSL04990
      REAL  ABS                                                    CSL05000
      DOUBLE PRECISION  DMAX1, CDABS                               CSL05010
C------------------------------------------------------------------CSL05020
C     USE TMINGAPS TO LABEL THE ISOLATED GOOD T-EIGENVALUES        CSL05030
C     THAT ARE VERY CLOSE TO SPURIOUS ONES.  ERROR ESTIMATES       CSL05040
C     WILL NOT BE COMPUTED FOR THESE T-EIGENVALUES.                CSL05050
C                                                                  CSL05060
C     ON ENTRY AND EXIT                                            CSL05070
C     VS CONTAINS THE COMPUTED DISTINCT EIGENVALUES OF T(1,MEV)    CSL05080
C     GR(K) = |VS(K)|, K = 1,NDIS, GR(K).LE.GR(K+1)               CSL05090
C     GG(K) = MIN(J.NE.K,|VS(K)-VS(J)|)   MINGAP                  CSL05100
C     MP CONTAINS THE CORRESPONDING T-MULTIPLICITIES               CSL05110
C     NDIS = NUMBER OF DISTINCT T-EIGENVALUES                      CSL05120
C     GAPTOL = RELATIVE GAP TOLERANCE SET IN MAIN                  CSL05130
C                                                                  CSL05140
C     ON EXIT                                                      CSL05150
C     MP(J) IS NOT CHANGED EXCEPT THAT  MP(J)=-1, IF MP(J)=1,      CSL05160
C     AND A SPURIOUS T-EIGENVALUE IS TOO CLOSE.                    CSL05170
C                                                                  CSL05180
C     IF MP(1)=-1 THAT SIMPLE GOOD T-EIGENVALUE WILL BE SKIPPED    CSL05190
C     IN THE SUBSEQUENT ERROR ESTIMATE COMPUTATIONS IN INVERR      CSL05200
C     THAT IS, WE COMPUTE ERROR ESTIMATES ONLY FOR THOSE GOOD      CSL05210
C     T-EIGENVALUES WITH MP(J)=1.                                  CSL05220
C------------------------------------------------------------------CSL05230
      ONE = 1.0D0                                                  CSL05240
      DGAP = SCALE1*SPUTOL                                         CSL05250
      NISO = 0                                                     CSL05260
      NG = 0                                                       CSL05270
      DO 40 J = 1,NDIS                                             CSL05280
      IF (MP(J).EQ.0) GO TO 40                                     CSL05290
      NG = NG+1                                                    CSL05300
      IF (MP(J).NE.1) GO TO 40                                     CSL05310
      TJ = GR(J)                                                   CSL05320
```

```
        TO = VS(J)                                           CSL05330
        TOL = DMAX1(DGAP,GAPTOL*TJ)                          CSL05340
C       TOL = DMAX1(ONE,TJ)*GAPTOL                           CSL05350
C       VS(J) IS NEXT SIMPLE GOOD T-EIGENVALUE               CSL05360
        NISO = NISO + 1                                      CSL05370
        IF (ABS(GG(J)).GT.TOL) GO TO 40                      CSL05380
        I = J                                                CSL05390
   10   I = I-1                                              CSL05400
        IF (I.LT.1) GO TO 20                                 CSL05410
        IF (TJ-GR(I).GT.TOL) GO TO 20                        CSL05420
        IF (MP(I).NE.0) GO TO 10                             CSL05430
        TEMP = CDABS(TO-VS(I))                               CSL05440
        IF (TEMP.GT.TOL) GO TO 10                            CSL05450
        MP(J) = -MP(J)                                       CSL05460
        NISO = NISO-1                                        CSL05470
        GO TO 40                                             CSL05480
   20   I = J                                                CSL05490
   30   I = I+1                                              CSL05500
        IF (I.GT.NDIS) GO TO 40                              CSL05510
        IF (GR(I)-TJ.GT.TOL) GO TO 40                        CSL05520
        IF (MP(I).NE.0) GO TO 30                             CSL05530
        TEMP = CDABS(TO-VS(I))                               CSL05540
        IF (TEMP.GT.TOL) GO TO 30                            CSL05550
        MP(J) = -MP(J)                                       CSL05560
        NISO = NISO-1                                        CSL05570
   40   CONTINUE                                             CSL05580
C                                                            CSL05590
C-----END OF ISOEV--------------------------------------------CSL05600
        RETURN                                               CSL05610
        END                                                  CSL05620
C---COMPEV---------------------------------------------------- CSL05630
C                                                            CSL05640
        SUBROUTINE COMPEV(ALPHA,BETA,V1,V2,VS,EVMAG,MULTOL,SPUTOL, CSL05650
       1MP,T2FLAG,MEV,NDIS,SAVTEV)                           CSL05660
C                                                            CSL05670
C       USES COMPLEX SYMMETRIC VERSION OF IMTQL1, CMTQL1, TO CSL05680
C       COMPUTE EIGENVALUES OF THE T-MATRIX T(1,MEV).        CSL05690
C                                                            CSL05700
C------------------------------------------------------------CSL05710
        COMPLEX*16 ALPHA(1),BETA(1),VS(1),V1(1),V2(1),EVAL,CTEMP CSL05720
        DOUBLE PRECISION EVMAG(1)                            CSL05730
        DOUBLE PRECISION TEMP,DGAP,TOL,DELMIN                CSL05740
        DOUBLE PRECISION MULTOL,SPUTOL,EVALR,EVALC           CSL05750
        INTEGER MP(1),T2FLAG(1),SAVTEV                       CSL05760
        DOUBLE PRECISION CDABS, DFLOAT                       CSL05770
C------------------------------------------------------------CSL05780
C                                                            CSL05790
        MEV1 = MEV - 1                                       CSL05800
C                                                            CSL05810
        IF (SAVTEV.GE.0) GO TO 40                            CSL05820
C                                                            CSL05830
        READ(10,10) MEV                                      CSL05840
   10   FORMAT(I6)                                           CSL05850
   20   FORMAT(20A4)                                         CSL05860
        MEV1 = MEV - 1                                       CSL05870
        READ(10,30) (VS(K), K = 1,MEV)                       CSL05880
   30   FORMAT(4Z20)                                         CSL05890
        READ(10,20) EXPLAN                                   CSL05900
        READ(10,20) EXPLAN                                   CSL05910
        READ(10,30) (V2(K), K = 1,MEV1)                      CSL05920
        GO TO 90                                             CSL05930
C                                                            CSL05940
   40   CONTINUE                                             CSL05950
C                                                            CSL05960
        DO 50 J = 1,MEV                                      CSL05970
        VS(J) = ALPHA(J)                                     CSL05980
```

```
   50 V1(J) = BETA(J)                                              CSL05990
C                                                                  CSL06000
      WRITE(6,60) MEV                                              CSL06010
   60 FORMAT(/' COMPUTE EIGENVALUES OF T(1,',I4,') USING CMTQL1'/) CSL06020
C                                                                  CSL06030
C-----------------------------------------------------------------CSL06040
      CALL CMTQL1(MEV,VS,V1,IERR)                                  CSL06050
C-----------------------------------------------------------------CSL06060
C                                                                  CSL06070
C     WRITE(6,70) IERR                                             CSL06080
   70 FORMAT(' T-EIGENVALUES VIA CMTQL1'/' IERR = ',I6/)           CSL06090
C                                                                  CSL06100
      IF (IERR.EQ.0) GO TO 90                                      CSL06110
C                                                                  CSL06120
      WRITE(6,80)                                                  CSL06130
   80 FORMAT(' ON RETURN FROM CMTQL1 ERROR FLAG WAS NOT ZERO'/)    CSL06140
      GO TO 410                                                    CSL06150
C                                                                  CSL06160
   90 CONTINUE                                                     CSL06170
C                                                                  CSL06180
C     T-EIGENVALUES ARE IN VS IN INCREASING ORDER OF MAGNITUDE     CSL06190
      DO 100 J = 1,MEV                                             CSL06200
  100 EVMAG(J) = CDABS(VS(J))                                      CSL06210
C                                                                  CSL06220
C     THE MAGNITUDES OF THE T-EIGENVALUES ARE IN EVMAG, IN ORDER OF CSL06230
C     INCREASING MAGNITUDE                                         CSL06240
C     WRITE(13,105) (EVMAG(J), J = 1,MEV)                          CSL06250
C 105 FORMAT(' MAGNITUDES OF T-EIGENVALUES'/(4E20.12))             CSL06260
C                                                                  CSL06270
      IF(SAVTEV.NE.1) GO TO 130                                    CSL06280
      WRITE(10,110) MEV                                            CSL06290
  110 FORMAT(I6,' = ORDER OF T-MATRIX, T-EIGVALS =')               CSL06300
      WRITE(10,120) (VS(J), J = 1,MEV)                             CSL06310
C 120 FORMAT(4Z20)                                                 CSL06320
  120 FORMAT(4E20.12)                                              CSL06330
C                                                                  CSL06340
C                                                                  CSL06350
  130 CONTINUE                                                     CSL06360
      MULTOL = MULTOL*EVMAG(MEV)                                   CSL06370
      SPUTOL = SPUTOL*EVMAG(MEV)                                   CSL06380
      TOL = 1000.0D0*SPUTOL                                        CSL06390
      WRITE(6,140) MULTOL,SPUTOL                                   CSL06400
  140 FORMAT(/' TOLERANCES USED IN T-MULTIPLICITY AND SPURIOUS TESTS =' CSL06410
     1 ,2E10.3/)                                                   CSL06420
C                                                                  CSL06430
C     T-MULTIPLICITY DETERMINATION                                 CSL06440
      J = 0                                                        CSL06450
      NDIS = 0                                                     CSL06460
      DO 150 I = 1,MEV                                             CSL06470
  150 T2FLAG(I) = 0                                                CSL06480
C                                                                  CSL06490
  160 J = J+1                                                      CSL06500
      IF (J.GT.MEV) GO TO 190                                      CSL06510
      IF (T2FLAG(J).EQ.1) GO TO 160                                CSL06520
      CTEMP = VS(J)                                                CSL06530
      EVAL = CTEMP                                                 CSL06540
      TEMP = EVMAG(J)                                              CSL06550
      NDIS = NDIS + 1                                              CSL06560
      INDEX = 1                                                    CSL06570
      T2FLAG(J) = 1                                                CSL06580
      I = J                                                        CSL06590
  170 I = I+1                                                      CSL06600
      IF (I.GT.MEV) GO TO 180                                      CSL06610
      IF (T2FLAG(I).EQ.1) GO TO 170                                CSL06620
      DGAP = EVMAG(I)-TEMP                                         CSL06630
      IF (DGAP.GT.MULTOL) GO TO 180                                CSL06640
      DGAP = CDABS(EVAL-VS(I))                                     CSL06650
```

```
         IF (DGAP.GT.MULTOL) GO TO 170                            CSL06660
C        T-MULTIPLICITY INCREASES                                 CSL06670
         INDEX = INDEX + 1                                        CSL06680
         CTEMP = CTEMP + VS(I)                                    CSL06690
         T2FLAG(I) = 1                                            CSL06700
         GO TO 170                                                CSL06710
C        T-MULTIPLICITY FOR VS(NDIS) HAS BEEN DETERMINED          CSL06720
     180 VS(NDIS) = CTEMP/DFLOAT(INDEX)                           CSL06730
         MP(NDIS) = INDEX                                         CSL06740
         GO TO 160                                                CSL06750
     190 CONTINUE                                                 CSL06760
C        T-MULTIPLICITY CALCULATION IS COMPLETE                   CSL06770
C                                                                 CSL06780
C        T(2,MEV) EIGENVALUE CALCULATION AND SPURIOUS TESTS       CSL06790
C                                                                 CSL06800
         IF (SAVTEV.LT.0) GO TO 240                               CSL06810
C                                                                 CSL06820
         WRITE(6,200) MEV1                                        CSL06830
     200 FORMAT(/' COMPUTE T(2,',I4,') EIGENVALUES'/)             CSL06840
C                                                                 CSL06850
         DO 210 J = 1,MEV1                                        CSL06860
         JP1 = J+1                                                CSL06870
         V2(J) = ALPHA(JP1)                                       CSL06880
     210 V1(J) = BETA(JP1)                                        CSL06890
C                                                                 CSL06900
C----------------------------------------------------------------CSL06910
         CALL CMTQL1(MEV1,V2,V1,IERR)                             CSL06920
C----------------------------------------------------------------CSL06930
C                                                                 CSL06940
C        WRITE(6,220) IERR                                        CSL06950
     220 FORMAT(' T2-HAT EIGENVALUES VIA CMTQL1'/' IERR = ',I6/)  CSL06960
C                                                                 CSL06970
         IF (IERR.EQ.0) GO TO 240                                 CSL06980
C                                                                 CSL06990
         WRITE(6,230)                                             CSL07000
     230 FORMAT(' ON RETURN FROM CMTQL1 ERROR FLAG WAS NOT ZERO'/)CSL07010
         GO TO 410                                                CSL07020
C                                                                 CSL07030
     240 CONTINUE                                                 CSL07040
C                                                                 CSL07050
         DO 250 J = 1,MEV1                                        CSL07060
     250 EVMAG(J) = CDABS(V2(J))                                  CSL07070
C                                                                 CSL07080
C        WRITE(13,255) (EVMAG(J), J = 1,MEV)                      CSL07090
C 255 FORMAT(/' MAGNITUDES OF T2 EIGENVALUES'/(4E20.12))          CSL07100
C                                                                 CSL07110
         IF(SAVTEV.NE.1) GO TO 270                                CSL07120
         WRITE(10,260) MEV1                                       CSL07130
     260 FORMAT(/I6,' = ORDER OF T2-HAT, T2EIGVALS = ')           CSL07140
         WRITE(10,120) (V2(J), J = 1,MEV1)                        CSL07150
     270 CONTINUE                                                 CSL07160
C                                                                 CSL07170
C        SPURIOUS TESTS                                           CSL07180
         DO 280 I = 1,MEV1                                        CSL07190
     280 T2FLAG(I) = 0                                            CSL07200
C                                                                 CSL07210
C        GO THROUGH THE EIGENVALUES OF T2-HAT.  FIND THE CLOSEST EIGENVALUECSL07220
C        OF T(1,MEV).  IF IT IS T-MULTIPLE GO ON.  IF IT IS SIMPLE DECLARE CSL07230
C        SPURIOUS WHENEVER DELMIN < SPUTOL BY SETTING MP(I) = 0  CSL07240
         J = 0                                                    CSL07250
     290 J = J+1                                                  CSL07260
         IF (J.GT.MEV1) GO TO 390                                 CSL07270
C                                                                 CSL07280
C        WRITE(14,300) J,V2(J)                                    CSL07290
     300 FORMAT('EIGENVALUE T2-HAT =', I6,2E22.14)                CSL07300
C                                                                 CSL07310
         TEMP = EVMAG(J)                                          CSL07320
```

```
      EVAL = V2(J)                                              CSL07330
      EVALR = TEMP + SPUTOL                                     CSL07340
      EVALC = TEMP - SPUTOL                                     CSL07350
      DELMIN = 2.D0*CDABS(VS(MEV))                              CSL07360
      IMIN = 0                                                  CSL07370
C     BACKWARD SEARCH                                           CSL07380
      I = J + 1                                                 CSL07390
  310 I = I - 1                                                 CSL07400
      IF(I.LT.1) GO TO 320                                      CSL07410
      IF(I.GT.NDIS) I = NDIS                                    CSL07420
C                                                               CSL07430
      TEMP = CDABS(VS(I))                                       CSL07440
      IF (TEMP.LT.EVALC) GO TO 320                              CSL07450
      IF(MP(I).EQ.0) GO TO 310                                  CSL07460
      DGAP = CDABS(VS(I) - EVAL)                                CSL07470
      IF (DGAP.GE.DELMIN) GO TO 310                             CSL07480
      DELMIN = DGAP                                             CSL07490
      IMIN = I                                                  CSL07500
C                                                               CSL07510
      GO TO 310                                                 CSL07520
C     FORWARD SEARCH                                            CSL07530
  320 I = J                                                     CSL07540
  330 I = I + 1                                                 CSL07550
      IF(I.GT.NDIS) GO TO 340                                   CSL07560
C                                                               CSL07570
      TEMP = CDABS(VS(I))                                       CSL07580
      IF (TEMP.GT.EVALR) GO TO 340                              CSL07590
      IF(MP(I).EQ.0) GO TO 330                                  CSL07600
      DGAP = CDABS(VS(I) - EVAL)                                CSL07610
      IF (DGAP.GE.DELMIN) GO TO 330                             CSL07620
      DELMIN = DGAP                                             CSL07630
      IMIN = I                                                  CSL07640
C                                                               CSL07650
      GO TO 330                                                 CSL07660
C                                                               CSL07670
  340 CONTINUE                                                  CSL07680
      IF(IMIN.EQ.0) GO TO 370                                   CSL07690
C     WRITE(14,350) IMIN, MP(IMIN),VS(IMIN),DELMIN,J            CSL07700
  350 FORMAT(/I6,' TH EVALUE, MP =',I3,' EVALUE =',2E22.13/     CSL07710
     1' MINDEL = ',E14.3,' OCCURS FOR',I6,' TH T2-HAT EVALUE')  CSL07720
      IF(DELMIN.GT.SPUTOL) GO TO 290                            CSL07730
      IF(MP(IMIN).GT.1)  GO TO 290                              CSL07740
      MP(IMIN) = 0                                              CSL07750
C     WRITE(14,360)                                             CSL07760
  360 FORMAT(' ABOVE T-EIGENVALUE IS SPURIOUS')                 CSL07770
      GO TO 290                                                 CSL07780
  370 CONTINUE                                                  CSL07790
      GO TO 290                                                 CSL07800
  390 CONTINUE                                                  CSL07810
C     END OF SPURIOUS TESTS                                     CSL07820
C                                                               CSL07830
      DO 400 J = 1,NDIS                                         CSL07840
  400 EVMAG(J) = CDABS(VS(J))                                   CSL07850
C                                                               CSL07860
      RETURN                                                    CSL07870
C-----END OF COMPEV----------------------------------------------CSL07880
  410 STOP                                                      CSL07890
      END                                                       CSL07900
C-----CMTQL1 (EIGENVALUES OF COMPLEX SYMMETRIC TRIDIAGONAL)------CSL07910
C                                                               CSL07920
      SUBROUTINE CMTQL1(N,D,E,IERR)                             CSL07930
C                                                               CSL07940
C---------------------------------------------------------------CSL07950
      INTEGER I,J,L,M,N,II,MML,IERR                             CSL07960
      COMPLEX*16 D(1),E(1),B,C,F,G,P,R,S,W,CZERO,CONE           CSL07970
      COMPLEX*16 CDSQRT,DCMPLX                                  CSL07980
      DOUBLE PRECISION MACHEP,EPS,TEMP,T0,T1,ZERO,HALF,ONE,TWO  CSL07990
```

```
      DOUBLE PRECISION CDABS,DSQRT                                   CSL08000
C----------------------------------------------------------------------CSL08010
      DATA MACHEP/Z3410000000000000/                                CSL08020
      EPS = 100.D0*MACHEP                                            CSL08030
C----------------------------------------------------------------------CSL08040
      ZERO = 0.0D0                                                   CSL08050
      HALF = 0.5D0                                                   CSL08060
      ONE  = 1.0D0                                                   CSL08070
      TWO  = 2.0D0                                                   CSL08080
      CZERO = DCMPLX(ZERO,ZERO)                                      CSL08090
      CONE  = DCMPLX(ONE,ZERO)                                       CSL08100
      IERR = 0                                                       CSL08110
      IF (N.EQ.1) GO TO 160                                          CSL08120
C                                                                    CSL08130
      DO 10 I = 2,N                                                  CSL08140
   10 E(I-1) = E(I)                                                  CSL08150
      E(N) = CZERO                                                   CSL08160
C                                                                    CSL08170
      DO 140 L = 1,N                                                 CSL08180
      J = 0                                                          CSL08190
C                                                                    CSL08200
C     DETERMINE FIRST NEGLIGIBLE SUBDIAGONAL ELEMENT IF ANY          CSL08210
   20 DO 30 M = L,N                                                  CSL08220
      IF (M.EQ.N) GO TO 40                                           CSL08230
      TEMP = CDABS(D(M)) + CDABS(D(M+1))                             CSL08240
      IF (CDABS(E(M)).LE.TEMP*MACHEP) GO TO 40                       CSL08250
   30 CONTINUE                                                       CSL08260
C                                                                    CSL08270
   40 P = D(L)                                                       CSL08280
C                                                                    CSL08290
      IF (M.EQ.L) GO TO 100                                          CSL08300
      IF (J.EQ.100) GO TO 150                                        CSL08310
      J = J+1                                                        CSL08320
C                                                                    CSL08330
C     FORM SHIFT AS EIGENVALUE OF (L,L+1) 2X2 CLOSEST TO D(L)        CSL08340
      G = (D(L+1) - P)*HALF                                          CSL08350
      T0 = CDABS(G)                                                  CSL08360
      T1 = CDABS(E(L))                                               CSL08370
      IF (T0.GT.T1) GO TO 50                                         CSL08380
      W = G/E(L)                                                     CSL08390
      R = CDSQRT(CONE + W**2)                                        CSL08400
      T0 = CDABS(W + R)                                              CSL08410
      T1 = CDABS(W - R)                                              CSL08420
      TEMP = ONE                                                     CSL08430
      IF (T1.GT.T0) TEMP = -ONE                                      CSL08440
      G = D(M) - P + E(L)/(W + TEMP*R)                               CSL08450
      GO TO 60                                                       CSL08460
   50 CONTINUE                                                       CSL08470
      W = E(L)/G                                                     CSL08480
      R = CDSQRT(CONE + W**2)                                        CSL08490
      T0 = CDABS(CONE + R)                                           CSL08500
      T1 = CDABS(CONE - R)                                           CSL08510
      TEMP = ONE                                                     CSL08520
      IF (T1.GT.T0) TEMP = -ONE                                      CSL08530
      G = D(M) - P + W*E(L)/(CONE + TEMP*R)                          CSL08540
   60 CONTINUE                                                       CSL08550
C                                                                    CSL08560
C     G IS SHIFTED D(M)                                              CSL08570
C     SPECIFY PARAMETERS FOR I = M-1 CASE, I = M-1,M-2,...,L         CSL08580
C                                                                    CSL08590
      S = CONE                                                       CSL08600
      C = -CONE                                                      CSL08610
      P = CZERO                                                      CSL08620
      MML = M - L                                                    CSL08630
C                                                                    CSL08640
      DO 90 II = 1,MML                                               CSL08650
```

```
      I = M - II                                               CSL08660
C                                                              CSL08670
C     FOR I<M-1 F=T(I+2,I), B=NEW E(I), AIM OF (I,I+1) TRANSFORMATION  CSL08680
C     IS TO ZERO OUT F                                        CSL08690
C                                                              CSL08700
      F =  S*E(I)                                              CSL08710
      B = -C*E(I)                                              CSL08720
      TO = CDABS(G)                                            CSL08730
      T1 = CDABS(F)                                            CSL08740
      IF (T1.GT.TO) GO TO 70                                   CSL08750
C     |G| >= |F|                                               CSL08760
      W = F/G                                                  CSL08770
      R = CDSQRT(CONE + W**2)                                  CSL08780
      E(I+1) = G*R                                             CSL08790
      C = CONE/R                                               CSL08800
      S = W*C                                                  CSL08810
      GO TO 80                                                 CSL08820
C     |F| > |G|                                                CSL08830
   70 CONTINUE                                                 CSL08840
      W = G/F                                                  CSL08850
      R = CDSQRT(CONE + W**2)                                  CSL08860
      E(I+1) = F*R                                             CSL08870
      S = CONE/R                                               CSL08880
      C = W*S                                                  CSL08890
   80 CONTINUE                                                 CSL08900
      TEMP = CDABS(W)**2 + ONE                                 CSL08910
      TO = DSQRT(TEMP)                                         CSL08920
      T1 = CDABS(R)                                            CSL08930
      IERR = -L                                                CSL08940
      IF (T1.LE.EPS*TO) GO TO 160                              CSL08950
      IERR = 0                                                 CSL08960
C                                                              CSL08970
C     C**2 + S**2 = CONE, -Q(I,I) = Q(I+1,I+1) = C, Q(I,I+1) = S  CSL08980
C     Q = Q-TRANSPOSE = Q-INVERSE   RR = CDSQRT(G**2 +F**2)   CSL08990
C     G = D(I+1) AFTER PREVIOUS TRANSFORMATION THEN G = NEW E(I)  CSL09000
C     NEW D(I) = D(I) - S*RR,   NEW D(I+1) = D(I+1) + S*RR    CSL09010
C     NEW E(I) = E(I) - C*RR,   NEW E(I+1) = RR,   P = S*RR   CSL09020
C                                                              CSL09030
      G = D(I+1) - P                                           CSL09040
      R = (D(I) - G)*S + TWO*C*B                               CSL09050
      P = S*R                                                  CSl.09060
      D(I+1) = G + P                                           CSL09070
      G = B - C*R                                              CSL09080
   90 CONTINUE                                                 CSL09090
C     END OF I LOOP                                            CSL09100
C                                                              CSL09110
C     UPDATE PARAMETERS FOR I = L CASE                         CSL09120
      D(L) = D(L) - P                                          CSL09130
      E(L) = G                                                 CSL09140
      E(M) = CZERO                                             CSL09150
      GO TO 20                                                 CSL09160
C                                                              CSL09170
C     ORDER EIGENVALUES   P = D(L)                             CSL09180
  100 IF (L.EQ.1) GO TO 120                                    CSL09190
      DO 110 II = 2,L                                          CSL09200
      I = L+2-II                                               CSL09210
      IF (CDABS(P).GE.CDABS(D(I-1))) GO TO 130                 CSL09220
      D(I) = D(I-1)                                            CSL09230
  110 CONTINUE                                                 CSL09240
C                                                              CSL09250
  120 I = 1                                                    CSL09260
C                                                              CSL09270
  130 D(I) = P                                                 CSL09280
C                                                              CSL09290
  140 CONTINUE                                                 CSL09300
      GO TO 160                                                CSL09310
C                                                              CSL09320
```

```
      150 IERR = L                                                     CSL09330
C-----END OF CMTQL1---------------------------------------------------CSL09340
      160 RETURN                                                       CSL09350
          END                                                          CSL09360
C                                                                      CSL09370
C-----COMGAP---------------------------------------------------------- CSL09380
C                                                                      CSL09390
          SUBROUTINE COMGAP(VC,VA,GG,MP,IND,M,IGAP,ITAG)               CSL09400
C                                                                      CSL09410
C------------------------------------------------------------------- CSL09420
          COMPLEX*16 VC(1),Z                                           CSL09430
          DOUBLE PRECISION  VA(1),TO,T1,TU,TK                          CSL09440
          REAL  GG(1),GTEMP                                            CSL09450
          INTEGER  MP(1),IND(1)                                        CSL09460
          REAL ABS                                                     CSL09470
          DOUBLE PRECISION CDABS                                       CSL09480
C--------------------------------------------------------------------- CSL09490
C     IF IGAP = 0 WE DO NOT ORDER EIGENVALUES BY INCREASING GAP SIZE   CSL09500
C     AND WE DO NOT WRITE GAP OUTPUT TO FILE 12                        CSL09510
C                                                                      CSL09520
C     VA(K) = |VC(K)|  VA(K) <= VA(K+1)                                CSL09530
C     GG(K) = MIN |VC(K)-VC(J)|  J .NE. K.                             CSL09540
C--------------------------------------------------------------------- CSL09550
          TU = VA(M) + VA(M)                                           CSL09560
          K = 0                                                        CSL09570
       10 K = K+1                                                      CSL09580
          IF (K.GT.M) GO TO 60                                         CSL09590
          INDEX = 0                                                    CSL09600
          T1 = TU                                                      CSL09610
          TK = VA(K)                                                   CSL09620
          Z = VC(K)                                                    CSL09630
          J = K                                                        CSL09640
C     BACKWARDS                                                        CSL09650
       20 J = J-1                                                      CSL09660
          IF (J.LT.1) GO TO 30                                         CSL09670
          TO = TK - VA(J)                                              CSL09680
          IF (TO.GT.T1) GO TO 30                                       CSL09690
          TO = CDABS(Z - VC(J))                                        CSL09700
          IF (T1.LE.TO) GO TO 20                                       CSL09710
          T1 = TO                                                      CSL09720
          INDEX = J                                                    CSL09730
          GO TO 20                                                     CSL09740
C     FORWARDS                                                         CSL09750
       30 J = K                                                        CSL09760
       40 J = J+1                                                      CSL09770
          IF (J.GT.M) GO TO 50                                         CSL09780
          TO = VA(J) - TK                                              CSL09790
          IF (TO.GT.T1) GO TO 50                                       CSL09800
          TO = CDABS(Z - VC(J))                                        CSL09810
          IF (T1.LE.TO) GO TO 40                                       CSL09820
          T1 = TO                                                      CSL09830
          INDEX = J                                                    CSL09840
          GO TO 40                                                     CSL09850
       50 IND(K) = INDEX                                               CSL09860
          GG(K) = T1                                                   CSL09870
          IF(ITAG.EQ.0) GO TO 10                                       CSL09880
          IF(MP(INDEX).EQ.0)  GG(K) = -GG(K)                           CSL09890
          GO TO 10                                                     CSL09900
C                                                                      CSL09910
       60 CONTINUE                                                     CSL09920
          IF (IGAP.EQ.0) GO TO 140                                     CSL09930
C                                                                      CSL09940
C     WRITE(12,70)                                                     CSL09950
       70 FORMAT(' MINGAPS FOR GOOD T-EIGENVALUES'/                    CSL09960
         1 1X,'EVNUM',1X,'NEIGH',15X,'R(EV)',15X,'I(EV)',4X,'MINGAP')  CSL09970
C     WRITE(12,80) (K,IND(K),VC(K),GG(K), K = 1,M)                     CSL09980
```

```
   80 FORMAT(2I6,2E20.12,E10.3)                                         CSL09990
C                                                                       CSL10000
C     ORDER VC G BY INCREASING MINGAP SIZE                              CSL10010
      DO 90 J = 1,M                                                     CSL10020
      IND(J) = J                                                        CSL10030
   90 CONTINUE                                                          CSL10040
C                                                                       CSL10050
      DO 110 K = 2,M                                                    CSL10060
      KM1 = K-1                                                         CSL10070
      DO 100 L = 1,KM1                                                  CSL10080
      KK = K-L                                                          CSL10090
      KP1 = KK+1                                                        CSL10100
      IF (ABS(GG(KP1)).GE.ABS(GG(KK))) GO TO 110                        CSL10110
      Z = VC(KK)                                                        CSL10120
      VC(KK) = VC(KP1)                                                  CSL10130
      VC(KP1) = Z                                                       CSL10140
      GTEMP = GG(KK)                                                    CSL10150
      GG(KK) = GG(KP1)                                                  CSL10160
      GG(KP1) = GTEMP                                                   CSL10170
      ITEMP = IND(KK)                                                   CSL10180
      IND(KK) = IND(KP1)                                                CSL10190
      IND(KP1) = ITEMP                                                  CSL10200
  100 CONTINUE                                                          CSL10210
  110 CONTINUE                                                          CSL10220
C                                                                       CSL10230
C     WRITE(12,120)                                                     CSL10240
  120 FORMAT(' T-EIGENVALUES ORDERED BY INCREASING MINGAP'/             CSL10250
     1 1X,'GAPNUM',1X,'EVNUM',15X,'R(EV)',15X,'I(EV)',4X,'MINGAP')      CSL10260
C                                                                       CSL10270
C     WRITE(12,130) (K,IND(K),VC(K),GG(K), K = 1,M)                     CSL10280
  130 FORMAT(I7,I6,2E20.12,E10.3)                                       CSL10290
C                                                                       CSL10300
  140 CONTINUE                                                          CSL10310
C-----END OF COMGAP-----------------------------------------------------CSL10320
      RETURN                                                            CSL10330
      END                                                               CSL10340
C                                                                       CSL10350
C-----START OF INVERM FOR TRIDIAGONAL COMPLEX SYMMETRIC MATRICES--------CSL10360
C                                                                       CSL10370
      SUBROUTINE INVERM(ALPHA,BETA,V1,V2,X1,ERROR,ERRORV,EPS,GR,GC,      CSL10380
     1INTERC,MEV,IT,IWRITE)                                             CSL10390
C                                                                       CSL10400
C----------------------------------------------------------------------CSL10410
      COMPLEX*16   ALPHA(1),BETA(1),V1(1),V2(1)                         CSL10420
      COMPLEX*16   X1,U,Z,TEMP,RATIO,BETAM,ZEROC                        CSL10430
      DOUBLE PRECISION   SUM,XU,NORM,TSUM,GSUM                          CSL10440
      DOUBLE PRECISION   EPS,EPS3,EPS4,ERROR,ERRORV,ZERO,ONE            CSL10450
      DOUBLE PRECISION GR(1),GC(1)                                      CSL10460
      INTEGER INTERC(1)                                                 CSL10470
      DOUBLE PRECISION DABS, DSQRT, DFLOAT, CDABS                       CSL10480
C     COMPLEX*16 DCMPLX                                                 CSL10490
C----------------------------------------------------------------------CSL10500
C                                                                       CSL10510
C     COMPUTES T-EIGENVECTORS FOR ISOLATED GOOD T-EIGENVALUES X1        CSL10520
C     USING INVERSE ITERATION ON T(1,MEV(X1)) SOLVING EQUATION          CSL10530
C     (T - X1*I)V2 = RIGHT-HAND SIDE (RANDOMLY-GENERATED) .             CSL10540
C     PROGRAM REFACTORS T- X1*I ON EACH ITERATION OF INVERSE ITERATION. CSL10550
C     TYPICALLY ONLY ONE ITERATION IS NEEDED PER T-EIGENVALUE X1.       CSL10560
C                                                                       CSL10570
C     IF IWRITE = 1 THEN THERE ARE EXTENDED WRITES TO FILE 6 (TERMINAL) CSL10580
C                                                                       CSL10590
C     ON ENTRY G CONTAINS A REAL*4 RANDOM VECTOR WHICH WAS GENERATED    CSL10600
C     IN MAIN PROGRAM.                                                  CSL10610
C                                                                       CSL10620
C     ON ENTRY AND EXIT                                                 CSL10630
C     MEV = ORDER OF T                                                  CSL10640
C     ALPHA, BETA CONTAIN THE DIAGONAL AND OFFDIAGONAL ENTRIES OF T.    CSL10650
```

```
C     EPS = 2. * MACHINE EPSILON                                        CSL10660
C                                                                       CSL10670
C     IN PROGRAM:                                                       CSL10680
C     ITER = MAXIMUM NUMBER STEPS ALLOWED FOR INVERSE ITERATION         CSL10690
C     ITER = IT ON ENTRY.                                               CSL10700
C     V1,V2 = WORK SPACES USED IN THE FACTORIZATION OF T(1,MEV).        CSL10710
C     V1 AND V2 MUST BE OF DIMENSION AT LEAST MEV.                      CSL10720
C                                                                       CSL10730
C     ON EXIT                                                           CSL10740
C     V2 = THE UNIT EIGENVECTOR OF T(1,MEV) CORRESPONDING TO X1.        CSL10750
C     ERROR = |V2(MEV)| = ERROR ESTIMATE FOR CORRESPONDING             CSL10760
C              RITZ VECTOR FOR X1.                                      CSL10770
C                                                                       CSL10780
C     ERRORV = || T*V2 - X1*V2 || = ERROR ESTIMATE ON T-EIGENVECTOR.    CSL10790
C     IF IT.GT.ITER THEN ERRORV = -ERRORV                               CSL10800
C     IT = NUMBER OF ITERATIONS ACTUALLY REQUIRED                       CSL10810
C----------------------------------------------------------------------CSL10820
C     INITIALIZATION AND PARAMETER SPECIFICATION                        CSL10830
      ONE  = 1.0D0                                                      CSL10840
      ZERO = 0.0D0                                                      CSL10850
      ZEROC = DCMPLX(ZERO,ZERO)                                         CSL10860
      ITER = IT                                                         CSL10870
      MP1 = MEV+1                                                       CSL10880
      MM1 = MEV-1                                                       CSL10890
      BETAM = BETA(MP1)                                                 CSL10900
      BETA(MP1) = ZEROC                                                 CSL10910
C                                                                       CSL10920
C     CALCULATE SCALE AND TOLERANCES                                    CSL10930
      TSUM = CDABS(ALPHA(1))                                            CSL10940
      DO 10 I = 2,MEV                                                   CSL10950
   10 TSUM = TSUM + CDABS(ALPHA(I)) + CDABS(BETA(I))                    CSL10960
C                                                                       CSL10970
      EPS3 = EPS*TSUM                                                   CSL10980
      EPS4 = DFLOAT(MEV)*EPS3                                           CSL10990
C                                                                       CSL11000
C     GENERATE SCALED RANDOM RIGHT-HAND SIDE                            CSL11010
      GSUM = ZERO                                                       CSL11020
      DO 20 I = 1,MEV                                                   CSL11030
   20 GSUM = GSUM + DABS(GR(I)) + DABS(GC(I))                           CSL11040
      GSUM = EPS4/GSUM                                                  CSL11050
C                                                                       CSL11060
C     INITIALIZE RIGHT HAND SIDE FOR INVERSE ITERATION                  CSL11070
      DO 30 I = 1,MEV                                                   CSL11080
      INTERC(I) = 0                                                     CSL11090
   30 V2(I) = GSUM*DCMPLX(GR(I),GC(I))                                  CSL11100
      IT = 1                                                            CSL11110
C                                                                       CSL11120
C     CALCULATE UNIT EIGENVECTOR OF T(1,MEV) FOR ISOLATED GOOD          CSL11130
C     T-EIGENVALUE X1.                                                  CSL11140
C                                                                       CSL11150
C     TRIANGULAR FACTORIZATION WITH NEAREST NEIGHBOR PIVOT              CSL11160
C     STRATEGY. INTERCHANGES ARE LABELLED BY SETTING INTERC(I)=0        CSL11170
C                                                                       CSL11180
   40 CONTINUE                                                          CSL11190
      U = ALPHA(1)-X1                                                   CSL11200
      Z = BETA(2)                                                       CSL11210
C                                                                       CSL11220
      DO 60 I=2,MEV                                                     CSL11230
      IF (CDABS(BETA(I)).GT.CDABS(U)) GO TO 50                          CSL11240
C     NO PIVOT INTERCHANGE                                              CSL11250
      V1(I-1) = Z/U                                                     CSL11260
      V2(I-1) = V2(I-1)/U                                               CSL11270
      V2(I) = V2(I)-BETA(I)*V2(I-1)                                     CSL11280
      RATIO = BETA(I)/U                                                 CSL11290
      U = ALPHA(I)-X1-Z*RATIO                                           CSL11300
      Z = BETA(I+1)                                                     CSL11310
```

```
      GO TO 60                                                    CSL11320
C     PIVOT INTERCHANGE                                           CSL11330
   50 CONTINUE                                                    CSL11340
      RATIO = U/BETA(I)                                           CSL11350
      INTERC(I) = 1                                               CSL11360
      V1(I-1) = ALPHA(I)-X1                                       CSL11370
      U = Z-RATIO*V1(I-1)                                         CSL11380
      Z = -RATIO*BETA(I+1)                                        CSL11390
      TEMP = V2(I-1)                                              CSL11400
      V2(I-1) = V2(I)                                             CSL11410
      V2(I) = TEMP-RATIO*V2(I)                                    CSL11420
   60 CONTINUE                                                    CSL11430
C                                                                 CSL11440
      IF (CDABS(U).EQ.ZERO) U= DCMPLX(EPS3,EPS3)                  CSL11450
C                                                                 CSL11460
C     SMALLNESS TEST AND DEFAULT VALUE FOR LAST COMPONENT         CSL11470
C     PIVOT(I-1) = |BETA(I)| FOR INTERCHANGE CASE                 CSL11480
C     (I-1,I+1) ELEMENT IN RIGHT FACTOR = BETA(I+1)               CSL11490
C     END OF FACTORIZATION AND FORWARD SUBSTITUTION               CSL11500
C                                                                 CSL11510
C     BACK SUBSTITUTION                                           CSL11520
      V2(MEV) = V2(MEV)/U                                         CSL11530
      DO 80 II = 1,MM1                                            CSL11540
      I = MEV-II                                                  CSL11550
      IF (INTERC(I+1).EQ.1) GO TO 70                              CSL11560
C     NO PIVOT INTERCHANGE                                        CSL11570
      V2(I) = V2(I)-V1(I)*V2(I+1)                                 CSL11580
      GO TO 80                                                    CSL11590
C     PIVOT INTERCHANGE                                           CSL11600
   70 V2(I) = (V2(I)-V1(I)*V2(I+1)-BETA(I+2)*V2(I+2))/BETA(I+1)   CSL11610
   80 CONTINUE                                                    CSL11620
C                                                                 CSL11630
C                                                                 CSL11640
C     TESTS FOR CONVERGENCE OF INVERSE ITERATION                  CSL11650
C     IF SUM |V2| COMPS. LE. 1 AND IT. LE. ITER DO ANOTHER INVIT STEP CSL11660
C                                                                 CSL11670
      NORM = CDABS(V2(MEV))                                       CSL11680
      DO 90 II = 1,MM1                                            CSL11690
      I = MEV-II                                                  CSL11700
   90 NORM = NORM+CDABS(V2(I))                                    CSL11710
C                                                                 CSL11720
C     IS DESIRED GROWTH IN VECTOR ACHIEVED ?                      CSL11730
C     IF NOT, DO ANOTHER INVERSE ITERATION STEP UNLESS NUMBER ALLOWED ISCSL11740
C     EXCEEDED.                                                   CSL11750
      IF (NORM.GE.ONE) GO TO 110                                  CSL11760
C                                                                 CSL11770
      IT=IT+1                                                     CSL11780
      IF (IT.GT.ITER) GO TO 110                                   CSL11790
C                                                                 CSL11800
      XU = EPS4/NORM                                              CSL11810
      DO 100 I=1,MEV                                              CSL11820
      INTERC(I) = 0                                               CSL11830
  100 V2(I) = V2(I)*XU                                            CSL11840
C                                                                 CSL11850
      GO TO 40                                                    CSL11860
C                                                                 CSL11870
C     NORMALIZE COMPUTED T-EIGENVECTOR : V2 = V2/||V2||           CSL11880
C                                                                 CSL11890
  110 CONTINUE                                                    CSL11900
C                                                         .       CSL11910
C---------------------------------------------------------------------CSL11920
      CALL CINPRD(V2,V2,SUM,MEV)                                  CSL11930
C---------------------------------------------------------------------CSL11940
C                                                                 CSL11950
      SUM = ONE/DSQRT(SUM)                                        CSL11960
      DO 120 II = 1,MEV                                           CSL11970
```

```
    120 V2(II) = SUM*V2(II)                                             CSL11980
C                                                                       CSL11990
C      SAVE ERROR ESTIMATE FOR LATER OUTPUT                             CSL12000
       ERROR = CDABS(V2(MEV))                                           CSL12010
C                                                                       CSL12020
C      GENERATE ERRORV = ||T*V2 - X1*V2||.                             CSL12030
C      LOOP IS BOTTOM UP BECAUSE LAST COMPONENTS MAY BE VERY SMALL      CSL12040
       V1(MEV) = ALPHA(MEV)*V2(MEV)+BETA(MEV)*V2(MEV-1)-X1*V2(MEV)      CSL12050
       DO 130 J = 2,MM1                                                 CSL12060
       JM = MP1 - J                                                     CSL12070
       V1(JM) = ALPHA(JM)*V2(JM) + BETA(JM)*V2(JM-1) + BETA(JM+1CSL12080
      1) - X1*V2(JM)                                                    CSL12090
    130 CONTINUE                                                        CSL12100
C                                                                       CSL12110
       V1(1) = ALPHA(1)*V2(1) + BETA(2)*V2(2) - X1*V2(1)               CSL12120
C                                                                       CSL12130
C-----------------------------------------------------------------CSL12140
       CALL CINPRD(V1,V1,ERRORV,MEV)                                    CSL12150
C-----------------------------------------------------------------CSL12160
C                                                                       CSL12170
       ERRORV = DSQRT(ERRORV)                                           CSL12180
       IF (IT.GT.ITER) ERRORV = -ERRORV                                 CSL12190
       IF (IWRITE.EQ.0) GO TO 150                                       CSL12200
C                                                                       CSL12210
C      FILE 6 (TERMINAL) OUTPUT OF ERROR ESTIMATES.                     CSL12220
       WRITE(6,140) MEV,X1,ERROR,ERRORV                                 CSL12230
    140 FORMAT(1X,'TSIZE',10X,'RE(GOODEV)',10X,'IM(GOODEV)',11X,'U(M)', CSL12240
      1 9X,'TERROR'/16,2E20.12,2E15.5)                                  CSL12250
C                                                                       CSL12260
C      RESTORE BETA(MEV+1) = BETAM                                      CSL12270
    150 CONTINUE                                                        CSL12280
       BETA(MP1) = BETAM                                                CSL12290
C-----END OF INVERM-----------------------------------------------CSL12300
       RETURN                                                           CSL12310
       END                                                              CSL12320
C                                                                       CSL12330
C-----START OF INNER PRODUCT ROUTINE------------------------------CSL12340
C                                                                       CSL12350
C      COMPUTES EUCLIDEAN INNER PRODUCT OF 2 COMPLEX VECTORS            CSL12360
C      SUMC = (V2-TRANSPOSE)*V1                                         CSL12370
C                                                                       CSL12380
       SUBROUTINE INPRDC(V2,V1,SUMC,N)                                  CSL12390
C                                                                       CSL12400
C-----------------------------------------------------------------CSL12410
       DOUBLE PRECISION  ZERO                                           CSL12420
       COMPLEX*16 V2(1),V1(1),SUMC                                      CSL12430
C-----------------------------------------------------------------CSL12440
C                                                                       CSL12450
       ZERO = 0.D0                                                      CSL12460
       SUMC = DCMPLX(ZERO,ZERO)                                         CSL12470
       DO 10 J=1,N                                                      CSL12480
     10 SUMC = SUMC + V2(J)*V1(J)                                       CSL12490
C                                                                       CSL12500
       RETURN                                                           CSL12510
C-----END OF EUCLIDEAN INNER PRODUCT SUBROUTINE-------------------CSL12520
       END                                                              CSL12530
C                                                                       CSL12540
C-----START OF HERMITIAN INNER PRODUCT ROUTINE-------------------CSL12550
C      COMPLEX INNER PRODUCT                                            CSL12560
C                                                                       CSL12570
       SUBROUTINE CINPRD(V2,V1,SUM,N)                                   CSL12580
C-----------------------------------------------------------------CSL12590
       DOUBLE PRECISION  ZERO,SUM                                       CSL12600
       COMPLEX*16 V2(1),V1(1),SUMC                                      CSL12610
C-----------------------------------------------------------------CSL12620
C      COMPUTES THE INNER PRODUCT OF THE CONJUGATE OF V2 WITH V1.       CSL12630
       ZERO = 0.D0                                                      CSL12640
```

```
      SUMC = DCMPLX(ZERO,ZERO)                                      CSL12650
      DO 10 J=1,N                                                   CSL12660
   10 SUMC = SUMC + DCONJG(V2(J))*V1(J)                             CSL12670
      SUM = DREAL(SUMC)                                             CSL12680
C                                                                   CSL12690
      RETURN                                                        CSL12700
C-----END OF COMPLEX INNER PRODUCT SUBROUTINE-------------------------CSL12710
      END                                                           CSL12720
```

SECTION 7.7   OPTIONAL PREPROCESSING PROGRAM, FILE DEFINITIONS, AND SAMPLE INPUT FILES

```
C-----LCCOMPAC-(STAND-ALONE PROGRAM)-------------------------------------LCC00010
C                                                                        LCC00020
C     THIS PROGRAM TRANSLATES A SPARSE, COMPLEX SYMMETRIC, N X N          LCC00030
C     MATRIX GIVEN AS I, J, A(I,J), INTO THE SPARSE MATRIX FORMAT         LCC00040
C     REQUIRED BY THE SAMPLE USPEC AND CMATV PROGRAMS PROVIDED            LCC00050
C     FOR USE WITH THE COMPLEX SYMMETRIC LANCZOS EIGENVALUE/EIGENVECTOR   LCC00060
C     PROCEDURES.  PROGRAM ASSUMES THAT THE MATRIX ENTRIES ARE            LCC00070
C     PRESENTED TO THE PROGRAM COLUMN BY COLUMN OR ROW BY ROW.           LCC00080
C                                                                        LCC00090
C     PORTABILITY:  THIS PROGRAM IS NOT PORTABLE BECAUSE OF ITS          LCC00100
C                   USE OF COMPLEX*16 VARIABLES.                          LCC00110
C                                                                        LCC00120
C-----------------------------------------------------------------------LCC00130
      COMPLEX*16  A(15000), AD(2000), ZEROC                              LCC00140
      DOUBLE PRECISION ZERO                                              LCC00150
      INTEGER IROW(15000),ICOL(15000)                                    LCC00160
C-----------------------------------------------------------------------LCC00170
C     INPUT FILE 7 CONTAINS THE SPARSE COMPLEX SYMMETRIC NXN MATRIX      LCC00180
C     STORED AS:                                                         LCC00190
C                                                                        LCC00200
C                 NZ,M,N,MATNO                                           LCC00210
C                 I(K)  J(K)  A(K)  K = 1,NZ                             LCC00220
C                                                                        LCC00230
C     WHERE NZ IS THE TOTAL NUMBER OF NONZEROS IN THE MATRIX A,          LCC00240
C     N IS THE ROW AND COLUMN DIMENSION OF A,                           LCC00250
C     AND A(K) ARE THE NONZERO ENTRIES STORED ROW BY ROW OR             LCC00260
C     COLUMN BY COLUMN.  PROGRAM READS THIS IN AS IROW(K) = I(K),       LCC00270
C     ICOL(K) = J(K), AND A(K) = A(K).                                   LCC00280
C                                                                        LCC00290
C     OUTPUT FILE = 8 CONTAINS THE A-MATRIX IN SPARSE FORMAT            LCC00300
C                                                                        LCC00310
C                 NZS,N,NZL,MATNO                                        LCC00320
C                 ICOL(K)    K = 1,NZL                                   LCC00330
C                 IROW(K)    K = 1,NZS                                   LCC00340
C                 AD(K)      K = 1,N                                     LCC00350
C                 A(K)       K = 1,NZS                                   LCC00360
C                                                                        LCC00370
C     WHERE N IS THE ORDER OF THE INPUT MATRIX A,                       LCC00380
C     NZ EQUALS THE NUMBER OF NONZERO ELEMENTS IN A WHICH ARE ON        LCC00390
C     OR BELOW THE MAIN DIAGONAL.  NZL EQUALS THE NUMBER OF THE         LCC00400
C     LAST COLUMN HAVING NONZEROES BELOW THE DIAGONAL IN A.             LCC00410
C     NZS EQUALS THE NUMBER OF NONZERO ELEMENTS BELOW THE MAIN          LCC00420
C     DIAGONAL.  AD(K), K=1,N, CONTAINS THE DIAGONAL ELEMENTS OF A.     LCC00430
C     A(K), K=1,NZS, CONTAINS THE KTH NONZERO SUB-DIAGONAL ELEMENT      LCC00440
C     OF THE INPUT MATRIX.  A IS STORED COLUMN BY COLUMN.              LCC00450
C     IROW(K), K=1,NZS, CONTAINS THE ROW INDEX OF THE  NONZERO         LCC00460
C     STRICTLY LOWER TRIANGULAR ELEMENT A(K).                          LCC00470
C     ICOL(K), K=1,NZL, EQUALS THE NUMBER OF STRICTLY LOWER            LCC00480
C     TRIANGULAR NONZEROES IN COLUMN K OF THE INPUT MATRIX.            LCC00490
C                                                                        LCC00500
C-----------------------------------------------------------------------LCC00510
      ZERO = 0.D0                                                        LCC00520
      ZEROC = DCMPLX(ZERO,ZERO)                                          LCC00530
C                                                                        LCC00540
      READ(7,10) NZ,N,MATNO,IIROW                                        LCC00550
   10 FORMAT(2I6,I8,I6)                                                  LCC00560
C                                                                        LCC00570
      WRITE(6,20) NZ,N,MATNO,IIROW                                       LCC00580
   20 FORMAT(I10,I6,I10,' = NO. NONZERO AIJ J.GE.I, ORDER OF A, MATNO'/  LCC00590
     1 I6,' = IIROW IF IIROW=0 ORDERING IS BY COLS IIROW=1 BY ROWS'/)    LCC00600
C                                                                        LCC00610
      DO 30 K = 1,N                                                      LCC00620
```

```
      30 AD(K) = ZEROC                                              LCC00630
C                                                                   LCC00640
         IF (IIROW.EQ.0) READ(7,40) (IROW(K),ICOL(K),A(K), K=1,NZ)  LCC00650
C                                                                   LCC00660
         IF (IIROW.EQ.1) READ(7,40) (ICOL(K),IROW(K),A(K), K=1,NZ)  LCC00670
      40 FORMAT(2I5,2E14.7)                                         LCC00680
C                                                                   LCC00690
         LCOUNT = 0                                                 LCC00700
         K = 1                                                      LCC00710
C                                                                   LCC00720
C        START OF A NEW COLUMN                                      LCC00730
      50 CONTINUE                                                   LCC00740
         J = ICOL(K)                                                LCC00750
         ICOL(J) = 0                                                LCC00760
      60 CONTINUE                                                   LCC00770
C                                                                   LCC00780
         IF (J.NE.IROW(K)) GO TO 70                                 LCC00790
C                                                                   LCC00800
C        DIAGONAL CASE                                             LCC00810
         AD(J) = A(K)                                               LCC00820
         GO TO 80                                                   LCC00830
C                                                                   LCC00840
C        SUB-DIAGONAL NONZERO                                      LCC00850
      70 CONTINUE                                                   LCC00860
         NZL = J                                                    LCC00870
         LCOUNT = LCOUNT + 1                                        LCC00880
         A(LCOUNT) = A(K)                                           LCC00890
         IROW(LCOUNT) = IROW(K)                                     LCC00900
         ICOL(J) = ICOL(J) + 1                                      LCC00910
C                                                                   LCC00920
      80 CONTINUE                                                   LCC00930
         K = K+1                                                    LCC00940
C                                                                   LCC00950
         IF(K.GT.NZ) GO TO 90                                       LCC00960
C                                                                   LCC00970
         IF(ICOL(K).GT.J) GO TO 50                                  LCC00980
C                                                                   LCC00990
         GO TO 60                                                   LCC01000
C                                                                   LCC01010
      90 CONTINUE                                                   LCC01020
         NZS = LCOUNT                                               LCC01030
C                                                                   LCC01040
         WRITE(8,100) NZS,N,NZL,MATNO                               LCC01050
         WRITE(6,100) NZS,N,NZL,MATNO                               LCC01060
     100 FORMAT(I10,2I6,I8,' = NZS N NZL MATNO')                    LCC01070
C                                                                   LCC01080
         WRITE(8,110) (ICOL(I), I=1,NZL)                            LCC01090
         WRITE(8,110) (IROW(K), K=1,NZS)                            LCC01100
     110 FORMAT(13I6)                                               LCC01110
C                                                                   LCC01120
         WRITE(8,120) (AD(K), K=1,N)                                LCC01130
         WRITE(8,120) (A(K), K=1,NZS)                               LCC01140
     120 FORMAT(4E19.10)                                            LCC01150
C                                                                   LCC01160
C-----END LCCOMPAC-------------------------------------------------LCC01170
         STOP                                                       LCC01180
         END                                                        LCC01190
```

-----INPUT/OUTPUT FILE DEFINITIONS FOR CSLEVAL AND CSLEVEC-----

Below is a listing of the input/output files which are accessed by the complex symmetric Lanczos eigenvalue program, CSLEVAL. Included also is a sample of the input file which CSLEVAL requires on file 5. The parameters in this file are supplied in free format. CSLEVAL computes eigenvalues of diagonalizable complex symmetric matrices.

SAMPLE DEFINITIONS OF THE INPUT/OUTPUT FILES FOR CSLEVAL

```
*CSLEVAL EXEC LANCZOS EIGENVALUE CALCULATION COMPLEX SYMMETRIC CASE
FI 06 TERM
FILEDEF  1 DISK &1       NHISTORY  A (RECFM F LRECL 80 BLOCK 80
FILEDEF  2 DISK &1       HISTORY   A (RECFM F LRECL 80 BLOCK 80
FILEDEF  3 DISK &1       GOODEV    A (RECFM F LRECL 80 BLOCK 80
FILEDEF  4 DISK &1       ERRINV    A (RECFM F LRECL 80 BLOCK 80
FILEDEF  5 DISK CSLEVAL  INPUT     A (RECFM F LRECL 80 BLOCK 80
FILEDEF  8 DISK &1       INPUT     A (RECFM F LRECL 80 BLOCK 80
FILEDEF 10 DISK &1       T-T2EVAL  A (RECFM F LRECL 80 BLOCK 80
FILEDEF 11 DISK &1       DISTINCT  A (RECFM F LRECL 80 BLOCK 80
LOAD   CSLEVAL   CSLESUB   CSLEMULT
```

SAMPLE INPUT FILE FOR CSLEVAL

```
 CSLEVAL INPUT LANCZOS EIGENVALUE COMPUTATION, NO REORTHOGONALIZATION
 OF A NONDEFECTIVE COMPLEX SYMMETRIC MATRIX.
 LINE 1    N      KMAX     NMEVS      MATNO
          528     2640       1       721830
 LINE 2   SVSEED     RHSEED      MXINIT
         49302312   5731029         5
 LINE 3   ISTART     ISTOP
            0          1
 LINE 4  IHIS  IDIST  SAVTEV  IWRITE (SAVE HIST.,DISTINCT EV,TEV,WRITE
          1      0      1       1
 LINE 5   RELTOL  (RELATIVE TOLERANCE IN 'COMBINING' GOODEV)
         .0000000001
 LINE 6   MB(1)   MB(2)   MB(3)   MB(4)    (ORDERS OF T(1,MEV) )
                   1056
C     NOTE THAT WHEN READING IN PREVIOUSLY COMPUTED EIGENVALUES
C     THE VALUE OF MB(1) MUST BE EQUAL TO THE SIZE AT WHICH
C     THOSE EIGENVALUES WERE COMPUTED AND KMAX MUST BE LISTED AS
C     LARGER THAN MB(1).
```

Below is a listing of the input/output files which are accessed by the complex symmetric Lanczos eigenvector program, CSLEVEC. Included also is a sample of the input file which CSLEVEC requires on file 5. The parameters in this file are supplied in free format. CSLEVEC computes eigenvectors for each of a user-specified subset of the eigenvalues computed by the companion program CSLEVAL.

SAMPLE DEFINITIONS OF THE INPUT/OUTPUT FILES FOR CSLEVEC

```
*CSLEVEC EXEC LANCZOS EIGENVECTOR PROGRAM COMPLEX SYMMETRIC CASE
FI 06 TERM
FILEDEF  2 DISK &1      HISTORY   A (RECFM F LRECL 80 BLOCK 80
FILEDEF  3 DISK &1      GOODEV    A (RECFM F LRECL 80 BLOCK 80
FILEDEF  4 DISK &1      ERRINV    A (RECFM F LRECL 80 BLOCK 80
FILEDEF  5 DISK CSLEVEC INPUT     A (RECFM F LRECL 80 BLOCK 80
FILEDEF  8 DISK &1      INPUT     A (RECFM F LRECL 80 BLOCK 80
FILEDEF  9 DISK &1      ERREST    A (RECFM F LRECL 80 BLOCK 80
FILEDEF 10 DISK &1      BOUNDS    A (RECFM F LRECL 80 BLOCK 80
FILEDEF 11 DISK &1      TEIGVECS  A (RECFM F LRECL 80 BLOCK 80
FILEDEF 12 DISK &1      RITZVECS  A (RECFM F LRECL 80 BLOCK 80
FILEDEF 13 DISK &1      PAIGE     A (RECFM F LRECL 80 BLOCK 80
LOAD  CSLEVEC  CSLESUB  CSLEMULT
```

SAMPLE INPUT FILE FOR CSLEVEC

```
 CSLEVEC EIGENVECTORS COMPLEX SYMMETRIC CASE NO REORTHOGONALIZATION
LINE 1  MDIMTV    MDIMRV  MBETA (MAX.DIMENSIONS,TVEC,RITVEC AND BETA
        10000     10000   2000
LINE 2     RELTOL
        .0000000001
LINE 3  MBOUND    NTVCON SVTVEC IREAD (FLAGS
           0         1      0      1
LINE 4  TVSTOP    LVCONT ERCONT  IWRITE (FLAGS
           0         1      1       1
LINE 5   RHSEED  (RANDOM GENERATOR SEED FOR STARTING VECTOR IN INVERM)
        45329517
LINE 6  MATNO       N
         100        100
```

CHAPTER 8

REAL SYMMETRIC MATRICES, 'BLOCK' LANCZOS CODE

SECTION 8.1   INTRODUCTION

The FORTRAN codes in this chapter address the question of using an iterative 'block' Lanczos procedure to compute a 'few' extreme eigenvalues and a basis for the corresponding invariant subspace of a given real symmetric matrix A.   An eigenvalue is extreme if it is one of the algebraically-smallest or the algebraically-largest eigenvalues.

For a given real symmetric matrix A, these codes compute the q algebraically-largest eigenvalues, λ_i, $1 \leq i \leq q$, of A and corresponding orthonormal real vectors $X_q \equiv \{x_1,...,x_q\}$ such that

$$AX_q = X_q A_q \quad \text{where} \quad A_q \equiv X_q^T A X_q. \tag{8.1.1}$$

Typically, $A_q = \Lambda_q$, a diagonal matrix whose nonzero entries are the eigenvalues λ_i.   The number q is small and specified by the user.

Real symmetric matrices are discussed in detail in Stewart [1973].   See Section 2.1 for a brief summary of the properties of real symmetric matrices which we use.   The Lanczos procedure included in this chapter is not a true block Lanczos procedure.   It is a hybrid Lanczos algorithm which combines ideas from the iterative block Lanczos procedures such as the one in Cullum and Donath [1974] and from the single-vector Lanczos procedure given in Chapter 2.

Several differences between the single-vector Lanczos codes in Chapter 2-5 and the iterative 'block' Lanczos codes should be stated explicitly.   The single-vector Lanczos codes do not have the capability of directly computing the A-multiplicities of the computed eigenvalues. The 'block' procedures however, will determine the true A-multiplicity of a given computed eigenvalue and compute a complete invariant subspace for such an eigenvalue, as long as the number of Lanczos vectors in the first block is large enough.   In order to determine A-multiplicities the single-vector codes have to do additional computation.   In some cases these multiplicities and a basis for the required eigenspace can be determined without too much additional computation.   This is true for example, whenever the desired eigenvalues replicate readily during the single-vector Lanczos computations.

The single-vector Lanczos procedures in Chapters 2-5 function in two stages.   First the eigenvalues of the matrix being considered are computed, and then a separate program is used to compute the corresponding desired eigenvectors.   The iterative 'block' Lanczos codes obtain approximations to the eigenvalues and to the eigenvectors simultaneously.   Both types of codes are restartable from pre-existing computations.   However, restarting has a different meaning for the two different types of codes.   In the single-vector codes, restarting means computing a larger Lanczos T-matrix, starting from a pre-existing smaller one. The eigenvalue and eigenvector computations are then repeated on the larger T-matrix. In the iterative block procedures, restarting means using the current approximations to the eigenvectors (or more correctly to a basis for the desired eigenspace), to initiate another iteration of the 'block' Lanczos procedure.

The single-vector Lanczos procedures in Chapters 2-7 are iterative only in the sense that one may consider several Lanczos T-matrices of different sizes before achieving the desired convergence. However, the 'block' procedure presented here is genuinely iterative. On each iteration a block version of the Lanczos recursion is used to generate a sequence of blocks of Lanczos vectors, simultaneously generating a 'small' real symmetric Lanczos T-matrix. The eigenvalues and eigenvectors of this small Lanczos matrix are computed and mapped into approximating eigenvectors for the given matrix using the Lanczos vectors. These approximate eigenvectors then become the starting block of Lanczos vectors for the next iteration of the block Lanczos procedure. This 'block' procedure is described in detail in Section 7.5 of Chapter 7 in Volume 1.

As we said earlier, the 'block' procedure included here is a hybrid of the single-vector and of the basic iterative block Lanczos procedures. This procedure is based upon a modification of the following basic block version of the Lanczos recursion

$$Q_{j+1}B_{j+1} = AQ_j - Q_jA_j - Q_{j-1}B_j^T \equiv P_j \qquad (8.1.2)$$

for $j=1,2,...,s$ where the coefficient matrices A_j and B_{j+1} are block analogs of the scalar coefficients in the single vector Lanczos recursion. In the standard block procedure,

$$A_j \equiv Q_j^T(AQ_j - Q_{j-1}B_j^T) \text{ and each } B_{j+1} \qquad (8.1.3)$$

is obtained by the Gram-Schmidt orthogonalization of the columns of P_j and $s<<n$, the order of the given A-matrix. Our single-vector Lanczos procedures do not use any reorthogonalization at any point in the computations. However, in our block procedures we require near-orthogonality of the Q-blocks. This orthogonality is maintained by incorporating reorthogonalization of the blocks generated within a given iteration, with respect to certain vectors in the first Lanczos block.

The sequence of 'blocks' generated on each iteration of this hybrid procedure has the property that the first Q-block, Q_1, contains at least as many vectors as the user is trying to compute; however, the second and succeeding blocks contain exactly one vector. The corresponding Lanczos T-matrices are not block tridiagonal. Each has a border of blocks occupying the first q rows and columns and is tridiagonal below these rows and columns.

The convergence of these procedures is monitored by the subroutine DIAGOM. Convergence requires reasonable gaps between the eigenvalues requested and the eigenvalues not being approximated by the block procedure. Typically, it is the ratio of these gaps to the spread, and the distribution of the A-eigenvalues over the A-spread which controls the rate of convergence. In particular, an iterative block Lanczos procedure may have difficulty with a matrix with evenly-distributed eigenvalues. Heuristics are incorporated which allow the number of vectors used in the first Lanczos block to vary. If the convergence stagnates the procedure will terminate to allow the user to intervene and reset the program parameters if desired.

BLEVAL, the main 'block' program for these real symmetric eigenelement computations, calls the subroutine LANCZS which on each iteration then calls the subroutine LANCII to generate a sequence of Q-blocks for that iteration. Subroutine LANCZS then calls the subroutine DIAGOM to diagonalize the Lanczos T-matrix generated on that iteration and to compute the updated approximations to the desired eigenspace. Convergence is checked and if it has not occurred, another iteration of the block Lanczos procedure is carried out.

In this 'block' procedure there is no identification or 'spurious' test for the eigenvalues of the Lanczos T-matrix. Since near-orthogonality of the Lanczos blocks is maintained, the q algebraically-largest eigenvalues of the T-matrices are approximations to the q algebraically-largest eigenvalues of the A-matrix being used in the recursions. This statement however, is not true for the other eigenvalues of these T-matrices because the orthogonality maintained is only with respect to the eigenspace which goes with the first q eigenvalues. The accuracy of the computed eigenvalues and eigenvectors is estimated on each iteration as part of the process of computing the second block of Lanczos vectors.

All computations are in double precision real arithmetic. The user must supply a subroutine USPEC which defines and initializes the A-matrix and a subroutine BMATV which computes Ax for any specified vector x. The small T-matrix eigenelement computations use two subroutines from the EISPACK Library [1976,1977], TRED2 and IMTQL2. If the q algebraically-smallest eigenvalues are required, then the user must supply the programs with a subroutine which computes -Ax rather that Ax. The user should refer to Chapter 7 in Volume 1 for more details on iterative block Lanczos procedures.

SECTION 8.2   DOCUMENTATION FOR PROGRAMS IN CHAPTERS 8 AND 9

```
C-----DOCUMENTATION FOR-----------------------------------------------------BLE00010
C     BLOCK LANCZOS EIGENVALUE/EIGENVECTOR PROGRAMS FOR              BLE00020
C     (1) REAL SYMMETRIC MATRICES                                   BLE00030
C     (2) FACTORED INVERSES OF REAL SYMMETRIC MATRICES              BLE00040
C                                                                   BLE00050
C     AUTHORS:   JANE CULLUM, W. E. DONATH AND RALPH A. WILLOUGHBY, BLE00060
C                IBM RESEARCH, YORKTOWN HEIGHTS, NY 10598.          BLE00070
C                PHONE: 914-945-2227                                BLE00080
C                                                                   BLE00090
C                                                                   BLE00100
C     REAL SYMMETRIC MATRICES:                                      BLE00110
C                                                                   BLE00120
C     GIVEN A REAL SYMMETRIC MATRIX A THE FILES BLEVAL, BLSUB AND   BLE00130
C     BLMULT CAN BE USED TO COMPUTE A FEW EXTREME EIGENVALUES       BLE00140
C     OF A, THAT IS THE ALGEBRAICALLY-LARGEST OR THE ALGEBRAICALLY- BLE00150
C     SMALLEST EIGENVALUES, AND A BASIS FOR THE CORRESPONDING       BLE00160
C     EIGENSPACE.                                                   BLE00170
C                                                                   BLE00180
C     FACTORED INVERSES OF REAL SYMMETRIC MATRICES:                 BLE00190
C                                                                   BLE00200
C     GIVEN A REAL SYMMETRIC MATRIX A, THE BLOCK PROCEDURE          BLE00210
C     CAN BE APPLIED TO AN ASSOCIATED B-MATRIX WHICH IS A           BLE00220
C     SCALED, SHIFTED AND PERMUTED VERSION OF A.  THAT IS,          BLE00230
C     B = SO*P*A*P' + SHIFT*I WHERE THE SCALE SO AND THE SHIFT      BLE00240
C     ARE CHOSEN BY THE USER TO PLACE THE DESIRED EIGENVALUES       BLE00250
C     AT THE EXTREME OF THE SPECTRUM OF B-INVERSE, AND THE          BLE00260
C     PERMUTATION P IS CHOSEN SO THAT THE SPARSITY OF THE A-MATRIX  BLE00270
C     IS PRESERVED IN THE SPARSITY OF THE FACTORIZATION OF B.       BLE00280
C     THE INVERSE BLOCK PROCEDURE REQUIRES A SUBROUTINE BLSOLV      BLE00290
C     THAT FOR A GIVEN VECTOR U, COMPUTES THE VECTOR V SUCH THAT    BLE00300
C     B*V = U, USING THE FACTORIZATION OF B.  THE SAMPLE BLSOLV     BLE00310
C     SUBROUTINE PROVIDED ASSUMES THAT THE B-MATRIX IS POSITIVE     BLE00320
C     DEFINITE AND THAT THE CHOLESKY FACTORS OF B ARE SUPPLIED      BLE00330
C     ON FILE 7.  HOWEVER, THE USER MAY REPLACE THIS SUBROUTINE     BLE00340
C     BY ONE THAT COMPUTES A MORE GENERAL FACTORIZATION             BLE00350
C     L*D*(L-TRANSPOSE) FOR AN INDEFINITE SYMMETRIC MATRIX.         BLE00360
C     THE BLOCK PROCEDURE USED IN THIS FASHION USES THE FILES       BLE00370
C     BLIEVAL, BLIMULT AND BLSUB.                                   BLE00380
C                                                                   BLE00390
C                                                                   BLE00400
C     ALGORITHM:                                                    BLE00410
C     THESE PROGRAMS USE A BLOCK FORM OF LANCZOS TRIDIAGONALIZATION BLE00420
C     WITH REORTHOGONALIZATION ONLY WITH RESPECT TO VECTORS         BLE00430
C     IN THE 1ST Q-BLOCK.  THE PROCEDURES ARE ITERATIVE, GENERATING BLE00440
C     ON EACH ITERATION A SMALL SYMMETRIC LANCZOS MATRIX, T.        BLE00450
C     THE EIGENVALUES AND EIGENVECTORS OF THE SMALL MATRIX ARE      BLE00460
C     COMPUTED USING SUBROUTINES FROM THE EISPACK LIBRARY.          BLE00470
C     THE RELEVANT SUBSET OF THE T-EIGENVECTORS IS THEN MAPPED      BLE00480
C     INTO THE LARGE N-SPACE CORRESPONDING TO THE MATRIX BEING      BLE00490
C     USED BY THE LANCZS SUBROUTINE, CONVERGENCE IS CHECKED,        BLE00500
C     AND IF CONVERGENCE OF THE DESIRED EIGENVALUES AND             BLE00510
C     EIGENVECTORS HAS NOT YET OCCURRED, THEN THE CURRENT           BLE00520
C     APPROXIMATIONS TO THE DESIRED EIGENSPACE ARE USED AS          BLE00530
C     STARTING VECTORS FOR THE NEXT ITERATION OF BLOCK LANCZOS.     BLE00540
C                                                                   BLE00550
C     USERS SHOULD NOTE THAT TYPICALLY IN THE BLOCK LANCZOS         BLE00560
C     PROCEDURES, IT IS THE RATIO OF THE GAPS TO THE SPREAD THAT    BLE00570
C     CONTROLS THE CONVERGENCE ALONG WITH HOW THE EIGENVALUES       BLE00580
C     ARE DISTRIBUTED OVER THAT SPREAD.  THE BIGGER THE GAPS        BLE00590
C     BETWEEN THE ONES BEING COMPUTED AND THE CLOSEST ONES NOT      BLE00600
C     BEING COMPUTED AND THE WEAKER THE SPREAD, THE FASTER THE      BLE00610
C     CONVERGENCE WILL BE.  WITHOUT DECENT GAPS THIS PROCEDURE      BLE00620
C     WILL NOT CONVERGE.  THE PROGRAMS CONTAIN CHECKS ON            BLE00630
C     THE ACTUAL RATE OF CONVERGENCE WHICH WILL CAUSE THE           BLE00640
```

```
C        PROCEDURE TO TERMINATE IF CONVERGENCE IS NOT OCCURRING          BLE00650
C        SUFFICIENTLY RAPIDLY.  THE USER MAY THEN CHANGE EITHER OR        BLE00660
C        BOTH THE MAXIMUM SIZE T-MATRIX ALLOWED AND THE NUMBER            BLE00670
C        OF VECTORS IN THE FIRST Q-BLOCK AND RERUN THE PROCEDURE          BLE00680
C        WITH THE CURRENT APPROXIMATION TO THE DESIRED EIGENSPACE         BLE00690
C        AS THE STARTING BLOCK OF VECTORS.                                BLE00700
C                                                                         BLE00710
C                                                                         BLE00720
C        THE IDEAS USED IN THESE PROGRAMS ARE DISCUSSED IN THE FOLLOWING  BLE00730
C        REFERENCES.                                                      BLE00740
C                                                                         BLE00750
C        1.  JANE CULLUM AND RALPH A. WILLOUGHBY, LANCZOS ALGORITHMS      BLE00760
C            FOR LARGE SYMMETRIC MATRICES,  PROGRESS IN                   BLE00770
C            SCIENTIFIC COMPUTING, EDITORS, G. GOLUB, H.O. KREISS,        BLE00780
C            S. ARBARBANEL, AND R. GLOWINSKI,  BIRKHAUSER BOSTON INC.,    BLE00790
C            CAMBRIDGE, MASSACHUSETTS, 1984.                              BLE00800
C                                                                         BLE00810
C        2.  JANE CULLUM AND W.E. DONATH, A BLOCK LANCZOS ALGORITHM       BLE00820
C            FOR COMPUTING THE Q ALGEBRAICALLY-LARGEST EIGENVALUES AND    BLE00830
C            A CORRESPONDING EIGENSPACE OF LARGE, SPARSE REAL SYMMETRIC   BLE00840
C            MATRICES, PROCEEDINGS OF THE 1974 IEEE CONFERENCE ON         BLE00850
C            DECISION AND CONTROL, PHOENIX, ARIZONA, PP.505-509, NOVEMBER BLE00860
C            1974.                                                        BLE00870
C                                                                         BLE00880
C        3.  JANE CULLUM, AN ACCELERATED 'BLOCK' LANCZOS ALGORITHM        BLE00890
C            FOR A FEW EXTREME EIGENVALUES OF A LARGE, SPARSE REAL        BLE00900
C            SYMMETRIC MATRIX.  IBM REPORT 1983.  PRESENTED AT THE        BLE00910
C            SPARSE MATRIX CONFERENCE, FAIRFIELD GLADE, TENNESSEE,        BLE00920
C            OCTOBER 1982.                                                BLE00930
C                                                                         BLE00940
C                                                                         BLE00950
C-----PORTABILITY---------------------------------------------------------BLE00960
C                                                                         BLE00970
C        PROGRAMS WERE TESTED FOR PORTABILITY USING THE PFORT VERIFIER.   BLE00980
C        FOR DETAILS OF THE VERIFIER SEE FOR EXAMPLE, B. G. RYDER AND     BLE00990
C        A. D. HALL, "THE PFORT VERIFIER", COMPUTING SCIENCE TECHNICAL    BLE01000
C        REPORT 12, BELL LABORATORIES, MURRAY HILL, NEW JERSEY 07974,     BLE01010
C        (REVISED), JANUARY 1981.                                         BLE01020
C                                                                         BLE01030
C        EXCEPT FOR THE FOLLOWING CONSTRUCTIONS WHICH CAN BE EASILY       BLE01040
C        MODIFIED BY THE USER TO MATCH THE PARTICULAR COMPUTER BEING      BLE01050
C        USED, THE PROGRAM STATEMENTS ARE PORTABLE.                       BLE01060
C                                                                         BLE01070
C        NONPORTABLE STATEMENTS.                                          BLE01080
C                                                                         BLE01090
C        IN BLEVAL, BLIEVAL (MAIN PROGRAMS)                               BLE01100
C            1.  DATA/MACHEP STATEMENT                                    BLE01110
C            2.  ALL READ(5,*) STATEMENTS (FREE FORMAT)                   BLE01120
C            3.  FORMAT(20A4) USED FOR THE EXPLANATORY HEADER ARRAY, EXPLANBLE01130
C            4.  FORMAT(4Z20) WHICH CAN BE USED TO WRITE LARGE VECTOR     BLE01140
C                FILES                                                    BLE01150
C            5.  THE COMMON BLOCK: LOOPS.                                 BLE01160
C        IN BLMULT, BLIMULT                                               BLE01170
C            1.  IN BMATV, BLSOLV, AND USPEC, THE ENTRIES WHICH           BLE01180
C                PASS THE STORAGE LOCATIONS OF THE ARRAYS DEFINING        BLE01190
C                THE USER-SPECIFIED MATRIX OR FACTORIZATION.              BLE01200
C        IN BLSUB                                                         BLE01210
C            1.  ALL STATEMENTS ARE PORTABLE EXCEPT THE ENTRY TO          BLE01220
C                SUBROUTINE LPERM WHICH PASSES THE PERMUTATION USED       BLE01230
C                TO OBTAIN THE B-MATRIX FROM SUBROUTINE USPEC.            BLE01240
C                SUBROUTINE LPERM IS USED ONLY IN CASE (2).               BLE01250
C                                                                         BLE01260
C                                                                         BLE01270
C-----MATRIX SPECIFICATION------------------------------------------------BLE01280
C                                                                         BLE01290
C        SUBROUTINE USPEC IS USED TO SPECIFY THE MATRIX WHICH THE BLOCK   BLE01300
C        LANCZOS PROCEDURE WILL USE.  IN CASE (1) THIS IS THE USER-       BLE01310
```

```
C       SPECIFIED A-MATRIX.   IN CASE (2) THE FACTORIZATION OF THE        BLE01320
C       ASSOCIATED B-MATRIX IS SPECIFIED.  SUBROUTINE USPEC HAS THE       BLE01330
C       CALLING SEQUENCE                                                  BLE01340
C                                                                         BLE01350
C            CALL USPEC(N,MATNO,NNZ,AVER)                                 BLE01360
C                                                                         BLE01370
C       WHERE N IS THE ORDER OF THE USER-SUPPLIED MATRIX A,               BLE01380
C       MATNO IS AN <= 8 DIGIT INTEGER USED AS A MATRIX AND               BLE01390
C       TEST IDENTIFICATION NUMBER, NNZ IS THE AVERAGE NUMBER             BLE01400
C       OF NONZERO ENTRIES IN EACH COLUMN, AND AVER IS THE                BLE01410
C       AVERAGE SIZE OF THE NONZERO ENTRIES IN THE MATRIX USED            BLE01420
C       BY LANCZS.  NOTE THAT NNZ AND AVER ARE DEFINED AS DOUBLE          BLE01430
C       PRECISION SCALARS.  THE MAIN PROGRAMS ASSUME THAT THEY            BLE01440
C       ARE COMPUTED IN USPEC.  THE USPEC SUBROUTINE                      BLE01450
C       DEFINES AND DIMENSIONS THE ARRAYS REQUIRED TO                     BLE01460
C       SPECIFY THE MATRIX THAT WILL BE USED BY THE LANCZS                BLE01470
C       SUBROUTINE AND INITIALIZES THESE ARRAYS.  THE STORAGE             BLE01480
C       LOCATIONS OF THESE ARRAYS ARE THEN PASSED TO THE                  BLE01490
C       SUBROUTINE BMATV IN CASE (1) AND TO THE SUBROUTINE BSOLV          BLE01500
C       IN CASE (2).  SAMPLE SUBROUTINES ARE INCLUDED FOR EACH            BLE01510
C       CASE.  CASE (1) ASSUMES THAT THE A-MATRIX IS STORED ON            BLE01520
C       FILE 8.  CASE (2) ASSUMES THAT THE FACTORIZATION OF THE           BLE01530
C       B-MATRIX IS STORED ON FILE 7.                                     BLE01540
C                                                                         BLE01550
C       IN CASE (1) :                                                     BLE01560
C       BMATV IS THE SUBROUTINE USED BY THE LANCZS SUBROUTINE             BLE01570
C       THAT GENERATES THE LANCZOS T-MATRICES.  SUBROUTINE                BLE01580
C       BMATV HAS THE CALLING SEQUENCE                                    BLE01590
C                                                                         BLE01600
C            CALL BMATV(W,U)                                              BLE01610
C                                                                         BLE01620
C       WHERE U AND W ARE DOUBLE PRECISION VECTORS.  FOR A GIVEN          BLE01630
C       W, BMATV CALCULATES U = A*W  FOR THE USER-SPECIFIED MATRIX A.     BLE01640
C       A SAMPLE BMATV IS INCLUDED FOR AN ARBITRARY SPARSE,               BLE01650
C       SYMMETRIC A-MATRIX STORED IN THE SPARSE FORMAT SPECIFIED          BLE01660
C       IN THE CORRESPONDING SAMPLE USPEC SUBROUTINE.                     BLE01670
C                                                                         BLE01680
C       IN CASE (2):                                                      BLE01690
C       THE LANCZOS T-MATRICES ARE GENERATED USING SPARSE MATRIX          BLE01700
C       INVERSION, USING THE SUBROUTINE BLSOLV.  THE CALLING              BLE01710
C       SEQUENCE OF BLSOLV IS                                             BLF01720
C                                                                         BLE01730
C            CALL BLSOLV(U,V)                                             BLE01740
C                                                                         BLE01750
C       WHERE U AND V ARE DOUBLE PRECISION VECTORS.  FOR A GIVEN V,       BLE01760
C       BLSOLV COMPUTES U = (B-INVERSE)*V USING A SPARSE                  BLE01770
C       FACTORIZATION OF THE B-MATRIX ASSOCIATED WITH THE USER-           BLE01780
C       SPECIFIED A-MATRIX.                                               BLE01790
C                                                                         BLE01800
C       THE FOLLOWING SPARSE MATRIX FORMAT IS USED TO STORE THE           BLE01810
C       MATRICES IN THE SAMPLE PROGRAMS:                                  BLE01820
C       ICOL(K), K = 1,NZL, NUMBER OF SUBDIAGONAL NONZEROS IN COLUMN K.   BLE01830
C       IROW(K), K = 1,NZS, ROW INDEX OF ASD(K).                         BLE01840
C       AD(K), K=1,N, CONTAINS THE DIAGONAL ELEMENTS OF THE A-MATRIX.     BLE01850
C       ASD(K), K=1,NZS  CONTAINS THE SUBDIAGONAL ELEMENTS OF A BY COLUMN.BLE01860
C       NZS = NUMBER OF NONZERO ELEMENTS BELOW THE DIAGONAL OF A          BLE01870
C       NZL = INDEX OF LAST COLUMN WITH NONZERO SUBDIAGONAL ENTRIES       BLE01880
C       N = ORDER OF THE A-MATRIX.                                        BLE01890
C                                                                         BLE01900
C       IN CASE (1) THE A-MATRIX IS STORED IN THIS FORMAT ON FILE 8.      BLE01910
C       IN CASE (2), IN THE SAMPLE USPEC PROVIDED WHICH IS ONLY           BLE01920
C       FOR POSITIVE DEFINITE B-MATRICES, THE SPARSE CHOLESKY FACTOR      BLE01930
C       OF B, L, IS STORED ON FILE 7 IN THE ABOVE SPARSE FORMAT           BLE01940
C       USING ARRAYS BD AND BSD.  IN CASE (2) THE OPTIONAL AUXILIARY      BLE01950
C       PROGRAMS PERMUT AND LORDER ALSO REQUIRE THE A-MATRIX;             BLE01960
C       HOWEVER, THE BLOCK LANCZOS PROCEDURE ONLY USES THE                BLE01970
C       FACTORIZATION OF THE B-MATRIX.                                    BLE01980
```

```
C                                                                      BLE01990
C                                                                      BLE02000
C-----MACHEP----------------------------------------------------------BLE02010
C                                                                      BLE02020
C                                                                      BLE02030
C     MACHEP IS A MACHINE DEPENDENT PARAMETER SPECIFYING THE RELATIVE   BLE02040
C     PRECISION OF THE FLOATING POINT ARITHMETIC USED.                  BLE02050
C     MACHEP = 2.2 * 10**-16 FOR DOUBLE PRECISION ARITHMETIC ON         BLE02060
C     IBM 370-3081.                                                     BLE02070
C                                                                      BLE02080
C     THE USER WILL HAVE TO RESET THIS PARAMETER TO                     BLE02090
C     THE CORRESPONDING VALUE FOR THE MACHINE BEING USED.  NOTE THAT    BLE02100
C     IF A MACHINE WITH A MACHINE EPSILON THAT IS MUCH LARGER THAN THE  BLE02110
C     VALUE GIVEN HERE IS BEING USED, THEN THERE COULD BE               BLE02120
C     PROBLEMS WITH THE TOLERANCES.                                     BLE02130
C                                                                      BLE02140
C                                                                      BLE02150
C-----SUBROUTINES AND FUNCTIONS USER MUST SUPPLY----------------------BLE02160
C                                                                      BLE02170
C                                                                      BLE02180
C     GENRAN,  FINPRO,  MASK,  USPEC,  AND                             BLE02190
C     CASE (1) BMATV:  CASE (2)  BLSOLV :                              BLE02200
C                                                                      BLE02210
C     GENRAN = COMPUTES K PSEUDO-RANDOM NUMBERS AND STORES THEM IN      BLE02220
C              THE REAL ARRAY, G.  THIS SUBROUTINE IS USED TO           BLE02230
C              GENERATE STARTING VECTORS FOR THE BLOCK LANCZOS          BLE02240
C              PROCEDURE.  CALLED FROM LANCZS SUBROUTINE.               BLE02250
C              USER CAN SUPPLY STARTING VECTORS FOR THE BLOCK           BLE02260
C              PROCEDURES.  ANY ADDITIONAL VECTORS REQUIRED ARE         BLE02270
C              GENERATED RANDOMLY BY GENRAN.  VECTORS SUPPLIED MUST      BLE02280
C              BE STORED ON FILE 10.  THE NUMBER OF SUCH VECTORS TO      BLE02290
C              BE READ IN IS SPECIFIED BY THE PARAMETER KSET.  THE       BLE02300
C              EXISTING CALLING SEQUENCE IS                             BLE02310
C                                                                      BLE02320
C                    CALL GENRAN(IIX,G,K).                             BLE02330
C                                                                      BLE02340
C              WHERE IIX =INTEGER SEED, G = REAL ARRAY WHOSE DIMENSION   BLE02350
C              MUST BE >= K.  K PSEUDO-RANDOM NUMBERS ARE GENERATED      BLE02360
C              AND PLACED IN G.                                        BLE02370
C                                                                      BLE02380
C     FINPRO = DOUBLE PRECISION FUNCTION WHICH COMPUTES THE INNER       BLE02390
C              PRODUCT OF 2 DOUBLE PRECISION VECTORS OF DIMENSION N.     BLE02400
C              EXISTING CALLING SEQUENCE IS                            BLE02410
C                                                                      BLE02420
C                    CALL FINPRO(N,V,J,W,K).                           BLE02430
C                                                                      BLE02440
C              COMPUTES THE INNER PRODUCT OF DIMENSION N OF THE VECTORS BLE02450
C              V AND W.  SUCCESSIVE COMPONENTS OF V AND OF W ARE STORED BLE02460
C              AT LOCATIONS THAT ARE ,RESPECTIVELY, J AND K UNITS APART.BLE02470
C                                                                      BLE02480
C     MASK = MASKS OVERFLOW AND UNDERFLOW.  OPTIONAL.                   BLE02490
C            USER MUST SUPPLY OR COMMENT OUT CALL.                      BLE02500
C                                                                      BLE02510
C     USPEC = DIMENSIONS AND INITIALIZES ARRAYS NEEDED TO SPECIFY       BLE02520
C             MATRIX USED BY LANCZS SUBROUTINE.  SEE MATRIX             BLE02530
C             SPECIFICATION SECTION.                                    BLE02540
C                                                                      BLE02550
C     BMATV = CASE (1) ONLY: COMPUTES MATRIX-VECTOR MULTIPLY FOR        BLE02560
C             USER-SUPPLIED A-MATRIX. SEE MATRIX SPECIFICATION SECTION. BLE02570
C                                                                      BLE02580
C     BLSOLV = CASE (2) ONLY: FOR GIVEN VECTOR V, COMPUTES U SUCH       BLE02590
C              B*U = V, GIVEN THE SPARSE FACTORIZTION OF THE B-MATRIX.  BLE02600
C                                                                      BLE02610
C                                                                      BLE02620
C-----PARAMETER CONTROLS----------------------------------------------BLE02630
C                                                                      BLE02640
C                                                                      BLE02650
```

```
C       PARAMETER CONTROLS ARE INTRODUCED TO CONTROL VARIOUS           BLE02660
C       ASPECTS OF THESE PROGRAMS.                                      BLE02670
C                                                                       BLE02680
C       THE FLAG EFLAG SPECIFIES THE NUMBER OF COMPUTATIONAL PHASES.    BLE02690
C                                                                       BLE02700
C       EFLAG  = (0,1)  MEANS                                           BLE02710
C                                                                       BLE02720
C               (0) PROGRAM TERMINATES AFTER COMPLETING PHASE 1         BLE02730
C                   COMPUTATIONS.                                       BLE02740
C                                                                       BLE02750
C               (1) PROGRAM COMPLETES BOTH PHASE 1 AND PHASE 2 OF       BLE02760
C                   THE COMPUTATIONS.                                   BLE02770
C                                                                       BLE02780
C       THE FLAG OFLAG CONTROLS THE ORTHOGONALITY CHECKS BETWEEN THE    BLE02790
C       JTH Q-BLOCK GENERATED AND THAT VECTOR IN THE 1ST Q-BLOCK THAT   BLE02800
C       IS GENERATING DESCENDANTS.  FOR SAFETY, OFLAG SHOULD BE 1.      BLE02810
C                                                                       BLE02820
C       OFLAG = (0,1)  MEANS                                            BLE02830
C                                                                       BLE02840
C               (0) NO ORTHOGONALITY CHECKS ARE MADE ON PHASE           BLE02850
C                   1 PORTION OF THE COMPUTATIONS.  ORTHOGONALITY       BLE02860
C                   CHECKS ARE ALWAYS MADE ON PHASE 2 PORTION.          BLE02870
C                                                                       BLE02880
C               (1) PROGRAM CHECKS ORTHOGONALITY OF GENERATED          BLE02890
C                   Q-BLOCKS W.R.T. THAT VECTOR IN THE 1ST Q-BLOCK      BLE02900
C                   THAT IS GENERATING DESCENDANTS IN BOTH PHASE        BLE02910
C                   1 AND PHASE 2 OF THE COMPUTATIONS.                  BLE02920
C                                                                       BLE02930
C       THE FLAG IWRITE DETERMINES THE AMOUNT OF OUTPUT TO FILE 6       BLE02940
C       DURING THE COMPUTATIONS                                         BLE02950
C                                                                       BLE02960
C       IWRITE = (0,1)   MEANS                                          BLE02970
C                                                                       BLE02980
C               (0) ABBREVIATED OUTPUT TO FILE 6.                       BLE02990
C                                                                       BLE03000
C               (1) ADDITIONAL COMMENTARY ON THE COMPUTATIONS IS        BLE03010
C                   PRINTED TO FILE 6.                                  BLE03020
C                                                                       BLE03030
C       THE PROGRAM ALWAYS WRITES A LIST OF THE COMPUTED EIGENVALUES    BLE03040
C       AND THE BASIS FOR THE CORRESPONDING EIGENSPACE TO FILE 15,      BLE03050
C       ALONG WITH ESTIMATES OF THE ERRORS IN THESE COMPUTED VALUES.    BLE03060
C                                                                       BLE03070
C-----INPUT/OUTPUT FILES----------------------------------------------BLE03080
C                                                                       BLE03090
C       ANY INPUT DATA OTHER THAN THE A-MATRIX, THE FACTORIZATION       BLE03100
C       OF THE B-MATRIX OR USER-SPECIFIED STARTING VECTORS SHOULD       BLE03110
C       BE STORED ON FILE 5.  SEE SAMPLE INPUT/OUTPUT FROM TYPICAL RUN. BLE03120
C       THE READ STATEMENTS IN THE GIVEN FORTRAN PROGRAM ASSUME THAT    BLE03130
C       THE DATA STORED ON FILE 5 IS IN FREE FORMAT.  USER SHOULD NOTE  BLE03140
C       THAT 'FREE FORMAT' IS NOT CLASSIFIED AS PORTABLE BY PFORT SO THAT BLE03150
C       THE USER MAY HAVE TO MODIFY THE READ STATEMENTS FROM FILE 5 TO  BLE03160
C       CONFORM TO WHAT IS PERMISSIBLE ON THE COMPUTER BEING USED.      BLE03170
C                                                                       BLE03180
C       FILE 6 WAS USED AS THE INTERACTIVE TERMINAL OUTPUT FILE.        BLE03190
C       THIS FILE PROVIDES A RUNNING ACCOUNT OF THE PROGRESS OF THE     BLE03200
C       COMPUTATIONS.  THE AMOUNT OF INFORMATION PRINTED OUT IS         BLE03210
C       CONTROLLED BY THE PARAMETER IWRITE.                            BLE03220
C                                                                       BLE03230
C DESCRIPTION OF OTHER I/O FILES                                        BLE03240
C                                                                       BLE03250
C FILE (K)    CONTAINS:                                                 BLE03260
C                                                                       BLE03270
C       (7)     INPUT FILE:                                             BLE03280
C               USED IN CASE (2).  CONTAINS THE FACTORIZATION           BLE03290
C               OF THE B-MATRIX.                                        BLE03300
C                                                                       BLE03310
C       (8)     INPUT FILE:                                             BLE03320
```

```
C                    USED IN CASE (1).  CONTAINS THE ARRAYS REQUIRED      BLE03330
C                    TO SPECIFY THE A-MATRIX.                             BLE03340
C                                                                         BLE03350
C     (10)    INPUT FILE:                                                 BLE03360
C                    CONTAINS USER-SUPPLIED STARTING VECTORS, IF ANY.     BLE03370
C                    TYPICALLY, THESE WOULD BE 1 OR MORE EIGENVECTOR      BLE03380
C                    APPROXIMATIONS OBTAINED DURING AN EARLIER RUN.       BLE03390
C                                                                         BLE03400
C     (13)    OUTPUT FILE:                                                BLE03410
C                    CONTAINS EXTRA EIGENVECTOR APPROXIMATIONS THAT       BLE03420
C                    WOULD OTHERWISE BE LOST UPON ANY REDUCTION IN THE    BLE03430
C                    SIZE OF THE 1ST Q-BLOCK.  IF AT ANY STAGE IN THE     BLE03440
C                    BLOCK PROCEDURE, THE SIZE OF THE 1ST Q-BLOCK IS      BLE03450
C                    REDUCED FROM KACT TO KACTN, THE Q-VECTORS FROM       BLE03460
C                    K = KACTN+1,KACT ARE WRITTEN TO FILE 13 FOR POSSIBLE BLE03470
C                    USE AS STARTING VECTORS IN A LATER RUN OF THE        BLE03480
C                    BLOCK LANCZOS PROCEDURE.                             BLE03490
C                                                                         BLE03500
C     (15)    OUTPUT FILE:                                                BLE03510
C                    CONTAINS COMPUTED EIGENVALUES AND CORRESPONDING      BLE03520
C                    COMPUTED EIGENSPACE AVAILABLE AT THE TIME OF         BLE03530
C                    TERMINATION OF THE BLOCK LANCZOS PROCEDURE.          BLE03540
C                                                                         BLE03550
C-----PARAMETERS SET BY THE BLOCK PROGRAMS----------------------------BLE03560
C                                                                         BLE03570
C                                                                         BLE03580
C     SPREC = TOLERANCE USED IN CHECKING ORTHOGONALITY BETWEEN            BLE03590
C             COMPUTED Q-BLOCKS AND THAT VECTOR IN THE FIRST             BLE03600
C             Q-BLOCK THAT IS GENERATING DESCENDANTS.  SEE COMMENTS       BLE03610
C             ON OFLAG.                                                   BLE03620
C                                                                         BLE03630
C-----USER-SPECIFIED PARAMETERS ---------------------------------------BLE03640
C                                                                         BLE03650
C                                                                         BLE03660
C     FOR BOTH CASES:                                                     BLE03670
C                                                                         BLE03680
C     N, MATNO = INTEGERS. SIZE OF USER-SPECIFIED MATRIX AND MATRIX       BLE03690
C                IDENTIFICATION NUMBER OF 8 OR FEWER DIGITS.              BLE03700
C                                                                         BLE03710
C     MDIMQ, MDIMTM = INTEGERS.  USER-SPECIFIED DIMENSIONS OF THE         BLE03720
C                     Q-ARRAY AND OF THE TM-ARRAY.  MDIMQ >= N*KMAX       BLE03730
C                     AND MDIMTM >= MXBLK**2.                             BLE03740
C                                                                         BLE03750
C     MAXIT,MAXIT2 = INTEGERS.  MAXIMUM NUMBER OF CALLS TO BMATV          BLE03760
C                    (CASE(1)) OR TO BLSOLV (CASE (2)) ALLOWED            BLE03770
C                    RESPECTIVELY, IN PHASE 1 AND IN PHASE 2.             BLE03780
C                                                                         BLE03790
C     RELTOL = DOUBLE PRECISION SCALAR.  RELATIVE TOLERANCE USED          BLE03800
C              TO COMPUTE CONVERGENCE CRITERION FOR PHASE 2 OF            BLE03810
C              THE BLOCK PROCEDURE.                                       BLE03820
C                                                                         BLE03830
C     SEED  = INTEGER.  SEED FOR RANDOM NUMBER GENERATOR.                 BLE03840
C             USED IN GENERATION OF STARTING VECTORS FOR                  BLE03850
C             THE BLOCK PROCEDURES.                                       BLE03860
C                                                                         BLE03870
C     KMAX  = INTEGER.  MXBLK = (KMAX - 1) IS MAXIMUM ALLOWED SIZE        BLE03880
C             FOR THE SMALL LANCZOS T-MATRICES.                           BLE03890
C                                                                         BLE03900
C     KM    = INTEGER.  NUMBER OF EIGENVALUES AND EIGENVECTORS            BLE03910
C             TO BE COMPUTED.                                             BLE03920
C                                                                         BLE03930
C     KACT  = INTEGER.  INITIAL NUMBER OF VECTORS IN THE 1ST Q-BLOCK.     BLE03940
C             IF THERE IS ANY POSSIBILITY THAT THE KM-TH DESIRED          BLE03950
C             EIGENVALUE IS MULTIPLE, AND THE USER NEEDS TO KNOW          BLE03960
C             THIS, THEN THE USER SHOULD SET KACT > KM.  OTHERWISE,       BLE03970
C             THIS PROGRAM WILL NOT BE ABLE TO DETERMINE THAT THAT        BLE03980
C             EIGENVALUE IS MULTIPLE UNLESS THE (KM-1)-TH AND KM-TH       BLE03990
```

```
C                    HAPPEN TO BE MULTIPLE.  IF IN FACT, THE KM-TH         BLE04000
C                    EIGENVALUE IS MULTIPLE AND THE USER NEEDS A BASIS FOR BLE04010
C                    THE CORRESPONDING EIGENSPACE, THEN THE PROCEDURE SHOULD BLE04020
C                    BE RERUN WITH THE EXISTING EIGENVECTORS APPROXIMATIONS BLE04030
C                    AS STARTING VECTORS AND A LARGER KACT TO GUARANTEE THAT BLE04040
C                    A COMPLETE BASIS FOR THAT EIGENSPACE HAS BEEN OBTAINED. BLE04050
C                                                                    BLE04060
C       KSET =  INTEGER.  NUMBER OF STARTING VECTORS SUPPLIED BY THE BLE04070
C                    THE USER.  THESE VECTORS SHOULD BE ON FILE 10.  BLE04080
C                                                                    BLE04090
C                                                                    BLE04100
C       NSTAG = INTEGER.  NUMBER OF THE ITERATION BEYOND WHICH THE   BLE04110
C                    CHANGE IN THE KM-TH RESIDUAL OVER THE PAST 10 ITERATIONS BLE04120
C                    IS MONITORED AND USED AS A MEASURE OF THE RATE OF BLE04130
C                    CONVERGENCE OF THE BLOCK PROCEDURE.             BLE04140
C                                                                    BLE04150
C       FRACT = DOUBLE PRECISION SCALAR.  EXPECTED OR HOPED FOR      BLE04160
C                    FRACTIONAL CHANGE IN THE KM-TH RESIDUAL OVER THE PAST BLE04170
C                    BLOCK LANCZOS ITERATIONS USED TO TEST FOR STAGNATION BLE04180
C                    OF CONVERGENCE.                                 BLE04190
C                                                                    BLE04200
C       NNZ   = DOUBLE PRECISION SCALAR.  AVERAGE NUMBER OF NONZERO  BLE04210
C                    ENTRIES PER ROW IN THE MATRIX USED IN THE LANCZOS BLE04220
C                    PROCEDURE.                                      BLE04230
C                                                                    BLE04240
C                                                                    BLE04250
C       AVER  = DOUBLE PRECISION SCALAR.  AVERAGE SIZE OF THE NONZERO BLE04260
C                    ENTRIES IN THE MATRIX USED IN THE LANCZOS PROCEDURE. BLE04270
C                                                                    BLE04280
C       CASE (2) ONLY:                                               BLE04290
C                                                                    BLE04300
C       SO, SHIFT = DOUBLE PRECISION SCALARS.  MATRIX USED BY LANCZS BLE04310
C                    SUBROUTINE IS B = SO*P*A*P' + SHIFT*I   WHERE P. BLE04320
C                    DENOTES A PERMUTATION MATRIX SELECTED TO PRESERVE BLE04330
C                    THE SPARSITY OF A IN THE FACTORIZATION OF B.    BLE04340
C                    SO AND SHIFT ARE CHOSEN BY THE USER SO THAT THE BLE04350
C                    DESIRED EIGENVALUES BECOME THE EXTREME EIGENVALUES BLE04360
C                    OF B-INVERSE.                                   BLE04370
C                                                                    BLE04380
C                                                                    BLE04390
C-----CONVERGENCE TEST-----------------------------------------------------BLE04400
C                                                                    BLE04410
C                                                                    BLE04420
C       THE CONVERGENCE TEST INCORPORATED IN THIS PROGRAM IS         BLE04430
C       BASED UPON THE FOLLOWING FACT:  GIVEN A REAL SYMMETRIC       BLE04440
C       MATRIX A, A VECTOR X OF NORM 1, AND A SCALAR EVAL            BLE04450
C       THEN THERE EXISTS AN EIGENVALUE AEVAL OF A SUCH THAT         BLE04460
C       DABS(AEVAL - EVAL) .LE. NORM(A*X - EVAL*X).  WITHIN          BLE04470
C       EACH ITERATION OF THE BLOCK LANCZOS PROCESS THESE TYPES      BLE04480
C       OF NORMS ARE COMPUTED IN THE PROCESS OF COMPUTING THE        BLE04490
C       2ND Q-BLOCK.                                                 BLE04500
C                                                                    BLE04510
C                                                                    BLE04520
C-----ARRAYS REQUIRED------------------------------------------------------BLE04530
C                                                                    BLE04540
C                                                                    BLE04550
C       Q(J)    = DOUBLE PRECISON ARRAY.  ITS DIMENSION MUST BE AT   BLE04560
C                    LEAST AS LARGE AS KMAX*N, WHERE N IS THE ORDER OF BLE04570
C                    THE GIVEN MATRIX, AND MXBLK = KMAX - 1 IS THE   BLE04580
C                    MAXIMUM SIZE T-MATRIX ALLOWED ON ANY GIVEN      BLE04590
C                    ITERATION.  THE COLUMNS OF Q HOLD THE LANCZOS   BLE04600
C                    VECTORS GENERATED ON EACH ITERATION OF BLOCK    BLE04610
C                    LANCZOS PLUS THERE MUST BE AN ADDITIONAL COLUMN BLE04620
C                    AVAILABLE FOR WORK SPACE.  THE FIRST KACT COLUMNS BLE04630
C                    OF Q CONTAIN THE CURRENT APPROXIMATING EIGENSPACE. BLE04640
C                                                                    BLE04650
C       E(J)    = DOUBLE PRECISION ARRAY.  ITS DIMENSION MUST BE AT  BLE04660
```

```
C                    LEAST MXBLK = KMAX - 1.  ON EACH ITERATION CONTAINS     BLE04670
C                    THE COMPUTED EIGENVALUES OF THE LANCZOS T-MATRIX.        BLE04680
C                                                                            BLE04690
C       TM(J)   = DOUBLE PRECISION ARRAY.  ITS DIMENSION MUST BE AT          BLE04700
C                    LEAST MXBLK**2 WHERE MXBLK = KMAX - 1.  CONTAINS         BLE04710
C                    THE LANCZOS T-MATRIX GENERATED ON EACH ITERATION         BLE04720
C                    AND THEN THE COMPUTED EIGENVECTORS OF THIS MATRIX.       BLE04730
C                    EISPACK SUBROUTINES ARE USED FOR THE SMALL              BLE04740
C                    EIGENELEMENT COMPUTATIONS.  EISPACK SUBROUTINE          BLE04750
C                    TRED2 IS USED TO REDUCE THE GIVEN T-MATRIX TO           BLE04760
C                    TRIDIAGONAL FORM.  THE EIGENELEMENT PROBLEM FOR THE      BLE04770
C                    TRIDIAGONAL MATRIX IS THEN SOLVED USING THE EISPACK      BLE04780
C                    SUBROUTINE IMTQL2.                                      BLE04790
C                                                                            BLE04800
C       EXPLAN(J) = REAL ARRAY.  ITS DIMENSION IS 20.  THIS ARRAY IS         BLE04810
C                   USED TO ALLOW EXPLANATORY COMMENTS IN THE INPUT FILES.   BLE04820
C                                                                            BLE04830
C       G(J) = REAL ARRAY.  ITS DIMENSION MUST BE >= N.  IT IS USED          BLE04840
C              FOR HOLDING THE PSEUDO-RANDOM NUMBERS USED TO GENERATE        BLE04850
C              ANY STARTING VECTORS NOT SUPPLIED BY THE USER.                BLE04860
C                                                                            BLE04870
C       RESIDL(J), = DOUBLE PRECISION ARRAYS.  DIMENSION >= MAXIMUM          BLE04880
C       RESIDK(J),   NUMBER OF ITERATIONS ALLOWED.  MAXIMUM IS              BLE04890
C                    CURRENTLY SET TO 100. USED TO MONITOR THE               BLE04900
C                    RATE OF CONVERGENCE.                                    BLE04910
C                                                                            BLE04920
C       TD(J), TOD(J), = DOUBLE PRECISION ARRAYS.  DIMENSION >= MXBLK.       BLE04930
C         SM(J)         WORK SPACES.                                         BLE04940
C                                                                            BLE04950
C       DESC(J), XLFT(J), = INTEGER ARRAYS.  DIMENSION >= MXBLK.             BLE04960
C         LEFT(J)            WORK SPACES.                                    BLE04970
C                                                                            BLE04980
C       DIR(2,J)  = 2-DIMENSIONAL INTEGER ARRAY.  COLUMN DIMENSION >=        BLE04990
C                    MXBLK, ROW DIMENSION 2.  KEEPS TRACK OF NUMBER          BLE05000
C                    OF VECTORS IN EACH QBLOCK.                              BLE05010
C                                                                            BLE05020
C       CASE (2) ONLY:                                                      BLE05030
C                                                                            BLE05040
C       IPR(J), IPT(J) = INTEGER ARRAYS.  EACH OF DIMENSION AT LEAST N.     BLE05050
C                    USED TO STORE THE REORDERING (IF ANY) OF               BLE05060
C                    THE GIVEN MATRIX.                                       BLE05070
C                                                                            BLE05080
C       OTHER ARRAYS                                                        BLE05090
C                                                                            BLE05100
C       THE USER IN THE SUBROUTINE USPEC MUST SPECIFY WHATEVER ARRAYS        BLE05110
C       ARE REQUIRED TO DEFINE THE MATRIX BEING USED BY LANCZS.              BLE05120
C                                                                            BLE05130
C                                                                            BLE05140
C-----SUBROUTINES INCLUDED---------------------------------------------------BLE05150
C                                                                            BLE05160
C                                                                            BLE05170
C       LANCZS = CONTAINS MAJOR LOOP FOR BLOCK LANCZOS PROCEDURES.           BLE05180
C                CALLED FROM MAIN PROGRAM, CALLS SUBROUTINE LANCII           BLE05190
C                TO GENERATE WITHIN A GIVEN ITERATION THE Q-BLOCKS           BLE05200
C                AND CORRESPONDING LANCZOS T-MATRICES.  THEN CALLS           BLE05210
C                SUBROUTINE DIAGOM TO COMPUTE THE EIGENELEMENTS              BLE05220
C                OF THE LANCZOS T-MATRIX AND TO MAP THE RELEVANT             BLE05230
C                T-EIGENVECTORS INTO RITZ VECTORS FOR THE A-MATRIX.          BLE05240
C                                                                            BLE05250
C       LANCII = ON EACH ITERATION OF BLOCK LANCZOS COMPUTES                 BLE05260
C                Q-SUBBLOCKS.                                                BLE05270
C                                                                            BLE05280
C       DIAGOM = CALLS EISPACK SUBROUTINES TO COMPUTE THE                    BLE05290
C                EIGENELEMENTS OF THE SMALL LANCZOS T-MATRICES               BLE05300
C                GENERATED ON EACH ITERATION OF BLOCK LANCZOS.               BLE05310
C                COMPUTES CORRESPONDING RITZ VECTORS FOR A-MATRIX.           BLE05320
C                MONITORS CONVERGENCE OF BLOCK LANCZOS PROCEDURE.            BLE05330
```

```
C                                                                     BLE05340
C      START   = GENERATES ANY REQUIRED STARTING VECTORS FOR 1ST      BLE05350
C                Q-BLOCK FOR FIRST ITERATION OF BLOCK LANCZOS.        BLE05360
C                                                                     BLE05370
C      ORTHOG  = GIVEN A SET OF Q-VECTORS, Q(J), J = MA,MB,           BLE05380
C                ORTHOGONALIZES THESE VECTORS W.R.T. THE Q-VECTORS    BLE05390
C                Q(J), J = 1,MA-1.                                    BLE05400
C                                                                     BLE05410
C      LPERM   = (USED IN CASE (2) ONLY) GIVEN A MATRIX B AND A       BLE05420
C                PERMUTATION P DEFINED IN THE VECTORS IPR AND IPT,    BLE05430
C                AND A VECTOR X COMPUTE EITHER (P-TRANSPOSE)*X OR PX. BLE05440
C                                                                     BLE05450
C      CASE (2) ONLY:                                                 BLE05460
C      FOR OPTIONAL PRELIMINARY PROCESSING:                          BLE05470
C                                                                     BLE05480
C      PERMUT  (STAND-ALONE PROGRAM):                                 BLE05490
C      USES THE NONZERO STRUCTURE OF A GIVEN MATRIX A.                BLE05500
C      CAN BE USED TO OBTAIN A REORDERING OF A THAT WILL PRESERVE     BLE05510
C      THE SPARSENESS OF A UNDER FACTORIZATION.  PERMUT CALLS         BLE05520
C      CALLS THE SPARSPAK PACKAGE, (A. GEORGE, J. LIU, E. NG,         BLE05530
C      U. WATERLOO).  SEE THE PERMUT FORTRAN CODE FOR DETAILS.        BLE05540
C                                                                     BLE05550
C      LORDER (STAND-ALONE PROGRAM) :                                 BLE05560
C      GIVEN A MATRIX C IN SPARSE FORMAT AND A PERMUTATION P,         BLE05570
C      COMPUTES THE REORDERED MATRIX B = P*C*P' AND WRITES IT         BLE05580
C      TO FILE 9 IN SPARSE FORMAT.  SEE THE LORDER FORTRAN CODE       BLE05590
C      FOR DETAILS.                                                   BLE05600
C                                                                     BLE05610
C      LFACT (STAND-ALONE PROGRAM) :                                  BLE05620
C      GIVEN A POSITIVE DEFINITE MATRIX B IN SPARSE FORMAT,           BLE05630
C      COMPUTES THE SPARSE CHOLESKY FACTOR L OF B AND WRITES IT       BLE05640
C      TO FILE 7 IN SPARSE FORMAT.  THUS, B = L*L'.                   BLE05650
C      SEE THE LFACT FORTRAN CODE FOR DETAILS.                        BLE05660
C                                                                     BLE05670
C      LTEST (STAND-ALONE MAIN PROGRAM ) :                            BLE05680
C            (USER MUST PROVIDE 3 SUBROUTINES)                        BLE05690
C      GIVEN THE FACTORIZATION OF A SPARSE MATRIX B, COMPUTES         BLE05700
C      THE SOLUTION OF THE EQUATION B*U = B*V1 FOR A KNOWN BUT        BLE05710
C      RANDOMLY-GENERATED VECTOR V1, SOLVING WITH AND WITHOUT ITERATIVE BLE05720
C      REFINEMENT TO OBTAIN A ROUGH CHECK ON THE NUMERICAL CONDITION  BLE05730
C      OF THE B-MATRIX.  THIS PROGRAM USES 3 USER-SUPPLIED SUBROUTINES BLE05740
C      CMATV, CMATS AND BLSOLV.  SEE THE LTEST FORTRAN CODE FOR DETAILS. BLE05750
C                                                                     BLE05760
C                                                                     BLE05770
C-----OTHER PROGRAMS PROVIDED------------------------------------------BLE05780
C                                                                     BLE05790
C                                                                     BLE05800
C      LECOMPAC  =  TRANSLATES A REAL SYMMETRIC MATRIX PROVIDED       BLE05810
C                   IN THE FORMAT  I, J, A(I,J) INTO THE SPARSE       BLE05820
C                   MATRIX FORMAT USED IN THE SAMPLE SUBROUTINES      BLE05830
C                   PROVIDED. IT ASSUMES THAT THE MATRIX              BLE05840
C                   ENTRIES ARE GIVEN EITHER COLUMN BY COLUMN OR      BLE05850
C                   ROW BY ROW.  THE DATA SET CREATED IS WRITTEN TO   BLE05860
C                   FILE 8.                                           BLE05870
C                                                                     BLE05880
C                                                                     BLE05890
C--------------------------------------------------------------------BLE05900
```

SECTION 8.3   MAIN PROGRAM, EIGENVALUE AND EIGENVECTOR CALCULATIONS

```
C-----BLEVAL (FEW EXTREME EIGENVALUES AND EIGENVECTORS)----------------BLE00010
C                    (REAL SYMMETRIC MATRICES)                         BLE00020
C                                                                      BLE00030
C        CONTAINS MAIN PROGRAM FOR COMPUTING A FEW OF THE ALGEBRAICALLY- BLE00040
C        LARGEST EIGENVALUES AND CORRESPONDING EIGENVECTORS OF A REAL  BLE00050
C        SYMMETRIC MATRIX, USING A BLOCK FORM OF LANCZOS TRIDIAGONALIZATIONBLE00060
C        WITH LIMITED REORTHOGONALIZATION.  PROCEDURE IS ITERATIVE.    BLE00070
C        PROCEDURE CAN BE USED TO COMPUTE THE ALGEBRAICALLY-SMALLEST   BLE00080
C        EIGENVALUES BY THE USER SUPPLYING -A*X RATHER THAN A*X, IN    BLE00090
C        WHICH CASE IT COMPUTES THE CORRESPONDING ALGEBRAICALLY-LARGEST BLE00100
C        EIGENVALUES OF -A.  IN THIS CASE THE SIGNS OF THE COMPUTED    BLE00110
C        EIGENVALUES ARE CHANGED PRIOR TO WRITING TO FILE 15 SO THAT   BLE00120
C        ON EXIT, FILE 15 CONTAINS THE ALGEBRAICALLY-SMALLEST EIGENVALUES BLE00130
C        OF A ALONG WITH THE CORRESPONDING EIGENVECTORS.               BLE00140
C                                                                      BLE00150
C        ITERATIVE 'BLOCK' LANCZOS PROCEDURE FOR WHICH ON EVERY        BLE00160
C        ITERATION, THE 2ND AND SUCCEEDING BLOCKS CONTAIN ONLY ONE     BLE00170
C        VECTOR WHICH IS SELECTED ON THE BASIS OF ITS EXPECTED INFLUENCE BLE00180
C        ON THE CONVERGENCE.   Q-BLOCKS GENERATED ON A GIVEN ITERATION BLE00190
C        ARE REORTHOGONALIZED ONLY W.R.T. THOSE VECTORS IN THE FIRST   BLE00200
C        Q-BLOCK THAT ARE NOT GENERATING DESCENDANTS ON THAT           BLE00210
C        ITERATION.                                                    BLE00220
C                                                                      BLE00230
C        PFORT VERIFIER IDENTIFIED THE FOLLOWING NONPORTABLE CONSTRUCTIONS:BLE00240
C        1.  DATA MACHEP DEFINITION                                    BLE00250
C        2.  FORMAT (20A4) USED FOR READING EXPLANATORY COMMENTS.      BLE00260
C        3.  FREE FORMAT (5,*), USED FOR PARAMETER INPUT FROM FILE 5.  BLE00270
C        4.  COMMON/LOOPS/ AS CONSTRUCTED IS NOT PORTABLE              BLE00280
C                                                                      BLE00290
C----------------------------------------------------------------------BLE00300
        DOUBLE PRECISION Q(44000),E(50),TM(2500),TOD(50),TD(50),EPSM,NNZ BLE00310
        DOUBLE PRECISION SM(100),ERRMAX,SPREC,MACHEP,AVER,RELTOL,ERRMAN BLE00320
        DOUBLE PRECISION EVAL, RESIDL(100), RESIDK(100), RESID, FRACT  BLE00330
        REAL EXPLAN(20),G(2000)                                        BLE00340
        INTEGER  DIR(2,100),DESC(100),LEFT(100),XLFT(100)              BLE00350
        INTEGER SEED,OFLAG,EFLAG                                       BLE00360
        COMMON/LOOPS/MAXIT,ITER                                        BLE00370
        COMMON /RANDOM/SEED                                            BLE00380
        COMMON/FLAGS/EFLAG,OFLAG                                       BLE00390
        DOUBLE PRECISION  DABS, DFLOAT                                 BLE00400
C----------------------------------------------------------------------BLE00410
        EXTERNAL BMATV                                                 BLE00420
        DATA MACHEP/Z3410000000000000/                                 BLE00430
C----------------------------------------------------------------------BLE00440
C                                                                      BLE00450
C        ARRAYS MUST DIMENSIONED AS FOLLOWS:                           BLE00460
C                                                                      BLE00470
C        1.  Q:   >=  KMAX*N                                           BLE00480
C        2.  G:   >= N                                                 BLE00490
C        3.  E:   >=  MXBLK                                            BLE00500
C        4.  TM:  >=  MXBLK**2                                         BLE00510
C        5.  TOD, TD, SM, DESC, LEFT, XLFT:  >= MXBLK                  BLE00520
C        6.  DIR: ROW DIMENSION = 2;  COLUMN DIMENSION >= MXBLK        BLE00530
C        7.  RESIDL, RESIDK:  >= MAXIMUM NUMBER OF ITERATIONS ALLOWED. BLE00540
C            PROGRAM CURRENTLY TERMINATES IF MORE THAN 100 ITERATIONS  BLE00550
C            ARE REQUESTED.  USED TO MONITOR CONVERGENCE.              BLE00560
C        8.  EXPLAN: DIMENSION = 20.                                   BLE00570
C                                                                      BLE00580
C----------------------------------------------------------------------BLE00590
C        OUTPUT HEADER                                                 BLE00600
        WRITE(6,10)                                                    BLE00610
     10 FORMAT(/' BLOCK LANCZOS PROCEDURE FOR REAL SYMMETRIC MATRICES'  BLE00620
```

```
      1 /' 2ND AND SUCCEEDING BLOCKS CONTAIN ONLY ONE VECTOR'//)    BLE00630
C                                                                    BLE00640
C     SET PROGRAM PARAMETERS                                         BLE00650
      EPSM = 2.D0*MACHEP                                             BLE00660
      SPREC = 1.D-5                                                  BLE00670
      MPMIN = -1000                                                  BLE00680
C                                                                    BLE00690
C     READ USER-SPECIFIED PARAMETERS FROM INPUT FILE 5 (FREE FORMAT) BLE00700
C                                                                    BLE00710
C     SELECT THE AMOUNT OF INTERMEDIATE OUTPUT DESIRED (IWRITE =0,1). BLE00720
C     IWRITE = 1 INCREASES THE AMOUNT OF INTERMEDIATE OUTPUT WRITTEN  BLE00730
C     TO FILE 6 ON EACH ITERATION OF THE BLOCK LANCZOS PROCEDURE.     BLE00740
      READ(5,20) EXPLAN                                              BLE00750
   20 FORMAT(20A4)                                                   BLE00760
      READ(5,*)  IWRITE                                              BLE00770
C                                                                    BLE00780
C     READ ORDER (N) OF MATRIX AND MATRIX IDENTIFICATION NUMBER (MATNO) BLE00790
      READ(5,20) EXPLAN                                              BLE00800
      READ(5,*) N,MATNO                                              BLE00810
C                                                                    BLE00820
C     READ USER-SPECIFIED DIMENSIONS OF Q-ARRAY (MDIMQ) AND OF THE    BLE00830
C     TM-ARRAY (MDIMTM).  READ MAXIMUM NUMBER (MAXIT) OF MATRIX-VECTOR BLE00840
C     MULTIPLIES ALLOWED IN PHASE 1.                                 BLE00850
      READ(5,20) EXPLAN                                              BLE00860
      READ(5,*) MDIMQ, MDIMTM, MAXIT                                 BLE00870
C                                                                    BLE00880
C     READ FLAGS:   EFLAG = (0,1).  EFLAG = 0, MEANS PROGRAM STOPS    BLE00890
C     AFTER COMPLETING PHASE 1 PORTION OF BLOCK LANCZOS PROCEDURE.    BLE00900
C     EFLAG = 1, MEANS PROGRAM COMPLETES BOTH PHASES BEFORE           BLE0091C
C     TERMINATING.                                                   BLE00920
C     OFLAG = (0,1).  OFLAG = 0, MEANS THAT IN PHASE 1 PORTION        BLE00930
C     OF THE COMPUTATION, THE PROGRAM DOES NO ORTHOGONALITY CHECKS    BLE00940
C     ON THE Q-BLOCKS GENERATED.  OFLAG = 1 MEANS THAT IN THE         BLE00950
C     PHASE 1 PORTION AND IN THE PHASE 2 PORTIONS OF THE COMPUTATIONS BLE00960
C     THE PROGRAM CHECKS THE ORTHOGONALITY OF THE Q-BLOCKS GENERATED  BLE00970
C     W.R.T. THAT VECTOR IN THE FIRST BLOCK THAT IS GENERATING        BLE00980
C     DESCENDANTS.  NOTE THAT IN PHASE 2, THE PROGRAM ALWAYS MAKES    BLE00990
C     THIS CHECK OF ORTHOGONALITY REGARDLESS OF THE VALUE OF OFLAG.   BLE01000
C     FOR SAFETY, OFLAG SHOULD ALWAYS BE SET TO 1, ALTHOUGH IN MANY   BLE01010
C     PROBLEMS THIS IS NOT NECESSARY.                                BLE01020
      READ(5,20) EXPLAN                                              BLE01030
      READ(5,*) EFLAG,OFLAG                                          BLE01040
C                                                                    BLE01050
C     READ SEED USED BY SUBROUTINE GENRAN TO OBTAIN THOSE STARTING    BLE01060
C     VECTORS WHICH ARE GENERATED RANDOMLY.                          BLE01070
      READ(5,20) EXPLAN                                              BLE01080
      READ(5,*) SEED                                                 BLE01090
C                                                                    BLE01100
C     SPECIFY MAXIMUM T-SIZE ALLOWED (KMAX-1); INITIAL SIZE OF        BLE01110
C     STARTING BLOCK (KACT);  NUMBER OF STARTING VECTORS SUPPLIED (KSET) BLE01120
C     SEE BLOCK LANCZOS HEADER FOR COMMENTS ON THE SIZE OF KACT.      BLE01130
      READ(5,20) EXPLAN                                              BLE01140
      READ(5,*) KMAX,KACT,KSET                                       BLE01150
C                                                                    BLE01160
C     SPECIFY NUMBER OF EXTREME EIGENVALUES AND EIGENVECTORS TO BE    BLE01170
C     COMPUTED (KM).  USER CAN SPECIFY THAT THE ALGEBRAICALLY-        BLE01180
C     SMALLEST EIGENVALUES ARE BEING COMPUTED BY SETTING KM < 0.      BLE01190
C     PROGRAM THEN ASSUMES THAT THE MATRIX-VECTOR MULTIPLY            BLE01200
C     SUBROUTINE WHICH THE USER HAS PROVIDED IS COMPUTING -A*X        BLE01210
C     INSTEAD OF A*X AND INTERNALLY IT COMPUTES THE |KM|              BLE01220
C     ALGEBRAICALLY-LARGEST EIGENVALUES OF -A.                       BLE01230
      READ(5,20) EXPLAN                                              BLE01240
      READ(5,*) KM                                                   BLE01250
      IF(KM.EQ.0) GO TO 490                                          BLE01260
      KML = IABS(KM)                                                 BLE01270
C                                                                    BLE01280
C     STAGNATION OF CONVERGENCE OF THE KM-TH EIGENVALUE WILL BE       BLE01290
```

```
C      TESTED AFTER NSTAG ITERATIONS.  CONVERGENCE WILL BE SAID TO      BLE01300
C      HAVE STAGNATED IF THE RATIO OF THE SQUARE OF THE CURRENT KM-TH   BLE01310
C      RESIDUAL TO THE SQUARE OF THE CORRESPONDING RESIDUAL OBTAINED    BLE01320
C      10 ITERATIONS EARLIER IS GREATER THAN FRACT.  NSTAG SHOULD BE    BLE01330
C      >= 25.  IN THE TESTS FRACT WAS SET TO .01.                       BLE01340
       READ(5,20) EXPLAN                                                BLE01350
       READ(5,*) NSTAG, FRACT                                           BLE01360
C                                                                       BLE01370
C      READ IN THE RELATIVE TOLERANCE (RELTOL) USED TO DETERMINE A      BLE01380
C      CONVERGENCE CRITERION FOR PHASE 2, AND THE MAXIMUM NUMBER (MAXIT2)BLE01390
C      OF MATRIX-VECTOR MULTIPLIES ALLOWED IN PHASE 2.                  BLE01400
       READ(5,20) EXPLAN                                                BLE01410
       IF(EFLAG.EQ.1)  READ(5,*)  RELTOL, MAXIT2                        BLE01420
C                                                .                      BLE01430
C      CONSISTENCY CHECKS                                               BLE01440
C      PROCEDURE REQUIRES ENOUGH ROOM IN Q-ARRAY FOR AT LEAST 2         BLE01450
C      BLOCKS OF SIZE KACT PLUS A WORKING VECTOR OF LENGTH N.           BLE01460
       MXBLK = KMAX -1                                                  BLE01470
       MXBLK2 = MXBLK*MXBLK                                             BLE01480
       IF(MDIMTM.LT.MXBLK2) GO TO 470                                   BLE01490
       NKMAX = N*KMAX                                                   BLE01500
       IF(MDIMQ.LT.NKMAX)   GO TO 510                                   BLE01510
       IF(KML.GT.KACT) GO TO 370                                        BLE01520
       IF(MXBLK.GT.N)  GO TO 390                                        BLE01530
       IF(2*KACT.GT.MXBLK)  GO TO 450                                   BLE01540
C                                                                       BLE01550
C----------------------------------------------------------------------BLE01560
C      DEFINE AND INITIALIZE THE ARRAYS FOR THE USER-SPECIFIED          BLE01570
C      A-MATRIX AND PASS THE STORAGE LOCATIONS OF THESE ARRAYS AND      BLE01580
C      OF ANY OTHER PARAMTERS NEEDED TO DEFINE THE MATRIX TO THE        BLE01590
C      MATRIX-VECTOR MULTIPLY SUBROUTINE BMATV.                         BLE01600
C                                                                       BLE01610
       CALL USPEC(N,MATNO,NNZ,AVER)                                     BLE01620
C                                                                       BLE01630
C----------------------------------------------------------------------BLE01640
C      MASK OVERFLOW AND UNDERFLOW                                      BLE01650
       CALL MASK                                                        BLE01660
C                                                                       BLE01670
C----------------------------------------------------------------------BLE01680
C      ARE THERE STARTING VECTORS TO READ IN FROM FILE 10 (KSET.NE.0) ? BLE01690
       IF(KSET.EQ.0) GO TO 70                                           BLE01700
C                                                                       BLE01710
       READ(10,30) NOLD,KACT                                            BLE01720
    30 FORMAT(I6,I4)                                                    BLE01730
       IF(NOLD.NE.N.OR.KSET.GT.KACT) GO TO 410                          BLE01740
       DO 50 J=1,KSET                                                   BLE01750
       READ(10,20) EXPLAN                                               BLE01760
       READ(10,40) EVAL,RESID                                           BLE01770
    40 FORMAT(E20.12,E13.4)                                             BLE01780
       READ(10,20) EXPLAN                                               BLE01790
       LINT= (J-1)*N + 1                                                BLE01800
       LFIN = J*N                                                       BLE01810
    50 READ(10,60) (Q(JL), JL = LINT,LFIN)                              BLE01820
    60 FORMAT(4E20.12)                                                  BLE01830
C                                                                       BLE01840
    70 CONTINUE                                                         BLE01850
C                                                                       BLE01860
C      WRITE TO A SUMMARY OF THE PARAMETERS FOR THIS RUN TO FILE 6      BLE01870
C                                                                       BLE01880
       MXBLK = KMAX - 1                                                 BLE01890
       WRITE(6,80)  N, NNZ, AVER, MATNO                                 BLE01900
    80 FORMAT(/6X,'ORDER OF MATRIX ',5X,'AVERAGE NONZEROES PER ROW'/    BLE01910
      1I15,E26.4/4X,'AVERAGE SIZE OF NONZERO ENTRIES',5X,'MATRIX ID'/   BLE01920
      1E25.4,12I/)                                                      BLE01930
C                                                                       BLE01940
       WRITE(6,90) MDIMQ, MDIMTM                                        BLE01950
    90 FORMAT(/18X,'USER-SPECIFIED'/2X,'MAX. DIMENSION Q-ARRAY',4X,'MAX. BLE01960
```

454

```
      1DIMENSION TM-ARRAY'/I16,I26/)                              BLE01970
C                                                                 BLE01980
      WRITE(6,100)  OFLAG, EFLAG                                  BLE01990
  100 FORMAT(/4X,'OFLAG',4X,'EFLAG'/I8,I9/)                       BLE02000
C                                                                 BLE02010
      IF(EFLAG.EQ.1)  WRITE(6,110) MAXIT,RELTOL,MAXIT2            BLE02020
  110 FORMAT(/4X,' MAXIT ',8X,' RELTOL ',6X,' MAXIT2 '/I10,E20.6,I12/)BLE02030
      IF(EFLAG.EQ.0)  WRITE(6,120) MAXIT                          BLE02040
  120 FORMAT(/4X,' MAXIT '/I10/)                                  BLE02050
C                                                                 BLE02060
      WRITE(6,130)  SEED                                          BLE02070
  130 FORMAT(/' SEED FOR RANDOM NUMBER GENERATOR'/I24/)           BLE02080
C                                                                 BLE02090
      IF(KM.GT.0)  WRITE(6,140) KML                               BLE02100
  140 FORMAT(/' COMPUTE THE',I3,' ALGEBRAICALLY-LARGEST EIGENVALUES AND BLE02110
     1CORRESPONDING VECTORS'/)                                    BLE02120
      IF(KM.LT.0)  WRITE(6,150) KML                               BLE02130
  150 FORMAT(/' COMPUTE THE',I3,' ALGEBRAICALLY-SMALLEST EIGENVALUES ANDBLE02140
     1 CORRESPONDING VECTORS'/' PROGRAM ASSUMES THAT USER IS PROVIDING -BLE02150
     1A*X INSTEAD OF A*X'/' AND COMPUTES THE ALGEBRAICALLY-LARGEST EIGENBLE02160
     1VALUES OF -A.'/' HOWEVER ON EXIT, FILE 15 CONTAINS THE ALGEBRAICALBLE02170
     1LY-SMALLEST EIGENVALUES OF'/' THE ORIGINAL A-MATRIX AND CORRESPONDBLE02180
     1ING EIGENVECTORS.'/)                                        BLE02190
      IF(KM.LT.0) KM = - KM                                       BLE02200
C                                                                 BLE02210
C     COMPUTE PHASE 1 CONVERGENCE TOLERANCE                       BLE02220
      IF(AVER.GE.1.)                                              BLE02230
     1ERRMAX = 2.D0*DFLOAT(N+1000)*NNZ*AVER*MACHEP                BLE02240
      IF(AVER.LT.1.)                                              BLE02250
     1ERRMAX = 2.D0*DFLOAT(N+1000)*NNZ*AVER**2*MACHEP             BLE02260
C                                                                 BLE02270
      WRITE(6,160) KACT,MXBLK,KSET                                BLE02280
  160 FORMAT(/' ON INITIAL ITERATIONS, THE FIRST BLOCK CONTAINS ',I3,' VBLE02290
     1ECTORS'/' HOWEVER THE SIZE OF THE FIRST BLOCK MAY CHANGE AS THE ITBLE02300
     1ERATIONS PROCEED'/' THE MAXIMUM SIZE T-MATRIX THAT CAN BE GENERATEBLE02310
     1D IS ',I4/' THE USER SUPPLIED ',I3,' STARTING VECTORS'/)    BLE02320
C                                                                 BLE02330
      WRITE(6,170)                                                BLE02340
  170 FORMAT(/' ITERATIVE PROCEDURE'/' PROCEDURE MONITORS THE SIZES OF TBLE02350
     1HE NORM(GRADIENTS)**2 ON EACH'/' ITERATION.  CONVERGENCE IS SAID  BLE02360
     1TO HAVE OCCURRED WHEN ALL'/' RELEVANT (NORMS)**2 ARE LESS THAN ERRBLE02370
     1MAX',E10.3/' TYPICALLY, PHASE 1 ERRMAX YIELDS SOMEWHAT LESS THAN'/BLE02380
     1' SINGLE PRECISION ACCURACY.  PHASE 2 REFINES THE VECTORS OBTAINEDBLE02390
     1'/' ON PHASE 1, ACCORDING TO THE ACCURACY SPECIFIED BY THE USER'/)BLE02400
C                                                                 BLE02410
      WRITE(6,180) ERRMAX                                         BLE02420
  180 FORMAT(//' PHASE 1 CONVERGENCE CRITERION, ERRMAX '/E22.3/)  BLE02430
C                                                                 BLE02440
C-----------------------------------------------------------------BLE02450
C     PASS STORAGE LOCATIONS OF VARIOUS ARRAYS TO LANCZS AND LANCI1 BLE02460
C     SUBROUTINES                                                 BLE02470
C                                                                 BLE02480
      CALL LANZP(DIR,DESC,SM,TM,TOD,TD,G,XLFT,LEFT,SPREC)         BLE02490
      CALL LANCP1(DIR,DESC,TM,SM,XLFT,LEFT)                       BLE02500
C                                                                 BLE02510
C-----------------------------------------------------------------BLE02520
C                                                                 BLE02530
C     ENTER PHASE 1 OF BLOCK LANCZOS PROCEDURE.  BLOCK PROCEDURE   BLE02540
C     HAS 2 POSSIBLE PHASES.  USER SPECIFIES PHASE 1 ONLY OR PHASE 1 BLE02550
C     AND PHASE 2 BY SETTING EFLAG = 0 OR 1, RESPECTIVELY. PHASE 1 BLE02560
C     COMPUTES VECTORS THAT MAY BE SOMEWHAT LESS ACCURATE THAN SINGLE BLE02570
C     PRECISION.  PHASE 2 TAKES THE VECTORS OBTAINED IN PHASE 1   BLE02580
C     AND ATTEMPTS TO REFINE THEM.  THE USER SPECIFIES THE DEGREE BLE02590
C     OF REFINEMENT DESIRED BY SETTING THE VALUES OF RELTOL AND MAXIT2. BLE02600
C     BOTH PHASES SHOULD BE USED.                                 BLE02610
      IPHASE = 1                                                  BLE02620
      NITER = 0                                                   BLE02630
```

```
      190 ITER = 0                                                  BLE02640
          RESIDL(1) = FRACT                                         BLE02650
          RESIDL(2) = NSTAG                                         BLE02660
    C                                                               BLE02670
    C-----------------------------------------------------------------BLE02680
    C     CALL INITIATES THE BLOCK LANCZOS PROCEDURE.               BLE02690
    C     ON RETURN EIGENVALUE APPROXIMATIONS ARE IN E(I), I=1,KACT BLE02700
    C     IN ALGEBRAICALLY DECREASING ORDER.  EIGENVECTOR APPROXIMATIONS BLE02710
    C     ARE IN FIRST N*KACT LOCATIONS IN THE Q-ARRAY.             3LE02720
    C                                                               BLE02730
          CALL LANCZS(BMATV,KML,KSET,KACT,MXBLK,N,Q,E,RESIDL,RESIDK,ERRMAX, BLE02740
         1 IPHASE,NITER,IWRITE)                                     BLE02750
    C                                                               BLE02760
    C-----------------------------------------------------------------BLE02770
    C                                                               BLE02780
          IF(IPHASE.EQ.MPMIN) WRITE(15,200) N,KACT                 BLE02790
      200 FORMAT(2I10,' PHASE 2 TERMINATED '/' PROGRAM INDICATES ACCURACY SPBLE02800
         1ECIFIED BY USER IS NOT ACHIEVABLE'/)                      BLE02810
    C                                                               BLE02820
          ITERA = IABS(ITER)                                        BLE02830
          IF(IWRITE.NE.MPMIN.AND.ITER.GT.0)  WRITE(6,210) IPHASE,ITERA BLE02840
      210 FORMAT(/1X,'PHASE COMPLETED',5X,' NUMBER MATRIX-VECTOR MULTIPLIES BLE02850
         1USED'/I10,I30)                                            BLE02860
    C                                                               BLE02870
          IF(IWRITE.EQ.MPMIN.OR.ITER.LT.0)  WRITE(6,220) IPHASE,ITERA BLE02880
      220 FORMAT(/1X,'PHASE TERMINATED',5X,' NUMBER MATRIX-VECTOR MULTIPLIESBLE02890
         1 USED'/I10,I30)                                           BLE02900
    C                                                               BLE02910
          IF(ITER.GT.0.AND.IWRITE.NE.MPMIN)  GO TO 250             BLE02920
    C                                                               BLE02930
          IF(ITER.LT.0)  WRITE(6,230)                               BLE02940
      230 FORMAT(//' SMALL EIGENVALUE SUBROUTINE DEFAULTED'/' BLOCK LANCZOS BLE02950
         1 PROCEDURE STOPS AFTER SAVING CURRENT EIGENVECTOR APPROXIMATIONS'/BLE02960
         1/)                                                        BLE02970
    C                                                               BLE02980
          WRITE(15,240)                                             BLE02990
          WRITE(6,240)                                              BLE03000
      240 FORMAT(//' BLOCK LANCZOS PROCEDURE TERMINATES WITHOUT CONVERGENCE BLE03010
         1'/' USER SHOULD EXAMINE OUTPUT TO DETERMINE REASONS FOR TERMINATIOBLE03020
         1N'//)                                                     BLE03030
    C                                                               BLE03040
    C     WRITE EIGENVALUE AND EIGENVECTOR APPROXIMATIONS CONTAINED IN BLE03050
    C     THE FIRST Q-BLOCK TO FILE 15                              BLE03060
    C                                                               BLE03070
      250 IF(IPHASE.EQ.1)  WRITE(15,260) N,KACT,SEED                BLE03080
      260 FORMAT(I6,I4,I12,' PHASE 1, ORDER A-MATRIX, SIZE OF Q(1), SEED') BLE03090
          IF(IPHASE.EQ.2)  WRITE(15,270) N,KACT,SEED                BLE03100
      270 FORMAT(I6,I4,I12,' PHASE 2, ORDER A-MATRIX, SIZE OF Q(1), SEED') BLE03110
    C                                                               BLE03120
          JJ=KACT                                                   BLE03130
          LINT = -N+1                                               BLE03140
          LFIN = 0                                                  BLE03150
          DO 290 J=1,KACT                                           BLE03160
          LINT = LINT + N                                           BLE03170
          LFIN = LFIN + N                                           BLE03180
          JJ=JJ+1                                                   BLE03190
    C                                                               BLE03200
    C     NOTE THAT RESIDUAL PRINTED OUT CORRESPONDS TO VALUE OBTAINED BLE03210
    C     PRIOR TO FINAL PROJECTION Q(1)-TRANSPOSE*AQ(1) DONE BEFORE BLE03220
    C     TERMINATION                                               BLE03230
    C                                                               BLE03240
          IF(KM.LT.0)   E(J) = -E(J)                                BLE03250
          WRITE(15,280) E(J), SM(JJ)                                BLE03260
      280 FORMAT(/E20.12,E13.4,'= EIGENVALUE, NORM(ERROR)**2,EIGENVECTOR='/)BLE03270
      290 WRITE(15,300) (Q(L), L=LINT,LFIN)                         BLE03280
          WRITE(15,310)                                             BLE03290
      300 FORMAT(4E20.12)                                           BLE03300
```

```
    310 FORMAT(/' ABOVE ARE COMPUTED APPROXIMATE EIGENVECTORS'/)              BLE03310
C                                                                            BLE03320
        IF(ITER.GT.MAXIT) WRITE(15,320) ITER,MAXIT                           BLE03330
    320 FORMAT(//' PROCEDURE TERMINATED BECAUSE NUMBER OF MATRIX-VECTOR MUBLE03340
       1LTIPLIES ',I6/' EXCEEDED MAXIMUM NUMBER ',I6,' ALLOWED'//)           BLE03350
C                                                                            BLE03360
        IF(ITER.LT.0) WRITE(15,330)                                          BLE03370
    330 FORMAT(//' USER BEWARE.  EIGENELEMENT COMPUTATIONS DEFAULTED BECAUBLE03380
       1SE'/'  EISPACK SUBROUTINE DEFAULTED.  EIGENVALUE AND EIGENVECTORBLE03390
       1 APPROXIMATIONS'/'  ABOVE WERE THOSE AVAILABLE AT THE TIME OF DEFBLE03400
       1AULT'/' SOMETHING IS SERIOUSLY WRONG.'//)                            BLE03410
C                                                                            BLE03420
C       CHECK FOR TERMINATION AFTER PHASE 1                                  BLE03430
C       ITER < 0 MEANS EISPACK SUBROUTINE DEFAULTED                          BLE03440
C       IPHASE = MPMIN MEANS THAT PHASE 2 TERMINATED DUE TO ORTHOGONALITY BLE03450
C       IWRITE = MPMIN MEANS THAT CONVERGENCE APPEARS TO HAVE STAGNATED     BLE03460
C       ITER > MAXIT MEANS MAXIMUM NUMBER OF MATRIX-VECTOR MULTIPLIES        BLE03470
C             ALLOWED BY USER WAS EXCEEDED                                   BLE03480
        IF(ITER.LT.0.OR.ITER.GT.MAXIT) GO TO 530                             BLE03490
        IF(IPHASE.EQ.MPMIN.OR.IWRITE.EQ.MPMIN) GO TO 530                     BLE03500
        IF(EFLAG.NE.1.OR.IPHASE.EQ.2) GO TO 530                              BLE03510
C                                                                            BLE03520
C       ENTER 2ND PHASE OF COMPUTATION TO ATTEMPT TO OBTAIN MORE            BLE03530
C       ACCURATE EIGENVECTOR APPROXIMATIONS.                                 BLE03540
C       USER CONTROLS THE SIZE OF THE ERROR TOLERANCE BY SPECIFYING          BLE03550
C       THE PARAMETER RELTOL.                                                BLE03560
C                                                                            BLE03570
        IPHASE = 2                                                           BLE03580
        MAXIT = MAXIT2                                                       BLE03590
        KSET = KACT                                                          BLE03600
C                                                                            BLE03610
C       ERROR TOLERANCE USES THE CONVERGED EIGENVALUE LARGEST IN            BLE03620
C       MAGNITUDE.                                                           BLE03630
        TD(1) = DABS(E(1))                                                   BLE03640
        IF(KML.EQ.1) GO TO 350                                               BLE03650
        DO 340 J = 2,KML                                                     BLE03660
    340 IF(DABS(E(J)).GT.TD(1))   TD(1) = DABS(E(J))                         BLE03670
    350 TD(1) = DMAX1(TD(1),1.D0)                                            BLE03680
        ERRMAN = RELTOL**2 * TD(1)**2                                        BLE03690
        IF(ERRMAN.GE.ERRMAX) GO TO 430                                       BLE03700
        ERRMAX = ERRMAN                                                      BLE03710
C                                                                            BLE03720
        WRITE(6,360) ERRMAX, MAXIT2                                          BLE03730
    360 FORMAT(//' ENTER PHASE 2 OF COMPUTATION'/' CONVERGENCE CRITERION IBLE03740
       1S REDUCED TO ',E13.4/' NO MORE THAN ',I5,' MATRIX VECTOR MULTIPLIEBLE03750
       1S WILL BE ALLOWED.'/' PROGRAM WILL TERMINATE IF BLOCK ORTHOGONALITYBLE03760
       1 PROBLEMS MATERIALIZE'/)                                             BLE03770
C                                                                            BLE03780
        GO TO 190                                                            BLE03790
C                                                                            BLE03800
C       INCONSISTENCIES IN THE DATA                                          BLE03810
C                                                                            BLE03820
    370 WRITE(6,380) KM,KACT                                                 BLE03830
    380 FORMAT(/' PROGRAM TERMINATES BECAUSE THE NUMBER OF EIGENELEMENTS    BLE03840
       1REQUESTED, KM =',I3/'  IS LARGER THAN THE SIZE OF THE FIRST Q BLOCBLE03850
       1K, KACT =',I3,' SPECIFIED'/'  USER MUST RESET KM OR KACT'/)         BLE03860
        GO TO 530                                                           BLE03870
C                                                                            BLE03880
    390 WRITE(6,400) KMAX,N                                                  BLE03890
    400 FORMAT(/' PROGRAM TERMINATES BECAUSE KMAX = ',I5,' IS TOO LARGE FOBLE03900
       1R THE SIZE, N = ',I5,', OF THE GIVEN MATRIX'/'  USER MUST DECREASEBLE03910
       1THE SIZE OF KMAX.'/)                                                 BLE03920
        GO TO 530                                                           BLE03930
C                                                                            BLE03940
    410 WRITE(6,420) NOLD,N,KACT,KSET                                        BLE03950
    420 FORMAT(/' PROGRAM TERMINATES BECAUSE FAULT OCCURRED IN READING IN  BLE03960
       1THE EIGENVECTOR APPROXIMATIONS'/'  EITHER THE SIZE MATRIX SPECIFIEDBLE03970
```

```
      1ON THE EIGENVECTOR FILE' ,I6/'  DID NOT MATCH THE SIZE SPECIFIED 'BLE03980
      1,I5,' IN THE PROGRAM OR THE NUMBER'/'    OF VECTORS IN FILE 10 = 'BLE03990
      1,I4,' IS LESS THAN THE NUMBER ',I3/'   USER SAID WERE THERE'/)   BLE04000
         GO TO 530                                                      BLE04010
C                                                                       BLE04020
  430 WRITE(6,440) ERRMAN, ERRMAX                                       BLE04030
  440 FORMAT(/' COMPUTED PHASE 2 CONVERGENCE CRITERION ',E13.4/' IS LARBLE04040
      1GER THAN PHASE 1 CRITERION ',E13.4/'  SO PROGRAM TERMINATES'/)   BLE04050
         GO TO 530                                                      BLE04060
C                                                                       BLE04070
  450 WRITE(6,460) KACT,MXBLK                                           BLE04080
  460 FORMAT(/' PROGRAM TERMINATES BECAUSE THERE IS NOT ENOUGH ROOM TO  BLE04090
      1GENERATE 2 BLOCKS','   BECAUSE KACT = ',I3,' AND MXBLK = ', I4/)  BLE04100
         GO TO 530                                                      BLE04110
C                                                                       BLE04120
C                                                                       BLE04130
  470 WRITE(6,480) MDIMTM, MXBLK                                        BLE04140
  480 FORMAT(/' PROGRAM TERMINATES BECAUSE THE DIMENSION ',I6,' OF THE TBLE04150
      1M ARRAY'/'  IS TOO SMALL FOR THE LARGEST T-MATRIX ALLOWED ',I4)  BLE04160
         GO TO 530                                                      BLE04170
C                                                                       BLE04180
  490 WRITE(6,500)                                                      BLE04190
  500 FORMAT(/' USER SPECIFIED NUMBER OF EIGENVALUES OF INTEREST AS 0'/'BLE04200
      1 PROGRAM TERMINATES FOR USER TO RESET KM TO DESIRED NONZERO VALUE'BLE04210
      1/)                                                               BLE04220
         GO TO 530                                                      BLE04230
C                                                                       BLE04240
  510 WRITE(6,520) MDIMQ, KMAX,N                                        BLE04250
  520 FORMAT(/' PROGRAM TERMINATES BECAUSE THE DIMENSION ',I6,' OF THE QBLE04260
      1-ARRAY'/'  IS TOO SMALL TO HOLD ',I5, ' VECTORS OF LENGTH ',I4)  BLE04270
         GO TO 530                                                      BLE04280
C                                                                       BLE04290
  530 CONTINUE                                                          BLE04300
C                                                                       BLE04310
      STOP                                                              BLE04320
C-----END OF MAIN PROGRAM FOR BLOCK LANCZOS PROCEDURE------------------BLE04330
      END                                                               BLE04340
```

SECTION 8.4   SAMPLE MATRIX-VECTOR MULTIPLY SUBROUTINES

```
C-----BLMULT--------------------------------------------------------BLM00010
C                                                                    BLM00020
C     CONTAINS SAMPLE USPEC AND BMATV SUBROUTINES FOR USE WITH       BLM00030
C     THE BLOCK LANCZOS PROCEDURE FOR REAL SYMMETRIC MATRICES.       BLM00040
C     PROGRAMS ARE USED WITH BLEVAL AND BLSUB FILES.                 BLM00050
C                                                                    BLM00060
C     NONPORTABLE CONSTRUCTIONS:                                     BLM00070
C     1.  THE ENTRY MECHANISM USED TO PASS THE STORAGE              BLM00080
C         LOCATIONS OF THE USER-SPECIFIED MATRIX FROM THE           BLM00090
C         SUBROUTINE USPEC TO THE MATRIX-VECTOR SUBROUTINE          BLM00100
C         BMATV.                                                     BLM00110
C     2.  IN THE SAMPLE USPEC AND BMATV SUBROUTINES FOR DIAGONAL    BLM00120
C         TEST MATRICES: FREE FORMAT (8,*) AND THE FORMAT (20A4).   BLM00130
C                                                                    BLM00140
C-----USPEC (GENERAL SYMMETRIC SPARSE MATRICES)---------------------BLM00150
C                                                                    BLM00160
C     SUBROUTINE USPEC(N,MATNO,NNZ,AVER)                            BLM00170
      SUBROUTINE GUSPEC(N,MATNO,NNZ,AVER)                           BLM00180
C                                                                    BLM00190
C------------------------------------------------------------------BLM00200
      DOUBLE PRECISION  ASD(10000),AD(5010),AVER,NNZ               BLM00210
      INTEGER  IROW(10000),ICOL(5010)                               BLM00220
C------------------------------------------------------------------BLM00230
C     USPEC DIMENSIONS AND INITIALIZES THE ARRAYS NEEDED TO DEFINE  BLM00240
C     THE USER-SPECIFIED MATRIX AND THEN PASSES THE STORAGE LOCATIONSBLM00250
C     OF THESE ARRAYS TO THE MULTIPLY SUBROUTINE BMATV.            BLM00260
C                                                                    BLM00270
C     MATRIX IS STORED IN FOLLOWING SPARSE MATRIX FORMAT:           BLM00280
C     N = ORDER OF A-MATRIX,                                         BLM00290
C     NZS = NUMBER OF NONZERO SUBDIAGONAL ENTRIES,                   BLM00300
C     NZL = INDEX OF LAST COLUMN CONTAINING NONZERO SUBDIAGONAL ENTRIES, BLM00310
C     ICOL(J), J=1,NZL IS THE NUMBER OF NONZERO SUBDIAGONAL ELEMENTS BLM00320
C             IN COLUMN J.                                          BLM00330
C     IROW(K), K = 1,NZS IS THE CORRESPONDING ROW INDEX FOR ASD(K). BLM00340
C     AD(I), I=1,N CONTAINS DIAGONAL ENTRIES (INCLUDING ANY 0       BLM00350
C            DIAGONAL ENTRIES).                                     BLM00360
C     ASD(K), K=1,NZS CONTAINS NONZERO SUBDIAGONAL ENTRIES, BY COLUMN BLM00370
C     FOR J > NZL THERE ARE NO NONZERO SUBDIAGONAL ELEMENTS IN COLUMN J. BLM00380
C     ICOL(J) = 0 IS ALLOWED                                        BLM00390
C                                                                    BLM00400
C------------------------------------------------------------------BLM00410
C     ARRAYS THAT DEFINE THE MATRIX ARE READ IN FROM FILE 8         BLM00420
C                                                                    BLM00430
      READ(8,10) NZS,NOLD,NZL,MATOLD                                BLM00440
   10 FORMAT(I10,2I6,I8)                                            BLM00450
C                                                                    BLM00460
      WRITE(6,20) NZS,NOLD,NZL,MATOLD                               BLM00470
   20 FORMAT(I10,2I6,I8,' = NZS,NOLD,NZL,MATOLD'/)                  BLM00480
C                                                                    BLM00490
C     TEST OF PARAMETER CORRECTNESS                                 BLM00500
      ITEMP = (NOLD-N)**2 + (MATNO-MATOLD)**2                       BLM00510
C                                                                    BLM00520
      IF(ITEMP.EQ.0) GO TO 40                                       BLM00530
C                                                                    BLM00540
      WRITE(6,30) NOLD,N,MATOLD,MATNO                               BLM00550
   30 FORMAT(/' PROGRAM TERMINATES BECAUSE EITHER THE SIZE ',I4,' OF THEBLM00560
     1 MATRIX'/' READ FROM FILE 8 DIFFERS FROM THE SIZE ',I4,' SPECIFIEDBLM00570
     1 BY'/' THE USER OR THE MATNO ',I8,' READ IN DIFFERS FROM THE MATNOBLM00580
     1 '/ I8,' SPECIFIED BY THE USER'/)                            BLM00590
      GO TO 100                                                     BLM00600
C                                                                    BLM00610
   40 CONTINUE                                                       BLM00620
C                                                                    BLM00630
```

```
C       NUMBER OF NONZERO SUBDIAGONAL ENTRIES IN EACH COLUMN IS READ    BLM00640
C       THEN THE CORRESPONDING ROW INDEX FOR EACH SUCH ENTRY IS READ    BLM00650
        READ(8,50) (ICOL(K), K=1,NZL)                                   BLM00660
        READ(8,50) (IROW(K), K=1,NZS)                                   BLM00670
     50 FORMAT(1316)                                                    BLM00680
C                                                                       BLM00690
C       DIAGONAL IS READ FIRST, THEN NONZERO BELOW DIAGONAL ENTRIES     BLM00700
        READ(8,60) (AD(K), K=1,N)                                       BLM00710
        READ(8,60) (ASD(K), K=1,NZS)                                    BLM00720
     60 FORMAT(4E19.10)                                                 BLM00730
C                                                                       BLM00740
C       COMPUTE NNZ, THE AVERAGE NUMBER OF NONZEROS PER COLUMN, AND     BLM00750
C       AVER, THE AVERAGE SIZE OF NONZERO ENTRIES.                      BLM00760
        ITCOL = 0                                                       BLM00770
        AVER = 0.DO                                                     BLM00780
        DO 70 K = 1,N                                                   BLM00790
        IF(DABS(AD(K)).EQ.0.DO)  GO TO 70                               BLM00800
        ITCOL = ITCOL + 1                                               BLM00810
        AVER = AVER + DABS(AD(K))                                       BLM00820
     70 CONTINUE                                                        BLM00830
        NTCOL = ITCOL                                                   BLM00840
        DO 80 K = 1,N                                                   BLM00850
     80 ITCOL = ITCOL + 2*ICOL(K)                                       BLM00860
        NNZ = DFLOAT(ITCOL)/DFLOAT(N)                                   BLM00870
        DO 90 K = 1,NZS                                                 BLM00880
     90 AVER = AVER + DABS(ASD(K))                                      BLM00890
        AVER = AVER/DFLOAT(NZS + NTCOL)                                 BLM00900
C                                                                       BLM00910
C-----------------------------------------------------------------------BLM00920
C       PASS STORAGE LOCATIONS OF ARRAYS THAT DEFINE THE MATRIX TO      BLM00930
C       THE MATRIX-VECTOR MULTIPLY SUBROUTINE BMATV                     BLM00940
C                                                                       BLM00950
        CALL BMATVE(ASD,AD,ICOL,IROW,N,NZL)                             BLM00960
C-----------------------------------------------------------------------BLM00970
C                                                                       BLM00980
        RETURN                                                          BLM00990
    100 STOP                                                            BLM01000
C-----END OF USPEC------------------------------------------------------BLM01010
        END                                                             BLM01020
C                                                                       BLM01030
C-----MATRIX-VECTOR MULTIPLY FOR REAL SPARSE SYMMETRIC MATRICES---------BLM01040
C                                                                       BLM01050
C       SUBROUTINE BMATV(W,U)                                           BLM01060
        SUBROUTINE GBMATV(W,U)                                          BLM01070
C                                                                       BLM01080
C-----------------------------------------------------------------------BLM01090
        DOUBLE PRECISION   U(1),W(1),ASD(1),AD(1)                       BLM01100
        INTEGER  IROW(1),ICOL(1)                                        BLM01110
        COMMON/LOOPS/MAXIT,ITER                                         BLM01120
C-----------------------------------------------------------------------BLM01130
C       SPARSE MATRIX-VECTOR MULTIPLY FOR LANCZS   U = A*W              BLM01140
C       SEE USPEC SUBROUTINE FOR DESCRIPTION OF THE ARRAYS THAT DEFINE  BLM01150
C       THE A-MATRIX                                                    BLM01160
C-----------------------------------------------------------------------BLM01170
C       INCREMENT THE A*W COUNTER                                       BLM01180
        ITER = ITER + 1                                                 BLM01190
C       COMPUTE THE DIAGONAL TERMS                                      BLM01200
        DO 10 I = 1,N                                                   BLM01210
     10 U(I) = AD(I)*W(I)                                               BLM01220
C                                                                       BLM01230
C       COMPUTE BY COLUMN                                               BLM01240
        LLAST = 0                                                       BLM01250
        DO 30 J = 1,NZL                                                 BLM01260
C                                                                       BLM01270
        IF (ICOL(J).EQ.0) GO TO 30                                      BLM01280
        LFIRST = LLAST + 1                                              BLM01290
```

```
      LLAST = LLAST + ICOL(J)                                  BLM01300
C                                                              BLM01310
      DO 20 L = LFIRST,LLAST                                   BLM01320
      I = IROW(L)                                              BLM01330
C                                                              BLM01340
      U(I) = U(I) + ASD(L)*W(J)                                BLM01350
      U(J) = U(J) + ASD(L)*W(I)                                BLM01360
C                                                              BLM01370
   20 CONTINUE                                                 BLM01380
C                                                              BLM01390
   30 CONTINUE                                                 BLM01400
C                                                              BLM01410
      RETURN                                                   BLM01420
C                                                              BLM01430
C------------------------------------------------------------BLM01440
C   STORAGE LOCATIONS OF ARRAYS ARE PASSED TO BMATV FROM USPEC BLM01450
C                                                              BLM01460
      ENTRY BMATVE(ASD,AD,ICOL,IROW,N,NZL)                     BLM01470
C------------------------------------------------------------BLM01480
C                                                              BLM01490
      RETURN                                                   BLM01500
C-----END OF BMATV-------------------------------------------BLM01510
      END                                                      BLM01520
C                                                              BLM01530
C-----MATRIX-VECTOR MULTIPLY FOR DIAGONAL TEST MATRICES------BLM01540
C   BMATV COMPUTES   U = (DIAGONAL MATRIX) * W                 BLM01550
C                                                              BLM01560
      SUBROUTINE BMATV(W,U)                                    BLM01570
C     SUBROUTINE DBMATV(W,U)                                   BLM01580
C                                                              BLM01590
C------------------------------------------------------------BLM01600
      DOUBLE PRECISION  W(1),U(1),D(1)                         BLM01610
      COMMON/LOOPS/MAXIT,ITER                                  BLM01620
C------------------------------------------------------------BLM01630
C   INCREMENT THE LOOP COUNTER                                 BLM01640
      ITER = ITER + 1                                          BLM01650
C                                                              BLM01660
      DO 10 I=1,N                                              BLM01670
   10 U(I)= D(I)*W(I)                                          BLM01680
C  10 U(I)= -D(I)*W(I)                                         BLM01690
C                                                              BLM01700
      RETURN                                                   BLM01710
C                                                              BLM01720
C------------------------------------------------------------BLM01730
      ENTRY MVDIAE(D,N)                                        BLM01740
C------------------------------------------------------------BLM01750
C                                                              BLM01760
      RETURN                                                   BLM01770
C-----END OF DIAGONAL TEST MATRIX  MULTIPLY-----------------BLM01780
      END                                                      BLM01790
C                                                              BLM01800
C-----START OF USPEC FOR DIAGONAL TEST MATRIX--------------BLM01810
C                                                              BLM01820
      SUBROUTINE USPEC(N,MATNO,NNZ,AVER)                       BLM01830
C     SUBROUTINE DUSPEC(N,MATNO,NNZ,AVER)                      BLM01840
C                                                              BLM01850
C------------------------------------------------------------BLM01860
      DOUBLE PRECISION  D(1000),SPACE,SHIFT,AVER,NNZ           BLM01870
      DOUBLE PRECISION  DABS, DFLOAT                           BLM01880
      REAL  EXPLAN(20)                                         BLM01890
C------------------------------------------------------------BLM01900
C                                                              BLM01910
      READ(8,10) EXPLAN                                        BLM01920
   10 FORMAT(20A4)                                             BLM01930
      READ(8,*) NOLD,NUNIF,SPACE,D(1),SHIFT                    BLM01940
      NNUNIF = NOLD - NUNIF                                    BLM01950
      WRITE(6,20) NOLD,SPACE,NNUNIF,D(1),SHIFT                 BLM01960
```

```
   20 FORMAT(/' DIAGONAL TEST MATRIX, SIZE = ',I4/' MOST ENTRIES ARE ',   BLM01970
     1E10.3,' UNITS APART.',I3,' ENTRIES'/' ARE IRREGULARLY SPACED. FIRSBLM01980
     1T ENTRY IS ',E10.3,' SHIFT = ',E10.3/)                            BLM01990
C                                                                       9LM02000
      IF(N.NE.NOLD) GO TO 100                                           BLM02010
C     COMPUTE THE UNIFORM PORTION OF THE SPECTRUM                       BLM02020
      DO 30 J=2,NUNIF                                                   BLM02030
   30 D(J) = D(1) - DFLOAT(J-1)*SPACE                                   BLM02040
      NUNIF1=NUNIF + 1                                                  BLM02050
      READ(8,10) EXPLAN                                                 BLM02060
      DO 40 J=NUNIF1,N                                                  BLM02070
   40 READ(8,*) D(J)                                                    BLM02080
C                                                                       BLM02090
      IF(SHIFT.EQ.0.) GO TO 60                                          BLM02100
      DO 50 J=1,N                                                       BLM02110
   50 D(J) = D(J) + SHIFT                                               BLM02120
C                                                                       BLM02130
C     PRINT OUT THE EIGENVALUES OF INTEREST                             BLM02140
   60 WRITE(6,70) (D(I), I=1,10 )                                       BLM02150
      NB = NUNIF - 2                                                    BLM02160
      WRITE(6,80) (D(I), I = NB,N)                                      BLM02170
   70 FORMAT(/' BLOCK LANCZOS TEST, 1ST 10 ENTRIES OF DIAGONAL TEST MATRBLM02180
     11X = '/(3E22.14))                                                 BLM02190
   80 FORMAT(/' MIDDLE UNIFORM PORTION OF MATRIX IS NOT PRINTED OUT'/    BLM02200
     1' END OF UNIFORM PLUS NONUNIFORM SECTION = '/(3E25.16))           BLM02210
C                                                                       BLM02220
C     DIAGONAL GENERATION COMPLETE                                      BLM02230
C     COMPUTE NNZ AND AVER                                              BLM02240
      NNZ = 1.D0                                                        BLM02250
      AVER = 0.D0                                                       BLM02260
      DO 90 K = 1,N                                                     BLM02270
   90 AVER = AVER + DABS(D(K))                                          BLM02280
      AVER = AVER/DFLOAT(N)                                             BLM02290
C                                                                       BLM02300
C-----------------------------------------------------------------------BLM02310
C     CALL ENTRY TO MATRIX-VECTOR MULTIPLY SUBROUTINE  TO PASS          BLM02320
C     STORAGE LOCATION OF D-ARRAY AND ORDER OF A-MATRIX.                BLM02330
C                                                                       BLM02340
      CALL MVDIAE(D,N)                                                  BLM02350
C-----------------------------------------------------------------------BLM02360
C                                                                       BLM02370
      RETURN                                                            BLM02380
  100 WRITE(6,110) NOLD,N                                               BLM02390
  110 FORMAT(' PROGRAM TERMINATES BECAUSE NOLD = ',I5,'DOES NOT EQUAL N BLM02400
     1 =',I5)                                                           BLM02410
C-----END OF USPEC SUBROUTINE FOR 'DIAGONAL' TEST MATRICES-------------BLM02420
      STOP                                                              BLM02430
      END                                                               BLM02440
```

SECTION 8.5   OTHER SUBROUTINES USED BY THE PROGRAMS IN CHAPTERS 8 AND 9

```
C-----BLSUB-------------------------------------------------------------BLS00010
C                                                                       BLS00020
C     PFORT VERIFIER IDENTIFIED THE FOLLOWING NONPORTABLE               BLS00030
C     CONSTRUCTIONS:                                                    BLS00040
C     1. ENTRY MECHANISMS USED TO PASS THE STORAGE LOCATIONS OF         BLS00050
C        SEVERAL ARRAYS FROM THE MAIN PROGRAM TO THE SUBROUTINES        BLS00060
C        LANCZS AND LANCI1.                                             BLS00070
C     2. COMMON BLOCK: LOOPS: USED IN LANCZS AND LANCI1.                BLS00080
C                                                                       BLS00090
C     SUBROUTINES:     LANCZS, LANCI1, ORTHOG, START, AND DIAGOM        BLS00100
C                      ARE USED WITH THE BLOCK LANCZOS PROGRAMS         BLS00110
C                      BLEVAL AND BLIEVAL. LPERM IS USED WITH BLIEVAL.  BLS00120
C                                                                       BLS00130
C                                                                       BLS00140
C-----LANCZS FOR BLOCK LANCZOS PROCEDURE--------------------------------BLS00150
C                                                                       BLS00160
C     ON EACH ITERATION CALLS LANCI1 SUBROUTINE TO GENERATE             BLS00170
C     THE Q-SUBBLOCKS AND THEN CALLS DIAGOM SUBROUTINE TO               BLS00180
C     DIAGONALIZE THE SMALL SYMMETRIC MATRIX WHICH IS THE PROJECTION    BLS00190
C     OF THE MATRIX BEING USED BY LANCZS ONTO THE SUBSPACE SPANNED      BLS00200
C     BY THESE Q-BLOCKS.                                                BLS00210
C                                                                       BLS00220
      SUBROUTINE LANCZS(MATVEC,KML,KSET,KACT,MXBLK,N,Q,E,RESIDL,        BLS00230
     1 RESIDK,ERRMAX,IPHASE,NITER,IWRITE)                               BLS00240
C                                                                       BLS00250
C---------------------------------------------------------------------- BLS00260
      DOUBLE PRECISION  E(1),Q(1),ERRMAX,SPREC,RESN,FRACT,RKM,SUM       BLS00270
      DOUBLE PRECISION  TM(1),SM(1),TD(1),TOD(1),RESIDL(1),RESIDK(1)    BLS00280
      REAL  G(1)                                                        BLS00290
      INTEGER  EFLAG,OFLAG,DIR(2,1),DESC(1),LEFT(1),XLFT(1)             BLS00300
      DOUBLE PRECISION FINPRO                                           BLS00310
      COMMON /LOOPS/MAXIT,ITER                                          BLS00320
      COMMON/FLAGS/EFLAG,OFLAG                                          BLS00330
      EXTERNAL MATVEC                                                   BLS00340
C---------------------------------------------------------------------- BLS00350
      KM = KML                                                          BLS00360
      MMT = MXBLK*MXBLK                                                 BLS00370
      MPMIN = -1000                                                     BLS00380
      IKACT = KACT + 10                                                 BLS00390
      FRACT = RESIDL(1)                                                 BLS00400
      NSTAG = RESIDL(2)                                                 BLS00410
      IORTHO = 0                                                        BLS00420
C                                                                       BLS00430
C     CONSTRUCT STARTING VECTORS                                        BLS00440
      IF(KSET.EQ.0) GO TO 10                                            BLS00450
C---------------------------------------------------------------------- BLS00460
      CALL ORTHOG(1,KSET,N,Q)                                           BLS00470
   10 CALL START (KSET+1,KACT,N,Q,G,ERRMAX)                            BLS00480
C---------------------------------------------------------------------- BLS00490
   20 CONTINUE                                                          BLS00500
C     INITIALIZE THE LANCZOS T-MATRIX.                                  BLS00510
      DO 30 J=1,MMT                                                     BLS00520
   30 TM(J)=0.D0                                                        BLS00530
C                                                                       BLS00540
C     INITIALIZE THE Q-BLOCK DIRECTORY                                  BLS00550
      DIR(1,1)=1                                                        BLS00560
      DIR(2,1)=KACT                                                     BLS00570
C                                                                       BLS00580
C     ORTHOGONALIZE THE STARTING VECTORS                               BLS00590
      IF(NITER.EQ.0) GO TO 40                                           BLS00600
C---------------------------------------------------------------------- BLS00610
      CALL ORTHOG(1,KACT,N,Q)                                           BLS00620
C---------------------------------------------------------------------- BLS00630
```

```
     40 CONTINUE                                                        BLS00640
C                                                                       BLS00650
C        GENERATE THE QSUBBLOCKS USED ON ITERATION NITER AND STORE IN   BLS00660
C        THE Q-ARRAY                                                    BLS00670
C                                                                       BLS00680
         DO 90 I=1,MXBLK                                                BLS00690
C                                                                       BLS00700
C---------------------------------------------------------------------BLS00710
         CALL LANCII(MATVEC,MXBLK,NITER,I,N,Q,KACT,KML,ERRMAX,RESN,RKM, 3LS00720
        1 IND,KACTN,IWRITE)                                             BLS00730
C---------------------------------------------------------------------BLS00740
C                                                                       BLS00750
C        HAS CONVERGENCE OCCURRED?                                      BLS00760
         II = I+1                                                       BLS00770
         IF (I.EQ.1.AND.DIR(2,I).EQ.DIR(2,II)) GO TO 140               BLS00780
C                                                                       BLS00790
C        WAS THERE ROOM FOR ANOTHER Q-BLOCK?                            BLS00800
         IF (DIR(2,II).LT.DIR(1,II))  GO TO 100                        BLS00810
C                                                                       BLS00820
C        IF OFLAG = 1 OR IPHASE = 2, CHECK THE ORTHOGONALITY OF         BLS00830
C        THE Q-SUBBLOCKS GENERATED WITH RESPECT TO THAT VECTOR          BLS00840
C        IN THE 1ST Q-BLOCK WHICH IS GENERATING DESCENDANTS.            BLS00850
C        IN PHASE 2 LOSSES IN ORTHOGONALITY ARE USED TO                 BLS00860
C        DETERMINE WHEN THE LIMITS ON THE ACHIEVABLE ACCURACY HAVE      BLS00870
C        BEEN REACHED.                                                  BLS00880
C                                                                       BLS00890
         IF(OFLAG.EQ.O.AND.IPHASE.EQ.1) GO TO 90                       BLS00900
C                                                                       BLS00910
         L1=DIR(1,II)                                                   BLS00920
         LL1 = (L1-1)*N + 1                                             BLS00930
         IND1 = (IND-1)*N + 1                                           BLS00940
C---------------------------------------------------------------------BLS00950
         SUM = FINPRO(N,Q(IND1),1,Q(LL1),1)                            BLS00960
C---------------------------------------------------------------------BLS00970
C                                                                       BLS00980
         IF(DABS(SUM).LT.SPREC)GO  TO 80                               BLS00990
C                                                                       BLS01000
         IF(IWRITE.EQ.1) WRITE(6,50) IND,L1,SUM,I                      BLS01010
     50 FORMAT(/' INNER PRODUCT OF VECTORS ',I3,' AND ',I3,' = ',E13.3/ BLS01020
        1' THIS VIOLATES ORTHOGONALITY TEST.  TERMINATE BLOCK GENERATION'BLS01030
        1/' WITH ',I3,'TH BLOCK '/)                                    BLS01040
C                                                                       BLS01050
C        ORTHOGONALITY TEST VIOLATED, TERMINATE BLOCK GENERATION        BLS01060
C        FOR THIS ITERATION.  IN PHASE 2 KEEP TRACK OF NUMBER OF        BLS01070
C        SUCH VIOLATIONS THAT LIMIT THE NUMBER OF BLOCKS TO < 10.       BLS01080
C        TERMINATE AFTER 3 SUCH VIOLATIONS IN PHASE 2.                  BLS01090
         IF(IPHASE.NE.1.AND.I.LT.IKACT)  IORTHO = IORTHO + 1           BLS01100
         IF(IORTHO.LT.3.AND.II.NE.2)  GO TO 70                         BLS01110
         WRITE(6,60)                                                    BLS01120
     60 FORMAT(/' THE ORTHOGONALITY TEST HAS FAILED THREE TIMES'/       BLS01130
        1' TERMINATE THE BLOCK PROCEDURE'/)                            BLS01140
         IPHASE = -1000                                                 BLS01150
C        BEFORE TERMINATING WRITE THE CURRENT EIGENVECTOR/EIGENVALUE    BLS01160
C        APPROXIMATIONS TO FILE 15                                      BLS01170
         GO TO 160                                                      BLS01180
C                                                                       BLS01190
C        TERMINATE THE Q-BLOCK GENERATION ON THIS ITERATION            BLS01200
     70 DIR(2,II)=DIR(2,I)                                              BLS01210
         GO TO 100                                                      BLS01220
C                                                                       BLS01230
     80 CONTINUE                                                        BLS01240
C                                                                       BLS01250
C        END OF ORTHOGONALITY TESTS                                     BLS01260
C                                                                       BLS01270
     90 CONTINUE                                                        BLS01280
C                                                                       BLS01290
```

```
C     END OF RECURSIVE Q-BLOCK GENERATION                               BLS01300
C                                                                       BLS01310
  100 CONTINUE                                                          BLS01320
      MM = DIR(2,II)                                                    BLS01330
      IF(IWRITE.EQ.1)  WRITE (6,110)  MM,I                             BLS01340
  110 FORMAT(' T-MATRIX IS OF ORDER ',I3, ' NUMBER OF BLOCKS = ',I3)   BLS01350
C                                                                       BLS01360
C-----------------------------------------------------------------------BLS01370
C     DIAGONALIZE THE PROJECTION MATRIX TM.  ON RETURN THE              BLS01380
C     UPDATED APPROXIMATIONS TO THE DESIRED EIGENVECTORS ARE IN THE     BLS01390
C     FIRST KACT COLUMNS OF THE Q-ARRAY.                                BLS01400
C     UPDATED EIGENVALUE APPROXIMATIONS ARE IN E.                       BLS01410
      TD(1) = RKM                                                       BLS01420
      TD(2) = FRACT                                                     BLS01430
      IERR = NSTAG                                                      BLS01440
C                                                                       BLS01450
      CALL DIAGOM(MXBLK,MM,TM,KACT,N,Q,E,RESIDL,RESIDK,                 BLS01460
     1 RESN,IND,KACTN,KM,TD,TOD,NITER,IERR,IWRITE)                      BLS01470
C-----------------------------------------------------------------------BLS01480
C                                                                       BLS01490
C     INCREMENT COUNTER FOR NUMBER OF BLOCK LANCZOS ITERATIONS          BLS01500
      NITER = NITER + 1                                                 BLS01510
C     IWRITE = MPMIN MEANS BLOCK LANCZOS PROCEDURE TERMINATED ABNORMALLYBLS01520
      IF(IWRITE.EQ.MPMIN)  GO TO 160                                    BLS01530
C     IERR .NE. 0 MEANS EISPACK SUBROUTINE DEFAULTED                    BLS01540
      IF(IERR.EQ.0) GO TO 130                                           BLS01550
      WRITE(6,120)                                                      BLS01560
  120 FORMAT(//' EISPACK SIGNALS TROUBLE IN SMALL IMTQL2 EIGENVALUE SUBROBLS01570
     1UTINE,'/' SO BLOCK LANCZOS PROGRAM TERMINATES'/)                  BLS01580
      ITER = -ITER                                                      BLS01590
C                                                                       BLS01600
      RETURN                                                            BLS01610
C                                                                       BLS01620
  130 IF (ITER.GE.MAXIT) GO TO 160                                      BLS01630
C                                                                       BLS01640
C     UPDATED APPROXIMATIONS WERE OBTAINED WITHOUT EXCEEDING            BLS01650
C     MAXIMUM NUMBER OF MATRIX-VECTOR MULTIPLIES SET BY THE USER.       BLS01660
C     CONTINUE BLOCK LANCZOS LOOP ITERATIONS                           BLS01670
C                                                                       BLS01680
      GO TO 20                                                          BLS01690
C                                                                       BLS01700
  140 WRITE(6,150)                                                      BLS01710
  150 FORMAT(//' BLOCK LANCZOS PROCEDURE CONVERGED'//)                  BLS01720
C                                                                       BLS01730
C     BLOCK LANCZOS PROCEDURE HAS CONVERGED.                            BLS01740
C     ATTEMPT TO IMPROVE THE APPROXIMATE EIGENVECTORS BY DIAGONALIZING  BLS01750
C     THE SMALL PROJECTION MATRIX OBTAINED BY USING ONLY THE            BLS01760
C     FIRST BLOCK IN Q-ARRAY.                                           BLS01770
C                                                                       BLS01780
  160 KACT2 = KACT*MXBLK                                                BLS01790
      DO 170 KK = 1,KACT2                                               BLS01800
  170 TM(KK) = 0.D0                                                     BLS01810
C-----------------------------------------------------------------------BLS01820
      CALL ORTHOG(1,KACT,N,Q)                                           BLS01830
C-----------------------------------------------------------------------BLS01840
      KKO = 1-N                                                         BLS01850
      KACTP1 = (KACT)*N + 1                                             BLS01860
      JJO = -MXBLK-1                                                    BLS01870
      DO 190 K=1,KACT                                                   BLS01880
      JJO = JJO + MXBLK + 1                                             BLS01890
      KKO = KKO + N                                                     BLS01900
C-----------------------------------------------------------------------BLS01910
      CALL MATVEC(Q(KKO),Q(KACTP1))                                     BLS01920
C-----------------------------------------------------------------------BLS01930
      LLO = (K-2)*N + 1                                                 BLS01940
      JJ = JJO                                                          BLS01950
      DO 180 L=K,KACT                                                   BLS01960
```

```
      LLO = LLO + N                                                 BLS01970
      JJ=JJ+1                                                       BLS01980
C-------------------------------------------------------------------BLS01990
      TM(JJ) = FINPRO(N,Q(LLO),1,Q(KACTP1),1)                       BLS02000
C-------------------------------------------------------------------BLS02010
  180 CONTINUE                                                      BLS02020
C                                                                   BLS02030
  190 CONTINUE                                                      BLS02040
C                                                                   BLS02050
C-------------------------------------------------------------------BLS02060
C     USE EISPACK SUBROUTINE TRED2 TO TRIDIAGONALIZE TM-MATRIX      BLS02070
C     TM = (1ST Q-BLOCK)-TRANSPOSE*A*(1ST Q-BLOCK).                 BLS02080
C     ON RETURN DIAGONAL ELEMENTS COMPUTED ARE IN TD, OFF-DIAGONAL  BLS02090
C     ELEMENTS ARE IN TOD, TRANSFORMATIONS USED ARE IN TM.         BLS02100
C     THEN USE EISPACK SUBROUTINE IMTQL2 TO DIAGONALIZE THE T-MATRIX.BLS02110
C     ON RETURN.  EIGENVALUES ARE IN TD IN ASCENDING ORDER.        BLS02120
C     CORRESPONDING EIGENVECTORS ARE IN TM.                        BLS02130
C                                                                   BLS02140
      CALL TRED2(MXBLK,KACT,TM,TD,TOD,TM)                           BLS02150
      CALL IMTQL2(MXBLK,KACT,TD,TOD,TM,IERR)                        BLS02160
C-------------------------------------------------------------------BLS02170
C                                                                   BLS02180
      IF(IERR.EQ.0)  GO TO 200                                      BLS02190
      WRITE(6,120)                                                  BLS02200
      ITER = -ITER                                                  BLS02210
C                                                                   BLS02220
      RETURN                                                        BLS02230
C                                                                   BLS02240
C     COMPUTE SUCCESSIVELY THE JTH-COMPONENTS OF THE RITZ VECTORS.  BLS02250
C     REORDER THE EIGENVALUES (AND EIGENVECTORS) SO THAT THEY       BLS02260
C     ARE IN ALGEBRAICALLY DECREASING ORDER.                       BLS02270
C                                                                   BLS02280
  200 DO 220 J=1,N                                                  BLS02290
      JJO = - MXBLK                                                 BLS02300
      JLO = -N + J                                                  BLS02310
      DO 210 K=1,KACT                                               BLS02320
      TOD(K)=0.D0                                                   BLS02330
      JJO = JJO + MXBLK                                             BLS02340
      JJ= JJO                                                       BLS02350
      JL = JLO                                                      BLS02360
      DO 210 L=1,KACT                                               BLS02370
      JJ=JJ+1                                                       BLS02380
      JL = JL + N                                                   BLS02390
  210 TOD(K)=TOD(K)+TM(JJ)*Q(JL)                                    BLS02400
      JK = JLO                                                      BLS02410
      DO 220 K=1,KACT                                               BLS02420
      JK = JK + N                                                   BLS02430
      KACTK = KACT - K + 1                                          BLS02440
      Q(JK)=TOD(KACTK)                                              BLS02450
  220 CONTINUE                                                      BLS02460
      DO 230 K=1,KACT                                               BLS02470
      KACTK = KACT - K + 1                                          BLS02480
  230 E(K)=TD(KACTK)                                                BLS02490
C                                                                   BLS02500
C     HAS CONVERGENCE OCCURRED?                                     BLS02510
      IF(I.EQ.1.AND.DIR(2,I).EQ.DIR(2,I+1))  GO TO 250             BLS02520
C                                                                   BLS02530
C     CONVERGENCE HAS NOT OCCURRED, PROCEDURE TERMINATED FOR SOME   BLS02540
C     OTHER REASON                                                  BLS02550
      WRITE(6,240)                                                  BLS02560
  240 FORMAT(//' BLOCK LANCZOS PROCEDURE TERMINATES WITHOUT CONVERGENCE'BLS02570
     1/' AFTER WRITING THE CURRENT EIGENVALUE AND EIGENVECTOR APPROXIMATBLS02580
     1IONS'/' TO FILE 15'/)                                        BLS02590
C                                                                   BLS02600
      RETURN                                                        BLS02610
C                                                                   BLS02620
  250 IF(IPHASE.EQ.1) WRITE(6,260) (E(K), K=1,KACT)                BLS02630
```

```
      IF(IPHASE.EQ.2) WRITE(6,270) (E(K), K=1,KACT)                 BLS02640
  260 FORMAT(/' AT END OF PHASE 1, COMPUTED EIGENVALUES ='/(4E20.12))  BLS02650
  270 FORMAT(/' AT END OF PHASE 2, COMPUTED EIGENVALUES ='/(4E20.12))  BLS02660
C                                                                   BLS02670
      RETURN                                                        BLS02680
C                                                                   BLS02690
C-----------------------------------------------------------------BLS02700
C     ENTRY RECEIVES STORAGE LOCATIONS OF SEVERAL OF THE ARRAYS    BLS02710
C     USED BY THE LANCZS SUBROUTINE.  THIS ALLOWS USER TO SPECIFY  BLS02720
C     THE DIMENSIONS OF THESE ARRAYS IN THE MAIN PROGRAM.          BLS02730
C                                                                   BLS02740
      ENTRY LANZP(DIR,DESC,SM,TM,TOD,TD,G,XLFT,LEFT,SPREC)         BLS02750
C-----------------------------------------------------------------BLS02760
C                                                                   BLS02770
C-----END OF LANCZS-----------------------------------------------BLS02780
      RETURN                                                        BLS02790
      END                                                           BLS02800
C                                                                   BLS02810
C-----START OF LANCI1---------------------------------------------BLS02820
C     GENERATES THE Q-SUBBLOCKS ON EACH ITERATION OF THE BLOCK LANCZOS BLS02830
C     PROCEDURE.                                                   BLS02840
C                                                                   BLS02850
      SUBROUTINE LANCI1(MATVEC,MXBLK,NITER,I,N,Q,KACT,KML,ERRMAX,  BLS02860
     1RESN,RKM,IND,KACTN,IWRITE)                                   BLS02870
C                                                                   BLS02880
C-----------------------------------------------------------------BLS02890
      DOUBLE PRECISION  Q(1),TM(1),S,SM(1),T,ERRMAX,SUM,RESN,RKM   BLS02900
      INTEGER  DIR(2,1),DESC(1),LEFT(1),XLFT(1)                    BLS02910
      DOUBLE PRECISION FINPRO, DSQRT                               BLS02920
      EXTERNAL MATVEC                                              BLS02930
C-----------------------------------------------------------------BLS02940
C     SIZE OF FIRST BLOCK CAN CHANGE.                              BLS02950
      IF(I.EQ.1) KACTN = KACT                                      BLS02960
C                                                                   BLS02970
C     XLFT(I+2) IS CUMULATIVE TOTAL OF VECTORS IN 1ST QBLOCK NOT   BLS02980
C     GENERATING DESCENDANTS.                                      BLS02990
C                                                                   BLS03000
      IF(I.GT.1) GO TO 10                                          BLS03010
      XLFT(1) = 0                                                  BLS03020
      XLFT(2) = 0                                                  BLS03030
   10 XLFT(I+2) = XLFT(I+1)                                        BLS03040
C                                                                   BLS03050
C     INITIALIZE THE DIRECTORY FOR NEXT QBLOCK Q(I+1)              BLS03060
C                                                                   BLS03070
      I2=DIR(2,I)                                                  BLS03080
      I1=DIR(1,I)                                                  BLS03090
      DIR(1,I+1)=I2+1                                              BLS03100
      DIR(2,I+1)=I2                                                BLS03110
C                                                                   BLS03120
C     IS THERE ROOM FOR ANOTHER QBLOCK?                            BLS03130
C                                                                   BLS03140
      MS = I2-I1+1                                                 BLS03150
      IF (MS+I2.LE.MXBLK) GO TO 70                                 BLS03160
C                                                                   BLS03170
C     NOT ENOUGH ROOM TO GENERATE ANOTHER BLOCK                    BLS03180
C     COMPLETE THE TM-MATRIX. NOTE THAT THE TM-MATRIX IS           BLS03190
C     DIMENSIONED AS (MXBLK,1) AND THE EISPACK SUBROUTINES         BLS03200
C     REQUIRE THE LOWER TRIANGULAR PART OF THIS MATRIX.            BLS03210
C                                                                   BLS03220
      I3=I2+1                                                      BLS03230
      JI30 = (I3-1)*N                                              BLS03240
      JI31 = JI30 + 1                                              BLS03250
      JK1 = (I1-2)*N + 1                                           BLS03260
      DO 60 K=I1,I2                                                BLS03270
      JK1 = JK1 + N                                                BLS03280
C-----------------------------------------------------------------BLS03290
```

```
      CALL MATVEC(Q(JK1),Q(J131))                                 BLS03300
C------------------------------------------------------------------BLS03310
C     COMPUTE LAST DIAGONAL BLOCK IN TM-MATRIX FOR THIS ITERATION  BLS03320
C                                                                  BLS03330
      JL1 = (K-2)*N + 1                                            BLS03340
      KK = (K-1)*MXBLK +  K - 1                                    BLS03350
   20 DO 30 L=K,I2                                                 BLS03360
      KK = KK + 1                                                  BLS03370
      JL1 = JL1 + N                                                BLS03380
C------------------------------------------------------------------BLS03390
      TM(KK) = FINPRO(N,Q(JL1),1,Q(J131),1)                        BLS03400
C------------------------------------------------------------------BLS03410
   30 CONTINUE                                                     BLS03420
C                                                                  BLS03430
C     COMPUTE ASSOCIATED CORRECTION TERMS IN TM-MATRIX.            BLS03440
      IF(XLFT(I).EQ.0) GO TO 50                                    BLS03450
      LUP = XLFT(I)                                                BLS03460
      DO 40 JJ = 1,LUP                                             BLS03470
      L= LEFT(JJ)                                                  BLS03480
      JL1 = (L-1)*N + 1                                            BLS03490
C------------------------------------------------------------------BLS03500
      SUM = FINPRO(N,Q(J131),1,Q(JL1),1)                           BLS03510
C------------------------------------------------------------------BLS03520
      KK = (L-1)*MXBLK + K                                         BLS03530
      TM(KK) = SUM + TM(KK)                                        BLS03540
   40 CONTINUE                                                     BLS03550
C                                                                  BLS03560
   50 CONTINUE                                                     BLS03570
C                                                                  BLS03580
   60 CONTINUE                                                     BLS03590
C                                                                  BLS03600
      RETURN                                                       BLS03610
C                                                                  BLS03620
C     ON EVERY BLOCK PASS THROUGH HERE TO GENERATE THE ITH-BLOCK   BLS03630
C     DIAGONAL ENTRY A(I) OF THE TM-MATRIX, EXCEPT THE LAST DIAGONAL BLS03640
C     BLOCK WHICH IS GENERATED ABOVE                               BLS03650
C                                                                  BLS03660
   70 CONTINUE                                                     BLS03670
C     COMPUTE (A-MATRIX)*(ITH-Q-BLOCK)                             BLS03680
      KA=I2                                                        BLS03690
      DO 80 K=I1,I2                                                BLS03700
      KA=KA+1                                                      BLS03710
      JKA1 = (KA-1)*N + 1                                          BLS03720
      JK1 = (K-1)*N + 1                                            BLS03730
C------------------------------------------------------------------BLS03740
      CALL MATVEC(Q(JK1),Q(JKA1))                                  BLS03750
C------------------------------------------------------------------BLS03760
      DESC(K)=KA                                                   BLS03770
   80 DESC(KA)=K                                                   BLS03780
C                                                                  BLS03790
C     COMPUTE (A-MATRIX)*(ITH-Q-BLOCK) - ((I-1)TH-Q-BLOCK)*B(I)-TRANS BLS03800
C     WHERE B(I) DENOTES THE ITH SUBDIAGONAL BLOCK                 BLS03810
C                                                                  BLS03820
      IF(I.EQ.1) GO TO 110                                         BLS03830
      J1 = DIR(1,I-1)                                              BLS03840
      J2 = DIR(2,I-1)                                              BLS03850
      DO 100 K=I1,I2                                               BLS03860
      KD=DESC(K)                                                   BLS03870
      JKD0 = (KD-1)*N                                              BLS03880
      KK = (J1-2)*MXBLK + K                                        BLS03890
      DO 90 L=J1,J2                                                BLS03900
      JL = (L-1)*N                                                 BLS03910
      KK = KK + MXBLK                                              BLS03920
      S=TM(KK)                                                     BLS03930
      JKD = JKD0                                                   BLS03940
      DO 90 J=1,N                                                  BLS03950
      JKD = JKD + 1                                                BLS03960
```

```
          JL = JL + 1                                               BLS03970
   90 Q(JKD) = Q(JKD) - S*Q(JL)                                     BLS03980
  100 CONTINUE                                                      BLS03990
          LINT = (KD-1)*N + 1                                       BLS04000
          LFIN = KD*N                                               BLS04010
C                                                                   BLS04020
C         COMPUTE A(I)                                              BLS04030
C                                                                   BLS04040
  110 DO 130 K=I1,I2                                                BLS04050
          KKMX = (K-1)*MXBLK                                        BLS04060
          KD=DESC(K)                                                BLS04070
          JKD1 = (KD-1)*N+ 1                                        BLS04080
          JL1 = (K-2)*N + 1                                         BLS04090
          DO 120 L=K,I2                                             BLS04100
          JL1 = JL1 + N                                             BLS04110
          KK = KKMX + L                                             BLS04120
C-------------------------------------------------------------------BLS04130
          TM(KK) = FINPRO(N,Q(JL1),1,Q(JKD1),1)                     BLS04140
C-------------------------------------------------------------------BLS04150
  120 CONTINUE                                                      BLS04160
  130 CONTINUE                                                      BLS04170
C                                                                   BLS04180
C         COMPUTE P(I) = P(I) - (ITH-Q-BLOCK)*A(I)                  BLS04190
C                                                                   BLS04200
          DO 170 K=I1,I2                                            BLS04210
          KKMX = (K-1)*MXBLK                                        BLS04220
          KD=DESC(K)                                                BLS04230
          JKD0 = (KD-1)*N                                           BLS04240
          JL = (I1-1)*N                                             BLS04250
          DO 140 L=I1,I2                                            BLS04260
          KK = KKMX + L                                             BLS04270
          IF(L.LT.K) KK=(L-1)*MXBLK + K                             BLS04280
          S=TM(KK)                                                  BLS04290
          JKD = JKD0                                                BLS04300
          DO 140 J=1,N                                              BLS04310
          JL = JL + 1                                               BLS04320
          JKD = JKD + 1                                             BLS04330
  140 Q(JKD) = Q(JKD) - S*Q(JL)                                     BLS04340
C                                                                   BLS04350
C         REORTHOGONALIZE THE BLOCK P(I) WITH RESPECT TO ALL VECTORS BLS04360
C         IN THE 1ST QBLOCK THAT ARE NOT CURRENTLY GENERATING ANY    BLS04370
C         DESCENDANTS.  NOTE THAT 2ND Q-BLOCK IS REORTHOGONALIZED    BLS04380
C         ELSEWHERE.                                                BLS04390
          IF(XLFT(I).EQ.0) GO TO 170                                BLS04400
          LUP = XLFT(I)                                             BLS04410
          DO 160 JJ = 1,LUP                                         BLS04420
          L= LEFT(JJ)                                               BLS04430
          JLO = (L-1)*N                                             BLS04440
          LLMX = (L-1)*MXBLK                                        BLS04450
          JL1 = JLO + 1                                             BLS04460
          JKD1 = JKD0 + 1                                           BLS04470
C-------------------------------------------------------------------BLS04480
          SUM = FINPRO(N,Q(JL1),1,Q(JKD1),1)                        BLS04490
C-------------------------------------------------------------------BLS04500
          JKD = JKD0                                                BLS04510
          JL = JLO                                                  BLS04520
          DO 150 J=1,N                                              BLS04530
          JKD = JKD + 1                                             BLS04540
          JL = JL + 1                                               BLS04550
  150 Q(JKD) = Q(JKD) - SUM* Q(JL)                                  BLS04560
          KK = LLMX + K                                             BLS04570
          TM(KK) = SUM + TM(KK)                                     BLS04580
C                                                                   BLS04590
  160 CONTINUE                                                      BLS04600
  170 CONTINUE                                                      BLS04610
C                                                                   BLS04620
C                                                                   BLS04630
```

```
C     GENERATE B(I+1)                                               BLS04640
C                                                                   BLS04650
      K1=DESC(I1)                                                   BLS04660
      K2=DESC(I2)                                                   BLS04670
      IFLAG=0                                                       BLS04680
C                                                                   BLS04690
C     COMPUTE NORMS                                                 BLS04700
C                                                                   BLS04710
  180 CONTINUE                                                      BLS04720
      JK1 = (K1-2)*N + 1                                            BLS04730
      DO 190 K=K1,K2                                                BLS04740
      JK1 = JK1 + N                                                 BLS04750
C-----------------------------------------------------------------BLS04760
      SM(K) = FINPRO(N,Q(JK1),1,Q(JK1),1)                           BLS04770
C-----------------------------------------------------------------BLS04780
  190 CONTINUE                                                      BLS04790
C                                                                   BLS04800
      IF(I.EQ.1.AND.K1.EQ.I2+1)  WRITE(6,200) NITER,               BLS04810
     1 (K,SM(K), K =K1,K2)                                          BLS04820
  200 FORMAT(//' ON ITERATION', I4,' NORM(GRADIENTS)**2 OF 1ST BLOCK = 'BLS04830
     1/5(I4,E12.3))                                                 BLS04840
C                                                                   BLS04850
C     TEST FOR CONVERGENCE OF BLOCK LANCZOS                         BLS04860
C                                                                   BLS04870
      IF(I.GT.1.OR.K1.GT.I2+1)  GO TO 250                           BLS04880
C                                                                   BLS04890
C     TEST THE FIRST KM OF THE EIGENVALUES FOR CONVERGENCE          BLS04900
      K2L = K1 + KML - 1                                            BLS04910
      RKM = SM(K2L)                                                 BLS04920
      DO 210 K = K1,K2L                                             BLS04930
      IF(SM(K).GT.ERRMAX )  GO TO 220                               BLS04940
  210 CONTINUE                                                      BLS04950
      GO TO 430                                                     BLS04960
C                                                                   BLS04970
C     CAN WE REDUCE KACT?  IF A SMALL RESIDUAL (GRADIENT) IS IDENTIFIED,BLS04980
C     SIZE OF 1ST BLOCK MAY BE REDUCED.                            BLS04990
  220 IF(KML.EQ.KACT) GO TO 250                                     BLS05000
      DO 230 K = K2L,K2                                             BLS05010
      IF(SM(K).GT.ERRMAX) GO TO 230                                 BLS05020
      KSAV = K                                                      BLS05030
      KACTN = KSAV - KACT                                           BLS05040
      GO TO 240                                                     BLS05050
C                                                                   BLS05060
  230 CONTINUE                                                      BLS05070
      GO TO 250                                                     BLS05080
C                                                                   BLS05090
  240 K2 = KSAV                                                     BLS05100
C                                                                   BLS05110
C     GENERATE THE TRANSPOSE OF B(I)                               BLS05120
C                                                                   BLS05130
  250 CONTINUE                                                      BLS05140
C                                                                   BLS05150
C     DETERMINE THE MAXIMAL NORM                                   BLS05160
      K=K1                                                          BLS05170
      S=SM(K)                                                       BLS05180
      DO 260 L=K1,K2                                                BLS05190
      IF (SM(L).LT.S) GOTO 260                                      BLS05200
      K=L                                                           BLS05210
      S=SM(L)                                                       BLS05220
  260 CONTINUE                                                      BLS05230
C     FOR 2ND QBLOCK, SAVE INDEX AND SIZE OF MAXIMAL NORM          BLS05240
      IF(I.GT.1) GO TO 270                                          BLS05250
      IND = K - KACT                                                BLS05260
      RESN = SM(K)                                                  BLS05270
C                                                                   BLS05280
  270 IF(S.LE.ERRMAX)GO TO 340                                      BLS05290
C                                                                   BLS05300
```

```
      IF(IFLAG.EQ.1) GO TO 340                                      BLS05310
C                                                                   BLS05320
      S=DSQRT(S)                                                    BLS05330
      JKO = (K-1)*N                                                 BLS05340
      JK = JKO                                                      BLS05350
      DO 280 J=1,N                                                  BLS05360
      JK = JK + 1                                                   BLS05370
  280 Q(JK)=Q(JK)/S                                                 BLS05380
      JLO = (K1-2)*N                                                BLS05390
      DO 310 L=K1,K2                                                BLS05400
      JLO = JLO + N                                                 BLS05410
      LL=(DESC(L) - 1)*MXBLK + K1                                   BLS05420
      IF (L.NE.K) GOTO 290                                          BLS05430
      TM(LL)=S                                                      BLS05440
      GO TO 310                                                     BLS05450
  290 JK1 = JKO + 1                                                 BLS05460
      JL1 = JLO + 1                                                 BLS05470
C-----------------------------------------------------------------BLS05480
      T = FINPRO(N,Q(JK1),1,Q(JL1),1)                               BLS05490
C-----------------------------------------------------------------BLS05500
      TM(LL)=T                                                      BLS05510
      JK = JKO                                                      BLS05520
      JL = JLO                                                      BLS05530
      DO 300 J=1,N                                                  BLS05540
      JK = JK + 1                                                   BLS05550
      JL = JL + 1                                                   BLS05560
  300 Q(JL) = Q(JL) - T*Q(JK)                                       BLS05570
  310 CONTINUE                                                      BLS05580
      IF (K.EQ.K1) GOTO 330                                         BLS05590
C                                                                   BLS05600
      JK1 = (K1-1)*N                                                BLS05610
      JK = JKO                                                      BLS05620
      DO 320 J=1,N                                                  BLS05630
      JK = JK + 1                                                   BLS05640
      JK1 = JK1 + 1                                                 BLS05650
      T=Q(JK1)                                                      BLS05660
      Q(JK1)=Q(JK)                                                  BLS05670
  320 Q(JK)=T                                                       BLS05680
      MA=DESC(K)                                                    BLS05690
      MB=DESC(K1)                                                   BLS05700
      DESC(K1)=MA                                                   BLS05710
      DESC(K)=MB                                                    BLS05720
      DESC(MA)=K1                                                   BLS05730
      DESC(MB)=K                                                    BLS05740
  330 CONTINUE                                                      BLS05750
C                                                                   BLS05760
      DIR(2,I+1)=K1                                                 BLS05770
C                                                                   BLS05780
      IFLAG=1                                                       BLS05790
C                                                                   BLS05800
      K1=K1+1                                                       BLS05810
      IF(I.EQ.1) GO TO 340                                          BLS05820
      IF (K1.LE.K2) GO TO 180                                       BLS05830
C     RETURN TO LANCZS                                              BLS05840
C                                                                   BLS05850
      RETURN                                                        BLS05860
C                                                                   BLS05870
C     IMPLICIT VECTOR DEFLATION                                     BLS05880
C                                                                   BLS05890
  340 CONTINUE                                                      BLS05900
      J= XLFT(I+2)                                                  BLS05910
      IF(K1.GT.K2) GO TO 360                                        BLS05920
      DO 350 L=K1,K2                                                BLS05930
      J = J+1                                                       BLS05940
  350 LEFT(J) = DESC(L)                                             BLS05950
  360 XLFT(I+2) = J                                                 BLS05960
C                                                                   BLS05970
```

```
C       FORCE REORTHGONALIZATION OF 2ND AND 3RD QBLOCKS W.R.T. THOSE       BLS05980
C       VECTORS IN 1ST QBLOCK THAT ARE NOT GENERATING DESCENDANTS          BLS05990
C       ON THIS ITERATION.                                                 BLS06000
        IF(I.GT.1)  GO TO 370                                              BLS06010
        XLFT(1) = XLFT(3)                                                  BLS06020
        XLFT(2) = XLFT(3)                                                  BLS06030
  370 IJJ = I + 2                                                          BLS06040
      IJJJ= XLFT(IJJ)                                                      BLS06050
C                                                                          BLS06060
        IF(IJJJ.EQ.0) GO TO 390                                           BLS06070
        IF(IWRITE.EQ.1)  WRITE(6,380) (LEFT(IJ),IJ= 1,IJJJ)               BLS06080
  380 FORMAT(' VECTORS NOT GENERATING DESCENDANTS ARE '/(1016))           BLS06090
C                                                                          BLS06100
  390 IF(I.EQ.1.AND.KML.GT.1) GO TO 400                                   BLS06110
C                                                                          BLS06120
      RETURN                                                               BLS06130
C                                                                          BLS06140
C       REORTHOGONALIZE 2ND QBLOCK W.R.T VECTORS IN 1ST BLOCK NOT          BLS06150
C       GENERATING DESCENDANTS                                             BLS06160
  400 IF(XLFT(I).EQ.0) RETURN                                             BLS06170
      LUP = XLFT(I)                                                        BLS06180
      KD = DIR(2,I+1)                                                      BLS06190
      JKDO = (KD-1)*N                                                      BLS06200
      DO 420 JJ = 1,LUP                                                    BLS06210
      L = LEFT(JJ)                                                         BLS06220
      JLO = (L-1)*N                                                        BLS06230
      JL1 = JLO + 1                                                        BLS06240
      JKD' = JKDO + 1                                                      BLS06250
C----------------------------------------------------------------------BLS06260
      SUM = FINPRO(N,Q(JKD1),1,Q(JL1),1)                                   BLS06270
C----------------------------------------------------------------------BLS06280
      JL = JLO                                                             BLS06290
      JKD = JKDO                                                           BLS06300
      DO 410 J=1,N                                                         BLS06310
      JL = JL + 1                                                          BLS06320
      JKD = JKD + 1                                                        BLS06330
  410 Q(JKD) = Q(JKD) - SUM *Q(JL)                                        BLS06340
  420 CONTINUE                                                             BLS06350
C                                                                          BLS06360
      RETURN                                                               BLS06370
C                                                                          BLS06380
C       EXIT IF CONVERGENCE OF DESIRED EIGENVECTORS IS CONFIRMED.          BLS06390
C                                                                          BLS06400
  430 CONTINUE                                                             BLS06410
      DO 440 L=K1,K2                                                       BLS06420
      M=DESC(L)                                                            BLS06430
  440 DESC(M)=0                                                            BLS06440
      DIR(2,2)=DIR(2,1)                                                    BLS06450
C                                                                          BLS06460
      WRITE(6,450) ERRMAX                                                  BLS06470
  450 FORMAT(/' CONVERGENCE OBSERVED, ALL RESIDUALS**2 .LT. ERRMAX = ',   BLS06480
     1 E20.12)                                                            BLS06490
C                                                                          BLS06500
      RETURN                                                               BLS06510
C                                                                          BLS06520
C----------------------------------------------------------------------BLS06530
C       ALLOWS PASSAGE OF LOCATIONS OF SOME OF THE ARRAYS USED BY LANCI1   BLS06540
C       SO THAT THESE ARRAYS CAN BE DIMENSIONED IN THE MAIN PROGRAM        BLS06550
C                                                                          BLS06560
      ENTRY LANCP1(DIR,DESC,TM,SM,XLFT,LEFT)                              BLS06570
C----------------------------------------------------------------------BLS06580
C                                                                          BLS06590
      RETURN                                                               BLS06600
C-----END OF LANCI1----------------------------------------------------BLS06610
      END                                                                  BLS06620
C                                                                          BLS06630
C-----ORTHOG-----------------------------------------------------------BLS06640
```

```
C     ORTHOGONALIZE COLUMNS M = MA,MB OF Q-ARRAY W.R.T COLUMNS M = 1,MB  BLS06650
C                                                                        BLS06660
      SUBROUTINE ORTHOG(MA,MB,N,Q)                                       BLS06670
C                                                                        BLS06680
C-----------------------------------------------------------------------BLS06690
      DOUBLE PRECISION Q(1), S                                           BLS06700
      DOUBLE PRECISION FINPRO, DSQRT                                     BLS06710
C-----------------------------------------------------------------------BLS06720
C     MAIN LOOP                                                          BLS06730
      DO 50 M = MA,MB                                                    BLS06740
      MMO = (M-1)*N                                                      BLS06750
      LLO = -N                                                           BLS06760
      DO 40 L = 1,M                                                      BLS06770
      LLO = LLO + N                                                      BLS06780
      LL = LLO + 1                                                       BLS06790
      MM = MMO + 1                                                       BLS06800
C-----------------------------------------------------------------------BLS06810
      S = FINPRO(N,Q(LL),1,Q(MM),1)                                      BLS06820
C-----------------------------------------------------------------------BLS06830
C                                                                        BLS06840
      IF (M.EQ.L) GO TO 20                                               BLS06850
C                                                                        BLS06860
      MM = MMO                                                           BLS06870
      LL = LLO                                                           BLS06880
      DO 10 I=1,N                                                        BLS06890
      LL = LL + 1                                                        BLS06900
      MM = MM + 1                                                        BLS06910
   10 Q(MM) = Q(MM) - S*Q(LL)                                            BLS06920
      GO TO 40                                                           BLS06930
C                                                                        BLS06940
   20 S = DSQRT(S)                                                       BLS06950
      MM = MMO                                                           BLS06960
      DO 30 I=1,N                                                        BLS06970
      MM = MM + 1                                                        BLS06980
   30 Q(MM) = Q(MM)/S                                                    BLS06990
C                                                                        BLS07000
   40 CONTINUE                                                           BLS07010
   50 CONTINUE                                                           BLS07020
C                                                                        BLS07030
      RETURN                                                             BLS07040
C-----END OF ORTHOG-----------------------------------------------------BLS07050
      END                                                                BLS07060
C                                                                        BLS07070
C-----START-------------------------------------------------------------BLS07080
C     GENERATES PSEUDO-RANDOM STARTING VECTORS.                          BLS07090
C                                                                        BLS07100
      SUBROUTINE START(KA,KB,N,Q,G,ERRMAX)                               BLS07110
C                                                                        BLS07120
C-----------------------------------------------------------------------BLS07130
      DOUBLE PRECISION Q(1), ERRMAX, S                                   BLS07140
      REAL  G(1)                                                         BLS07150
      COMMON/RANDOM/IIX                                                  BLS07160
      DOUBLE PRECISION FINPRO, DSQRT                                     BLS07170
C-----------------------------------------------------------------------BLS07180
      IF(KA.GT.KB) RETURN                                                BLS07190
C                                                                        BLS07200
      IIL = IIX                                                          BLS07210
      DO 110 K = KA,KB                                                   BLS07220
      KKO = (K-1)*N                                                      BLS07230
C                                                                        BLS07240
C-----------------------------------------------------------------------BLS07250
      CALL GENRAN(IIL,G,N)                                               BLS07260
C-----------------------------------------------------------------------BLS07270
C                                                                        BLS07280
      KK = KKO                                                           BLS07290
      DO 10 I = 1,N                                                      BLS07300
      KK = KK + 1                                                        BLS07310
```

```
   10 Q(KK) = G(I)                                                      BLS07320
      LLO = -N                                                          BLS07330
   20 DO 70 L=1,K                                                       BLS07340
      LLO = LLO + N                                                     BLS07350
      LL = LLO + 1                                                      BLS07360
      KK = KKO + 1                                                      BLS07370
C-----------------------------------------------------------------------BLS07380
      S = FINPRO(N,Q(LL),1,Q(KK),1)                                     BLS07390
C-----------------------------------------------------------------------BLS07400
C                                                                       BLS07410
      IF (K.EQ.L) GO TO 40                                              BLS07420
C                                                                       BLS07430
      LL = LLO                                                          BLS07440
      KK = KKO                                                          BLS07450
      DO 30 I=1,N                                                       BLS07460
      LL = LL + 1                                                       BLS07470
      KK = KK + 1                                                       BLS07480
   30 Q(KK) = Q(KK) - S*Q(LL)                                           BLS07490
      GO TO 70                                                          BLS07500
C                                                                       BLS07510
   40 S = DSQRT(S)                                                      BLS07520
      IF(S.LE.ERRMAX) GO TO 80                                          BLS07530
      KK = KKO                                                          BLS07540
      DO 50 I=1,N                                                       BLS07550
      KK = KK + 1                                                       BLS07560
   50 Q(KK) = Q(KK)/S                                                   BLS07570
C                                                                       BLS07580
      WRITE(6,60) K                                                     BLS07590
   60 FORMAT(I6,' TH STARTING VECTOR IS GENERATED RANDOMLY')            BLS07600
C                                                                       BLS07610
   70 CONTINUE                                                          BLS07620
      GO TO 110                                                         BLS07630
C                                                                       BLS07640
C-----------------------------------------------------------------------BLS07650
   80 CALL GENRAN(IIX,G,N)                                              BLS07660
C-----------------------------------------------------------------------BLS07670
C                                                                       BLS07680
      WRITE(6,90) K                                                     BLS07690
   90 FORMAT(/I6,' TH RANDOM VECTOR REJECTED, GENERATE ANOTHER'/)       BLS07700
C                                                                       BLS07710
      KK = KKO                                                          BLS07720
      DO 100 I = 1,N                                                    BLS07730
      KK = KK + 1                                                       BLS07740
  100 Q(KK) = G(I)                                                      BLS07750
      GO TO 20                                                          BLS07760
C                                                                       BLS07770
  110 CONTINUE                                                          BLS07780
      RETURN                                                            BLS07790
C-----END OF START------------------------------------------------------BLS07800
      END                                                               BLS07810
C                                                                       BLS07820
C-----START OF DIAGOM---------------------------------------------------BLS07830
C     DIAGOM CALLS THE EISPACK SUBROUTINES TRED2 AND IMTQL2 TO          BLS07840
C     DIAGONALIZE THE SMALL SYMMETRIC MATRICES GENERATED AT EACH        BLS07850
C     ITERATION OF BLOCK LANCZOS.                                       BLS07860
C                                                                       BLS07870
      SUBROUTINE DIAGOM(MXBLK,MM,TM,KACT,N,Q,E,RESID,RESK,RESN,IND,     BLS07880
     1 KACTN,KM,TD,TOD,NITER,IERR,IWRITE)                               BLS07890
C                                                                       BLS07900
C-----------------------------------------------------------------------BLS07910
      DOUBLE PRECISION   TM(MXBLK,1),Q(1),E(1),TD(1),TOD(1),RESID(1)    BLS07920
      DOUBLE PRECISION   RESK(1),RESN,RATIO,FRACT,RKM,EMAX,SPREAD,EGAP  BLS07930
      DOUBLE PRECISION   DABS,DFLOAT,DMAX1                              BLS07940
C-----------------------------------------------------------------------BLS07950
      IF(NITER.GE.100) GO TO 270                                        BLS07960
      RKM = TD(1)                                                       BLS07970
      FRACT = TD(2)                                                     BLS07980
```

```
        NSTAG = IERR                                                    BLS07990
        KWANT = KACT                                                    BLS08000
C                                                                       BLS08010
C       STORE KM-TH RESIDUALS**2 FOR CHECK ON STAGNATION OF CONVERGENCE BLS09020
        NITER1 = NITER + 1                                              BLS08030
        RESK(NITER1) = RKM                                              BLS08040
        IF(NITER.LE.NSTAG) GO TO 10                                     BLS08050
C       TEST FOR STAGNATION                                             BLS08060
        NITERM = NITER - 10                                             BLS08070
        RATIO = RKM / RESK(NITERM)                                      BLS08080
        IF(RATIO.GT.FRACT) GO TO 250                                    BLS08090
C                                                                       BLS08100
     10 CONTINUE                                                        BLS08110
C                                                                       BLS08120
C       TEST GAPS TO DETERMINE IF SIZE OF 1ST Q-BLOCK CAN BE REDUCED    BLS08130
        IF(NITER.EQ.0) GO TO 40                                         BLS08140
        IF(KM.EQ.KACT.OR.NITER.LT.10) GO TO 30                          BLS08150
        KACT1 = KACT - 1                                                BLS08160
        DO 20 K = KM,KACT1                                              BLS08170
        RATIO =  DABS(E(K+1) - E(K))                                    BLS08180
        IF(RATIO.LT.25*EGAP) GO TO 20                                   BLS08190
        KACT = K                                                        BLS08200
        GO TO 40                                                        BLS08210
     20 CONTINUE                                                        BLS08220
C                                                                       BLS08230
C       IF KACT.NE.KACTN, THEN SUBROUTINE LANCII IDENTIFIED A VERY      BLS08240
C       SMALL RESIDUAL FOR SOME E(J), J>= KM.                           BLS08250
     30 IF(KACT.EQ.KACTN)  GO TO 50                                     BLS08260
        RATIO = DABS(E(KACTN+1) - E(KACTN))                             BLS08270
        IF(RATIO.LE.EGAP)  GO TO 50                                     BLS08280
        KACT = KACTN                                                    BLS08290
     40 ICOUNT = 1                                                      BLS08300
        INDEXP = IND                                                    BLS08310
        RESID(1) = RESN                                                 BLS08320
        GO TO 80                                                        BLS08330
C                                                                       BLS08340
     50 CONTINUE                                                        BLS08350
        IF(IND.NE.INDEXP) GO TO 70                                      BLS08360
C       INDEX OF VECTOR OF MAXIMUM NORM IS SAME AS ON PREVIOUS ITERATION BLS08370
        ICOUNT = ICOUNT + 1                                             BLS08380
        IF(ICOUNT.LE.5) GO TO 60                                        BLS08390
        ITEST = ICOUNT - 4                                              BLS08400
        RATIO = RESID(ITEST)/RESN                                       BLS08410
        IF(DABS(RATIO).GT.10.D0) GO TO 60                               BLS08420
C                                                                       BLS08430
C       CONVERGENCE STAGNATED, ADD NEXT RITZ VECTOR IN THE CHAIN        BLS08440
C       TO THE 1ST Q-BLOCK AND RESET THE FLAGS THAT KEEP TRACK OF       BLS08450
C       CONVERGENCE.                                                    BLS08460
        INDEXP = IND                                                    BLS08470
        ICOUNT = 0                                                      BLS08480
        KACT = KACT + 1                                                 BLS08490
        KWANT = KACT                                                    BLS08500
C       CHECK THAT THERE IS ENOUGH ROOM TO ENLARGE THE 1ST QBLOCK       BLS08510
        IF(2*KACT.GT.MXBLK) GO TO 230                                   BLS08520
        GO TO 80                                                        BLS08530
C                                                                       BLS08540
     60 RESID(ICOUNT) = RESN                                            BLS08550
        INDEXP = IND                                                    BLS08560
        GO TO 80                                                        BLS08570
C                                                                       BLS08580
     70 ICOUNT = 1                                                      BLS08590
        RESID(1) = RESN                                                 BLS08600
        INDEXP = IND                                                    BLS08610
C                                                                       BLS08620
C-----------------------------------------------------------------------BLS08630
C       USE EISPACK SUBROUTINES TO DIAGONALIZE THE SMALL TM-MATRIX.     BLS08640
C                                                                       BLS08650
```

```
    80 CALL TRED2(MXBLK,MM,TM,TD,TOD,TM)                                  BLS08660
       CALL IMTQL2(MXBLK,MM,TD,TOD,TM,IERR)                              BLS08670
C-----------------------------------------------------------------------BLS08680
       IF(IERR.EQ.0) GO TO 90                                           BLS08690
       RETURN                                                           BLS08700
C                                                                       BLS08710
C    SELECT RELEVANT EIGENVALUES AND EIGENVECTORS OF THE T-MATRIX.      BLS08720
    90 CONTINUE                                                         BLS08730
C                                                                       BLS08740
C    IMTQL2 RETURNS EIGENVALUES (AND CORRESPONDING EIGENVECTORS) IN     BLS08750
C    ALGEBRAICALLY-ASCENDING ORDER.  REARRANGE TO DESCENDING ORDER.     BLS08760
C                                                                       BLS08770
       DO 100 L=1,MM                                                    BLS08780
       MML = MM-L+1                                                     BLS08790
   100 E(L) = TD(MML)                                                   BLS08800
C                                                                       BLS08810
   110 WRITE(6,120) KACT, (E(J), J=1,KACT)                             BLS08820
   120 FORMAT(' COMPUTED',I4,' ALGEBRAICALLY-LARGEST EIGENVALUES'/(4E20.1BLS08830
      12))                                                              BLS08840
C                                                                       BLS08850
C    COMPUTE ESTIMATE MAXIMUM EIGENVALUE AND OF SPREAD                  BLS08860
       IF(NITER.GT.1) GO TO 140                                         BLS08870
       EMAX = DMAX1(DABS(E(1)),DABS(E(MM)))                             BLS08880
       SPREAD = DABS(E(1) - E(MM))                                      BLS08890
       EGAP = SPREAD/DFLOAT(N)                                          BLS08900
       IF(NITER.EQ.1)  WRITE(6,130) EMAX,SPREAD,EGAP                    BLS08910
   130 FORMAT(/4X,'ESTIMATED NORM OF MATRIX',4X,'ESTIMATED SPREAD',6X,'SPBLS08920
      1READ*(SIZE)*(-1)'/E28.4,E20.4,E24.3)                             BLS08930
   140 CONTINUE                                                         BLS08940
C                                                                       BLS08950
C    COMPUTE RITZ VECTORS                                              BLS08960
       DO 180 I=1,N                                                     BLS08970
       DO 150 KK=1,KWANT                                                BLS08980
       TOD(KK)=0.D0                                                     BLS08990
       K = MM - KK + 1                                                  BLS09000
       IL = - N + I                                                     BLS09010
       DO 150 L = 1,MM                                                  BLS09020
       IL = IL + N                                                      BLS09030
   150 TOD(KK) = TOD(KK) + TM(L,K)*Q(IL)                                BLS09040
       IKK = -N + I                                                     BLS09050
   160 DO 170 KK=1,KACT                                                 BLS09060
       IKK = IKK + N                                                    BLS09070
   170 Q(IKK)=TOD(KK)                                                   BLS09080
   180 CONTINUE                                                         BLS09090
C                                                                       BLS09100
C    ON FILE 13 SAVE ANY EXTRA VECTORS NO LONGER NEEDED IN 1ST Q-BLOCK  BLS09110
       IF(KWANT.EQ.KACT) GO TO 290                                      BLS09120
       K1 = KACT + 1                                                    BLS09130
       K2 = KWANT                                                       BLS09140
       DUMMY = 100.                                                     BLS09150
       DO 190 K = K1,K2                                                 BLS09160
       LINT = (K-1)*N + 1                                               BLS09170
       LFIN = K*N                                                       BLS09180
       WRITE(13,210) E(K),DUMMY,K                                       BLS09190
       WRITE(13,220) (Q(L), L=LINT,LFIN)                                BLS09200
   190 CONTINUE                                                         BLS09210
       KDELTA = KWANT - KACT                                            BLS09220
       WRITE(13,200) KDELTA                                             BLS09230
   200 FORMAT(/' ABOVE ARE ',I3,' VECTORS STRIPPED FROM A 1ST Q-BLOCK'/  BLS09240
      1' DURING A BLOCK LANZCOS RUN WHICH COULD BE USED AS STARTING VECTOBLS09250
      1RS'/' IN A LATER RUN IF THE USER DECIDES THAT THESE EIGENVALUES SHBLS09260
      1OULD'/' BE COMPUTED AFTER ALL.  FORMAT USED IN THE SAME AS WAS USEBLS09270
      1D'/' IN THE CORRESPONDING BLSTARTV FILE'/)                       BLS09280
   210 FORMAT(/E20.12,E13.4,I6,' = EVAL,DUMMY,EVAL NUMBER,EVEC='/)       BLS09290
   220 FORMAT(4E20.12)                                                  BLS09300
```

```
          GO TO 290                                                   BLS09310
C                                                                     BLS09320
C     DEFAULT, SIZE OF 1ST Q-BLOCK TOO LARGE FOR MXBLK                BLS09330
  230 IWRITE = -1000                                                  BLS09340
      WRITE(6,240) KACT,MXBLK                                         BLS09350
      WRITE(15,240) KACT,MXBLK                                        BLS09360
  240 FORMAT(//' BLOCK LANCZOS PROCEDURE TRIED TO INCREASE THE SIZE OF 1BLS09370
     1ST QBLOCK'/' TO ',I3,' BUT THIS IS NOT FEASIBLE BECAUSE TWICE THISBLS09380
     1 SIZE'/' IS G.T. MXBLK WHICH EQUALS ',I4/' USER CAN RERUN PROGRAM BLS09390
     1WITH LARGER MXBLK'/)                                            BLS09400
      GO TO 290                                                       BLS09410
C                                                                     BLS09420
C     DEFAULT, CONVERGENCE RATE IS TOO SLOW                           BLS09430
  250 IWRITE = -1000                                                  BLS09440
      WRITE(6,260) NITER,RATIO,FRACT                                  BLS09450
      WRITE(15,260) NITER,RATIO,FRACT                                 BLS09460
  260 FORMAT(//' ON ITERATION ',I3,' CONVERGENCE APPEARS TO BE STAGNATEDBLS09470
     1'/' RATIO OF SQUARE OF CURRENT KM-TH RESIDUAL TO CORRESPONDING SQUBLS09480
     1ARE'/' 10 ITERATIONS EARLIER IS ',E10.3,' COMPARED TO '/        BLS09490
     1' USER-SPECIFIED RATIO ',E10.3,'. THEREFORE, PROGRAM TERMINATES'/'BLS09500
     1 USER SHOULD LOOK AT THE OUTPUT. IF CONVERGENCE HAS STAGNATED, USEBLS09510
     1R'/' CAN EITHER INCREASE KACT OR KMAX OR RESET THE STAGNATION PARABLS09520
     1METERS'/' NSTAG AND FRACT, AND RESTART THE BLOCK PROCEDURE USING TBLS09530
     1HE'/' CURRENT EIGENVECTOR APPROXIMATIONS AS STARTING VECTORS'/)   BLS09540
      GO TO 290                                                       BLS09550
C                                                                     BLS09560
  270 IWRITE = -1000                                                  BLS09570
      WRITE(6,280)                                                    BLS09580
      WRITE(15,280)                                                   BLS09590
  280 FORMAT(//' SOMETHING IS SERIOUSLY WRONG.  NUMBER OF ITERATIONS IS BLS09600
     1EXCESSIVE'/' PROGRAM TERMINATES FOR USER TO DECIDE WHAT TO DO'/'  BLS09610
     1ALTERNATIVES INCLUDE INCREASING KACT OR KMAX OR BOTH, AND RESTARTIBLS09620
     1NG'/' USING THE CURRENT APPROXIMATIONS AS STARTING VECTORS'//)    BLS09630
C                                                                     BLS09640
  290 CONTINUE                                                        BLS09650
      RETURN                                                          BLS09660
C-----END OF DIAGOM-------------------------------------------------BLS09670
      END                                                             BLS09680
C-----LPERM PERMUTES VECTORS--------------------------------------- BLS09690
C                                                                     BLS09700
      SUBROUTINE LPERM(W,U,IPERM)                                     BLS09710
C                                                                     BLS09720
C-----------------------------------------------------------------BLS09730
      DOUBLE PRECISION  U(1),W(1)                                     BLS09740
      INTEGER  IPR(1),IPT(1)                                          BLS09750
C-----------------------------------------------------------------BLS09760
C     SUBROUTINE HAS 2 BRANCHES:  IPERM = 1,   CALCULATES            BLS09770
C     U = P*W  WHERE  P  IS THE PERMUTATION REPRESENTED BY IPR       BLS09780
C     LET J = IPR(K) THEN U(K) = W(J), K = 1,N. WE SET W(K)=U(K), K=1,N BLS09790
C     IPERM = 2, USING THE PERMUTATION IPT (P-TRANSPOSE) U = P'*W, W=U BLS09800
C     LET J = IPT(K) THEN U(K) = W(J), K=1,N. WE SET W(K) = U(K), K=1,N BLS09810
C-----------------------------------------------------------------BLS09820
C                                                                     BLS09830
      IF(IPERM.EQ.2) GO TO 10                                         BLS09840
C     IPERM = 1                                                       BLS09850
      DO 20 K = 1,N                                                   BLS09860
      J = IPR(K)                                                      BLS09870
   20 U(K) = W(J)                                                     BLS09880
      DO 30 K = 1,N                                                   BLS09890
   30 W(K) = U(K)                                                     BLS09900
      GO TO 60                                                        BLS09910
C     IPERM = 2                                                       BLS09920
   10 DO 40  K = 1,N                                                  BLS09930
      J = IPT(K)                                                      BLS09940
   40 U(K) = W(J)                                                     BLS09950
      DO 50 K = 1,N                                                   BLS09960
   50 W(K) = U(K)                                                     BLS09970
```

```
      60 CONTINUE                                                      BLS09980
C                                                                      BLS09990
      RETURN                                                           BLS10000
C                                                                      BLS10010
C---------------------------------------------------------------------BLS10020
      ENTRY LPERME(IPR,IPT,N)                                          BLS10030
C---------------------------------------------------------------------BLS10040
C                                                                      BLS10050
C-----END OF LPERM-------------------------------------------          BLS10060
      RETURN                                                           BLS10070
      END                                                              BLS10080
```

SECTION 8.6   FILE DEFINITIONS AND SAMPLE INPUT FILE

Below is a listing of the input/output files which are accessed by the
real symmetric block Lanczos eigenvalue/eigenvector program, BLEVAL.
BLEVAL computes a few extreme eigenvalues and corresponding eigenvectors
of a real symmetric matrix. Also below is a sample of the input file
which BLEVAL requires on file 5. The parameters in this file are
supplied in free format.

```
SAMPLE DEFINITIONS OF THE INPUT/OUTPUT FILES FOR BLEVAL
-------------------------------------------------------------------
*BLEVAL EXEC
FI 06 TERM
FILEDEF  5 DISK BLEVAL     INPUT     A (RECFM F LRECL 80 BLOCK 80
FILEDEF  8 DISK &1         INPUT     A (RECFM F LRECL 80 BLOCK 80
FILEDEF 10 DISK &1         BLSTARTV  A (RECFM F LRECL 80 BLOCK 80
FILEDEF 13 DISK &1         BLEXTRAV  A (RECFM F LRECL 80 BLOCK 80
FILEDEF 15 DISK &1         BLEIGVEC  A (RECFM F LRECL 80 BLOCK 80
*IMTQL2 AND TRED2 ARE 2 EISPACK LIBRARY SUBROUTINES
LOAD  BLEVAL  BLSUB  BLMULT  IMTQL2  TRED2
-------------------------------------------------------------------
```

```
SAMPLE INPUT FILE FOR BLEVAL
-------------------------------------------------------------------
LINE 1 IWRITE (SPECIFY MESSAGE LEVEL TO FILE 6: 1 MEANS DETAILED
            1
LINE 2    N       MATNO  (SIZE OF A-MATRIX, MATRIX IDENT. NUMBER
        528         528
LINE 3 MDIMQ         MDIMTM       MAXIT (DIMS. Q, TM, MAX A*X'S
        40000        2500          1000
LINE 4 EFLAG   OFLAG (EFLAG=(0,1) 1=2PHASES. OFLAG: 1=ORTHOG CHECK
          1       1
LINE 5 SEED      (STARTING VECTOR SEED, RANDOM NUMBER GENERATOR
       3482736
LINE 6 KMAX   KACT  KSET (MAX T SIZE +1,SIZE 1ST BLOCK,VECS SUPPLIED
        21      4      0
LINE 7 KM   (NUMB. EVS FOR ALG-LARGEST, -(NUMB. EVS) FOR ALG-SMALLEST
        4
LINE 8 NSTAG   FRACT (NO. ITNS BEFORE TEST CONVERGENCE, TEST FRACTION
         25     .01
LINE 9 RELTOL    MAXIT2 (PHASE 2, CONVERGE. TOL. , NO. A*X'S ALLOWED
       .00000001    1000
-------------------------------------------------------------------
```

CHAPTER 9

FACTORED INVERSES, REAL SYMMETRIC BLOCK LANCZOS

SECTION 9.1   INTRODUCTION

The FORTRAN codes in this chapter address the question of using an iterative block Lanczos procedure to compute a 'few' eigenvalues and a basis for the corresponding eigenspace of a real symmetric matrix A by computing a few extreme eigenvalues and a corresponding basis for the inverse of a real symmetric matrix B obtained from A by scaling, shifting and permuting A. For a given real symmetric matrix A, the codes consider the inverse of a matrix B where

$$B \equiv PCP^T \text{ with } C \equiv S0*A + SHIFT*I, \qquad (9.1.1)$$

the scale S0 and the shift SHIFT are specified by the user, and the permutation matrix P is chosen so that for a sparse matrix A (or C), the resulting factorization of the associated B matrix is also sparse. An eigenvalue is extreme if it is one of the algebraically-smallest or the algebraically-largest eigenvalues in the eigenvalue spectrum.

Specifically, for a given real symmetric matrix A and associated B-matrix as defined in Eqn(9.1.1), the codes in this chapter compute the q algebraically-largest eigenvalues, λ_i, $1 \le i \le q$, of B^{-1} and corresponding orthonormal real vectors $X_q \equiv \{x_1,...,x_q\}$ such that

$$B^{-1}X_q = X_q A_q, \text{ where } A_q \equiv X_q^T A X_q. \qquad (9.1.2)$$

Typically, $A_q = \Lambda_q$, a diagonal matrix whose nonzero entries are the eigenvalues λ_i. The number q is small and specified by the user.

Real symmetric matrices and factorizations of real symmetric matrices are discussed in Stewart [1973]. See also Bunch and Kaufman [1977] and George and Liu [1981]. Chapter 2, Section 2.1 contains a brief summary of the properties of real symmetric matrices which we use in these codes.

The Lanczos code contained in this chapter is a simple modification of the hybrid 'block' Lanczos procedure given in Chapter 8 to handle the factored inverse of the B-matrix given in Eqn(9.1.1). Therefore please see Chapter 8, Section 8.1, for comments about this procedure and for comments regarding the differences between iterative block Lanczos procedures and single-vector Lanczos procedures.

BLIEVAL is the main 'block' program for the factored inverse version of the 'block' Lanczos codes in Chapter 8. BLIEVAL uses the same subroutines as the real symmetric codes in Chapter 8, with the exception of the user-supplied subroutines. The user must supply a subroutine USPEC which defines and initializes the matrix which is to be used by the LANCZS and LANCI1 subroutines. In the factored inverse case, USPEC specifies the factorization of the particular B-matrix being used. These Lanczos programs do not require the A-matrix. However, the user must supply the scalars S0 and SHIFT, and the permutation P (if any). The user must also supply a subroutine BLSOLV which solves the system of equations Bu = x for any given vector x.

The sample USPEC and BLSOLV subroutines provided assume that the B-matrix being used is positive definite and that the Cholesky factors of B

$$B = LL^T \text{ where L is a lower triangular matrix,} \qquad (9.1.3)$$

are used for the matrix-vector multiply, $B^{-1}x$, for any given vector x. However, the user may replace these subroutines by subroutines which define and use a more general factorization. These Lanczos codes only require that the BLSOLV subroutine solves the system Bu = x, rapidly and accurately.

All computations are in double precision real arithmetic. On each iteration, the accuracy of the computed eigenvectors is checked in the process of computing the second block of Lanczos vectors on that iteration. Note that the eigenvectors of B^{-1} are simple permutations of the eigenvectors of A. These permutations are undone prior to the termination of the block procedure. The corresponding eigenvalues of A are obtained from those of B^{-1} by a simple scalar transformation which is included in the codes. The eigenelement computations for the small Lanczos matrices use two subroutines from the EISPACK Library [1976,1977], TRED2 and IMTQL2.

Several optional preprocessing programs are provided, PERMUT, LORDER, LFACT, and LTEST. Listings for these programs are given in Chapter 4. PERMUT calls the SPARSPAK Library [1979] to attempt to identify a reordering or permutation P of a given matrix A for which sparseness is preserved under factorization of the permuted matrix. LORDER takes a given matrix C and permutation P and computes the sparse matrix format for the permuted matrix, $B \equiv PCP^T$. LFACT computes the Cholesky factors of a given positive definite matrix. LTEST performs a very crude check on the numerical condition of the matrix supplied to it, by solving a system of equations with and without iterative refinement, LINPACK [1979].

The usefulness of this code for computing a few interior eigenvalues of a given real symmetric matrix is dubious. For such an application one would have to select a shift SHIFT that places the desired eigenvalues of the A-matrix on the extreme of the spectrum of the associated matrix B^{-1} and is chosen so that the B-matrix is well-conditioned numerically. This is not a trivial task. The user should refer to Chapter 7 of Volume 1 of this book for more details on iterative block Lanczos procedures.

SECTION 9.2   MAIN PROGRAM, EIGENVALUE AND EIGENVECTOR COMPUTATIONS

```
C-----BLIEVAL (FEW EXTREME EIGENVALUES AND EIGENVECTORS)---------------BLI00010
C            (USING FACTORED INVERSE OF A REAL SYMMETRIC MATRIX)       BLI00020
C                                                                      BLI00030
C      CONTAINS MAIN PROGRAM FOR COMPUTING A FEW EIGENVALUES           BLI00040
C      AND CORRESPONDING EIGENVECTORS OF A REAL SYMMETRIC MATRIX       BLI00050
C      BY COMPUTING A FEW OF THE ALGEBRAICALLY-LARGEST OR              BLI00060
C      ALGEBRAICALLY-SMALLEST EIGENVALUES OF THE INVERSE OF A SCALED,  BLI00070
C      SHIFTED, AND PERMUTED VERSION B OF THE ORIGINAL A-MATRIX        BLI00080
C      USING A BLOCK FORM OF LANCZOS TRIDIAGONALIZATION WITH LIMITED   BLI00090
C      REORTHOGONALIZATION.  THIS BLOCK PROCEDURE IS ITERATIVE AND     BLI00100
C      REQUIRES A SUBROUTINE BLSOLV THAT FOR ANY GIVEN VECTOR W        BLI00110
C      COMPUTES U SUCH THAT B*U = W.   THE SAMPLE BLSOLV SUBROUTINES   BLI00120
C      PROVIDED FOR SPARSE MATRICES ARE ONLY FOR THE CASE THAT B IS    BLI00130
C      POSITIVE DEFINITE AND USE THE CHOLESKY FACTORS OF B.  HOWEVER,  BLI00140
C      THE USER COULD REPLACE THESE BY A SUBROUTINE WHICH COMPUTES     BLI00150
C      FOR AN INDEFINITE MATRIX THE FACTORIZATION L*D*(L-TRANSPOSE).   BLI00160
C                                                                      BLI00170
C      THIS BLOCK PROCEDURE COMPUTES THE ALGEBRAICALLY-LARGEST         BLI00180
C      EIGENVALUES OF THE INVERSE OF THE B-MATRIX, UNLESS THE USER     BLI00190
C      SUPPLIES -(B-INVERSE)*X RATHER THAN (B-INVERSE)*X, IN WHICH     BLI00200
C      CASE IT COMPUTES THE CORRESPONDING ALGEBRAICALLY-SMALLEST       BLI00210
C      EIGENVALUES OF (B-INVERSE) BY COMPUTING THE ALGEBRAICALLY-      BLI00220
C      LARGEST EIGENVALUES OF -(B-INVERSE).  IN THIS CASE THE SIGNS    BLI00230
C      OF THE COMPUTED EIGENVALUES ARE CHANGED PRIOR TO WRITING TO     BLI00240
C      FILE 15 SO THAT ON EXIT, FILE 15 CONTAINS THE ALGEBRAICALLY-    BLI00250
C      SMALLEST EIGENVALUES OF B-INVERSE ALONG WITH THE CORRESPONDING  BLI00260
C      EIGENVALUES OF THE ORIGINAL A-MATRIX AND CORRESPONDING          BLI00270
C      EIGENVECTORS.  THE MATRIX B = SO*P*A*P' + SHIFT*I WHERE THE     BLI00280
C      SCALE SO AND SHIFT ARE READ IN THIS PROGRAM, AND THE            BLI00290
C      PERMUTATION P IS DEFINED IN THE CORRESPONDING USPEC SUBROUTINE. BLI00300
C      THE PROGRAM ASSUMES THAT THE FACTORIZATION READ IN USPEC        BLI00310
C      CORRESPONDS TO THE SO, SHIFT AND PERMUTATION READ IN.  THE SO   BLI00320
C      AND SHIFT ARE CHOSEN SO THAT THE DESIRED EIGENVALUES ARE AT     BLI00330
C      THE EXTREME OF THE SPECTRUM OF B-INVERSE.                       BLI00340
C                                                                      BLI00350
C      THIS IS AN ITERATIVE 'BLOCK' LANCZOS PROCEDURE FOR WHICH ON     BLI00360
C      EVERY ITERATION, THE 2ND AND SUCCEEDING BLOCKS CONTAIN ONLY ONE BLI00370
C      VECTOR WHICH IS SELECTED ON THE BASIS OF ITS EXPECTED INFLUENCE BLI00380
C      ON THE CONVERGENCE.   Q-BLOCKS GENERATED ON A GIVEN ITERATION   BLI00390
C      ARE REORTHOGONALIZED ONLY W.R.T. THOSE VECTORS IN THE FIRST     BLI00400
C      Q-BLOCK WHICH ARE NOT ALLOWED TO GENERATE DESCENDANTS ON        BLI00410
C      THAT ITERATION.                                                 BLI00420
C                                                                      BLI00430
C      PFORT VERIFIER IDENTIFIED THE FOLLOWING NONPORTABLE CONSTRUCTIONS:BLI00440
C      1.  DATA MACHEP DEFINITION                                      BLI00450
C      2.  FORMAT (20A4) USED FOR READING EXPLANATORY COMMENTS.        BLI00460
C      3.  FREE FORMAT (5,*), USED FOR PARAMETER INPUT FROM FILE 5.    BLI00470
C      4.  COMMON/LOOPS/ AS CONSTRUCTED IS NOT PORTABLE                BLI00480
C                                                                      BLI00490
C----------------------------------------------------------------------BLI00500
       DOUBLE PRECISION Q(44000),E(50),TM(2500),TOD(50),TD(50),EPSM,NNZ BLI00510
       DOUBLE PRECISION SM(100),ERRMAX,SPREC,MACHEP,AVER,RELTOL,ERRMAN  BLI00520
       DOUBLE PRECISION  EVAL, RESIDL(100), RESIDK(100), RESID, FRACT   BLI00530
       DOUBLE PRECISION SO,SHIFT                                        BLI00540
       REAL EXPLAN(20),G(2000)                                         BLI00550
       INTEGER   DIR(2,100),DESC(100),LEFT(100),XLFT(100)              BLI00560
       INTEGER SEED,OFLAG,EFLAG                                        BLI00570
       COMMON/LOOPS/MAXIT,ITER                                         BLI00580
       COMMON /RANDOM/SEED                                             BLI00590
       COMMON/FLAGS/EFLAG,OFLAG                                        BLI00600
       DOUBLE PRECISION  DABS, DFLOAT                                  BLI00610
C----------------------------------------------------------------------BLI00620
       EXTERNAL BLSOLV                                                 BLI00630
```

```
      DATA MACHEP/Z3410000000000000/                                    BLI00640
C--------------------------------------------------------------------BLI00650
C                                                                       BLI00660
C     ARRAYS MUST DIMENSIONED AS FOLLOWS:                               BLI00670
C                                                                       BLI00680
C     1.  Q:   >=  KMAX*N                                               BLI00690
C     2.  G:   >=  N                                                    BLI00700
C     3.  E:   >=  MXBLK                                                BLI00710
C     4.  TM:  >=  MXBLK**2                                             BLI00720
C     5.  TOD, TD, SM, DESC, LEFT, XLFT:   >= MXBLK                     BLI00730
C     6.  DIR:  ROW DIMENSION = 2;   COLUMN DIMENSION >= MXBLK          BLI00740
C     7.  RESIDL, RESIDK:  >= MAXIMUM NUMBER OF ITERATIONS ALLOWED.     BLI00750
C         PROGRAM CURRENTLY TERMINATES IF MORE THAN 100 ITERATIONS      BLI00760
C         ARE REQUESTED.  USED TO MONITOR CONVERGENCE. SEE SUBROUTINE   BLI00770
C         DIAGOM.                                                       BLI00780
C     8.  EXPLAN:  DIMENSION = 20.                                      BLI00790
C                                                                       BLI00800
C--------------------------------------------------------------------BLI00810
C     OUTPUT HEADER                                                     BLI00820
      WRITE(6,10)                                                       BLI00830
   10 FORMAT(/' BLOCK LANCZOS PROCEDURE, USES FACTORED INVERSE OF A USERBLI00840
     1-SPECIFIED MATRIX'/' 2ND AND SUCCEEDING BLOCKS GENERATED ON EACH BBLI00850
     1LOCK ITERATION '/' CONTAIN ONLY ONE VECTOR'//)                    BLI00860
C                                                                       BLI00870
C     SET PROGRAM PARAMETERS                                            BLI00880
      EPSM = 2.DO*MACHEP                                                BLI00890
      SPREC = 1.D-5                                                     BLI00900
      MPMIN = -1000                                                     BLI00910
C                                                                       BLI00920
C     READ USER-SPECIFIED PARAMETERS FROM INPUT FILE 5 (FREE FORMAT)    BLI00930
C                                                                       BLI00940
C     SELECT THE AMOUNT OF INTERMEDIATE OUTPUT DESIRED (IWRITE =0,1).   BLI00950
C     IWRITE = 1 INCREASES THE AMOUNT OF INTERMEDIATE OUTPUT WRITTEN    BLI00960
C     TO FILE 6 ON EACH ITERATION OF THE BLOCK LANCZOS PROCEDURE.       BLI00970
      READ(5,20) EXPLAN                                                 BLI00980
   20 FORMAT(20A4)                                                      BLI00990
      READ(5,*)  IWRITE                                                 BLI01000
C                                                                       BLI01010
C     READ ORDER (N) OF MATRIX AND MATRIX IDENTIFICATION NUMBER (MATNO) BLI01020
C     READ SCALE (SO) AND SHIFT (SHIFT) APPLIED TO MATRIX AND           BLI01030
C     FLAG JPERM.  JPERM = (0,1):  JPERM = 1 MEANS MATRIX HAS BEEN      BLI01040
C     PERMUTED                                                          BLI01050
      READ(5,20) EXPLAN                                                 BLI01060
      READ(5,*) N,MATNO,SO,SHIFT,JPERM                                  BLI01070
C                                                                       BLI01080
C     READ USER-SPECIFIED DIMENSIONS OF Q-ARRAY (MDIMQ) AND OF THE      BLI01090
C     TM-ARRAY (MDIMTM).  READ MAXIMUM NUMBER (MAXIT) OF CALLS TO THE   BLI01100
C     BLSOLV SUBROUTINE ALLOWED IN PHASE 1.                            BLI01110
      READ(5,20) EXPLAN                                                 BLI01120
      READ(5,*) MDIMQ, MDIMTM, MAXIT                                    BLI01130
C                                                                       BLI01140
C     READ FLAGS:  EFLAG = (0,1).  EFLAG = 0, MEANS PROGRAM STOPS       BLI01150
C     AFTER COMPLETING PHASE 1 PORTION OF BLOCK LANCZOS PROCEDURE.      BLI01160
C     EFLAG = 1, MEANS PROGRAM COMPLETES BOTH PHASES BEFORE             BLI01170
C     TERMINATING.                                                      BLI01180
C     OFLAG = (0,1).  OFLAG = 0, MEANS THAT IN PHASE 1 PORTION          BLI01190
C     OF THE COMPUTATION, THE PROGRAM DOES NO ORTHOGONALITY CHECKS      BLI01200
C     ON THE Q-BLOCKS GENERATED.  OFLAG = 1 MEANS THAT IN THE           BLI01210
C     PHASE 1 PORTION AND IN THE PHASE 2 PORTIONS OF THE COMPUTATIONS   BLI01220
C     THE PROGRAM CHECKS THE ORTHOGONALITY OF THE Q-BLOCKS GENERATED    BLI01230
C     W.R.T. THAT VECTOR IN THE FIRST BLOCK THAT IS GENERATING          BLI01240
C     DESCENDANTS.  NOTE THAT IN PHASE 2, THE PROGRAM ALWAYS MAKES      BLI01250
C     THIS CHECK OF ORTHOGONALITY REGARDLESS OF THE VALUE OF OFLAG.     BLI01260
C     FOR SAFETY, OFLAG SHOULD ALWAYS BE SET TO 1, ALTHOUGH FOR MANY    BLI01270
C     PROBLEMS THIS IS NOT NECESSARY.                                   BLI01280
      READ(5,20) EXPLAN                                                 BLI01290
```

```
      READ(5,*) EFLAG,OFLAG                                        BLI01300
C                                                                 BLI01310
C     READ SEED USED BY SUBROUTINE GENRAN TO OBTAIN THOSE STARTING BLI01320
C     VECTORS WHICH ARE GENERATED RANDOMLY.                       BLI01330
      READ(5,20) EXPLAN                                           BLI01340
      READ(5,*) SEED                                              BLI01350
C                                                                 BLI01360
C     SPECIFY MAXIMUM T-SIZE ALLOWED (KMAX-1); INITIAL SIZE OF    BLI01370
C     STARTING BLOCK (KACT);  NUMBER OF STARTING VECTORS SUPPLIED (KSET)BLI01380
C     SEE BLOCK LANCZOS HEADER FOR COMMENTS REGARDING THE SIZE OF KACT. BLI01390
      READ(5,20) EXPLAN                                           BLI01400
      READ(5,*) KMAX,KACT,KSET                                    BLI01410
C                                                                 BLI01420
C     SPECIFY NUMBER (KM) OF EXTREME EIGENVALUES AND EIGENVECTORS BLI01430
C     OF B-INVERSE TO BE COMPUTED.  THE BLOCK PROCEDURE WORKS WITH THE BLI01440
C     INVERSE OF THE MATRIX B = SO*P*A*P' + SHIFT*I, USING A      BLI01450
C     FACTORIZATION OF B.  TO INDICATE THAT THE ALGEBRAICALLY-    BLI01460
C     SMALLEST EIGENVALUES OF B-INVERSE ARE BEING COMPUTED SET KM < 0. BLI01470
C     IF KM < 0, THE PROGRAM ASSUMES THAT BLSOLV SUBROUTINE WHICH BLI01480
C     THE USER HAS PROVIDED IS COMPUTING -(B-INVERSE)*X           BLI01490
C     INSTEAD OF (B-INVERSE)*X AND INTERNALLY IT COMPUTES THE |KM| BLI01500
C     ALGEBRAICALLY-LARGEST EIGENVALUES OF -(B-INVERSE).          BLI01510
      READ(5,20) EXPLAN                                           BLI01520
      READ(5,*) KM                                                BLI01530
      IF(KM.EQ.0) GO TO 540                                       BLI01540
      KML = IABS(KM)                                              BLI01550
C                                                                 BLI01560
C     STAGNATION OF CONVERGENCE OF THE KM-TH EIGENVALUE WILL BE   BLI01570
C     TESTED AFTER NSTAG ITERATIONS.  CONVERGENCE WILL BE SAID TO BLI01580
C     HAVE STAGNATED IF THE RATIO OF THE SQUARE OF THE CURRENT KM-TH BLI01590
C     RESIDUAL TO THE SQUARE OF THE CORRESPONDING RESIDUAL OBTAINED BLI01600
C     10 ITERATIONS EARLIER IS GREATER THAN FRACT.  NSTAG SHOULD BE BLI01610
C     >= 25.  FRACT WAS SET EQUAL TO .01 IN THE TESTS.            BLI01620
      READ(5,20) EXPLAN                                           BLI01630
      READ(5,*) NSTAG, FRACT                                      BLI01640
C                                                                 BLI01650
C     READ IN THE RELATIVE TOLERANCE (RELTOL) USED TO DETERMINE A BLI01660
C     CONVERGENCE CRITERION FOR PHASE 2, AND THE MAXIMUM NUMBER (MAXIT2)BLI01670
C     OF CALLS TO SUBROUTINE BLSOLV ALLOWED IN PHASE 2.          BLI01680
      READ(5,20) EXPLAN                                           BLI01690
      IF(EFLAG.EQ.1)  READ(5,*)  RELTOL, MAXIT2                   BLI01700
C                                                                 BLI01710
C     CONSISTENCY CHECKS                                          BLI01720
C     PROCEDURE REQUIRES ENOUGH ROOM IN THE Q-ARRAY FOR AT LEAST 2 BLI01730
C     BLOCKS OF SIZE KACT PLUS A WORKING VECTOR OF LENGTH N.      BLI01740
      MXBLK = KMAX -1                                             BLI01750
      MXBLK2 = MXBLK*MXBLK                                        BLI01760
      IF(MDIMTM.LT.MXBLK2) GO TO 520                              BL!01770
      NKMAX = N*KMAX                                              BLI01780
      IF(MDIMQ.LT.NKMAX)   GO TO 560                              BLI01790
      IF(KML.GT.KACT) GO TO 420                                   BLI01800
      IF(MXBLK.GT.N)  GO TO 440                                   BLI01810
      IF(2*KACT.GT.MXBLK)  GO TO 500                              BLI01820
C                                                                 BLI01830
C-------------------------------------------------------------------BLI01840
C     DEFINE AND INITIALIZE THE ARRAYS NEEDED TO DEFINE THE       BLI01850
C     FACTORIZATION OF THE B-MATRIX.  PASS THE STORAGE LOCATIONS  BLI01860
C     OF THESE ARRAYS TO THE SUBROUTINE BLSOLV.                   BLI01870
C                                                                 BLI01880
      CALL USPEC(N,MATNO,NNZ,AVER)                                BLI01890
C                                                                 BLI01900
C-------------------------------------------------------------------BLI01910
C     MASK OVERFLOW AND UNDERFLOW                                 BLI01920
      CALL MASK                                                   BLI01930
C                                                                 BLI01940
C-------------------------------------------------------------------BLI01950
C     ARE THERE STARTING VECTORS TO READ IN FROM FILE 10 (KSET.NE.0) ? BLI01960
```

```
      IF(KSET.EQ.0) GO TO 70                                          BLI01970
C                                                                     BLI01980
      READ(10,30) NOLD,KACT                                           BLI01990
   30 FORMAT(I6,I4)                                                   BLI02000
      IF(NOLD.NE.N.OR.KSET.GT.KACT) GO TO 460                         BLI02010
      DO 50 J=1,KSET                                                  BLI02020
      READ(10,20) EXPLAN                                              BLI02030
      READ(10,40) EVAL,RESID                                          BLI02040
   40 FORMAT(E20.12,E13.4)                                            BLI02050
      READ(10,20) EXPLAN                                              BLI02060
      LINT= (J-1)*N + 1                                               BLI02070
      LFIN = J*N                                                      BLI02080
   50 READ(10,60) (Q(JL), JL = LINT,LFIN)                             BLI02090
   60 FORMAT(4E20.12)                                                 BLI02100
C                                                                     BLI02110
   70 CONTINUE                                                        BLI02120
C                                                                     BLI02130
C     WRITE TO A SUMMARY OF THE PARAMETERS FOR THIS RUN TO FILE 6     BLI02140
C                                                                     BLI02150
      MXBLK = KMAX - 1                                                BLI02160
      WRITE(6,80) N, NNZ, AVER, MATNO                                 BLI02170
   80 FORMAT(/4X,'ORDER OF B-MATRIX ',5X,'AVERAGE NUMBER NONZEROES PER RBLI02180
     10W IN FACTOR'/                                                  BLI02190
     1I15,E47.4/3X,'CRUDE ESTIMATE OF SIZE NONZERO ENTRIES',5X,'MATRIX IBLI02200
     1D'/E31.4,I21/)                                                  BLI02210
C                                                                     BLI02220
      WRITE(6,90) SO, SHIFT                                           BLI02230
   90 FORMAT(/4X,'SCALE USED ON A-MATRIX',5X,'SHIFT USED ON A-MATRIX'/ BLI02240
     1E26.4,E27.4/)                                                   BLI02250
C                                                                     BLI02260
      WRITE(6,100) MDIMQ, MDIMTM                                      BLI02270
  100 FORMAT(/18X,'USER-SPECIFIED'/2X,'MAX. DIMENSION Q-ARRAY',4X,'MAX. BLI02280
     1DIMENSION TM-ARRAY'/I16,I26/)                                   BLI02290
C                                                                     BLI02300
      WRITE(6,110) OFLAG, EFLAG                                       BLI02310
  110 FORMAT(/4X,'OFLAG',4X,'EFLAG'/I8,I9/)                           BLI02320
C                                                                     BLI02330
      IF(OFLAG.EQ.1) WRITE(6,120) SPREC                              BLI02340
  120 FORMAT(/4X,'ORTHOGONALITY TEST TOLERANCE'/E25.2)                BLI02350
C                                                                     BLI02360
      IF(EFLAG.EQ.1) WRITE(6,130) MAXIT,RELTOL,MAXIT2                 BLI02370
  130 FORMAT(/4X,' MAXIT ',8X,' RELTOL ',6X,' MAXIT2 '/I10,E20.6,I12/) BLI02380
      IF(EFLAG.EQ.0) WRITE(6,140) MAXIT                               BLI02390
  140 FORMAT(/4X,' MAXIT '/I10/)                                      BLI02400
C                                                                     BLI02410
      WRITE(6,150) SEED                                               BLI02420
  150 FORMAT(/' SEED FOR RANDOM NUMBER GENERATOR'/I24/)               BLI02430
C                                                                     BLI02440
      IF(KM.GT.0) WRITE(6,160) KML                                    BLI02450
  160 FORMAT(/' COMPUTE THE',I3,' ALGEBRAICALLY-LARGEST EIGENVALUES AND BLI02460
     1CORRESPONDING VECTORS'/' OF THE INVERSE OF B = (SO*P*A*P-TRANS + BLI02470
     1HIFT*I)'/)                                                      BLI02480
      IF(KM.LT.0) WRITE(6,170) KML                                    BLI02490
  170 FORMAT(/' COMPUTE THE',I3,' ALGEBRAICALLY-SMALLEST EIGENVALUES ANDBLI02500
     1 CORRESPONDING VECTORS'/' OF THE INVERSE OF THE MATRIX B = (SO*P*ABLI02510
     1*P-TRANS + SHIFT*I).'/' PROGRAM ASSUMES THAT USER IS PROVIDING -(BBLI02520
     1-INVERSE)*X INSTEAD OF (B-INVERSE)*X'/' AND COMPUTES THE ALGEBRAICBLI02530
     1ALLY-LARGEST EIGENVALUES OF -(B-INVERSE).'/' HOWEVER ON EXIT, FILEBLI02540
     1 15 CONTAINS THE ALGEBRAICALLY-SMALLEST EIGENVALUES'/' OF B-INVERSBLI02550
     1E, THE CORRESPONDING EIGENVALUES OF THE ORIGINAL A-MATRIX'/' AND TBLI02560
     1HE CORRESPONDING EIGENVECTORS OF A.'/)                          BLI02570
C                                                                     BLI02580
C     NOTE THAT THE ESTIMATE FOR AVER IN THE INVERSE CASE IS VERY CRUDE BLI02590
C     COMPUTE PHASE 1 CONVERGENCE TOLERANCE                           BLI02600
      IF(AVER.GE.1.)                                                  BLI02610
     1ERRMAX = 2.D0*DFLOAT(N+1000)*NNZ*AVER*MACHEP                    BLI02620
      IF(AVER.LT.1.)                                                  BLI02630
```

```
     1ERRMAX = 2.DO*DFLOAT(N+1000)*NNZ*AVER**2*MACHEP              BLI02640
C                                                                 BLI02650
     WRITE(6,180) KACT,MXBLK,KSET                                 BLI02660
 180 FORMAT(/' ON INITIAL ITERATIONS, THE FIRST BLOCK CONTAINS ',I3,' VBLI02670
    1ECTORS'/' HOWEVER THE SIZE OF THE FIRST BLOCK MAY CHANGE AS THE ITBLI02680
    1ERATIONS PROCEED'/' THE MAXIMUM SIZE T-MATRIX THAT CAN BE GENERATEBLI02690
    1D IS ',I4/' THE USER SUPPLIED ',I3,' STARTING VECTORS'/)      BLI02700
C                                                                 BLI02710
     WRITE(6,190)                                                 3LI02720
 190 FORMAT(/' ITERATIVE PROCEDURE'/' PROCEDURE MONITORS THE SIZES OF TBLI02730
    1HE NORM(GRADIENTS)**2 ON EACH'/' ITERATION.  CONVERGENCE IS SAID  BLI02740
    1TO HAVE OCCURRED WHEN ALL'/' RELEVANT (NORMS)**2 ARE LESS THAN ERRBLI02750
    1MAX',E10.3/' PHASE 1 ERRMAX MAY YIELD SOMEWHAT LESS THAN SINGLE PRBLI02760
    1ECISION ACCURACY.'/'  PHASE 2 REFINES THE VECTORS OBTAINED ON PHASBLI02770
    1E 1, ACCORDING TO'/' THE ACCURACY SPECIFIED BY THE USER'/)    BLI02780
C                                                                 BLI02790
     WRITE(6,200) ERRMAX                                          BLI02800
 200 FORMAT(//' PHASE 1 CONVERGENCE CRITERION, ERRMAX '/E22.3/)   BLI02810
C                                                                 BLI02820
C------------------------------------------------------------------BLI02830
C    PASS STORAGE LOCATIONS OF VARIOUS ARRAYS TO LANCZS AND LANCI1 BLI02840
C    SUBROUTINES                                                  BLI02850
C                                                                 BLI02860
     CALL LANZP(DIR,DESC,SM,TM,TOD,TD,G,XLFT,LEFT,SPREC)          BLI02870
     CALL LANCPI(DIR,DESC,TM,SM,XLFT,LEFT)                        BLI02880
C                                                                 BLI02890
C------------------------------------------------------------------BLI02900
C                                                                 BLI02910
C    ENTER PHASE 1 OF BLOCK LANCZOS PROCEDURE.  BLOCK PROCEDURE    BLI02920
C    HAS 2 POSSIBLE PHASES.  USER SPECIFIES PHASE 1 ONLY OR PHASE 1 BLI02930
C    AND PHASE 2 BY SETTING EFLAG = 0 OR 1, RESPECTIVELY. PHASE 1  BLI02940
C    COMPUTES VECTORS THAT ARE USUALLY ACCURATE TO SINGLE PRECISION. BLI02950
C    PHASE 2 TAKES THE VECTORS OBTAINED IN PHASE 1 AND REFINES THEM. BLI02960
C    THE USER SPECIFIES THE DEGREE OF REFINEMENT DESIRED BY SELECTING BLI02970
C    THE VALUE OF RELTOL AND MAXIT2.  BOTH PHASES SHOULD BE USED.  BLI02980
     IPHASE = 1                                                   BLI02990
     NITER = 0                                                    BLI03000
 210 ITER = 0                                                     BLI03010
     RESIDL(1) = FRACT                                            BLI03020
     RESIDL(2) = NSTAG                                            BLI03030
C                                                                 BLI03040
C------------------------------------------------------------------BLI03050
C    CALL INITIATES THE BLOCK LANCZOS PROCEDURE.                  BLI03060
C    ON RETURN EIGENVALUE APPROXIMATIONS ARE IN E(I),            BLI03070
C    I = 1,KACT, IN ALGEBRAICALLY DECREASING ORDER.  CORRESPONDING BLI03080
C    EIGENVECTOR APPROXIMATIONS ARE IN FIRST N*KACT LOCATIONS IN  BLI03090
C    THE Q-ARRAY.                                                 BLI03100
C                                                                 BLI03110
     CALL LANCZS(BLSOLV,KML,KSET,KACT,MXBLK,N,Q,E,RESIDL,RESIDK,ERRMAX,BLI03120
    1 IPHASE,NITER,IWRITE)                                        BLI03130
C                                                                 BLI03140
C------------------------------------------------------------------BLI03150
C                                                                 BLI03160
     IF(IPHASE.EQ.MPMIN) WRITE(15,220) N,KACT                    BLI03170
 220 FORMAT(2I10,' PHASE 2 TERMINATED '/' PROGRAM INDICATES ACCURACY SPBLI03180
    1ECIFIED BY USER IS NOT ACHIEVABLE'/)                         BLI03190
C                                                                 BLI03200
     ITERA = IABS(ITER)                                           BLI03210
     IF(IWRITE.NE.MPMIN.AND.ITER.GT.0)  WRITE(6,230) IPHASE,ITERA BLI03220
 230 FORMAT(/1X,'PHASE COMPLETED',5X,' NUMBER CALLS TO BLSOLV SUBROUTINBLI03230
    1E USED'/I10,I32)                                             BLI03240
C                                                                 BLI03250
     IF(IWRITE.EQ.MPMIN.OR.ITER.LT.0) WRITE(6,240) IPHASE,ITERA  BLI03260
 240 FORMAT(/1X,'PHASE TERMINATED',5X,' NUMBER CALLS TO BLSOLV SUBROUTIBLI03270
    1NE USED'/I10,I32)                                            BLI03280
C                                                                 BLI03290
```

```
      IF(ITER.GT.O.AND.IWRITE.NE.MPMIN)  GO TO 270              BLI03300
C                                                               BLI03310
      IF(ITER.LT.0)  WRITE(6,250)                               BLI03320
  250 FORMAT(//' SMALL EIGENVALUE SUBROUTINE DEFAULTED'/' BLOCK LANCZOS BLI03330
     1 PROCEDURE STOPS AFTER SAVING CURRENT EIGENVECTOR APPROXIMATIONS'/BLI03340
     1/)                                                        BLI03350
C                                                               BLI03360
      WRITE(15,260)                                             BLI03370
      WRITE(6,260)                                              BLI03380
  260 FORMAT(//' BLOCK LANCZOS PROCEDURE TERMINATES WITHOUT CONVERGENCE BLI03390
     1'/' USER SHOULD EXAMINE OUTPUT TO DETERMINE REASONS FOR TERMINATIOBLI03400
     1N'//)                                                     BLI03410
C                                                               BLI03420
C     WRITE EIGENVALUE AND EIGENVECTOR APPROXIMATIONS CONTAINED IN      BLI03430
C     THE FIRST Q-BLOCK TO FILE 15                             BLI03440
C                                                               BLI03450
  270 IF(IPHASE.EQ.1)  WRITE(15,280) N,KACT,SEED                BLI03460
  280 FORMAT(I6,I4,I12,' PHASE 1, ORDER A-MATRIX, SIZE OF Q(1), SEED')  BLI03470
      IF(IPHASE.EQ.2)  WRITE(15,290) N,KACT,SEED                BLI03480
  290 FORMAT(I6,I4,I12,' PHASE 2, ORDER A-MATRIX, SIZE OF Q(1), SEED')  BLI03490
C                                                               BLI03500
C     PERMUTE THE EIGENVECTORS IF NECESSARY                     BLI03510
      IF(JPERM.EQ.0) GO TO 310                                  BLI03520
      LINT = -N + 1                                             BLI03530
      KACT1 = KACT*N + 1                                        BLI03540
      DO 300 J = 1,KACT                                         BLI03550
      LINT = LINT + N                                           BLI03560
C-----------------------------------------------------------------------BLI03570
      IPERM = 2                                                 BLI03580
      CALL LPERM(Q(LINT),Q(KACT1),IPERM)                        BLI03590
C-----------------------------------------------------------------------BLI03600
  300 CONTINUE                                                  BLI03610
C                                                               BLI03620
C     COMPUTE THE EIGENVALUES OF THE A-MATRIX                   BLI03630
  310 DO 320 J = 1,KACT                                         BLI03640
      IF(KM.LT.0)   E(J) = -E(J)                                BLI03650
      TD(J) = 1.DO/E(J)                                         BLI03660
  320 TD(J) = (TD(J) - SHIFT)/SO                                BLI03670
C                                                               BLI03680
C     NOTE THAT RESIDUAL PRINTED OUT CORRESPONDS TO VALUE OBTAINED      BLI03690
C     PRIOR TO FINAL PROJECTION Q(1)-TRANSPOSE*AQ(1) DONE BEFORE        BLI03700
C     TERMINATION                                               BLI03710
      JJ=KACT                                                   BLI03720
      LINT = -N + 1                                             BLI03730
      LFIN = 0                                                  BLI03740
      DO 340 J=1,KACT                                           BLI03750
      LINT = LINT + N                                           BLI03760
      LFIN = LFIN + N                                           BLI03770
      JJ=JJ+1                                                   BLI03780
C                                                               BLI03790
C     NOTE THAT RESIDUAL PRINTED OUT CORRESPONDS TO VALUE OBTAINED      BLI03800
C     PRIOR TO FINAL PROJECTION Q(1)-TRANSPOSE*(B-INVERS)*Q(1) DONE     BLI03810
C     BEFORE TERMINATION                                        BLI03820
C                                                               BLI03830
      WRITE(15,330) E(J), SM(JJ),TD(J)                          BLI03840
  330 FORMAT(/E20.12,E13.4,E20.12,'BI-EVAL,ER**2,A-EVAL,A-EVEC'/)       BLI03850
  340 WRITE(15,350) (Q(L), L=LINT,LFIN)                         BLI03860
      WRITE(15,360)                                             BLI03870
  350 FORMAT(4E20.12)                                           BLI03880
  360 FORMAT(/' ABOVE ARE COMPUTED APPROXIMATE EIGENVECTORS'/)  BLI03890
C                                                               BLI03900
      IF(ITER.GT.MAXIT) WRITE(15,370) ITER,MAXIT                BLI03910
  370 FORMAT(//' PROCEDURE TERMINATED BECAUSE NUMBER OF CALLS TO BLSOLV BLI03920
     1 SUBROUTINE',I6/' EXCEEDED MAXIMUM NUMBER ',I6,' ALLOWED'//)      BLI03930
C                                                               BLI03940
      IF(ITER.LT.0) WRITE(15,380)                               BLI03950
  380 FORMAT(//' USER BEWARE.  EIGENELEMENT COMPUTATIONS DEFAULTED BECAUBLI03960
```

```
                ISE'/'   EISPACK SUBROUTINE DEFAULTED.   EIGENVALUE AND EIGENVECTORBLI03970
             1 APPROXIMATIONS'/'   ABOVE WERE THOSE AVAILABLE AT THE TIME OF DEFBLI03980
             1AULT'/'  SOMETHING IS SERIOUSLY WRONG.'//)                      BLI03990
C                                                                            BLI04000
C         CHECK FOR TERMINATION AFTER PHASE 1                                BLI04010
C         ITER < 0 MEANS EISPACK SUBROUTINE DEFAULTED                        BLI04020
C         IPHASE = MPMIN MEANS THAT PHASE 2 TERMINATED DUE TO ORTHOGONALITY BLI04030
C         IWRITE = MPMIN MEANS THAT CONVERGENCE APPEARS TO HAVE STAGNATED    BLI04040
C         ITER > MAXIT MEANS MAXIMUM NUMBER OF CALLS TO BLSOLV               BLI04050
C                 ALLOWED BY USER WAS EXCEEDED                               BLI04060
             IF(ITER.LT.0.OR.ITER.GT.MAXIT) GO TO 580                        BLI04070
             IF(IPHASE.EQ.MPMIN.OR.IWRITE.EQ.MPMIN) GO TO 580                BLI04080
             IF(EFLAG.NE.1.OR.IPHASE.EQ.2) GO TO 580                         BLI04090
C                                                                            BLI04100
C         ENTER 2ND PHASE OF COMPUTATION TO ATTEMPT TO OBTAIN MORE           BLI04110
C         ACCURATE EIGENVECTOR APPROXIMATIONS.                              BLI04120
C         USER CONTROLS THE SIZE OF THE ERROR TOLERANCE BY SPECIFYING        BLI04130
C         THE PARAMETER RELTOL.                                             BLI04140
C                                                                            BLI04150
             IPHASE = 2                                                      BLI04160
             MAXIT = MAXIT2                                                  BLI04170
             KSET = KACT                                                     BLI04180
C                                                                            BLI04190
C         ERROR TOLERANCE USES THE CONVERGED EIGENVALUE LARGEST IN           BLI04200
C         MAGNITUDE.                                                         BLI04210
             TD(1) = DABS(E(1))                                              BLI04220
             IF(KML.EQ.1) GO TO 400                                          BLI04230
             DO 390 J = 2,KML                                                BLI04240
         390 IF(DABS(E(J)).GT.TD(1))     TD(1) = DABS(E(J))                  BLI04250
         400 TD(1) = DMAX1(TD(1),1.D0)                                       BLI04260
             ERRMAN = RELTOL**2 * TD(1)**2                                   BLI04270
             IF(ERRMAN.GE.ERRMAX) GO TO 480                                  BLI04280
             ERRMAX = ERRMAN                                                 BLI04290
C                                                                            BLI04300
             WRITE(6,410) ERRMAX, MAXIT2                                     BLI04310
         410 FORMAT(//' ENTER PHASE 2 OF COMPUTATION'/' CONVERGENCE CRITERION IBLI04320
             1S REDUCED TO ',E13.4/' NO MORE THAN ',I5,' CALLS TO SUBROUTINE BLSBLI04330
             1OLV WILL BE ALLOWED.'/' PROGRAM WILL TERMINATE IF BLOCK ORTHGONALIBLI04340
             1TY PROBLEMS MATERIALIZE'/)                                     BLI04350
C                                                                            BLI04360
             GO TO 210                                                       BLI04370
C                                                                            BLI04380
C         INCONSISTENCIES IN THE DATA                                        BLI04390
C                                                                            BLI04400
         420 WRITE(6,430) KM,KACT                                            BLI04410
         430 FORMAT(/' PROGRAM TERMINATES BECAUSE THE NUMBER OF EIGENELEMENTS BLI04420
             1REQUESTED, KM =',I3/'  IS LARGER THAN THE SIZE OF THE FIRST Q BLOCBLI04430
             1K, KACT =',I3,' SPECIFIED'/'  USER MUST RESET KM OR KACT'/)    BLI04440
             GO TO 580                                                       BLI04450
C                                                                            BLI04460
         440 WRITE(6,450) KMAX,N                                             BLI04470
         450 FORMAT(/' PROGRAM TERMINATES BECAUSE KMAX = ',I5,' IS TOO LARGE FOBLI04480
             1R THE SIZE, N = ',I5,', OF THE GIVEN MATRIX'/'  USER MUST DECREASEBLI04490
             1THE SIZE OF KMAX.'/)                                           BLI04500
             GO TO 580                                                       BLI04510
C                                                                            BLI04520
         460 WRITE(6,470) NOLD,N,KACT,KSET                                   BLI04530
         470 FORMAT(/' PROGRAM TERMINATES BECAUSE FAULT OCCURRED IN READING IN BLI04540
             1THE EIGENVECTOR APPROXIMATIONS'/' EITHER THE SIZE MATRIX SPECIFIEDBLI04550
             1ON THE EIGENVECTOR FILE' ,I6/' DID NOT MATCH THE SIZE SPECIFIED 'BLI04560
             1,I5,' IN THE PROGRAM OR THE NUMBER'/'   OF VECTORS IN FILE 10 = 'BLI04570
             1,I4,' IS LESS THAN THE NUMBER ',I3/'  USER SAID WERE THERE'/)   BLI04580
             GO TO 580                                                       BLI04590
C                                                                            BLI04600
         480 WRITE(6,490) ERRMAN, ERRMAX                                     BLI04610
         490 FORMAT(/' COMPUTED PHASE 2 CONVERGENCE CRITERION ',E13.4/' IS LARBLI04620
             1GER THAN PHASE 1 CRITERION ',E13.4/'  SO PROGRAM TERMINATES'/)  BLI04630
```

```
      GO TO 580                                                 BLI04640
C                                                               BLI04650
  500 WRITE(6,510)  KACT,MXBLK                                  BLI04660
  510 FORMAT(/' PROGRAM TERMINATES BECAUSE THERE IS NOT ENOUGH ROOM TO  BLI04670
     1GENERATE 2 BLOCKS','   BECAUSE KACT = ',I3,' AND MXBLK = ', I4/)   BLI04680
      GO TO 580                                                 BLI04690
C                                                               BLI04700
C                                                               BLI04710
  520 WRITE(6,530)  MDIMTM, MXBLK                               BLI04720
  530 FORMAT(/' PROGRAM TERMINATES BECAUSE THE DIMENSION ',I6,' OF THE TBLI04730
     1M ARRAY'/'   IS TOO SMALL FOR THE LARGEST T-MATRIX ALLOWED ',I4)   BLI04740
      GO TO 580                                                 BLI04750
C                                                               BLI04760
  540 WRITE(6,550)                                              BLI04770
  550 FORMAT(/' USER SPECIFIED NUMBER OF EIGENVALUES OF INTEREST AS 0'/'BLI04780
     1 PROGRAM TERMINATES FOR USER TO RESET KM TO DESIRED NONZERO VALUE'BLI04790
     1/)                                                        BLI04800
      GO TO 580                                                 BLI04810
C                                                               BLI04820
  560 WRITE(6,570)  MDIMQ, KMAX,N                               BLI04830
  570 FORMAT(/' PROGRAM TERMINATES BECAUSE THE DIMENSION ',I6,' OF THE QBLI04840
     1-ARRAY'/'   IS TOO SMALL TO HOLD ',I5, ' VECTORS OF LENGTH ',I4)   BLI04850
      GO TO 580                                                 BLI04860
C                                                               BLI04870
  580 CONTINUE                                                  BLI04880
C                                                               BLI04890
      STOP                                                      BLI04900
C-----END OF MAIN PROGRAM FOR INVERSE BLOCK LANCZOS PROCEDURE-----------BLI04910
      END                                                       BLI04920
```

SECTION 9.3   SAMPLE MATRIX-VECTOR MULTIPLY SUBROUTINES

```
C---BLIMULT-(INVERSES OF REAL SYMMETRIC MATRICES)----------------------BLI00010
C                                                                      BLI00020
C      CONTAINS SUBROUTINES  LANCZS AND SAMPLE USPEC AND BLSOLV        BLI00030
C      USED BY THE VERSION OF THE BLOCK LANCZOS ALGORITHMS FOR         BLI00040
C      FACTORED INVERSES OF REAL SYMMETRIC MATRICES, BLIVAL.           BLI00050
C      NOTE THAT SAMPLE BLSOLV FOR SPARSE MATRICES ASSUMES THAT        BLI00060
C      B-MATRIX IS POSITIVE DEFINITE AND USES CHOLESKY FACTORS.        BLI00070
C      HOWEVER, THE USER CAN DIRECTLY REPLACE THAT SUBROUTINE BY       BLI00080
C      A SUBROUTINE FOR INDEFINITE MATRICES THAT COMPUTES THE          BLI00090
C      GENERALIZED FACTORIZATION L*D*(L-TRANSPOSE).                    BLI00100
C                                                                      BLI00110
C      NONPORTABLE CONSTRUCTIONS:                                      BLI00120
C      1.   THE ENTRY MECHANISM USED TO PASS THE STORAGE LOCATIONS     BLI00130
C           OF THE FACTORIZATION OF THE MATRIX THAT WILL BE USED       BLI00140
C           BY THE LANCZS SUBROUTINE TO THE SUBROUTINE BLSOLV.         BLI00150
C      2.   IN THE SAMPLE USPEC AND BLSOLV SUBROUTINES PROVIDED:       BLI00160
C           THE FREE FORMAT (7,*), THE FORMAT (20A4) USED FOR          BLI00170
C           READING EXPLANATORY COMMENTS IN THE MATRIX SPECIFICATION   BLI00180
C           FILES, AND THE HEX FORMAT (4Z20) USED IN THE USPECS.       BLI00190
C      3.   THE COMMON BLOCK: LOOPS                                    BLI00200
C                                                                      BLI00210
C-----USPEC FOR FACTORED INVERSES OF REAL SYMMETRIC MATRICES-----------BLI00220
C                                                                      BLI00230
       SUBROUTINE CUSPEC(N,MATNO,NNZ,AVER)                             BLI00240
C      SUBROUTINE USPEC(N,MATNO,NNZ,AVER)                              BLI00250
C                                                                      BLI00260
C---------------------------------------------------------------------BLI00270
       DOUBLE PRECISION  BD(2200),BSD(10000),NNZ,AVER                  BLI00280
       INTEGER  KCOL(2200),KROW(10000),IPR(2200),IPT(2200)             BLI00290
C---------------------------------------------------------------------BLI00300
C      THIS SAMPLE SUBROUTINE ASSUMES THAT B IS POSITIVE DEFINITE      BLI00310
C      USER COULD REPLACE BY SIMILAR SUBROUTINE FOR GENERAL FACTORIZATIONBLI00320
C      DIMENSIONS ARRAYS NEEDED TO DEFINE CHOLESKY FACTOR OF B-MATRIX, BLI00330
C      READS CHOLESKY FACTOR FROM FILE 7, AND THEN PASSES STORAGE      BLI00340
C      LOCATIONS OF THESE ARRAYS TO THE B-MATRIX SOLVE SUBROUTINE BLSOLV.BLI00350
C                                                                      BLI00360
C      HERE WE HAVE B = P*C*P' = L*L'  WHERE  C = SO*A + SHIFT*I.      BLI00370
C      P IS A PERMUTATION MATRIX DEFINED BY THE VECTOR MAPS IPR AND IPT.BLI00380
C      THE ITH ROW OF B CORRESPONDS TO THE JTH ROW OF C (A) WHERE      BLI00390
C      J = IPR(I) AND I = IPT(J).                                      BLI00400
C                                                                      BLI00410
C      THE B-CHOLESKY FACTOR IS STORED IN THE FOLLOWING SPARSE FORMAT: BLI00420
C      N = ORDER OF THE B-MATRIX.                                      BLI00430
C      NZT = NUMBER OF NONZERO SUBDIAGONAL ENTRIES IN THE CHOLESKY     BLI00440
C           FACTOR, L.                                                 BLI00450
C      KCOL(J), J=1,N IS THE NUMBER OF NONZERO SUBDIAGONAL ELEMENTS IN BLI00460
C           COLUMN J OF L.                                             BLI00470
C      KROW(K), K=1,NZT IS THE ROW INDEX FOR CORRESPONDING ENTRY BSD(K).BLI00480
C      BD(J), J = 1,N CONTAINS THE DIAGONAL ENTRIES OF L.             BLI00490
C      BSD(K), K =1,NZT CONTAINS THE NONZERO SUBDIAGONAL ENTRIES OF L  BLI00500
C      JPERM = (0,1):  1 MEANS CHOLEKSY FACTOR CORRESPONDS TO          BLI00510
C                      PERMUTED C. O MEANS NO PERMUTATION WAS USED.    BLI00520
C---------------------------------------------------------------------BLI00530
C      READ CHOLESKY FACTOR FROM FILE 7.  MUST BE STORED              BLI00540
C      IN SPARSE MATRIX FORMAT.                                        BLI00550
       READ(7,10) NZT,NOLD,NZL,MATOLD,JPERM                            BLI00560
   10 FORMAT(I10,2I6,I8,I6)                                            BLI00570
C                                                                      BLI00580
       WRITE(6,20) NZT,NZL,N,NOLD,MATOLD,JPERM                         BLI00590
   20 FORMAT(' HEADER, CHOLESKY FACTOR FILE'/                          BLI00600
     1 3X,'NZT',3X,'NZL',5X,'N',2X,'NOLD',2X,'MATOLD',1X,'JPERM'/      BLI00610
     1 4I6,I8,I6/)                                                     BLI00620
C                                                                      BLI00630
```

```
      IF (N.NE.NOLD.OR.MATNO.NE.MATOLD) GO TO 100                   BLI00640
C                                                                   BLI00650
      READ(7,30) (KCOL(K), K = 1,NZL)                               BLI00660
      READ(7,30) (KROW(K), K = 1,NZT)                               BLI00670
   30 FORMAT(13I6)                                                  BLI00680
      READ(7,40) (BD(K), K = 1,N)                                   BLI00690
      READ(7,40) (BSD(K), K = 1,NZT)                                BLI00700
   40 FORMAT(4Z20)                                                  BLI00710
C  20 FORMAT(3E25.16)                                               BLI00720
C                                                                   BLI00730
C     DOES CHOLESKY FACTOR CORRESPOND TO PERMUTED B?                BLI00740
      IF(JPERM.EQ.0)  GO TO 60                                      BLI00750
      READ(7,30) (IPR(K), K = 1,N)                                  BLI00760
C                                                                   BLI00770
      DO 50 K = 1,N                                                 BLI00780
      J = IPR(K)                                                    BLI00790
   50 IPT(J) = K                                                    BLI00800
C------------------------------------------------------------------BLI00810
      CALL LPERME(IPR,IPT,N)                                        BLI00820
C------------------------------------------------------------------BLI00830
   60 CONTINUE                                                      BLI00840
C                                                                   BLI00850
C     COMPUTE NNZ, THE AVERAGE NUMBER OF NONZEROS PER COLUMN, AND   BLI00860
C     AVER, THE AVERAGE SIZE OF NONZERO ENTRIES IN THE FACTORS      BLI00870
C     OF THE B-MATRIX.  FROM THIS, ESTIMATE (TOO CRUDELY) THE       BLI00880
C     AVERAGE FOR B-INVERSE AS AVER = 1/AVER.                       BLI00890
      ITCOL = 0                                                     BLI00900
      AVER = 0.D0                                                   BLI00910
      DO 70 K = 1,N                                                 BLI00920
      IF(DABS(BD(K)).EQ.0.D0)  GO TO 70                             BLI00930
      ITCOL = ITCOL + 1                                             BLI00940
      AVER = AVER + DABS(BD(K))                                     BLI00950
   70 CONTINUE                                                      BLI00960
      NTCOL = ITCOL                                                 BLI00970
      DO 80 K = 1,N                                                 BLI00980
   80 ITCOL = ITCOL + 2*KCOL(K)                                     BLI00990
      NNZ = DFLOAT(ITCOL)/DFLOAT(N)                                 BLI01000
      DO 90 K = 1,NZS                                               BLI01010
   90 AVER = AVER + DABS(BSD(K))                                    BLI01020
      AVER = AVER/DFLOAT(NZS + NTCOL)                               BLI01030
      AVER = 1.D0/AVER                                              BLI01040
C                                                                   BL'01050
C------------------------------------------------------------------BLI01060
C     PASS STORAGE LOCATIONS OF FACTORS TO INVERSION SUBROUTINE BLSOLV BLI01070
      CALL BSOLVE(BSD,BD,KCOL,KROW,N,NZT,NZL)                       BLI01080
C------------------------------------------------------------------BLI01090
C                                                                   BLI01100
      GO TO 120                                                     BLI01110
C                                                                   BLI01120
  100 CONTINUE                                                      BLI01130
C     DEFAULT EXIT                                                  BLI01140
      WRITE(6,110)                                                  BLI01150
  110 FORMAT(/' TERMINATE.  PARAMETERS IN CHOLESKY FACTOR FILE'/    BLI01160
     1' DO NOT AGREE WITH THOSE SPECIFIED BY THE USER'/)            BLI01170
      STOP                                                          BLI01180
C                                                                   BLI01190
  120 CONTINUE                                                      BLI01200
C-----END OF USPEC-------------------------------------------------BLI01210
      RETURN                                                        BLI01220
      END                                                           BLI01230
C                                                                   BLI01240
C-----BLSOLV-(FACTORED INVERSES OF REAL SYMMETRIC MATRICES)--------BLI01250
C                                                                   BLI01260
C     SUBROUTINE BLSOLV(V,U)                                        BLI01270
      SUBROUTINE CBSOLV(V,U)                                        BLI01280
C                                                                   BLI01290
C------------------------------------------------------------------BLI01300
```

```
      DOUBLE PRECISION  BD(1),BSD(1),U(1),V(1),TEMP,ZERO,ONE           BLI01310
      INTEGER  KCOL(1),KROW(1)                                         BLI01320
      COMMON/LOOPS/MAXIT,ITER                                          BLI01330
C-----------------------------------------------------------------------BLI01340
      ITER = ITER + 1                                                  BLI01350
      ZERO = 0.0D0                                                     BLI01360
      ONE = 1.0D0                                                      BLI01370
C     SOLVE B*U = V FOR U WHERE  B = L*L'                              BLI01380
C     SET U = V. FIRST SOLVE L*U = U FOR U, THEN SOLVE L'*U = U FOR U  BLI01390
      KL = 0                                                           BLI01400
      DO 10 K = 1,N                                                    BLI01410
   10 U(K) = V(K)                                                      BLI01420
      DO 30 K = 1,N                                                    BLI01430
      TEMP = U(K)/BD(K)                                                BLI01440
      U(K) = TEMP                                                      BLI01450
      IF (KCOL(K).EQ.0.OR.K.EQ.N) GO TO 30                             BLI01460
      KF = KL + 1                                                      BLI01470
      KL = KL + KCOL(K)                                                BLI01480
      DO 20 KK = KF,KL                                                 BLI01490
      KR = KROW(KK)                                                    BLI01500
   20 U(KR) = U(KR) - TEMP*BSD(KK)                                     BLI01510
   30 CONTINUE                                                         BLI01520
      NP1 = N+1                                                        BLI01530
      KF = NZT + 1                                                     BLI01540
      DO 50 K = 1,N                                                    BLI01550
      L = NP1 - K                                                      BLI01560
      TEMP = U(L)                                                      BLI01570
      IF (KCOL(L).EQ.0.OR.L.EQ.N) GO TO 50                             BLI01580
      KL = KF - 1                                                      BLI01590
      KF = KF - KCOL(L)                                                BLI01600
      DO 40 LL = KF,KL                                                 BLI01610
      LR = KROW(LL)                                                    BLI01620
   40 TEMP = TEMP - BSD(LL)*U(LR)                                      BLI01630
   50 U(L) = TEMP/BD(L)                                                BLI01640
   60 CONTINUE                                                         BLI01650
C                                                                      BLI01660
      RETURN                                                           BLI01670
C                                                                      BLI01680
C-----------------------------------------------------------------------BLI01690
      ENTRY BSOLVE(BSD,BD,KCOL,KROW,N,NZT,NZL)                         BLI01700
C-----------------------------------------------------------------------BLI01710
C                                                                      BLI01720
      RETURN                                                           BLI01730
C-----END OF BLSOLV---------------------------------------------------BLI01740
      END                                                              BLI01750
C                                                                      BLI01760
C-----SUBROUTINES FOR DIAGONAL TEST MATRICES-------------------------BLI01770
C     BLSOLV AND USPEC SUBROUTINES FOR DIAGONAL TEST MATRICES          BLI01780
C                                                                      BLI01790
C-------BLSOLV DIAGONAL TEST MATRIX----------------------------------BLI01800
C                                                                      BLI01810
C     SUBROUTINE DBSOLV(V,U)                                           BLI01820
      SUBROUTINE BLSOLV(V,U)                                           BLI01830
C                                                                      BLI01840
C-----------------------------------------------------------------------BLI01850
      DOUBLE PRECISION  V(1),U(1),D(1)                                 BLI01860
      COMMON/LOOPS/MAXIT,ITER                                          BLI01870
C-----------------------------------------------------------------------BLI01880
      ITER = ITER + 1                                                  BLI01890
   10 DO 20 I=1,N                                                      BLI01900
   20 U(I)= V(I)/D(I)                                                  BLI01910
C  20 U(I)= -V(I)/D(I)                                                 BLI01920
C                                                                      BLI01930
   30 CONTINUE                                                         BLI01940
      RETURN                                                           BLI01950
C                                                                      BLI01960
```

```
C-----------------------------------------------------------------BLI01970
C      BELOW ENTRY IS FOR A DIAGONAL TEST MATRIX               BLI01980
       ENTRY DSOLVE(D,N)                                       BLI01990
C-----------------------------------------------------------------BLI02000
       RETURN                                                  BLI02010
C-----END OF 'DIAGONAL' TEST MATRIX  BLSOLV------------------------BLI02020
       END                                                     BLI02030
C                                                              BLI02040
C-----START OF USPEC FOR DIAGONAL TEST MATRIX---------------------BLI02050
C                                                              BLI02060
       SUBROUTINE USPEC(N,MATNO,NNZ,AVER)                      BLI02070
C      SUBROUTINE DUSPEC(N,MATNO,NNZ,AVER)                     BLI02080
C                                                              BLI02090
C-----------------------------------------------------------------BLI02100
       DOUBLE PRECISION  D(1000),DI(1000),SHIFT,SPACE,NNZ,AVER BLI02110
       DOUBLE PRECISION  DABS, DFLOAT                          BLI02120
       REAL  EXPLAN(20)                                        BLI02130
C-----------------------------------------------------------------BLI02140
C                                                              BLI02150
       READ(7,10) EXPLAN                                       BLI02160
    10 FORMAT(20A4)                                            BLI02170
       READ(7,*) NOLD,NUNIF,SPACE,D(1),SHIFT                   BLI02180
       NNUNIF = NOLD - NUNIF                                   BLI02190
       WRITE(6,20) NOLD,SPACE,NNUNIF,D(1),SHIFT                BLI02200
    20 FORMAT(/' DIAGONAL TEST MATRIX, SIZE = ',I4/' IS THE INVERSE OF MABLI02210
      1TRIX WITH MOST ENTRIES',E10.3/' UNITS APART AND WITH ',I3,' ENTRIEBLI02220
      1S IRREGULARLY SPACED'/' FIRST ENTRY WAS ',E13.4,' SHIFT = ',E10.3 BLI02230
      1/)                                                      BLI02240
C                                                              BLI02250
       IF(N.NE.NOLD) GO TO 120                                 BLI02260
C      COMPUTE THE UNIFORM PORTION OF THE SPECTRUM             BLI02270
       DO 30 J=2,NUNIF                                         BLI02280
    30 D(J) = D(1) - DFLOAT(J-1)*SPACE                         BLI02290
       NUNIF1=NUNIF + 1                                        BLI02300
       READ(7,10)  EXPLAN                                      BLI02310
       DO 40 J=NUNIF1,N                                        BLI02320
    40 READ(7,*) D(J)                                          BLI02330
       NB = NUNIF - 2                                          BLI02340
C                                                              BLI02350
       IF(SHIFT.EQ.0.) GO TO 60                                BLI02360
       DO 50 J=1,N                                             BLI02370
    50 D(J) = D(J) + SHIFT                                     BLI02380
C                                                              BLI02390
C      COMPUTE EIGENVALUES OF INVERSE FOR PRINTOUT ONLY        BLI02400
    60 DO 70 J = 1,N                                           BLI02410
    70 DI(J) = 1.D0/D(J)                                       BLI02420
       WRITE(6,80) (J,DI(J), J=1,N )                           BLI02430
    80 FORMAT(/' INVERSE BLOCK LANCZOS TEST, LANCZS USES INVERSE OF GIVENBLI02440
      1MATRIX'/' ENTRIES OF INVERSE OF DIAGONAL TEST MATRIX = '/(I4,E20.1BLI02450
      12,I4,E20.12,I4,E20.12))                                 BLI02460
C                                                              BLI02470
C      DIAGONAL GENERATION COMPLETE                            BLI02480
C                                                              BLI02490
C      COMPUTE NNZ AND AVER                                    BLI02500
       NNZ = 1.D0                                              BLI02510
       AVER = 0.D0                                             BLI02520
       DO 90 K = 1,N                                           BLI02530
    90 AVER = AVER + DABS(DI(K))                               BLI02540
       AVER = AVER/DFLOAT(N)                                   BLI02550
       AVER = 1.D0/AVER                                        BLI02560
C                                                              BLI02570
C      COMPUTE THE GAPS                                        BLI02580
       N1 = N-1                                                BLI02590
       DO 100 K = 1,N1                                         BLI02600
   100 DI(K) = DI(K+1) - DI(K)                                 BLI02610
       WRITE(6,110) (K,DI(K), K=1,N1)                          BLI02620
   110 FORMAT(/' GAPS BETWEEN EIGENVALUES'/(I4,E13.4,I4,E13.4,I4,E13.4,I4BLI02630
```

```
      1,E13.4))                                                   BLI02640
C                                                                 BLI02650
C-----------------------------------------------------------------BLI02660
C      PASS STORAGE LOCATIONS OF D AND N TO DSOLV SUBROUTINE      BLI02670
       CALL DSOLVE(D,N)                                           BLI02680
C-----------------------------------------------------------------BLI02690
C                                                                 BLI02700
       RETURN                                                     BLI02710
  120 WRITE(6,130) NOLD,N                                         BLI02720
  130 FORMAT(' PROGRAM TERMINATES BECAUSE NOLD = ',I5,'DOES NOT EQUAL N BLI02730
      1 =',I5)                                                    BLI02740
C-----END OF USPEC SUBROUTINE FOR 'DIAGONAL' TEST MATRICES-------------BLI02750
       STOP                                                       BLI02760
       END                                                        BLI02770
```

SECTION 9.4   FILE DEFINITIONS AND SAMPLE INPUT FILE

Below is a listing of the input/output files which are accessed by the
real symmetric block Lanczos eigenvalue/eigenvector program, BLIEVAL.
BLIEVAL calculates a few extreme eigenvalues and corresponding eigenvectors
of the inverse of a real symmetric matrix (scaled and shifted). Also
below is a sample of the input file which BLIEVAL requires on file 5.
The parameters in this file are supplied in free format.

```
SAMPLE DEFINITIONS OF THE INPUT/OUTPUT FILES FOR BLIEVAL
-----------------------------------------------------------------
*BLIEVAL EXEC
FI 06 TERM
FILEDEF   5 DISK BLIEVAL   INPUT     A (RECFM F LRECL 80 BLOCK 80
FILEDEF   8 DISK &1        INPUT     A (RECFM F LRECL 80 BLOCK 80
FILEDEF  10 DISK &1        BLSTARTV  A (RECFM F LRECL 80 BLOCK 80
FILEDEF  13 DISK &1        BLEXTRAV  A (RECFM F LRⅧCL 80 BLOCK 80
FILEDEF  15 DISK &1        BLEIGVEC  A (RECFM F LRⅬCL 80 BLOCK 80
*IMTQL2 AND TRED2 ARE 2 EISPACK LIBRARY SUBROUTINES
LOAD  BLIEVAL  BLSUB  BLIMULT  IMTQL2  TRED2
-----------------------------------------------------------------
```

```
SAMPLE INPUT FILE FOR BLIEVAL
-------------------------------------------------------------------
 LINE 1 IWRITE (SPECIFY MESSAGE LEVEL TO FILE 6: 1 MEANS DETAILED
          1
 LINE 2 N      MATNO     SO     SHIFT    JPERM (SIZE,ID,SCALE,SHIFT,PERM?
        1250    1250     1.      0.        0
 LINE 3 MDIMQ         MDIMTM        MAXIT (DIMS. Q, TM, MAX A*X'S
        40000         2500          1000
 LINE 4 EFLAG   OFLAG ( EFLAG=(0,1) 1=2PHASES. OFLAG: 1=ORTHOG CHECK
          1       1
 LINE 5 SEED      (STARTING VECTOR SEED, RANDOM NUMBER GENERATOR
        3482736
 LINE 6 KMAX   KACT   KSET (MAX T SIZE +1,SIZE 1ST BLOCK,VECTORS SUPPLIED
         31     3      0
 LINE 7 KM (NUMB. EVS FOR ALG-LARGEST, -(NUMB. EVS) FOR ALG-SMALLEST
         3
 LINE 8 NSTAG  FRACT (NO. ITNS BEFORE TEST CONVERGENCE, TEST FRACTION
         25     .05
 LINE 9 RELTOL         MAXIT2 (PHASE 2,CONVERGE.TOL.,NO. A*X'S ALLOWED
        .00000001       1000
-------------------------------------------------------------------
```

REFERENCES[a]

Björck, Å (1967). Solving linear least squares problems by Gram-Schmidt orthogonalization. *BIT* 7, 1-21.

Bunch, J. R., and Kaufman, L. (1977). Some stable methods for calculating inertia and solving symmetric linear systems. *Math. Comp.* **31**, 163-179.

Cullum, J., and Donath, W. E. (1974). A block Lanczos algorithm for computing the q algebraically largest eigenvalues and a corresponding eigenspace for large, sparse symmetric matrices. *Proc. 1974 IEEE Conference on Decision and Control,* 505-509.

Cullum, J., and Willoughby, R. A. (1984). Lanczos Algorithms, Large Symmetric Matrices: Volume 1, Theory. Progress in Scientific Computing Series, Birkhauser, Boston.

EISPACK Guide (1976). Matrix Eigensystem Routines - EISPACK Guide. Smith, B. T., Boyle, J. M., Garbow, B. S., Ikebe, Y., Klema, V. C., and Moler, C. B. *Lecture Notes in Computer Science* **6**, Second Edition, Springer, New York.

EISPACK Guide (1977). Matrix Eigensystem Routines - EISPACK Guide Extension. Garbow, B. S., Boyle, J. M., Dongarra, J. J., and Moler, C. B. *Lecture Notes in Computer Science* **51**, Second Edition, Springer, New York.

George, A., and Liu, J. W. H. (1981). Computer Solution of Large Sparse Positive Definite Systems. Prentice-Hall, Englewood Cliffs, NJ.

Golub, G. H., and Kahan, W. (1965). Calculating the singular values and pseudoinverse of a matrix. *SIAM J. Numer. Anal.* **2**, 205-224.

Golub, G. H., Luk, F. T., and Overton, M. L. (1981). A block Lanczos method for computing the singular values and corresponding singular vectors of a matrix. *ACM TOMS* **7**, 149-169.

Lanczos, C. (1961). Linear Differential Operators. Van Nostrand, New York.

LINPACK User's Guide. (1979). Dongarra, J. J., Bunch, J. R., Moler, C. B., and Stewart, G. W. SIAM Press, Philadelphia.

Moro, G., and Freed, J. H. (1981). Calculation of ESR spectra and related Fokker-Planck forms by the use of the Lanczos algorithm. *J. Chem. Phys.* **74**, 3757-3773.

Paige, C. C. (1971). The computation of eigenvalues and eigenvectors of very large sparse matrices. *Ph.D. Thesis,* U. London.

Paige, C. C. (1972). Computational variants of the Lanczos method for the eigenproblem. *J. Inst. Math. Appl.* **10**, 373-381.

Paige, C. C. (1976). Error analysis of the Lanczos algorithms for tridiagonalizing a symmetric matrix. *J. Inst. Math. Appl.* **18**, 341-349.

Paige, C. C. (1980). Accuracy and effectiveness of the Lanczos algorithm for the symmetric eigenproblem. *Linear Algebra Appl.* **34**, 235-258.

PFORT Verifier (1981). Ryder, B. G., and Hall, A. D. Bell Laboratory Technical Report 12.

SPARSPAK User Guide (1979). George, A., Liu, J. W. H., and Ng, E. CS Dept. Tech. Report, U. Waterloo.

Stewart, G. W. (1973). Introduction to Matrix Computations. Academic Press, New York.

Wilkinson, J. H. (1965). The Algebraic Eigenvalue Problem. Oxford University Press, New York.

---

[a] Extended set of references is contained in Volume 1.

Index

FACTORED INVERSES, REAL SYMMETRIC MATRICES

SINGLE-VECTOR CODES

REAL, SYMMETRIC GENERALIZED PROBLEMS

SINGLE-VECTOR CODES

REAL RECTANGULAR MATRICES

SINGLE-VECTOR CODES

NONDEFECTIVE, COMPLEX SYMMETRIC MATRICES

SINGLE-VECTOR CODES